INSTRUCTOR'S SOLUTIONS MANUAL

HEIDI A. HOWARD

Florida Community College at Jacksonville

to accompany

TRIGONOMETRY

EIGHTH EDITION

Margaret L. Lial

American River College

John Hornsby

University of New Orleans

David I. Schneider

University of Maryland

Boston San Francisco New York
London Toronto Sydney Tokyo Singapore Madrid
Mexico City Munich Paris Cape Town Hong Kong Montreal

Reproduced by Pearson Addison-Wesley from electronic files supplied by the author.

Publishing as Pearson Addison-Wesley, 75 Arlington Street, Boston, MA 02116

ISBN 0-321-22737-9

1 2 3 4 5 6 BB 08 07 06 05 04

CONTENTS

5 TRIGONOMETRIC IDENTITIES

6 INVERSE CIRCULAR FUNCTIONS AND TRIGONOMETRIC EQUATIONS

7 APPLICATIONS OF TRIGONOMETRY AND VECTORS

8 COMPLEX NUMBERS, POLAR EQUATIONS, AND PARAMETRIC EQUATIONS

I wish to thank Faiz Al-Rubaee (a never-ending resource),
John Samons (a consummate professional), and Amanda Nunley
(a budding mathematician) for all their contributions to this supplement.
I also want to acknowledge the endless efforts of Joanne Ha,
Sandra Scholten, Sheila Spinney, and
Joe Vetere (aka Mr. Bender).
Thanks to you all.

Chapter 1
TRIGONOMETRIC FUNCTIONS

Section 1.1: Angles

1. Answers will vary.

2. $1°$ represents $\frac{1}{360}$ of a rotation. Therefore, $45°$ represents $\frac{45}{360} = \frac{1}{8}$ of a complete rotation.

3. Let $x =$ the measure of the angle.
If the angle is its own complement, then we have the following.

$$x + x = 90$$
$$2x = 90$$
$$x = 45$$

A $45°$ angle is its own complement.

4. Let $x =$ the measure of the angle.
If the angle is its own supplement, then we have the following.

$$x + x = 180$$
$$2x = 180$$
$$x = 90.$$

A $90°$ angle is its own supplement.

5. $30°$

(a) $90° - 30° = 60°$

(b) $180° - 30° = 150°$

6. $60°$

(a) $90° - 60° = 30°$

(b) $180° - 60° = 120°$

7. $45°$

(a) $90° - 45° = 45°$

(b) $180° - 45° = 135°$

8. $18°$

(a) $90° - 18° = 72°$

(b) $180° - 18° = 162°$

9. $54°$

(a) $90° - 54° = 36°$

(b) $180° - 54° = 126°$

10. $89°$

(a) $90° - 89° = 1° °$

(b) $180° - 89° = 91°$

11. $\frac{25\text{ minutes}}{60\text{ minutes}} = \frac{x}{360°} \Rightarrow x = \frac{25}{60}(360) = 25(6) = 150°$

12. Since the minute hand is $\frac{3}{4}$ the way around, the hour hand is $\frac{3}{4}$ of the way between the 1 and 2. Thus, the hour hand is located 8.75 minutes past 12. The minute hand is 15 minutes before the 12. The smaller angle formed by the hands of the clock can be found by solving the proportion $\frac{(15+8.75)\text{ minutes}}{60\text{ minutes}} = \frac{x}{360°}$.

$$\frac{(15+8.75)\text{ minutes}}{60\text{ minutes}} = \frac{x}{360°} \Rightarrow \frac{23.75}{60} = \frac{x}{360} \Rightarrow x = \frac{23.75}{60}(360) = 23.75(6) = 142.5°$$

13. The two angles form a straight angle.

$$7x + 11x = 180 \Rightarrow 18x = 180 \Rightarrow x = 10$$

The measures of the two angles are $(7x)° = [7(10)]° = 70°$ and $(11x)° = [11(10)]° = 110°$.

14. The two angles form a right angle.

$$4y+2y=90 \Rightarrow 6y=90 \Rightarrow y=15$$

The two angles have measures of $(4y)° = [4(15)]° = 60°$ and $(2y)° = [2(15)]° = 30°$.

15. The two angles form a right angle.

$$(5k+5)+(3k+5)=90 \Rightarrow 8k+10=90 \Rightarrow 8k=80 \Rightarrow k=10$$

The measures of the two angles are $(5k+5)° = [5(10)+5]° = (50+5)° = 55°$ and $(3k+5)° = [3(10)+5]° = (30+5)° = 35°$.

16. The sum of the measures of two supplementary angles is 180°.

$$(10m+7)+(7m+3)=180 \Rightarrow 17m+10=180 \Rightarrow 17m=170 \Rightarrow m=10$$

Since $(10m+7)° = [10(10)+7]° = (100+7)° = 107°$ and $(7m+3)° = [7(10)+3]° = (70+3)° = 73°$, the angle measures are 107° and 73°.

17. The sum of the measures of two supplementary angles is 180°.

$$(6x-4)+(8x-12)=180 \Rightarrow 14x-16=180 \Rightarrow 14x=196 \Rightarrow x=14$$

Since $(6x-4)° = [6(14)-4]° = (84-4)° = 80°$ and $(8x-12)° = [8(14)-12]° = (112-12)° = 100°$, the angle measures are 80° and 100°.

18. The sum of the measures of two complementary angles is 90°.

$$(9z+6)+3z=90 \Rightarrow 12z+6=90 \Rightarrow 12z=84 \Rightarrow z=7$$

Since $(9z+6)° = [9(7)+6]° = (63+6)° = 69°$ and $(3z)° = [3(7)]° = 21°$, the angle measures are 69° and 21°.

19. If an angle measures x degrees and two angles are complementary if their sum is 90°, then the complement of an angle of $x°$ is $(90 - x)°$.

20. If an angle measures x degrees and two angles are supplementary if their sum is 180°, then the supplement of an angle of $x°$ is $(180 - x)°$.

21. The first negative angle coterminal with x between 0° and 60° is $(x - 360)°$.

22. The first positive angle coterminal with x between 0° and –60° is $(x + 360)°$.

23.

$$\begin{array}{r} 62° \ 18' \\ +21° \ 41' \\ \hline 83° \ 59' \end{array}$$

24.

$$\begin{array}{r} 75° \ 15' \\ +\ 83° \ 32' \\ \hline 158° \ 47' \end{array}$$

25. $71°18' - 47°29' = 70°78' - 47°29'$

$$\begin{array}{r} 70° \ 78' \\ -47° \ 29' \\ \hline 23° \ 49' \end{array}$$

26. $47^\circ 23' - 73^\circ 48' = -(73^\circ 48' - 47^\circ 23')$

Since $\begin{array}{r} 73^\circ 48' \\ \underline{-47^\circ 23'} \\ 26^\circ 25' \end{array}$, we have $47^\circ 23' - 73^\circ 48' = -(73^\circ 48' - 47^\circ 23') = -26^\circ 25'$.

27. $90^\circ - 51^\circ 28' = 89^\circ 60' - 51^\circ 28'$

$$\begin{array}{r} 89^\circ 60' \\ \underline{-51^\circ 28'} \\ 38^\circ 32' \end{array}$$

28. $180^\circ - 124^\circ 51' = 179^\circ 60' - 124^\circ 51'$

$$\begin{array}{r} 179^\circ 60' \\ \underline{-124^\circ 51'} \\ 55^\circ \ 9' \end{array}$$

29. $90^\circ - 72^\circ 58' 11'' = 89^\circ 59' 60'' - 72^\circ 58' 11''$

$$\begin{array}{r} 89^\circ 59' 60'' \\ \underline{-72^\circ 58' 11''} \\ 17^\circ \ 1' 49'' \end{array}$$

30. $90^\circ - 36^\circ 18' 47'' = 89^\circ 59' 60'' - 36^\circ 18' 47''$

$$\begin{array}{r} 89^\circ 59' 60'' \\ \underline{-36^\circ 18' 47''} \\ 53^\circ 41' 13'' \end{array}$$

31. $20^\circ 54' = 20^\circ + \frac{54}{60}^\circ = 20^\circ + .900^\circ = 20.900^\circ$

32. $38^\circ 42' = 38^\circ + \frac{42}{60}^\circ = 38.700^\circ$

33. $91^\circ 35' 54'' = 91^\circ + \frac{35}{60}^\circ + \frac{54}{3600}^\circ$
$\approx 91^\circ + .5833^\circ + .0150^\circ \approx 91.598^\circ$

34. $34^\circ 51' 35'' = 34^\circ + \frac{51}{60}^\circ + \frac{35}{3600}^\circ$
$\approx 34^\circ + .8500^\circ + .0097^\circ$
$\approx 34.860^\circ$

35 $274^\circ 18' 59'' = 274^\circ + \frac{18}{60}^\circ + \frac{59}{3600}^\circ$
$\approx 274^\circ + .3000^\circ + .0164^\circ$
$\approx 274.316^\circ$

36. $165^\circ 51' 9'' = 165^\circ + \frac{51}{60}^\circ + \frac{9}{3600}^\circ$
$= 165^\circ + .8500^\circ + .0025^\circ$
$= 165.853^\circ$

37. $31.4296^\circ = 31^\circ + .4296^\circ = 31^\circ + .4296(60')$
$= 31^\circ + 25.776' = 31^\circ + 25' + .776'$
$= 31^\circ + 25' + .776(60'')$
$= 31^\circ 25' 46.56'' \approx 31^\circ 25' 47''$

38. $59.0854^\circ = 59^\circ + .0854^\circ = 59^\circ + .0854(60')$
$= 59^\circ + 5.124' = 59^\circ + 5' + .124'$
$= 59^\circ + 5' + .124(60'')$
$= 59^\circ \ 5' \ 7.44'' \approx 59^\circ 5' 7''$

39. $89.9004^\circ = 89^\circ + .9004^\circ = 89^\circ + .9004(60')$
$= 89^\circ + 54.024' = 89^\circ + 54' + .024'$
$= 89^\circ + 54' + .024(60'')$
$= 89^\circ 54' 1.44'' \approx 89^\circ 54' 1''$

40. $102.3771^\circ = 102^\circ + .3771^\circ = 102^\circ + .3771(60') = 102^\circ + 22.626' = 102^\circ + 22' + .626'$
$= 120^\circ + 22' + .626(60'') = 102^\circ \ 22' \ 37.56'' \approx 102^\circ \ 22' \ 38''$

41. $178.5994^\circ = 178^\circ + .5994^\circ = 178^\circ + .5994(60') = 178^\circ + 35.964' = 178^\circ + 35' + .964'$
$= 178^\circ + 35' + .964(60'') = 178^\circ 35' 57.84'' \approx 178^\circ 35' 58''$

42. $122.6853^\circ = 122^\circ + .6853^\circ = 122^\circ + .6853(60') = 122^\circ + 41.118' = 122^\circ + 41' + .118'$
$= 122^\circ + 41' + .118(60'') = 122^\circ \ 41' \ 7.08'' \approx 122^\circ \ 41' \ 7''$

43. Answers will vary.

44. $1.21 \text{ hr} = 1 \text{ hr} + .21 \text{ hr} = 1 \text{ hr} + .21(60 \text{ min}) = 1 \text{ hr} + 12.6 \text{ min}$
$= 1 \text{ hr} + 12 \text{ min} + .6 \text{ min} = 1 \text{ hr} + 12 \text{ min} + .6(60 \text{ sec})$
$= 1 \text{ hr, } 12 \text{ min, } 36 \text{ sec}$

Additional answers will vary.

45. $-40°$ is coterminal with $360° + (-40°) = 320°$.

46. $-98°$ is coterminal with $360° + (-98°) = 262°$.

47. $-125°$ is coterminal with $360° + (-125°) = 235°$.

48. $-203°$ is coterminal with $360° + (-203°) = 157°$.

49. $539°$ is coterminal with $539° - 360° = 179°$.

50. $699°$ is coterminal with $699° - 360° = 339°$.

51. $850°$ is coterminal with $850° - 2(360°) = 850° - 720° = 130°$.

52. $1000°$ is coterminal with $1000° - 2 \cdot 360° = 1000° - 720° = 280°$.

53. $30°$
A coterminal angle can be obtained by adding an integer multiple of 360°.
$30° + n \cdot 360°$

54. $45°$
A coterminal angle can be obtained by adding an integer multiple of 360°.
$45° + n \cdot 360°$

55. $135°$
A coterminal angle can be obtained by adding an integer multiple of 360°.
$135° + n \cdot 360°$

56. $270°$
A coterminal angle can be obtained by adding an integer multiple of 360°.
$270° + n \cdot 360°$

57. $-90°$
A coterminal angle can be obtained by adding an integer multiple of 360°.
$-90° + n \cdot 360°$

58. $-135°$
A coterminal angle can be obtained by adding integer multiple of 360°.
$-135° + n \cdot 360°$

59. The answers to Exercises 56 and 57 give the same set of angles since $-90°$ is coterminal with $-90° + 360° = 270°$.

60. **A.** $360° + r°$ is coterminal with $r°$ because you are adding an integer multiple of 360° to $r°$, $r° + 1 \cdot 360°$.

B. $r° - 360°$ is coterminal with $r°$ because you are adding an integer multiple of 360° to $r°$, $r° + (-1) \cdot 360°$.

C. $360° - r°$ is not coterminal with $r°$ because you are not adding an integer multiple of 360° to $r°$.
$360° - r° \neq r° + n \cdot 360°$ for an integer value n.

D. $r° + 180°$ is not coterminal with $r°$ because you are not adding an integer multiple of 360° to $r°$.
$r° + 180° \neq r° + n \cdot 360°$ for an integer value n. You are adding $\frac{1}{2}(360°)$.

Choices C and D are not coterminal with $r°$.

For Exercises 61 – 68, angles other than those given are possible.

61.

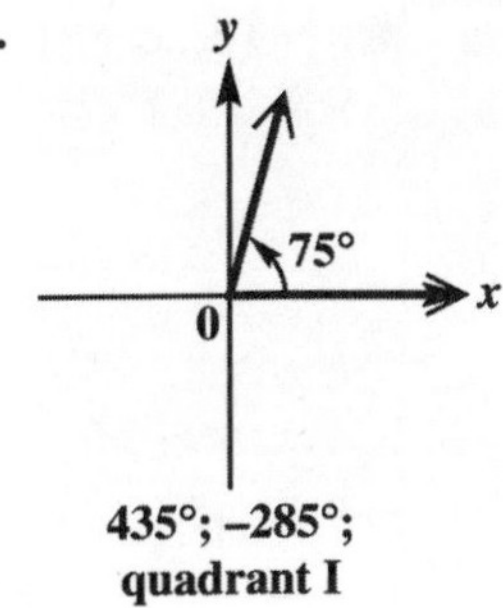

75° is coterminal with 75° + 360° = 435° and 75° – 360° = –285°. These angles are in quadrant I.

62.

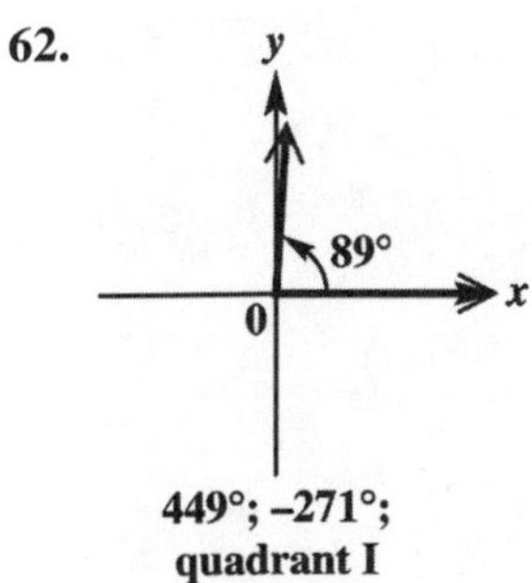

89° is coterminal with 89° + 360° = 449° and 89° – 360° = –271°. These angles are in quadrant I.

63.

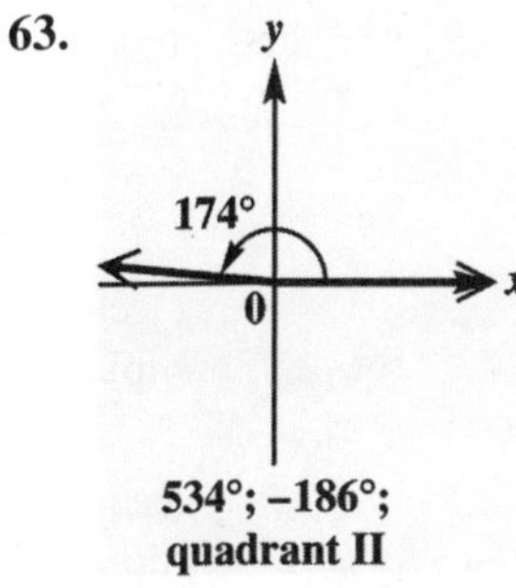

174° is coterminal with 174° + 360° = 534° and 174° – 360° = –186°. These angles are in quadrant II.

64.

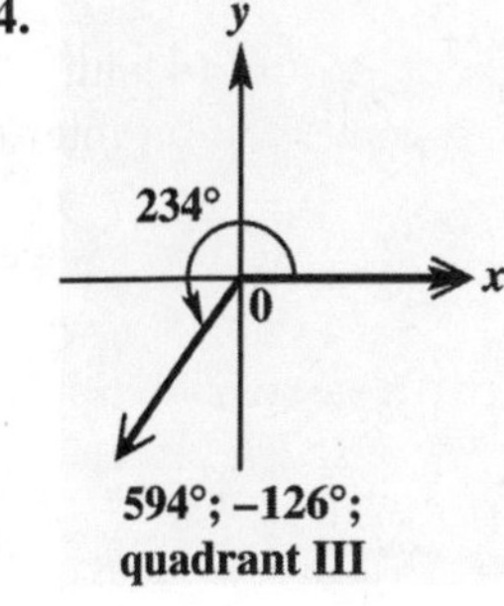

234° is coterminal with 234° + 360° = 594° and 234° – 360° = –126°. These angles are in quadrant III.

65.

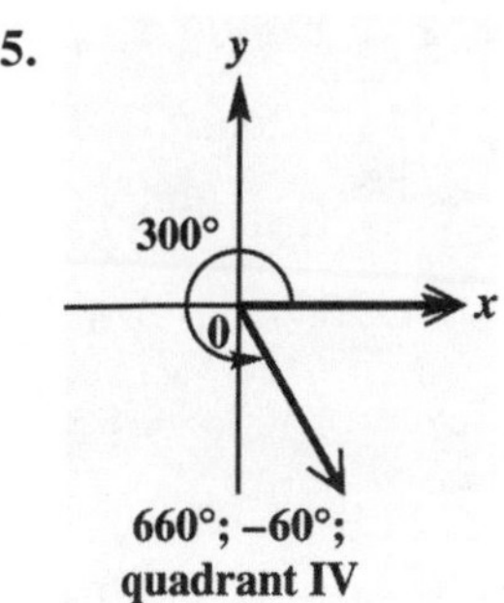

300° is coterminal with 300° + 360° = 660° and 300° – 360° = –60°. These angles are in quadrant IV.

66.

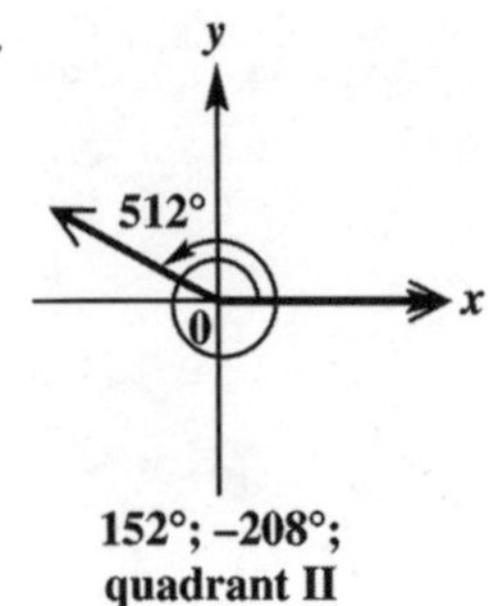

$512°$ is coterminal with $512° - 360° = 152°$ and $512° - 2 \cdot 360° = -208°$. These angles are in quadrant II.

67.

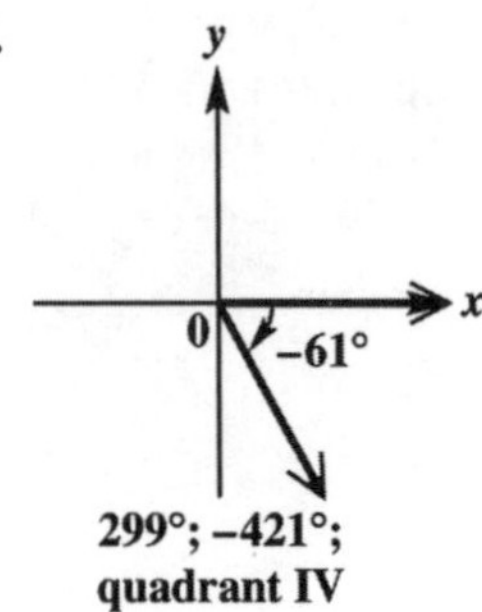

$-61°$ is coterminal with $-61° + 360° = 299°$ and $-61° - 360° = -421°$. These angles are in quadrant IV.

68.

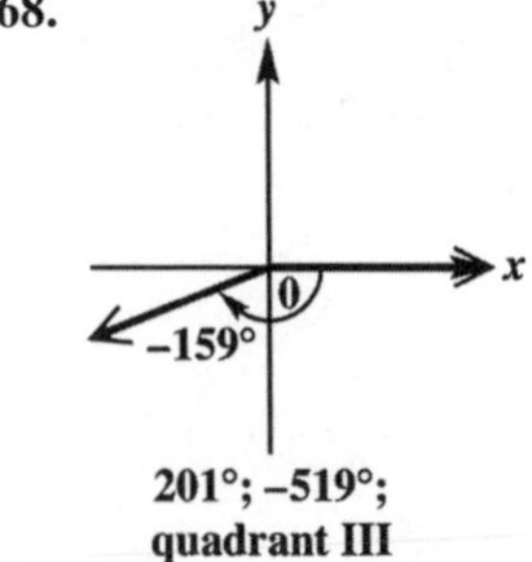

$-159°$ is coterminal with $-159° + 360° = 201°$ and $-159° - 360° = -519°$. These angles are in quadrant III.

69.

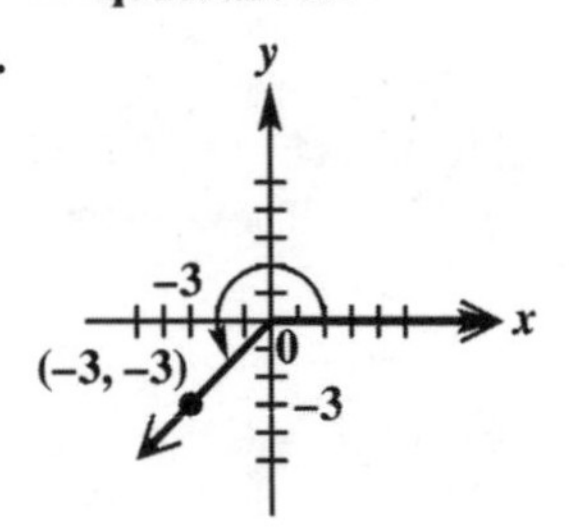

Points: $(0,0)$ and $(-3,-3)$

$$
\begin{aligned}
r &= \sqrt{(-3-0)^2 + (-3-0)^2} \\
&= \sqrt{(-3)^2 + (-3)^2} \\
&= \sqrt{9+9} \\
&= \sqrt{18} \\
&= \sqrt{9 \cdot 2} \\
&= 3\sqrt{2}
\end{aligned}
$$

70.

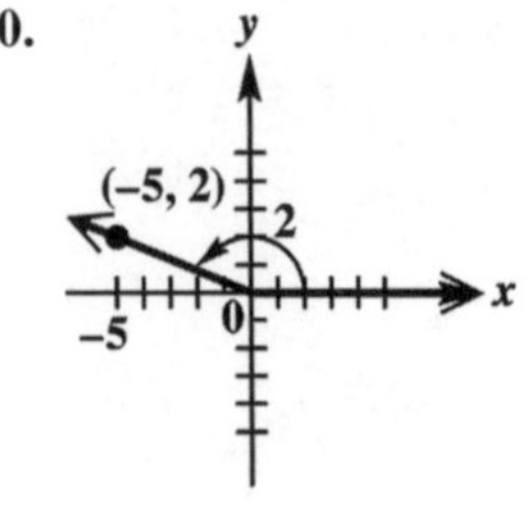

Points: $(0,0)$ and $(-5,2)$

$$
\begin{aligned}
r &= \sqrt{(-5-0)^2 + (2-0)^2} \\
&= \sqrt{(-5)^2 + 2^2} \\
&= \sqrt{25+4} \\
&= \sqrt{29}
\end{aligned}
$$

71.

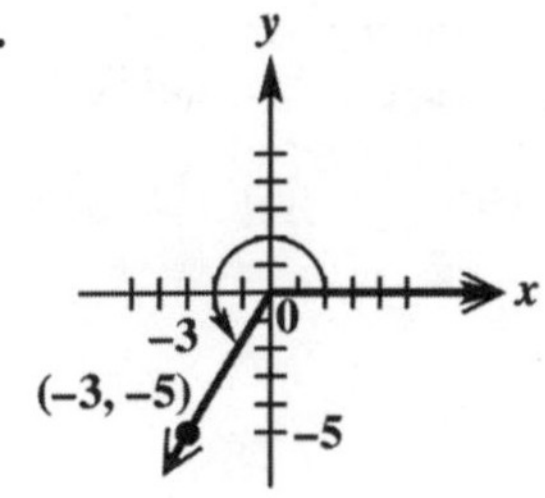

Points: $(0,0)$ and $(-3,-5)$

$$r=\sqrt{(-3-0)^2+(-5-0)^2}$$
$$=\sqrt{(-3)^2+(-5)^2}$$
$$=\sqrt{9+25}$$
$$=\sqrt{34}$$

72.

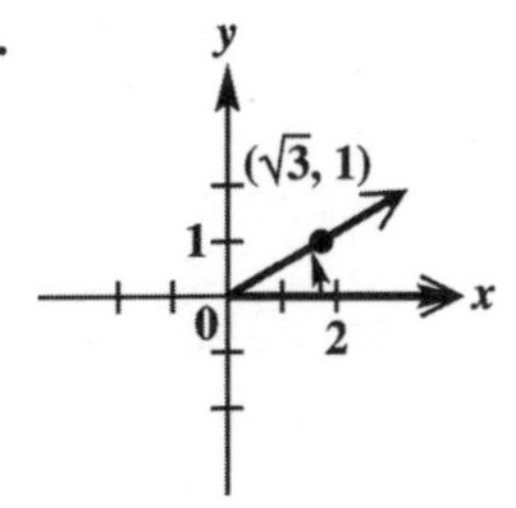

Points: $(0,0)$ and $\left(\sqrt{3},1\right)$

$$r=\sqrt{\left(\sqrt{3}-0\right)^2+(1-0)^2}$$
$$=\sqrt{\left(\sqrt{3}\right)^2+1^2}$$
$$=\sqrt{3+1}$$
$$=\sqrt{4}=2$$

73.

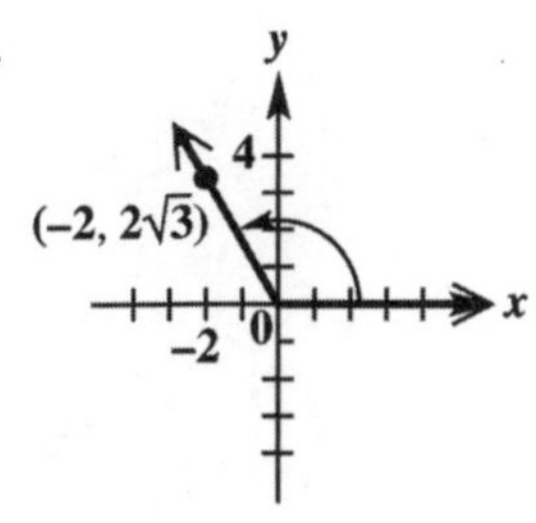

Points: $(0,0)$ and $\left(-2,2\sqrt{3}\right)$

$$r=\sqrt{(-2-0)^2+\left(2\sqrt{3}-0\right)^2}$$
$$=\sqrt{(-2)^2+\left(2\sqrt{3}\right)^2}$$
$$=\sqrt{4+12}$$
$$=\sqrt{16}=4$$

74.

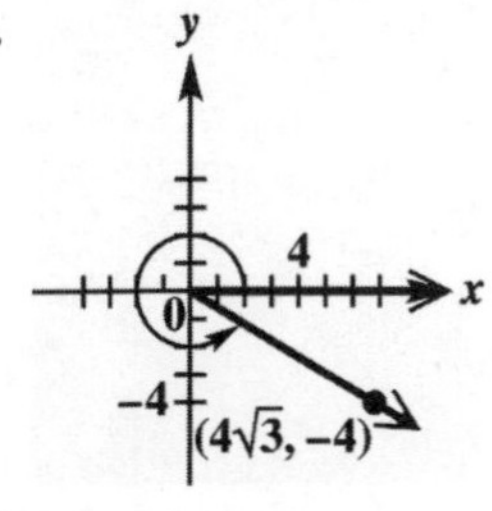

Points: $(0,0)$ and $\left(4\sqrt{3},-4\right)$

$$r=\sqrt{\left(4\sqrt{3}-0\right)^2+(-4-0)^2}$$
$$=\sqrt{\left(4\sqrt{3}\right)^2+(-4)^2}$$
$$=\sqrt{48+16}$$
$$=\sqrt{64}=8$$

75. 45 revolutions per min $=\frac{45}{60}$ revolution per sec $=\frac{3}{4}$ revolution per sec

A turntable will make $\frac{3}{4}$ revolution in 1 sec.

76. 90 revolutions per min $=\frac{90}{60}$ revolutions per sec $=1.5$ revolutions per sec

A windmill will make 1.5 revolutions in 1 sec.

77. 600 rotations per min $=\frac{600}{60}$ rotations per sec $=10$ rotations per sec $=5$ rotations per $\frac{1}{2}$ sec
$=5(360°)$ per $\frac{1}{2}$ sec $=1800°$ per $\frac{1}{2}$ sec

A point on the edge of the tire will move 1800° in $\frac{1}{2}$ sec.

78. If the propeller rotates 1000 times per minute, then it rotates $\frac{1000}{60} = 16\frac{2}{3}$ times per sec. Each rotation is $360°$, so the total number of degrees a point rotates in 1 sec is $(360°)(16\frac{2}{3}) = 6000°$.

79. $75°$ per min $= 75°(60)$ per hr $= 4500°$ per hr $= \frac{4500°}{360°}$ rotations per hr $= 12.5$ rotations per hr

The pulley makes 12.5 rotations in 1 hr.

80. First, convert $74.25°$ to degrees and minutes. Find the difference between this measurement and $74°\,20'$.

$$74.25° = 74° + .25(60') = 74°\,15' \quad \text{and} \quad \begin{array}{r} 74°\,20' \\ \underline{-74°\,15'} \\ 5' \end{array}$$

Next, convert $74°20'$ to decimal degrees. Find the difference between this measurement and $74.25°$, rounded to the nearest hundredth of a degree.

$$74°\,20' = 74° + \tfrac{20}{60}° \approx 74.333° \quad \text{and} \quad \begin{array}{r} 74.333° \\ \underline{-74.250°} \\ .083° \approx .08° \end{array}$$

The difference in measurements is $5'$ to the nearest minute or $.08°$ to the nearest hundredth of a degree.

81. The earth rotates $360°$ in 24 hr. $360°$ is equal to $360(60') = 21{,}600'$.

$$\frac{24\,\text{hr}}{21{,}600'} = \frac{x}{1'} \Rightarrow x = \frac{24}{21{,}600}\,\text{hr} = \frac{24}{21{,}600}(60\text{ min}) = \frac{1}{15}\text{ min} = \frac{1}{15}(60\text{ sec}) = 4\text{ sec}$$

It should take the motor 4 sec to rotate the telescope through an angle of 1 min.

82. Since we have five central angles that comprise a full circle, we find have the following.

$$5(2\theta) = 360° \Rightarrow 10\theta = 360° \Rightarrow \theta = 36°$$

The angle of each point of the five-pointed star measures $36°$.

Section 1.2: Angle Relationships and Similar Triangles

1. In geometry, opposite angles are called vertical angles.

2. In Figure 14, since we have two parallel lines and a transversal, the measures of angles 1, 4, 5 and 8 are all the same. Also, the measures of angles 2, 3, 6, and 7 are the same. If you know one of the angles, say angle 1, then you also know the measures of angles 4, 5, and 8. Since angles 1 and 2 form a straight angle, you know that the sum of their measures is $180°$. Thus, you can determine the measure of angle 2. Since the measure of angle 2 is the same as measures of angles 3, 6, and 8, you know the measures of all eight angles.

3. The two indicated angles are vertical angles. Hence, their measures are equal.

$$5x - 129 = 2x - 21 \Rightarrow 3x - 129 = -21 \Rightarrow 3x = 108 \Rightarrow x = 36$$

Since $5x - 129 = 5(36) - 129 = 180 - 129 = 51$ and $2x - 21 = 2(36) - 21 = 72 - 21 = 51$, both angles measure $51°$.

4. The two indicated angles are vertical angles. Hence, their measures are equal.

$$11x - 37 = 7x + 27 \Rightarrow 4x - 37 = 27 \Rightarrow 4x = 64 \Rightarrow x = 16$$

Since $11x - 37 = 11(16) - 37 = 176 - 37 = 139$ and $7x + 27 = 7(16) + 27 = 112 + 27 = 139$, both angles measure $139°$.

5. The three angles are the interior angles of a triangle. Hence, the sum of their measures is 180°.

$$x+(x+20)+(210-3x)=180 \Rightarrow -x+230=180 \Rightarrow -x=-50 \Rightarrow x=50$$

Since $x+20=50+20=70$ and $210-3x=210-3(50)=210-150=60,$ the three angles measure 50°, 70°, and 60°.

6. The three angles are the interior angles of a triangle. Hence, the sum of their measures is 180°.

$$(10x-20)+(x+15)+(x+5)=180 \Rightarrow 12x=180 \Rightarrow x=15$$

Since $10x-20=10(15)-20=150-20=130,$ $x+15=15+15=30,$ and $x+5=15+5=20,$ the three angles measure 130°, 30°, and 20°.

7. The three angles are the interior angles of a triangle. Hence, the sum of their measures is 180°.

$$(x-30)+(2x-120)+(\tfrac{1}{2}x+15)=180 \Rightarrow \tfrac{7}{2}x-135=180 \Rightarrow \tfrac{7}{2}x=315 \Rightarrow x=\tfrac{2}{7}(315)=90$$

Since $x-30=90-30=60,$ $\frac{1}{2}x+15=\frac{1}{2}(90)+15=45+15=60,$ and $2x-120=2(90)-120=$ $180-120=60,$ all three angles measure 60°.

8. The three angles are the interior angles of a triangle. Hence, the sum of their measures is 180°.

$$x+x+x=180 \Rightarrow 3x=180 \Rightarrow x=60$$

All three angles measure 60°.

9. The two angles are supplementary since they form a straight angle. Hence, their sum is 180°.

$$(3x+5)+(5x+15)=180 \Rightarrow 8x+20=180 \Rightarrow 8x=160 \Rightarrow x=20$$

Since $3x+5=3(20)+5=60+5=65$ and $5x+15=5(20)+15=100+15=115,$ the two angles measure 65° and 115°.

10. The two angles are complementary since they form a right angle. Hence, the sum of their measures is 90°.

$$(5x-1)+2x=90 \Rightarrow 7x-1=90 \Rightarrow 7x=91 \Rightarrow x=13$$

Since $5x-1=5(13)-1=65-1=64$ and $2x=2(13)=26,$ the two angles measure 64° and 26°.

11. Since the two angles are alternate interior angles, their measures are equal.

$$2x-5=x+22 \Rightarrow x-5=22 \Rightarrow x=27$$

Since $2x-5=2(27)-5=54-5=49$ and $x+22=27+22=49,$ both angles measure 49°.

12. Since the two angles are alternate exterior angles, their measures are equal.

$$2x+61=6x-51 \Rightarrow 61=4x-51 \Rightarrow 112=4x \Rightarrow 28=x$$

Since $2x+61=2(28)+61=56+61=117$ and $6x-51=6(28)-51=168-51=117,$ both angles measure 117°

13. Since the two angles are interior angles on the same side of the transversal, the sum of their measures is 180°.

$$(x+1)+(4x-56)=180 \Rightarrow 5x-55=180 \Rightarrow 5x=235 \Rightarrow x=47$$

Since $x+1=47+1=48$ and $4x-56=4(47)-56=188-56=132,$ the angles measure 48° and 132°.

14. Since the two angles are alternate exterior angles, their measures are equal.

$$10x+11=15x-54 \Rightarrow 11=5x-54 \Rightarrow 65=5x \Rightarrow 13=x$$

Since $10x+11=10(13)+11=130+11=141$ and $15x-54=15(13)-54=195-54=141$, both angles measure $141°$.

15. Let x = the measure of the third angle.

$$37°+57°+x=180°$$
$$89°+x=180°$$
$$x=91°$$

The third angle of the triangle measures $91°$.

16. Let x = the measure of the third angle.

$$x+29°+104°=180°$$
$$x+133°=180°$$
$$x=47°$$

The third angle of the triangle measures $47°$.

17. Let x = the measure of the third angle.

$$147°12'+30°19'+x=180°$$
$$177°31'+x=180°$$
$$x=180°-177°31'$$
$$x=179°60'-177°31'$$
$$x=2°29'$$

The third angle of the triangle measures $2°29'$.

18. Let x = the measure of the third angle.

$$x+136°50'+41°38'=180°$$
$$x+178°28'=180°$$
$$x=180°-178°28'$$
$$x=179°60'-178°28'$$
$$x=1°32'$$

The third angle of the triangle measures $1°32'$.

19. Let x = the measure of the third angle.

$$74.2°+80.4°+x=180°$$
$$154.6°+x=180°$$
$$x=25.4°$$

The third angle of the triangle measures $25.4°$.

20. Let x = the measure of the third angle.

$$29.6°+49.7°+x=180°$$
$$79.3+x=180°$$
$$x=100.7°$$

The third angle of the triangle measures $100.7°$.

21. A triangle cannot have angles of measures $85°$ and $100°$. The sum of the measures of these two angles is $85° + 100° = 185°$, which exceeds $180°$.

22. A triangle cannot have two obtuse angles. Since an obtuse angle measures between $90°$ and $180°$, the sum of two obtuse angles would be between $180°$ and $360°$, which exceeds $180°$.

23. Angle 1 and the $55°$ angle are vertical angles, which have the same measure, so angle 1 measures $55°$. Angle 5 and the $120°$ angle are interior angles on the same side of transversal, which means they are supplements. Thus, the measure of angle 5 is $180°-120°=60°$. Since angles 3 and 5 are vertical angles, the measure of angle 3 is also $60°$. Since angles 1, 2, and 3 form a line, the sum of their measures is $180°$.

$$\text{measure angle 1} + \text{measure angle 2} + \text{measure angle 3} = 180°$$
$$55° + \text{measure angle 2} + 60° = 180°$$
$$115° + \text{measure angle 2} = 180°$$
$$\text{measure angle 2} = 65°$$

Since angles 2 and 4 are vertical angles, the measure of angle 4 is $65°$. Angle 6 and the $120°$ angle are vertical angles, so the measure of angle 6 is $120°$. Angles 6 and 8 are supplements, so the sum of their measures is $180°$.

Continued on next page

23. (continued)

$$\begin{aligned} \text{measure angle 6} + \text{measure angle 8} &= 180^\circ \\ 120^\circ + \text{measure angle 8} &= 180^\circ \\ \text{measure angle 8} &= 60^\circ \end{aligned}$$

Since angles 7 and 8 are vertical angles, the measure of angle 7 is 60°.

Thus, the angle measurements given in numerical order (angle 1, angle 2, etc.) are as follows.

55°, 65°, 60°, 65°, 60°, 120°, 60°, 60°, 55°, and 55°.

24. The sum of the measures of the two angles is 180°. Thus, we have the following.

$$(x+2y)+11x=180 \Rightarrow 12x+2y=180 \Rightarrow 6x+y=90 \text{ (1)}$$

Since it is given that $x+y=40$, we can solve this equation for y (or x) and substitute into equation 1.

$$y=40-x \text{ (2)}$$

$$6x+y=90 \Rightarrow 6x+(40-x)=90 \Rightarrow 5x+40=90 \Rightarrow 5x=50 \Rightarrow x=10$$

Substitute 10 for x in equation 1 in order to solve for y.

$$y=40-x \Rightarrow y=40-10=30$$

Using the values of x and y, we can find the measure of the two angles.

Since $x+2y=10+2(30)=10+60=70$ and $11x=11(10)=110$, the measure of the angles are 70° and 110°.

25. The triangle has a right angle, but each side has a different measure. The triangle is a right triangle and a scalene triangle.

26. The triangle has one obtuse angle and three unequal sides, so it is obtuse and scalene.

27. The triangle has three acute angles and three equal sides, so it is acute and equilateral.

28. The triangle has two equal sides and all angles are acute, so it is acute and isosceles.

29. The triangle has a right angle and three unequal sides, so it is right and scalene.

30. The triangle has one obtuse angle and two equal sides, so it is obtuse and isosceles.

31. The triangle has a right angle and two equal sides, so it is right and isosceles.

32. The triangle has a right angle with three unequal sides, so it is right and scalene.

33. The triangle has one obtuse angle and three unequal sides, so it is obtuse and scalene.

34. This triangle has three equal sides and all angles are acute, so it is acute and equilateral.

35. The triangle has three acute angles and two equal sides, so it is acute and isosceles.

36. This triangle has a right angle with three unequal sides, so it is right and scalene.

37. – 39. Answers will vary.

40. Connect the right end of the semi-circle to the point where the arc crosses the semi-circle. Since the setting of the compass has never changed, the triangle is equilateral. Therefore, each of its angles measures 60°.

41. Corresponding angles are A and P, B and Q, C and R; Corresponding sides are AC and PR, BC and QR, AB and PQ

42. Corresponding angles are A and P, C and R, B and Q; Corresponding sides are AC and PR, CB and RQ, AB and PQ

43. Corresponding angles are A and C, E and D, ABE and CBD; Corresponding sides are EB and DB, AB and CB, AE and CD

44. Corresponding angles are H and F, K and E, HGK and FGE; Corresponding sides are HK and FE, GK and GE, HG and FG

45. Since angle Q corresponds to angle A, the measure of angle Q is 42°.

Since angles A, B, and C are interior angles of a triangle, the sum of their measures is 180°.

$$\begin{aligned} \text{measure angle } A + \text{measure angle } B + \text{measure angle } C &= 180^\circ \\ 42^\circ + \text{ measure angle } B + 90^\circ &= 180^\circ \\ 132^\circ + \text{ measure angle } B &= 180^\circ \\ \text{measure angle } B &= 48^\circ \end{aligned}$$

Since angle R corresponds to angle B, the measure of angle R is 48°.

46. Since angle M corresponds to angle B, the measure of angle M is 46°. Since angle P corresponds to angle C, the measure of angle P is 78°.

Since angles A, B, and C are interior angles of a triangle, the sum of their measures is 180°.

$$\begin{aligned} \text{measure angle } A + \text{measure angle } B + \text{measure angle } C &= 180^\circ \\ \text{measure angle } A + 46^\circ + 78^\circ &= 180^\circ \\ \text{measure angle } A \ + 124^\circ &= 180^\circ \\ \text{measure angle } A &= 56^\circ \end{aligned}$$

Since angle N corresponds to angle A, the measure of angle N is 56°.

47. Since angle B corresponds to angle K, the measure of angle B is 106°.

Since angles A, B, and C are interior angles of a triangle, the sum of their measures is 180°.

$$\begin{aligned} \text{measure angle } A + \text{measure angle } B + \text{measure angle } C &= 180^\circ \\ \text{measure angle } A + 106^\circ + 30^\circ &= 180^\circ \\ \text{measure angle } A \ + 136^\circ &= 180^\circ \\ \text{measure angle } A &= 44^\circ \end{aligned}$$

Since angle M corresponds to angle A, the measure of angle M is 44°.

48. Since angle Y corresponds to angle V, the measure of angle Y is 28°.

Since angle T corresponds to angle X, the measure of angle T is 74°.

Since angles X, Y, and Z are interior angles of a triangle, the sum of their measures is 180°.

$$\begin{aligned} \text{measure angle } X + \text{measure angle } Y + \text{measure angle } Z &= 180^\circ \\ 74^\circ + 28^\circ + \text{measure angle } Z &= 180^\circ \\ 102^\circ + \text{measure angle } Z &= 180^\circ \\ \text{measure angle } Z &= 78^\circ \end{aligned}$$

Since angle W corresponds to angle Z, the measure of angle W is 78°.

49. Since angles X, Y, and Z are interior angles of a triangle, the sum of their measures is 180°.

$$\begin{aligned} \text{measure angle } X + \text{measure angle } Y + \text{measure angle } Z &= 180^\circ \\ \text{measure angle } X + 90^\circ + 38^\circ + &= 180^\circ \\ \text{measure angle } X + 128^\circ &= 180^\circ \\ \text{measure angle } X &= 52^\circ \end{aligned}$$

Since angle M corresponds to angle X, the measure of angle M is 52°.

50. Since angle T corresponds to angle P, the measure of angle T is 20°.
Since angle V corresponds to angle Q, the measure of angle V is 64°.
Since angles P, Q, and R are interior angles of a triangle, the sum of their measures is 180°.

$$\begin{aligned} \text{measure angle } P + \text{measure angle } Q + \text{measure angle } R &= 180^\circ \\ 20^\circ + 64^\circ + \text{measure angle } R &= 180^\circ \\ 84^\circ + \text{measure angle } R &= 180^\circ \\ \text{measure angle } R &= 96^\circ \end{aligned}$$

Since angle U corresponds to angle R, the measure of angle U is 96°.

In Exercises 51-56, corresponding sides of similar triangles are proportional. Other proportions are possible in solving these exercises.

51. $\frac{25}{10} = \frac{a}{8} \Rightarrow \frac{5}{2} = \frac{a}{8} \Rightarrow 5(8) = 2a \Rightarrow 40 = 2a \Rightarrow 20 = a$
$\frac{25}{10} = \frac{b}{6} \Rightarrow \frac{5}{2} = \frac{b}{6} \Rightarrow 5(6) = 2b \Rightarrow 30 = 2b \Rightarrow 15 = b$

52. $\frac{a}{10} = \frac{75}{25} \Rightarrow \frac{a}{10} = \frac{3}{1} \Rightarrow a = 30$ and $\frac{b}{20} = \frac{75}{25} \Rightarrow \frac{b}{20} = \frac{3}{1} \Rightarrow b = 60$

53. $\frac{6}{12} = \frac{a}{12} \Rightarrow 6(12) = 12a \Rightarrow 72 = 12a \Rightarrow 6 = a$
$\frac{b}{15} = \frac{1}{2} \Rightarrow 2b = (1)15 \Rightarrow 2b = 15 \Rightarrow b = \frac{15}{2} = 7\frac{1}{2}$

54. $\frac{a}{6} = \frac{3}{9} \Rightarrow \frac{a}{6} = \frac{1}{3} \Rightarrow 3a = 6(1) \Rightarrow 3a = 6 \Rightarrow a = 2$

55. $\frac{x}{4} = \frac{9}{6} \Rightarrow \frac{x}{4} = \frac{3}{2} \Rightarrow 2x = 4(3) \Rightarrow 2x = 12 \Rightarrow x = 6$

56. $\frac{m}{12} = \frac{21}{14} \Rightarrow \frac{m}{12} = \frac{3}{2} \Rightarrow 2m = 12(3) \Rightarrow 2m = 36 \Rightarrow m = 18$

57. Let x = the height of the tree.
The triangle formed by the tree and its shadow is similar to the triangle formed by the stick and its shadow.

$$\begin{aligned} \frac{x}{2} &= \frac{45}{3} \\ \frac{x}{2} &= \frac{15}{1} \\ x &= 30 \end{aligned}$$

The tree is 30 m high.

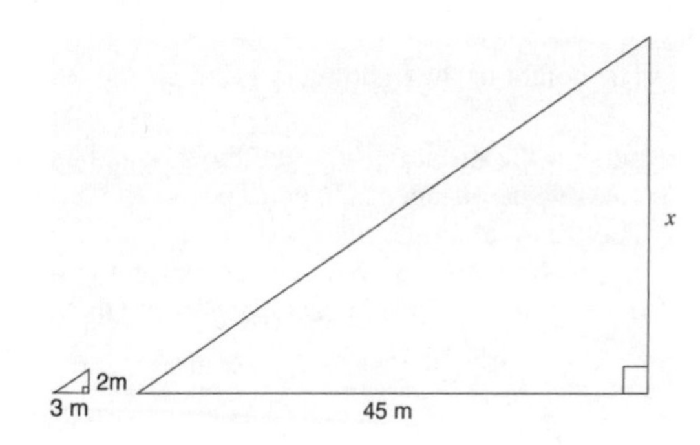

58. Let x = the height of the tower.
The triangle formed by the lookout tower and its shadow is similar to the triangle formed by the truck and its shadow.

$$\frac{x}{180}=\frac{9}{15}\Rightarrow\frac{x}{180}=\frac{3}{5}$$

$$5x=180(3)\Rightarrow 5x=540\Rightarrow x=108$$

The height of the tower is 108 ft.

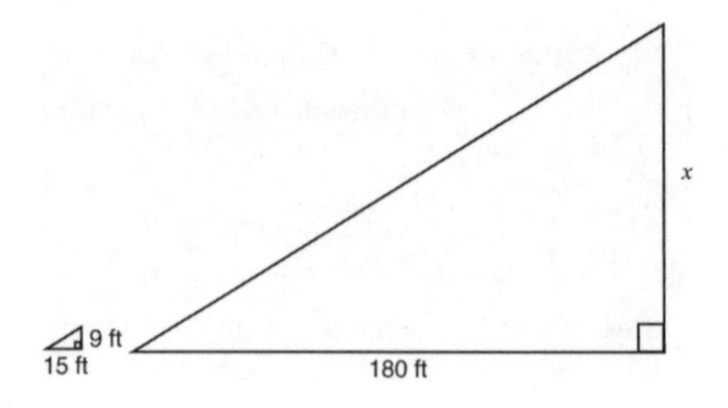

59. Let x = the middle side of the actual triangle (in meters);
y = the longest side of the actual triangle (in meters).
The triangles in the photograph and the piece of land are similar. The shortest side on the land corresponds to the shortest side on the photograph. We can set up the following proportions. Notice in each proportion, the units of the variable will be meters.

$$\frac{400\text{ m}}{4\text{ cm}}=\frac{x}{5\text{ cm}}\Rightarrow\frac{100}{1}=\frac{x}{5}\Rightarrow 500=x \text{ and } \frac{400\text{ m}}{4\text{ cm}}=\frac{y}{7\text{ cm}}\Rightarrow\frac{100}{1}=\frac{y}{7}\Rightarrow 700=y$$

The other two sides are 500 m and 700 m long.

60. Let x = the height of the lighthouse.

$$\frac{x}{1.75}=\frac{28}{3.5}\Rightarrow 3.5x=1.75(28)\Rightarrow 3.5x=49\Rightarrow x=14$$

The lighthouse is 14 m tall.

61. Let x = the height of the building.
The triangle formed by the house and its shadow is similar to the triangle formed by the building and its shadow.

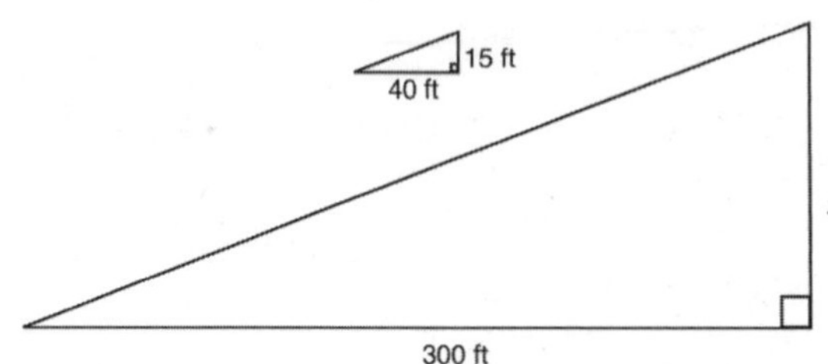

$$\frac{15}{40}=\frac{x}{300}\Rightarrow 15(300)=40x\Rightarrow 4500=40x\Rightarrow 112.5=x$$

The height of the building is 112.5 ft.

62. Let x = the distance between Phoenix and Tucson (in km);
y = the distance between Tucson and Yuma (in km).
The triangles in the map and on land are similar. The distance between Phoenix and Tucson on land corresponds to the distance between Phoenix and Tucson on the map. We can set up the following proportions. Notice in each proportion, the units of the variable will be km.

$$\frac{x}{8\text{ cm}}=\frac{230\text{ km}}{12\text{ cm}}\Rightarrow 12x=8(230)\Rightarrow 12x=1840\Rightarrow x\approx 153.3$$

$$\frac{y}{17\text{ cm}}=\frac{230\text{ km}}{12\text{ cm}}\Rightarrow 12y=17(230)\Rightarrow 12y=3910\Rightarrow y\approx 325.8$$

The distance from Phoenix to Tucson is about 153.3 km. The distance from Tucson to Yuma is about 325.8 km.

63. $\frac{x}{50}=\frac{100+120}{100}\Rightarrow\frac{x}{50}=\frac{220}{100}\Rightarrow\frac{x}{50}=\frac{11}{5}\Rightarrow 5x=550\Rightarrow x=110$

64. $\frac{y}{60}=\frac{40}{40+160}\Rightarrow\frac{y}{60}=\frac{40}{200}\Rightarrow\frac{y}{60}=\frac{1}{5}\Rightarrow 5y=60\Rightarrow y=12$

65. $\frac{c}{100}=\frac{10+90}{90}\Rightarrow\frac{c}{100}=\frac{100}{90}\Rightarrow\frac{c}{100}=\frac{10}{9}\Rightarrow 9c=100(10)\Rightarrow 9c=1000\Rightarrow c=\frac{1000}{9}\approx 111.1$

66. $\frac{m}{80}=\frac{75+5}{75}\Rightarrow\frac{m}{80}=\frac{80}{75}\Rightarrow\frac{m}{80}=\frac{16}{15}\Rightarrow 15m=80(16)\Rightarrow 15m=1280\Rightarrow m=\frac{1280}{15}=\frac{256}{3}\approx 85.3$

67. In these two similar quadrilaterals, the largest of the three shortest sides of the first quadrilateral (32 cm) corresponds to the smaller of the two longest sides of the second quadrilateral (48 cm). The following diagram is not drawn to scale.

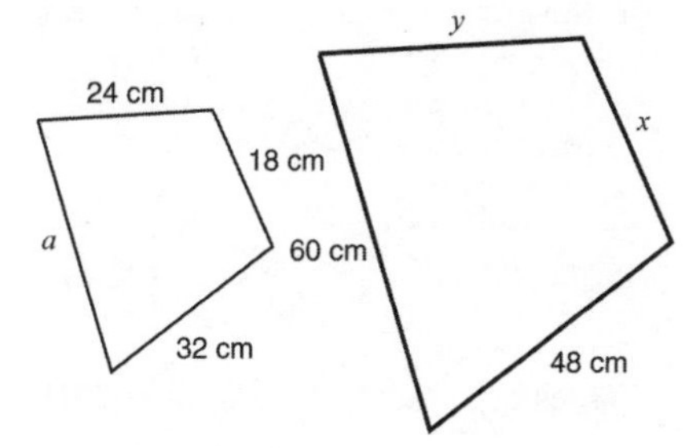

Let a = the length of the longest side of the first quadrilateral;

x = the length of the shortest side of the second quadrilateral;

y = the other unknown length.

Corresponding sides are in proportion.

$$\frac{a}{60}=\frac{32}{48}\Rightarrow\frac{a}{60}=\frac{2}{3}\Rightarrow 3a=60(2)\Rightarrow 3a=120\Rightarrow a=40\text{ cm}$$
$$\frac{24}{y}=\frac{32}{48}\Rightarrow\frac{24}{y}=\frac{2}{3}\Rightarrow 24(3)=2y\Rightarrow 2y=72\Rightarrow y=36\text{ cm}$$
$$\frac{18}{x}=\frac{32}{48}\Rightarrow\frac{18}{x}=\frac{2}{3}\Rightarrow 18(3)=2x\Rightarrow 2x=54\Rightarrow x=27\text{ cm}$$

68. Let x = length of the entire body carved into the mountain.

$$\frac{\frac{3}{4}}{60}=\frac{6\frac{1}{3}}{x}$$
$$\frac{\frac{3}{4}}{60}=\frac{\frac{19}{3}}{x}$$
$$\frac{3}{4}x=60\left(\frac{19}{3}\right)$$
$$\frac{3}{4}x=380\Rightarrow x=\frac{1520}{3}=506\frac{2}{3}$$

Abraham Lincoln's body would be $506\frac{2}{3}$ ft tall.

69. Since the two triangles are similar, the corresponding sides are in proportion.

$$\frac{x-5}{15}=\frac{5}{10+5}\Rightarrow\frac{x-5}{15}=\frac{5}{15}\Rightarrow\frac{x-5}{15}=\frac{1}{3}\Rightarrow 3(x-5)=15\Rightarrow 3x-15=15\Rightarrow 3x=30\Rightarrow x=10$$

$$\frac{x-2y}{(x-2y)+(x+y)}=\frac{5}{10+5}\Rightarrow\frac{x-2y}{2x-y}=\frac{5}{15}$$

$$\frac{x-2y}{2x-y}=\frac{1}{3}\Rightarrow 3(x-2y)=2x-y\Rightarrow 3x-6y=2x-y\Rightarrow x=5y$$

Substituting 10 for x we have, $x=5y\Rightarrow 10=5y\Rightarrow y=2$.

Therefore, $x = 10$ and $y = 2$.

70. Since the two triangles are similar, the corresponding angles have the same measure. Note: The angle that measures $(10y+8)°$ and the angle that measures $58°$ are also vertical angles.

$$10y+8=58\Rightarrow 10y=50\Rightarrow y=5$$

Since the two triangles are similar, the corresponding sides are in proportion.

$$\frac{x-y}{x+y}=\frac{6}{18}\Rightarrow\frac{x-y}{x+y}=\frac{1}{3}\Rightarrow 3(x-y)=x+y\Rightarrow 3x-3y=x+y\Rightarrow 2x=4y\Rightarrow x=2y$$

Substituting 5 for y we have, $x=2y\Rightarrow x=2(5)=10$

Therefore, $x = 10$ and $y = 5$

71. **(a)** Thumb covers about 2 arc degrees or about 120 arc minutes. This would be $\frac{31}{120}$ or approximately $\frac{1}{4}$ of the thumb would cover the moon

(b) 20 + 10 = 30 arc degrees

72. **(a)** Let D_s be the Mars-Sun Distance, and d_s the diameter of the sun, D_m the Mars-Phobos distance, and d_m the diameter of the Phobos. Then, by similar triangles

$$\frac{D_s}{D_m}=\frac{d_s}{d_m}\Rightarrow D_m=\frac{D_s d_m}{d_s}=\frac{142{,}000{,}000\times 17.4}{865{,}000}\approx 2856 \text{ mi}$$

(b) No. Phobos does not come close enough to the surface of Mars.

73. **(a)** Let D_s be the Jupiter-Sun distance, and d_s the diameter of the sun, D_m the Jupiter-Ganymede distance, and d_m the diameter of the Ganymede. Then, by similar triangles

$$\frac{D_s}{D_m}=\frac{d_s}{d_m}\Rightarrow D_m=\frac{D_s d_m}{d_s}\Rightarrow D_m=\frac{484{,}000{,}000\times 3270}{865{,}000}\approx 1{,}830{,}000 \text{ mi}$$

(b) Yes; Ganymede is usually less than half this distance from Jupiter.

Section 1.3: Trigonometric Functions

1.

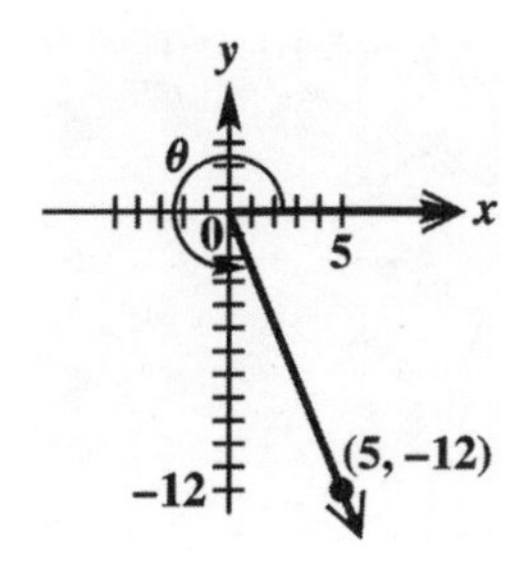

2.

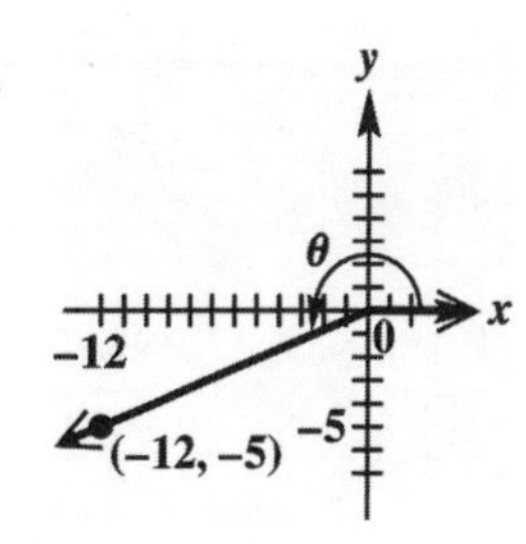

3.

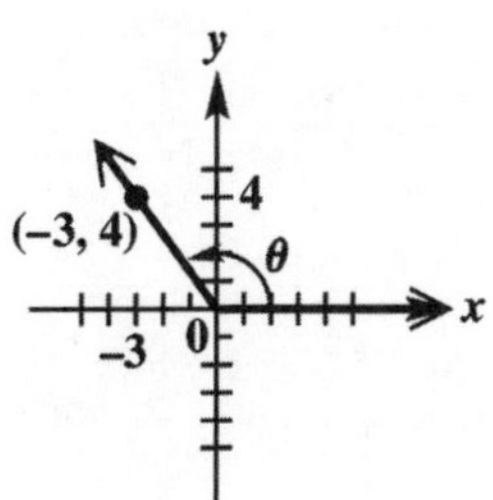

4.

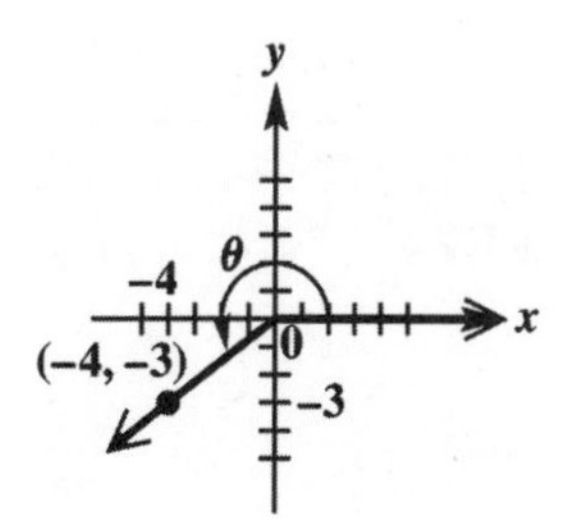

5. $(-3,4)$

$x=-3$, $y=4$ and $r=\sqrt{x^2+y^2}=\sqrt{(-3)^2+4^2}=\sqrt{9+16}=\sqrt{25}=5$

$\sin\theta=\frac{y}{r}=\frac{4}{5}$

$\cos\theta=\frac{x}{r}=\frac{-3}{5}=-\frac{3}{5}$

$\tan\theta=\frac{y}{x}=\frac{4}{-3}=-\frac{4}{3}$

$\cot\theta=\frac{x}{y}=\frac{-3}{4}=-\frac{3}{4}$

$\sec\theta=\frac{r}{x}=\frac{5}{-3}=-\frac{5}{3}$

$\csc\theta=\frac{r}{y}=\frac{5}{4}$

6. $(-4,-3)$

$x=-4$, $y=-3$, and $r=\sqrt{x^2+y^2}=\sqrt{(-4)^2+(-3)^2}=\sqrt{16+9}=\sqrt{25}=5$

$\sin\theta=\frac{y}{r}=\frac{-3}{5}=-\frac{3}{5}$

$\cos\theta=\frac{x}{r}=\frac{-4}{5}=-\frac{4}{5}$

$\tan\theta=\frac{y}{x}=\frac{-3}{-4}=\frac{3}{4}$

$\cot\theta=\frac{x}{y}=\frac{-4}{-3}=\frac{4}{3}$

$\sec\theta=\frac{r}{x}=\frac{5}{-4}=-\frac{5}{4}$

$\csc\theta=\frac{r}{y}=\frac{5}{-3}=-\frac{5}{3}$

7. $(0,2)$

$x=0$, $y=2$, and $r=\sqrt{x^2+y^2}=\sqrt{0^2+2^2}=\sqrt{0+4}=\sqrt{4}=2$

$\sin\theta=\frac{y}{r}=\frac{2}{2}=1$

$\cos\theta=\frac{x}{r}=\frac{0}{2}=0$

$\tan\theta=\frac{y}{x}=\frac{2}{0}$ undefined

$\cot\theta=\frac{x}{y}=\frac{0}{2}=0$

$\sec\theta=\frac{r}{x}=\frac{2}{0}$ undefined

$\csc\theta=\frac{r}{y}=\frac{2}{2}=1$

8. $(-4,0)$

$x=-4$, $y=2$, and $r=\sqrt{x^2+y^2}=\sqrt{(-4)^2+0^2}=\sqrt{16+0}=\sqrt{16}=4$

$\sin\theta=\frac{y}{r}=\frac{0}{4}=0$

$\cos\theta=\frac{x}{r}=\frac{-4}{4}=-1$

$\tan\theta=\frac{y}{x}=\frac{0}{-4}=0$

$\cot\theta=\frac{x}{y}=\frac{-4}{0}$ undefined

$\sec\theta=\frac{r}{x}=\frac{4}{-4}=-1$

$\csc\theta=\frac{r}{y}=\frac{4}{0}$ undefined

9. $\left(1,\sqrt{3}\right)$

$x=1,\ y=\sqrt{3}$, and $r=\sqrt{x^2+y^2}=\sqrt{1^2+\left(\sqrt{3}\right)^2}=\sqrt{1+3}=\sqrt{4}=2$

$\sin\theta=\frac{y}{r}=\frac{\sqrt{3}}{2}$

$\cos\theta=\frac{x}{r}=\frac{1}{2}$

$\tan\theta=\frac{y}{x}=\frac{\sqrt{3}}{1}=\sqrt{3}$

$\cot\theta=\frac{x}{y}=\frac{1}{\sqrt{3}}=\frac{1}{\sqrt{3}}\cdot\frac{\sqrt{3}}{\sqrt{3}}=\frac{\sqrt{3}}{3}$

$\sec\theta=\frac{r}{x}=\frac{2}{1}=2$

$\csc\theta=\frac{r}{y}=\frac{2}{\sqrt{3}}=\frac{2}{\sqrt{3}}\cdot\frac{\sqrt{3}}{\sqrt{3}}=\frac{2\sqrt{3}}{3}$

10. $\left(-2\sqrt{3},-2\right)$

$x=-2\sqrt{3},\ y=-2$, and $r=\sqrt{x^2+y^2}=\sqrt{\left(-2\sqrt{3}\right)^2+(-2)^2}=\sqrt{12+4}=\sqrt{16}=4$

$\sin\theta=\frac{y}{r}=\frac{-2}{4}=-\frac{1}{2}$

$\cos\theta=\frac{x}{r}=\frac{-2\sqrt{3}}{4}=-\frac{\sqrt{3}}{2}$

$\tan\theta=\frac{y}{x}=\frac{-2}{-2\sqrt{3}}=\frac{1}{\sqrt{3}}=\frac{\sqrt{3}}{3}$

$\cot\theta=\frac{x}{y}=\frac{-2\sqrt{3}}{-2}=\sqrt{3}$

$\sec\theta=\frac{r}{x}=\frac{4}{-2\sqrt{3}}=-\frac{2}{\sqrt{3}}=-\frac{2\sqrt{3}}{3}$

$\csc\theta=\frac{r}{y}=\frac{4}{-2}=-2$

11. $(-2,0)$

$x=-2,\ y=0$, and $r=\sqrt{x^2+y^2}=\sqrt{(-2)^2+0^2}=\sqrt{4+0}=\sqrt{4}=2$

$\sin\theta=\frac{y}{r}=\frac{0}{2}=0$

$\cos\theta=\frac{x}{r}=\frac{-2}{2}=-1$

$\tan\theta=\frac{y}{x}=\frac{0}{-2}=0$

$\cot\theta=\frac{x}{y}=\frac{-2}{0}$ undefined

$\sec\theta=\frac{r}{x}=\frac{2}{-2}=-1$

$\csc\theta=\frac{r}{y}=\frac{2}{0}$ undefined

12. (3,–4)

$x=3,\ \ y=-4,$ and $r=\sqrt{x^2+y^2}=\sqrt{3^2+(-4)^2}=\sqrt{9+16}=\sqrt{25}=5$

$\sin\theta=\frac{y}{r}=\frac{-4}{5}=-\frac{4}{5}$

$\cos\theta=\frac{x}{r}=\frac{3}{5}$

$\tan\theta=\frac{y}{x}=\frac{-4}{3}=-\frac{4}{3}$

$\cot\theta=\frac{x}{y}=\frac{3}{-4}=-\frac{3}{4}$

$\sec\theta=\frac{r}{x}=\frac{5}{3}$

$\csc\theta=\frac{r}{y}=\frac{5}{-4}=-\frac{5}{4}$

13. Answers will vary.
For any nonquadrantal angle θ, a point on the terminal side of θ will be of the form (x,y) where $x,y\neq 0$. Now $\sin\theta=\frac{y}{r}$ and $\csc\theta=\frac{r}{y}$ both exist and are simply reciprocals of each other, and hence will have the same sign.

14. Answers will vary.
For any point on the terminal side of a nonquadrantal θ, a right triangle can be formed by connecting the point to the origin and connecting the point to the x-axis. The value of r can be interpreted as the length of the hypotenuse of this right triangle.

15. Since the cot θ is undefined, and $\cot\theta=\frac{x}{y}$, where (x,y) is a point on the terminal side of θ, $y=0$ and x can be any nonzero number. Therefore, tan θ = 0.

16. If a point on the terminal side of θ lies in quadrant III, it will be of the form (x,y) where $x<0$ and $y<0$. Since $r>0$, the signs of the six trigonometric function values of θ will be as follows.

$\sin\theta=\frac{y}{r}=\frac{negative}{positive}=negative$

$\cos\theta=\frac{x}{r}=\frac{negative}{positive}=negative$

$\tan\theta=\frac{y}{x}=\frac{negative}{negative}=positive$

$\cot\theta=\frac{x}{y}=\frac{negative}{negative}=positive$

$\sec\theta=\frac{r}{x}=\frac{positive}{negative}=negative$

$\csc\theta=\frac{r}{y}=\frac{positive}{negative}=negative$

In Exercises 17–24, $r=\sqrt{x^2+y^2}$, which is positive.

17. In quadrant II, x is negative, so $\frac{x}{r}$ is negative.

18. In quadrant III, y is negative, so $\frac{y}{r}$ is negative.

19. In quadrant IV, x is positive and y is negative, so $\frac{y}{x}$ is negative.

20. In quadrant IV, x is positive and y is negative, so $\frac{x}{y}$ is negative.

21. In quadrant II, y is positive so, $\frac{y}{r}$ is positive.

22. In quadrant III, x is negative so, $\frac{x}{r}$ is negative.

23. In quadrant IV, x is positive so, $\frac{x}{r}$ is positive.

24. In quadrant IV, y is negative so, $\frac{y}{r}$ is positive.

25. Since $x \geq 0$, the graph of the line $2x + y = 0$ is shown to the right of the y-axis. A point on this line is $(1, -2)$ since $2(1) + (-2) = 0$. The corresponding value of r is $r = \sqrt{1^2 + (-2)^2} = \sqrt{1+4} = \sqrt{5}$.

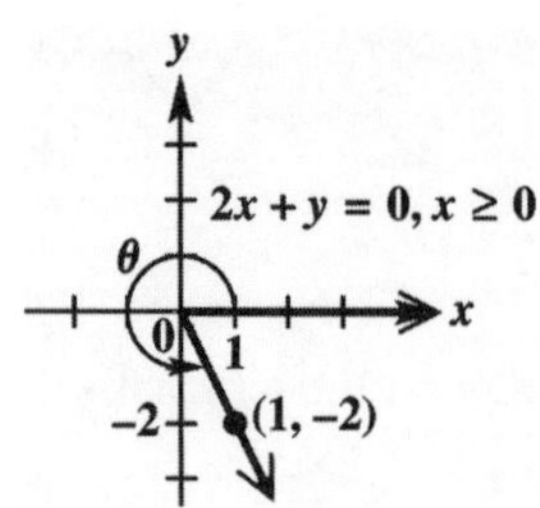

$\sin\theta = \frac{y}{r} = \frac{-2}{\sqrt{5}} = -\frac{2}{\sqrt{5}} \cdot \frac{\sqrt{5}}{\sqrt{5}} = -\frac{2\sqrt{5}}{5}$

$\cos\theta = \frac{x}{r} = \frac{1}{\sqrt{5}} = \frac{1}{\sqrt{5}} \cdot \frac{\sqrt{5}}{\sqrt{5}} = \frac{\sqrt{5}}{5}$

$\tan\theta = \frac{y}{x} = \frac{-2}{1} = -2$

$\cot\theta = \frac{x}{y} = \frac{1}{-2} = -\frac{1}{2}$

$\sec\theta = \frac{r}{x} = \frac{\sqrt{5}}{1} = \sqrt{5}$

$\csc\theta = \frac{r}{y} = \frac{\sqrt{5}}{-2} = -\frac{\sqrt{5}}{2}$

26. Since $x \geq 0$, the graph of the line $3x + 5y = 0$ is shown to the right of the y-axis. A point on this graph is $(5, -3)$ since $3(5) + 5(-3) = 0$. The corresponding value of r is $r = \sqrt{5^2 + (-3)^2} = \sqrt{25+9} = \sqrt{34}$.

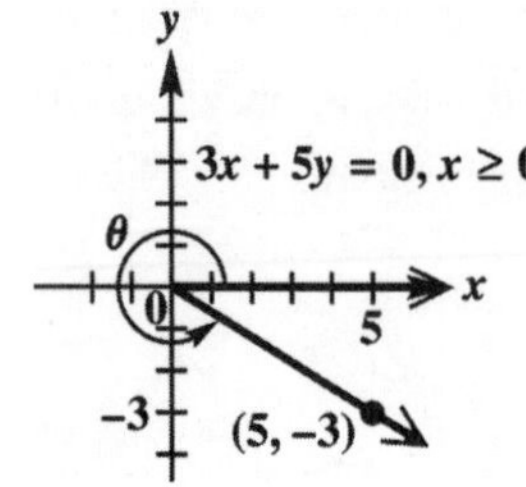

$\sin\theta = \frac{y}{r} = \frac{-3}{\sqrt{34}} = -\frac{3}{\sqrt{34}} \cdot \frac{\sqrt{34}}{\sqrt{34}} = -\frac{3\sqrt{34}}{34}$

$\cos\theta = \frac{x}{r} = \frac{5}{\sqrt{34}} = \frac{5}{\sqrt{34}} \cdot \frac{\sqrt{34}}{\sqrt{34}} = \frac{5\sqrt{34}}{34}$

$\tan\theta = \frac{y}{x} = \frac{-3}{5} = -\frac{3}{5}$

$\cot\theta = \frac{x}{y} = \frac{5}{-3} = -\frac{5}{3}$

$\sec\theta = \frac{r}{x} = \frac{\sqrt{34}}{5}$

$\csc\theta = \frac{r}{y} = \frac{\sqrt{34}}{-3} = -\frac{\sqrt{34}}{3}$

27. Since $x \le 0$, the graph of the line $-6x - y = 0$ is shown to the left of the y-axis. A point on this graph is $(-1,6)$ since $-6(-1)-6=0$. The corresponding value of r is $r=\sqrt{(-1)^2+6^2}=\sqrt{1+36}=\sqrt{37}$.

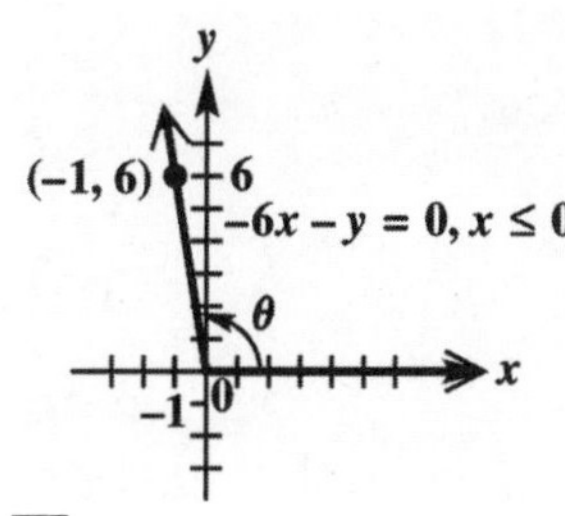

$$\sin\theta=\frac{y}{r}=\frac{6}{\sqrt{37}}=\frac{6}{\sqrt{37}}\cdot\frac{\sqrt{37}}{\sqrt{37}}=\frac{6\sqrt{37}}{37}$$
$$\cos\theta=\frac{x}{r}=\frac{-1}{\sqrt{37}}=-\frac{1}{\sqrt{37}}\cdot\frac{\sqrt{37}}{\sqrt{37}}=-\frac{\sqrt{37}}{37}$$
$$\tan\theta=\frac{y}{x}=\frac{6}{-1}=-6$$
$$\cot\theta=\frac{x}{y}=\frac{-1}{6}=-\frac{1}{6}$$
$$\sec\theta=\frac{r}{x}=\frac{\sqrt{37}}{-1}=-\sqrt{37}$$
$$\csc\theta=\frac{r}{y}=\frac{\sqrt{37}}{6}$$

28. Since $x \le 0$, the graph of the line $-5x-3y=0$ is shown to the left of the y-axis. A point on this line is $(-3,5)$ since $-5(-3)-3(5)=0$. The corresponding value of r is $r=\sqrt{(-3)^2+5^2}=\sqrt{9+25}=\sqrt{34}$.

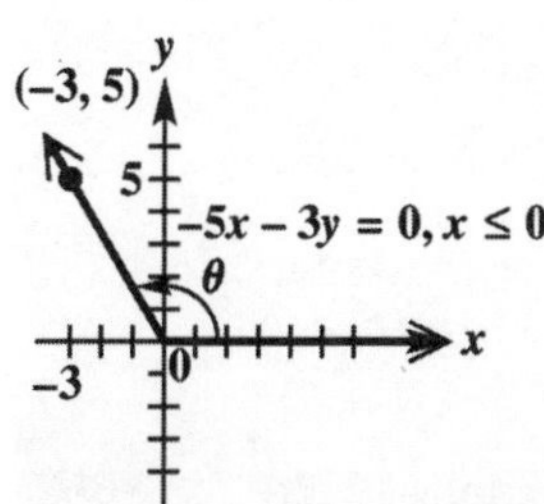

$$\sin\theta=\frac{y}{r}=\frac{5}{\sqrt{34}}=\frac{5}{\sqrt{34}}\cdot\frac{\sqrt{34}}{\sqrt{34}}=\frac{5\sqrt{34}}{34}$$
$$\cos\theta=\frac{x}{r}=\frac{-3}{\sqrt{34}}=-\frac{3}{\sqrt{34}}\cdot\frac{\sqrt{34}}{\sqrt{34}}=-\frac{3\sqrt{34}}{34}$$
$$\tan\theta=\frac{y}{x}=\frac{5}{-3}=-\frac{5}{3}$$
$$\cot\theta=\frac{x}{y}=\frac{-3}{5}=-\frac{3}{5}$$
$$\sec\theta=\frac{r}{x}=\frac{\sqrt{34}}{-3}=-\frac{\sqrt{34}}{3}$$
$$\csc\theta=\frac{r}{y}=\frac{\sqrt{34}}{5}$$

29. Since $x \le 0$, the graph of the line $-4x+7y=0$ is shown to the left of the y-axis. A point on this line is $(-7,-4)$ since $-4(-7)+7(-4)=0$. The corresponding value of r is
$r=\sqrt{(-7)^2+(-4)^2}=\sqrt{49+16}=\sqrt{65}$.

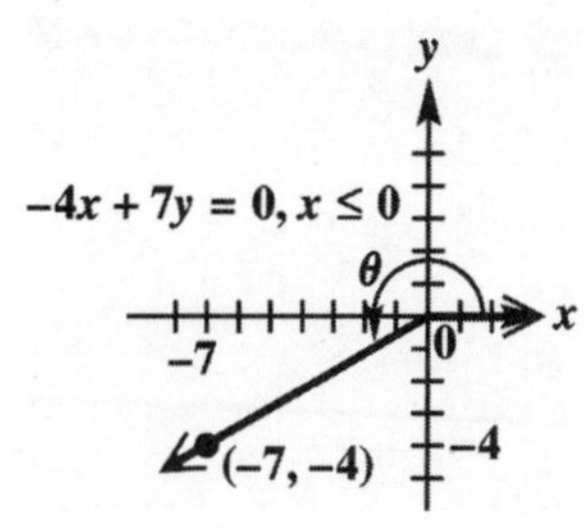

Continued on next page

29. (continued)

$\sin\theta = \frac{y}{r} = \frac{5}{\sqrt{34}} = \frac{5}{\sqrt{34}}\cdot\frac{\sqrt{34}}{\sqrt{34}} = -\frac{4\sqrt{65}}{65}$

$\cos\theta = \frac{x}{r} = \frac{-3}{\sqrt{34}} = -\frac{3}{\sqrt{34}}\cdot\frac{\sqrt{34}}{\sqrt{34}} = -\frac{7\sqrt{65}}{65}$

$\tan\theta = \frac{y}{x} = \frac{-4}{-7} = \frac{4}{7}$

$\cot\theta = \frac{x}{y} = \frac{-7}{-4} = \frac{7}{4}$

$\sec\theta = \frac{r}{x} = \frac{\sqrt{65}}{-7} = -\frac{\sqrt{65}}{7}$

$\csc\theta = \frac{r}{y} = \frac{\sqrt{65}}{-4} = -\frac{\sqrt{65}}{4}$

30. Since $x \ge 0$, the graph of the line $6x - 5y = 0$ is shown to the right of the *y*-axis. A point on this line is $(5,6)$ since $6(5) - 5(6) = 0$. The corresponding value of *r* is $r = \sqrt{5^2 + 6^2} = \sqrt{25+36} = \sqrt{61}$.

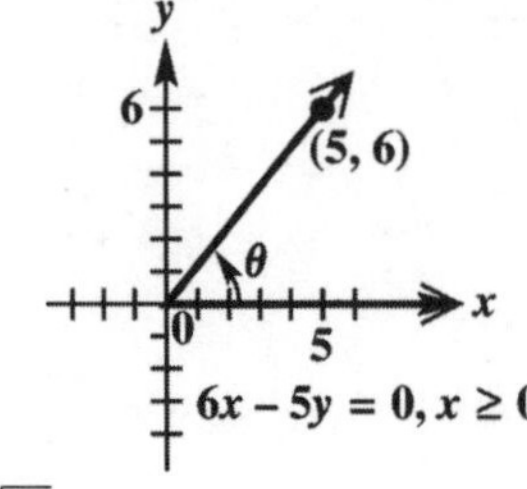

$\sin\theta = \frac{y}{r} = \frac{6}{\sqrt{61}} = \frac{6}{\sqrt{61}}\cdot\frac{\sqrt{61}}{\sqrt{61}} = \frac{6\sqrt{61}}{61}$

$\cos\theta = \frac{x}{r} = \frac{5}{\sqrt{61}} = \frac{5}{\sqrt{61}}\cdot\frac{\sqrt{61}}{\sqrt{61}} = \frac{5\sqrt{61}}{61}$

$\tan\theta = \frac{y}{x} = \frac{6}{5}$

$\cot\theta = \frac{x}{y} = \frac{5}{6}$

$\sec\theta = \frac{r}{x} = \frac{\sqrt{61}}{5}$

$\csc\theta = \frac{r}{y} = \frac{\sqrt{61}}{6}$

31. The quadrantal angle $\theta = 450°$ is coterminal with $450° - 360° = 90°$. A point on the terminal side of θ is of the form $(0, y)$, where *y* is a positive real number. The corresponding value of *r* is $r = \sqrt{0^2 + y^2} = y$. The six trigonometric function values of θ are as follows.

$\sin\theta = \frac{y}{r} = \frac{y}{y} = 1$

$\cos\theta = \frac{x}{r} = \frac{0}{y} = 0$

$\tan\theta = \frac{y}{x} = \frac{y}{0}$ undefined

$\cot\theta = \frac{x}{y} = \frac{0}{y} = 0$

$\sec\theta = \frac{r}{x} = \frac{y}{0}$ undefined

$\csc\theta = \frac{r}{y} = \frac{y}{y} = 1$

32. To find the point on the line $x+2y=0$, where $x \geq 0$, we can first use the fact that a circle centered at the origin with radius 1 has the equation $x^2+y^2=1$. (1)

Solving $x+2y=0$ for x we have, $x=-2y$. In order for $x \geq 0$, we must have $y \leq 0$. Substituting $-2y$ for x in equation 1 and solving for y, we have the following.

$$x^2+y^2=1 \Rightarrow (-2y)^2+y^2=1 \Rightarrow 4y^2+y^2=1 \Rightarrow 5y^2=1 \Rightarrow y^2=\frac{1}{5}$$

$$y=-\sqrt{\frac{1}{5}} \text{ since } y \leq 0$$

$$y=-\frac{1}{\sqrt{5}} \Rightarrow y=-\frac{1}{\sqrt{5}}\cdot\frac{\sqrt{5}}{\sqrt{5}} \Rightarrow y=-\frac{\sqrt{5}}{5}$$

Substituting $y=-\frac{\sqrt{5}}{5}$ into $x=-2y$ we have, $x=-2\left(-\frac{\sqrt{5}}{5}\right)=\frac{2\sqrt{5}}{5}$.

A point on the terminal side of θ which intersects the circle of radius 1 is $\left(\frac{2\sqrt{5}}{5},-\frac{\sqrt{5}}{5}\right)$, where y is a negative real number. The corresponding value of r is 1. The six trigonometric function values of θ are as follows.

$$\sin\theta=\frac{y}{r}=\frac{-\frac{\sqrt{5}}{5}}{1}=-\frac{\sqrt{5}}{5}$$

$$\cos\theta=\frac{x}{r}=\frac{\frac{2\sqrt{5}}{5}}{1}=\frac{2\sqrt{5}}{5}$$

$$\tan\theta=\frac{y}{x}=\frac{\frac{\sqrt{5}}{5}}{-\frac{2\sqrt{5}}{5}}=\frac{\sqrt{5}}{5}\cdot\left(-\frac{5}{2\sqrt{5}}\right)=-\frac{1}{2}$$

$$\cot\theta=\frac{x}{y}=\frac{-\frac{2\sqrt{5}}{5}}{\frac{\sqrt{5}}{5}}=\left(-\frac{2\sqrt{5}}{5}\right)\cdot\left(\frac{5}{\sqrt{5}}\right)=-2$$

$$\sec\theta=\frac{r}{x}=\frac{1}{\frac{2\sqrt{5}}{5}}=1\cdot\frac{5}{2\sqrt{5}}=\frac{5}{2\sqrt{5}}\cdot\frac{\sqrt{5}}{\sqrt{5}}=\frac{\sqrt{5}}{2}$$

$$\csc\theta=\frac{r}{y}=\frac{1}{-\frac{\sqrt{5}}{5}}=1\cdot\left(-\frac{5}{\sqrt{5}}\right)=-\frac{5}{\sqrt{5}}\cdot\frac{\sqrt{5}}{\sqrt{5}}=-\sqrt{5}$$

The six trigonometric function values are the same as those obtained in Example 3.

For Exercises 33 – 42, all angles are quadrantal angles. For $0°$ we will choose the point $(1,0)$, so $x=1$, $y=0$, and $r=1$. For $90°$ we will choose the point $(0,1)$, so $x=0$, $y=1$, and $r=1$. For $180°$ we will choose the point $(-1,0)$, so $x=-1$, $y=0$, and $r=1$. Finally, for $270°$ we will choose the point $(0,-1)$, so $x=0$, $y=-1$, and $r=1$.

33. $\cos 90°+3\sin 270°$

$$\cos 90°=\frac{x}{r}=\frac{0}{1}=0$$

$$\sin 270°=\frac{y}{r}=\frac{-1}{1}=-1$$

$$\cos 90°+3\sin 270°=0+3(-1)=-3$$

34. $\tan 0°-6\sin 90°$

$$\tan 0°=\frac{y}{x}=\frac{0}{1}=0$$

$$\sin 90°=\frac{y}{r}=\frac{1}{1}=1$$

$$\tan 0°-6\sin 90°=0-6(1)=-6$$

35. $3\sec 180° - 5\tan 360°$

$\sec 180° = \frac{r}{x} = \frac{1}{-1} = -1$ and $\tan 360° = \tan 0° = \frac{y}{x} = \frac{0}{1} = 0$

$3\sec 180° - 5\tan 360° = 3(-1) - 5(0) = -3 - 0 = -3$

36. $4\csc 270° + 3\cos 180°$

$\csc 270° = \frac{r}{y} = \frac{1}{-1} = -1$ and $\cos 180° = \frac{x}{r} = \frac{-1}{1} = -1$

$4\csc 270° + 3\cos 180° = 4(-1) + 3(-1) = -4 + (-3) = -7$

37. $\tan 360° + 4\sin 180° + 5\cos^2 180°$

$\tan 360° = \tan 0° = \frac{y}{x} = \frac{0}{1} = 0$, $\sin 180° = \frac{y}{r} = \frac{0}{1} = 0$, and $\cos 180° = \frac{x}{r} = \frac{-1}{1} = -1$

$\tan 360° + 4\sin 180° + 5\cos^2 180° = 0 + 4(0) + 5(-1)^2 = 0 + 0 + 5(1) = 5$

38. $2\sec 0° + 4\cot^2 90° + \cos 360°$

$\sec 0° = \frac{r}{x} = \frac{1}{1} = 1$, $\cot 90° = \frac{x}{y} = \frac{0}{1} = 0$, and $\cos 360° = \cos 0° = \frac{x}{r} = \frac{1}{1} = 1$

$2\sec 0° + 4\cot^2 90° + \cos 360° = 2(1) + 4(0)^2 + 1 = 2 + 4(0) + 1 = 2 + 0 + 1 = 3$

39. $\sin^2 180° + \cos^2 180°$

$\sin 180° = \frac{y}{r} = \frac{0}{1} = 0$ and $\cos 180° = \frac{x}{r} = \frac{-1}{1} = -1$

$\sin^2 180° + \cos^2 180° = 0^2 + (-1)^2 = 0 + 1 = 1$

40. $\sin^2 360° + \cos^2 360°$

$\sin 360° = \sin 0° = \frac{y}{r} = \frac{0}{1} = 0$ and $\cos 360° = \cos 0° = \frac{x}{r} = \frac{1}{1} = 1$

$\sin^2 360° + \cos^2 360° = 0^2 + 1^2 = 0 + 1 = 1$

41. $\sec^2 180° - 3\sin^2 360° + 2\cos 180°$

$\sec 180° = \frac{r}{x} = \frac{1}{-1} = -1$, $\sin 360° = \sin 0° = \frac{y}{r} = \frac{0}{1} = 0$, and $\cos 180° = \frac{x}{r} = \frac{-1}{1} = -1$

$\sec^2 180° - 3\sin^2 360° + 2\cos 180° = (-1)^2 - 3(0) + 2(-1) = 1 - 0 - 2 = -1$

42. $5\sin^2 90° + 2\cos^2 270° - 7\tan^2 360°$

$\sin 90° = \frac{y}{r} = \frac{1}{1} = 1$, $\cos 270° = \frac{x}{r} = \frac{0}{1} = 0$, and $\tan 360° = \tan 0° = \frac{y}{x} = \frac{0}{1} = 0$

$5\sin^2 90° + 2\cos^2 270° - 7\tan^2 360° = 5(1)^2 + 2(0)^2 - 7(0)^2 = 5(1) + 2(0) - 7(0) = 5 + 0 - 0 = 5$

43. $\cos\left[(2n+1)\cdot 90°\right]$

This angle is a quadrantal angle whose terminal side lies on either the positive part of the y-axis or the negative part of the y-axis. Any point on these terminal sides would have the form $(0,k)$, where k is any real number, $k \neq 0$.

$$\cos\left[(2n+1)\cdot 90°\right]=\frac{x}{r}=\frac{0}{\sqrt{0^2+k^2}}=\frac{0}{\sqrt{k^2}}=\frac{0}{|k|}=0$$

44. $\sin\left[n\cdot 180°\right]$

This angle is a quadrantal angle whose terminal side lies on either the positive part of the x-axis or the negative part of the x-axis. Any point on these terminal sides would have the form $(k,0)$, where k is any real number, $k \neq 0$.

$$\sin\left[n\cdot 180°\right]=\frac{y}{r}=\frac{0}{\sqrt{0^2+k^2}}=\frac{0}{\sqrt{k^2}}=\frac{0}{|k|}=0$$

45. $\tan\left[n\cdot 180°\right]$

The angle is a quadrantal angle whose terminal side lies on either the positive part of the x-axis or the negative part of the x-axis. Any point on these terminal sides would have the form $(k, 0)$, where k is any real number, $k \neq 0$.

$$\tan\left[n\cdot 180°\right]=\frac{y}{x}=\frac{0}{k}=0$$

46. $\tan\left[(2n+1)\cdot 90°\right]$

This angle is a quadrantal angle whose terminal side lies on either the positive part of the y-axis or the negative part of the y-axis. Any point on these terminal sides would have the form $(0,k)$, where k is any real number, $k \neq 0$.

$$\tan\left[(2n+1)\cdot 90°\right]=\frac{y}{x}=\frac{k}{0} \quad \text{undefined}$$

47. Using a calculator, sin 15° = .258819045 and cos 75° = .258819045. We can conjecture that the sines and cosines of complementary angles are equal. Trying another pair of complementary angles we obtain $\sin 30° = \cos 60° = .5$. Therefore, our conjecture appears to be true.

48. Using a calculator, tan 25° = .466307658 and cot 65° = .466307658. We can conjecture that the tangent and cotangent of complementary angles are equal. Trying another pair of complementary angles we obtain $\tan 45° = .\cot 45° = 1$. Therefore, our conjecture appears to be true.

49. Using a calculator, sin 10° = .173648178 and sin(–10°) = – .173648178. We can conjecture that the sines of an angle and its negative are opposites of each other. Using a circle, an angle θ having the point (x, y) on its terminal side has a corresponding angle $-\theta$ with a point $(x, -y)$ on its terminal side. From the definition of sine, $\sin(-\theta)=\frac{-y}{r}=-\frac{y}{r}$ and $\sin\theta=\frac{y}{r}$. The sines are negatives of each other.

50. Using a calculator, cos 20° = .93969262 and cos (–20°) = .93969262. We can conjecture that the cosines of an angle and its negative are equal. Using a circle, and angle θ having the point (x, y) on its terminal side has a corresponding angle $-\theta$ with point $(x, -y)$ on its terminal side. From the definition of cosine, $\cos(-\theta)=\frac{x}{r}$ and $\cos\theta=\frac{x}{r}$. The cosines are equal.

51. When $\sin\theta = \cos\theta$, then $\frac{y}{r} = \frac{x}{r}$. As a result, the two values of θ must lie along the line $y = x$, which occurs when θ =45° and θ =225°.

52. When $\tan\theta = 1$, we have $\frac{y}{x} = 1$ or $y = x$. The two values of θ must lie along the line $y = x$, which occurs, so θ =45° and θ =225°.

53. Since $\cos\theta = .8$ occurs when x = .8 and r = 1 and $\theta \approx 36.87°$, $\cos\theta = -.8$ occurs when $x = -.8$ and r = 1. As a result, $\theta = 180° - 36.87° = 143.13°$ and $\theta = 180° + 36.87° = 216.87°$.

54. Since $\sin\theta = .8$ occurs when y = .8 and r = 1 and $\theta \approx 53.13°$, $\sin\theta = -.8$ occurs when $y = -.8$ and r = 1. As a result, $\theta = 180° - 53.13° = 306.87°$ and $\theta = 180° + 53.13° = 233.13°$.

In Exercises 55 – 60, make sure your calculator is in the modes indicated in the instructions.

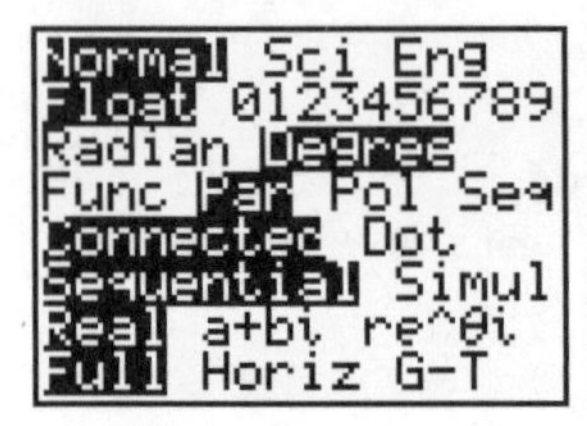

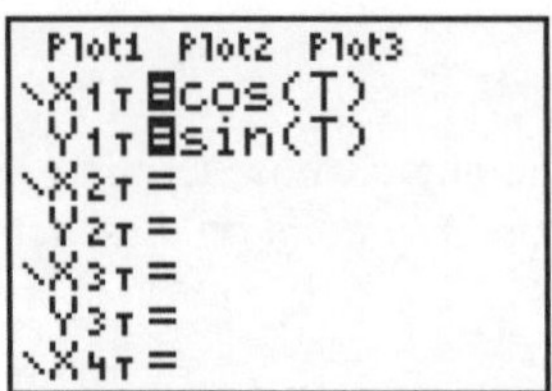

55. Use the TRACE feature to move around the circle in quadrant I to find $T = 20$. We see that cos 20° is about .940, and sin 20° is about .342.

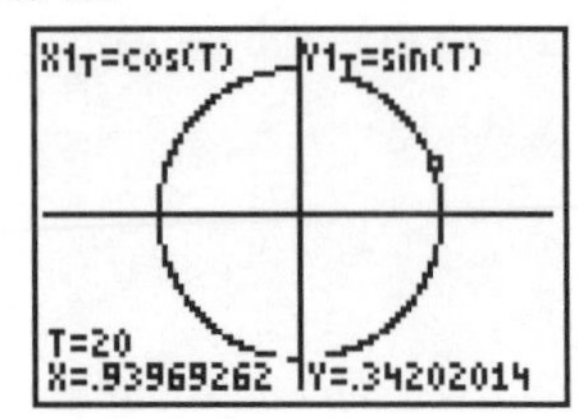

56. Use the TRACE feature to move around the circle in quadrant I. We see that cos 40° ≈ .766 so, T = 40°.

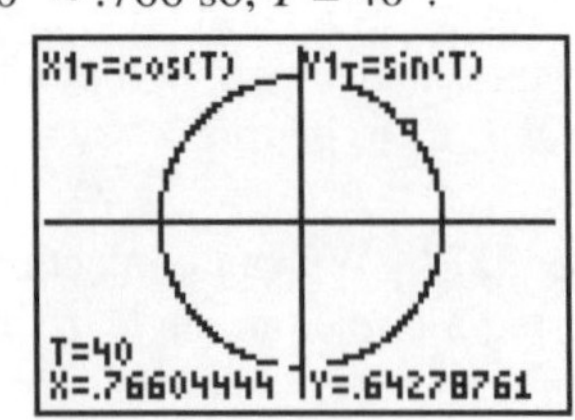

57. Use the TRACE feature to move around the circle in quadrant I. We see that sin 35° ≈ .574, so T =35°.

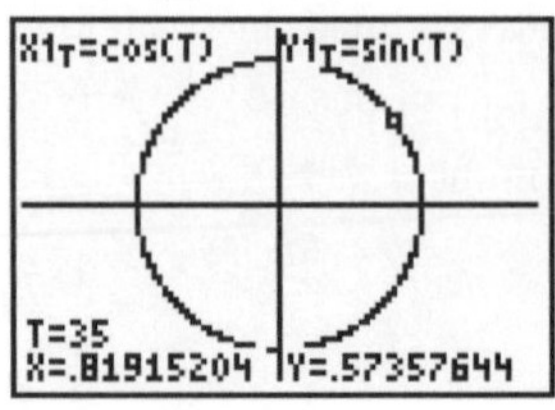

58. Use the TRACE feature to move around the circle in quadrant I. We see that cos 45° = sin 45°, so T = 45°.

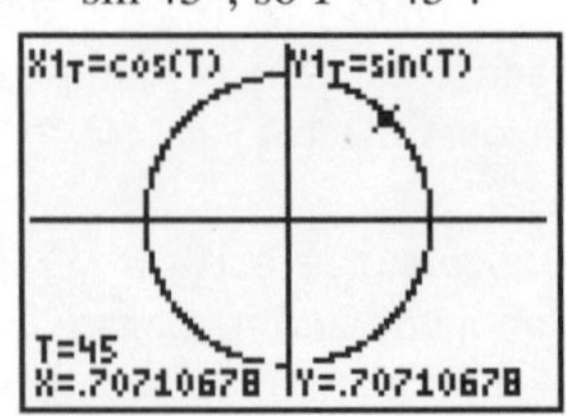

59. As T increases from 0° to 90°, the cosine decreases and the sine increases.

60. As T increases from 90° to 180°, the cosine decreases and the sine decreases.

61. The area of a triangle is $\frac{1}{2}bh$. The area of one of the six equilateral triangles is the following.

$$\frac{1}{2}(x)\left(\frac{\sqrt{3}}{2}x\right)=\left(\frac{1}{2}x^2\right)\left(\frac{\sqrt{3}}{2}\right)$$

Since $\sin\theta=\frac{\frac{\sqrt{3}}{2}x}{x}=\frac{\sqrt{3}}{2}$, we can substitute $\sin\theta$ for $\frac{\sqrt{3}}{2}$ in the area of a triangle, which gives $\frac{1}{2}x^2\sin\theta$. Thus, the area of the hexagon is $6\left(\frac{1}{2}x^2\sin\theta\right)=3x^2\sin\theta$.

62. **(a)** From the figure in the text and the definition of $\tan\theta$, we can see that $\tan\theta=\frac{y}{x}$.

(b) Solving for x we have, $\tan\theta=\frac{y}{x}\Rightarrow x\tan\theta=y\Rightarrow x=\frac{y}{\tan\theta}$.

Section 1.4: Using the Definitions of the Trigonometric Functions

1. 1 is its own reciprocal.
$\sin\theta=\csc\theta=1$
$\sin 90°=\csc 90°=1$
Therefore, $\theta=90°$.

2. –1 is its own reciprocal.
$\cos\theta=\sec\theta=-1$
$\cos 180°=\sec 180°=-1$
Therefore, $\theta=180°$.

3. $\cos\theta=\frac{1}{\sec\theta}=\frac{1}{-2.5}=-.4$

4. $\cot\theta=\frac{1}{\tan\theta}=\frac{1}{-\frac{1}{5}}=-5$

5. $\sin\theta=\frac{1}{\csc\theta}=\frac{1}{3}$

6. $\sec\theta=\frac{1}{\cos\theta}=\frac{1}{-\frac{\sqrt{7}}{7}}=-\frac{7}{\sqrt{7}}\cdot\frac{\sqrt{7}}{\sqrt{7}}=-\sqrt{7}$

7. $\sin\theta=\frac{1}{\csc\theta}=\frac{1}{\sqrt{15}}=\frac{1}{\sqrt{15}}\cdot\frac{\sqrt{15}}{\sqrt{15}}=\frac{\sqrt{15}}{15}$

8. $\tan\theta=\frac{1}{\cot\theta}=\frac{1}{-\frac{\sqrt{5}}{3}}=-\frac{3}{\sqrt{5}}$
$=-\frac{3}{\sqrt{5}}\cdot\frac{\sqrt{5}}{\sqrt{5}}=-\frac{3\sqrt{5}}{5}$

9. $\sin\theta=\frac{1}{\csc\theta}=\frac{1}{1.42716327}\approx .70069071$

10. $\cos\theta=\frac{1}{\sec\theta}=\frac{1}{9.80425133}\approx .10199657$

11. No. Since $\sin\theta=\frac{1}{\csc\theta}$, if $\sin\theta>0$ then its reciprocal, $\csc\theta$, must also be greater than zero.

12. Since $-1\le\cos\theta\le 1$, it is not possible to have $\cos\theta=\frac{3}{2}$.

13. Since $\tan 90°$ is undefined, it does not have a reciprocal.

14. Since $\tan\theta=\frac{1}{\cot\theta}$, multiply both sides by $\cot\theta$ to obtain $\tan\theta\cot\theta=1$. Now divide both sides by $\tan\theta$ to obtain $\cot\theta=\frac{1}{\tan\theta}$.

15. $\tan\theta = \frac{1}{\cot\theta} = \frac{1}{-3} = -\frac{1}{3}$

16. $\tan\theta = \frac{1}{\cot\theta} = \frac{1}{2}$

17. $\tan\theta = \frac{1}{\cot\theta} = \frac{1}{\frac{\sqrt{6}}{12}} = \frac{12}{\sqrt{6}}$

$= \frac{12}{\sqrt{6}} \cdot \frac{\sqrt{6}}{\sqrt{6}} = \frac{12\sqrt{6}}{6} = 2\sqrt{6}$

18. $\tan\theta = \frac{1}{\cot\theta} = \frac{1}{\frac{\sqrt{3}}{3}} = \frac{3}{\sqrt{3}} = \frac{3}{\sqrt{3}} \cdot \frac{\sqrt{3}}{\sqrt{3}} = \sqrt{3}$

19. $\tan\theta = \frac{1}{\cot\theta} = \frac{1}{.4} = 2.5$

20. $\tan\theta = \frac{1}{\cot\theta} = \frac{1}{-.01} = -100$

21. $\tan(3\theta - 4°) = \frac{1}{\cot(5\theta - 8°)} \Rightarrow \tan(3\theta - 4°) = \tan(5\theta - 8°)$

The second equation is true if $3\theta - 4° = 5\theta - 8°$, so solving this equation will give a value (but not the only value) for which the given equation is true.

$$3\theta - 4° = 5\theta - 8° \Rightarrow 4° = 2\theta \Rightarrow \theta = 2°$$

22. $\sec(2\theta + 6°)\cos(5\theta + 3°) = 1 \Rightarrow \cos(5\theta + 3°) = \frac{1}{\sec(2\theta + 6°)} \Rightarrow \cos(5\theta + 3°) = \cos(2\theta + 6°)$

The third equation is true if $5\theta + 3° = 2\theta + 6°$, so solving this equation will give a value (but not the only value) for which the given equation is true.

$$5\theta + 3° = 2\theta + 6° \Rightarrow 3\theta = 3° \Rightarrow \theta = 1°$$

23. $\sin(4\theta + 2°)\csc(3\theta + 5°) = 1 \Rightarrow \sin(4\theta + 2°) = \frac{1}{\csc(3\theta + 5°)} \Rightarrow \sin(4\theta + 2°) = \sin(3\theta + 5°)$

The third equation is true if $4\theta + 2° = 3\theta + 5°$, so solving this equation will give a value (but not the only value) for which the given equation is true.

$$4\theta + 2° = 3\theta + 5° \Rightarrow \theta = 3°$$

24. $\cos(6A + 5°) = \frac{1}{\sec(4A + 15°)} \Rightarrow \cos(6A + 5°) = \cos(4A + 15°)$

The second equation is true if $6A + 5° = 4A + 15°$, so solving this equation will give a value (but not the only value) for which the given equation is true.

$$6A + 5° = 4A + 15° \Rightarrow 2A = 10° \Rightarrow A = 5°$$

25. In quadrant II, the cosine is negative and the sine is positive.

26. An angle of $1294°$ must lie in quadrant III since the sine is negative and the tangent is positive.

27. $\sin\theta > 0$ implies θ is in quadrant I or II. $\cos\theta < 0$ implies θ is in quadrant II or III. Thus, θ must be in quadrant II.

28. $\cos\theta > 0$ implies θ is in quadrant I or IV. $\tan\theta > 0$ implies θ is in quadrant I or III. Thus, θ must be in quadrant I.

29. $\tan\theta > 0$ implies θ is in quadrant I or III. $\cot\theta > 0$ also implies θ is in quadrant I or III. Thus, θ is in quadrant I or III.

30. $\tan\theta < 0$ implies θ is in quadrant II or IV. $\cot\theta < 0$ also implies θ is in quadrant II or IV. Thus, θ is in quadrant II or IV.

31. $\cos\theta < 0$ implies θ is in quadrant II or III.

32. $\tan\theta > 0$ implies θ is in quadrant I or III.

33. 74° is in quadrant I.

$\sin 74° : +, \cos 74° : +, \tan 74° : +$

34. 298° is in quadrant IV

$\sin 298° : -, \cos 298° : +, \tan 298° : -$

35. 129° is in quadrant II.
$\sin 129° : +, \cos 129° : -, \tan 129° : -$

36. 183° is in quadrant III.
$\sin 183° : -, \cos 183° : -, \tan 183° : +$

37. Since $406° - 360° = 46°$, 406° is in quadrant I

$\sin 406° : +, \cos 406° : +, \tan 406° : +$

38. Since $412° - 360° = 52°$, 412° is in quadrant I.
$\sin 52° : +, \cos 52° : +, \tan 52° : +$

39. Since $-82° + 360° = 278°$, $-82°$ is in quadrant IV.
$\sin(-82°) : -, \cos(-82°) : +, \tan(-82°) : -$

40. Since $-121° + 360° = 239°, -121°$ is in quadrant III.
$\sin 74° : -, \cos 74° : -, \tan 74° : +$

41. $\tan 30° = \dfrac{\sin 30°}{\cos 30°}$

$\cos 30° < 1$, so $\tan 30° = \dfrac{\sin 30°}{\cos 30°} > \sin 30°$; that is, tan 30° is greater than sin 30°.

42. $\sin 21° > \sin 20°$
$\sin\theta$ increases as θ increases from 0° to 90°, so sin 21° is greater than sin 20°.

43. $\sec 33° = \dfrac{1}{\cos 33°}$; since $\cos 33° < 1, \dfrac{1}{\cos 33°} > \cos 33°$.
$\sin 33° < \cos 33° < \sec 33°$.
Therefore, sec 33° is greater than sin 33°.

44. $\cos 5° < 1$ so $(\cos 5°)^2 = \cos^2 5° < \cos 5°$

Therefore, $\cos 5°$ is greater than $\cos^2 5°$.

45. $\cos 26° > \cos 27°$

$\cos\theta$ decreases as θ goes from 0° to 90°, so $\cos 26°$ is greater than $\cos 27°$.

46. $\cot 2° = \dfrac{\cos 2°}{\sin 2°}$

$\sin 2° < 1$, so $\dfrac{1}{\sin 2°} > 1$. Since $\cos 2° > 0$, we have $\dfrac{\cos 2°}{\sin 2°} > \cos 2°$ by multiplying both sides by cos 2°. Thus, $\cot 2° > \cos 2°$.

47. Since $-1 \le \sin\theta \le 1$ for all θ and 2 is not in this range, $\sin\theta = 2$ is impossible.

48. Since $-1 \le \cos\theta \le 1$ for all θ and -1.001 is not in this range, $\cos\theta = -1.001$ is impossible.

49. $\tan\theta$ can take on all values. Thus, $\tan\theta = .92$ is possible.

50. $\cot\theta$ can take on all values. Thus, $\cot\theta = -12.1$ is possible.

51. $\sec\theta = 1$ is possible since $\sec\theta \le -1$ or $\sec\theta \ge 1$.

52. $\tan\theta$ can take on all values. Thus, $\tan\theta = 1$ is possible.

53. $\sin\theta = \frac{1}{2}$ is possible because $-1 \le \sin\theta \le 1$. Furthermore, when $\sin\theta = \frac{1}{2}$, $\csc\theta = \frac{1}{\sin\theta} = \frac{1}{\frac{1}{2}} = 2.$

$\sin\theta = \frac{1}{2}$ and $\csc\theta = 2$ is possible.

54. $\tan\theta = 2$ is possible, but when $\tan\theta = 2$, $\cot\theta = \frac{1}{\tan\theta} = \frac{1}{2}$. $\tan\theta = 2$ and $\cot\theta = \frac{1}{2}$ is impossible.

55. Find $\tan\theta$ is $\sec\theta = 3$, with θ in quadrant IV.
Start with the identity $\tan^2\alpha + 1 = \sec^2\alpha$, and replace $\sec\theta$ with 3.

$$\tan^2\alpha + 1 = 3^2 \Rightarrow \tan^2\alpha + 1 = 9 \Rightarrow \tan^2\alpha = 8 \Rightarrow \tan\alpha = \pm\sqrt{8} = \pm 2\sqrt{2}$$

Since θ is in quadrant IV, $\tan\theta < 0$, so $\tan\theta = -2\sqrt{2}$.

56. Find $\sin\theta$, if $\cos\theta = -\frac{1}{4}$, with θ in quadrant II.

Start with the identity $\sin^2\theta + \cos^2\theta = 1$. Replace $\cos\theta$ with $-\frac{1}{4}$.

$$\sin^2\theta + \left(-\frac{1}{4}\right)^2 = 1 \Rightarrow \sin^2\theta + \frac{1}{16} = 1 \Rightarrow \sin^2\theta = \frac{15}{16} \Rightarrow \sin\theta = \pm\sqrt{\frac{15}{16}} = \pm\frac{\sqrt{15}}{4}$$

Since θ is in quadrant II, $\sin\theta > 0$, so $\sin\theta = \frac{\sqrt{15}}{4}$.

57. Find $\csc\theta$, if $\cot\theta = -\frac{1}{2}$ and θ is in quadrant IV.

Start with the identity $1 + \cot^2\theta = \csc^2\theta$. Replace $\cot\theta$ with $-\frac{1}{2}$.

$$\csc^2\theta = 1 + \left(-\frac{1}{2}\right)^2 \Rightarrow \csc^2\theta = 1 + \frac{1}{4} \Rightarrow \csc^2\theta = \frac{5}{4} \Rightarrow \csc\theta = \pm\sqrt{\frac{5}{4}} = \pm\frac{\sqrt{5}}{2}$$

Since θ is in quadrant IV, $\csc\theta < 0$, so $\csc\theta = -\frac{\sqrt{5}}{2}$.

58. Find $\sec\theta$, if $\tan\theta = \frac{\sqrt{7}}{3}$, and θ is in quadrant III.

Start with the identity $\tan^2\theta + 1 = \sec^2\theta$. Replace $\tan\theta$ with $\frac{\sqrt{7}}{3}$.

$$\sec^2\theta = \left(\frac{\sqrt{7}}{3}\right)^2 + 1 \Rightarrow \sec^2\theta = \frac{7}{9} + \frac{9}{9} \Rightarrow \sec^2\theta = \frac{16}{9} \Rightarrow \sec\theta = \pm\sqrt{\frac{16}{9}} = \pm\frac{4}{3}$$

Since θ is in quadrant III, $\sec\theta < 0$, so $\sec\theta = -\frac{4}{3}$.

59. Find $\cos\theta$, if $\csc\theta = -4$ with θ in quadrant III.

Start with the identity $\sin\theta = \frac{1}{\csc\theta}$. We have $\sin\theta = \frac{1}{\csc\theta} = \frac{1}{-4} = -\frac{1}{4}$. Now use the identity

$\sin^2\theta + \cos^2\theta = 1$ and replace $\sin\theta$ with $-\frac{1}{4}$.

$$\left(-\frac{1}{4}\right)^2 + \cos^2\theta = 1 \Rightarrow \frac{1}{16} + \cos^2\theta = 1 \Rightarrow \cos^2\theta = \frac{15}{16} \Rightarrow \cos\theta = \pm\sqrt{\frac{15}{16}} = \pm\frac{\sqrt{15}}{4}$$

Since θ is in quadrant III, $\cos\theta < 0$, so $\cos\theta = -\frac{\sqrt{15}}{4}$.

60. Find $\sin\theta$, if $\sec\theta = 2$, with θ in quadrant IV.

Start with the identity $\cos\theta = \frac{1}{\sec\theta}$ We have $\cos\theta = \frac{1}{\sec\theta} = \frac{1}{2}$. Now use the identity

$\sin^2\theta + \cos^2\theta = 1$ and replace $\cos\theta$ with $\frac{1}{2}$.

$$\sin^2\theta + \left(\frac{1}{2}\right)^2 = 1 \Rightarrow \sin^2\theta + \frac{1}{4} = 1 \Rightarrow \sin^2\theta = \frac{3}{4} \Rightarrow \sin\theta = \pm\sqrt{\frac{3}{4}} = \pm\frac{\sqrt{3}}{2}$$

Since θ is in quadrant IV, $\sin\theta < 0$; therefore, $\sin\theta = -\frac{\sqrt{3}}{2}$.

61. Find $\cot\theta$, if $\csc\theta = -3.5891420$, with θ in quadrant III.

Start with the identity $1 + \cot^2\theta = \csc^2\theta$. Replace $\csc\theta$ with -3.5891420.

$$1 + \cot^2\theta = (-3.5891420)^2 \Rightarrow 1 + \cot^2\theta \approx 12.8819403$$
$$\cot^2\theta \approx 11.8819403 \Rightarrow \cot\theta \approx \pm 3.44701905$$

Since θ is in quadrant III, $\cot\theta > 0$; therefore, $\cot\theta \approx 3.44701905$.

62. Find $\tan\theta$, if $\sin\theta = .49268329$, with θ in quadrant II.

Start with the identity $\sin^2\theta + \cos^2\theta = 1$. Replace $\sin\theta$ with $.49268329$.

$$(.49268329)^2 + \cos^2\theta = 1 \Rightarrow .24273682 + \cos^2\theta \approx 1$$
$$\cos^2\theta \approx .75726318 \Rightarrow \cos\theta \approx \pm .87020870$$

Since θ is in quadrant II, $\cos\theta < 0$; therefore, $\cos\theta \approx -.87020870$. Since $\tan\theta = \frac{\sin\theta}{\cos\theta}$, we have

$\tan\theta \approx = \frac{.49268329}{-.87020870} \approx -.56616682$.

63. We need to check if $\sin^2\theta + \cos^2\theta = 1$. Substituting $-.6$ for $\cos\theta$ and $.8$ for $\sin\theta$ we have the following.

$$\sin^2\theta + \cos^2\theta = 1$$
$$(.8)^2 + (-.6)^2 \stackrel{?}{=} 1$$
$$.64 + .36 \stackrel{?}{=} 1$$
$$1 = 1 \text{ True}$$

Yes, there is an angle θ for which $\cos\theta = -.6$ and $\sin\theta = .8$.

For Exercises 64–71, remember that r is always positive.

64. $\cos\theta = -\frac{3}{5} = \frac{-3}{5}$, with θ in quadrant III

$\cos\theta = \frac{x}{r}$ and θ in quadrant III, so let $x = -3$, $r = 5$.

$$x^2 + y^2 = r^2$$
$$(-3)^2 + y^2 = 5^2$$
$$9 + y^2 = 25$$
$$y^2 = 16$$
$$y = \pm\sqrt{16}$$
$$y = \pm 4$$

θ is in quadrant III, so $y = -4$.

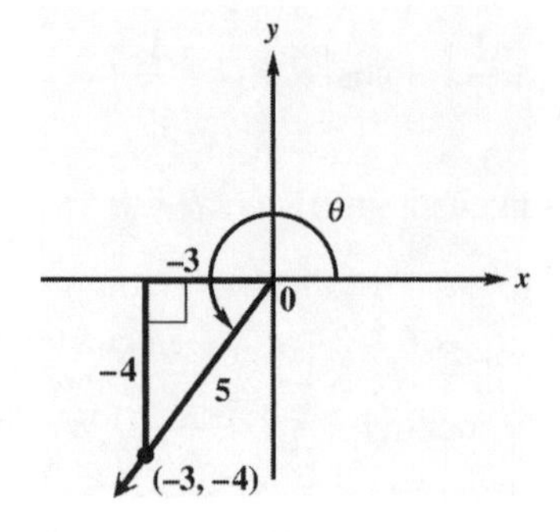

$\sin\theta = \frac{y}{r} = \frac{-4}{5} = -\frac{4}{5}$

$\cos\theta = \frac{x}{r} = \frac{-3}{5} = -\frac{3}{5}$

$\tan\theta = \frac{y}{x} = \frac{-4}{-3} = \frac{4}{3}$

$\cot\theta = \frac{x}{y} = \frac{-3}{-4} = \frac{3}{4}$

$\sec\theta = \frac{r}{x} = \frac{5}{-3} = -\frac{5}{3}$

$\csc\theta = \frac{r}{y} = \frac{5}{-4} = -\frac{5}{4}$

65. $\tan\theta = -\frac{15}{8} = \frac{15}{-8}$, with θ in quadrant II

$\tan\theta = \frac{y}{x}$ and θ is in quadrant II, so let $y = 15$ and $x = -8$.

$$x^2 + y^2 = r^2$$
$$(-8)^2 + 15^2 = r^2$$
$$64 + 225 = r^2$$
$$289 = r^2$$
$$r = 17, \text{ since } r \text{ is positive}$$

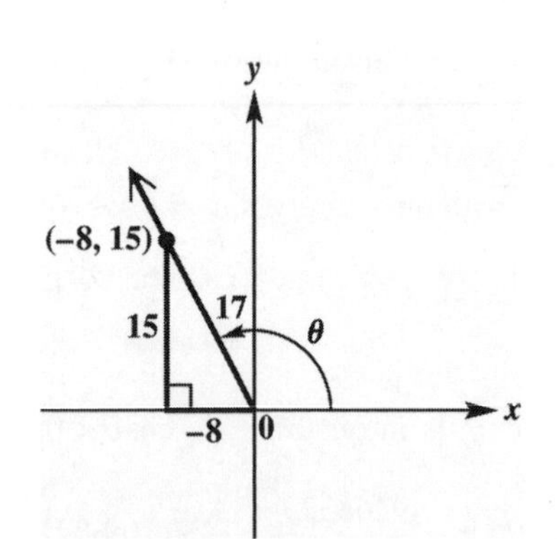

$\sin\theta = \frac{y}{r} = \frac{15}{17}$

$\cos\theta = \frac{x}{r} = \frac{-8}{17} = -\frac{8}{17}$

$\tan\theta = \frac{y}{x} = \frac{15}{-8} = -\frac{15}{8}$

$\cot\theta = \frac{x}{y} = \frac{-8}{15} = -\frac{8}{15}$

$\sec\theta = \frac{r}{x} = \frac{17}{-8} = -\frac{17}{8}$

$\csc\theta = \frac{r}{y} = \frac{17}{15}$

66. $\tan\theta = \sqrt{3} = \frac{-\sqrt{3}}{-1}$

$\tan\theta = \frac{y}{x}$ and θ is in quadrant III, so let $y = -\sqrt{3}$ and $x = -1$.

$$x^2 + y^2 = r^2$$
$$(-1)^2 + \left(-\sqrt{3}\right)^2 = r^2$$
$$1 + 3 = r^2$$
$$4 = r^2$$

$r = 2$, since r is positive

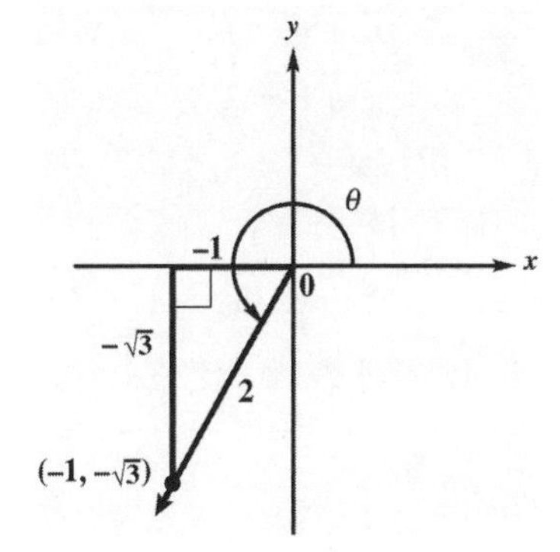

$$\sin\theta = \frac{y}{r} = \frac{-\sqrt{3}}{2} = -\frac{\sqrt{3}}{2}$$
$$\cos\theta = \frac{x}{r} = \frac{-1}{2} = -\frac{1}{2}$$
$$\tan\theta = \frac{y}{x} = \frac{-\sqrt{3}}{-1} = \sqrt{3}$$
$$\cot\theta = \frac{x}{y} = \frac{-1}{-\sqrt{3}} = \frac{1}{\sqrt{3}} = \frac{1}{\sqrt{3}}\cdot\frac{\sqrt{3}}{\sqrt{3}} = \frac{\sqrt{3}}{3}$$
$$\sec\theta = \frac{r}{x} = \frac{2}{-1} = -2$$
$$\csc\theta = \frac{r}{y} = \frac{2}{-\sqrt{3}} = -\frac{2}{\sqrt{3}}\cdot\frac{\sqrt{3}}{\sqrt{3}} = -\frac{2\sqrt{3}}{3}$$

67. $\sin\theta = \frac{\sqrt{5}}{7}$, with θ in quadrant I

$\sin\theta = \frac{y}{r}$ and θ in quadrant I, so let $y = \sqrt{5}, r = 7$.

$$x^2 + y^2 = r^2$$
$$x^2 + \left(\sqrt{5}\right)^2 = 7^2$$
$$x^2 + 5 = 49$$
$$x^2 = 44$$
$$x = \pm\sqrt{44}$$
$$x = \pm 2\sqrt{11}$$

θ is in quadrant I, so $x = 2\sqrt{11}$.

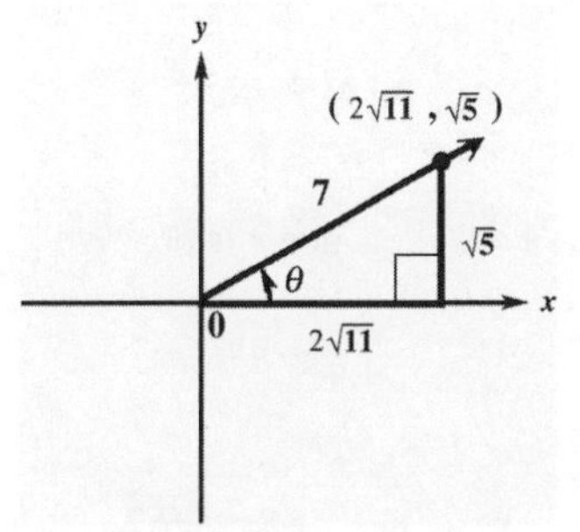

Drawing not to scale

$$\sin\theta = \frac{y}{r} = \frac{\sqrt{5}}{7}$$
$$\cos\theta = \frac{x}{r} = \frac{2\sqrt{11}}{7}$$
$$\tan\theta = \frac{y}{x} = \frac{\sqrt{5}}{2\sqrt{11}} = \frac{\sqrt{5}}{2\sqrt{11}}\cdot\frac{\sqrt{11}}{\sqrt{11}} = \frac{\sqrt{55}}{22}$$
$$\cot\theta = \frac{x}{y} = \frac{2\sqrt{11}}{\sqrt{5}} = \frac{2\sqrt{11}}{\sqrt{5}}\cdot\frac{\sqrt{5}}{\sqrt{5}} = \frac{2\sqrt{55}}{5}$$
$$\sec\theta = \frac{r}{x} = \frac{7}{2\sqrt{11}} = \frac{7}{2\sqrt{11}}\cdot\frac{\sqrt{11}}{\sqrt{11}} = \frac{7\sqrt{11}}{22}$$
$$\csc\theta = \frac{r}{y} = \frac{7}{\sqrt{5}} = \frac{7}{\sqrt{5}}\cdot\frac{\sqrt{5}}{\sqrt{5}} = \frac{7\sqrt{5}}{5}$$

68. $\csc\theta = 2 = \frac{2}{1}$, with θ in quadrant II

$\csc\theta = \frac{r}{y}$ and θ in quadrant II, so let $y = 1$, $r = 2$.

$$x^2 + y^2 = r^2$$
$$x^2 + 1^2 = 2^2$$
$$x^2 + 1 = 4$$
$$x^2 = 3$$
$$x = \pm\sqrt{3}$$

θ is in quadrant II, so $x = -\sqrt{3}$.

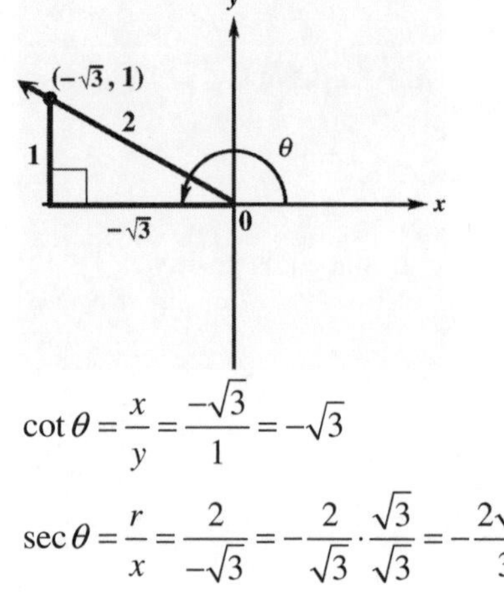

$$\sin\theta = \frac{y}{r} = \frac{1}{2}$$
$$\cos\theta = \frac{x}{r} = \frac{-\sqrt{3}}{2} = -\frac{\sqrt{3}}{2}$$
$$\tan\theta = \frac{y}{x} = \frac{1}{-\sqrt{3}} = -\frac{1}{\sqrt{3}} \cdot \frac{\sqrt{3}}{\sqrt{3}} = -\frac{\sqrt{3}}{3}$$

$$\cot\theta = \frac{x}{y} = \frac{-\sqrt{3}}{1} = -\sqrt{3}$$
$$\sec\theta = \frac{r}{x} = \frac{2}{-\sqrt{3}} = -\frac{2}{\sqrt{3}} \cdot \frac{\sqrt{3}}{\sqrt{3}} = -\frac{2\sqrt{3}}{3}$$
$$\csc\theta = \frac{r}{y} = \frac{2}{1} = 2$$

69. $\cot\theta = \frac{\sqrt{3}}{8}$, with θ in quadrant I

$\cot\theta = \frac{x}{y}$ and θ in quadrant I, so let $x = \sqrt{3}$, $y = 8$.

$$x^2 + y^2 = r^2$$
$$\left(\sqrt{3}\right)^2 + 8^2 = r^2$$
$$3 + 64 = r^2$$
$$67 = r^2$$

$r = \sqrt{67}$, since r is positive

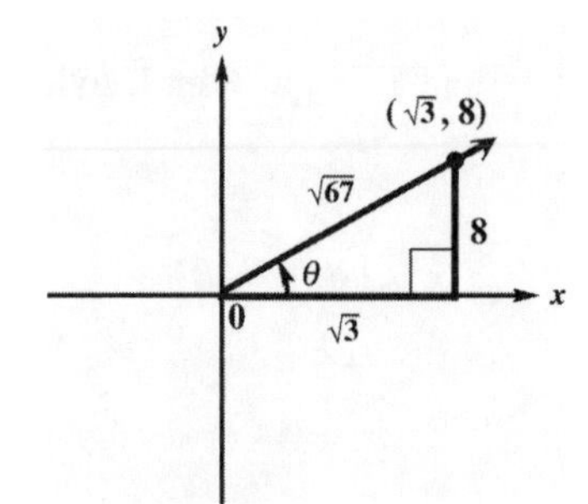

Drawing not to scale

$$\sin\theta = \frac{y}{r} = \frac{8}{\sqrt{67}} = \frac{8}{\sqrt{67}} \cdot \frac{\sqrt{67}}{\sqrt{67}} = \frac{8\sqrt{67}}{67}$$
$$\cos\theta = \frac{x}{r} = \frac{\sqrt{3}}{\sqrt{67}} = \frac{\sqrt{3}}{\sqrt{67}} \cdot \frac{\sqrt{67}}{\sqrt{67}} = \frac{\sqrt{201}}{67}$$
$$\tan\theta = \frac{y}{x} = \frac{8}{\sqrt{3}} = \frac{8}{\sqrt{3}} \cdot \frac{\sqrt{3}}{\sqrt{3}} = \frac{8\sqrt{3}}{3}$$

$$\cot\theta = \frac{x}{y} = \frac{\sqrt{3}}{8}$$
$$\sec\theta = \frac{r}{x} = \frac{\sqrt{67}}{\sqrt{3}} = \frac{\sqrt{67}}{\sqrt{3}} \cdot \frac{\sqrt{3}}{\sqrt{3}} = \frac{\sqrt{201}}{3}$$
$$\csc\theta = \frac{r}{y} = \frac{\sqrt{67}}{8}$$

70. $\cot\theta = -1.49586 = \dfrac{1.49586}{-1}$, with θ in quadrant IV

$\cot\theta = \dfrac{x}{y}$ and θ is in quadrant IV, let $x = 1.49586$ and $y = -1$.

$$x^2 + y^2 = r^2$$
$$(1.49586)^2 + (-1)^2 = r^2$$
$$2.23760 + 1 \approx r^2$$
$$3.23760 \approx r^2$$
$r \approx 1.79933$, since r is positive

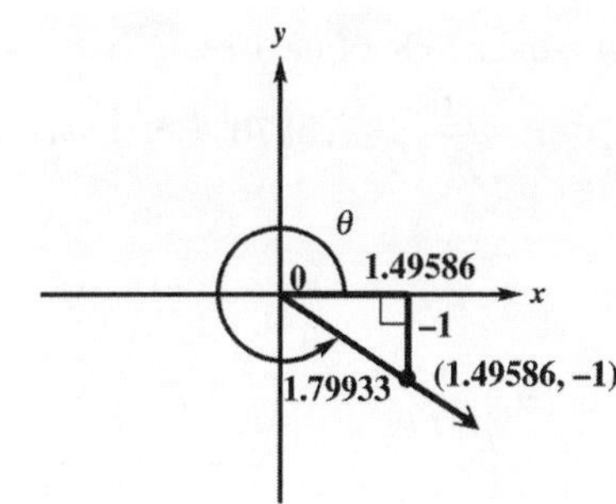

$\sin\theta = \dfrac{y}{r} \approx \dfrac{-1}{1.79933} \approx -.555762$

$\cos\theta = \dfrac{x}{r} \approx \dfrac{1.49586}{1.79933} \approx .831343$

$\tan\theta = \dfrac{y}{x} \approx \dfrac{-1}{1.49586} \approx -.668512$

$\cot\theta = \dfrac{x}{y} \approx \dfrac{1.49586}{-1} \approx -1.49586$

$\sec\theta = \dfrac{r}{x} \approx \dfrac{1.79933}{1.49586} \approx 1.20287$

$\csc\theta = \dfrac{r}{y} \approx \dfrac{1.79933}{-1} \approx -1.79933$

71. $\sin\theta = .164215 = \dfrac{.164215}{1}$, with θ in quadrant II

$\sin\theta = \dfrac{y}{r}$ and θ is in quadrant II, let $y = .164215$ and $r = 1$.

$$x^2 + y^2 = r^2$$
$$x^2 + (.164215)^2 = 1^2$$
$$x^2 + .026966 = 1$$
$$x^2 = .973034$$
$$x \approx \pm\sqrt{.973034}$$
$$x \approx \pm .986425$$
θ is in quadrant II, so, $x = -.986425$.

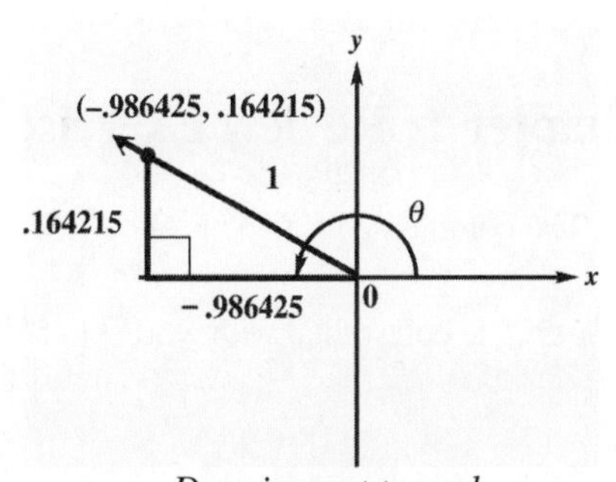

Drawing not to scale

$\sin\theta = \dfrac{y}{r} = \dfrac{.164215}{1} = .164215$

$\cos\theta = \dfrac{x}{r} \approx \dfrac{-.986425}{1} = -.986425$

$\tan\theta = \dfrac{y}{x} \approx \dfrac{.164215}{-.986425} \approx -.166475$

$\cot\theta = \dfrac{x}{y} \approx \dfrac{-.986425}{.164215} \approx -6.00691$

$\sec\theta = \dfrac{r}{x} \approx \dfrac{1}{-.986425} \approx -1.01376$

$\csc\theta = \dfrac{r}{y} = \dfrac{1}{.164215} \approx 6.08958$

72. $x^2 + y^2 = r^2 \Rightarrow \dfrac{x^2 + y^2}{y^2} = \dfrac{r^2}{y^2} \Rightarrow \dfrac{x^2}{y^2} + \dfrac{y^2}{y^2} = \dfrac{r^2}{y^2} \Rightarrow \left(\dfrac{x}{y}\right)^2 + 1 = \left(\dfrac{r}{y}\right)^2 \Rightarrow 1 + \left(\dfrac{x}{y}\right)^2 = \left(\dfrac{r}{y}\right)^2$

Since $\cot\theta = \dfrac{x}{y}$ and $\csc\theta = \dfrac{r}{y}$, we have $1 + (\cot\theta)^2 = (\csc\theta)^2$ or $1 + \cot^2\theta = \csc^2\theta$.

73. $\dfrac{\cos\theta}{\sin\theta} = \dfrac{\frac{x}{r}}{\frac{y}{r}} = \dfrac{x}{r} \div \dfrac{y}{r} = \dfrac{x}{r} \cdot \dfrac{r}{y} = \dfrac{x}{y} = \cot\theta$

74. The statement is false. For example, $\sin 30° + \cos 30° \approx .5 + .8660 = 1.3660 \neq 1$.

75. The statement is false since $-1 \le \sin\theta \le 1$ for all θ.

76. Let h = the height of the tree.

$$\cos 70° = \frac{50\text{ ft}}{h} \Rightarrow h\cos 70° = 50 \Rightarrow h = \frac{50}{\cos 70°} = \frac{50}{0.3420} \approx 146 \text{ feet}$$

77. **(a)** $r^2 = 5^2 + 12^2 \Rightarrow r^2 = 25 + 144 \Rightarrow r^2 = 169 \Rightarrow r = \sqrt{169} = 13$ prism diopters

(b) $\tan\theta = \frac{5}{12}$

78. $90° < \theta < 180° \Rightarrow 180° < 2\theta < 360°$, so 2θ lies in quadrants III and IV. Thus, $\sin 2\theta$ is negative.

79. $90° < \theta < 180° \Rightarrow 45° < \frac{\theta}{2} < 90°$, so $\frac{\theta}{2}$ lies in quadrant I. Thus, $\tan\frac{\theta}{2}$ is positive.

80. $90° < \theta < 180° \Rightarrow 270° < \theta + 180° < 360°$, so $\theta + 180°$ lies in quadrant IV. Thus, $\cot(\theta + 180°)$ is negative.

81. $90° < \theta < 180° \Rightarrow -90° > -\theta > -180° \Rightarrow -180° < -\theta < -90°$, so $-\theta$ lies in quadrant III ($-180°$ is coterminal with $180°$ and $-90°$ is coterminal with $270°$.). Thus, $\cos(-\theta)$ is negative.

82. Answers will vary.

Chapter 1: Review Exercises

1. The complement of 35° is $90° - 35° = 55°$. The supplement of 35° is $180° - 35° = 145°$.

2. $-51°$ is coterminal with $360° + (-51°) = 309°$.

3. $-174° + 360° = 186°$

4. 792° is coterminal with 792° - 2(360°) = 72°

5. Let n represent any integer. Any angle coterminal with 270° would be $270° + n \cdot 360°$.

6. 320 rotations per min $= \frac{320}{60}$ rotations per sec $= \frac{16}{3}$ rotations per sec
$= \frac{32}{9}$ rotations per $\frac{2}{3}$ sec $= \frac{32}{9}(360°)$ per $\frac{2}{3}$ sec $= 1280°$ per $\frac{2}{3}$ sec

A point on the edge of the pulley will move 1280° in $\frac{2}{3}$ sec.

7. 650 rotations per min $= \frac{650}{60}$ rotations per sec $= \frac{65}{6}$ rotations per sec = 26 rotations per 2.4 sec
$= 26(360°)$ per 2.4 sec $= 9360°$ per 2.4 sec

A point on the edge of the propeller will rotate 9360° in 2.4 sec.

8. $47°\, 25'\, 11'' = 47° + \frac{25}{60}° + \frac{11}{3600}° \approx 47° + .4167° + .0031° \approx 47.420°$

9. $119°\,8'\,3'' = 119° + \frac{8}{60}° + \frac{3}{3600}° \approx 119° + .1333° + .0008° \approx 119.134°$

10. $-61.5034° = -(61.5034°) = -[61° + .5034(60')] = -[61° + 30.204']$

$$= -[61° + 30' + .204'] = -[61° + 30' + .204(60'')]$$

$$= -[61° + 30' + 12.24''] = -61°30'12.24'' \approx -61°30'12''$$

11. $275.1005° = 275° + .1005(60') = 275° + 6.03' = 275° + 6' + .03'$

$$= 275° + 6' + .03(60'') = 275°6'1.8'' \approx 275°6'2''$$

12. The two indicated angles are vertical angles. Hence, their measures are equal.

$$\begin{aligned} 9x + 4 &= 12x - 14 \\ 4 &= 3x - 14 \\ 18 &= 3x \\ 6 &= x \end{aligned}$$

Since $9x + 4 = 9(6) + 4 = 54 + 4 = 58$ and $12x - 14 = 12(6) - 14 = 72 - 14 = 58$, both angles measure 58°.

13. The three angles are the interior angles of a triangle. Hence, the sum of their measures is 180°.

$$4x + 6x + 8x = 180° \Rightarrow 18x = 180° \Rightarrow x = 10°$$

Since $4x = 4(10) = 40$, $6x = 6(10) = 60$, and $8x = 8(10) = 80$, the three angles measure 40°, 60°, and 80°.

14. Since a line has 180°, the angle supplementary to β is $180 - \beta$. The sum of the angles of a triangle is 180°, so

$$\theta + \alpha + (180 - \beta) = 180° \Rightarrow \theta = 180° - \alpha - 180° + \beta \Rightarrow \theta = -\alpha + \beta \Rightarrow \theta = \beta - \alpha.$$

15. Assuming *PQ* and *BA* are parallel, ΔPCQ is similar to ΔACB since the measure of angle *PCQ* is equal to the measure of *ACB* (vertical angles). Since corresponding sides of similar triangles are proportional, we can write $\frac{PQ}{BA} = \frac{PC}{AC}$. Thus, we have the following.

$$\frac{PQ}{BA} = \frac{PC}{AC} \Rightarrow \frac{1.25\text{mm}}{BA} = \frac{150\text{mm}}{30\text{km}}$$

Solving for *BA*, we have $\frac{1.25}{BA} = \frac{150}{30} \Rightarrow \frac{1.25}{BA} = \frac{5}{1} \Rightarrow 1.25 = 5BA \Rightarrow BA = \frac{1.25}{5} = .25\,\text{km}.$

16. Since angle *Z* corresponds to angle *T*, the measure of angle *Z* is 32°.

Since angle *V* corresponds to angle *X*, the measure of angle *V* is 41°.

Since angles *X*, *Y*, and *Z* are interior angles of a triangle, the sum of their measures is 180°.

$$\begin{aligned} \text{measure angle } X + \text{measure angle } Y + \text{measure angle } Z &= 180° \\ 41° + \text{measure angle } Y + 32° &= 180° \\ 73° + \text{measure angle } Y &= 180° \\ \text{measure angle } Y &= 107° \end{aligned}$$

Since angle *U* corresponds to angle *Y*, the measure of angle *U* is 107°.

17. Since angle R corresponds to angle P, the measure of angle R is 82°.

Since angle M corresponds to angle S, the measure of angle M is 86°.

Since angle N corresponds to angle Q, the measure of angle N is 12°.

Note: $12° + 82° + 86° = 180°$.

18. $\frac{m}{30} = \frac{75}{50} \Rightarrow \frac{m}{30} = \frac{3}{2} \Rightarrow 2m = 90 \Rightarrow m = 45$

$\frac{n}{40} = \frac{75}{50} \Rightarrow \frac{n}{40} = \frac{3}{2} \Rightarrow 2n = 120 \Rightarrow n = 60$

19. Since the large triangle is equilateral, the small triangle is also equilateral. Thus, $p = q = 7$.

20. $\frac{6}{r} = \frac{7}{11+7} \Rightarrow \frac{6}{r} = \frac{7}{18} \Rightarrow 7r = 108 \Rightarrow r = \frac{108}{7} \approx 15.4$

21. $\frac{k}{6} = \frac{12+9}{9} \Rightarrow \frac{k}{6} = \frac{21}{9} \Rightarrow \frac{k}{6} = \frac{7}{3} \Rightarrow 3k = 42 \Rightarrow k = 14$

22. If two triangles are similar, their corresponding sides are <u>proportional</u> and the measures of their corresponding angles are <u>equal</u>.

23. Let x = the shadow of the 30-ft tree.

$$\frac{20}{8} = \frac{30}{x} \Rightarrow \frac{5}{2} = \frac{30}{x} \Rightarrow 5x = 60 \Rightarrow x = 12 \text{ ft}$$

24. $x = -3,\ y = -3$ and $r = \sqrt{x^2 + y^2} = \sqrt{(-3)^2 + (-3)^2} = \sqrt{9+9} = \sqrt{18} = 3\sqrt{2}$

$\sin\theta = \frac{y}{r} = -\frac{3}{3\sqrt{2}} = -\frac{1}{\sqrt{2}} \cdot \frac{\sqrt{2}}{\sqrt{2}} = -\frac{\sqrt{2}}{2}$

$\cos\theta = \frac{x}{r} = \frac{-3}{3\sqrt{2}} = -\frac{1}{\sqrt{2}} = -\frac{1}{\sqrt{2}} \cdot \frac{\sqrt{2}}{\sqrt{2}} = -\frac{\sqrt{2}}{2}$

$\tan\theta = \frac{y}{x} = \frac{-3}{-3} = 1$

$\cot\theta = \frac{x}{y} = \frac{-3}{-3} = 1$

$\sec\theta = \frac{r}{x} = \frac{3\sqrt{2}}{-3} = -\sqrt{2}$

$\csc\theta = \frac{r}{y} = \frac{3\sqrt{2}}{-3} = -\sqrt{2}$

25. $x = 1,\ y = -\sqrt{3}$ and $r = \sqrt{x^2 + y^2} = \sqrt{1^2 + \left(-\sqrt{3}\right)^2} = \sqrt{1+3} = \sqrt{4} = 2$

$\sin\theta = \frac{y}{r} = \frac{-\sqrt{3}}{2} = -\frac{\sqrt{3}}{2}$

$\cos\theta = \frac{x}{r} = \frac{1}{2}$

$\tan\theta = \frac{y}{x} = \frac{-\sqrt{3}}{1} = -\sqrt{3}$

$\cot\theta = \frac{x}{y} = \frac{1}{-\sqrt{3}} = -\frac{1}{\sqrt{3}} \cdot \frac{\sqrt{3}}{\sqrt{3}} = -\frac{\sqrt{3}}{3}$

$\sec\theta = \frac{r}{x} = \frac{2}{1} = 2$

$\csc\theta = \frac{r}{y} = \frac{2}{-\sqrt{3}} = -\frac{2}{\sqrt{3}} \cdot \frac{\sqrt{3}}{\sqrt{3}} = -\frac{2\sqrt{3}}{3}$

26. For the quadrantal angle, $180°$, we will choose the point $(-1,0)$, so $x=-1$, $y=0$, and $r=1$.

$\sin 180° = \frac{y}{r} = \frac{0}{1} = 0$

$\cos 180° = \frac{-1}{1} = -1$

$\tan 180° = \frac{y}{x} = \frac{0}{-1} = 0$

$\cot 180° = \frac{x}{y} = \frac{-1}{0}$ undefined

$\sec 180° = \frac{r}{x} = \frac{1}{-1} = -1$

$\csc 180° = \frac{r}{y} = \frac{1}{0}$ undefined

27. $(3,-4)$

$x = 3$, $y = -4$ and $r = \sqrt{3^2 + (-4)^2} = \sqrt{9+16} = \sqrt{25} = 5$

$\sin\theta = \frac{y}{r} = \frac{-4}{5} = -\frac{4}{5}$

$\cos\theta = \frac{x}{r} = \frac{3}{5}$

$\tan\theta = \frac{y}{x} = \frac{-4}{3} = -\frac{4}{3}$

$\cot\theta = \frac{x}{y} = \frac{3}{-4} = -\frac{3}{4}$

$\sec\theta = \frac{r}{x} = \frac{5}{3}$

$\csc\theta = \frac{r}{y} = \frac{5}{-4} = -\frac{5}{4}$

28. $(9,-2)$

$x=9$, $y=-2$, and $r = \sqrt{9^2 + (-2)^2} = \sqrt{81+4} = \sqrt{85}$

$\sin\theta = \frac{y}{r} = \frac{-2}{\sqrt{85}} = -\frac{2}{\sqrt{85}} \cdot \frac{\sqrt{85}}{\sqrt{85}} = -\frac{2\sqrt{85}}{85}$

$\cos\theta = \frac{x}{r} = \frac{9}{\sqrt{85}} = \frac{9}{\sqrt{85}} \cdot \frac{\sqrt{85}}{\sqrt{85}} = \frac{9\sqrt{85}}{85}$

$\tan\theta = \frac{y}{x} = \frac{-2}{9} = -\frac{2}{9}$

$\cot\theta = \frac{x}{y} = \frac{9}{-2} = -\frac{9}{2}$

$\sec\theta = \frac{r}{x} = \frac{\sqrt{85}}{9}$

$\csc\theta = \frac{r}{y} = \frac{\sqrt{85}}{-2} = -\frac{\sqrt{85}}{2}$

29. $(-8,15)$

$x=-8$, $y=15$, and $r = \sqrt{(-8)^2 + 15^2} = \sqrt{64+225} = \sqrt{289} = 17$

$\sin\theta = \frac{y}{r} = \frac{15}{17}$

$\cos\theta = \frac{x}{r} = -\frac{8}{17}$

$\tan\theta = \frac{y}{x} = \frac{15}{-8} = -\frac{15}{8}$

$\cot\theta = \frac{x}{y} = -\frac{8}{15}$

$\sec\theta = \frac{r}{x} = \frac{17}{-8} = -\frac{17}{8}$

$\csc\theta = \frac{r}{y} = \frac{17}{15}$

30. $(1,-5)$

$x=1,\ y=-5,$ and $r=\sqrt{1^2+5^2}=\sqrt{1+26}=\sqrt{26}$

$\sin\theta=\frac{y}{r}=\frac{-5}{\sqrt{26}}=-\frac{5}{\sqrt{26}}\cdot\frac{\sqrt{26}}{\sqrt{26}}=-\frac{5\sqrt{26}}{26}$

$\cos\theta=\frac{x}{r}=\frac{1}{\sqrt{26}}=\frac{1}{\sqrt{26}}\cdot\frac{\sqrt{26}}{\sqrt{26}}=\frac{\sqrt{26}}{26}$

$\tan\theta=\frac{y}{x}=\frac{-5}{1}=-5$

$\cot\theta=\frac{x}{y}=-\frac{1}{5}$

$\sec\theta=\frac{r}{x}=\frac{\sqrt{26}}{1}=\sqrt{26}$

$\csc\theta=\frac{r}{y}=\frac{\sqrt{26}}{-5}=-\frac{\sqrt{26}}{5}$

31. $\left(6\sqrt{3},-6\right)$

$x=6\sqrt{3},\ y=-6,$ and $r=\sqrt{\left(6\sqrt{3}\right)^2+(-6)^2}=\sqrt{108+36}=\sqrt{144}=12$

$\sin\theta=\frac{y}{r}=\frac{-6}{12}=-\frac{1}{2}$

$\cos\theta=\frac{x}{r}=\frac{6\sqrt{3}}{12}=\frac{\sqrt{3}}{2}$

$\tan\theta=\frac{y}{x}=\frac{-6}{6\sqrt{3}}=-\frac{1}{\sqrt{3}}=-\frac{1}{\sqrt{3}}\cdot\frac{\sqrt{3}}{\sqrt{3}}=-\frac{\sqrt{3}}{3}$

$\cot\theta=\frac{x}{y}=-\frac{6\sqrt{3}}{6}=-\sqrt{3}$

$\sec\theta=\frac{r}{x}=\frac{12}{6\sqrt{3}}=\frac{2}{\sqrt{3}}=\frac{2}{\sqrt{3}}\cdot\frac{\sqrt{3}}{\sqrt{3}}=\frac{2\sqrt{3}}{3}$

$\csc\theta=\frac{r}{y}=\frac{12}{-6}=-2$

32. $\left(-2\sqrt{2},\ 2\sqrt{2}\right)$

$x=-2\sqrt{2},\ y=2\sqrt{2},$ and $r=\sqrt{\left(-2\sqrt{2}\right)^2+\left(2\sqrt{2}\right)^2}=\sqrt{8+8}=\sqrt{16}=4$

$\sin\theta=\frac{y}{r}=\frac{2\sqrt{2}}{4}=\frac{\sqrt{2}}{2}$

$\cos\theta=\frac{x}{r}=\frac{-2\sqrt{2}}{4}=-\frac{\sqrt{2}}{2}$

$\tan\theta=\frac{y}{x}=\frac{2\sqrt{2}}{-2\sqrt{2}}=-1$

$\cot\theta=\frac{x}{y}=\frac{-2\sqrt{2}}{2\sqrt{2}}=-1$

$\sec\theta=\frac{r}{x}=\frac{4}{-2\sqrt{2}}=-\frac{2}{\sqrt{2}}=-\frac{2}{\sqrt{2}}\cdot\frac{\sqrt{2}}{\sqrt{2}}=-\sqrt{2}$

$\csc\theta=\frac{r}{y}=\frac{4}{2\sqrt{2}}=\frac{2}{\sqrt{2}}=\frac{2}{\sqrt{2}}\cdot\frac{\sqrt{2}}{\sqrt{2}}=\sqrt{2}$

33. If the terminal side of a quadrantal angle lies along the y-axis, a point on the terminal side would be of the form $(0, k)$, where k is a real number, $k\neq 0$.

$\sin\theta=\frac{y}{r}=\frac{k}{r}$

$\cos\theta=\frac{x}{r}=\frac{0}{r}=0$

$\tan\theta=\frac{y}{x}=\frac{k}{0}$, undefined

$\cot\theta=\frac{x}{y}=\frac{0}{k}=0$

$\sec\theta=\frac{r}{x}=\frac{r}{0}$, undefined

$\csc\theta=\frac{r}{y}=\frac{r}{k}$

The tangent and secant are undefined.

34. Since the terminal side of the angle is defined by $5x-3y=0, x\ge 0$, a point on this terminal side would be $(3,5)$ since $5(3)-3(5)=0$. The corresponding value of r is $r=\sqrt{3^2+5^2}=\sqrt{9+25}=\sqrt{34}$.

$$\sin\theta=\frac{y}{r}=\frac{5}{\sqrt{34}}=\frac{5}{\sqrt{34}}\cdot\frac{\sqrt{34}}{\sqrt{34}}=\frac{5\sqrt{34}}{34}\qquad \cot\theta=\frac{x}{y}=\frac{3}{5}$$

$$\cos\theta=\frac{x}{r}=\frac{3}{\sqrt{34}}=\frac{3}{\sqrt{34}}\cdot\frac{\sqrt{34}}{\sqrt{34}}=\frac{3\sqrt{34}}{34}\qquad \sec\theta=\frac{r}{x}=\frac{\sqrt{34}}{3}$$

$$\tan\theta=\frac{y}{x}=\frac{5}{3}\qquad \csc\theta=\frac{r}{y}=\frac{\sqrt{34}}{5}$$

35.

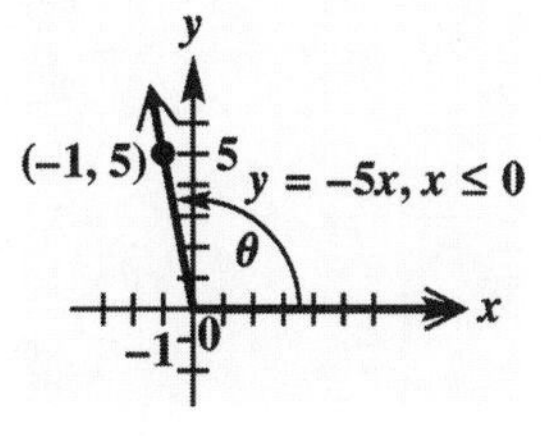

36. Using the point (–1, 5), the corresponding value of r is $r=\sqrt{(-1)^2+(5)^2}=\sqrt{1+25}=\sqrt{26}$.

$$\sin\theta=\frac{y}{r}=\frac{5}{\sqrt{26}}=\frac{5}{\sqrt{26}}\cdot\frac{\sqrt{26}}{\sqrt{26}}=\frac{5\sqrt{26}}{26}$$

$$\cos\theta=\frac{x}{r}=\frac{-1}{\sqrt{26}}=-\frac{1}{\sqrt{26}}\cdot\frac{\sqrt{26}}{\sqrt{26}}=-\frac{\sqrt{26}}{26}$$

$$\tan\theta=\frac{y}{x}=\frac{5}{-1}=-5$$

For Exercises 37 – 38, all angles are quadrantal angles. For $90°$ we will choose the point $(0,1)$, so $x=0,\ y=1$, and $r=1$. For $180°$ we will choose the point $(-1,0)$, so $x=-1,\ y=0$, and $r=1$. Finally, for $270°$ we will choose the point $(0,-1)$, so $x=0,\ y=-1$, and $r=1$.

37. $4\sec 180°-2\sin^2 270°$

$$\sec 180°=\frac{r}{x}=\frac{1}{-1}=-1 \text{ and } \sin 270°=\frac{y}{r}=\frac{-1}{1}=-1$$

$$4\sec 180°-2\sin^2 270°=4(-1)-2(-1)^2=-4-2(1)=-4-2=-6$$

38. $-\cot^2 90°+4\sin 270°-3\tan 180°$

$$\cot 90°=\frac{x}{y}=\frac{0}{1}=0,\ \sin 270°=\frac{y}{r}=\frac{-1}{1}=-1, \text{ and } \tan 180°=\frac{y}{x}=\frac{0}{-1}=0$$

$$-\cot^2 90°+4\sin 270°-3\tan 180°=-(0^2)+4(-1)-3(0)=0+(-4)-0=-4$$

39. **(a)** For any angle θ, $\sec\theta\ge 1$ or $\sec\theta\le -1$. Therefore, $\sec\theta=-\frac{2}{3}$ is impossible.

(b) $\tan\theta$ can take on all values. Therefore, $\tan\theta=1.4$ is possible.

(c) For any angle θ, $\csc\theta\ge 1$ or $\csc\theta\le -1$. Therefore, $\csc\theta=5$ is possible.

40. $\sin\theta = \frac{\sqrt{3}}{5}$ and $\cos\theta < 0$.

$\sin\theta = \frac{y}{r}$, so let $y = \sqrt{3}, r = 5$.

$$x^2 + y^2 = r^2 \Rightarrow x^2 + \left(\sqrt{3}\right)^2 = 5^2 \Rightarrow x^2 + 3 = 25 \Rightarrow x^2 = 22 \Rightarrow x = \pm\sqrt{22}$$

$\cos\theta < 0$, so $x = -\sqrt{22}$.

$\sin\theta = \frac{y}{r} = \frac{\sqrt{3}}{5}$

$\cos\theta = \frac{x}{r} = \frac{-\sqrt{22}}{5} = -\frac{\sqrt{22}}{5}$

$\tan\theta = \frac{y}{x} = \frac{\sqrt{3}}{-\sqrt{22}} = -\frac{\sqrt{3}}{\sqrt{22}} \cdot \frac{\sqrt{22}}{\sqrt{22}} = -\frac{\sqrt{66}}{22}$

$\cot\theta = \frac{x}{y} = \frac{-\sqrt{22}}{\sqrt{3}} = -\frac{\sqrt{22}}{\sqrt{3}} \cdot \frac{\sqrt{3}}{\sqrt{3}} = -\frac{\sqrt{66}}{3}$

$\sec\theta = \frac{r}{x} = \frac{5}{-\sqrt{22}} = -\frac{5}{\sqrt{22}} \cdot \frac{\sqrt{22}}{\sqrt{22}} = -\frac{5\sqrt{22}}{22}$

$\csc\theta = \frac{r}{y} = \frac{5}{\sqrt{3}} = \frac{5}{\sqrt{3}} \cdot \frac{\sqrt{3}}{\sqrt{3}} = \frac{5\sqrt{3}}{3}$

41. $\cos\theta = -\frac{5}{8} = \frac{-5}{8}$, with θ in quadrant III

$\cos\theta = \frac{x}{r}$ and θ in quadrant III, so let $x = -5$, $r = 8$.

$$x^2 + y^2 = r^2 \Rightarrow (-5)^2 + y^2 = 8^2 \Rightarrow 25 + y^2 = 64 \Rightarrow y^2 = 39 \Rightarrow y = \pm\sqrt{39}$$

θ is in quadrant III, so $y = -\sqrt{39}$.

$\sin\theta = \frac{y}{r} = \frac{-\sqrt{39}}{8} = -\frac{\sqrt{39}}{8}$

$\cos\theta = \frac{x}{r} = -\frac{5}{8}$

$\tan\theta = \frac{y}{x} = \frac{-\sqrt{39}}{-5} = \frac{\sqrt{39}}{5}$

$\cot\theta = \frac{x}{y} = \frac{-5}{-\sqrt{39}} = \frac{5}{\sqrt{39}} \cdot \frac{\sqrt{39}}{\sqrt{39}} = \frac{5\sqrt{39}}{39}$

$\sec\theta = \frac{r}{x} = \frac{8}{-5} = -\frac{8}{5}$

$\csc\theta = \frac{r}{y} = \frac{8}{-\sqrt{39}} = -\frac{8}{\sqrt{39}}$

$= -\frac{8}{\sqrt{39}} \cdot \frac{\sqrt{39}}{\sqrt{39}} - \frac{8\sqrt{39}}{39}$

42. $\tan\alpha = 2 = \frac{-2}{-1}$, with α in quadrant III.

$\tan\alpha = \frac{y}{x}$, so let $y = -2$ and $x = -1$.

$$x^2 + y^2 = r^2 \Rightarrow (-1)^2 + (-2)^2 = r^2 \Rightarrow 1 + 4 = r^2 \Rightarrow 5 = r^2 \Rightarrow r = \sqrt{5}, \text{ since } r \text{ is positive}$$

$\sin\alpha = \frac{y}{r} = \frac{-2}{\sqrt{5}} = -\frac{2}{\sqrt{5}} \cdot \frac{\sqrt{5}}{\sqrt{5}} = -\frac{2\sqrt{5}}{5}$

$\cos\alpha = \frac{x}{r} = \frac{-1}{\sqrt{5}} = -\frac{1}{\sqrt{5}} \cdot \frac{\sqrt{5}}{\sqrt{5}} = -\frac{\sqrt{5}}{5}$

$\tan\alpha = \frac{y}{x} = \frac{-2}{-1} = 2$

$\cot\alpha = \frac{x}{y} = \frac{-1}{-2} = \frac{1}{2}$

$\sec\alpha = \frac{r}{x} = \frac{\sqrt{5}}{-1} = -\sqrt{5}$

$\csc\alpha = \frac{r}{y} = \frac{\sqrt{5}}{-2} = -\frac{\sqrt{5}}{2}$

43. $\sec\beta = -\sqrt{5} = \frac{\sqrt{5}}{-1}$, with β in quadrant II.

$\sec\beta = \frac{r}{x}$, so let $r = \sqrt{5}$ and $x = -1$.

$$x^2 + y^2 = r^2 \Rightarrow (-1)^2 + y^2 = \left(\sqrt{5}\right)^2 \Rightarrow 1 + y^2 = 5 \Rightarrow y^2 = 4 \Rightarrow y = \pm 2$$

β is in quadrant II, so $y = 2$.

$$\sin\beta = \frac{y}{r} = \frac{2}{\sqrt{5}} = \frac{2}{\sqrt{5}} \cdot \frac{\sqrt{5}}{\sqrt{5}} = \frac{2\sqrt{5}}{5}$$

$$\cos\beta = \frac{x}{r} = \frac{-1}{\sqrt{5}} = -\frac{1}{\sqrt{5}} \cdot \frac{\sqrt{5}}{\sqrt{5}} = -\frac{\sqrt{5}}{5}$$

$$\tan\beta = \frac{y}{x} = \frac{2}{-1} = -2$$

$$\cot\beta = \frac{x}{y} = \frac{-1}{2} = -\frac{1}{2}$$

$$\sec\beta = \frac{r}{x} = \frac{\sqrt{5}}{-1} = -\sqrt{5}$$

$$\csc\beta = \frac{r}{y} = \frac{\sqrt{5}}{2}$$

44. $\sin\theta = -\frac{2}{5} = \frac{-2}{5}$, with θ in quadrant III.

$\sin\theta = \frac{y}{r}$, so let $r = 5$ and $y = -2$.

$$x^2 + y^2 = r^2 \Rightarrow x^2 + (-2)^2 = 5^2 \Rightarrow x^2 + 4 = 25 \Rightarrow x^2 = 21 \Rightarrow x = \pm\sqrt{21}$$

θ is in quadrant III, so $x = -\sqrt{21}$.

$$\sin\theta = \frac{y}{r} = \frac{-2}{5} = -\frac{2}{5}$$

$$\cos\theta = \frac{x}{r} = \frac{-\sqrt{21}}{5} = -\frac{\sqrt{21}}{5}$$

$$\tan\theta = \frac{y}{x} = \frac{-2}{-\sqrt{21}} = \frac{2}{\sqrt{21}} \cdot \frac{\sqrt{21}}{\sqrt{21}} = \frac{2\sqrt{21}}{21}$$

$$\cot\theta = \frac{x}{y} = \frac{-\sqrt{21}}{-2} = \frac{\sqrt{21}}{2}$$

$$\sec\theta = \frac{r}{x} = \frac{5}{-\sqrt{21}} = -\frac{5}{\sqrt{21}} \cdot \frac{\sqrt{21}}{\sqrt{21}} = -\frac{5\sqrt{21}}{21}$$

$$\csc\theta = \frac{r}{y} = \frac{5}{-2} = -\frac{5}{2}$$

45. $\sec\alpha = \frac{5}{4}$, with θ in quadrant IV.

$\sec\alpha = \frac{r}{x}$, so let $r = 5$ and $x = 4$.

$$x^2 + y^2 = r^2 \Rightarrow 4^2 + y^2 = 5^2 \Rightarrow 16 + y^2 = 25 \Rightarrow y^2 = 9 \Rightarrow y = \pm 3$$

α is in quadrant IV, so $y = -3$.

$$\sin\alpha = \frac{y}{r} = \frac{-3}{5} = -\frac{3}{5}$$

$$\cos\alpha = \frac{x}{r} = \frac{4}{5}$$

$$\tan\alpha = \frac{y}{x} = \frac{-3}{4} = -\frac{3}{4}$$

$$\cot\alpha = \frac{x}{y} = \frac{4}{-3} = -\frac{4}{3}$$

$$\sec\alpha = \frac{r}{x} = \frac{5}{4}$$

$$\csc\alpha = \frac{r}{y} = \frac{5}{-3} = -\frac{5}{3}$$

46. The sine is negative in quadrants III and IV. The cosine is positive in quadrants I and IV. Since $\sin\theta < 0$ and $\cos\theta > 0, \theta$ lies in quadrant IV. Since $\tan\theta = \frac{\sin\theta}{\cos\theta}$ and $\sin\theta < 0$ and $\cos\theta > 0$, the sign of $\tan\theta$ is negative.

47. Since $\cot\theta = \frac{1}{\tan\theta}$, use the calculator to find $\frac{1}{1.6778490}$ or 1.6778490^{-1}. The result would be .5960011.

48. $\dfrac{360°}{26{,}000 \text{ yr}} = \dfrac{9}{650} \dfrac{\text{deg}}{\text{yr}} = \dfrac{9}{650}(60) \dfrac{\text{min}}{\text{yr}} = \dfrac{54}{13} \dfrac{\text{min}}{\text{yr}} = \dfrac{54}{13}(60)\dfrac{\text{sec}}{\text{yr}} = \dfrac{648}{13} \text{ sec / yr} \approx 50 \text{ sec/yr}$

49. Let x be the depth of the crater Autolycus. We can set up and solve the following proportion.

$$\frac{x}{11{,}000} = \frac{1.3}{1.5} \Rightarrow 1.5x = 1.3(11{,}000) \Rightarrow 1.5x = 14{,}300 \Rightarrow x = \frac{14{,}300}{1.5} \approx 9500$$

The depth of the crater Autolycus is about 9500 ft.

50. Let x be the height of Bradley. We can set up and solve the following proportion.

$$\frac{x}{21{,}000} = \frac{1.8}{2.8} \Rightarrow 2.8x = 1.8(21{,}000) \Rightarrow 2.8x = 37{,}800 \Rightarrow x = \frac{37{,}800}{2.8} = 13{,}500$$

Bradley is 13,500 ft tall.

Chapter 1: Test

1. $74°\, 17'\, 54'' = 74° + \frac{17}{60}° + \frac{54}{3600}° \approx 74° + .2833° + .0150° \approx 74.2983°$

2. $360° + (-157°) = 203°$

3. 450 rotations per min $= \frac{450}{60}$ rotations per sec $= \frac{15}{2}$ rotations per sec $= \frac{15}{2}(360°)$ per sec $= 2700°$ per sec

A point on the edge of the tire will move $2700°$ in 1 sec.

4. The sum of the measures of the interior angles of a triangle is $180°$.

$$60° + y + 90° = 180° \Rightarrow 150° + y = 180° \Rightarrow y = 30°$$

Also, since the measures of x and y sum to be the third angle of a triangle, we have the following.

$$30° + (x + y) + 90° = 180°$$

Substitute $30°$ for y and solve for x.

$$30° + (x + y) + 90° = 180° \Rightarrow 30° + (x + 30°) + 90° = 180° \Rightarrow x + 150° = 180° \Rightarrow x = 30°$$

Thus, $x = y = 30°$.

5. Since we have similar triangles, we can set up and solve the following proportion.

$$\frac{20}{30} = \frac{x}{100 - x} \Rightarrow 20(100 - x) = 30x \Rightarrow 2000 - 20x = 30x \Rightarrow 2000 = 50x \Rightarrow x = 40$$

The lifeguard will enter the water 40 yd east of his original position.

6. $(2, -5)$

$x = 2,\ y = -5,$ and $r = \sqrt{x^2 + y^2} = \sqrt{2^2 + (-5)^2} = \sqrt{4 + 25} = \sqrt{29}$

$$\sin\theta = \frac{y}{r} = \frac{-5}{\sqrt{29}} = -\frac{5}{\sqrt{29}} \cdot \frac{\sqrt{29}}{\sqrt{29}} = -\frac{5\sqrt{29}}{29}$$

$$\cos\theta = \frac{x}{r} = \frac{2}{\sqrt{29}} = \frac{2}{\sqrt{29}} \cdot \frac{\sqrt{29}}{\sqrt{29}} = \frac{2\sqrt{29}}{29}$$

$$\tan\theta = \frac{y}{x} = \frac{-5}{2} = -\frac{5}{2}$$

7. If $\cos\theta < 0$, then θ is in quadrant II or III. If $\cot\theta > 0$, then θ is in quadrant I or III. Therefore, θ terminates in quadrant III.

8. $\cos\theta = \frac{4}{5}$, with θ in quadrant IV

$\cos\theta = \frac{x}{r}$ so let $x = 4$, $r = 5$.

$$x^2 + y^2 = r^2 \Rightarrow 4^2 + y^2 = 5^2 \Rightarrow 16 + y^2 = 25 \Rightarrow y^2 = 9 \Rightarrow y = \pm 3$$

θ is in quadrant IV, so $y = -3$.

$$\cos\theta = \frac{x}{r} = \frac{4}{5} \qquad \cot\theta = \frac{x}{y} = \frac{4}{-3} = -\frac{4}{3}$$

$$\sin\theta = \frac{y}{r} = \frac{-3}{5} = -\frac{3}{5} \qquad \sec\theta = \frac{r}{x} = \frac{5}{4}$$

$$\tan\theta = \frac{y}{x} = \frac{-3}{4} = -\frac{3}{4} \qquad \csc\theta = \frac{r}{y} = \frac{5}{-3} = -\frac{5}{3}$$

9. No; when $\sin\theta$ and $\cos\theta$ are both negative for the same value of θ, $\tan\theta = \frac{\sin\theta}{\cos\theta}$ is positive.

10. (a) If θ is in the interval $(90°, 180°)$, then $90° < \theta < 180° \Rightarrow 45° < \frac{\theta}{2} < 90°$. Thus, $\frac{\theta}{2}$ lies in quadrant I and $\cos\frac{\theta}{2}$ is positive.

(b) If θ is in the interval $(90°, 180°)$ then $90° < \theta < 180° \Rightarrow -90° < \theta - 180° < 0°$. Thus, $\theta - 180°$ lies in quadrant IV and $\cot(\theta - 180°)$ is negative.

11. $\cos\theta = \frac{1}{\sec\theta} = \frac{1}{-10} = -\frac{1}{10}$ or $-.1$

12. $\cot\theta = \frac{1}{\tan\theta}$; $\frac{1}{\tan\theta} = \frac{1}{\frac{\sin\theta}{\cos\theta}} = 1 \div \frac{\sin\theta}{\cos\theta} = 1 \cdot \frac{\cos\theta}{\sin\theta} = \frac{\cos\theta}{\sin\theta}$;

$1 + \cot^2\theta = \csc^2\theta \Rightarrow \cot^2\theta = \csc^2\theta - 1 \Rightarrow \cot\theta = \pm\sqrt{\csc^2\theta - 1}$

Chapter 1: Quantitative Reasoning

Step 1: The ratios of corresponding sides of similar triangles CAG and HAD are equal and $HD = 1$.

Step 2: The ratios of corresponding sides of similar triangles AGE and ADB are equal.

Step 3: $EF = BD = 1$

Step 4: From Steps 1 – 3, $CG = \frac{AG}{AD} = \frac{EG}{EF} = \frac{EG}{1} = EG$.

The height of the tree (in feet) is (approximately) the number of paces.

Chapter 2

ACUTE ANGLES AND RIGHT TRIANGLES

Section 2.1: Trigonometric Functions of Acute Angles

1. $\sin A = \frac{\text{side opposite}}{\text{hypotenuse}} = \frac{21}{29}$

$\cos A = \frac{\text{side adjacent}}{\text{hypotenuse}} = \frac{20}{29}$

$\tan A = \frac{\text{side opposite}}{\text{side adjacent}} = \frac{21}{20}$

2. $\sin A = \frac{\text{side opposite}}{\text{hypotenuse}} = \frac{45}{53}$

$\cos A = \frac{\text{side adjacent}}{\text{hypotenuse}} = \frac{28}{53}$

$\tan A = \frac{\text{side opposite}}{\text{side adjacent}} = \frac{45}{28}$

3. $\sin A = \frac{\text{side opposite}}{\text{hypotenuse}} = \frac{n}{p}$

$\cos A = \frac{\text{side adjacent}}{\text{hypotenuse}} = \frac{m}{p}$

$\tan A = \frac{\text{side opposite}}{\text{side adjacent}} = \frac{n}{m}$

4. $\sin A = \frac{\text{side opposite}}{\text{hypotenuse}} = \frac{k}{z}$

$\cos A = \frac{\text{side adjacent}}{\text{hypotenuse}} = \frac{y}{z}$

$\tan A = \frac{\text{side opposite}}{\text{side adjacent}} = \frac{k}{y}$

For Exercises 5–10, refer to the Function Values of Special Angles chart on page 50 of your text.

5. C; $\sin 30° = \frac{1}{2}$

6. H; $\cos 45° = \frac{\sqrt{2}}{2}$

7. B; $\tan 45° = 1$

8. G; $\sec 60° = \frac{1}{\cos 60°} = \frac{1}{\frac{1}{2}} = 2$

9. E; $\csc 60° = \frac{1}{\sin 60°} = \frac{1}{\frac{\sqrt{3}}{2}} = \frac{2}{\sqrt{3}} = \frac{2}{\sqrt{3}} \cdot \frac{\sqrt{3}}{\sqrt{3}} = \frac{2\sqrt{3}}{3}$

10. A; $\cot 30° = \frac{\cos 30°}{\sin 30°} = \frac{\frac{\sqrt{3}}{2}}{\frac{1}{2}} = \frac{\sqrt{3}}{2} \cdot \frac{2}{1} = \sqrt{3}$

11. $a = 5, b = 12$

$c^2 = a^2 + b^2 \Rightarrow c^2 = 5^2 + 12^2 \Rightarrow c^2 = 25 + 144 \Rightarrow c^2 = 169 \Rightarrow c = 13$

$\sin B = \frac{\text{side opposite}}{\text{hypotenuse}} = \frac{b}{c} = \frac{12}{13}$

$\cos B = \frac{\text{side adjacent}}{\text{hypotenuse}} = \frac{a}{c} = \frac{5}{13}$

$\tan B = \frac{\text{side opposite}}{\text{side adjacent}} = \frac{b}{a} = \frac{12}{5}$

$\cot B = \frac{\text{side adjacent}}{\text{side opposite}} = \frac{a}{b} = \frac{5}{12}$

$\sec B = \frac{\text{hypotenuse}}{\text{side adjacent}} = \frac{c}{a} = \frac{13}{5}$

$\csc B = \frac{\text{hypotenuse}}{\text{side opposite}} = \frac{c}{b} = \frac{13}{12}$

12. $a = 3, b = 5$

$c^2 = a^2 + b^2 \Rightarrow c^2 = 3^2 + 5^2 \Rightarrow c^2 = 9 + 25 \Rightarrow c^2 = 34 \Rightarrow c = \sqrt{34}$

$\sin B = \frac{\text{side opposite}}{\text{hypotenuse}} = \frac{b}{c} = \frac{5}{\sqrt{34}} = \frac{5}{\sqrt{34}} \cdot \frac{\sqrt{34}}{\sqrt{34}} = \frac{5\sqrt{34}}{34}$

$\cos B = \frac{\text{side adjacent}}{\text{hypotenuse}} = \frac{a}{c} = \frac{3}{\sqrt{34}} = \frac{3}{\sqrt{34}} \cdot \frac{\sqrt{34}}{\sqrt{34}} = \frac{3\sqrt{34}}{34}$

$\tan B = \frac{\text{side opposite}}{\text{side adjacent}} = \frac{b}{a} = \frac{5}{3}$

$\cot B = \frac{\text{side adjacent}}{\text{side opposite}} = \frac{a}{b} = \frac{3}{5}$

$\sec B = \frac{\text{hypotenuse}}{\text{side adjacent}} = \frac{c}{a} = \frac{\sqrt{34}}{3}$

$\csc B = \frac{\text{hypotenuse}}{\text{side opposite}} = \frac{c}{b} = \frac{\sqrt{34}}{5}$

13. $a = 6, c = 7$

$c^2 = a^2 + b^2 \Rightarrow 7^2 = 6^2 + b^2 \Rightarrow 49 = 36 + b^2 \Rightarrow b^2 = 13 \Rightarrow b = \sqrt{13}$

$\sin B = \frac{\text{side opposite}}{\text{hypotenuse}} = \frac{b}{c} = \frac{\sqrt{13}}{7}$

$\cos B = \frac{\text{side adjacent}}{\text{hypotenuse}} = \frac{a}{c} = \frac{6}{7}$

$\tan B = \frac{\text{side opposite}}{\text{side adjacent}} = \frac{b}{a} = \frac{\sqrt{13}}{6}$

$\cot B = \frac{\text{side adjacent}}{\text{side opposite}} = \frac{a}{b} = \frac{6}{\sqrt{13}} = \frac{6}{\sqrt{13}} \cdot \frac{\sqrt{13}}{\sqrt{13}} = \frac{6\sqrt{13}}{13}$

$\sec B = \frac{\text{hypotenuse}}{\text{side adjacent}} = \frac{c}{a} = \frac{7}{6}$

$\csc B = \frac{\text{hypotenuse}}{\text{side opposite}} = \frac{c}{b} = \frac{7}{\sqrt{13}} = \frac{7}{\sqrt{13}} \cdot \frac{\sqrt{13}}{\sqrt{13}} = \frac{7\sqrt{13}}{13}$

14. $b = 7, c = 12$

$c^2 = a^2 + b^2 \Rightarrow 12^2 = a^2 + 7^2 \Rightarrow 144 = a^2 + 49 \Rightarrow a^2 = 95 \Rightarrow a = \sqrt{95}$

$\sin B = \frac{\text{side opposite}}{\text{hypotenuse}} = \frac{b}{c} = \frac{7}{12}$

$\cos B = \frac{\text{side adjacent}}{\text{hypotenuse}} = \frac{a}{c} = \frac{\sqrt{95}}{12}$

$\tan B = \frac{\text{side opposite}}{\text{side adjacent}} = \frac{b}{a} = \frac{7}{\sqrt{95}} = \frac{7}{\sqrt{95}} \cdot \frac{\sqrt{95}}{\sqrt{95}} = \frac{7\sqrt{95}}{95}$

$\cot B = \frac{\text{side adjacent}}{\text{side opposite}} = \frac{a}{b} = \frac{\sqrt{95}}{7}$

$\sec B = \frac{\text{hypotenuse}}{\text{side adjacent}} = \frac{c}{a} = \frac{12}{\sqrt{95}} = \frac{12}{\sqrt{95}} \cdot \frac{\sqrt{95}}{\sqrt{95}} = \frac{12\sqrt{95}}{95}$

$\csc B = \frac{\text{hypotenuse}}{\text{side opposite}} = \frac{c}{b} = \frac{12}{7}$

15. $\sin\theta = \cos(90° - \theta)$; $\cos\theta = \sin(90° - \theta)$;
$\tan\theta = \cot(90° - \theta)$; $\cot\theta = \tan(90° - \theta)$;
$\sec\theta = \csc(90° - \theta)$; $\csc\theta = \sec(90° - \theta)$

16. $\cot 73° = \tan(90° - 73°) = \tan 17°$

17. $\sec 39° = \csc(90° - 39°) = \csc 51°$

18. $\cos(\alpha + 20°) = \sin[90° - (\alpha + 20°)]$
$= \sin(90° - \alpha - 20°)$
$= \sin(70° - \alpha)$

19. $\cot(\theta - 10°) = \tan[90° - (\theta - 10°)]$
$= \tan(90° - \theta + 10°)$
$= \tan(100° - \theta)$

20. $\tan 25.4° = \cot(90° - 25.4°) = \cot 64.6°$

21. $\sin 38.7° = \cos(90° - 38.7°) = \cos 51.3°$

22. Using $A = 50°, 102°, 248°,$ and $-26°,$ we see that $\sin(90° - A)$ and $\cos A$ yield the same values.

```
sin(90-50)
      .6427876097
cos(50)
      .6427876097
```

```
sin(90-102)
     -.2079116908
cos(102)
     -.2079116908
```

```
sin(90-248)
     -.3746065934
cos(248)
     -.3746065934
```

```
sin(90--26)
      .8987940463
cos(-26)
      .8987940463
```

23. $\tan \alpha = \cot(\alpha + 10°)$

Since tangent and cotangent are cofunctions, this equation is true if the sum of the angles is 90°.

$$\alpha + (\alpha + 10°) = 90° \Rightarrow 2\alpha + 10° = 90° \Rightarrow 2\alpha = 80° \Rightarrow \alpha = 40°$$

24. $\cos\theta = \sin 2\theta$

Since sine and cosine are cofunctions, this equation is true if the sum of the angles is 90°.

$$\theta + 2\theta = 90° \Rightarrow 3\theta = 90° \Rightarrow \theta = 30°$$

25. $\sin(2\theta + 10°) = \cos(3\theta - 20°)$

Since sine and cosine are cofunctions, this equation is true if the sum of the angles is 90°.

$$(2\theta + 10°) + (3\theta - 20°) = 90° \Rightarrow 5\theta - 10° = 90° \Rightarrow 5\theta = 100° \Rightarrow \theta = 20°$$

26. $\sec(\beta + 10°) = \csc(2\beta + 20°)$

Since secant and cosecant are cofunctions, this equation is true if the sum of the angles is 90°.

$$(\beta + 10°) + (2\beta + 20°) = 90° \Rightarrow 3\beta + 30° = 90° \Rightarrow 3\beta = 60° \Rightarrow \beta = 20°$$

27. $\tan(3B + 4°) = \cot(5B - 10°)$

Since tangent and cotangent are cofunctions, this equation is true if the sum of the angles is 90°.

$$(3B + 4°) + (5B - 10°) = 90° \Rightarrow 8B - 6° = 90° \Rightarrow 8B = 96° \Rightarrow B = 12°$$

28. $\cot(5\theta + 2°) = \tan(2\theta + 4°)$

Since tangent and cotangent are cofunctions, this equation is true if the sum of the angles is 90°.

$$(5\theta + 2°) + (2\theta + 4°) = 90° \Rightarrow 7\theta + 6° = 90° \Rightarrow 7\theta = 84° \Rightarrow \theta = 12°$$

29. $\sin 50° > \sin 40°$

In the interval from 0° to 90°, as the angle increases, so does the sine of the angle, so $\sin 50° > \sin 40°$ is true.

30. $\tan 28° \leq \tan 40°$

$\tan\theta$ increases as θ increases from 0° to 90°. Since 40°>28°, tan 40° > tan 28°. Therefore, the given statement is true.

31. $\sin 46° < \cos 46°$

$\sin\theta$ increases as θ increases from 0° to 90°. Since $46° > 44°$, $\sin 46° > \sin 44°$ and $\sin 44° = \cos 46°$, we have $\sin 46° > \cos 46°$. Thus, the statement is false.

32. $\cos 28° < \sin 28°$

$\cos\theta$ decreases as θ increases from 0° to 90°. Since $28° < 62°$, $\cos 28° > \cos 62°$ and $\cos 62° = \sin 28°$, we have $\cos 28° > \sin 28°$. Thus, the statement is false.

33. $\tan 41° < \cot 41°$

$\tan\theta$ increases as θ increases from 0° to 90°. Since $49° > 41°$, $\tan 49° > \tan 41°$. Since $\tan 49° = \cot 41°$, we have $\cot 41° > \tan 41°$. Therefore, the statement is true.

34. $\cot 30° < \tan 40°$

$\tan(90° - 30°) = \tan 60° = \cot 30°$

In the interval from 0° to 90°, the tangent increases so, $\tan 60° > \tan 40°$. Therefore, $\cot 30° > \tan 40°$, and the statement is false.

For Exercises 35 – 40, refer to the following figure (Figure 6 from page 49 of your text).

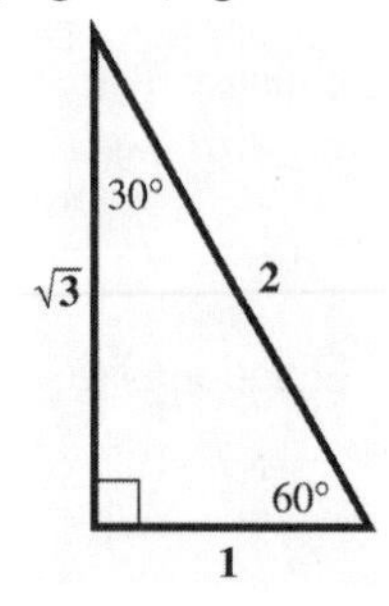

35. $\tan 30° = \dfrac{\text{side opposite}}{\text{side adjacent}} = \dfrac{1}{\sqrt{3}} = \dfrac{1}{\sqrt{3}} \cdot \dfrac{\sqrt{3}}{\sqrt{3}} = \dfrac{\sqrt{3}}{3}$

36. $\cot 30° = \dfrac{\text{side adjacent}}{\text{side opposite}} = \dfrac{\sqrt{3}}{1} = \sqrt{3}$

37. $\sin 30° = \dfrac{\text{side opposite}}{\text{hypotenuse}} = \dfrac{1}{2}$

38. $\cos 30° = \dfrac{\text{side adjacent}}{\text{hypotenuse}} = \dfrac{\sqrt{3}}{2}$

39. $\sec 30° = \dfrac{\text{hypotenuse}}{\text{side adjacent}} = \dfrac{2}{\sqrt{3}} = \dfrac{2}{\sqrt{3}} \cdot \dfrac{\sqrt{3}}{\sqrt{3}} = \dfrac{2\sqrt{3}}{3}$

40. $\csc 30° = \dfrac{\text{hypotenuse}}{\text{side opposite}} = \dfrac{2}{1} = 2$

For Exercises 41 – 44, refer to the following figure (Figure 7 from page 50 of your text).

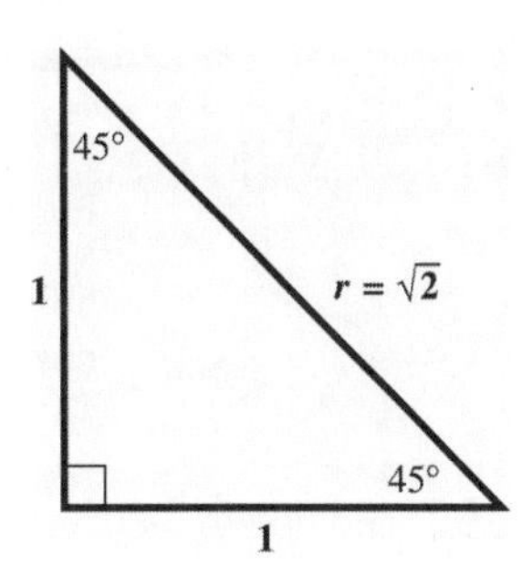

41. $\csc 45° = \dfrac{\text{hypotenuse}}{\text{side opposite}} = \dfrac{\sqrt{2}}{1} = \sqrt{2}$

42. $\sec 45° = \dfrac{\text{hypotenuse}}{\text{side adjacent}} = \dfrac{\sqrt{2}}{1} = \sqrt{2}$

43. $\cos 45° = \dfrac{\text{side adjacent}}{\text{hypotenuse}} = \dfrac{1}{\sqrt{2}}$

$= \dfrac{1}{\sqrt{2}} \cdot \dfrac{\sqrt{2}}{\sqrt{2}} = \dfrac{\sqrt{2}}{2}$

44. $\cot 45° = \dfrac{\text{side adjacent}}{\text{side opposite}} = \dfrac{\sqrt{2}}{\sqrt{2}} = 1$

45.

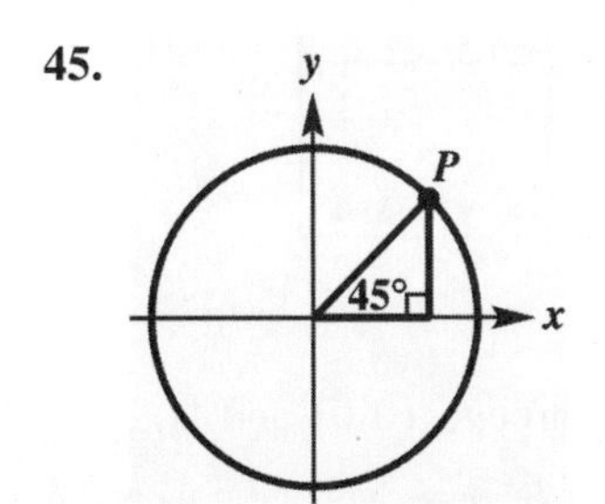

46. $\sin 45° = \dfrac{y}{4} \Rightarrow y = 4\sin 45° = 4 \cdot \dfrac{\sqrt{2}}{2} = 2\sqrt{2}$

$\cos 45° = \dfrac{x}{4} \Rightarrow x = 4\cos 45° = 4 \cdot \dfrac{\sqrt{2}}{2} = 2\sqrt{2}$

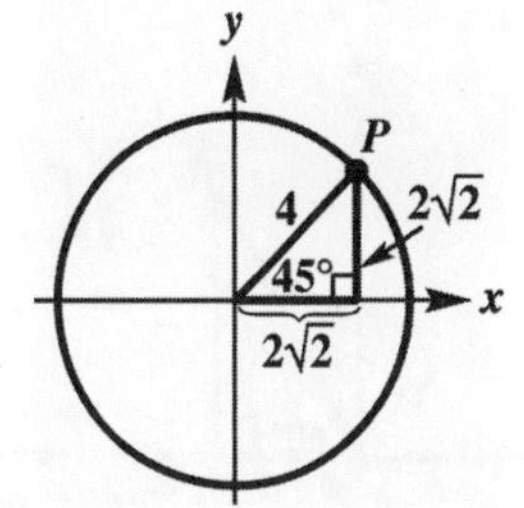

47. The legs of the right triangle provide the coordinates of *P*. *P* is $\left(2\sqrt{2}, 2\sqrt{2}\right)$.

48.

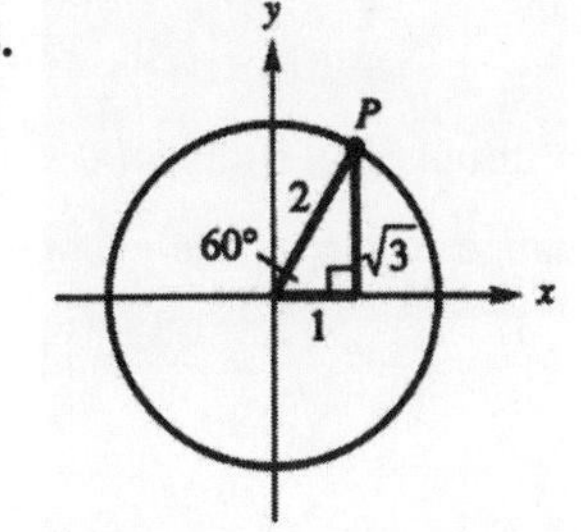

$\sin 60° = \dfrac{y}{2} \Rightarrow y = 2\sin 60° = 2 \cdot \dfrac{\sqrt{3}}{2} = \sqrt{3}$ and $\cos 60° = \dfrac{x}{2} \Rightarrow x = 2\cos 60° = 2 \cdot \dfrac{1}{2} = 1$

The legs of the right triangle provide the coordinates of *P*. *P* is $\left(1, \sqrt{3}\right)$.

49. Y_1 is $\sin x$ and Y_2 is $\tan x$.

$\sin 0° = 0$	$\tan 0° = 0$
$\sin 30° = .5$	$\tan 30° \approx .57735$
$\sin 45° \approx .70711$	$\tan 45° = 1$
$\sin 60° \approx .86603$	$\tan 60° = 1.7321$
$\sin 90° = 1$	$\tan 90°$: undefined

50. Y_1 is $\cos x$ and Y_2 is $\csc x$.

$\cos 0° = 1$	$\csc 0°$: undefined
$\cos 30° \approx .86603$	$\csc 30° = 2$
$\cos 45° \approx .70711$	$\csc 45° \approx 1.4142$
$\cos 60° = .5$	$\csc 60° \approx 1.1547$
$\cos 90° = 0$	$\csc 90° = 1$

51. Since $\sin 60° = \frac{\sqrt{3}}{2}$ and 60° is between 0° and 90°, A = 60°.

52. .7071067812 is a rational approximation for the exact value $\frac{\sqrt{2}}{2}$ (an irrational value).

53. The point of intersection is (.70710678, .70710678). This corresponds to the point $\left(\frac{\sqrt{2}}{2}, \frac{\sqrt{2}}{2}\right)$.

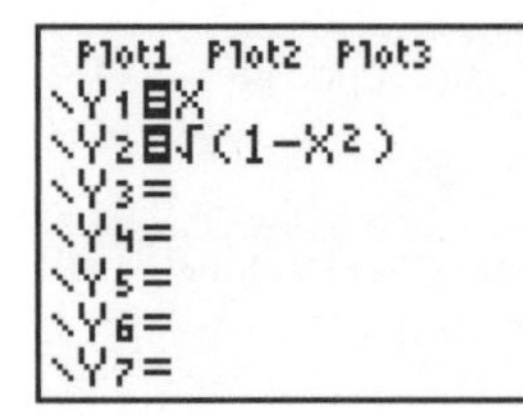

WINDOW
Xmin=-1.8
Xmax=1.8
Xscl=1
Ymin=-1.2
Ymax=1.2
Yscl=1
Xres=1

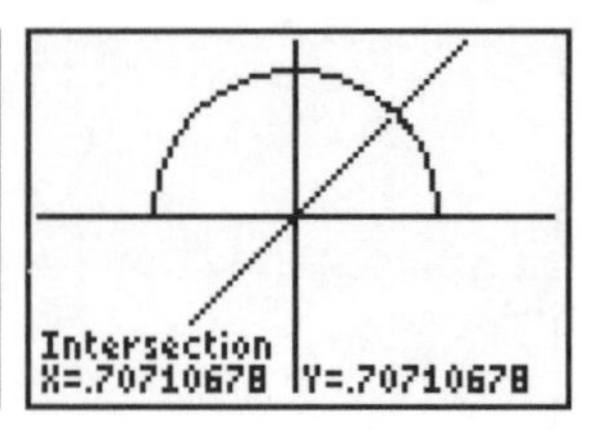

These coordinates are the sine and cosine of 45°.

54.

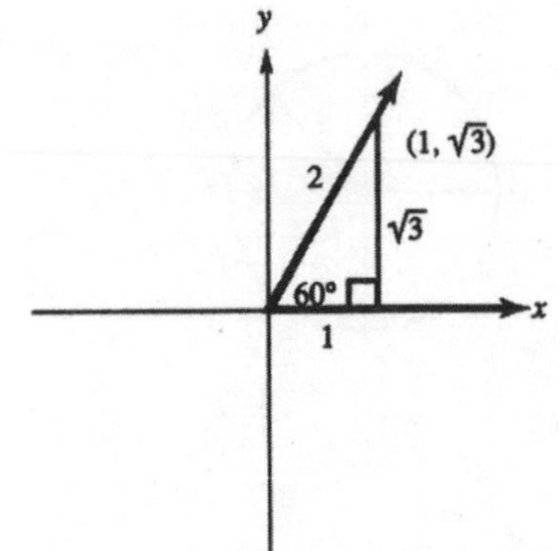

The line passes through $(0,0)$ and $(1,\sqrt{3})$. The slope is change in y over the change in x. Thus, $m = \frac{\sqrt{3}}{1} = \sqrt{3}$ and the equation of the line is $y = \sqrt{3}x$.

55.

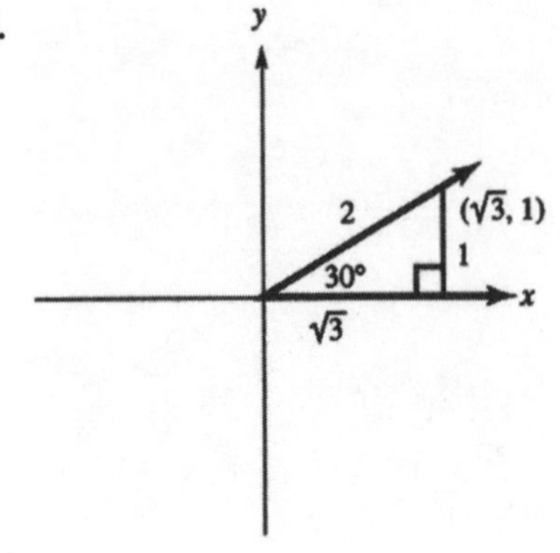

The line passes through (0, 0) and $(\sqrt{3}, 1)$. The slope is change in y over the change in x. Thus, $m = \frac{1}{\sqrt{3}} = \frac{1}{\sqrt{3}} \cdot \frac{\sqrt{3}}{\sqrt{3}} = \frac{\sqrt{3}}{3}$ and the equation of the line is $y = \frac{\sqrt{3}}{3}x$.

56. One point on the line $y=\frac{\sqrt{3}}{3}x$, is the origin $(0,0)$. Let (x, y) be any other point on this line. Then, by the definition of slope, $m=\frac{y-0}{x-0}=\frac{y}{x}=\frac{\sqrt{3}}{3}$, but also, by the definition of tangent, $\tan\theta=\frac{\sqrt{3}}{3}$. Because $\tan 30°=\frac{\sqrt{3}}{3}$, the line $y=\frac{\sqrt{3}}{3}x$ makes a 30° angle with the positive x-axis. (See Exercise 55).

57. One point on the line $y=\sqrt{3}x$ is the origin $(0,0)$. Let (x, y) be any other point on this line. Then, by the definition of slope, $m=\frac{y-0}{x-0}=\frac{y}{x}=\sqrt{3}$, but also, by the definition of tangent, $\tan\theta=\sqrt{3}$. Because $\tan 60°=\sqrt{3}$, the line $y=\sqrt{3}x$ makes a 60° angle with the positive x-axis (See Exercise 54).

58. (a) The diagonal forms two isosceles right triangles. Each angle formed by a side of the square and the diagonal measures 45°.

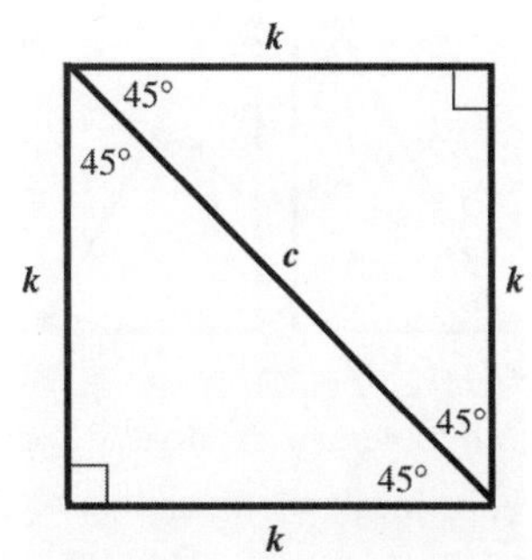

(b) By the Pythagorean theorem, $k^2+k^2=c^2 \Rightarrow 2k^2=c^2 \Rightarrow c=\sqrt{2k^2} \Rightarrow c=k\sqrt{2}$. The length of the diagonal is $k\sqrt{2}$.

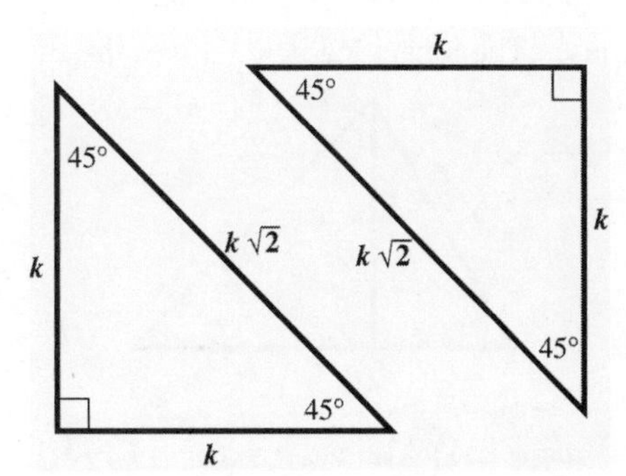

(c) In a $45°-45°$ right triangle, the hypotenuse has a length that is $\underline{\sqrt{2}}$ times as long as either leg.

59. **(a)** Each of angles of the equilateral triangle has measure of $\frac{1}{3}(180°) = 60°$.

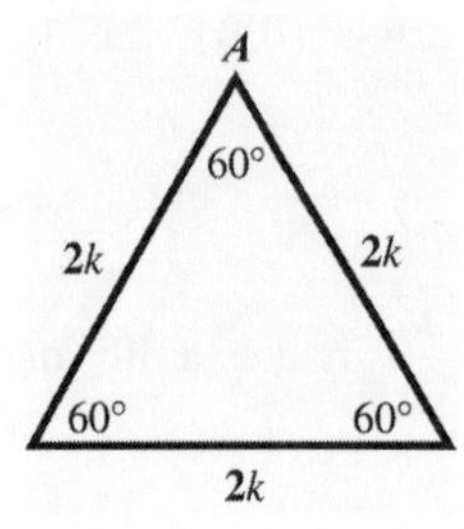

(b) The perpendicular bisects the opposite side so the length of each side opposite each 30° angle is *k*.

(c) Let *x* equal the length of the perpendicular and apply the Pythagorean theorem.

$$x^2 + k^2 = (2k)^2 \Rightarrow x^2 + k^2 = 4k^2 \Rightarrow x^2 = 3k^2 \Rightarrow x = k\sqrt{3}$$

The length of the perpendicular is $k\sqrt{3}$.

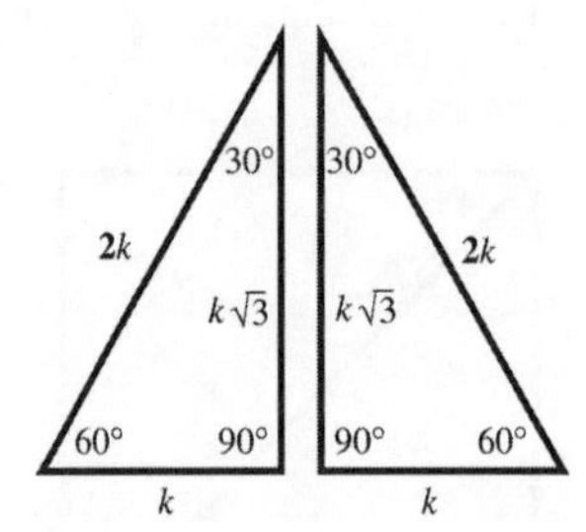

(d) In a 30° - 60° right triangle, the hypotenuse is always 2 times as long as the shorter leg, and the longer leg has a length that is $\sqrt{3}$ times as long as that of the shorter leg. Also, the shorter leg is opposite the 30° angle, and the longer leg is opposite the 60° angle.

60. Apply the relationships between the lengths of the sides of a $30°-60°$ right triangle first to the triangle on the left to find the values of *a* and *b*. In the $30°-60°$ right triangle, the side opposite the $30°$ angle is $\frac{1}{2}$ the length of the hypotenuse. The longer leg is $\sqrt{3}$ times the shorter leg.

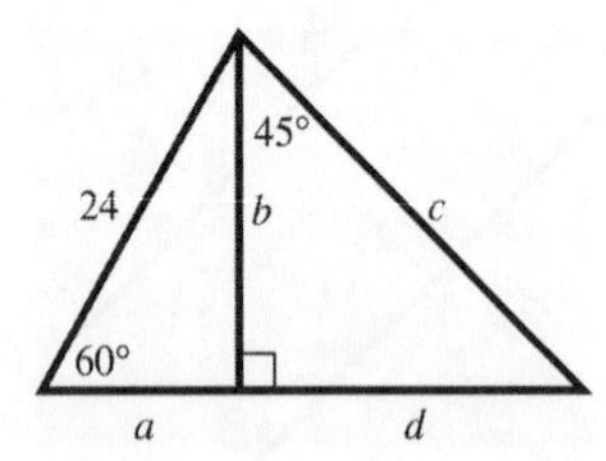

$$a = \frac{1}{2}(24) = 12 \text{ and } b = a\sqrt{3} = 12\sqrt{3}$$

Apply the relationships between the lengths of the sides of a $45°-45°$ right triangle next to the triangle on the right to find the values of *d* and *c*. In the $45°-45°$ right triangle, the sides opposite the $45°$ angles measure the same. The hypotenuse is $\sqrt{2}$ times the measure of a leg.

$$d = b = 12\sqrt{3} \text{ and } c = d\sqrt{2} = \left(12\sqrt{3}\right)\left(\sqrt{2}\right) = 12\sqrt{6}$$

61. Apply the relationships between the lengths of the sides of a $30°-60°$ right triangle first to the triangle on the left to find the values of y and x, and then to the triangle on the right to find the values of z and w. In the $30°-60°$ right triangle, the side opposite the $30°$ angle is $\frac{1}{2}$ the length of the hypotenuse. The longer leg is $\sqrt{3}$ times the shorter leg.

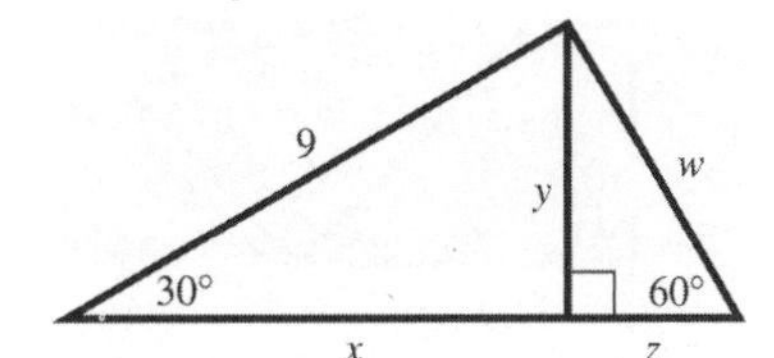

Thus, we have the following.

$$y=\frac{1}{2}(9)=\frac{9}{2} \text{ and } x=y\sqrt{3}=\frac{9\sqrt{3}}{2}$$

$$y=z\sqrt{3}, \text{ so } z=\frac{y}{\sqrt{3}}=\frac{\frac{9}{2}}{\sqrt{3}}=\frac{9\sqrt{3}}{6}=\frac{3\sqrt{3}}{2}, \text{ and } w=2z, \text{ so } w=2\left(\frac{3\sqrt{3}}{2}\right)=3\sqrt{3}$$

62. Apply the relationships between the lengths of the sides of a $30°-60°$ right triangle first to the triangle on the right to find the values of m and a. In the $30°-60°$ right triangle, the side opposite the $60°$ angle is $\sqrt{3}$ times as long as the side opposite to the $30°$ angle. The length of the hypotenuse is 2 times as long as the shorter leg (opposite the $30°$ angle).

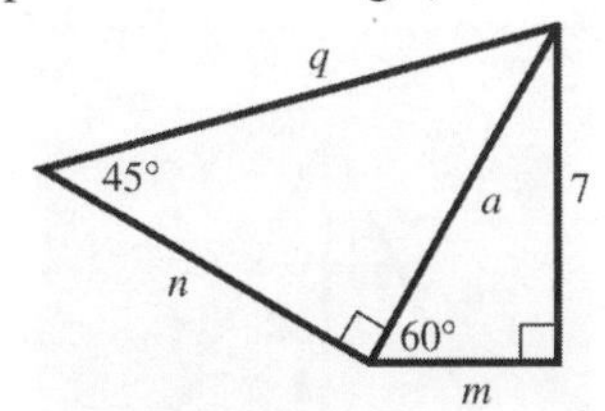

Thus, we have the following.

$$7=m\sqrt{3}\Rightarrow m=\frac{7}{\sqrt{3}}=\frac{7}{\sqrt{3}}\cdot\frac{\sqrt{3}}{\sqrt{3}}=\frac{7\sqrt{3}}{3} \text{ and } a=2m\Rightarrow a=2\left(\frac{7\sqrt{3}}{3}\right)=\frac{14\sqrt{3}}{3}$$

Apply the relationships between the lengths of the sides of a $45°-45°$ right triangle next to the triangle on the left to find the values of n and q. In the $45°-45°$ right triangle, the sides opposite the $45°$ angles measure the same. The hypotenuse is $\sqrt{2}$ times the measure of a leg.

Thus, we have the following.

$$n=a=\frac{14\sqrt{3}}{3} \text{ and } q=n\sqrt{2}=\left(\frac{14\sqrt{3}}{3}\right)\sqrt{2}=\frac{14\sqrt{6}}{3}$$

63. Apply the relationships between the lengths of the sides of a $45°-45°$ right triangle to the triangle on the left to find the values of *p* and *r*. In the $45°-45°$ right triangle, the sides opposite the $45°$ angles measure the same. The hypotenuse is $\sqrt{2}$ times the measure of a leg.

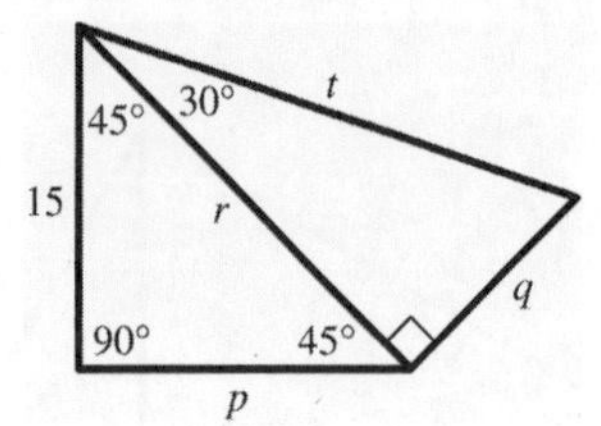

Thus, we have the following.

$$p = 15 \text{ and } r = p\sqrt{2} = 15\sqrt{2}$$

Apply the relationships between the lengths of the sides of a $30°-60°$ right triangle next to the triangle on the right to find the values of *q* and *t*. In the $30°-60°$ right triangle, the side opposite the $60°$ angle is $\sqrt{3}$ times as long as the side opposite to the $30°$ angle. The length of the hypotenuse is 2 times as long as the shorter leg (opposite the $30°$ angle).

Thus, we have the following.

$$r = q\sqrt{3} \Rightarrow q = \frac{r}{\sqrt{3}} = \frac{15\sqrt{2}}{\sqrt{3}} = \frac{15\sqrt{2}}{\sqrt{3}} \cdot \frac{\sqrt{3}}{\sqrt{3}} = 5\sqrt{6} \text{ and } t = 2q = 2\left(5\sqrt{6}\right) = 10\sqrt{6}$$

64. Let *h* be the height of the equilateral triangle. *h* bisects the base, *s*, and forms two 30°–60° right triangles.

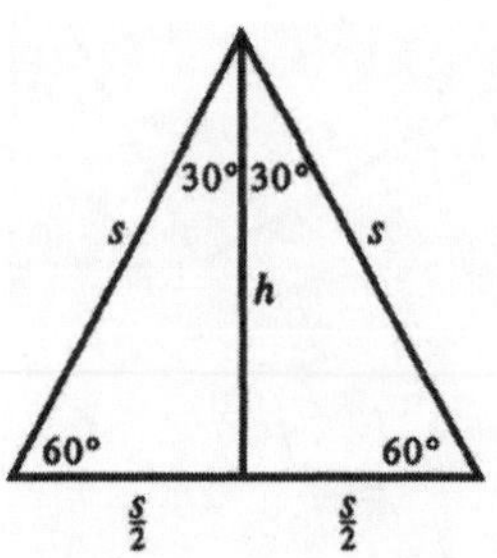

The formula for the area of a triangle is $A = \frac{1}{2}bh.$ In this triangle, $b = s$. The height *h* of the triangle is the side opposite the 60° angle in either 30°–60° right triangle. The side opposite the 30° angle is $\frac{s}{2}$.

The height is $\sqrt{3} \cdot \frac{s}{2} \cdot = \frac{s\sqrt{3}}{2}$ So the area of the entire triangle is $A = \frac{1}{2}s\left(\frac{s\sqrt{3}}{2}\right) = \frac{s^2\sqrt{3}}{4}.$

65. Since $A = \frac{1}{2}bh,$ we have $A = \frac{1}{2} \cdot s \cdot s = \frac{1}{2}s^2$ or $A = \frac{s^2}{2}.$

66. Yes, the third angle can be found by subtracting the given acute angle from 90°, and the remaining two sides can be found using a trigonometric function involving the known angle and side.

67. Answers will vary.

Section 2.2: Trigonometric Functions of Non-Acute Angles

1. C; $180° - 98° = 82°$
(98° is in quadrant II)

2. F; $212° - 180° = 32°$
(212° is in quadrant III)

3. A; $-135° + 360° = 225°$ and
$225° - 180° = 45°$
(225° is in quadrant III)

4. B; $-60° + 360° = 300°$ and
$360° - 300° = 60°$
(300° is in quadrant IV)

5. D; $750° - 2 \cdot 360° = 30°$
(30° is in quadrant I)

6. B; $480° - 360° = 120°$ and
$180° - 120° = 60°$
(120° is in quadrant II)

7. 2 is a good choice for r because in a $30° - 60°$ right triangle, the hypotenuse is twice the length of the shorter side (the side opposite to the $30°$ angle). By choosing 2, one avoids introducing a fraction (or decimal) when determining the length of the shorter side. Choosing any even positive integer for r would have this result; however, 2 is the most convenient value.

8. – 9. Answers will vary.

	θ	$\sin\theta$	$\cos\theta$	$\tan\theta$	$\cot\theta$	$\sec\theta$	$\csc\theta$
10.	30°	$\frac{1}{2}$	$\frac{\sqrt{3}}{2}$	$\frac{\sqrt{3}}{3}$	$\sqrt{3}$	$\frac{2\sqrt{3}}{3}$	2
11.	45°	$\frac{\sqrt{2}}{2}$	$\frac{\sqrt{2}}{2}$	1	1	$\sqrt{2}$	$\sqrt{2}$
12.	60°	$\frac{\sqrt{3}}{2}$	$\frac{1}{2}$	$\sqrt{3}$	$\frac{\sqrt{3}}{3}$	2	$\frac{2\sqrt{3}}{3}$
13.	120°	$\frac{\sqrt{3}}{2}$	$\cos 120° = -\cos 60° = -\frac{1}{2}$	$-\sqrt{3}$	$\cot 120° = -\cot 60° = -\frac{\sqrt{3}}{3}$	$\sec 120° = -\sec 60° = -2$	$\frac{2\sqrt{3}}{3}$
14.	135°	$\frac{\sqrt{2}}{2}$	$-\frac{\sqrt{2}}{2}$	$\tan 135° = -\tan 45° = -1$	$\cot 135° = -\cot 45° = -1$	$-\sqrt{2}$	$\sqrt{2}$
15.	150°	$\sin 150° = \sin 30° = \frac{1}{2}$	$-\frac{\sqrt{3}}{2}$	$-\frac{\sqrt{3}}{3}$	$\cot 150° = -\cot 30° = -\sqrt{3}$	$\sec 150° = -\sec 30° = -\frac{2\sqrt{3}}{3}$	2
16.	210°	$-\frac{1}{2}$	$\cos 210° = -\cos 30° = -\frac{\sqrt{3}}{2}$	$\frac{\sqrt{3}}{3}$	$\sqrt{3}$	$\sec 210° = -\sec 30° = -\frac{2\sqrt{3}}{3}$	−2
17.	240°	$-\frac{\sqrt{3}}{2}$	$-\frac{1}{2}$	$\tan 240° = \tan 60° = \sqrt{3}$	$\cot 240° = \cot 60° = \frac{\sqrt{3}}{3}$	−2	$-\frac{2\sqrt{3}}{3}$

18. To find the reference angle for 300°, sketch this angle in standard position.

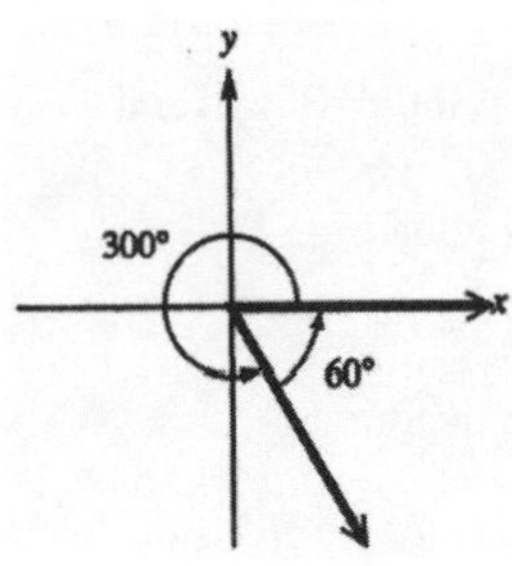

The reference angle is $360° - 300° = 60°$. Since 300° lies in quadrant IV, the sine, tangent, cotangent, and cosecant are negative.

$$\sin 300° = -\sin 60° = -\frac{\sqrt{3}}{2}$$
$$\cos 300° = \cos 60° = \frac{1}{2}$$
$$\tan 300° = -\tan 60° = -\sqrt{3}$$
$$\cot 300° = -\cot 60° = -\frac{\sqrt{3}}{3}$$
$$\sec 300° = \sec 60° = 2$$
$$\csc 300° = -\csc 60° = -\frac{2\sqrt{3}}{3}$$

19. To find the reference angle for 315°, sketch this angle in standard position.

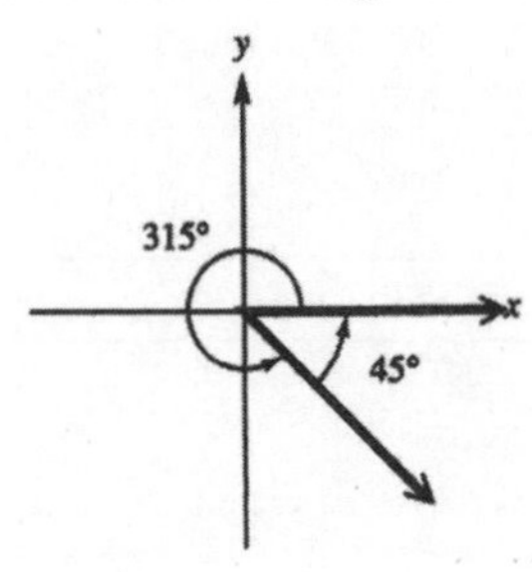

The reference angle is $360° - 315° = 45°$. Since 315° lies in quadrant IV, the sine, tangent, cotangent, and cosecant are negative.

$$\sin 315° = -\sin 45° = -\frac{\sqrt{2}}{2}$$
$$\cos 315° = \cos 45° = \frac{\sqrt{2}}{2}$$
$$\tan 315° = -\tan 45° = -1$$
$$\cot 315° = -\cot 45° = -1$$
$$\sec 315° = \sec 45° = \sqrt{2}$$
$$\csc 315° = -\csc 45° = -\sqrt{2}$$

20. To find the reference angle for 405°, sketch this angle in standard position.

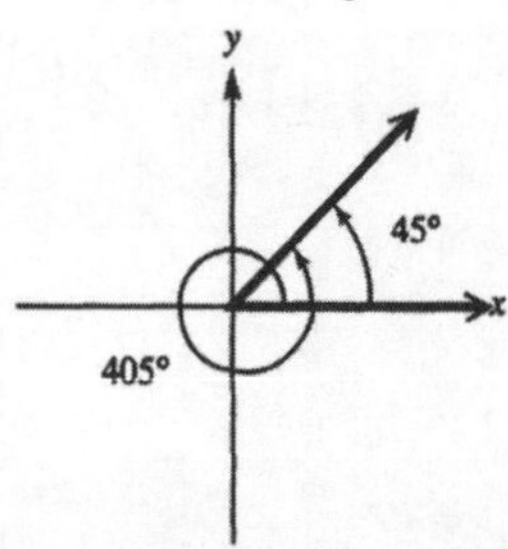

The reference angle for 405° is $405° - 360° = 45°$. Because 405° lies in quadrant I, the values of all of its trigonometric functions will be positive, so these values will be identical to the trigonometric function values for 45°. See the Function Values of Special Angles table that follows Example 5 in Section 2.1 on page 50.)

$$\sin 405° = \sin 45° = \frac{\sqrt{2}}{2}$$
$$\cos 405° = \cos 45° = \frac{\sqrt{2}}{2}$$
$$\tan 405° = \tan 45° = 1$$
$$\cot 405° = \cot 45° = 1$$
$$\sec 405° = \sec 45° = \sqrt{2}$$
$$\csc 405° = \csc 45° = \sqrt{2}$$

21. To find the reference angle for $-300°$, sketch this angle in standard position.

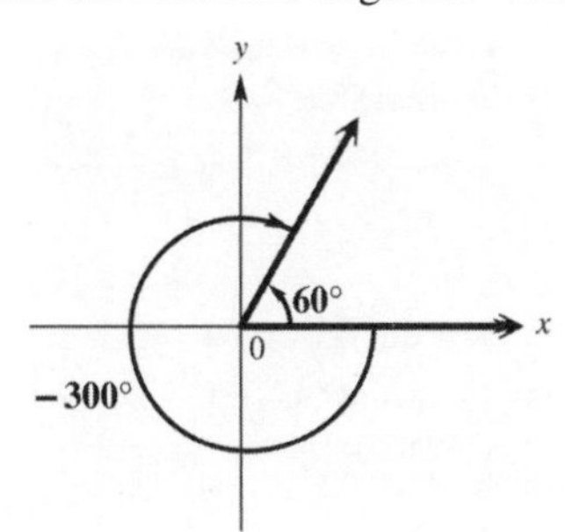

The reference angle for $-300°$ is $-300° + 360° = 60°$. Because $-300°$ lies in quadrant I, the values of all of its trigonometric functions will be positive, so these values will be identical to the trigonometric function values for 60°. See the Function Values of Special Angles table that follows Example 5 in Section 2.1 on page 50.)

$$\sin(-300°) = \sin 60° = \frac{\sqrt{3}}{2}$$
$$\cos(-300°) = \cos 60° = \frac{1}{2}$$
$$\tan(-300°) = \tan 60° = \sqrt{3}$$
$$\cot(-300°) = \cot 60° = \frac{\sqrt{3}}{3}$$
$$\sec(-300°) = \sec 60° = 2$$
$$\csc(-300°) = \csc 60° = \frac{2\sqrt{3}}{3}$$

22. To find the reference angle for $420°$, sketch this angle in standard position.

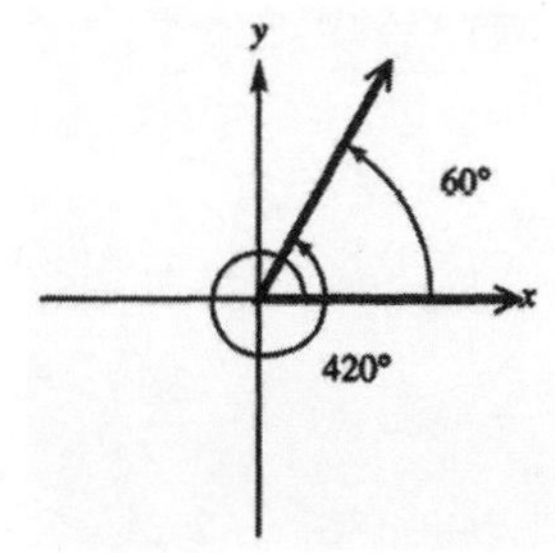

The reference angle for $420°$ is $420° - 360° = 60°$. Because $420°$ lies in quadrant I, the values of all of its trigonometric functions will be positive, so these values will be identical to the trigonometric function values for 60°. See the Function Values of Special Angles table that follows Example 5 in Section 2.1 on page 50.)

$$\sin 420° = \sin 60° = \frac{\sqrt{3}}{2}$$
$$\cos 420° = \cos 60° = \frac{1}{2}$$
$$\tan 420° = \tan 60° = \sqrt{3}$$
$$\cot 420° = \cot 60° = \frac{\sqrt{3}}{3}$$
$$\sec 420° = \sec 60° = 2$$
$$\csc 420° = \csc 60° = \frac{2\sqrt{3}}{3}$$

23. To find the reference angle for $480°$, sketch this angle in standard position.

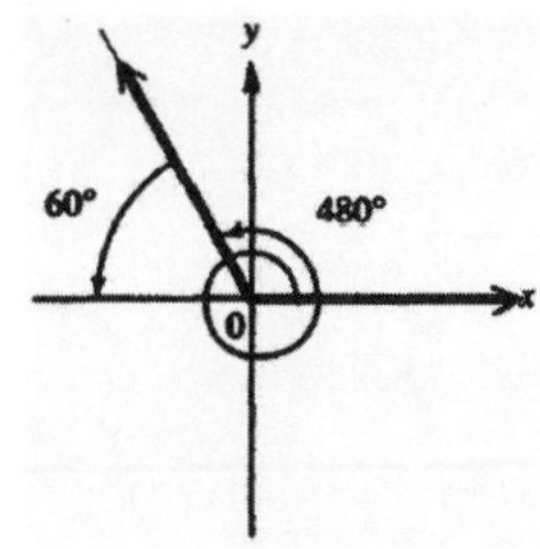

$480°$ is coterminal with $480° - 360° = 120°$. The reference angle is $180° - 120° = 60°$. Since $480°$ lies in quadrant II, the cosine, tangent, cotangent, and secant are negative.

$$\sin 480° = \sin 60° = \frac{\sqrt{3}}{2}$$
$$\cos 480° = \cos 60° = -\frac{1}{2}$$
$$\tan 480° = \tan 60° = -\sqrt{3}$$
$$\cot 480° = \cot 60° = -\frac{\sqrt{3}}{3}$$
$$\sec 480° = \sec 60° = -2$$
$$\csc 480° = \csc 60° = \frac{2\sqrt{3}}{3}$$

24. To find the reference angle for 495°, sketch this angle in standard position.

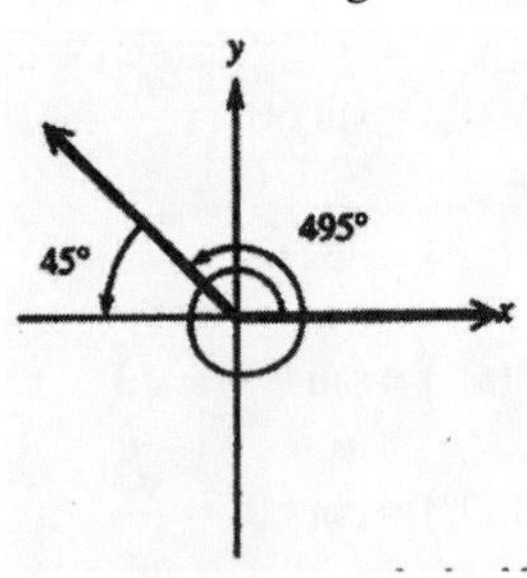

495° is coterminal with $495° - 360° = 135°$. The reference angle is $180° - 135° = 45°$. Since 495° lies in quadrant II, the cosine, tangent, cotangent, and secant are negative.

$$\sin 495° = \sin 45° = \frac{\sqrt{2}}{2}$$

$$\cos 495° = -\cos 45° = -\frac{\sqrt{2}}{2}$$

$$\tan 495° = -\tan 45° = -1$$

$$\cot 495° = -\cot 45° = -1$$

$$\sec 495° = -\sec 45° = -\sqrt{2}$$

$$\csc 495° = \csc 45° = \sqrt{2}$$

25. To find the reference angle for 570°, sketch this angle in standard position.

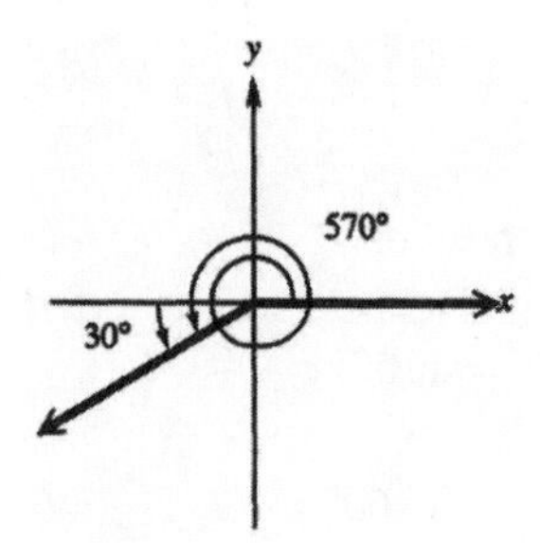

570° is coterminal with $570° - 360° = 210°$. The reference angle is $210° - 180° = 30°$. Since 570° lies in quadrant III, the sine, cosine, secant, and cosecant are negative.

$$\sin 570° = -\sin 30° = -\frac{1}{2}$$

$$\cos 570° = -\cos 30° = -\frac{\sqrt{3}}{2}$$

$$\tan 570° = \tan 30° = \frac{\sqrt{3}}{3}$$

$$\cot 570° = \cot 30° = \sqrt{3}$$

$$\sec 570° = -\sec 30° = -\frac{2\sqrt{3}}{3}$$

$$\csc 570° = -\csc 30° = -2$$

26. To find the reference angle for 750°, sketch this angle in standard position.

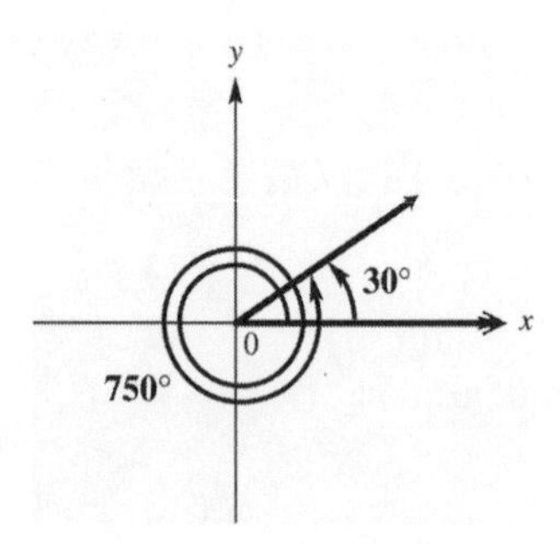

750° is coterminal with 30° because $750° - 2 \cdot 360° = 750° - 720° = 30°$. Since 750° lies in quadrant I, the values of all of its trigonometric functions will be positive, so these values will be identical to the trigonometric function values for 30°.

$$\sin 750° = \sin 30° = \frac{1}{2}$$

$$\cos 750° = \cos 30° = \frac{\sqrt{3}}{2}$$

$$\tan 750° = \tan 30° = \frac{\sqrt{3}}{3}$$

$$\cot 750° = \cot 30° = \sqrt{3}$$

$$\sec 750° = \sec 30° = \frac{2\sqrt{3}}{3}$$

$$\csc 750° = \csc 30° = 2$$

27. $1305°$ is coterminal with $1305° - 3 \cdot 360° = 1305° - 1080° = 225°$. This angle lies in quadrant III and the reference angle is $225° - 180° = 45°$. Since $1305°$ lies in quadrant III, the sine, cosine, secant, and cosecant are negative.

$\sin 1305° = -\sin 45° = -\frac{\sqrt{2}}{2}$

$\cos 1305° = -\cos 45° = -\frac{\sqrt{2}}{2}$

$\tan 1305° = \tan 45° = 1$

$\cot 1305° = \cot 45° = 1$

$\sec 1305° = -\sec 45° = -\sqrt{2}$

$\csc 1305° = -\csc 45° = -\sqrt{2}$

28. $1500°$ is coterminal with $1500° - 4 \cdot 360° = 1500° - 1440° = 60°$. Because $420°$ lies in quadrant I, the values of all of its trigonometric functions will be positive, so these values will be identical to the trigonometric function values for $60°$.

$\sin 1500° = \sin 60° = \frac{\sqrt{3}}{2}$

$\cos 1500° = \cos 60° = \frac{1}{2}$

$\tan 1500° = \tan 60° = \sqrt{3}$

$\cot 1500° = \cot 60° = \frac{\sqrt{3}}{3}$

$\sec 1500° = \sec 60° = 2$

$\csc 1500° = \csc 60° = \frac{2\sqrt{3}}{3}$

29. $2670°$ is coterminal with $2670° - 7 \cdot 360° = 2670° - 2520° = 150°$. The reference angle is $180° - 150° = 30°$. Since $2670°$ lies in quadrant II, the cosine, tangent, cotangent, and secant are negative.

$\sin 2670° = \sin 30° = \frac{1}{2}$

$\cos 2670° = -\cos 30° = -\frac{\sqrt{3}}{2}$

$\tan 2670° = -\tan 30° = -\frac{\sqrt{3}}{3}$

$\cot 2670° = -\cot 30° = -\sqrt{3}$

$\sec 2670° = -\sec 30° = -\frac{2\sqrt{3}}{3}$

$\csc 2670° = \csc 30° = 2$

30. $-390°$ is coterminal with $-390° + 2 \cdot 360° = -390° + 720° = 330°$. The reference angle is $360° - 330° = 30°$. Since $-390°$ lies in quadrant IV, the sine, tangent, cotangent, and cosecant are negative.

$\sin(-390°) = -\sin 30° = -\frac{1}{2}$

$\cos(-390°) = \cos 30° = \frac{\sqrt{3}}{2}$

$\tan(-390°) = -\tan 30° = -\frac{\sqrt{3}}{3}$

$\cot(-390°) = -\cot 30° = -\sqrt{3}$

$\sec(-390°) = \sec 30° = \frac{2\sqrt{3}}{3}$

$\csc(-390°) = -\csc 30° = -2$

31. $-510°$ is coterminal with $-510° + 2 \cdot 360° = -510° + 720° = 210°$. The reference angle is $210° - 180° = 30°$. Since $-510°$ lies in quadrant III, the sine, cosine, and secant and cosecant are negative.

$\sin(-510°) = -\sin 30° = -\frac{1}{2}$

$\cos(-510°) = -\cos 30° = -\frac{\sqrt{3}}{2}$

$\tan(-510°) = \tan 30° = \frac{\sqrt{3}}{3}$

$\cot(-510°) = \cot 30° = \sqrt{3}$

$\sec(-510°) = -\sec 30° = -\frac{2\sqrt{3}}{3}$

$\csc(-510°) = -\csc 30° = -2$

32. $-1020°$ is coterminal with $-1020° + 3 \cdot 360° = -1020° + 1080° = 60°$. Because $-1020°$ lies in quadrant I, the values of all of its trigonometric functions will be positive, so these values will be identical to the trigonometric function values for $60°$.

$$\sin(-1020°) = \sin 60° = \frac{\sqrt{3}}{2}$$
$$\cos(-1020°) = \cos 60° = \frac{1}{2}$$
$$\tan(-1020°) = \tan 60° = \sqrt{3}$$
$$\cot(-1020°) = \cot 60° = \frac{\sqrt{3}}{3}$$
$$\sec(-1020°) = \sec 60° = 2$$
$$\csc(-1020°) = \csc 60° = \frac{2\sqrt{3}}{3}$$

33. $-1290°$ is coterminal with $-1290° + 4 \cdot 360° = -1290° + 1440° = 150°$. This angle lies in quadrant II and the reference angle is $180° - 150° = 30°$. Since $-1290°$ lies in quadrant II, the cosine, tangent, cotangent, and secant are negative.

$$\sin(-1290°) = \sin 30° = \frac{1}{2}$$
$$\cos(-1290°) = -\cos 30° = -\frac{\sqrt{3}}{2}$$
$$\tan(-1290°) = -\tan 30° = -\frac{\sqrt{3}}{3}$$
$$\cot(-1290°) = -\cot 30° = -\sqrt{3}$$
$$\sec(-1290°) = -\sec 30° = -\frac{2\sqrt{3}}{3}$$
$$\csc(-1290°) = \csc 30° = 2$$

34. Since $1305°$ is coterminal with an angle of $1305° - 3 \cdot 360° = 1305° - 1080° = 225°$, it lies in quadrant III. Its reference angle is $225° - 180° = 45°$. Since the sine is negative in quadrant III, we have $\sin 1305° = -\sin 45° = -\frac{\sqrt{2}}{2}$.

35. Since $-510°$ is coterminal with an angle of $-510° + 2 \cdot 360° = -510° + 720° = 210°$, it lies in quadrant III. Its reference angle is $210° - 180° = 30°$. Since the cosine is negative in quadrant III, we have $\cos(-510°) = -\cos 30° = -\frac{\sqrt{3}}{2}$.

36. Since $-1020°$ is coterminal with an angle of $-1020° + 3 \cdot 360° = -1020° + 1080° = 60°$, it lies in quadrant I. Because $-1020°$ lies in quadrant I, the values of all of its trigonometric functions will be positive, so $\tan(-1020°) = \tan 60° = \sqrt{3}$.

37. Since $1500°$ is coterminal with an angle of $1500° - 4 \cdot 360° = 1500° - 1440° = 60°$, it lies in quadrant I. Because $1500°$ lies in quadrant I, the values of all of its trigonometric functions will be positive, so $\sin 1500° = \sin 60° = \frac{\sqrt{3}}{2}$.

38. $\sin 30° + \sin 60° \stackrel{?}{=} \sin(30° + 60°)$

Evaluate each side to determine whether this statement is true or false.

$$\sin 30° + \sin 60° = \frac{1}{2} + \frac{\sqrt{3}}{2} = \frac{1+\sqrt{3}}{2} \text{ and } \sin(30° + 60°) = \sin 90° = 1$$

Since $\frac{1+\sqrt{3}}{2} \neq 1$, the given statement is false.

39. $\sin(30° + 60°) \stackrel{?}{=} \sin 30° \cdot \cos 60° + \sin 60° \cdot \cos 30°$

Evaluate each side to determine whether this equation is true or false.

$$\sin(30° + 60°) = \sin 90° = 1 \text{ and } \sin 30° \cdot \cos 60° + \sin 60° \cdot \cos 30° = \frac{1}{2} \cdot \frac{1}{2} + \frac{\sqrt{3}}{2} \cdot \frac{\sqrt{3}}{2} = \frac{1}{4} + \frac{3}{4} = 1$$

Since, 1 = 1, the statement is true.

40. $\cos 60° \stackrel{?}{=} 2\cos^2 30° - 1$

Evaluate each side to determine whether this statement is true or false.

$$\cos 60° = \frac{1}{2} \text{ and } 2\cos^2 30° - 1 = 2\left(\frac{\sqrt{3}}{2}\right)^2 - 1 = 2\left(\frac{3}{4}\right) - 1 = \frac{3}{2} - 1 = \frac{1}{2}$$

Since $\frac{1}{2} = \frac{1}{2}$, the statement is true.

41. $\cos 60° \stackrel{?}{=} 2\cos 30°$

Evaluate each side to determine whether this statement is true or false.

$$\cos 60° = \frac{1}{2} \text{ and } 2\cos 30° = 2\left(\frac{\sqrt{3}}{2}\right) = \sqrt{3}$$

Since $\frac{1}{2} \neq \sqrt{3}$, the statement is false.

42. $\sin 120° \stackrel{?}{=} \sin 150° - \sin 30°$

Evaluate each side to determine whether this statement is true or false.

$$\sin 120° = \frac{\sqrt{3}}{2} \text{ and } \sin 150° - \sin 30° = \frac{1}{2} - \frac{1}{2} = 0$$

Since $\frac{\sqrt{3}}{2} \neq 0$, the statement is false.

43. $\sin 120° \stackrel{?}{=} \sin 180° \cdot \cos 60° - \sin 60° \cdot \cos 180°$

Evaluate each side to determine whether this statement is true or false.

$$\sin 120° = \frac{\sqrt{3}}{2} \text{ and } \sin 180° \cdot \cos 60° - \sin 60° \cdot \cos 180° = 0\left(\frac{1}{2}\right) - \left(\frac{\sqrt{3}}{2}\right)(-1) = 0 - \left(-\frac{\sqrt{3}}{2}\right) = \frac{\sqrt{3}}{2}$$

Since $\frac{\sqrt{3}}{2} = \frac{\sqrt{3}}{2}$, the statement is true.

44. $225°$ is in quadrant III, so the reference angle is $225° - 180° = 45°$.

$$\cos 45° = \frac{x}{r} \Rightarrow x = r\cos 45° = 10 \cdot \frac{\sqrt{2}}{2} = 5\sqrt{2} \text{ and } \sin 45° = \frac{y}{r} \Rightarrow y = r\sin 45° = 10 \cdot \frac{\sqrt{2}}{2} = 5\sqrt{2}$$

Since $225°$ is in quadrant III, both the x and y coordinate will be negative. The coordinates of P are: $\left(-5\sqrt{2}, -5\sqrt{2}\right)$.

45. $150°$ is in quadrant III, so the reference angle is $180° - 150° = 30°$.

$$\cos 30° = \frac{x}{r} \Rightarrow x = r\cos 30° = 6 \cdot \frac{\sqrt{3}}{2} = 3\sqrt{3} \text{ and } \sin 30° = \frac{y}{r} \Rightarrow y = r\sin 30° = 6 \cdot \frac{1}{2} = 3$$

Since $150°$ is in quadrant II, x will be negative and y will be positive. The coordinated of P are: $\left(-3\sqrt{3}, 3\right)$.

46. For every angle θ, $\sin^2\theta+\cos^2\theta=1$. Since $(-.8)^2+(.6)^2=.64+.36=1$, there is an angle θ for which $\cos\theta=.6$ and $\sin\theta=-.8$. Since $\cos\theta>0$ and $\sin\theta<0$, it is an angle that lies in quadrant IV.

47. For every angle θ, $\sin^2\theta+\cos^2\theta=1$. Since $\left(\frac{3}{4}\right)^2+\left(\frac{2}{3}\right)^2=\frac{9}{16}+\frac{4}{9}=\frac{145}{144}\neq 1$, there is no angle θ for which $\cos\theta=\frac{2}{3}$ and $\sin\theta=\frac{3}{4}$.

48. If θ is in the interval $(90°,180°)$, then $90°<\theta<180°\Rightarrow 45°<\frac{\theta}{2}<90°$. Thus, $\frac{\theta}{2}$ is a quadrant I angle and $\sin\frac{\theta}{2}$ is positive.

49. If θ is in the interval $(90°,180°)$, then $90°<\theta<180°\Rightarrow 45°<\frac{\theta}{2}<90°$. Thus, $\frac{\theta}{2}$ is a quadrant I angle and $\cos\frac{\theta}{2}$ is positive.

50. If θ is in the interval $(90°,180°)$, then $90°<\theta<180°\Rightarrow 270°<\theta+180°<360°$. Thus, $\theta+180°$ is a quadrant IV angle and $\cot(\theta+180°)$ is negative.

51. If θ is in the interval $(90°,180°)$, then $90°<\theta<180°\Rightarrow 270°<\theta+180°<360°$. Thus, $\theta+180°$ is a quadrant IV angle and $\sec(\theta+180°)$ is positive.

52. If θ is in the interval $(90°,180°)$, then $90°<\theta<180°\Rightarrow -90°>-\theta>-180°\Rightarrow -180°<-\theta<-90°$. Since $-180°$ is coterminal with $-180°+360°=180°$ and $-90°$ is coterminal with $-90°+360°=270°$, $-\theta$ is a quadrant III angle and $\cos(-\theta)$ is negative.

53. If θ is in the interval $(90°,180°)$, then $90°<\theta<180°\Rightarrow -90°>-\theta>-180°\Rightarrow -180°<-\theta<-90°$. Since $-180°$ is coterminal with $-180°+360°=180°$ and $-90°$ is coterminal with $-90°+360°=270°$, $-\theta$ is a quadrant III angle and $\sin(-\theta)$ is negative.

54. θ and $\theta+n\cdot 360°$ are coterminal angles, so the sine of each of these will result in the same value.

55. θ and $\theta+n\cdot 360°$ are coterminal angles, so the cosine of each of these will result in the same value.

56. The reference angle for $115°$ is $180°-115°=65°$. Since $115°$ is in quadrant II the cosine is negative. Cos θ decreases on the interval $(90°,180°)$ from 0 to –1. Therefore, $\cos 115°$ is closest to $-.4$.

57. The reference angle for $115°$ is $180°-115°=65°$. Since $115°$ is in quadrant II the sine is positive. Sin θ decreases on the interval $(90°,180°)$ from 1 to 0. Therefore, $\sin 115°$ is closest to .9.

58. When θ = 45°, $\sin\theta=\cos\theta=\frac{\sqrt{2}}{2}$. Sine and cosine are opposites in quadrants II and IV. Thus, $180°-\theta=180°-45°=135°$ in quadrant II and $360°-\theta=360°-45°=315°$ in quadrant IV.

59. When $\theta=45°$, $\sin\theta=\cos\theta=\frac{\sqrt{2}}{2}$. Sine and cosine are both positive in quadrant I and both negative in quadrant III. Since $\theta+180°=45°+180°=225°$, 45° is the quadrant I angle and 225° is the quadrant III angle.

60. $L = \dfrac{(\theta_2 - \theta_1) S^2}{200(h + S \tan\alpha)}$

(a) Substitute $h = 1.9$ ft, $\alpha = .9°$, $\theta_1 = -.3°$, $\theta_2 = 4°$, and $S = 336$ ft.

$$L = \frac{[4-(-3)]336^2}{200(1.9 + 336\tan .9°)} \approx 550\text{ft}$$

(b) Substitute $h = 1.9$ ft, $\alpha = 1.5°$, $\theta_1 = -.3°$, $\theta_2 = 4°$, and $S = 336$ ft.

$$L = \frac{[4-(-3)]336^2}{200(1.9 + 336\tan 1.5°)} \approx 369\text{ft}$$

(c) Answers will vary.

61. $\sin\theta = \dfrac{1}{2}$

Since $\sin\theta$ is positive, θ must lie in quadrants I or II. Since one angle, namely $30°$, lies in quadrant I, that angle is also the reference angle, θ'. The angle in quadrant II will be $180° - \theta' = 180° - 30° = 150°$.

62. $\cos\theta = \dfrac{\sqrt{3}}{2}$

Since $\cos\theta$ is positive, θ must lie in quadrants I or IV. Since one angle, namely $30°$, lies in quadrant I, that angle is also the reference angle, θ'. The angle in quadrant IV will be $360° - \theta' = 360° - 30° = 330°$.

63. $\tan\theta = -\sqrt{3}$

Since $\tan\theta$ is negative, θ must lie in quadrants II or IV. Since the absolute value of $\tan\theta$ is $\sqrt{3}$, the reference angle, θ' must be $60°$. The quadrant II angle θ equals $180° - \theta' = 180° - 60° = 120°$, and the quadrant IV angle θ equals $360° - \theta' = 360° - 60° = 300°$.

64. $\sec\theta = -\sqrt{2}$

Since $\sec\theta$ is negative, θ must lie in quadrants II or III. Since the absolute value of $\sec\theta$ is $\sqrt{2}$, the reference angle, θ' must be $45°$. The quadrant II angle θ equals $180° - \theta' = 180° - 45° = 135°$, and the quadrant III angle θ equals $180° + \theta' = 180° + 45° = 225°$.

65. $\cos\theta = \dfrac{\sqrt{2}}{2}$

Since $\cos\theta$ is positive, θ must lie in quadrants I or IV. Since one angle, namely $45°$, lies in quadrant I, that angle is also the reference angle, θ'. The angle in quadrant IV will be $360° - \theta' = 360° - 45° = 315°$.

66. $\cot\theta = -\dfrac{\sqrt{3}}{3}$

Since $\cot\theta$ is negative, θ must lie in quadrants II or IV. Since the absolute value of $\cot\theta$ is $\dfrac{\sqrt{3}}{3}$, the reference angle, θ' must be $60°$. The quadrant II angle θ equals $180° - \theta' = 180° - 60° = 120°$, and the quadrant IV angle θ equals $360° - \theta' = 360° - 60° = 300°$.

Section 2.3: Finding Trigonometric Function Values Using a Calculator

1. The CAUTION at the beginning of this section verifying that a calculator is in degree mode by finding <u>sin</u> 90°. If the calculator is in degree mode, the display should be <u>1</u>.

2. When a scientific or graphing calculator is used to find a trigonometric function value, in most cases the result is an <u>approximate</u> value.

3. To find values of the cotangent, secant, and cosecant functions with a calculator, it is necessary to find the <u>reciprocal</u> of the <u>reciprocal</u> function value.

4. The reciprocal is used <u>before</u> the inverse function key when finding the angle, but <u>after</u> the function key when finding the trigonometric function value.

For Exercises 5 – 21, be sure your calculator is in degree mode. If your calculator accepts angles in degrees, minutes, and seconds, it is not necessary to change angles to decimal degrees. Keystroke sequences may vary on the type and/or model of calculator being used. Screens shown will be from a TI-83 Plus calculator. To obtain the degree (°) and (′) symbols, go to the ANGLE menu (2nd APPS).

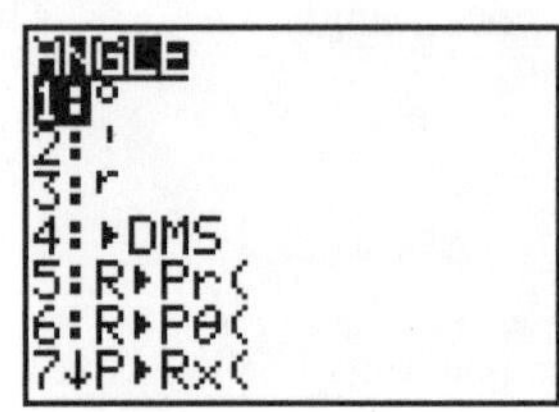

For Exercises 5 – 15, the calculation for decimal degrees is indicated for calculators that do not accept degree, minutes, and seconds.

5. $\sin 38°42' \approx .6252427$

$38°\,42' = \left(38 + \frac{42}{60}\right)° = 38.7°$

6. $\cot 41°24' \approx 1.1342773$

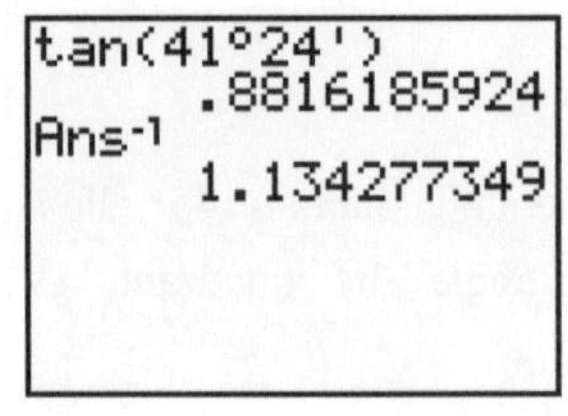

$42°\,24' = \left(41 + \frac{24}{60}\right)° = 41.4°$

7. $\sec 13°\,15' \approx 1.0273488$

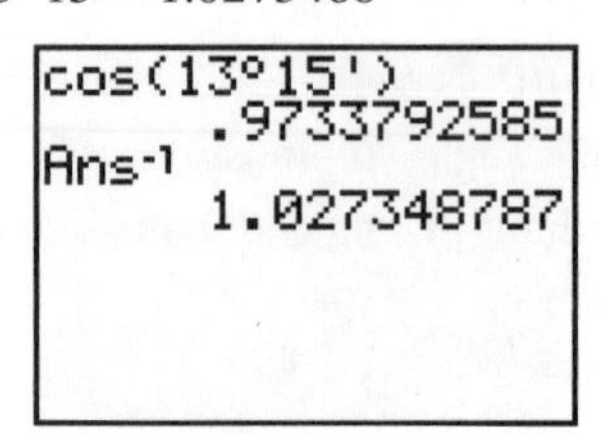

$13°15' = \left(13 + \frac{15}{60}\right)° = 13.25°$

8. $\csc 145°\,45' \approx 1.7768146$

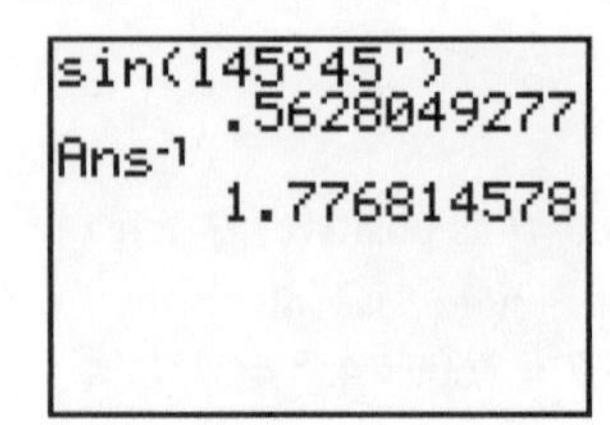

$145°45' = \left(145 + \frac{45}{60}\right)° = 145.75°$

9. $\cot 183°48' \approx 15.055723$

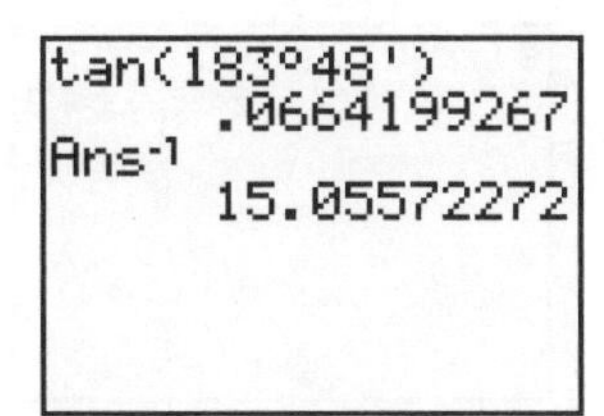

$183°48' = \left(183 + \frac{48}{60}\right)° = 183.8°$

10. $\cos 421°30' \approx .4771588$

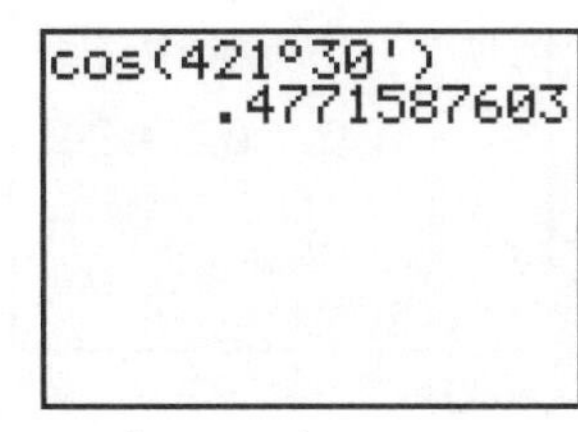

$421°30' = \left(421 + \frac{30}{60}\right)° = 421.5°$

11. $\sec 312°12' \approx 1.4887142$

```
cos(312°12')
            .6717205893
Ans-1
             1.48871423
```

$312°12' = \left(312 + \frac{12}{60}\right)° = 312.2°$

12. $\tan(-80°6') \approx -5.7297416$

```
tan(-80°6')
           -5.729741647
```

$-80°6' = -\left(80 + \frac{6}{60}\right)° = -80.1°$

13. $\sin(-317°36') \approx .6743024$

```
sin(-317°36')
            .6743023876
```

$-317°36' = -\left(317 + \frac{36}{60}\right)° = -317.6°$

14. $\cot(-512°20') \approx 1.9074147$

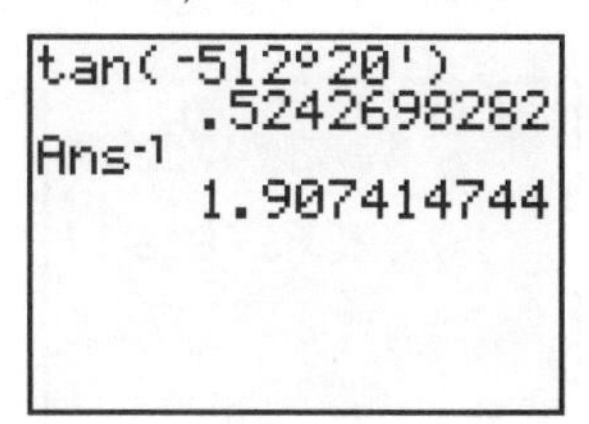

$-512°20' = -\left(512 + \frac{20}{60}\right)° \approx -512.3333333°$

15. $\cos(-15') \approx .9999905$

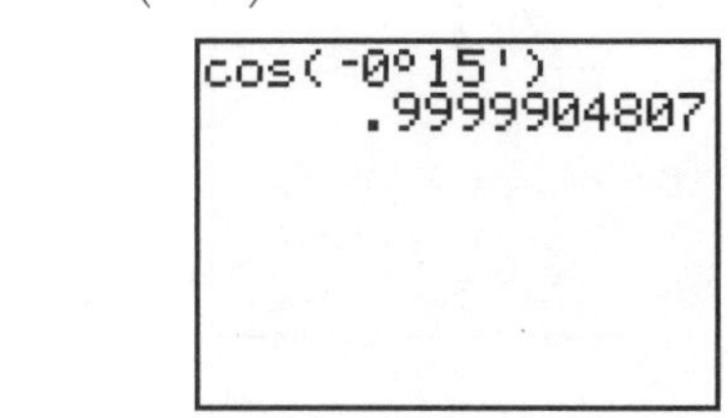

$-15' = -\frac{15}{60}' = -.25°$

16. $\frac{1}{\sec 14.8°} = \cos 14.8° \approx .9668234$

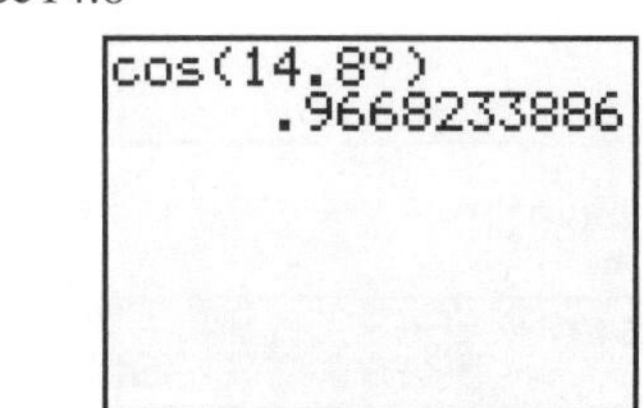

17. $\frac{1}{\cot 23.4°} = \tan 23.4° \approx .4327386$

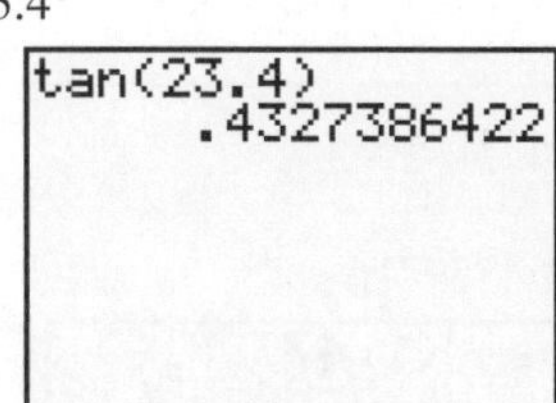

18. $\frac{\sin 33°}{\cos 33°} = \tan 33° \approx .6494076$

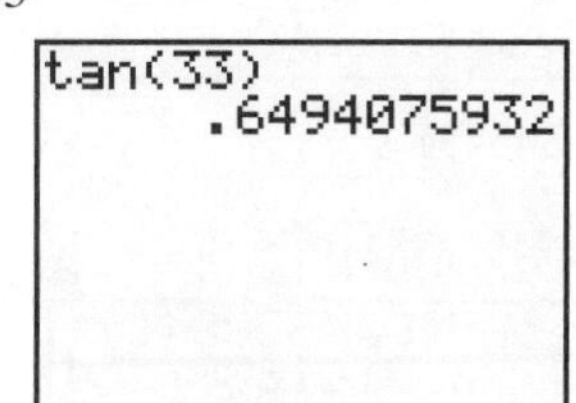

19. $\dfrac{\cos 77°}{\sin 77°} = \cot 77° \approx .2308682$

```
tan(77)
     4.331475874
Ans-1
     .2308681911
```

20. $\cos(90° - 3.69°) = \sin(3.69°) \approx .0643581$

```
sin(3.69)
     .0643581381
```

21. $\cot(90° - 4.72°) = \tan 4.72° \approx .0825664$

```
tan(4.72)
     .0825664011
```

22. $\sin\theta = .84802194$

```
sin-1(.84802194)
     57.99717206
```

$\theta \approx 57.997172°$

23. $\tan\theta = 1.4739716$

```
tan-1(1.4739716)
     55.84549629
```

$\theta \approx 55.845496°$

24. $\tan\theta = 6.4358841$

```
tan-1(6.4358841)
     81.16807334
```

$\theta \approx 81.168073°$

25. $\sin\theta = .27843196$

```
sin-1(.27843196)
     16.16664145
```

$\theta \approx 16.166641°$

26. $\sec\theta = 1.1606249$

```
1/1.1606249
     .8616048131
cos-1(Ans)
     30.50274845
```

$\theta \approx 30.502748°$

27. $\cot\theta = 1.2575516$

```
1/1.2575516
     .7951959983
tan-1(Ans)
     38.49157974
```

$\theta \approx 38.491580°$

28. $\csc\theta = 1.3861147$

```
1/1.3861147
     .7214410178
sin-1(Ans)
     46.17358205
```

$\theta \approx 46.173582°$

29. $\sec\theta = 2.7496222$

```
1/2.7496222
     .3636863275
cos-1(Ans)
     68.6732406
```

$\theta \approx 68.673241°$

30. A common mistake is to have the calculator in radian mode, when it should be in degree mode (and vice verse).

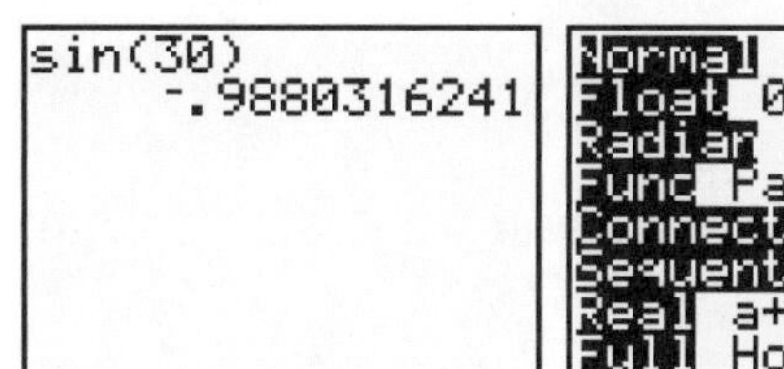

31. If the calculator allowed an angle θ where $0° \le \theta < 360°$, then one would need to find an angle within this interval that is coterminal with $2000°$ by subtracting a multiple of $360°$, i.e. $2000° - 5 \cdot 360° = 2000° - 1800° = 200°$. If the calculator had a higher restriction on evaluating angles (such as $0 \le \theta < 90°$) then one would need to use reference angles.

32. $\tan A = 1.482560969$

Find the arctan of 1.482560969.

```
tan⁻¹(1.482560969
)
          56.00000001
tan(56)
          1.482560969
```

$A \approx 56°$

33. $\sin^{-1} A = 22$

Find the sine of $22°$.

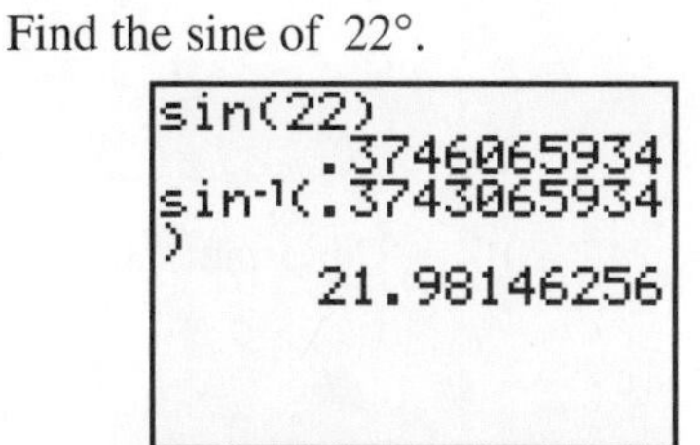

$A \approx .3746065934°$

34. $\sin 35° \cos 55° + \cos 35° \sin 55° = 1$

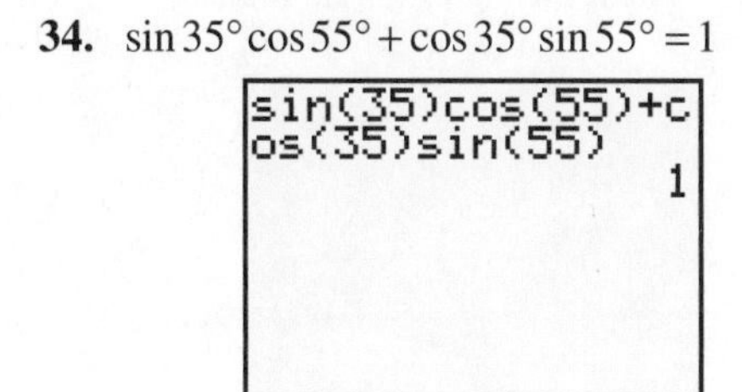

35. $\cos 100° \cos 80° - \sin 100° \sin 80° = -1$

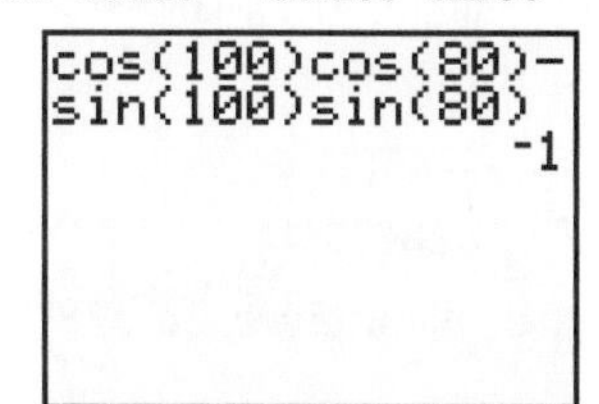

36. $\cos 75°29' \cos 14°31' - \sin 75°29' \sin 14°31' = 0$

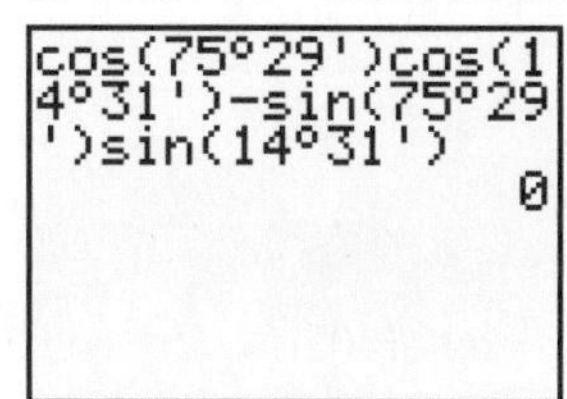

37. $\sin 28°14' \cos 61°46' + \cos 28°14' \sin 61°46' = 1$

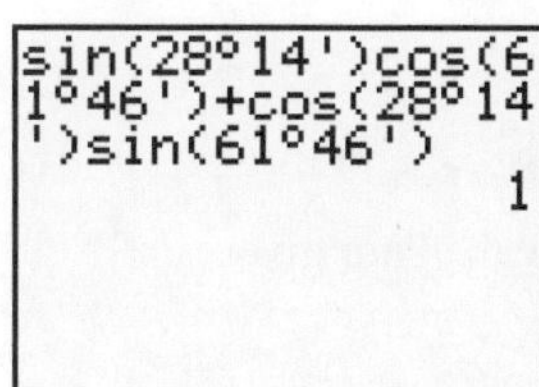

38. For Auto A, calculate $70 \cdot \cos 10° \approx 68.94$. Auto A's reading is approximately 68.94 mph.
For Auto B, calculate $70 \cdot \cos 20° \approx 65.78$. Auto B's reading is approximately 65.78 mph.

39. The figure for this exercise indicates a right triangle. Because we are not considering the time involved in detecting the speed of the car, we will consider the speeds as sides of the right triangle. Given angle θ, $\cos\theta = \dfrac{r}{a}$. Thus, the speed that the radar detects is $r = a\cos\theta$.

40. $\cos 40° \overset{?}{=} 2\cos 20°$

Using a calculator gives $\cos 40° \approx .76604444$ and $2\cos 20° \approx 1.87938524$. Thus, the statement is false.

41. $\sin 10° + \sin 10° \overset{?}{=} \sin 20°$

Using a calculator gives $\sin 10° + \sin 10° \approx .34729636$ and $\sin 20° \approx .34202014$. Thus, the statement is false.

42. $\cos 70° \overset{?}{=} 2\cos^2 35° - 1$

Using a calculator gives $\cos 70° \approx .34202014$ and $2\cos^2 35° - 1 \approx .34202014$. Thus, the statement is true.

43. $\sin 50° \overset{?}{=} 2\sin 25° \cos 25°$

Using a calculator gives $\sin 50° \approx .76604444$ and $2\sin 25° \cos 25° \approx .76604444$. Thus, the statement is true.

44. $2\cos 38°22' \overset{?}{=} \cos 76°44'$

Using a calculator gives $2\cos 38°22' \approx 1.56810939$ and $\cos 76°44' \approx .22948353$. Thus, the statement is false.

45. $\cos 40° \overset{?}{=} 1 - 2\sin^2 80°$

Using a calculator gives $\cos 40° \approx .76604444$ and $1 - 2\sin^2 80° \approx -.93969262$. Thus, the statement is false.

46. $\frac{1}{2}\sin 40° \overset{?}{=} \sin\frac{1}{2}(40°)$

Using a calculator gives $\frac{1}{2}\sin 40° \approx .32139380$ and $\sin\frac{1}{2}(40°) \approx .34202014$ Thus, the statement is false.

47. $\sin 39°48' + \cos 39°48' \overset{?}{=} 1$

Using a calculator gives $\sin 39°48' + \cos 39°48' \approx 1.40839322 \neq 1$. Thus, the statement is false.

48. $F = W \sin\theta$

$F = 2400 \sin(-2.4°) \approx -100.5$ lb

F is negative because the car is traveling downhill.

49. $F = W \sin\theta$

$F = 2100 \sin 1.8° \approx 65.96$ lb

50. $F = W \sin\theta$

$$-145 = W \sin(-3°) \Rightarrow \frac{-145}{\sin(-3°)} = W \Rightarrow W \approx 2771 \text{ lb}$$

51. $F = W\sin\theta$

$$-130 = 2600\sin\theta \Rightarrow \frac{-130}{2600} = \sin\theta \Rightarrow -.05 = \sin\theta \Rightarrow \theta = \sin^{-1}(-.05) \approx -2.87°$$

52. $F = W\sin\theta$

$F = 2200\sin 2° \approx 76.77889275$ lb

$F = 2000\sin 2.2° \approx 76.77561818$ lb

The 2200-lb car on a 2° uphill grade has the greater grade resistance.

53. $F = W\sin\theta$

$$150 = 3000\sin\theta \Rightarrow \frac{150}{3000} = \sin\theta \Rightarrow .05 = \sin\theta \Rightarrow \theta = \sin^{-1}.05 \approx 2.87°$$

54.

θ	$\sin\theta$	$\tan\theta$	$\frac{\pi\theta}{180}$
0°	.0000	.0000	.0000
.5°	.0087	.0087	.0087
1°	.0175	.0175	.0175
1.5°	.0262	.0262	.0262
2°	.0349	.0349	.0349
2.5°	.0436	.0437	.0436
3°	.0523	.0524	.0524
3.5°	.0610	.0612	.0611
4°	.0698	.0699	.0698

(a) From the table we see that if θ is small, $\sin\theta \approx \tan\theta \approx \frac{\pi\theta}{180}$.

(b) $F = W\sin\theta \approx W\tan\theta \approx \frac{W\pi\theta}{180}$

(c) $\tan\theta = \frac{4}{100} = .04$

$F \approx W\tan\theta = 2000(.04) = 80\text{lb}$

(d) Use $F \approx \frac{W\pi\theta}{180}$ from part (b)

Let $\theta = 3.75$ and $W = 1800$.

$$F \approx \frac{1800\pi(3.75)}{180} \approx 117.81 \text{ lb}$$

55. $R = \frac{V^2}{g(f + \tan\theta)}$

(a) Since 45 mph = 66 ft/sec, $V = 66$, $\theta = 3°$, $g = 32.2$, and $f = .14$, we have the following.

$$R = \frac{V^2}{g(f + \tan\theta)} = \frac{66^2}{32.2(.14 + \tan 3°)} \approx 703 \text{ ft}$$

Continued on next page

55. (continued)

(b) Since there are 5280 ft in one mile and 3600 sec in one min, we have the following.

$$70 \text{ mph} = 70 \text{ mph} \cdot 1 \text{ hr}/\ 3600 \text{ sec} \cdot 5280 \text{ ft}/1 \text{ mi} = 102\frac{2}{3} \text{ ft per sec} \approx 102.67 \text{ ft per sec}$$

Since $V = 102.67$, $\theta = 3°$, $g = 32.2$, and $f = .14$, we have the following.

$$R = \frac{V^2}{g(f + \tan\theta)} \approx \frac{102.67^2}{32.2(.14 + \tan 3°)} \approx 1701 \text{ ft}$$

(c) Intuitively, increasing θ would make it easier to negotiate the curve at a higher speed much like is done at a race track. Mathematically, a larger value of θ (acute) will lead to a larger value for $\tan\theta$. If $\tan\theta$ increases, then the ratio determining R will *decrease*. Thus, the radius can be smaller and the curve sharper if θ is increased.

$$R = \frac{V^2}{g(f + \tan\theta)} = \frac{66^2}{32.2(.14 + \tan 4°)} \approx 644 \text{ ft and } R = \frac{V^2}{g(f + \tan\theta)} \approx \frac{102.67^2}{32.2(.14 + \tan 4°)} \approx 1559 \text{ ft}$$

As predicted, both values are less.

56. From Exercise 55, $R = \dfrac{V^2}{g(f + \tan\theta)}$. Solving for V we have the following.

$$R = \frac{V^2}{g(f + \tan\theta)} \Rightarrow V^2 = Rg(f + \tan\theta) \Rightarrow V = \sqrt{Rg(f + \tan\theta)}$$

Since $R = 1150$, $\theta = 2.1°$, $g = 32.2$, and $f = .14$, we have the following.

$$V = \sqrt{Rg(f + \tan\theta)} = \sqrt{1150(32.2)(.14 + \tan 2.1°)} \approx 80.9 \text{ ft/sec}$$

Since 80.9 ft/sec · 3600 sec/hr · 1 mi/5280 ft ≈ 55 mph, it should have a 55 mph speed limit.

57. (a) $\theta_1 = 46°$, $\theta_2 = 31°$, $c_1 = 3 \times 10^8$ m per sec

$$\frac{c_1}{c_2} = \frac{\sin\theta_1}{\sin\theta_2} \Rightarrow c_2 = \frac{c_1 \sin\theta_2}{\sin\theta_1} \Rightarrow c_2 = \frac{(3\times10^8)(\sin 31°)}{\sin 46°} \approx 2\times10^8$$

Since c_1 is only given to one significant digit, c_2 can only be given to one significant digit. The speed of light in the second medium is about 2×10^8 m per sec.

(b) $\theta_1 = 39°$, $\theta_2 = 28°$, $c_1 = 3\times10^8$ m per sec

$$\frac{c_1}{c_2} = \frac{\sin\theta_1}{\sin\theta_2} \Rightarrow c_2 = \frac{c_1 \sin\theta_2}{\sin\theta_1} \Rightarrow c_2 = \frac{(3\times10^8)(\sin 28°)}{\sin 39°} \approx 2\times10^8$$

Since c_1 is only given to one significant digit, c_2 can only be given to one significant digit. The speed of light in the second medium is about 2×10^8 m per sec.

58. (a) $\theta_1 = 40°, c_2 = 1.5\times10^8$ m per sec, and $c_1 = 3\times10^8$ m per sec

$$\frac{c_1}{c_2} = \frac{\sin\theta_1}{\sin\theta_2} \Rightarrow \sin\theta_2 = \frac{c_2 \sin\theta_1}{c_1} \Rightarrow \sin\theta_2 = \frac{(1.5\times10^8)(\sin 40°)}{3\times10^8} \Rightarrow \theta_2 = \sin^{-1}\left[\frac{(1.5\times10^8)(\sin 40°)}{3\times10^8}\right] \approx 19°$$

(b) $\theta_1 = 62°$, $c_2 = 2.6\times10^8$ m per sec and $c_1 = 3\times10^8$ m per sec

$$\frac{c_1}{c_2} = \frac{\sin\theta_1}{\sin\theta_2} \Rightarrow \sin\theta_2 = \frac{c_2 \sin\theta_1}{c_1} \Rightarrow \sin\theta_2 = \frac{(2.6\times10^8)(\sin 62°)}{3\times10^8} \Rightarrow \theta_2 = \sin^{-1}\left[\frac{(2.6\times10^8)(\sin 62°)}{3\times10^8}\right] \approx 50°$$

59. $\theta_1 = 90°$, $c_1 = 3\times10^8$ m per sec, and $c_2 = 2.254\times10^8$

$$\frac{c_1}{c_2} = \frac{\sin\theta_1}{\sin\theta_2} \Rightarrow \sin\theta_2 = \frac{c_2 \sin\theta_1}{c_1}$$

$$\sin\theta_2 = \frac{(2.254\times10^8)(\sin 90°)}{3\times10^8} = \frac{2.254\times10^8(1)}{3\times10^8} = \frac{2.254}{3} \Rightarrow \theta_2 = \sin^{-1}\left(\frac{2.254}{3}\right) \approx 48.7°$$

60. $\theta_1 = 90° - 29.6° = 60.4°$, $c_1 = 3\times10^8$ m per sec, and $c_2 = 2.254\times10^8$ $c_1 = 3\times10^8$ m per sec

$$\frac{c_1}{c_2} = \frac{\sin\theta_1}{\sin\theta_2} \Rightarrow \sin\theta_2 = \frac{c_2 \sin\theta_1}{c_1}$$

$$\sin\theta_2 = \frac{(2.254\times10^8)(\sin 60.4°)}{3\times10^8} = \frac{2.254}{3}(\sin 60.4°) \Rightarrow \theta_2 = \sin^{-1}\left(\frac{2.254}{3}(\sin 60.4°)\right) \approx 40.8°$$

Light from the object is refracted at an angle of 40.8° from the vertical. Light from the horizon is refracted at an angle of 48.7° from the vertical. Therefore, the fish thinks the object lies at an angle of 48.7° – 40.8° = 7.9° above the horizon.

61. (a) Let $V_1 = 55$ mph $= 55$ mph $\cdot$ 1 hr/ 3600 sec $\cdot$ 5280 ft/1 mi $= 80\frac{2}{3}$ ft per sec ≈ 80.67 ft per sec, and V_2=30 mph $= 30$ mph $\cdot$ 1 hr/ 3600 sec $\cdot$ 5280 ft/1 mi $= 44$ ft per sec. Also, let $\theta = 3.5°$, $K_1 = .4$, and $K_2 = .02$.

$$D = \frac{1.05(V_1^2 - V_2^2)}{64.4(K_1 + K_2 + \sin\theta)} = \frac{1.05(80.67^2 - 44^2)}{64.4(.4 + .02 + \sin 3.5°)} \approx 155 \text{ ft}$$

(b) Let $V_1 \approx 80.67$ ft per sec, $V_2 = 44$ ft per sec, $\theta = -2°$, $K_1 = .4$, and $K_2 = .02$.

$$D = \frac{1.05(V_1^2 - V_2^2)}{64.4(K_1 + K_2 + \sin\theta)} = \frac{1.05(80.67^2 - 44^2)}{64.4[.4 + .02 + \sin(-2°)]} \approx 194 \text{ ft}$$

62. Using the values for K_1 and K_2 from Exercise 61, determine V_2 when $D = 200$, $\theta = -3.5°$, V_1=90 mph $= 30$ mph $\cdot$ 1 hr/ 3600 sec $\cdot$ 5280 ft/1 mi $= 132$ ft per sec.

$$D = \frac{1.05(V_1^2 - V_2^2)}{64.4(K_1 + K_2 + \sin\theta)} \Rightarrow 200 = \frac{1.05(132^2 - V_2^2)}{64.4[.4 + .02 + \sin(-3.5°)]}$$

$$200 = \frac{1.05(132^2) - 1.05V_2^2}{23.12} \Rightarrow 200(23.12) = 18{,}295.2 - 1.05V_2^2 \Rightarrow 4624 = 18{,}295.2 - 1.05V_2^2$$

$$-13{,}671.2 = -1.05V_2^2 \Rightarrow V_2^2 = \frac{-13{,}671.2}{-1.05} \Rightarrow V_2^2 = 13020.19048 \Rightarrow V_2 \approx 114.106$$

$V_2 \approx 114$ ft/sec $\cdot$ 3600 sec/hr $\cdot$ 1 mi/5280 ft ≈ 78 mph

Section 2.4: Solving Right Triangles

Connections (page 72)
Steps 1 and 2 compare to his second step. The first part of Step 3 (solving the equation) compares to his third step, and the last part of Step 3 (checking) compares to his fourth step.

Exercises

1. 20,385.5 to 20,386.5

2. 28,999.5 to 29,000.5

3. 8958.5 to 8959.5

4. Answers will vary.
No; the number of points scored will be a whole number.

5. Answer will vary.
It would be cumbersome to write 2 as 2.00 or 2.000, for example, if the measurements had 3 or 4 significant digits (depending on the problem). In the formula, it is understood that 2 is an exact value. Since the radius measurement, 54.98 cm, has four significant digits, an appropriate answer would be 345.4 cm.

6. 23.0 ft indicates 3 significant digits and 23.00 ft indicates four significant digits.

7. If h is the actual height of a building and the height is measured as 58.6 ft, then $|h - 58.6| \leq \underline{.05}$.

8. If w is the actual weight of a car and the weight is measured as 15.00×10^2 lb, then $|w - 1500| \leq \underline{.5}$.

9. $A = 36°20'$, $c = 964$ m

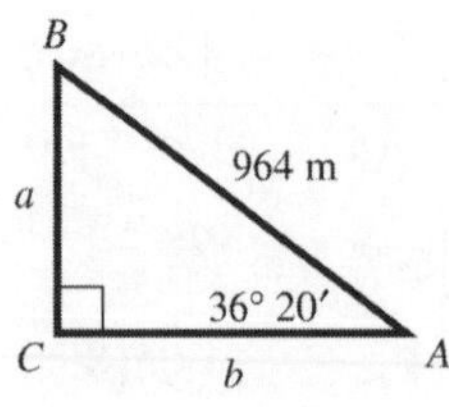

$$A + B = 90° \Rightarrow B = 90° - A \Rightarrow B = 90° - 36°20' = 89°60' - 36°20' = 53°40'$$

$$\sin A = \frac{a}{c} \Rightarrow \sin 36°20' = \frac{a}{964} \Rightarrow a = 964 \sin 36°20' \approx 571 \text{ m} \text{ (rounded to three significant digits)}$$

$$\cos A = \frac{b}{c} \Rightarrow \cos 36°20' = \frac{b}{964} \Rightarrow b = 964 \cos 36°20' \approx 777 \text{ m} \text{ (rounded to three significant digits)}$$

10. $A = 31°40'$, $a = 35.9$ km

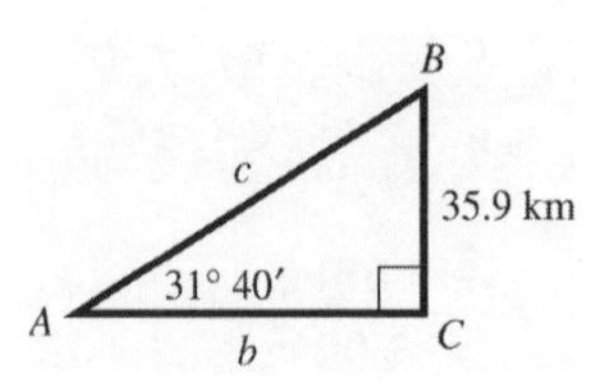

$$A + B = 90° \Rightarrow B = 90° - A \Rightarrow B = 90° - 31°40' = 89°60' - 31°40' = 58°20'$$

$$\sin A = \frac{a}{c} \Rightarrow \sin 31°40' = \frac{35.9}{c} \Rightarrow c = \frac{35.9}{\sin 31°40'} \approx 68.4 \text{ km (rounded to three significant digits)}$$

$$\tan A = \frac{a}{b} \Rightarrow \tan 31°40' = \frac{35.9}{b} \Rightarrow b = \frac{35.9}{\tan 31°40'} \approx 58.2 \text{ km (rounded to three significant digits)}$$

11. $N = 51.2°$, $m = 124$ m

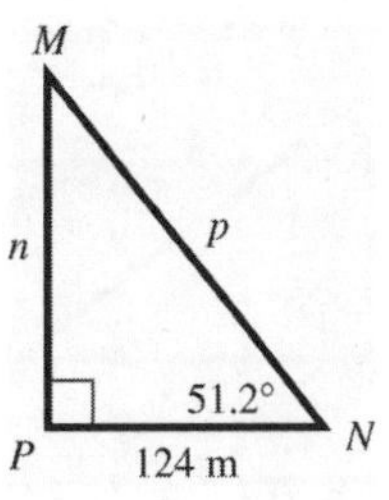

$$M + N = 90° \Rightarrow M = 90° - N \Rightarrow M = 90° - 51.2° = 38.8°$$

$\tan N = \dfrac{n}{m} \Rightarrow \tan 51.2° = \dfrac{n}{124} \Rightarrow n = 124 \tan 51.2° \approx 154$ m (rounded to three significant digits)

$\cos N = \dfrac{m}{p} \Rightarrow \cos 51.2° = \dfrac{124}{p} \Rightarrow p = \dfrac{124}{\cos 51.2°} \approx 198$ m (rounded to three significant digits)

12. $X = 47.8°$, $z = 89.6$ cm

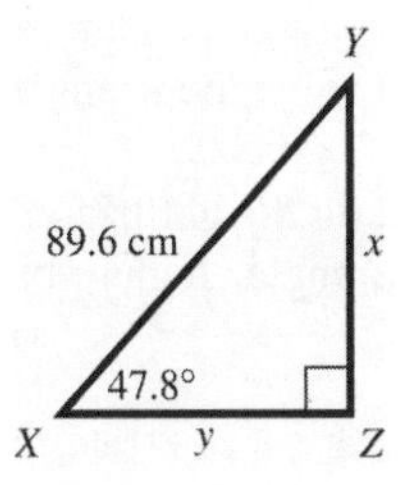

$$Y + X = 90° \Rightarrow Y = 90° - X \Rightarrow Y = 90° - 47.8° = 42.2°$$

$\sin X = \dfrac{x}{z} \Rightarrow \sin 47.8° = \dfrac{x}{89.6} \Rightarrow x = 89.6 \sin 47.8° \approx 66.4$ cm (rounded to three significant digits)

$\cos X = \dfrac{y}{z} \Rightarrow \cos 47.8° = \dfrac{y}{89.6} \Rightarrow y = 89.6 \cos 47.8° \approx 60.2$ cm (rounded to three significant digits)

13. $B = 42.0892°$, $b = 56.851$ cm

$$A + B = 90° \Rightarrow A = 90° - B \Rightarrow A = 90° - 42.0892° = 47.9108°$$

$\sin B = \dfrac{b}{c} \Rightarrow \sin 42.0892° = \dfrac{56.851}{c} \Rightarrow c = \dfrac{56.851}{\sin 42.0892°} \approx 84.816$ cm (rounded to five significant digits)

$\tan B = \dfrac{b}{a} \Rightarrow \tan 42.0892° = \dfrac{56.851}{a} \Rightarrow a = \dfrac{56.851}{\tan 42.0892°} \approx 62.942$ cm (rounded to five significant digits)

14. $B = 68.5142°,\ c = 3579.42$ m

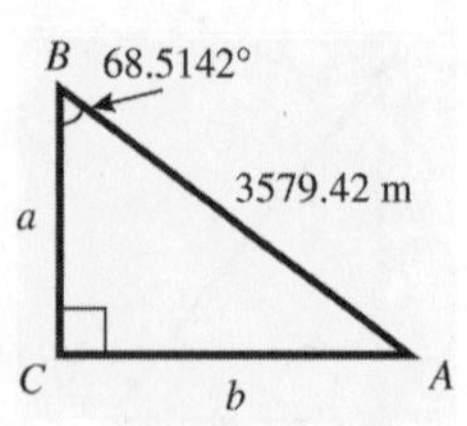

$A + B = 90° \Rightarrow A = 90° - B \Rightarrow A = 90° - 68.5142° = 21.4858°$

$\sin B = \frac{b}{c} \Rightarrow \sin 68.5142° = \frac{b}{3579.42} \Rightarrow b = 3579.42 \sin 68.5142° \approx 3330.68$ m (rounded to six significant digits)

$\cos B = \frac{a}{c} \Rightarrow \cos 68.5142° = \frac{a}{3579.42} \Rightarrow a = 3579.42 \cos 68.5142° \approx 1311.04$ m (rounded to six significant digits)

15. No; You need to have at least one side to solve the triangle.

16. If we are given an acute angle and a side in a right triangle, the unknown part of the triangle requiring the least work to find is the other acute angle. It may be found by subtracting the given acute angle from 90°.

17. Answers will vary.

If you know one acute angle, the other acute angle may be found by subtracting the given acute angle from 90°. If you know one of the sides, then choose two of the trigonometric ratios involving sine, cosine or tangent that involve the known side in order to find the two unknown sides.

18. Answers will vary.

If you know the lengths of two sides, you can set up a trigonometric ratio to solve for one of the acute angles. The other acute angle may be found by subtracting the calculated acute angle from 90°. With either of the two acute angles that have been determined, you can set up a trigonometric ratio along with one of the known sides to solve for the missing side.

19. $A = 28.00°$, $c = 17.4$ ft

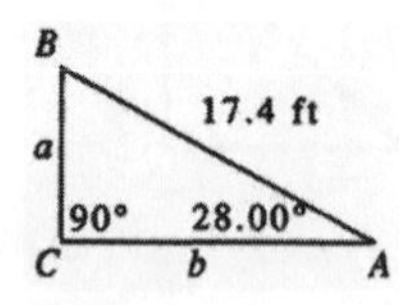

$A + B = 90° \Rightarrow B = 90° - A \Rightarrow B = 90° - 28.00° = 62.00°$

$\sin A = \frac{a}{c} \Rightarrow \sin 28.00° = \frac{a}{17.4} \Rightarrow a = 17.4 \sin 28.00° \approx 8.17$ ft (rounded to three significant digits)

$\cos A = \frac{b}{c} \Rightarrow \cos 28.00° = \frac{b}{17.4} \Rightarrow b = 17.4 \cos 28.00° \approx 15.4$ ft (rounded to three significant digits)

20. $B = 46.00°$, $c = 29.7$ m

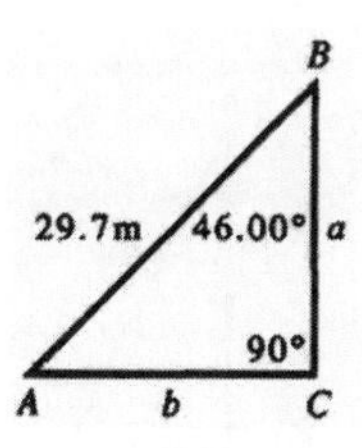

$$A + B = 90° \Rightarrow A = 90° - B \Rightarrow A = 90° - 46.00° = 44.00°$$

$$\cos B = \frac{a}{c} \Rightarrow \cos 46.00° = \frac{a}{29.7} \Rightarrow a = 29.7 \cos 46.00° \approx 20.6 \text{ m} \text{ (rounded to three significant digits)}$$

$$\sin B = \frac{b}{c} \Rightarrow \sin 46.00° = \frac{b}{29.7} \Rightarrow b = 29.7 \sin 46.00° \approx 21.4 \text{ m} \text{ (rounded to three significant digits)}$$

21. Solve the right triangle with $B = 73.00°$, $b = 128$ in. and $C = 90°$

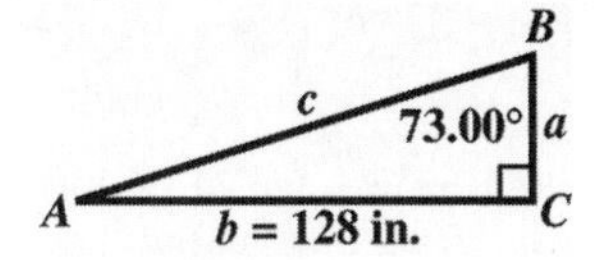

$$A + B = 90° \Rightarrow A = 90° - B \Rightarrow A = 90° - 73.00° = 17.00°$$

$$\tan B° = \frac{b}{a} \Rightarrow \tan 73.00° = \frac{128}{a} \Rightarrow a = \frac{128}{\tan 73.00°} \Rightarrow a = 39.1 \text{ in (rounded to three significant digits)}$$

$$\sin B° = \frac{b}{c} \Rightarrow \sin 73.00° = \frac{128}{c} \Rightarrow c = \frac{128}{\sin 73.00°} \Rightarrow c = 134 \text{ in (rounded to three significant digits)}$$

22. $A = 61.00°$, $b = 39.2$ cm

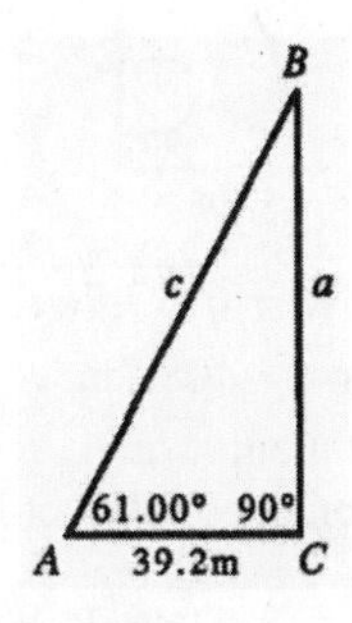

$$A + B = 90° \Rightarrow B = 90° - A \Rightarrow B = 90° - 61.00° = 29.00°$$

$$\tan A = \frac{a}{b} \Rightarrow \tan 61.00° = \frac{a}{39.2} \Rightarrow a = 39.2 \tan 61.00 \approx 70.7 \text{ cm (rounded to three significant digits)}$$

$$\cos A = \frac{b}{c} \Rightarrow \cos 61.00° = \frac{39.2}{c} \Rightarrow c = \frac{39.2}{\cos 61.00°} \approx 80.9 \text{ cm (rounded to three significant digits)}$$

23. $a = 76.4$ yd, $b = 39.3$ yd

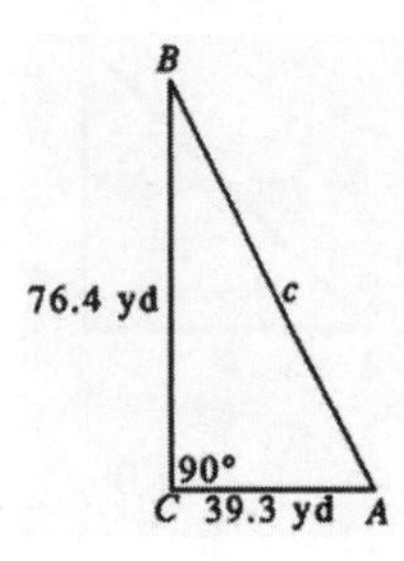

$$c^2 = a^2 + b^2 \Rightarrow c = \sqrt{a^2 + b^2} = \sqrt{(76.4)^2 + (39.3)^2} = \sqrt{5836.96 + 1544.49} = \sqrt{7381.45} \approx 85.9 \text{ yd}$$

(rounded to three significant digits)

We will determine the measurements of both A and B by using the sides of the right triangle. In practice, once you find one of the measurements, subtract it from $90°$ to find the other.

$$\tan A = \frac{a}{b} \Rightarrow \tan A = \frac{76.4}{39.3} \approx 1.944020356 \Rightarrow A \approx \tan^{-1}(1.944020356) \approx 62.8° \approx 62°50'$$

$$\tan B = \frac{b}{a} \Rightarrow \tan B = \frac{39.3}{76.4} \approx .5143979058 \Rightarrow B \approx \tan^{-1}(.5143979058) \approx 27.2° \approx 27°10'$$

24. $a = 958$ m, $b = 489$ m

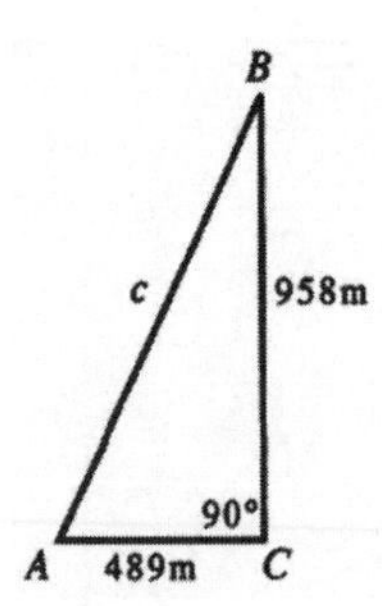

$$c^2 = a^2 + b^2 \Rightarrow c = \sqrt{a^2 + b^2} = \sqrt{958^2 + 489^2} = \sqrt{917{,}764 + 239{,}121} = \sqrt{1{,}156{,}885} \approx 1075.565887$$

If we round to three significant digits, then $c \approx 1080$ m.

We will determine the measurements of both A and B by using the sides of the right triangle. In practice, once you find one of the measurements, subtract it from $90°$ to find the other.

$$\tan A = \frac{a}{b} \Rightarrow \tan A = \frac{958}{489} \approx 1.959100204 \Rightarrow A \approx \tan^{-1}(1.959100204) \approx 63.0° \approx 63°00'$$

$$\tan B = \frac{b}{a} \Rightarrow \tan B = \frac{489}{958} \approx .5104384134 \Rightarrow B \approx \tan^{-1}(.5104384134) \approx 27.0° \approx 27°00'$$

25. $a = 18.9$ cm, $c = 46.3$ cm

$$c^2 = a^2 + b^2 \Rightarrow c^2 - a^2 = b^2 \Rightarrow b = \sqrt{c^2 - a^2} \Rightarrow b = \sqrt{(46.3)^2 - (18.9)^2}$$

$b = \sqrt{2143.69 - 357.21} = \sqrt{1786.48} \approx 42.3$ cm (rounded to three significant digits)

$$\sin A = \frac{a}{c} \Rightarrow \sin A = \frac{18.9}{46.3} \approx .4082073434 \Rightarrow A = \sin^{-1}(.4082073434) \approx 24.1° \approx 24°10'$$

$$\cos B = \frac{a}{c} \Rightarrow \cos B = \frac{18.9}{46.3} \approx .4082073434 \Rightarrow B = \cos^{-1}(.4082073434) \approx 65.9° \approx 65°50'$$

26. $b = 219$ m, $c = 647$ m

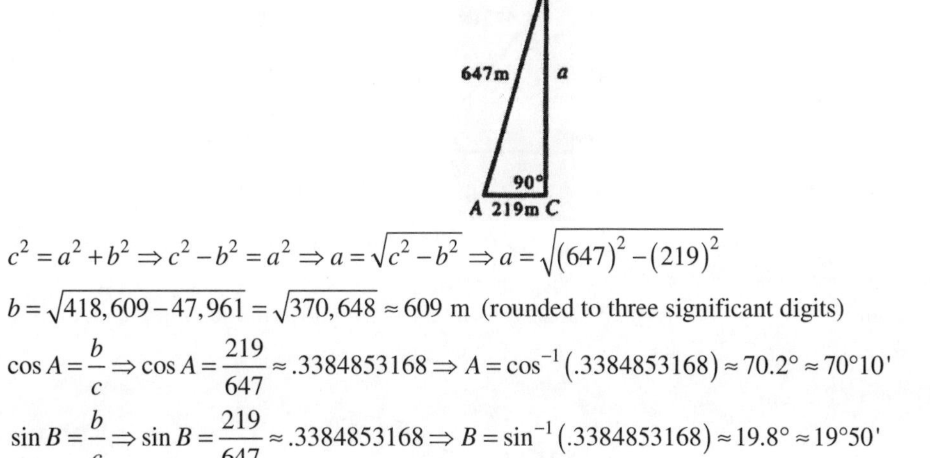

$$c^2 = a^2 + b^2 \Rightarrow c^2 - b^2 = a^2 \Rightarrow a = \sqrt{c^2 - b^2} \Rightarrow a = \sqrt{(647)^2 - (219)^2}$$

$b = \sqrt{418{,}609 - 47{,}961} = \sqrt{370{,}648} \approx 609$ m (rounded to three significant digits)

$$\cos A = \frac{b}{c} \Rightarrow \cos A = \frac{219}{647} \approx .3384853168 \Rightarrow A = \cos^{-1}(.3384853168) \approx 70.2° \approx 70°10'$$

$$\sin B = \frac{b}{c} \Rightarrow \sin B = \frac{219}{647} \approx .3384853168 \Rightarrow B = \sin^{-1}(.3384853168) \approx 19.8° \approx 19°50'$$

27. $A = 53°24'$, $c = 387.1$ ft

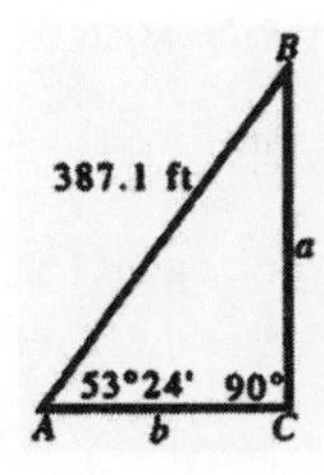

$$A + B = 90° \Rightarrow B = 90° - A \Rightarrow B = 90° - 53°24' = 89°60' - 53°24° = 36°36'$$

$$\sin A = \frac{a}{c} \Rightarrow \sin 53°24' = \frac{a}{387.1} \Rightarrow a = 387.1 \sin 53°24' \approx 310.8 \text{ ft (rounded to four significant digits)}$$

$$\cos A = \frac{b}{c} \Rightarrow \cos 53°24' = \frac{b}{387.1} \Rightarrow b = 387.1 \cos 53°24' \approx 230.8 \text{ ft (rounded to four significant digits)}$$

28. $A = 13°47'$, $c = 1285$ m

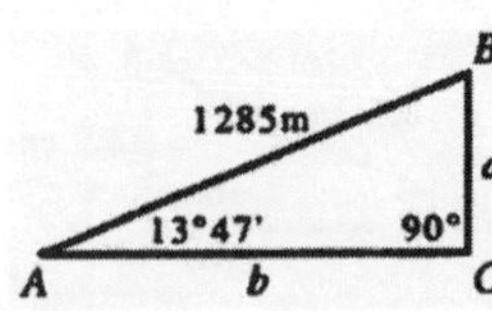

$A + B = 90° \Rightarrow B = 90° - A \Rightarrow B = 90° - 13°47' = 89°60' - 13°47' = 76°13'$

$\sin A = \frac{a}{c} \Rightarrow \sin 13°47' = \frac{a}{1285} \Rightarrow a = 1285 \sin 13°47' \Rightarrow a \approx 306.2$ m (rounded to four significant digits)

$\cos A = \frac{b}{c} \Rightarrow \cos 13°47' = \frac{b}{1285} \Rightarrow b = 1285 \cos 13°47' \approx 1248$ m (rounded to four significant digits)

29. $B = 39°9'$, $c = .6231$ m

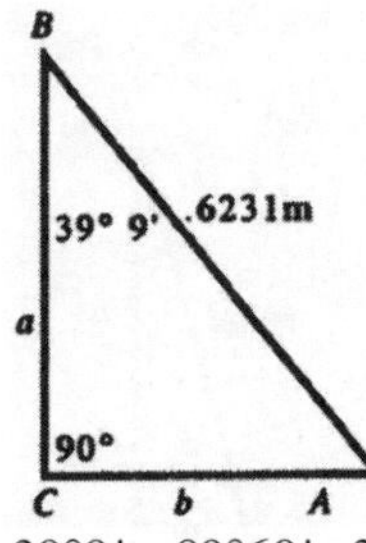

$A + B = 90° \Rightarrow A = 90° - B \Rightarrow A = 90° - 39°9' = 89°60' - 39°9' = 50°51'$

$\cos B = \frac{a}{c} \Rightarrow \cos 39°9' = \frac{a}{.6231} \Rightarrow a = .6231 \cos 39°9' \approx .4832$ m (rounded to four significant digits)

$\sin B = \frac{b}{c} \Rightarrow \sin 39°9' = \frac{b}{.6231} \Rightarrow b = .6231 \sin 39°9' \approx .3934$ m (rounded to four significant digits)

30. $B = 82°51'$, $c = 4.825$ cm

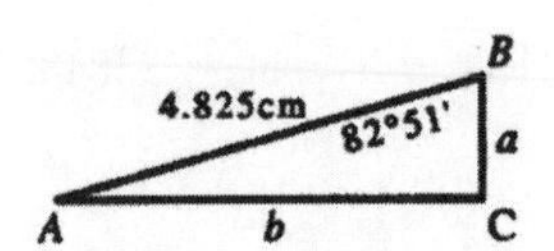

$A + B = 90° \Rightarrow A = 90° - B \Rightarrow B = 90° - 82°51' = 89°60' - 82°51' = 7°9'$

$\sin B = \frac{b}{c} \Rightarrow \sin 82°51' = \frac{b}{4.825} \Rightarrow b = 4.825 \sin 82°51' \approx 4.787$ cm (rounded to three significant digits)

$\cos B = \frac{a}{c} \Rightarrow \cos 82°51' = \frac{a}{4.825} \Rightarrow a = 4.825 \cos 82°51'' \approx .6006$ cm (rounded to three significant digits)

31. The angle of elevation from X to Y is 90° whenever Y is directly above X.

32. The angle of elevation from X to Y is the acute angle formed by ray XY and a horizontal ray with endpoint at X. Therefore, the angle of elevation cannot be more than 90°.

33. Answers will vary.
The angle of elevation and the angle of depression are measured between the line of sight and a horizontal line. So, in the diagram, lines AD and CB are both horizontal. Hence, they are parallel. The line formed by AB is a transversal and angles DAB and ABC are alternate interior angle and thus have the same measure.

34. The angle of depression is measured between the line of sight and a horizontal line. This angle is measured between the line of sight and a vertical line.

35. $\sin 43°50' = \dfrac{d}{13.5} \Rightarrow d = 13.5 \sin 43°50' \approx 9.3496000$

The ladder goes up the wall 9.35 m. (rounded to three significant digits)

36. $T = 32°\ 10'$ and $S = 57°50'$

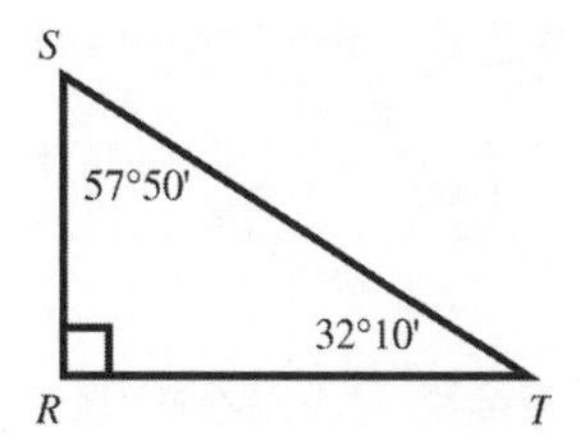

Since $S + T = 32°\ 10' + 57°50' = 89°60' = 90°$, triangle RST is a right triangle. Thus, we have

$\tan 32°10' = \dfrac{RS}{53.1} \Rightarrow RS = 53.1 \tan 32°10' \approx 33.395727.$

The distance across the lake is 33.4 m. (rounded to three significant digits)

37. Let x represent the horizontal distance between the two buildings and y represent the height of the portion of the building across the street that is higher than the window.

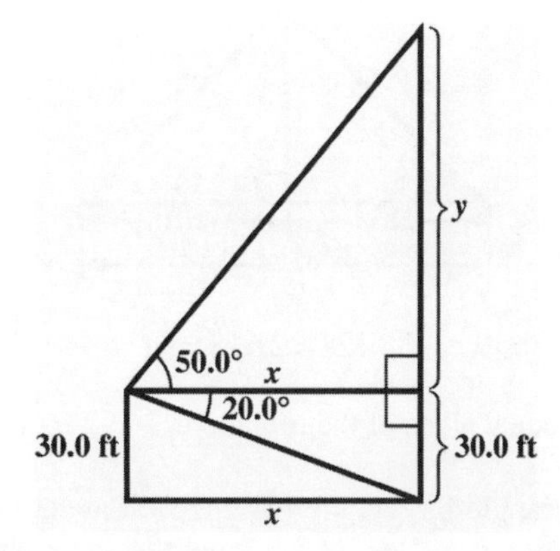

We have the following.

$$\tan 20.0° = \frac{30.0}{x} \Rightarrow x = \frac{30.3}{\tan 20.0°} \approx 82.4$$

$$\tan 50.0° = \frac{y}{x} \Rightarrow y = x \tan 50.0° = \left(\frac{30.0}{\tan 20.0°}\right) \tan 50.0°$$

$$\text{height} = y + 30.0 = \left(\frac{30.0}{\tan 20.0°}\right) \tan 50.0° + 30.0 \approx 128.2295$$

The height of the building across the street is about 128 ft. (rounded to three significant digits)

38. Let x = the diameter of the sun.

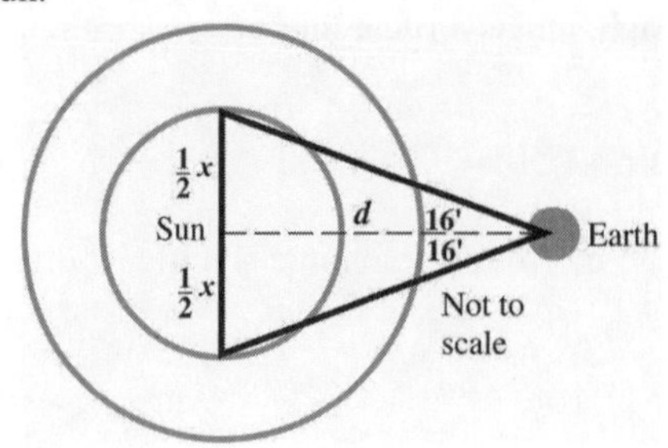

Since the included angle is $32'$, $\frac{1}{2}(32')=16'$. We will use this angle, d, and half of the diameter to set up the following equation.

$$\frac{\frac{1}{2}x}{92,919,800}=\tan 16' \Rightarrow x=2(92,919,800)(\tan 16')\approx 864,943.0189$$

The diameter of the sun is about 865,000 mi. (rounded to three significant digits)

39. The altitude of an isosceles triangle bisects the base as well as the angle opposite the base. The two right triangles formed have interior angles, which have the same measure angle. The lengths of the corresponding sides also have the same measure. Since the altitude bisects the base, each leg (base) of the right triangles is $\frac{42.36}{2}=21.18$ in.

Let x = the length of each of the two equal sides of the isosceles triangle.

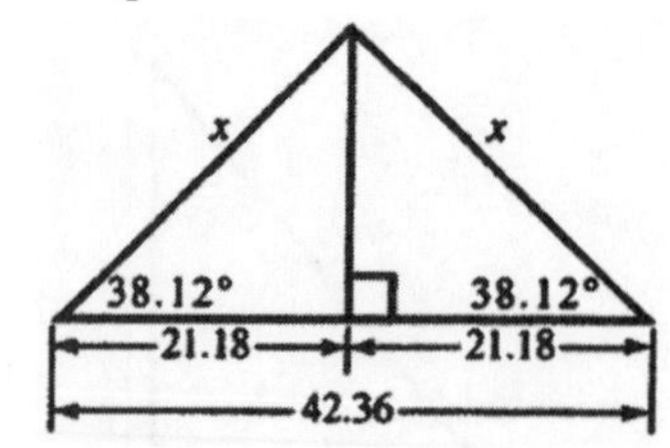

$$\cos 38.12°=\frac{21.18}{x}\Rightarrow x\cos 38.12°=21.18\Rightarrow x=\frac{21.18}{\cos 38.12°}\approx 26.921918$$

The length of each of the two equal sides of the triangle is 26.92 in. (rounded to four significant digits)

40. The altitude of an isosceles triangle bisects the base as well as the angle opposite the base. The two right triangles formed have interior angles, which have the same measure angle. The lengths of the corresponding sides also have the same measure. Since the altitude bisects the base, each leg (base) of the right triangles are $\frac{184.2}{2}=92.10$ cm. Each angle opposite to the base of the right triangles measures $\frac{1}{2}(68°44')=34°22'$.

Let h = the altitude.

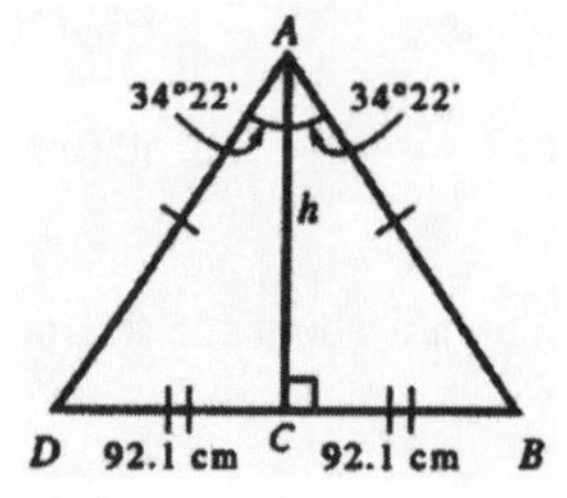

In triangle ABC, $\tan 34°22'=\frac{92.10}{h}\Rightarrow h\tan 34°22'=92.10\Rightarrow h=\frac{92.10}{\tan 34°22'}\approx 134.67667.$

The altitude of the triangle is 134.7 cm. (rounded to four significant digits)

41. In order to find the angle of elevation, θ, we need to first find the length of the diagonal of the square base. The diagonal forms two isosceles right triangles. Each angle formed by a side of the square and the diagonal measures 45°.

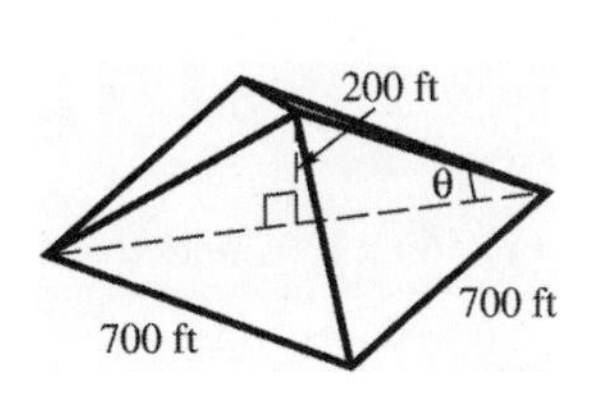

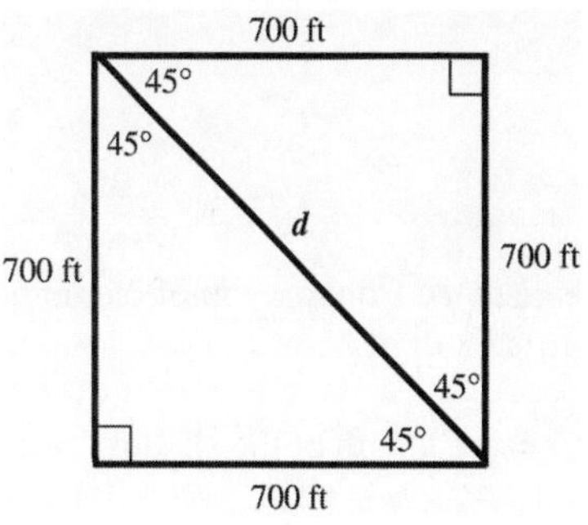

By the Pythagorean theorem, $700^2 + 700^2 = d^2 \Rightarrow 2 \cdot 700^2 = d^2 \Rightarrow d = \sqrt{2 \cdot 700^2} \Rightarrow d = 700\sqrt{2}$. Thus, length of the diagonal is $700\sqrt{2}$ ft. To to find the angle, θ, we consider the following isosceles triangle.

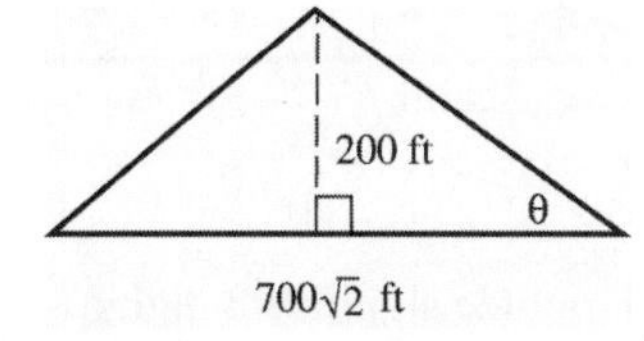

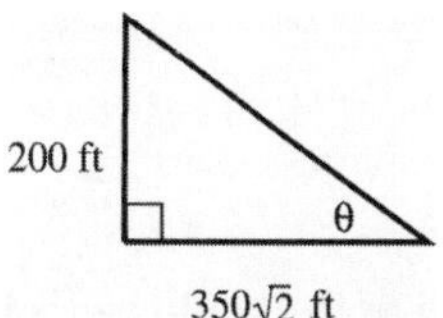

The height of the pyramid bisects the base of this triangle and forms two right triangles. We can use one of these triangles to find the angle of elevation, θ.

$$\tan\theta = \frac{200}{350\sqrt{2}} \approx .4040610178 \Rightarrow \theta \approx \tan^{-1}(.4040610178) \approx 22.0017$$

Rounding this figure to two significant digits, we have $\theta \approx 22°$.

42. Let y = the height of the spotlight (this measurement starts 6 feet above ground)

We have the following.

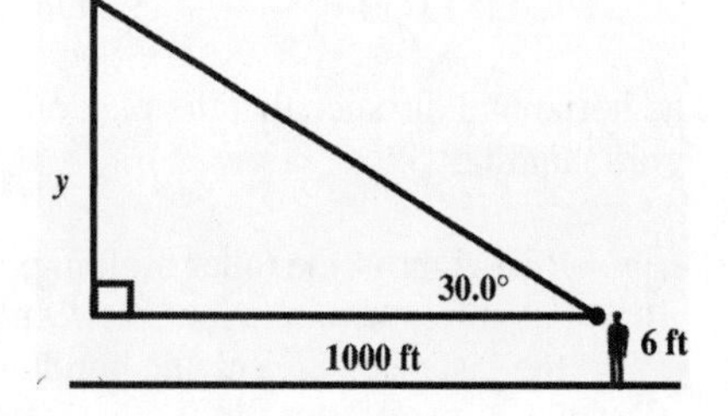

$$\tan 30.0° = \frac{y}{1000}$$
$$y = 1000 \cdot \tan 30.0° \approx 577.3502$$

Rounding this figure to three significant digits, we have $y \approx 577$.

However, the observer's eye-height is 6 feet from the ground, so the cloud ceiling is $577 + 6 = 583$ ft.

43. Let h represent the height of the tower.

In triangle ABC we have the following.

$$\tan 34.6° = \frac{h}{40.6}$$
$$h = 40.6 \tan 34.6° \approx 28.0081$$

The height of the tower is 28.0 m. (rounded to three significant digits)

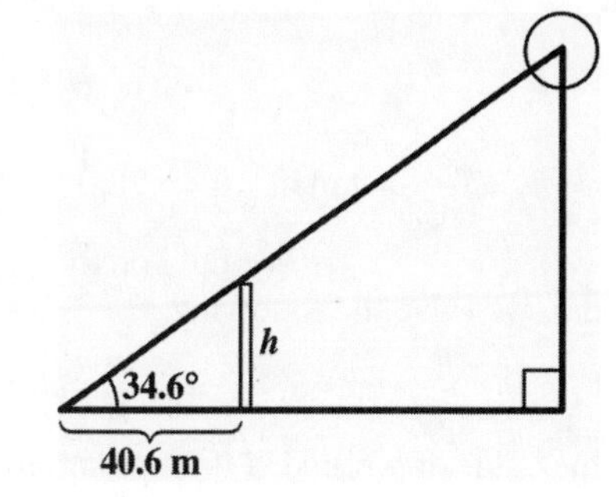

44. Let d = the distance from the top B of the building to the point on the ground A.

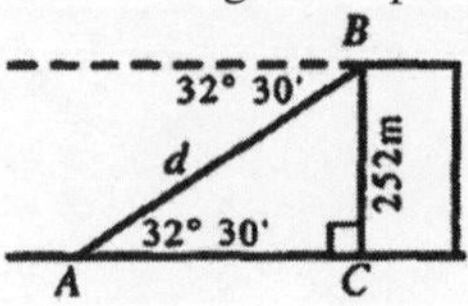

In triangle ABC, $\sin 32°30' = \dfrac{252}{d} \Rightarrow d = \dfrac{252}{\sin 32°30'} \approx 469.0121.$
The distance from the top of the building to the point on the ground is 469 m. (rounded to three significant digits)

45. Let x = the length of the shadow cast by Diane Carr.

$$\tan 23.4° = \frac{5.75}{x}$$
$$x \tan 23.4° = 5.75$$
$$x = \frac{5.75}{\tan 23.4°} \approx 13.2875$$

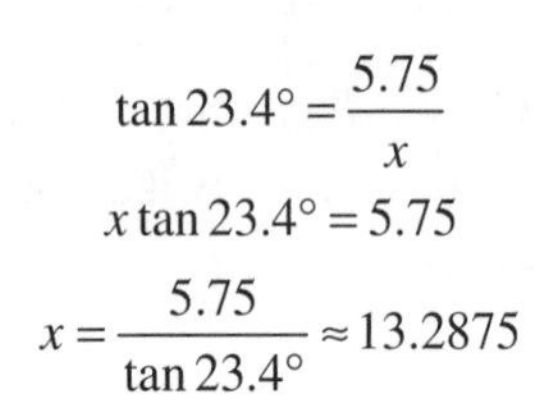

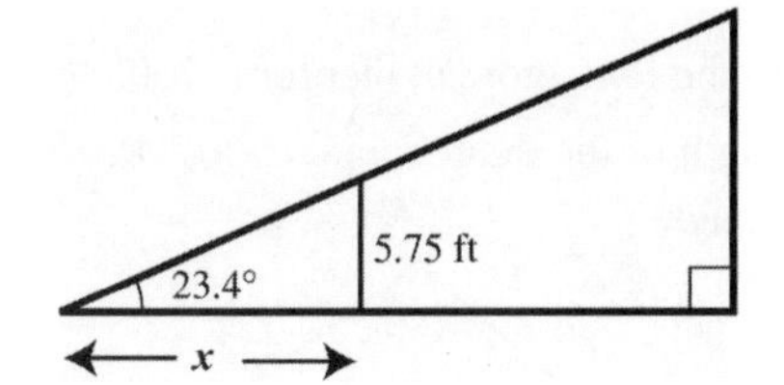

The length of the shadow cast by Diane Carr is 13.3 ft. (rounded to three significant digits)

46. Let x = the horizontal distance that the plan must fly to be directly over the tree.

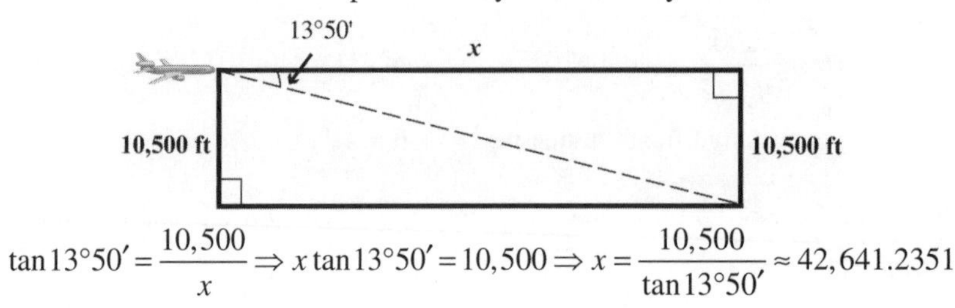

$$\tan 13°50' = \frac{10{,}500}{x} \Rightarrow x \tan 13°50' = 10{,}500 \Rightarrow x = \frac{10{,}500}{\tan 13°50'} \approx 42{,}641.2351$$

The horizontal distance that the plan must fly to be directly over the tree is 42,600 ft. (rounded to three significant digits)

47. Let x = the height of the taller building;
h = the difference in height between the shorter and taller buildings;
d = the distance between the buildings along the ground.

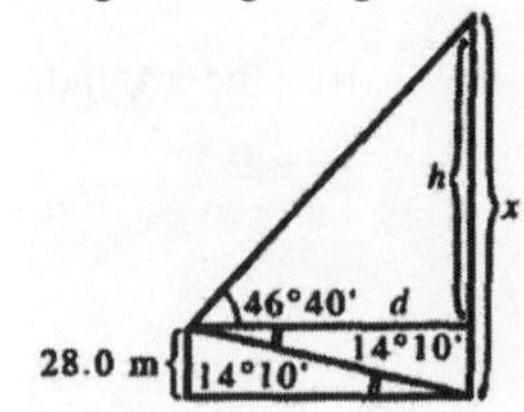

$$\frac{28.0}{d} = \tan 14°10' \Rightarrow 28.0 = d \tan 14°10' \Rightarrow d = \frac{28.0}{\tan 14°10'} \approx 110.9262493 \text{ m}$$

(We hold on to these digits for the intermediate steps.)

To find h, we solve the following.

$$\frac{h}{d} = \tan 46°40' \Rightarrow h = d \tan 46°40' \approx (110.9262493)\tan 46°40' \approx 117.5749$$

Thus, the value of h rounded to three significant digits is 118 m.
Since $x = h + 28.0 = 118 + 28.0 \approx 146\text{m}$, the height of the taller building is 146 m.

48. Let $\theta =$ the angle of depression.

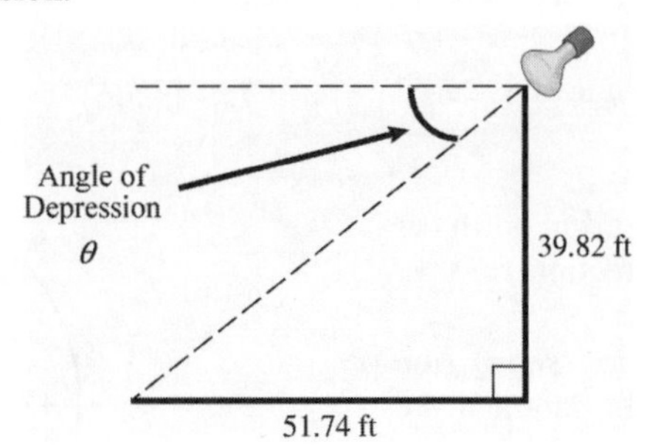

$$\tan\theta = \frac{39.82}{51.74} \approx .7696173174 \Rightarrow \theta = \tan^{-1}(.7696173174) \Rightarrow \theta \approx 37.58° \approx 37°35'$$

49. **(a)** Let x = the height of the peak above 14,545 ft.

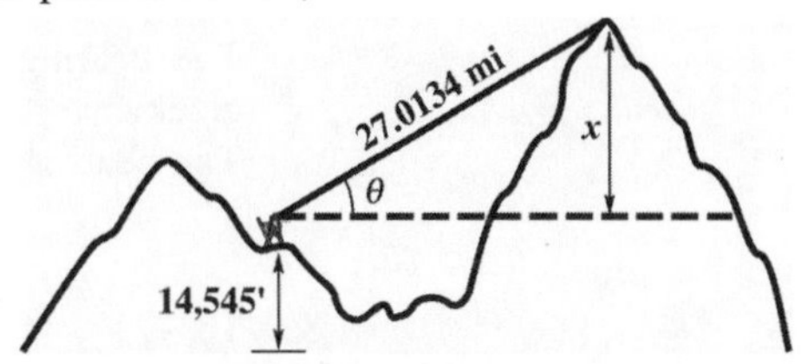

Since the diagonal of the right triangle formed is in miles, we must first convert this measurement to feet. Since there are 5280 ft in one mile, we have the length of the diagonal is $27.0134(5280) = 142,630.752$. To find the value of x, we solve the following.

$$\sin 5.82° = \frac{x}{142,630.752} \Rightarrow x = 142,630.752 \sin 5.82° \approx 14,463.2674$$

Thus, the value of x rounded to five significant digits is 14,463 ft. Thus, the total height is about 14,545 + 14,463 = 29,008 ft.

(b) The curvature of the earth would make the peak appear shorter than it actually is. Initially the surveyors did not think Mt. Everest was the tallest peak in the Himalayas. It did not look like the tallest peak because it was farther away than the other large peaks.

50. Let x = the distance from the assigned target.

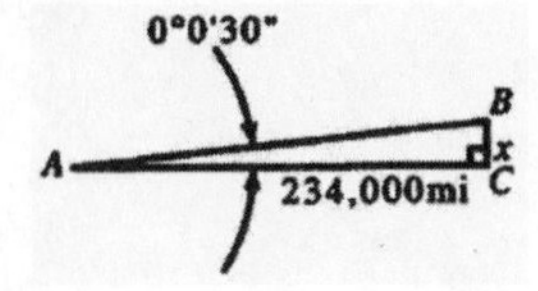

In triangle ABC, we have the following.

$$\tan 0°0'30'' = \frac{x}{234,000} \Rightarrow x = 234,000 \tan 0°0'30'' \approx 34.0339$$

The distance from the assigned target is 34.0 mi. (rounded to three significant digits)

Section 2.5: Further Applications of Right Triangles

1. It should be shown as an angle measured clockwise from due north.

2. It should be shown measured from north (or south) in the east (or west) direction.

3. A sketch is important to show the relationships among the given data and the unknowns.

4. The angle of elevation (or depression) from X to Y is measured from the horizontal line through X to the ray XY.

5. $(-4,0)$

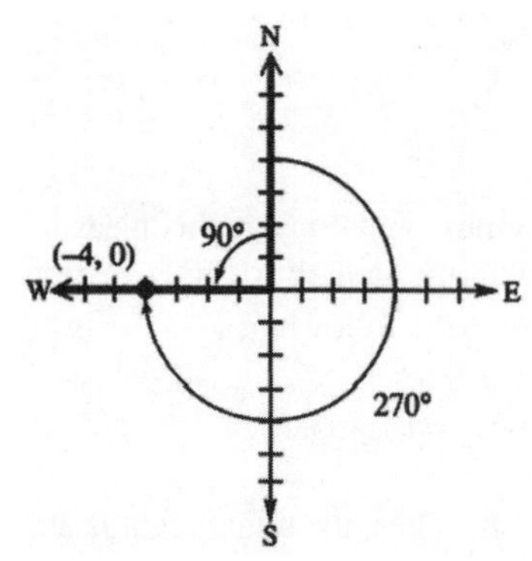

The bearing of the airplane measured in a clockwise direction from due north is 270°. The bearing can also be expressed as N 90° W, or S 90° W.

6. $(-3,-3)$

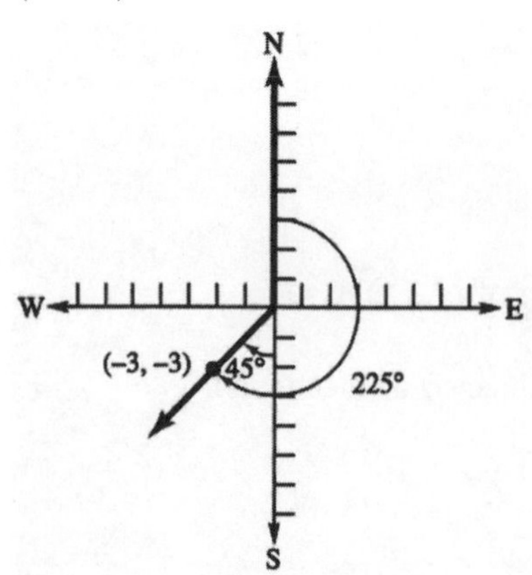

The bearing of the airplane measured in a clockwise direction from due north is 225°. The bearing can also be expressed as S 45° W.

7. $(-5,5)$

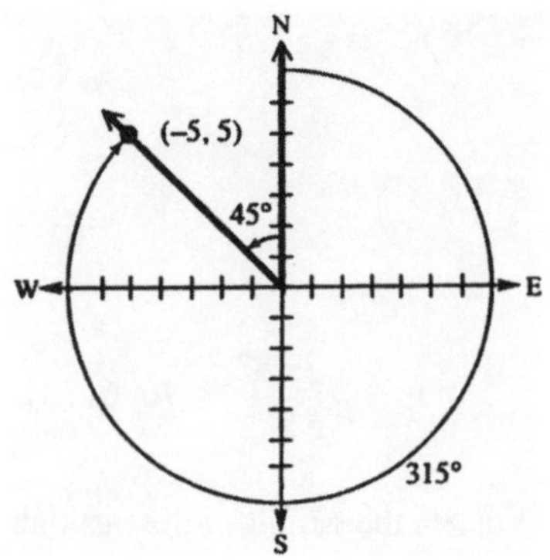

The bearing of the airplane measured in a clockwise direction from due north is 315°. The bearing can also be expressed as N 45° W.

8. $(0,-2)$

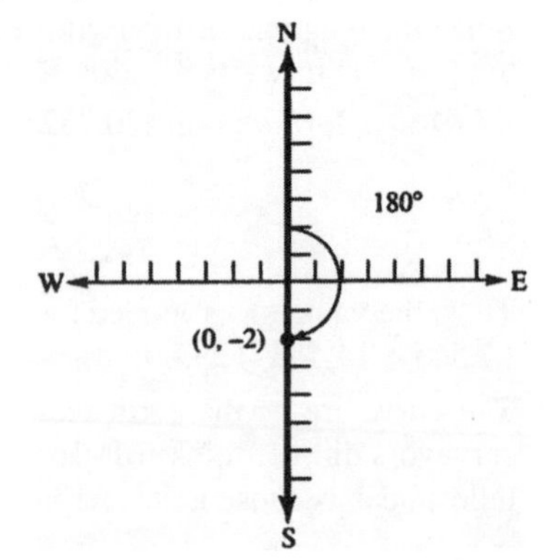

The bearing of the airplane measured in a clockwise direction from due north is 180°. The bearing can also be expressed as S 0° E or S 0° W.

9. All points whose bearing from the origin is $240°$ lie in quadrant III.

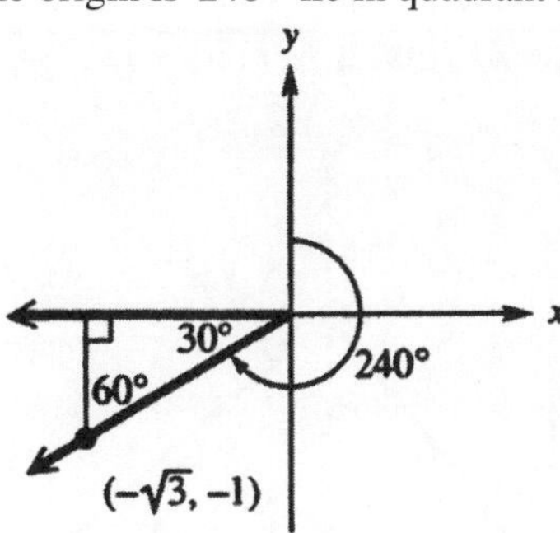

The reference angle, θ', is 30°. For any point, (x, y), on the ray $\frac{x}{r} = -\cos\theta'$ and $\frac{y}{r} = -\sin\theta'$, where r is the distance the point is from the origin. If we let $r = 2$, then we have the following.

$$\frac{x}{r} = -\cos\theta' \Rightarrow x = -r\cos\theta' = -2\cos 30° = -2\cdot\tfrac{\sqrt{3}}{2} = -\sqrt{3}$$

$$\frac{y}{r} = -\sin\theta' \Rightarrow y = -r\sin\theta' = -2\sin 30° = -2\cdot\tfrac{1}{2} = -1$$

Thus, a point on the ray is $\left(-\sqrt{3}, -1\right)$. Since the ray contains the origin, the equation is of the form $y = mx$. Substituting the point $\left(-\sqrt{3}, -1\right)$, we have $-1 = m\left(-\sqrt{3}\right) \Rightarrow m = \frac{-1}{-\sqrt{3}} = \frac{1}{\sqrt{3}}\cdot\frac{\sqrt{3}}{\sqrt{3}} = \frac{\sqrt{3}}{3}$. Thus, the equation of the ray is $y = \frac{\sqrt{3}}{3}x,\ x \le 0$ since the ray lies in quadrant III.

10. All points whose bearing from the origin is $150°$ lie in quadrant IV.

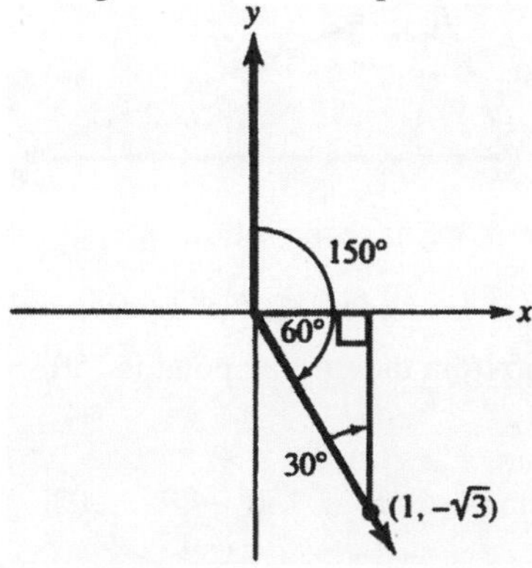

The reference angle, θ', is 60°. For any point, (x, y), on the ray $\frac{x}{r} = \cos\theta'$ and $\frac{y}{r} = -\sin\theta'$, where r is the distance the point is from the origin. If we let $r = 2$, then we have the following.

$$\frac{x}{r} = \cos\theta' \Rightarrow x = r\cos\theta' = 2\cos 60° = 2\cdot\tfrac{1}{2} = 1$$

$$\frac{y}{r} = -\sin\theta' \Rightarrow y = -r\sin\theta' = -2\sin 60° = -2\cdot\tfrac{\sqrt{3}}{2} = -\sqrt{3}$$

Thus, a point on the ray is $\left(1, -\sqrt{3}\right)$. Since the ray contains the origin, the equation is of the form $y = mx$. Substituting the point $\left(1, -\sqrt{3}\right)$, we have $-\sqrt{3} = m(1) \Rightarrow m = -\sqrt{3}$. Thus, the equation of the ray is $y = -\sqrt{3}x,\ x \ge 0$ since the ray lies in quadrant IV.

11. Let x = the distance the plane is from its starting point.
In the figure, the measure of angle ACB is $40^\circ + (180^\circ - 130^\circ) = 40^\circ + 50^\circ = 90^\circ$. Therefore, triangle ACB is a right triangle.

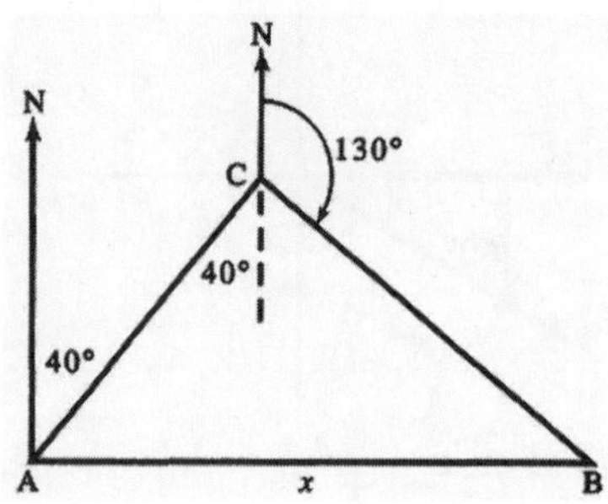

Since $d = rt$, the distance traveled in 1.5 hr is (1.5 hr)(110 mph) = 165 mi. The distance traveled in 1.3 hr is (1.3 hr)(110 mph) = 143 mi.
Using the Pythagorean theorem, we have the following.

$$x^2 = 165^2 + 143^2 \Rightarrow x^2 = 27{,}225 + 20{,}449 \Rightarrow x^2 = 47{,}674 \Rightarrow x \approx 218.3438$$

The plane is 220 mi from its starting point. (rounded to two significant digits)

12. Let x = the distance from the starting point.
In the figure, the measure of angle ACB is $27^\circ + (180^\circ - 117^\circ) = 27^\circ + 63^\circ = 90^\circ$. Therefore, triangle ACB is a right triangle.

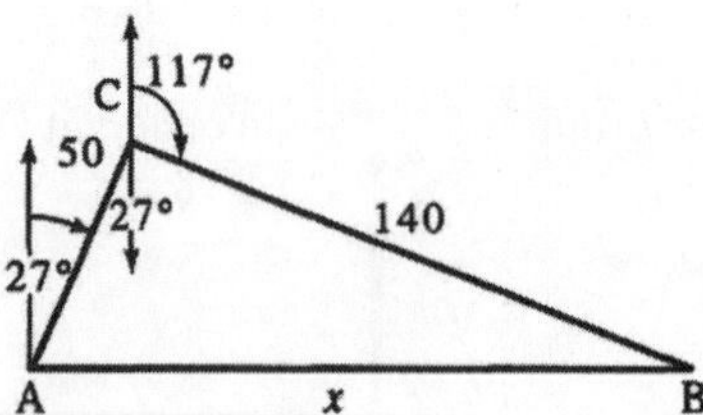

Applying the Pythagorean theorem, we have the following.

$$x^2 = 50^2 + 140^2 \Rightarrow x^2 = 2500 + 19{,}600 \Rightarrow x^2 = 22{,}100 \Rightarrow x = \sqrt{22{,}100} \approx 148.6607$$

The distance of the end of the trip from the starting point is 150 km. (rounded to two significant digits)

13. Let x = distance the ships are apart.
In the figure, the measure of angle CAB is $130^\circ - 40^\circ = 90^\circ$. Therefore, triangle CAB is a right triangle.

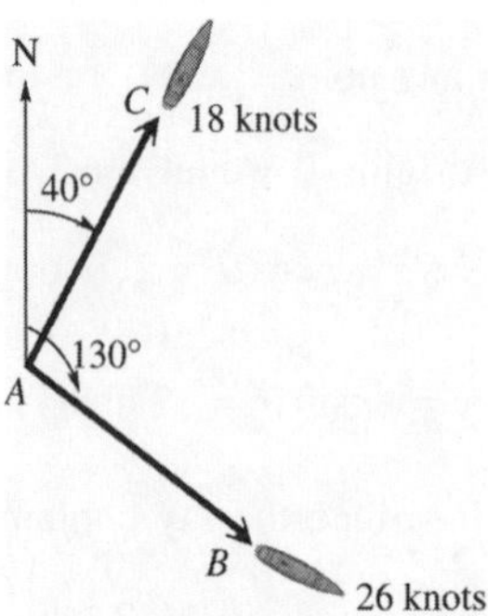

Since $d = rt$, the distance traveled by the first ship in 1.5 hr is (1.5 hr)(18 knots) = 27 nautical mi and the second ship is (1.5hr)(26 knots) = 39 nautical mi.
Applying the Pythagorean theorem, we have the following.

$$x^2 = 27^2 + 39^2 \Rightarrow x^2 = 729 + 1521 \Rightarrow x^2 = 2250 \Rightarrow x = \sqrt{2250} \approx 47.4342$$

The ships are 47 nautical mi apart. (rounded to 2 significant digits)

14. Let C = the location of the ship;
c = the distance between the lighthouses.

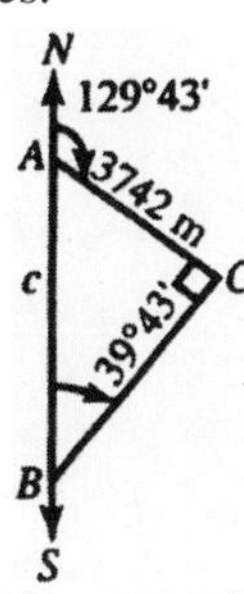

The measure of angle BAC is $180° - 129°43' = 179°60' - 129°43' = 50°17'$.
Since $50°17' + 39°43' = 89°60' = 90°$, we have a right triangle and can get set up and solve the following equation.

$$\sin 39°43 = \frac{3742}{c} \Rightarrow c \sin 39°43 = 3742 \Rightarrow c = \frac{3742}{\sin 39°43} \approx 5856.1020$$

The distance between the lighthouses is 5856 m. (rounded to four significant digits)

15. Draw triangle WDG with W representing Winston-Salem, D representing Danville, and G representing Goldsboro. Name any point X on the line due south from D.

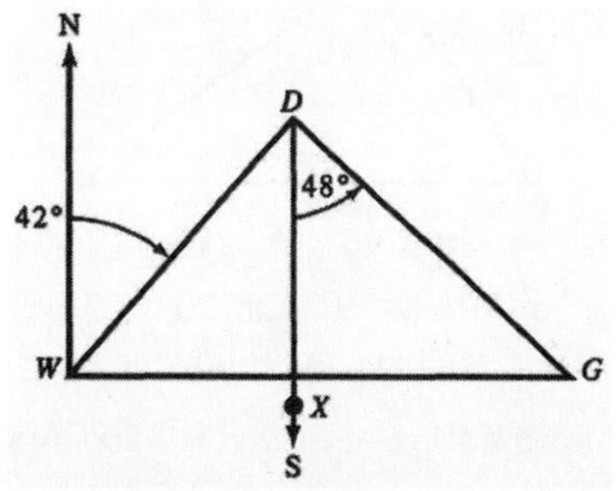

Since the bearing from W to D is 42° (equivalent to N 42° E), angle WDX measures 42°. Since angle XDG measures 48°, the measure of angle D is 42° + 48° = 90°. Thus, triangle WDG is a right triangle. Using $d = rt$ and the Pythagorean theorem, we have the following.

$$WG = \sqrt{(WD)^2 + (DG)^2} = \sqrt{[60(1)]^2 + [60(1.8)]^2}$$

$$WG = \sqrt{60^2 + 108^2} = \sqrt{3600 + 11{,}664} = \sqrt{15{,}264} \approx 123.5476$$

The distance from Winston-Salem to Goldsboro is 120 mi. (rounded to two significant digits)

16. Let x = the distance from Atlanta to Augusta.

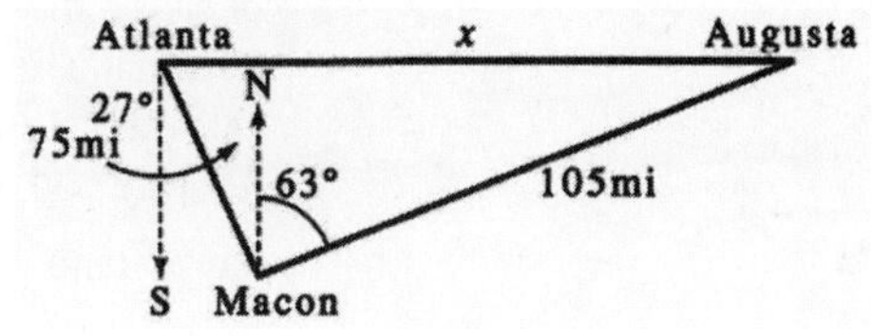

The line from Atlanta to Macon makes an angle of $27° + 63° = 90°$, with the line from Macon to Augusta. Since $d = rt$, the distance from Atlanta to Macon is $60\left(1\frac{1}{4}\right) = 75$ mi. The distance from Macon to Augusta is $60\left(1\frac{3}{4}\right) = 105$ mi.

Use the Pythagorean theorem to find x, we have the following.

$$x^2 = 75^2 + 105^2 \Rightarrow x^2 = 5635 + 11{,}025 \Rightarrow x^2 = 16{,}650 \approx 129.0349$$

The distance from Atlanta to Augusta is 130 mi. (rounded to two significant digits)

17. Let x = distance between the two ships.

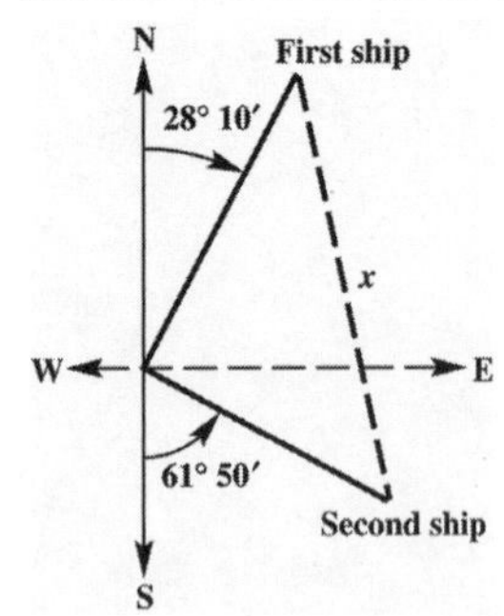

The angle between the bearings of the ships is $180° - (28°10' + 61°50') = 90°$. The triangle formed is a right triangle. The distance traveled at 24.0 mph is (4 hr) (24.0 mph) = 96 mi. The distance traveled at 28.0 mph is (4 hr)(28.0 mph) = 112 mi.

Applying the Pythagorean theorem we have the following.

$$x^2 = 96^2 + 112^2 \Rightarrow x^2 = 9216 + 12{,}544 \Rightarrow x^2 = 21{,}760 \Rightarrow x = \sqrt{21{,}760} \approx 147.5127$$

The ships are 148 mi apart. (rounded to three significant digits)

18. Let C = the location of the transmitter;
a = the distance of the transmitter from B.

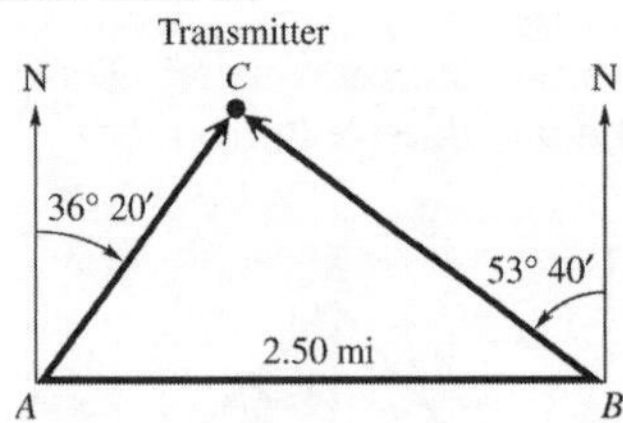

The measure of angle CBA is $90° - 53°40' = 89°60' - 53°40' = 36°20'$.
The measure of angle CAB $90° - 36°20' = 89°60' - 36°20' = 53°40'$.
Since $A + B = 90°$, so $C = 90°$. Thus, we have the following.

$$\sin A = \frac{a}{2.50} \Rightarrow \sin 53°40' = \frac{a}{2.50} \Rightarrow a = 2.50 \sin 53°40' \approx 2.0140$$

The distance of the transmitter from B is 2.01 mi. (rounded to 3 significant digits)

19. Solve the equation $ax = b + cx$ for x in terms of a, b, c.

$$ax = b + cx \Rightarrow ax - cx = b \Rightarrow x(a-c) = b \Rightarrow x = \frac{b}{a-c}$$

20. Suppose we have a line that has x-intercept a and y-intercept b. Assume for the following diagram that a and b are both positive. This is not a necessary condition, but it makes the visualization easier.

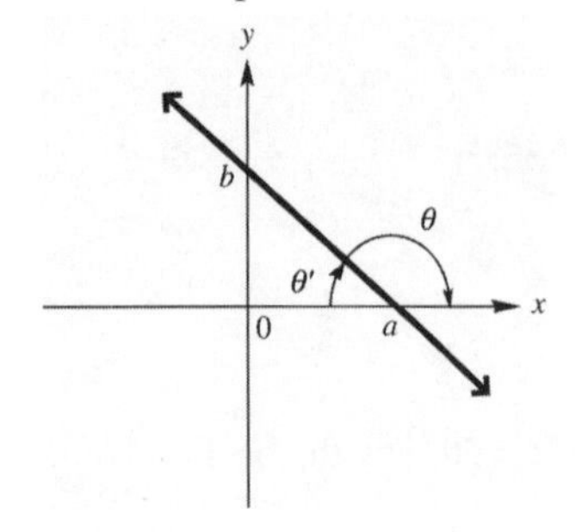

Now $\tan\theta = -\tan(180° - \theta)$. This is because the angle represented by $180° - \theta$ terminates in quadrant II if $0° < \theta < 90°$. If $90° < \theta < 180°$, then the angle represented by $180° - \theta$ terminates in quadrant I. Thus, $\tan\theta$ and $\tan(180° - \theta)$ are opposite in sign.

Clearly, the slope of the line is $m = -\frac{b}{a}$. and $\tan\theta = -\tan(180° - \theta) = -\tan\theta' = -\frac{b}{a}$.

Thus, $m = -\frac{b}{a} = -\tan\theta$.

The point-slope form of the equation of a line is $y - y_1 = m(x - x_1)$. Substituting $-\tan\theta$ for m into $y - y_1 = m(x - x_1)$, we have $y - y_1 = (\tan\theta)(x - x_1)$. If the line passes through $(a, 0)$, then therefore have $y - 0 = (\tan\theta)(x - a)$ or $y = (\tan\theta)(x - a)$.

21. Using the equation $y = (\tan\theta)(x-a)$ where $(a,0)$ is a point on the line and θ is the angle the line makes with the x-axis, $y = (\tan 35°)(x-25)$.

22. Using the equation $y = (\tan\theta)(x-a)$ where $(a,0)$ is a point on the line and θ is the angle the line makes with the x-axis, $y = (\tan 15°)(x-5)$.

For Exercises 23 and 24, we will provide both the algebraic and graphing calculator solutions.

23. Algebraic Solution:
Let x = the side adjacent to 49.2° in the smaller triangle.

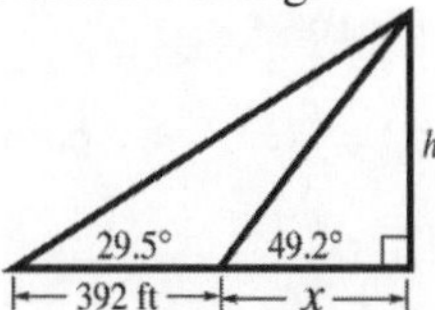

In the larger right triangle, we have $\tan 29.5° = \dfrac{h}{392+x} \Rightarrow h = (392+x)\tan 29.5°$.

In the smaller right triangle, we have $\tan 49.2° = \dfrac{h}{x} \Rightarrow h = x\tan 49.2°$.

Substitute the first expression for h in this equation, and solve for x.

$$(392+x)\tan 29.5° = x\tan 49.2°$$
$$392\tan 29.5° + x\tan 29.5° = x\tan 49.2°$$
$$392\tan 29.5° = x\tan 49.2° - x\tan 29.5°$$
$$392\tan 29.5 = x(\tan 49.2° - \tan 29.5°)$$
$$\frac{392\tan 29.5°}{\tan 49.2° - \tan 29.5°} = x$$

Then substitute this value for x in the equation for the smaller triangle to obtain the following.

$$h = x\tan 49.2° = \frac{392\tan 29.5°}{\tan 49.2° - \tan 29.5°}\tan 49.2° \approx 433.4762$$

Graphing Calculator Solution:

The first line considered is $y = (\tan 29.5°)x$ and the second is $y = (\tan 49.2°)(x-392)$.

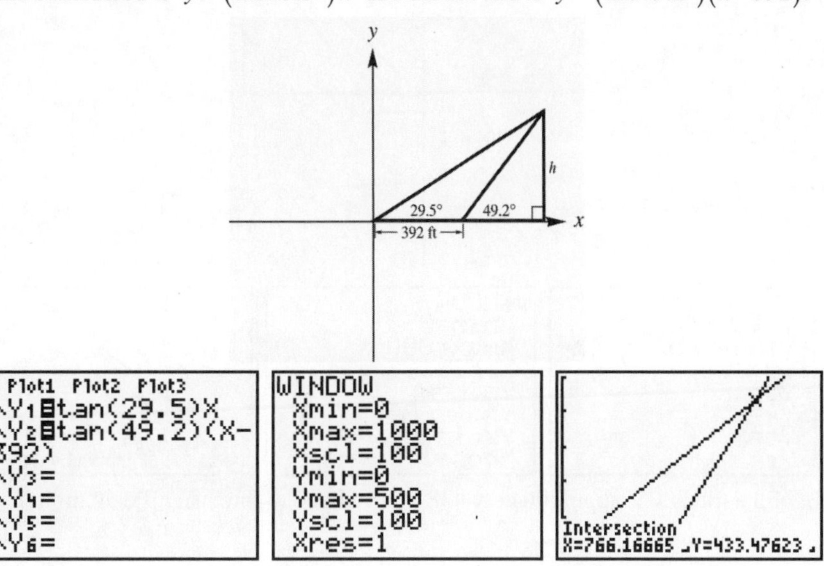

The height of the triangle is 433 ft. (rounded to three significant digits)

24. Algebraic Solution:
Let x = the side adjacent to 52.5° in the smaller triangle.

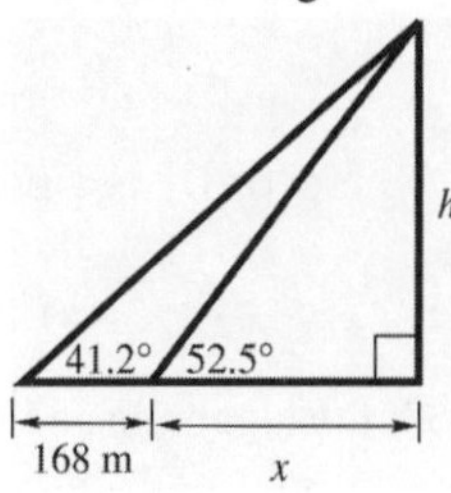

In the larger triangle, we have $\tan 41.2° = \dfrac{h}{168+x} \Rightarrow h = (168+x)\tan 41.2°$.

In the smaller triangle, we have $\tan 52.5° = \dfrac{h}{x} \Rightarrow h = x \tan 52.5°$.

Substitute for h in this equation and solve for x.

$$\begin{aligned}(168+x)\tan 41.2° &= x\tan 52.5°\\ 168\tan 41.2° + x\tan 41.2° &= x\tan 52.5°\\ 168\tan 41.2° &= x\tan 52.5° - x\tan 41.2°\\ 168\tan 41.2° &= x(\tan 52.5° - \tan 41.2°)\\ \frac{168\tan 41.2°}{\tan 52.5° - \tan 41.2°} &= x\end{aligned}$$

Substituting for x in the equation for the smaller triangle, we have the following.

$$h = x\tan 52.5° \Rightarrow h = \frac{168\tan 41.2°\tan 52.5°}{\tan 52.5° - \tan 41.2°} \approx 448.0432$$

Graphing Calculator Solution:

The first line considered is $y = (\tan 41.2°)x$ and the second is $y = (\tan 52.5°)(x-168)$.

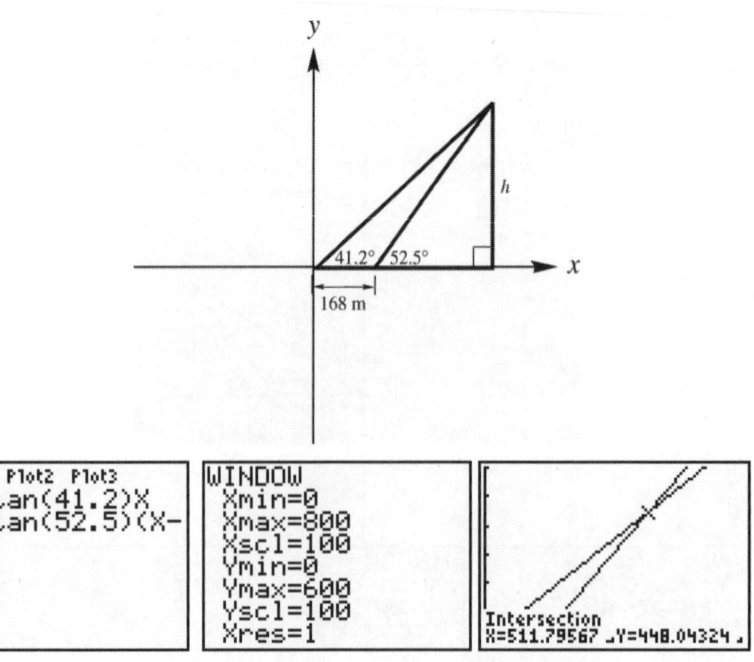

The height of the triangle is approximately 448 m. (rounded to three significant digits)

25. Let x = the distance from the closer point on the ground to the base of height h of the pyramid.

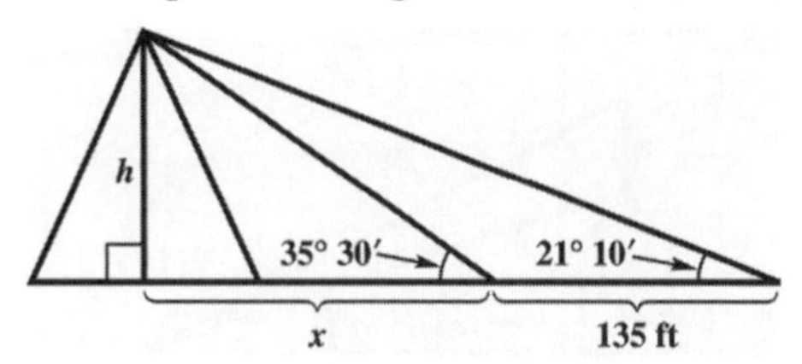

In the larger right triangle, we have $\tan 21°10' = \frac{h}{135+x} \Rightarrow h = (135+x)\tan 21°10'$.

In the smaller right triangle, we have $\tan 35°30' = \frac{h}{x} \Rightarrow h = x\tan 35°30'$.

Substitute for h in this equation, and solve for x to obtain the following.

$$(135+x)\tan 21°10' = x\tan 35°30'$$
$$135\tan 21°10' + x\tan 21°10' = x\tan 35°30'$$
$$135\tan 21°10' = x\tan 35°30' - x\tan 21°10'$$
$$135\tan 21°10' = x(\tan 35°30' - \tan 21°10')$$
$$\frac{135\tan 21°10'}{\tan 35°30' - \tan 21°10'} = x$$

Substitute for x in the equation for the smaller triangle.

$$h = \frac{135\tan 21°10'}{\tan 35°30' - \tan 21°10'}\tan 35°30' \approx 114.3427$$

The height of the pyramid is 114 ft. (rounded to three significant digits)

26. Let x = the distance traveled by the whale as it approaches the tower;
y = the distance from the tower to the whale as it turns.

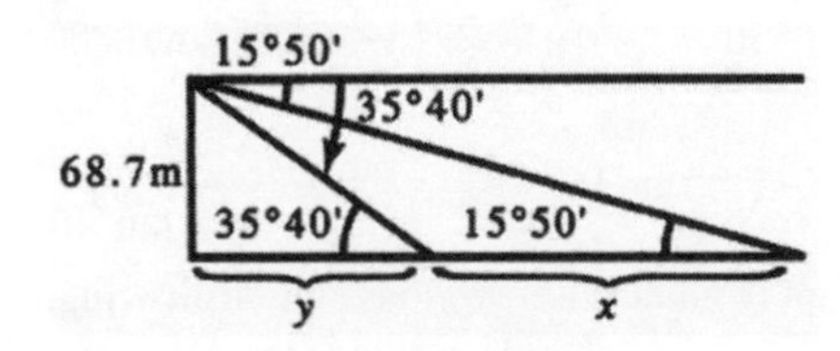

$$\frac{68.7}{y} = \tan 35°40' \Rightarrow 68.7\, y\tan 35°40' \Rightarrow y = \frac{68.7}{\tan 35°40'}$$

and

$$\frac{68.7}{x+y} = \tan 15°50' \Rightarrow 68.7 = (x+y)\tan 15°50' \Rightarrow x+y = \frac{68.7}{\tan 15°50'}$$

$$x = \frac{68.7}{\tan 15°50'} - y \Rightarrow x = \frac{68.7}{\tan 15°50'} - \frac{68.7}{\tan 35°40'} \approx 146.5190$$

The whale traveled 147 m as it approached the lighthouse. (rounded to three significant digits)

27. Let x = the height of the antenna;
h = the height of the house.

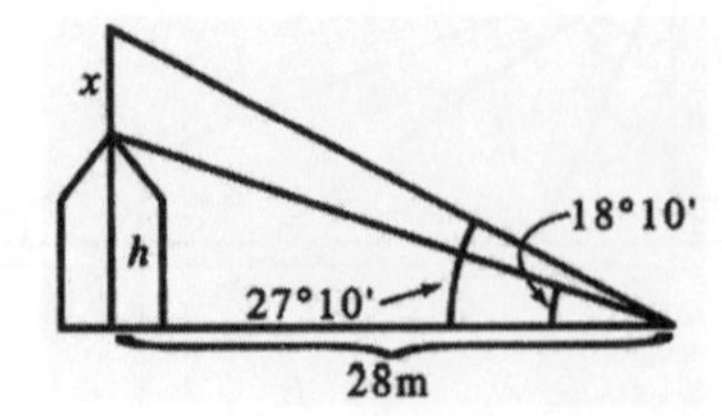

In the smaller right triangle, we have $\tan 18°10' = \frac{h}{28} \Rightarrow h = 28\tan 18°10'$.

In the larger right triangle, we have the following.

$$\tan 27°10' = \frac{x+h}{28} \Rightarrow x+h = 28\tan 27°10' \Rightarrow x = 28\tan 27°10' - h$$
$$x = 28\tan 27°10' - 28\tan 18°10' \approx 5.1816$$

The height of the antenna is 5.18 m. (rounded to three significant digits)

28. Let x = the height of Mt. Whitney above the level of the road;
y = the distance shown in the figure below.

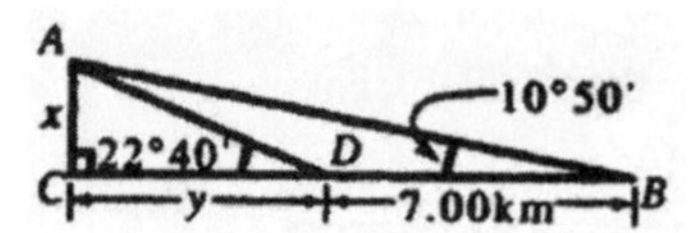

In triangle ADC, $\tan 22°40' = \frac{x}{y} \Rightarrow y\tan 22°40' = x \Rightarrow y = \frac{x}{\tan 22°40'}$. (1)

In triangle ABC, we have the following.

$$\tan 10°50' = \frac{x}{y+7.00} \Rightarrow (y+7.00)\tan 10°50' = x$$
$$y\tan 10°50' + 7.00\tan 10°50' = x \Rightarrow y = \frac{x - 7.00\tan 10°50'}{\tan 10°50'} \quad (2)$$

Setting equations 1 and 2 equal to each other, we have the following.

$$\frac{x}{\tan 22°40'} = \frac{x - 7.00\tan 10°50'}{\tan 10°50'}$$
$$x\tan 10°50' = x\tan 22°40' - 7.00(\tan 10°50')(\tan 22°40')$$
$$7.00(\tan 10°50')(\tan 22°40') = x\tan 22°40' - x\tan 10°50'$$
$$7.00(\tan 10°50')(\tan 22°40') = x(\tan 22°40' - \tan 10°50')$$
$$\frac{7.00(\tan 10°50')(\tan 22°40')}{\tan 22°40' - \tan 10°50'} = x$$
$$x \approx 2.4725.$$

The height of the top of Mt. Whitney above road level is 2.47 km. (rounded to three significant digits)

29. **(a)** From the figure in the text, $d = \frac{b}{2}\cot\frac{\alpha}{2} + \frac{b}{2}\cot\frac{\beta}{2} \Rightarrow d = \frac{b}{2}\left(\cot\frac{\alpha}{2} + \cot\frac{\beta}{2}\right)$.

(b) Using the result of part a, let $\alpha = 37'48"$, $\beta = 42'3"$, and $b = 2.000$.

$$d = \frac{b}{2}\left(\cot\frac{\alpha}{2} + \cot\frac{\beta}{2}\right) \Rightarrow d = \frac{2}{2}\left(\cot\frac{37'48"}{2} + \cot\frac{42'3"}{2}\right)$$

$$d \approx \cot .315° + \cot .3504166667° = \frac{1}{\tan .315°} + \frac{1}{\tan .3504166667°} \approx 345.3951$$

The distance between the two points P and Q is 345.3951 cm. (rounded)

30. Let h = the minimum height above the surface of the earth so a pilot at A can see an object on the horizon at C.

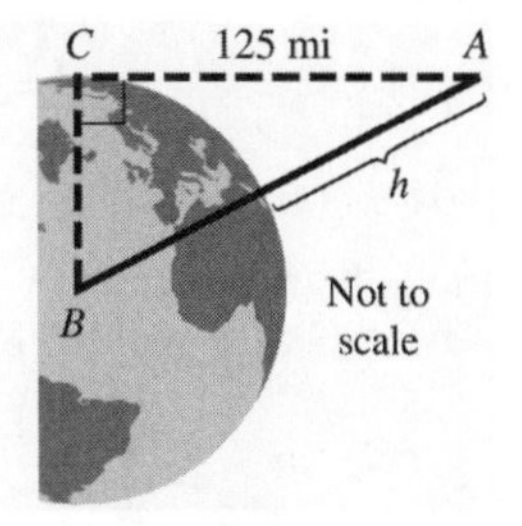

Using the Pythagorean theorem, we have the following.

$$\left(4.00\times10^3 + h\right)^2 = \left(4.00\times10^3\right)^2 + 125^2$$

$$(4000 + h)^2 = 4000^2 + 125^2$$

$$(4000 + h)^2 = 16{,}000{,}000 + 15{,}625$$

$$(4000 + h)^2 = 16{,}015{,}625$$

$$4000 + h = \sqrt{16{,}015{,}625}$$

$$h = \sqrt{16{,}015{,}625} - 4000 \approx 4001.9526 - 4000 = 1.9526$$

The minimum height above the surface of the earth would be 1.95 mi. (rounded to 3 significant digits)

31. Let x = the minimum distance that a plant needing full sun can be placed from the fence.

Fence
4.65 ft
23° 20′
Plant
x

$$\tan 23°20' = \frac{4.65}{x} \Rightarrow x\tan 23°20' = 4.65 \Rightarrow x = \frac{4.65}{\tan 23°20'} \approx 10.7799$$

The minimum distance is 10.8 ft. (rounded to three significant digits)

32. $\tan A = \frac{1.0837}{1.4923} \approx .7261944649 \Rightarrow A \approx \tan^{-1}(.7261944649) \approx 35.987° \approx 35°59.2' \approx 35°59'10''$

$\tan B = \frac{1.4923}{1.0837} \approx 1.377041617 \Rightarrow B \approx \tan^{-1}(1.377041617) \approx 54.013° \approx 54°00.8' \approx 54°00'50''$

33. **(a)** If $\theta = 37°$, then $\frac{\theta}{2} = \frac{37°}{2} = 18.5°$.

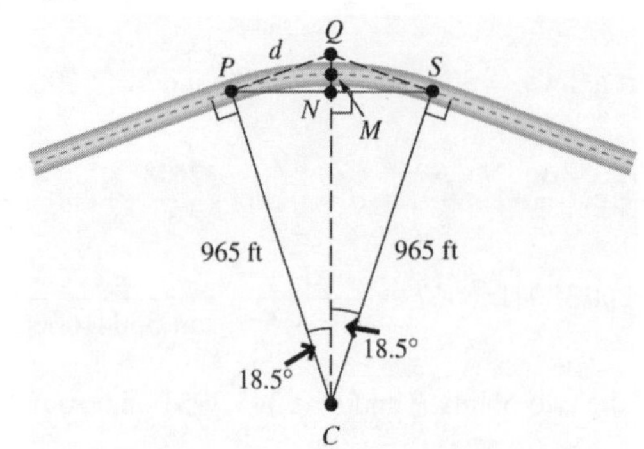

To find the distance between P and Q, d, we first note that angle QPC is a right angle. Hence, triangle QPC is a right triangle and we can solve the following.

$$\tan 18.5° = \frac{d}{965} \Rightarrow d = 965 \tan 18.5° \approx 322.8845$$

The distance between P and Q, is 323 ft. (rounded to three significant digits)

(b) Since we are dealing with a circle, the distance between M and C is R. If we let x be the distance from N to M, then the distance from C to N will be $R - x$.

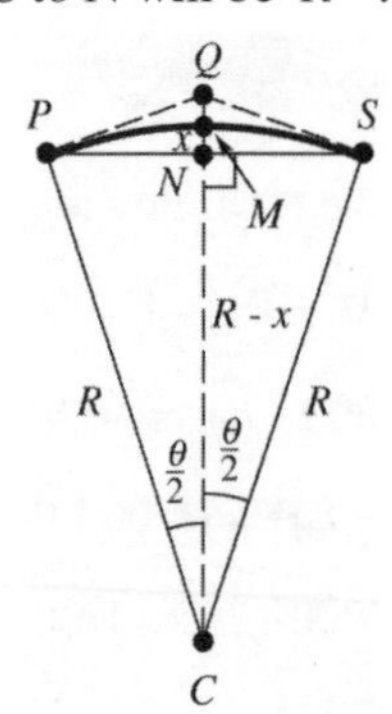

Since triangle CNP is a right triangle, we can set up the following equation.

$$\cos\frac{\theta}{2} = \frac{R-x}{R} \Rightarrow R\cos\frac{\theta}{2} = R - x \Rightarrow x = R - R\cos\frac{\theta}{2} \Rightarrow x = R\left(1 - \cos\frac{\theta}{2}\right)$$

34. Let y = the common hypotenuse of the two right triangles.

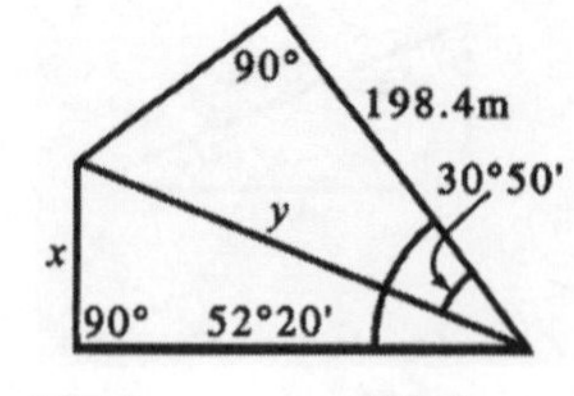

$$\cos 30°50' = \frac{198.4}{y} \Rightarrow y = \frac{198.4}{\cos 30°50'} \approx 231.0571948$$

To find x, first find the angle opposite x in the right triangle by find the following.

$$52°20' - 30°50' = 51°80' - 30°50' = 21°30'$$

$$\sin 21°30' = \frac{x}{y} \Rightarrow \sin 21°30' \approx \frac{x}{231.0571948} \Rightarrow x \approx 231.0571948 \sin 21°30' \approx 84.6827$$

The length x is approximate 84.7 m. (rounded)

35. (a)

$$\theta \approx \frac{57.3S}{R} = \frac{57.3(336)}{600} = 32.088°$$

$$d = R\left(1 - \cos\frac{\theta}{2}\right) = 600(1 - \cos 16.044°) \approx 23.3702 \text{ ft}$$

The distance is 23.4 ft. (rounded to three significant digits)

(b)

$$\theta \approx \frac{57.3S}{R} = \frac{57.3(485)}{600} = 46.3175°$$

$$d = R\left(1 - \cos\frac{\theta}{2}\right) = 600(1 - \cos 23.15875°) \approx 48.3488$$

The distance is 48.3 ft. (rounded to three significant digits)

(c) The faster the speed, the more land needs to be cleared on the inside of the curve.

Chapter 2: Review Exercises

1.
$$\sin A = \frac{\text{side opposite}}{\text{hypotenuse}} = \frac{60}{61} \qquad \cot A = \frac{\text{side adjacent}}{\text{side opposite}} = \frac{11}{60}$$

$$\cos A = \frac{\text{side adjacent}}{\text{hypotenuse}} = \frac{11}{61} \qquad \sec A = \frac{\text{hypotenuse}}{\text{side adjacent}} = \frac{61}{11}$$

$$\tan A = \frac{\text{side opposite}}{\text{side adjacent}} = \frac{60}{11} \qquad \csc A = \frac{\text{hypotenuse}}{\text{side opposite}} = \frac{61}{60}$$

2.
$$\sin A = \frac{\text{side opposite}}{\text{hypotenuse}} = \frac{40}{58} = \frac{20}{29} \qquad \cot A = \frac{\text{side adjacent}}{\text{side opposite}} = \frac{42}{40} = \frac{21}{20}$$

$$\cos A = \frac{\text{side adjacent}}{\text{hypotenuse}} = \frac{42}{58} = \frac{21}{29} \qquad \sec A = \frac{\text{hypotenuse}}{\text{side adjacent}} = \frac{58}{42} = \frac{29}{21}$$

$$\tan A = \frac{\text{side opposite}}{\text{side adjacent}} = \frac{40}{42} = \frac{20}{21} \qquad \csc A = \frac{\text{hypotenuse}}{\text{side opposite}} = \frac{58}{40} = \frac{29}{20}$$

3. $\sin 4\beta = \cos 5\beta$

Since sine and cosine are cofunctions, we have the following.

$$4\beta + 5\beta = 90° \Rightarrow 9\beta = 90° \Rightarrow \beta = 10°$$

4. $\sec(2\theta + 10°) = \csc(4\theta + 20°)$

Since secant and cosecant are cofunctions, we have the following.

$$(2\theta + 10°) + (4\theta + 20°) = 90° \Rightarrow 6\theta + 30° = 90° \Rightarrow 6\theta = 60° \Rightarrow \theta = 10°$$

5. $\tan(5x + 11°) = \cot(6x + 2°)$

Since tangent and cotangent are cofunctions, we have the following.

$$(5x + 11°) + (6x + 2°) = 90° \Rightarrow 11x + 13° = 90° \Rightarrow 11x = 77° \Rightarrow x = 7°$$

6. $\cos\left(\frac{3\theta}{5}+11°\right)=\sin\left(\frac{7\theta}{10}+40°\right)$

Since sine and cosine are cofunctions, we have the following.

$$\left(\frac{3\theta}{5}+11°\right)+\left(\frac{7\theta}{10}+40°\right)=90° \Rightarrow \frac{6\theta}{10}+\frac{7\theta}{10}+51°=90° \Rightarrow \frac{13}{10}\theta+51°=90°$$

$$\frac{13}{10}\theta=39° \Rightarrow \theta=\frac{10}{13}(39°)=30°$$

7. $\sin 46° < \sin 58°$

$\sin\theta$ increases as θ increases from 0° to 90°. Since $58° > 46°$, we have $\sin 58°$ is greater than $\sin 46°$. Thus, the statement is true.

8. $\cos 47° < \cos 58°$

$\cos\theta$ decreases as θ increases from 0° to 90°. Since $47° < 58°$, we have $\cos 47°$ is greater than $\cos 58°$. Thus, the statement is false.

9. $\sec 48° \geq \cos 42°$

Since 48° and 42° are in quadrant I, $\sec 48°$ and $\cos 42°$ are both positive. Since $0 < \sin 42° < 1$, $\frac{1}{\sin 42°} = \csc 42° > 1$. Moreover, $0 < \cos 42° < 1$. Thus, $\sec 48° \geq \cos 42°$ and the statement is true.

10. $\sin 22° \geq \csc 68°$

Since 22° and 68° are in quadrant I, $\sin 22°$ and $\csc 68°$ are both positive. Since $0 < \sin 68° < 1$, $\frac{1}{\sin 68°} = \csc 68° > 1$. Moreover, $0 < \sin 22° < 1$. Thus, $\sin 22° < \csc 68°$ and the statement is false.

11. The measures of angles *A* and *B* sum to be 90°, and are complementary angles. Since sine and cosine are cofunctions, we have $\sin B = \cos(90° - B) = \cos A$.

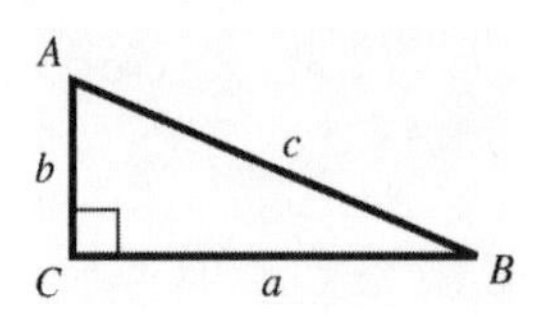

12. 120°

This angle lies in quadrant II, so the reference angle is 180° – 120° = 60°. Since 120° is in quadrant II, the cosine, tangent, cotangent and secant are negative.

$\sin 120° = \sin 60° = \frac{\sqrt{3}}{2}$

$\cos 120° = -\cos 60° = -\frac{1}{2}$

$\tan 120° = -\tan 60° = -\sqrt{3}$

$\cot 120° = -\cot 60° = -\frac{\sqrt{3}}{3}$

$\sec 120° = -\sec 60° = -2$

$\csc 120° = \csc 60° = \frac{2\sqrt{3}}{3}$

13. 300°

This angle lies in quadrant IV, so the reference angle is 360° – 300° = 60°. Since 300° is in quadrant IV, the sine, tangent, cotangent and cosecant are negative.

$\sin 300° = -\sin 60° = -\frac{\sqrt{3}}{2}$

$\cos 300° = \cos 60° = \frac{1}{2}$

$\tan 300° = -\tan 60° = -\sqrt{3}$

$\cot 300° = -\cot 60° = -\frac{\sqrt{3}}{3}$

$\sec 300° = \sec 60° = 2$

$\csc 300° = -\csc 60° = -\frac{2\sqrt{3}}{3}$

14. –225°

–225° is coterminal with –225° + 360° = 135°. This angle lies in quadrant II. The reference angle is $180° - 135° = 45°$. Since –225° is in quadrant II, the cosine, tangent, cotangent, and secant are negative.

$\sin(-225°) = \sin 45° = \frac{\sqrt{2}}{2}$

$\cos(-225°) = -\cos 45° = -\frac{\sqrt{2}}{2}$

$\tan(-225°) = -\tan 45° = -1$

$\cot(-225°) = -\cot 45° = -1$

$\sec(-225°) = -\sec 45° = -\sqrt{2}$

$\csc(-225°) = \csc 45° = \sqrt{2}$

15. –390° is coterminal with $-390° + 2 \cdot 360° = -390° + 720° = 330°$. This angle lies in quadrant IV. The reference angle is $360° - 330° = 30°$. Since –390° is in quadrant IV, the sine, tangent, cotangent, and cosecant are negative.

$\sin(-390°) = -\sin 30° = -\frac{1}{2}$

$\cos(-390°) = \cos 30° = \frac{\sqrt{3}}{2}$

$\tan(-390°) = -\tan 30° = -\frac{\sqrt{3}}{3}$

$\cot(-390°) = -\cot 30° = -\sqrt{3}$

$\sec(-390°) = \sec 30° = \frac{2\sqrt{3}}{3}$

$\csc(-390°) = -\csc 30° = -2$

16. $\sin\theta = -\frac{1}{2}$

Since $\sin\theta$ is negative, θ must lie in quadrants III or IV. Since the absolute value of $\sin\theta$ is $\frac{1}{2}$, the reference angle, θ', must be 30°. The angle in quadrant III will be $180° + \theta' = 180° + 30° = 210°$. The angle in quadrant IV will be $360° - \theta' = 360° - 30° = 330°$.

17. $\cos\theta = -\frac{1}{2}$

Since $\cos\theta$ is negative, θ must lie in quadrants II or III. Since the absolute value of $\cos\theta$ is $\frac{1}{2}$, the reference angle, θ', must be 60°. The quadrant II angle θ equals $180° - \theta' = 180° - 60° = 120°$, and the quadrant III angle θ equals $180° + \theta' = 180° + 60° = 240°$.

18. $\cot\theta = -1$

Since $\cot\theta$ is negative, θ must lie in quadrants II or IV. Since the absolute value of $\cot\theta$ is 1, the reference angle, θ', must be 45°. The quadrant II angle θ equals $180° - \theta' = 180° - 45° = 135°$, and the quadrant IV angle θ equals $360° - \theta' = 360° - 45° = 315°$.

19. $\sec\theta = -\dfrac{2\sqrt{3}}{3}$

Since $\sec\theta$ is negative, θ must lie in quadrants II or III. Since the absolute value of $\sec\theta$ is $\dfrac{2\sqrt{3}}{3}$, the reference angle, θ', must be 30°. The quadrant II angle θ equals $180° - \theta' = 180° - 30° = 150°$, and the quadrant III angle θ equals $180° + \theta' = 180° + 30° = 210°$.

20. $\cos 60° + 2\sin^2 30° = \dfrac{1}{2} + 2\left(\dfrac{1}{2}\right)^2 = \dfrac{1}{2} + 2\left(\dfrac{1}{4}\right) = \dfrac{1}{2} + \dfrac{1}{2} = 1$

21. $\tan^2 120° - 2\cot 240° = \left(-\sqrt{3}\right)^2 - 2\left(\dfrac{\sqrt{3}}{3}\right) = 3 - \dfrac{2\sqrt{3}}{3}$

22. $\sec^2 300° - 2\cos^2 150° + \tan 45° = 2^2 - 2\left(-\dfrac{\sqrt{3}}{2}\right)^2 + 1 = 4 - 2\left(\dfrac{3}{4}\right) + 1 = 4 - \dfrac{3}{2} + 1 = \dfrac{7}{2}$

23. **(a)** $(-3,-3)$

Given the point (x, y), we need to determine the distance from the origin, r.

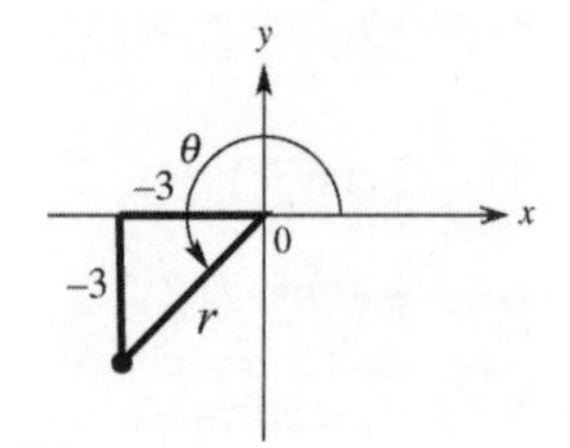

$$r = \sqrt{x^2 + y^2}$$
$$r = \sqrt{(-3)^2 + (-3)^2}$$
$$r = \sqrt{9+9}$$
$$r = \sqrt{18}$$
$$r = 3\sqrt{2}$$

$$\sin\theta = \frac{y}{r} = \frac{-3}{3\sqrt{2}} = -\frac{1}{\sqrt{2}} = -\frac{1}{\sqrt{2}} \cdot \frac{\sqrt{2}}{\sqrt{2}} = -\frac{\sqrt{2}}{2};\ \cos\theta = \frac{x}{r} = \frac{-3}{3\sqrt{2}} = -\frac{1}{\sqrt{2}} = -\frac{1}{\sqrt{2}} \cdot \frac{\sqrt{2}}{\sqrt{2}} = -\frac{\sqrt{2}}{2}$$

$$\tan\theta = \frac{y}{x} = \frac{-3}{-3} = 1$$

(b) $\left(1, -\sqrt{3}\right)$

Given the point (x, y), we need to determine the distance from the origin, r.

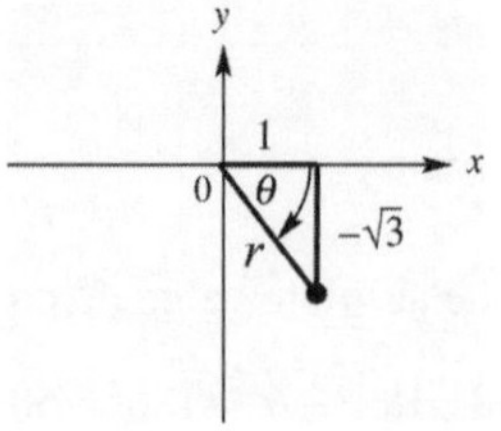

$$r = \sqrt{x^2 + y^2}$$
$$r = \sqrt{1^2 + \left(-\sqrt{3}\right)^2}$$
$$r = \sqrt{1+3}$$
$$r = \sqrt{4}$$
$$r = 2$$

$$\sin\theta = \frac{y}{r} = \frac{-\sqrt{3}}{2} = -\frac{\sqrt{3}}{2};\ \cos\theta = \frac{x}{r} = \frac{1}{2};\ \tan\theta = \frac{y}{x} = \frac{-\sqrt{3}}{1} = -\sqrt{3}$$

For the remainder of the exercises in this section, be sure your calculator is in degree mode. If your calculator accepts angles in degrees, minutes, and seconds, it is not necessary to change angles to decimal degrees. Keystroke sequences may vary on the type and/or model of calculator being used. Screens shown will be from a TI-83 Plus calculator. To obtain the degree (°) and (′) symbols, go to the ANGLE menu (2nd APPS).

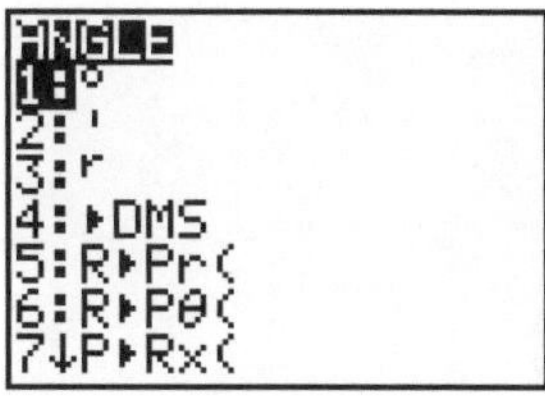

For Exercises 24, 25, and 27, the calculation for decimal degrees is indicated for calculators that do not accept degree, minutes, and seconds.

24. $\sin 72°30' \approx .95371695$

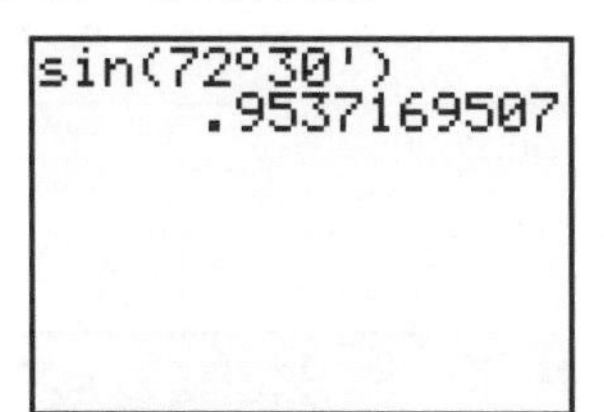

$72°30' = \left(72 + \frac{30}{60}\right)° = 72.5°$

25. $\sec 222°30' \approx -1.3563417$

cos(222°30')
-.7372773368
Ans⁻¹
-1.356341705

$222°30' = \left(222 + \frac{30}{60}\right)° = 222.5°$

26. $\cot 305.6° \approx -.71592968$

tan(305.6)
-1.39678522
Ans⁻¹
-.7159296831

27. $\csc 78°21' \approx 1.0210339$

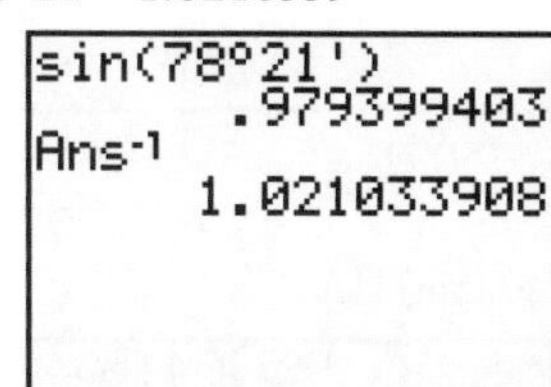

$78°21' = \left(78 + \frac{21}{60}\right)° = 78.35°$

28. $\sec 58.9041° \approx 1.9362132$

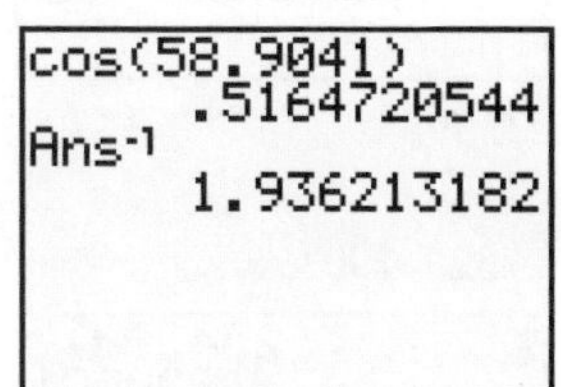

29. $\tan 11.7689° \approx .20834446$

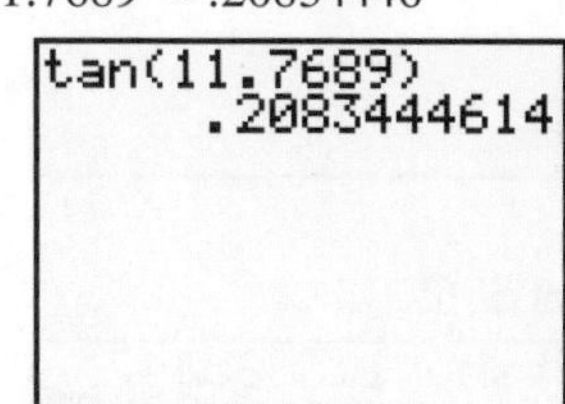

30. If $\theta = 135°$, $\theta = 45°$.
If $\theta = 45°$, $\theta = 45°$.
If $\theta = 300°$, $\theta = 60°$.
If $\theta = 140°$, $\theta = 40°$.

Of these reference angles, 40° is the only one which is not a special angle, so D, tan 140°, is the only one which cannot be determined exactly using the methods of this chapter.

31. $\sin\theta = .8254121$

$\theta \approx 55.673870°$

32. $\cot\theta = 1.1249386$

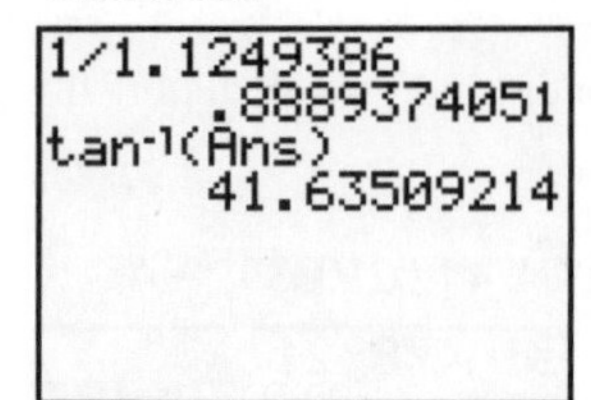

$\theta \approx 41.635092°$

33. $\cos\theta = .97540415$

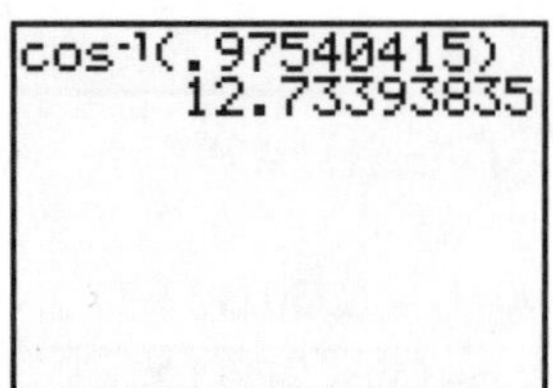

$\theta \approx 12.733938°$

34. $\sec\theta = 1.2637891$

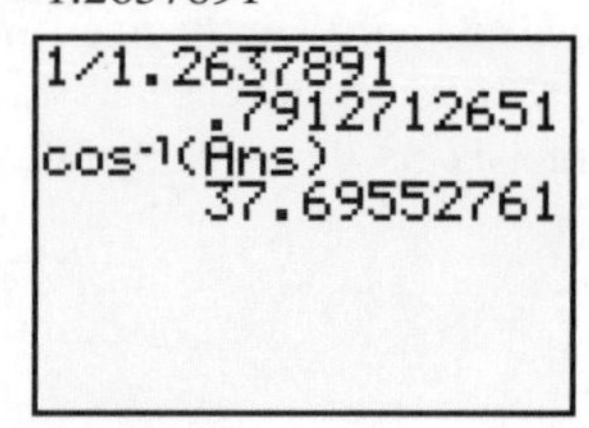

$\theta \approx 37.695528$

35. $\tan\theta = 1.9633124$

$\theta \approx 63.008286°$

36. $\csc\theta = 9.5670466$

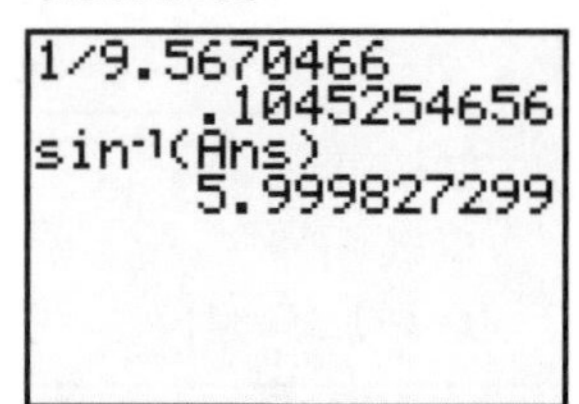

$\theta \approx 5.9998273°$

37. $\sin\theta = .73254290$

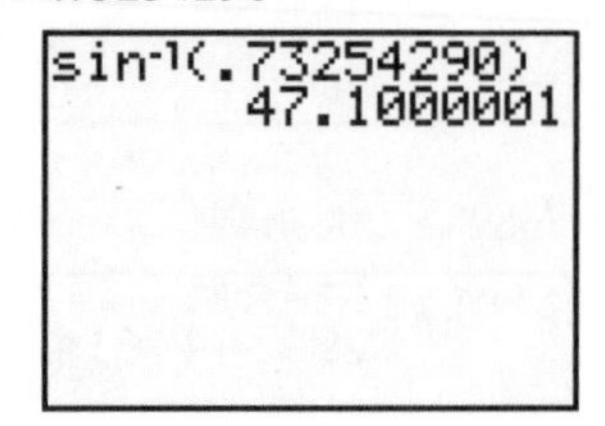

Since $\sin\theta$ is positive, there will be one angle in quadrant I and one angle in quadrant II. If θ' is the reference angle, then the two angles are θ' and $180° - \theta'$. Thus, the quadrant I angle is approximately equal to $47.1°$, and the quadrant II angle is $180° - 47.1° = 132.9°$.

38. $\tan\theta = 1.3865342$

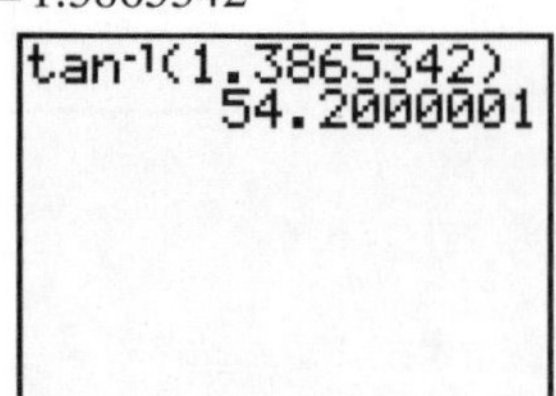

Since $\tan\theta$ is positive, there will be one angle in quadrant I and one angle in quadrant III. If θ' is the reference angle, then the two angles are θ' and $180° + \theta'$. Thus, the quadrant I angle is approximately equal to $54.2°$, and the quadrant III angle is $180° + 54.2° = 234.2°$.

39. $\sin 50° + \sin 40° \overset{?}{=} \sin 90°$

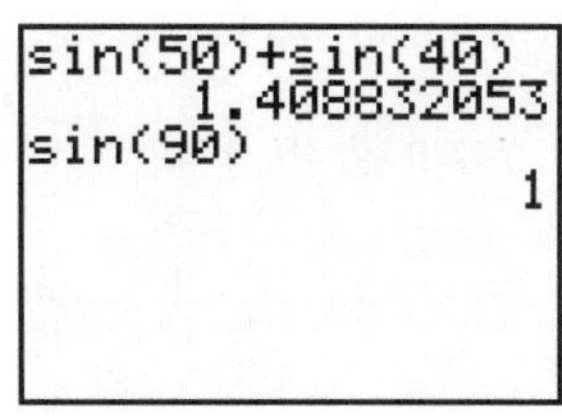

Since $\sin 50° + \sin 40° \approx 1.408832053$ and $\sin 90° = 1$, the statement is false.

40. $\cos 210° \overset{?}{=} \cos 180° \cdot \cos 30° - \sin 180° \cdot \sin 30°$

Since $\cos 210° = -\cos 30° = -\frac{\sqrt{3}}{2}$ and $\cos 180° \cdot \cos 30° - \sin 180° \cdot \sin 30° = (-1)\left(\frac{\sqrt{3}}{2}\right) - (0)\left(\frac{1}{2}\right) = -\frac{\sqrt{3}}{2}$, the statement is true.

41. $\sin 240° \overset{?}{=} 2\sin 120° \cos 120°$

Since $\sin 240° = -\sin 60° = -\frac{\sqrt{3}}{2}$ and $2\sin 120° \cos 120° = 2(\sin 60°)(-\cos 60°) = 2\left(\frac{\sqrt{3}}{2}\right)\left(\frac{1}{2}\right) = -\frac{\sqrt{3}}{2}$, the statement is true.

42. $\sin 42° + \sin 42° \overset{?}{=} \sin 84°$

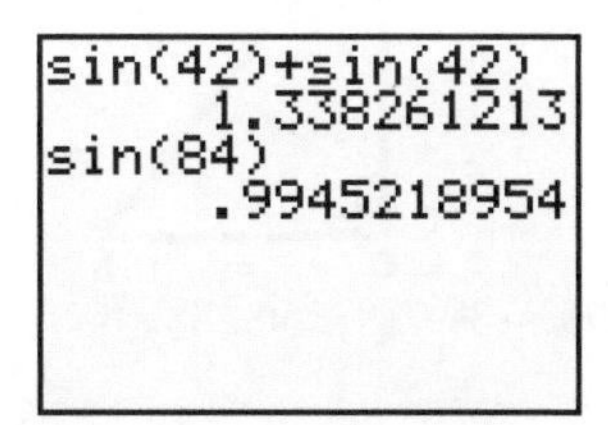

Using a calculator, we have $\sin 42° + \sin 42° = 1.338261213$ and $\sin 84° = .9945218954$. Thus, the statement is false.

43. No, $\cot 25° = \dfrac{1}{\tan 25°} \neq \tan^{-1} 25°$.

44. $\theta = 2976°$

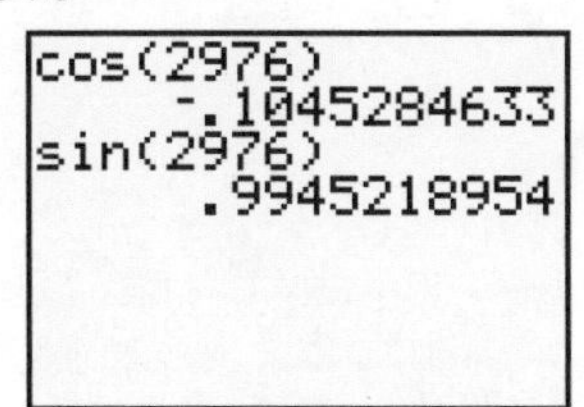

Since cosine is negative and sine is positive, the angle θ is in quadrant II.

45. $\theta = 1997°$

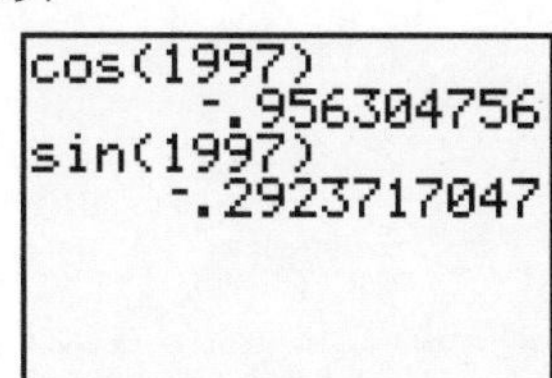

Since sine and cosine are both negative, the angle θ is in quadrant III.

46. $\theta = 4000°$

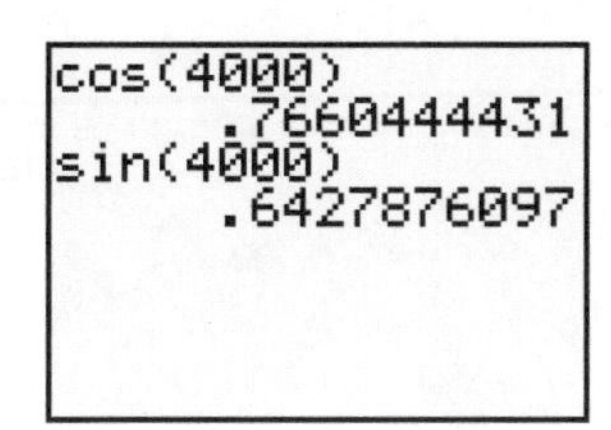

Since sine and cosine are both positive, the angle θ is in quadrant I.

47. $A = 58° \ 30', \ c = 748$

$A + B = 90° \Rightarrow B = 90° - A \Rightarrow B = 90° - 58°30' = 89°60' - 58°30' = 31°30'$

$\sin A = \dfrac{a}{c} \Rightarrow \sin 58°30' = \dfrac{a}{748} \Rightarrow a = 748 \sin 58°30' \approx 638$ (rounded to three significant digits)

$\cos A = \dfrac{b}{c} \Rightarrow \cos 58°30' = \dfrac{b}{748} \Rightarrow b = 748 \cos 58°30' \approx 391$ (rounded to three significant digits)

48. $a = 129.70, \ b = 368.10$

$\tan A = \dfrac{129.70}{368.10} \approx .3523499049 \Rightarrow A \approx \tan^{-1}(.3523499049) \approx 19.41° \approx 19°25'$

$\tan B = \dfrac{368.10}{129.70} \approx 2.838087895 \Rightarrow B \approx \tan^{-1}(2.838087895) \approx 70.59° \approx 70°35'$

Note: $A + B = 90°$

$c = \sqrt{a^2 + b^2} = \sqrt{129.70^2 + 368.10^2} = \sqrt{16,822.09 + 135,497.61} = \sqrt{152,319.7} \approx 390.28$ (rounded to five significant digits)

49. $A = 39.72°, \ b = 38.97$ m

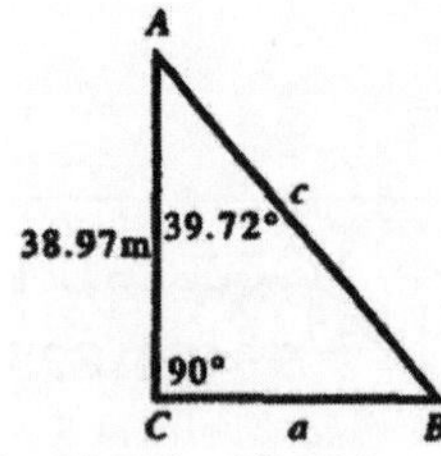

$A + B = 90° \Rightarrow B = 90° - A \Rightarrow B = 90° - 39.72° = 50.28°$

$\tan A = \dfrac{a}{b} \Rightarrow \tan 39.72° = \dfrac{a}{38.97} \Rightarrow a = 38.97 \tan 39.72° \approx 32.38$ m (rounded to four significant digits)

$\cos A = \dfrac{b}{c} \Rightarrow \cos 39.72° = \dfrac{38.97}{c} \Rightarrow c \cos 39.72° = 38.97 \Rightarrow c = \dfrac{38.97}{\cos 39.72°} \approx 50.66$ m (rounded to five significant digits)

50. $B = 47°53', \ b = 298.6$ m

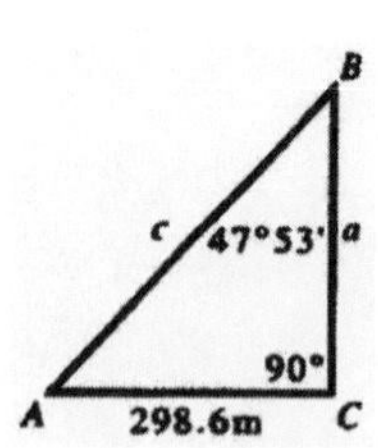

$A + B = 90° \Rightarrow A = 90° - B \Rightarrow A = 90° - 47°53' = 89°60' - 47°53' = 42°7'$

$\tan B = \dfrac{b}{a} \Rightarrow \tan 47°53' = \dfrac{298.6}{a} \Rightarrow a \tan 47°53' = 298.6 \Rightarrow a = \dfrac{298.6}{\tan 47°53'} \approx 270.0$ m (rounded to four significant digits)

$\sin B = \dfrac{b}{c} \Rightarrow \sin 47°53' = \dfrac{298.6}{c} \Rightarrow c \sin 47°53' = 298.6 \Rightarrow c = \dfrac{298.6}{\sin 47°53'} \approx 402.5$ m (rounded to four significant digits)

51. Let x = height of the tower.

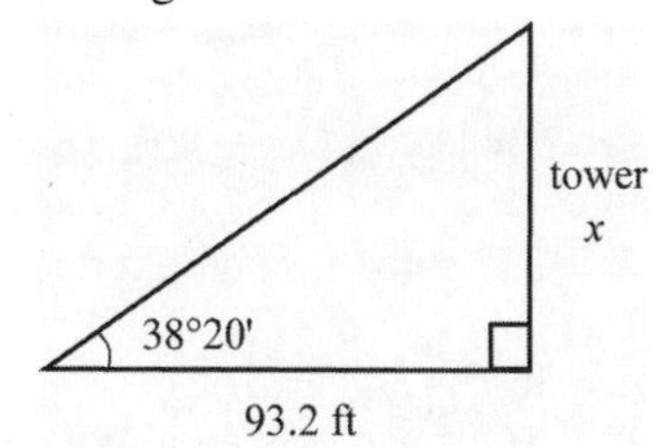

$$\tan 38°20' = \frac{x}{93.2}$$
$$x = 93.2 \tan 38°20'$$
$$x \approx 73.6930$$

The height of the tower is 73.7 ft. (rounded to three significant digits)

52. Let h = height of the tower.

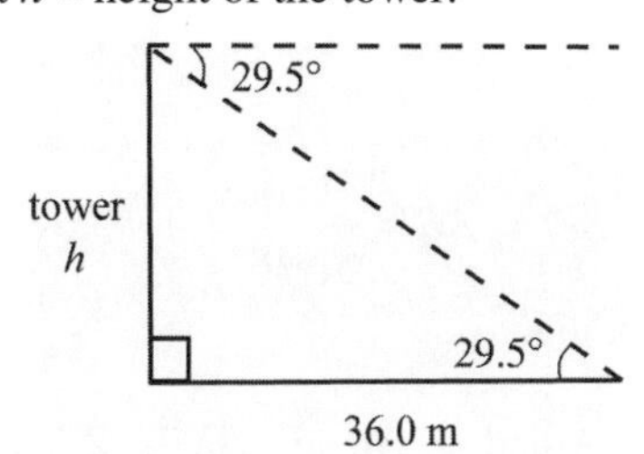

$$\tan 29.5° = \frac{h}{36.0}$$
$$h = 36.0 \tan 29.5°$$
$$h \approx 20.3678$$

The height of the tower is 20.4 m. (rounded to three significant digits)

53. Let x = length of the diagonal.

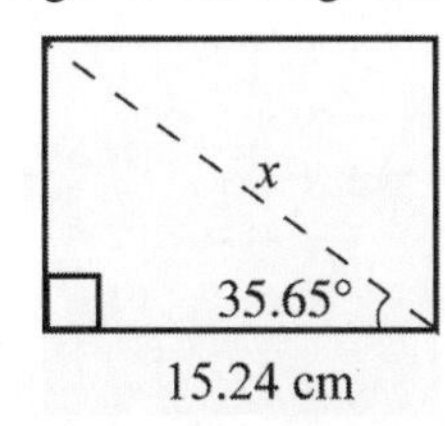

$$\cos 35.65° = \frac{15.24}{x}$$
$$x \cos 35.65° = 15.24$$
$$x = \frac{15.24}{\cos 35.65°}$$
$$x \approx 18.7548$$

The length of the diagonal of the rectangle is 18.75 cm. (rounded to three significant digits)

54. Let x = the length of the equal sides of an isosceles triangle.

Divide the isosceles triangle into two congruent right triangles.

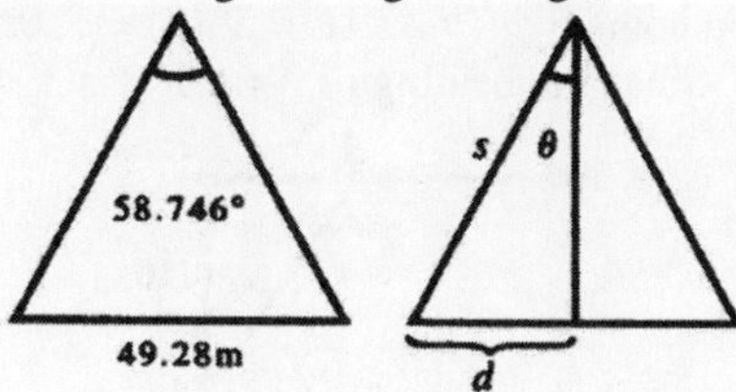

$$d = \frac{1}{2}(49.28) = 24.64 \text{ and } \theta = \frac{1}{2}(58.746°) = 29.373°$$

and

$$\sin\theta = \frac{d}{s} \Rightarrow \sin 29.373° = \frac{24.64}{s} \Rightarrow \sin 29.373° = \frac{24.64}{s} \Rightarrow s \sin 29.373° = 24.64$$
$$s = \frac{24.64}{\sin 29.373°} \Rightarrow s \approx 50.2352$$

Each side is 50.24 m long. (rounded to 4 significant digits)

55. Draw triangle ABC and extend the north-south lines to a point X south of A and S to a point Y, north of C.

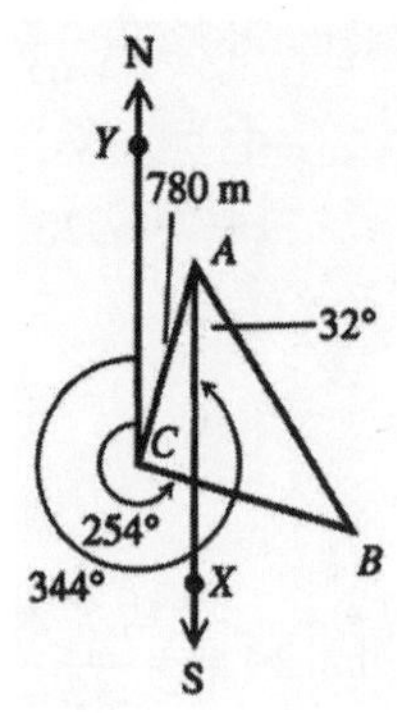

Angle ACB = 344° – 254° = 90°, so ABC is a right triangle.
Angle BAX = 32° since it is an alternate interior angle to 32°.
Angle YCA = 360° – 344° = 16°
Angle XAC = 16° since it is an alternate interior angle to angle YCA.
Angle BAC = 32° + 16° = 48°.

In triangle ABC, $\cos A = \frac{AC}{AB} \Rightarrow \cos 48° = \frac{780}{AB} \Rightarrow AB\cos 48° = 780 \Rightarrow AB = \frac{780}{\cos 48°} \approx 1165.6917.$

The distance from A to B is 1200 m. (rounded to two significant digits)

56. Draw triangle ABC and extend north-south lines from points A and B. Angle ABX is 55° (alternate interior angles of parallel lines cut by a transversal have the same measure) so Angle ABC is 55° + 35° = 90°.

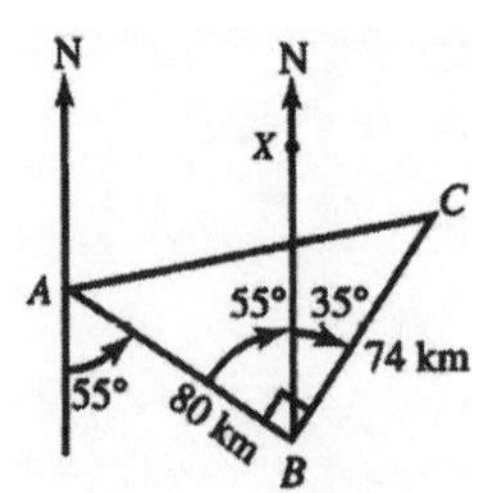

Since angle ABC is a right angle, use the Pythagorean theorem to find the distance from A to C.

$$(AC)^2 = 80^2 + 74^2 \Rightarrow (AC)^2 = 6400 + 5476 \Rightarrow (AC)^2 = 11,876 \Rightarrow AC = \sqrt{11,876} \approx 108.9771$$

It is 110 km from A to C. (rounded to two significant digits)

57. Suppose A is the car heading south at 55 mph, B is the car heading west, and point C is the intersection from which they start. After two hours by $d = rt$, $AC = 55(2) = 110$. Since angle ACB is a right angle, triangle ACB is a right triangle. Since the bearing of A from B is 324°, angle CAB = 360° – 324° = 36°.

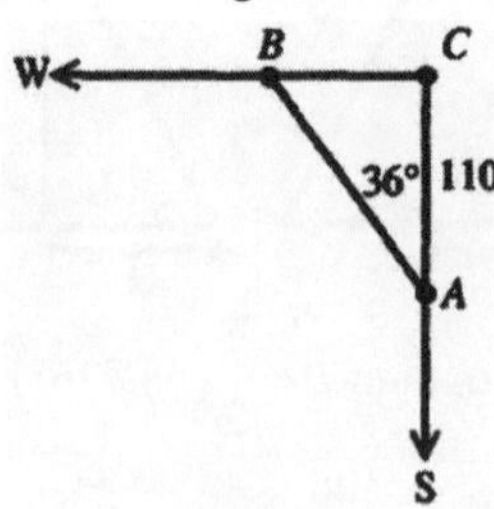

$$\cos CAB = \frac{AC}{AB} \Rightarrow \cos 36° = \frac{110}{AB} \Rightarrow AB\cos 36° = 110 \Rightarrow AB = \frac{110}{\cos 36°} \approx 135.9675$$

There are 140 mi apart. (rounded to two significant digits)

58. Let x = the leg opposite angle A.

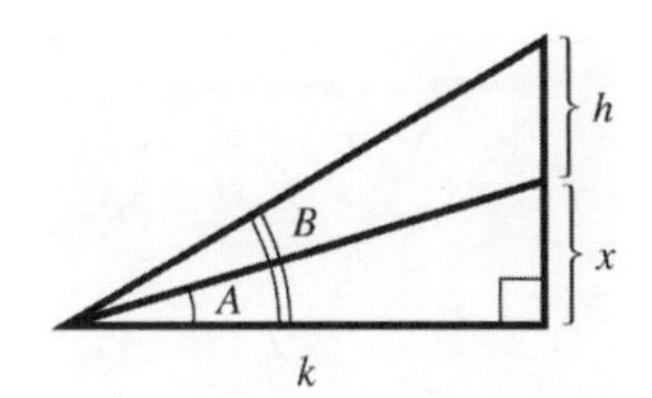

$$\tan A = \frac{x}{k} \Rightarrow x = k\tan A \text{ and } \tan B = \frac{h+x}{k} \Rightarrow x = k\tan B - h$$

Therefore, we have the following.

$$k\tan A = k\tan B - h \Rightarrow h = k\tan B - k\tan A \Rightarrow h = k\left(\tan B - \tan A\right)$$

59. – 60. Answers will vary.

61. $h = R\left(\dfrac{1}{\cos\left(\frac{180T}{P}\right)} - 1\right)$

(a) Let R = 3955 mi, T = 25 min, P = 140 min.

$$h = R\left(\frac{1}{\cos\left(\frac{180T}{P}\right)} - 1\right) \Rightarrow h = 3955\left(\frac{1}{\cos\left(\frac{180\cdot 25}{140}\right)} - 1\right) \approx 715.9424$$

The height of the satellite is approximately 716 mi.

(b) Let R = 3955 mi, T = 30 min, P = 140 min.

$$h = R\left(\frac{1}{\cos\left(\frac{180T}{P}\right)} - 1\right) \Rightarrow h = 3955\left(\frac{1}{\cos\left(\frac{180\cdot 30}{140}\right)} - 1\right) \approx 1103.6349$$

The height of the satellite is approximately 1104 mi.

62. **(a)** From the figure we see that, $\sin\theta = \dfrac{x_Q - x_P}{d} \Rightarrow x_Q = x_P + d\sin\theta$. Similarly, we have

$\cos\theta = \dfrac{y_Q - y_P}{d} \Rightarrow y_Q = y_P + d\cos\theta.$

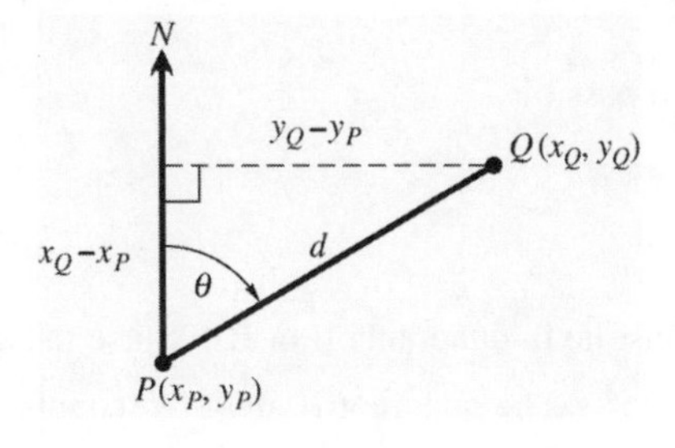

(b) Let $(x_P, y_P) = (123.62, 337.95)$, $\theta = 17°\ 19'\ 22''$, and $d = 193.86$.

$x_Q = x_P + d\sin\theta \Rightarrow x_Q = 123.62 + 193.86\sin 17°19'22'' \approx 181.3427$

$y_Q = y_P + d\cos\theta \Rightarrow = 337.95 + 193.86\cos 17°19'22'' \approx 523.0170$

The coordinates of Q are (181.34, 523.02). (rounded to five significant digits)

Chapter 2: Test

1. $\sin A = \frac{\text{side opposite}}{\text{hypotenuse}} = \frac{12}{13}$

$\cos A = \frac{\text{side adjacent}}{\text{hypotenuse}} = \frac{5}{13}$

$\tan A = \frac{\text{side opposite}}{\text{side adjacent}} = \frac{12}{5}$

$\cot A = \frac{\text{side adjacent}}{\text{side opposite}} = \frac{5}{12}$

$\sec A = \frac{\text{hypotenuse}}{\text{side adjacent}} = \frac{13}{5}$

$\csc A = \frac{\text{hypotenuse}}{\text{side opposite}} = \frac{13}{12}$

2. Apply the relationships between the lengths of the sides of a $30°-60°$ right triangle first to the triangle on the right to find the values of y and w. In the $30°-60°$ right triangle, the side opposite the $60°$ angle is $\sqrt{3}$ times as long as the side opposite to the $30°$ angle. The length of the hypotenuse is 2 times as long as the shorter leg (opposite the $30°$ angle).

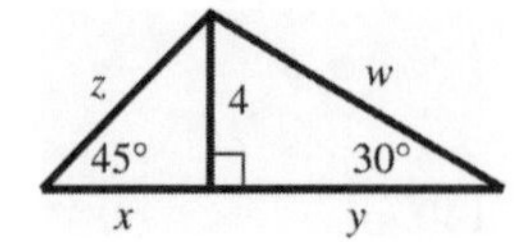

Thus, we have the following.

$$y = 4\sqrt{3} \text{ and } w = 2(4) = 8$$

Apply the relationships between the lengths of the sides of a $45°-45°$ right triangle next to the triangle on the left to find the values of x and z. In the $45°-45°$ right triangle, the sides opposite the $45°$ angles measure the same. The hypotenuse is $\sqrt{2}$ times the measure of a leg.

Thus, we have the following.

$$x = 4 \text{ and } z = 4\sqrt{2}$$

3. $\sin(B+15°) = \cos(2B+30°)$

Since sine and cosine are cofunctions, the equation is true when the following holds.

$$(B+15°)+(2B+30°) = 90°$$
$$3B+45° = 90°$$
$$3B = 45°$$
$$B = 15°$$

This is one solution; others are possible.

4. $\sin\theta = .27843196$

$\theta \approx 16.16664145°$

5. $\cos\theta = -\frac{\sqrt{2}}{2}$

Since $\cos\theta$ is negative, θ must lie in quadrants II or III. Since the absolute value of $\cos\theta$ is $\frac{\sqrt{2}}{2}$, the reference angle, θ', must be $45°$. The quadrant II angle θ equals $180° - \theta' = 180° - 45° = 135°$, and the quadrant III angle θ equals $180° + \theta' = 180° + 45° = 225°$.

6. $\tan\theta = 1.6778490$

Since $\cot\theta = \frac{1}{\tan\theta} = (\tan\theta)^{-1}$, we can use division or the inverse key (multiplicative inverse).

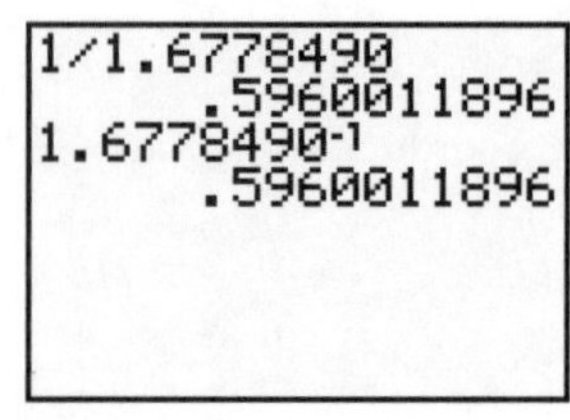

$\cot\theta \approx .5960011896$

7. **(a)** $\sin 24° < \sin 48°$

Sine increases from 0 to 1 in the interval $0° \le \theta \le 90°$. Therefore, $\sin 24°$ is less than $\sin 48°$ and the statement is true.

(b) $\cos 24° < \cos 48°$

Cosine decreases from 1 to 0 in the interval $0° \le \theta \le 90°$ Therefore, $\cos 24°$ is not less than $\cos 48°$ and the statement is false.

(c) $\tan 24° < \tan 48°$

Tangent increases in the interval $0° \le \theta \le 90°$. Therefore, $\tan 24°$ is less than $\tan 48°$ and the statement is true.

8. $\cot(-750°)$

$-750°$ is coterminal with $-750° + 3 \cdot 360° = -750° + 1080° = 330°$, which is in quadrant IV. The cotangent is negative in quadrant IV and the reference angle is $360° - 330° = 30°$.

$$\cot(-750°) = -\cot 30° = -\sqrt{3}$$

9. **(a)** $\sin 78°21' \approx .97939940$

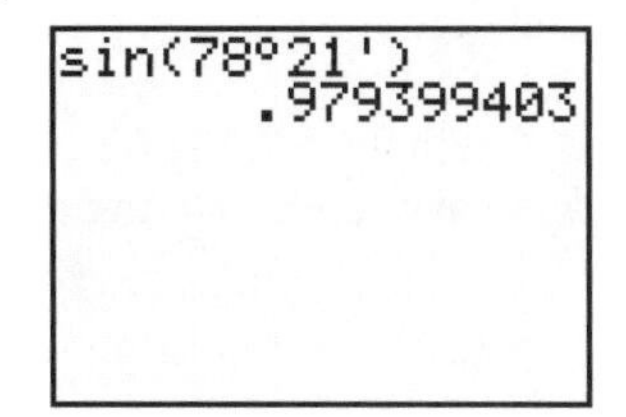

$78°21' = \left(78 + \frac{21}{60}\right)° = 78.35°$

(b) $\tan 117.689° \approx -1.9056082$

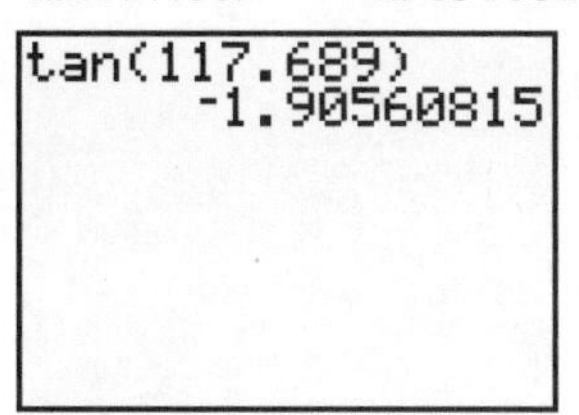

(c) $\sec 58.9041° \approx 1.9362132$

```
cos(58.9041)
          .5164720544
Ans-1
          1.936213182
```

10. $A = 58°30'$, $a = 748$

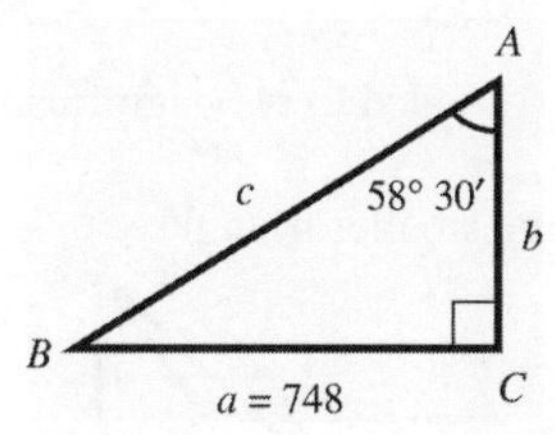

$A + B = 90° \Rightarrow B = 90° - A \Rightarrow B = 90° - 58°30' = 89°60' - 58°30' = 31°30'$

$\tan A = \dfrac{a}{b} \Rightarrow \tan 58°30' = \dfrac{748}{b} \Rightarrow b \tan 58°30' = 748 \Rightarrow b = \dfrac{748}{\tan 58°30'} \approx 458$ (rounded to three significant digits)

$\sin A = \dfrac{a}{c} \Rightarrow \sin 58°30' = \dfrac{748}{c} \Rightarrow c \sin 58°30' = 748 \Rightarrow c = \dfrac{748}{\sin 58°30'} \approx 877$ (rounded to three significant digits)

11. Let $\theta =$ the measure of the angle that the guy wire makes with the ground.

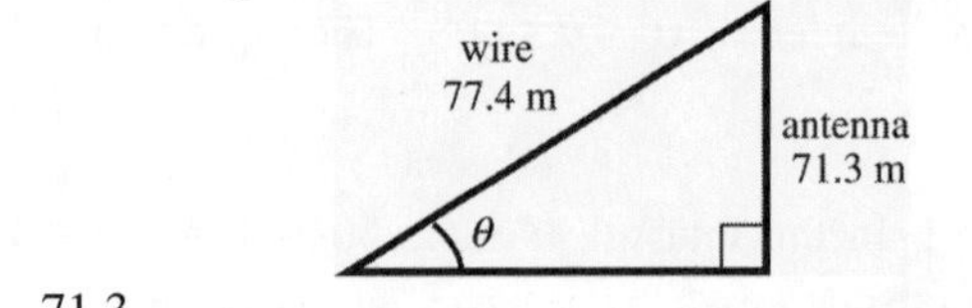

$$\sin\theta = \frac{71.3}{77.4} \approx .9211886305 \Rightarrow \theta \approx \sin^{-1}(.9211886305) \approx 61.1° \approx 61°10'$$

12. Let $x =$ the height of the flagpole.

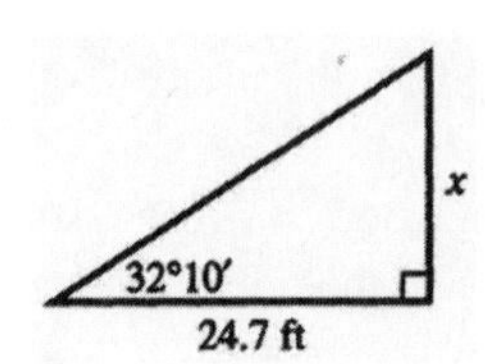

$$\tan 32°10' = \frac{x}{24.7} \Rightarrow x = 24.7\tan 32°10' \approx 15.5344$$

The flagpole is approximately 15.5 ft high. (rounded to three significant digits)

13. Draw triangle ACB and extend north-south lines from points A and C. Angle ACD is 62° (alternate interior angles of parallel lines cut by a transversal have the same measure), so Angle ACB is 62° + 28° = 90°.

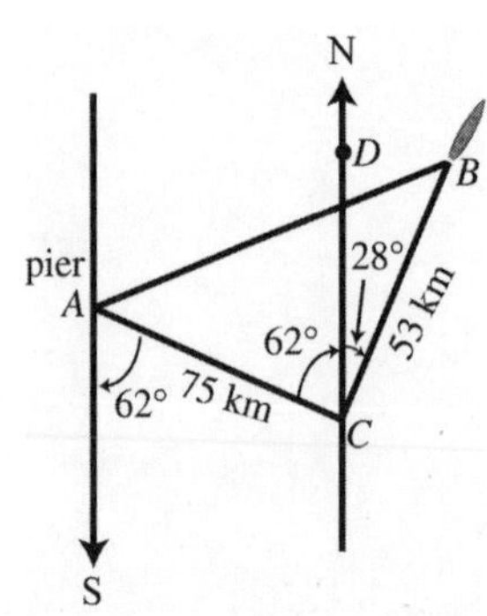

Since angle ACB is a right angle, use the Pythagorean theorem to find the distance from A to B.

$$(AB)^2 = 75^2 + 53^2 \Rightarrow (AB)^2 = 5625 + 2809 \Rightarrow (AB)^2 = 8434 \Rightarrow AB = \sqrt{8434} \approx 91.8368$$

It is 92 km from the pier to the boat. (rounded to two significant digits)

14. Let $x =$ the side adjacent to 52.5° in the smaller triangle.

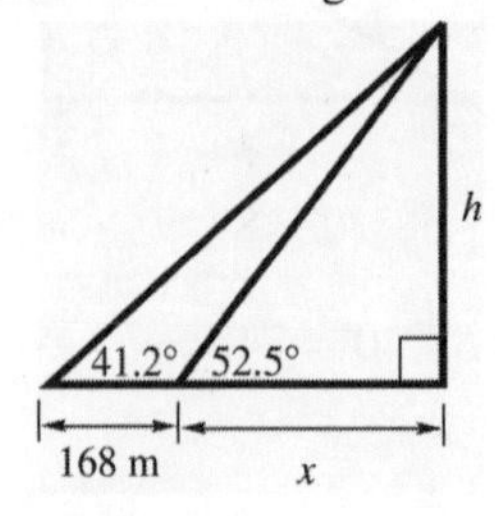

In the larger triangle, we have $\tan 41.2° = \frac{h}{168 + x} \Rightarrow h = (168 + x)\tan 41.2°$.

In the smaller triangle, we have $\tan 52.5° = \frac{h}{x} \Rightarrow h = x \tan 52.5°$.

Continued on next page

14. (continued)

Substitute for h in this equation and solve for x.

$$\begin{aligned}(168+x)\tan 41.2° &= x\tan 52.5°\\ 168\tan 41.2° + x\tan 41.2° &= x\tan 52.5°\\ 168\tan 41.2° &= x\tan 52.5° - x\tan 41.2°\\ 168\tan 41.2° &= x(\tan 52.5° - \tan 41.2°)\\ \frac{168\tan 41.2°}{\tan 52.5° - \tan 41.2°} &= x\end{aligned}$$

Substituting for x in the equation for the smaller triangle, we have the following.

$$h = x\tan 52.5° \Rightarrow h = \frac{168\tan 41.2° \tan 52.5°}{\tan 52.5° - \tan 41.2°} \approx 448.0432$$

The height of the triangle is approximately 448 m. (rounded to three significant digits)

Chapter 2: Quantitative Reasoning

$$D = \frac{v^2 \sin\theta\cos\theta + v\cos\theta\sqrt{(v\sin\theta)^2 + 64h}}{32}$$

All answers are rounded to four significant digits.

1. Since $v = 44$ ft per sec and $h = 7$ ft, we have $D = \dfrac{44^2 \sin\theta\cos\theta + 44\cos\theta\sqrt{(44\sin\theta)^2 + 64\cdot 7}}{32}$.

If $\theta = 40°$, $D = \dfrac{1936\sin 40\cos 40 + 44\cos 40\sqrt{(44\sin 40)^2 + 448}}{32} \approx 67.00$ ft.

If $\theta = 42°$, $D = \dfrac{1936\sin 42\cos 42 + 44\cos 42\sqrt{(44\sin 42)^2 + 448}}{32} \approx 67.14$ ft.

If $\theta = 45°$, $D = \dfrac{1936\sin 45\cos 45 + 44\cos 45\sqrt{(44\sin 45)^2 + 448}}{32} \approx 66.84$ ft.

As θ increases, D increases and then decreases.

2. Since $h = 7$ ft and $\theta = 42°$, we have $D = \dfrac{v^2 \sin 42\cos 42 + v\cos 42\sqrt{(v\sin 42)^2 + 64h}}{32}$.

If $v = 43$, $D = \dfrac{43^2 \sin 42\cos 42 + 43\cos 42\sqrt{(43\sin 42)^2 + 448}}{32} \approx 64.40$ ft.

If $v = 44$, $D = \dfrac{44^2 \sin 42\cos 42 + 44\cos 42\sqrt{(44\sin 42)^2 + 448}}{32} \approx 67.14$ ft.

If $v = 45$, $D = \dfrac{45^2 \sin 42\cos 42 + 45\cos 42\sqrt{(45\sin 42)^2 + 448}}{32} \approx 69.93$ ft.

As v increases, D increases.

3. The velocity affects the distance more. The shot-putter should concentrate on achieving as large a value of v as possible.

Chapter 3

RADIAN MEASURE AND CIRCULAR FUNCTIONS

Section 3.1: Radian Measure

1. Since θ is in quadrant I, $0<\theta<\frac{\pi}{2}$. Since $\frac{\pi}{2}\approx 1.57$, 1 is the only integer value in the interval. Thus, the radian measure of θ is 1 radian.

2. Since θ is in quadrant II, $\frac{\pi}{2}<\theta<\pi$. Since $\frac{\pi}{2}\approx 1.57$ and $\pi\approx 3.14$, 2 and 3 are the only integers in the interval. Since θ is closer to $\frac{\pi}{2}$, the radian measure of θ is 2 radians.

3. Since θ is in quadrant II, $\frac{\pi}{2}<\theta<\pi$. Since $\frac{\pi}{2}\approx 1.57$ and $\pi\approx 3.14$, 2 and 3 are the only integers in the interval. Since θ is closer to π, the radian measure of θ is 3 radians.

4. Since θ is an angle in quadrant IV drawn in a clockwise direction, $-\frac{\pi}{2}<\theta<0$. Also $-\frac{\pi}{2}\approx -1.57$, and -1 is the only integer in the interval. Thus, the radian measure of θ is -1 radian.

5. $60^\circ = 60\left(\frac{\pi}{180}\text{ radian}\right)=\frac{\pi}{3}\text{ radians}$

6. $30^\circ = 30\left(\frac{\pi}{180}\text{radian}\right)=\frac{\pi}{6}\text{ radians}$

7. $90^\circ = 90\left(\frac{\pi}{180}\text{ radian}\right)=\frac{\pi}{2}\text{ radians}$

8. $120^\circ = 120\left(\frac{\pi}{180}\text{radian}\right)=\frac{2\pi}{3}\text{ radians}$

9. $150^\circ = 150\left(\frac{\pi}{180}\text{ radian}\right)=\frac{5\pi}{6}\text{ radians}$

10. $270^\circ = 270\left(\frac{\pi}{180}\text{ radian}\right)=\frac{3\pi}{2}\text{ radians}$

11. $300^\circ = 300\left(\frac{\pi}{180}\text{radian}\right)=\frac{5\pi}{3}\text{ radians}$

12. $315^\circ = 315\left(\frac{\pi}{180}\text{ radian}\right)=\frac{7\pi}{4}\text{ radians}$

13. $450^\circ = 450\left(\frac{\pi}{180}\text{ radian}\right)=\frac{5\pi}{2}\text{ radians}$

14. $480^\circ = 480\left(\frac{\pi}{180}\text{ radian}\right)=\frac{8\pi}{3}\text{ radians}$

15. Multiply the degree measure by $\frac{\pi}{180}$ radian and reduce. Your answer will be in radians. Leave the answer as a multiple of π, unless otherwise directed.

16. Multiply the radian measure by $\frac{180^\circ}{\pi}$ and reduce. Your answer will be in degrees.

17. – 20. Answers will vary.

21. $\frac{\pi}{3}=\frac{\pi}{3}\left(\frac{180^\circ}{\pi}\right)=60^\circ$

22. $\frac{8\pi}{3}=\frac{8\pi}{3}\left(\frac{180^\circ}{\pi}\right)=480^\circ$

23. $\frac{7\pi}{4}=\frac{7\pi}{4}\left(\frac{180^\circ}{\pi}\right)=315^\circ$

24. $\frac{2\pi}{3}=\frac{2\pi}{3}\left(\frac{180^\circ}{\pi}\right)=120^\circ$

25. $\frac{11\pi}{6}=\frac{11\pi}{6}\left(\frac{180^\circ}{\pi}\right)=330^\circ$

26. $\frac{15\pi}{4}=\frac{15\pi}{4}\left(\frac{180^\circ}{\pi}\right)=675^\circ$

27. $-\frac{\pi}{6}=-\frac{\pi}{6}\left(\frac{180^\circ}{\pi}\right)=-30^\circ$

28. $\frac{8\pi}{5}=\frac{8\pi}{5}\left(\frac{180^\circ}{\pi}\right)=288^\circ$

29. $\frac{7\pi}{10}=\frac{7\pi}{10}\left(\frac{180^\circ}{\pi}\right)=126^\circ$

30. $\frac{11\pi}{15}=\frac{11\pi}{15}\left(\frac{180^\circ}{\pi}\right)=132^\circ$

31. $\frac{4\pi}{15}=\frac{4\pi}{15}\left(\frac{180^\circ}{\pi}\right)=48^\circ$

32. $\frac{7\pi}{20}=\frac{7\pi}{20}\left(\frac{180^\circ}{\pi}\right)=63^\circ$

33. $\frac{17\pi}{20}=\frac{17\pi}{20}\left(\frac{180^\circ}{\pi}\right)=153^\circ$

34. $\frac{11\pi}{30}=\frac{11\pi}{30}\left(\frac{180^\circ}{\pi}\right)=66^\circ$

35. $39^\circ=39\left(\frac{\pi}{180}\text{ radian}\right)\approx .68\text{ radian}$

36. $74^\circ=74\left(\frac{\pi}{180}\text{ radian}\right)\approx 1.29\text{ radians}$

37. $42.5^\circ=42.5\left(\frac{\pi}{180}\text{ radian}\right)\approx .742\text{ radians}$

38. $264.9^\circ=264.9\left(\frac{\pi}{180}\text{ radian}\right)$
$\approx 4.623\text{ radians}$

39. $139^\circ 10'=\left(139+\frac{10}{60}\right)^\circ\approx 139.1666667\left(\frac{\pi}{180}\text{ radian}\right)\approx 2.43\text{ radians}$

40. $174^\circ 50'=\left(174+\frac{50}{60}\right)^\circ\approx 174.8333333\left(\frac{\pi}{180}\text{ radian}\right)\approx 3.05\text{ radians}$

41. $64.29^\circ=64.29\left(\frac{\pi}{180}\text{ radian}\right)\approx 1.122\text{ radians}$

42. $85.04^\circ=85.04\left(\frac{\pi}{180}\text{ radian}\right)=1.484\text{ radians}$

43. $56^\circ 25'=\left(56+\frac{25}{60}\right)^\circ\approx 56.41666667\left(\frac{\pi}{180}\text{ radian}\right)\approx .9847\text{ radian}$

44. $122^\circ 37'=\left(122+\frac{37}{60}\right)^\circ\approx 122.6166667\left(\frac{\pi}{180}\text{ radian}\right)\approx 2.140\text{ radians}$

45. $47.6925^\circ=47.6925\left(\frac{\pi}{180}\text{ radian}\right)\approx .832391\text{ radian}$

46. $23.0143° = 23.0143\left(\dfrac{\pi}{180}\text{ radian}\right) \approx .401675\text{ radian}$

47. $2\text{ radians} = 2\left(\dfrac{180°}{\pi}\right) \approx 114.591559°$
$= 114° + .591559(60')$
$\approx 114° + 35.49354' \approx 114°35'$

48. $5\text{ radians} = 5\left(\dfrac{180°}{\pi}\right) \approx 286.4788976°$
$= 286° + .4788976(60')$
$\approx 286° + 28.733856' \approx 286°29'$

49. $1.74\text{ radians} = 1.74\left(\dfrac{180°}{\pi}\right) \approx 99.69465635° = 99° + .69465635(60') \approx 99° + 41.679381' \approx 99°42'$

50. $3.06\ \text{ radians} = 3.06\ \left(\dfrac{180°}{\pi}\right) \approx 175.3250853° = 175° + .3250853(60') \approx 175° + 19.505118' \approx 175°20'$

51. $.3417\ \text{ radian} = .3417\ \left(\dfrac{180°}{\pi}\right) \approx 19.57796786° = 99° + .57796786(60') \approx 99° + 34.678072' \approx 19°35'$

52. $9.84763\text{ radians} = 9.84763\left(\dfrac{180°}{\pi}\right) \approx 564.2276372° = 564° + .2276372(60')$
$\approx 564° + 13.658232' \approx 564°14'$

53. $5.01095\text{ radians} = 5.01095\left(\dfrac{180°}{\pi}\right) \approx 287.1062864° = 287° + .1062864(60')$
$\approx 287° + 6.377184' \approx 287°6'$

54. $-3.47189\text{ radians} = -3.47189\left(\dfrac{180°}{\pi}\right) \approx -198.9246439° = -\left[198° + .9246439(60')\right]$
$\approx -[198° + 55.478634'] \approx -198°55'$

55. Without the degree symbol on the 30, it is assumed that 30 is measured in radians. Thus, the approximate value of $\sin 30$ is $-.98803$, not $\frac{1}{2}$.

56. Answers will vary.

In Exercises 57 – 74, we will start by substituting 180° for π as is done in Example 3.

57. $\sin\dfrac{\pi}{3} = \sin\left(\dfrac{1}{3}\cdot 180°\right) = \sin 60° = \dfrac{\sqrt{3}}{2}$

58. $\cos\dfrac{\pi}{6} = \cos\left(\dfrac{1}{6}\cdot 180°\right) = \cos 30° = \dfrac{\sqrt{3}}{2}$

59. $\tan\dfrac{\pi}{4} = \tan\left(\dfrac{1}{4}\cdot 180°\right) = \tan 45° = 1$

60. $\cot\dfrac{\pi}{3} = \cot\left(\dfrac{1}{3}\cdot 180°\right) = \cot 60° = \dfrac{\sqrt{3}}{3}$

61. $\sec\dfrac{\pi}{6} = \sec\left(\dfrac{1}{6}\cdot 180°\right) = \sec 30° = \dfrac{2\sqrt{3}}{3}$

62. $\csc\dfrac{\pi}{4} = \csc\left(\dfrac{1}{4}\cdot 180°\right) = \csc 45° = \sqrt{2}$

63. $\sin\dfrac{\pi}{2} = \sin\left(\dfrac{1}{2}\cdot 180°\right) = \sin 90° = 1$

64. $\csc\dfrac{\pi}{2} = \csc\left(\dfrac{1}{2}\cdot 180°\right) = \csc 90° = 1$

65. $\tan\frac{5\pi}{3} = \tan\left(\frac{5}{3}\cdot 180°\right) = \tan 300° = -\tan 60° = -\sqrt{3}$

66. $\cot\frac{2\pi}{3} = \cot\left(\frac{2}{3}\cdot 180°\right) = \cot 120° = -\cot 60° - \frac{\sqrt{3}}{3}$

67. $\sin\frac{5\pi}{6} = \sin\left(\frac{5}{6}\cdot 180°\right) = \sin 150° = \sin 30° = \frac{1}{2}$

68. $\tan\frac{5\pi}{6} = \tan\left(\frac{5}{6}\cdot 180°\right) = \tan 150° = -\tan 30° = -\frac{\sqrt{3}}{3}$

69. $\cos 3\pi = \cos(3\cdot 180°) = \cos 540° = \cos(540 - 360°) = \cos(180°) = -1$

70. $\sec\pi = \sec 180° = -1$

71. $\sin\left(-\frac{8\pi}{3}\right) = \sin\left(-\frac{8}{3}\cdot 180°\right) = \sin(-480°) = \sin(-480° + 2\cdot 360°) = \sin(240°) = -\sin(60°) = -\frac{\sqrt{3}}{2}$

72. $\cot\left(-\frac{2\pi}{3}\right) = \cot\left(-\frac{2}{3}\cdot 180°\right) = \cot(-120°) = \frac{\sqrt{3}}{3}$

73. $\sin\left(-\frac{7\pi}{6}\right) = \sin\left(-\frac{7}{6}\cdot 180°\right) = \sin(-210°) = \frac{1}{2}$

74. $\cos\left(-\frac{\pi}{6}\right) = \cos\left(-\frac{1}{6}\cdot 180°\right) = \cos(-30°) = \frac{\sqrt{3}}{2}$

75. We begin the calculations with the blank next to $30°$, and then proceed counterclockwise from there.

$30° = 30\left(\frac{\pi}{180}\text{ radian}\right) = \frac{\pi}{6}\text{ radian};$ $\quad \frac{\pi}{4}\text{ radian} = \frac{\pi}{4}\left(\frac{180°}{\pi}\right) = 45°;$

$60° = 60\left(\frac{\pi}{180}\text{ radian}\right) = \frac{\pi}{3}\text{ radians};$ $\quad \frac{2\pi}{3}\text{ radians} = \frac{2\pi}{3}\left(\frac{180°}{\pi}\right) = 120°;$

$\frac{3\pi}{4}\text{ radians} = \frac{3\pi}{4}\left(\frac{180°}{\pi}\right) = 135°;$ $\quad 150° = 150\left(\frac{\pi}{180}\text{ radian}\right) = \frac{5\pi}{6}\text{ radians};$

$180° = 180\left(\frac{\pi}{180}\text{ radian}\right) = \pi\text{ radians};$ $\quad 210° = 210\left(\frac{\pi}{180}\text{ radian}\right) = \frac{7\pi}{6}\text{ radians};$

$225° = 225\left(\frac{\pi}{180}\text{ radian}\right) = \frac{5\pi}{4}\text{ radians};$ $\quad \frac{4\pi}{3}\text{ radians} = \frac{4\pi}{3}\left(\frac{180°}{\pi}\right) = 240°;$

$\frac{5\pi}{3}\text{ radians} = \frac{5\pi}{3}\left(\frac{180°}{\pi}\right) = 300°;$ $\quad 315° = 315\left(\frac{\pi}{180}\text{ radian}\right) = \frac{7\pi}{4}\text{ radians};$

$330° = 330\left(\frac{\pi}{180}\text{ radian}\right) = \frac{11\pi}{6}\text{ radians}$

76. **(a)** $\frac{9}{10}\cdot\frac{\pi}{180}=\frac{\pi}{200}$

(b) For degrees we have, $3.5\times.9°=3.15°$. For radians we have, $3.5\times\frac{\pi}{200}\approx.055$ radian.

77. **(a)** In 24 hr, the hour hand will rotate twice around the clock. Since one complete rotation measures 2π radians, the two rotations will measure $2(2\pi)=4\pi$ radians.

(b) In 4 hr, the hour hand will rotate $\frac{4}{12}=\frac{1}{3}$ of the way around the clock, which will measure $\frac{1}{3}(2\pi)=\frac{2\pi}{3}$ radians.

78. In each rotation, the pulley would rotate 2π radians.

(a) In 8 rotations, the pulley would turn $8(2\pi)=16\pi$ radians.

(b) In 30 rotations, the pulley would turn $30(2\pi)=60\pi$ radians

79. In each rotation around earth, the space vehicle would rotate 2π radians.

(a) In 2.5 orbits, the space vehicle travels $2.5(2\pi)=5\pi$ radians

(b) In $\frac{4}{3}$ of an orbit, the space vehicle travels $\frac{4}{3}(2\pi)=\frac{8\pi}{3}$ radians.

Section 3.2: Applications of Radian Measure

Connections (page 102)
Answers will vary.
The longitude at Greenwich is $0°$.

Exercises

1. $r=4,\ \theta=\frac{\pi}{2}$

$s=r\theta=4\left(\frac{\pi}{2}\right)=2\pi$

2. $r=12,\ \theta=\frac{\pi}{3}$

$s=r\theta=12\left(\frac{\pi}{3}\right)=4\pi$

3. $s=6\pi,\ \theta=\frac{3\pi}{4}$

$s=r\theta\Rightarrow r=\frac{s}{\theta}=\frac{6\pi}{\frac{3\pi}{4}}=6\pi\cdot\frac{4}{3\pi}=8$

4. $s=3\pi,\ \theta=\frac{\pi}{2}$

$s=r\theta\Rightarrow r=\frac{s}{\theta}=\frac{3\pi}{\frac{\pi}{2}}=3\pi\cdot\frac{2}{\pi}=6$

5. $r=3,\ s=3$

$s=r\theta\Rightarrow\theta=\frac{s}{r}=\frac{3}{3}=1$

6. $s=6,\ r=4$

$s=r\theta\Rightarrow\theta=\frac{s}{r}=\frac{6}{4}=\frac{3}{2}$ or 1.5

7. $r=12.3$ cm, $\theta=\frac{2\pi}{3}$ radians

$s=r\theta=12.3\left(\frac{2\pi}{3}\right)=8.2\pi\approx25.8$ cm

8. $r=.892$ cm, $\theta=\frac{11\pi}{10}$ radians

$s=r\theta=.892\left(\frac{11\pi}{10}\right)=.9812\pi$ cm ≈3.08 cm

9. $r = 4.82$ m, $\theta = 60°$

Converting θ to radians, we have $\theta = 60° = 60\left(\frac{\pi}{180} \text{ radian}\right) = \frac{\pi}{3}$ radians.

Thus, the arc is $s = r\theta = 4.82\left(\frac{\pi}{3}\right) = \frac{4.82\pi}{3} \approx 5.05$ m. (rounded to three significant digits)

10. $r = 71.9$ cm, $\theta = 135°$

Converting θ to radians, we have $\theta = 135° = 135\left(\frac{\pi}{180} \text{ radian}\right) = \frac{3\pi}{4}$ radians.

Thus, the arc is $s = r\theta = 71.9\left(\frac{3\pi}{4}\right)\text{cm} = \frac{215.7\pi}{4}$ cm ≈ 169 cm. (rounded to three significant digits)

11. The formula for arc length is $s = r\theta$. Substituting $2r$ for r we obtain $s = (2r)\theta = 2(r\theta)$. The length of the arc is doubled.

12. Recall that π radians = 180°. If θ is measured in degrees, then the formula becomes the following.

$$s = r \cdot \frac{\theta\pi}{180} \Rightarrow s = \frac{\pi r\theta}{180}$$

For Exercises 13 – 18, note that since 6400 has two significant digits and the angles are given to the nearest degree, we can have only two significant digits in the answers.

13. 9° N, 40° N

$\theta = 40° - 9° = 31° = 31\left(\frac{\pi}{180} \text{ radian}\right) = \frac{31\pi}{180}$ radian and $s = r\theta = 6400\left(\frac{31\pi}{180}\right) \approx 3500$ km

14. 36° N, 49°N

$\theta = 49° - 36° = 13° = 13\left(\frac{\pi}{180} \text{ radian}\right) = \frac{13\pi}{180}$ radian and $s = r\theta = 6400\left(\frac{13\pi}{180}\right) \approx 1500$ km

15. 41° N, 12° S
12° S = –12° N

$\theta = 41° - (-12°) = 53° = 53\left(\frac{\pi}{180} \text{ radian}\right) = \frac{53\pi}{180}$ radian and $s = r\theta = 6400\left(\frac{53\pi}{180}\right) \approx 5900$ km

16. 45° N, 34° S
34° S = –34° N

$\theta = 45° - (-34°) = 79° = 79\left(\frac{\pi}{180} \text{ radian}\right) = \frac{79\pi}{180}$ radians and $s = r\theta = 6400\left(\frac{79\pi}{180}\right) \approx 8800$ km

17. $r = 6400$ km, $s = 1200$ km

$s = r\theta \Rightarrow 1200 = 6400\theta \Rightarrow \theta = \frac{1200}{6400} = \frac{3}{16}$

Converting $\frac{3}{16}$ radian to degrees we have, $\theta = \frac{3}{16}\left(\frac{180°}{\pi}\right) \approx 11°$.
The north-south distance between the two cities is 11°.
Let x = the latitude of Madison.

$$x - 33° = 11° \Rightarrow x = 44° \text{ N}$$

The latitude of Madison is 44° N.

18. $r = 6400$ km, $s = 1100$ km

$s = r\theta \Rightarrow 1100 = 6400\theta \Rightarrow \theta = \frac{1100}{6400} = \frac{11}{64}$

Converting $\frac{11}{64}$ radian to degrees we have $\theta = \frac{11}{64} \cdot \frac{180°}{\pi} \approx 10°$.

The north-south distance between the two cities is 10°.
Let x = the latitude of Toronto.

$$x - 33° = 10° \Rightarrow x = 43° \text{ N}$$

The latitude of Toronto is 43° N

19. **(a)** The number of inches lifted is the arc length in a circle with $r = 9.27$ in. and $\theta = 71°50'$.

$$71°50' = \left(71 + \tfrac{50}{60}\right)\left(\frac{\pi}{180°}\right) \text{ and } s = r\theta \Rightarrow s = 9.27\left(71 + \tfrac{50}{60}\right)\left(\frac{\pi}{180°}\right) \approx 11.6221$$

The weight will rise 11.6 in. (rounded to three significant digits)

(b) When the weight is raised 6 in., we have the following.

$$s = r\theta \Rightarrow 6 = 9.27\theta \Rightarrow \theta = \frac{6}{9.27} \text{radian} = \frac{6}{9.27}\left(\frac{180°}{\pi}\right) \approx 37.0846° = 37° + .0846(60') \approx 37°5'$$

The pulley must be rotated through $37°5'$.

20. To find the radius of the pulley, first convert 51.6° to radians.

$$\theta = 51.6° = 51.6\left(\frac{\pi}{180} \text{ radians}\right) = \frac{51.6\pi}{180} \text{ radians}$$

Now substitute this value of θ and $s = 11.4$ cm into the equation $s = r\theta$ and solve for r.

$$s = r\theta \Rightarrow 11.4 = r\left(\frac{51.6\pi}{180}\right) \Rightarrow r = 11.4 \cdot \frac{180}{51.6\pi} \approx 12.6584$$

The radius of the pulley is 12.7 cm. (rounded to three significant digits)

21. A rotation of $\theta = 60.0\left(\frac{\pi}{180} \text{ radian}\right) = \frac{\pi}{3}$ radians on the smaller wheel moves through an arc length of

$s = r\theta = 5.23\left(\frac{\pi}{3}\right) \approx 5.4768$ cm. (holding on to more digits for the intermediate steps)

Since both wheels move together, the larger wheel moves 5.48 cm, which rotates it through an angle θ, where $5.4768 = 8.16\theta \Rightarrow \theta = \frac{5.4768}{8.16} = .671$ radian $= .671\left(\frac{180°}{\pi}\right) \approx 38.5°$.

The larger wheel rotates through 38.5°.

22. The arc length s represents the distance traveled by a point on the rim of a wheel. Since the two wheels rotate together, s will be the same for both wheels.

For the smaller wheel, $\theta = 80° = 80\left(\frac{\pi}{180}\right) = \frac{4\pi}{9}$ radians and $s = r\theta = 11.7\left(\frac{4\pi}{9}\right) \approx 16.3363$ cm.

For the larger wheel, $\theta = 50° = 50\left(\frac{\pi}{180} \text{ radian}\right) = \frac{5\pi}{18}$ radian. Thus, we can solve the following.

$$s = r\theta \Rightarrow 16.3363 = r\left(\frac{5\pi}{18}\right) \Rightarrow r = 16.3363 \cdot \frac{18}{5\pi} \approx 18.720$$

The radius of the larger wheel is 18.7 cm. (rounded to 3 significant digits)

23. A rotation of $\theta = 180\left(\frac{\pi}{180} \text{ radian}\right) = \pi$ radians. The chain moves a distance equal to half the arc length of the larger gear. So, for the large gear and pedal, $s = r\theta \Rightarrow 4.72\pi$. Thus, the chain moves 4.72π in. The small gear rotates through an angle as follows.

$$\theta = \frac{s}{r} \Rightarrow \theta = \frac{4.72\pi}{1.38} \approx 3.42\pi$$

θ for the wheel and θ for the small gear are the same, or 3.42π. So, for the wheel, we have the following.

$$s = r\theta \Rightarrow r = 13.6(3.42\pi) \approx 146.12$$

The bicycle will move 146 in. (rounded to three significant digits)

24. **(a)** In one hour, the truck travels 55 mi. Since the radius is given in in., convert 55 mi to in.

$$s = 55 \text{ mi} = 55(5280) \text{ ft} = 290,400 \text{ ft} = 290,400(12) \text{ in.} = 3,484,800 \text{ in.}$$

Solving for the radius, we have the following.

$$s = r\theta \Rightarrow 3,484,800 \text{ in.} = (14 \text{ in.})\theta \Rightarrow \theta = \frac{3,484,800}{14} \approx 248,914.29 \text{ radians}$$

Each rotation is 2π radians. Thus, we have the following.

$$\frac{\theta}{2\pi} = \frac{248,914.29}{2\pi} \approx 39,615.94$$

Thus, the number of rotations is 39,616. (rounded to the nearest whole rotation)

(b) Find s for the 16-in. wheel.

$$s = r\theta \Rightarrow s \approx (16 \text{ in.})(248,914.29) = 3,982,628.64 \text{ in.}$$

$$(3,982,628.64 \text{ in.})(1 \text{ ft}/12 \text{ in})(1 \text{ mi}/5280 \text{ ft}) \approx 62.9 \text{ mi}$$

The truck with the 16-in. tires has gone 62.9 mi in one hour, so its speed is 62.9 mph. Yes, the driver deserves a ticket.

25. Let t = the length of the train.
t is approximately the arc length subtended by $3° \, 20'$. First convert $\theta = 3°20'$ to radians.

$$\theta = 3°20' = \left(3 + \tfrac{20}{60}\right)° = 3\tfrac{1}{3}° = \left(3\tfrac{1}{3}\right)\left(\frac{\pi}{180} \text{ radian}\right) = \left(\frac{10}{3}\right)\left(\frac{\pi}{180} \text{ radian}\right) = \frac{\pi}{54} \text{ radian}$$

The length of the train is $t = r\theta \Rightarrow t = 3.5\left(\frac{\pi}{54}\right) \approx .20 \text{ km}$ long. (rounded to two significant digits)

26. Let r = the distance of the boat.
The height of the mast, 32 ft, is approximately the arc length subtended by $2° \, 10'$. First convert $\theta = 2°10'$ to radians.

$$\theta = 2°10' = \left(2 + \tfrac{10}{60}\right)° = 2\tfrac{1}{6}° = \left(2\tfrac{1}{6}\right)\left(\frac{\pi}{180} \text{ radian}\right) = \left(\frac{13}{6}\right)\left(\frac{\pi}{180} \text{ radian}\right) = \frac{13\pi}{1080} \text{ radian}$$

We must now find the radius, r.

$$s = r\theta \Rightarrow r = \frac{s}{\theta} \Rightarrow r = \frac{32}{\frac{13\pi}{1080}} = 32 \cdot \frac{1080}{13\pi} \approx 846.2146$$

The boat is about 850 ft away. (rounded to two significant digits)

27. $r = 6, \theta = \dfrac{\pi}{3}$

$$A = \frac{1}{2}r^2\theta \Rightarrow A = \frac{1}{2}(6)^2\left(\frac{\pi}{3}\right) = \frac{1}{2}(36)\left(\frac{\pi}{3}\right) = 6\pi$$

28. $r = 8, \theta = \dfrac{\pi}{2}$

$$A = \frac{1}{2}r^2\theta \Rightarrow A = \frac{1}{2}(8)^2\left(\frac{\pi}{2}\right) = \frac{1}{2}(64)\left(\frac{\pi}{2}\right) = 16\pi$$

29. $A = 3$ sq units, $r = 2$

$$A = \frac{1}{2}r^2\theta \Rightarrow 3 = \frac{1}{2}(2)^2\theta \Rightarrow 3 = \frac{1}{2}(4)\theta \Rightarrow 3 = 2\theta \Rightarrow \theta = \frac{3}{2} = 1.5 \text{ radians}$$

30. $A = 8$ sq units, $r = 4$

$$A = \frac{1}{2}r^2\theta \Rightarrow 8 = \frac{1}{2}(4)^2\theta \Rightarrow 8 = \frac{1}{2}(16)\theta \Rightarrow 8 = 8\theta \Rightarrow \theta = 1 \text{ radian}$$

In Exercises 31 – 38, we will be rounding to the nearest tenth.

31. $r = 29.2$ m, $\theta = \dfrac{5\pi}{6}$ radians

$$A = \frac{1}{2}r^2\theta \Rightarrow A = \frac{1}{2}(29.2)^2\left(\frac{5\pi}{6}\right) = \frac{1}{2}(852.64)\left(\frac{5\pi}{6}\right) \approx 1116.1032$$

The area of the sector is 1116.1 m^2. (1120 m^2 rounded to three significant digits)

32. $r = 59.8$ km, $\theta = \dfrac{2\pi}{3}$ radians

$$A = \frac{1}{2}r^2\theta \Rightarrow A = \frac{1}{2}(59.8)^2\left(\frac{2\pi}{3}\right) = \frac{1}{2}(3576.04)\left(\frac{2\pi}{3}\right) \approx 3744.8203$$

The area of the sector is 3744.8 km^2. (3740 km^2 rounded to three significant digits)

33. $r = 30.0$ ft, $\theta = \dfrac{\pi}{2}$ radians

$$A = \frac{1}{2}r^2\theta \Rightarrow A = \frac{1}{2}(30.0)^2\left(\frac{\pi}{2}\right) = \frac{1}{2}(900)\left(\frac{\pi}{2}\right) = 225\pi \approx 706.8583$$

The area of the sector is 706.9 ft^2. (707 ft^2 rounded to three significant digits)

34. $r = 90.0$ yd, $\theta = \dfrac{5\pi}{6}$ radians

$$A = \frac{1}{2}r^2\theta \Rightarrow A = \frac{1}{2}(90.0)^2\left(\frac{5\pi}{6}\right) = \frac{1}{2}(8100)\left(\frac{2\pi}{3}\right) = 3375\pi \approx 10{,}602.8752$$

The area of the sector is $10{,}602.9 \text{ yd}^2$. ($10{,}600 \text{ yd}^2$ rounded to three significant digits)

35. $r = 12.7$ cm, $\theta = 81°$

The formula $A = \frac{1}{2}r^2\theta$ requires that θ be measured in radians. Converting 81° to radians, we have $\theta = 81\left(\frac{\pi}{180}\text{ radian}\right) = \frac{9\pi}{20}$ radians. Since $A = \frac{1}{2}(12.7)^2\left(\frac{9\pi}{20}\right) = \frac{1}{2}(161.29)\left(\frac{9\pi}{20}\right) \approx 114.0092$, the area of the sector is 114.0 cm^2. (114 cm^2 rounded to three significant digits)

36. $r = 18.3$ m, $\theta = 125°$

The formula $A = \frac{1}{2}r^2\theta$ requires that θ be measured in radians. Converting 125° to radians, we have $\theta = 125\left(\frac{\pi}{180}\text{ radian}\right) = \frac{25\pi}{36}$ radians. Since $A = \frac{1}{2}(18.3)^2\left(\frac{25\pi}{36}\right) = \frac{1}{2}(334.89)\left(\frac{25\pi}{36}\right) \approx 365.3083$, the area of the sector is 365.3 m^2. (365 m^2 rounded to three significant digits)

37. $r = 40.0$ mi, $\theta = 135°$

The formula $A = \frac{1}{2}r^2\theta$ requires that θ be measured in radians. Converting 135° to radians, we have $\theta = 135\left(\frac{\pi}{180}\text{ radian}\right) = \frac{3\pi}{4}$ radians. Since $A = \frac{1}{2}(40.0)^2\left(\frac{3\pi}{4}\right) = \frac{1}{2}(1600)\left(\frac{3\pi}{4}\right) = 600\pi \approx 1884.9556$, the area of the sector is 1885.0 mi^2. (1880 mi^2 rounded to three significant digits)

38. $r = 90.0$ km, $\theta = 270°$

The formula $A = \frac{1}{2}r^2\theta$ requires that θ be measured in radians. Converting 270° to radians, we have $\theta = 270\left(\frac{\pi}{180}\text{ radian}\right) = \frac{3\pi}{2}$ radians. Since $A = \frac{1}{2}(90.0)^2\left(\frac{3\pi}{2}\right) = \frac{1}{2}(8100)\left(\frac{3\pi}{2}\right) = 6075\pi \approx 19{,}085.1754$, the area of the sector is $19{,}085.2\text{ km}^2$. ($19{,}100\text{ km}^2$ rounded to three significant digits)

39. $A = 16\text{ in.}^2$, $r = 3.0$ in.

$A = \frac{1}{2}r^2\theta \Rightarrow 16 = \frac{1}{2}(3)^2\theta \Rightarrow 16 = \frac{9}{2}\theta \Rightarrow \theta = 16 \cdot \frac{2}{9} = \frac{32}{9} \approx 3.6$ radians (rounded to two significant digits)

40. $A = 64\text{ m}^2$, $\theta = \frac{\pi}{6}$ radian

$A = \frac{1}{2}r^2\theta \Rightarrow 64 = \frac{1}{2}r^2\left(\frac{\pi}{6}\right) \Rightarrow 64 = \frac{\pi}{12}r^2 \Rightarrow r^2 = 64 \cdot \frac{12}{\pi} \Rightarrow r^2 = \frac{768}{\pi} \Rightarrow r = \sqrt{\frac{768}{\pi}} \approx 16$ m (rounded to two significant digits)

41. **(a)** The central angle in degrees measures $\frac{360°}{27} = 13\tfrac{1}{3}°$. Converting to radians, we have the following.

$$13\tfrac{1}{3}° = \left(13\tfrac{1}{3}\right)\left(\frac{\pi}{180}\text{ radian}\right) = \left(\frac{40}{3}\right)\left(\frac{\pi}{180}\text{ radian}\right) = \frac{2\pi}{27}\text{ radian}$$

(b) Since $C = 2\pi r$, and $r = 76$ ft, we have $C = 2\pi(76) = 152\pi \approx 477.5221$. The circumference is 480 ft. (rounded to two significant digits)

Continued on next page

41. (continued)

(c) Since $r = 76$ ft and $\theta = \frac{2\pi}{27}$, we have $s = r\theta = 76\left(\frac{2\pi}{27}\right) = \frac{152\pi}{27} \approx 17.6860.$ Thus, the length of the arc is 17.7 ft. (rounded to three significant digits)

Note: If this measurement is approximated to be $\frac{160}{9}$, then the approximated value would be 17.8 ft, rounded to three significant digits.

(d) Area of sector with $r = 76$ ft and $\theta = \frac{2\pi}{27}$ is as follows.

$$A = \frac{1}{2}r^2\theta \Rightarrow A = \frac{1}{2}\left(76^2\right)\frac{2\pi}{27} = \frac{1}{2}(5776)\frac{2\pi}{27} = \frac{5776\pi}{27} \approx 672.0681 \approx 672 \text{ ft}^2$$

42. The area cleaned is the area of the sector "wiped" by the total area and blade minus the area "wiped" by the arm only. We must first convert 95° to radians.

$$95° = (95)\left(\frac{\pi}{180} \text{ radian}\right) = \frac{19\pi}{36} \text{ radians}$$

Since 10 − 7 = 3, the arm was 3 in. long. Thus, we have the following.

$$A_{\text{arm only}} = \frac{1}{2}(3)^2\left(\frac{19\pi}{36}\right) = \frac{1}{2}(9)\left(\frac{19\pi}{36}\right) = \frac{19\pi}{8} \approx 7.4613 \text{ in.}^2$$

and

$$A_{\text{arm and blade}} = \frac{1}{2}(10)^2\left(\frac{19\pi}{36}\right) = \frac{1}{2}(100)\left(\frac{19\pi}{36}\right) = \frac{475\pi}{18} \approx 82.9031 \text{ in.}^2$$

Since $82.9031 - 7.4613 = 75.4418$, the area of the region cleaned was about 75.4 in.^{2}.

43. (a)

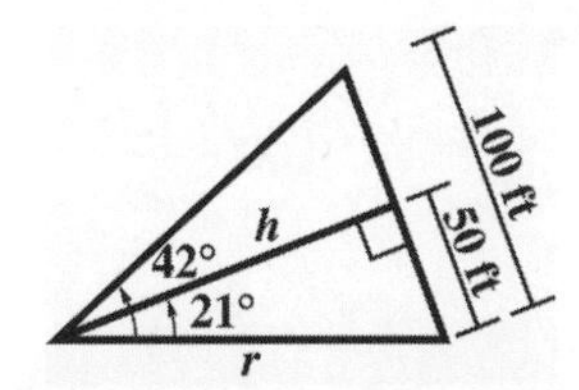

The triangle formed by the central angle and the chord is isosceles. Therefore, the bisector of the central angle is also the perpendicular bisector of the chord.

$$\sin 21° = \frac{50}{r} \Rightarrow r = \frac{50}{\sin 21°} \approx 140 \text{ ft}$$

(b) $r = \frac{50}{\sin 21°}$; $\theta = 42°$

Converting θ to radians, we have $42\left(\frac{\pi}{180} \text{ radian}\right) = \frac{7\pi}{30}$ radian. Solving for the arc length, we have $s = r\theta \Rightarrow s = \frac{50}{\sin 21°} \cdot \frac{7\pi}{30} = \frac{35\pi}{3\sin 21°} \approx 102$ ft.

Continued on next page

43. (continued)

(c)

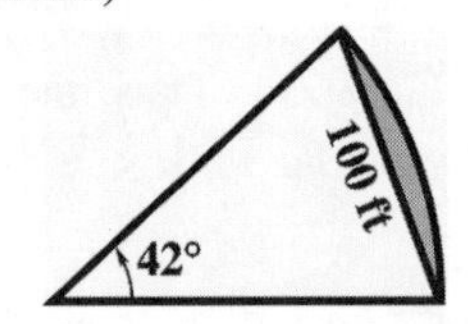

The area of the portion of the circle can be found by subtracting the area of the triangle from the area of the sector. From the figure in part (a), we have $\tan 21° = \frac{50}{h}$ so $h = \frac{50}{\tan 21°}$.

$$A_{\text{sector}} = \frac{1}{2}r^2\theta \Rightarrow A_{\text{sector}} = \frac{1}{2}\left(\frac{50}{\sin 21°}\right)^2\left(\frac{7\pi}{30}\right) \approx 7135 \text{ ft}^2$$

and

$$A_{\text{triangle}} = \frac{1}{2}bh \Rightarrow A_{\text{triangle}} = \frac{1}{2}(100)\left(\frac{50}{\tan 21°}\right) \approx 6513 \text{ ft}^2$$

The area bounded by the arc and the chord is 7135 – 6513 = 622 ft^2.

44. (a) Since the area of a circle is $A = \pi r^2$, we have the following.

$$950{,}000 = \pi r^2 \Rightarrow r^2 = \frac{950{,}000}{\pi} \Rightarrow r = \sqrt{\frac{950{,}000}{\pi}} \approx 549.9040$$

Thus, the radius is 550 m. (rounded to two significant digits)

(b) Converting $\theta = 35°$ to radians, we have $35\left(\frac{\pi}{180} \text{ radian}\right) = \frac{7\pi}{36}$ radian.

Since the area of a sector is $A = \frac{1}{2}r^2\theta$, we have the following.

$$A = \frac{1}{2}r^2\theta \Rightarrow 950{,}000 = \frac{1}{2}r^2\left(\frac{7\pi}{36}\right) \Rightarrow 950{,}000 = r^2\left(\frac{7\pi}{72}\right) \Rightarrow r^2 = 950{,}000 \cdot \frac{72}{7\pi}$$

$$r^2 = \frac{68{,}400{,}000}{7\pi} \Rightarrow r = \sqrt{\frac{68{,}400{,}000}{7\pi}} \approx 1763.6163$$

Thus, the radius is 1800 m. (rounded to two significant digits)

45. Use the Pythagorean theorem to find the hypotenuse of the triangle, which is also the radius of the sector of the circle.

$$r^2 = 30^2 + 40^2 \Rightarrow r^2 = 900 + 1600 \Rightarrow r^2 = 2500 \Rightarrow r = 50$$

The total area of the lot is the sum of the areas of the triangle and the sector.

Converting $\theta = 60°$ to radians, we have $60\left(\frac{\pi}{180} \text{ radian}\right) = \frac{\pi}{3}$ radians.

$$A_{\text{triangle}} = \frac{1}{2}bh = \frac{1}{2}(30)(40) = 600 \text{ yd}^2$$

and

$$A_{\text{sector}} = \frac{1}{2}r^2\theta = \frac{1}{2}(50)^2\left(\frac{\pi}{3}\right) = \frac{1}{2}(2500)\left(\frac{\pi}{3}\right) = \frac{1250\pi}{3} \text{ yd}^2$$

Total area $A_{\text{triangle}} + A_{\text{sector}} = 600 + \frac{1250\pi}{3} \approx 1908.9969$ or 1900 yd^2, rounded to two significant digits.

46. Converting $\theta = 1' = \left(\frac{1}{60}\right)^\circ$ to radians, we have $\frac{1}{60}\left(\frac{\pi}{180} \text{ radian}\right) = \frac{\pi}{10,800}$ radian. Solving for the arc length, we have $s = r\theta \Rightarrow s = 3963 \cdot \frac{\pi}{10,800} = \frac{11\pi}{30} \approx 1.1519$. Thus, there are approximately 1.15 statute miles in 1 nautical mile. (rounded to two decimal places)

47. Converting $\theta = 7^\circ 12' = \left(7 + \frac{12}{60}\right)^\circ = 7.2^\circ$ to radians, we have $7.2\left(\frac{\pi}{180} \text{ radian}\right) = \frac{7.2\pi}{180} = \frac{\pi}{25}$ radian. Solving for the radius with the arc length formula, we have the following.

$$s = r\theta \Rightarrow 496 = r \cdot \frac{\pi}{25} \Rightarrow r = 496 \cdot \frac{25}{\pi} = \frac{12,400}{\pi} \approx 3947.0426$$

Thus, the radius is approximately 3947 mi. (rounded to the nearest mile)

Using this approximate radius, we can find the circumference of the Earth. Since $C = 2\pi r$, we have $C \approx 2\pi(3947) \approx 24,799.7324$. Thus, the approximate circumference is 24,800 mi. (rounded to three significant digits)

48. The central angle in degrees measures .517°. Converting to radians, we have the following.

$$.517^\circ = (.517)\left(\frac{\pi}{180} \text{ radian}\right) = \frac{.517\pi}{180} \text{ radian}$$

$$s = r\theta \Rightarrow s = 238,900\left(\frac{.517\pi}{180}\right) \approx 2155.6788$$

Recall, from page 105 of your text (above Exercises 25 – 26), for very small central angles, there is little difference between the arc and the inscribed chord. Thus, the diameter of the moon is approximately 2156 mi. (rounded to four significant digits)

49. Since L is the arc length, we have $L = r\theta$. Thus, $r = \frac{L}{\theta}$.

50. Since $\cos\frac{\theta}{2} = \frac{h}{r}$, $h = r\cos\frac{\theta}{2}$.

51. $d = r - h \Rightarrow d = r - r\cos\frac{\theta}{2} \Rightarrow d = r\left(1 - \cos\frac{\theta}{2}\right)$

52. Since $r = \frac{L}{\theta}$, we have $d = r\left(1 - \cos\frac{\theta}{2}\right) \Rightarrow d = \frac{L}{\theta}\left(1 - \cos\frac{\theta}{2}\right)$.

53. If we let $r' = 2r$, then $A_{\text{sector}} = \frac{1}{2}(r')^2\theta = \frac{1}{2}(2r)^2\theta = \frac{1}{2}(4r^2)\theta = 4\left(\frac{1}{2}r^2\theta\right)$. Thus, the area, $\frac{1}{2}r^2\theta$, is quadrupled.

54. $A_{\text{sector}} = \frac{1}{2}r^2\theta$, where θ is in radians. Thus, $A_{\text{sector}} = \frac{1}{2}r^2\frac{\theta}{180} = \frac{\pi r^2\theta}{360}$, θ is in degrees.

55. The base area is $A_{\text{sector}} = \frac{1}{2}r^2\theta$. Thus, the volume is $V = \frac{1}{2}r^2\theta h$ or $V = \frac{r^2\theta h}{2}$, where θ is in radians.

56. To find the base area, we need to find the area of the outer sector and subtract from it the area of the inner sector (the "missing" sector).

$$A_{\text{outer sector}} = \frac{1}{2}r_1^2\theta \text{ and } A_{\text{inner sector}} = \frac{1}{2}r_2^2\theta$$

$$A_{\text{outer sector}} - A_{\text{inner sector}} = \frac{1}{2}r_1^2\theta - \frac{1}{2}r_2^2\theta = \frac{1}{2}\theta\left(r_1^2 - r_2^2\right)$$

Thus, the volume is $V = \frac{1}{2}\theta\left(r_1^2 - r_2^2\right)h$, where θ is in radians.

Section 3.3: The Unit Circle and Circular Functions

Connections (page 113)

$PQ = y = \frac{y}{1} = \sin\theta$; $OQ = x = \frac{x}{1} = \cos\theta$; $AB = \frac{AB}{1} = \frac{AB}{AO} = \frac{y}{x} = \tan\theta$ (by similar triangles)

Exercises

1. An angle of $\theta = \frac{\pi}{2}$ radians intersects the unit circle at the point $(0,1)$.

(a) $\sin\theta = y = 1$

(b) $\cos\theta = x = 0$

(c) $\tan\theta = \frac{y}{x} = \frac{1}{0}$; undefined

2. An angle of $\theta = \pi$ radians intersects the unit circle at the point $(-1,0)$.

(a) $\sin\theta = y = 0$

(b) $\cos\theta = x = -1$

(c) $\tan\theta = \frac{y}{x} = \frac{0}{-1} = 0$

3. An angle of $\theta = 2\pi$ radians intersects the unit circle at the point $(1,0)$.

(a) $\sin\theta = y = 0$

(b) $\cos\theta = x = 1$

(c) $\tan\theta = \frac{y}{x} = \frac{0}{1} = 0$

4. An angle of $\theta = 3\pi$ radians intersects the unit circle at the point $(-1,0)$.

(a) $\sin\theta = y = 0$

(b) $\cos\theta = x = -1$

(c) $\tan\theta = \frac{y}{x} = \frac{0}{-1} = 0$

5. An angle of $\theta = -\pi$ radians intersects the unit circle at the point $(-1,0)$.

(a) $\sin\theta = y = 0$

(b) $\cos\theta = x = -1$

(c) $\tan\theta = \frac{y}{x} = \frac{0}{-1} = 0$

6. An angle of $\theta = -\frac{3\pi}{2}$ radians intersects the unit circle at the point $(0,1)$.

(a) $\sin\theta = y = 1$

(b) $\cos\theta = x = 0$

(c) $\tan\theta = \frac{y}{x} = \frac{1}{0}$; undefined

For Exercises 7 – 22, use the following copy of Figure 12 on page 109 of your text.

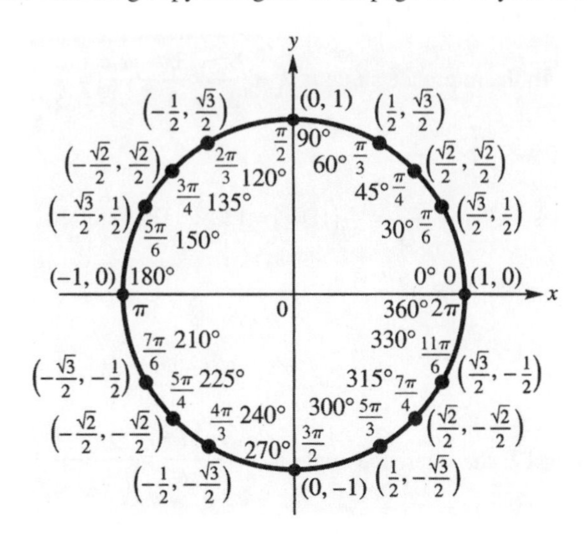

7. Find $\sin\frac{7\pi}{6}$.

Since $\frac{7\pi}{6}$ is in quadrant III, the reference angle is $\frac{7\pi}{6}-\pi=\frac{7\pi}{6}-\frac{6\pi}{6}=\frac{\pi}{6}$. In quadrant III, the sine is negative. Thus, $\sin\frac{7\pi}{6}=-\sin\frac{\pi}{6}=-\frac{1}{2}$.

Converting $\frac{7\pi}{6}$ to degrees we have, $\frac{7\pi}{6}=\frac{7}{6}(180°)=210°$. The reference angle is $210°-180°=30°$. Thus, $\sin\frac{7\pi}{6}=\sin 210°=-\sin 30°=-\frac{1}{2}$.

8. Find $\cos\frac{5\pi}{3}$.

Since $\frac{5\pi}{3}$ is in quadrant IV, the reference angle is $2\pi-\frac{5\pi}{3}=\frac{6\pi}{3}-\frac{5\pi}{3}=\frac{\pi}{3}$. In quadrant IV, the cosine is positive. Thus, $\cos\frac{5\pi}{3}=\cos\frac{\pi}{3}=\frac{1}{2}$.

Converting $\frac{5\pi}{3}$ to degrees we have, $\frac{5\pi}{3}=\frac{5}{3}(180°)=300°$. The reference angle is $360°-300°=60°$. Thus, $\cos\frac{5\pi}{3}=\cos 300°=\cos 60°=\frac{1}{2}$.

9. Find $\tan\frac{3\pi}{4}$.

Since $\frac{3\pi}{4}$ is in quadrant II, the reference angle is $\pi-\frac{3\pi}{4}=\frac{4\pi}{4}-\frac{3\pi}{4}=\frac{\pi}{4}$. In quadrant II, the tangent is negative. Thus, $\tan\frac{3\pi}{4}=-\tan\frac{\pi}{4}=-1$.

Converting $\frac{3\pi}{4}$ to degrees we have, $\frac{3\pi}{4}=\frac{3}{4}(180°)=135°$. The reference angle is $180°-135°=45°$. Thus, $\tan\frac{3\pi}{4}=\tan 135°=-\tan 45°=-1$.

10. Find $\sec\frac{2\pi}{3}$.

Since $\frac{2\pi}{3}$ is in quadrant II, the reference angle is $\pi-\frac{2\pi}{3}=\frac{3\pi}{3}-\frac{2\pi}{3}=\frac{\pi}{3}$. In quadrant II, the secant is negative. Thus, $\sec\frac{2\pi}{3}=-\sec\frac{\pi}{3}=-2$.

Converting $\frac{2\pi}{3}$ to degrees we have, $\frac{2\pi}{3}=\frac{2}{3}(180°)=120°$. The reference angle is $180°-120°=60°$. Thus, $\sec\frac{2\pi}{3}=\sec 120°=-\sec 60°=-2$.

11. Find $\csc\frac{11\pi}{6}$.

Since $\frac{11\pi}{6}$ is in quadrant IV, the reference angle is $2\pi-\frac{11\pi}{6}=\frac{12\pi}{6}-\frac{11\pi}{6}=\frac{\pi}{6}$. In quadrant IV, the cosecant is negative. Thus, $\csc\frac{11\pi}{6}=-\csc\frac{\pi}{6}=-2$.

Converting $\frac{11\pi}{6}$ to degrees we have, $\frac{11\pi}{6}=\frac{11}{6}(180°)=330°$. The reference angle is $360°-330°=30°$. Thus, $\csc\frac{11\pi}{6}=\csc 330°=-\csc 30°=-2$.

12. Find $\cot\frac{5\pi}{6}$.

Since $\frac{5\pi}{6}$ is in quadrant II, the reference angle is $\pi-\frac{5\pi}{6}=\frac{6\pi}{6}-\frac{5\pi}{6}=\frac{\pi}{6}$. In quadrant II, the cotangent is negative. Thus, $\cot\frac{5\pi}{6}=-\cot\frac{\pi}{6}=-\sqrt{3}$.

Converting $\frac{5\pi}{6}$ to degrees we have, $\frac{5\pi}{6}=\frac{5}{6}(180°)=150°$. The reference angle is $180°-150°=30°$. Thus, $\cot\frac{5\pi}{6}=\cot 150°=-\cot 30°=-\sqrt{3}$.

13. Find $\cos\left(-\frac{4\pi}{3}\right)$.

$-\frac{4\pi}{3}$ is coterminal with $-\frac{4\pi}{3}+2\pi=-\frac{4\pi}{3}+\frac{6\pi}{3}=\frac{2\pi}{3}$. Since $\frac{2\pi}{3}$ is in quadrant II, the reference angle is $\pi-\frac{2\pi}{3}=\frac{3\pi}{3}-\frac{2\pi}{3}=\frac{\pi}{3}$. In quadrant II, the cosine is negative. Thus, $\cos\left(-\frac{4\pi}{3}\right)=\cos\frac{2\pi}{3}=-\cos\frac{\pi}{3}=-\frac{1}{2}$.

Converting $\frac{2\pi}{3}$ to degrees we have, $\frac{2\pi}{3}=\frac{2}{3}(180°)=120°$. The reference angle is $180°-120°=60°$. Thus, $\cos\left(-\frac{4\pi}{3}\right)=\cos\frac{2\pi}{3}=\cos 120°=-\cos 60°=-\frac{1}{2}$.

14. Find $\tan\frac{17\pi}{3}$.

$\frac{17\pi}{3}$ is coterminal with $\frac{17\pi}{3}-2(2\pi)=\frac{17\pi}{3}-4\pi=\frac{17\pi}{3}-\frac{12\pi}{3}=\frac{5\pi}{3}$. Since $\frac{5\pi}{3}$ is in quadrant IV, the reference angle is $2\pi-\frac{5\pi}{3}=\frac{6\pi}{3}-\frac{5\pi}{3}=\frac{\pi}{3}$. In quadrant IV, the tangent is negative. Thus, $\tan\frac{17\pi}{3}=\tan\frac{5\pi}{3}=-\tan\frac{\pi}{3}=-\sqrt{3}$.

Converting $\frac{5\pi}{3}$ to degrees we have, $\frac{5\pi}{3}=\frac{5}{3}(180°)=300°$. The reference angle is $360°-300°=60°$. Thus, $\tan\frac{5\pi}{3}=\tan 300°=-\tan 60°=-\sqrt{3}$.

15. Find $\cos\frac{7\pi}{4}$

Since $\frac{7\pi}{4}$ is in quadrant IV, the reference angle is $2\pi-\frac{7\pi}{4}=\frac{8\pi}{4}-\frac{7\pi}{4}=\frac{\pi}{4}$. In quadrant IV, the cosine is positive. Thus, $\cos\frac{7\pi}{4}=\cos\frac{\pi}{4}=\frac{\sqrt{2}}{2}$.

Converting $\frac{7\pi}{4}$ to degrees we have, $\frac{7\pi}{4}=\frac{7}{4}(180°)=315°$. The reference angle is $360°-315°=45°$. Thus, $\cos\frac{7\pi}{4}=\cos 315°=\cos 45°=\frac{\sqrt{2}}{2}$.

16. Find $\sec\frac{5\pi}{4}$

Since $\frac{5\pi}{4}$ is in quadrant III, the reference angle is $\frac{5\pi}{4}-\pi=\frac{5\pi}{4}-\frac{4\pi}{4}=\frac{\pi}{4}$. In quadrant III, the secant is negative. Thus, $\sec\frac{5\pi}{4}=-\sec\frac{\pi}{4}=-\sqrt{2}$.

Converting $\frac{5\pi}{4}$ to degrees we have, $\frac{5\pi}{4}=\frac{5}{4}(180°)=225°$. The reference angle is $225°-180°=45°$. Thus, $\sec\frac{5\pi}{4}=\sec 225°=-\sec 45°=-\sqrt{2}$.

17. Find $\sin\left(-\frac{4\pi}{3}\right)$

$-\frac{4\pi}{3}$ is coterminal with $-\frac{4\pi}{3}+2\pi=-\frac{4\pi}{3}+\frac{6\pi}{3}=\frac{2\pi}{3}$. Since $\frac{2\pi}{3}$ is in quadrant II, the reference angle is $\pi-\frac{2\pi}{3}=\frac{3\pi}{3}-\frac{2\pi}{3}=\frac{\pi}{3}$. In quadrant II, the sine is positive. Thus, $\sin\left(-\frac{4\pi}{3}\right)=\sin\frac{2\pi}{3}=\sin\frac{\pi}{3}=\frac{\sqrt{3}}{2}$.

Converting $\frac{2\pi}{3}$ to degrees we have, $\frac{2\pi}{3}=\frac{2}{3}(180°)=120°$. The reference angle is $180°-120°=60°$. Thus, $\sin\left(-\frac{4\pi}{3}\right)=\sin\frac{2\pi}{3}=\sin 120°=\sin 60°=\frac{\sqrt{3}}{2}$.

18. Find $\sin\left(-\frac{5\pi}{6}\right)$

$-\frac{5\pi}{6}$ is coterminal with $-\frac{5\pi}{6}+2\pi=-\frac{5\pi}{6}+\frac{12\pi}{6}=\frac{7\pi}{6}$. Since $\frac{7\pi}{6}$ is in quadrant III, the reference angle is $\frac{7\pi}{6}-\pi=\frac{7\pi}{6}-\frac{6\pi}{6}=\frac{\pi}{6}$. In quadrant III, the sine is negative. Thus, $\sin\left(-\frac{5\pi}{6}\right)=\sin\left(\frac{7\pi}{6}\right)=-\sin\frac{\pi}{6}=-\frac{1}{2}$.

Converting $\frac{7\pi}{6}$ to degrees we have, $\frac{7\pi}{6}=\frac{7}{6}(180°)=210°$. The reference angle is $210°-180°=30°$. Thus, $\sin\left(-\frac{5\pi}{6}\right)=\sin\frac{7\pi}{6}=\sin 210°=-\sin 30°=-\frac{1}{2}$.

19. Find $\sec\frac{23\pi}{6}$.

$\frac{23\pi}{6}$ is coterminal with $\frac{23\pi}{6}-2\pi=\frac{23\pi}{6}-\frac{12\pi}{6}=\frac{11\pi}{6}$. Since $\frac{11\pi}{6}$ is in quadrant IV, the reference angle is $2\pi-\frac{11\pi}{6}=\frac{12\pi}{6}-\frac{11\pi}{6}=\frac{\pi}{6}$. In quadrant IV, the secant is positive. Thus,

$\sec\frac{23\pi}{6}=\sec\frac{11\pi}{6}=\sec\frac{\pi}{6}=\frac{2\sqrt{3}}{3}$.

Converting $\frac{11\pi}{6}$ to degrees we have, $\frac{11\pi}{6}=\frac{11}{6}(180°)=330°$. The reference angle is $360°-330°=30°$. Thus, $\sec\frac{23\pi}{6}=\sec\frac{11\pi}{6}=\sec 330°=\sec 30°=\frac{2\sqrt{3}}{3}$.

20. Find $\csc\frac{13\pi}{3}$.

$\frac{13\pi}{3}$ is coterminal with $\frac{13\pi}{3}-2(2\pi)=\frac{13\pi}{3}-4\pi=\frac{13\pi}{3}-\frac{12\pi}{3}=\frac{\pi}{3}$. Since $\frac{\pi}{3}$ is in quadrant I, we have $\csc\frac{13\pi}{3}=\csc\frac{\pi}{3}=\frac{2\sqrt{3}}{3}$

Converting $\frac{\pi}{3}$ to degrees, we have $\frac{\pi}{3}=\frac{1}{3}(180°)=60°$. Thus, $\csc\frac{\pi}{3}=\csc 60°=\frac{2\sqrt{3}}{3}$.

21. Find $\tan\frac{5\pi}{6}$.

Since $\frac{5\pi}{6}$ is in quadrant II, the reference angle is $\pi-\frac{5\pi}{6}=\frac{6\pi}{6}-\frac{5\pi}{6}=\frac{\pi}{6}$. In quadrant II, the tangent is negative. Thus, $\tan\frac{5\pi}{6}=-\tan\frac{\pi}{6}=-\frac{\sqrt{3}}{3}$.

Converting $\frac{5\pi}{6}$ to degrees we have, $\frac{5\pi}{6}=\frac{5}{6}(180°)=150°$. The reference angle is $180°-150°=30°$. Thus, $\tan\frac{5\pi}{6}=\tan 150°=-\tan 30°=-\frac{\sqrt{3}}{3}$.

22. Find $\cos\frac{3\pi}{4}$.

Since $\frac{3\pi}{4}$ is in quadrant II, the reference angle is $\pi-\frac{3\pi}{4}=\frac{4\pi}{4}-\frac{3\pi}{4}=\frac{\pi}{4}$. In quadrant II, the cosine is negative. Thus, $\cos\frac{3\pi}{4}=-\cos\frac{\pi}{4}=-\frac{\sqrt{2}}{2}$.

Converting $\frac{3\pi}{4}$ to degrees we have, $\frac{3\pi}{4}=\frac{3}{4}(180°)=135°$. The reference angle is $180°-135°=45°$.

Thus, $\tan\frac{3\pi}{4}=\cos 135°=-\cos 45°=-\frac{\sqrt{2}}{2}$.

For Exercises 23 – 34, 49 – 54, and 61 – 70, your calculator must be set in radian mode. Keystroke sequences may vary based on the type and/or model of calculator being used. As in Example 3, we will set the calculator to show four decimal digits.

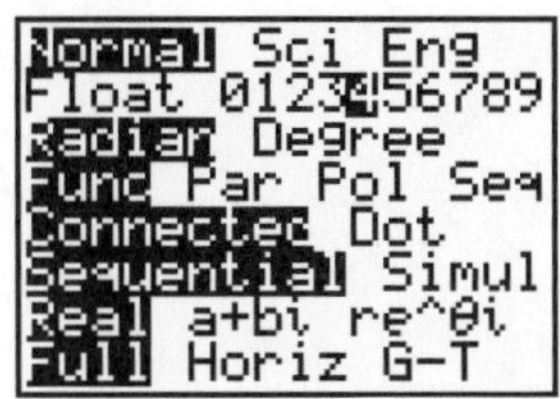

23. $\sin .6109 \approx .5736$

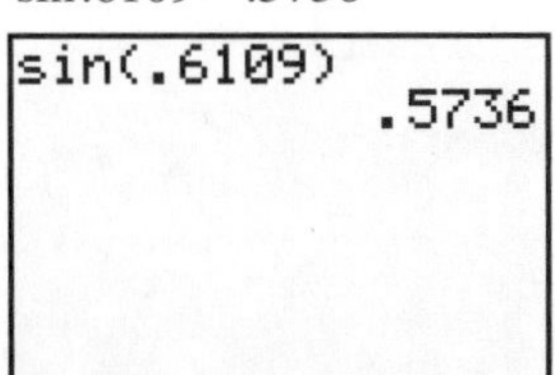

24. $\sin .8203 \approx .7314$

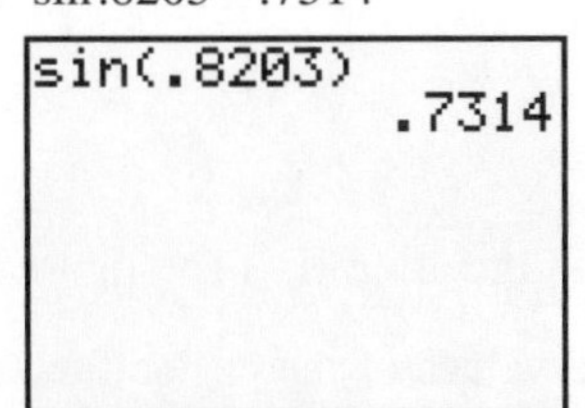

25. $\cos(-1.1519) \approx .4068$

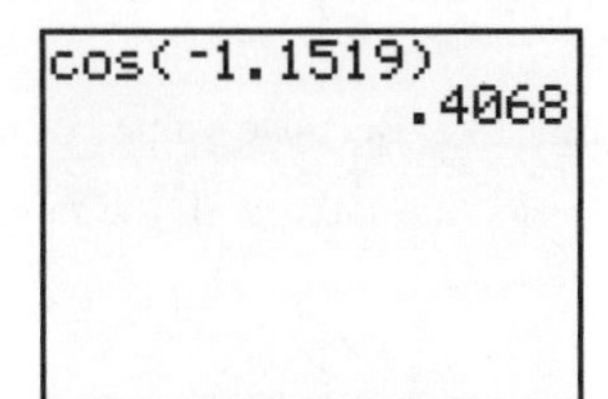

26. $\cos(-5.2825) \approx .5397$

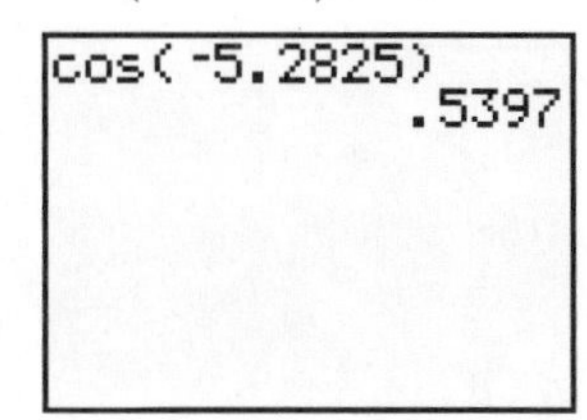

27. $\tan 4.0203 \approx 1.2065$ t

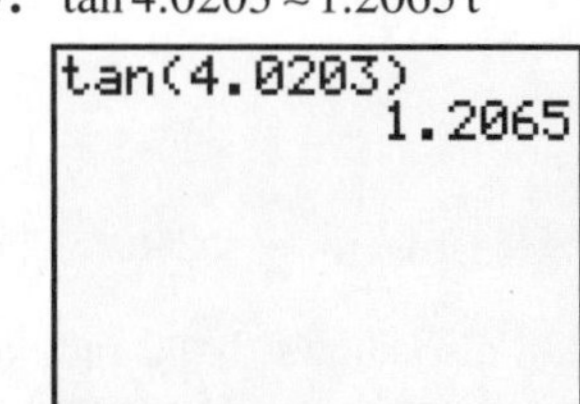

28. $\tan 6.4752 \approx .1944$

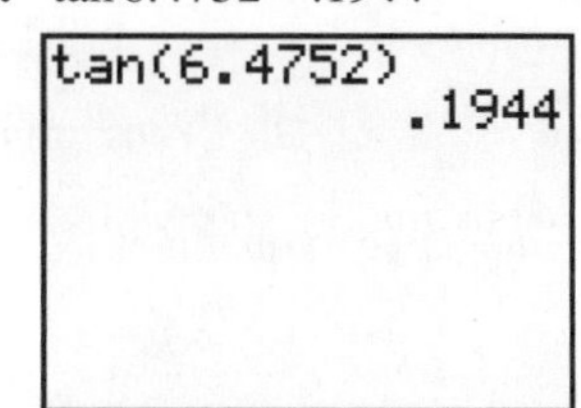

29. $\csc(-9.4946) \approx 14.3338$

```
sin(-9.4946)
               .0698
Ans-1
             14.3338
```

30. $\csc 1.3875 \approx 1.0170$

```
sin(1.3875)
               .9832
Ans-1
              1.0170
```

31. $\sec 2.8440 \approx -1.0460$

```
cos(2.8440)
              -.9560
Ans-1
             -1.0460
```

32. $\sec(-8.3429) \approx -2.1291$

```
cos(-8.3429)
              -.4697
Ans-1
             -2.1291
```

33. $\cot 6.0301 \approx -3.8665$

```
tan(6.0301)
              -.2586
Ans-1
             -3.8665
```

34. $\cot 3.8426 \approx 1.1848$

```
tan(3.8426)
               .8440
Ans-1
              1.1848
```

35. $\cos .8 \approx .7$

36. $\sin 4 \approx -.75$

37. $x = -.65$ when $\theta \approx 4$ radians.

38. $y = -.95$ when $\theta \approx 4.4$ radians.

39. $\cos 2$

$\frac{\pi}{2} \approx 1.57$ and $\pi \approx 3.14$, so $\frac{\pi}{2} < 2 < \pi$. Thus, an angle of 2 radians is in quadrant II. (The figure for Exercises 35 – 38 also shows that 2 radians is in quadrant II.) Because values of the cosine function are negative in quadrant II, cos 2 is negative.

40. $\sin(-1)$

$-\frac{\pi}{2} \approx -1.57$, so $-\frac{\pi}{2} < -1 < 0$. Thus, an angle of -1 radian is in quadrant IV. Because values of the sine function are negative in quadrant IV, $\sin(-1)$ is negative.

41. $\sin 5$

$\frac{3\pi}{2} \approx 4.71$ and $2\pi \approx 6.28$, so $\frac{3\pi}{2} < 5 < 2\pi$. Thus, an angle of 5 radians is in quadrant IV. (The figure for Exercises 35 – 38 also shows that 5 radians is in quadrant IV.) Because values of the sine function are negative in quadrant IV, sin 5 is negative.

42. $\cos 6$

$\frac{3\pi}{2} \approx 4.71$ and so $\frac{3\pi}{2} < 6 < 2\pi$. Thus, an angle of 6 radians is in quadrant IV. (The figure for Exercises 35 – 38 also shows that 6 radians is in quadrant IV.) Because values of the cosine function are positive in quadrant IV, cos 6 is positive.

43. $\tan 6.29$

$2\pi \approx 6.28$ and $2\pi + \frac{\pi}{2} = \frac{4\pi}{2} + \frac{\pi}{2} = \frac{5\pi}{2} \approx 7.85$, so $2\pi < 6.29 < \frac{5\pi}{2}$. Notice that 2π is coterminal with 0 and $\frac{5\pi}{2}$ is coterminal with $\frac{\pi}{2}$. Thus, an angle of 6.29 radians is in quadrant I. Because values of the tangent function are positive in quadrant I, $\tan 6.29$ is positive.

44. $\tan(-6.29)$

$-2\pi - \frac{\pi}{2} = -\frac{4\pi}{2} - \frac{\pi}{2} = -\frac{5\pi}{2} \approx -7.85$ and $-2\pi \approx -6.28$, so $-\frac{5\pi}{2} < -6.29 < -2\pi$. Notice that $\frac{5\pi}{2}$ is coterminal with $\frac{3\pi}{2}$ and -2π is coterminal with 0. Thus, an angle of -6.29 is in quadrant IV. Because values of the tangent function are negative in quadrant IV, $\tan(-6.29)$ is negative.

45. $\sin\theta = y = \frac{\sqrt{2}}{2}$

$\cos\theta = x = \frac{\sqrt{2}}{2}$

$\tan\theta = \frac{y}{x} = \frac{\frac{\sqrt{2}}{2}}{\frac{\sqrt{2}}{2}} = 1$

$\cot\theta = \frac{x}{y} = \frac{\frac{\sqrt{2}}{2}}{\frac{\sqrt{2}}{2}} = 1$

$\sec\theta = \frac{1}{x} = \frac{1}{\frac{\sqrt{2}}{2}} = \frac{2}{\sqrt{2}} = \frac{2}{\sqrt{2}} \cdot \frac{\sqrt{2}}{\sqrt{2}} = \sqrt{2}$

$\csc\theta = \frac{1}{y} = \frac{1}{\frac{\sqrt{2}}{2}} = \frac{2}{\sqrt{2}} = \frac{2}{\sqrt{2}} \cdot \frac{\sqrt{2}}{\sqrt{2}} = \sqrt{2}$

46. $\sin\theta = y = \frac{8}{17}$

$\cos\theta = x = -\frac{15}{17}$

$\tan\theta = \frac{y}{x} = \frac{\frac{8}{17}}{-\frac{15}{17}} = \frac{8}{17}\left(-\frac{17}{15}\right) = -\frac{8}{15}$

$\cot\theta = \frac{x}{y} = \frac{-\frac{15}{17}}{\frac{8}{17}} = -\frac{15}{17}\left(\frac{17}{8}\right) = -\frac{15}{8}$

$\sec\theta = \frac{1}{x} = \frac{1}{-\frac{15}{17}} = -\frac{17}{15}$

$\csc\theta = \frac{1}{y} = \frac{1}{\frac{8}{17}} = \frac{17}{8}$

47. $\sin\theta = y = -\frac{12}{13}$

$\cos\theta = x = \frac{5}{13}$

$\tan\theta = \frac{y}{x} = \frac{-\frac{12}{13}}{\frac{5}{13}} = -\frac{12}{13}\left(\frac{13}{5}\right) = -\frac{12}{5}$

$\cot\theta = \frac{x}{y} = \frac{\frac{5}{13}}{-\frac{12}{13}} = \frac{5}{13}\left(-\frac{13}{12}\right) = -\frac{5}{12}$

$\sec\theta = \frac{1}{x} = \frac{1}{\frac{5}{13}} = \frac{13}{5}$

$\csc\theta = \frac{1}{y} = \frac{1}{-\frac{12}{13}} = -\frac{13}{12}$

48. $\sin\theta = y = -\dfrac{1}{2}$

$\cos\theta = x = -\dfrac{\sqrt{3}}{2}$

$\tan\theta = \dfrac{y}{x} = \dfrac{-\frac{1}{2}}{-\frac{\sqrt{3}}{2}} = -\dfrac{1}{2}\left(-\dfrac{2}{\sqrt{3}}\right)$

$= \dfrac{1}{\sqrt{3}} = \dfrac{1}{\sqrt{3}} \cdot \dfrac{\sqrt{3}}{\sqrt{3}} = \dfrac{\sqrt{3}}{3}$

$\cot\theta = \dfrac{x}{y} = \dfrac{-\frac{\sqrt{3}}{2}}{-\frac{1}{2}} = -\dfrac{\sqrt{3}}{2}\left(-\dfrac{2}{1}\right) = \sqrt{3}$

$\sec\theta = \dfrac{1}{x} = \dfrac{1}{-\frac{\sqrt{3}}{2}} = -\dfrac{2}{\sqrt{3}}$

$= -\dfrac{2}{\sqrt{3}} \cdot \dfrac{\sqrt{3}}{\sqrt{3}} = -\dfrac{2\sqrt{3}}{3}$

$\csc\theta = \dfrac{1}{y} = \dfrac{1}{-\frac{1}{2}} = 1 \cdot \left(-\dfrac{2}{1}\right) = -2$

49. $\tan s = .2126$

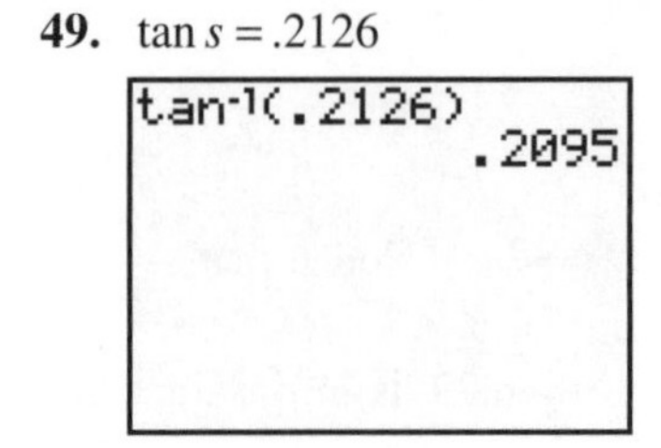

$s \approx .2095$

50. $\cos s = .7826$

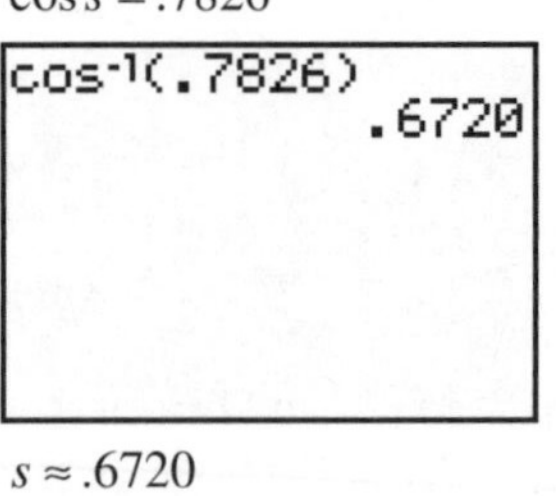

$s \approx .6720$

51. $\sin s = .9918$

```
sin⁻¹(.9918)
            1.4426
```

$s \approx 1.4426$

52. $\cot s = .2994$

```
1/.2994
            3.3400
tan⁻¹(Ans)
            1.2799
```

$s \approx 1.2799$

53. $\sec s = 1.0806$

```
1/1.0806
             .9254
cos⁻¹(Ans)
             .3887
```

$s \approx .3887$

54. $\csc s = 1.0219$

```
1/1.0219
             .9786
sin⁻¹(Ans)
            1.3634
```

$s \approx 1.3634$

55. $\left[\dfrac{\pi}{2}, \pi\right]$; $\sin s = \dfrac{1}{2}$

Recall that $\sin\dfrac{\pi}{6} = \dfrac{1}{2}$ and in quadrant II, sin s is positive. Therefore, $\sin\left(\pi - \dfrac{\pi}{6}\right) = \sin\dfrac{5\pi}{6} = \dfrac{1}{2}$, and thus, $s = \dfrac{5\pi}{6}$.

56. $\left[\frac{\pi}{2}, \pi\right]$; $\cos s = -\frac{1}{2}$

Recall that $\cos\frac{\pi}{3} = \frac{1}{2}$ and in quadrant II, cos s is negative. Therefore, $\cos\left(\pi - \frac{\pi}{3}\right) = \cos\frac{2\pi}{3} = -\frac{1}{2}$, and thus, $s = \frac{2\pi}{3}$.

57. $\left[\pi, \frac{3\pi}{2}\right]$; $\tan s = \sqrt{3}$

Recall that $\tan\frac{\pi}{3} = \sqrt{3}$ and in quadrant III, tan s is positive. Therefore, $\tan\left(\pi + \frac{\pi}{3}\right) = \tan\frac{4\pi}{3} = \sqrt{3}$, and thus, $s = \frac{4\pi}{3}$.

58. $\left[\pi, \frac{3\pi}{2}\right]$; $\sin s = -\frac{1}{2}$

Recall that $\sin\frac{\pi}{6} = \frac{1}{2}$ and in quadrant III, sin s is negative. Therefore, $\sin\left(\pi + \frac{\pi}{6}\right) = \sin\frac{7\pi}{6} = -\frac{1}{2}$, and thus, $s = \frac{7\pi}{6}$.

59. $\left[\frac{3\pi}{2}, 2\pi\right]$; $\tan s = -1$

Recall that $\tan\frac{\pi}{4} = 1$ and in quadrant IV, tan s is negative. Therefore, $\tan\left(2\pi - \frac{\pi}{4}\right) = \tan\frac{7\pi}{4} = -1$, and thus, $s = \frac{7\pi}{4}$.

60. $\left[\frac{3\pi}{2}, 2\pi\right]$; $\cos s = \frac{\sqrt{3}}{2}$

Recall that $\cos\frac{\pi}{6} = \frac{\sqrt{3}}{2}$ and in quadrant IV, cos s is positive. Therefore, $\cos\left(2\pi - \frac{\pi}{6}\right) = \cos\frac{11\pi}{6} = \frac{\sqrt{3}}{2}$, and thus, $s = \frac{11\pi}{6}$.

61. s = the length of an arc on the unit circle = 2.5

$x = \cos s \Rightarrow x = \cos 2.5$

$y = \sin s \Rightarrow y = \sin 2.5$

$(-.8011, .5985)$

```
cos(2.5)
            -.8011
sin(2.5)
             .5985
```

62. s = the length of an arc on the unit circle = 3.4

$x = \cos s \Rightarrow x = \cos 3.4$

$y = \sin s \Rightarrow y = \sin 3.4$

$(-.9668, -.2555)$

```
cos(3.4)
              -.9668
sin(3.4)
              -.2555
```

63. $s = -7.4$

$x = \cos s \Rightarrow x = \cos(-7.4)$

$y = \sin s \Rightarrow y = \sin(-7.4)$

$(.4385, -.8987)$

```
cos(-7.4)
               .4385
sin(-7.4)
              -.8987
```

64. $s = -3.9$

$x = \cos s \Rightarrow x = \cos(-3.9)$

$y = \sin s \Rightarrow y = \sin(-3.9)$

$(-.7259, .6878)$

```
cos(-3.9)
              -.7259
sin(-3.9)
               .6878
```

65. $s = 51$

```
cos(51)
               .7422
sin(51)
               .6702
```

Since cosine and sine are both positive, an angle of 51 radians lies in quadrant I.

66. $s = 49$

```
cos(49)
               .3006
sin(49)
              -.9538
```

Since cosine is positive and sine is negative, an angle of 49 radians lies in quadrant IV.

67. $s = 65$

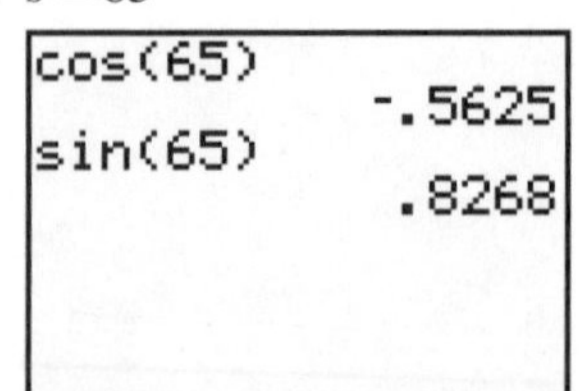

Since cosine is negative and sine is positive, an angle of 65 radians lies in quadrant II.

68. $s = 79$

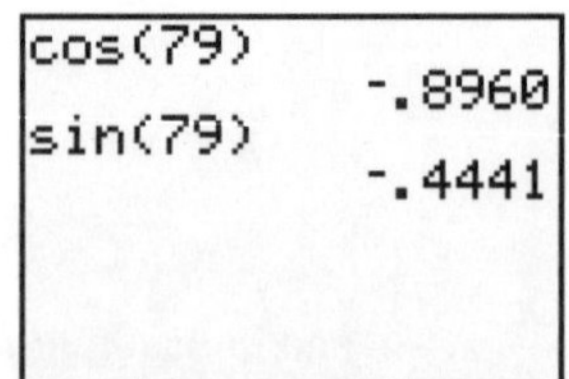

Since cosine and sine are both negative, an angle of 79 radians lies in quadrant III.

69. Since $x = \cos s$, $\cos s = .55319149$. Also, since $y = \sin s$, $\sin s = .83305413$.

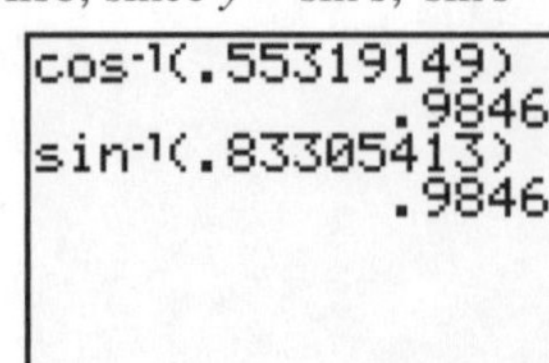

$s \approx .9846$

70. Since $x = \cos s$, $\cos s = -.9361702$.
Also, since $y = \sin s$, $\sin s = .35154709$. To determine s using sine, we must subtract $\sin^{-1}(.35154709)$ from π because s lies in quadrant II.

```
cos⁻¹(-.9361702)
              2.7824
π-sin⁻¹(.35154709
)
              2.7824
```

$s \approx 2.782$

71. To solve this problem make sure your calculator is in radian mode; $23.44° \approx .4091$ radians and $44.88° \approx .7833$ radians.

The shortest day can be found as follows.

$$\cos(.1309H) = -\tan D \tan L$$
$$\cos(.1309H) = -\tan(-.4091)\tan(.7833)$$
$$\cos(.1309H) \approx .4317$$
$$.1309H \approx \cos^{-1}(.4317)$$
$$.1309H \approx 1.1244$$
$$H \approx 8.6 \text{ hr}$$

The longest day can be found as follows.

$$\cos(.1309H) = -\tan D \tan L$$
$$\cos(.1309H) = -\tan(.4091)\tan(.7833)$$
$$\cos(.1309H) \approx -.4317$$
$$.1309H \approx 2.0172$$
$$H \approx 15.4 \text{ hr}$$

72. **(a)** New Orleans has a latitude of $L = 30°$ or .5236 radians. Since the day and time have not changed, $D \approx -.1425$ and $\omega \approx .7854$

$$\sin\theta = \cos D \cos L \cos\omega + \sin D \sin L$$
$$\sin\theta = \cos(-.1425)\cos(.5236)\cos(.7854) + \sin(-.1425)\sin(.5236)$$
$$\sin\theta \approx .5352$$

Thus, $\theta \approx \sin^{-1}(.5352) \approx .5647$ radians or 32.4°.

(b) Answers will vary.

73. $t = 60 - 30\cos\dfrac{x\pi}{6}$

(a) January: $x = 0$; $t = 60 - 30\cos\dfrac{0\cdot\pi}{6} = 60 - 30\cos 0 = 60 - 30(1) = 60 - 30 = 30°$

(b) April: $x = 3$; $t = 60 - 30\cos\dfrac{3\pi}{6} = 60 - 30\cos\dfrac{\pi}{2} = 60 - 30(0) = 60 - 0 = 60°$

(c) May: $x = 4$; $t = 60 - 30\cos\dfrac{4\pi}{6} = 60 - 30\cos\dfrac{2\pi}{3} = 60 - 30\left(-\dfrac{1}{2}\right) = 60 + 15 = 75°$

(d) June: $x = 5$; $t = 60 - 30\cos\dfrac{5\pi}{6} = 60 - 30\left(-\dfrac{\sqrt{3}}{2}\right) = 60 + 15\sqrt{3} \approx 86°$

(e) August: $x = 7$; $t = 60 - 30\cos\dfrac{7\pi}{6} = 60 - 30\left(-\dfrac{\sqrt{3}}{2}\right) = 60 + 15\sqrt{3} \approx 86°$

(f) October: $x = 9$; $t = 60 - 30\cos\dfrac{9\pi}{6} = 60 - 30\cos\dfrac{3\pi}{2} = 60 - 30(0) = 60 - 0 = 60°$

74. $T(x) = 37\sin\left[\frac{2\pi}{365}(x-101)\right]+25$

(a) March 1 (day 60)

$$T(60) = 37\sin\left[\frac{2\pi}{365}(60-101)\right]+25 \approx 1°F$$

(b) April 1 (day 91)

$$T(91) = 37\sin\left[\frac{2\pi}{365}(91-101)\right]+25 \approx 19°F$$

(c) Day 150

$$T(150) = 37\sin\left[\frac{2\pi}{365}(150-101)\right]+25 \approx 53°F$$

(d) June 15 is day 166 $(31+28+31+30+31+15=166)$.

$$T(166) = 37\sin\left[\frac{2\pi}{365}(166-101)\right]+25 \approx 58°F$$

(e) September 1 is day 244 $(31+28+31+30+31+30+31+31+1=244)$.

$$T(244) = 37\sin\left[\frac{2\pi}{365}(244-101)\right]+25 \approx 48°F$$

(f) October 31 is day 304 $(31+28+31+30+31+30+31+31+30+31=304)$.

$$T(304) = 37\sin\left[\frac{2\pi}{365}(304-101)\right]+25 \approx 12°F$$

Section 3:4: Linear and Angular Speed

1. The circumference of the unit circle is 2π.
$\omega = 1$ radian per sec, $\theta = 2\pi$ radians

$$\omega = \frac{\theta}{t} \Rightarrow 1 = \frac{2\pi}{t} \Rightarrow t = 2\pi \text{ sec}$$

2. The circumference of the unit circle is 2π.
$v = 1$ unit per sec, $s = 2\pi$ units

$$v = \frac{s}{t} \Rightarrow 1 = \frac{2\pi}{t} \Rightarrow t = 2\pi \text{ sec}$$

3. $r = 20$ cm, $\omega = \frac{\pi}{12}$ radian per sec, $t = 6$ sec

(a) $\omega = \frac{\theta}{t} \Rightarrow \frac{\pi}{12} = \frac{\theta}{6} \Rightarrow \theta = 6 \cdot \frac{\pi}{12} = \frac{\pi}{2}$ radians

(b) $s = r\theta \Rightarrow s = 20 \cdot \frac{\pi}{2} = 10\pi$ cm

(c) $v = \frac{r\theta}{t} \Rightarrow v = \frac{20 \cdot \frac{\pi}{2}}{6} = \frac{10\pi}{6} = \frac{5\pi}{3}$ cm per sec

4. $r = 30$ cm, $\omega = \frac{\pi}{10}$ radian per sec, $t = 4$ sec

(a) $\omega = \frac{\theta}{t} \Rightarrow \frac{\pi}{10} = \frac{\theta}{4} \Rightarrow \theta = 4 \cdot \frac{\pi}{10} = \frac{2\pi}{5}$ radians

(b) $s = r\theta \Rightarrow s = 30 \cdot \frac{2\pi}{5} = 12\pi$ cm

(c) $v = \frac{r\theta}{t} \Rightarrow v = \frac{30 \cdot \frac{2\pi}{5}}{4} = \frac{12\pi}{4} = 3\pi$ cm per sec

5. $\omega = \frac{2\pi}{3}$ radians per sec, $t = 3$ sec

$\omega = \frac{\theta}{t} \Rightarrow \frac{2\pi}{3} = \frac{\theta}{3} \Rightarrow \theta = 2\pi$ radians

6. $\omega = \frac{\pi}{4}$ radian per min, $t = 5$

$\omega = \frac{\theta}{t} \Rightarrow \frac{\pi}{4} = \frac{\theta}{5} \Rightarrow \theta = 5\left(\frac{\pi}{4}\right) = \frac{5\pi}{4}$ radians

7. $\theta = \frac{3\pi}{4}$ radians, $t = 8$ sec

$\omega = \frac{\theta}{t} \Rightarrow \theta = \frac{\frac{3\pi}{4}}{8} = \frac{3\pi}{4} \cdot \frac{1}{8} = \frac{3\pi}{32}$ radian per sec

8. $\theta = \frac{2\pi}{5}$ radians, $t = 10$ sec

$\omega = \frac{\theta}{t} \Rightarrow \omega = \frac{\frac{2\pi}{5}}{10} = \frac{2\pi}{5} \cdot \frac{1}{10} = \frac{\pi}{25}$ radian per sec

9. $\theta = \frac{2\pi}{9}$ radian, $\omega = \frac{5\pi}{27}$ radian per min

$\omega = \frac{\theta}{t} \Rightarrow \frac{5\pi}{27} = \frac{\frac{2\pi}{9}}{t} \Rightarrow \frac{5\pi}{27} = \frac{2\pi}{9t} \Rightarrow 45\pi t = 54\pi \Rightarrow t = \frac{54\pi}{45\pi} = \frac{6}{5}$ min

10. $\theta = \frac{3\pi}{8}$ radians, $\omega = \frac{\pi}{24}$ radian per min

$\omega = \frac{\theta}{t} \Rightarrow \frac{\pi}{24} = \frac{\frac{3\pi}{8}}{t} \Rightarrow \frac{\pi}{24} = \frac{3\pi}{8t} \Rightarrow 8\pi t = 72\pi \Rightarrow t = \frac{72\pi}{8\pi} = 9$ min

11. $\theta = 3.871142$ radians, $t = 21.4693$sec

$\omega = \frac{\theta}{t} \Rightarrow \omega = \frac{3.871142}{21.4693} \approx .180311$ radian per sec

12. $\omega = .90674$ radian per min, $t = 11.876$ min

$\omega = \frac{\theta}{t} \Rightarrow .90674 = \frac{\theta}{11.876} \Rightarrow \theta = (.90674)(11.876) \approx 10.768$ radians

13. $r = 12$ m, $\omega = \frac{2\pi}{3}$ radians per sec

$v = r\omega \Rightarrow v = 12\left(\frac{2\pi}{3}\right) = 8\pi$ m per sec

14. $r = 8$ cm, $\omega = \frac{9\pi}{5}$ radians per sec

$v = r\omega \Rightarrow v = 8\left(\frac{9\pi}{5}\right) = \frac{72\pi}{5}$ cm per sec

15. $v = 9$ m per sec, $r = 5$ m

$v = r\omega \Rightarrow 9 = 5\omega \Rightarrow \omega = \frac{9}{5}$ radians per sec

16. $v = 18$ ft per sec, $r = 3$ ft

$v = r\omega \Rightarrow 18 = 3\omega \Rightarrow \omega = 6$ radians per sec

17. $v = 107.692$ m per sec, $r = 58.7413$ m

$v = r\omega \Rightarrow 107.692 = 58.7413\omega \Rightarrow \omega = \frac{107.692}{58.7413} \approx 1.83333$ radians per sec

18. $r = 24.93215$ cm, $\omega = .372914$ radian per sec

$v = r\omega \Rightarrow v = (24.93215)(.372914) \approx 9.29755$ cm per sec

19. $r = 6$ cm, $\omega = \frac{\pi}{3}$ radians per sec, $t = 9$ sec

$s = r\omega t \Rightarrow s = 6\left(\frac{\pi}{3}\right)(9) = 18\pi$ cm

20. $r = 9$ yd, $\omega = \frac{2\pi}{5}$ radians per sec, $t = 12$ sec

$s = r\omega t \Rightarrow s = 9\left(\frac{2\pi}{5}\right)(12) = \frac{216\pi}{5}$ yd

21. $s = 6\pi$ cm, $r = 2$ cm, $\omega = \frac{\pi}{4}$ radian per sec

$s = r\omega t \Rightarrow 6\pi = 2\left(\frac{\pi}{4}\right)t \Rightarrow 6\pi = \left(\frac{\pi}{2}\right)t \Rightarrow t = 6\pi\left(\frac{2}{\pi}\right) = 12$ sec

22. $s = \frac{12\pi}{5}$ m, $r = \frac{3}{2}$ m, $\omega = \frac{2\pi}{5}$ radians per sec

$s = r\omega t \Rightarrow \frac{12\pi}{5} = \frac{3}{2}\left(\frac{2\pi}{5}\right)t \Rightarrow \frac{12\pi}{5} = \left(\frac{3\pi}{5}\right)t \Rightarrow t = \frac{12\pi}{5}\left(\frac{5}{3\pi}\right) = 4$ sec

23. $s = \frac{3\pi}{4}$ km, $r = 2$ km, $t = 4$ sec

$s = r\omega t \Rightarrow \frac{3\pi}{4} = 2(\omega)4 \Rightarrow \frac{3\pi}{4} = 8\omega \Rightarrow \omega = \frac{3\pi}{4} \cdot \frac{1}{8} = \frac{3\pi}{32}$ radian per sec

24. $s = \frac{8\pi}{9}$ m, $r = \frac{4}{3}$ m, $t = 12$ sec

$s = r\omega t \Rightarrow \frac{8\pi}{9} = \frac{4}{3}(\omega)12 \Rightarrow \frac{8\pi}{9} = 16\omega \Rightarrow \omega = \frac{8\pi}{9} \cdot \frac{1}{16} = \frac{\pi}{18}$ radian per sec

25. The hour hand of a clock moves through an angle of 2π radians (one complete revolution) in 12 hours, so $\omega = \frac{\theta}{t} = \frac{2\pi}{12} = \frac{\pi}{6}$ radian per hr.

26. The line makes 300 revolutions per minute. Each revolution is 2π radians, so we have the following.

$$\omega = 2\pi(300) = 600\pi \text{ radians per min}$$

27. The minute hand makes one revolution per hour. Each revolution is 2π radians. Thus, we have $\omega = 2\pi(1) = 2\pi$ radians per hr. Since there are 60 min in 1 hr, $\omega = \frac{2\pi}{60} = \frac{\pi}{30}$ radian per min.

28. The second hand of a clock makes one revolution per minute. Each revolution is 2π radians, so $\omega = 2\pi(1) = 2\pi$ radians per min. Since there are 60 sec in 1 min, $\omega = \frac{2\pi}{60} = \frac{\pi}{30}$ radian per sec.

29. The minute hand of a clock moves through an angle of 2π radians in 60 min, and at the tip of the minute hand, $r = 7$ cm, so we have the following.

$$v = \frac{r\theta}{t} \Rightarrow v = \frac{7(2\pi)}{60} = \frac{7\pi}{30} \text{ cm per min}$$

30. The second hand makes one revolution per minute. Each revolution is 2π radians, and at the tip of the second hand, $r = 28$ mm, so we have $v = r\omega \Rightarrow v = 28(2\pi) = 56\pi$ mm per min. Since there are 60 sec in 1 min, $v = \frac{56\pi}{60} = \frac{14\pi}{15}$ mm per sec.

31. The flywheel making 42 rotations per min turns through an angle $42(2\pi) = 84\pi$ radians in 1 min with $r = 2$m. Thus, we have the following.

$$v = \frac{r\theta}{t} \Rightarrow v = \frac{2(84\pi)}{1} = 168\pi \text{ m per min}$$

32. The point on the tread of the tire is rotating 35 times per min. Each rotation is 2π radians. Thus, we have $\omega = 35(2\pi) = 70\pi$ radians per min. Since $v = r\omega$, we have $v = 18(70\pi) = 1260\pi$ cm per min.

33. At 500 rotations per min, the propeller turns through an angle of $\theta = 500(2\pi) = 1000\pi$ radians in 1 min with $r = \frac{3}{2} = 1.5$ m, we have $v = \frac{r\theta}{t} \Rightarrow v = \frac{1.5(1000\pi)}{1} = 1500\pi$ m per min.

34. The point on the edge of the gyroscope is rotating 680 times per min. Each rotation is 2π radians. Since $\omega = 680(2\pi) = 1360\pi$ radians per min, we have the following.

$$v = r\omega \Rightarrow v = 83(1360\pi) = 112{,}880\pi \text{ cm per min}$$

35. At 200 revolutions per minute, the bicycle tire is moving 200 (2π) = 400π radians per min. This is the angular velocity ω. The linear velocity of the bicycle is $v = r\omega = 13(400\pi) = 5200\pi$ in. per min. To convert this to miles per hour, we find the following.

$$v = \frac{5200\pi \text{ in.}}{\text{min}} \cdot \frac{60 \text{ min}}{\text{hr}} \cdot \frac{1 \text{ ft}}{12 \text{ in.}} \cdot \frac{1 \text{ mi}}{5280 \text{ ft}} \approx 15.5 \text{ mph}$$

36. Mars will make one full rotation (of 2π radians) during the course of one day. Thus, we have the following.

$$2\pi \text{ radians}\left(\frac{1 \text{ hr}}{0.2552 \text{ radian}}\right) \approx 24.62 \text{ hr}$$

37. (a) $\theta = \frac{1}{365}(2\pi) = \frac{2\pi}{365}$ radian

(b) $\omega = \frac{2\pi}{365}$ radian per day $= \frac{2\pi}{365} \cdot \frac{1}{24}$ radian per hr $= \frac{\pi}{4380}$ radian per hr

(c) $v = r\omega \Rightarrow v = (93{,}000{,}000)\left(\frac{\pi}{4380}\right) \approx 66{,}700$ mph

38. (a) The earth completes one revolution per day, so it turns through $\theta = 2\pi$ radians in time $t = 1$ day $= 24$ hr. Thus, we have the following.

$$\omega = \frac{\theta}{t} \Rightarrow \omega = \frac{2\pi}{1} = 2\pi \text{ radians per day} = \omega = \frac{2\pi}{24} \text{ radian per hr} = \frac{\pi}{12} \text{ radian per hr}$$

(b) At the poles, $r = 0$, so $v = r\omega = 0$.

(c) At the equator, $r = 6400$ km. Thus, we have the following.

$$v = r\omega \Rightarrow v = 6400(2\pi) = 12{,}800\pi \text{ km per day} = \frac{12{,}800\pi}{24} \text{ km per hr} \approx 533\pi \text{ km per hr}$$

(d) Salem rotates about the axis in a circle of radius r at an angular velocity $\omega = 2\pi$ radians per day.

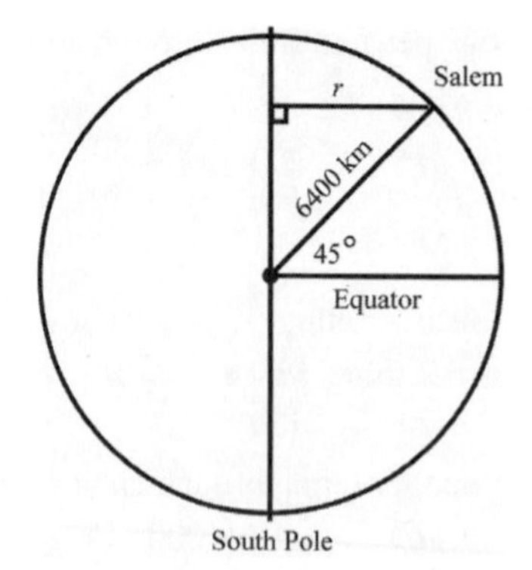

$$\sin 45° = \frac{r}{6400} \Rightarrow r = 6400 \sin 45° = 6400\left(\frac{\sqrt{2}}{2}\right) = 3200\sqrt{2} \text{ km}$$

and

$$v = r\omega \Rightarrow v = 3200\sqrt{2}(2\pi) \approx 9050\pi \text{ km per day} \approx 28{,}000 \text{ km per day}$$

$$v = \frac{9050\pi}{24} \text{ km per hr} \approx 377\pi \text{ km per hr} \approx 1200 \text{ km per hr}$$

39. (a) Since the 56 cm belt goes around in 18 sec, we have the following.

$$v = r\omega \Rightarrow \frac{56}{18} = (12.96)\omega \Rightarrow \frac{28}{9} = (12.96)\omega \Rightarrow \omega = \frac{\frac{28}{9}}{12.96} \approx .24 \text{ radian per sec}$$

(b) Since $s = 56$ cm of belt go around in $t = 18$ sec, the linear velocity is as follows.

$$v = \frac{s}{t} \Rightarrow v = \frac{56}{18} = \frac{28}{9} \approx 3.11 \text{ cm per sec}$$

40. The larger pulley rotates 25 times in 36 sec or $\frac{25}{36}$ times per sec. Thus, its angular velocity is $\omega = \frac{25}{36}(2\pi) = \frac{25\pi}{18}$ radians per sec. The linear velocity of the belt is the following.

$$v = r\omega \Rightarrow v = 15\left(\frac{25\pi}{18}\right) = \frac{125\pi}{6} \text{ cm per sec}$$

To find the angular velocity of the smaller pulley, use $v = \frac{125\pi}{6}$ cm per sec and $r = 8$ cm.

$$v = r\omega \Rightarrow \frac{125\pi}{6} = 8\omega \Rightarrow \omega = \frac{125\pi}{6}\left(\frac{1}{8}\right) = \frac{125\pi}{48} \text{ radians per sec}$$

41. $\omega = (152)(2\pi) = 304\pi$ radians per min $= \frac{304\pi}{60}$ radians per sec $= \frac{76\pi}{15}$ radians per sec

$$v = r\omega \Rightarrow 59.4 = r\left(\frac{76\pi}{15}\right) \Rightarrow r = 59.4\left(\frac{15}{76\pi}\right) \approx 3.73 \text{ cm}$$

42. Let s = the length of the track on the arc.

First, converting $40°$ to radians, we have $40° = 40\left(\frac{\pi}{180}\text{radian}\right) = \frac{2\pi}{9}\text{radian}.$

The length of track, s, is the arc length when $\theta = \frac{2\pi}{9}$ and $r = 1800$ ft.

$$s = r\theta \Rightarrow s = 1800\left(\frac{2\pi}{9}\right) = 400\pi \text{ ft}$$

Expressing the velocity $v = 30$ mph in ft per sec, we have the following.

$$v = 30 \text{ mph} = \frac{30}{3600} \text{ mi per sec} = \frac{(30)5280}{3600} \text{ ft per sec} = 44 \text{ ft per sec}$$

Since $v = 30 \text{ mph} = \frac{30}{3600}$ mi per sec $= \frac{(30)5280}{3600}$ ft per sec $= 44$ ft per sec, we have the following.

$$v = \frac{s}{t} \Rightarrow 44 = \frac{400\pi}{t} \Rightarrow 44t = 400\pi \Rightarrow t = \frac{400\pi}{44} = \frac{100\pi}{11} \approx 29 \text{ sec}$$

43. In one minute, the propeller makes 5000 revolutions. Each revolution is 2π radians, so we have $5000(2\pi) = 10{,}000\pi$ radians per min. Since there are 60 sec in a minute, we have the following.

$$\omega = \frac{10{,}000\pi}{60} = \frac{500\pi}{3} \approx 523.6 \text{ radians per sec}$$

44. $r = 5\text{ft};\ \omega = 25$ radians per sec

$v = r\omega \Rightarrow v = 5(25) = 125$ ft per sec

Chapter 3: Review Exercises

1. An angle of $1°$ is $\frac{1}{360}$ of the way around a circle. 1 radian is the measurement of an angle who vertex is at the center of a circle and intercepts an arc equal in length to the radius of the circle. 1 radian is $\frac{1}{2\pi}$ of the way around this circle. Since $\frac{1}{360} < \frac{1}{2\pi}$, the angular measurement of 1 radian is larger than $1°$.

2. **(a)** Since $\frac{\pi}{2} \approx 1.57$ and $\pi \approx 3.14$, we have $\frac{\pi}{2} < 3 < \pi$. Thus, the terminal side is in quadrant II.

(b) Since $\pi \approx 3.14$ and $\frac{3\pi}{2} \approx 4.71$, we have $\pi < 4 < \frac{3\pi}{2}$. Thus, the terminal side is in quadrant III.

(c) Since $-\frac{\pi}{2} \approx -1.57$ and $-\pi \approx -3.14$, we have $-\frac{\pi}{2} > -2 > -\pi$. Thus, the terminal side is in quadrant III.

(d) Since $2\pi \approx 6.28$ and $\frac{5\pi}{2} \approx 7.85$. we have $2\pi < 7 < \frac{5\pi}{2}$. Thus, the terminal side is in quadrant I.

3. To find a coterminal angle, add or subtract multiples of 2π. Three of the many possible answers are $1+2\pi$, $1+4\pi$, and $1+6\pi$.

4. To find a coterminal angle, add or subtract multiples of 2π. Since n represents any integer, the expression $\frac{\pi}{6} + 2n\pi$ would generate all coterminal angles with an angle of $\frac{\pi}{6}$ radian.

5. $45° = 45\left(\frac{\pi}{180}\text{ radian}\right) = \frac{\pi}{4}\text{ radian}$

6. $120° = 120\left(\frac{\pi}{180}\text{ radian}\right) = \frac{2\pi}{3}\text{ radians}$

7. $175° = 175\left(\frac{\pi}{180}\text{ radian}\right) = \frac{35\pi}{36}\text{ radians}$

8. $330° = 330\left(\frac{\pi}{180}\text{ radian}\right) = \frac{11\pi}{6}\text{ radians}$

9. $800° = 800\left(\frac{\pi}{180}\text{ radian}\right) = \frac{40\pi}{9}\text{ radians}$

10. $1020° = 1020\left(\frac{\pi}{180}\text{ radian}\right) = \frac{17\pi}{3}\text{ radians}$

11. $\frac{5\pi}{4} = \frac{5\pi}{4}\left(\frac{180°}{\pi}\right) = 225°$

12. $\frac{9\pi}{10} = \frac{9\pi}{10}\left(\frac{180°}{\pi}\right) = 162°$

13. $\frac{8\pi}{3} = \frac{8\pi}{3}\left(\frac{180°}{\pi}\right) = 480°$

14. $-\frac{6\pi}{5} = -\frac{6\pi}{5}\left(\frac{180°}{\pi}\right) = -216°$

15. $-\frac{11\pi}{18} = -\frac{11\pi}{18}\left(\frac{180°}{\pi}\right) = -110°$

16. $\frac{21\pi}{5} = \frac{21\pi}{5}\left(\frac{180°}{\pi}\right) = 756°$

17. Since $\frac{15}{60} = \frac{1}{4}$ rotation, we have $\theta = \frac{1}{4}(2\pi) = \frac{\pi}{2}$. Thus, $s = r\theta \Rightarrow s = 2\left(\frac{\pi}{2}\right) = \pi$ in.

18. Since $\frac{20}{60} = \frac{1}{3}$ rotation, we have $\theta = \frac{1}{3}(2\pi) = \frac{2\pi}{3}$. Thus, $s = r\theta \Rightarrow s = 2\left(\frac{2\pi}{3}\right) = \frac{4\pi}{3}$ in.

19. Since $\theta = 3(2\pi) = 6\pi,$ we have $s = r\theta \Rightarrow s = 2(6\pi) = 12\pi$ in.

20. Since $\theta = 10.5(2\pi) = 21\pi,$ we have $s = r\theta \Rightarrow s = 2(21\pi) = 42\pi$ in.

21. $r = 15.2$ cm, $\theta = \frac{3\pi}{4}$

$s = r\theta \Rightarrow s = 15.2\left(\frac{3\pi}{4}\right) = 11.4\pi \approx 35.8$ cm

22. $r = 11.4$ cm, $\theta = .769$

$s = r\theta \Rightarrow s = (11.4)(.769) \approx 8.77$ cm

23. $r = 8.973$ cm, $\theta = 49.06°$

First convert $\theta = 49.06°$ to radians.

$$\theta = 49.06° = 49.06\left(\frac{\pi}{180}\right) = \frac{49.06\pi}{180}$$

$$s = r\theta \Rightarrow s = 8.973\left(\frac{49.06\pi}{180}\right) \approx 7.683 \text{ cm}$$

24. $r = 28.69, \theta = \frac{7\pi}{4}$

$$A = \frac{1}{2}r^2\theta \Rightarrow A = \frac{1}{2}(28.69)^2\left(\frac{7\pi}{4}\right) = \frac{1}{2}(823.1161)\left(\frac{7\pi}{4}\right) \approx 2263 \text{ in.}^2$$

25. $r = 38.0\text{m}, \theta = 21°40'$

First convert $\theta = 21°40'$ to radians.

$$\theta = 21°40' = \left(21 + \tfrac{40}{60}\right)\left(\frac{\pi}{180}\right) = \left(21\tfrac{2}{3}\right)\left(\frac{\pi}{180}\right) = \frac{65}{3}\left(\frac{\pi}{180}\right) = \frac{13\pi}{108}$$

$$A = \frac{1}{2}r^2\theta \Rightarrow A = \frac{1}{2}(38.0)^2\left(\frac{13\pi}{108}\right) = \frac{1}{2}(1444)\left(\frac{13\pi}{108}\right) \approx 273\text{m}^2$$

26. Because the central angle is very small, the arc length is approximately equal to the length of the inscribed chord. (See directions for Exercises 25–26 in Section 3.2.)
Let h = height of tree, r = 2000, and $\theta = 1°10'$.

First, convert $\theta = 1°10'$ to radians.

$$\theta = 1°10' = \left(1 + \tfrac{10}{60}\right)° = \left(1\tfrac{1}{6}\right)° = \frac{7}{6}\left(\frac{\pi}{180}\text{radian}\right) = \frac{7\pi}{1080}\text{radian}$$

Thus, we have the following.

$$h \approx r\theta \Rightarrow h \approx 2000\left(\frac{7\pi}{1080}\right) \approx 41 \text{ yd}$$

27. 28°N, 12°S
12°N = −12°S

$\theta = 28° - (-12°) = 40° = 40\left(\frac{\pi}{180}\text{ radian}\right) = \frac{2\pi}{9}$ radian and $s = r\theta = 6400\left(\frac{2\pi}{9}\right) \approx 4500\text{km}$ (rounded to two significant digits)

28. 72°E, 35°W
35°W = −35°E

$\theta = 75° - (-35°) = 110° = 110\left(\frac{\pi}{180}\text{ radian}\right) = \frac{11\pi}{18}$ radian and $s = r\theta = 6400\left(\frac{11\pi}{18}\right) \approx 12{,}000\text{km}$
(rounded to two significant digits)

29. $s = 1.5,\ r = 2$

$$s = r\theta \Rightarrow 1.5 = 2\theta \Rightarrow \theta = \frac{1.5}{2} = \frac{3}{4}$$

$$A = \frac{1}{2}r^2\theta \Rightarrow A = \frac{1}{2}(2)^2\left(\frac{3}{4}\right) = \frac{1}{2}(4)\left(\frac{3}{4}\right) = \frac{3}{2} \text{ or } 1.5 \text{ sq units}$$

30. $s = 4,\ r = 8$

$$s = r\theta \Rightarrow 4 = 8\theta \Rightarrow \theta = \frac{4}{8} = \frac{1}{2}$$

$$A = \frac{1}{2}r^2\theta \Rightarrow A = \frac{1}{2}(8)^2\left(\frac{1}{2}\right) = \frac{1}{2}(64)\left(\frac{1}{2}\right) = 16 \text{ sq units}$$

31. (a) The hour hand of a clock moves through an angle of 2π radians in 12 hr, so we have the following.

$$\omega = \frac{2\pi}{12} = \frac{\pi}{6} \text{ radian per hr}$$

Since $\omega = \frac{\theta}{t} \Rightarrow \theta = t\omega$, in 2 hr the angle would be $2\left(\frac{\pi}{6}\right)$ or $\frac{\pi}{3}$ radians.

(b) The distance s the tip of the hour hand travels during the time period from 1 o'clock to 3 o'clock, is the arc length when $\theta = \frac{\pi}{3}$ and $r = 6$in. Thus, $s = r\theta \Rightarrow s = 6\left(\frac{\pi}{3}\right) = 2\pi$ in.

32. Answers will vary.

33. $\tan\frac{\pi}{3} = \sqrt{3}$

Converting $\frac{\pi}{3}$ to degrees we have, $\frac{\pi}{3} = \frac{1}{3}(180°) = 60°$.

$\tan\frac{\pi}{3} = \tan 60° = \sqrt{3}$

34. Find $\cos\frac{2\pi}{3}$.

Since $\frac{2\pi}{3}$ is in quadrant II, the reference angle is $\pi - \frac{2\pi}{3} = \frac{3\pi}{3} - \frac{2\pi}{3} = \frac{\pi}{3}$. In quadrant II, the cosine is negative. Thus, $\cos\frac{2\pi}{3} = -\cos\frac{\pi}{3} = -\frac{1}{2}$.

Converting $\frac{2\pi}{3}$ to degrees, we have $\frac{2\pi}{3} = \frac{2}{3}(180°) = 120°$. The reference angle is $180° - 120° = 60°$.

Thus, $\cos\frac{2\pi}{3} = \cos 120° = -\cos 60° = -\frac{1}{2}$.

35. Find $\sin\left(-\frac{5\pi}{6}\right)$.

$-\frac{5\pi}{6}$ is coterminal with $-\frac{5\pi}{6}+2\pi=-\frac{5\pi}{6}+\frac{12\pi}{6}=\frac{7\pi}{6}$. Since $\frac{7\pi}{6}$ is in quadrant III, the reference angle is $\frac{7\pi}{6}-\pi=\frac{7\pi}{6}-\frac{6\pi}{6}=\frac{\pi}{6}$. In quadrant III, the sine is negative. Thus, $\sin\left(-\frac{5\pi}{6}\right)=\sin\frac{7\pi}{6}=-\sin\frac{\pi}{6}=-\frac{1}{2}$.

Converting $\frac{7\pi}{6}$ to degrees, we have $\frac{7\pi}{6}=\frac{7}{6}(180°)=210°$. The reference angle is $210°-180°=30°$.

Thus, $\sin\left(-\frac{5\pi}{6}\right)=\sin\frac{7\pi}{6}=\sin 210°=-\sin 30°=-\frac{1}{2}$.

36. Find $\tan\left(-\frac{7\pi}{3}\right)$.

$-\frac{7\pi}{3}$ is coterminal with $-\frac{7\pi}{3}+2(2\pi)=-\frac{7\pi}{3}+4\pi=-\frac{7\pi}{3}+\frac{12\pi}{3}=\frac{5\pi}{3}$. Since $\frac{5\pi}{3}$ is in quadrant IV, the reference angle is $2\pi-\frac{5\pi}{3}=\frac{6\pi}{3}-\frac{5\pi}{3}=\frac{\pi}{3}$. In quadrant IV, the tangent is negative. Thus, $\tan\left(-\frac{7\pi}{3}\right)=\tan\frac{5\pi}{3}=-\tan\frac{\pi}{3}=-\sqrt{3}$.

Converting $\frac{5\pi}{3}$ to degrees, we have $\frac{5\pi}{3}=\frac{5}{3}(180°)=300°$. The reference angle is $360°-300°=60°$.

Thus, $\tan\left(-\frac{7\pi}{3}\right)=\tan\frac{5\pi}{3}=\tan 300°=-\tan 60°=-\sqrt{3}$.

37. Find $\csc\left(-\frac{11\pi}{6}\right)$.

$-\frac{11\pi}{6}$ is coterminal with $-\frac{11\pi}{6}+2\pi=-\frac{11\pi}{6}+\frac{12\pi}{6}=\frac{\pi}{6}$. Since $\frac{\pi}{6}$ is in quadrant I, we have $\csc\left(-\frac{11\pi}{6}\right)=\csc\frac{\pi}{6}=2$.

Converting $\frac{\pi}{6}$ to degrees, we have $\frac{\pi}{6}=\frac{1}{6}(180°)=30°$. Thus, $\csc\left(-\frac{11\pi}{6}\right)=\csc\frac{\pi}{6}=\csc 30°=2$.

38. Find $\cot\left(-\frac{17\pi}{3}\right)$.

$-\frac{17\pi}{3}$ is coterminal with $-\frac{17\pi}{3}+3(2\pi)=-\frac{17\pi}{3}+6\pi=-\frac{17\pi}{3}+\frac{18\pi}{3}=\frac{\pi}{3}$. Since $\frac{\pi}{3}$ is in quadrant I, we have $\cot\left(-\frac{17\pi}{3}\right)=\cot\frac{\pi}{3}=\frac{\sqrt{3}}{3}$.

Converting $\frac{\pi}{3}$ to degrees, we have $\frac{\pi}{3}=\frac{1}{3}(180°)=60°$. Thus, $\cot\left(-\frac{17\pi}{3}\right)=\cot\frac{\pi}{3}=\cot 60°=\frac{\sqrt{3}}{3}$.

39. Since $0<1<\frac{\pi}{2}$, sin 1 and cos 1 are both positive. Thus, tan 1 is positive. Also, since $\frac{\pi}{2}<2<\pi$, sin 2 is positive but cos 2 is negative. Thus, tan 2 is negative. Therefore, tan 1 > tan 2.

40. Since $0<1<\frac{\pi}{2}$, sin 1 and cos 1 are both positive. Also, $\cos 1<1\Rightarrow\frac{1}{\cos 1}>1$. Since $\sin 1>0$, we have $\frac{1}{\cos 1}\cdot\sin 1>1\cdot\sin 1\Rightarrow\frac{\sin 1}{\cos 1}>\sin 1\Rightarrow\tan 1>\sin 1$.

41. Since $\frac{\pi}{2}<2<\pi$, sin 2 is positive and cos 2 is negative. Thus, sin 2 > cos 2.

42. Since $\sin 0=0, \cos(\sin 0)=\cos 0=1$. Also, since $\cos 0=1, \sin(\cos 0)=\sin 1$. Since $0<1<\frac{\pi}{2}$, $0<\sin 1<1$. Thus, $\cos(\sin 0)>\sin(\cos 0)$.

43. $\sin 1.0472\approx .8660$

```
sin(1.0472)
          .8660
```

44. $\tan\ 1.2275\approx 2.7976$

```
tan(1.2275)
         2.7976
```

45. $\cos(-.2443)\approx .9703$

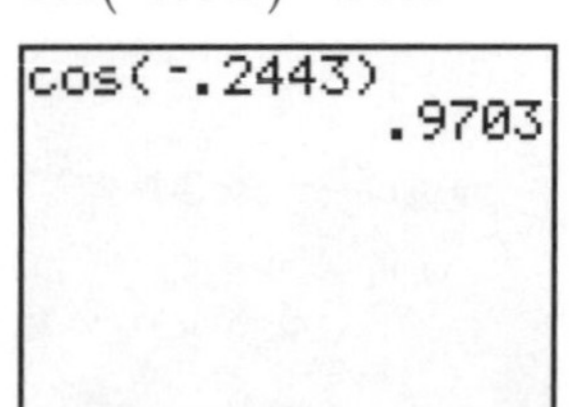

46. $\cot 3.0543\approx -11.4266$

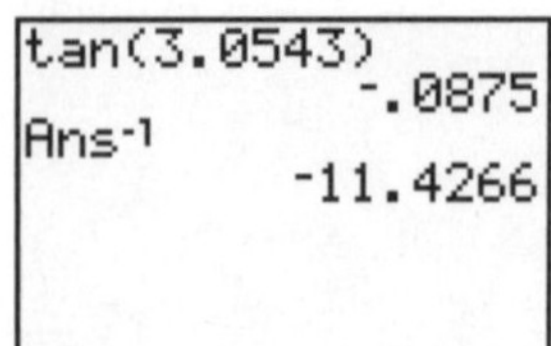

47. $\sec 7.3159\approx 1.9513$

```
cos(7.3159)
          .5125
Ans-1
         1.9513
```

48. $\csc 4.8386\approx -1.0080$

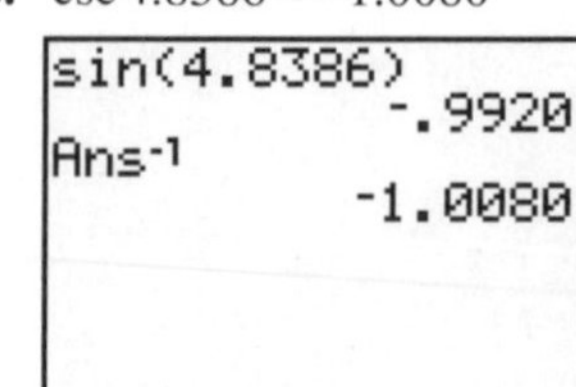

49. $\cos s=.9250$

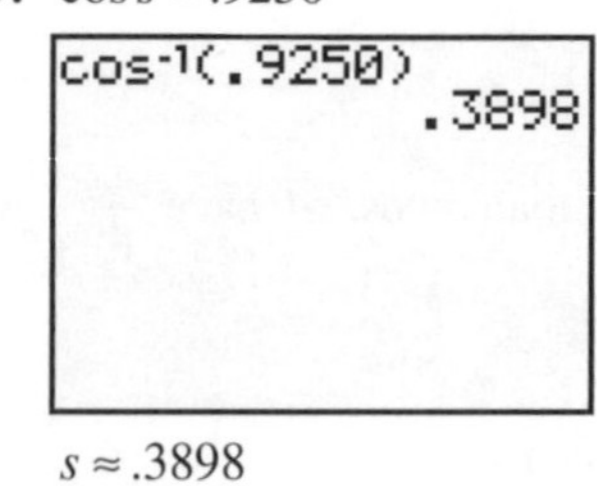

$s\approx .3898$

50. $\tan s=4.0112$

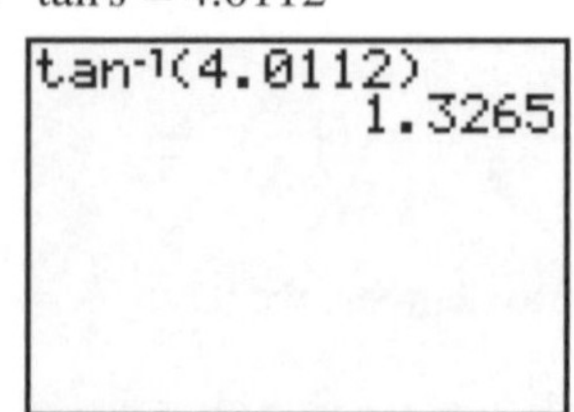

$s\approx 1.3265$

51. $\sin s = .4924$

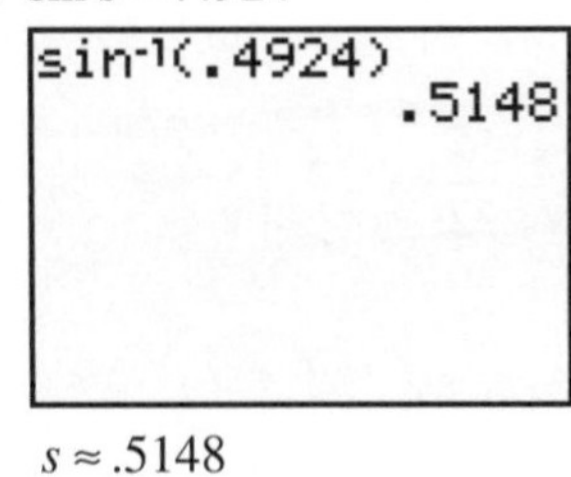

$s \approx .5148$

52. $\csc s = 1.2361$

```
1/1.2361
                .8090
sin⁻¹(Ans)
                .9424
```

$s \approx .9424$

53. $\cot s = .5022$

```
1/.5022
               1.9912
tan⁻¹(Ans)
               1.1054
```

$s \approx 1.1054$

54. $\sec s = 4.5600$

```
1/4.5600
                .2193
cos⁻¹(Ans)
               1.3497
```

$s \approx 1.3497$

55. $\left[0, \frac{\pi}{2}\right], \cos s = \frac{\sqrt{2}}{2}$

Because $\cos s = \frac{\sqrt{2}}{2}$, the reference angle for s must be $\frac{\pi}{4}$ since $\cos\frac{\pi}{4} = \frac{\sqrt{2}}{2}$. For s to be in the interval $\left[0, \frac{\pi}{2}\right]$, s must be the reference angle. Therefore $s = \frac{\pi}{4}$.

56. $\left[\frac{\pi}{2}, \pi\right]$, $\tan s = -\sqrt{3}$

Because $\tan s = -\sqrt{3}$, the reference angle for s must be $\frac{\pi}{3}$ since $\tan\frac{\pi}{3} = \sqrt{3}$. For s to be in the interval $\left[\frac{\pi}{2}, \pi\right]$, we must subtract the reference angle from π. Therefore, $s = \pi - \frac{\pi}{3} = \frac{2\pi}{3}$.

57. $\left[\pi, \frac{3\pi}{2}\right]$, $\sec s = -\frac{2\sqrt{3}}{3}$

Because $\sec s = -\frac{2\sqrt{3}}{3}$, the reference angle for s must be $\frac{\pi}{6}$ since $\sec\frac{\pi}{6} = \frac{2\sqrt{3}}{3}$. For s to be in the interval $\left[\pi, \frac{3\pi}{2}\right]$, we must add the reference angle to π. Therefore, $s = \pi + \frac{\pi}{6} = \frac{7\pi}{6}$.

58. $\left[\frac{3\pi}{2}, 2\pi\right]$, $\sin s = -\frac{1}{2}$

Because $\sin s = -\frac{1}{2}$, the reference angle for s must be $\frac{\pi}{6}$ since $\sin\frac{\pi}{6} = -\frac{1}{2}$. For s to be in the interval $\left[\frac{3\pi}{2}, 2\pi\right]$, we must subtract the reference angle from 2π. Therefore, $s = 2\pi - \frac{\pi}{6} = \frac{11\pi}{6}$.

59. $\theta=\frac{5\pi}{12},\ \omega=\frac{8\pi}{9}$ radians per sec

$$\omega=\frac{\theta}{t}\Rightarrow\frac{8\pi}{9}=\frac{\frac{5\pi}{12}}{t}\Rightarrow\frac{8\pi}{9}=\frac{5\pi}{12t}\Rightarrow 96\pi t=45\pi\Rightarrow t=\frac{45\pi}{96\pi}=\frac{15}{32}\text{sec}$$

60. $t=12$ sec, $\omega=9$ radians per sec

$$\omega=\frac{\theta}{t}\Rightarrow 9=\frac{\theta}{12}\Rightarrow\theta=108\text{ radians}$$

61. $t=8$ sec, $\theta=\frac{2\pi}{5}$ radians

$$\omega=\frac{\theta}{t}\Rightarrow\omega=\frac{\frac{2\pi}{5}}{8}=\frac{2\pi}{5}\left(\frac{1}{8}\right)=\frac{\pi}{20}\text{ radian per sec}$$

62. $s=\frac{12\pi}{25}$ ft, $r=\frac{3}{5}$ ft, $t=15$ sec

$$v=\frac{s}{t}\Rightarrow v=\frac{\frac{12\pi}{25}}{15}=\frac{12\pi}{25}\left(\frac{1}{15}\right)=\frac{12\pi}{375}$$

$$v=r\omega\Rightarrow\frac{12\pi}{375}=\frac{3}{5}\omega\Rightarrow\omega=\frac{5}{3}\left(\frac{12\pi}{375}\right)=\frac{4\pi}{75}\text{ radian per sec}$$

63. $r=11.46\text{cm},\ \omega=4.283$ radians per sec, $t=5.813$ sec

$s=r\theta=r\omega t=(11.46)(4.283)(5.813)\approx 285.3\text{ cm}$

64. The flywheel is rotating 90 times per sec or $90(2\pi)=180\pi$ radians per sec. Since $r=7$ cm, we have the following.

$$v=r\omega\Rightarrow v=7(180\pi)=1260\pi\text{ cm per sec}$$

65. Since $t=30$ sec and $\theta=\frac{5\pi}{6}$, radians we have $\omega=\frac{\theta}{t}\Rightarrow\omega=\frac{\frac{5\pi}{6}}{30}=\frac{5\pi}{6}\cdot\frac{1}{30}=\frac{\pi}{36}$ radian per sec.

66. $F(t)=\frac{1}{2}(1-\cos t)$

(a) $F(0)=\frac{1}{2}(1-\cos 0)=\frac{1}{2}(1-1)=\frac{1}{2}(0)=0$; The face of the moon is not visible.

(b) $F\left(\frac{\pi}{2}\right)=\frac{1}{2}\left(1-\cos\frac{\pi}{2}\right)=\frac{1}{2}(1-0)=\frac{1}{2}(1)=\frac{1}{2}$; Half the face of the moon is visible.

(c) $F(\pi)=\frac{1}{2}(1-\cos\pi)=\frac{1}{2}[1-(-1)]=\frac{1}{2}(2)=1$; The face of the moon is completely visible.

(d) $F\left(\frac{3\pi}{2}\right)=\frac{1}{2}\left(1-\cos\frac{3\pi}{2}\right)=\frac{1}{2}(1-0)=\frac{1}{2}(1)=\frac{1}{2}$; Half the face of the moon is visible.

67. **(a)** Because alternate interior angles of parallel lines with transversal have the same measure, the measure of angle *ABC* is equal to the measure of angle *BCD* (the angle of elevation). Moreover, triangle BAC is a right triangle and we can write the relation $\sin\theta = \frac{h}{d}$.

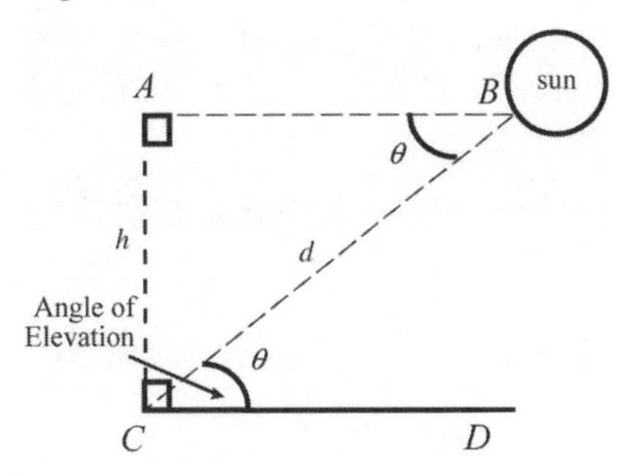

Solving for d we have, $\sin\theta = \frac{h}{d} \Rightarrow d\sin\theta = h \Rightarrow d = h\cdot\frac{1}{\sin\theta} \Rightarrow d = h\csc\theta.$

(b) Since $d = h\csc\theta$, we substitute $2h$ for *d* and solve.

$$2h = h\csc\theta \Rightarrow 2 = \csc\theta \Rightarrow \sin\theta = \frac{1}{2} \Rightarrow \theta = \frac{\pi}{6}$$

d is double *h* when the sun is $\frac{\pi}{6}$ radians (30°) above the horizon.

(c) $\csc\frac{\pi}{2} = 1$ and $\csc\frac{\pi}{3} = \frac{2\sqrt{3}}{3} \approx 1.15$

When the sun is lower in the sky $\left(\theta = \frac{\pi}{3}\right)$, sunlight is filtered by more atmosphere. There is less ultraviolet light reaching the earth's surface, and therefore, there is less likelihood of becoming sunburned. In this case, sunlight passes through 15% more atmosphere.

68. **(a)** $\sin\theta = \frac{y}{2.625} \Rightarrow y = 2.625\sin\theta$

(b) $\cos\theta = \frac{u}{2.625} \Rightarrow u = 2.625\cos\theta$

(c) $\cos\alpha = \frac{s}{10.5} \Rightarrow s = 10.5\cos\alpha$

(d) $x = u + s \Rightarrow x = 2.625\cos\theta + 10.5\cos\alpha$

(e) The maximum velocity of 21.6 mph occurs when $\theta = 4.94$ radians.

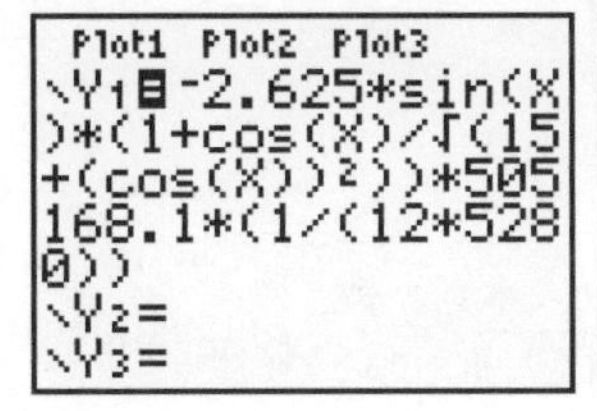

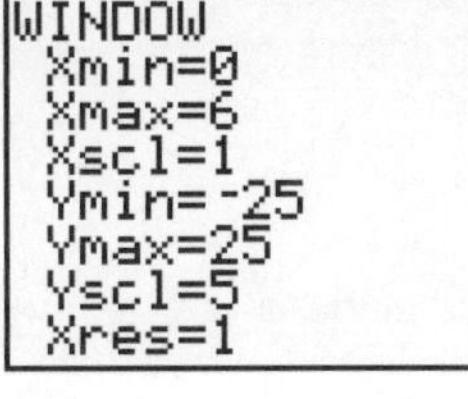

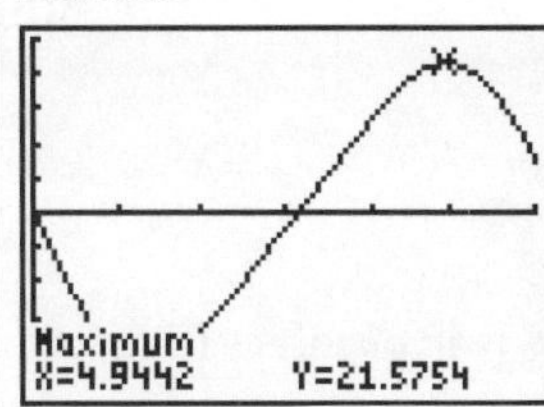

Chapter 3: Test

1. $120° = 120\left(\dfrac{\pi}{180}\text{ radian}\right) = \dfrac{2\pi}{3}\text{ radians}$

2. $-45° = -45\left(\dfrac{\pi}{180}\text{ radian}\right) = -\dfrac{\pi}{4}\text{ radian}$

3. $5° = 5\left(\dfrac{\pi}{180}\text{ radian}\right) = \dfrac{\pi}{36} \approx .09\text{ radian}$

4. $\dfrac{3\pi}{4} = \dfrac{3\pi}{4}\left(\dfrac{180°}{\pi}\right) = 135°$

5. $-\dfrac{7\pi}{6} = -\dfrac{7\pi}{6}\left(\dfrac{180°}{\pi}\right) = -210°$

6. $4 = 4\left(\dfrac{180°}{\pi}\right) \approx 229.18°$

7. $r = 150\text{ cm},\ s = 200\text{ cm}$

(a) $s = r\theta \Rightarrow 200 = 150\theta \Rightarrow \theta = \dfrac{200}{150} = \dfrac{4}{3}$

(b) $A = \dfrac{1}{2}r^2\theta \Rightarrow A = \dfrac{1}{2}(150)^2\left(\dfrac{4}{3}\right) = \dfrac{1}{2}(22{,}500)\left(\dfrac{4}{3}\right) = 15{,}000\text{ cm}^2$

8. $r = \dfrac{1}{2}\text{ in.},\ s = 1\text{ in.}$

$s = r\theta \Rightarrow 1 = \dfrac{1}{2}\theta \Rightarrow \theta = 2\text{ radians}$

For Exercises 9 – 14, use the following copy of the figure on pages 109 and 123 of your text.

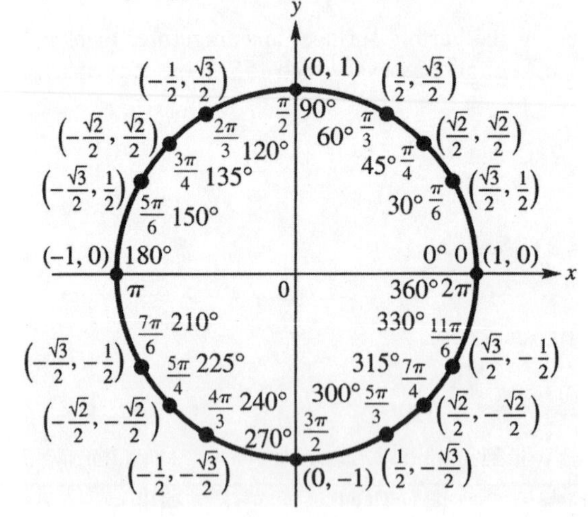

9. Find $\sin\dfrac{3\pi}{4}$.

Since $\dfrac{3\pi}{4}$ is in quadrant II, the reference angle is $\pi - \dfrac{3\pi}{4} = \dfrac{4\pi}{4} - \dfrac{3\pi}{4} = \dfrac{\pi}{4}$. In quadrant II, the sine is positive. Thus, $\sin\dfrac{3\pi}{4} = \sin\dfrac{\pi}{4} = \dfrac{\sqrt{2}}{2}$.

Converting $\dfrac{3\pi}{4}$ to degrees, we have $\dfrac{3\pi}{4} = \dfrac{3}{4}(180°) = 135°$. The reference angle is $180° - 135° = 45°$.

Thus, $\sin\dfrac{3\pi}{4} = \sin 135° = \sin 45° = \dfrac{\sqrt{2}}{2}$.

10. Find $\cos\left(-\frac{7\pi}{6}\right)$.

$-\frac{7\pi}{6}$ is coterminal with $-\frac{7\pi}{6}+2\pi=-\frac{7\pi}{6}+\frac{12\pi}{6}=\frac{5\pi}{6}$. Since $\frac{5\pi}{6}$ is in quadrant II, the reference angle is $\pi-\frac{5\pi}{6}=\frac{6\pi}{6}-\frac{5\pi}{6}=\frac{\pi}{6}$. In quadrant II, the cosine is negative. Thus, $\cos\left(-\frac{7\pi}{6}\right)=\cos\frac{5\pi}{6}=-\cos\frac{\pi}{6}=-\frac{\sqrt{3}}{2}$.

Converting $\frac{5\pi}{6}$ to degrees, we have $\frac{5\pi}{6}=\frac{5}{6}(180°)=150°$. The reference angle is $180°-150°=30°$.

Thus, $\cos\left(-\frac{7\pi}{6}\right)=\cos\frac{5\pi}{6}=\cos 150°=-\cos 30°=-\frac{\sqrt{3}}{2}$.

11. $\tan\frac{3\pi}{2}=\tan 270°$ is undefined.

12. Find $\sec\frac{8\pi}{3}$

$\frac{8\pi}{3}$ is coterminal with $\frac{8\pi}{3}-2\pi=\frac{8\pi}{3}-\frac{6\pi}{3}=\frac{2\pi}{3}$. Since $\frac{2\pi}{3}$ is in quadrant II, the reference angle is $\pi-\frac{2\pi}{3}=\frac{3\pi}{3}-\frac{2\pi}{3}=\frac{\pi}{3}$. In quadrant II, the secant is negative. Thus, $\sec\frac{8\pi}{3}=\sec\frac{2\pi}{3}=-\sec\frac{\pi}{3}=-2$.

Converting $\frac{2\pi}{3}$ to degrees, we have $\frac{2\pi}{3}=\frac{2}{3}(180°)=120°$. The reference angle is $180°-120°=60°$.

Thus, $\sec\frac{8\pi}{3}=\sec\frac{2\pi}{3}=\sec 120°=-\sec 60°=-2$.

13. $\tan\pi=\tan 180°=0$

14. $\cos\frac{3\pi}{2}=\cos 270°=0$

15. $s=\frac{7\pi}{6}$

Since $\frac{7\pi}{6}$ is in quadrant III, the reference angle is $\frac{7\pi}{6}-\pi=\frac{7\pi}{6}-\frac{6\pi}{6}=\frac{\pi}{6}$. In quadrant III, the sine and cosine are negative.

$\sin\frac{7\pi}{6}=-\sin\frac{\pi}{6}=-\frac{1}{2}$

$\cos\frac{7\pi}{6}=-\cos\frac{\pi}{6}=-\frac{\sqrt{3}}{2}$

$\tan\frac{7\pi}{6}=\tan\frac{\pi}{6}=\frac{\sqrt{3}}{3}$

16. For any point, (x,y), on the unit circle, we have $\tan s=\frac{y}{x}$ and $\sec s=\frac{1}{x}$. Thus, for any point on the unit circle where $x=0$, these two functions will be undefined. This occurs at the following.

$$s=\ldots,-\frac{5\pi}{2},-\frac{3\pi}{2},-\frac{\pi}{2},\frac{\pi}{2},\frac{3\pi}{2},\frac{5\pi}{2},\ldots$$

Therefore, the domains of the tangent and secant functions are $\{s\,|\,s\neq(2n+1)\frac{\pi}{2}$, where n is any integer$\}$.

17. (a) $\sin s = .8258$

```
sin⁻¹(.8258)
           .9716
```

$s \approx .9716$

(b) Since $\cos\frac{\pi}{3} = \frac{1}{2}$ and $0 \le \frac{\pi}{3} \le \frac{\pi}{2}$,

$$s = \frac{\pi}{3}.$$

18. (a) The speed of ray OP is $\omega = \frac{\pi}{12}$ radian per sec. Since $\omega = \frac{\theta}{t}$, then in 8 sec we have the following.

$$\omega = \frac{\theta}{t} \Rightarrow \frac{\pi}{12} = \frac{\theta}{8} \Rightarrow \theta = \frac{8\pi}{12} = \frac{2\pi}{3} \text{ radians}$$

(b) From part (a), P generates an angle of $\frac{2\pi}{3}$ radians in 8 sec. The distance traveled by P along the circle is $s = r\theta \Rightarrow s = 60\left(\frac{2\pi}{3}\right) = 40\pi$ cm.

(c) $v = \frac{s}{t} \Rightarrow \frac{40\pi}{8} = 5\pi$ cm per sec.

19. $r = 483{,}600{,}000$ mi

In 11.64 years, Jupiter travels 2π radians.

$$\omega = \frac{\theta}{t} \Rightarrow \omega = \frac{2\pi}{11.64}$$

$$v = r\omega \Rightarrow v = 483{,}600{,}000\left(\frac{2\pi}{11.64}\right) \approx 261{,}043{,}678.2 \text{ miles per year}$$

Next, convert this to miles per second.

$$v = \left(\frac{261{,}043{,}678.2 \text{ miles}}{\text{year}}\right)\left(\frac{1 \text{ year}}{365 \text{ days}}\right)\left(\frac{1 \text{ day}}{24 \text{ hours}}\right)\left(\frac{1 \text{ hour}}{60 \text{ minutes}}\right)\left(\frac{1 \text{ minute}}{60 \text{ seconds}}\right) \approx 8.278 \text{ mi per sec}$$

20. (a) Suppose the person takes a seat at point A. When they travel $\frac{\pi}{2}$ radians, they are 50 ft above the ground. When they travel $\frac{\pi}{6}$ more radians, we can let x be the additional vertical distance traveled.

$$\sin\frac{\pi}{6} = \frac{x}{50} \Rightarrow x = 50\sin\frac{\pi}{6} = 50 \cdot \frac{1}{2} = 25$$

Thus, an additional 25 ft above ground it traveled, for a total of 75 ft above ground.

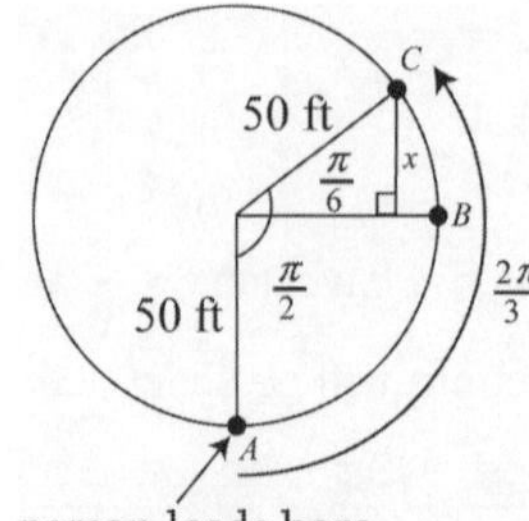

person loads here

(b) Ferris wheel goes $\frac{2\pi}{3}$ radians per 30 sec or $\frac{2\pi}{90}$ radian per sec.

Chapter 3: Quantitative Reasoning

1. Triangle *RQP* is similar to triangle *RMO* because angle *R* = angle *R* and angle *Q* = angle *M*.

2. $\frac{r}{c} = \frac{\frac{c}{2}}{b} \Rightarrow \frac{r}{c} = \frac{c}{2b} \Rightarrow r = \frac{c^2}{2b}$

3. $a^2 + b^2 = c^2$; Thus, we have $r = \frac{a^2 + b^2}{2b}$.

4. In the diagram, $a = 1.4$ in. and $b = .2$ in. Since $r = \frac{a^2 + b^2}{2b}$, we have the following.

$$r = \frac{(1.4)^2 + (.2)^2}{2(.2)} = \frac{1.96 + .04}{.4} = \frac{2.00}{.4} = 5$$

Thus, the radius is 5 in.

Chapter 4
GRAPHS OF THE CIRCULAR FUNCTIONS

Section 4.1: Graphs of the Sine and Cosine Functions

Connections (page 140)

1. $X = -0.4161468,\ Y = 0.90929743,\ X$ is cos2 and Y is sin2

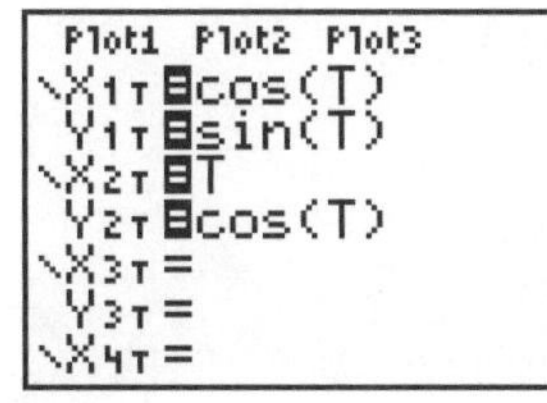

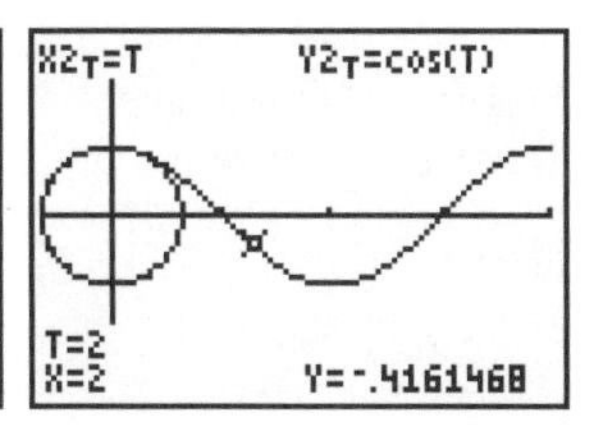

2. $X = 1.9, Y = 0.94630009;$

$\sin 1.9 = 0.94630009$

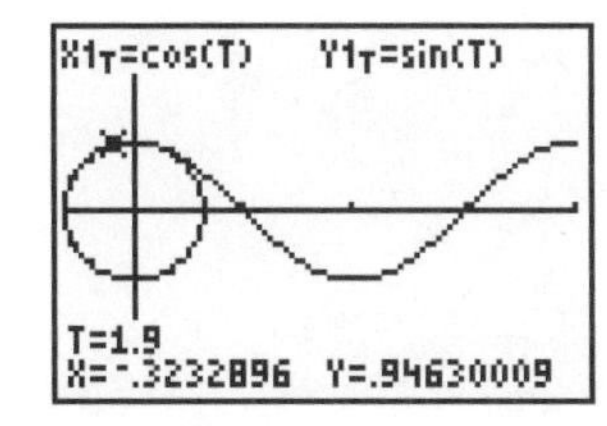

3. $X = 1.9, Y = -0.3232896;$

$\cos 1.9 = -0.3232896$

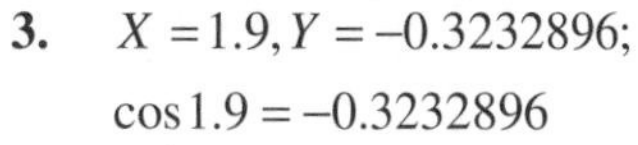

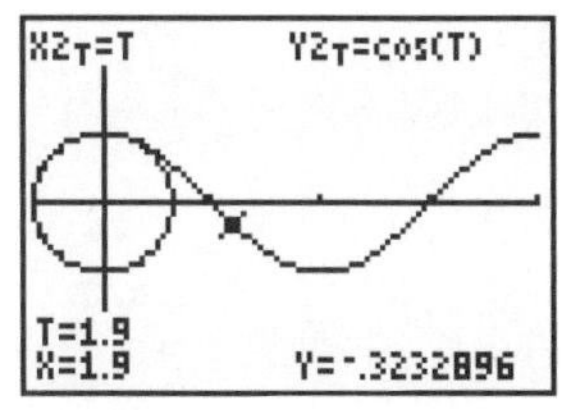

Exercises

1. $y = \sin x$

The graph is a sinusoidal curve with amplitude 1 and period 2π.

Since sin 0 = 0, the point $(0,0)$ is on the graph. This matches with graph G.

2. $y = \cos x$

The graph is a sinusoidal curve with amplitude 1 and period 2π.

Since $\cos 0 = 1$, the point $(0,1)$ is on the graph. This matches with graph A.

3. $y = -\sin x$

The graph is a sinusoidal curve with amplitude 1 and period 2π.

Because $a = -1$, the graph is a reflection of $y = \sin x$ in the x-axis. This matches with graph E.

4. $y = -\cos x$

The graph is a sinusoidal curve with amplitude 1 and period 2π.

Because $a = -1$, the graph is a reflection of $y = \cos x$ in the x-axis. This matches with graph D.

5. $y = \sin 2x$

The graph is a sinusoidal curve with a period of π and an amplitude of 1.

Since $\sin(2\cdot 0) = \sin 0 = 0$, the point $(0,0)$ is on the graph. This matches with graph B.

6. $y = \cos 2x$

The graph is a sinusoidal curve with an amplitude of 1 and a period of π.

Since $\cos(2\cdot 0) = \cos 0 = 1$, the point $(0,1)$ is on the graph. This matches with graph H.

7. $y = 2\sin x$

The graph is a sinusoidal curve with amplitude 2 and period 2π.

Since $2\sin 0 = 2\cdot 0 = 0$, the point $(0,0)$ is on the graph. This matches with graph F.

8. $y = 2\cos x$
The graph is a sinusoidal curve with period 2π and amplitude 2. Since $2\cos 0 = 2 \cdot 1 = 2$, the point $(0, 2)$ is on the graph. This matches with graph C.

9. $y = 2\cos x$
Amplitude: $|2| = 2$

x	0	$\frac{\pi}{2}$	π	$\frac{3\pi}{2}$	2π
$\cos x$	1	0	-1	0	1
$2\cos x$	2	0	-2	0	2

This table gives five values for graphing one period of the function. Repeat this cycle for the interval $[-2\pi, 0]$.

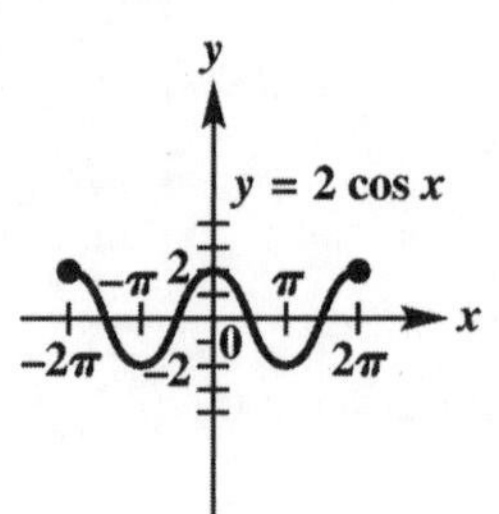

10. $y = 3\sin x$
Amplitude: $|3| = 3$

x	0	$\frac{\pi}{2}$	π	$\frac{3\pi}{2}$	2π
$\sin x$	0	1	0	-1	0
$3\sin x$	0	3	0	-3	0

This table gives five values for graphing one period of $y = 3\sin x$. Repeat this cycle for the interval $[-2\pi, 0]$.

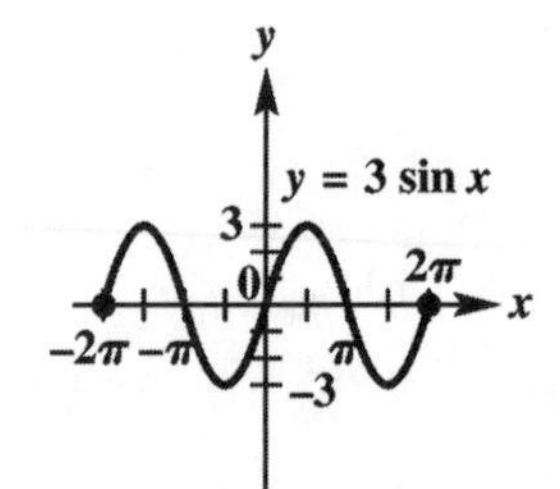

11. $y = \frac{2}{3}\sin x$

Amplitude: $\left|\frac{2}{3}\right| = \frac{2}{3}$

x	0	$\frac{\pi}{2}$	π	$\frac{3\pi}{2}$	2π
$\sin x$	0	1	0	-1	0
$\frac{2}{3}\sin x$	0	$\frac{2}{3} \approx .7$	0	$-\frac{2}{3} \approx -.7$	0

This table gives five values for graphing one period of $y = \frac{2}{3}\sin x$. Repeat this cycle for the interval $[-2\pi, 0]$.

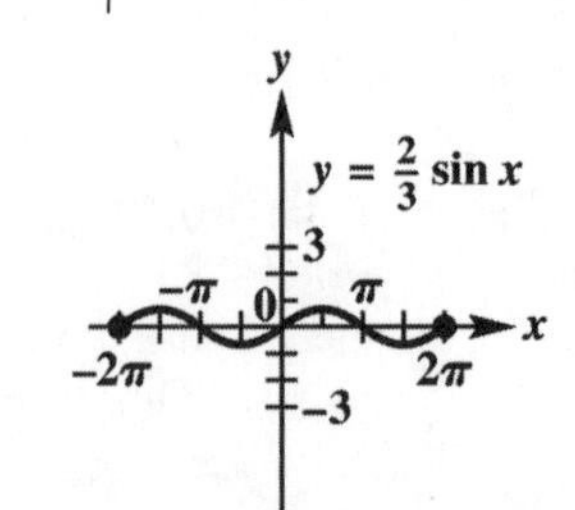

12. $y = \frac{3}{4}\cos x$

Amplitude: $\left|\frac{3}{4}\right| = \frac{3}{4}$

x	0	$\frac{\pi}{2}$	π	$\frac{3\pi}{2}$	2π
$\cos x$	1	0	-1	0	1
$\frac{3}{4}\cos x$	$\frac{3}{4}$	0	$-\frac{3}{4}$	0	$\frac{3}{4}$

This table gives five values for graphing one period of $y = \frac{3}{4}\cos x$. Repeat this cycle for the interval $[-2\pi, 0]$.

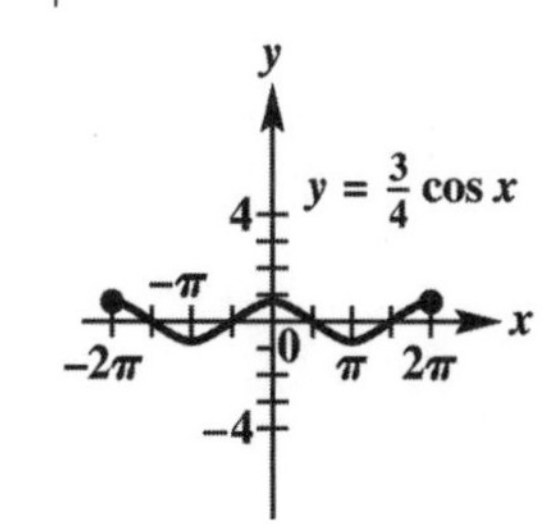

13. $y = -\cos x$

Amplitude: $|-1| = 1$

x	0	$\frac{\pi}{2}$	π	$\frac{3\pi}{2}$	2π
$\cos x$	1	0	-1	0	1
$-\cos x$	-1	0	1	0	-1

This table gives five values for graphing one period of $y = -\cos x$. Repeat this cycle for the interval $[-2\pi, 0]$.

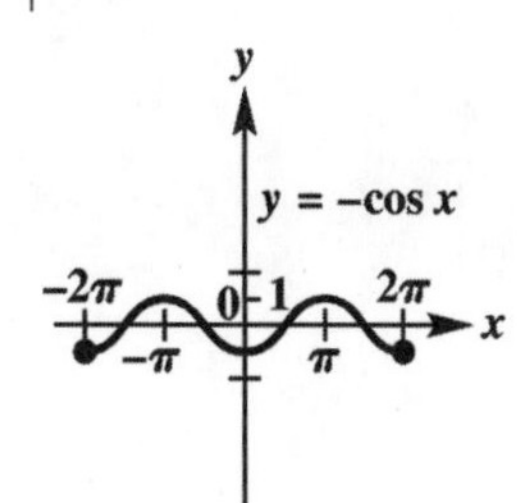

14. $y = -\sin x$

Amplitude: $|-1| = 1$

x	0	$\frac{\pi}{2}$	π	$\frac{3\pi}{2}$	2π
$\sin x$	0	1	0	-1	0
$-\sin x$	0	-1	0	1	0

This table gives five values for graphing one period of $y = -\sin x$. Repeat this cycle for the interval $[-2\pi, 0]$.

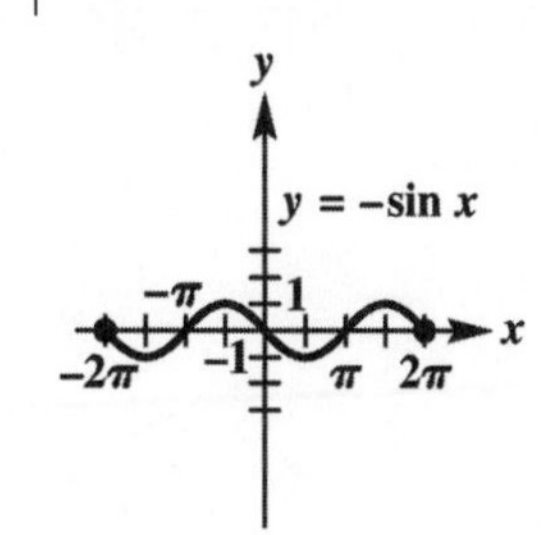

15. $y = -2 \sin x$

Amplitude: $|-2| = 2$

x	0	$\frac{\pi}{2}$	π	$\frac{3\pi}{2}$	2π
$\sin x$	0	1	0	−1	0
$-2 \sin x$	0	−2	0	2	0

This table gives five values for graphing one period of $y = -2\sin x$. Repeat this cycle for the interval $[-2\pi, 0]$.

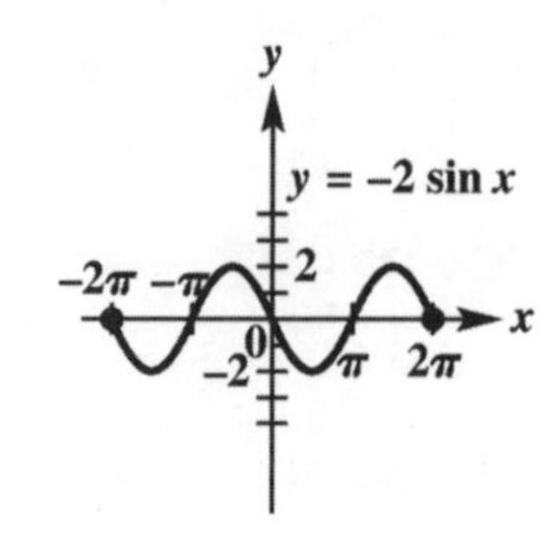

16. $y = -3 \cos x$

Amplitude: $|-3| = 3$

x	0	$\frac{\pi}{2}$	π	$\frac{3\pi}{2}$	2π
$\cos x$	1	0	−1	0	1
$-3 \cos x$	−3	0	3	0	−3

This table gives five values for graphing one period of $y = -3\cos x$. Repeat this cycle for the interval $[-2\pi, 0]$.

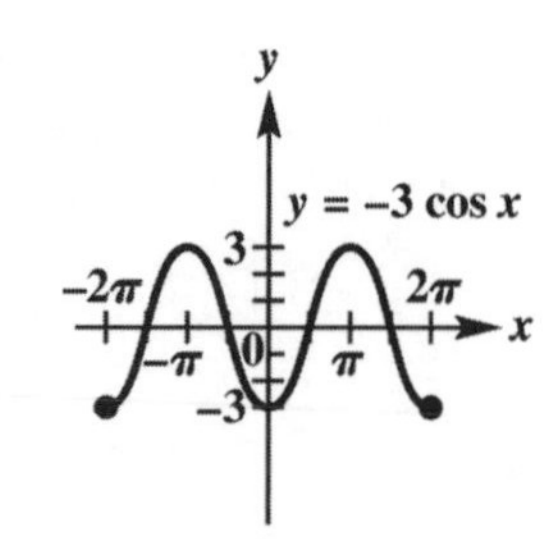

17. $y = \sin \frac{1}{2} x$

Period: $\frac{2\pi}{\frac{1}{2}} = 4\pi$ and Amplitude: $|1| = 1$

Divide the interval $[0, 4\pi]$ into four equal parts to get x-values that will yield minimum and maximum points and x-intercepts. Then make a table. Repeat this cycle for the interval $[-4\pi, 0]$.

x	0	π	2π	3π	4π
$\frac{1}{2}x$	0	$\frac{\pi}{2}$	π	$\frac{3\pi}{2}$	2π
$\sin \frac{1}{2}x$	0	1	0	−1	0

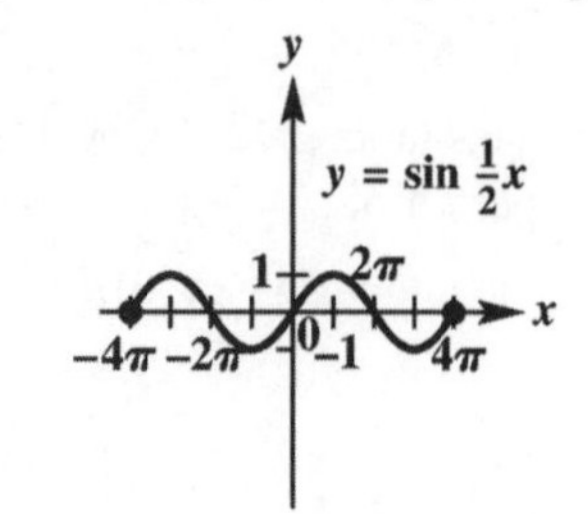

18. $y = \sin\frac{2}{3}x$

Period: $\frac{2\pi}{\frac{2}{3}} = 2\pi \cdot \frac{3}{2} = 3\pi$ and Amplitude: $|1| = 1$

Divide the interval $[0, 3\pi]$ into four equal parts to get the x-values that will yield minimum and maximum points and x-intercepts. Then make a table. Repeat this cycle for the interval $[-3\pi, 0]$.

x	0	$\frac{3\pi}{4}$	$\frac{3\pi}{2}$	$\frac{9\pi}{4}$	3π
$\frac{2}{3}x$	0	$\frac{\pi}{2}$	π	$\frac{3\pi}{2}$	2π
$\sin\frac{2}{3}x$	0	1	0	-1	0

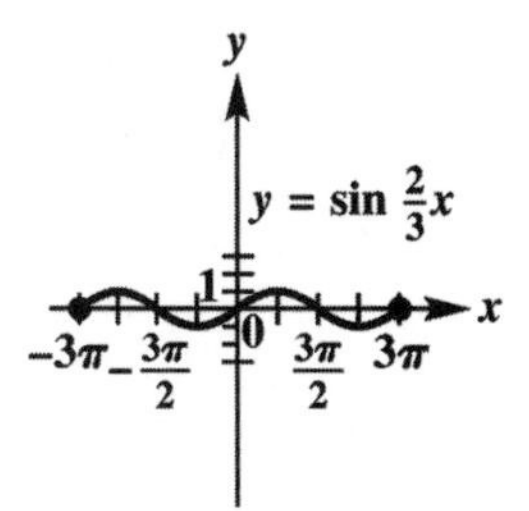

19. $y = \cos\frac{3}{4}x$

Period: $\frac{2\pi}{\frac{3}{4}} = 2\pi \cdot \frac{4}{3} = \frac{8\pi}{3}$ and Amplitude: $|1| = 1$

Divide the interval $\left[0, \frac{8\pi}{3}\right]$ into four equal parts to get the x-values that will yield minimum and maximum points and x-intercepts. Then make a table. Repeat this cycle for the interval $\left[-\frac{8\pi}{3}, 0\right]$.

x	0	$\frac{2\pi}{3}$	$\frac{4\pi}{3}$	2π	$\frac{8\pi}{3}$
$\frac{3}{4}x$	0	$\frac{\pi}{2}$	π	$\frac{3\pi}{2}$	2π
$\cos\frac{3}{4}x$	1	0	-1	0	1

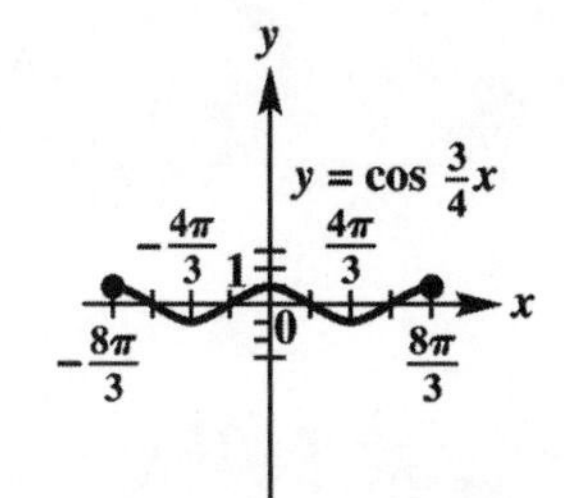

20. $y = \cos\frac{1}{3}x$

Period: $\frac{2\pi}{\frac{1}{3}} = 2\pi \cdot \frac{3}{1} = 6\pi$ and Amplitude: $|1| = 1$

Divide the interval $[0, 6\pi]$ into four equal parts to get the x-values that will yield minimum and maximum points and x-intercepts. Then make a table. Repeat this cycle for the interval $[-6\pi, 0]$.

x	0	$\frac{3\pi}{2}$	3π	$\frac{9\pi}{2}$	6π
$\frac{1}{3}x$	0	$\frac{\pi}{2}$	π	$\frac{3\pi}{2}$	2π
$\cos\frac{1}{3}x$	1	0	-1	0	1

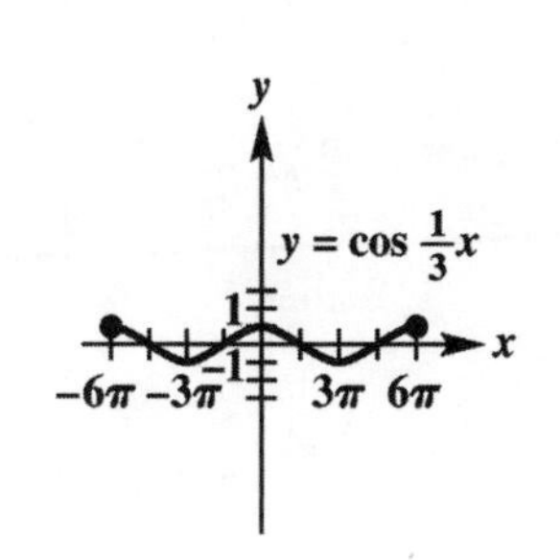

21. $y = \sin 3x$

Period: $\frac{2\pi}{3}$ and Amplitude: $|1| = 1$

Divide the interval $\left[0, \frac{2\pi}{3}\right]$ into four equal parts to get the x-values that will yield minimum and maximum points and x-intercepts. Then make a table. Repeat this cycle for the interval $\left[-\frac{2\pi}{3}, 0\right]$.

x	0	$\frac{\pi}{6}$	$\frac{\pi}{3}$	$\frac{\pi}{2}$	$\frac{2\pi}{3}$
$3x$	0	$\frac{\pi}{2}$	π	$\frac{3\pi}{2}$	2π
$\sin 3x$	0	1	0	-1	0

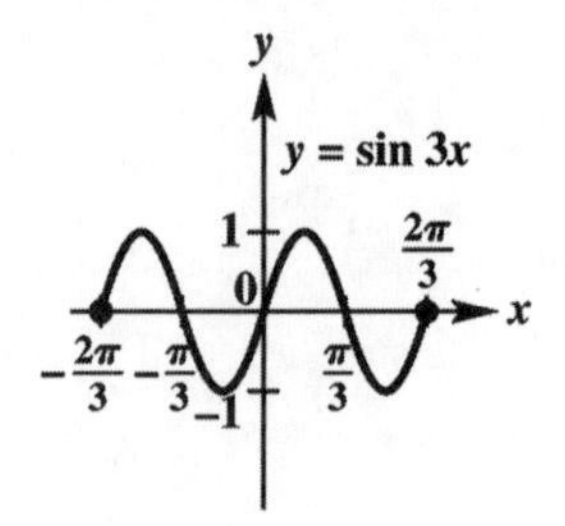

22. $y = \cos 2x$

Period: $\frac{2\pi}{2} = \pi$ and Amplitude: $|1| = 1$

Divide the interval $[0, \pi]$ into four equal parts to get the x-values that will yield minimum and maximum points and x-intercepts. Then make a table. Repeat this cycle for the interval $[-\pi, 0]$.

x	0	$\frac{\pi}{4}$	$\frac{\pi}{2}$	$\frac{3\pi}{4}$	π
$2x$	0	$\frac{\pi}{2}$	π	$\frac{3\pi}{2}$	2π
$\cos 2x$	1	0	-1	0	1

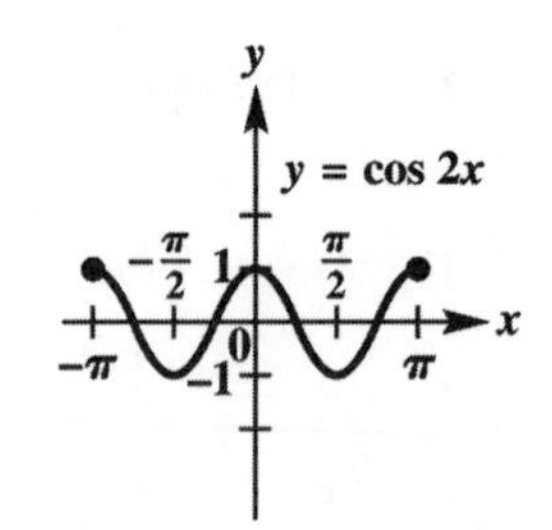

23. $y = 2\sin\frac{1}{4}x$

Period: $\frac{2\pi}{\frac{1}{4}} = 2\pi \cdot \frac{4}{1} = 8\pi$ and Amplitude: $|2| = 2$

Divide the interval $[0, 8\pi]$ into four equal parts to get the x-values that will yield minimum and maximum points and x-intercepts. Then make a table. Repeat this cycle for the interval $[-8\pi, 0]$.

x	0	2π	4π	6π	8π
$\frac{1}{4}x$	0	$\frac{\pi}{2}$	π	$\frac{3\pi}{2}$	2π
$\sin\frac{1}{4}x$	0	1	0	-1	0
$2\sin\frac{1}{4}x$	0	2	0	-2	0

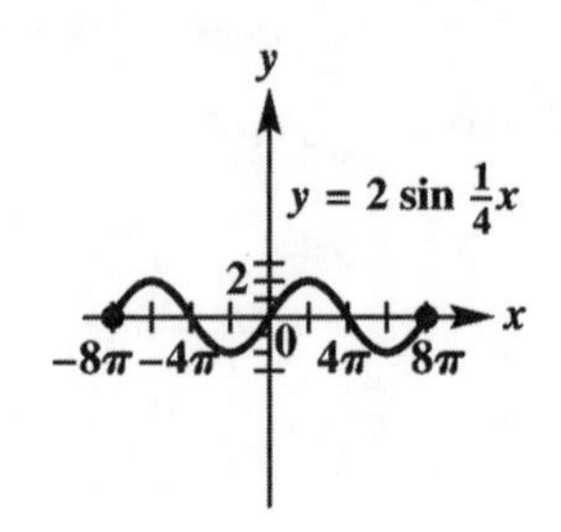

24. $y = 3\sin 2x$

Period: $\frac{2\pi}{2} = \pi$ and Amplitude: $|3| = 3$

Divide the interval $[0, \pi]$ into four equal parts to get the x-values that will yield minimum and maximum points and x-intercepts. Then make a table. Repeat this cycle for the interval $[-\pi, 0]$.

x	0	$\frac{\pi}{4}$	$\frac{\pi}{2}$	$\frac{3\pi}{4}$	π
$2x$	0	$\frac{\pi}{2}$	π	$\frac{3\pi}{2}$	2π
$\sin 2x$	0	1	0	−1	0
$3\sin 2x$	0	3	0	−3	0

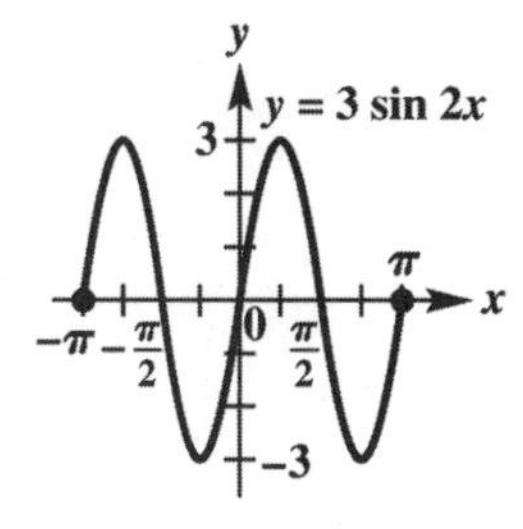

25. $y = -2\cos 3x$

Period: $\frac{2\pi}{3}$ and Amplitude: $|-2| = 2$

Divide the interval $\left[0, \frac{2\pi}{3}\right]$ into four equal parts to get the x-values that will yield minimum and maximum points and x-intercepts. Then make a table. Repeat this cycle for the interval $\left[-\frac{2\pi}{3}, 0\right]$.

x	0	$\frac{\pi}{6}$	$\frac{\pi}{3}$	$\frac{\pi}{2}$	$\frac{2\pi}{3}$
$3x$	0	$\frac{\pi}{2}$	π	$\frac{3\pi}{2}$	2π
$\cos 3x$	−1	0	1	0	−1
$-2\cos 3x$	−2	0	2	0	−2

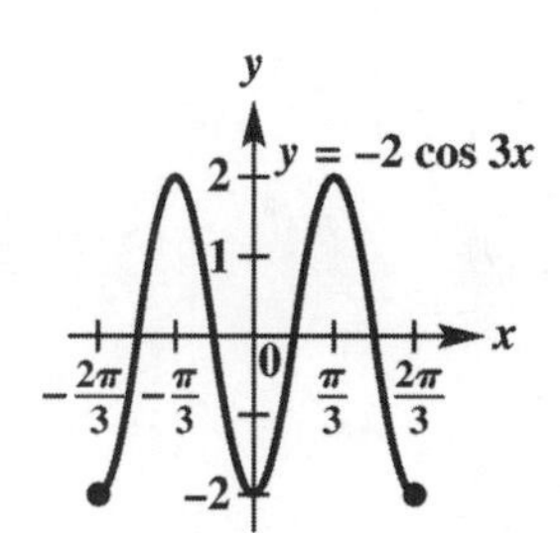

26. $y = -5\cos 2x$

Period: $\frac{2\pi}{2} = \pi$ and Amplitude: $|-5| = 5$

Divide the interval $[0, \pi]$ into four equal parts to get the x-values that will yield minimum and maximum points and x-intercepts. Then make a table. Repeat this cycle for the interval $[-\pi, 0]$.

x	0	$\frac{\pi}{4}$	$\frac{\pi}{2}$	$\frac{3\pi}{4}$	π
$2x$	0	$\frac{\pi}{2}$	π	$\frac{3\pi}{2}$	2π
$\cos 2x$	1	0	−1	0	1
$-5\cos 2x$	−5	0	5	0	−5

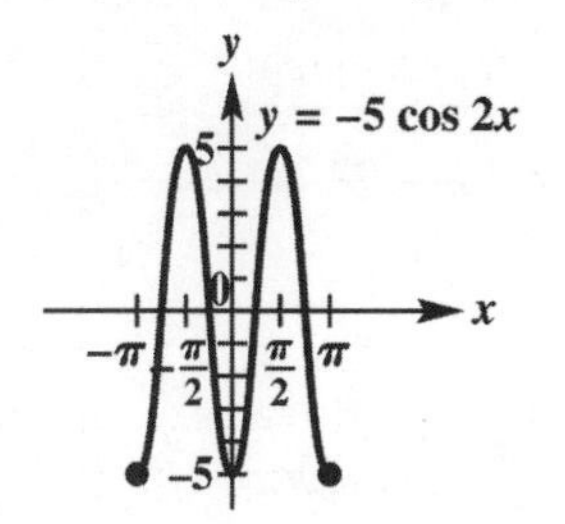

27. $y = \cos \pi x$

Period: $\frac{2\pi}{\pi} = 2$ and Amplitude: $|1| = 1$

Divide the interval $[0, 2]$ into four equal parts to get the x-values that will yield minimum and maximum points and x-intercepts. Then make a table. Repeat this cycle for the interval $[-2, 0]$.

x	0	$\frac{1}{2}$	1	$\frac{3}{2}$	2
πx	0	$\frac{\pi}{2}$	π	$\frac{3\pi}{2}$	2π
$\cos \pi x$	1	0	-1	0	1

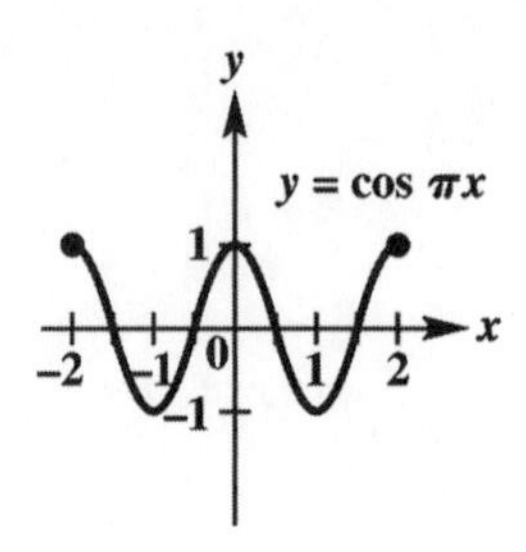

28. $y = -\sin \pi x$

Period: $\frac{2\pi}{\pi} = 2$ and Amplitude: $|-1| = 1$

Divide the interval $[0, 2]$ into four equal parts to get the x-values that will yield minimum and maximum points and x-intercepts. Then make a table. Repeat this cycle for the interval $[-2, 0]$.

x	0	$\frac{1}{2}$	1	$\frac{3}{2}$	2
πx	0	$\frac{\pi}{2}$	π	$\frac{3\pi}{2}$	2π
$\sin \pi x$	0	1	0	-1	0
$-\sin \pi x$	0	-1	0	1	0

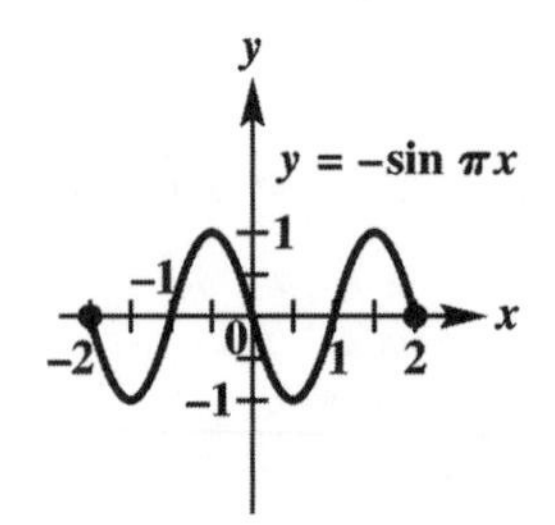

29. $y = -2\sin 2\pi x$

Period: $\frac{2\pi}{2\pi} = 1$ and Amplitude: $|-2| = 2$

Divide the interval $[0, 1]$ into four equal parts to get the x-values that will yield minimum and maximum points and x-intercepts. Then make a table. Repeat this cycle for the interval $[-1, 0]$.

x	0	$\frac{1}{4}$	$\frac{1}{2}$	$\frac{3}{4}$	1
$2\pi x$	0	$\frac{\pi}{2}$	π	$\frac{3\pi}{2}$	2π
$\sin 2\pi x$	0	1	0	-1	0
$-2\sin 2\pi x$	0	-2	0	2	0

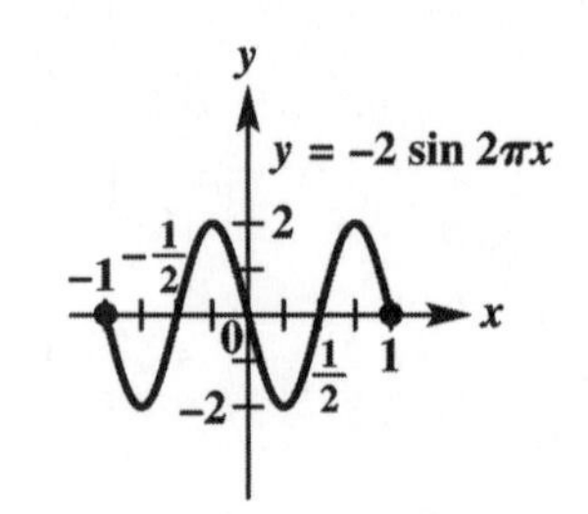

30. $y = 3\cos 2\pi x$

Period: $\frac{2\pi}{2\pi} = 1$ and Amplitude: $|3| = 3$

Divide the interval $[0,1]$ into four equal parts to get the x-values that will yield minimum and maximum points and x-intercepts. Then make a table. Repeat this cycle for the interval $[-1,0]$.

x	0	$\frac{1}{4}$	$\frac{1}{2}$	$\frac{3}{4}$	1
$2\pi x$	0	$\frac{\pi}{2}$	π	$\frac{3\pi}{2}$	2π
$\cos 2\pi x$	1	0	-1	0	1
$3\cos 2\pi x$	3	0	-3	0	3

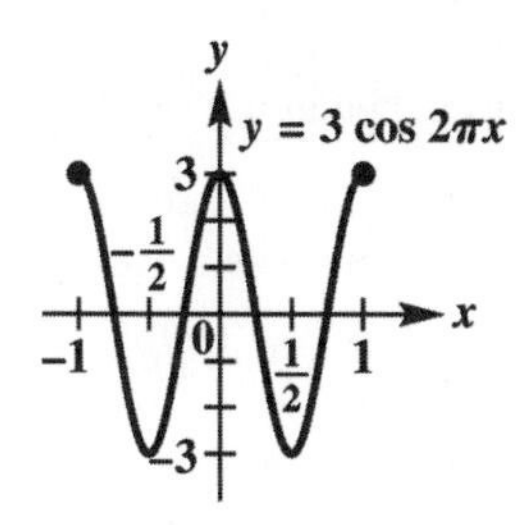

31. $y = \frac{1}{2}\cos\frac{\pi}{2}x$

Period: $\frac{2\pi}{\frac{\pi}{2}} = 2\pi \cdot \frac{2}{\pi} = 4$ and Amplitude: $\left|\frac{1}{2}\right| = \frac{1}{2}$

Divide the interval $[0,4]$ into four equal parts to get the x-values that will yield minimum and maximum points and x-intercepts. Then make a table. Repeat this cycle for the interval $[-4,0]$.

x	0	1	2	3	4
$\frac{\pi}{2}x$	0	$\frac{\pi}{2}$	π	$\frac{3\pi}{2}$	2π
$\cos\frac{\pi}{2}x$	1	0	-1	0	1
$\frac{1}{2}\cos\frac{\pi}{2}x$	$\frac{1}{2}$	0	$-\frac{1}{2}$	0	$\frac{1}{2}$

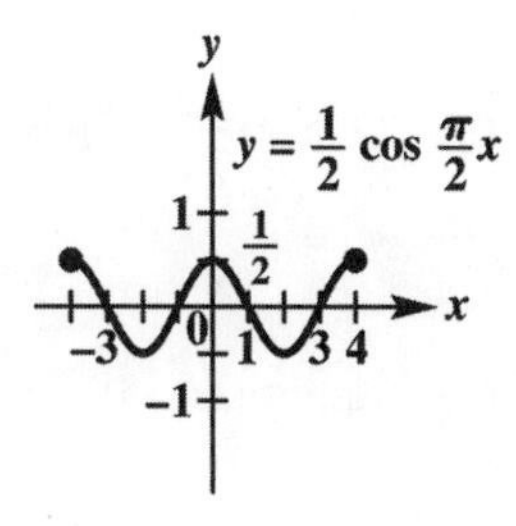

32. $y = -\frac{2}{3}\sin\frac{\pi}{4}x$

Period: $\frac{2\pi}{\frac{\pi}{4}} = 2\pi \cdot \frac{4}{\pi} = 8$ and Amplitude: $\left|-\frac{2}{3}\right| = \frac{2}{3}$

Divide the interval $[0,8]$ into four equal parts to get the x-values that will yield minimum and maximum points and x-intercepts. Then make a table. Repeat this cycle for the interval $[-8,0]$.

x	0	2	4	6	8
$\frac{\pi}{4}x$	0	$\frac{\pi}{2}$	π	$\frac{3\pi}{2}$	2π
$\sin\frac{\pi}{4}x$	0	1	0	-1	0
$-\frac{2}{3}\sin\frac{\pi}{4}x$	0	$-\frac{2}{3}$	0	$\frac{2}{3}$	0

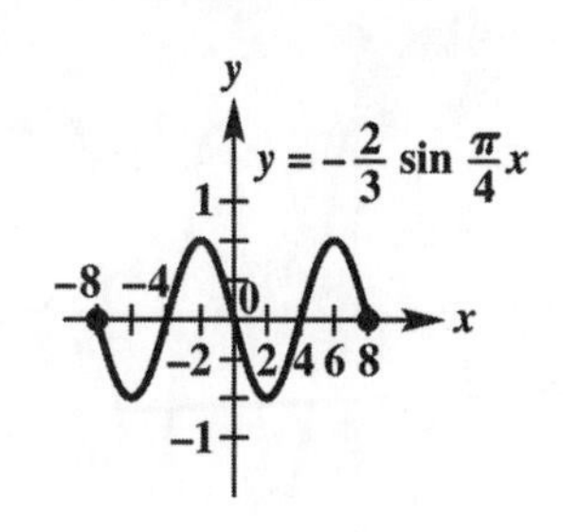

33. **(a)** The highest temperature is 80°; the lowest is 50°.

(b) The amplitude is $\frac{1}{2}(80° - 50°) = \frac{1}{2}(30°) = 15°$.

(c) The period is about 35,000 yr.

(d) The trend of the temperature now is downward.

34. **(a)** The amplitude is $\frac{1}{2}(120 - 80) = \frac{1}{2}(40) = 20$.

(b) Since the period is .8 sec, there are $\frac{1}{.8} = 1.25$ beats per sec and the pulse rate is $60(1.25) = 75$ beats per min.

35. **(a)** The latest time that the animals begin their evening activity is 8:00 P.M., the earliest time is 4:00 P.M. So, $4{:}00 \le y \le 8{:}00$. Since there is a difference of 4 hr in these times, the amplitude is $\frac{1}{2}(4) = 2$ hr.

(b) The length of this period is 1 yr.

36. **(a)** The amplitude of the graph is $\frac{1}{3}$; the period is $\frac{3}{2}$. Since $\frac{2\pi}{k} = \frac{3}{2}$, $k = \frac{4\pi}{3}$. The equation is $y = \frac{1}{3}\sin\frac{4\pi t}{3}$.

(b) It takes $\frac{3}{2}$ sec for a complete movement of the arm.

37. $E = 5 \cos 120\pi r$

(a) Amplitude: $|5| = 5$ and Period: $\frac{2\pi}{120\pi} = \frac{1}{60}$ sec

(b) Since the period is $\frac{1}{60}$, one cycle is completed in $\frac{1}{60}$ sec. Therefore, in 1 sec, 60 cycles are completed.

(c) $t = 0,\ E = 5\cos 120\pi(0) = 5\cos 0 = 5(1) = 5$

$t = .03,\ E = 5\cos 120\pi(.03) = 5\cos 3.6\pi \approx 1.545$

$t = .06,\ E = 5\cos 120\pi(.06) = 5\cos 7.2\pi \approx -4.045$

$t = .09,\ E = 5\cos 120\pi(.09) = 5\cos 10.8\pi 0 \approx -4.045$

$t = .12,\ E = 5\cos 120\pi(.12) = 5\cos 14.4\pi \approx 1.545$

(d)

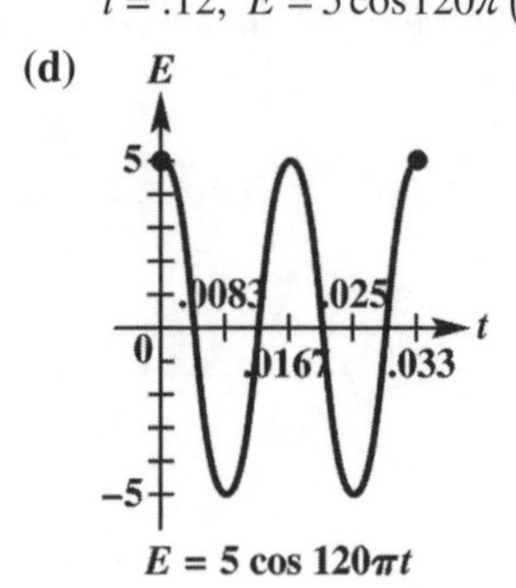

38. $E = 3.8 \cos 40\pi t$

(a) Amplitude: 3.8 and Period: $\frac{2\pi}{40\pi} = \frac{1}{20}$

(b) Frequency $= \frac{1}{\text{period}}$ = number of cycles per second = 20

(c) $t = .02,\ E = 3.8\cos 40\pi(.02) = 3.8\cos .8\pi \approx -3.074$

$t = .04,\ E = 3.8\cos 40\pi(.04) = 3.8\cos 1.6\pi \approx 1.174$

$t = .08,\ E = 3.8\cos 40\pi(.08) = 3.8\cos 3.2\pi \approx -3.074$

$t = .12,\ E = 3.8\cos 40\pi(.12) = 3.8\cos 4.8\pi \approx -3.074$

$t = .14,\ E = 3.8\cos 40\pi(.14) = 3.8\cos 5.6\pi \approx 1.174$

(d)

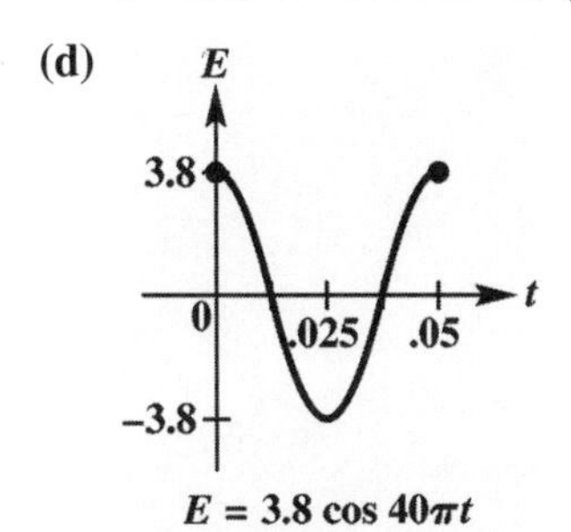

39. **(a)** The graph has a general upward trend along with small annual oscillations.

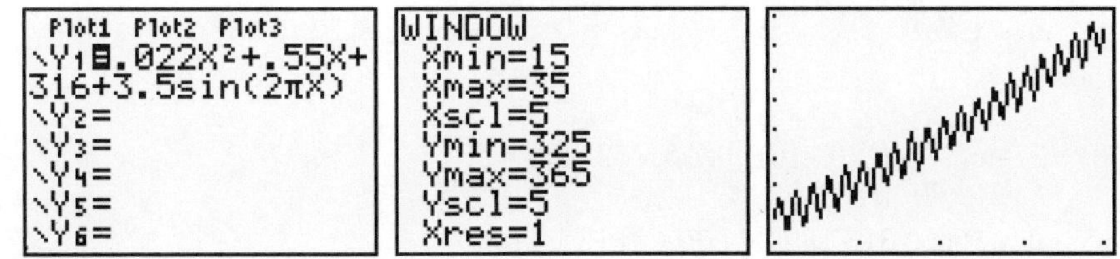

(b) The seasonal variations are caused by the term $3.5\sin 2\pi x$. The maximums will occur when $2\pi x = \frac{\pi}{2} + 2n\pi$, where n is an integer. Since x cannot be negative, n cannot be negative. This is equivalent to the following.

$$2\pi x = \frac{\pi}{2} + 2n\pi,\ n = 0, 1, 2, \ldots \Rightarrow 2x = \frac{1}{2} + 2n,\ n = 0, 1, 2, \ldots \Rightarrow x = \frac{1}{4} + n,\ n = 0, 1, 2, \ldots$$

$$x = \frac{4n+1}{4},\ n = 0, 1, 2, \ldots \Rightarrow x = \frac{1}{4}, \frac{5}{4}, \frac{9}{4}, \ldots$$

Since x is in years, $x = \frac{1}{4}$ corresponds to April when the seasonal carbon dioxide levels are maximum.

The minimums will occur when $2\pi x = \frac{3\pi}{2} + 2n\pi$, where n is an integer. Since x cannot be negative, n cannot be negative. This is equivalent to the following.

$$2\pi x = \frac{3\pi}{2} + 2n\pi,\ n = 0, 1, 2, \ldots \Rightarrow 2x = \frac{3}{2} + 2n,\ n = 0, 1, 2, \ldots \Rightarrow x = \frac{3}{4} + n,\ n = 0, 1, 2, \ldots$$

$$x = \frac{4n+3}{4},\ n = 0, 1, 2, \ldots \Rightarrow x = \frac{3}{4}, \frac{7}{4}, \frac{11}{4}, \ldots$$

This is $\frac{1}{2}$ yr later, which corresponds to October.

40. **(a)** The graph of C has a general upward trend similar to L (in Exercise 39) except that both the carbon dioxide levels and the seasonal oscillations are larger for C than L.

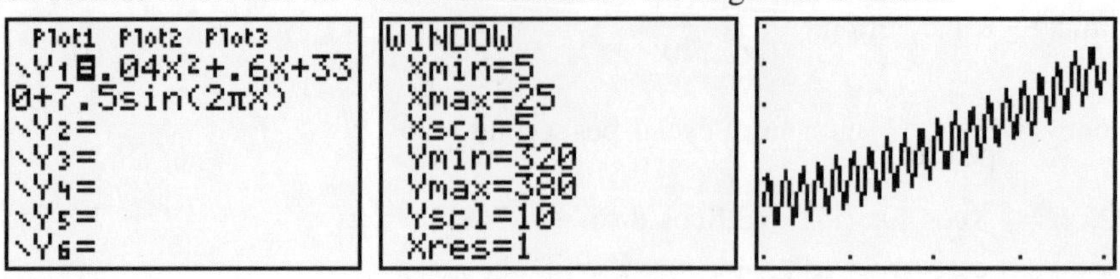

(b) Answers will vary.

(c) To solve this problem, horizontally translate the graph of C a distance of 1970 units to the right. The new C function can be written as the following.

$C(x) = .04(x-1970)^2 + .6(x-1970) + 330 + 7.5\sin\left[2\pi(x-1970)\right]$, where x is the actual year

This function would now be valid for $1970 \le x \le 1995$.

41. $T(x) = 37\sin\left[\frac{2\pi}{365}(x-101)\right] + 25$

(a) March 1 (day 60)

$$T(60) = 37\sin\left[\frac{2\pi}{365}(60-101)\right] + 25 \approx 1.001° \approx 1°$$

(b) April 1 (day 91)

$$T(91) = 37\sin\left[\frac{2\pi}{365}(91-101)\right] + 25 \approx 18.662° \approx 19°$$

(c) Day 150

$$T(150) = 37\sin\left[\frac{2\pi}{365}(150-101)\right] + 25 \approx 52.638° \approx 53°$$

(d) June 15 is day 166. $(31+28+31+30+31+15=166)$

$$T(166) = 37\sin\left[\frac{2\pi}{365}(166-101)\right] + 25 \approx 58.286° \approx 58°$$

(e) September 1 is day 244. $(31+28+31+30+31+30+31+31+1=244)$

$$T(244) = 37\sin\left[\frac{2\pi}{365}(244-101)\right] + 25 \approx 48.264° \approx 48°$$

(f) October 31 is day 304. $(31+28+31+30+31+30+31+31+30+31=304)$

$$T(304) = 37\sin\left[\frac{2\pi}{365}(304-101)\right] + 25 \approx 12.212° \approx 12°$$

42. $\Delta S = .034(1367)\sin\left[\frac{2\pi(82.5-N)}{365.25}\right]$

(a) $N = 80,\ \Delta S = .034(1367)\sin\left[\frac{2\pi(82.5-80)}{365.25}\right] \approx 1.998$ watts per m^2.

(b) $N = 1268,\ \Delta S = .034(1367)\sin\left[\frac{2\pi(82.5-1268)}{365.25}\right] \approx -46.461$ watts per m^2.

Continued on next page

42. (continued)

(c) Since the greatest value the sine function can be is 1, the maximum of

$$\Delta S = .034(1367)\sin\left[\frac{2\pi(82.5-N)}{365.25}\right] \text{ is } .034(1367) = 46.478 \text{ watts per m}^2.$$

(d) $\Delta S = .034(1367)\sin\left[\frac{2\pi(82.5-N)}{365.25}\right] = 0$ when $\sin\left[\frac{2\pi(82.5-N)}{365.25}\right] = 0.$ This can occur when $\frac{2\pi(82.5-N)}{365.25} = 0$ or $N = 82.5$. Other answers are possible. Since N represents a day number, which should be a natural number, we might interpret day 82.5 as noon on the 82nd day.

43. The graph repeats each day, so the period is 24 hours.

44. Approximately $\frac{1}{2}(2.6-.2) = \frac{1}{2}(2.4) = 1.2$

45. Approximately 6 P.M., approximately 0.2 feet.

46. The low tide occurred at 6 P.M. + 1:19, or 7:19 P.M., the height was approximately $.2 - .2 = 0$ ft.

47. Approximately 2 A.M.; approximately 2.6 feet.

48. The high tide occurred at 2 A.M. + 1:18, or 3:18 A.M.; the height was approximately $2.6 - .2 = 2.4$ ft.

49. $-1 \le y \le 1$
Amplitude: 1
Period: 8 squares $= 8(30°) = 240°$ or $\frac{4\pi}{3}$

50. $-1 \le y \le 1$
Amplitude: 1
Period: 4 squares $= 4(30°) = 120°$ or $\frac{2\pi}{3}$

51. No, we can't say $\sin bx = b\sin x$. If b is not zero, then the period of $y = \sin bx$ is $\frac{2\pi}{|b|}$ and the amplitude is 1. The period of $y = b\sin x$ is 2π and the amplitude is $|b|$.

52. No, we can't say $\cos bx = b\cos x$. If b is not zero, then the period of $y = \cos bx$ is $\frac{2\pi}{|b|}$ and the amplitude is 1. The period of $y = b\cos x$ is 2π and the amplitude is $|b|$.

53. The functions are graphed in the following window.

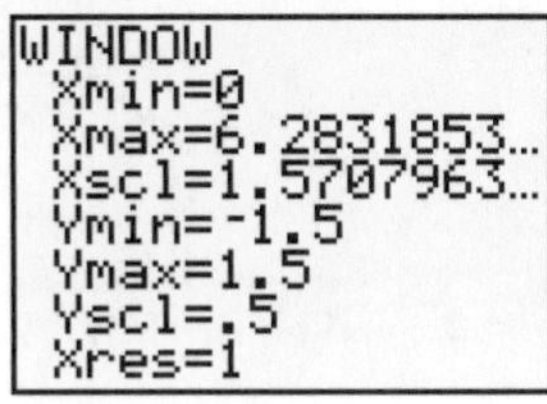

$y = \sin(3x)$ shows 3 cycles

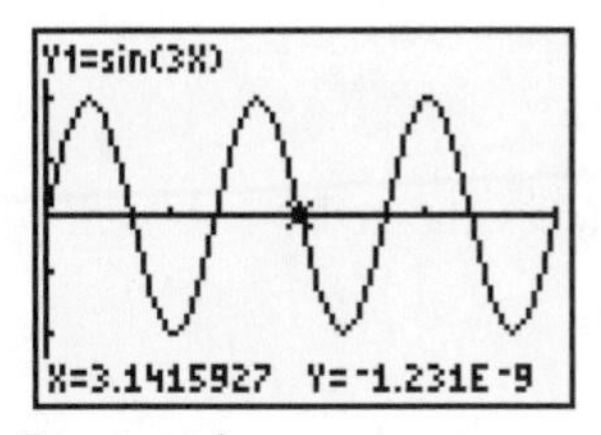

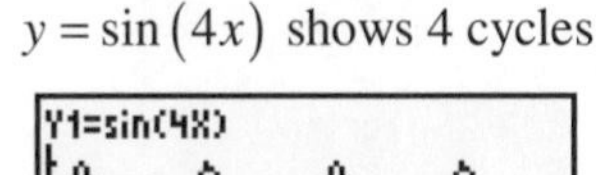

$y = \sin(4x)$ shows 4 cycles

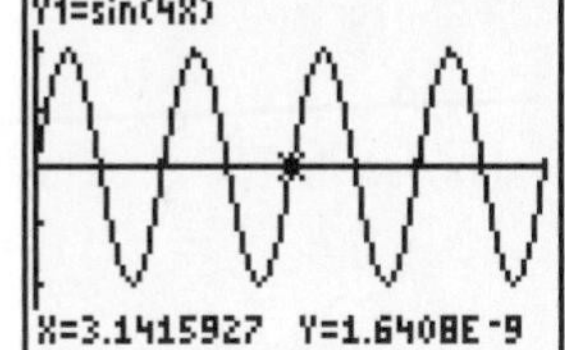

$y = \sin(5x)$ shows 5 cycles

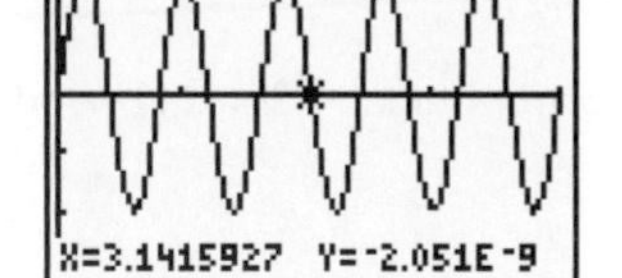

Continued on next page

53. (continued)

$y=\sin\left(\frac{1}{2}x\right)$ shows $\frac{1}{2}$ cycle

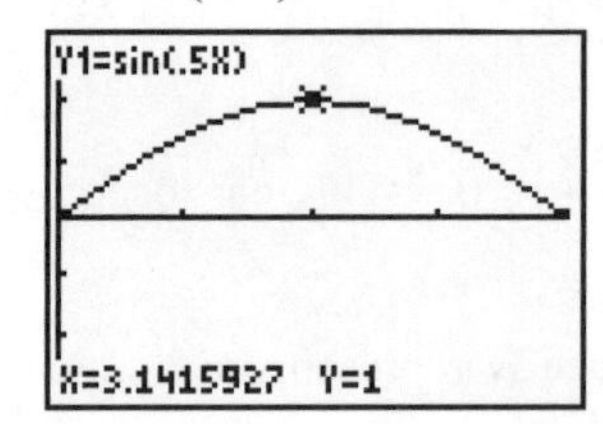

$y=\sin\left(\frac{3}{2}x\right)$ shows $\frac{3}{2}$ cycles

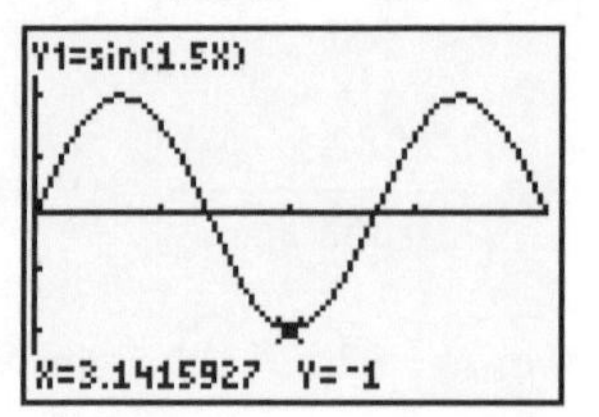

Section 4.2: Translations of the Graphs of the Sine and Cosine Functions

1. $y=\sin\left(x-\frac{\pi}{4}\right)$

The graph is a sinusoidal curve $y = \sin x$ shifted $\frac{\pi}{4}$ units to the right. This matches graph D.

2. $y=\sin\left(x+\frac{\pi}{4}\right)=\sin\left[x-\left(-\frac{\pi}{4}\right)\right]$

The graph is a sinusoidal curve $y = \sin x$ shifted $\frac{\pi}{4}$ units to the left. This matches graph G.

3. $y=\cos\left(x-\frac{\pi}{4}\right)$

The graph is a sinusoidal curve $y=\cos x$ shifted $\frac{\pi}{4}$ units to the right. This matches graph H.

4. $y=\cos\left(x+\frac{\pi}{4}\right)=\cos\left[x-\left(-\frac{\pi}{4}\right)\right]$

The graph is a sinusoidal curve $y=\cos x$ shifted $\frac{\pi}{4}$ units to the left. This matches graph A.

5. $y = 1 + \sin x$
The graph is a sinusoidal curve $y = \sin x$ translated vertically 1 unit up. This matches graph B.

6. $y = -1 + \sin x$
The graph is a sinusoidal curve $y = \sin x$ translated vertically 1 unit down. This matches graph E.

7. $y = 1 + \cos x$
The graph is a sinusoidal curve $y = \cos x$ translated vertically 1 unit up. This matches graph F.

8. $y = -1 + \cos x$
The graph is a sinusoidal curve $y = \cos x$ translated vertically 1 unit down. This matches graph C.

9. $y=3\sin(2x-4)=3\sin\left[2(x-2)\right]$

The amplitude = $|3|=3$, period = $\frac{2\pi}{2}=\pi$, and phase shift = 2. This matches choice B.

10. $y=2\sin(3x-4)=2\sin\left[3\left(x-\frac{4}{3}\right)\right]$

The amplitude = $|2|=2$, period = $\frac{2\pi}{3}$, and phase shift = $\frac{4}{3}$. This matches choice D.

11. $y = 4\sin(3x-2) = 4\sin\left[3\left(x-\frac{2}{3}\right)\right]$

The amplitude = $|4| = 4$, period = $\frac{2\pi}{3}$, and phase shift = $\frac{2}{3}$. This matches choice C.

12. $y = 2\sin(4x-3) = 2\sin\left[4\left(x-\frac{3}{4}\right)\right]$

The amplitude = $|2| = 2$, period = $\frac{2\pi}{4} = \frac{\pi}{2}$, and phase shift = $\frac{3}{4}$. This matches choice A.

13. If the graph of $y = \cos x$ is translated $\frac{\pi}{2}$ units horizontally to the right, it will coincide with the graph of $y = \sin x$.

14. If the graph of $y = \sin x$ is translated $\frac{\pi}{2}$ units horizontally to the left, it will coincide with the graph of $y = \cos x$.

15. $y = 2\sin(x-\pi)$

amplitude: $|2| = 2$; period: $\frac{2\pi}{1} = 2\pi$;
There is no vertical translation.
The phase shift is π units to the right.

16. $y = \frac{2}{3}\sin\left(x+\frac{\pi}{2}\right) = \frac{2}{3}\sin\left[x-\left(-\frac{\pi}{2}\right)\right]$

amplitude: $\left|\frac{2}{3}\right| = \frac{2}{3}$; period: $\frac{2\pi}{1} = 2\pi$;
There is no vertical translation.
The phase shift is $\frac{\pi}{2}$ units to the left.

17. $y = 4\cos\left(\frac{x}{2}+\frac{\pi}{2}\right) = 4\cos\frac{1}{2}\left[x-(-\pi)\right]$

amplitude: $|4| = 4$;
period: $\frac{2\pi}{\frac{1}{2}} = 2\pi\cdot\frac{2}{1} = 4\pi$;
There is no vertical translation.
The phase shift is π units to the left.

18. $y = -\cos\pi\left(x-\frac{1}{3}\right)$

amplitude: $|-1| = 1$; period: $\frac{2\pi}{\pi} = 2$;
There is no vertical translation.
The phase shift is $\frac{1}{3}$ unit to the right.

19. $y = 3\cos\frac{\pi}{2}\left(x-\frac{1}{2}\right)$

amplitude: $|3| = 3$;
period: $\frac{2\pi}{\frac{\pi}{2}} = 2\pi\cdot\frac{2}{\pi} = 4$;
There is no vertical translation.
The phase shift is $\frac{1}{2}$ unit to the right.

20. $y = \frac{1}{2}\sin\left(\frac{x}{2}+\pi\right) = \frac{1}{2}\sin\frac{1}{2}\left[x-(-2\pi)\right]$

amplitude: $\left|\frac{1}{2}\right| = \frac{1}{2}$;
period: $\frac{2\pi}{\frac{1}{2}} = 2\pi\cdot\frac{2}{1} = 4\pi$;
There is no vertical translation.
The phase shift is 2π units to the left.

21. $y = 2 - \sin\left(3x - \frac{\pi}{5}\right) = -\sin 3\left(x - \frac{\pi}{15}\right) + 2$

amplitude: $|-1| = 1$; period: $\frac{2\pi}{3}$;

The vertical translation is 2 units up.

The phase shift is $\frac{\pi}{15}$ unit to the right.

22. $y = -1 + \frac{1}{2}\cos(2x - 3\pi)$

$= \frac{1}{2}\cos 2\left(x - \frac{3\pi}{2}\right) - 1$

amplitude: $\left|\frac{1}{2}\right| = \frac{1}{2}$; period: $\frac{2\pi}{2} = \pi$;

The vertical translation is 1 unit down.

The phase shift is $\frac{3\pi}{2}$ units to the right.

23. $y = \cos\left(x - \frac{\pi}{2}\right)$

Step 1 Find the interval whose length is $\frac{2\pi}{b}$.

$$0 \le x - \frac{\pi}{2} \le 2\pi \Rightarrow 0 + \frac{\pi}{2} \le x \le 2\pi + \frac{\pi}{2} \Rightarrow \frac{\pi}{2} \le x \le \frac{5\pi}{2}$$

Step 2 Divide the period into four equal parts to get the following x-values.

$$\frac{\pi}{2},\ \pi,\ \frac{3\pi}{2},\ 2\pi,\ \frac{5\pi}{2}$$

Step 3 Evaluate the function for each of the five x-values.

x	$\frac{\pi}{2}$	π	$\frac{3\pi}{2}$	2π	$\frac{5\pi}{2}$
$x - \frac{\pi}{2}$	0	$\frac{\pi}{2}$	π	$\frac{3\pi}{2}$	2π
$\cos\left(x - \frac{\pi}{2}\right)$	1	0	−1	0	1

Steps 4 and 5 Plot the points found in the table and join them with a sinusoidal curve. By graphing an additional period to the right, we obtain the following graph.

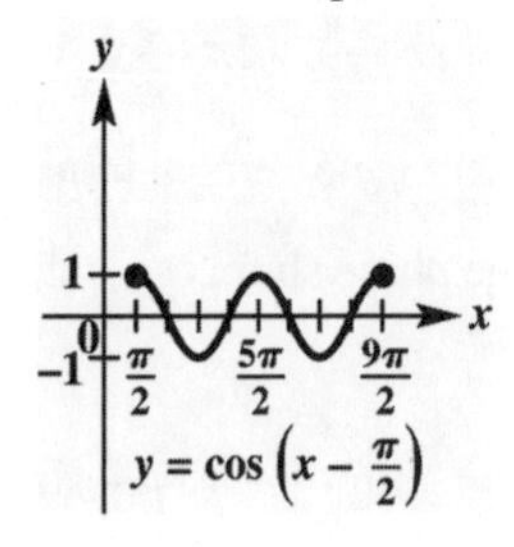

The amplitude is 1.

The period is 2π.

There is no vertical translation.

The phase shift is $\frac{\pi}{2}$ units to the right.

24. $y = \sin\left(x - \frac{\pi}{4}\right)$

Step 1 Find the interval whose length is $\frac{2\pi}{b}$.

$$0 \le x - \frac{\pi}{4} \le 2\pi \Rightarrow 0 + \frac{\pi}{4} \le x \le 2\pi + \frac{\pi}{4} \Rightarrow \frac{\pi}{4} \le x \le \frac{9\pi}{4}$$

Continued on next page

24. (continued)

Step 2 Divide the period into four equal parts to get the following x-values.

$$\frac{\pi}{4}, \frac{3\pi}{4}, \frac{5\pi}{4}, \frac{7\pi}{4}, \frac{9\pi}{4}$$

Step 3 Evaluate the function for each of the five x-values

x	$\frac{\pi}{4}$	$\frac{3\pi}{4}$	$\frac{5\pi}{4}$	$\frac{7\pi}{4}$	$\frac{9\pi}{4}$
$x-\frac{\pi}{4}$	0	$\frac{\pi}{2}$	π	$\frac{3\pi}{2}$	2π
$\sin\left(x-\frac{\pi}{4}\right)$	0	1	0	−1	0

Steps 4 and 5 Plot the points found in the table and join them with a sinusoidal curve. By graphing an additional period to the right, we obtain the following graph.

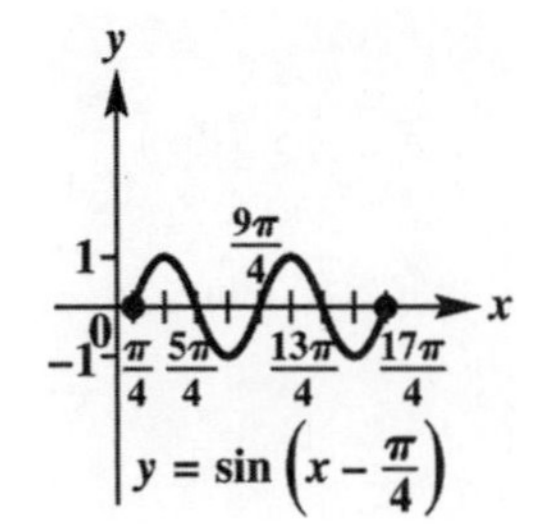

The amplitude is 1.

The period is 2π.

There is no vertical translation.

The phase shift is $\frac{\pi}{4}$ unit to the right.

25. $y=\sin\left(x+\frac{\pi}{4}\right)$

Step 1 Find the interval whose length is $\frac{2\pi}{b}$.

$$0\le x+\frac{\pi}{4}\le 2\pi \Rightarrow 0-\frac{\pi}{4}\le x\le 2\pi-\frac{\pi}{4}\Rightarrow -\frac{\pi}{4}\le x\le \frac{7\pi}{4}$$

Step 2 Divide the period into four equal parts to get the following x-values.

$$-\frac{\pi}{4}, \frac{\pi}{4}, \frac{3\pi}{4}, \frac{5\pi}{4}, \frac{7\pi}{4}$$

Step 3 Evaluate the function for each of the five x-values.

x	$-\frac{\pi}{4}$	$\frac{\pi}{4}$	$\frac{3\pi}{4}$	$\frac{5\pi}{4}$	$\frac{7\pi}{4}$
$x+\frac{\pi}{4}$	0	$\frac{\pi}{2}$	π	$\frac{3\pi}{2}$	2π
$\sin\left(x+\frac{\pi}{4}\right)$	0	1	0	−1	0

Continued on next page

25. (continued)

Steps 4 and 5 Plot the points found in the table and join then with a sinusoidal curve. By graphing an additional period to the right, we obtain the following graph.

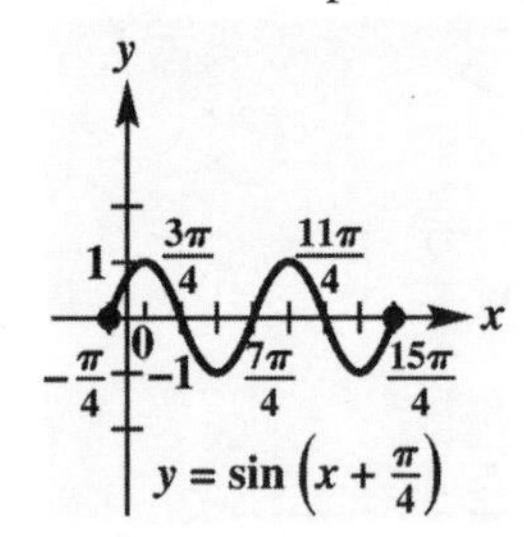

The amplitude is 1.

The period is 2π.

There is no vertical translation.

The phase shift is $\frac{\pi}{4}$ unit to the left.

26. $y = \cos\left(x - \frac{\pi}{3}\right)$

Step 1 Find the interval whose length is $\frac{2\pi}{b}$.

$$0 \le x - \frac{\pi}{3} \le 2\pi \Rightarrow 0 + \frac{\pi}{3} \le x \le 2\pi + \frac{\pi}{3} \Rightarrow \frac{\pi}{3} \le x \le \frac{7\pi}{3}$$

Step 2 Divide the period into four equal parts to get the following x-values.

$$\frac{\pi}{3}, \frac{5\pi}{6}, \frac{4\pi}{3}, \frac{11\pi}{6}, \frac{7\pi}{3}$$

Step 3 Evaluate the function for each of the five x-values.

x	$\frac{\pi}{3}$	$\frac{5\pi}{6}$	$\frac{4\pi}{3}$	$\frac{11\pi}{6}$	$\frac{7\pi}{3}$
$x - \frac{\pi}{3}$	0	$\frac{\pi}{2}$	π	$\frac{3\pi}{2}$	2π
$\cos\left(x - \frac{\pi}{3}\right)$	1	0	−1	0	1

Steps 4 and 5 Plot the points found in the table and join them with a sinusoidal curve. By graphing an additional period to the right, we obtain the following graph.

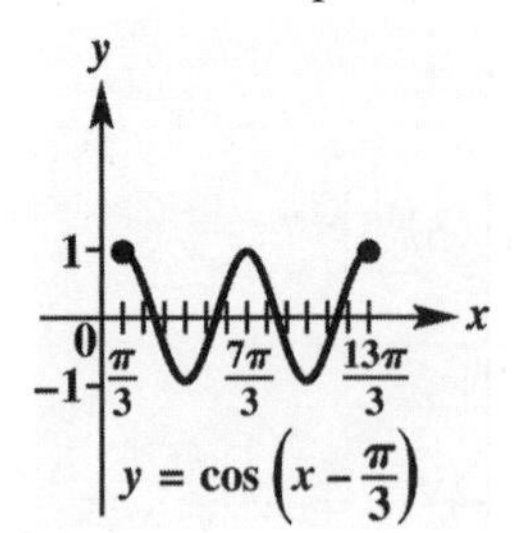

The amplitude is 1.

The period is 2π.

There is no vertical translation.

The phase shift is $\frac{\pi}{3}$ units to the right.

27. $y = 2\cos\left(x - \frac{\pi}{3}\right)$

Step 1 Find the interval whose length is $\frac{2\pi}{b}$.

$$0 \le x - \frac{\pi}{3} \le 2\pi \Rightarrow 0 + \frac{\pi}{3} \le x \le 2\pi + \frac{\pi}{3} \Rightarrow \frac{\pi}{3} \le x \le \frac{7\pi}{3}$$

Continued on next page

27. (continued)

Step 2 Divide the period into four equal parts to get the following x-values.

$$\frac{\pi}{3}, \frac{5\pi}{6}, \frac{4\pi}{3}, \frac{11\pi}{6}, \frac{7\pi}{3}$$

Step 3 Evaluate the function for each of the five x-values.

x	$\frac{\pi}{3}$	$\frac{5\pi}{6}$	$\frac{4\pi}{3}$	$\frac{11\pi}{6}$	$\frac{7\pi}{3}$
$x-\frac{\pi}{3}$	0	$\frac{\pi}{2}$	π	$\frac{3\pi}{2}$	2π
$\cos\left(x-\frac{\pi}{3}\right)$	1	0	−1	0	1
$2\cos\left(x-\frac{\pi}{3}\right)$	2	0	−2	0	2

Steps 4 and 5 Plot the points found in the table and join them with a sinusoidal curve. By graphing an additional period to the right, we obtain the following graph.

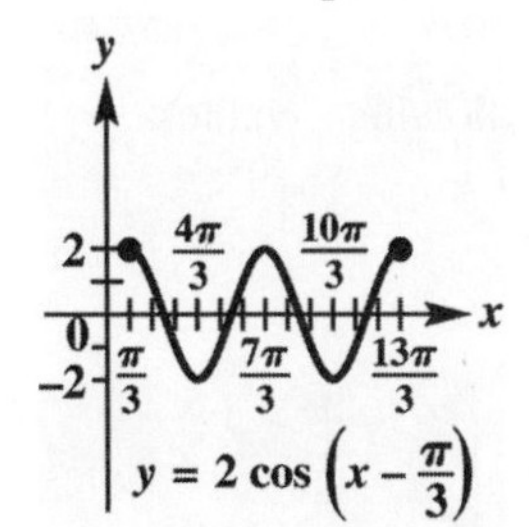

The amplitude is 2.

The period is 2π.

There is no vertical translation.

The phase shift is $\frac{\pi}{3}$ units to the right.

28. $y = 3\sin\left(x-\frac{3\pi}{2}\right)$

Step 1 Find the interval whose length is $\frac{2\pi}{b}$.

$$0 \le x-\frac{3\pi}{2} \le 2\pi \Rightarrow 0+\frac{3\pi}{2} \le x \le 2\pi+\frac{3\pi}{2} \Rightarrow \frac{3\pi}{2} \le x \le \frac{7\pi}{2}$$

Step 2 Divide the period into four equal parts to get the following x-values.

$$\frac{3\pi}{2}, 2\pi, \frac{5\pi}{2}, 3\pi, \frac{7\pi}{2}$$

Step 3 Evaluate the function for each of the five x-values.

x	$\frac{3\pi}{2}$	2π	$\frac{5\pi}{2}$	3π	$\frac{7\pi}{2}$
$x-\frac{3\pi}{2}$	0	$\frac{\pi}{2}$	π	$\frac{3\pi}{2}$	2π
$\sin\left(x-\frac{3\pi}{2}\right)$	0	1	0	−1	0
$3\sin\left(x-\frac{3\pi}{2}\right)$	0	3	0	−3	0

Continued on next page

28. (continued)

Steps 4 and 5 Plot the points found in the table and join them with a sinusoidal curve. By graphing an additional period to the right, we obtain the following graph.

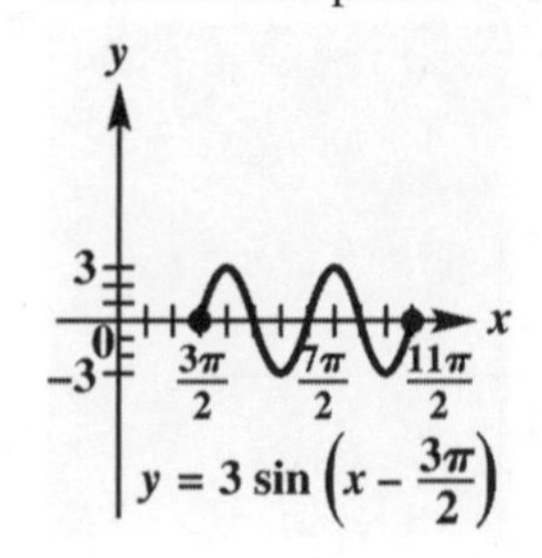

The amplitude is 3.

The period is 2π.

There is no vertical translation.

The phase shift is $\frac{3\pi}{2}$ units to the right.

29. $y=\frac{3}{2}\sin 2\left(x+\frac{\pi}{4}\right)$

Step 1 Find the interval whose length is $\frac{2\pi}{b}$.

$$0\le 2\left(x+\frac{\pi}{4}\right)\le 2\pi \Rightarrow 0\le x+\frac{\pi}{4}\le\frac{2\pi}{2}\Rightarrow 0\le x+\frac{\pi}{4}\le\pi\Rightarrow -\frac{\pi}{4}\le x\le\frac{3\pi}{4}$$

Step 2 Divide the period into four equal parts to get the following x-values.

$$-\frac{\pi}{4},\ 0,\ \frac{\pi}{4},\ \frac{\pi}{2},\ \frac{3\pi}{4}$$

Step 3 Evaluate the function for each of the five x-values

x	$-\frac{\pi}{4}$	0	$\frac{\pi}{4}$	$\frac{\pi}{2}$	$\frac{3\pi}{4}$
$2\left(x+\frac{\pi}{4}\right)$	0	$\frac{\pi}{2}$	π	$\frac{3\pi}{2}$	2π
$\sin 2\left(x+\frac{\pi}{4}\right)$	0	1	0	-1	0
$\frac{3}{2}\sin 2\left(x+\frac{\pi}{4}\right)$	0	$\frac{3}{2}$	0	$-\frac{3}{2}$	0

Steps 4 and 5 Plot the points found in the table and join them with a sinusoidal curve.

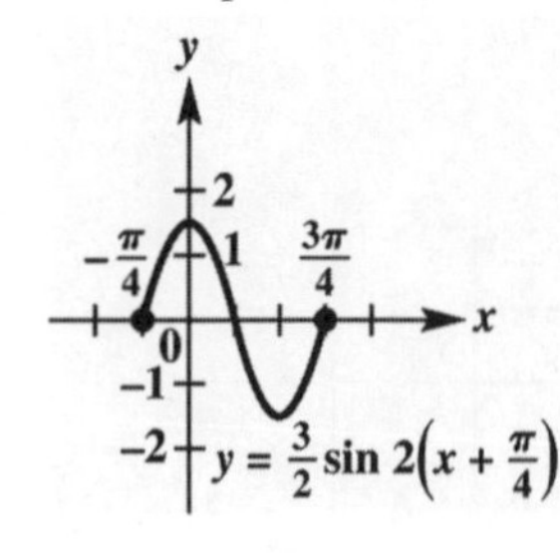

The amplitude is $\frac{3}{2}$.

The period is $\frac{2\pi}{2}$, which is π.

There is no vertical translation.

The phase shift is $\frac{\pi}{4}$ unit to the left.

30. $y = -\frac{1}{2}\cos 4\left(x + \frac{\pi}{2}\right)$

Step 1 Find the interval whose length is $\frac{2\pi}{b}$.

$$0 \le 4\left(x + \frac{\pi}{2}\right) \le 2\pi \Rightarrow 0 \le x + \frac{\pi}{2} \le \frac{\pi}{2} \Rightarrow 0 - \frac{\pi}{2} \le x \le \frac{\pi}{2} - \frac{\pi}{2} \Rightarrow -\frac{\pi}{2} \le x \le 0$$

Step 2 Divide the period into four equal parts to get the following x-values.

$$-\frac{\pi}{2}, -\frac{3\pi}{8}, -\frac{\pi}{4}, -\frac{\pi}{8}, 0$$

Step 3 Evaluate the function for each of the five x-values.

x	$-\frac{\pi}{2}$	$-\frac{3\pi}{8}$	$-\frac{\pi}{4}$	$-\frac{\pi}{8}$	0
$4\left(x + \frac{\pi}{2}\right)$	0	$\frac{\pi}{2}$	π	$\frac{3\pi}{2}$	2π
$\cos 4\left(x + \frac{\pi}{2}\right)$	1	0	-1	0	1
$-\frac{1}{2}\cos 4\left(x + \frac{\pi}{2}\right)$	$-\frac{1}{2}$	0	$\frac{1}{2}$	0	$-\frac{1}{2}$

Steps 4 and 5 Plot the points found in the table and join them with a sinusoidal curve.

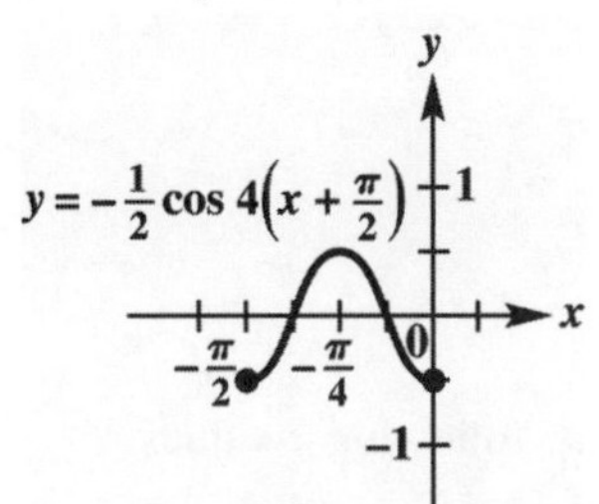

The amplitude is $\left|-\frac{1}{2}\right|$, which is $\frac{1}{2}$.

The period is $\frac{2\pi}{4}$, which is $\frac{\pi}{2}$.

There is no vertical translation.

The phase shift is $\frac{\pi}{2}$ units to the left.

31. $y = -4\sin(2x - \pi) = -4\sin 2\left(x - \frac{\pi}{2}\right)$

Step 1 Find the interval whose length is $\frac{2\pi}{b}$.

$$0 \le 2\left(x - \frac{\pi}{2}\right) \le 2\pi \Rightarrow 0 \le x - \frac{\pi}{2} \le \frac{2\pi}{2} \Rightarrow 0 \le x - \frac{\pi}{2} \le \pi \Rightarrow \frac{\pi}{2} \le x \le \frac{3\pi}{2}$$

Step 2 Divide the period into four equal parts to get the following x-values.

$$\frac{\pi}{2}, \frac{3\pi}{4}, \pi, \frac{5\pi}{4}, \frac{3\pi}{2}$$

Continued on next page

31. (continued)

Step 3 Evaluate the function for each of the five x-values

x	$\frac{\pi}{2}$	$\frac{3\pi}{4}$	π	$\frac{5\pi}{4}$	$\frac{3\pi}{2}$
$2\left(x-\frac{\pi}{2}\right)$	0	$\frac{\pi}{2}$	π	$\frac{3\pi}{2}$	2π
$\sin 2\left(x-\frac{\pi}{2}\right)$	0	1	0	-1	0
$-4\sin 2\left(x-\frac{\pi}{2}\right)$	0	-4	0	4	0

Steps 4 and 5 Plot the points found in the table and join them with a sinusoidal curve.

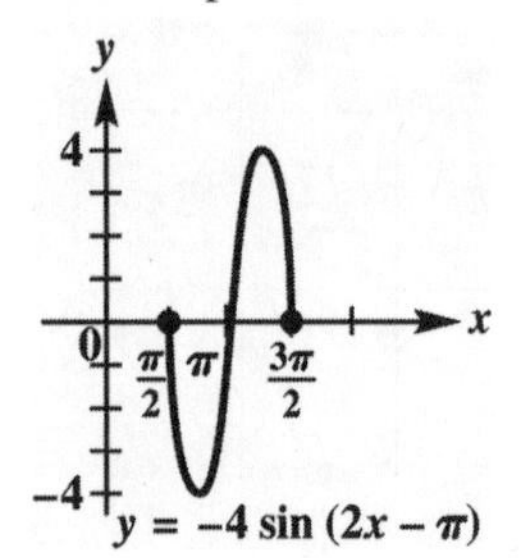

The amplitude is $|-4|$, which is 4.

The period is $\frac{2\pi}{2}$, which is π.

There is no vertical translation.

The phase shift is $\frac{\pi}{2}$ units to the right.

32. $y = 3\cos(4x+\pi) = 3\cos 4\left(x+\frac{\pi}{4}\right)$

Step 1 Find the interval whose length is $\frac{2\pi}{b}$.

$$0 \le 4\left(x+\frac{\pi}{4}\right) \le 2\pi \Rightarrow 0 \le x+\frac{\pi}{4} \le \frac{\pi}{2} \Rightarrow 0-\frac{\pi}{4} \le x \le \frac{\pi}{2}-\frac{\pi}{4} \Rightarrow -\frac{\pi}{4} \le x \le \frac{\pi}{4}$$

Step 2 Divide the period into four equal parts to get the following x-values.

$$-\frac{\pi}{4},\ -\frac{\pi}{8},\ 0,\ \frac{\pi}{8},\ \frac{\pi}{4}$$

Step 3 Evaluate the function for each of the five x-values.

x	$-\frac{\pi}{4}$	$-\frac{\pi}{8}$	0	$\frac{\pi}{8}$	$\frac{\pi}{4}$
$4\left(x+\frac{\pi}{4}\right)$	0	$\frac{\pi}{2}$	π	$\frac{3\pi}{2}$	2π
$\cos 4\left(x+\frac{\pi}{4}\right)$	1	0	-1	0	1
$3\cos 4\left(x+\frac{\pi}{4}\right)$	3	0	-3	0	3

Continued on next page

32. (continued)

Steps 4 and 5 Plot the points found in the table and join them with a sinusoidal curve.

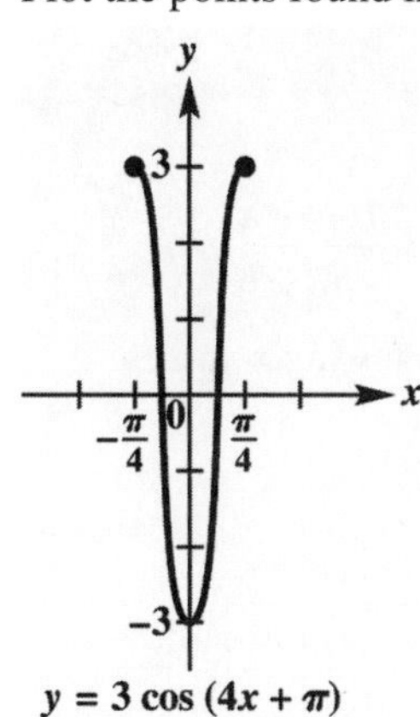

The amplitude is 3.

The period is $\frac{2\pi}{4}$, which is $\frac{\pi}{2}$.

There is no vertical translation.

The phase shift is $\frac{\pi}{4}$ unit to the left.

33. $y = \frac{1}{2}\cos\left(\frac{1}{2}x - \frac{\pi}{4}\right) = \frac{1}{2}\cos\frac{1}{2}\left(x - \frac{\pi}{2}\right)$

Step 1 Find the interval whose length is $\frac{2\pi}{b}$.

$$0 \le \frac{1}{2}\left(x - \frac{\pi}{2}\right) \le 2\pi \Rightarrow 0 \le x - \frac{\pi}{2} \le 4\pi \Rightarrow \frac{\pi}{2} \le x \le \frac{8\pi}{2} + \frac{\pi}{2} \Rightarrow \frac{\pi}{2} \le x \le \frac{9\pi}{2}$$

Step 2 Divide the period into four equal parts to get the following x-values.

$$\frac{\pi}{2}, \frac{11\pi}{4}, 5\pi, \frac{19\pi}{4}, \frac{9\pi}{2}$$

Step 3 Evaluate the function for each of the five x-values.

x	$\frac{\pi}{2}$	$\frac{11\pi}{4}$	5π	$\frac{19\pi}{4}$	$\frac{9\pi}{2}$
$\frac{1}{2}\left(x - \frac{\pi}{2}\right)$	0	$\frac{\pi}{2}$	π	$\frac{3\pi}{2}$	2π
$\cos\frac{1}{2}\left(x - \frac{\pi}{2}\right)$	1	0	−1	0	1
$\frac{1}{2}\cos\frac{1}{2}\left(x - \frac{\pi}{2}\right)$	$\frac{1}{2}$	0	$-\frac{1}{2}$	0	$\frac{1}{2}$

Steps 4 and 5 Plot the points found in the table and join them with a sinusoidal curve.

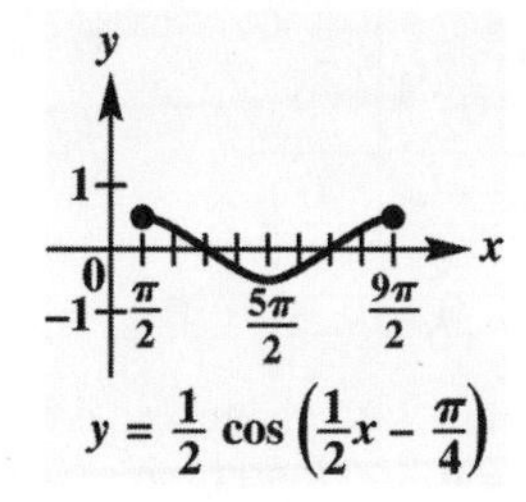

The amplitude is $\frac{1}{2}$.

The period is $\frac{2\pi}{\frac{1}{2}}$, which is 4π.

This is no vertical translation.

The phase shift is $\frac{\pi}{2}$ units to the right.

34. $y=-\frac{1}{4}\sin\left(\frac{3}{4}x+\frac{\pi}{8}\right)=-\frac{1}{4}\sin\frac{3}{4}\left(x+\frac{\pi}{6}\right)$

Step 1 Find the interval whose length is $\frac{2\pi}{b}$.

$$0\le\frac{3}{4}\left(x+\frac{\pi}{6}\right)\le 2\pi\Rightarrow 0\le x+\frac{\pi}{6}\le\frac{8\pi}{3}\Rightarrow -\frac{\pi}{6}\le x\le\frac{16\pi}{6}-\frac{\pi}{6}\Rightarrow -\frac{\pi}{6}\le x\le\frac{15\pi}{6}$$

Step 2 Divide the period into four equal parts to get the following x-values.

$$-\frac{\pi}{6},\ \frac{\pi}{2},\ \frac{7\pi}{6},\ \frac{11\pi}{6},\ \frac{15\pi}{6}$$

Step 3 Evaluate the function for each of the five x-values.

x	$-\frac{\pi}{6}$	$\frac{\pi}{2}$	$\frac{7\pi}{6}$	$\frac{11\pi}{6}$	$\frac{15\pi}{6}$
$\frac{3}{4}\left(x+\frac{\pi}{6}\right)$	0	$\frac{\pi}{2}$	π	$\frac{3\pi}{2}$	2π
$\sin\frac{3}{4}\left(x+\frac{\pi}{6}\right)$	0	1	0	-1	0
$-\frac{1}{4}\sin\frac{3}{4}\left(x+\frac{\pi}{6}\right)$	0	$-\frac{1}{4}$	0	$\frac{1}{4}$	0

Steps 4 and 5 Plot the points found in the table and join them with a sinusoidal curve.

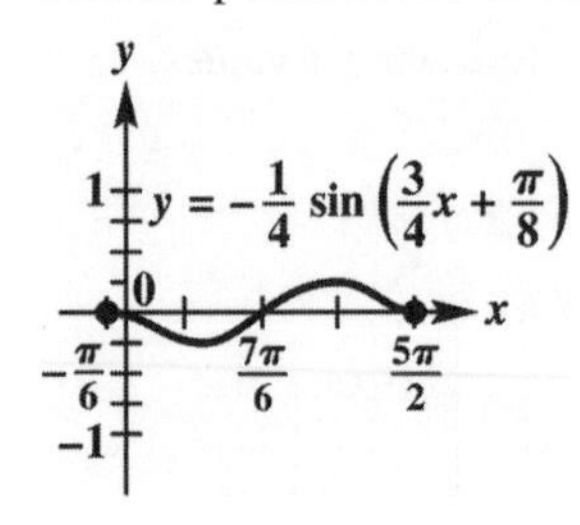

The amplitude is $\left|-\frac{1}{4}\right|$, which is $\frac{1}{4}$.

The period is $\frac{2\pi}{\frac{3}{4}}$, which is $\frac{8\pi}{3}$.

This is no vertical translation.

The phase shift if $\frac{\pi}{6}$ units to the left.

35. $y=-3+2\sin x$

Step 1 The period is 2π.

Step 2 Divide the period into four equal parts to get the following x-values.

$$0,\ \frac{\pi}{2},\ \pi,\ \frac{3\pi}{2},\ 2\pi$$

Step 3 Evaluate the function for each of the five x-values.

x	0	$\frac{\pi}{2}$	π	$\frac{3\pi}{2}$	2π
$\sin x$	0	1	0	-1	0
$2\sin x$	0	2	0	-2	0
$-3+2\sin x$	-3	-1	-3	-5	-3

Steps 4 and 5 Plot the points found in the table and join them with a sinusoidal curve. By graphing an additional period to the left, we obtain the following graph.

The amplitude is $|2|$, which is 2.

The vertical translation is 3 units down.

There is no phase shift.

36. $y = 2 - 3\cos x$

Step 1 The period is 2π.

Step 2 Divide the period into four equal parts to get the following x-values.

$$0,\ \frac{\pi}{2},\ \pi,\ \frac{3\pi}{2},\ 2\pi$$

Step 3 Evaluate the function for each of the five x-values.

x	0	$\frac{\pi}{2}$	π	$\frac{3\pi}{2}$	2π
$\cos x$	1	0	−1	0	1
$-3\cos x$	−3	0	3	0	−3
$2-3\cos x$	−1	2	5	2	−1

Steps 4 and 5 Plot the points found in the table and join them with a sinusoidal curve. By graphing an additional period to the left, we obtain the following graph.

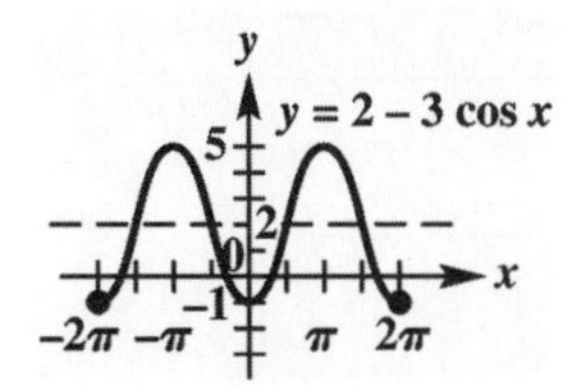

The amplitude is $|-3|$, which is 3.

The vertical translation is 2 units up.

There is no phase shift..

37. $y = -1 - 2\cos 5x$

Step 1 Find the interval whose length is $\frac{2\pi}{b}$.

$$0 \le 5x \le 2\pi \Rightarrow 0 \le x \le \frac{2\pi}{5}$$

Step 2 Divide the period into four equal parts to get the following x-values.

$$0,\ \frac{\pi}{10},\ \frac{\pi}{5},\ \frac{3\pi}{10},\ \frac{2\pi}{5}$$

Step 3 Evaluate the function for each of the five x-values.

x	0	$\frac{\pi}{10}$	$\frac{\pi}{5}$	$\frac{3\pi}{10}$	$\frac{2\pi}{5}$
$5x$	0	$\frac{\pi}{2}$	π	$\frac{3\pi}{2}$	2π
$\cos 5x$	1	0	−1	0	1
$-2\cos 5x$	−2	0	2	0	−2
$-1-2\cos 5x$	−3	−1	1	−1	−3

Steps 4 and 5 Plot the points found in the table and join them with a sinusoidal curve. By graphing an additional period to the left, we obtain the following graph.

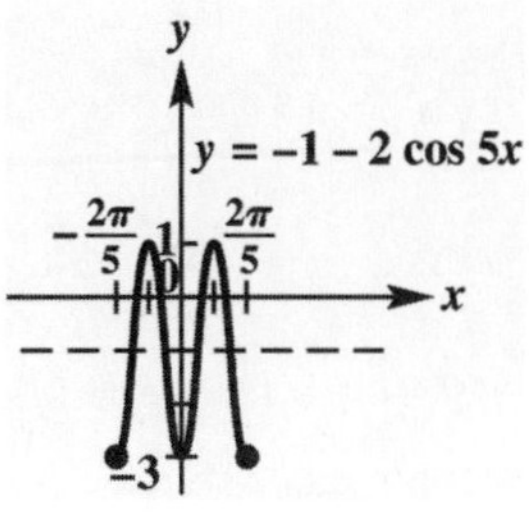

The period is $\frac{2\pi}{5}$.

The amplitude is $|-2|$, which is 2.

The vertical translation is 1 unit down.

There is no phase shift.

38. $y = 1 - \frac{2}{3}\sin\frac{3}{4}x$

Step 1 Find the interval whose length is $\frac{2\pi}{b}$.

$$0 \le \frac{3}{4}x \le 2\pi \Rightarrow 0 \le x < \frac{4}{3}\cdot 2\pi \Rightarrow 0 \le x \le \frac{8\pi}{3}$$

Step 2 Divide the period into four equal parts to get the following x-values.

$$0,\ \frac{2\pi}{3},\ \frac{4\pi}{3},\ 2\pi,\ \frac{8\pi}{3}$$

Step 3 Evaluate the function for each of the five x-values.

x	0	$\frac{2\pi}{3}$	$\frac{4\pi}{3}$	2π	$\frac{8\pi}{3}$
$\frac{3}{4}x$	0	$\frac{\pi}{2}$	π	$\frac{3\pi}{2}$	2π
$\sin\frac{3}{4}x$	0	1	0	-1	0
$-\frac{2}{3}\sin\frac{3}{4}x$	0	$-\frac{2}{3}$	0	$\frac{2}{3}$	0
$1-\frac{2}{3}\sin\frac{3}{4}x$	1	$\frac{1}{3}$	1	$\frac{5}{3}$	1

Steps 4 and 5 Plot the points found in the table and join them with a sinusoidal curve. By graphing an additional period to the left, we obtain the following graph.

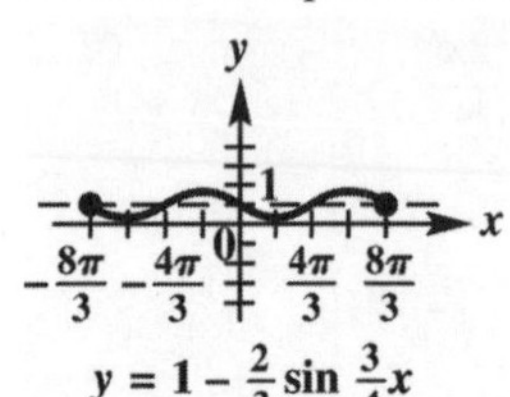

The amplitude is $\left|-\frac{2}{3}\right|$, which is $\frac{2}{3}$.

The period is $\frac{2\pi}{\frac{3}{4}}$, which is $\frac{8\pi}{3}$.

The vertical translation is 1 unit up.

There is no phase shift.

39. $y = 1 - 2\cos\frac{1}{2}x$

Step 1 Find the interval whose length is $\frac{2\pi}{b}$.

$$0 \le \frac{1}{2}x \le 2\pi \Rightarrow 0 \le x \le 4\pi$$

Step 2 Divide the period into four equal parts to get the following x-values.

$$0,\ \pi,\ 2\pi,\ 3\pi,\ 4\pi$$

Continued on next page

39. (continued)

Step 3 Evaluate the function for each of the five x-values.

x	0	π	2π	3π	4π
$\frac{1}{2}x$	0	$\frac{\pi}{2}$	π	$\frac{3\pi}{2}$	2π
$\cos\frac{1}{2}x$	1	0	-1	0	1
$-2\cos\frac{1}{2}x$	-2	0	2	0	-2
$1-2\cos\frac{1}{2}x$	-1	1	3	1	-1

Steps 4 and 5 Plot the points found in the table and join them with a sinusoidal curve. By graphing an additional period to the left, we obtain the following graph.

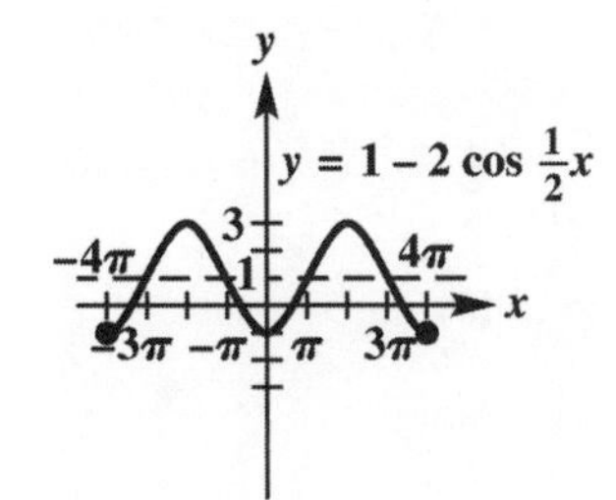

The amplitude is $|-2|$, which is 2.

The period is $\frac{2\pi}{\frac{1}{2}}$, which is 4π.

The vertical translation is 1 unit up. There is no phase shift.

40. $y=-3+3\sin\frac{1}{2}x$

Step 1 Find the interval whose length is $\frac{2\pi}{b}$.

$$0\le\frac{1}{2}x\le 2\pi \Rightarrow 0\le x\le 4\pi$$

Step 2 Divide the period into four equal parts to get the following x-values.

$$0,\ \pi,\ 2\pi,\ 3\pi,\ 4\pi$$

Step 3 Evaluate the function for each of the five x-values.

x	0	π	2π	3π	4π
$\frac{1}{2}x$	0	$\frac{\pi}{2}$	π	$\frac{3\pi}{2}$	2π
$\sin\frac{1}{2}x$	0	1	0	-1	0
$3\sin\frac{1}{2}x$	0	3	0	-3	0
$-3+3\sin\frac{1}{2}x$	-3	0	-3	-6	-3

Continued on next page

40. (continued)

Steps 4 and 5 Plot the points found in the table and join them with a sinusoidal curve. By graphing an additional period to the left, we obtain the following graph.

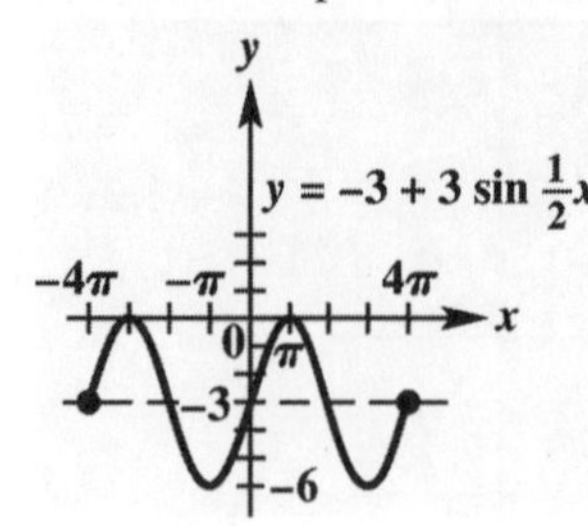

The amplitude is $|3|$, which is 3.

The period is $\frac{2\pi}{\frac{1}{2}}$, which is 4π.

The vertical translation is 3 units down.

There is no phase shift.

41. $y = -2 + \frac{1}{2}\sin 3x$

Step 1 Find the interval whose length is $\frac{2\pi}{b}$.

$$0 \le 3x \le 2\pi \Rightarrow 0 \le x \le \frac{2\pi}{3}$$

Step 2 Divide the period into four equal parts to get the following x-values.

$$0, \frac{\pi}{6}, \frac{\pi}{3}, \frac{\pi}{2}, \frac{2\pi}{3}$$

Step 3 Evaluate the function for each of the five x-values.

x	0	$\frac{\pi}{6}$	$\frac{\pi}{3}$	$\frac{\pi}{2}$	$\frac{2\pi}{3}$
$3x$	0	$\frac{\pi}{2}$	π	$\frac{3\pi}{2}$	2π
$\sin 3x$	0	1	0	-1	0
$\frac{1}{2}\sin 3x$	0	$\frac{1}{2}$	0	$-\frac{1}{2}$	0
$-2+\frac{1}{2}\sin 3x$	-2	$-\frac{3}{2}$	-2	$-\frac{5}{2}$	-2

Steps 4 and 5 Plot the points found in the table and join them with a sinusoidal curve. By graphing an additional period to the left, we obtain the following graph.

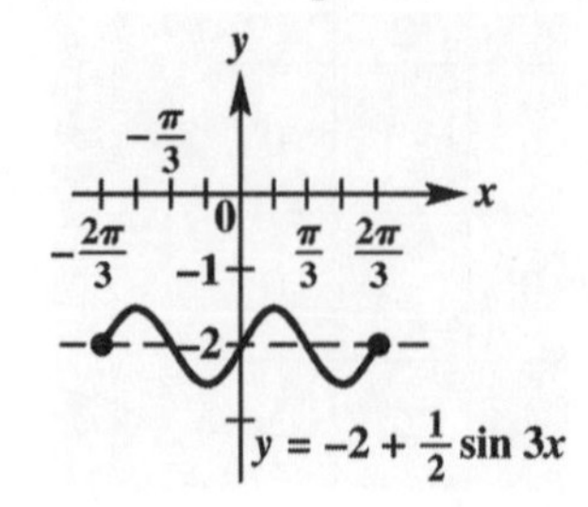

The amplitude is $\left|\frac{1}{2}\right|$, which is $\frac{1}{2}$.

The period is $\frac{2\pi}{3}$.

The vertical translation is 2 units down.

There is no phase shift.

42. $y=1+\frac{2}{3}\cos\frac{1}{2}x$

Step 1 Find the interval whose length is $\frac{2\pi}{b}$.

$$0\le\frac{1}{2}x\le 2\pi\Rightarrow 0\le x\le 4\pi$$

Step 2 Divide the period into four equal parts to get the following x-values.

$$0,\ \pi,\ 2\pi,\ 3\pi,\ 4\pi$$

Step 3 Evaluate the function for each of the five x-values.

x	0	π	2π	3π	4π
$\frac{1}{2}x$	0	$\frac{\pi}{2}$	π	$\frac{3\pi}{2}$	2π
$\cos\frac{1}{2}x$	1	0	-1	0	1
$\frac{2}{3}\cos\frac{1}{2}x$	$\frac{2}{3}$	0	$-\frac{2}{3}$	0	$\frac{2}{3}$
$1+\frac{2}{3}\cos\frac{1}{2}x$	$\frac{5}{3}$	1	$\frac{1}{3}$	1	$\frac{5}{3}$

Steps 4 and 5 Plot the points found in the table and join them with a sinusoidal curve. By graphing an additional period to the left, we obtain the following graph.

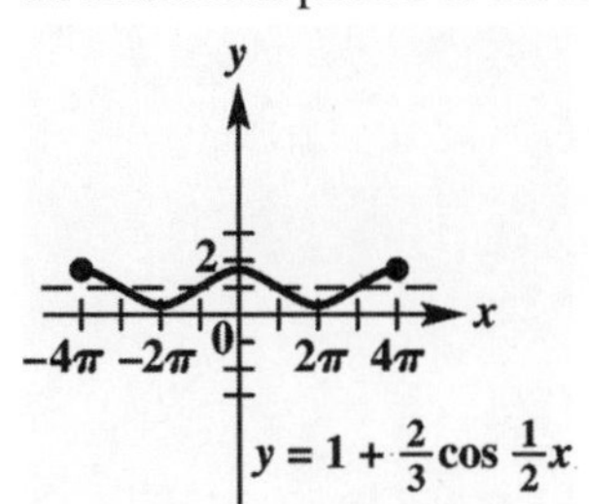

The amplitude is $\left|\frac{2}{3}\right|$, which is $\frac{2}{3}$.

The period is 4π.

The vertical translation is 1 unit up.

There is no phase shift.

43. $y=-3+2\sin\left(x+\frac{\pi}{2}\right)$

Step 1 Find the interval whose length is $\frac{2\pi}{b}$.

$$0\le x+\frac{\pi}{2}\le 2\pi\Rightarrow\Rightarrow 0-\frac{\pi}{2}\le x\le 2\pi-\frac{\pi}{2}\Rightarrow -\frac{\pi}{2}\le x\le\frac{3\pi}{2}$$

Step 2 Divide the period into four equal parts to get the following x-values.

$$-\frac{\pi}{2},\ 0,\ \frac{\pi}{2},\ \pi,\ \frac{3\pi}{2}$$

Continued on next page

43. (continued)

Step 3 Evaluate the function for each of the five x-values

x	$-\frac{\pi}{2}$	0	$\frac{\pi}{2}$	π	$\frac{3\pi}{2}$
$x+\frac{\pi}{2}$	0	$\frac{\pi}{2}$	π	$\frac{3\pi}{2}$	2π
$\sin\left(x+\frac{\pi}{2}\right)$	0	1	0	−1	0
$2\sin\left(x+\frac{\pi}{2}\right)$	0	2	0	−2	0
$-3+2\sin\left(x+\frac{\pi}{2}\right)$	−3	−1	−3	−5	−3

Steps 4 and 5 Plot the points found in the table and join them with a sinusoidal curve.

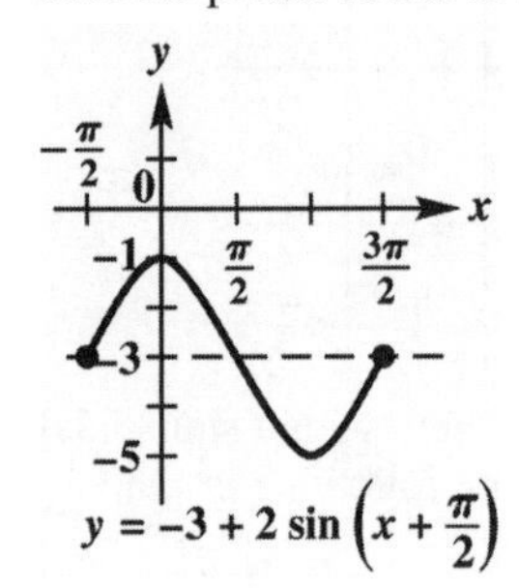

The amplitude is $|2|$, which is 2.

The period is 2π.

The vertical translation is 3 units down.

The phase shift is $\frac{\pi}{2}$ units to the left.

44. $y = 4 - 3\cos(x - \pi)$

Step 1 Find the interval whose length is $\frac{2\pi}{b}$.

$$0 \le x - \pi \le 2\pi \Rightarrow \pi \le x \le 3\pi$$

Step 2 Divide the period into four equal parts to get the following x-values.

$$\pi, \frac{3\pi}{2}, 2\pi, \frac{5\pi}{2}, 3\pi$$

Step 3 Evaluate the function for each of the five x-values.

x	π	$\frac{3\pi}{2}$	2π	$\frac{5\pi}{2}$	3π
$x-\pi$	0	$\frac{\pi}{2}$	π	$\frac{3\pi}{2}$	2π
$\cos(x-\pi)$	1	0	−1	0	1
$-3\cos(x-\pi)$	−3	0	3	0	−3
$4-3\cos(x-\pi)$	1	4	7	4	1

Continued on next page

44. (continued)

Steps 4 and 5 Plot the points found in the table and join then with a sinusoidal curve.

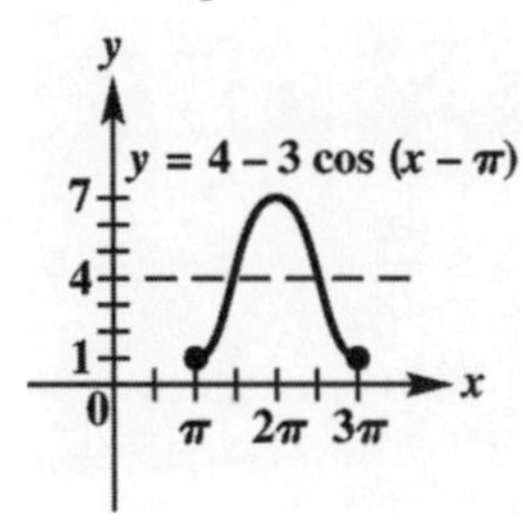

The amplitude is $|-3|$, which is 3.

The period is 2π.

The vertical translation is 4 units up.

The phase shift is π units to the right.

45. $y=\frac{1}{2}+\sin 2\left(x+\frac{\pi}{4}\right)$

Step 1 Find the interval whose length is $\frac{2\pi}{b}$.

$$0\le 2\left(x+\frac{\pi}{4}\right)\le 2\pi \Rightarrow 0\le x+\frac{\pi}{4}\le \frac{2\pi}{2} \Rightarrow 0\le x+\frac{\pi}{4}\le \pi \Rightarrow -\frac{\pi}{4}\le x\le \frac{3\pi}{4}$$

Step 2 Divide the period into four equal parts to get the following x-values.

$$-\frac{\pi}{4},\ 0,\ \frac{\pi}{4},\ \frac{\pi}{2},\ \frac{3\pi}{4}$$

Step 3 Evaluate the function for each of the five x-values.

x	$-\frac{\pi}{4}$	0	$\frac{\pi}{4}$	$\frac{\pi}{2}$	$\frac{3\pi}{4}$
$2\left(x+\frac{\pi}{4}\right)$	0	$\frac{\pi}{2}$	π	$\frac{3\pi}{2}$	2π
$\sin 2\left(x+\frac{\pi}{4}\right)$	0	1	0	-1	0
$\frac{1}{2}+\sin 2\left(x+\frac{\pi}{4}\right)$	$\frac{1}{2}$	$\frac{3}{2}$	$\frac{1}{2}$	$-\frac{1}{2}$	$\frac{1}{2}$

Steps 4 and 5 Plot the points found in the table and join them with a sinusoidal curve.

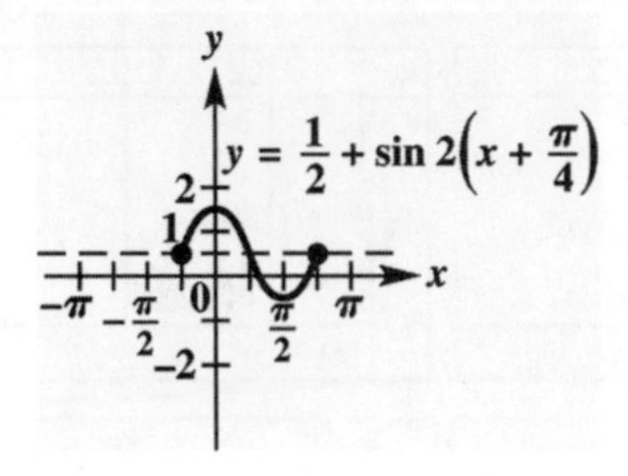

The amplitude is $|1|$, which is 1.

The period is $\frac{2\pi}{2}$, which is π.

The vertical translation is $\frac{1}{2}$ unit up.

The phase shift is $\frac{\pi}{4}$ units to the left.

46. $y = -\frac{5}{2} + \cos 3\left(x - \frac{\pi}{6}\right)$

Step 1 Find the interval whose length is $\frac{2\pi}{b}$.

$$0 \le 3\left(x - \frac{\pi}{6}\right) \le 2\pi \Rightarrow 0 \le x - \frac{\pi}{6} \le \frac{2\pi}{3} \Rightarrow \frac{\pi}{6} \le x \le \frac{4\pi}{6} + \frac{\pi}{6} \Rightarrow \frac{\pi}{6} \le x \le \frac{5\pi}{6}$$

Step 2 Divide the period into four equal parts to get the following x-values.

$$\frac{\pi}{6}, \frac{\pi}{3}, \frac{\pi}{2}, \frac{2\pi}{3}, \frac{5\pi}{6}$$

Step 3 Evaluate the function for each of the five x-values.

x	$\frac{\pi}{6}$	$\frac{\pi}{3}$	$\frac{\pi}{2}$	$\frac{2\pi}{3}$	$\frac{5\pi}{6}$
$3\left(x - \frac{\pi}{6}\right)$	0	$\frac{\pi}{2}$	π	$\frac{3\pi}{2}$	2π
$\cos 3\left(x - \frac{\pi}{6}\right)$	1	0	-1	0	1
$-\frac{5}{2} + \cos 3\left(x - \frac{\pi}{6}\right)$	$-\frac{3}{2}$	$-\frac{5}{2}$	$-\frac{7}{2}$	$-\frac{5}{2}$	$-\frac{3}{2}$

Steps 4 and 5 Plot the points found in the table and join them with a sinusoidal curve.

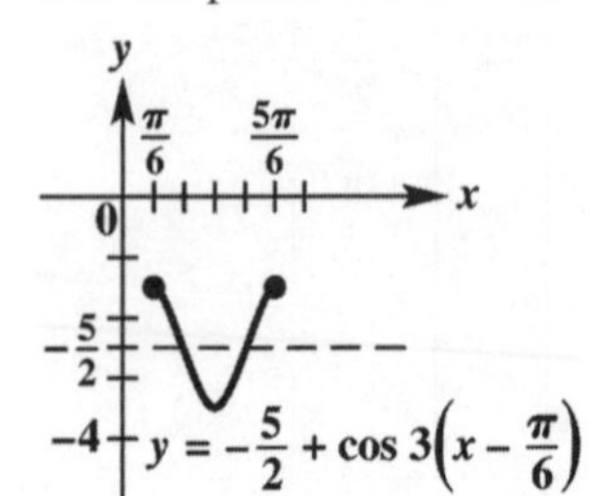

The amplitude is $|1|$, which is 1.

The period is $\frac{2\pi}{3}$.

The vertical translation is $\frac{5}{2}$ units down.

The phase shift is $\frac{\pi}{6}$ units to the right.

47. **(a)** Let January correspond to $x = 1$, February to $x = 2$, … , and December of the second year to $x = 24$. Yes, the data appears to outline the graph of a translated sine graph.

L1	L2	L3
1	36	------
2	39	
3	43	
4	48	
5	55	
6	59	
7	64	

L2(1)=36

L1	L2	L3
7	64	
8	63	
9	57	
10	50	
11	43	
12	39	
13	36	

L2(7) =64

L1	L2	L3
13	36	
14	39	
15	43	
16	48	
17	55	
18	59	
19	64	

L2(19) =64

L1	L2	L3
18	59	
19	64	
20	63	
21	57	
22	50	
23	43	
24	39	

L2(24)=39

WINDOW
Xmin=1
Xmax=25
Xscl=5
Ymin=30
Ymax=70
Yscl=5
Xres=1

Continued on next page

47. (continued)

(b) The sine graph is vertically centered around the line $y = 50$. This line represents the average yearly temperature in Vancouver of 50°F. (This is also the actual average yearly temperature.)

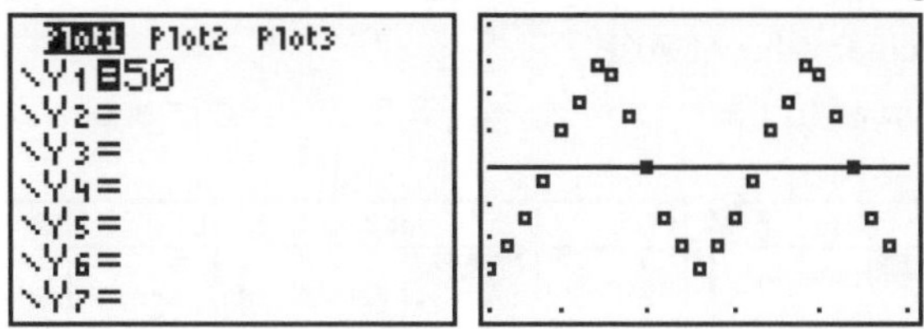

(c) The amplitude of the sine graph is approximately 14 since the average monthly high is 64, the average monthly low is 36, and $\frac{1}{2}(64-36)=\frac{1}{2}(28)=14.$ The period is 12 since the temperature cycles every twelve months. Let $b=\frac{2\pi}{12}=\frac{\pi}{6}.$ One way to determine the phase shift is to use the following technique. The minimum temperature occurs in January. Thus, when $x = 1$, $b(x-d)$ must equal $\left(-\frac{\pi}{2}\right)+2\pi n,$ where n is an integer, since the sine function is minimum at these values. Solving for d, we have the following.

$$\frac{\pi}{6}(1-d)=-\frac{\pi}{2}\Rightarrow 1-d=\frac{6}{\pi}\left(-\frac{\pi}{2}\right)\Rightarrow 1-d=-3\Rightarrow -d=-4\Rightarrow d=4$$

This can be used as a first approximation.

(d) Let $f(x) = a \sin b(x-d) + c$. Since the amplitude is 14, let $a = 14$. The period is equal to 1 yr or 12 mo, so $b=\frac{\pi}{6}.$ The average of the maximum and minimum temperatures is as follows.

$$\frac{1}{2}(64+36)=\frac{1}{2}(100)=50$$

Let the vertical translation be $c = 50$.
Since the phase shift is approximately 4, it can be adjusted slightly to give a better visual fit. Try 4.2. Since the phase shift is 4.2, let $d = 4.2$. Thus, $f(x)=14\sin\left[\frac{\pi}{6}(x-4.2)\right]+50.$

(e) Plotting the data with $f(x)=14\sin\left[\frac{\pi}{6}(x-4.2)\right]+50$ on the same coordinate axes gives a good fit.

(f)

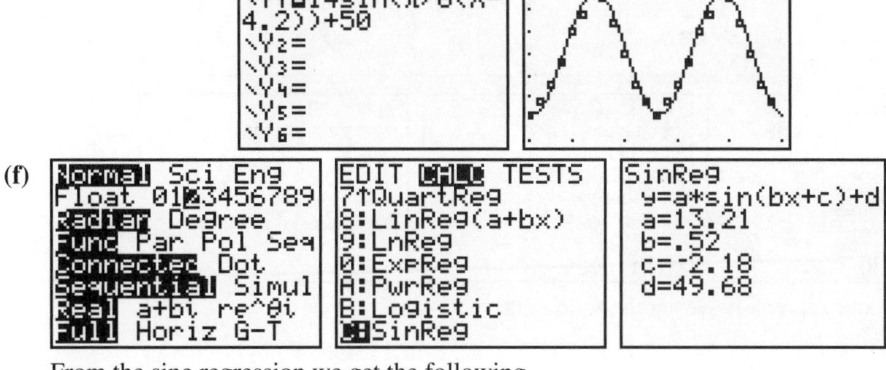

From the sine regression we get the following.

$$y=13.21\sin(.52x-2.18)+49.68=13.21\sin\left[.52(x-4.19)\right]+49.68$$

48. (a) We can predict the average yearly temperature by finding the mean of the average monthly temperatures: $\frac{51+55+63+67+77+86+50+90+84+71+59+52}{12}=\frac{845}{12}\approx 70.4°\text{F}$, which is very close to the actual value of 70°F.

(b) Let January correspond to $x=1$, February to $x=2$, … , and December of the second year to $x=24$.

(c) Let the amplitude a be $\frac{1}{2}(90-51)=\frac{1}{2}(39)=19.5$. Since the period is 12, let $b=\frac{\pi}{6}$. Let $c=\frac{1}{2}(90+51)=\frac{1}{2}(141)=70.5$. The minimum temperature occurs in January. Thus, when $x=1$, $b(x-d)$ must equal an odd multiple of π since the cosine function is minimum at these values. Solving for d, we have the following.

$$\frac{\pi}{6}(1-d)=-\pi\Rightarrow 1-d=\frac{6}{\pi}(-\pi)\Rightarrow 1-d=-6\Rightarrow -d=-7\Rightarrow d=7$$

d can be adjusted slightly to give a better visual fit. Try $d=7.2$. Thus, we have the following.

$$f(x)=a\cos b(x-d)+c=19.5\cos\left[\frac{\pi}{6}(x-7.2)\right]+70.5$$

(d) Plotting the data with $f(x)=19.5\cos\left[\frac{\pi}{6}(x-7.2)\right]+70.5$ on the same coordinate axes give a good fit.

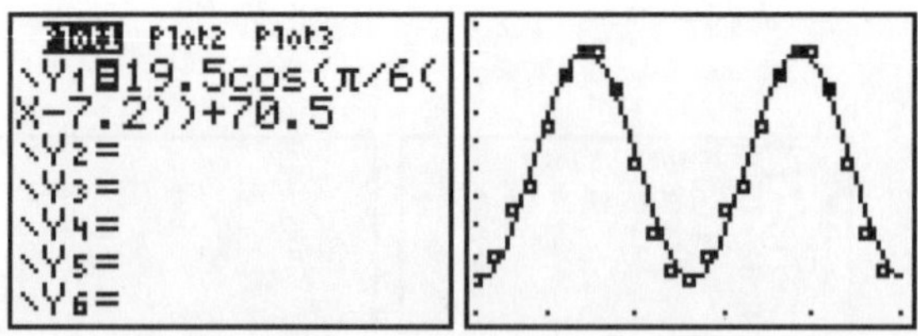

(e)

From the sine regression we get the following.

$$y=19.72\sin(.52x-2.17)+70.47=19.72\sin\left[.52(x-4.17)\right]+70.47$$

Section 4.3: Graphs of the Other Circular Functions

1. $y = -\csc x$
The graph is the reflection of the graph of $y = \csc x$ about the x-axis. This matches with graph B.

2. $y = -\sec x$
The graph is the reflection of the graph of $y = \sec x$ about the x-axis. This matches with graph C.

3. $y = -\tan x$
The graph is the reflection of the graph of $y = \tan x$ about the x-axis. This matches with graph E.

4. $y = -\cot x$
The graph is the reflection of the graph of $y = \cot x$ about the x-axis. This matches with graph A.

5. $y = \tan\left(x - \frac{\pi}{4}\right)$

The graph is the graph of $y = \tan x$ shifted $\frac{\pi}{4}$ units to the right. This matches with graph D.

6. $y = \cot\left(x - \frac{\pi}{4}\right)$

The graph is the graph of $y = \cot x$ shifted $\frac{\pi}{4}$ units to the right. This matches with graph F.

7. $y = 3\sec\frac{1}{4}x$

Step 1 Graph the corresponding reciprocal function $y = 3\cos\frac{1}{4}x$. The period is $\frac{2\pi}{\frac{1}{4}} = 2\pi \cdot \frac{4}{1} = 8\pi$ and its amplitude is $|3| = 3$. One period is in the interval $0 \le x \le 8\pi$. Dividing the interval into four equal parts gives us the following key points.

$$(0,1),\ (2\pi,0),\ (4\pi,-1),\ (6\pi,0),\ (8\pi,1)$$

Step 2 The vertical asymptotes of $y = \sec\frac{1}{4}x$ are at the x-intercepts of $y = \cos\frac{1}{4}x$, which are $x = 2\pi$ and $x = 6\pi$. Continuing this pattern to the left, we also have a vertical asymptote of $x = -2\pi$.

Step 3 Sketch the graph.

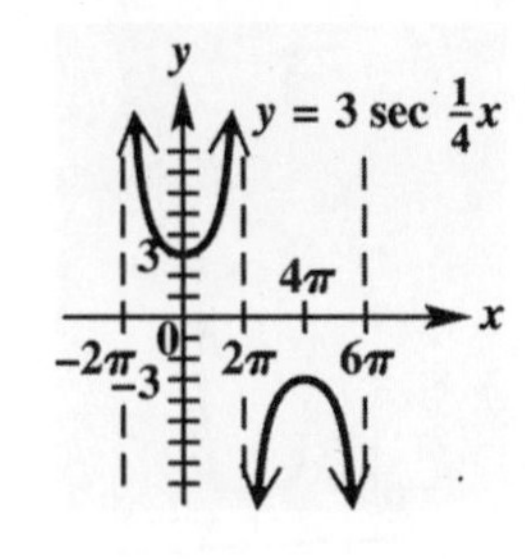

8. $y = -2\sec\frac{1}{2}x$

Step 1 Graph the corresponding reciprocal function $y = -2\cos\frac{1}{2}x$. The period is $\frac{2\pi}{\frac{1}{2}} = 2\pi \cdot \frac{2}{1} = 4\pi$ and its amplitude is $|-2| = 2$. One period is in the interval $0 \le x \le 4\pi$. Dividing the interval into four equal parts gives us the following key points.

$$(0,-2),\ (\pi,0),\ (2\pi,2),\ (3\pi,0),\ (4\pi,-2)$$

Step 2 The vertical asymptotes of $y = -2\sec\frac{1}{2}x$ are at the x-intercepts of $y = -2\cos\frac{1}{2}x$, which are $x = \pi$ and $x = 3\pi$. Continuing this pattern to the left, we also have a vertical asymptote of $x = -\pi$.

Step 3 Sketch the graph.

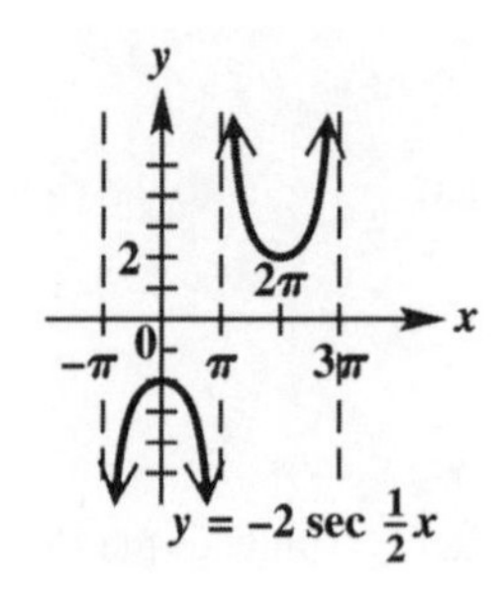

9. $y = -\frac{1}{2}\csc\left(x + \frac{\pi}{2}\right)$

Step 1 Graph the corresponding reciprocal function $y = -\frac{1}{2}\sin\left(x + \frac{\pi}{2}\right)$. The period is 2π and its amplitude is $\left|-\frac{1}{2}\right| = \frac{1}{2}$. One period is in the interval $-\frac{\pi}{2} \le x \le \frac{3\pi}{2}$. Dividing the interval into four equal parts gives us the following key points.

$$\left(-\frac{\pi}{2},0\right),\ \left(0,-\frac{1}{2}\right),\ \left(\frac{\pi}{2},0\right),\ \left(\pi,\frac{1}{2}\right),\ \left(\frac{3\pi}{2},0\right)$$

Step 2 The vertical asymptotes of $y = -\frac{1}{2}\csc\left(x + \frac{\pi}{2}\right)$ are at the x-intercepts of $y = -\frac{1}{2}\sin\left(x + \frac{\pi}{2}\right)$, which are $x = -\frac{\pi}{2}$, $x = \frac{\pi}{2}$, and $x = \frac{3\pi}{2}$.

Step 3 Sketch the graph.

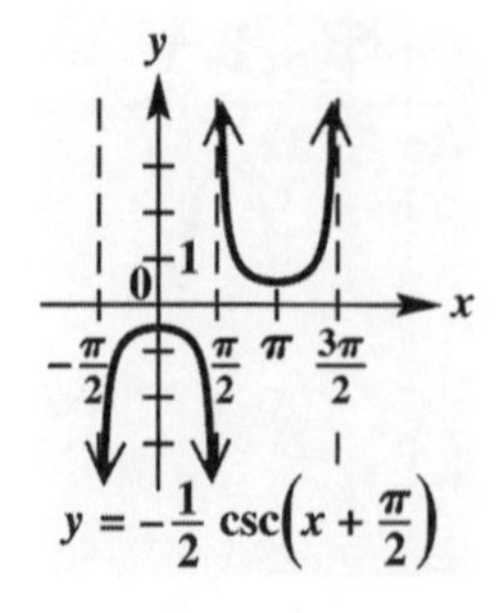

10. $y = \frac{1}{2}\csc\left(x - \frac{\pi}{2}\right)$

Step 1 Graph the corresponding reciprocal function $y = \frac{1}{2}\sin\left(x - \frac{\pi}{2}\right)$. The period is 2π and its amplitude is $\left|\frac{1}{2}\right| = \frac{1}{2}$. One period is in the interval $\frac{\pi}{2} \le x \le \frac{5\pi}{2}$. Dividing the interval into four equal parts us the following key points.

$$\left(\frac{\pi}{2},0\right),\ \left(\pi,\frac{1}{2}\right),\ \left(\frac{3\pi}{2},0\right),\ \left(2\pi,-\frac{1}{2}\right),\ \left(\frac{5\pi}{2},0\right)$$

Step 2 The vertical asymptotes of $y = \frac{1}{2}\csc\left(x - \frac{\pi}{2}\right)$ are at the x-intercepts of $y = \frac{1}{2}\sin\left(x - \frac{\pi}{2}\right)$, which are $x = \frac{\pi}{2}$, $x = \frac{3\pi}{2}$, and $x = \frac{5\pi}{2}$.

Step 3 Sketch the graph.

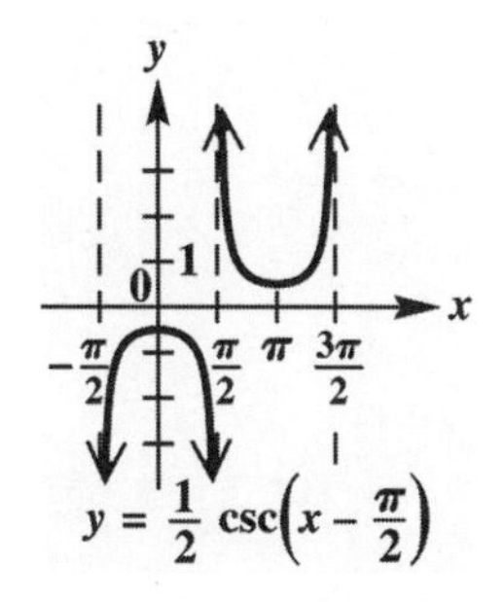

11. $y = \csc\left(x - \frac{\pi}{4}\right)$

Step 1 Graph the corresponding reciprocal function $y = \sin\left(x - \frac{\pi}{4}\right)$ The period is 2π and its amplitude is $|1| = 1$. One period is in the interval $\frac{\pi}{4} \le x \le \frac{9\pi}{4}$. Dividing the interval into four equal parts gives us the following key points.

$$\left(\frac{\pi}{4},0\right),\ \left(\frac{3\pi}{4},1\right),\ \left(\frac{5\pi}{4},0\right),\ \left(\frac{7\pi}{4},-1\right),\ \left(\frac{9\pi}{4},0\right)$$

Step 2 The vertical asymptotes of $y = \csc\left(x - \frac{\pi}{4}\right)$ are at the x-intercepts of $y = \sin\left(x - \frac{\pi}{4}\right)$, which are $x = \frac{\pi}{4}$, $x = \frac{5\pi}{4}$, and $x = \frac{9\pi}{4}$.

Step 3 Sketch the graph.

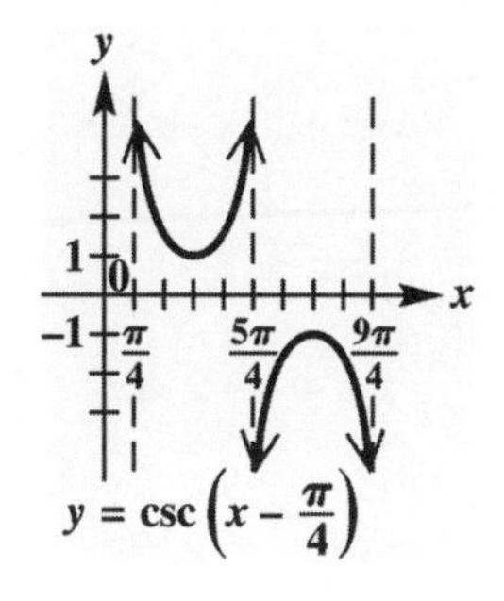

12. $y = \sec\left(x + \frac{3\pi}{4}\right)$

Step 1 Graph the corresponding reciprocal function $y = \cos\left(x + \frac{3\pi}{4}\right)$. The period is 2π and its amplitude is $|1| = 1$. One period is in the interval $-\frac{3\pi}{4} \le x \le \frac{5\pi}{4}$. Dividing the interval into four equal parts gives us the following key points.

$$\left(-\frac{3\pi}{4}, 1\right), \left(-\frac{\pi}{4}, 0\right), \left(\frac{\pi}{4}, -1\right), \left(\frac{3\pi}{4}, 0\right), \left(\frac{5\pi}{4}, 1\right)$$

Step 2 The vertical asymptotes of $y = \sec\left(x + \frac{3\pi}{4}\right)$ are at the x-intercepts of $y = \cos\left(x + \frac{3\pi}{4}\right)$, which are $x = -\frac{\pi}{4}$ and $x = \frac{3\pi}{4}$. Continuing this pattern to the right, we also have a vertical asymptote of $x = \frac{7\pi}{4}$.

Step 3 Sketch the graph.

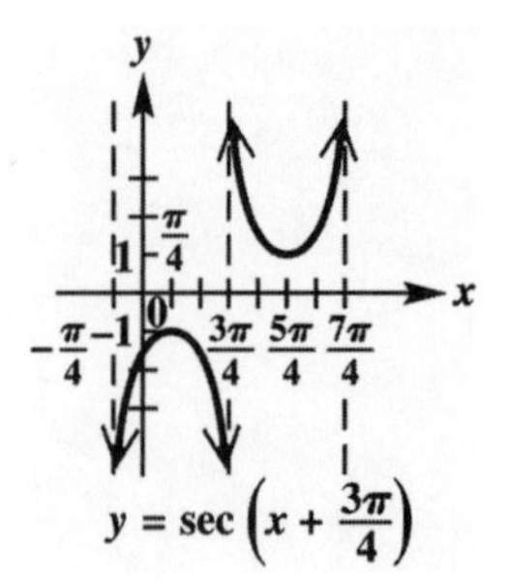

13. $y = \sec\left(x + \frac{\pi}{4}\right)$

Step 1 Graph the corresponding reciprocal function $y = \cos\left(x + \frac{\pi}{4}\right)$. The period is 2π and its amplitude is $|1| = 1$. One period is in the interval $-\frac{\pi}{4} \le x \le \frac{7\pi}{4}$. Dividing the interval into four equal parts gives us the following key points.

$$\left(-\frac{\pi}{4}, 1\right), \left(\frac{\pi}{4}, 0\right), \left(\frac{3\pi}{4}, -1\right), \left(\frac{5\pi}{4}, 0\right), \left(\frac{7\pi}{4}, 1\right)$$

Step 2 The vertical asymptotes of $y = \sec\left(x + \frac{\pi}{4}\right)$ are at the x-intercepts of $y = \cos\left(x + \frac{\pi}{4}\right)$, which are $x = \frac{\pi}{4}$ and $x = \frac{5\pi}{4}$. Continuing this pattern to the right, we also have a vertical asymptote of $x = \frac{9\pi}{4}$.

Step 3 Sketch the graph.

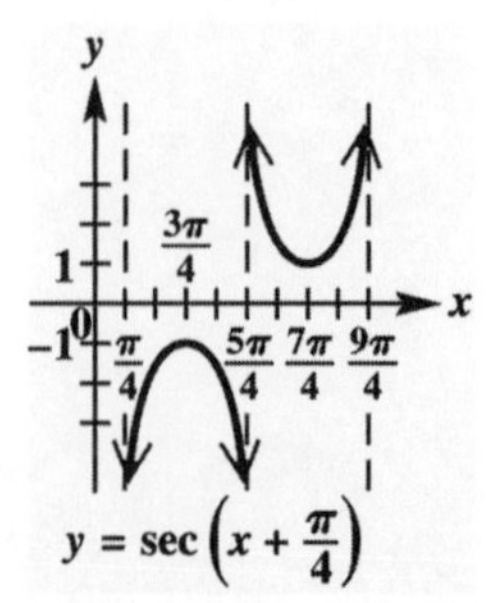

14. $y = \csc\left(x + \frac{\pi}{3}\right)$

Step 1 Graph the corresponding reciprocal function $y = \sin\left(x + \frac{\pi}{3}\right)$. The period is 2π and its amplitude is 1. One period is in the interval $-\frac{\pi}{3} \le x \le \frac{5\pi}{3}$. Dividing the interval into four equal parts gives us the following key points.

$$\left(-\frac{\pi}{3},0\right),\ \left(\frac{\pi}{6},1\right),\ \left(\frac{2\pi}{3},0\right),\ \left(\frac{7\pi}{6},-1\right),\ \left(\frac{5\pi}{3},0\right)$$

Step 2 The vertical asymptotes of $y = \csc\left(x + \frac{\pi}{3}\right)$ are at the x-intercepts of $y = \sin\left(x + \frac{\pi}{3}\right)$, which are $x = -\frac{\pi}{3}$, $x = \frac{2\pi}{3}$, and $x = \frac{5\pi}{3}$.

Step 3 Sketch the graph.

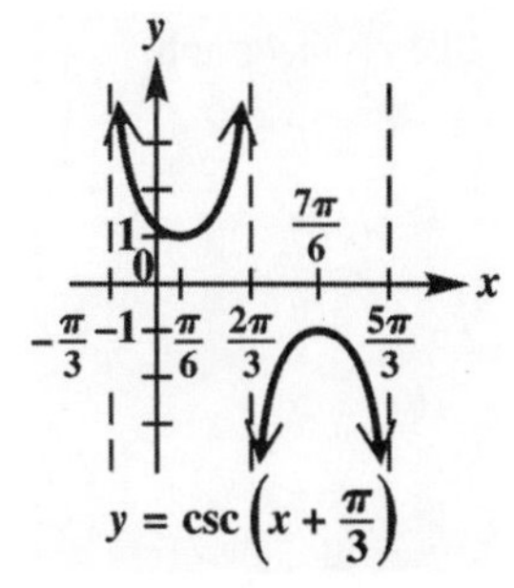

15. $y = \sec\left(\frac{1}{2}x + \frac{\pi}{3}\right) = \sec\frac{1}{2}\left(x + \frac{2\pi}{3}\right)$

Step 1 Graph the corresponding reciprocal function $y = \cos\frac{1}{2}\left(x + \frac{2\pi}{3}\right)$. The period is $\frac{2\pi}{\frac{1}{2}} = 2\pi \cdot \frac{2}{1} = 4\pi$ and its amplitude is $\left|\frac{1}{2}\right| = \frac{1}{2}$. One period is in the interval $-\frac{2\pi}{3} \le x \le \frac{10\pi}{3}$. Dividing the interval into four equal parts gives us the following key points.

$$\left(-\frac{2\pi}{3},\frac{1}{2}\right),\ \left(\frac{\pi}{3},0\right),\ \left(\frac{4\pi}{3},-\frac{1}{2}\right),\ \left(\frac{7\pi}{3},0\right),\ \left(\frac{10\pi}{3},\frac{1}{2}\right)$$

Step 2 The vertical asymptotes of $y = \sec\frac{1}{2}\left(x + \frac{2\pi}{3}\right)$ are at the x-intercepts of $y = \cos\frac{1}{2}\left(x + \frac{2\pi}{3}\right)$, which are $x = \frac{\pi}{3}$ and $x = \frac{7\pi}{3}$. Continuing this pattern to the right, we also have a vertical asymptote of $x = \frac{13\pi}{3}$.

Step 3 Sketch the graph.

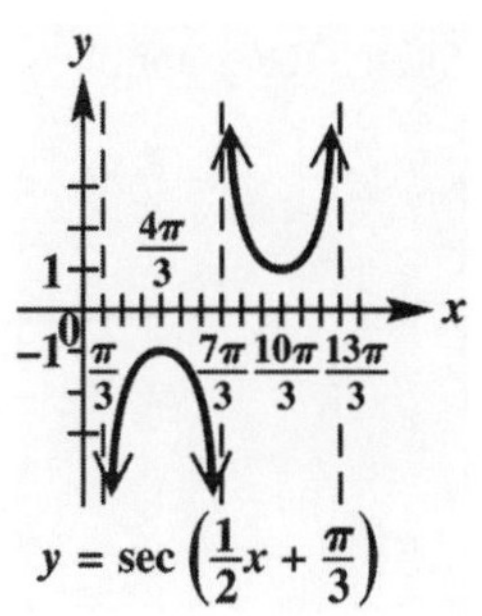

16. $y=\csc\left(\frac{1}{2}x-\frac{\pi}{4}\right)=\csc\frac{1}{2}\left(x-\frac{\pi}{2}\right)$

Step 1 Graph the corresponding reciprocal function $y=\sin\frac{1}{2}\left(x-\frac{\pi}{2}\right)$. The period is $\frac{2\pi}{\frac{1}{2}}=2\pi\cdot\frac{2}{1}=4\pi$ and its amplitude is $|1|=1$. One period is in the interval $\frac{\pi}{2}\le x\le\frac{9\pi}{2}$. Dividing the interval into four equal parts gives us the following key points.

$$\left(\frac{\pi}{2},0\right),\ \left(\frac{3\pi}{2},1\right),\ \left(\frac{5\pi}{2},0\right),\ \left(\frac{7\pi}{2},-1\right),\ \left(\frac{9\pi}{2},0\right)$$

Step 2 The vertical asymptotes of $y=\csc\frac{1}{2}\left(x-\frac{\pi}{2}\right)$ are at the x-intercepts of $y=\sin\frac{1}{2}\left(x-\frac{\pi}{2}\right)$, which are $x=\frac{\pi}{2}$, $x=\frac{5\pi}{2}$, and $x=\frac{9\pi}{2}$.

Step 3 Sketch the graph.

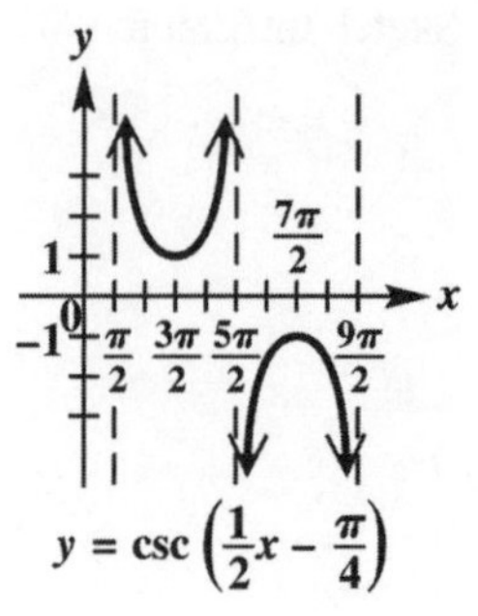

17. $y=2+3\sec(2x-\pi)=2+3\sec 2\left(x-\frac{\pi}{2}\right)$

Step 1 Graph the corresponding reciprocal function $y=2+3\cos 2\left(x-\frac{\pi}{2}\right)$. The period is π and its amplitude is $|3|=3$. One period is in the interval $\frac{\pi}{2}\le x\le\frac{3\pi}{2}$. Dividing the interval into four equal parts gives us the following key points.

$$\left(\frac{\pi}{2},5\right),\ \left(\frac{3\pi}{4},2\right),\ (\pi,-1),\ \left(\frac{5\pi}{4},2\right),\ \left(\frac{3\pi}{2},5\right)$$

Step 2 The vertical asymptotes of $y=2+3\sec 2\left(x-\frac{\pi}{2}\right)$ are at the x-intercepts of $y=3\cos 2\left(x-\frac{\pi}{2}\right)$, which are $x=\frac{3\pi}{4}$ and $x=\frac{5\pi}{4}$. Continuing this pattern to the left, we also have a vertical asymptote of $x=\frac{\pi}{4}$.

Step 3 Sketch the graph.

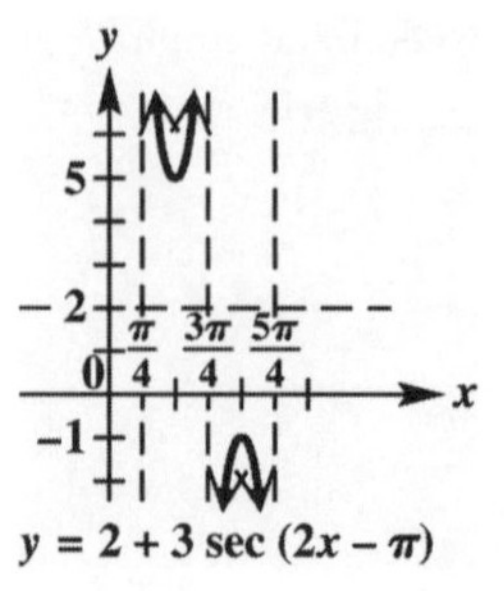

18. $y = 1 - 2\csc\left(x + \frac{\pi}{2}\right)$

Step 1 Graph the corresponding reciprocal function $y = 1 - 2\sin\left(x + \frac{\pi}{2}\right)$. The period is 2π and its amplitude is $|-2| = 2$. One period is in the interval $-\frac{\pi}{2} \le x \le \frac{3\pi}{2}$. Dividing the interval into four equal parts gives us the following key points.

$$\left(-\frac{\pi}{2}, 0\right),\ (0, -1),\ \left(\frac{\pi}{2}, 1\right),\ (\pi, 3),\ \left(\frac{3\pi}{2}, 1\right)$$

Step 2 The vertical asymptotes of $y = 1 - 2\csc\left(x + \frac{\pi}{2}\right)$ are at the x-intercepts of $y = -2\sin\left(x + \frac{\pi}{2}\right)$, which are $x = -\frac{\pi}{2}$, $x = \frac{\pi}{2}$, and $x = \frac{3\pi}{2}$.

Step 3 Sketch the graph.

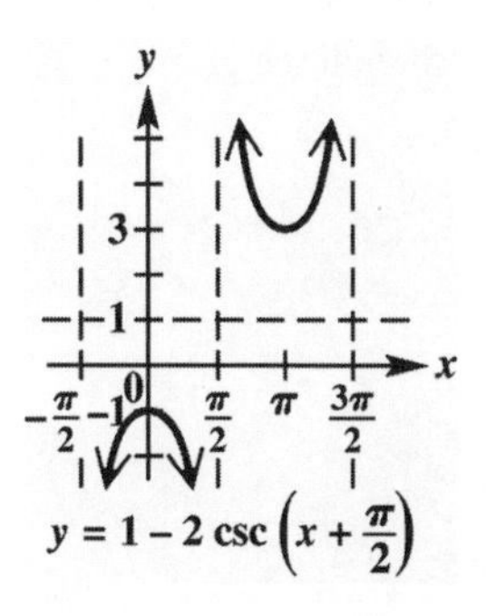

19. $y = 1 - \frac{1}{2}\csc\left(x - \frac{3\pi}{4}\right)$

Step 1 Graph the corresponding reciprocal function $y = 1 - \frac{1}{2}\sin\left(x - \frac{3\pi}{4}\right)$. The period is 2π and its amplitude is $\frac{1}{2}$. One period is in the interval $\frac{3\pi}{4} \le x \le \frac{11\pi}{4}$. Dividing the interval into four equal parts gives us the following key points.

$$\left(\frac{3\pi}{4}, 1\right),\ \left(\frac{5\pi}{4}, \frac{1}{2}\right),\ \left(\frac{7\pi}{4}, 1\right),\ \left(\frac{9\pi}{4}, \frac{3}{2}\right),\ \left(\frac{11\pi}{4}, 1\right)$$

Step 2 The vertical asymptotes of $y = 1 - \frac{1}{2}\csc\left(x - \frac{3\pi}{4}\right)$ are at the x-intercepts of $y = -\frac{1}{2}\sin\left(x - \frac{3\pi}{4}\right)$, which are $x = \frac{3\pi}{4}$, $x = \frac{7\pi}{4}$, and $x = \frac{11\pi}{4}$.

Step 3 Sketch the graph.

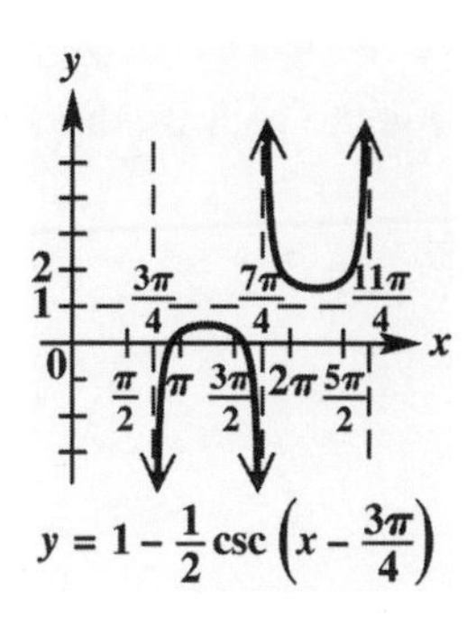

20. $y=2+\frac{1}{4}\sec\left(\frac{1}{2}x-\pi\right)=2+\frac{1}{4}\sec\frac{1}{2}(x-2\pi)$

Step 1 Graph the corresponding reciprocal function $y=2+\frac{1}{4}\cos\frac{1}{2}(x-2\pi)$. The period is $\frac{2\pi}{\frac{1}{2}}=2\pi\cdot\frac{2}{1}=4\pi$ and its amplitude is $\left|\frac{1}{4}\right|=\frac{1}{4}$.. One period is in the interval $2\pi\le x\le 6\pi$. Dividing the interval into four equal parts gives us the following key points.

$$\left(2\pi,\frac{9}{4}\right),\ (3\pi,2),\ \left(4\pi,\frac{7}{4}\right),\ (5\pi,2),\ \left(6\pi,\frac{9}{4}\right)$$

Step 2 The vertical asymptotes of $y=2+\frac{1}{4}\sec\frac{1}{2}(x-2\pi)$ are at the x-intercepts of $y=\frac{1}{4}\cos\frac{1}{2}(x-2\pi)$, which are $x=3\pi$ and $x=5\pi$. Continuing this pattern to the left, we also have a vertical asymptote of $x=-\pi$.

Step 3 Sketch the graph.

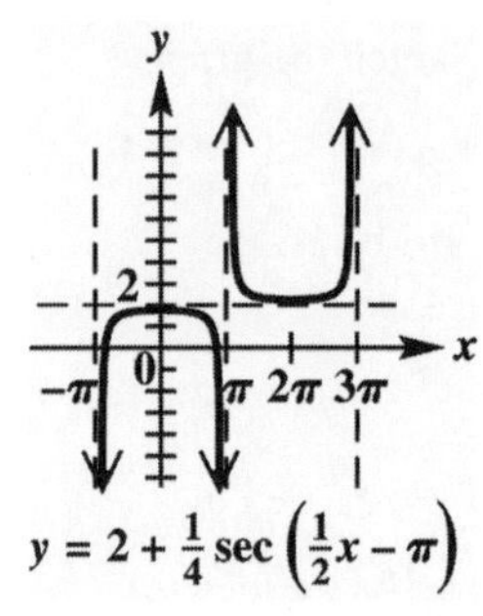

21. $y=\tan 4x$

Step 1 Find the period and locate the vertical asymptotes. The period of tangent is $\frac{\pi}{b}$, so the period for this function is $\frac{\pi}{4}$. Tangent has asymptotes of the form $bx=-\frac{\pi}{2}$ and $bx=\frac{\pi}{2}$. Therefore, the asymptotes for $y=\tan 4x$ are as follows.

$$4x=-\frac{\pi}{2}\Rightarrow x=-\frac{\pi}{8}\text{ and }4x=\frac{\pi}{2}\Rightarrow x=\frac{\pi}{8}.$$

Step 2 Sketch the two vertical asymptotes found in Step 1.

Step 3 Divide the interval into four equal parts.

$$-\frac{\pi}{8},\ -\frac{\pi}{16},\ 0,\ \frac{\pi}{16},\ \frac{\pi}{8}$$

Step 4 Finding the first-quarter point, midpoint, and third-quarter point, we have the following.

$$\left(-\frac{\pi}{16},-1\right),\ (0,0),\ \left(\frac{\pi}{16},1\right)$$

Step 5 Join the points with a smooth curve.

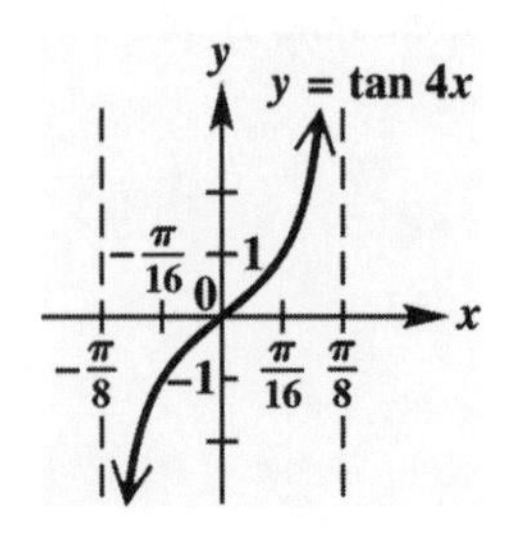

22. $y = \tan \frac{1}{2}x$

Step 1 Find the period and locate the vertical asymptotes. The period of tangent is $\frac{\pi}{b}$, so the period for this function is 2π. Tangent has asymptotes of the form $bx = -\frac{\pi}{2}$ and $bx = \frac{\pi}{2}$. Therefore, the asymptotes for $y = \tan \frac{1}{2}x$ are as follows.

$$\frac{1}{2}x = -\frac{\pi}{2} \Rightarrow x = -\pi \text{ and } \frac{1}{2}x = \frac{\pi}{2} \Rightarrow x = \pi$$

Step 2 Sketch the two vertical asymptotes found in Step 1.

Step 3 Divide the interval into four equal parts.

$$-\pi, -\frac{\pi}{2}, 0, \frac{\pi}{2}, \pi$$

Step 4 Finding the first-quarter point, midpoint, and third-quarter point, we have the following.

$$\left(-\frac{\pi}{2}, -1\right), (0,0), \left(\frac{\pi}{2}, 1\right)$$

Step 5 Join the points with a smooth curve.

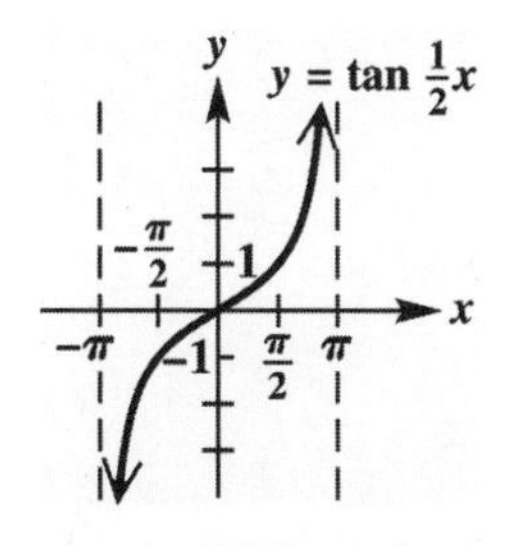

23. $y = 2 \tan x$

Step 1 Find the period and locate the vertical asymptotes. The period of tangent is $\frac{\pi}{b}$, so the period for this function is π. Tangent has asymptotes of the form $bx = -\frac{\pi}{2}$ and $bx = \frac{\pi}{2}$. Therefore, the asymptotes for $y = 2 \tan x$ are $x = -\frac{\pi}{2}$ and $x = \frac{\pi}{2}$.

Step 2 Sketch the two vertical asymptotes found in Step 1.

Step 3 Divide the interval into four equal parts.

$$-\frac{\pi}{2}, -\frac{\pi}{4}, 0, \frac{\pi}{4}, \frac{\pi}{2}$$

Step 4 Finding the first-quarter point, midpoint, and third-quarter point, we have the following.

$$\left(-\frac{\pi}{4}, -2\right), (0,0), \left(\frac{\pi}{4}, 2\right)$$

Step 5 Join the points with a smooth curve. The graph is "stretched" because $a = 2$ and $|2| > 1$.

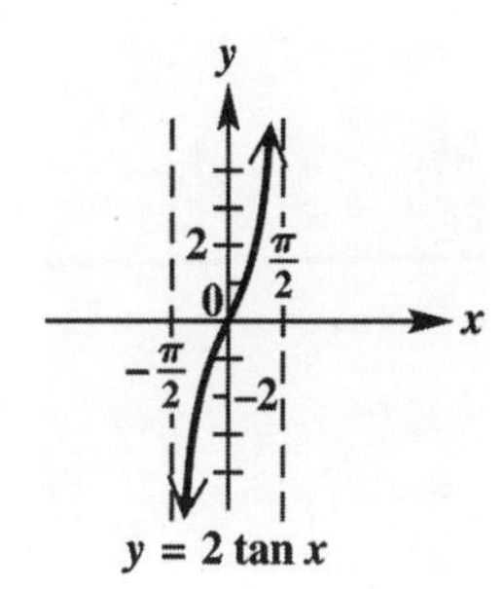

24. $y = 2 \cot x$

Step 1 Find the period and locate the vertical asymptotes. The period of cotangent is $\frac{\pi}{b}$, so the period for this function is π. Cotangent has asymptotes of the form $bx = 0$ and $bx = \pi$. The asymptotes for $y = 2 \cot x$ are $x = 0$ and $x = \pi$.

Step 2 Sketch the two vertical asymptotes found in Step 1.

Step 3 Divide the interval into four equal parts.

$$0, \frac{\pi}{4}, \frac{\pi}{2}, \frac{3\pi}{4}, \pi$$

Step 4 Finding the first-quarter point, midpoint, and third-quarter point, we have the following.

$$\left(\frac{\pi}{4}, 2\right), \left(\frac{\pi}{2}, 0\right), \left(\frac{3\pi}{4}, -2\right)$$

Step 5 Join the points with a smooth curve. The graph is "stretched" because $a = 2$ and $|2| > 1$.

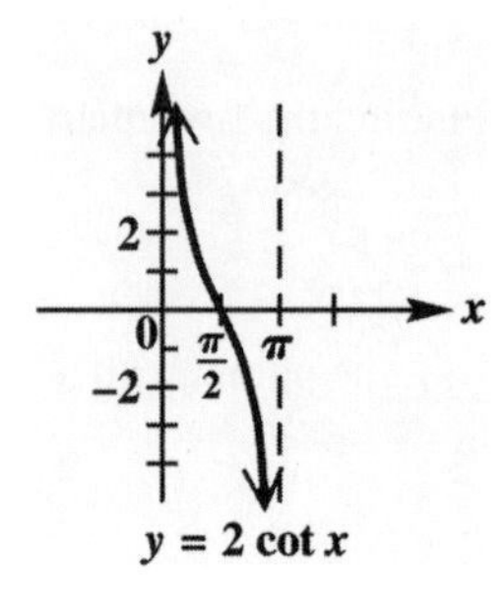

25. $y = 2 \tan \frac{1}{4} x$

Step 1 Find the period and locate the vertical asymptotes. The period of tangent is $\frac{\pi}{b}$, so the period for this function is 4π. Tangent has asymptotes of the form $bx = -\frac{\pi}{2}$ and $bx = \frac{\pi}{2}$.

Therefore, the asymptotes for $y = 2 \tan \frac{1}{4} x$ are as follows.

$$\frac{1}{4} x = -\frac{\pi}{2} \Rightarrow x = -2\pi \text{ and } \frac{1}{4} x = \frac{\pi}{2} \Rightarrow x = 2\pi$$

Step 2 Sketch the two vertical asymptotes found in Step 1.

Step 3 Divide the interval into four equal parts.

$$-2\pi, -\pi, 0, \pi, 2\pi$$

Step 4 Finding the first-quarter point, midpoint, and third-quarter point, we have the following.

$$(-\pi, -2), (0,0), (\pi, 2)$$

Step 5 Join the points with a smooth curve.

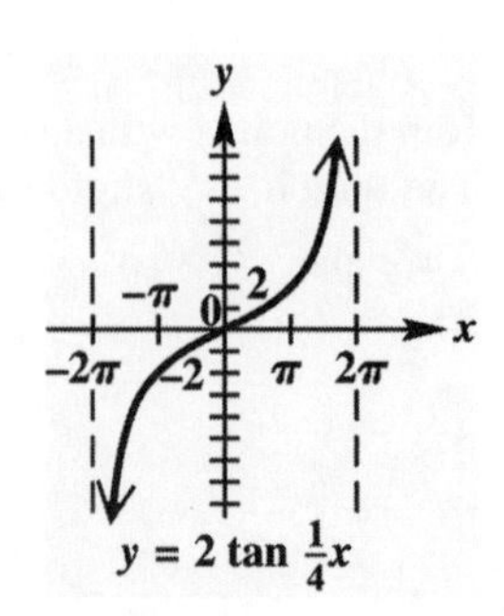

26. $y = \frac{1}{2}\cot x$

Step 1 Find the period and locate the vertical asymptotes. The period of cotangent is $\frac{\pi}{b}$, so the period for this function is π. Cotangent has asymptotes of the form $bx = 0$ and $bx = \pi$. The asymptotes for $y = \frac{1}{2}\cot x$ are $x = 0$ and $x = \pi$.

Step 2 Sketch the two vertical asymptotes found in Step 1.

Step 3 Divide the interval into four equal parts.

$$0, \frac{\pi}{4}, \frac{\pi}{2}, \frac{3\pi}{4}, \pi$$

Step 4 Finding the first-quarter point, midpoint, and third-quarter point, we have the following.

$$\left(\frac{\pi}{4}, \frac{1}{2}\right), \left(\frac{\pi}{2}, 0\right), \left(\frac{3\pi}{4}, -\frac{1}{2}\right)$$

Step 5 Join the points with a smooth curve. The graph is "compressed" because $a = \frac{1}{2}$ and $\left|\frac{1}{2}\right| < 1$.

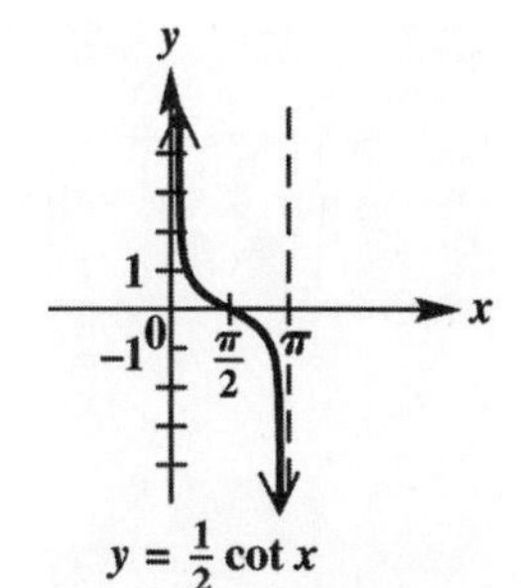

27. $y = \cot 3x$

Step 1 Find the period and locate the vertical asymptotes. The period of cotangent is $\frac{\pi}{b}$, so the period for this function is $\frac{\pi}{3}$. Cotangent has asymptotes of the form $bx = 0$ and $bx = \pi$. The asymptotes for $y = \cot 3x$ are as follows.

$$3x = 0 \Rightarrow x = 0 \text{ and } 3x = \pi \Rightarrow x = \frac{\pi}{3}$$

Step 2 Sketch the two vertical asymptotes found in Step 1.

Step 3 Divide the interval into four equal parts.

$$0, \frac{\pi}{12}, \frac{\pi}{6}, \frac{\pi}{4}, \frac{\pi}{3}$$

Step 4 Finding the first-quarter point, midpoint, and third-quarter point, we have the following.

$$\left(\frac{\pi}{12}, 1\right), \left(\frac{\pi}{6}, 0\right), \left(\frac{\pi}{4}, -1\right)$$

Step 5 Join the points with a smooth curve.

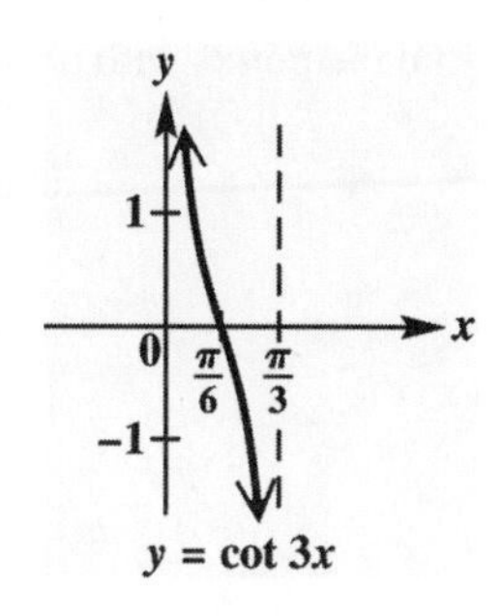

28. $y = -\cot \frac{1}{2}x$

Step 1 Find the period and locate the vertical asymptotes. The period of cotangent is $\frac{\pi}{b}$, so the period for this function is 2π. Cotangent has asymptotes of the form $bx = 0$ and $bx = \pi$. The asymptotes for $y = -\cot \frac{1}{2}x$ are as follows.

$$\frac{1}{2}x = 0 \Rightarrow x = 0 \text{ and } \frac{1}{2}x = \pi \Rightarrow x = 2\pi$$

Step 2 Sketch the two vertical asymptotes found in Step 1.

Step 3 Divide the interval into four equal parts.

$$0, \frac{\pi}{2}, \pi, \frac{3\pi}{2}, 2\pi$$

Step 4 Finding the first-quarter point, midpoint, and third-quarter point, we have the following.

$$\left(\frac{\pi}{2}, -1\right), \ (\pi, 0), \ \left(\frac{3\pi}{2}, 1\right)$$

Step 5 Join the points with a smooth curve. The graph is the reflection of the graph of $y = \cot \frac{1}{2}x$ about the x-axis.

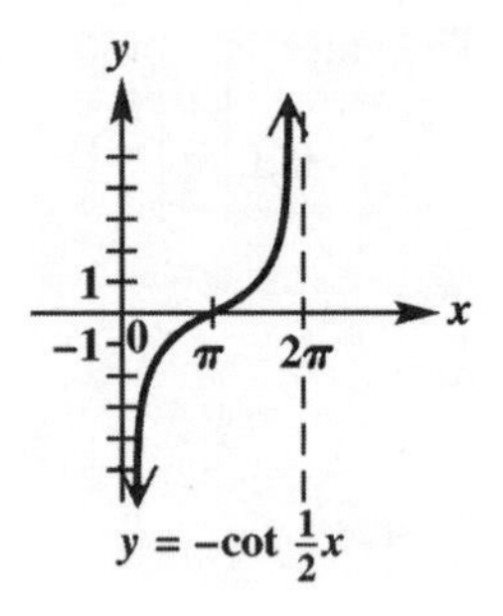

29. $y = -2\tan \frac{1}{4}x$

Step 1 Find the period and locate the vertical asymptotes. The period of tangent is $\frac{\pi}{b}$, so the period for this function is 4π. Tangent has asymptotes of the form $bx = -\frac{\pi}{2}$ and $bx = \frac{\pi}{2}$. Therefore, the asymptotes for $y = -2\tan \frac{1}{4}x$ are as follows.

$$\frac{1}{4}x = -\frac{\pi}{2} \Rightarrow x = -2\pi \text{ and } \frac{1}{4}x = \frac{\pi}{2} \Rightarrow x = 2\pi.$$

Step 2 Sketch the two vertical asymptotes found in Step 1.

Step 3 Divide the interval into four equal parts.

$$-2\pi, -\pi, 0, \pi, 2\pi$$

Step 4 Finding the first-quarter point, midpoint, and third-quarter point, we have the following.

$$(-\pi, 2), \ (0,0), \ (\pi, -2)$$

Step 5 Join the points with a smooth curve.

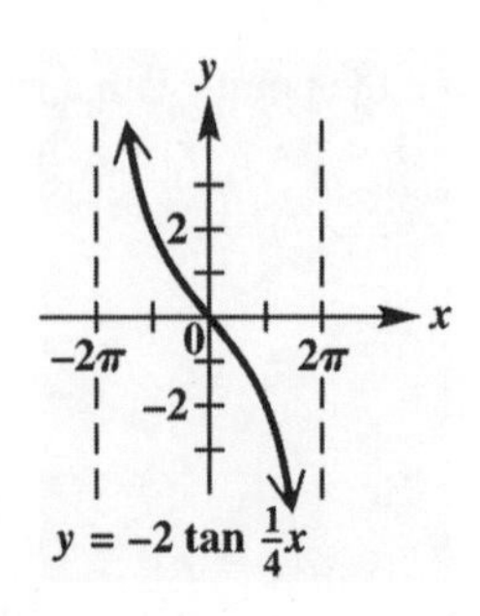

30. $y = 3\tan\frac{1}{2}x$

Step 1 Find the period and locate the vertical asymptotes. The period of tangent is $\frac{\pi}{b}$, so the period for this function is 2π. Tangent has asymptotes of the form $bx = -\frac{\pi}{2}$ and $bx = \frac{\pi}{2}$. Therefore, the asymptotes for $y = 3\tan\frac{1}{2}x$ are as follows.

$$\frac{1}{2}x = -\frac{\pi}{2} \Rightarrow x = -\pi \text{ and } \frac{1}{2}x = \frac{\pi}{2} \Rightarrow x = \pi.$$

Step 2 Sketch the two vertical asymptotes found in Step 1.

Step 3 Divide the interval into four equal parts.

$$-\pi, -\frac{\pi}{2}, 0, \frac{\pi}{2}, \pi$$

Step 4 Finding the first-quarter point, midpoint, and third-quarter point, we have the following.

$$\left(-\frac{\pi}{2}, -3\right), \ (0,0), \ \left(\frac{\pi}{2}, 3\right)$$

Step 5 Join the points with a smooth curve.

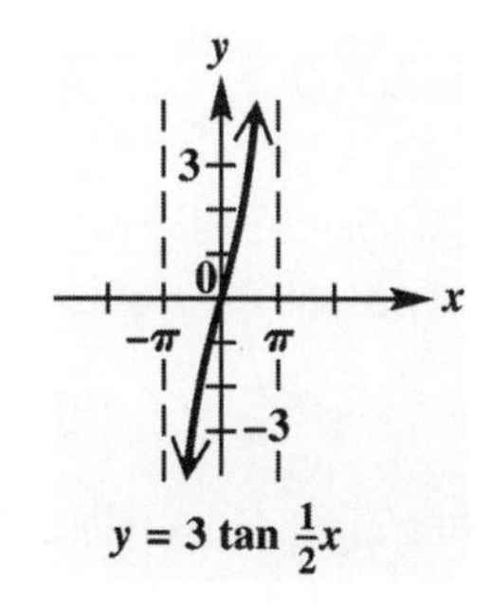

31. $y = \frac{1}{2}\cot 4x$

Step 1 Find the period and locate the vertical asymptotes. The period of cotangent is $\frac{\pi}{b}$, so the period for this function is $\frac{\pi}{4}$. Cotangent has asymptotes of the form $bx = 0$ and $bx = \pi$. The asymptotes for $y = \frac{1}{2}\cot 4x$ are as follows.

$$4x = 0 \Rightarrow x = 0 \text{ and } 4x = \pi \Rightarrow x = \frac{\pi}{4}$$

Step 2 Sketch the two vertical asymptotes found in Step 1.

Step 3 Divide the interval into four equal parts.

$$0, \frac{\pi}{16}, \frac{\pi}{8}, \frac{3\pi}{16}, \frac{\pi}{4}$$

Step 4 Finding the first-quarter point, midpoint, and third-quarter point, we have the following.

$$\left(\frac{\pi}{16}, \frac{1}{2}\right), \left(\frac{\pi}{8}, 0\right), \left(\frac{3\pi}{16}, -\frac{1}{2}\right)$$

Continued on next page

31. (continued)

Step 5 Join the points with a smooth curve.

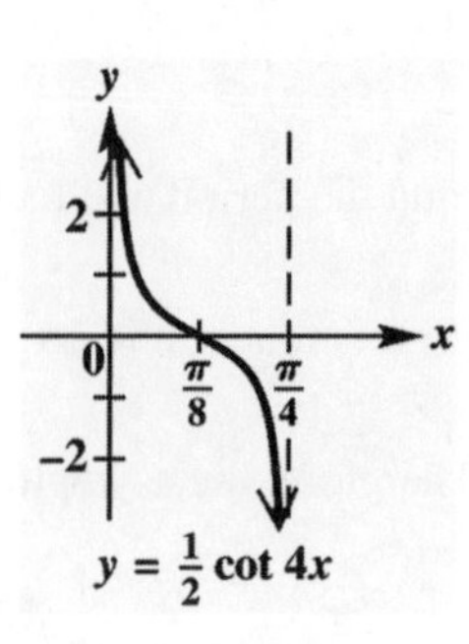

32. $y = -\frac{1}{2}\cot 2x$

Step 1 Find the period and locate the vertical asymptotes. The period of cotangent is $\frac{\pi}{b}$, so the period for this function is $\frac{\pi}{2}$. Cotangent has asymptotes of the form $bx = 0$ and $bx = \pi$.

The asymptotes for $y = -\frac{1}{2}\cot 2x$ are as follows.

$$2x = 0 \Rightarrow x = 0 \text{ and } 2x = \pi \Rightarrow x = \frac{\pi}{2}$$

Step 2 Sketch the two vertical asymptotes found in Step 1.

Step 3 Divide the interval into four equal parts.

$$0, \frac{\pi}{8}, \frac{\pi}{4}, \frac{3\pi}{8}, \frac{\pi}{2}$$

Step 4 Finding the first-quarter point, midpoint, and third-quarter point, we have the following.

$$\left(\frac{\pi}{8}, -\frac{1}{2}\right), \left(\frac{\pi}{4}, 0\right), \left(\frac{3\pi}{8}, \frac{1}{2}\right)$$

Step 5 Join the points with a smooth curve.

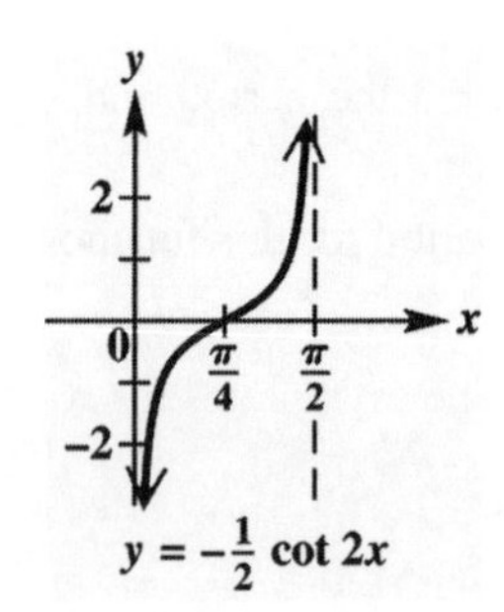

33. $y = \tan(2x - \pi) = \tan 2\left(x - \frac{\pi}{2}\right)$

Period: $\frac{\pi}{b} = \frac{\pi}{2}$

Vertical translation: none

Phase shift (horizontal translation): $\frac{\pi}{2}$ units to the right

Because the function is to be graphed over a two-period interval, locate three adjacent vertical asymptotes. Because asymptotes of the graph $y = \tan x$ occur at $-\frac{\pi}{2}$, $\frac{\pi}{2}$, and $\frac{3\pi}{2}$, the following equations can be solved to locate asymptotes.

$$2\left(x - \frac{\pi}{2}\right) = -\frac{\pi}{2},\ 2\left(x - \frac{\pi}{2}\right) = \frac{\pi}{2}, \text{ and } 2\left(x - \frac{\pi}{2}\right) = \frac{3\pi}{2}$$

Solve each of these equations.

$$2\left(x - \frac{\pi}{2}\right) = -\frac{\pi}{2} \Rightarrow x - \frac{\pi}{2} = -\frac{\pi}{4} \Rightarrow x = \frac{\pi}{4}$$

$$2\left(x - \frac{\pi}{2}\right) = \frac{\pi}{2} \Rightarrow x - \frac{\pi}{2} = \frac{\pi}{4} \Rightarrow x = \frac{3\pi}{4}$$

$$2\left(x - \frac{\pi}{2}\right) = \frac{3\pi}{2} \Rightarrow x - \frac{\pi}{2} = \frac{3\pi}{4} \Rightarrow x = \frac{5\pi}{4}$$

Divide the interval $\left(\frac{\pi}{4}, \frac{3\pi}{4}\right)$ into four equal parts to obtain the following key x-values.

$$\text{first-quarter value: } \frac{3\pi}{8}\ \text{ middle value: } \frac{\pi}{2};\ \text{third-quarter value: } \frac{5\pi}{8}$$

Evaluating the given function at these three key x-values gives the following points.

$$\left(\frac{3\pi}{8}, -1\right),\ \left(\frac{\pi}{2}, 0\right),\ \left(\frac{5\pi}{8}, 1\right)$$

Connect these points with a smooth curve and continue to graph to approach the asymptote $x = \frac{\pi}{4}$ and $x = \frac{3\pi}{4}$ to complete one period of the graph. Sketch the identical curve between the asymptotes $x = \frac{3\pi}{4}$ and $x = \frac{5\pi}{4}$ to complete a second period of the graph.

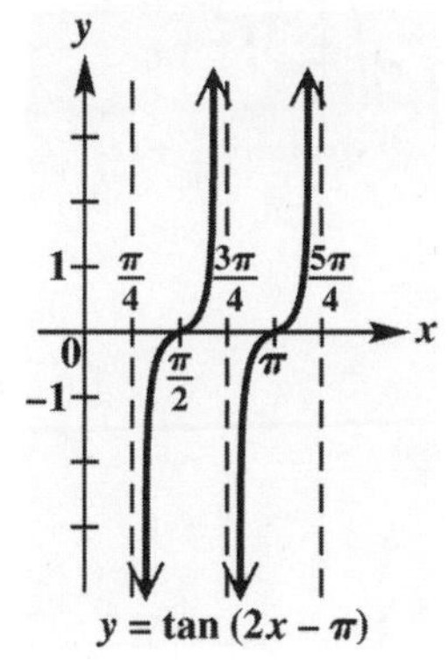

34. $y = \tan\left(\frac{x}{2} + \pi\right) = \tan\frac{1}{2}(x + 2\pi)$

Period: $\frac{\pi}{b} = \frac{\pi}{\frac{1}{2}} = 2\pi$

Vertical translation: none

Phase shift (horizontal translation): 2π units to the left

Because the function is to be graphed over a two-period interval, locate three adjacent vertical asymptotes. Because asymptotes of the graph $y = \tan x$ occur at $\frac{\pi}{2}$, $\frac{3\pi}{2}$, and $\frac{5\pi}{2}$, the following equations can be solved to locate asymptotes.

$$\frac{1}{2}(x + 2\pi) = \frac{\pi}{2},\ \frac{1}{2}(x + 2\pi) = \frac{3\pi}{2},\ \text{and}\ \frac{1}{2}(x + 2\pi) = \frac{5\pi}{2}$$

Solve each of these equations.

$$\frac{1}{2}(x + 2\pi) = \frac{\pi}{2} \Rightarrow x + 2\pi = \pi \Rightarrow x = -\pi$$

$$\frac{1}{2}(x + 2\pi) = \frac{3\pi}{2} \Rightarrow x + 2\pi = 3\pi \Rightarrow x = \pi$$

$$\frac{1}{2}(x + 2\pi) = \frac{5\pi}{2} \Rightarrow x + 2\pi = 5\pi \Rightarrow x = 3\pi$$

Divide the interval $(\pi, 3\pi)$ into four equal parts to obtain the following key x-values.

first-quarter value: $\frac{3\pi}{2}$; middle value: 2π; third-quarter value: $\frac{5\pi}{2}$

Evaluating the given function at these three key x-values gives the following points.

$$\left(\frac{3\pi}{2}, -1\right),\ (2\pi, 0),\ \left(\frac{5\pi}{2}, 1\right)$$

Connect these points with a smooth curve and continue to graph to approach the asymptote $x = \pi$ and $x = 3\pi$ to complete one period of the graph. Sketch the identical curve between the asymptotes $x = -\pi$ and $x = \pi$ to complete a second period of the graph.

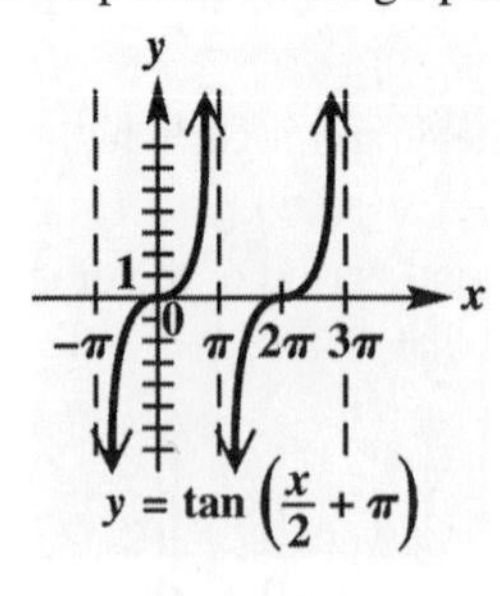

35. $y=\cot\left(3x+\frac{\pi}{4}\right)=\cot 3\left(x+\frac{\pi}{12}\right)$

Period: $\frac{\pi}{b}=\frac{\pi}{3}$

Vertical translation: none

Phase shift (horizontal translation): $\frac{\pi}{12}$ unit to the left

Because the function is to be graphed over a two-period interval, locate three adjacent vertical asymptotes. Because asymptotes of the graph $y=\cot x$ occur at multiples of π, the following equations can be solved to locate asymptotes.

$$3\left(x+\frac{\pi}{12}\right)=0,\ 3\left(x+\frac{\pi}{12}\right)=\pi,\text{ and } 3\left(x+\frac{\pi}{12}\right)=2\pi$$

Solve each of these equations.

$$3\left(x+\frac{\pi}{12}\right)=0\Rightarrow x+\frac{\pi}{12}=0\Rightarrow x=-\frac{\pi}{12}$$

$$3\left(x+\frac{\pi}{12}\right)=\pi\Rightarrow x+\frac{\pi}{12}=\frac{\pi}{3}\Rightarrow x=\frac{\pi}{3}-\frac{\pi}{12}=\frac{\pi}{4}$$

$$3\left(x+\frac{\pi}{12}\right)=2\pi\Rightarrow x+\frac{\pi}{12}=\frac{2\pi}{3}\Rightarrow x=\frac{2\pi}{3}-\frac{\pi}{12}\Rightarrow x=\frac{7\pi}{12}$$

Divide the interval $\left(\frac{\pi}{4},\frac{7\pi}{12}\right)$ into four equal parts to obtain the following key x-values.

$$\text{first-quarter value: } \frac{\pi}{3};\ \text{middle value: } \frac{5\pi}{12};\ \text{third-quarter value: } \frac{\pi}{2}$$

Evaluating the given function at these three key x-values gives the following points.

$$\left(\frac{\pi}{3},1\right),\ \left(\frac{5\pi}{12},0\right),\ \left(\frac{\pi}{2},-1\right)$$

Connect these points with a smooth curve and continue to graph to approach the asymptote $x=\frac{\pi}{4}$ and $x=\frac{7\pi}{12}$ to complete one period of the graph. Sketch the identical curve between the asymptotes $x=-\frac{\pi}{12}$ and $x=\frac{\pi}{4}$ to complete a second period of the graph.

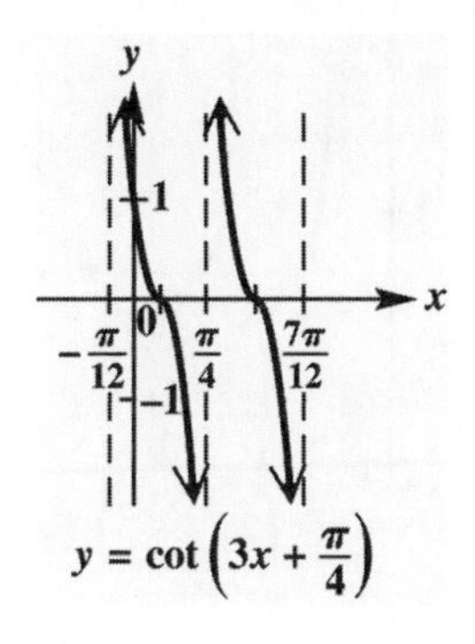

36. $y = \cot\left(2x - \frac{3\pi}{2}\right) = \cot 2\left(x - \frac{3\pi}{4}\right)$

Period: $\frac{\pi}{b} = \frac{\pi}{2}$

Vertical translation: none

Phase shift (horizontal translation): $\frac{3\pi}{4}$ units to the right

Because the function is to be graphed over a two-period interval, locate three adjacent vertical asymptotes. Because asymptotes of the graph $y = \cot x$ occur at multiples of π, the following equations can be solved to locate asymptotes.

$$2\left(x - \frac{3\pi}{4}\right) = -\pi,\ 2\left(x - \frac{3\pi}{4}\right) = 0, \text{ and } 2\left(x - \frac{3\pi}{4}\right) = \pi$$

Solve each of these equations.

$$2\left(x - \frac{3\pi}{4}\right) = -\pi \Rightarrow x - \frac{3\pi}{4} = -\frac{\pi}{2} \Rightarrow x = -\frac{\pi}{2} + \frac{3\pi}{4} \Rightarrow x = \frac{\pi}{4}$$

$$2\left(x - \frac{3\pi}{4}\right) = 0 \Rightarrow x - \frac{3\pi}{4} = 0 \Rightarrow x = \frac{3\pi}{4}$$

$$2\left(x - \frac{3\pi}{4}\right) = \pi \Rightarrow x - \frac{3\pi}{4} = \frac{\pi}{2} \Rightarrow x = \frac{\pi}{2} + \frac{3\pi}{4} = \frac{5\pi}{4}$$

Divide the interval $\left(\frac{\pi}{4}, \frac{3\pi}{4}\right)$ into four equal parts to obtain the following key x-values.

first-quarter value: $\frac{3\pi}{8}$; middle value: $\frac{\pi}{2}$; third-quarter value: $\frac{5\pi}{8}$

Evaluating the given function at these three key x-values gives the following points.

$$\left(\frac{3\pi}{8}, 1\right), \left(\frac{\pi}{2}, 0\right), \left(\frac{5\pi}{8}, -1\right)$$

Connect these points with a smooth curve and continue to graph to approach the asymptote $x = \frac{\pi}{4}$ and $x = \frac{3\pi}{4}$ to complete one period of the graph. Sketch the identical curve between the asymptotes $x = \frac{3\pi}{4}$ and $x = \frac{5\pi}{4}$ to complete a second period of the graph.

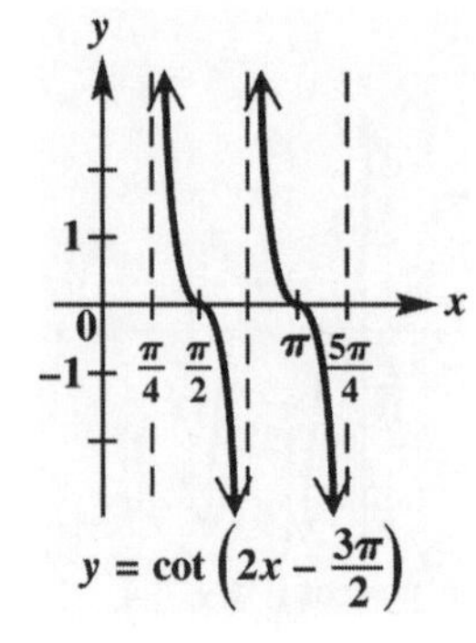

37. $y = 1 + \tan x$
This is the graph of $y = \tan x$ translated vertically 1 unit up.

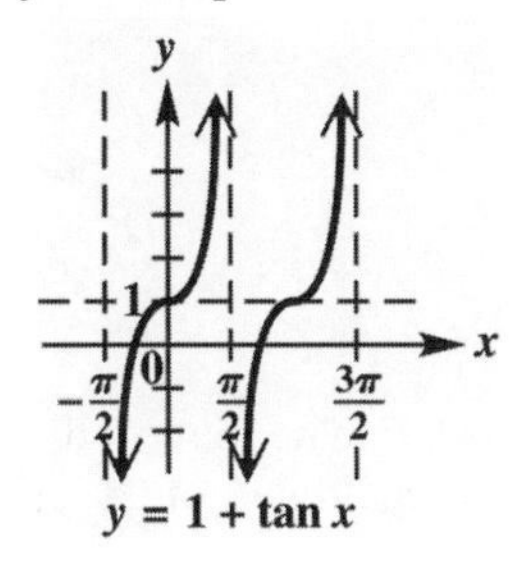

38. $y = 1 - \tan x$

This is the graph of $y = \tan x$, reflected over the x-axis and then translated vertically 1 unit up.

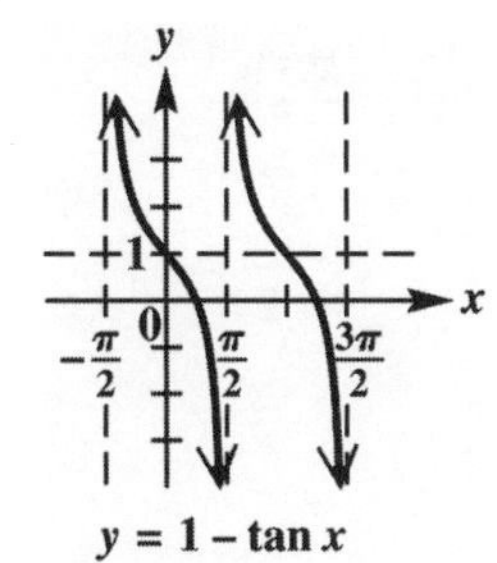

39. $y = 1 - \cot x$
This is the graph of $y = \cot x$ reflected about the x-axis and then translated vertically 1 unit up.

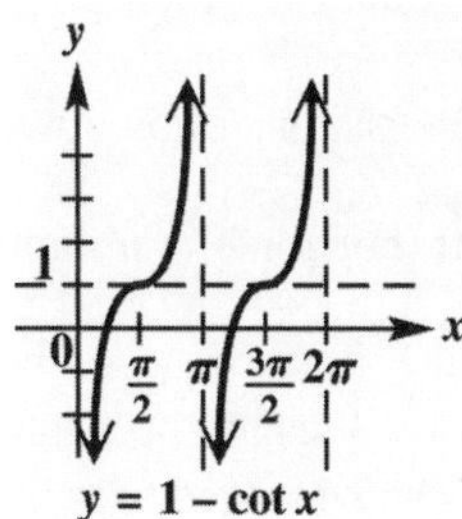

40. $y = -2 - \cot x$
This is the graph of $y = \cot x$ reflected about the x-axis and then translated vertically 2 units down.

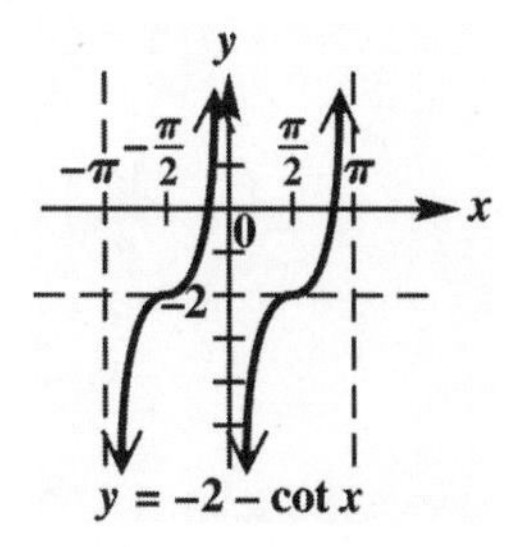

41. $y = -1 + 2 \tan x$
This is the graph of $y = 2 \tan x$ translated vertically 1 unit down.

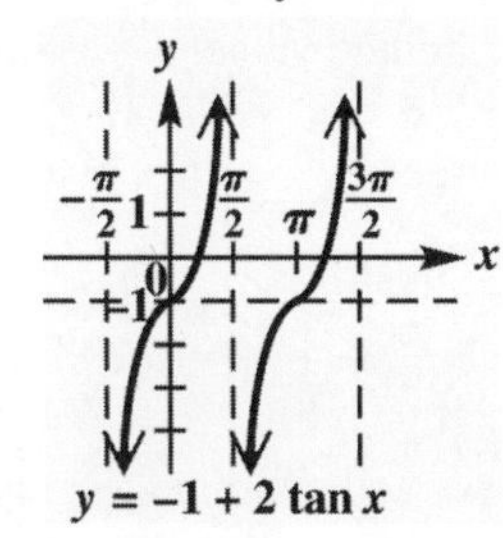

42. $y = 3 + \frac{1}{2} \tan x$

This is the graph of $y = \frac{1}{2} \tan x$ translated vertically 3 units up.

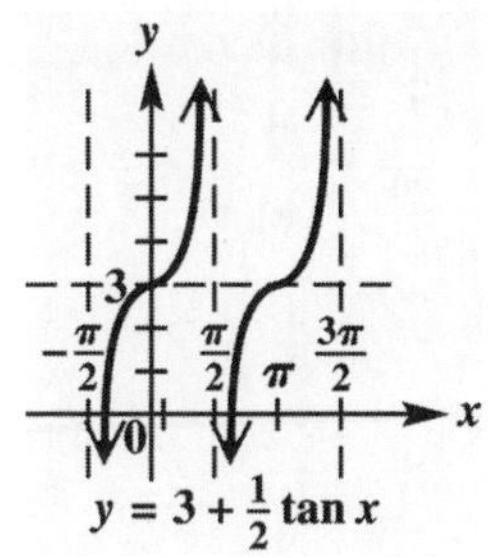

43. $y=-1+\frac{1}{2}\cot(2x-3\pi)=-1+\frac{1}{2}\cot 2\left(x-\frac{3\pi}{2}\right)$

Period: $\frac{\pi}{b}=\frac{\pi}{2}$.

Vertical translation: 1 unit down

Phase shift (horizontal translation): $\frac{3\pi}{2}$ units to the right

Because the function is to be graphed over a two-period interval, locate three adjacent vertical asymptotes. Because asymptotes of the graph $y=\cot x$ occur at multiples of π, the following equations can be solved to locate asymptotes.

$$2\left(x-\frac{3\pi}{2}\right)=-2\pi,\ 2\left(x-\frac{3\pi}{2}\right)=-\pi,\text{ and }2\left(x-\frac{3\pi}{2}\right)=0$$

Solve each of these equations.

$$2\left(x-\frac{3\pi}{2}\right)=-2\pi\Rightarrow x-\frac{3\pi}{2}=-\pi\Rightarrow x=-\pi+\frac{3\pi}{2}=\frac{\pi}{2}$$

$$2\left(x-\frac{3\pi}{2}\right)=-\pi\Rightarrow x-\frac{3\pi}{2}=-\frac{\pi}{2}\Rightarrow x=-\frac{\pi}{2}+\frac{3\pi}{2}\Rightarrow x=\frac{2\pi}{2}=\pi$$

$$2\left(x-\frac{3\pi}{2}\right)=0\Rightarrow x-\frac{3\pi}{2}=0\Rightarrow x=\frac{3\pi}{2}$$

Divide the interval $\left(\frac{\pi}{2},\pi\right)$ into four equal parts to obtain the following key x-values.

first-quarter value: $\frac{5\pi}{8}$; middle value: $\frac{3\pi}{4}$; third-quarter value: $\frac{7\pi}{8}$

Evaluating the given function at these three key x-values gives the following points.

$$\left(\frac{5\pi}{8},-\frac{1}{2}\right),\ \left(\frac{3\pi}{4},-1\right),\ \left(\frac{7\pi}{8},-\frac{3}{2}\right)$$

Connect these points with a smooth curve and continue to graph to approach the asymptote $x=\frac{\pi}{2}$ and $x=\pi$ to complete one period of the graph. Sketch the identical curve between the asymptotes $x=\pi$ and $x=\frac{3\pi}{2}$ to complete a second period of the graph.

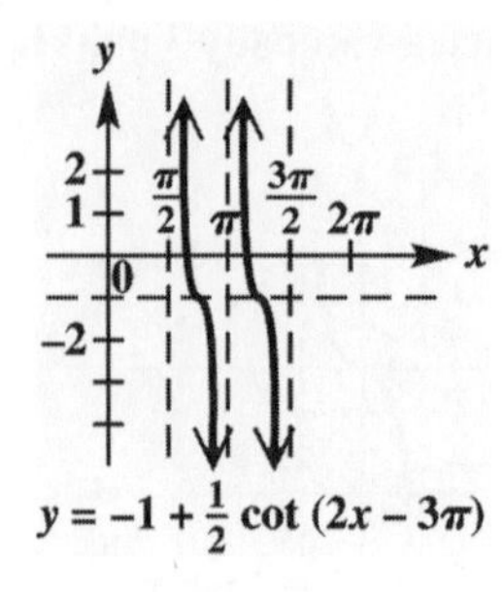

44. $y=-2+3\tan(4x+\pi)=-2+3\tan 4\left(x+\frac{\pi}{4}\right)$

Period: $\frac{\pi}{b}=\frac{\pi}{4}$

Vertical translation: 2 units down

Phase shift (horizontal translation): $\frac{\pi}{4}$ unit to the left

Because the function is to be graphed over a two-period interval, locate three adjacent vertical asymptotes. Because asymptotes of the graph $y=\tan x$ occur at $-\frac{\pi}{2}$, $\frac{\pi}{2}$, and $\frac{3\pi}{2}$, the following equations can be solved to locate asymptotes.

$$4\left(x+\frac{\pi}{4}\right)=-\frac{\pi}{2},\ 4\left(x+\frac{\pi}{4}\right)=\frac{\pi}{2},\text{ and } 4\left(x+\frac{\pi}{4}\right)=\frac{3\pi}{2}$$

Solve each of these equations.

$$4\left(x+\frac{\pi}{4}\right)=-\frac{\pi}{2}\Rightarrow x+\frac{\pi}{4}=-\frac{\pi}{8}\Rightarrow x=-\frac{3\pi}{8}$$

$$4\left(x+\frac{\pi}{4}\right)=\frac{\pi}{2}\Rightarrow x+\frac{\pi}{4}=\frac{\pi}{8}\Rightarrow x=-\frac{\pi}{8}$$

$$4\left(x+\frac{\pi}{4}\right)=\frac{3\pi}{2}\Rightarrow x+\frac{\pi}{4}=\frac{3\pi}{8}\Rightarrow x=\frac{\pi}{8}$$

Divide the interval $\left(-\frac{3\pi}{8},-\frac{\pi}{8}\right)$ into four equal parts to obtain the following key x-values.

first-quarter value: $-\frac{5\pi}{16}$; middle value: $-\frac{\pi}{4}$; third-quarter value: $-\frac{3\pi}{16}$

Evaluating the given function at these three key x-values gives the following points.

$$\left(-\frac{5\pi}{16},-5\right),\left(-\frac{\pi}{4},-2\right),\left(-\frac{3\pi}{16},1\right)$$

Connect these points with a smooth curve and continue to graph to approach the asymptote $x=-\frac{3\pi}{8}$ and $x=-\frac{\pi}{8}$ to complete one period of the graph. Sketch the identical curve between the asymptotes $x=-\frac{\pi}{8}$ and $x=\frac{\pi}{8}$ to complete a second period of the graph.

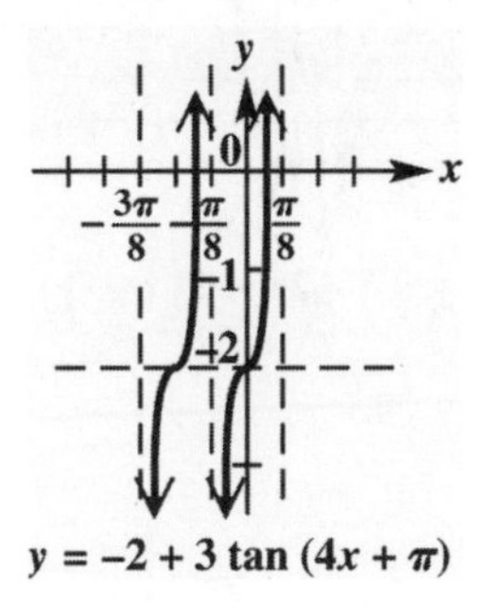

$y=-2+3\tan(4x+\pi)$

45. $y=1-2\cot 2\left(x+\frac{\pi}{2}\right)$

Period: $\frac{\pi}{b}=\frac{\pi}{2}$

Vertical translation: 1 unit up

Phase shift (horizontal translation): $\frac{\pi}{2}$ unit to the left

Because the function is to be graphed over a two-period interval, locate three adjacent vertical asymptotes. Because asymptotes of the graph $y=\cot x$ occur at multiples of π, the following equations can be solved to locate asymptotes.

$$2\left(x+\frac{\pi}{2}\right)=0,\ 2\left(x+\frac{\pi}{2}\right)=\pi, \text{ and } 2\left(x+\frac{\pi}{2}\right)=2\pi$$

Solve each of these equations.

$$2\left(x+\frac{\pi}{2}\right)=0\Rightarrow x+\frac{\pi}{2}=0\Rightarrow x=0-\frac{\pi}{2}=-\frac{\pi}{2}$$

$$2\left(x+\frac{\pi}{2}\right)=\pi\Rightarrow x+\frac{\pi}{2}=\frac{\pi}{2}\Rightarrow x=\frac{\pi}{2}-\frac{\pi}{2}\Rightarrow x=0$$

$$2\left(x+\frac{\pi}{2}\right)=2\pi\Rightarrow x+\frac{\pi}{2}=\pi\Rightarrow x=\pi-\frac{\pi}{2}=\frac{\pi}{2}$$

Divide the interval $\left(0,\frac{\pi}{2}\right)$ into four equal parts to obtain the following key x-values.

first-quarter value: $\frac{\pi}{8}$; middle value: $\frac{\pi}{4}$; third-quarter value: $\frac{3\pi}{8}$

Evaluating the given function at these three key x-values gives the following points.

$$\left(\frac{\pi}{8},-1\right),\ \left(\frac{\pi}{4},1\right),\ \left(\frac{3\pi}{8},3\right)$$

Connect these points with a smooth curve and continue to graph to approach the asymptote $x=0$ and $x=\frac{\pi}{2}$ to complete one period of the graph. Sketch the identical curve between the asymptotes $x=-\frac{\pi}{2}$ and $x=0$ to complete a second period of the graph.

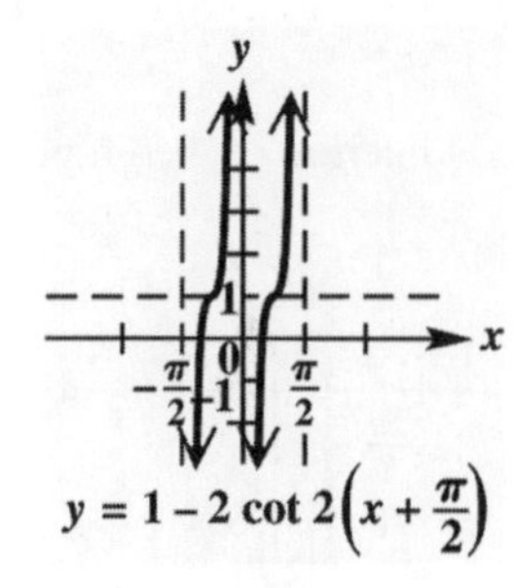

46. $y = \frac{2}{3}\tan\left(\frac{3}{4}x - \pi\right) - 2 = -2 + \frac{2}{3}\tan\frac{3}{4}\left(x - \frac{4\pi}{3}\right)$

Period is $\frac{\pi}{b} = \frac{\pi}{\frac{3}{4}} = \frac{4}{3}\pi = \frac{4\pi}{3}$

Vertical translation: 2 units down

Phase shift (horizontal translation): $\frac{4\pi}{3}$ units to the right

Because the function is to be graphed over a two-period interval, locate three adjacent vertical asymptotes. Because asymptotes of the graph $y = \tan x$ occur at $-\frac{\pi}{2}$, $\frac{\pi}{2}$, and $\frac{3\pi}{2}$, the following equations can be solved to locate asymptotes.

$$\frac{3}{4}\left(x - \frac{4\pi}{3}\right) = -\frac{\pi}{2},\ \frac{3}{4}\left(x - \frac{4\pi}{3}\right) = \frac{\pi}{2},\text{ and } \frac{3}{4}\left(x - \frac{4\pi}{3}\right) = \frac{3\pi}{2}$$

Solve each of these equations.

$$\frac{3}{4}\left(x - \frac{4\pi}{3}\right) = -\frac{\pi}{2} \Rightarrow x - \frac{4\pi}{3} = -\frac{2\pi}{3} \Rightarrow x = \frac{2\pi}{3}$$

$$\frac{3}{4}\left(x - \frac{4\pi}{3}\right) = \frac{\pi}{2} \Rightarrow x - \frac{4\pi}{3} = \frac{2\pi}{3} \Rightarrow x = 2\pi$$

$$\frac{3}{4}\left(x - \frac{4\pi}{3}\right) = \frac{3\pi}{2} \Rightarrow x - \frac{4\pi}{3} = 2\pi \Rightarrow x = \frac{10\pi}{3}$$

Divide the interval $\left(\frac{2\pi}{3}, 2\pi\right)$ into four equal parts to obtain the following key x-values.

first-quarter value: π; middle value: $\frac{4\pi}{3}$; third-quarter value: $\frac{5\pi}{3}$

Evaluating the given function at these three key x-values gives the following points.

$$\left(\pi, -\frac{8}{3}\right), \left(\frac{4\pi}{3}, -2\right), \left(\frac{5\pi}{3}, -\frac{4}{3}\right)$$

Connect these points with a smooth curve and continue to graph to approach the asymptote $x = \frac{2\pi}{3}$ and $x = 2\pi$ to complete one period of the graph. Sketch the identical curve between the asymptotes $x = 2\pi$ and $x = \frac{10\pi}{3}$ to complete a second period of the graph.

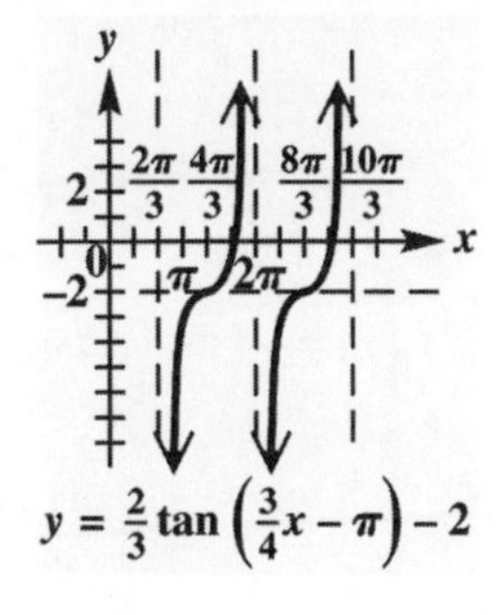

47. True; $\frac{\pi}{2}$ is the smallest positive value where $\cos\frac{\pi}{2}=0$. Since $\tan\frac{\pi}{2}=\frac{\sin\frac{\pi}{2}}{\cos\frac{\pi}{2}}$, $\frac{\pi}{2}$ is the smallest positive value where the tangent function is undefined. Thus, $k=\frac{\pi}{2}$ is the smallest positive value for which $x=k$ is an asymptote for the tangent function.

48. False; $\cot\frac{\pi}{2}=\frac{\cos\frac{\pi}{2}}{\sin\frac{\pi}{2}}=\frac{0}{1}=0$. The smallest such number is π.

49. True; since $\tan x=\frac{\sin x}{\cos x}$ and $\sec x=\frac{1}{\cos x}$, the tangent and secant functions will be undefined at the same values.

50. False; secant values are undefined when $x=\frac{\pi}{2}+n\pi$, while cosecant values are undefined when $x=n\pi$.

51. False; $\tan(-x)=\frac{\sin(-x)}{\cos(-x)}=\frac{-\sin x}{\cos x}=-\tan x$ (since $\sin x$ is odd and $\cos x$ is even) for all x in the domain. Moreover, if $\tan(-x)=\tan x$, then the graph would be symmetric about the y-axis, which it is not.

52. True; $\sec(-x)=\frac{1}{\cos(-x)}=\frac{1}{\cos(x)}=\sec(x)$ (since $\cos x$ is even) for all x in the domain. Moreover, if $\sec(-x)=\sec x$, then the graph would be symmetric about the y-axis, which it is.

53. The function $\tan x$ has a period of π, so it repeats four times over the interval $(-2\pi,2\pi]$. Since its range is $(-\infty,\infty)$, $\tan x=c$ has four solutions for every value of c.

54. None; $\cos x\le 1$ for all x, so $\frac{1}{\cos x}\ge 1$ and $\sec x\ge 1$. Since $\sec x\ge 1$, $\sec x$ has no values in the interval $(-1,1)$.

55. The domain of the tangent function, $y=\tan x$, is $\left\{x \,\middle|\, x\ne\frac{\pi}{2}+n\pi, \text{ where } n \text{ is any integer}\right\}$, and the range is $(-\infty,\infty)$. For the function $f(x)=-4\tan(2x+\pi)=-4\tan 2\left(x+\frac{\pi}{2}\right)$, the period is $\frac{\pi}{2}$. Therefore, the domain is $\left\{x \,\middle|\, x\ne\frac{\pi}{4}+\frac{\pi}{2}n, \text{ where } n \text{ is any integer}\right\}$. This can also be written as $\left\{x \,\middle|\, x\ne(2n+1)\frac{\pi}{4}, \text{ where } n \text{ is any integer}\right\}$. The range remains $(-\infty,\infty)$.

56. The domain of the cosecant function, $y = \csc x$, is $\{x \mid x \neq n\pi$, where n is any integer$\}$, and the range is $(-\infty, -1] \cup [1, \infty)$. For the function $g(x) = -2\csc(4x+\pi) = -2\csc 4\left(x+\frac{\pi}{4}\right)$, the period is $\frac{\pi}{2}$. Therefore, the domain is $\left\{x \,\middle|\, x \neq \frac{n\pi}{4}, \text{ where } n \text{ is any integer}\right\}$. The range becomes $(-\infty, -2] \cup [2, \infty)$ since $a = -2$.

57. $d = 4 \tan 2\pi t$

(a) $d = 4\tan 2\pi(1) = 4\tan 2\pi \approx 4(0) = 0$ m

(b) $d = 4\tan 2\pi(.4) = 4\tan .8\pi \approx 4(-.7265) \approx -2.9$ m

(c) $d = 4\tan 2\pi(.8) = 4\tan 1.6\pi \approx 4(-3.0777) \approx -12.3$ m

(d) $d = 4\tan 2\pi(1.2) = 4\tan 2.4\pi \approx 4(3.0777) \approx 12.3$ m

(e) $t = .25$ leads to $\tan \frac{\pi}{2}$, which is undefined.

58. $a = 4\left|\sec 2\pi t\right|$

(a) $t = 0$

$a = 4\left|\sec 0\right| = 4\left|1\right| = 4(1) = 4$ m

(b) $t = .86$

$a = 4\left|\sec 2\pi(.86)\right| \approx 4\left|1.5688\right| = 4(1.5688) \approx 6.3$ m

(c) $t = 1.24$

$a = 4\left|\sec 2\pi(1.24)\right| \approx 4\left|15.9260\right| = 4(15.9260) \approx 63.7$ m

59. Answers will vary.

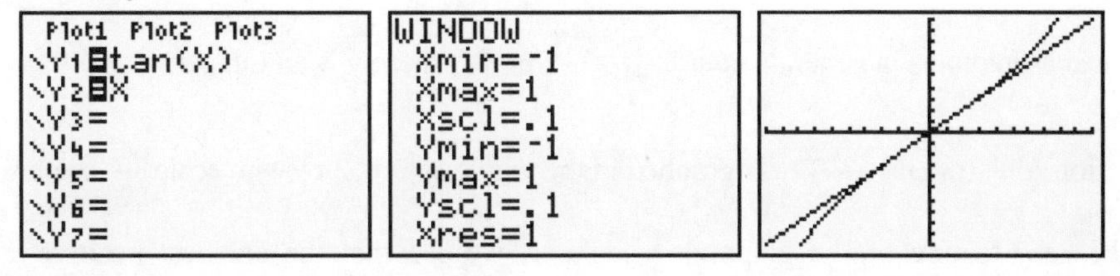

60. Answers will vary.
No, these portions are not actually parabolas.

61. Graph the functions.

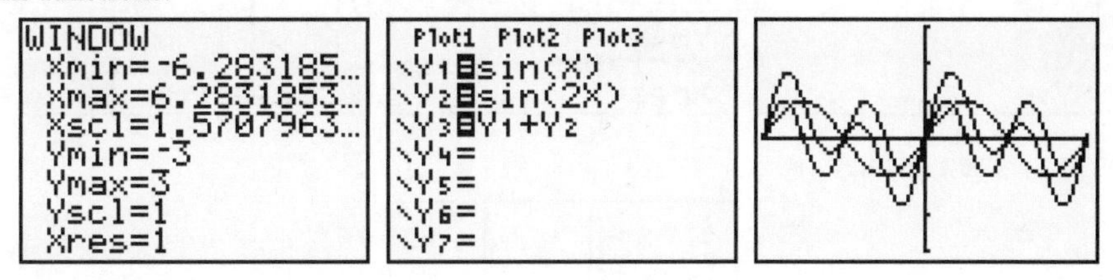

Continued on next page

61. (continued)

Notice that $Y_1\left(\frac{\pi}{6}\right)+Y_2\left(\frac{\pi}{6}\right)\approx .5+.8660254=1.3660254=Y_3\left(\frac{\pi}{6}\right)=\left(Y_1+Y_2\right)\left(\frac{\pi}{6}\right)$.

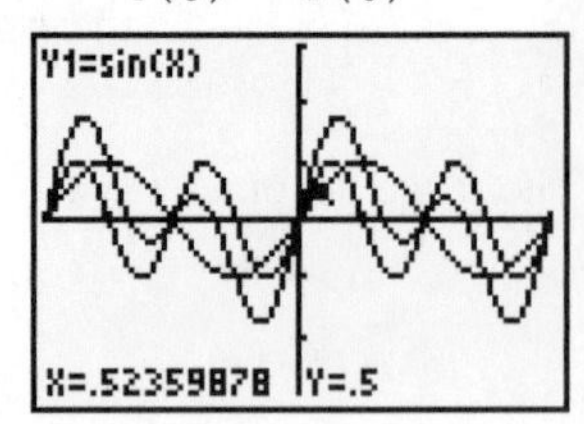

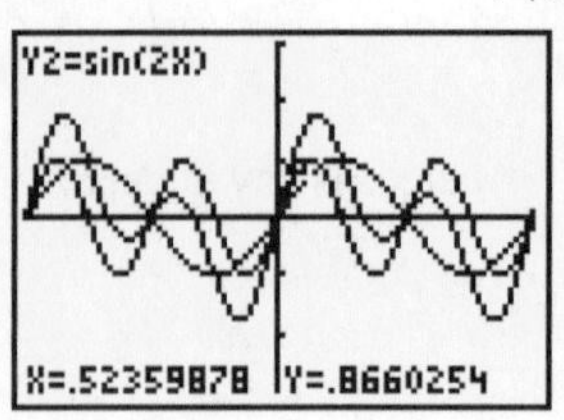

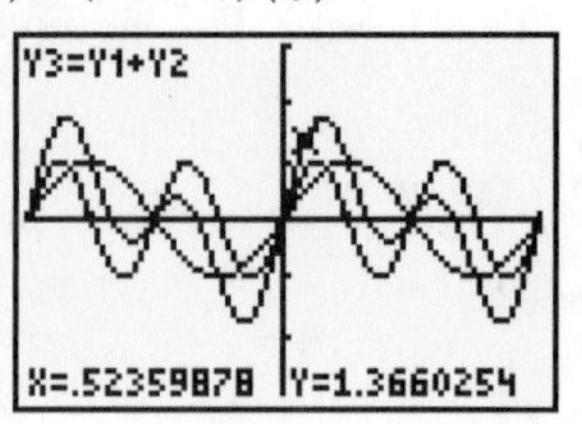

62. Graph the functions.

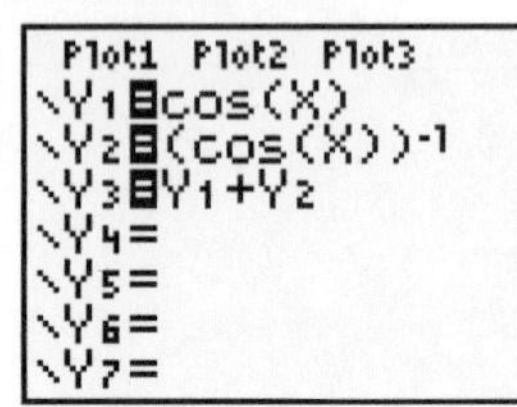

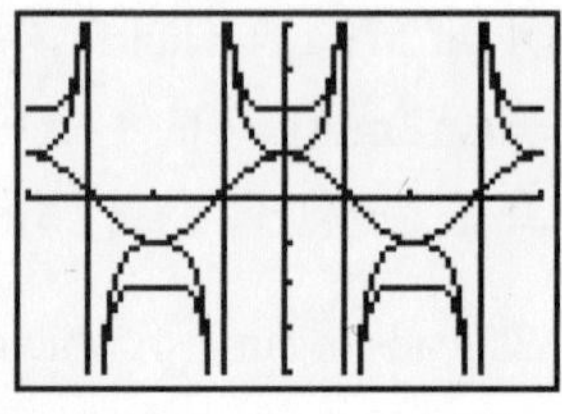

Notice that $Y_1\left(\frac{\pi}{6}\right)+Y_2\left(\frac{\pi}{6}\right)\approx .8660254+1.1547005=2.0207259=Y_3\left(\frac{\pi}{6}\right)=\left(Y_1+Y_2\right)\left(\frac{\pi}{6}\right)$.

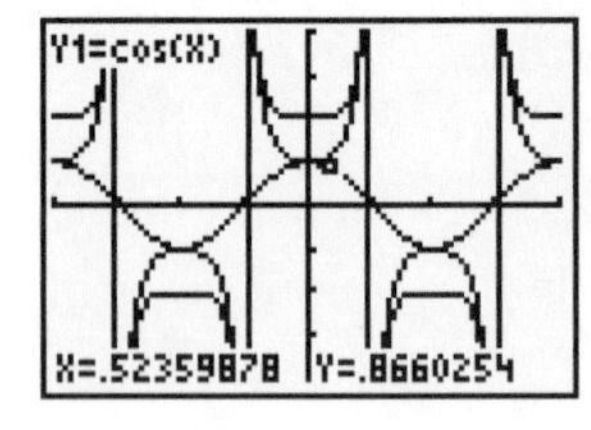

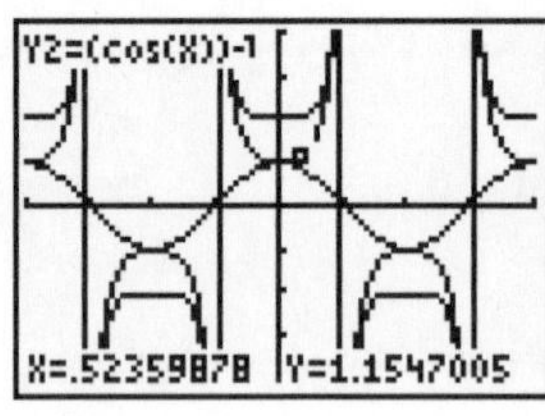

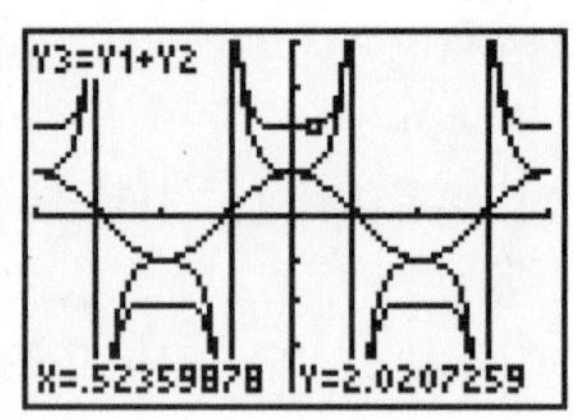

63. π

64. $\pi = x-\frac{\pi}{4} \Rightarrow x=\pi+\frac{\pi}{4}=\frac{5\pi}{4}$

65. The vertical asymptotes in general occur at $x=\frac{5\pi}{4}+n\pi$, where n is an integer.

66. The function $y=-2-\cot\left(x-\frac{\pi}{4}\right)$ is graphed in the window $[-\pi,2\pi]$ with scale $\frac{\pi}{2}$ (with respect to x).

Using the zero feature of the graphing calculator, we see that the smallest positive x-intercept is approximately .3217505544.

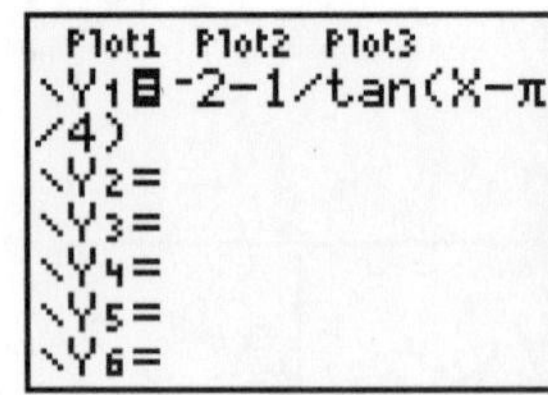

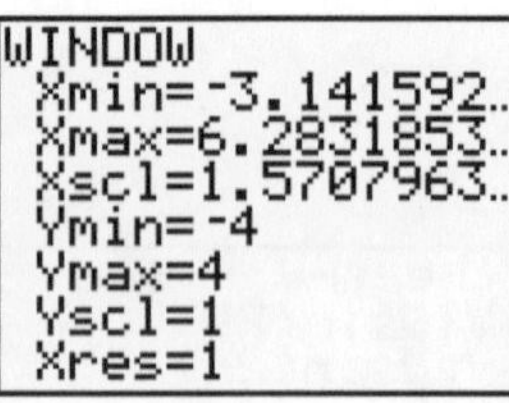

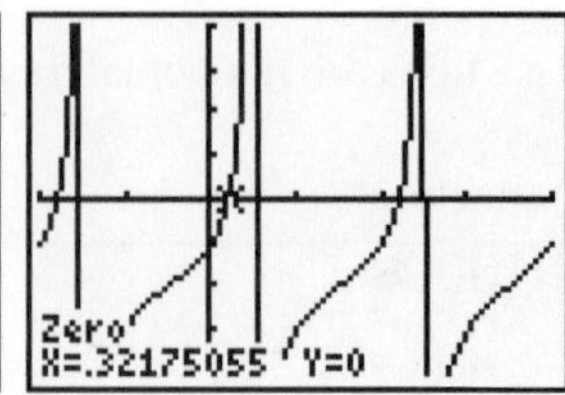

67. .3217505544 + $\pi \approx$ 3.463343208

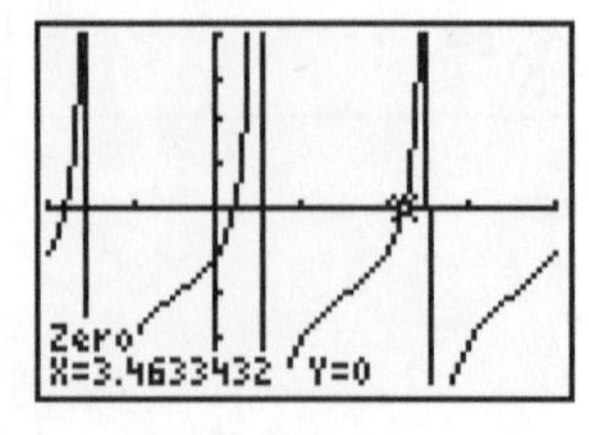

68. $\{x \mid x=.3217505544+n\pi,$ where n is an integer$\}$

Summary Exercises on Graphing Circular Functions

1. $y = 2\sin \pi x$

Period: $\frac{2\pi}{\pi} = 2$ and Amplitude: $|2| = 2$

Divide the interval $[0, 2]$ into four equal parts to get the x-values that will yield minimum and maximum points and x-intercepts. Then make a table.

x	0	$\frac{1}{2}$	1	$\frac{3}{2}$	2
πx	0	$\frac{\pi}{2}$	π	$\frac{3\pi}{2}$	2π
$\sin \pi x$	0	1	0	-1	0
$2\sin \pi x$	0	2	0	-2	0

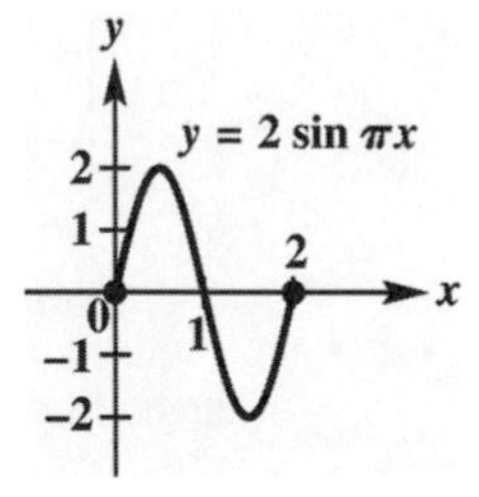

2. $y = 4\cos(1.5x) = y = 4\cos\left(\frac{3}{2}x\right)$

Period: $\frac{2\pi}{\frac{3}{2}} = 2\pi \cdot \frac{2}{3} = \frac{4\pi}{3}$ and Amplitude: $|4| = 4$

Divide the interval $\left[0, \frac{4\pi}{3}\right]$ into four equal parts to get the x-values that will yield minimum and maximum points and x-intercepts. Then make a table.

x	0	$\frac{\pi}{3}$	$\frac{2\pi}{3}$	π	$\frac{4\pi}{3}$
$\frac{3}{2}x$	0	$\frac{\pi}{2}$	π	$\frac{3\pi}{2}$	2π
$\cos\frac{3}{2}x$	1	0	-1	0	1
$4\cos\frac{3}{2}x$	4	0	-4	0	4

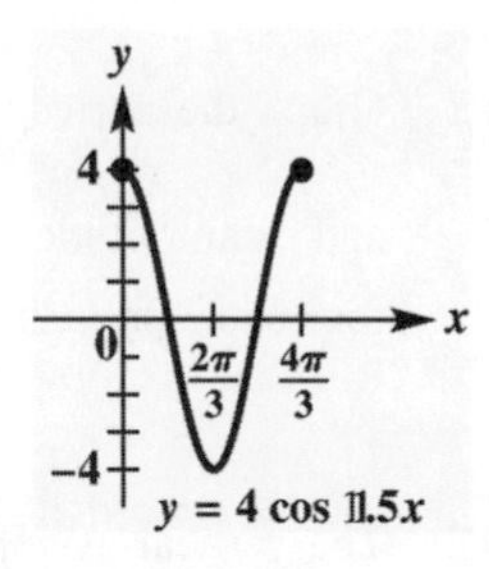

3. $y = -2 + .5\cos\frac{\pi}{4}x$

Step 1 To find the interval whose length is $\frac{2\pi}{b}$.

$$0 \le \frac{\pi}{4}x \le 2\pi \Rightarrow 0 \le x \le 8$$

Step 2 Divide the period into four equal parts to get the following x-values.

$$0, 2, 4, 6, 8$$

Continued on next page

3. (continued)

Step 3 Evaluate the function for each of the five x-values.

x	0	2	4	6	8
$\frac{\pi}{4}x$	0	$\frac{\pi}{2}$	π	$\frac{3\pi}{2}$	2π
$\cos\frac{\pi}{4}x$	1	0	−1	0	1
$\frac{1}{2}\cos\frac{\pi}{4}x$	$\frac{1}{2}$	0	$-\frac{1}{2}$	0	$\frac{1}{2}$
$-2+\frac{1}{2}\cos\frac{\pi}{4}x$	$-\frac{3}{2}$	−2	$-\frac{5}{2}$	−2	$-\frac{3}{2}$

Steps 4 and 5 Plot the points found in the table and join them with a sinusoidal curve. By graphing an additional period to the left, we obtain the following graph.

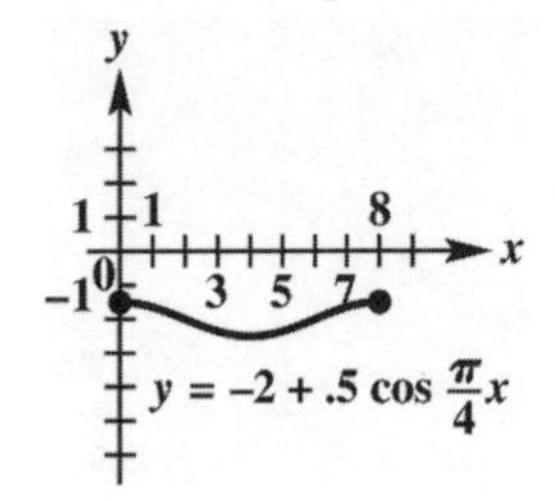

The amplitude is $|.5| = \left|\frac{1}{2}\right|$, which is $\frac{1}{2}$.

The period is 8.

The vertical translation is 2 units down.

There is no phase shift.

4. $y = 3\sec\frac{\pi x}{2}$

Step 1 Graph the corresponding reciprocal function $y = 3\cos\frac{\pi x}{2}$ The period is $\frac{2\pi}{\frac{\pi}{2}} = 2\pi \cdot \frac{2}{\pi} = 4$ and its amplitude is $|3| = 3$. One period is in the interval $0 \le x \le 4$. Dividing the interval into four equal parts gives us the following key points.

$$(0,3),\ (1,0),\ (2,-3),\ (3,0),\ (4,3)$$

Step 2 The vertical asymptotes of $y = 3\sec\frac{\pi x}{2}$ are at the x-intercepts of $y = 3\cos\frac{\pi x}{2}$, which are $x = 1$ and $x = 3$. Continuing this pattern to the left, we also have a vertical asymptote of $x = -1$.

Step 3 Sketch the graph.

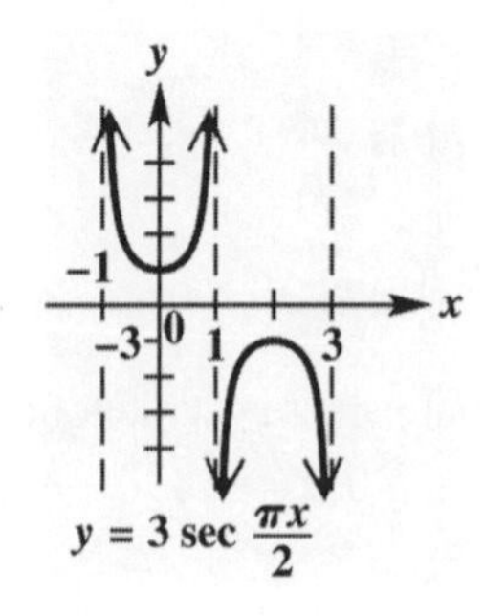

5. $y = -4\csc .5x = -4\csc\frac{1}{2}x$

Step 1 Graph the corresponding reciprocal function $y = -4\sin\frac{1}{2}x$ The period is $\frac{2\pi}{\frac{1}{2}} = 2\pi\cdot\frac{2}{1} = 4\pi$ and its amplitude is $|-4| = 4$. One period is in the interval $0 \le x \le 4\pi$. Dividing the interval into four equal parts gives us the following key points.

$$(0,0),\ (\pi,-4),\ (2\pi,0),\ (3\pi,4),\ (4\pi,0)$$

Step 2 The vertical asymptotes of $y = -4\csc\frac{1}{2}x$ are at the x-intercepts of $y = -4\sin\frac{1}{2}x$, which are $x = 0,\ x = 2\pi,$ and $x = 4\pi.$

Step 3 Sketch the graph.

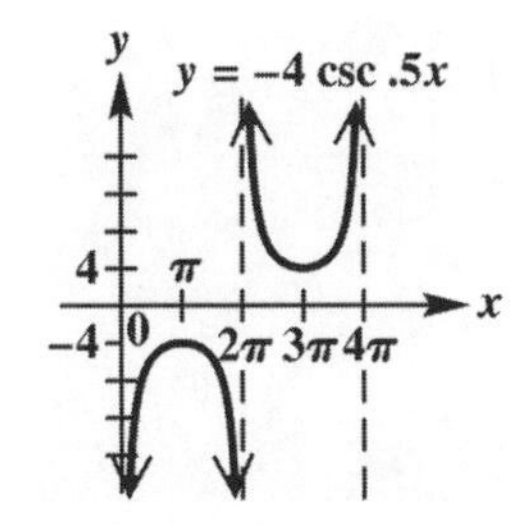

6. $y = 3\tan\left(\frac{\pi x}{2} + \pi\right) = 3\tan\frac{\pi}{2}(x+2)$

Step 1 Find the period and locate the vertical asymptotes. The period of tangent is $\frac{\pi}{b}$, so the period for this function is $\frac{\pi}{\frac{\pi}{2}} = \pi\cdot\frac{2}{\pi} = 2$. Tangent has asymptotes of the form $bx = -\frac{\pi}{2}$ and $bx = \frac{\pi}{2}$.

The asymptotes for $y = 3\tan\frac{\pi}{2}(x+2)$ are as follows.

$$\frac{\pi}{2}(x+2) = -\frac{\pi}{2} \Rightarrow x+2 = -1 \Rightarrow x = -3 \text{ and } \frac{\pi}{2}(x+2) = \frac{\pi}{2} \Rightarrow x+2 = 1 \Rightarrow x = -1$$

Continuing this pattern we see that $x = 1$ is also a vertical asymptote.

. *Step 2* Sketch the vertical asymptotes, $x = -1$ and $x = 1$.

Step 3 Divide the interval into four equal parts.

$$-1, -\frac{1}{2}, 0, \frac{1}{2}, 1$$

Step 4 Finding the first-quarter point, midpoint, and third-quarter point, we have the following.

$$\left(-\frac{1}{2},-3\right),\ (0,0),\ \left(\frac{1}{2},3\right)$$

Step 5 Join the points with a smooth curve.

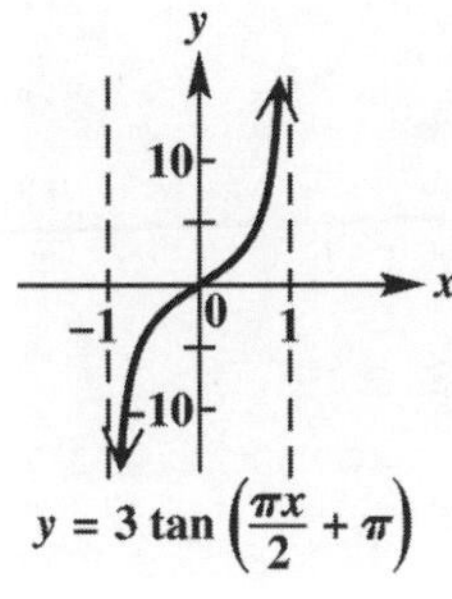

7. $y = -5\sin\frac{x}{3}$

Period: $\frac{2\pi}{\frac{1}{3}} = 2\pi \cdot \frac{3}{1} = 6\pi$ and Amplitude: $|-5| = 5$

Divide the interval $[0, 6\pi]$ into four equal parts to get the x-values that will yield minimum and maximum points and x-intercepts. Then make a table. Repeat this cycle for the interval $[-6\pi, 0]$.

x	0	$\frac{3\pi}{2}$	3π	$\frac{9\pi}{2}$	6π
$\frac{x}{3}$	0	$\frac{\pi}{2}$	π	$\frac{3\pi}{2}$	2π
$\sin\frac{x}{3}$	0	1	0	-1	0
$-5\sin\frac{x}{3}$	0	-5	0	5	0

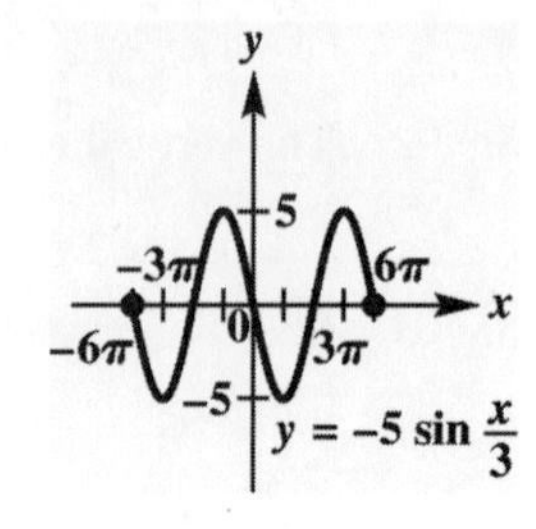

8. $y = 10\cos\left(\frac{x}{4} + \frac{\pi}{2}\right) = 10\cos\frac{1}{4}(x + 2\pi)$

Step 1 Find the interval whose length is $\frac{2\pi}{b}$.

$$0 \le \frac{1}{4}(x + 2\pi) \le 2\pi \Rightarrow 0 \le x + 2\pi \le 8\pi \Rightarrow -2\pi \le x \le 6\pi$$

Step 2 Divide the period into four equal parts to get the following x-values.

$$-2\pi,\ 0,\ 2\pi,\ 4\pi,\ 6\pi$$

Step 3 Evaluate the function for each of the five x-values.

x	-2π	0	2π	4π	6π
$\frac{1}{4}(x+2\pi)$	0	$\frac{\pi}{2}$	π	$\frac{3\pi}{2}$	2π
$\cos\frac{1}{4}(x+2\pi)$	1	0	-1	0	1
$10\cos\frac{1}{4}(x+2\pi)$	10	0	-10	0	10

Steps 4 and 5 Plot the points found in the table and join them with a sinusoidal curve. By graphing an additional period to the left, we obtain the following graph.

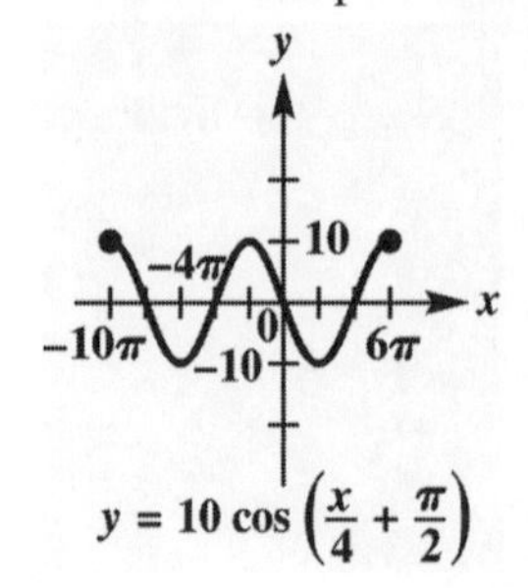

The amplitude is 10.

The period is 8π.

There is no vertical translation.

The phase shift is 2π units to the left.

9. $y = 3 - 4\sin(2.5x + \pi) = 3 - 4\sin\left(\frac{5}{2}x + \pi\right) = 3 - 4\sin\frac{5}{2}\left(x + \frac{2\pi}{5}\right)$

Step 1 Find the interval whose length is $\frac{2\pi}{b}$.

$$0 \le \frac{5}{2}\left(x + \frac{2\pi}{5}\right) \le 2\pi \Rightarrow 0 \le x + \frac{2\pi}{5} \le \frac{4\pi}{5} \Rightarrow -\frac{2\pi}{5} \le x \le \frac{2\pi}{5}$$

Step 2 Divide the period into four equal parts to get the following x-values.

$$-\frac{2\pi}{5},\ -\frac{\pi}{5},\ 0,\ \frac{\pi}{2},\ \frac{2\pi}{5}$$

Step 3 Evaluate the function for each of the five x-values

x	$-\frac{2\pi}{5}$	$-\frac{\pi}{5}$	0	$\frac{\pi}{5}$	$\frac{2\pi}{5}$
$\frac{5}{2}\left(x + \frac{2\pi}{5}\right)$	0	$\frac{\pi}{2}$	π	$\frac{3\pi}{2}$	2π
$\sin\frac{5}{2}\left(x + \frac{2\pi}{5}\right)$	0	1	0	-1	0
$-4\sin\frac{5}{2}\left(x + \frac{2\pi}{5}\right)$	0	-4	0	4	0
$3 - 4\sin\frac{5}{2}\left(x + \frac{2\pi}{5}\right)$	3	-1	3	7	3

Steps 4 and 5 Plot the points found in the table and join them with a sinusoidal curve. By graphing an additional period to the right, we obtain the following graph.

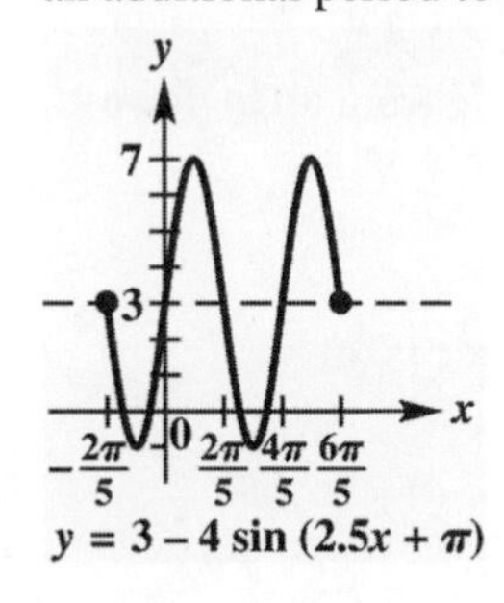

y = 3 − 4 sin (2.5x + π)

The amplitude is $|-4|$, which is 4.

The period is $\frac{2\pi}{\frac{5}{2}}$, which is $\frac{4\pi}{5}$.

The vertical translation is 3 units up.

The phase shift is $\frac{2\pi}{5}$ units to the left.

10. $y = 2 - \sec\left[\pi(x-3)\right]$

Step 1 Graph the corresponding reciprocal function $y = 2 - \cos\left[\pi(x-3)\right]$. The period is 2 and its amplitude is $|-1| = 1$. One period is in the interval $\frac{3}{2} \le x \le \frac{7}{2}$. Dividing the interval into four equal parts gives us the following key points.

$$\left(\frac{3}{2}, 2\right),\ (2,3),\ \left(\frac{5}{2}, 2\right),\ (3,1),\ \left(\frac{7}{2}, 2\right)$$

Continued on next page

10. (continued)

Step 2 The vertical asymptotes of $y = 2 - \sec\left[\pi(x-3)\right]$ are at the x-intercepts of $y = -\cos\left[\pi(x-3)\right]$, which are $x = \frac{3}{2}$, $x = \frac{5}{2}$, and $x = \frac{7}{2}$. Continuing this pattern to the left, we also have a vertical asymptote of $x = \frac{1}{2}$ and $x = -\frac{1}{2}$.

Step 3 Sketch the graph.

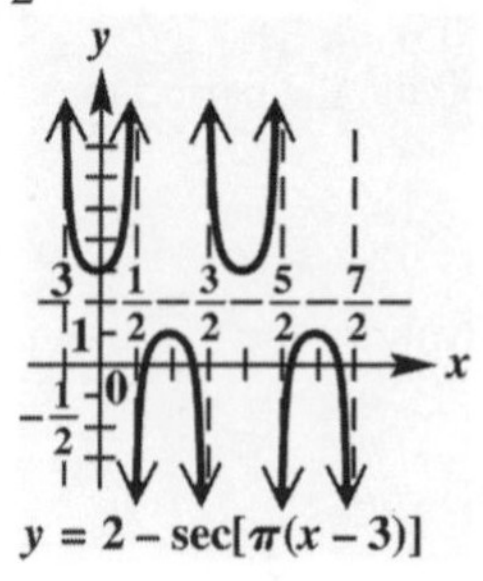

$y = 2 - \sec[\pi(x-3)]$

Section 4.4: Harmonic Motion

1. $s(0) = 2$ in.; $P = .5$ sec

(a) Given $s(t) = a\cos\omega t$, the period is $\frac{2\pi}{\omega}$ and the amplitude is $|a|$.

$$P = .5 \text{ sec} \Rightarrow .5 = \frac{2\pi}{\omega} \Rightarrow \frac{1}{2} = \frac{2\pi}{\omega} \Rightarrow \omega = 4\pi$$

$$s(0) = 2 = a\cos\left[\omega(0)\right] \Rightarrow 2 = a\cos 0 \Rightarrow 2 = a(1) \Rightarrow a = 2$$

Thus, $s(t) = 2\cos 4\pi t$.

(b) Since $s(1) = 2\cos\left[4\pi(1)\right] = 2\cos 4\pi = 2(1) = 2$, the weight is neither moving upward nor downward. At $t = 1$, the motion of the weight is changing from up to down.

2. $s(0) = 5$ in.; $P = 1.5$ sec

(a) Given $s(t) = a\cos\omega t$, the period is $\frac{2\pi}{\omega}$ and the amplitude is $|a|$.

$$P = 1.5 \text{ sec} \Rightarrow 1.5 = \frac{2\pi}{\omega} \Rightarrow \frac{3}{2} = \frac{2\pi}{\omega} \Rightarrow 3\omega = 4\pi \Rightarrow \omega = \frac{4\pi}{3}$$

$$s(0) = 5 = a\cos\left[\omega(0)\right] \Rightarrow 5 = a\cos 0 \Rightarrow 5 = a(1) \Rightarrow a = 5$$

Thus, $s(t) = 5\cos\frac{4\pi}{3}t$.

(b) Since $s(1) = 5\cos\left[\frac{4\pi}{3}(1)\right] = 5\cos\frac{4\pi}{3} = 5\left(-\frac{1}{2}\right) = -\frac{5}{2} = -2.5$, the weight is moving upward.

3. $s(0)=-3$ in.; $P=.8$ sec

(a) Given $s(t)=a\cos\omega t$, the period is $\frac{2\pi}{\omega}$ and the amplitude is $|a|$.

$$P=.8\text{ sec}\Rightarrow .8=\frac{2\pi}{\omega}\Rightarrow \frac{4}{5}=\frac{2\pi}{\omega}\Rightarrow 4\omega=10\pi\Rightarrow \omega=\frac{10\pi}{4}=2.5\pi$$

$$s(0)=-3=a\cos\left[\omega(0)\right]\Rightarrow -3=a\cos 0\Rightarrow -3=a(1)\Rightarrow a=-3$$

Thus, $s(t)=-3\cos 2.5\pi t$.

(b) Since $s(1)=-3\cos\left[2.5\pi(1)\right]=-3\cos\frac{5\pi}{2}=-3(0)=0$, the weight is moving upward.

4. $s(0)=-4$ in.; $P=1.2$ sec

(a) Given $s(t)=a\cos\omega t$, the period is $\frac{2\pi}{\omega}$ and the amplitude is $|a|$.

$$P=1.2\text{ sec}\Rightarrow 1.2=\frac{2\pi}{\omega}\Rightarrow \frac{6}{5}=\frac{2\pi}{\omega}\Rightarrow 6\omega=10\pi\Rightarrow \omega=\frac{10\pi}{6}=\frac{5\pi}{3}$$

$$s(0)=-4=a\cos\left[\omega(0)\right]\Rightarrow -4=a\cos 0\Rightarrow -4=a(1)\Rightarrow a=-4$$

Thus, $s(t)=-4\cos\frac{5\pi}{3}t$.

(b) Since $s(1)=-4\cos\left[\frac{5\pi}{3}(1)\right]=-4\cos\frac{5\pi}{3}=-4\left(\frac{1}{2}\right)=-2$, the weight is moving downward.

5. Since frequency is $\frac{\omega}{2\pi}$, we have $27.5=\frac{\omega}{2\pi}\Rightarrow \omega=55\pi$. Since $s(0)=.21$, $.21=a\cos\left[\omega(0)\right]\Rightarrow$ $.21=a\cos 0\Rightarrow .21=a(1)\Rightarrow a=.21$. Thus, $s(t)=.21\cos 55\pi t$.

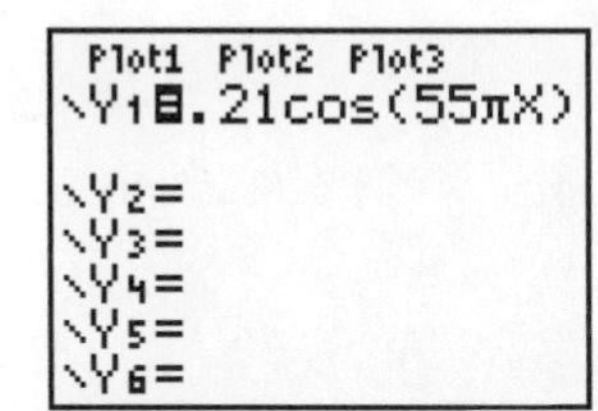

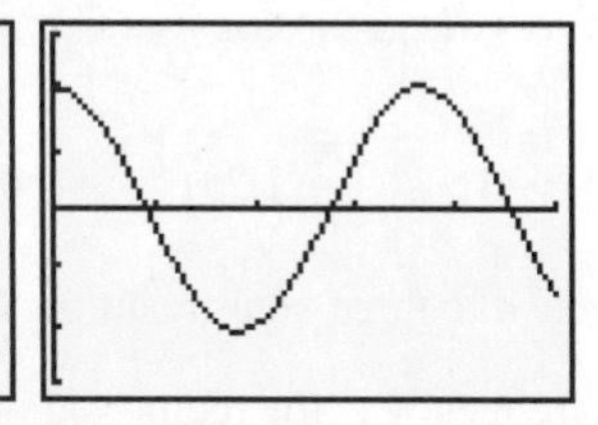

6. Since frequency is $\frac{\omega}{2\pi}$, we have $110=\frac{\omega}{2\pi}\Rightarrow \omega=220\pi$. Since $s(0)=.11$, $.11=a\cos\left[\omega(0)\right]\Rightarrow .11=a\cos 0\Rightarrow .11=a(1)\Rightarrow a=.11$. Thus, $s(t)=.11\cos 220\pi t$.

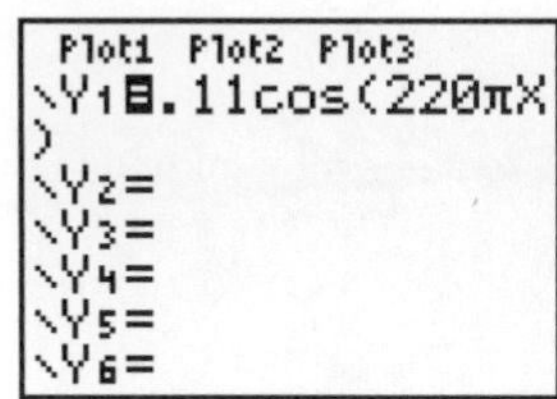

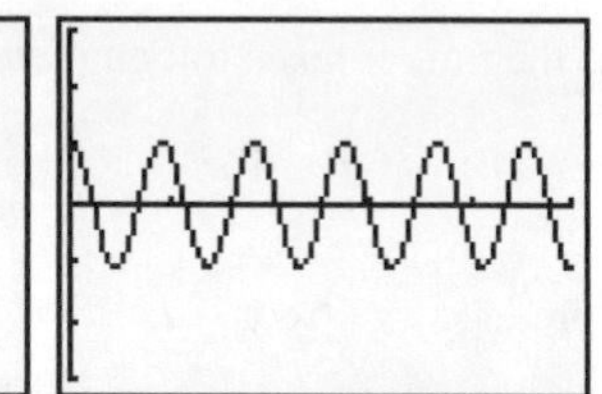

7. Since frequency is $\frac{\omega}{2\pi}$, we have $55=\frac{\omega}{2\pi}\Rightarrow\omega=110\pi$. Since $s(0)=.14$, $.14=a\cos[\omega(0)]\Rightarrow$ $.14=a\cos 0\Rightarrow .14=a(1)\Rightarrow a=.14$. Thus, $s(t)=.14\cos 110\pi t$.

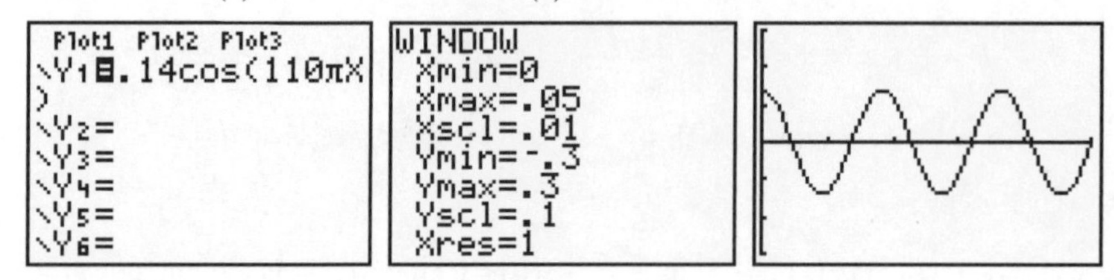

8. Since frequency is $\frac{\omega}{2\pi}$, we have $220=\frac{\omega}{2\pi}\Rightarrow\omega=440\pi$. Since $s(0)=.06$, $.06=a\cos[\omega(0)]\Rightarrow$ $.06=a\cos 0\Rightarrow .06=a(1)\Rightarrow a=.06$. Thus, $s(t)=.06\cos 440\pi t$.

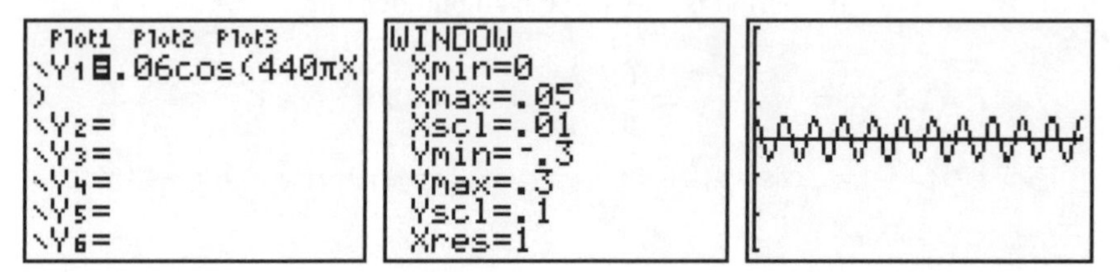

9. **(a)** Since the object is pulled down 4 units, $s(0)=-4$. Thus, we have the following.

$$s(0)=-4=a\cos[\omega(0)]\Rightarrow -4=a\cos 0\Rightarrow -4=a(1)\Rightarrow a=-4$$

Since the time it takes to complete one oscillation is 3 sec, $P=3$ sec.

$$P=3\text{ sec}\Rightarrow 3=\frac{2\pi}{\omega}\Rightarrow 3=\frac{2\pi}{\omega}\Rightarrow 3\omega=2\pi\Rightarrow\omega=\frac{2\pi}{3}$$

Therefore, $s(t)=-4\cos\frac{2\pi}{3}t$.

(b) $s(1)=-4\cos\left[\frac{2\pi}{3}(1.25)\right]=-4\cos\left[\frac{2\pi}{3}\left(\frac{5}{4}\right)\right]=-4\cos\frac{5\pi}{6}=-4\left(-\frac{\sqrt{3}}{2}\right)=2\sqrt{3}\approx 3.46$ units (rounded to three significant digits)

(c) The frequency is the reciprocal of the period, or $\frac{1}{3}$.

10. **(a)** Since the object is pulled down 6 units, $s(0)=-6$. Thus, we have the following.

$$s(0)=-6=a\cos[\omega(0)]\Rightarrow -6=a\cos 0\Rightarrow -6=a(1)\Rightarrow a=-6$$

Since the time it takes to complete one oscillation is 4 sec, $P=4$ sec.

$$P=4\text{ sec}\Rightarrow 4=\frac{2\pi}{\omega}\Rightarrow 4=\frac{2\pi}{\omega}\Rightarrow 4\omega=2\pi\Rightarrow\omega=\frac{2\pi}{4}=\frac{\pi}{2}$$

Therefore, $s(t)=-6\cos\frac{\pi}{2}t$.

(b) $s(1)=-6\cos\left[\frac{\pi}{2}(1.25)\right]=-6\cos\left[\frac{\pi}{2}\left(\frac{5}{4}\right)\right]=-6\cos\frac{5\pi}{8}\approx 2.30$ units (rounded to three significant digits)

(c) The frequency is the reciprocal of the period, or $\frac{1}{4}$.

11. **(a)** $a = 2, \omega = 2$

$s(t) = a\sin\omega t \Rightarrow s(t) = 2\sin 2t$

amplitude $= |a| = |2| = 2$; period $= \frac{2\pi}{\omega} = \frac{2\pi}{2} = \pi$; frequency $= \frac{\omega}{2\pi} = \frac{1}{\pi}$

(b) $a = 2,\ \omega = 4$

$s(t) = a\sin\omega t \Rightarrow s(t) = 2\sin 4t$

amplitude $= |a| = |2| = 2$; period $= \frac{2\pi}{\omega} = \frac{2\pi}{4} = \frac{\pi}{2}$; frequency $= \frac{\omega}{2\pi} = \frac{4}{2\pi} = \frac{2}{\pi}$

12. $P = 2\pi\sqrt{\frac{L}{32}}; L = \frac{1}{2}$ ft

The period is $P = 2\pi\sqrt{\frac{L}{32}} \Rightarrow P = 2\pi\sqrt{\frac{\frac{1}{2}}{32}} = 2\pi\sqrt{\frac{1}{64}} = 2\pi\cdot\frac{1}{8} = \frac{\pi}{4}$ sec.

The frequency is the reciprocal of the period, or $\frac{4}{\pi}$.

13. Since $P = 2\pi\sqrt{\frac{L}{32}},\ 1 = 2\pi\sqrt{\frac{L}{32}} \Rightarrow \frac{1}{2\pi} = \sqrt{\frac{L}{32}} \Rightarrow \frac{1}{(2\pi)^2} = \frac{L}{32} \Rightarrow \frac{1}{4\pi^2} = \frac{L}{32} \Rightarrow \frac{32}{4\pi^2} = L \Rightarrow L = \frac{8}{\pi^2}$ ft.

14. $s(t) = a\sin\sqrt{\frac{k}{m}}t;\ k = 4$; $P = 1$ sec

A period of 1 sec is produced when $\frac{2\pi}{\sqrt{\frac{k}{m}}} = 1$. Since $k = 4$, we solve the following.

$$\frac{2\pi}{\sqrt{\frac{k}{m}}} = 1 \Rightarrow \frac{2\pi}{\sqrt{\frac{4}{m}}} = 1 \Rightarrow 2\pi = \sqrt{\frac{4}{m}} \Rightarrow 4\pi^2 = \frac{4}{m} \Rightarrow 4\pi^2 m = 4 \Rightarrow m = \frac{1}{\pi^2}$$

15. $s(t) = -4\cos 8\pi t$

(a) The maximum of $s(t) = -4\cos 8\pi t$ is $|-4| = 4$ in.

(b) In order for $s(t) = -4\cos 8\pi t$ to reach its maximum, $y = \cos 8\pi t$ needs to be at a minimum. This occurs after $8\pi t = \pi \Rightarrow t = \frac{\pi}{8\pi} = \frac{1}{8}$ sec.

(c) Since $s(t) = -4\cos 8\pi t$ and $s(t) = a\cos\omega t$, $\omega = 8\pi$. Therefore, frequency $= \frac{\omega}{2\pi} = \frac{8\pi}{2\pi} = 4$ cycles per sec. The period is the reciprocal of the frequency, or $\frac{1}{4}$ sec.

16. $k = 2,\ m = 1$

(a) Since the spring is stretched $\frac{1}{2}$ ft, amplitude $= a = \frac{1}{2}$. From Exercise 14 we have, $s(t) = a \sin \sqrt{\frac{k}{m}} t$. Thus, $s(t) = a \sin \sqrt{\frac{2}{1}} t = a \sin \sqrt{2} t$. Since $s(t) = a \sin \omega t$, we have $\omega = \sqrt{2}$. This yields the following.

$$\text{period} = \frac{2\pi}{\omega} = \frac{2\pi}{\sqrt{2}} = \sqrt{2}\,\pi \text{ and frequency } = \frac{\omega}{2\pi} = \frac{\sqrt{2}}{2\pi}$$

(b) $s(t) = a \sin \omega t \Rightarrow s(t) = \frac{1}{2} \sin \sqrt{2}\, t$

17. $s(t) = -5 \cos 4\pi t$, $a = |-5| = 5,\ \omega = 4\pi$

(a) maximum height = amplitude = $a = |-5| = 5$ in.

(b) frequency $= \frac{\omega}{2\pi} = \frac{4\pi}{2\pi} = 2$ cycles per sec; period $= \frac{2\pi}{\omega} = \frac{1}{2}$ sec

(c) $s(t) = -5 \cos 4\pi t = 5 \Rightarrow \cos 4\pi t = -1 \Rightarrow 4\pi t = \pi \Rightarrow t = \frac{1}{4}$

The weight first reaches its maximum height after $\frac{1}{4}$ sec.

(d) Since $s(1.3) = -5 \cos\left[4\pi(1.3)\right] = -5 \cos 5.2\pi \approx 4$, after 1.3 sec, the weight is about 4 in. above the equilibrium position.

18. $s(t) = -4 \cos 10t,\ a = -4,\ \omega = 10$

(a) maximum height = amplitude = $a = |-4| = 4$ in.

(b) frequency $= \frac{\omega}{2\pi} = \frac{10}{2\pi} = \frac{5}{\pi}$ cycles per sec; period $= \frac{2\pi}{\omega} = \frac{2\pi}{10} = \frac{\pi}{5}$ sec

(c) $s(t) = -4 \cos 10t = 4 \Rightarrow \cos 10t = -1 \Rightarrow 10t = \pi \Rightarrow t = \frac{\pi}{10}$

The weight first reaches its maximum height after $\frac{\pi}{10}$ sec.

(d) $s(1.466) = -4 \cos(10 \cdot 1.466) = -4 \cos(14.66) \approx 2$

After 1.466 sec, the weight is about 2 in. above the equilibrium position.

19. $a = -3$

(a) We will use a model of the form $s(t) = a \cos \omega t$ with $a = -3$.

Since $s(0) = -3 \cos\left[\omega(0)\right] = -3 \cos 0 = -3(1) = -3$,

Using a cosine function rather than a sine function will avoid the need for a phase shift.

Since the frequency $= \frac{6}{\pi}$ cycles per sec, by definition, $\frac{\omega}{2\pi} = \frac{6}{\pi} \Rightarrow \omega\pi = 12\pi \Rightarrow \omega = 12$.

Therefore, a model for the position of the weight at time *t* seconds is $s(t) = -3 \cos 12t$.

(b) The period is the reciprocal of the frequency, or $\frac{\pi}{6}$ sec.

20. $a = -2$

(a) period: $\frac{2\pi}{\omega} = \frac{1}{3} \Rightarrow 6\pi = \omega$

$s(t) = a\cos\omega t \Rightarrow s(t) = -2\cos 6\pi t$

(b) frequency $= \frac{\omega}{2\pi} = \frac{6\pi}{2\pi} = 3$ cycles per sec

For Exercises 21 – 22, we have the following.

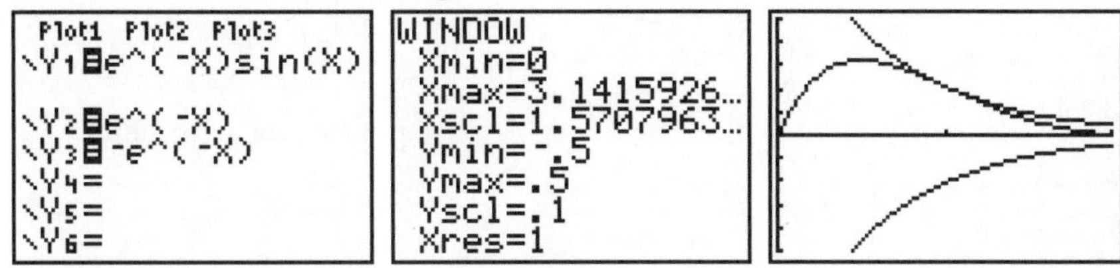

21. Since $e^{-t} \neq 0$, we have $e^{-t}\sin t = 0 \Rightarrow \sin t = 0 \Rightarrow t = 0,\ \pi$. The x-intercepts of Y_1 are the same as these of $\sin x$.

22. Since $Y_1 = Y_2 \Rightarrow e^{-t}\sin t = e^{-t} \Rightarrow \sin t = 1 \Rightarrow t = \frac{\pi}{2}$, the intersection occurs when $\sin t$ is at a maximum, that is, when $t = \frac{\pi}{2}$. Thus, the point of intersection is $\left(\frac{\pi}{2}, e^{-\pi/2}\right)$

Since $Y_1 = Y_3 \Rightarrow e^{-t}\sin t = -e^{-t} \Rightarrow \sin t = -1 \Rightarrow x = \frac{3\pi}{2}$, the intersection occurs when $\sin t$ is at a minimum but the minimum, value of $\sin t$ does not occur in $[0, \pi]$.

Chapter 4: Review Exercises

1. B; The amplitude is $|4| = 4$ and period is $\frac{2\pi}{2} = \pi$.

2. D; The amplitude is $|-3| = 3$, but the period is $\frac{2\pi}{\frac{1}{2}} = 4\pi$. All other statements are true.

3. The range for $\sin x$ and $\cos x$ is $[-1,1]$. The range for $\tan x$ and $\cot x$ is $(-\infty,\infty)$. Since $\frac{1}{2}$ falls in these intervals, those trigonometric functions can attain the value $\frac{1}{2}$.

4. The range for $\sec x$ and $\csc x$ is $(-\infty,-1]\cup[1,\infty)$. The range for $\tan x$ and $\cot x$ is $(-\infty,\infty)$. Since 2 falls in these intervals, those trigonometric functions can attain the value 2.

5. $y = 2\sin x$
Amplitude: 2
Period: 2π
Vertical translation: none
Phase shift: none

6. $y = \tan 3x$
Amplitude: not applicable
Period: $\frac{\pi}{3}$
Vertical translation: none
Phase shift: none

7. $y = -\frac{1}{2}\cos 3x$

Amplitude: $\left|-\frac{1}{2}\right| = \frac{1}{2}$

Period: $\frac{2\pi}{3}$

Vertical translation: none
Phase shift: none

8. $y = 2\sin 5x$

Amplitude: $|2| = 2$

Period: $\frac{2\pi}{5}$

Vertical translation: none
Phase shift: none

9. $y = 1 + 2\sin\frac{1}{4}x$

Amplitude: $|2| = 2$

Period: $\frac{2\pi}{\frac{1}{4}} = 8\pi$

Vertical translation: up 1 unit
Phase shift: none

10. $y = 3 - \frac{1}{4}\cos\frac{2}{3}x$

Amplitude: $\left|-\frac{1}{4}\right| = \frac{1}{4}$

Period: $\frac{2\pi}{\frac{2}{3}} = 3\pi$

Vertical translation: up 3 units
Phase shift: none

11. $y = 3\cos\left(x + \frac{\pi}{2}\right) = 3\cos\left[x - \left(-\frac{\pi}{2}\right)\right]$

Amplitude: $|3| = 3$

Period: 2π

Vertical translation: none

Phase shift: $\frac{\pi}{2}$ units to the left

12. $y = -\sin\left(x - \frac{3\pi}{4}\right)$

Amplitude: $|-1| = 1$

Period: 2π

Vertical translation: none

Phase shift: $\frac{3\pi}{4}$ units to the right

13. $y = \frac{1}{2}\csc\left(2x - \frac{\pi}{4}\right) = \frac{1}{2}\csc 2\left(x - \frac{\pi}{8}\right)$

Amplitude: not applicable

Period: $\frac{2\pi}{2} = \pi$

Vertical translation: none

Phase shift: $\frac{\pi}{8}$ unit to the right

14. $y = 2\sec(\pi x - 2\pi) = y = 2\sec\pi(x-2)$

Amplitude: not applicable

Period: $\frac{2\pi}{\pi} = 2$

Vertical translation: none
Phase shift: 2 units to the right

15. $y = \frac{1}{3}\tan\left(3x - \frac{\pi}{3}\right) = \frac{1}{3}\tan 3\left(x - \frac{\pi}{9}\right)$

Amplitude: not applicable

Period: $\frac{\pi}{3}$

Vertical translation: none

Phase shift: $\frac{\pi}{9}$ units to the right

16. $y = \cot\left(\frac{x}{2} + \frac{3\pi}{4}\right) = \cot\frac{1}{2}\left(x + \frac{3\pi}{2}\right)$

$= \cot\frac{1}{2}\left[x - \left(-\frac{3\pi}{2}\right)\right]$

Amplitude: not applicable

Period: $\frac{\pi}{\frac{1}{2}} = 2\pi$

Vertical translation: none

Phase shift: $\frac{3\pi}{2}$ units to the left

17. The tangent function has a period of π and x-intercepts at integral multiples of π.

18. The sine function has a period of 2π and passes through the origin.

19. The cosine function has a period of 2π and has the value 0 when $x = \frac{\pi}{2}$.

20. The cosecant function has a period of 2π and is not defined at integral multiples of π.

21. The cotangent function has a period of π and decreases on the interval $(0, \pi)$.

22. The secant function has a period of 2π and vertical asymptotes at odd multiples of $\frac{\pi}{2}$; that is, at $x = (2n+1)\frac{\pi}{2}$, where n is an integer.

23. – 24. Answers will vary.

25. $y = 3\sin x$

Period: 2π and Amplitude: $|3| = 3$

Divide the interval $[0, 2\pi]$ into four equal parts to get the x-values that will yield minimum and maximum points and x-intercepts. Then make a table.

x	0	$\frac{\pi}{2}$	π	$\frac{3\pi}{2}$	2π
$\sin x$	0	1	0	-1	0
$3\sin x$	0	3	0	-3	0

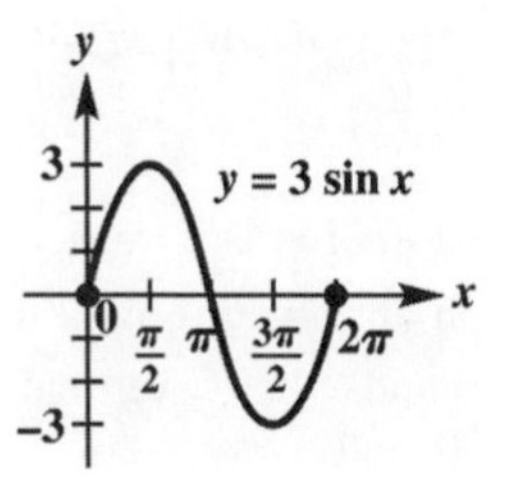

26. $y = \frac{1}{2}\sec x$

Step 1 Graph the corresponding reciprocal function $y = \frac{1}{2}\cos x$ The period is 2π and its amplitude is $\left|\frac{1}{2}\right| = \frac{1}{2}$. One period is in the interval $0 \le x \le 2\pi$. Dividing the interval into four equal parts gives us the following key points.

$$\left(0, \frac{1}{2}\right), \left(\frac{\pi}{2}, 0\right), \left(\pi, -\frac{1}{2}\right), \left(\frac{3\pi}{2}, 0\right), \left(2\pi, \frac{1}{2}\right)$$

Step 2 The vertical asymptotes of $y = \frac{1}{2}\sec x$ are at the x-intercepts of $y = \frac{1}{2}\cos x$, which are $x = \frac{\pi}{2}$ and $x = \frac{3\pi}{2}$. Continuing this pattern to the left, we also have a vertical asymptote of $x = -\frac{\pi}{2}$.

Step 3 Sketch the graph.

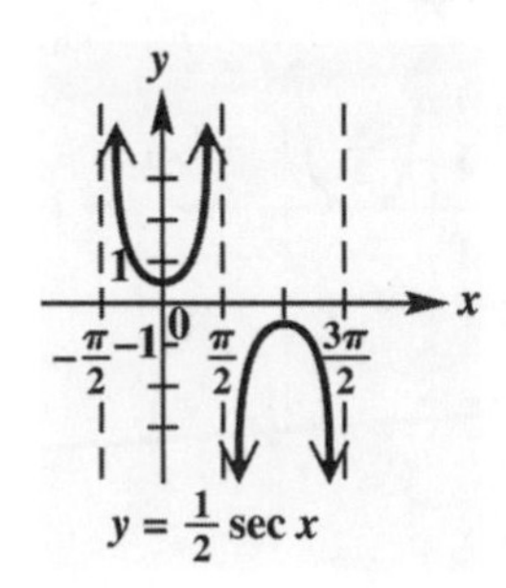

27. $y=-\tan x$ is a reflection of the graph of $y=\tan x$ over the x-axis. The period is π and has vertical asymptotes of $x=-\frac{\pi}{2}$ and $x=\frac{\pi}{2}$.

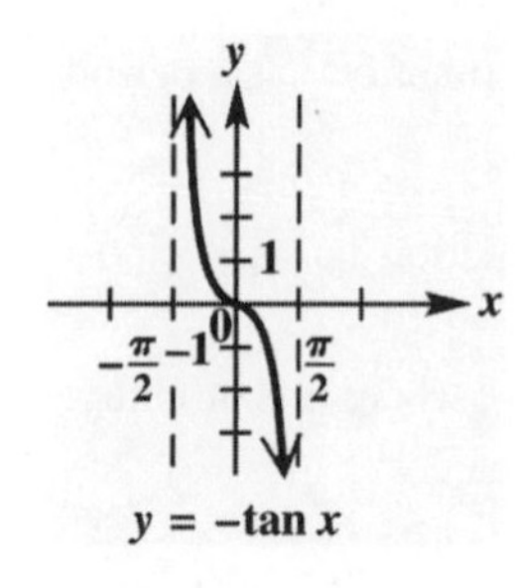

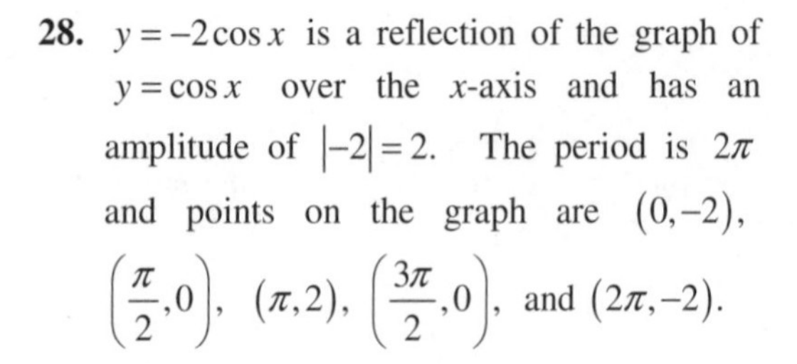

28. $y=-2\cos x$ is a reflection of the graph of $y=\cos x$ over the x-axis and has an amplitude of $|-2|=2$. The period is 2π and points on the graph are $(0,-2)$, $\left(\frac{\pi}{2},0\right)$, $(\pi,2)$, $\left(\frac{3\pi}{2},0\right)$, and $(2\pi,-2)$.

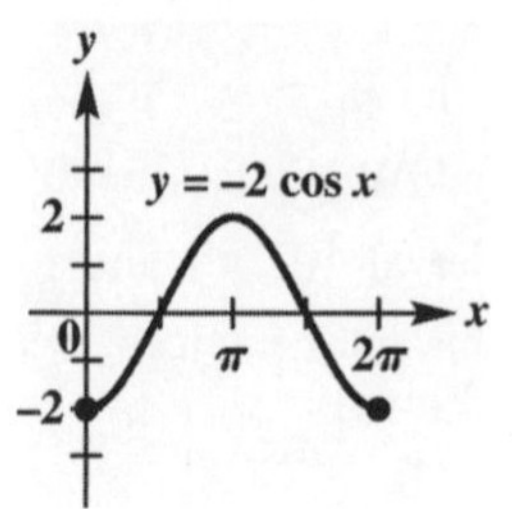

29. $y=2+\cot x$ is a vertical translation of the graph of $y=\cot x$ up 2 units. The period is π and has vertical asymptotes of $x=0$ and $x=\pi$.

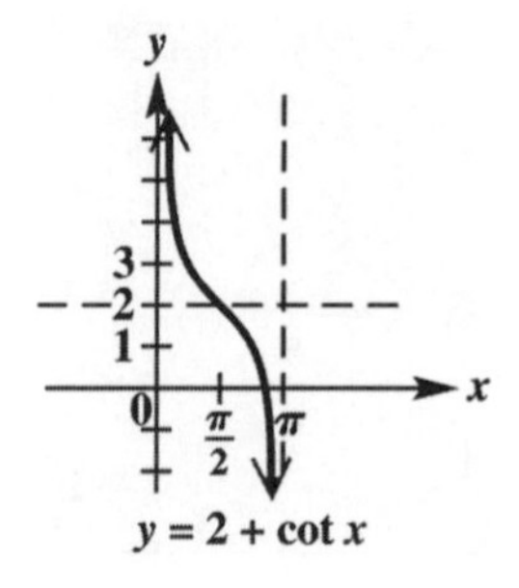

30. $y=-1+\csc x$ is a vertical translation of the graph of $y=\csc x$ down 1 unit. The period is 2π and has vertical asymptotes of $x=0$, $x=\pi$, and $x=2\pi$.

Step 1 Graph the corresponding reciprocal function $y=-1+\sin x$ The period is 2π and its amplitude is $|1|=1$. One period is in the interval $0\le x\le 2\pi$. Dividing the interval into four equal parts gives us the following key points.

$$(0,-1),\ \left(\frac{\pi}{2},-1\right),\ (\pi,-1),\ \left(\frac{3\pi}{2},-2\right),\ (2\pi,-1)$$

Step 2 The vertical asymptotes of $y=-1+\csc x$ are at the x-intercepts of $y=\sin x$, which are $x=0$, $x=\pi$, and $x=2\pi$.

Step 3 Sketch the graph.

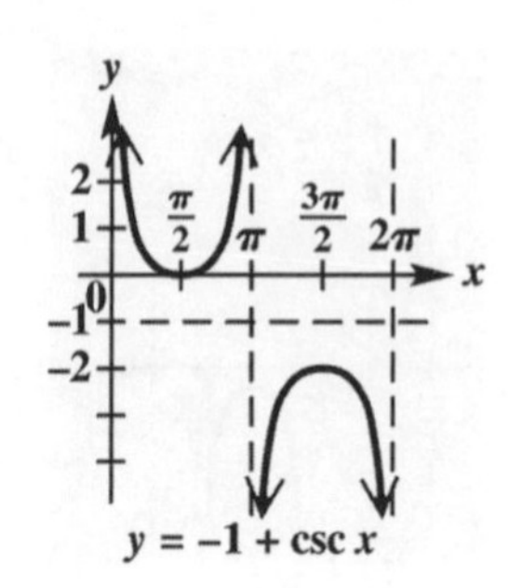

31. $y = \sin 2x$

Period: $\frac{2\pi}{2} = \pi$ and Amplitude: $|1| = 1$

Divide the interval $[0, \pi]$ into four equal parts to get the x-values that will yield minimum and maximum points and x-intercepts. Then make a table.

x	0	$\frac{\pi}{4}$	$\frac{\pi}{2}$	$\frac{3\pi}{4}$	π
$2x$	0	$\frac{\pi}{2}$	π	$\frac{3\pi}{2}$	2π
$\sin 2x$	0	1	0	-1	0

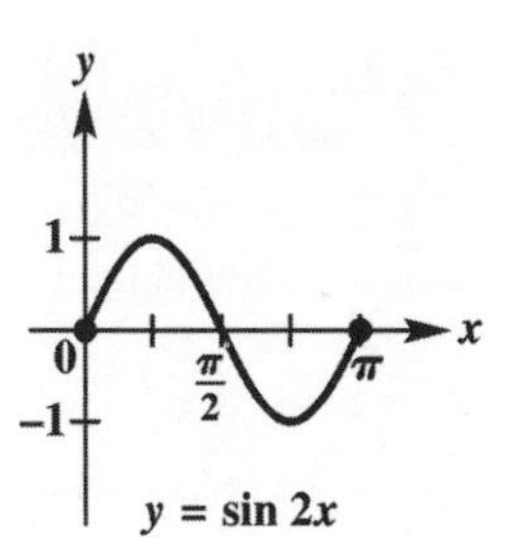

32. $y = \tan 3x$

Step 1 Find the period and locate the vertical asymptotes. The period of tangent is $\frac{\pi}{b}$, so the period for this function is $\frac{\pi}{3}$. Tangent has asymptotes of the form $bx = -\frac{\pi}{2}$ and $bx = \frac{\pi}{2}$. The asymptotes for $y = \tan 3x$ are as follows.

$$3x = -\frac{\pi}{2} \Rightarrow x = -\frac{\pi}{6} \text{ and } 3x = \frac{\pi}{2} \Rightarrow x = \frac{\pi}{6}$$

Step 2 Sketch the vertical asymptotes found in Step 1.

Step 3 Divide the interval into four equal parts.

$$-\frac{\pi}{6}, -\frac{\pi}{12}, 0, \frac{\pi}{12}, \frac{\pi}{6}$$

Step 4 Finding the first-quarter point, midpoint, and third-quarter point, we have the following.

$$\left(-\frac{\pi}{12}, -1\right), (0,0), \left(\frac{\pi}{12}, 1\right)$$

Step 5 Join the points with a smooth curve.

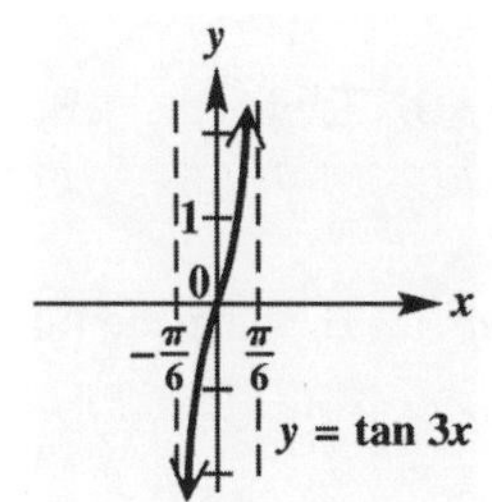

33. $y = 3 \cos 2x$

Period: $\frac{2\pi}{2} = \pi$ and Amplitude: $|3| = 3$

Divide the interval $[0, \pi]$ into four equal parts to get the x-values that will yield minimum and maximum points and x-intercepts. Then make a table.

x	0	$\frac{\pi}{4}$	$\frac{\pi}{2}$	$\frac{3\pi}{4}$	π
$2x$	0	$\frac{\pi}{2}$	π	$\frac{3\pi}{2}$	2π
$\cos 2x$	1	0	-1	0	1
$3\cos 2x$	3	0	-3	0	3

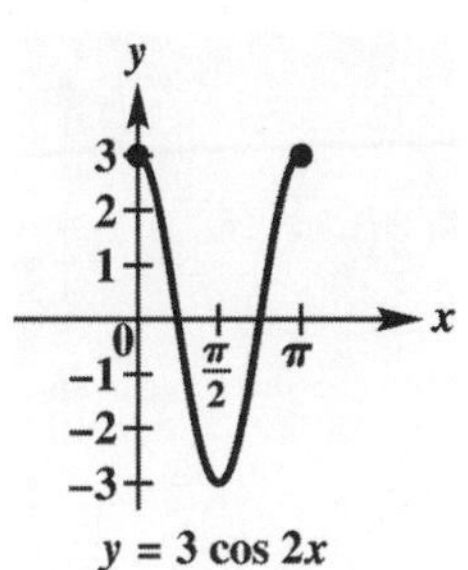

34. $y = \frac{1}{2}\cot 3x$

Step 1 Find the period and locate the vertical asymptotes. The period of cotangent is $\frac{\pi}{b}$, so the period for this function is $\frac{\pi}{3}$. Cotangent has asymptotes of the form $bx = 0$ and $bx = \pi$.

The asymptotes for $y = \frac{1}{2}\cot 3x$ are $x = 0$ and $x = \frac{\pi}{3}$.

Step 2 Sketch the two vertical asymptotes found in Step 1.

Step 3 Divide the interval into four equal parts.

$$0, \frac{\pi}{12}, \frac{\pi}{6}, \frac{\pi}{4}, \frac{\pi}{3}$$

Step 4 Finding the first-quarter point, midpoint, and third-quarter point, we have the following.

$$\left(\frac{\pi}{12}, \frac{1}{2}\right), \left(\frac{\pi}{6}, 0\right), \left(\frac{\pi}{4}, -\frac{1}{2}\right)$$

Step 5 Join the points with a smooth curve. The graph is "shrunk" because $a = \frac{1}{2}$ and $\left|\frac{1}{2}\right| < 1$.

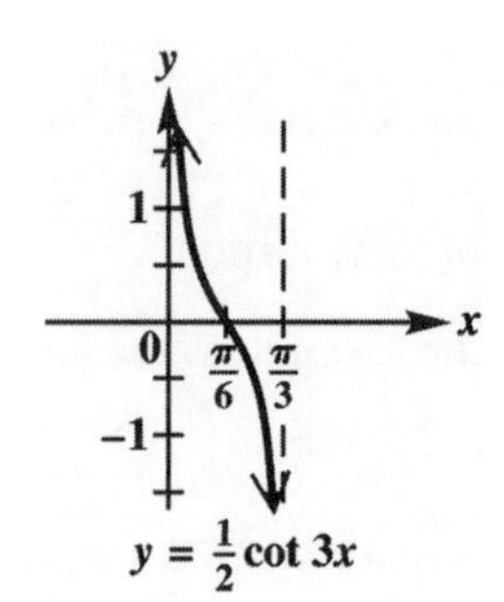

35. $y = \cos\left(x - \frac{\pi}{4}\right)$

Step 1 Find the interval whose length is $\frac{2\pi}{b}$.

$$0 \le x - \frac{\pi}{4} \le 2\pi \Rightarrow 0 + \frac{\pi}{4} \le x \le 2\pi + \frac{\pi}{4} \Rightarrow \frac{\pi}{4} \le x \le \frac{9\pi}{4}$$

Step 2 Divide the period into four equal parts to get the following x-values.

$$\frac{\pi}{4},\ \pi,\ \frac{3\pi}{2},\ 2\pi,\ \frac{9\pi}{4}$$

Step 3 Evaluate the function for each of the five x-values.

x	$\frac{\pi}{4}$	$\frac{3\pi}{4}$	$\frac{5\pi}{4}$	$\frac{7\pi}{4}$	$\frac{9\pi}{4}$
$x - \frac{\pi}{4}$	0	$\frac{\pi}{2}$	π	$\frac{3\pi}{2}$	2π
$\cos\left(x - \frac{\pi}{4}\right)$	1	0	-1	0	1

Steps 4 and 5 Plot the points found in the table and join them with a sinusoidal curve.

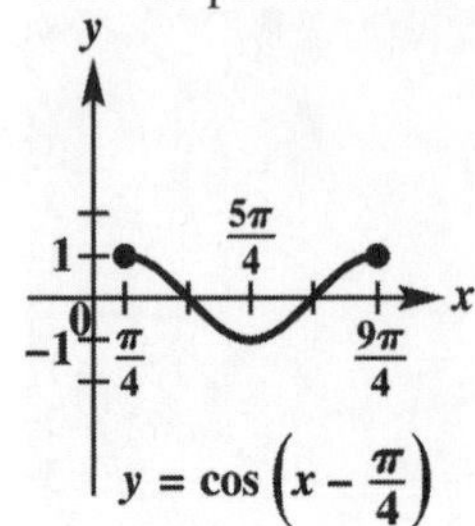

The amplitude is 1.

The period is 2π.

There is no vertical translation.

The phase shift is $\frac{\pi}{4}$ units to the right.

36. $y = \tan\left(x - \frac{\pi}{2}\right)$

Period: π

Vertical translation: none

Phase shift (horizontal translation): $\frac{\pi}{2}$ units to the right

Because the function is to be graphed over a two-period interval, locate three adjacent vertical asymptotes. Because asymptotes of the graph $y = \tan x$ occur at $-\frac{\pi}{2}$, and $\frac{\pi}{2}$, the following equations can be solved to locate asymptotes.

$$x - \frac{\pi}{2} = -\frac{\pi}{2} \Rightarrow x = 0 \text{ and } x - \frac{\pi}{2} = \frac{\pi}{2} \Rightarrow x = \pi$$

Divide the interval $(0, \pi)$ into four equal parts to obtain the following key x-values.

$$\text{first-quarter value: } \frac{\pi}{4}\text{; middle value: } \frac{\pi}{2}\text{; third-quarter value: } \frac{3\pi}{4}$$

Evaluating the given function at these three key x-values gives the following points.

$$\left(\frac{\pi}{4}, -1\right), \left(\frac{\pi}{2}, 0\right), \left(\frac{3\pi}{4}, 1\right)$$

Connect these points with a smooth curve and continue to graph to approach the asymptote $x = 0$ and $x = \pi$ to complete one period of the graph.

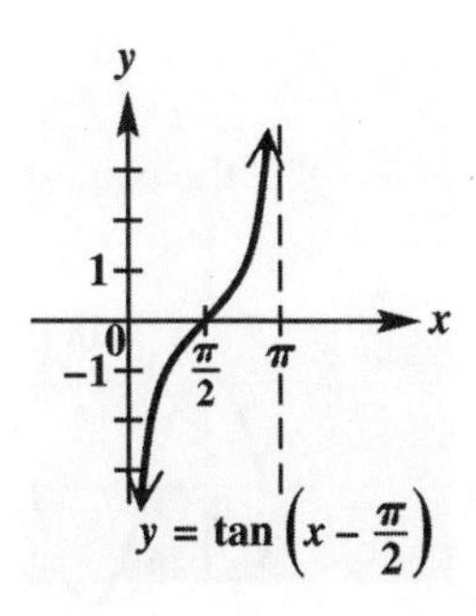

37. $y = \sec\left(2x + \frac{\pi}{3}\right) = \sec 2\left(x + \frac{\pi}{6}\right)$

Step 1 Graph the corresponding reciprocal function $\cos 2\left(x + \frac{\pi}{6}\right)$. The period is $\frac{2\pi}{2} = \pi$ and its amplitude is $|2| = 2$. One period is in the interval $\frac{\pi}{12} \le x \le \frac{13\pi}{12}$. Dividing the interval into four equal parts gives us the following key points.

$$\left(\frac{\pi}{12}, 0\right), \left(\frac{\pi}{3}, -1\right), \left(\frac{7\pi}{12}, 0\right), \left(\frac{5\pi}{6}, 1\right), \left(\frac{13\pi}{12}, 0\right)$$

Step 2 The vertical asymptotes of $y = \sec 2\left(x + \frac{\pi}{6}\right)$ are at the x-intercepts of $y = \cos 2\left(x + \frac{\pi}{6}\right)$, which are $x = \frac{\pi}{12}$, $x = \frac{7\pi}{12}$, and $x = \frac{13\pi}{12}$.

Step 3 Sketch the graph.

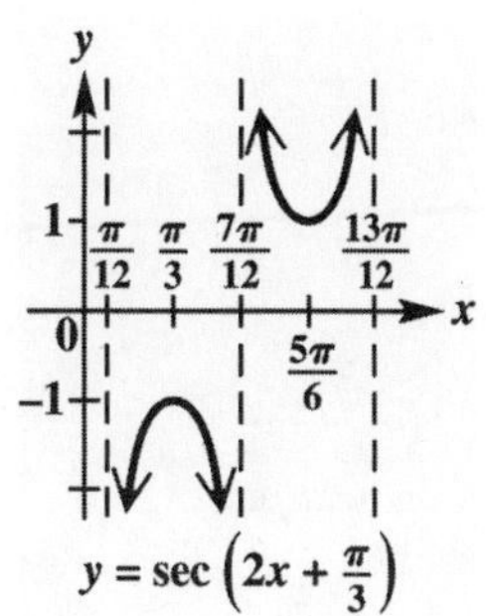

38. $y=\sin\left(3x+\frac{\pi}{2}\right)=\sin 3\left(x+\frac{\pi}{6}\right)$

Step 1 Find the interval whose length is $\frac{2\pi}{b}$.

$$0\le 3\left(x+\frac{\pi}{6}\right)\le 2\pi \Rightarrow 0\le x+\frac{\pi}{6}\le\frac{2\pi}{3}\Rightarrow -\frac{\pi}{6}\le x\le\frac{\pi}{2}$$

Step 2 Divide the period into four equal parts to get the following x-values.

$$-\frac{\pi}{6},\ 0,\ \frac{\pi}{6},\ \frac{\pi}{3},\ \frac{\pi}{2}$$

Step 3 Evaluate the function for each of the five x-values

x	$-\frac{\pi}{6}$	0	$\frac{\pi}{6}$	$\frac{\pi}{3}$	$\frac{\pi}{2}$
$3\left(x+\frac{\pi}{6}\right)$	0	$\frac{\pi}{2}$	π	$\frac{3\pi}{2}$	2π
$\sin 3\left(x+\frac{\pi}{6}\right)$	0	1	0	-1	0

Steps 4 and 5 Plot the points found in the table and join them with a sinusoidal curve.

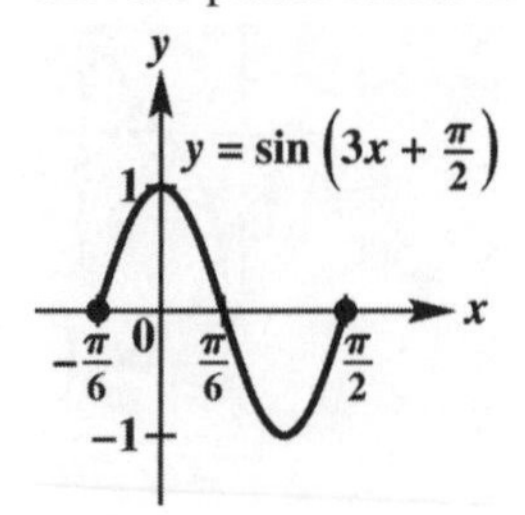

The amplitude is 1.

The period is $\frac{2\pi}{3}$.

There is no vertical translation.

The phase shift is $\frac{\pi}{6}$ unit to the left.

39. $y = 1 + 2\cos 3x$

Step 1 Find the interval whose length is $\frac{2\pi}{b}$.

$$0\le 3x\le 2\pi \Rightarrow 0\le x\le\frac{2\pi}{3}$$

Step 2 Divide the period into four equal parts to get the following x-values.

$$0,\ \frac{\pi}{6},\ \frac{\pi}{3},\ \frac{\pi}{2},\ \frac{2\pi}{3}$$

Step 3 Evaluate the function for each of the five x-values.

x	0	$\frac{\pi}{6}$	$\frac{\pi}{3}$	$\frac{\pi}{2}$	$\frac{2\pi}{3}$
$3x$	0	$\frac{\pi}{2}$	π	$\frac{3\pi}{2}$	2π
$\cos 3x$	1	0	-1	0	1
$2\cos 3x$	2	0	-2	0	2
$1+2\cos 3x$	3	1	-1	1	3

Continued on next page

39. (continued)

Steps 4 and 5 Plot the points found in the table and join them with a sinusoidal curve. By graphing an additional period to the left, we obtain the following graph.

The period is $\frac{2\pi}{3}$.

The amplitude is $|2|$, which is 2.

The vertical translation is 1 unit up.

There is no phase shift.

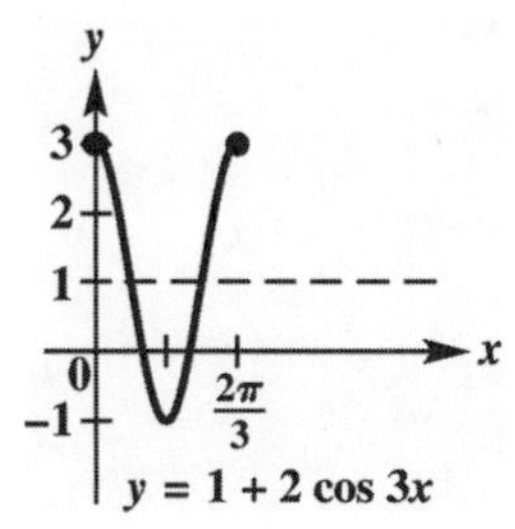

40. $y = -1 - 3\sin 2x$

Step 1 Find the interval whose length is $\frac{2\pi}{b}$.

$$0 \le 2x \le 2\pi \Rightarrow 0 \le x \le \pi$$

Step 2 Divide the period into four equal parts to get the following x-values.

$$0, \frac{\pi}{4}, \frac{\pi}{2}, \frac{3\pi}{4}, \pi$$

Step 3 Evaluate the function for each of the five x-values.

x	0	$\frac{\pi}{4}$	$\frac{\pi}{2}$	$\frac{3\pi}{4}$	π
$2x$	0	$\frac{\pi}{2}$	π	$\frac{3\pi}{2}$	2π
$\sin 2x$	0	1	0	−1	0
$-3\sin 2x$	0	−3	0	3	0
$-1-3\sin 2x$	−1	−4	−1	2	−1

Steps 4 and 5 Plot the points found in the table and join them with a sinusoidal curve. By graphing an additional period to the left, we obtain the following graph.

The amplitude is $|-3|$, which is 3.

The period is $\frac{2\pi}{2} = \pi$.

The vertical translation is 1 unit down.

There is no phase shift.

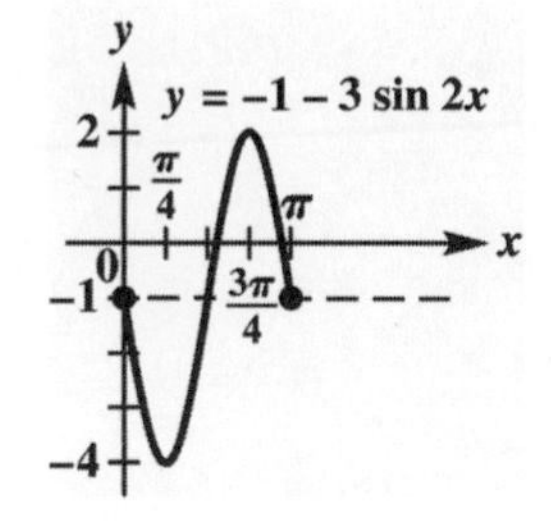

41. $y = 2\sin \pi x$

Period: $\frac{2\pi}{\pi} = 2$ and Amplitude: $|2| = 2$

Divide the interval $[0, 2]$ into four equal parts to get the x-values that will yield minimum and maximum points and x-intercepts. Then make a table.

x	0	$\frac{1}{2}$	1	$\frac{3}{2}$	2
πx	0	$\frac{\pi}{2}$	π	$\frac{3\pi}{2}$	2π
$\sin \pi x$	0	1	0	-1	0
$2\sin \pi x$	0	2	0	-2	0

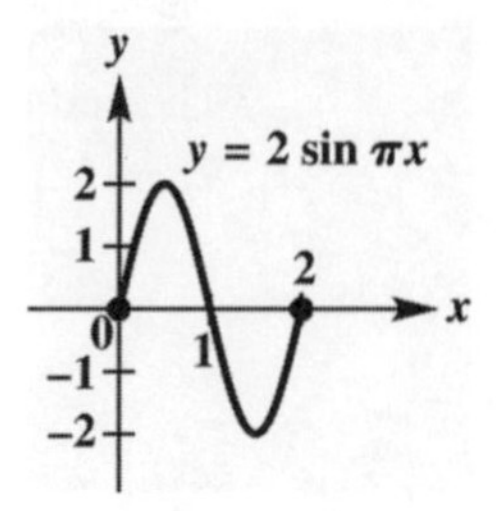

42. $y = -\frac{1}{2}\cos(\pi x - \pi) = -\frac{1}{2}\cos \pi(x-1)$

Step 1 Find the interval whose length is $\frac{2\pi}{b}$.

$$0 \le \pi(x-1) \le 2\pi \Rightarrow 0 \le x-1 \le 2 \Rightarrow 1 \le x \le 3$$

Step 2 Divide the period into four equal parts to get the following x-values.

$$1, \frac{3}{2}, 2, \frac{5}{2}, 3$$

Step 3 Evaluate the function for each of the five x-values.

x	1	$\frac{3}{2}$	2	$\frac{5}{2}$	3
$\pi(x-1)$	0	$\frac{\pi}{2}$	π	$\frac{3\pi}{2}$	2π
$\cos \pi(x-1)$	1	0	-1	0	1
$-\frac{1}{2}\cos \pi(x-1)$	$-\frac{1}{2}$	0	$\frac{1}{2}$	0	$-\frac{1}{2}$

Steps 4 and 5 Plot the points found in the table and join then with a sinusoidal curve.

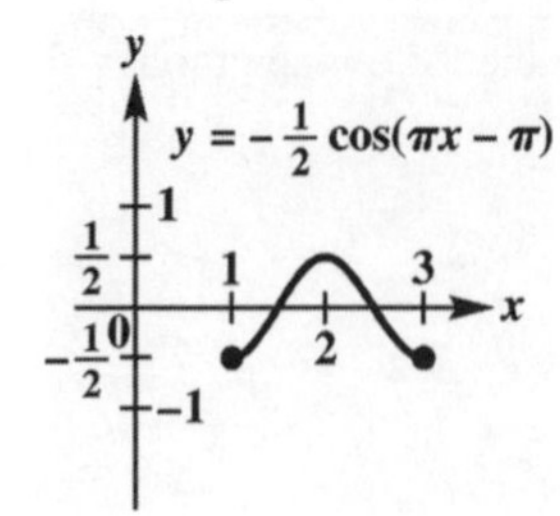

The amplitude is $\left|-\frac{1}{2}\right| = \frac{1}{2}$.

The period is $\frac{2\pi}{\pi}$, which is 2.

This is no vertical translation.

The phase shift is 1 unit to the right.

43. – 44. Answers will vary.

45. **(a)** The shorter leg of the right triangle has length $h_2 - h_1$. Thus, we have the following.

$$\cot\theta = \frac{d}{h_2 - h_1} \Rightarrow d = (h_2 - h_1)\cot\theta$$

(b) When $h_2 = 55$ and $h_1 = 5$, $d = (55-5)\cot\theta = 50\cot\theta$.

The period is π, but the graph wanted is d for $0<\theta<\frac{\pi}{2}$. The asymptote is the line $\theta = 0$. Also, when $\theta = \frac{\pi}{4}, d = 50\cot\frac{\pi}{4} = 50(1) = 50.$

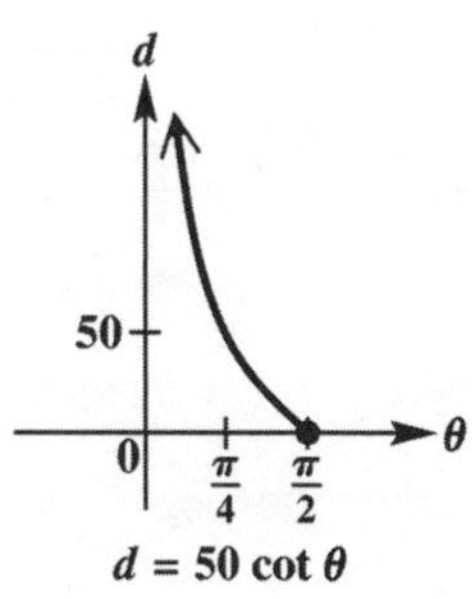

$d = 50\cot\theta$

46. **(a)** Since $27 - 14.7 = 12.3$ and $14.7 - 2.4 = 12.3$, the time between high tides is 12.3 hr.

(b) Since $2.6 - 1.4 = 1.2$, the difference in water levels between high tide and low tide is 1.2 ft.

(c) Since $f(x) = .6\cos[.511(x-2.4)] + 2$, we have the following.

$$f(10) = .6\cos[.511(10-2.4)] + 2 = .6\cos(3.8836) + 2 \approx 1.56 \text{ ft}$$

47. $t = 60 - 30\cos\frac{x\pi}{6}$

(a) For January, $x = 0$. Thus, $t = 60 - 30\cos\frac{0\cdot\pi}{6} = 60 - 30\cos 0 = 60 - 30(1) = 60 - 30 = 30°$.

(b) For April, $x = 3$. Thus, $t = 60 - 30\cos\frac{3\pi}{6} = 60 - 30\cos\frac{\pi}{2} = 60 - 30(0) = 60 - 0 = 60°$.

(c) For May, $x = 4$. Thus, $t = 60 - 30\cos\frac{4\pi}{6} = 60 - 30\cos\frac{2\pi}{3} = 60 - 30\left(-\frac{1}{2}\right) = 60 + 15 = 75°$.

(d) For June, $x = 5$. Thus, $t = 60 - 30\cos\frac{5\pi}{6} = 60 - 30\left(-\frac{\sqrt{3}}{2}\right) = 60 + 15\sqrt{3} \approx 86°$.

(e) For August, $x = 7$. Thus, $t = 60 - 30\cos\frac{7\pi}{6} = 60 - 30\left(-\frac{\sqrt{3}}{2}\right) = 60 + 15\sqrt{3} \approx 86°$.

(f) For October, $x = 9$. Thus, $t = 60 - 30\cos\frac{9\pi}{6} = 60 - 30\cos\frac{3\pi}{2} = 60 - 30(0) = 60 - 0 = 60°$.

48. **(a)** Let January correspond to x = 1, February to x = 2, … , and December of the second year to x = 24.

L1	L2	L3	2
1	25	------	
2	28		
3	36		
4	48		
5	61		
6	72		
7	74		

L2(1)=25

L1	L2	L3	2
7	74		
8	75		
9	66		
10	55		
11	39		
12	28		
13	25		

L2(7) =74

L1	L2	L3	2
13	25		
14	28		
15	36		
16	48		
17	61		
18	72		
19	74		

L2(19) =74

Continued on next page

48. (a) (continued)

L1	L2	L3
18	72	
19	74	
20	75	
21	66	
22	55	
23	39	
24	28	

L2(24)=28

WINDOW
Xmin=1
Xmax=25
Xscl=5
Ymin=20
Ymax=80
Yscl=10
Xres=1

(b) The amplitude of the sine graph is approximately 25 since the average monthly high is 75, the average monthly low is 25, and $\frac{1}{2}(75-25)=\frac{1}{2}(50)=25$. The period is 12 since the temperature cycles every twelve months. Let $b=\frac{2\pi}{12}=\frac{\pi}{6}$. One way to determine the phase shift is to use the following technique. The minimum temperature occurs in January. Thus, when $x = 1$, $b(x-d)$ must equal $\left(-\frac{\pi}{2}\right)+2\pi n$, where n is an integer, since the sine function is minimum at these values. Solving for d, we have the following.

$$\frac{\pi}{6}(1-d)=-\frac{\pi}{2}\Rightarrow 1-d=\frac{6}{\pi}\left(-\frac{\pi}{2}\right)\Rightarrow 1-d=-3\Rightarrow -d=-4\Rightarrow d=4$$

This can be used as a first approximation.

Let $f(x) = a \sin b(x-d) + c$. Since the amplitude is 25, let $a = 25$. The period is equal to 1 yr or 12 mo, so $b=\frac{\pi}{6}$. The average of the maximum and minimum temperatures is as follows.

$$\frac{1}{2}(75+25)=\frac{1}{2}(100)=50$$

Let the vertical translation be $c = 50$.

Since the phase shift is approximately 4, it can be adjusted slightly to give a better visual fit. Try 4.2. Since the phase shift is 4.2, let $d = 4.2$. Thus, $f(x)=25\sin\left[\frac{\pi}{6}(x-4.2)\right]+50.$

(c) See part b.

(d) Plotting the data with $f(x)=25\sin\left[\frac{\pi}{6}(x-4.2)\right]+50$ on the same coordinate axes gives an excellent fit.

(e)

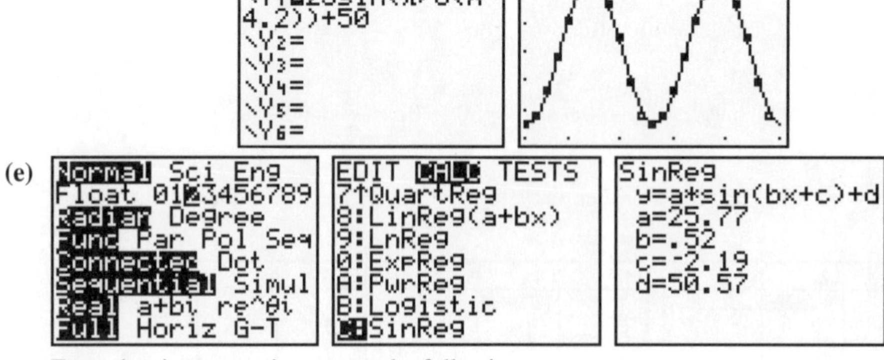

From the sine regression we get the following.

$$y=25.77\sin(.52x-2.19)+50.57 \text{ or } y=25.77\sin\left[.52(x-4.21)\right]+50.57$$

49. $P(t) = 7(1-\cos 2\pi t)(t+10)+100e^{.2t}$

(a) January 1, base year $t = 0$

$$P(0)=7(1-\cos 0)(10)+100e^{0}=7(1-1)(10)+100(1)=7(0)(10)+100=0+100=100$$

(b) July 1, base year $t = .5$

$$P(.5)=7(1-\cos\pi)(.5+10)+100e^{.2(.5)}$$
$$=7[1-(-1)](10.5)+100e^{.1}=7(2)(10.5)+100e^{.1}=147+100e^{.1}\approx 258$$

(c) January 1, following year $t = 1$

$$P(1)=7(1-\cos 2\pi)(1+10)+100e^{.2}$$
$$=7(1-1)(1+10)+100e^{.2}=7(0)(11)+100e^{.2}=0+100e^{.2}=100e^{.2}\approx 122$$

(d) July 1, following year $t = 1.5$

$$P(1.5)=7(1-\cos 3\pi)(1.5+10)+100e^{.2(1.5)}$$
$$=7[1-(-1)](11.5)+100e^{.3}=7(2)(11.5)+100e^{.3}=161++100e^{.3}\approx 296$$

50. **(a)** From the graph, one period is about 20 yr.

(b) The population of hares fluctuates between a maximum of about 150,000 and a minimum of about 5000.

51. $s(t)=4\sin\pi t$

$a=4,\ \omega=\pi$

amplitude $=|a|=4$

period $=\dfrac{2\pi}{\omega}=\dfrac{2\pi}{\pi}=2$

frequency $=\dfrac{\omega}{2\pi}=\dfrac{\pi}{2\pi}=\dfrac{1}{2}$

52. $s(t)=3\cos 2t$

$a=3,\ \omega=2$

amplitude $=|a|=3$

period $=\dfrac{2\pi}{\omega}=\dfrac{2\pi}{2}=\pi$

frequency $=\dfrac{\omega}{2\pi}=\dfrac{2}{2\pi}=\dfrac{1}{\pi}$

53. The frequency is the number of cycles in one unit of time.

$$s(1.5)=4\sin 1.5\pi=4\sin\frac{3\pi}{2}=4(-1)=-4;\ s(2)=4\sin 2\pi=4(0)=0$$

$$s(3.25)=4\sin 3.25\pi=4\sin\frac{13\pi}{4}=4\sin\frac{5\pi}{4}=4\left(-\frac{\sqrt{2}}{2}\right)=-2\sqrt{2}$$

54. The period is the time to complete one cycle. The amplitude is the maximum distance (on either side) from the initial point.

Chapter 4: Test

1. $y=3-6\sin\left(2x+\dfrac{\pi}{2}\right)=3-6\sin 2\left(x+\dfrac{\pi}{4}\right)=3-6\sin 2\left[x-\left(-\dfrac{\pi}{4}\right)\right]$

(a) $\dfrac{2\pi}{2}=\pi$

(b) 6

(c) $[-3,9]$

(d) The y-intercept occurs when $x=0$.

$$-6\sin\left(2\cdot 0+\frac{\pi}{2}\right)+3=-6\sin\left(0+\frac{\pi}{2}\right)+3=-6\sin\left(\frac{\pi}{2}\right)+3=-6(1)+3=-6+3=-3$$

(e) $\dfrac{\pi}{4}$ unit to the left $\left(\text{that is, } -\dfrac{\pi}{4}\right)$

2. $y = -\cos 2x$

Period: $\frac{2\pi}{2} = \pi$ and Amplitude: $|-1| = 1$

Divide the interval $[0, \pi]$ into four equal parts to get the x-values that will yield minimum and maximum points and x-intercepts. Then make a table. Repeat this cycle for the interval $[-\pi, 0]$.

x	0	$\frac{\pi}{4}$	$\frac{\pi}{2}$	$\frac{3\pi}{4}$	π
$2x$	0	$\frac{\pi}{2}$	π	$\frac{3\pi}{2}$	2π
$\cos 2x$	1	0	–1	0	1
$-\cos 2x$	–1	0	1	0	–1

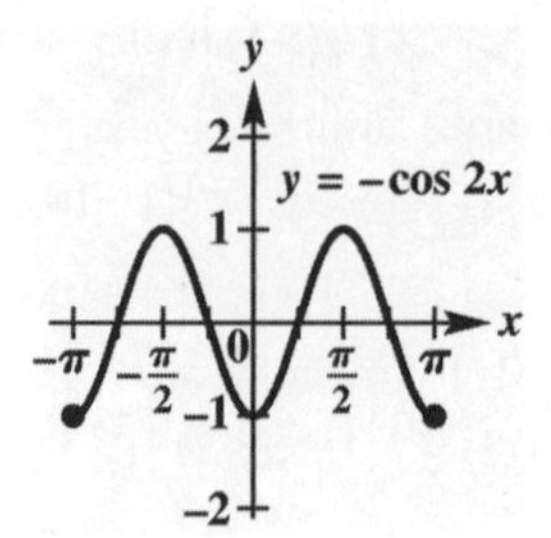

3. $y = -\csc 2x$

Step 1 Graph the corresponding reciprocal function $y = -\sin 2x$ The period is $\frac{2\pi}{2} = \pi$ and its amplitude is $|-1| = 1$. One period is in the interval $0 \le x \le \pi$. Dividing the interval into four equal parts gives us the following key points.

$$(0,0),\ \left(\frac{\pi}{2},-1\right),\ \left(\frac{\pi}{2},0\right),\ \left(\frac{3\pi}{4},1\right),\ (\pi,0)$$

Step 2 The vertical asymptotes of $y = -\csc 2x$ are at the x-intercepts of $y = -\sin 2x$, which are $x = 0, x = \frac{\pi}{2}$, and $x = \pi$. Continuing this pattern to the right, we also have a vertical asymptotes of $x = \frac{3\pi}{2}$ and $x = 2\pi$.

Step 3 Sketch the graph.

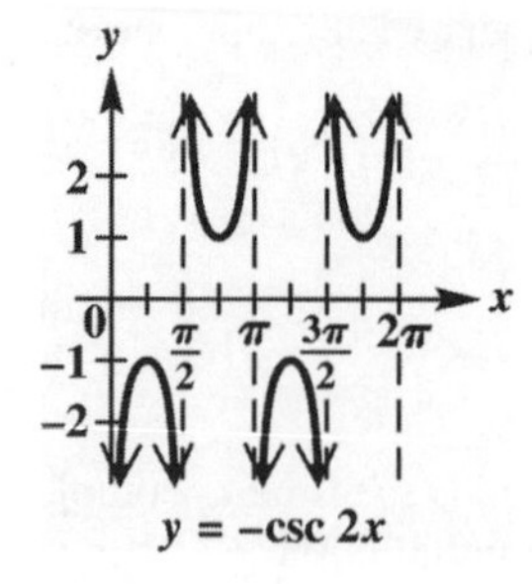

4. $y = \tan\left(x - \frac{\pi}{2}\right)$

Period: π

Vertical translation: none

Phase shift (horizontal translation): $\frac{\pi}{2}$ units to the right

Because the function is to be graphed over a two-period interval, locate three adjacent vertical asymptotes. Because asymptotes of the graph $y = \tan x$ occur at $-\frac{\pi}{2}$, and $\frac{\pi}{2}$, the following equations can be solved to locate asymptotes.

$$x - \frac{\pi}{2} = -\frac{\pi}{2} \Rightarrow x = 0 \text{ and } x - \frac{\pi}{2} = \frac{\pi}{2} \Rightarrow x = \pi$$

Continued on next page

4. (continued)

Divide the interval $(0,\pi)$ into four equal parts to obtain the following key x-values.

$$\text{first-quarter value: } \frac{\pi}{4};\ \text{middle value: } \frac{\pi}{2};\ \text{third-quarter value: } \frac{3\pi}{4}$$

Evaluating the given function at these three key x-values gives the following points.

$$\left(\frac{\pi}{4},-1\right),\ \left(\frac{\pi}{2},0\right),\ \left(\frac{3\pi}{4},1\right)$$

Connect these points with a smooth curve and continue to graph to approach the asymptote $x=0$ and $x=\pi$ to complete one period of the graph. Repeat this cycle for the interval $[-\pi,0]$.

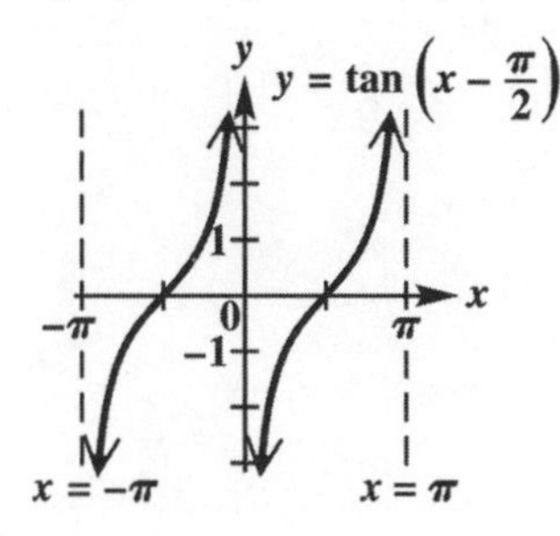

5. $y=-1+2\sin(x+\pi)$

Step 1 Find the interval whose length is $\frac{2\pi}{b}$.

$$0\le x+\pi\le 2\pi \Rightarrow\Rightarrow -\pi\le x\le\pi$$

Step 2 Divide the period into four equal parts to get the following x-values.

$$-\pi,\ -\frac{\pi}{2},\ 0,\ \frac{\pi}{2},\ \pi$$

Step 3 Evaluate the function for each of the five x-values

x	$-\frac{\pi}{2}$	$-\frac{\pi}{2}$	0	$\frac{\pi}{2}$	π
$x+\pi$	0	$\frac{\pi}{2}$	π	$\frac{3\pi}{2}$	2π
$\sin(x+\pi)$	0	1	0	−1	0
$2\sin(x+\pi)$	0	2	0	−2	0
$-1+2\sin(x+\pi)$	−1	1	−1	−3	−1

Steps 4 and 5 Plot the points found in the table and join them with a sinusoidal curve. Repeat this cycle for the interval $[-\pi,0]$.

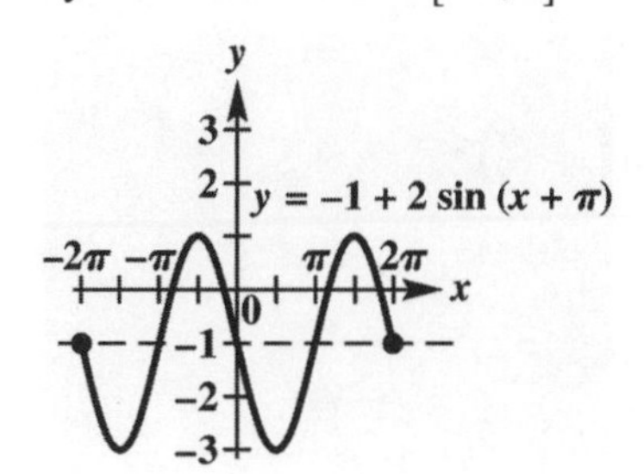

The amplitude is $|2|$, which is 2.

The period is 2π.

The vertical translation is 1 unit down.

The phase shift is π units to the left.

6. $y=-2-\cot\left(x-\frac{\pi}{2}\right)$

Period: $\frac{\pi}{b}=\frac{\pi}{1}=\pi$

Vertical translation: 2 units down

Phase shift (horizontal translation): $\frac{\pi}{2}$ units to the right

Because the function is to be graphed over a two-period interval, locate three adjacent vertical asymptotes. Because asymptotes of the graph $y=\cot x$ occur at multiples of π, the following equations can be solved to locate asymptotes.

$$x-\frac{\pi}{2}=-\pi,\ x-\frac{\pi}{2}=0, \text{ and } x-\frac{\pi}{2}=\pi$$

Solve each of these equations.

$$x-\frac{\pi}{2}=-\pi \Rightarrow x=-\frac{\pi}{2}$$

$$x-\frac{\pi}{2}=0 \Rightarrow x=\frac{\pi}{2}$$

$$x-\frac{\pi}{2}=\pi \Rightarrow x=\frac{3\pi}{2}$$

Divide the interval $\left(-\frac{\pi}{2},\frac{\pi}{2}\right)$ into four equal parts to obtain the following key x-values.

$$\text{first-quarter value: } -\frac{\pi}{4};\ \text{middle value: } 0\,;\ \text{third-quarter value: } \frac{\pi}{4}$$

Evaluating the given function at these three key x-values gives the following points.

$$\left(-\frac{\pi}{4},-3\right),\ (0,-2),\ \left(\frac{\pi}{4},-1\right)$$

Connect these points with a smooth curve and continue to graph to approach the asymptote $x=-\frac{\pi}{2}$ and $x=\frac{\pi}{2}$ to complete one period of the graph. Sketch the identical curve between the asymptotes $x=\frac{\pi}{2}$ and $x=\frac{3\pi}{2}$ to complete a second period of the graph.

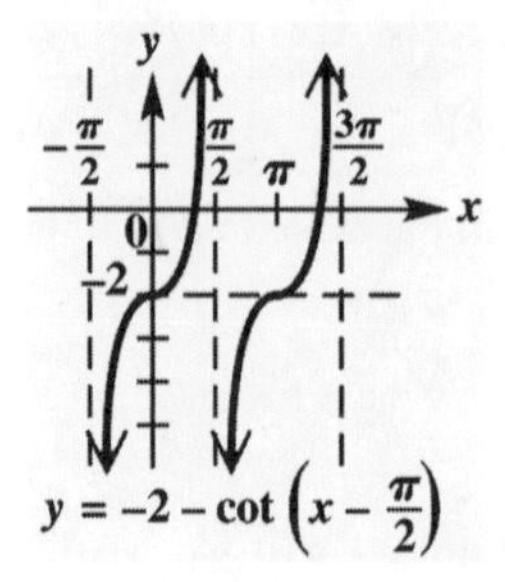

7. $y = 3\csc \pi x$

Step 1 Graph the corresponding reciprocal function $y = 3\sin \pi x$. The period is $\frac{2\pi}{\pi} = 2$ and its amplitude is $|3| = 3$. One period is in the interval $0 \le x \le 2$. Dividing the interval into four equal parts gives us the following key points.

$$(0,0),\ \left(\frac{1}{2},1\right),\ (1,0),\ \left(\frac{3}{2},-1\right),\ (2,0)$$

Step 2 The vertical asymptotes of $y = 3\csc \pi x$ are at the x-intercepts of $y = 3\sin \pi x$, which are $x = 0$, $x = 1$, and $x = 2$.

Step 3 Sketch the graph. Repeat this cycle for the interval $[-2,0]$.

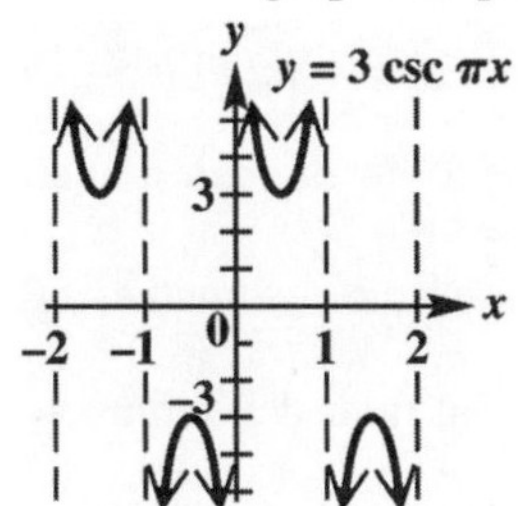

Period: 2
Amplitude: Not applicable
Phase shift: None
Vertical translation: None

8. **(a)** $f(x) = 17.5\sin\left[\frac{\pi}{6}(x-4)\right] + 67.5$

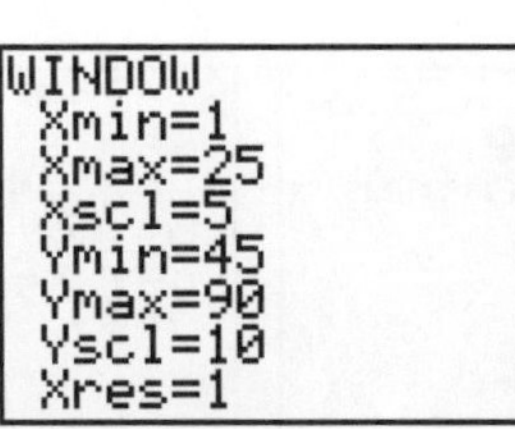

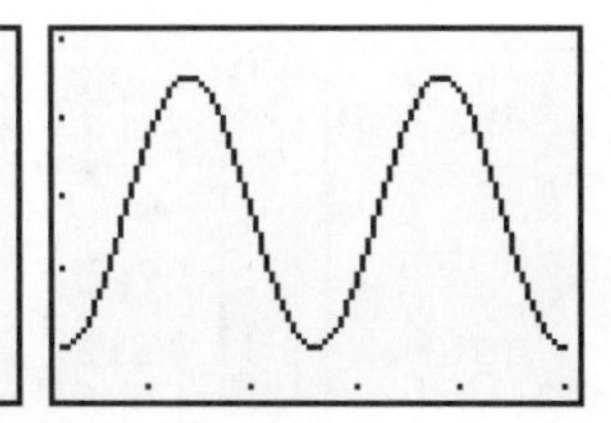

(b) Amplitude: 17.5; Period: $\frac{2\pi}{\frac{\pi}{6}} = 2\pi \cdot \frac{6}{\pi} = 12$;

Phase shift: 4 units to the right; Vertical translation: 67.5 units up

(c) For the month of December, $x = 12$.

$$f(12) = 17.5\sin\left[\frac{\pi}{6}(12-4)\right] + 67.5 = 17.5\sin\left(\frac{4\pi}{3}\right) + 67.5 = 17.5\left(-\frac{\sqrt{3}}{2}\right) + 67.5 \approx 52°$$

(d) A minimum of 50 occurring at $x = 13 = 12 + 1$ implies 50°F in January.

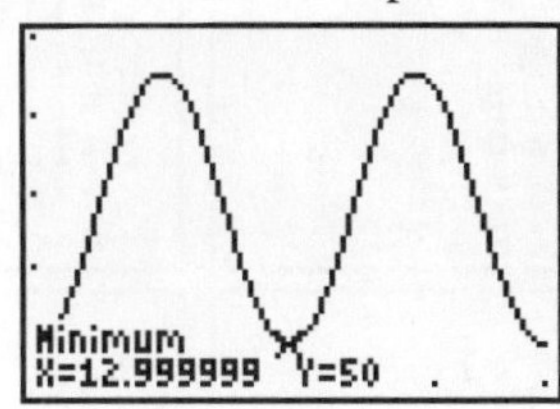

A maximum of 85 occurring at $x = 7$ and $x = 19 = 12 + 7$ implies 85°F in July.

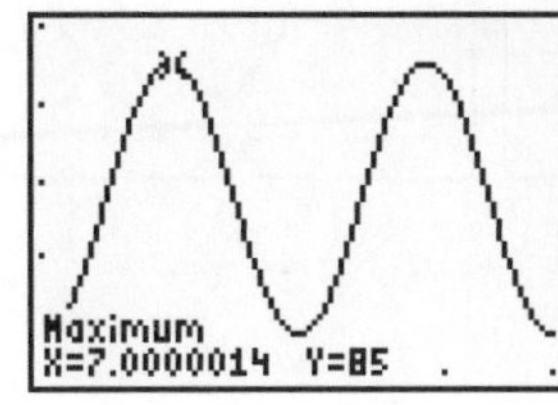

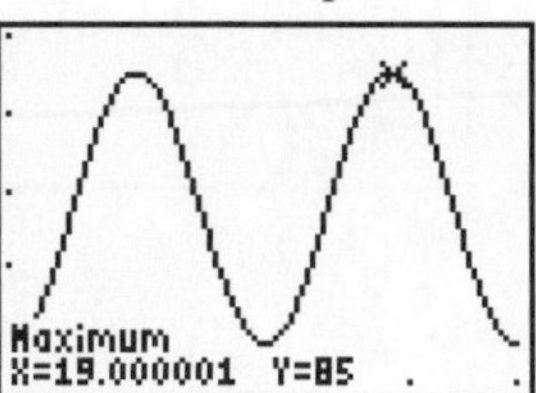

(e) Approximately 67.5° would be an average yearly temperature. This is the vertical translation.

9. $s(t) = -4\cos 8\pi t$, $a = |-4| = 4$, $\omega = 8\pi$

(a) maximum height = amplitude = $a = |-4| = 4$ in.

(b) $s(t) = -4\cos 8\pi t = 4 \Rightarrow \cos 8\pi t = -1 \Rightarrow 8\pi t = \pi \Rightarrow t = \frac{1}{8}$

The weight first reaches its maximum height after $\frac{1}{8}$ sec.

(c) frequency $= \frac{\omega}{2\pi} = \frac{8\pi}{2\pi} = 4$ cycles per sec; period $= \frac{2\pi}{\omega} = \frac{2\pi}{8\pi} = \frac{1}{4}$ sec

10. The function $y = \sin x$ and $y = \cos x$ both have all real numbers as their domains. The functions $f(x) = \tan x = \frac{\sin x}{\cos x}$ and $f(x) = \sec x = \frac{1}{\cos x}$ both have $y = \cos x$ in their denominators. Therefore, both the tangent and secant functions have the same restrictions on their domains.

The functions $y = \sin x$ and $y = \cos x$ both have all real numbers as their domains. The functions $f(x) = \cot x = \frac{\cos x}{\sin x}$ and $f(x) = \csc x = \frac{1}{\sin x}$ both have $y = \sin x$ in their denominators. Therefore, both the cotangent and cosecant functions have the same restrictions on their domains.

Chapter 4: Quantitative Reasoning

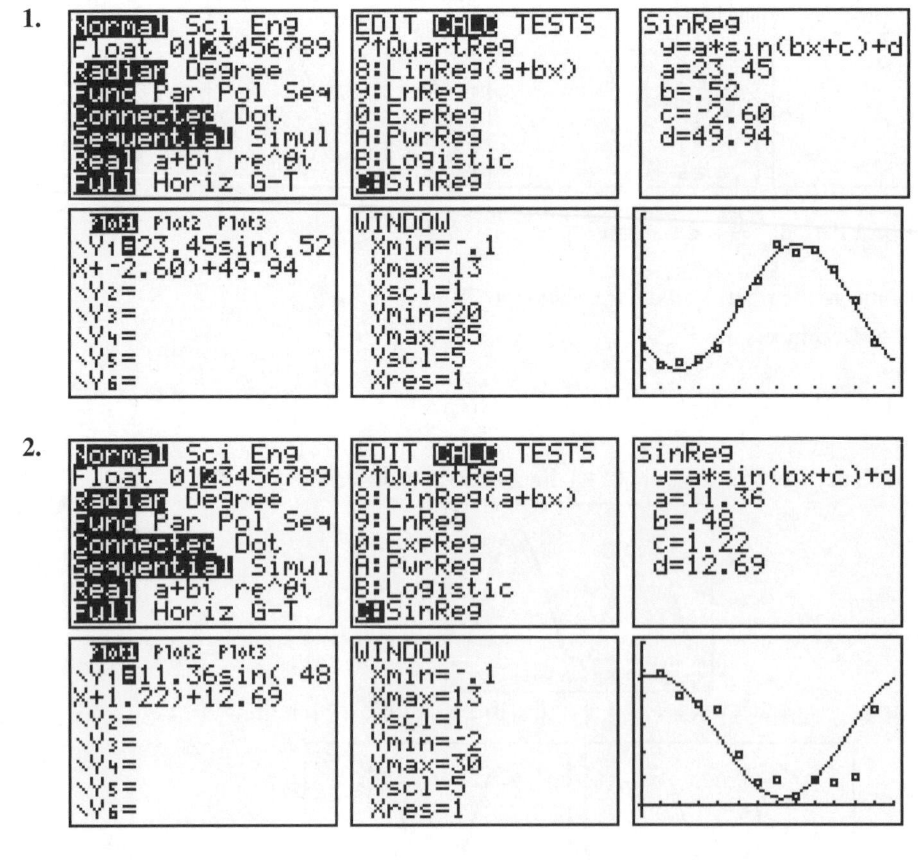

3. Answers will vary.

Chapter 5
TRIGONOMETRIC IDENTITIES

Section 5.1: Fundamental Identities

1. By a negative-angle identity, $\tan(-x) = -\tan x$. Since $\tan x = 2.6$, $\tan(-x) = -\tan x = -2.6$. Thus, we have the following.

$$\text{If } \tan x = 2.6, \text{ then } \tan(-x) = \underline{-2.6}.$$

2. By a negative angle identity, $\cos(-x) = \cos x$. Since $\cos x = -.65$, $\cos(-x) = \cos x = -.65$. Thus, we have the following.

$$\text{If } \cos x = -.65, \text{ then } \cos(-x) = \underline{-.65}.$$

3. By a reciprocal identity, $\cot x = \dfrac{1}{\tan x}$. Since $\tan x = 1.6$, $\cot x = \dfrac{1}{1.6} = .625$. Thus, we have the following.

$$\text{If } \tan x = 1.6, \text{ then } \cot x = \underline{.625}.$$

4. By a quotient identity, $\tan x = \dfrac{\sin x}{\cos x}$. Since $\cos x = .8$ and $\sin x = .6$, $\tan x = \dfrac{\sin x}{\cos x} = \dfrac{.6}{.8} = .75$. Also by a negative angle identity, $\tan(-x) = -\tan x$, so $\tan(-x) = -.75$. Thus, we have the following.

$$\text{If } \cos x = .8 \text{ and } \sin x = .6, \text{ then } \tan(-x) = \underline{-.75}.$$

5. $\cos s = \dfrac{3}{4}$, s is in quadrant I.

An identity that relates sine and cosine is $\sin^2 s + \cos^2 s = 1$.

$$\sin^2 s + \cos^2 s = 1 \Rightarrow \sin^2 s + \left(\frac{3}{4}\right)^2 = 1 \Rightarrow \sin^2 s + \frac{9}{16} = 1 \Rightarrow \sin^2 s = 1 - \frac{9}{16} = \frac{7}{16} \Rightarrow \sin s = \pm\frac{\sqrt{7}}{4}$$

Since s is in quadrant I, $\sin s = \dfrac{\sqrt{7}}{4}$.

6. $\cot s = -\dfrac{1}{3}$, s in quadrant IV

We will use the identity $1 + \cot^2 s = \csc^2 s$ since $\sin s = \dfrac{1}{\csc s}$.

$$1 + \cot^2 s = \csc^2 s \Rightarrow 1 + \left(-\frac{1}{3}\right)^2 = \csc^2 s \Rightarrow 1 + \frac{1}{9} = \csc^2 s \Rightarrow \frac{10}{9} = \csc^2 s \Rightarrow \csc s = \pm\frac{\sqrt{10}}{3}$$

Since s is in quadrant IV, $\csc s < 0$, so $\csc s = -\dfrac{\sqrt{10}}{3}$. Thus, we have the following.

$$\sin s = \frac{1}{\csc s} = -\frac{3}{\sqrt{10}} = -\frac{3}{\sqrt{10}} \cdot \frac{\sqrt{10}}{\sqrt{10}} = -\frac{3\sqrt{10}}{10}$$

7. $\cos(-s) = \frac{\sqrt{5}}{5}, \quad \tan s < 0$

Since $\cos(-s) = \frac{\sqrt{5}}{5}$, we have $\cos s = \frac{\sqrt{5}}{5}$ by a negative angle identity.

An identity that relates sine and cosine is $\sin^2 s + \cos^2 s = 1$.

$$\sin^2 s + \cos^2 s = 1 \Rightarrow \sin^2 s + \left(\frac{\sqrt{5}}{5}\right)^2 = 1 \Rightarrow \sin^2 s + \frac{5}{25} = 1 \Rightarrow \sin^2 s + \frac{1}{5} = 1$$

$$\sin^2 s = 1 - \frac{1}{5} = \frac{4}{5} \Rightarrow \sin s = \pm\frac{2}{\sqrt{5}} = \pm\frac{2}{\sqrt{5}} \cdot \frac{\sqrt{5}}{\sqrt{5}} = \pm\frac{2\sqrt{5}}{5}$$

Since $\tan s < 0$ and $\cos s > 0$, s is in quadrant IV, so $\sin s < 0$. Thus, $\sin s = -\frac{2\sqrt{5}}{5}$.

8. $\tan s = -\frac{\sqrt{7}}{2}, \sec s > 0$

$$\tan^2 s + 1 = \sec^2 s \Rightarrow \left(-\frac{\sqrt{7}}{2}\right)^2 + 1 = \sec^2 s \Rightarrow \frac{7}{4} + 1 = \sec^2 s \Rightarrow \frac{11}{4} = \sec^2 s \Rightarrow \sec s = \pm\frac{\sqrt{11}}{2}$$

Since $\sec s > 0$, $\sec s = \frac{\sqrt{11}}{2}$. Also, since $\cos s = \frac{1}{\sec s}$, $\cos s = \frac{1}{\frac{\sqrt{11}}{2}} = \frac{2}{\sqrt{11}}$.

Now, using the identity $\sin^2 s + \cos^2 s = 1$, we have the following.

$$\sin^2 s + \left(\frac{2}{\sqrt{11}}\right)^2 = 1 \Rightarrow \sin^2 s + \frac{4}{11} = 1 \Rightarrow \sin^2 s = 1 - \frac{4}{11} = \frac{7}{11}$$

$$\sin s = \pm\sqrt{\frac{7}{11}} = \pm\frac{\sqrt{7}}{\sqrt{11}} = \pm\frac{\sqrt{7}}{\sqrt{11}} \cdot \frac{\sqrt{11}}{\sqrt{11}} = \pm\frac{\sqrt{77}}{11}$$

Since $\tan s < 0$ and $\sec s > 0$, s is in quadrant IV, so $\sin s < 0$. Thus, $\sin s = -\sqrt{\frac{7}{11}} = -\frac{\sqrt{77}}{11}$.

9. $\sec s = \frac{11}{4}, \tan s < 0$

Since $\cos s = \frac{1}{\sec s}, \cos s = \frac{1}{\frac{11}{4}} = \frac{4}{11}$. Using the identity $\sin^2 s + \cos^2 s = 1$, we have the following.

$$\sin^2 s + \cos^2 s = 1 \Rightarrow \sin^2 s + \left(\frac{4}{11}\right)^2 = 1 \Rightarrow \sin^2 s + \frac{16}{121} = 1$$

$$\sin^2 s = 1 - \frac{16}{121} \Rightarrow \sin^2 s = \frac{105}{121} \Rightarrow \sin s = \pm\frac{\sqrt{105}}{11}$$

Since $\tan s < 0$ and $\sec s > 0$, s is in quadrant IV, so $\sin s < 0$. Thus, $\sin s = -\frac{\sqrt{105}}{11}$.

10. $\csc s = -\frac{8}{5}$

Since $\sin s = \frac{1}{\csc s}$, $\sin s = \frac{1}{-\frac{8}{5}} = -\frac{5}{8}$.

11. The quadrants are given so that one can determine which sign $(+ \text{ or } -)$ $\sin s$ will take. Since $\sin s = \frac{1}{\csc s}$, the sign of $\sin s$ will be the same as $\csc s$.

12. $\sin(-x) = \underline{-\sin x}$

13. Since $f(-x) = \sin(-x) = -\sin x = -f(x)$, $f(x) = \sin x$ is odd.

14. $\cos(-x) = \underline{\cos x}$

15. Since $f(-x) = \cos(-x) = \cos x = f(x)$, $f(x) = \cos x$ is even

16. $\tan(-x) = \underline{-\tan x}$

17. Since $f(-x) = \tan(-x) = -\tan x = -f(x)$, $f(x) = \tan x$ is odd.

18. This is the graph of $f(x) = \sec x$. It is symmetric about the *y*-axis. Moreover, since $f(-x) = \sec(-x) = \dfrac{1}{\cos(-x)} = \dfrac{1}{\cos x} = \sec x = f(x)$, $f(-x) = f(x)$.

19. This is the graph of $f(x) = \csc x$. It is symmetric about the origin. Moreover, since $f(-x) = \csc(-x) = \dfrac{1}{\sin(-x)} = \dfrac{1}{-\sin x} = -\csc x = -f(x)$, $f(-x) = -f(x)$.

20. This is the graph of $f(x) = \cot x$. It is symmetric about the origin. Moreover, since $f(-x) = \cot(-x) = \dfrac{\cos(-x)}{\sin(-x)} = \dfrac{\cos x}{-\sin x} = -\dfrac{\cos x}{\sin x} = -\cot x = -f(x)$, $f(-x) = -f(x)$.

21. $\sin\theta = \dfrac{2}{3}, \theta$ in quadrant II

Since θ is in quadrant II, the sine and cosecant function values are positive. The cosine, tangent, cotangent, and secant function values are negative.

$$\sin^2\theta + \cos^2\theta = 1 \Rightarrow \cos^2\theta = 1 - \sin^2\theta = 1 - \left(\frac{2}{3}\right)^2 = 1 - \frac{4}{9} = \frac{5}{9} \Rightarrow \cos\theta = -\frac{\sqrt{5}}{3}, \text{ since } \cos\theta < 0$$

$$\tan\theta = \frac{\sin\theta}{\cos\theta} = \frac{\frac{2}{3}}{-\frac{\sqrt{5}}{3}} = -\frac{2}{\sqrt{5}} = -\frac{2}{\sqrt{5}} \cdot \frac{\sqrt{5}}{\sqrt{5}} = -\frac{2\sqrt{5}}{5}$$

$$\cot\theta = \frac{1}{\tan\theta} = \frac{1}{-\frac{2}{\sqrt{5}}} = -\frac{\sqrt{5}}{2}$$

$$\sec\theta = \frac{1}{\cos\theta} = \frac{1}{-\frac{\sqrt{5}}{3}} = -\frac{3}{\sqrt{5}} = -\frac{3}{\sqrt{5}} \cdot \frac{\sqrt{5}}{\sqrt{5}} = -\frac{3\sqrt{5}}{5}$$

$$\csc\theta = \frac{1}{\sin\theta} = \frac{1}{\frac{2}{3}} = \frac{3}{2}$$

22. $\cos\theta = \frac{1}{5}, \theta$ in quadrant I

Since θ is in quadrant I, all the function values are positive.

$$\sin^2\theta + \cos^2\theta = 1 \Rightarrow \sin^2\theta = 1 - \cos^2\theta = 1 - \left(\frac{1}{5}\right)^2 = 1 - \frac{1}{25} = \frac{24}{25}$$

$$\sin\theta = \frac{\sqrt{24}}{25} = \frac{2\sqrt{6}}{5}, \text{ since } \sin\theta > 0$$

$$\tan\theta = \frac{\sin\theta}{\cos\theta} = \frac{\frac{2\sqrt{6}}{5}}{\frac{1}{5}} = 2\sqrt{6}$$

$$\cot\theta = \frac{1}{\tan\theta} = \frac{1}{2\sqrt{6}} = \frac{1}{2\sqrt{6}} \cdot \frac{\sqrt{6}}{\sqrt{6}} = \frac{\sqrt{6}}{12}$$

$$\sec\theta = \frac{1}{\cos\theta} = \frac{1}{\frac{1}{5}} = 5$$

$$\csc\theta = \frac{1}{\sin\theta} = \frac{1}{\frac{2\sqrt{6}}{5}} = \frac{5}{2\sqrt{6}} = \frac{5}{2\sqrt{6}} \cdot \frac{\sqrt{6}}{\sqrt{6}} = \frac{5\sqrt{6}}{12}$$

23. $\tan\theta = -\frac{1}{4}, \theta$ in quadrant IV

Since θ is in quadrant IV, the cosine and secant function values are positive. The sine, tangent, cotangent, and cosecant function values are negative.

$$\sec^2\theta = 1 + \tan^2\theta = 1 + \left(-\frac{1}{4}\right)^2 = 1 + \frac{1}{16} = \frac{17}{16} \Rightarrow \sec\theta = \frac{\sqrt{17}}{4}, \text{ since } \sec\theta > 0$$

$$\cos\theta = \frac{1}{\sec\theta} = \frac{1}{\frac{\sqrt{17}}{4}} = \frac{4}{\sqrt{17}} = \frac{4}{\sqrt{17}} \cdot \frac{\sqrt{17}}{\sqrt{17}} = \frac{4\sqrt{17}}{17}$$

$$\sin^2\theta + \cos^2\theta = 1 \Rightarrow \sin^2\theta = 1 - \cos^2\theta = 1 - \left(\frac{4}{\sqrt{17}}\right)^2$$

$$\sin^2\theta = 1 - \frac{16}{17} = \frac{1}{17} \Rightarrow \sin\theta = -\frac{1}{\sqrt{17}} = -\frac{1}{\sqrt{17}} \cdot \frac{\sqrt{17}}{\sqrt{17}} = -\frac{\sqrt{17}}{17}, \text{ since } \sin\theta < 0$$

$$\csc\theta = \frac{1}{\sin\theta} = \frac{1}{-\frac{1}{\sqrt{17}}} = -\sqrt{17}$$

$$\cot\theta = \frac{1}{\tan\theta} = \frac{1}{-\frac{1}{4}} = -4$$

24. $\csc\theta = -\frac{5}{2}$, θ in quadrant III

Since θ is in quadrant III, the tangent, and cotangent function values are positive. The sine, cosine, cosecant, and secant function values are negative.

$$\sin\theta = \frac{1}{\csc\theta} = \frac{1}{-\frac{5}{2}} = -\frac{2}{5}$$

$$\cos^2\theta = 1 - \sin^2\theta = 1 - \left(-\frac{2}{5}\right)^2 = 1 - \frac{4}{25} = \frac{21}{25} \Rightarrow \cos\theta = -\frac{\sqrt{21}}{5}, \text{ since } \cos\theta < 0$$

$$\tan\theta = \frac{\sin\theta}{\cos\theta} = \frac{-\frac{2}{5}}{-\frac{\sqrt{21}}{5}} = \frac{2}{\sqrt{21}} = \frac{2}{\sqrt{21}} \cdot \frac{\sqrt{21}}{\sqrt{21}} = \frac{2\sqrt{21}}{21}$$

$$\cot\theta = \frac{1}{\tan\theta} = \frac{1}{\frac{2}{\sqrt{21}}} = \frac{\sqrt{21}}{2}$$

$$\sec\theta = \frac{1}{\cos\theta} = \frac{1}{-\frac{\sqrt{21}}{5}} = -\frac{5}{\sqrt{21}} = -\frac{5}{\sqrt{21}} \cdot \frac{\sqrt{21}}{\sqrt{21}} = -\frac{5\sqrt{21}}{21}$$

25. $\cot\theta = \frac{4}{3}, \sin\theta > 0$

Since $\cot\theta > 0$ and $\sin\theta > 0$, θ is in quadrant I, so all the function values are positive.

$$\tan = \frac{1}{\cot\theta} = \frac{1}{\frac{4}{3}} = \frac{3}{4}$$

$$\sec^2\theta = 1 + \tan^2\theta = 1 + \left(\frac{3}{4}\right)^2 = 1 + \frac{9}{16} = \frac{25}{16} \Rightarrow \sec\theta = \frac{5}{4}, \text{ since } \sec\theta > 0$$

$$\cos\theta = \frac{1}{\sec\theta} = \frac{1}{\frac{5}{4}} = \frac{4}{5}$$

$$\sin^2\theta = 1 - \cos^2\theta = 1 - \left(\frac{4}{5}\right)^2 = 1 - \frac{16}{25} = \frac{9}{25} \Rightarrow \sin\theta = \frac{3}{5}, \text{ since } \sin\theta > 0$$

$$\csc\theta = \frac{1}{\sin\theta} = \frac{1}{\frac{3}{5}} = \frac{5}{3}$$

26. $\sin\theta = -\frac{4}{5}, \cos\theta < 0$

Since $\sin\theta < 0$ and $\cos\theta<0$, θ is in quadrant III. Since θ is in quadrant III, the tangent, and cotangent function values are positive. The cosecant and secant function values are negative.

$$\cos^2\theta = 1-\sin^2\theta = 1-\left(-\frac{4}{5}\right)^2 = 1-\frac{16}{25} = \frac{9}{25} \Rightarrow \cos\theta = -\frac{3}{5}, \text{ since } \cos\theta < 0$$

$$\tan\theta = \frac{\sin\theta}{\cos\theta} = \frac{-\frac{4}{5}}{-\frac{3}{5}} = \frac{4}{3}$$

$$\cot\theta = \frac{1}{\tan\theta} = \frac{1}{\frac{4}{3}} = \frac{3}{4}$$

$$\sec\theta = \frac{1}{\cos\theta} = \frac{1}{-\frac{3}{5}} = -\frac{5}{3}$$

$$\csc\theta = \frac{1}{\sin\theta} = \frac{1}{-\frac{4}{5}} = -\frac{5}{4}$$

27. $\sec\theta = \frac{4}{3}, \sin\theta < 0$

Since $\sec\theta > 0$ and $\sin\theta < 0$, θ is in quadrant IV. Since θ is in quadrant IV, the cosine function value is positive. The tangent, cotangent, and cosecant function values are negative.

$$\cos\theta = \frac{1}{\sec\theta} = \frac{1}{\frac{4}{3}} = \frac{3}{4}$$

$$\sin^2\theta = 1-\cos^2\theta = 1-\left(\frac{3}{4}\right)^2 = 1-\frac{9}{16} = \frac{7}{16} \Rightarrow \sin\theta = -\frac{\sqrt{7}}{4}, \text{ since } \sin\theta < 0$$

$$\tan\theta = \frac{\sin\theta}{\cos\theta} = \frac{-\frac{\sqrt{7}}{4}}{\frac{3}{4}} = -\frac{\sqrt{7}}{3}$$

$$\cot\theta = \frac{1}{\tan\theta} = -\frac{1}{-\frac{\sqrt{7}}{3}} = -\frac{3}{\sqrt{7}}\cdot\frac{\sqrt{7}}{\sqrt{7}} = -\frac{3\sqrt{7}}{7}$$

$$\csc\theta = \frac{1}{\sin\theta} = -\frac{1}{-\frac{\sqrt{7}}{4}} = -\frac{4}{\sqrt{7}}\cdot\frac{\sqrt{7}}{\sqrt{7}} = -\frac{4\sqrt{7}}{7}$$

28. $\cos\theta = -\frac{1}{4}, \sin\theta > 0$

Since $\cos\theta < 0$ and $\sin\theta > 0$, θ is in quadrant II. Since θ is in quadrant II, the cosecant function value is positive. The tangent, cotangent, and secant function values are negative.

$$\sin^2\theta = 1 - \cos^2\theta = 1 - \left(-\frac{1}{4}\right)^2 = 1 - \frac{1}{16} = \frac{15}{16} \Rightarrow \sin\theta = \frac{\sqrt{15}}{4}, \text{ since } \sin\theta < 0$$

$$\tan\theta = \frac{\sin\theta}{\cos\theta} = \frac{\frac{\sqrt{15}}{4}}{-\frac{1}{4}} = -\sqrt{15}$$

$$\cot\theta = \frac{1}{\tan\theta} = \frac{1}{-\sqrt{15}} = -\frac{1}{\sqrt{15}} \cdot \frac{\sqrt{15}}{\sqrt{15}} = -\frac{\sqrt{15}}{15}$$

$$\sec\theta = \frac{1}{\cos\theta} = \frac{1}{-\frac{1}{4}} = -4$$

$$\csc\theta = \frac{1}{\sin\theta} = \frac{1}{\frac{\sqrt{15}}{4}} = \frac{4}{\sqrt{15}} = \frac{4}{\sqrt{15}} \cdot \frac{\sqrt{15}}{\sqrt{15}} = \frac{4\sqrt{15}}{15}$$

29. Since $\frac{\cos x}{\sin x} = \cot x$, choose expression B.

30. Since $\tan x = \frac{\sin x}{\cos x}$, choose expression D.

31. Since $\cos(-x) = \cos x$, choose expression E.

32. Since $\tan^2 x + 1 = \sec^2 x$, choose expression C.

33. Since $1 = \sin^2 x + \cos^2 x$, choose expression A.

34. Since $-\tan x \cos x = -\frac{\sin x}{\cos x} \cdot \cos x = -\sin x = \sin(-x)$, choose expression C.

35. Since $\sec^2 x - 1 = \tan^2 x = \frac{\sin^2 x}{\cos^2 x}$, choose expression A.

36. Since $\frac{\sec x}{\csc x} = \frac{\frac{1}{\cos x}}{\frac{1}{\sin x}} = \frac{\sin x}{\cos x} = \tan x$, choose expression E.

37. Since $1 + \sin^2 x = (\csc^2 x - \cot^2 x) + \sin^2 x$, choose expression D.

38. Since $\cos^2 x = \frac{1}{\sec^2 x}$, choose expression B.

39. It is incorrect to state $1+\cot^2=\csc^2$. Cotangent and cosecant are functions of some variable such as $\theta, x,$ or t. An acceptable statement would be $1+\cot^2\theta=\csc^2\theta$.

40. In general, it is false that $\sqrt{x^2+y^2}=x+y$. Stating $\sin^2\theta+\cos^2\theta=1$ implies $\sin\theta+\cos\theta=1$ is a false statement.

41. Find $\sin\theta$ if $\cos\theta=\dfrac{x}{x+1}$.

Since $\sin^2\theta+\cos^2\theta=1$ and $\cos\theta=\dfrac{x}{x+1}$, we have the following.

$$\sin^2\theta=1-\cos^2\theta=1-\left(\frac{x}{x+1}\right)^2=1-\frac{x^2}{(x+1)^2}=\frac{(x+1)^2-x^2}{(x+1)^2}=\frac{x^2+2x+1-x^2}{(x+1)^2}=\frac{2x+1}{(x+1)^2}$$

Thus, $\sin\theta=\dfrac{\pm\sqrt{2x+1}}{x+1}$.

42. Find $\tan\alpha$ if $\sec\alpha=\dfrac{p+4}{p}$.

Since $\tan^2\alpha+1=\sec^2\alpha$ and $\sec\alpha=\dfrac{p+4}{p}$, we have the following.

$$\tan^2\alpha=\sec^2\alpha-1=\frac{(p+4)^2}{p^2}-1=\frac{p^2+8p+16}{p^2}-\frac{p^2}{p^2}=\frac{8p+16}{p^2}=\frac{4(2p+4)}{p^2}$$

Thus, $\tan\alpha=\dfrac{\pm2\sqrt{2p+4}}{p}$.

43. $\sin^2 x+\cos^2 x=1\Rightarrow\sin^2 x=1-\cos^2 x\Rightarrow\sin x=\pm\sqrt{1-\cos^2 x}$

44. $\cot^2 x+1=\csc^2 x\Rightarrow\cot^2 x=\csc^2 x-1\Rightarrow\cot x=\pm\sqrt{\csc^2 x-1}=\pm\sqrt{\dfrac{1}{\sin^2 x}-1}=\dfrac{\pm\sqrt{1-\sin^2 x}}{\sin x}$

45. $\tan^2 x+1=\sec^2 x\Rightarrow\tan^2 x=\sec^2 x-1\Rightarrow\tan x=\pm\sqrt{\sec^2-1}$

46. $\cot^2 x+1=\csc^2 x\Rightarrow\cot^2 x=\csc^2 x-1\Rightarrow\cot x=\pm\sqrt{\csc^2-1}$

47. $\csc x=\dfrac{1}{\sin x}\Rightarrow\csc x=\dfrac{1}{\pm\sqrt{1-\cos^2 x}}=\dfrac{\pm1}{\sqrt{1-\cos^2 x}}\cdot\dfrac{\sqrt{1-\cos^2 x}}{\sqrt{1-\cos^2 x}}=\dfrac{\pm\sqrt{1-\cos^2 x}}{1-\cos^2 x}$

48. $\sec x=\dfrac{1}{\cos x}\Rightarrow\sec x=\dfrac{1}{\pm\sqrt{1-\sin^2 x}}=\dfrac{1}{\pm\sqrt{1-\sin^2 x}}\cdot\dfrac{\sqrt{1-\sin^2 x}}{\sqrt{1-\sin^2 x}}=\dfrac{\pm\sqrt{1-\sin^2 x}}{1-\sin^2 x}$

49. $\cot\theta\sin\theta=\dfrac{\cos\theta}{\sin\theta}\cdot\sin\theta=\cos\theta$

50. $\sec\theta\cot\theta\sin\theta=\dfrac{1}{\cos\theta}\cdot\dfrac{\cos\theta}{\sin\theta}\cdot\dfrac{\sin\theta}{1}$
$=\dfrac{\sin\theta\cos\theta}{\cos\theta\sin\theta}=1$

51. $\cos\theta\csc\theta = \cos\theta \cdot \dfrac{1}{\sin\theta} = \dfrac{\cos\theta}{\sin\theta} = \cot\theta$

52. $\cot^2\theta\left(1+\tan^2\theta\right) = \dfrac{\cos^2\theta}{\sin^2\theta}\left(\sec^2\theta\right) = \dfrac{\cos^2\theta}{\sin^2\theta}\left(\dfrac{1}{\cos^2\theta}\right) = \dfrac{1}{\sin^2\theta} = \csc^2\theta$

53. $\sin^2\theta\left(\csc^2\theta - 1\right) = \sin^2\theta\left(\dfrac{1}{\sin^2\theta} - 1\right) = \dfrac{\sin^2\theta}{\sin^2\theta} - \sin^2\theta = 1 - \sin^2\theta = \cos^2\theta$

54. $\left(\sec\theta - 1\right)\left(\sec\theta + 1\right) = \sec^2\theta - 1 = \tan^2\theta$

55.
$$\begin{aligned}(1-\cos\theta)(1+\sec\theta) &= 1+\sec\theta-\cos\theta-\cos\theta\sec\theta = 1+\sec\theta-\cos\theta-\cos\theta\left(\frac{1}{\cos\theta}\right)\\ &= 1+\sec\theta-\cos\theta-1 = \sec\theta-\cos\theta\end{aligned}$$

56. $\dfrac{\cos\theta+\sin\theta}{\sin\theta} = \dfrac{\cos\theta}{\sin\theta} + \dfrac{\sin\theta}{\sin\theta} = \cot\theta + 1$

57. $\dfrac{\cos^2\theta - \sin^2\theta}{\sin\theta\cos\theta} = \dfrac{\cos^2\theta}{\sin\theta\cos\theta} - \dfrac{\sin^2\theta}{\sin\theta\cos\theta} = \dfrac{\cos\theta}{\sin\theta} - \dfrac{\sin\theta}{\cos\theta} = \cot\theta - \tan\theta$

58. $\dfrac{1-\sin^2\theta}{1+\cot^2\theta} = \dfrac{\cos^2\theta}{\csc^2\theta} = \dfrac{\cos^2\theta}{\frac{1}{\sin^2\theta}} = \sin^2\theta\cos^2\theta$

59. $\sec\theta - \cos\theta = \dfrac{1}{\cos\theta} - \cos\theta = \dfrac{1}{\cos\theta} - \dfrac{\cos^2\theta}{\cos\theta} = \dfrac{1-\cos^2\theta}{\cos\theta} = \dfrac{\sin^2\theta}{\cos\theta} = \dfrac{\sin\theta}{\cos\theta}\cdot\sin\theta = \tan\theta\sin\theta$

60.
$$\begin{aligned}(\sec\theta+\csc\theta)(\cos\theta-\sin\theta) &= \left(\frac{1}{\cos\theta}+\frac{1}{\sin\theta}\right)(\cos\theta-\sin\theta)\\ &= \frac{1}{\cos\theta}(\cos\theta) - \frac{1}{\cos\theta}(\sin\theta) + \frac{1}{\sin\theta}(\cos\theta) - \frac{1}{\sin\theta}(\sin\theta)\\ &= 1 - \frac{\sin\theta}{\cos\theta} + \frac{\cos\theta}{\sin\theta} - 1 = -\tan\theta + \cot\theta = \cot\theta - \tan\theta\end{aligned}$$

61. $\sin\theta\left(\csc\theta - \sin\theta\right) = \sin\theta\csc\theta - \sin^2\theta = \sin\theta\cdot\dfrac{1}{\sin\theta} - \sin^2\theta = 1 - \sin^2\theta = \cos^2\theta$

62. $\dfrac{1+\tan^2\theta}{1+\cot^2\theta} = \dfrac{\sec^2\theta}{\csc^2\theta} = \dfrac{\frac{1}{\cos^2\theta}}{\frac{1}{\sin^2\theta}} = \dfrac{1}{\cos^2\theta}\cdot\dfrac{\sin^2\theta}{1} = \dfrac{\sin^2\theta}{\cos^2\theta} = \tan^2\theta$

63. $\sin^2\theta + \tan^2\theta + \cos^2\theta = \left(\sin^2\theta + \cos^2\theta\right) + \tan^2\theta = 1 + \tan^2\theta = \sec^2\theta$

64. $\dfrac{\tan(-\theta)}{\sec\theta} = \dfrac{-\tan\theta}{\frac{1}{\cos\theta}} = -\dfrac{\sin\theta}{\cos\theta}\cdot\dfrac{\cos\theta}{1} = -\sin\theta$

65. Since $\cos x = \frac{1}{5}$, which is positive, x is in quadrant I or quadrant IV.

$$\sin x = \pm\sqrt{1-\cos^2 x} = \pm\sqrt{1-\left(\frac{1}{5}\right)^2} = \pm\sqrt{\frac{24}{25}} = \pm\frac{\sqrt{24}}{5} = \pm\frac{2\sqrt{6}}{5}$$

$$\tan x = \frac{\sin x}{\cos x} = \frac{\pm\frac{2\sqrt{6}}{5}}{\frac{1}{5}} = \pm 2\sqrt{6}$$

$$\sec x = \frac{1}{\cos x} = \frac{1}{\frac{1}{5}} = 5$$

Quadrant I:

$$\frac{\sec x - \tan x}{\sin x} = \frac{5-2\sqrt{6}}{\frac{2\sqrt{6}}{5}} = \frac{25-10\sqrt{6}}{2\sqrt{6}} = \frac{25-10\sqrt{6}}{2\sqrt{6}}\cdot\frac{\sqrt{6}}{\sqrt{6}} = \frac{25\sqrt{6}-60}{12}$$

Quadrant IV:

$$\frac{\sec x - \tan x}{\sin x} = \frac{5-\left(-2\sqrt{6}\right)}{-\frac{2\sqrt{6}}{5}} = \frac{25+10\sqrt{6}}{-2\sqrt{6}} = \frac{25+10\sqrt{6}}{-2\sqrt{6}}\cdot\frac{-\sqrt{6}}{-\sqrt{6}} = \frac{-25\sqrt{6}-60}{12}$$

66. Since $\csc x = -3$, which is negative, x is in quadrant III or quadrant IV.

Quadrant III:

$$\sin x = \frac{1}{\csc x} = -\frac{1}{3} \Rightarrow \cos x = -\sqrt{1-\sin^2 x} = -\sqrt{1-\left(-\frac{1}{3}\right)^2} = -\sqrt{1-\frac{1}{9}} = -\sqrt{\frac{8}{9}} = -\frac{2\sqrt{2}}{3}$$

$$\sec x = \frac{1}{\cos x} = -\frac{3}{2\sqrt{2}}$$

$$\frac{\sin x + \cos x}{\sec x} = \frac{-\frac{1}{3}-\frac{2\sqrt{2}}{3}}{-\frac{3}{2\sqrt{2}}} = \left(\frac{-1-2\sqrt{2}}{3}\right)\left(-\frac{2\sqrt{2}}{3}\right) = \frac{2\sqrt{2}+8}{9}$$

Quadrant IV:

$$\sin x = \frac{1}{\csc x} = -\frac{1}{3} \Rightarrow \cos x = \sqrt{1-\sin^2 x} = \sqrt{1-\left(-\frac{1}{3}\right)^2} = \sqrt{1-\frac{1}{9}} = \sqrt{\frac{8}{9}} = \frac{2\sqrt{2}}{3}$$

$$\sec x = \frac{3}{2\sqrt{2}}$$

$$\frac{\sin x + \cos x}{\sec x} = \frac{-\frac{1}{3}+\frac{2\sqrt{2}}{3}}{\frac{3}{2\sqrt{2}}} = \left(\frac{-1+2\sqrt{2}}{3}\right)\left(\frac{2\sqrt{2}}{3}\right) = \frac{-2\sqrt{2}+8}{9}$$

67. $y = \sin(-2x) \Rightarrow y = -\sin(2x)$

68. It is the negative of $\sin(2x)$.

69. $y = \cos(-4x) \Rightarrow y = \cos(4x)$

70. It is the same function.

71. **(a)** $y=\sin(-4x)\Rightarrow y=-\sin(4x)$

(b) $y=\cos(-2x)\Rightarrow y=\cos(2x)$

(c) $y=-5\sin(-3x)\Rightarrow y=-5[-\sin(3x)]\Rightarrow y=5\sin(3x)$

In Exercises 72 – 76, the functions are graphed in the following window.

```
ZOOM MEMORY        WINDOW
1:ZBox              Xmin=-6.152285…
2:Zoom In           Xmax=6.1522856…
3:Zoom Out          Xscl=1.5707963…
4:ZDecimal          Ymin=-4
5:ZSquare           Ymax=4
6:ZStandard         Yscl=1
7:ZTrig             Xres=1
```

72. The equation $\cos 2x=1-2\sin^2 x$ is an identity.

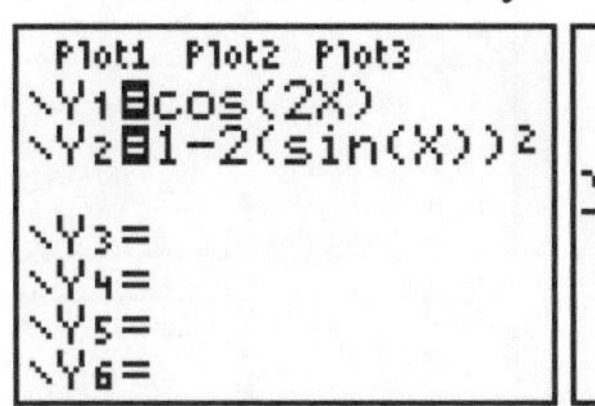

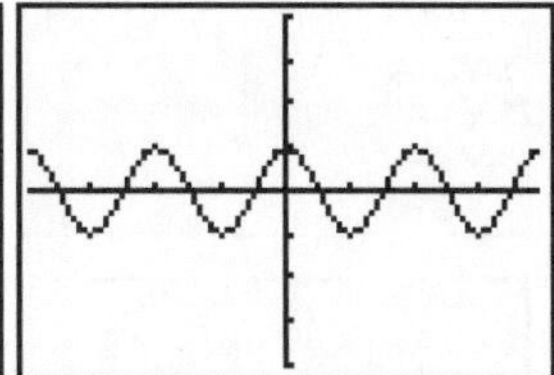

73. The equation $2\sin s=\sin 2s$ is not an identity.

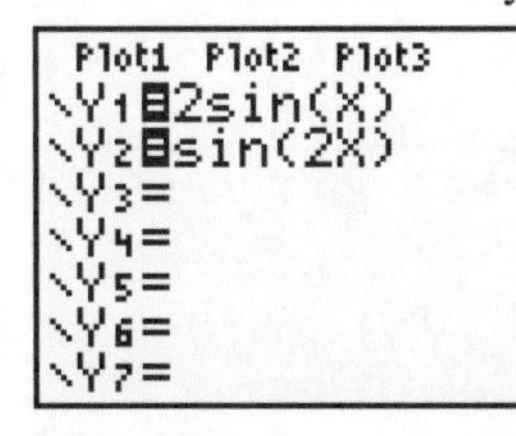

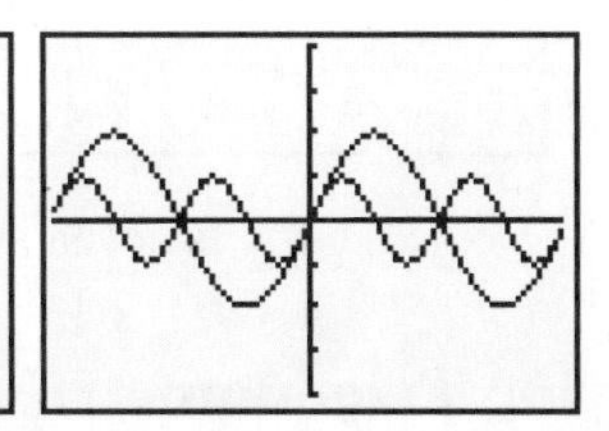

74. The equation $\sin x=\sqrt{1-\cos^2 x}$ is not an identity.

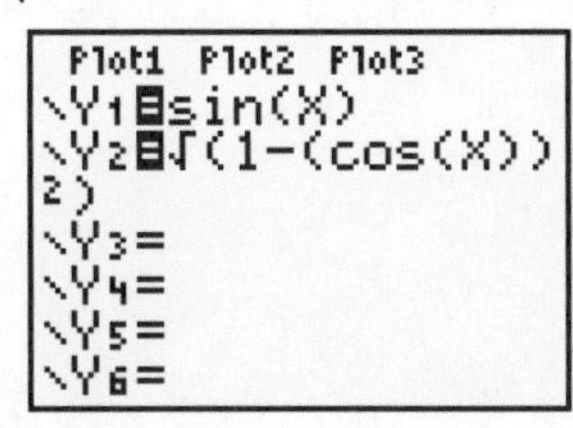

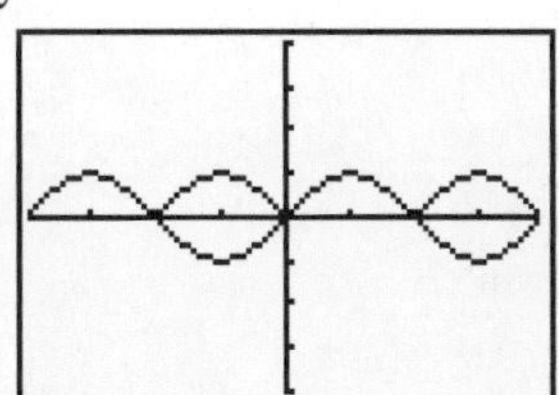

75. The equation $\cos 2x=\cos^2 x-\sin^2 x$ is an identity.

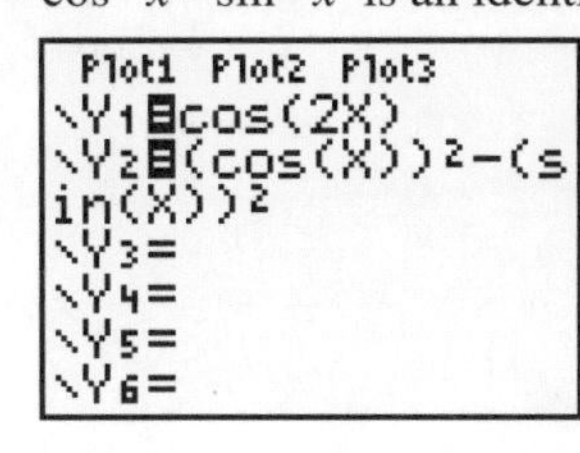

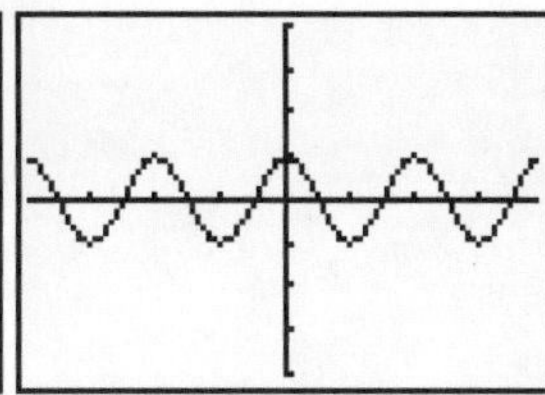

76. Does $\cos(x-y)=\cos x-\cos y$?

If it does, then the graphs of $Y_1=\cos(x-y)$ and $Y_2=\cos x-\cos y$ for specific values of y will overlap. We will graph 3 cases: $y=\frac{\pi}{4}$, $y=\frac{\pi}{2}$, and $y=\pi$.

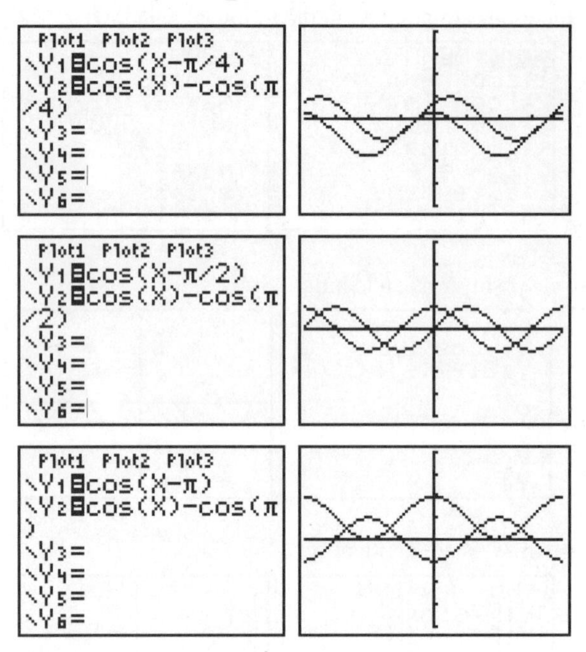

The equation $\cos(x-y)=\cos x-\cos y$ is not an identity.

Section 5.2: Verifying Trigonometric Identities

1. $\cot\theta+\dfrac{1}{\cot\theta}=\cot\theta+\tan\theta$

$$=\frac{\cos\theta}{\sin\theta}+\frac{\sin\theta}{\cos\theta}$$

$$=\frac{\cos^2\theta+\sin^2\theta}{\sin\theta\cos\theta}$$

$$=\frac{1}{\sin\theta\cos\theta}\text{ or }\csc\theta\sec\theta$$

2. $\dfrac{\sec x}{\csc x}+\dfrac{\csc x}{\sec x}=\dfrac{\frac{1}{\cos x}}{\frac{1}{\sin x}}+\dfrac{\frac{1}{\sin x}}{\frac{1}{\cos x}}$

$$\frac{\sin x}{\cos x}+\frac{\cos x}{\sin x}$$

$$=\frac{\sin^2 x+\cos^2 x}{\sin x\cos x}$$

$$=\frac{1}{\sin x\cos x}\text{ or }\csc x\sec s$$

3. $\tan s\,(\cot s+\csc s)=\dfrac{\sin s}{\cos s}\left(\dfrac{\cos s}{\sin s}+\dfrac{1}{\sin s}\right)$

$$=1+\frac{1}{\cos s}$$

$$=1+\sec s$$

4. $\cos\beta(\sec\beta+\csc\beta)$

$$=\cos\beta\left(\frac{1}{\cos\beta}+\frac{1}{\sin\beta}\right)$$

$$=1+\cot\beta$$

5. $\dfrac{1}{\csc^2\theta}+\dfrac{1}{\sec^2\theta}=\sin^2\theta+\cos^2\theta=1$

6. $$\frac{1}{\sin\alpha-1}-\frac{1}{\sin\alpha+1}=\frac{\sin\alpha+1}{(\sin\alpha-1)(\sin\alpha+1)}-\frac{\sin\alpha-1}{(\sin\alpha-1)(\sin\alpha+1)}=\frac{(\sin\alpha+1)-(\sin\alpha-1)}{(\sin\alpha-1)(\sin\alpha+1)}$$
$$=\frac{\sin\alpha+1-\sin\alpha+1}{(\sin\alpha-1)(\sin\alpha+1)}=\frac{2}{\sin^2\alpha-1}=-\frac{2}{\cos^2\alpha}\text{ or }-2\sec^2\alpha$$

7. $$\frac{\cos x}{\sec x}+\frac{\sin x}{\csc x}=\frac{\cos x}{\frac{1}{\cos x}}+\frac{\sin x}{\frac{1}{\sin x}}=\cos x\left(\frac{\cos x}{1}\right)+\sin x\left(\frac{\sin x}{1}\right)=\cos^2 x+\sin^2 x=1$$

8. $$\frac{\cos\theta}{\sin\theta}+\frac{\sin\theta}{1+\cos\theta}=\frac{\cos\theta(1+\cos\theta)}{\sin\theta(1+\cos\theta)}+\frac{\sin\theta\cdot\sin\theta}{\sin\theta(1+\cos\theta)}$$
$$=\frac{\cos\theta+\cos^2\theta}{\sin\theta(1+\cos\theta)}+\frac{\sin^2\theta}{\sin\theta(1+\cos\theta)}=\frac{\cos\theta+\cos^2\theta+\sin^2\theta}{\sin\theta(1+\cos\theta)}$$
$$=\frac{\cos\theta+(\cos^2\theta+\sin^2\theta)}{\sin\theta(1+\cos\theta)}=\frac{\cos\theta+1}{\sin\theta(1+\cos\theta)}=\frac{1}{\sin\theta}\text{ or }\csc\theta$$

9. $$(1+\sin t)^2+\cos^2 t=1+2\sin t+\sin^2 t+\cos^2 t$$
$$=1+2\sin t+(\sin^2 t+\cos^2 t)=1+2\sin t+1=2+2\sin t$$

10. $$(1+\tan s)^2-2\tan s=1+2\tan s+\tan^2 s-2\tan s=1+\tan^2 s=\sec^2 s$$

11. $$\frac{1}{1+\cos x}-\frac{1}{1-\cos x}=\frac{1-\cos x}{(1+\cos x)(1-\cos x)}-\frac{1+\cos x}{(1+\cos x)(1-\cos x)}=\frac{(1-\cos x)-(1+\cos x)}{(1+\cos x)(1-\cos x)}$$
$$=\frac{1-\cos x-1-\cos x}{1-\cos^2 x}=-\frac{2\cos x}{\sin^2 x}$$
or $$-\frac{2\cos x}{\sin^2 x}=-\frac{2\cos x}{\sin x\sin x}=-2\left(\frac{\cos x}{\sin x}\right)\left(\frac{1}{\sin x}\right)=-2\cot x\csc x$$

12. $$(\sin\alpha-\cos\alpha)^2=\sin^2\alpha-2\sin\alpha\cos\alpha+\cos^2\alpha=(\sin^2\alpha+\cos^2\alpha)-2\sin\alpha\cos\alpha=1-2\sin\alpha\cos\alpha$$

13. $$\sin^2\theta-1=(\sin\theta+1)(\sin\theta-1)$$

14. $$\sec^2\theta-1=(\sec\theta+1)(\sec\theta-1)$$

15. $$(\sin x+1)^2-(\sin x-1)^2=[(\sin x+1)+(\sin x-1)][(\sin x+1)-(\sin x-1)]$$
$$=(\sin x+1+\sin x-1)(\sin x+1-\sin x+1)=(2\sin x)(2)=4\sin x$$

16. $$(\tan x+\cot x)^2-(\tan x-\cot x)^2=[(\tan x+\cot x)+(\tan x-\cot x)][(\tan x+\cot x)-(\tan x-\cot x)]$$
$$=(\tan x+\cot x+\tan x-\cot x)(\tan x+\cot x-\tan x+\cot x)$$
$$=(2\tan x)(2\cot x)=4\cdot\frac{\sin x}{\cos x}\cdot\frac{\cos x}{\sin x}=4$$

17. $2\sin^2 x + 3\sin x + 1$
Let $a = \sin x$.
$2\sin^2 x + 3\sin x + 1 = 2a^2 + 3a + 1 = (2a+1)(a+1) = (2\sin x + 1)(\sin x + 1)$

18. $4\tan^2 \beta + \tan\beta - 3$
Let $a = \tan\beta$.
$4\tan^2 \beta + \tan\beta - 3 = 4a^2 + a - 3 = (4a-3)(a+1) = (4\tan\beta - 3)(\tan\beta + 1)$

19. $\cos^4 x + 2\cos^2 x + 1$
Let $\cos^2 x = a$.
$\cos^4 x + 2\cos^2 x + 1 = a^2 + 2a + 1 = (a+1)^2 = (\cos^2 x + 1)^2$

20. $\cot^4 x + 3\cot^2 x + 2$
Let $\cot^2 x = a$.
$$\begin{aligned}\cot^4 x + 3\cot^2 x + 2 &= a^2 + 3a + 2 = (a+2)(a+1)\\ &= (\cot^2 x + 2)(\cot^2 x + 1) = (\cot^2 x + 2)(\csc^2 x) = \csc^2 x(\cot^2 x + 2)\end{aligned}$$

21. $\sin^3 x - \cos^3 x$
Let $\sin x = a$ and $\cos x = b$.
$$\begin{aligned}\sin^3 x - \cos^3 x &= a^3 - b^3 = (a-b)(a^2 + ab + b^2) = (\sin x - \cos x)(\sin^2 x + \sin x\cos x + \cos^2 x)\\ &= (\sin x - \cos x)\left[(\sin^2 x + \cos^2 x) + \sin x\cos x\right] = (\sin x - \cos x)(1 + \sin x\cos x)\end{aligned}$$

22. $\sin^3 \alpha + \cos^3 \alpha$
Let $\sin x = a$ and $\cos x = b$.
$$\begin{aligned}\sin^3 x + \cos^3 x &= a^3 + b^3 = (a+b)(a^2 - ab + b^2) = (\sin x + \cos x)(\sin^2 x - \sin x\cos x + \cos^2 x)\\ &= (\sin x + \cos x)\left[(\sin^2 x + \cos^2 x) - \sin x\cos x\right] = (\sin x + \cos x)(1 - \sin x\cos x)\end{aligned}$$

23. $\tan\theta\cos\theta = \dfrac{\sin\theta}{\cos\theta}\cos\theta = \sin\theta$

24. $\cot\alpha\sin\alpha = \dfrac{\cos\alpha}{\sin\alpha}\cdot\sin\alpha = \cos\alpha$

25. $\sec r\cos r = \dfrac{1}{\cos r}\cdot\cos r = 1$

26. $\cot t\tan t = \dfrac{\cos t}{\sin t}\cdot\dfrac{\sin t}{\cos t} = 1$

27. $\dfrac{\sin\beta\tan\beta}{\cos\beta} = \tan\beta\tan\beta = \tan^2\beta$

28.
$$\begin{aligned}\frac{\csc\theta\sec\theta}{\cot\theta} &= \frac{\frac{1}{\sin\theta}\cdot\frac{1}{\cos\theta}}{\frac{\cos\theta}{\sin\theta}}\\ &= \frac{1}{\sin\theta}\cdot\frac{1}{\cos\theta}\cdot\frac{\sin\theta}{\cos\theta}\\ &= \frac{1}{\cos^2\theta} = \sec^2\theta\end{aligned}$$

29. $\sec^2 x - 1 = \dfrac{1}{\cos^2 x} - 1 = \dfrac{1}{\cos^2 x} - \dfrac{\cos^2 x}{\cos^2 x} = \dfrac{1-\cos^2 x}{\cos^2 x} = \dfrac{\sin^2 x}{\cos^2 x} = \tan^2 x$

30. $\csc^2 t - 1 = \cot^2 t$

31. $\frac{\sin^2 x}{\cos^2 x} + \sin x \csc x = \tan^2 x + \sin x \cdot \frac{1}{\sin x} = \tan^2 x + 1 = \sec^2 x$

32. $\frac{1}{\tan^2 \alpha} + \cot \alpha \tan \alpha = \cot^2 \alpha + \frac{\cos \alpha}{\sin \alpha} \cdot \frac{\sin \alpha}{\cos \alpha} = \cot^2 \alpha + 1 = \csc^2 \alpha$

33. Verify $\frac{\cot \theta}{\csc \theta} = \cos \theta.$

$$\frac{\cot \theta}{\csc \theta} = \frac{\frac{\cos \theta}{\sin \theta}}{\frac{1}{\sin \theta}} = \frac{\cos \theta}{\sin \theta} \cdot \frac{\sin \theta}{1} = \cos \theta$$

34. Verify $\frac{\tan \alpha}{\sec \alpha} = \sin \alpha.$

$$\frac{\tan \alpha}{\sec \alpha} = \frac{\frac{\sin \alpha}{\cos \alpha}}{\frac{1}{\cos \alpha}} = \frac{\sin \alpha}{\cos \alpha} \cdot \cos \alpha = \sin \alpha$$

35. Verify $\frac{1 - \sin^2 \beta}{\cos \beta} = \cos \beta.$

$$\frac{1 - \sin^2 \beta}{\cos \beta} = \frac{\cos^2 \beta}{\cos \beta} = \cos \beta$$

36. Verify $\frac{\tan^2 \alpha + 1}{\sec \alpha} = \sec \alpha$.

$$\frac{\tan^2 \alpha + 1}{\sec \alpha} = \frac{\sec^2 \alpha}{\sec \alpha} = \sec \alpha$$

37. Verify $\cos^2 \theta (\tan^2 \theta + 1) = 1.$

$$\cos^2 \theta (\tan^2 \theta + 1) = \cos^2 \theta \left(\frac{\sin^2 \theta}{\cos^2 \theta} + 1 \right)$$

$$= \cos^2 \theta \left(\frac{\sin^2 \theta}{\cos^2 \theta} + \frac{\cos^2 \theta}{\cos^2 \theta} \right) = \cos^2 \theta \left(\frac{\sin^2 \theta + \cos^2 \theta}{\cos^2 \theta} \right) = \cos^2 \theta \left(\frac{1}{\cos^2 \theta} \right) = 1$$

38. Verify $\sin^2 \beta (1 + \cot^2 \beta) = 1.$

$$\sin^2 \beta (1 + \cot^2 \beta) = \sin^2 \beta \csc^2 \beta = \sin^2 \beta \cdot \frac{1}{\sin^2 \beta} = 1$$

39. Verify $\cot s + \tan s = \sec s \csc s.$

$$\cot s + \tan s = \frac{\cos s}{\sin s} + \frac{\sin s}{\cos s}$$

$$= \frac{\cos^2 s}{\sin s \cos s} + \frac{\sin^2 s}{\sin s \cos s} = \frac{\cos^2 s + \sin^2 s}{\cos s \sin s} = \frac{1}{\cos s \sin s} = \frac{1}{\cos s} \cdot \frac{1}{\sin s} = \sec s \csc s$$

40. Verify $\sin^2 \alpha + \tan^2 \alpha + \cos^2 \alpha = \sec^2 \alpha.$

$$\sin^2 \alpha + \tan^2 \alpha + \cos^2 \alpha = \left(\sin^2 \alpha + \cos^2 \alpha \right) + \tan^2 \alpha = 1 + \tan^2 \alpha = \sec^2 \alpha$$

41. Verify $\frac{\cos \alpha}{\sec \alpha} + \frac{\sin \alpha}{\csc \alpha} = \sec^2 \alpha - \tan^2 \alpha.$

Working with the left side, we have $\frac{\cos \alpha}{\sec \alpha} + \frac{\sin \alpha}{\csc \alpha} = \frac{\cos \alpha}{\frac{1}{\cos \alpha}} + \frac{\sin \alpha}{\frac{1}{\sin \alpha}} = \cos^2 \alpha + \sin^2 \alpha = 1.$

Working with the right side, we have $\sec^2 \alpha - \tan^2 \alpha = 1.$

Since $\frac{\cos\alpha}{\sec\alpha}+\frac{\sin\alpha}{\csc\alpha}=1=\sec^2\alpha-\tan^2\alpha$, the statement has been verified.

42. Verify $\frac{\sin^2\theta}{\cos\theta}=\sec\theta-\cos\theta$.

$$\frac{\sin^2\theta}{\cos\theta}=\frac{1-\cos^2\theta}{\cos\theta}=\frac{1}{\cos\theta}-\frac{\cos^2\theta}{\cos\theta}=\sec\theta-\cos\theta$$

43. Verify $\sin^4\theta-\cos^4\theta=2\sin^2\theta-1$.

$$\sin^4\theta-\cos^4\theta=\left(\sin^2\theta+\cos^2\theta\right)\left(\sin^2\theta-\cos^2\theta\right)=1\cdot\left(\sin^2\theta-\cos^2\theta\right)=\sin^2\theta-\cos^2\theta$$
$$=\sin^2\theta-\left(1-\sin^2\theta\right)=\sin^2\theta-1+\sin^2\theta=2\sin^2\theta-1$$

44. Verify $\frac{\cos\theta}{\sin\theta\cot\theta}=1$.

$$\frac{\cos\theta}{\sin\theta\cot\theta}=\frac{\cos\theta}{\sin\theta\cdot\frac{\cos\theta}{\sin\theta}}=\frac{\cos\theta}{\cos\theta}=1$$

45. Verify $\left(1-\cos^2\alpha\right)\left(1+\cos^2\alpha\right)=2\sin^2\alpha-\sin^4\alpha$.

$$\left(1-\cos^2\alpha\right)\left(1+\cos^2\alpha\right)=\sin^2\alpha\left(1+\cos^2\alpha\right)=\sin^2\alpha\left(2-\sin^2\alpha\right)=2\sin^2\alpha-\sin^4\alpha$$

46. Verify $\tan^2\alpha\sin^2\alpha=\tan^2\alpha+\cos^2\alpha-1$.

Work with the left side.

$$\tan^2\alpha\sin^2\alpha=\tan^2\alpha\left(1-\cos^2\alpha\right)=\tan^2\alpha-\tan^2\alpha\cos^2\alpha=\tan^2\alpha-\sin^2\alpha$$

Now work with the right side.

$$\tan^2\alpha+\cos^2\alpha-1=\tan^2\alpha-\left(1-\cos^2\alpha\right)=\tan^2\alpha-\sin^2\alpha$$

Since $\tan^2\alpha+\cos^2\alpha-1=\tan^2\alpha-\sin^2\alpha=\tan^2\alpha+\cos^2\alpha-1$, the statement has been verified.

47. Verify $\frac{\cos\theta+1}{\tan^2\theta}=\frac{\cos\theta}{\sec\theta-1}$.

Work with the left side.

$$\frac{\cos\theta+1}{\tan^2\theta}=\frac{\cos\theta+1}{\sec^2\theta-1}=\frac{\cos\theta+1}{\frac{1}{\cos^2\theta}-1}$$
$$=\frac{(\cos\theta+1)\cos^2\theta}{\left(\frac{1}{\cos^2\theta}-1\right)\cos^2\theta}=\frac{\cos^2\theta(\cos\theta+1)}{1-\cos^2\theta}=\frac{\cos^2\theta(\cos\theta+1)}{(1+\cos\theta)(1-\cos\theta)}=\frac{\cos^2\theta}{1-\cos\theta}$$

Now work with the right side.

$$\frac{\cos\theta}{\sec\theta-1}=\frac{\cos\theta}{\frac{1}{\cos\theta}-1}=\frac{\cos\theta}{\frac{1}{\cos\theta}}\cdot\frac{\cos\theta}{\cos\theta}=\frac{\cos^2\theta}{1-\cos\theta}$$

Since $\frac{\cos\theta+1}{\tan^2\theta}=\frac{\cos^2\theta}{1-\cos\theta}=\frac{\cos\theta}{\sec\theta-1}$, the statement has been verified.

48. Verify $\dfrac{(\sec\theta-\tan\theta)^2+1}{\sec\theta\csc\theta-\tan\theta\csc\theta}=2\tan\theta.$

$$\frac{(\sec\theta-\tan\theta)^2+1}{\sec\theta\csc\theta-\tan\theta\csc\theta}=\frac{\sec^2\theta-2\sec\theta\tan\theta+\tan^2\theta+1}{\csc\theta(\sec\theta-\tan\theta)}$$

$$=\frac{\sec^2\theta-2\sec\theta\tan\theta+(\tan^2\theta+1)}{\csc\theta(\sec\theta-\tan\theta)}=\frac{\sec^2\theta-2\sec\theta\tan\theta+\sec^2\theta}{\csc\theta(\sec\theta-\tan\theta)}$$

$$=\frac{2\sec^2\theta-2\sec\theta\tan\theta}{\csc\theta(\sec\theta-\tan\theta)}=\frac{2\sec\theta(\sec\theta-\tan\theta)}{\csc\theta(\sec\theta-\tan\theta)}$$

$$=\frac{2\sec\theta}{\csc\theta}=2\cdot\frac{\sin\theta}{\cos\theta}=2\tan\theta$$

49. Verify $\dfrac{1}{1-\sin\theta}+\dfrac{1}{1+\sin\theta}=2\sec^2\theta.$

$$\frac{1}{1-\sin\theta}+\frac{1}{1+\sin\theta}=\frac{1+\sin\theta}{(1+\sin\theta)(1-\sin\theta)}+\frac{1-\sin\theta}{(1+\sin\theta)(1-\sin\theta)}$$

$$=\frac{(1+\sin\theta)+(1-\sin\theta)}{(1+\sin\theta)(1-\sin\theta)}=\frac{1+\sin\theta+1-\sin\theta}{(1+\sin\theta)(1-\sin\theta)}=\frac{2}{1-\sin^2\theta}=\frac{2}{\cos^2\theta}=2\sec^2\theta$$

50. Verify $\dfrac{1}{\sec\alpha-\tan\alpha}=\sec\alpha+\tan\alpha.$

$$\frac{1}{\sec\alpha-\tan\alpha}=\frac{1}{\sec\alpha-\tan\alpha}\cdot\frac{\sec\alpha+\tan\alpha}{\sec\alpha+\tan\alpha}=\frac{\sec\alpha+\tan\alpha}{\sec^2\alpha-\tan^2\alpha}$$

$$=\frac{\sec\alpha+\tan\alpha}{\frac{1}{\cos^2\alpha}-\frac{\sin^2\alpha}{\cos^2\alpha}}=\frac{\sec\alpha+\tan\alpha}{\frac{1-\sin^2\alpha}{\cos^2\alpha}}=\frac{\sec\alpha+\tan\alpha}{\frac{\cos^2\alpha}{\cos^2\alpha}}=\frac{\sec\alpha+\tan\alpha}{1}=\sec\alpha+\tan\alpha$$

51. Verify $\dfrac{\tan s}{1+\cos s}+\dfrac{\sin s}{1-\cos s}=\cot s+\sec s\csc s.$

$$\frac{\tan s}{1+\cos s}+\frac{\sin s}{1-\cos s}=\frac{\tan s(1-\cos s)}{(1+\cos s)(1-\cos s)}+\frac{\sin s(1+\cos s)}{(1+\cos s)(1-\cos s)}$$

$$=\frac{\tan s(1-\cos s)+\sin s(1+\cos s)}{(1+\cos s)(1-\cos s)}$$

$$=\frac{\tan s-\sin s+\sin s+\sin s\cos s}{1-\cos^2 s}$$

$$=\frac{\tan s+\sin s\cos s}{\sin^2 s}=\frac{\tan s}{\sin^2 s}+\frac{\sin s\cos s}{\sin^2 s}$$

$$=\tan s\cdot\frac{1}{\sin^2 s}+\frac{\cos s}{\sin s}$$

$$=\frac{\sin s}{\cos s}\cdot\frac{1}{\sin^2 s}+\cot s$$

$$=\frac{1}{\cos s}\cdot\frac{1}{\sin s}+\cot s$$

$$=\sec s\csc s+\cot s$$

52. Verify $\frac{1-\cos x}{1+\cos x}=(\cot x-\csc x)^2$.

Work with the left side.

$$\frac{1-\cos x}{1+\cos x}=\frac{(1-\cos x)(1-\cos x)}{(1+\cos x)(1-\cos x)}=\frac{1-2\cos x+\cos^2 x}{1-\cos^2 x}=\frac{1-2\cos x+\cos^2 x}{\sin^2 x}$$

Work with the right side.

$$(\cot x-\csc x)^2=\left(\frac{\cos x}{\sin x}-\frac{1}{\sin x}\right)^2=\left(\frac{\cos x-1}{\sin x}\right)^2=\frac{\cos^2 x-2\cos x+1}{\sin^2 x}$$

Since $\frac{1-\cos x}{1+\cos x}=\frac{\cos^2 x-2\cos x+1}{\sin^2 x}=(\cot x-\csc x)^2$, the statement has been verified.

53. Verify $\frac{\cot\alpha+1}{\cot\alpha-1}=\frac{1+\tan\alpha}{1-\tan\alpha}$.

$$\frac{\cot\alpha+1}{\cot\alpha-1}=\frac{\frac{\cos\alpha}{\sin\alpha}+1}{\frac{\cos\alpha}{\sin\alpha}-1}=\frac{\frac{\cos\alpha}{\sin\alpha}+1}{\frac{\cos\alpha}{\sin\alpha}-1}\cdot\frac{\sin\alpha}{\sin\alpha}$$

$$=\frac{\cos\alpha+\sin\alpha}{\cos\alpha-\sin\alpha}=\frac{\cos\alpha+\sin\alpha}{\cos\alpha-\sin\alpha}\cdot\frac{\frac{1}{\cos\alpha}}{\frac{1}{\cos\alpha}}=\frac{\frac{\cos\alpha}{\cos\alpha}+\frac{\sin\alpha}{\cos\alpha}}{\frac{\cos\alpha}{\cos\alpha}-\frac{\sin\alpha}{\cos\alpha}}=\frac{1+\tan\alpha}{1-\tan\alpha}$$

54. Verify $\frac{1}{\tan\alpha-\sec\alpha}+\frac{1}{\tan\alpha+\sec\alpha}=-2\tan\alpha$.

$$\frac{1}{\tan\alpha-\sec\alpha}+\frac{1}{\tan\alpha+\sec\alpha}=\frac{\tan\alpha+\sec\alpha}{(\tan\alpha+\sec\alpha)(\tan\alpha-\sec\alpha)}+\frac{\tan\alpha-\sec\alpha}{(\tan\alpha+\sec\alpha)(\tan\alpha-\sec\alpha)}$$

$$=\frac{\tan\alpha+\sec\alpha+\tan\alpha-\sec\alpha}{(\tan\alpha+\sec\alpha)(\tan\alpha-\sec\alpha)}=\frac{2\tan\alpha}{\tan^2\alpha-\sec^2\alpha}$$

$$=\frac{2\tan\alpha}{\tan^2\alpha-(\tan^2\alpha+1)}=\frac{2\tan\alpha}{\tan^2\alpha-\tan^2\alpha-1}=\frac{2\tan\alpha}{-1}=-2\tan\alpha$$

55. Verify $\sin^2\alpha\sec^2\alpha+\sin^2\alpha\csc^2\alpha=\sec^2\alpha$.

$$\sin^2\alpha\sec^2\alpha+\sin^2\alpha\csc^2\alpha=\sin^2\alpha\cdot\frac{1}{\cos^2\alpha}+\sin^2\alpha\cdot\frac{1}{\sin^2\alpha}=\frac{\sin^2\alpha}{\cos^2\alpha}+1=\tan^2\alpha+1=\sec^2\alpha$$

56. Verify $\frac{\csc\theta+\cot\theta}{\tan\theta+\sin\theta}=\cot\theta\csc\theta$.

$$\frac{\csc\theta+\cot\theta}{\tan\theta+\sin\theta}=\frac{\frac{1}{\sin\theta}+\frac{\cos\theta}{\sin\theta}}{\frac{\sin\theta}{\cos\theta}+\sin\theta}=\frac{\frac{1+\cos\theta}{\sin\theta}}{\sin\theta\left(\frac{1}{\cos\theta}+1\right)}\cdot\frac{\sin\theta\cos\theta}{\sin\theta\cos\theta}$$

$$=\frac{(1+\cos\theta)\cos\theta}{\sin^2\theta(1+\cos\theta)}=\frac{\cos\theta}{\sin^2\theta}=\frac{1}{\sin\theta}\cdot\frac{\cos\theta}{\sin\theta}=\csc\theta\cot\theta$$

57. Verify $\sec^4 x - \sec^2 x = \tan^4 x + \tan^2 x$.
Simplify left side.

$$\sec^4 x - \sec^2 x = \sec^2 x\left(\sec^2 x - 1\right) = \sec^2 x \tan^2 x = \tan^2 x \sec^2 x$$

Simplify right side.

$$\tan^4 x + \tan^2 x = \tan^2 x\left(\tan^2 x + 1\right) = \tan^2 x \sec^2 x$$

Since $\sec^4 x - \sec^2 x = \tan^2 x \sec^2 x = \tan^4 x + \tan^2 x,$ the statement has been verified.

58. Verify $\dfrac{1-\sin\theta}{1+\sin\theta} = \sec^2\theta - 2\sec\theta\tan\theta + \tan^2\theta$.
Working with the right side, we have the following.

$$\sec^2\theta - 2\sec\theta\tan\theta + \tan^2\theta = \frac{1}{\cos^2\theta} - 2\cdot\frac{1}{\cos\theta}\cdot\frac{\sin\theta}{\cos\theta} + \frac{\sin^2\theta}{\cos^2\theta} = \frac{1-2\sin\theta+\sin^2\theta}{\cos^2\theta}$$

$$= \frac{(1-\sin\theta)^2}{1-\sin^2\theta} = \frac{(1-\sin\theta)^2}{(1+\sin\theta)(1-\sin\theta)} = \frac{1-\sin\theta}{1+\sin\theta}$$

59. Verify $\sin\theta + \cos\theta = \dfrac{\sin\theta}{1-\dfrac{\cos\theta}{\sin\theta}} + \dfrac{\cos\theta}{1-\dfrac{\sin\theta}{\cos\theta}}$.
Working with the right side, we have the following.

$$\frac{\sin\theta}{1-\dfrac{\cos\theta}{\sin\theta}} + \frac{\cos\theta}{1-\dfrac{\sin\theta}{\cos\theta}} = \frac{\sin\theta}{1-\dfrac{\cos\theta}{\sin\theta}}\cdot\frac{\sin\theta}{\sin\theta} + \frac{\cos\theta}{1-\dfrac{\sin\theta}{\cos\theta}}\cdot\frac{\cos\theta}{\cos\theta} = \frac{\sin^2\theta}{\sin\theta-\cos\theta} + \frac{\cos^2\theta}{\cos\theta-\sin\theta}$$

$$= \frac{\sin^2\theta}{\sin\theta-\cos\theta} + \frac{\cos^2\theta}{-(\sin\theta-\cos\theta)} = \frac{\sin^2\theta}{\sin\theta-\cos\theta} - \frac{\cos^2\theta}{\sin\theta-\cos\theta}$$

$$= \frac{\sin^2\theta-\cos^2\theta}{\sin\theta-\cos\theta} = \frac{(\sin\theta+\cos\theta)(\sin\theta-\cos\theta)}{\sin\theta-\cos\theta} = \sin\theta+\cos\theta$$

60. Verify $\dfrac{\sin\theta}{1-\cos\theta} - \dfrac{\sin\theta\cos\theta}{1+\cos\theta} = \csc\theta\left(1+\cos^2\theta\right)$.

$$\frac{\sin\theta}{1-\cos\theta} - \frac{\sin\theta\cos\theta}{1+\cos\theta} = \frac{\sin\theta(1+\cos\theta)}{(1+\cos\theta)(1-\cos\theta)} - \frac{\sin\theta\cos\theta(1-\cos\theta)}{(1+\cos\theta)(1-\cos\theta)}$$

$$= \frac{\sin\theta(1+\cos\theta) - \sin\theta\cos\theta(1-\cos\theta)}{(1+\cos\theta)(1-\cos\theta)}$$

$$= \frac{\sin\theta + \sin\theta\cos\theta - \sin\theta\cos\theta + \sin\theta\cos^2\theta}{1-\cos^2\theta}$$

$$= \frac{\sin\theta + \sin\theta\cos^2\theta}{\sin^2\theta}$$

$$= \frac{1+\cos^2\theta}{\sin\theta}$$

$$= \frac{1}{\sin\theta}\left(1+\cos^2\theta\right)$$

$$= \csc\theta\left(1+\cos^2\theta\right)$$

61. Verify $\frac{\sec^4 s-\tan^4 s}{\sec^2 s+\tan^2 s}=\sec^2 s-\tan^2 s$.

$$\frac{\sec^4 s-\tan^4 s}{\sec^2 s+\tan^2 s}=\frac{\left(\sec^2 s+\tan^2 s\right)\left(\sec^2 s-\tan^2 s\right)}{\sec^2 s+\tan^2 s}=\sec^2 s-\tan^2 s$$

62. Verify $\frac{\cot^2 t-1}{1+\cot^2 t}=1-2\sin^2 t$.

$$\frac{\cot^2 t-1}{1+\cot^2 t}=\frac{\frac{\cos^2 t}{\sin^2 t}-1}{1+\frac{\cos^2 t}{\sin^2 t}}=\frac{\frac{\cos^2 t}{\sin^2 t}-1}{1+\frac{\cos^2 t}{\sin^2 t}}\cdot\frac{\sin^2 t}{\sin^2 t}=\frac{\cos^2 t-\sin^2 t}{\sin^2 t+\cos^2 t}$$

$$=\frac{\cos^2 t-\sin^2 t}{1}=\cos^2 t-\sin^2 t=\left(1-\sin^2 t\right)-\sin^2 t=1-2\sin^2 t$$

63. Verify $\frac{\tan^2 t-1}{\sec^2 t}=\frac{\tan t-\cot t}{\tan t+\cot t}$.

Working with the right side, we have the following.

$$\frac{\tan t-\cot t}{\tan t+\cot t}=\frac{\tan t-\frac{1}{\tan t}}{\tan t+\frac{1}{\tan t}}=\frac{\tan t-\frac{1}{\tan t}}{\tan t+\frac{1}{\tan t}}\cdot\frac{\tan t}{\tan t}=\frac{\tan^2 t-1}{\tan^2 t+1}=\frac{\tan^2 t-1}{\sec^2 t}$$

64. Verify $(1+\sin x+\cos x)^2=2(1+\sin x)(1+\cos x)$.

Working with the right side, we have the following.

$$\begin{aligned}2(1+\sin x)(1+\cos x)&=2(1+\sin x+\cos x+\sin x\cos x)\\&=2+2\sin x+2\cos x+2\sin x\cos x\end{aligned}$$

Working with the left side, we have the following.

$$\begin{aligned}(1+\sin x+\cos x)^2&=1+\sin x+\cos x+\sin x+\sin^2 x+\sin x\cos x+\cos x+\sin x\cos x+\cos^2 x\\&=2+2\sin x+2\cos x+2\sin x\cos x\end{aligned}$$

Since $(1+\sin x+\cos x)^2=2+2\sin x+2\cos x+2\sin x\cos x=2(1+\sin x)(1+\cos x)$, the statement has been verified.

65. Verify $\frac{1+\cos x}{1-\cos x}-\frac{1-\cos x}{1+\cos x}=4\cot x\csc x$.

$$\begin{aligned}\frac{1+\cos x}{1-\cos x}-\frac{1-\cos x}{1+\cos x}&=\frac{(1+\cos x)^2}{(1+\cos x)(1-\cos x)}-\frac{(1-\cos x)^2}{(1+\cos x)(1-\cos x)}\\&=\frac{1+2\cos x+\cos^2 x}{(1+\cos x)(1-\cos x)}-\frac{1-2\cos x+\cos^2 x}{(1+\cos x)(1-\cos x)}\\&=\frac{1+2\cos x+\cos^2 x-1+2\cos x-\cos^2 x}{(1+\cos x)(1-\cos x)}\\&=\frac{4\cos x}{1-\cos^2 x}=\frac{4\cos x}{\sin^2 x}=4\cdot\frac{\cos x}{\sin x}\cdot\frac{1}{\sin x}=4\cot x\csc x\end{aligned}$$

66. Verify $\left(\sec\alpha-\tan\alpha\right)^2=\dfrac{1-\sin\alpha}{1+\sin\alpha}$.

$$\left(\sec\alpha-\tan\alpha\right)^2=\sec^2\alpha-2\sec\alpha\tan\alpha+\tan^2\alpha=\frac{1}{\cos^2\alpha}-2\cdot\frac{1}{\cos\alpha}\cdot\frac{\sin\alpha}{\cos\alpha}+\frac{\sin^2\alpha}{\cos^2\alpha}$$

$$=\frac{1-2\sin\alpha+\sin^2\alpha}{\cos^2\alpha}=\frac{\left(1-\sin\alpha\right)^2}{1-\sin^2\alpha}=\frac{\left(1-\sin\alpha\right)^2}{\left(1-\sin\alpha\right)\left(1+\sin\alpha\right)}=\frac{1-\sin\alpha}{1+\sin\alpha}$$

67. Verify $\left(\sec\alpha+\csc\alpha\right)\left(\cos\alpha-\sin\alpha\right)=\cot\alpha-\tan\alpha$.

$$\left(\sec\alpha+\csc\alpha\right)\left(\cos\alpha-\sin\alpha\right)=\left(\frac{1}{\cos\alpha}+\frac{1}{\sin\alpha}\right)\left(\cos\alpha-\sin\alpha\right)$$

$$=\frac{\cos\alpha}{\cos\alpha}-\frac{\sin\alpha}{\cos\alpha}+\frac{\cos\alpha}{\sin\alpha}-\frac{\sin\alpha}{\sin\alpha}=1-\tan\alpha+\cot\alpha-1=\cot\alpha-\tan\alpha$$

68. Verify $\dfrac{\sin^4\alpha-\cos^4\alpha}{\sin^2\alpha-\cos^2\alpha}=1$.

$$\frac{\sin^4\alpha-\cos^4\alpha}{\sin^2\alpha-\cos^2\alpha}=\frac{\left(\sin^2\alpha+\cos^2\alpha\right)\left(\sin^2\alpha-\cos^2\alpha\right)}{\sin^2\alpha-\cos^2\alpha}=\sin^2\alpha+\cos^2\alpha=1$$

69. One example does not verify a statement to be true. However, if you find one example where it shows the statement is false, then that one example is enough to state that the statement is false. For example, if the student let $\theta=45°$ (or $\frac{\pi}{4}$ radians), then the student would have observed $\cos 45°+\sin 45°=\frac{\sqrt{2}}{2}+\frac{\sqrt{2}}{2}=\frac{2\sqrt{2}}{2}=\sqrt{2}\neq 1$. This would show that $\cos\theta+\sin\theta=1$ is not an identity.

70. Answers will vary.
No, if an equation has an infinite number of solutions then it does not necessarily imply that one has an identity. An example of this is the equation $\sin x=\cos x$. There are infinitely many values of x in which this statement is true, but it does not imply that $\sin x=\cos x$ is an identity.

In Exercises 71 – 78, the functions are graphed in the following window.

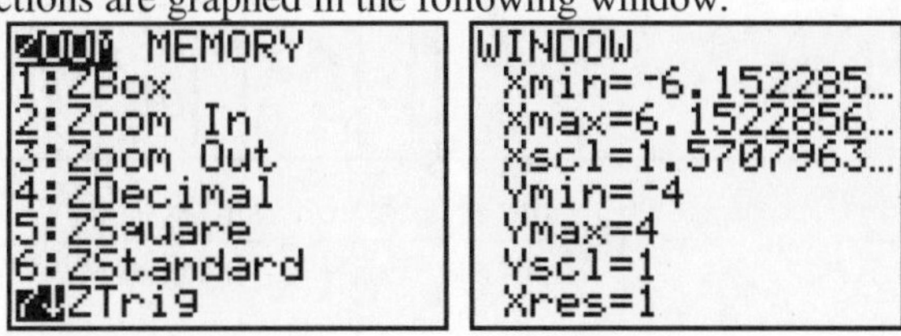

71. $\left(\sec\theta+\tan\theta\right)\left(1-\sin\theta\right)$ appears to be equivalent to $\cos\theta$.

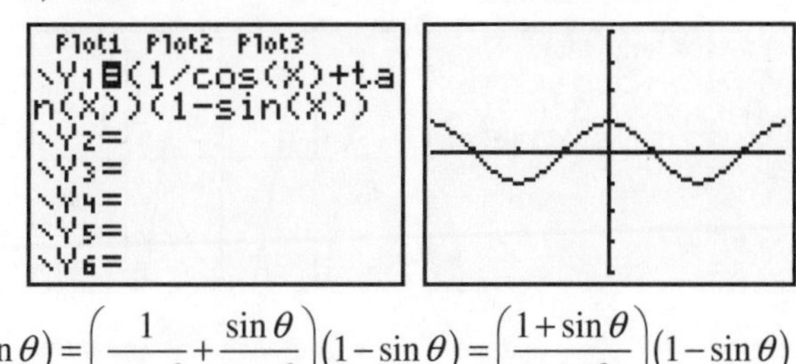

$$\left(\sec\theta+\tan\theta\right)\left(1-\sin\theta\right)=\left(\frac{1}{\cos\theta}+\frac{\sin\theta}{\cos\theta}\right)\left(1-\sin\theta\right)=\left(\frac{1+\sin\theta}{\cos\theta}\right)\left(1-\sin\theta\right)$$

$$=\frac{\left(1+\sin\theta\right)\left(1-\sin\theta\right)}{\cos\theta}=\frac{1-\sin^2\theta}{\cos\theta}=\frac{\cos^2\theta}{\cos\theta}=\cos\theta$$

72. $(\csc\theta+\cot\theta)(\sec\theta-1)$ appears to be equivalent to $\tan\theta$.

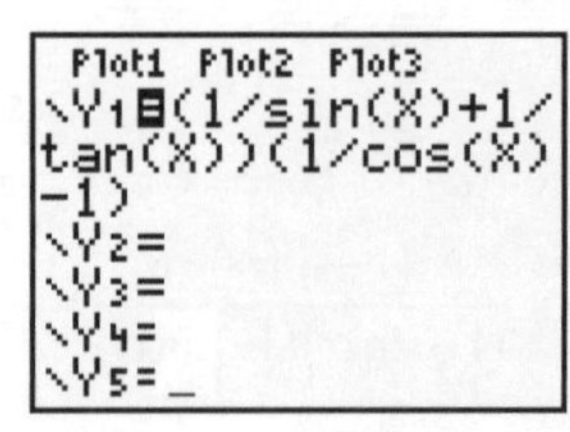

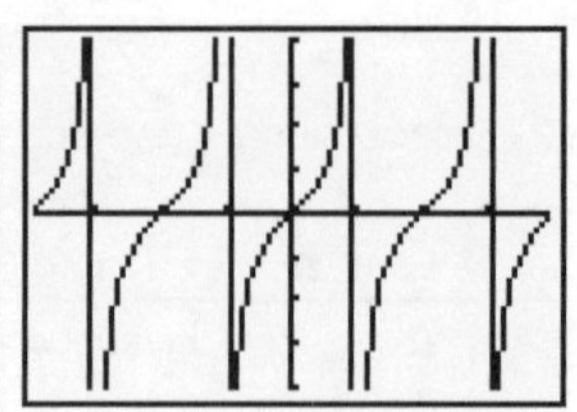

$$(\csc\theta+\cot\theta)(\sec\theta-1)=\left(\frac{1}{\sin\theta}+\frac{\cos\theta}{\sin\theta}\right)\left(\frac{1}{\cos\theta}-1\right)=\left(\frac{1+\cos\theta}{\sin\theta}\right)\left(\frac{1-\cos\theta}{\cos\theta}\right)$$

$$=\frac{1-\cos^2\theta}{\sin\theta\cos\theta}=\frac{\sin^2\theta}{\sin\theta\cos\theta}=\frac{\sin\theta}{\cos\theta}=\tan\theta$$

73. $\dfrac{\cos\theta+1}{\sin\theta+\tan\theta}$ appears to be equivalent to $\cot\theta$.

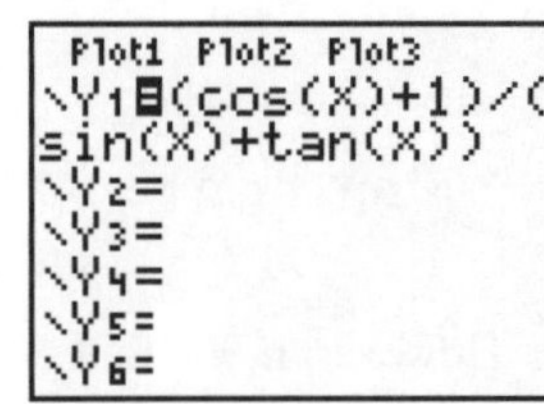

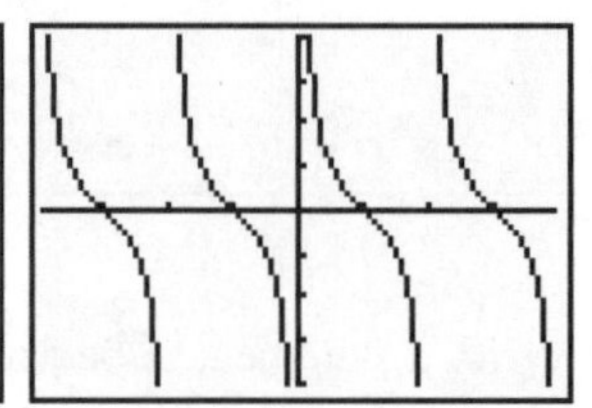

$$\frac{\cos\theta+1}{\sin\theta+\tan\theta}=\frac{1+\cos\theta}{\sin\theta+\frac{\sin\theta}{\cos\theta}}=\frac{1+\cos\theta}{\sin\theta\left(1+\frac{1}{\cos\theta}\right)}$$

$$=\frac{1+\cos\theta}{\sin\theta\left(1+\frac{1}{\cos\theta}\right)}\cdot\frac{\cos\theta}{\cos\theta}=\frac{(1+\cos\theta)\cos\theta}{\sin\theta(\cos\theta+1)}=\frac{\cos\theta}{\sin\theta}=\cot\theta$$

74. $\tan\theta\sin\theta+\cos\theta$ appears to be equivalent to $\sec\theta$.

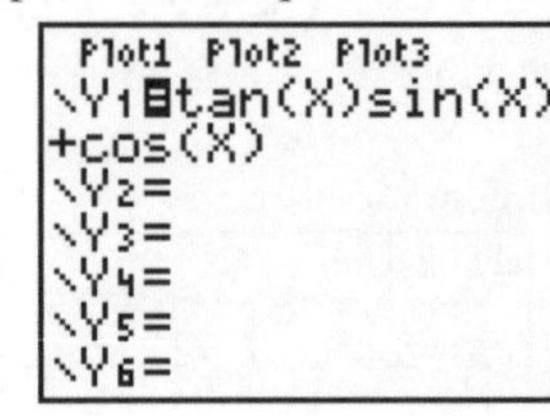

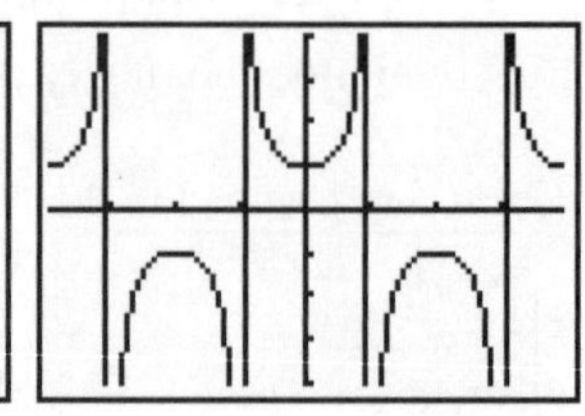

$$\tan\theta\sin\theta+\cos\theta=\frac{\sin\theta}{\cos\theta}\cdot\sin\theta+\cos\theta=\frac{\sin^2\theta}{\cos\theta}+\frac{\cos^2\theta}{\cos\theta}=\frac{\sin^2\theta+\cos^2\theta}{\cos\theta}=\frac{1}{\cos\theta}=\sec\theta$$

75. Is $\dfrac{2+5\cos s}{\sin s}=2\csc s+5\cot s$ an identity?

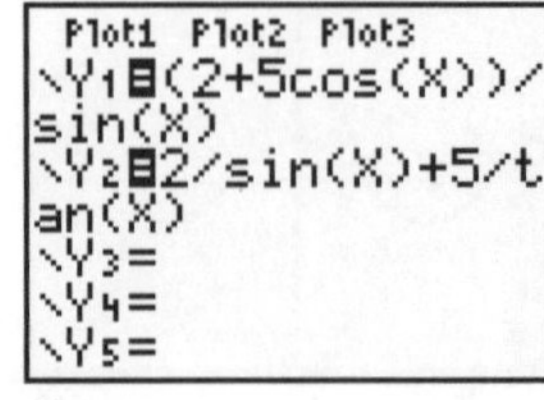

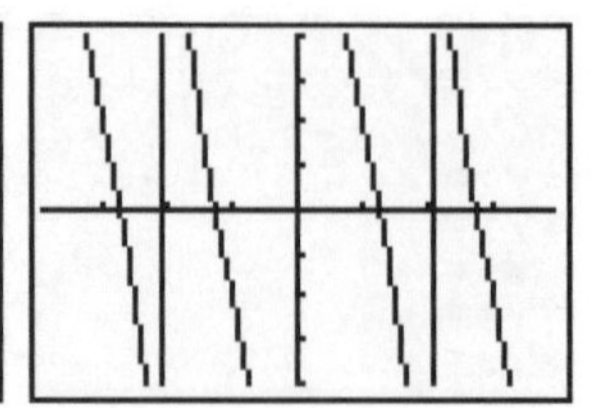

The graphs of $y=\dfrac{2+5\cos x}{\sin x}$ and $y=2\csc x+5\cot x$ appear to be the same. The given equation may be an identity. Since $\dfrac{2+5\cos s}{\sin s}=\dfrac{2}{\sin s}+\dfrac{5\cos s}{\sin s}=2\csc s+5\cot s$, the given statement is an identity.

76. Is $1+\cot^2 s=\dfrac{\sec^2 s}{\sec^2 s-1}$ an identity?

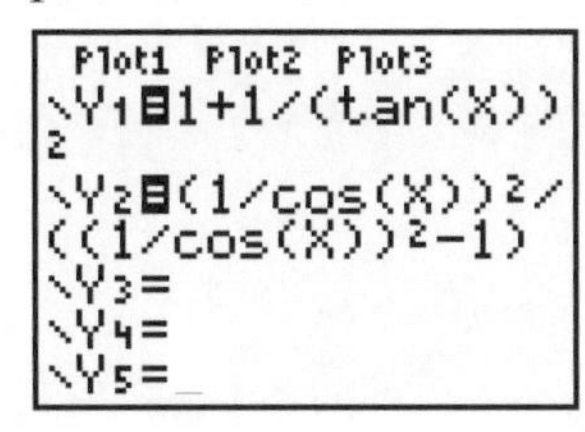

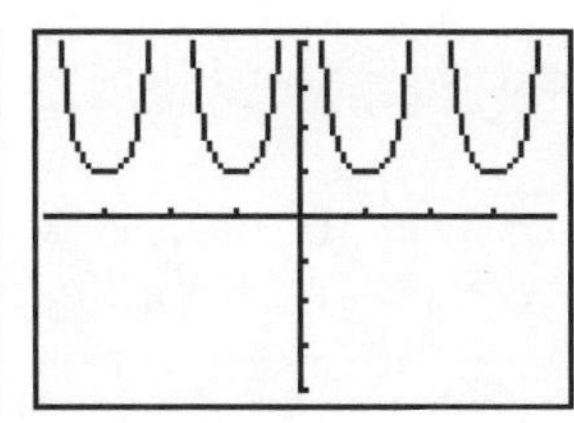

The graphs of $y=1+\cot^2 x$ and $y=\dfrac{\sec^2 x}{\sec^2 x-1}$ appear to be the same. The given equation may be an identity. Since $\dfrac{\sec^2 s}{\sec^2 s-1}=\dfrac{\sec^2 s}{\tan^2 s}=\dfrac{1}{\cos^2 s}\cdot\dfrac{\cos^2 s}{\sin^2 s}=\dfrac{1}{\sin^2 s}=\csc^2 s=1+\cot^2 s,$ the given statement is an identity.

77. Is $\dfrac{\tan s-\cot s}{\tan s+\cot s}=2\sin^2 s$ an identity?

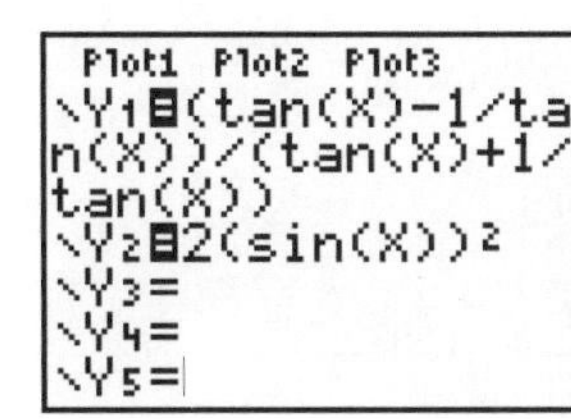

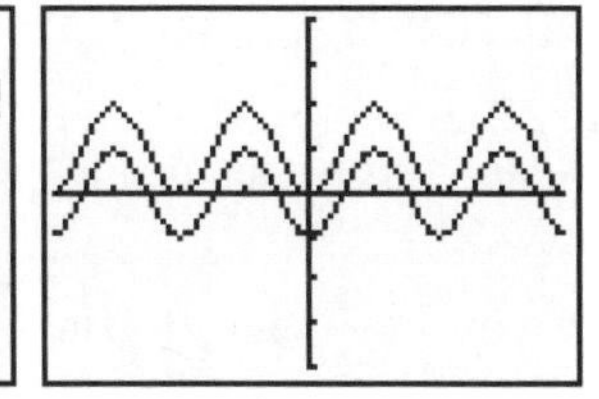

The graphs of $y=\dfrac{\tan x-\cot x}{\tan x+\cot x}$ and $y=2\sin^2 x$ are not the same. The given statement is not an identity.

78. Is $\dfrac{1}{1+\sin s}+\dfrac{1}{1-\sin s}=\sec^2 s$ an identity?

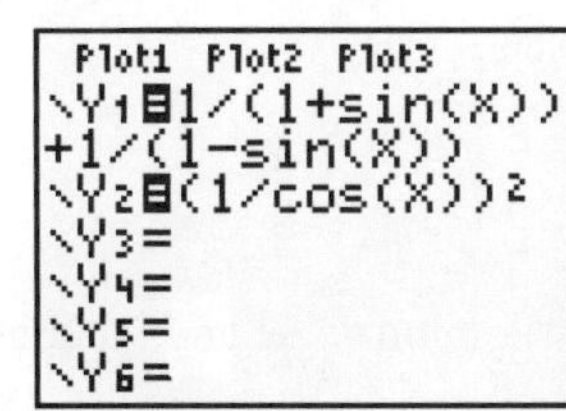

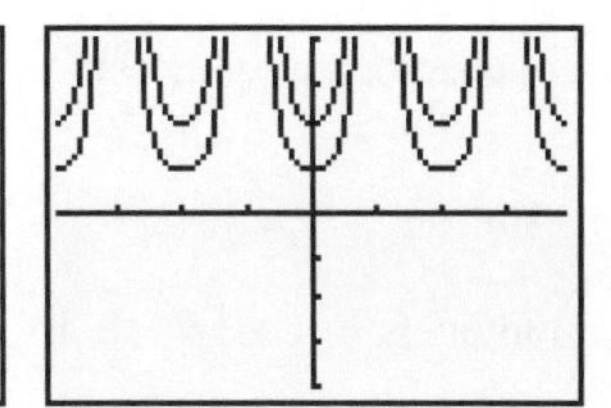

The graphs of $y=\dfrac{1}{1+\sin x}+\dfrac{1}{1-\sin x}$ and $y=\sec^2 x$ are not the same. The given statement is not an identity.

79. Show that $\sin(\csc s)=1$ is not an identity.

We need to find only one value for which the statement is false. Let $s = 2$. Use a calculator to find that sin (csc 2) $\approx$ .891094, which is not equal to 1.

$\sin(\csc s)=1$ does not hold true for *all* real numbers s. Thus, it is not an identity.

80. Show that $\sqrt{\cos^2 s} = \cos s$ is not an identity.

Let $s = \frac{\pi}{3}$. We have $\cos\frac{\pi}{3} = \frac{1}{2}$ and $\sqrt{\cos^2\frac{\pi}{3}} = \sqrt{\left(\frac{1}{2}\right)^2} = \sqrt{\frac{1}{4}} = \frac{1}{2}$. But let $s = \frac{2\pi}{3}$. We have,

$\cos s = -\frac{1}{2}$ and $\sqrt{\cos^2\frac{2\pi}{3}} = \sqrt{\left(-\frac{1}{2}\right)^2} = \sqrt{\frac{1}{4}} = \frac{1}{2}$.

$\sqrt{\cos^2 s} = \cos s$ does not hold true for *all* real numbers s. Thus, it is not an identity.

81. Show that $\csc t = \sqrt{1+\cot^2 t}$ is not an identity.

Let $t = \frac{\pi}{4}$. We have $\csc\frac{\pi}{4} = \sqrt{2}$ and $\sqrt{1+\cot^2\frac{\pi}{4}} = \sqrt{1+1^2} = \sqrt{1+1} = \sqrt{2}$. But let $t = -\frac{\pi}{4}$. We have,

$\csc -\frac{\pi}{4} = -\sqrt{2}$ and $\sqrt{1+\cot^2\left(-\frac{\pi}{4}\right)} = \sqrt{1+(-1)^2} = \sqrt{1+1} = \sqrt{2}$.

$\csc t = \sqrt{1+\cot^2 t}$ does not hold true for *all* real numbers s. Thus, it is not an identity.

82. Show that $\cos t = \sqrt{1-\sin^2 t}$ is not an identity.

Let $t = \frac{\pi}{3}$. We have $\cos\frac{\pi}{3} = \frac{1}{2}$ and $\sqrt{1-\sin^2\frac{\pi}{3}} = \sqrt{1-\left(\frac{\sqrt{3}}{2}\right)^2} = \sqrt{1-\frac{3}{4}} = \sqrt{\frac{1}{4}} = \frac{1}{2}$. But let $t = \frac{2\pi}{3}$.

We have, . $\cos\frac{2\pi}{3} = -\frac{1}{2}$ and $\sqrt{1-\sin^2\frac{2\pi}{3}} = \sqrt{1-\left(\frac{\sqrt{3}}{2}\right)^2} = \sqrt{1-\frac{3}{4}} = \sqrt{\frac{1}{4}} = \frac{1}{2}$.

$\cos t = \sqrt{1-\sin^2 t}$ does not hold true for *all* real numbers s. Thus, it is not an identity.

83. $\sin x = \sqrt{1-\cos^2 x}$ is a true statement when $\sin x \geq 0$.

84. **(a)** $I = k\cos^2\theta = k\left(1-\sin^2\theta\right)$

(b) For $\theta = 2\pi n$ for all integers n, $\cos^2\theta = 1$, its maximum value and I attains a maximum value of k.

85. **(a)** $P = ky^2$ and $y = 4\cos(2\pi t)$

$$P = ky^2 = k\left[4\cos(2\pi t)\right]^2 = k\left[16\cos^2(2\pi t)\right] = 16k\cos^2(2\pi t)$$

(b) $P = 16k\cos^2(2\pi t) = 16k\left[1-\sin^2(2\pi t)\right]$

86. **(a)** The sum of L and C equals 3.

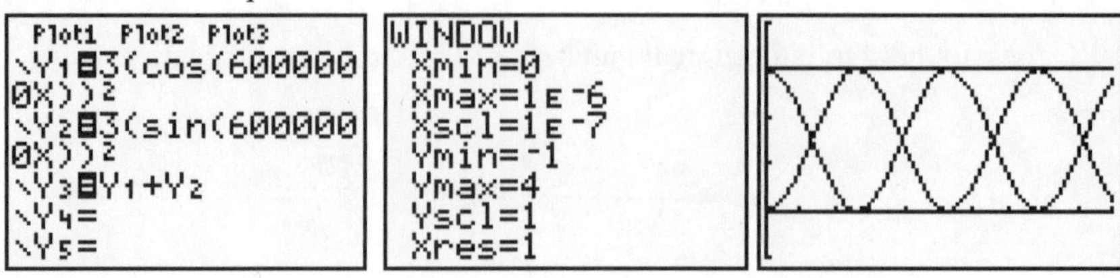

Continued on next page

86. (continued)

(b) Let $Y_1 = L(t), Y_2 = C(t)$, and $Y_3 = E(t)$ $Y_3 = 3$ for all inputs.

X	Y2	Y3
0	0	3
1E-7	.95646	3
2E-7	2.6061	3
3E-7	2.8451	3
4E-7	1.3688	3
5E-7	.05974	3
6E-7	.58747	3

Y3=Y1+Y2

(c) $$\begin{aligned} E(t) &= L(t)+C(t) \\ &= 3\cos^2(6{,}000{,}000t)+3\sin^2(6{,}000{,}000t) \\ &= 3\left[\cos^2(6{,}000{,}000t)+\sin^2(6{,}000{,}000t)\right] \\ &= 3\cdot 1 = 3 \end{aligned}$$

Section 5.3: Sum and Difference Identities for Cosine

1. Since $\cos(x+y) = \cos x \cos y - \sin x \sin y$, the correct choice is F.

2. Since $\cos(x-y) = \cos x \cos y + \sin x \sin y$, the correct choice is A.

3. Since $\cos(90° - x) = \cos 90° \cos x - \sin 90° \sin x = (0)\cos x + (1)\sin x = 0 + \sin x = \sin x$, the correct choice is E. Also, cosine is the cofunction of sine; hence, $\cos(90° - x) = \sin x$.

4. Since cosine is the cofunction of sine, $\sin(90° - x) = \cos x$. the correct choice is B.

5. $\cos 75° = \cos(30° + 45°) = \cos 30° \cos 45° - \sin 30° \sin 45° = \frac{\sqrt{3}}{2}\cdot\frac{\sqrt{2}}{2} - \frac{1}{2}\cdot\frac{\sqrt{2}}{2} = \frac{\sqrt{6}}{4} - \frac{\sqrt{2}}{4} = \frac{\sqrt{6}-\sqrt{2}}{4}$

6. $\cos(-15°) = \cos(30° - 45°) = \cos 30° \cos 45° + \sin 30° \sin 45° = \frac{\sqrt{3}}{2}\cdot\frac{\sqrt{2}}{2} + \frac{1}{2}\cdot\frac{\sqrt{2}}{2} = \frac{\sqrt{6}}{4} + \frac{\sqrt{2}}{4} = \frac{\sqrt{6}+\sqrt{2}}{4}$

7. $\cos 105° = \cos(60° + 45°) = \cos 60° \cos 45° - \sin 60° \sin 45° = \frac{1}{2}\cdot\frac{\sqrt{2}}{2} - \frac{\sqrt{3}}{2}\cdot\frac{\sqrt{2}}{2} = \frac{\sqrt{2}}{4} - \frac{\sqrt{6}}{4} = \frac{\sqrt{2}-\sqrt{6}}{4}$

8. $$\begin{aligned} \cos(-105°) &= \cos\left[-60° + (-45°)\right] = \cos(-60°)\cos(-45°) - \sin(-60°)\sin(-45°) \\ &= \frac{1}{2}\cdot\frac{\sqrt{2}}{2} - \left(-\frac{\sqrt{3}}{2}\right)\left(-\frac{\sqrt{2}}{2}\right) = \frac{\sqrt{2}}{4} - \frac{\sqrt{6}}{4} = \frac{\sqrt{2}-\sqrt{6}}{4} \end{aligned}$$

9. $$\begin{aligned} \cos\left(\frac{7\pi}{12}\right) &= \cos\left(\frac{4\pi}{12} + \frac{3\pi}{12}\right) = \cos\left(\frac{\pi}{3} + \frac{\pi}{4}\right) \\ &= \cos\frac{\pi}{3}\cos\frac{\pi}{4} - \sin\frac{\pi}{3}\sin\frac{\pi}{4} = \frac{1}{2}\cdot\frac{\sqrt{2}}{2} - \frac{\sqrt{3}}{2}\cdot\frac{\sqrt{2}}{2} = \frac{\sqrt{2}}{4} - \frac{\sqrt{6}}{4} = \frac{\sqrt{2}-\sqrt{6}}{4} \end{aligned}$$

10. $\cos\left(-\frac{\pi}{12}\right)=\cos\left(\frac{2\pi}{12}-\frac{3\pi}{12}\right)=\cos\left(\frac{\pi}{6}-\frac{\pi}{4}\right)$

$$=\cos\frac{\pi}{6}\cos\frac{\pi}{4}+\sin\frac{\pi}{6}\sin\frac{\pi}{4}=\frac{\sqrt{3}}{2}\cdot\frac{\sqrt{2}}{2}+\frac{1}{2}\cdot\frac{\sqrt{2}}{2}=\frac{\sqrt{6}}{4}+\frac{\sqrt{2}}{4}=\frac{\sqrt{6}+\sqrt{2}}{4}$$

11. $\cos 40°\cos 50°-\sin 40°\sin 50°=\cos(40°+50°)=\cos 90°=0$

12. $\cos\frac{7\pi}{9}\cos\frac{2\pi}{9}-\sin\frac{7\pi}{9}\sin\frac{2\pi}{9}=\cos\left(\frac{7\pi}{9}+\frac{2\pi}{9}\right)=\cos\pi=-1$

13. The answer to Exercise 11 is 0. Using a calculator to evaluate $\cos 40°\cos 50°-\sin 40°\sin 50°$ also gives a value of 0.

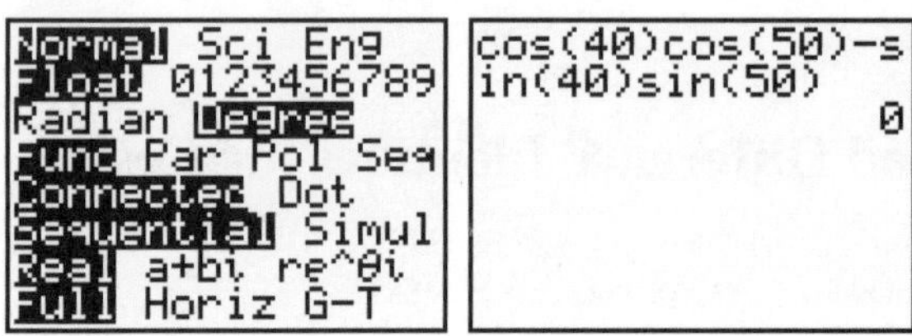

14. The answer to Exercise 12 is -1. Using a calculator to evaluate $\cos\frac{7\pi}{9}\cos\frac{2\pi}{9}-\sin\frac{7\pi}{9}\sin\frac{2\pi}{9}$ also gives a value of -1.

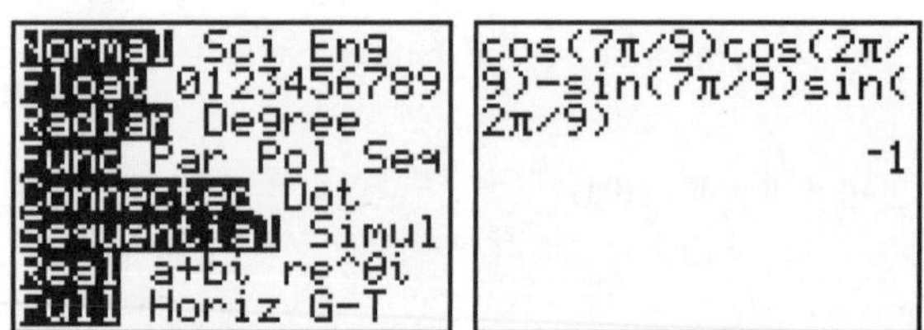

15. $\tan 87° = \cot(90° - 87°) = \cot 3°$

16. $\sin 15° = \cos(90° - 15°) = \cos 75°$

17. $\cos\frac{\pi}{12}=\sin\left(\frac{\pi}{2}-\frac{\pi}{12}\right)=\sin\frac{5\pi}{12}$

18. $\sin\frac{2\pi}{5}=\cos\left(\frac{\pi}{2}-\frac{2\pi}{5}\right)=\cos\frac{\pi}{10}$

19. $\csc(-14°24')=\sec\left[90°-(-14°24')\right]=\sec(90°+14°24')=\sec 104°24'$

20. $\sin 142°14'=\cos(90°-142°14')=\cos\left[-(142°14'-90°)\right]=\cos(-52°14')$

21. $\sin\frac{5\pi}{8}=\cos\left(\frac{\pi}{2}-\frac{5\pi}{8}\right)=\cos\left(\frac{4\pi}{8}-\frac{5\pi}{8}\right)=\cos\left(-\frac{\pi}{8}\right)$

22. $\cot\frac{9\pi}{10}=\tan\left(\frac{\pi}{2}-\frac{9\pi}{10}\right)=\tan\left(\frac{5\pi}{10}-\frac{9\pi}{10}\right)=\tan\left(-\frac{4\pi}{10}\right)=\tan\left(-\frac{2\pi}{5}\right)$

23. $\sec 146°42'=\csc(90°-146°42')=\csc\left[-(146°42'-90°)\right]=\csc(-56°42')$

24. $\tan 174°3' = \cot(90° - 174°3') = \cot[-(174°3'° - 90)] = \cot(-84°3')$

25. $\cot 176.9814° = \tan(90° - 176.9814°) = \tan[-(176.9814° - 90°)] = \tan(-86.9814°)$

26. $\sin 98.0142° = \cos(90° - 98.0142°) = \cos[-(98.0142° - 90°)] = \cos(-8.0142°)$

27. Since $\frac{\pi}{6} = \frac{\pi}{2} - \frac{\pi}{3}$, $\cot\frac{\pi}{3} = \tan\frac{\pi}{6}$.

$\cot\frac{\pi}{3} = \underline{\tan}\,\frac{\pi}{6}$

28. Since $-\frac{\pi}{6} = \frac{\pi}{2} - \frac{2\pi}{3}$, $\sin\frac{2\pi}{3} = \cos\left(\frac{\pi}{2} - \frac{2\pi}{3}\right) = \cos\left(-\frac{\pi}{6}\right)$.

$\sin\frac{2\pi}{3} = \underline{\cos}\left(-\frac{\pi}{6}\right)$

29. Since $90° - 57° = 33°$, $\sin 57° = \cos 33°$.

$\sin 57° = \underline{\cos} 33°$

30. Since $90° - 18° = 72°$, $\cot 18° = \tan 72°$.

$\underline{\tan} 72° = \cot 18°$

31. Since $90° - 70° = 20°$ and $\sin x = \frac{1}{\csc x}$, $\cos 70° = \frac{1}{\csc 20°}$.

$\cos 70° = \frac{1}{\underline{\csc} 20°}$

32. Since $90° - 24° = 66°$ and $\cot x = \frac{1}{\tan x}$, $\tan 24° = \frac{1}{\tan 66°}$.

$\tan 24° = \frac{1}{\underline{\tan} 66°}$

33. $\tan\theta = \cot(45° + 2\theta)$
Since $\tan\theta = \cot(90° - \theta)$, $90° - \theta = 45° + 2\theta \Rightarrow 90° = 45° + 3\theta \Rightarrow 3\theta = 45° \Rightarrow \theta = 15°$.

34. $\sin\theta = \cos(2\theta - 10°)$

Since $\sin\theta = \cos(90° - \theta)$, $90° - \theta = 2\theta - 10° \Rightarrow 90° = 3\theta - 10° \Rightarrow 3\theta = 100° \Rightarrow \theta = \frac{100°}{3}$.

35. $\sec\theta = \csc\left(\frac{\theta}{2} + 20°\right)$

By a cofunction identity, $\sec\theta = \csc(90° - \theta)$. Thus, $\csc\left(\frac{\theta}{2} + 20°\right) = \csc(90° - \theta)$ and we can solve the following.

$$\csc\left(\frac{\theta}{2} + 20°\right) = \csc(90° - \theta) \Rightarrow \frac{\theta}{2} + 20° = 90° - \theta \Rightarrow \frac{3\theta}{2} + 20° = 90° \Rightarrow \frac{3\theta}{2} = 70° \Rightarrow \theta = \frac{2}{3}(70°) = \frac{140°}{3}$$

36. $\cos\theta = \sin(3\theta + 10°)$

By a cofunction identity, $\cos\theta = \sin(90° - \theta)$. Thus, $\sin(90° - \theta) = \sin(3\theta + 10°)$ and we can solve the following.

$$\sin(90° - \theta) = \sin(3\theta + 10°) \Rightarrow 90° - \theta = 3\theta + 10° \Rightarrow 90° = 4\theta + 10° \Rightarrow 4\theta = 80° \Rightarrow \theta = 20°$$

37. $\sin(3\theta - 15°) = \cos(\theta + 25°)$

Since $\sin\theta = \cos(90° - \theta)$, we have the following.

$$\sin(3\theta - 15°) = \cos[90° - (3\theta - 15°)] = \cos(90° - 3\theta + 15°) = \cos(105° - 3\theta)$$

We can next solve the equation $\cos(105° - 3\theta) = \cos(\theta + 25°)$.

$$\cos(105° - 3\theta) = \cos(\theta + 25°) \Rightarrow 105° - 3\theta = \theta + 25° \Rightarrow 105° = 4\theta + 25° \Rightarrow 4\theta = 80° \Rightarrow \theta = 20°$$

38. $\cot(\theta - 10°) = \tan(2\theta + 20°)$

Since $\cot\theta = \tan(90° - \theta)$, we have the following.

$$\cot(\theta - 10°) = \tan[90° - (\theta - 10°)] = \tan(90° - \theta + 10°) = \tan(100° - \theta)$$

We can next solve the equation $\tan(100° - \theta) = \tan(2\theta + 20°)$.

$$\tan(100° - \theta) = \tan(2\theta + 20°) \Rightarrow 100° - \theta = 2\theta + 20° \Rightarrow 100° = 3\theta + 20° \Rightarrow 3\theta = 80° \Rightarrow \theta = \frac{80°}{3}$$

39. $\cos(0° - \theta) = \cos 0° \cos\theta + \sin\theta \sin 0° = (1)\cos\theta + (0)\sin\theta = \cos\theta + 0 = \cos\theta$

40. $\cos(90° - \theta) = \cos 90° \cos\theta + \sin 90° \sin\theta = (0)\cos\theta + (1)\sin\theta = 0 + \sin\theta = \sin\theta$

41. $\cos(180° - \theta) = \cos 180° \cos\theta + \sin 180° \sin\theta = (-1)\cos\theta + (0)\sin\theta = -\cos\theta + 0 = -\cos\theta$

42. $\cos(270° - \theta) = \cos 270° \cos\theta + \sin 270° \sin\theta = (0)\cos\theta + (-1)\sin\theta = 0 - \sin\theta = -\sin\theta$

43. $\cos(0° + \theta) = \cos 0° \cos\theta - \sin\theta \sin 0° = (1)\cos\theta - (0)\sin\theta = \cos\theta - 0 = \cos\theta$

44. $\cos(90° + \theta) = \cos 90° \cos\theta - \sin 90° \sin\theta = (0)\cos\theta - (1)\sin\theta = 0 - \sin\theta = -\sin\theta$

45. $\cos(180° + \theta) = \cos 180° \cos\theta - \sin 180° \sin\theta = (-1)\cos\theta - (0)\sin\theta = -\cos\theta - 0 = -\cos\theta$

46. $\cos(270° + \theta) = \cos 270° \cos\theta - \sin 270° \sin\theta = (0)\cos\theta - (-1)\sin\theta = 0 + \sin\theta = \sin\theta$

47. $\cos s = -\frac{1}{5}$ and $\sin t = \frac{3}{5}$, s and t are in quadrant II.

Since $\cos s = \frac{x}{r}$, we have $\cos s = -\frac{1}{5} = \frac{-1}{5}$. Thus, we let $x = -1$ and $r = 5$. Substituting into the Pythagorean theorem, we get the following.

$$(-1)^2 + y^2 = 5^2 \Rightarrow 1 + y^2 = 25 \Rightarrow y^2 = 24 \Rightarrow y = \sqrt{24}, \text{ since } \sin s > 0$$

Thus, $\sin s = \frac{y}{r} = \frac{\sqrt{24}}{5}$.

Since $\sin t = \frac{y}{r}$, we have $\sin t = \frac{3}{5}$. Thus, we let $y = 3$ and $r = 5$. Substituting into the Pythagorean theorem, we get the following.

$$x^2 + 3^2 = 5^2 \Rightarrow x^2 + 9 = 25 \Rightarrow x^2 = 16 \Rightarrow x = -4, \text{ since } \cos t < 0$$

Thus, $\cos t = \frac{x}{r} = \frac{-4}{5} = -\frac{4}{5}$.

$$\cos(s+t) = \cos s \cos t - \sin s \sin t = \left(-\frac{1}{5}\right)\left(-\frac{4}{5}\right) - \left(\frac{\sqrt{24}}{5}\right)\left(\frac{3}{5}\right) = \frac{4}{25} - \frac{3\sqrt{24}}{25} = \frac{4}{25} - \frac{6\sqrt{6}}{25} = \frac{4-6\sqrt{6}}{25}$$

$$\cos(s-t) = \cos s \cos t + \sin s \sin t = \left(-\frac{1}{5}\right)\left(-\frac{4}{5}\right) + \left(\frac{\sqrt{24}}{5}\right)\left(\frac{3}{5}\right) = \frac{4}{25} + \frac{3\sqrt{24}}{25} = \frac{4+6\sqrt{6}}{25}$$

48. $\sin s = \frac{2}{3}$ and $\sin t = -\frac{1}{3}$, s is in quadrant II and t is in quadrant IV

Since $\sin s = \frac{y}{r}$, we have $\sin s = \frac{2}{3}$. Thus, we let $y = 2$ and $r = 3$. Substituting into the Pythagorean theorem, we get the following.

$$x^2 + 2^2 = 3^2 \Rightarrow x^2 + 4 = 9 \Rightarrow x^2 = 5 \Rightarrow x = -\sqrt{5}, \text{ since } \cos s < 0$$

Thus, $\cos s = \frac{x}{r} = \frac{-\sqrt{5}}{3} = -\frac{\sqrt{5}}{3}$.

Since $\sin t = \frac{y}{r}$, we have $\sin t = -\frac{1}{3} = \frac{-1}{3}$. Thus, we let $y = -1$ and $r = 3$. Substituting into the Pythagorean theorem, we get the following.

$$x^2 + (-1)^2 = 3^2 \Rightarrow x^2 + 1 = 9 \Rightarrow x^2 = 8 \Rightarrow x = \sqrt{8} = 2\sqrt{2}, \text{ since } \cos t > 0$$

Thus, $\cos t = \frac{x}{r} = \frac{2\sqrt{2}}{3}$.

$$\cos(s+t) = \cos s \cos t - \sin s \sin t = \left(-\frac{\sqrt{5}}{3}\right)\left(\frac{2\sqrt{2}}{3}\right) - \left(\frac{2}{3}\right)\left(-\frac{1}{3}\right) = \frac{-2\sqrt{10}}{9} + \frac{2}{9} = \frac{-2\sqrt{10}+2}{9}$$

$$\cos(s-t) = \cos s \cos t + \sin s \sin t = \left(-\frac{\sqrt{5}}{3}\right)\left(\frac{2\sqrt{2}}{3}\right) + \left(\frac{2}{3}\right)\left(-\frac{1}{3}\right) = \frac{-2\sqrt{10}}{9} - \frac{2}{9} = \frac{-2\sqrt{10}-2}{9}$$

49. $\sin s=\frac{3}{5}$ and $\sin t=-\frac{12}{13}$, s is in quadrant I and t is in quadrant III

Since $\sin s=\frac{y}{r}$, we have $\sin s=\frac{3}{5}$. Thus, we let $y=3$ and $r=5$. Substituting into the Pythagorean theorem, we get the following.

$$x^2+3^2=5^2 \Rightarrow x^2+9=25 \Rightarrow x^2=16 \Rightarrow x=4, \text{ since } \cos s>0$$

Thus, $\cos s=\frac{x}{r}=\frac{4}{5}$.

Since $\sin t=\frac{y}{r}$, we have $\sin t=-\frac{12}{13}=\frac{-12}{13}$. Thus, we let $y=-12$ and $r=13$. Substituting into the Pythagorean theorem, we get the following.

$$x^2+(-12)^2=13^2 \Rightarrow x^2+144=169 \Rightarrow x^2=25 \Rightarrow x=-5, \text{ since } \cos t<0$$

Thus, $\cos t=\frac{x}{r}=\frac{-5}{13}=-\frac{5}{13}$.

$$\cos(s+t)=\cos s\cos t-\sin s\sin t=\left(\frac{4}{5}\right)\left(-\frac{5}{13}\right)-\left(\frac{3}{5}\right)\left(-\frac{12}{13}\right)=-\frac{20}{65}+\frac{36}{65}=\frac{16}{65}$$

$$\cos(s-t)=\cos s\cos t+\sin s\sin t=\left(\frac{4}{5}\right)\left(-\frac{5}{13}\right)+\left(\frac{3}{5}\right)\left(-\frac{12}{13}\right)=-\frac{20}{65}-\frac{36}{65}=-\frac{56}{65}$$

50. $\cos s=-\frac{8}{17}$ and $\cos t=-\frac{3}{5}$, s and t are in quadrant III

Since $\cos s=\frac{x}{r}$, we have $\cos s=-\frac{8}{17}=\frac{-8}{17}$. Thus, we let $x=-8$ and $r=17$. Substituting into the Pythagorean theorem, we get the following.

$$(-8)^2+y^2=17^2 \Rightarrow 64+y^2=289 \Rightarrow y^2=225 \Rightarrow y=-15, \text{ since } \sin s<0$$

Thus, $\sin s=\frac{y}{r}=\frac{-15}{17}=-\frac{15}{17}$.

Since $\cos t=\frac{x}{r}$, we have $\cos t=-\frac{3}{5}=\frac{-3}{5}$. Thus, we let $x=-3$ and $r=5$. Substituting into the Pythagorean theorem, we get the following.

$$(-3)^2+y^2=5^2 \Rightarrow 9+y^2=25 \Rightarrow y^2=16 \Rightarrow y=-4, \text{ since } \sin t<0$$

Thus, $\sin s=\frac{y}{r}=\frac{-4}{5}=-\frac{4}{5}$.

$$\cos(s+t)=\cos s\cos t-\sin s\sin t=\left(-\frac{8}{17}\right)\left(-\frac{3}{5}\right)-\left(-\frac{15}{17}\right)\left(-\frac{4}{5}\right)=\frac{24}{85}-\frac{60}{85}=-\frac{36}{85}$$

$$\cos(s-t)=\cos s\cos t+\sin s\sin t=\left(-\frac{8}{17}\right)\left(-\frac{3}{5}\right)+\left(-\frac{15}{17}\right)\left(-\frac{4}{5}\right)=\frac{24}{85}+\frac{60}{85}=\frac{84}{85}$$

51. $\sin s = \frac{\sqrt{5}}{7}$ and $\sin t = \frac{\sqrt{6}}{8}$, s and t are in quadrant I.

Since $\sin s = \frac{y}{r}$, we have $\sin s = \frac{\sqrt{5}}{7}$. Thus, we let $y = \sqrt{5}$ and $r = 7$. Substituting into the Pythagorean theorem, we get the following.

$$x^2 + \left(\sqrt{5}\right)^2 = 7^2 \Rightarrow x^2 + 5 = 49 \Rightarrow x^2 = 44 \Rightarrow x = \sqrt{44}, \text{ since } \cos s > 0$$

Thus, $\cos s = \frac{x}{r} = \frac{\sqrt{44}}{7}$.

Since $\sin t = \frac{y}{r}$, we have $\sin t = \frac{\sqrt{6}}{8}$. Thus, we let $y = \sqrt{6}$ and $r = 8$. Substituting into the Pythagorean theorem, we get the following.

$$x^2 + \left(\sqrt{6}\right)^2 = 8^2 \Rightarrow x^2 + 6 = 64 \Rightarrow x^2 = 58 \Rightarrow x = \sqrt{58}, \text{ since } \cos t > 0$$

Thus, $\cos t = \frac{x}{r} = \frac{\sqrt{58}}{8}$.

$$\cos(s+t) = \cos s \cos t - \sin s \sin t = \left(\frac{\sqrt{44}}{7}\right)\left(\frac{\sqrt{58}}{8}\right) - \left(\frac{\sqrt{5}}{7}\right)\left(\frac{\sqrt{6}}{8}\right) = \frac{2\sqrt{638}}{56} - \frac{\sqrt{30}}{56} = \frac{2\sqrt{638} - \sqrt{30}}{56}$$

$$\cos(s-t) = \cos s \cos t + \sin s \sin t = \left(\frac{\sqrt{44}}{7}\right)\left(\frac{\sqrt{58}}{8}\right) + \left(\frac{\sqrt{5}}{7}\right)\left(\frac{\sqrt{6}}{8}\right) = \frac{2\sqrt{638}}{56} + \frac{\sqrt{30}}{56} = \frac{2\sqrt{638} + \sqrt{30}}{56}$$

52. $\cos s = \frac{\sqrt{2}}{4}$ and $\sin t = -\frac{\sqrt{5}}{6}$, s and t are in quadrant IV.

Since $\cos s = \frac{x}{r}$, we have $\cos s = \frac{\sqrt{2}}{4}$. Thus, we let $x = \sqrt{2}$ and $r = 4$. Substituting into the Pythagorean theorem, we get the following.

$$\left(\sqrt{2}\right)^2 + y^2 = 4^2 \Rightarrow 2 + y^2 = 16 \Rightarrow y^2 = 14 \Rightarrow y = -\sqrt{14}, \text{ since } \sin s < 0$$

Thus, $\sin s = \frac{y}{r} = \frac{-\sqrt{14}}{4} = -\frac{\sqrt{14}}{4}$.

Since $\sin t = \frac{y}{r}$, we have $\sin t = -\frac{\sqrt{5}}{6} = \frac{-\sqrt{5}}{6}$. Thus, we let $y = -\sqrt{5}$ and $r = 6$. Substituting into the Pythagorean theorem, we get the following.

$$x^2 + \left(-\sqrt{5}\right)^2 = 6^2 \Rightarrow x^2 + 5 = 36 \Rightarrow x^2 = 31 \Rightarrow x = \sqrt{31}, \text{ since } \cos t > 0$$

Thus, $\cos t = \frac{x}{r} = \frac{\sqrt{31}}{6}$.

$$\cos(s+t) = \cos s \cos t - \sin s \sin t = \left(\frac{\sqrt{2}}{4}\right)\left(\frac{\sqrt{31}}{6}\right) - \left(-\frac{\sqrt{14}}{4}\right)\left(-\frac{\sqrt{5}}{6}\right) = \frac{\sqrt{62}}{24} - \frac{\sqrt{70}}{24} = \frac{\sqrt{62} - \sqrt{70}}{24}$$

$$\cos(s-t) = \cos s \cos t + \sin s \sin t = \left(\frac{\sqrt{2}}{4}\right)\left(\frac{\sqrt{31}}{6}\right) + \left(-\frac{\sqrt{14}}{4}\right)\left(-\frac{\sqrt{5}}{6}\right) = \frac{\sqrt{62}}{24} + \frac{\sqrt{70}}{24} = \frac{\sqrt{62} + \sqrt{70}}{24}$$

53. True or false: $\cos 42° = \cos(30° + 12°)$

Since 42° = 30° +12°, the given statement is true.

54. True or false: $\cos(-24°) = \cos 16° - \cos 40°$

Since $\cos(-24°) = \cos(16° - 40°) = \cos 16° \cos 40° + \sin 16° \sin 40° \neq \cos 16° - \cos 40°$, the given statement is false.

55. True or false: $\cos 74° = \cos 60° \cos 14° + \sin 60° \sin 14°$

Since $\cos 74° = \cos(60° + 14°) = \cos 60° \cos 14° - \sin 60° \sin 14° \neq \cos 60° \cos 14° + \sin 60° \sin 14°$, the given statement is false.

56. True or false: $\cos 140° = \cos 60° \cos 80° - \sin 60° \sin 80°$

Since $\cos 140° = \cos(60° + 80°) = \cos 60° \cos 80° - \sin 60° \sin 80°$, the given statement is true.

57. True or false: $\cos\frac{\pi}{3} = \cos\frac{\pi}{12}\cos\frac{\pi}{4} - \sin\frac{\pi}{12}\sin\frac{\pi}{4}$.

Since $\frac{\pi}{3} = \frac{4\pi}{12} = \frac{\pi}{12} + \frac{3\pi}{12} = \frac{\pi}{12} + \frac{\pi}{4}$, $\cos\frac{\pi}{3} = \cos\left(\frac{\pi}{12} + \frac{\pi}{4}\right) = \cos\frac{\pi}{12}\cos\frac{\pi}{4} - \sin\frac{\pi}{12}\sin\frac{\pi}{4}$.

The given statement is true.

58. True or false: $\cos\frac{2\pi}{3} = \cos\frac{11\pi}{12}\cos\frac{\pi}{4} + \sin\frac{11\pi}{12}\sin\frac{\pi}{4}$

Since $\frac{2\pi}{3} = \frac{8\pi}{12} = \frac{11\pi}{12} - \frac{3\pi}{12} = \frac{11\pi}{12} - \frac{\pi}{4}$, $\cos\frac{2\pi}{3} = \cos\left(\frac{11\pi}{12} - \frac{\pi}{4}\right) = \cos\frac{11\pi}{12}\cos\frac{\pi}{4} + \sin\frac{11\pi}{12}\sin\frac{\pi}{4}$.

The given statement is true.

59. True or false: $\cos 70° \cos 20° - \sin 70° \sin 20° = 0$

Since $\cos 70° \cos 20° - \sin 70° \sin 20° = \cos(70° + 20°) = \cos 90° = 0$, the given statement is true.

60. True or false: $\cos 85° \cos 40° + \sin 85° \sin 40° = \frac{\sqrt{2}}{2}$

Since $\cos 85° \cos 40° + \sin 85° \sin 40° = \cos(85° - 40°) = \cos 45° = \frac{\sqrt{2}}{2}$, the given statement is true.

61. True or false: $\tan\left(\theta - \frac{\pi}{2}\right) = \cot\theta$

Since $\tan\left(\theta - \frac{\pi}{2}\right) = -\tan\left[-\left(\theta - \frac{\pi}{2}\right)\right] = -\tan\left(\frac{\pi}{2} - \theta\right) = -\cot\theta \neq \cot\theta$, the given statement is false.

62. True or false: $\sin\left(\theta - \frac{\pi}{2}\right) = \cos\theta$

Since $\sin\left(\theta - \frac{\pi}{2}\right) = -\sin\left[-\left(\theta - \frac{\pi}{2}\right)\right] = -\sin\left(\frac{\pi}{2} - \theta\right) = -\cos\theta \neq \cos\theta$, the given statement is false.

63. Verify $\cos\left(\dfrac{\pi}{2}+x\right)=-\sin x$.

$$\cos\left(\frac{\pi}{2}+x\right)=\cos\frac{\pi}{2}\cos x-\sin\frac{\pi}{2}\sin x=(0)\cos x-(1)\sin x=0-\sin x=-\sin x$$

64. Verify $\sec(\pi-x)=-\sec x$

$$\sec(\pi-x)=\frac{1}{\cos(\pi-x)}=\frac{1}{\cos\pi\cos x+\sin\pi\sin x}$$
$$=\frac{1}{(-1)\cos x+(0)\sin x}=\frac{1}{-\cos x+0}=\frac{1}{-\cos x}=-\frac{1}{\cos x}=-\sec x$$

65. Verify $\cos 2x=\cos^2 x-\sin^2 x$

$$\cos 2x=\cos(x+x)=\cos x\cos x-\sin x\sin x=\cos^2 x-\sin^2 x$$

66. Verify $1+\cos 2x-\cos^2 x=\cos^2 x$.

From Exercise 65, $\cos 2x=\cos^2 x-\sin^2 x$.

$$1+\cos 2x-\cos^2 x=1+\left(\cos^2 x-\sin^2 x\right)-\cos^2 x=1-\sin^2 x=\cos^2 x$$

67. $\cos 195°=\cos(180°+15°)$
$$=\cos 180°\cos 15°-\sin 180°\sin 15°=(-1)\cos 15°-(0)\sin 15°=-\cos 15°-0=-\cos 15°$$

68. $-\cos 15°=-\cos(45°-30°)$
$$=-(\cos 45°\cos 30°+\sin 45°\sin 30°)=-\left(\frac{\sqrt{2}}{2}\cdot\frac{\sqrt{3}}{2}+\frac{\sqrt{2}}{2}\cdot\frac{1}{2}\right)=-\left(\frac{\sqrt{6}}{4}+\frac{\sqrt{2}}{4}\right)=\frac{-\sqrt{6}-\sqrt{2}}{4}$$

69. $\cos 195°=-\cos 15°=\dfrac{-\sqrt{6}-\sqrt{2}}{4}$

70. (a) $\cos 255°=\cos(180°+75°)=\cos 180°\cos 75°-\sin 180°\sin 75°=(-1)\cos 75°-(0)\sin 75°$
$$=-\cos 75°=-\cos(45°+30°)=-(\cos 45°\cos 30°-\sin 45°\sin 30°)$$
$$=-\left(\frac{\sqrt{2}}{2}\cdot\frac{\sqrt{3}}{2}-\frac{\sqrt{2}}{2}\cdot\frac{1}{2}\right)=-\left(\frac{\sqrt{6}}{4}-\frac{\sqrt{2}}{4}\right)=-\left(\frac{\sqrt{6}-\sqrt{2}}{4}\right)=\frac{\sqrt{2}-\sqrt{6}}{4}$$

(b) $\cos\dfrac{11\pi}{12}=\cos\left(\dfrac{12\pi}{12}-\dfrac{\pi}{12}\right)=\cos\left(\pi-\dfrac{\pi}{12}\right)=\cos\pi\cos\dfrac{\pi}{12}+\sin\pi\sin\dfrac{\pi}{12}=(-1)\cos\dfrac{\pi}{12}+(0)\sin\dfrac{\pi}{12}$
$$=-\cos\frac{\pi}{12}=-\cos\left(\frac{3\pi}{12}-\frac{2\pi}{12}\right)=-\cos\left(\frac{\pi}{4}-\frac{\pi}{6}\right)=-\left(\cos\frac{\pi}{4}\cos\frac{\pi}{6}+\sin\frac{\pi}{4}\sin\frac{\pi}{6}\right)$$
$$=-\left(\frac{\sqrt{2}}{2}\cdot\frac{\sqrt{3}}{2}+\frac{\sqrt{2}}{2}\cdot\frac{1}{2}\right)=-\left(\frac{\sqrt{6}}{4}+\frac{\sqrt{2}}{4}\right)=-\left(\frac{\sqrt{6}+\sqrt{2}}{4}\right)=\frac{-\sqrt{6}-\sqrt{2}}{4}$$

71. (a) Since there are 60 cycles per sec, the number of cycles in .05 sec is given by the following.

$$(.05 \text{ sec})(60 \text{ cycles per sec}) = 3 \text{ cycles}$$

(b) Since $V = 163 \sin \omega t$ and the maximum value of $\sin \omega t$ is 1, the maximum voltage is 163. Since $V = 163 \sin \omega t$ and the minimum value of $\sin \omega t$ is -1, the minimum voltage is -163. Therefore, the voltage is not always equal to 115.

72. (a) Graph 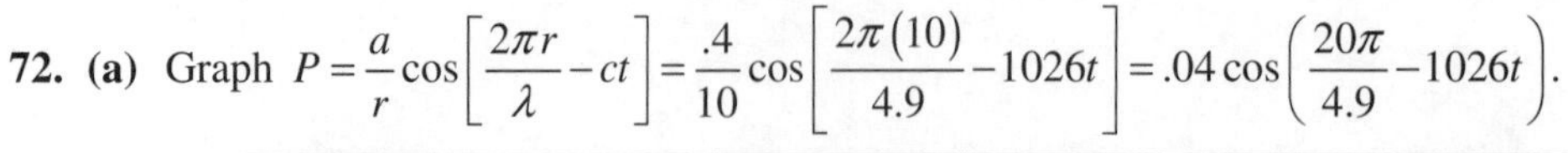 $P = \frac{a}{r}\cos\left[\frac{2\pi r}{\lambda} - ct\right] = \frac{.4}{10}\cos\left[\frac{2\pi(10)}{4.9} - 1026t\right] = .04\cos\left(\frac{20\pi}{4.9} - 1026t\right)$.

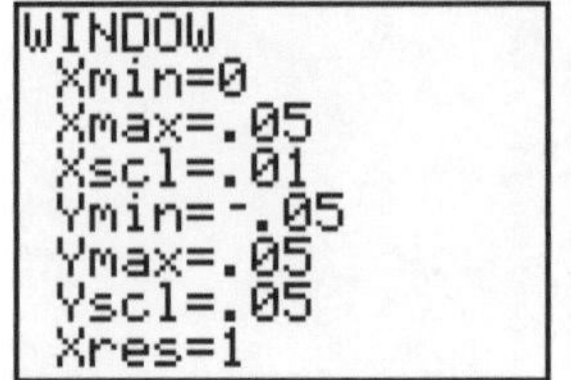

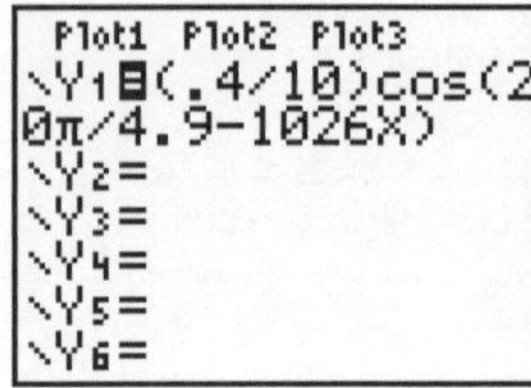

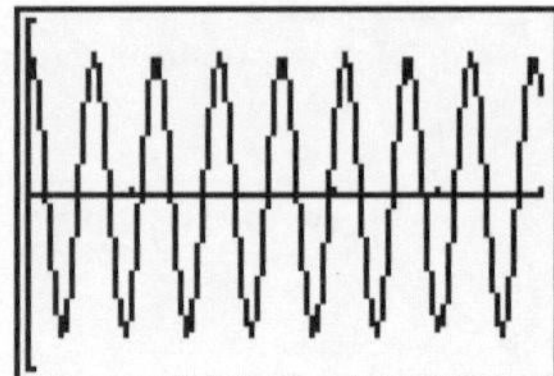

The pressure P is oscillating.

(b) Graph $P = \frac{a}{r}\cos\left[\frac{2\pi r}{\lambda} - ct\right] = \frac{3}{r}\cos\left[\frac{2\pi(r)}{4.9} - 1026(10)\right] = \frac{3}{r}\cos\left(\frac{2\pi r}{4.9} - 10,260\right)$.

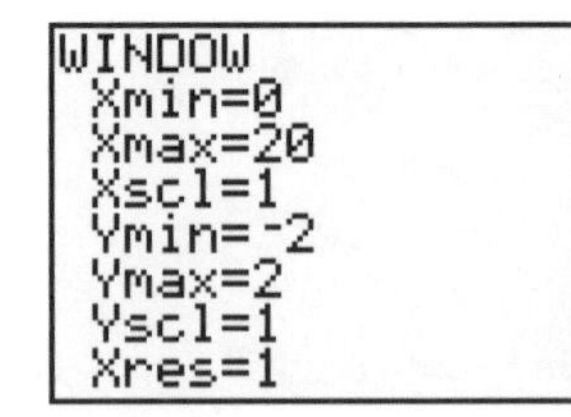

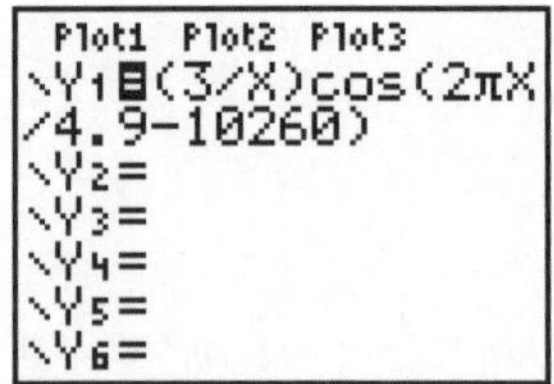

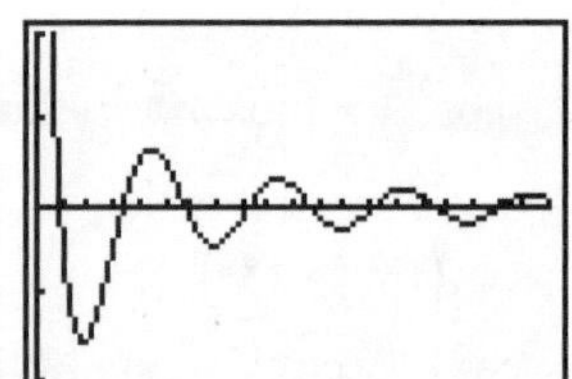

The pressure oscillates, and amplitude decreases as r increases.

(c) $P = \frac{a}{r}\cos\left[\frac{2\pi r}{\lambda} - ct\right]$

Let $r = n\lambda$.

$$P = \frac{a}{r}\cos\left[\frac{2\pi r}{\lambda} - ct\right] = \frac{a}{n\lambda}\cos\left[\frac{2\pi n\lambda}{\lambda} - ct\right] = \frac{a}{n\lambda}\cos[2\pi n - ct]$$

$$= \frac{a}{n\lambda}\left[\cos(2\pi n)\cos(ct) + \sin(2\pi n)\sin(ct)\right] = \frac{a}{n\lambda}\left[(1)\cos(ct) + (0)\sin(ct)\right]$$

$$= \frac{a}{n\lambda}\left[\cos(ct) + 0\right] = \frac{a}{n\lambda}\cos(ct)$$

Section 5.4: Sum and Difference Identities for Sine and Tangent

1. – 2. Answers will vary.

3. Since we have the following, the correct choice is C.

$$\sin 15° = \sin(45° - 30°) = \sin 45° \cos 30° - \cos 45° \sin 30° = \frac{\sqrt{2}}{2}\cdot\frac{\sqrt{3}}{2} - \frac{\sqrt{2}}{2}\cdot\frac{1}{2} = \frac{\sqrt{6}}{4} - \frac{\sqrt{2}}{4} = \frac{\sqrt{6} - \sqrt{2}}{4}$$

4. Since we have the following, the correct choice is A.

$$\sin 105° = \sin(60° + 45°) = \sin 60° \cos 45° + \cos 60° \sin 45° = \frac{\sqrt{3}}{2}\cdot\frac{\sqrt{2}}{2} + \frac{1}{2}\cdot\frac{\sqrt{2}}{2} = \frac{\sqrt{6}}{4} + \frac{\sqrt{2}}{4} = \frac{\sqrt{6} + \sqrt{2}}{4}$$

5. Since we have the following, the correct choice is E.

$$\tan 15° = \tan(60° - 45°) = \frac{\tan 60° - \tan 45°}{1 + \tan 60° \tan 45°} = \frac{\sqrt{3} - 1}{1 + \sqrt{3}(1)} = \frac{\sqrt{3} - 1}{1 + \sqrt{3}} \cdot \frac{1 - \sqrt{3}}{1 - \sqrt{3}}$$

$$= \frac{\sqrt{3} - 3 - 1 + \sqrt{3}}{1 - 3} = \frac{-4 + 2\sqrt{3}}{-2} = 2 - \sqrt{3}$$

6. Since we have the following, the correct choice is F.

$$\tan 105° = \tan(60° + 45°) = \frac{\tan 60° + \tan 45°}{1 - \tan 60° \tan 45°} = \frac{\sqrt{3} + 1}{1 - \sqrt{3}(1)} = \frac{\sqrt{3} + 1}{1 - \sqrt{3}} \cdot \frac{1 + \sqrt{3}}{1 + \sqrt{3}}$$

$$= \frac{(1 + \sqrt{3})^2}{1^2 - (\sqrt{3})^2} = \frac{1 + 2\sqrt{3} + 3}{1 - 3} = \frac{4 + 2\sqrt{3}}{-2} = -2 - \sqrt{3}$$

7. Since we have the following, the correct choice is B.

$$\sin(-105°) = \sin(45° - 150°) = \sin 45° \cos 150° - \cos 45° \sin 150°$$

$$= \left(\frac{\sqrt{2}}{2}\right)\left(-\frac{\sqrt{3}}{2}\right) - \left(\frac{\sqrt{2}}{2}\right)\left(\frac{1}{2}\right) = -\frac{\sqrt{6}}{4} - \frac{\sqrt{2}}{4} = \frac{-\sqrt{6} - \sqrt{2}}{4}$$

8. Since we have the following, the correct choice is D.

$$\tan(-105°) = -\tan 105° = -\tan(60° + 45°) = -\frac{\tan 60° + \tan 45°}{1 - \tan 60° \tan 45°} = -\frac{\sqrt{3} + 1}{1 - \sqrt{3}(1)} = -\frac{\sqrt{3} + 1}{1 - \sqrt{3}} \cdot \frac{1 + \sqrt{3}}{1 + \sqrt{3}}$$

$$= -\frac{(1 + \sqrt{3})^2}{1^2 - (\sqrt{3})^2} = -\frac{1 + 2\sqrt{3} + 3}{1 - 3} = -\frac{4 + 2\sqrt{3}}{-2} = 2 + \sqrt{3}$$

9. $$\sin\frac{5\pi}{12} = \sin\left(\frac{\pi}{4} + \frac{\pi}{6}\right) = \sin\frac{\pi}{4}\cos\frac{\pi}{6} + \cos\frac{\pi}{4}\sin\frac{\pi}{6} = \frac{\sqrt{2}}{2} \cdot \frac{\sqrt{3}}{2} + \frac{\sqrt{2}}{2} \cdot \frac{1}{2} = \frac{\sqrt{6}}{4} + \frac{\sqrt{2}}{4} = \frac{\sqrt{6} + \sqrt{2}}{4}$$

10. $$\tan\frac{5\pi}{12} = \tan\left(\frac{\pi}{6} + \frac{\pi}{4}\right) = \frac{\tan\frac{\pi}{6} + \tan\frac{\pi}{4}}{1 - \tan\frac{\pi}{6}\tan\frac{\pi}{4}} = \frac{\frac{\sqrt{3}}{3} + 1}{1 - \frac{\sqrt{3}}{3}} = \frac{\frac{\sqrt{3}}{3} + 1}{1 - \frac{\sqrt{3}}{3}} \cdot \frac{3}{3} = \frac{\sqrt{3} + 3}{3 - \sqrt{3}}$$

$$= \frac{\sqrt{3} + 3}{3 - \sqrt{3}} \cdot \frac{3 + \sqrt{3}}{3 + \sqrt{3}} = \frac{(3 + \sqrt{3})^2}{3^2 - (\sqrt{3})^2} = \frac{9 + 6\sqrt{3} + 3}{9 - 3} = \frac{12 + 6\sqrt{3}}{6} = 2 + \sqrt{3}$$

11. $$\tan\frac{\pi}{12} = \tan\left(\frac{\pi}{4} - \frac{\pi}{6}\right) = \frac{\tan\frac{\pi}{4} - \tan\frac{\pi}{6}}{1 + \tan\frac{\pi}{4}\tan\frac{\pi}{6}} = \frac{1 - \frac{\sqrt{3}}{3}}{1 + \frac{\sqrt{3}}{3}} = \frac{1 - \frac{\sqrt{3}}{3}}{1 + \frac{\sqrt{3}}{3}} \cdot \frac{3}{3} = \frac{3 - \sqrt{3}}{3 + \sqrt{3}}$$

$$= \frac{3 - \sqrt{3}}{3 + \sqrt{3}} \cdot \frac{3 - \sqrt{3}}{3 - \sqrt{3}} = \frac{(3 - \sqrt{3})^2}{3^2 - (\sqrt{3})^2} = \frac{9 - 6\sqrt{3} + 3}{9 - 3} = \frac{12 - 6\sqrt{3}}{6} = 2 - \sqrt{3}$$

12. $\sin\frac{\pi}{12}=\sin\left(\frac{\pi}{4}-\frac{\pi}{6}\right)=\sin\frac{\pi}{4}\cos\frac{\pi}{6}-\cos\frac{\pi}{4}\sin\frac{\pi}{6}=\frac{\sqrt{2}}{2}\cdot\frac{\sqrt{3}}{2}-\frac{\sqrt{2}}{2}\cdot\frac{1}{2}=\frac{\sqrt{6}}{4}-\frac{\sqrt{2}}{4}=\frac{\sqrt{6}-\sqrt{2}}{4}$

13. $\sin\left(-\frac{7\pi}{12}\right)=\sin\left(-\frac{\pi}{3}-\frac{\pi}{4}\right)=\sin\left(-\frac{\pi}{3}\right)\cos\frac{\pi}{4}-\cos\left(-\frac{\pi}{3}\right)\sin\frac{\pi}{4}=-\sin\frac{\pi}{3}\cos\frac{\pi}{4}-\cos\frac{\pi}{3}\sin\frac{\pi}{4}$

$$=-\frac{\sqrt{3}}{2}\frac{\sqrt{2}}{2}\cdot-\frac{1}{2}\cdot\frac{\sqrt{2}}{2}=-\frac{\sqrt{6}}{4}-\frac{\sqrt{2}}{4}=\frac{-\sqrt{6}-\sqrt{2}}{4}$$

14. $\tan\left(-\frac{7\pi}{12}\right)=\tan\left[-\frac{\pi}{4}+\left(-\frac{\pi}{3}\right)\right]=\frac{\tan\left(-\frac{\pi}{4}\right)-\tan\left(-\frac{\pi}{3}\right)}{1+\tan\left(-\frac{\pi}{4}\right)\tan\left(-\frac{\pi}{3}\right)}=\frac{-1+\left(-\sqrt{3}\right)}{1-(-1)\left(-\sqrt{3}\right)}$

$$=\frac{-1-\sqrt{3}}{1-\sqrt{3}}=\frac{-1-\sqrt{3}}{1-\sqrt{3}}\cdot\frac{1+\sqrt{3}}{1+\sqrt{3}}=\frac{-1-\sqrt{3}-\sqrt{3}-3}{1^2-\left(\sqrt{3}\right)^2}=\frac{-4-2\sqrt{3}}{1-3}=\frac{-4-2\sqrt{3}}{-2}=2+\sqrt{3}$$

15. $\sin 76° \cos 31° - \cos 76° \sin 31° = \sin(76° - 31°) = \sin 45° = \frac{\sqrt{2}}{2}$

16. $\sin 40° \cos 50° + \cos 40° \sin 50° = \sin(40° + 50°) = \sin 90° = 1$

17. $\frac{\tan 80°+\tan 55°}{1-\tan 80° \tan 55°}=\tan\left(80°+55°\right)=\tan 135°=-1$

18. $\frac{\tan 80°-\tan\left(-55°\right)}{1+\tan 80° \tan\left(-55°\right)}=\tan\left[80°-\left(-55°\right)\right]=\tan 135°=-1$

19. $\frac{\tan 100°+\tan 80°}{1-\tan 100° \tan 80°}=\tan\left(100°+80°\right)=\tan 180°=0$

20. $\sin 100° \cos 10° - \cos 100° \sin 10° = \sin\left(100°-10°\right)=\sin 90°=1$

21. $\sin\frac{\pi}{5}\cos\frac{3\pi}{10}-\cos\frac{\pi}{5}\sin\frac{3\pi}{10}=\sin\left(\frac{\pi}{5}+\frac{3\pi}{10}\right)=\sin\left(\frac{2\pi}{10}+\frac{3\pi}{10}\right)=\sin\frac{5\pi}{10}=\sin\frac{\pi}{2}=1$

22. $\frac{\tan\frac{5\pi}{12}+\tan\frac{\pi}{4}}{1-\tan\frac{5\pi}{12}\tan\frac{\pi}{4}}=\tan\left(\frac{5\pi}{12}+\frac{\pi}{4}\right)$

$$=\tan\frac{8\pi}{12}$$

$$=\tan\frac{2\pi}{3}$$

$$=-\sqrt{3}$$

23. $\cos\left(30°+\theta\right)=\cos 30° \cos\theta-\sin 30° \sin\theta$

$$=\frac{\sqrt{3}}{2}\cos\theta-\frac{1}{2}\sin\theta$$

$$=\frac{1}{2}\left(\sqrt{3}\cos\theta-\sin\theta\right)$$

$$=\frac{\sqrt{3}\cos\theta-\sin\theta}{2}$$

24. $\cos(45° - \theta) = \cos 45° \cos\theta + \sin 45° \sin\theta$
$$= \frac{\sqrt{2}}{2}\cos\theta + \frac{\sqrt{2}}{2}\sin\theta$$
$$= \frac{\sqrt{2}(\cos\theta + \sin\theta)}{2}$$

25. $\cos(60° + \theta) = \cos 60° \cos\theta - \sin 60° \sin\theta$
$$= \frac{1}{2}\cos\theta - \frac{\sqrt{3}}{2}\sin\theta$$
$$= \frac{1}{2}\left(\cos\theta - \sqrt{3}\sin\theta\right)$$
$$= \frac{\cos\theta - \sqrt{3}\sin\theta}{2}$$

26. $\cos(\theta - 30°) = \cos\theta\cos 30° + \sin\theta\sin 30°$
$$= \frac{\sqrt{3}}{2}\cos\theta + \frac{1}{2}\sin\theta$$
$$= \frac{\sqrt{3}\cos\theta + \sin\theta}{2}$$

27. $\cos\left(\frac{3\pi}{4} - x\right) = \cos\frac{3\pi}{4}\cos x + \sin\frac{3\pi}{4}\sin x$
$$= -\frac{\sqrt{2}}{2}\cos x + \frac{\sqrt{2}}{2}\sin x$$
$$= \frac{\sqrt{2}}{2}(-\cos x + \sin x)$$
$$= \frac{\sqrt{2}(\sin x - \cos x)}{2}$$

28. $\sin(45° + \theta) = \sin 45° \cos\theta + \sin\theta\cos 45°$
$$= \frac{\sqrt{2}}{2}\sin\theta + \frac{\sqrt{2}}{2}\cos\theta$$
$$= \frac{\sqrt{2}}{2}(\sin\theta + \cos\theta)$$
$$= \frac{\sqrt{2}(\sin\theta + \cos\theta)}{2}$$

29. $\tan(\theta + 30°) = \frac{\tan\theta + \tan 30°}{1 - \tan\theta\tan 30°}$
$$= \frac{\tan\theta + \frac{1}{\sqrt{3}}}{1 - \frac{1}{\sqrt{3}}\tan\theta}$$
$$= \frac{\sqrt{3}\tan\theta + 1}{\sqrt{3} - \tan\theta}$$

30. $\tan\left(\frac{\pi}{4} + x\right) = \frac{\tan\frac{\pi}{4} + \tan x}{1 - \tan\frac{\pi}{4}\tan x}$
$$= \frac{1 + \tan x}{1 - \tan x}$$

31. $\sin\left(\frac{\pi}{4} + x\right) = \sin\frac{\pi}{4}\cos x + \cos\frac{\pi}{4}\sin x$
$$= \frac{\sqrt{2}}{2}\cos x + \frac{\sqrt{2}}{2}\sin x$$
$$= \frac{\sqrt{2}(\cos x + \sin x)}{2}$$

32. $\sin(180° - \theta) = \sin 180° \cos\theta - \cos 180° \sin\theta = (0)(\cos\theta) - (-1)(\sin\theta) = 0 + \sin\theta = \sin\theta$

33. $\sin(270° - \theta) = \sin 270° \cos\theta - \cos 270° \sin\theta = (-1)\cos\theta - (0)\sin\theta = -\cos\theta - 0 = -\cos\theta$

34. $\tan(180° + \theta) = \frac{\tan 180° + \tan\theta}{1 - \tan 180° \tan\theta}$
$$= \frac{0 + \tan\theta}{1 - 0}$$
$$= \tan\theta$$

35. $\tan(360° - \theta) = \frac{\tan 360° - \tan\theta}{1 + \tan 360° \tan\theta}$
$$= \frac{0 - \tan\theta}{1 + 0}$$
$$= -\tan\theta$$

36. $\sin(\pi + \theta) = \sin\pi\cos\theta + \cos\pi\sin\theta$
$$= 0 \cdot \cos\theta + (-1)\sin\theta$$
$$= -\sin\theta$$

37. $\tan(\pi - \theta) = \frac{\tan\pi - \tan\theta}{1 + \tan\pi\tan\theta} = \frac{0 - \tan\theta}{1 + 0(\tan\theta)}$
$$= -\tan\theta$$

38. To follow Example 2 to find $\tan(270° - \theta)$, we would need to use the tangent of a difference formula, $\tan(A - B) = \dfrac{\tan A - \tan B}{1 + \tan A \tan B}$. In this formula, $A = 270°$ and $\tan 270°$ is undefined.

39. Since $\tan 65.902° \tan 24.098° = \cot(90 - 65.902°) \tan 24.098° = \cot 24.098° \tan 24.098° = 1$, the denominator of $\dfrac{\tan 65.902° + \tan 24.098°}{1 - \tan 65.902° \tan 24.098°}$ becomes zero, which is undefined. Also,
$\dfrac{\tan 65.902° + \tan 24.098°}{1 - \tan 65.902° \tan 24.098°} = \tan(65.902° + 24.098°) = \tan 90°$, which is undefined.

40. If A, B, and C are angles of a triangle, then $A + B + C = 180°$. Therefore, we have the following.

$$\sin(A + B + C) = \sin 180° = 0$$

41. $\cos s = \dfrac{3}{5}$, $\sin t = \dfrac{5}{13}$, and s and t are in quadrant I

First, find $\sin s$, $\tan s$, $\cos t$, and $\tan t$. Since s and t are in quadrant I, all are positive.

$$\sin^2 s + \cos^2 s = 1 \Rightarrow \sin^2 s + \left(\frac{3}{5}\right)^2 = 1 \Rightarrow \sin^2 s + \frac{9}{25} = 1 \Rightarrow \sin^2 s = \frac{16}{25} \Rightarrow \sin s = \frac{4}{5}, \text{ since } \sin s > 0$$

$$\tan s = \frac{\sin s}{\cos s} = \frac{\frac{4}{5}}{\frac{3}{5}} = \frac{4}{5} \cdot \frac{5}{3} = \frac{4}{3}$$

$$\sin^2 t + \cos^2 t = 1 \Rightarrow \left(\frac{5}{13}\right)^2 + \cos^2 t = 1 \Rightarrow \frac{25}{169} + \cos^2 t = 1 \Rightarrow \cos^2 t = \frac{144}{169} \Rightarrow \cos t = \frac{12}{13}, \text{ since } \cos t > 0$$

$$\tan t = \frac{\frac{5}{13}}{\frac{12}{13}} = \frac{5}{13} \cdot \frac{13}{12} = \frac{5}{12}$$

(a) $\sin(s + t) = \sin s \cos t + \cos s \sin t = \dfrac{4}{5} \cdot \dfrac{12}{13} + \dfrac{5}{13} \cdot \dfrac{3}{5} = \dfrac{63}{65}$

(b) $\tan(s + t) = \dfrac{\tan s + \tan t}{1 - \tan s \tan t} = \dfrac{\frac{4}{3} + \frac{5}{12}}{1 - \left(\frac{4}{3}\right)\left(\frac{5}{12}\right)} = \dfrac{\frac{4}{3} + \frac{5}{12}}{1 - \left(\frac{4}{3}\right)\left(\frac{5}{12}\right)} \cdot \dfrac{36}{36} = \dfrac{48 + 15}{36 - 20} = \dfrac{63}{16}$

(c) To find the quadrant of $s + t$, notice that $\sin(s + t) > 0$, which implies $s + t$ is in quadrant I or II. $\tan(s + t) > 0$, which implies $s + t$ is in quadrant I or III. Therefore, $s + t$ is in quadrant I.

42. $\cos s = -\frac{1}{5}$, $\sin t = \frac{3}{5}$, s and t are in quadrant II

First, find sin s, tan s, cos t, and tan t. Since s and t are in quadrant II, sin s is positive. Tan s, cos t, and tan t are all negative.

$$\sin^2 s + \cos^2 s = 1 \Rightarrow \sin^2 s + \left(-\frac{1}{5}\right)^2 = 1 \Rightarrow \sin^2 s + \frac{1}{25} = 1$$

$$\Rightarrow \sin^2 s = \frac{24}{25} \Rightarrow \sin s = \frac{\sqrt{24}}{5} = \frac{2\sqrt{6}}{5}, \text{ since } \sin s > 0$$

$$\sin^2 t + \cos^2 t = 1 \Rightarrow \left(\frac{3}{5}\right)^2 + \cos^2 t = 1 \Rightarrow \frac{9}{25} + \cos^2 t = 1 \Rightarrow \cos^2 t = \frac{16}{25} \Rightarrow \cos t = -\frac{4}{5}, \text{ since } \cos t < 0$$

$$\tan s = \frac{\sin s}{\cos s} = \frac{\frac{2\sqrt{6}}{5}}{-\frac{1}{5}} = \frac{2\sqrt{6}}{5} \cdot \left(-\frac{5}{1}\right) = -2\sqrt{6};\ \tan t = \frac{\frac{3}{5}}{-\frac{4}{5}} = \frac{3}{5} \cdot \left(-\frac{5}{4}\right) = -\frac{3}{4}$$

(a) $$\sin(s+t) = \sin s \cos t + \cos s \sin t = \left(\frac{2\sqrt{6}}{5}\right)\left(-\frac{4}{5}\right) + \left(-\frac{1}{5}\right)\left(\frac{3}{5}\right) = \frac{-8\sqrt{6}-3}{25}$$

(b) Different forms of tan $(s + t)$ will be obtained depending on whether tan s and tan t are written with rationalized denominators.

$$\tan(s+t) = \frac{\tan s + \tan t}{1 - \tan s \tan t} = \frac{-2\sqrt{6} + \left(-\frac{3}{4}\right)}{1 - \left(-2\sqrt{6}\right)\left(-\frac{3}{4}\right)} = \frac{\frac{-8\sqrt{6}-3}{4}}{\frac{4-6\sqrt{6}}{4}} = \frac{-8\sqrt{6}-3}{4-6\sqrt{6}}$$

(c) To find the quadrant of $s + t$, notice from the preceding that $\sin(s+t) = \frac{-8\sqrt{6}-3}{25} < 0$ and $\tan(s+t) = \frac{-8\sqrt{6}-3}{4-6\sqrt{6}} > 0$. The sine is negative in quadrants III and IV, the tangent is positive in quadrants I and III. Therefore, $s + t$ is in quadrant III.

43. $\sin s = \frac{2}{3}$, $\sin t = -\frac{1}{3}$, s is in quadrant II, and t is in quadrant IV

In order to substitute into sum and difference identities, we need to find the values of $\cos s$ and $\cos t$, and also the values of $\tan s$ and $\tan t$. Because s is in quadrant II, the values of both $\cos s$ and $\tan s$ will be negative. Because t is in quadrant IV, $\cos t$ will be positive, while $\tan t$ will be negative.

$$\cos s = \sqrt{1 - \sin^2 s} = -\sqrt{1 - \left(\frac{2}{3}\right)^2} = -\frac{\sqrt{5}}{3}$$

$$\cos t = \sqrt{1 - \sin^2 t} = \sqrt{1 - \left(-\frac{1}{3}\right)^2} = \frac{\sqrt{8}}{3} = \frac{2\sqrt{2}}{3}$$

$$\tan s = \frac{\frac{2}{3}}{-\frac{\sqrt{5}}{3}} = -\frac{2}{\sqrt{5}} = -\frac{2\sqrt{5}}{5};\ \tan t = \frac{-\frac{1}{3}}{\frac{2\sqrt{2}}{3}} = -\frac{1}{2\sqrt{2}} = -\frac{\sqrt{2}}{4}$$

Continued on next page

43. (continued)

(a) $\sin(s+t)=\frac{2}{3}\left(\frac{2\sqrt{2}}{3}\right)+\left(-\frac{\sqrt{5}}{3}\right)\left(-\frac{1}{3}\right)=\frac{4\sqrt{2}}{9}+\frac{\sqrt{5}}{9}=\frac{4\sqrt{2}+\sqrt{5}}{9}$

(b) Different forms of tan $(s + t)$ will be obtained depending on whether tan s and tan t are written with rationalized denominators.

$$\tan(s+t)=\frac{-\frac{2\sqrt{5}}{5}+\left(-\frac{\sqrt{2}}{4}\right)}{1-\left(-\frac{2\sqrt{5}}{5}\right)\left(-\frac{\sqrt{2}}{4}\right)}=\frac{-\frac{2\sqrt{5}}{5}+\left(-\frac{\sqrt{2}}{4}\right)}{1-\left(-\frac{2\sqrt{5}}{5}\right)\left(-\frac{\sqrt{2}}{4}\right)}\cdot\frac{20}{20}=\frac{-8\sqrt{5}-5\sqrt{2}}{20-2\sqrt{10}}$$

(c) From parts (a) and (b), $\sin(s + t) > 0$ and $\tan(s + t) < 0$. The only quadrant in which values of sine are positive and values of tangent are negative is quadrant II.

44. $\sin s=\frac{3}{5}$, $\sin t=-\frac{12}{13}$, s is in quadrant I, and t is in quadrant III

In order to substitute into sum and difference identities, we need to find the values of $\cos s$ and $\cos t$, and also the values of $\tan s$ and $\tan t$. Because s is in quadrant I, the values of both $\cos s$ and $\tan s$ will be positive. Because t is in quadrant III the value $\cos t$ will be negative, while $\tan t$ will be positive.

$$\cos s=\sqrt{1-\sin^2 s}=\sqrt{1-\left(\frac{3}{5}\right)^2}=\sqrt{1-\frac{9}{25}}=\sqrt{\frac{16}{25}}=\frac{4}{5}$$

$$\cos t=\sqrt{1-\sin^2 t}=-\sqrt{1-\left(-\frac{12}{13}\right)^2}=-\sqrt{1-\frac{144}{169}}=-\sqrt{\frac{25}{169}}=-\frac{5}{13}$$

$$\tan s=\frac{\frac{3}{5}}{\frac{4}{5}}=\frac{3}{4};\ \tan t=\frac{-\frac{12}{13}}{-\frac{5}{13}}=\frac{12}{5}$$

(a) $\sin(s+t)=\left(\frac{3}{5}\right)\left(-\frac{5}{13}\right)+\left(\frac{4}{5}\right)\left(-\frac{12}{13}\right)=\frac{-15}{65}+\frac{-48}{65}=-\frac{63}{65}$

(b) $\tan(s+t)=\dfrac{\frac{3}{4}+\frac{12}{5}}{1-\left(\frac{3}{4}\right)\left(\frac{12}{5}\right)}=\dfrac{\frac{3}{4}+\frac{12}{5}}{1-\left(\frac{3}{4}\right)\left(\frac{12}{5}\right)}\cdot\frac{20}{20}=\frac{15+48}{20-36}=-\frac{63}{16}$

(c) From parts (a) and (b), $\sin(s + t) < 0$ and $\tan(s + t) < 0$. The only quadrant in which values of sine and tangent are negative is quadrant IV.

45. $\cos s = -\frac{8}{17}, \cos t = -\frac{3}{5}$, and s and t are in quadrant III

In order to substitute into sum and difference identities, we need to find the values of $\sin s$ and $\sin t$, and also the values of $\tan s$ and $\tan t$. Because s and t are both in quadrant III, the values of $\sin s$ and $\sin t$ will be negative, while $\tan s$ and $\tan t$ will be positive.

$$\sin s = \sqrt{1-\cos^2 s} = -\sqrt{1-\left(-\frac{8}{17}\right)^2} = -\sqrt{1-\frac{64}{289}} = -\sqrt{\frac{225}{289}} = -\frac{15}{17}$$

$$\sin t = \sqrt{1-\cos^2 t} = -\sqrt{1-\left(-\frac{3}{5}\right)^2} = -\sqrt{1-\frac{9}{25}} = -\sqrt{\frac{16}{25}} = -\frac{4}{5}$$

$$\tan s = \frac{-\frac{15}{17}}{-\frac{8}{17}} = \frac{15}{8};\ \tan t = \frac{-\frac{4}{5}}{-\frac{3}{5}} = \frac{4}{3}$$

(a) $\sin(s+t) = \left(-\frac{15}{17}\right)\left(-\frac{3}{5}\right) + \left(-\frac{4}{5}\right)\left(-\frac{8}{17}\right) = \frac{45}{85} + \frac{32}{85} = \frac{77}{85}$

(b) $\tan(s+t) = \dfrac{\frac{15}{8}+\frac{4}{3}}{1-\left(\frac{15}{8}\right)\left(\frac{4}{3}\right)} = \dfrac{\frac{15}{8}+\frac{4}{3}}{1-\left(\frac{15}{8}\right)\left(\frac{4}{3}\right)} \cdot \frac{24}{24} = \frac{45+32}{24-60} = \frac{77}{-36} = -\frac{77}{36}$

(c) From parts (a) and (b), $\sin(s + t) > 0$ and $\tan(s + t) < 0$. The only quadrant in which values of sine are positive and values of tangent are negative is quadrant II.

46. $\cos s = -\frac{15}{17}$, $\sin t = \frac{4}{5}$, s is in quadrant II, and t is in quadrant I

In order to substitute into sum and difference identities, we need to find the values of $\sin s$ and $\cos t$, and also the values of $\tan s$ and $\tan t$. Because s is in quadrant II, the value of $\sin s$ is positive, while the value of $\tan s$ is negative. Because t is in quadrant I, both the values of $\cos t$ and $\tan t$ are positive.

$$\sin s = \sqrt{1-\cos^2 s} = -\sqrt{1-\left(-\frac{15}{17}\right)^2} = -\sqrt{1-\frac{225}{289}} = \sqrt{\frac{64}{289}} = \frac{8}{17}$$

$$\cos t = \sqrt{1-\sin^2 t} = \sqrt{1-\left(\frac{4}{5}\right)^2} = \sqrt{1-\frac{16}{25}} = \sqrt{\frac{9}{25}} = \frac{3}{5}$$

$$\tan s = \frac{\frac{8}{17}}{-\frac{15}{17}} = -\frac{8}{15};\ \tan t = \frac{\frac{4}{5}}{\frac{3}{5}} = \frac{4}{3}$$

(a) $\sin(s+t) = \left(\frac{8}{17}\right)\left(\frac{3}{5}\right) + \left(-\frac{15}{17}\right)\left(\frac{4}{5}\right) = \frac{24}{85} + \left(-\frac{60}{85}\right) = -\frac{36}{85}$

(b) $\tan(s+t) = \dfrac{-\frac{8}{15}+\frac{4}{3}}{1-\left(-\frac{8}{15}\right)\left(\frac{4}{3}\right)} = \dfrac{-\frac{8}{15}+\frac{4}{3}}{1-\left(-\frac{8}{15}\right)\left(\frac{4}{3}\right)} \cdot \frac{45}{45} = \frac{-24+60}{45+32} = \frac{36}{77}$

(c) From parts (a) and (b), $\sin(s + t) < 0$ and $\tan(s + t) > 0$. The only quadrant in which values of sine are negative and values of tangent are positive is quadrant III.

47. $\sin 165° = \sin(180° - 15°) = \sin 180° \cos 15° - \cos 180° \sin 15°$

$= (0)\cos 15° - (-1)\sin 15° = 0 + \sin 15° = \sin 15° = \sin(45° - 30°)$

$= \sin 45° \cos 30° - \cos 45° \sin 30° = \frac{\sqrt{2}}{2} \cdot \frac{\sqrt{3}}{2} - \frac{\sqrt{2}}{2} \cdot \frac{1}{2} = \frac{\sqrt{6}}{4} - \frac{\sqrt{2}}{4} = \frac{\sqrt{6} - \sqrt{2}}{4}$

48. $\tan 165° = \tan(180° - 15°) = \frac{\tan 180° - \tan 15°}{1 + \tan 180° \tan 15°} = \frac{0 - \tan 15°}{1 + 0 \cdot \tan 15°} = -\tan 15°$

Now use a difference identity to find $\tan 15°$.

$$\tan 15° = \tan(45° - 30°) = \frac{\tan 45° - \tan 30°}{1 + \tan 45° \tan 30°} = \frac{1 - \frac{\sqrt{3}}{3}}{1 + 1 \cdot \frac{\sqrt{3}}{3}} = \frac{3 - \sqrt{3}}{3 + \sqrt{3}}$$

$$= \frac{3 - \sqrt{3}}{3 + \sqrt{3}} \cdot \frac{3 - \sqrt{3}}{3 - \sqrt{3}} = \frac{9 - 3\sqrt{3} - 3\sqrt{3} + 3}{9 - 3} = \frac{12 - 6\sqrt{3}}{6} = 2 - \sqrt{3}$$

Thus, $\tan 165° = -\tan 15° = -(2 - \sqrt{3}) = -2 + \sqrt{3}.$

49. $\sin 255° = \sin(270° - 15°) = \sin 270° \cos 15° - \cos 270° \sin 15°$

$= (-1)\cos 15° - (0)\sin 15° = -\cos 15° - 0 = -\cos 15°$

Now use a difference identity to find $\cos 15°$.

$\cos 15° = \cos(45° - 30°) = (\cos 45° \cos 30° + \sin 45° \sin 30°)$

$= \frac{\sqrt{2}}{2} \cdot \frac{\sqrt{3}}{2} + \frac{\sqrt{2}}{2} \cdot \frac{1}{2} = \frac{\sqrt{6}}{4} + \frac{\sqrt{2}}{4} = \frac{\sqrt{6} + \sqrt{2}}{4}$

Thus, $\sin 255° = -\cos 15° = -\left(\frac{\sqrt{6} + \sqrt{2}}{4}\right) = \frac{-\sqrt{6} - \sqrt{2}}{4}.$

50. $\tan 285° = \tan(360° - 75°) = \frac{\tan 360° - \tan 75°}{1 + \tan 360° \tan 75°} = \frac{0 - \tan 75°}{1 + 0 \cdot \tan 75°} = \frac{-\tan 75°}{1 + 0} = -\tan 75°$

Now use a difference identity to find $\tan 75°$.

$$\tan 75° = \tan(30° + 45°) = \frac{\frac{\sqrt{3}}{3} + 1}{1 - \frac{\sqrt{3}}{3} \cdot 1}$$

$$= \frac{\sqrt{3} + 3}{3 - \sqrt{3}} = \frac{\sqrt{3} + 3}{3 - \sqrt{3}} \cdot \frac{\sqrt{3} + 3}{3 + \sqrt{3}} = \frac{3 + 3\sqrt{3} + 3\sqrt{3} + 9}{9 - 3} = \frac{12 + 6\sqrt{3}}{6} = 2 + \sqrt{3}$$

Thus, $\tan 285° = -\tan 75° = -(2 + \sqrt{3}) = -2 - \sqrt{3}.$

51. $\tan\frac{11\pi}{12}=\tan\left(\pi-\frac{\pi}{12}\right)=\frac{\tan\pi-\tan\frac{\pi}{12}}{1+\tan\pi\tan\frac{\pi}{12}}=-\tan\frac{\pi}{12}$

Now use a difference identity to find $\tan\frac{\pi}{12}$.

$$\tan\frac{\pi}{12}=\tan\left(\frac{\pi}{4}-\frac{\pi}{6}\right)=\frac{\tan\frac{\pi}{4}-\tan\frac{\pi}{6}}{1+\tan\frac{\pi}{4}\tan\frac{\pi}{6}}=\frac{1-\frac{\sqrt{3}}{3}}{1+1\cdot\frac{\sqrt{3}}{3}}=\frac{1-\frac{\sqrt{3}}{3}}{1+\frac{\sqrt{3}}{3}}=\frac{\frac{3-\sqrt{3}}{3}}{\frac{3+\sqrt{3}}{3}}$$

$$=\frac{3-\sqrt{3}}{3+\sqrt{3}}=\frac{3-\sqrt{3}}{3+\sqrt{3}}\cdot\frac{3-\sqrt{3}}{3-\sqrt{3}}=\frac{9-6\sqrt{3}+3}{9-3}=\frac{12-6\sqrt{3}}{6}=\frac{6\left(2-\sqrt{3}\right)}{6}=2-\sqrt{3}$$

Thus, $\tan\frac{11\pi}{12}=-\tan\frac{\pi}{12}=-\left(2-\sqrt{3}\right)=-2+\sqrt{3}.$

52. $\sin\left(-\frac{13\pi}{12}\right)=-\sin\left(\frac{13\pi}{12}\right)=-\sin\left(\pi+\frac{\pi}{12}\right)=-\left(\sin\pi\cos\frac{\pi}{12}+\cos\pi\sin\frac{\pi}{12}\right)$

$$=-\left[(0)\cos\frac{\pi}{12}+(-1)\sin\frac{\pi}{12}\right]=-\left(0-\sin\frac{\pi}{12}\right)=-\left(-\sin\frac{\pi}{12}\right)=\sin\frac{\pi}{12}$$

$$=\sin\left(\frac{\pi}{4}-\frac{\pi}{6}\right)=\sin\frac{\pi}{4}\cos\frac{\pi}{6}-\cos\frac{\pi}{4}\sin\frac{\pi}{6}=\frac{\sqrt{2}}{2}\cdot\frac{\sqrt{3}}{2}-\frac{\sqrt{2}}{2}\cdot\frac{1}{2}=\frac{\sqrt{6}-\sqrt{2}}{4}$$

The graphs in Exercises 53 –56 are shown in the following window. This window can be obtained through the Zoom Trig feature.

```
WINDOW
 Xmin=-6.152285...
 Xmax=6.1522856...
 Xscl=1.5707963...
 Ymin=-4
 Ymax=4
 Yscl=1
 Xres=1
```

```
ZOOM MEMORY
1:ZBox
2:Zoom In
3:Zoom Out
4:ZDecimal
5:ZSquare
6:ZStandard
7:ZTrig
```

53. $\sin\left(\frac{\pi}{2}+x\right)$ appears to be equivalent to $\cos x$.

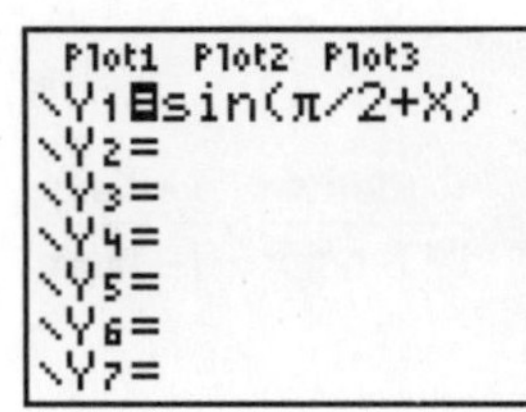

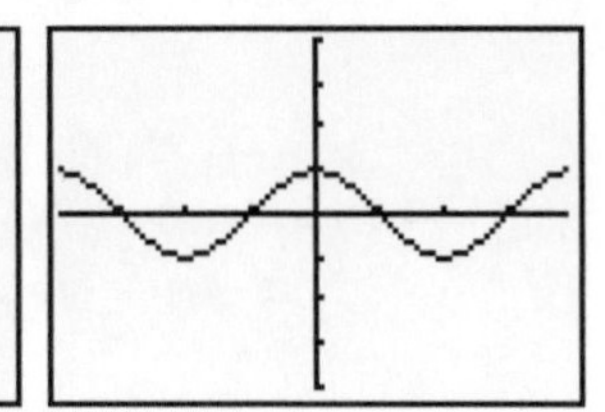

$$\sin\left(\frac{\pi}{2}+x\right)=\sin\frac{\pi}{2}\cos x+\sin x\cos\frac{\pi}{2}=1\cdot\cos x+\sin x\cdot 0=\cos x+0=\cos x$$

54. $\sin\left(\frac{3\pi}{2}+x\right)$ appears to be equivalent to $-\cos x$.

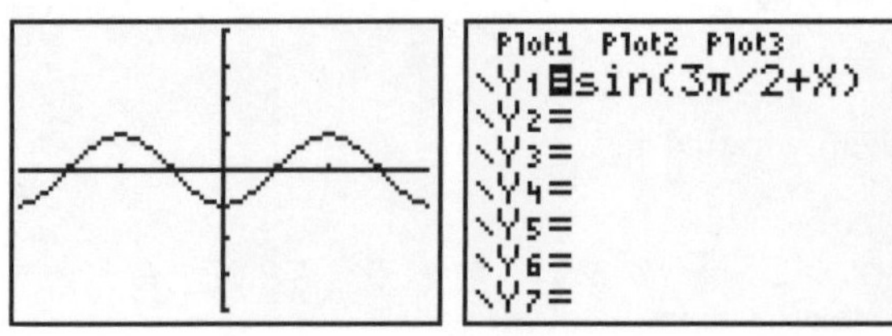

$$\sin\left(\frac{3\pi}{2}+x\right)=\sin\frac{3\pi}{2}\cos x+\cos\frac{3\pi}{2}\sin x=-1\cdot\cos x+0\cdot\sin x=-\cos x+0=-\cos x$$

55. $\tan\left(\frac{\pi}{2}+x\right)$ appears to be equivalent to $-\cot x$.

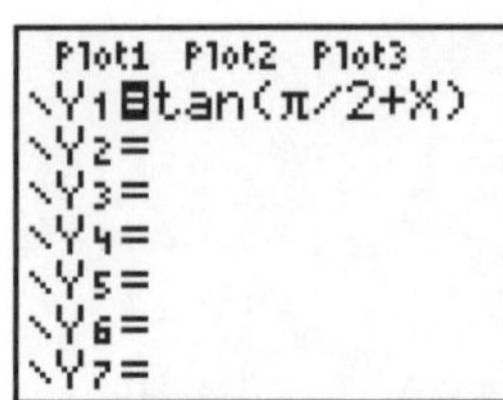

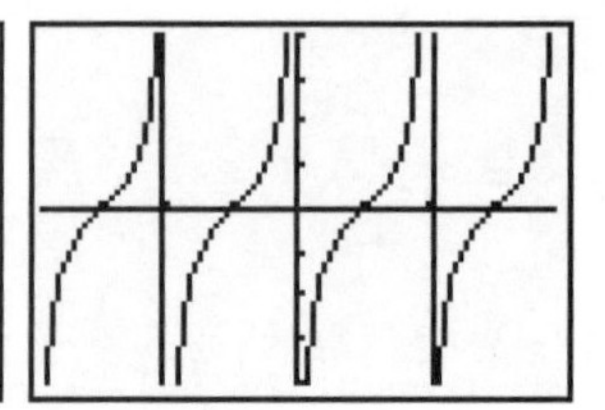

$$\tan\left(\frac{\pi}{2}+x\right)=\frac{\sin\left(\frac{\pi}{2}+x\right)}{\cos\left(\frac{\pi}{2}+x\right)}=\frac{\sin\frac{\pi}{2}\cos x+\cos\frac{\pi}{2}\sin x}{\cos\frac{\pi}{2}\cos x-\sin\frac{\pi}{2}\sin x}$$

$$=\frac{1\cdot\cos x+0\cdot\sin x}{0\cdot\cos x-1\cdot\sin x}=\frac{\cos x+0}{0-\sin x}=\frac{\cos x}{-\sin x}=-\frac{\cos x}{\sin x}=-\cot x$$

56. $\frac{1+\tan x}{1-\tan x}$ appears to be equivalent to $\tan\left(\frac{\pi}{4}+x\right)$.

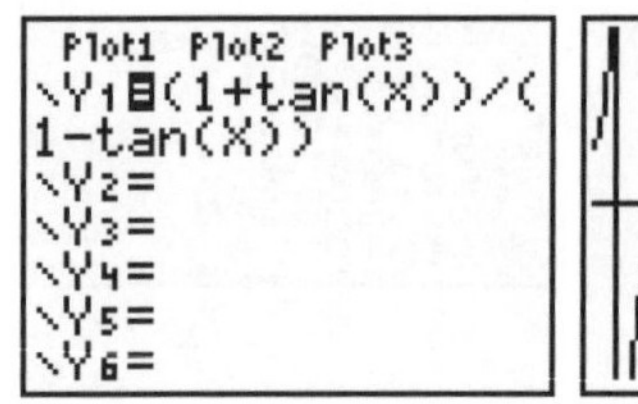

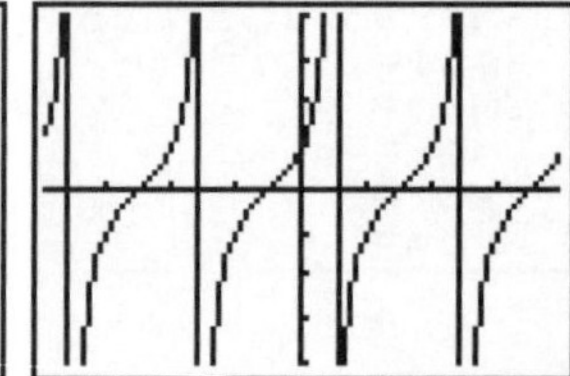

Working with $\tan\left(\frac{\pi}{4}+x\right)$ we have the following.

$$\tan\left(\frac{\pi}{4}+x\right)=\frac{\tan\frac{\pi}{4}+\tan x}{1-\tan\frac{\pi}{4}\tan x}=\frac{1+\tan x}{1-1\cdot\tan x}=\frac{1+\tan x}{1-\tan x}$$

57. Verify $\sin 2x=2\sin x\cos x$ is an identity.

$$\sin 2x=\sin(x+x)=\sin x\cos x+\sin x\cos x=2\sin x\cos x$$

58. Verify $\sin(x+y)+\sin(x-y)=2\sin x\cos y$ is an identity.

$$\sin(x+y)+\sin(x-y)=(\sin x\cos y+\cos x\sin y)+(\sin x\cos y-\cos x\sin y)$$
$$=2\sin x\cos y$$

59. Verify $\sin(210°+x)-\cos(120°+x)=0$ is an identity.

$$\begin{aligned}\sin(210°+x)-\cos(120°+x)&=(\sin 210°\cos x+\cos 210°\sin x)-(\cos 120°\cos x-\sin 120°\sin x)\\&=-\frac{1}{2}\cos x-\frac{\sqrt{3}}{2}\sin x-\left(-\frac{1}{2}\cos x-\frac{\sqrt{3}}{2}\sin x\right)\\&=-\frac{1}{2}\cos x-\frac{\sqrt{3}}{2}\sin x+\frac{1}{2}\cos x+\frac{\sqrt{3}}{2}\sin x=0\end{aligned}$$

60. Verify $\tan(x-y)-\tan(y-x)=\dfrac{2(\tan x-\tan y)}{1+\tan x\tan y}$ is an identity.

$$\begin{aligned}\tan(x-y)-\tan(y-x)&=\frac{\tan x-\tan y}{1+\tan x\tan y}-\frac{\tan y-\tan x}{1+\tan y\tan x}\\&=\frac{\tan x-\tan y-\tan y+\tan x}{1+\tan x\tan y}=\frac{2(\tan x-\tan y)}{1+\tan x\tan y}\end{aligned}$$

61. Verify $\dfrac{\cos(\alpha-\beta)}{\cos\alpha\sin\beta}=\tan\alpha+\cot\beta$ is an identity.

$$\frac{\cos(\alpha-\beta)}{\cos\alpha\sin\beta}=\frac{\cos\alpha\cos\beta+\sin\alpha\sin\beta}{\cos\alpha\sin\beta}=\frac{\cos\alpha\cos\beta}{\cos\alpha\sin\beta}+\frac{\sin\alpha\sin\beta}{\cos\alpha\sin\beta}=\frac{\cos\beta}{\sin\beta}+\frac{\sin\alpha}{\cos\alpha}=\cot\beta+\tan\alpha$$

62. Verify $\dfrac{\sin(s+t)}{\cos s\cos t}=\tan s+\tan t$ is an identity.

$$\frac{\sin(s+t)}{\cos s\cos t}=\frac{\sin s\cos t+\cos s\sin t}{\cos s\cos t}=\frac{\sin s\cos t}{\cos s\cos t}+\frac{\cos s\sin t}{\cos s\cos t}=\frac{\sin s}{\cos s}+\frac{\sin t}{\cos t}=\tan s+\tan t$$

63. Verify that $\dfrac{\sin(x-y)}{\sin(x+y)}=\dfrac{\tan x-\tan y}{\tan x+\tan y}$ is an identity.

$$\frac{\sin(x-y)}{\sin(x+y)}=\frac{\sin x\cos y-\cos x\sin y}{\sin x\cos y+\cos x\sin y}=\frac{\frac{\sin x\cos y}{\cos x\cos y}-\frac{\cos x\sin y}{\cos x\cos y}}{\frac{\sin x\cos y}{\cos x\cos y}+\frac{\cos x\sin y}{\cos x\cos y}}$$

$$=\frac{\frac{\sin x}{\cos x}\cdot\frac{\cos y}{\cos y}-\frac{\cos x}{\cos x}\cdot\frac{\sin y}{\cos y}}{\frac{\sin x}{\cos x}\cdot\frac{\cos y}{\cos y}+\frac{\cos x}{\cos x}\cdot\frac{\sin y}{\cos y}}=\frac{\frac{\sin x}{\cos x}\cdot 1-1\cdot\frac{\sin y}{\cos y}}{\frac{\sin x}{\cos x}\cdot 1+1\cdot\frac{\sin y}{\cos y}}=\frac{\frac{\sin x}{\cos x}-\frac{\sin y}{\cos y}}{\frac{\sin x}{\cos x}+\frac{\sin y}{\cos y}}=\frac{\tan x-\tan y}{\tan x+\tan y}$$

64. Verify $\dfrac{\sin(x-y)}{\cos(x-y)}=\dfrac{\cot x+\cot y}{1+\cot x\cot y}$. is an identity.

Working with the right side we have the following.

$$\frac{\cot x+\cot y}{1+\cot x\cot y}=\frac{\frac{\cos x}{\sin x}+\frac{\cos y}{\sin y}}{1+\frac{\cos x}{\sin x}\cdot\frac{\cos y}{\sin y}}=\frac{\frac{\cos x}{\sin x}+\frac{\cos y}{\sin y}}{1+\frac{\cos x}{\sin x}\cdot\frac{\cos y}{\sin y}}\cdot\frac{\sin x\sin y}{\sin x\sin y}=\frac{\cos x\sin y+\sin x\cos y}{\sin x\sin y+\cos x\cos y}=\frac{\sin(x+y)}{\cos(x-y)}$$

65. Verify $\frac{\sin(s-t)}{\sin t}+\frac{\cos(s-t)}{\cos t}=\frac{\sin s}{\sin t\cos t}$ is an identity.

$$\begin{aligned}\frac{\sin(s-t)}{\sin t}+\frac{\cos(s-t)}{\cos t}&=\frac{\sin s\cos t-\sin t\cos s}{\sin t}+\frac{\cos s\cos t+\sin s\sin t}{\cos t}\\&=\frac{\sin s\cos^2 t-\sin t\cos t\cos s}{\sin t\cos t}+\frac{\sin t\cos t\cos s+\sin^2 t\sin s}{\sin t\cos t}\\&=\frac{\sin s\cos^2 t+\sin s\sin^2 t}{\sin t\cos t}=\frac{\sin s\left(\cos^2 t+\sin^2 t\right)}{\sin t\cos t}=\frac{\sin s}{\sin t\cos t}\end{aligned}$$

66. Verify $\frac{\tan(\alpha+\beta)-\tan\beta}{1+\tan(\alpha+\beta)\tan\beta}=\tan\alpha$ is an identity.

$$\frac{\tan(\alpha+\beta)-\tan\beta}{1+\tan(\alpha+\beta)\tan\beta}=\tan\left[(\alpha+\beta)-\beta\right]=\tan\alpha$$

67. Since angle β and angle ABC are supplementary, the measure of angle ABC is $180°-\beta$.

68. $\alpha+(180°-\beta)+\theta=180°\Rightarrow\alpha-\beta+\theta=0\Rightarrow\theta=\beta-\alpha$

69. $\tan\theta=\tan(\beta-\alpha)=\frac{\tan\beta-\tan\alpha}{1+\tan\beta\tan\alpha}$

70. Since $\tan\theta=\frac{\tan\beta-\tan\alpha}{1+\tan\beta\tan\alpha}$, if we let $\tan\alpha=m_1$ and $\tan\alpha=m_2$, we have $\tan\theta=\frac{m_1-m_2}{1+m_1m_2}$.

71. $x+y=9, 2x+y=-1$

Change the equation to slope-intercept form to find their slopes.

$$x+y=9\Rightarrow y=-x+9\Rightarrow m_1=-1 \text{ and } 2x+y=-1\Rightarrow y=-2x-1\Rightarrow m_2=-2$$

$$\tan\theta=\frac{m_1-m_2}{1+m_1m_2}=\frac{-1-(-2)}{1+(-1)(-2)}=\frac{1}{1+2}=\frac{1}{3}\Rightarrow\theta=\tan^{-1}\frac{1}{3}\approx 18.4°$$

72. $5x-2y+4=0, 3x+5y=6$

Change the equation to slope-intercept form to find their slopes.

$$5x-2y+4=0\Rightarrow -2y=-5x-4\Rightarrow y=\frac{5}{2}x+2\Rightarrow m_1=\frac{5}{2}$$

$$3x+5y=6\Rightarrow 5y=-3x+6\Rightarrow y=-\frac{3}{5}x+\frac{6}{5}\Rightarrow m_2=-\frac{3}{5}$$

$$\tan\theta=\frac{m_1-m_2}{1+m_1m_2}=\frac{-\frac{3}{5}-\frac{5}{2}}{1+\left(-\frac{3}{5}\right)\left(\frac{5}{2}\right)}=\frac{-\frac{3}{5}-\frac{5}{2}}{1+\left(-\frac{3}{5}\right)\left(\frac{5}{2}\right)}\cdot\frac{10}{10}=\frac{-6-25}{10-15}=\frac{-31}{-5}=\frac{31}{5}=6.2$$

$\theta=\tan^{-1}6.2\approx 80.8°$

73. **(a)** $F = \dfrac{.6W \sin(\theta + 90°)}{\sin 12°} = \dfrac{.6(170)\sin(30+90)°}{\sin 12°} = \dfrac{102 \sin 120°}{\sin 12°} \approx 425 \text{ lb}$

(This is a good reason why people frequently have back problems.)

(b) $F = \dfrac{.6W \sin(\theta + 90°)}{\sin 12°} = \dfrac{.6W(\sin\theta \cos 90° + \sin 90° \cos\theta)}{\sin 12°} = \dfrac{.6W(\sin\theta \cdot 0 + 1 \cdot \cos\theta)}{\sin 12°}$

$= \dfrac{.6W(0 + \cos\theta)}{\sin 12°} = \dfrac{.6}{\sin 12°} W \cos\theta \approx 2.9W \cos\theta$

(c) F will be maximum when $\cos\theta = 1$ or $\theta = 0°$. ($\theta = 0°$ corresponds to the back being horizontal which gives a maximum force on the back muscles. This agrees with intuition since stress on the back increases as one bends farther until the back is parallel with the ground.)

74. **(a)** $F = \dfrac{.6W \sin(\theta + 90°)}{\sin 12°}, W = 200, \theta = 45°$

$F = \dfrac{.6(200)\sin(45+90)°}{\sin 12°} = \dfrac{120 \sin 135°}{\sin 12°} \approx 408 \text{ lb}$

(b) The calculator should be in degree mode.

Graph $y = \dfrac{.6(200)\sin(x + 90°)}{\sin 12°} = \dfrac{120 \sin(x + 90°)}{\sin 12°}$ and $y = 400$ on the same screen and find the point of intersection.

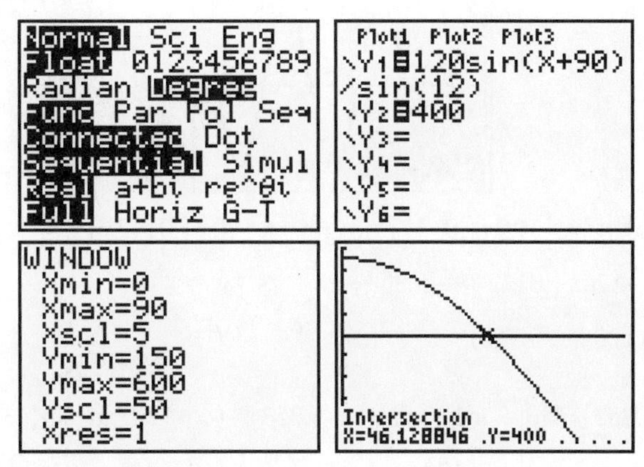

A force of 400 lb is exerted when $\theta \approx 46.1°$.

75. $e = 20\sin\left(\dfrac{\pi t}{4} - \dfrac{\pi}{2}\right) = 20\left(\sin\dfrac{\pi t}{4}\cos\dfrac{\pi}{2} - \cos\dfrac{\pi t}{4}\sin\dfrac{\pi}{2}\right)$

$= 20\left[\left(\sin\dfrac{\pi t}{4}\right)(0) - \left(\cos\dfrac{\pi t}{4}\right)(1)\right] = 20\left(0 - \cos\dfrac{\pi t}{4}\right) = -20\cos\dfrac{\pi t}{4}$

76. (a) The calculator should be in radian mode.

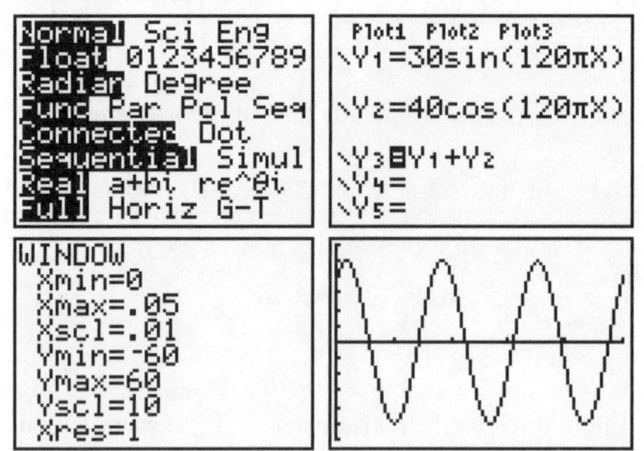

(b) Using the maximum and minimum features on your calculator, the amplitude appears to be 50. So, let $a = 50$.

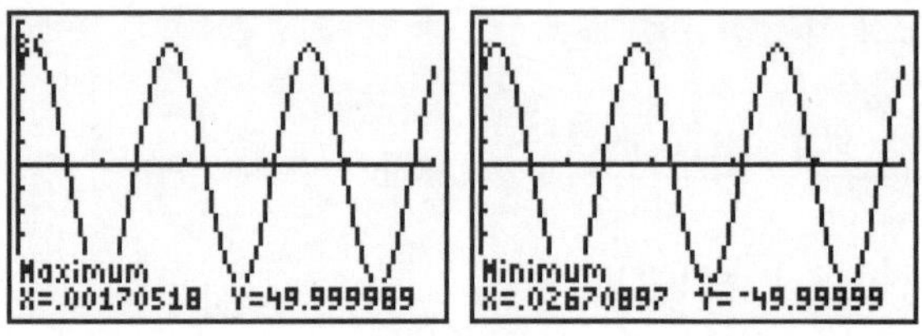

Estimate the phase shift by approximating the first t-intercept where the graph of V is increasing. This is located at $t \approx .0142$.

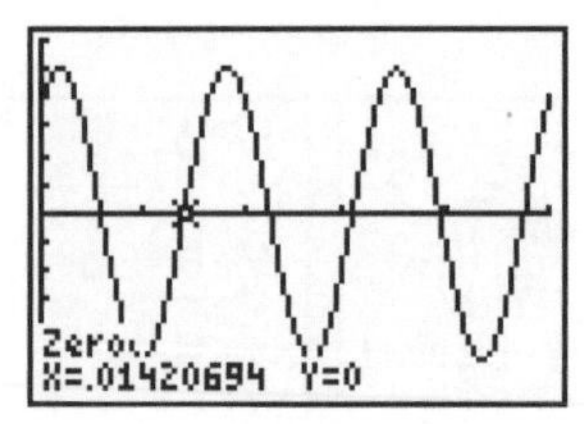

$$\sin(120\pi t+\phi)=0 \Rightarrow \sin\left[120\pi(.0142)+\phi\right]=0 \Rightarrow 120\pi(.0142)+\phi=0$$

$$\phi=-120\pi(.0142)\approx -5.353$$

Thus, $V = 50\sin(120\pi t - 5.353)$.

(c)
$$\begin{aligned}50\sin(120\pi t-5.353)&=50\left[(\sin 120\pi t)(\cos 5.353)-(\cos 120\pi t)(\sin 5.353)\right]\\&\approx 50\left[(\sin 120\pi t)(.5977)-(\cos 120\pi t)(-.8017)\right]\\&\approx 29.89\sin 120\pi t+40.09\cos 120\pi t\\&\approx 30\sin 120\pi t+40\cos 120\pi t\end{aligned}$$

Section 5.5: Double-Angle Identities

1. Since $2\cos^2 15°-1=\cos(2\cdot 15°)=\cos 30°=\dfrac{\sqrt{3}}{2}$, the correct choice is C.

2. Since $\dfrac{2\tan 15°}{1-\tan^2 15°}=\tan(2\cdot 15°)=\tan 30°=\dfrac{\sqrt{3}}{3}$, the correct choice is E.

3. Since $2\sin 22.5^\circ \cos 22.5^\circ = \sin(2\cdot 22.5^\circ) = \sin 45^\circ = \frac{\sqrt{2}}{2}$, the correct choice is B.

4. Since $\cos^2 \frac{\pi}{6} - \sin^2 \frac{\pi}{6} = \cos\left(2\cdot\frac{\pi}{6}\right) = \cos\frac{\pi}{3} = \frac{1}{2}$, the correct choice is A.

5. Since $2\sin\frac{\pi}{3}\cos\frac{\pi}{3} = \sin\left(2\cdot\frac{\pi}{3}\right) = \sin\frac{2\pi}{3} = \frac{\sqrt{3}}{2}$, the correct choice is C.

6. Since $\dfrac{2\tan\frac{\pi}{3}}{1-\tan^2\frac{\pi}{3}} = \tan\left(2\cdot\frac{\pi}{3}\right) = \tan\frac{2\pi}{3} = -\sqrt{3}$, the correct choice is D.

7. $\cos 2\theta = \frac{3}{5}, \theta$ is in quadrant I.

$$\cos 2\theta = 2\cos^2\theta - 1 \Rightarrow \frac{3}{5} = 2\cos^2\theta - 1 \Rightarrow 2\cos^2\theta = \frac{3}{5} + 1 = \frac{8}{5} \Rightarrow \cos^2\theta = \frac{8}{10} = \frac{4}{5}$$

Since θ is in quadrant I, $\cos\theta > 0$. Thus, $\cos\theta = \sqrt{\frac{4}{5}} = \frac{2}{\sqrt{5}} = \frac{2\sqrt{5}}{5}$.

Since θ is in quadrant I, $\sin\theta > 0$.

$$\sin\theta = \sqrt{1-\cos^2\theta} = \sqrt{1-\left(\frac{2\sqrt{5}}{5}\right)^2} = \sqrt{1-\frac{20}{25}} = \sqrt{\frac{5}{25}} = \sqrt{\frac{1}{5}} = \frac{1}{\sqrt{5}} = \frac{1}{\sqrt{5}}\cdot\frac{\sqrt{5}}{\sqrt{5}} = \frac{\sqrt{5}}{5}$$

8. $\cos 2\theta = \frac{3}{4}, \theta$ is in quadrant III.

$$\cos 2\theta = 2\cos^2\theta - 1 \Rightarrow \frac{3}{4} = 2\cos^2\theta - 1 \Rightarrow 2\cos^2\theta = \frac{7}{4} \Rightarrow \cos^2\theta = \frac{7}{8}$$

Since θ is in quadrant III, $\cos\theta < 0$.

$$\cos\theta = -\sqrt{\frac{7}{8}} = -\frac{\sqrt{7}}{2\sqrt{2}} = -\frac{\sqrt{7}}{2\sqrt{2}}\cdot\frac{\sqrt{2}}{\sqrt{2}} = -\frac{\sqrt{14}}{4}$$

Since θ is in quadrant III, $\sin\theta < 0$.

$$\sin\theta = -\sqrt{1-\cos^2\theta} = -\sqrt{1-\frac{7}{8}} = -\sqrt{\frac{1}{8}} = -\frac{1}{2\sqrt{2}} = -\frac{1}{2\sqrt{2}}\cdot\frac{\sqrt{2}}{\sqrt{2}} = -\frac{\sqrt{2}}{4}$$

9. $\cos 2\theta = -\frac{5}{12}, \frac{\pi}{2} < \theta < \pi$

$$\cos 2\theta = 2\cos^2\theta - 1 \Rightarrow 2\cos^2\theta = \cos 2\theta + 1 = -\frac{5}{12} + 1 = \frac{7}{12} \Rightarrow \cos^2\theta = \frac{7}{24}$$

Since $\frac{\pi}{2} < \theta < \pi, \cos\theta < 0$. Thus, $\cos\theta = -\sqrt{\frac{7}{24}} = -\frac{\sqrt{7}}{\sqrt{24}} = -\frac{\sqrt{7}}{2\sqrt{6}} = -\frac{\sqrt{7}}{2\sqrt{6}} \cdot \frac{\sqrt{6}}{\sqrt{6}} = -\frac{\sqrt{42}}{12}$.

$$\sin^2\theta = 1 - \cos^2\theta = 1 - \left(-\sqrt{\frac{7}{24}}\right)^2 = 1 - \frac{7}{24} = \frac{17}{24}$$

Since $\frac{\pi}{2} < \theta < \pi, \sin\theta > 0$. Thus, $\sin\theta = \sqrt{\frac{17}{24}} = \frac{\sqrt{17}}{\sqrt{24}} = \frac{\sqrt{17}}{2\sqrt{6}} = \frac{\sqrt{17}}{2\sqrt{6}} \cdot \frac{\sqrt{6}}{\sqrt{6}} = \frac{\sqrt{102}}{12}$.

10. $\cos 2x = \frac{2}{3}, \frac{\pi}{2} < x < \pi$

$$\cos 2x = 2\cos^2 x - 1 \Rightarrow \frac{2}{3} = 2\cos^2 x - 1 \Rightarrow \cos^2 x = \frac{5}{6}$$

Since x is in quadrant II, $\cos x < 0$.

$$\cos x = -\sqrt{\frac{5}{6}} = -\frac{\sqrt{5}}{\sqrt{6}} = -\frac{\sqrt{5}}{\sqrt{6}} \cdot \frac{\sqrt{6}}{\sqrt{6}} = -\frac{\sqrt{30}}{6}$$

Since x is in quadrant II, $\sin x > 0$.

$$\sin t = \sqrt{1 - \cos^2 t} = \sqrt{1 - \left(-\sqrt{\frac{5}{6}}\right)^2} = \sqrt{1 - \frac{5}{6}} = \sqrt{\frac{1}{6}} = \frac{1}{\sqrt{6}} = \frac{1}{\sqrt{6}} \cdot \frac{\sqrt{6}}{\sqrt{6}} = \frac{\sqrt{6}}{6}$$

11. $\sin\theta = \frac{2}{5}, \cos\theta < 0$

$$\cos 2\theta = 1 - 2\sin^2\theta \Rightarrow \cos 2\theta = 1 - 2\left(\frac{2}{5}\right)^2 \; 1 - 2 \cdot \frac{4}{25} = 1 - \frac{8}{25} = \frac{17}{25}$$

$$\cos^2 2\theta + \sin^2 2\theta = 1 \Rightarrow \sin^2 2\theta = 1 - \cos^2 2\theta \Rightarrow \sin^2 2\theta = 1 - \left(\frac{17}{25}\right)^2 = 1 - \frac{289}{625} = \frac{336}{625}$$

Since $\cos\theta < 0$, $\sin 2\theta < 0$ because $\sin 2\theta = 2\sin\theta\cos\theta < 0$ and $\sin\theta > 0$.

$$\sin 2\theta = -\sqrt{\frac{336}{625}} = -\frac{\sqrt{336}}{25} = -\frac{4\sqrt{21}}{25}$$

12. $\cos\theta = -\frac{12}{13}, \sin\theta > 0$

$$\sin\theta = \sqrt{1 - \cos^2\theta} \Rightarrow \sin\theta = \sqrt{1 - \left(-\frac{12}{13}\right)^2} = \sqrt{1 - \frac{144}{169}} = \sqrt{\frac{25}{169}} = \frac{5}{13}$$

$$\sin 2\theta = 2\sin\theta\cos\theta = 2\left(\frac{5}{13}\right)\left(-\frac{12}{13}\right) = -\frac{120}{169}$$

$$\cos 2\theta = 2\cos^2\theta - 1 = 2\left(-\frac{12}{13}\right)^2 - 1 = 2 \cdot \frac{144}{169} - 1 = \frac{288}{169} - 1 = \frac{119}{169}$$

13. $\tan x = 2,\ \cos x > 0$

$$\tan 2x = \frac{2\tan x}{1-\tan^2 x} = \frac{2(2)}{1-2^2} = \frac{4}{1-4^2} = -\frac{4}{3}$$

Since both tan x and cos x are positive, x must be in quadrant I. Since $0° < x < 90°$, then $0° < 2x < 180°$. Thus, $2x$ must be in either quadrant I or quadrant II. Hence, $\sec 2x$ could be positive or negative.

$$\sec^2 2x = 1 + \tan^2 2x = 1 + \left(-\frac{4}{3}\right) = 1 + \frac{16}{9} = \frac{25}{9}$$

$$\sec 2x = \pm\frac{5}{3} \Rightarrow \cos 2x = \frac{1}{\sec 2x} = \pm\frac{3}{5}$$

$$\cos^2 2x + \sin^2 2x = 1 \Rightarrow \sin^2 2x = 1 - \cos^2 2x \Rightarrow \sin^2 2x = 1 - \left(\pm\frac{3}{5}\right)^2 \Rightarrow \sin^2 2x = 1 - \frac{9}{25} = \frac{16}{25}$$

In quadrants I and II, $\sin 2x > 0$. Thus, we have $\sin 2x = \sqrt{\frac{16}{25}} = \frac{4}{5}$.

14. $\tan x = \frac{5}{3}, \sin x < 0$

$$\sec^2 x = 1 + \tan^2 x = 1 + \left(\frac{5}{3}\right)^2 = 1 + \frac{25}{9} = \frac{34}{9}$$

Since $\sin x < 0$, and $\tan x > 0$, x is in quadrant III, so cos x and sec x are negative.

$$\sec x = -\frac{\sqrt{34}}{3}$$

$$\cos x = -\frac{3}{\sqrt{34}} = -\frac{3}{\sqrt{34}} \cdot \frac{\sqrt{34}}{\sqrt{34}} = -\frac{3\sqrt{34}}{34}$$

$$\sin\theta = -\sqrt{1-\cos^2\theta} = -\sqrt{1-\left(-\frac{3\sqrt{34}}{34}\right)} = -\sqrt{1-\frac{306}{1156}} = -\sqrt{\frac{850}{1156}} = -\frac{5\sqrt{34}}{34}$$

$$\cos 2x = 2\cos^2 x - 1 = 2\left(\frac{9}{34}\right) - 1 = \frac{9}{17} - 1 = -\frac{8}{17}$$

$$\sin 2x = 2\sin x \cos x = 2\left(\frac{5\sqrt{34}}{34}\right)\left(\frac{3\sqrt{34}}{34}\right) = \frac{15}{17}$$

15. $\sin\theta = -\frac{\sqrt{5}}{7}, \cos\theta > 0$

$$\cos^2\theta = 1 - \sin^2\theta = 1 - \left(-\frac{\sqrt{5}}{7}\right)^2 = 1 - \frac{5}{49} = \frac{44}{49}$$

Since $\cos\theta > 0$, $\cos\theta = \sqrt{\frac{44}{49}} = \frac{\sqrt{44}}{7} = \frac{2\sqrt{11}}{7}$.

$$\cos 2\theta = 1 - 2\sin^2\theta = 1 - 2\left(-\frac{\sqrt{5}}{7}\right)^2 = 1 - 2\cdot\frac{5}{49} = 1 - \frac{10}{49} = \frac{39}{49}$$

$$\sin 2\theta = 2\sin\theta\cos\theta = 2\left(-\frac{\sqrt{5}}{7}\right)\left(\frac{2\sqrt{11}}{7}\right) = -\frac{4\sqrt{55}}{49}$$

16. $\cos\theta = \frac{\sqrt{3}}{5}, \sin\theta > 0$

$$\sin\theta = \sqrt{1-\cos^2\theta} = \sqrt{1-\left(\frac{\sqrt{3}}{5}\right)^2} = \sqrt{1-\frac{3}{25}} = \sqrt{\frac{22}{25}} = \frac{\sqrt{22}}{5}$$

$$\cos 2\theta = 2\cos^2\theta - 1 = 2\left(\frac{\sqrt{3}}{5}\right)^2 - 1 = 2\cdot\frac{3}{25} - 1 = \frac{6}{25} - 1 = -\frac{19}{25}$$

$$\sin 2\theta = 2\sin\theta\cos\theta = 2\left(\frac{\sqrt{22}}{5}\right)\left(\frac{\sqrt{3}}{5}\right) = \frac{2\sqrt{66}}{25}$$

17. $\cos^2 15° - \sin^2 15° = \cos\left[2(15°)\right] = \cos 30° = \frac{\sqrt{3}}{2}$

18. $\frac{2\tan 15°}{1-\tan^2 15°} = \frac{\tan 15° + \tan 15°}{1-\tan^2 15°} = \tan 2(15°) = \tan 30° = \frac{\sqrt{3}}{3}$

19. $1 - 2\sin^2 15° = \cos\left[2(15°)\right] = \cos 30° = \frac{\sqrt{3}}{2}$

20. $1 - 2\sin^2 22\frac{1}{2}^\circ = \cos 2\left(22\frac{1}{2}^\circ\right) = \cos 45° = \frac{\sqrt{2}}{2}$

21. $2\cos^2 67\frac{1}{2}^\circ - 1 = \cos^2 67\frac{1}{2}^\circ - \sin^2 67\frac{1}{2}^\circ = \cos 2\left(67\frac{1}{2}^\circ\right) = \cos 135° = -\frac{\sqrt{2}}{2}$

22. $\cos^2\frac{\pi}{8} - \frac{1}{2} = \frac{1}{2}\left(2\cos^2\frac{\pi}{8} - 1\right) = \frac{1}{2}\left[\cos\left(2\cdot\frac{\pi}{8}\right)\right] = \frac{1}{2}\cos\frac{\pi}{4} = \frac{1}{2}\left(\frac{\sqrt{2}}{2}\right) = \frac{\sqrt{2}}{4}$

23. $\frac{\tan 51°}{1-\tan^2 51°}$

Since $\frac{2\tan A}{1-\tan^2 A} = \tan 2A$, we have the following.

$$\frac{1}{2}\left(\frac{2\tan A}{1-\tan^2 A}\right) = \frac{1}{2}\tan 2A \Rightarrow \frac{\tan A}{1-\tan^2 A} = \frac{1}{2}\tan 2A \Rightarrow \frac{\tan 51°}{1-\tan^2 51°} = \frac{1}{2}\tan\left[2(51°)\right] = \frac{1}{2}\tan(102°)$$

24. $\frac{\tan 34°}{2(1-\tan^2 34°)} = \frac{1}{4}\left(\frac{2\tan 34°}{1-\tan^2 34°}\right) = \frac{1}{4}\tan\left[2(34°)\right] = \frac{1}{4}\tan 68°$

25. $\frac{1}{4} - \frac{1}{2}\sin^2 47.1° = \frac{1}{4}(1 - 2\sin^2 47.1°) = \frac{1}{4}\cos\left[2(47.1°)\right] = \frac{1}{4}\cos 94.2°$

26. $\frac{1}{8}\sin 29.5°\cos 29.5° = \frac{1}{16}(2\sin 29.5°\cos 29.5°) = \frac{1}{16}\sin\left[2(29.5°)\right] = \frac{1}{16}\sin 59°$

27. $\sin^2 \frac{2\pi}{5} - \cos^2 \frac{2\pi}{5} = -\left(\cos^2 \frac{2\pi}{5} - \sin^2 \frac{2\pi}{5}\right)$

Since $\cos^2 A - \sin^2 A = -\cos 2A$, $-\left(\cos^2 \frac{2\pi}{5} - \sin^2 \frac{2\pi}{5}\right) = -\cos\left(2 \cdot \frac{2\pi}{5}\right) = -\cos \frac{4\pi}{5}$.

28. $\cos^2 2x - \sin^2 2x = \cos\left[2(2x)\right] = \cos 4x$

29. $\cos 3x = \cos(2x + x) = \cos 2x \cos x - \sin 2x \sin x = \left(1 - 2\sin^2 x\right)\cos x - (2\sin x \cos x)\sin x$

$= \cos x - 2\sin^2 x \cos x - 2\sin^2 x \cos x = \cos x - 4\sin^2 x \cos x = \cos x\left(1 - 4\sin^2 x\right)$

$= \cos x\left[1 - 4\left(1 - \cos^2 x\right)\right] = \cos x\left(-3 + 4\cos^2 x\right) = -3\cos x + 4\cos^3 x$

30. $\sin 4x = \sin\left[2(2x)\right] = 2\sin 2x \cos 2x = 2(2\sin x \cos x)\left(\cos^2 x - \sin^2 x\right) = 4\sin x \cos^3 x - 4\sin^3 x \cos x$

31. $\tan 3x = \tan(2x + x) = \dfrac{\tan 2x + \tan x}{1 - \tan 2x \tan x} = \dfrac{\dfrac{2\tan x}{1 - \tan^2 x} + \tan x}{1 - \dfrac{2\tan x}{1 - \tan^2 x} \cdot \tan x} = \dfrac{\dfrac{2\tan x}{1 - \tan^2 x} + \dfrac{\left(1 - \tan^2 x\right)\tan x}{1 - \tan^2 x}}{\dfrac{1 - \tan^2 x}{1 - \tan^2 x} - \dfrac{2\tan^2 x}{1 - \tan^2 x}}$

$= \dfrac{\dfrac{2\tan x}{1 - \tan^2 x} + \dfrac{\tan x - \tan^3 x}{1 - \tan^2 x}}{\dfrac{1 - \tan^2 x}{1 - \tan^2 x} - \dfrac{2\tan^2 x}{1 - \tan^2 x}} \cdot \dfrac{1 - \tan^2 x}{1 - \tan^2 x} = \dfrac{2\tan x + \tan x - \tan^3 x}{1 - \tan^2 x - 2\tan^2 x} = \dfrac{3\tan x - \tan^3 x}{1 - 3\tan^2 x}$

32. $\cos 4x = \cos\left[2(2x)\right] = 2\cos^2 2x - 1 = 2\left(2\cos^2 x - 1\right)^2 - 1$

$= 2\left(4\cos^4 x - 4\cos^2 x + 1\right) - 1 = 8\cos^4 x - 8\cos^2 x + 2 - 1 = 8\cos^4 x - 8\cos^2 x + 1$

Exercises 33 – 36 are graphed in the following window.

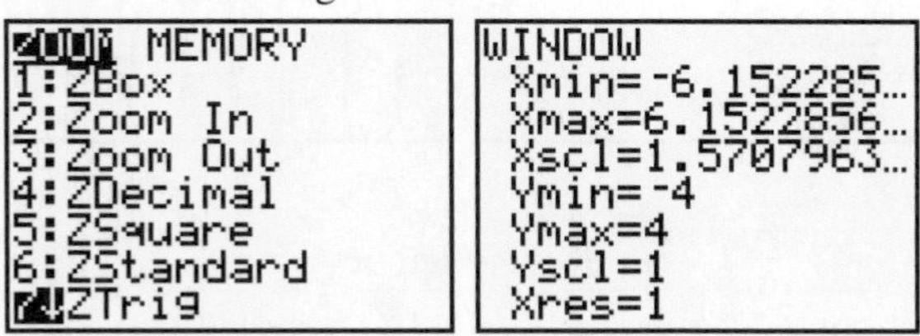

33. $\cos^4 x - \sin^4 x$ appears to be equivalent to cos 2*x*.

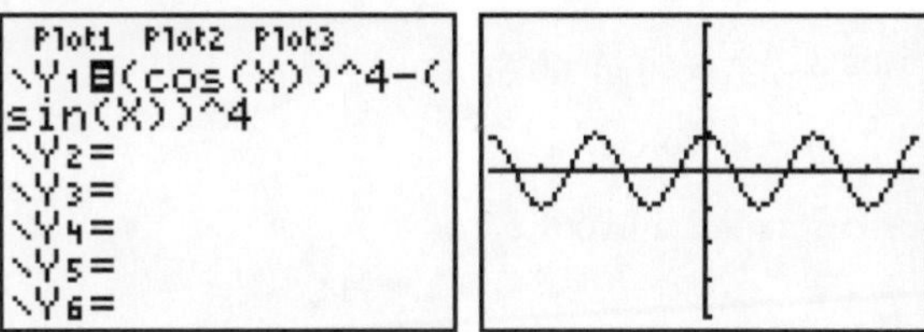

$\cos^4 x - \sin^4 x = \left(\cos^2 x + \sin^2 x\right)\left(\cos^2 x - \sin^2 x\right) = 1 \cdot \cos 2x = \cos 2x$

34. $\dfrac{4\tan x\cos^2 x-2\tan x}{1-\tan^2 x}$ appears to be equivalent to $\sin 2x$.

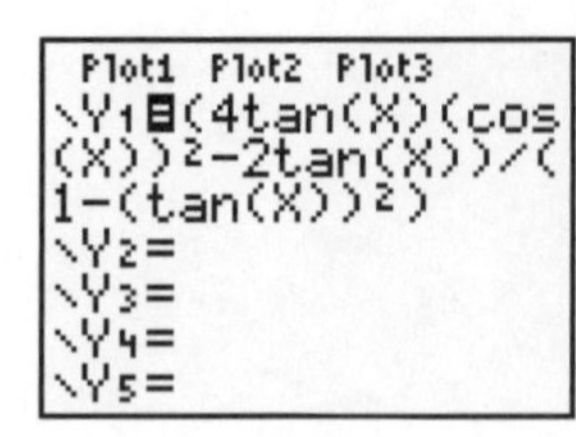

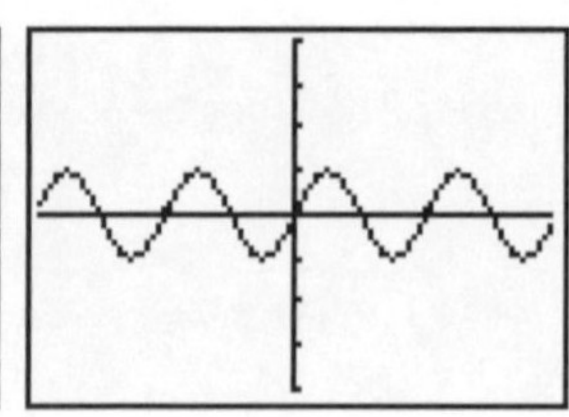

$$\frac{4\tan x\cos^2 x-2\tan x}{1-\tan^2 x}=\frac{2\tan x\left(2\cos^2 x-1\right)}{1-\tan^2 x}$$
$$=\frac{2\tan x}{1-\tan^2 x}\left(2\cos^2 x-1\right)=\tan 2x\cos 2x=\frac{\sin 2x}{\cos 2x}\cdot\cos 2x=\sin 2x$$

35. $\dfrac{2\tan x}{2-\sec^2 x}$ appears to be equivalent to $\tan 2x$.

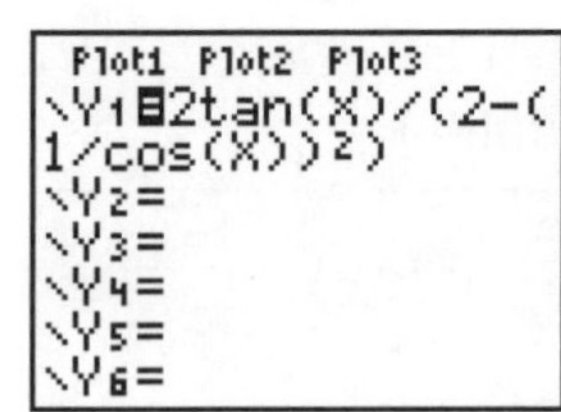

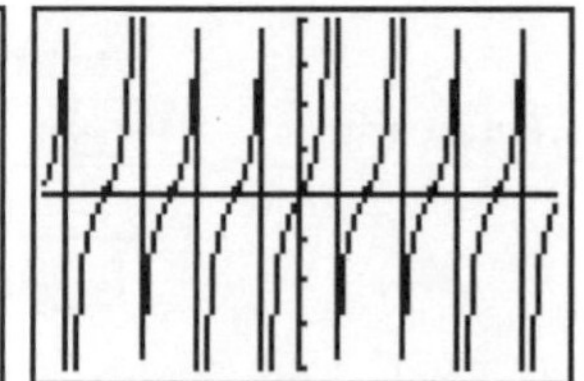

$$\frac{2\tan x}{2-\sec^2 x}=\frac{2\tan x}{1-\left(\sec^2 x-1\right)}=\frac{2\tan x}{1-\tan^2 x}=\tan 2x$$

36. $\dfrac{\cot^2 x-1}{2\cot x}$ appears to be equivalent to $\cot 2x$.

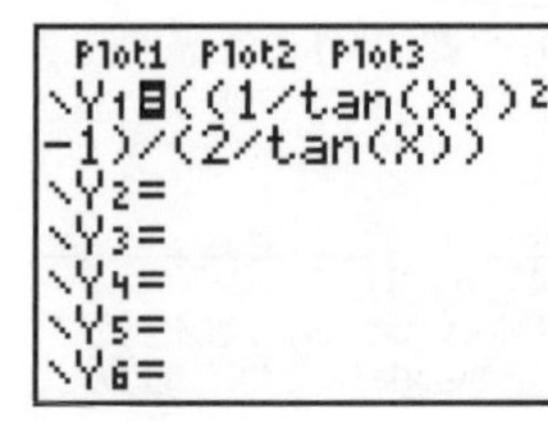

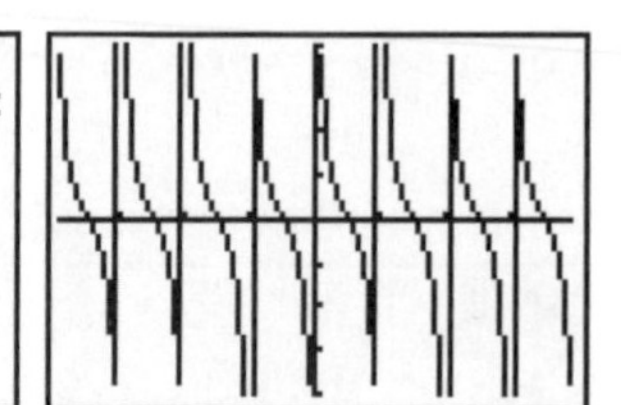

$$\frac{\cot^2 x-1}{2\cot x}=\frac{\frac{1}{\tan^2 x}-1}{\frac{2}{\tan x}}=\frac{\frac{1}{\tan^2 x}-1}{\frac{2}{\tan x}}\cdot\frac{\tan^2 x}{\tan^2 x}=\frac{1-\tan^2 x}{2\tan x}=\frac{1}{\frac{2\tan x}{1-\tan^2 x}}=\frac{1}{\tan 2x}=\cot 2x$$

37. Verify $(\sin x+\cos x)^2=\sin 2x+1$ is an identity.

$$\begin{aligned}(\sin x+\cos x)^2&=\sin^2 x+2\sin x\cos x+\cos^2 x\\&=\left(\sin^2 x+\cos^2 x\right)+2\sin x\cos x\\&=1+\sin 2x\end{aligned}$$

38. Verify $\sec 2x = \dfrac{\sec^2 x + \sec^4 x}{2 + \sec^2 x - \sec^4 x}$ is an identity.

Work with the right side.

$$\frac{\sec^2 x + \sec^4 x}{2 + \sec^2 x - \sec^4 x} = \frac{\frac{1}{\cos^2 x} + \frac{1}{\cos^4 x}}{2 + \frac{1}{\cos^2 x} - \frac{1}{\cos^4 x}} = \frac{\frac{1}{\cos^2 x} + \frac{1}{\cos^4 x}}{2 + \frac{1}{\cos^2 x} - \frac{1}{\cos^4 x}} \cdot \frac{\cos^4 x}{\cos^4 x} = \frac{\cos^2 x + 1}{2\cos^4 x + \cos^2 x - 1}$$

$$= \frac{\cos^2 x + 1}{\left(2\cos^2 x - 1\right)\left(\cos^2 x + 1\right)} = \frac{1}{2\cos^2 x - 1} = \frac{1}{\cos 2x} = \sec 2x$$

39. Verify $\tan 8k - \tan 8k \tan^2 4k = 2 \tan 4k$ is an identity.

$$\tan 8k - \tan 8k \tan^2 4k = \tan 8k\left(1 - \tan^2 4k\right) = \frac{2 \tan 4k}{1 - \tan^2 4k}\left(1 - \tan^2 4k\right) = 2 \tan 4k$$

40. Verify $\sin 2x = \dfrac{2 \tan x}{1 + \tan^2 x}$ is an identity.

Work with the right side.

$$\frac{2 \tan x}{1 + \tan^2 x} = \frac{2 \cdot \frac{\sin x}{\cos x}}{1 + \frac{\sin^2 x}{\cos^2 x}} = \frac{2 \cdot \frac{\sin x}{\cos x}}{1 + \frac{\sin^2 x}{\cos^2 x}} \cdot \frac{\cos^2 x}{\cos^2 x} = \frac{2 \sin x \cos x}{\cos^2 x + \sin^2 x} = 2 \sin x \cos x = \sin 2x$$

41. Verify $\cos 2y = \dfrac{2 - \sec^2 y}{\sec^2 y}$ is an identity.

Work with the right side.

$$\frac{2 - \sec^2 y}{\sec^2 y} = \frac{2 - \frac{1}{\cos^2 y}}{\frac{1}{\cos^2 y}} = \frac{2 - \frac{1}{\cos^2 y}}{\frac{1}{\cos^2 y}} \cdot \frac{\cos^2 y}{\cos^2 y} = \frac{2\cos^2 y - 1}{1} = \cos 2y$$

42. Verify $-\tan 2\theta = \dfrac{2 \tan \theta}{\sec^2 \theta - 2}$ is an identity.

Work with the right side.

$$\frac{2 \tan \theta}{\sec^2 \theta - 2} = \frac{2 \tan \theta}{\left(1 + \tan^2 \theta\right) - 2} = \frac{2 \tan \theta}{\tan^2 \theta - 1} = -\tan 2\theta$$

43. Verify that $\sin 4x = 4 \sin x \cos x \cos 2x$ is an identity.

$$\sin 4x = \sin 2\left(2x\right) = 2 \sin 2x \cos 2x = 2\left(2 \sin x \cos x\right) \cos 2x = 4 \sin x \cos x \cos 2x$$

44. Verify $\dfrac{1 + \cos 2x}{\sin 2x} = \cot x$ is an identity.

$$\frac{1 + \cos 2x}{\sin 2x} = \frac{1 + \left(2\cos^2 x - 1\right)}{2 \sin x \cos x} = \frac{2\cos^2 x}{2 \sin x \cos x} = \frac{\cos x}{\sin x} = \cot x$$

45. Verify $\tan(\theta-45°)+\tan(\theta+45°)=2\tan 2\theta$ is an identity.

$$\tan(\theta-45°)+\tan(\theta+45°)=\frac{\tan\theta-\tan 45°}{1+\tan\theta\tan 45°}+\frac{\tan\theta+\tan 45°}{1-\tan\theta\tan 45°}=\frac{\tan\theta-1}{1+\tan\theta}+\frac{\tan\theta+1}{1-\tan\theta}$$
$$=\frac{\tan\theta-1}{\tan\theta+1}-\frac{\tan\theta+1}{\tan\theta-1}=\frac{\tan\theta-1}{\tan\theta+1}\cdot\frac{\tan\theta-1}{\tan\theta-1}-\frac{\tan\theta+1}{\tan\theta-1}\cdot\frac{\tan\theta+1}{\tan\theta+1}$$
$$=\frac{(\tan\theta-1)^2}{(\tan\theta+1)(\tan\theta-1)}-\frac{(\tan\theta+1)^2}{(\tan\theta+1)(\tan\theta-1)}$$
$$=\frac{\tan^2\theta-2\tan\theta+1}{(\tan\theta+1)(\tan\theta-1)}-\frac{\tan^2\theta+2\tan\theta+1}{(\tan\theta+1)(\tan\theta-1)}$$
$$=\frac{(\tan^2\theta-2\tan\theta+1)-(\tan^2\theta+2\tan\theta+1)}{(\tan\theta+1)(\tan\theta-1)}$$
$$=\frac{\tan^2\theta-2\tan\theta+1-\tan^2\theta-2\tan\theta-1}{(\tan\theta+1)(\tan\theta-1)}$$
$$=\frac{-4\tan\theta}{\tan^2\theta-1}=\frac{4\tan\theta}{1-\tan^2\theta}=\frac{2(2\tan\theta)}{1-\tan^2\theta}=2\tan 2\theta$$

46. Verify $\cot 4\theta=\frac{1-\tan^2 2\theta}{2\tan 2\theta}$ is an identity.

$$\cot 4\theta=\frac{1}{\tan 4\theta}=\frac{1}{\frac{2\tan 2\theta}{1-\tan^2 2\theta}}=\frac{1}{\frac{2\tan 2\theta}{1-\tan^2 2\theta}}\cdot\frac{1-\tan^2 2\theta}{1-\tan^2 2\theta}=\frac{1-\tan^2 2\theta}{2\tan 2\theta}$$

47. Verify $\frac{2\cos 2\theta}{\sin 2\theta}=\cot\theta-\tan\theta$ is an identity.

Work with the right side.

$$\cot\theta-\tan\theta=\frac{\cos\theta}{\sin\theta}-\frac{\sin\theta}{\cos\theta}$$
$$=\frac{\cos\theta}{\sin\theta}\cdot\frac{\cos\theta}{\cos\theta}-\frac{\sin\theta}{\cos\theta}\cdot\frac{\sin\theta}{\sin\theta}$$
$$=\frac{\cos^2\theta-\sin^2\theta}{\sin\theta\cos\theta}$$
$$=\frac{2(\cos^2\theta-\sin^2\theta)}{2\sin\theta\cos\theta}=\frac{2\cos 2\theta}{\sin 2\theta}$$

48. Verify $\sin 4x=4\sin x\cos x-8\sin^3 x\cos x$ is an identity.

$$\sin 4x=\sin 2(2x)=2\sin 2x\cos 2x=2(2\sin x\cos x)(1-2\sin^2 x)=4\sin x\cos x-8\sin^3 x\cos x$$

49. Verify $\sin 2A\cos 2A=\sin 2A-4\sin^3 A\cos A$ is an identity.

$$\sin 2A\cos 2A=(2\sin A\cos A)(1-2\sin^2 A)=2\sin A\cos A-4\sin^3 A\cos A=\sin 2A-4\sin^3 A\cos A$$

50. Verify $\cos 2x = \dfrac{1-\tan^2 x}{1+\tan^2 x}$ is an identity.

Work with the right side.

$$\frac{1-\tan^2 x}{1+\tan^2 x} = \frac{1-\frac{\sin^2 x}{\cos^2 x}}{1+\frac{\sin^2 x}{\cos^2 x}} = \frac{1-\frac{\sin^2 x}{\cos^2 x}}{1+\frac{\sin^2 x}{\cos^2 x}} \cdot \frac{\cos^2 x}{\cos^2 x} = \frac{\cos^2 x - \sin^2 x}{\cos^2 x + \sin^2 x} = \frac{\cos^2 x - \sin^2 x}{1} = \cos^2 x - \sin^2 x = \cos 2x$$

51. Verify $\tan s + \cot s = 2\csc 2s$ is an identity.

$$\tan s + \cot s = \frac{\sin s}{\cos s}\cdot\frac{\sin s}{\sin s} + \frac{\cos s}{\sin s}\cdot\frac{\cos s}{\cos s} = \frac{\sin^2 s + \cos^2 s}{\cos s \sin s} = \frac{1}{\cos s \sin s} = \frac{2}{2\cos s \sin s} = \frac{2}{\sin 2s} = 2\csc 2s$$

52. Verify $\dfrac{\cot A - \tan A}{\cot A + \tan A} = \cos 2A$ is an identity.

$$\frac{\cot A - \tan A}{\cot A + \tan A} = \frac{\frac{\cos A}{\sin A} - \frac{\sin A}{\cos A}}{\frac{\cos A}{\sin A} + \frac{\sin A}{\cos A}} = \frac{\frac{\cos A}{\sin A} - \frac{\sin A}{\cos A}}{\frac{\cos A}{\sin A} + \frac{\sin A}{\cos A}} \cdot \frac{\sin A \cos A}{\sin A \cos A}$$

$$= \frac{\cos^2 A - \sin^2 A}{\cos^2 A + \sin^2 A} = \frac{\cos^2 A - \sin^2 A}{1} = \cos^2 A - \sin^2 A = \cos 2A$$

53. Verify $1 + \tan x \tan 2x = \sec 2x$ is an identity.

$$1 + \tan x \tan 2x = 1 + \tan x\left(\frac{2\tan x}{1-\tan^2 x}\right) = 1 + \frac{2\tan^2 x}{1-\tan^2 x} = \frac{1-\tan^2 x + 2\tan^2 x}{1-\tan^2 x} = \frac{1+\tan^2 x}{1-\tan^2 x}$$

$$= \frac{1+\frac{\sin^2 x}{\cos^2 x}}{1-\frac{\sin^2 x}{\cos^2 x}} = \frac{1+\frac{\sin^2 x}{\cos^2 x}}{1-\frac{\sin^2 x}{\cos^2 x}} \cdot \frac{\cos^2 x}{\cos^2 x} = \frac{\cos^2 x + \sin^2 x}{\cos^2 x - \sin^2 x} = \frac{1}{\cos^2 x - \sin^2 x} = \frac{1}{\cos 2x} = \sec 2x$$

54. Verify $\cot\theta \tan(\theta+\pi) - \sin(\pi-\theta)\cos\left(\frac{\pi}{2}-\theta\right) = \cos^2\theta$ is an identity.

$$\begin{aligned}
\cot\theta \tan(\theta+\pi) - \sin(\pi-\theta)\cos\left(\frac{\pi}{2}-\theta\right) &= \cot\theta \cdot \frac{\tan\theta + \tan\pi}{1-\tan\theta\tan\pi} \\
&\quad -(\sin\pi\cos\theta - \cos\pi\sin\theta)\left(\cos\frac{\pi}{2}\cos\theta + \sin\frac{\pi}{2}\sin\theta\right) \\
&= \cot\theta \cdot \frac{\tan\theta + 0}{1-\tan\theta(0)} \\
&\quad -(0\cdot\cos\theta - (-1)\sin\theta)(0\cdot\cos\theta + 1\cdot\sin\theta) \\
&= \cot\theta \cdot \frac{\tan\theta}{1-0} - (0+\sin\theta)(0+\sin\theta) \\
&= \cot\theta\cdot\tan\theta - \sin^2\theta = 1 - \sin^2\theta = \cos^2\theta
\end{aligned}$$

55. $2\sin 58^\circ \cos 102^\circ = 2\left(\frac{1}{2}\left[\sin(58^\circ + 102^\circ) + \sin(58^\circ - 102^\circ)\right]\right) = \sin 160^\circ + \sin(-44^\circ) = \sin 160^\circ - \sin 44^\circ$

56. $2\cos 85°\sin 140° = 2\left(\frac{1}{2}\left[\sin(85°+140°)-\sin(85°-140°)\right]\right) = \sin 225° - \sin(-55°) = \sin 225° + \sin 55°$

57. $5\cos 3x\cos 2x = 5\left(\frac{1}{2}\left[\cos(3x+2x)+\cos(3x-2x)\right]\right) = \frac{5}{2}(\cos 5x+\cos x) = \frac{5}{2}\cos 5x + \frac{5}{2}\cos x$

58. $\sin 4x\sin 5x = \frac{1}{2}\left[\cos(4x-5x)-\cos(4x+5x)\right] = \frac{1}{2}\cos(-x) - \frac{1}{2}\cos 9x = \frac{1}{2}\cos x - \frac{1}{2}\cos 9x$

59. $\cos 4x - \cos 2x = -2\sin\left(\frac{4x+2x}{2}\right)\sin\left(\frac{4x-2x}{2}\right) = -2\sin\frac{6x}{2}\sin\frac{2x}{2} = -2\sin 3x\sin x$

60. $\cos 5x + \cos 8x = 2\cos\left(\frac{5x+8x}{2}\right)\cos\left(\frac{5x-8x}{2}\right) = 2\cos\frac{13x}{2}\cos\frac{-3x}{2} = 2\cos 6.5x\cos 1.5x$

61. $\sin 25° + \sin(-48°) = 2\sin\left(\frac{25°+(-48°)}{2}\right)\cos\left(\frac{25°-(-48°)}{2}\right)$

$$= 2\sin\frac{-23°}{2}\cos\frac{73°}{2} = 2\sin(-11.5°)\cos 36.5° = -2\sin 11.5°\cos 36.5°$$

62. $\sin 102° - \sin 95° = 2\cos\left(\frac{102°+95°}{2}\right)\sin\left(\frac{102°-95°}{2}\right) = 2\cos\frac{197°}{2}\sin\frac{7°}{2} = 2\cos 98.5°\sin 3.5°$

63. $\cos 4x + \cos 8x = 2\cos\left(\frac{4x+8x}{2}\right)\cos\left(\frac{4x-8x}{2}\right)$

$$= 2\cos\frac{12x}{2}\cos\frac{-4x}{2} = 2\cos 6x\cos(-2x) = 2\cos 6x\cos 2x$$

64. $\sin 9x - \sin 3x = 2\cos\left(\frac{9x+3x}{2}\right)\sin\left(\frac{9x-3x}{2}\right) = 2\cos\frac{12x}{2}\sin\frac{6x}{2} = 2\cos 6x\sin 3x$

65. From Example 6, $W = \frac{(163\sin 120\pi t)^2}{15} \Rightarrow W \approx 1771.3(\sin 120\pi t)^2$.

Thus, we have the following.

$$\begin{aligned} 1771.3(\sin 120\pi t)^2 &= 1771.3\sin 120\pi t\cdot\sin 120\pi t \\ &= (1771.3)\left(\frac{1}{2}\right)\left[\cos(120\pi t-120\pi t) - \cos(120\pi t+120\pi t)\right] \\ &= 885.6(\cos 0 - \cos 240\pi t) = 885.6(1-\cos 240\pi t) \\ &= -885.6\cos 240\pi t + 885.6 \end{aligned}$$

If we compare this to $W = a\cos(\omega t)+c$, then $a = -885.6$, $c = 885.6$, and $\omega = 240\pi$.

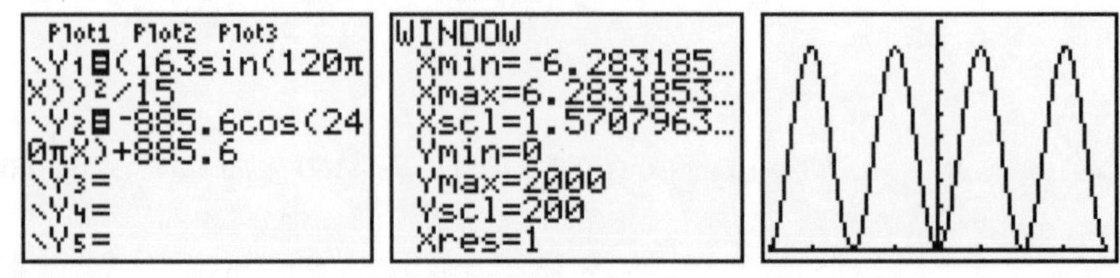

66. **(a)** Graph $W = VI = \left[163\sin(120\pi t)\right]\left[1.23\sin(120\pi t)\right]$ over the interval $0 \le t \le .05$.

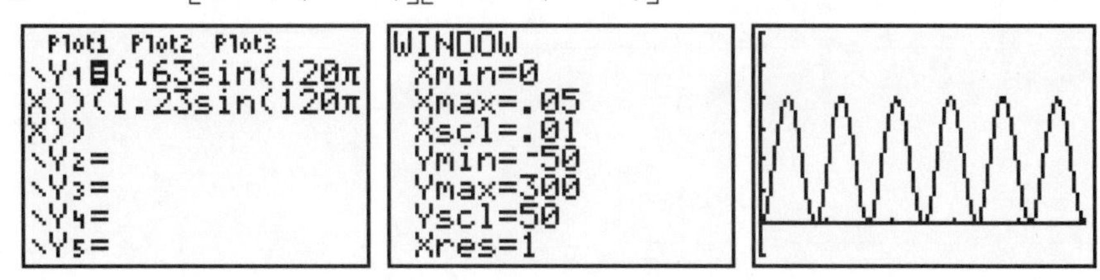

(b) The minimum wattage is 0 and the maximum wattage occurs whenever $\sin(120\pi t) = 1$. This would be 163(1.23) = 200.49 watts.

(c) $\left[163\sin(120\pi t)\right]\left[1.23\sin(120\pi t)\right] = 200.49\sin^2(120\pi t)$

$$= 200.49\left[\frac{1}{2}(1 - \cos 240\pi t)\right] = -100.245\cos 240\pi t + 100.245$$

Then $a = -100.245$, $\omega = 240\pi$, and $c = 100.245$.

(d) The graphs of $W = \left[163\sin(120\pi t)\right]\left[1.23\sin(120\pi t)\right]$ and $W = -100.245\cos 240\pi t + 100.245$ are the same.

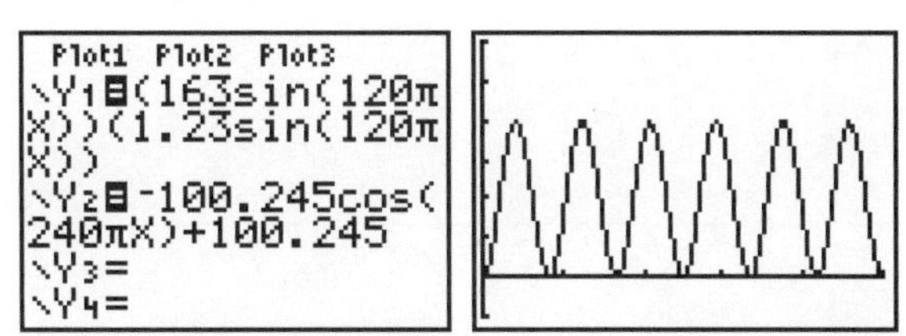

(e) The graph of W is vertically centered about the line $y = 100.245$. An estimate for the average wattage consumed is 100.245 watts. (For sinusoidal current, the average wattage consumed by an electrical device will be equal to half of the peak wattage.)

Section 5.6: Half-Angle Identities

1. Since 195° is in quadrant III and since the sine is negative in quadrant III, use the negative square root.

2. Since 58° is in quadrant I and since the cosine is positive in quadrant I, use the positive square root.

3. Since 225° is in quadrant III and since the tangent is positive in quadrant III, use the positive square root.

4. Since −10° is in quadrant IV and since the sine is negative in quadrant IV, use the negative square root.

5. Since $\sin 15° = \sqrt{\frac{1 - \cos 30°}{2}} = \sqrt{\frac{1 - \frac{\sqrt{3}}{2}}{2}} = \sqrt{\frac{2 - \sqrt{3}}{2 \cdot 2}} = \frac{\sqrt{2 - \sqrt{3}}}{\sqrt{4}} = \frac{\sqrt{2 - \sqrt{3}}}{2}$, the correct choice is C.

6. Since $\tan 15° = \frac{1 - \cos 30°}{\sin 30°} = \frac{1 - \frac{\sqrt{3}}{2}}{\frac{1}{2}} = \frac{1 - \frac{\sqrt{3}}{2}}{\frac{1}{2}} \cdot \frac{2}{2} = \frac{2 - \sqrt{3}}{1} = 2 - \sqrt{3}$, the correct choice is A.

7. Since $\cos\frac{\pi}{8}=\sqrt{\frac{1+\cos\frac{\pi}{4}}{2}}=\sqrt{\frac{1+\frac{\sqrt{2}}{2}}{2}}=\sqrt{\frac{1+\frac{\sqrt{2}}{2}}{2}\cdot\frac{2}{2}}=\frac{\sqrt{2+\sqrt{2}}}{\sqrt{4}}=\frac{\sqrt{2+\sqrt{2}}}{2}$, the correct choice is D.

8. Since $\tan\left(-\frac{\pi}{8}\right)=\frac{1-\cos\left(-\frac{\pi}{4}\right)}{\sin\left(-\frac{\pi}{4}\right)}=\frac{1-\cos\frac{7\pi}{4}}{\sin\frac{7\pi}{4}}$

$$=\frac{1-\frac{\sqrt{2}}{2}}{-\frac{\sqrt{2}}{2}}=\frac{1-\frac{\sqrt{2}}{2}}{-\frac{\sqrt{2}}{2}}\cdot\frac{2}{2}=\frac{2-\sqrt{2}}{-\sqrt{2}}\cdot\frac{\sqrt{2}}{\sqrt{2}}=\frac{2\sqrt{2}-2}{-2}=\frac{2-2\sqrt{2}}{2}=1-\sqrt{2},$$

the correct choice is E.

9. Since $\tan 67.5^\circ=\frac{1-\cos 135^\circ}{\sin 135^\circ}$

$$=\frac{1-\left(-\frac{\sqrt{2}}{2}\right)}{\frac{\sqrt{2}}{2}}=\frac{1-\left(-\frac{\sqrt{2}}{2}\right)}{\frac{\sqrt{2}}{2}}\cdot\frac{2}{2}=\frac{2+\sqrt{2}}{\sqrt{2}}=\frac{2+\sqrt{2}}{\sqrt{2}}\cdot\frac{\sqrt{2}}{\sqrt{2}}=\frac{2\sqrt{2}+2}{2}=1+\sqrt{2},$$

the correct choice is F.

10. Since $\cos 67.5^\circ=\sqrt{\frac{1-\cos 135^\circ}{2}}$

$$=\sqrt{\frac{1+\left(-\frac{\sqrt{2}}{2}\right)}{2}}=\sqrt{\frac{1+\left(-\frac{\sqrt{2}}{2}\right)}{2}\cdot\frac{2}{2}}=\sqrt{\frac{2-\sqrt{2}}{2}\cdot\frac{1}{2}}=\frac{\sqrt{2-\sqrt{2}}}{\sqrt{4}}=\frac{\sqrt{2-\sqrt{2}}}{2},$$

the correct choice is B.

11. $\sin 67.5^\circ=\sin\left(\frac{135^\circ}{2}\right)$

Since 67.5° is in quadrant I, $\sin 67.5^\circ > 0$.

$$\sin 67.5^\circ=\sqrt{\frac{1-\cos 135^\circ}{2}}=\sqrt{\frac{1-(-\cos 45^\circ)}{2}}=\sqrt{\frac{1+\frac{\sqrt{2}}{2}}{2}}=\sqrt{\frac{1+\frac{\sqrt{2}}{2}}{2}\cdot\frac{2}{2}}=\frac{\sqrt{2+\sqrt{2}}}{\sqrt{4}}=\frac{\sqrt{2+\sqrt{2}}}{2}$$

12. $\sin 195° = \sin\left(\frac{390°}{2}\right)$

Since 195° is in quadrant III, sin 195° < 0.

$$\sin 195° = -\sqrt{\frac{1-\cos 390°}{2}} = -\sqrt{\frac{1-\cos 30°}{2}} = -\sqrt{\frac{1-\frac{\sqrt{3}}{2}}{2}} = -\sqrt{\frac{1-\frac{\sqrt{3}}{2}}{2}\cdot\frac{2}{2}} = -\frac{\sqrt{2-\sqrt{3}}}{\sqrt{4}} = \frac{\sqrt{2-\sqrt{3}}}{2}$$

13. $\cos 195° = \cos\left(\frac{390°}{2}\right)$

Since 195° is in quadrant III, cos 195° < 0.

$$\cos 195° = -\sqrt{\frac{1+\cos 390°}{2}} = -\sqrt{\frac{1+\cos 30°}{2}} = -\sqrt{\frac{1+\frac{\sqrt{3}}{2}}{2}} = -\sqrt{\frac{2+\sqrt{3}}{4}} = \frac{-\sqrt{2+\sqrt{3}}}{2}$$

14. $$\tan 195° = \tan\left(\frac{390°}{2}\right) = \frac{\sin 390°}{1+\cos 390°} = \frac{\sin 30°}{1+\cos 30°} = \frac{\frac{1}{2}}{1+\frac{\sqrt{3}}{2}} = \frac{\frac{1}{2}}{1+\frac{\sqrt{3}}{2}}\cdot\frac{2}{2} = \frac{1}{2+\sqrt{3}}$$

$$= \frac{1}{2+\sqrt{3}}\cdot\frac{2-\sqrt{3}}{2-\sqrt{3}} = \frac{2-\sqrt{3}}{4-3} = \frac{2-\sqrt{3}}{1} = 2-\sqrt{3}$$

15. $\cos 165° = \cos\left(\frac{330°}{2}\right)$

Since 165° is in quadrant II, cos 165° < 0.

$$\cos 165° = -\sqrt{\frac{1+\cos 330°}{2}} = -\sqrt{\frac{1+\cos 30°}{2}} = -\sqrt{\frac{1+\frac{\sqrt{3}}{2}}{2}} = -\sqrt{\frac{2+\sqrt{3}}{4}} = \frac{-\sqrt{2+\sqrt{3}}}{2}$$

16. $\sin 165° = \sin\left(\frac{330°}{2}\right)$

Since 165° is in quadrant II, sin 165° > 0.

$$\sin 165° = \sqrt{\frac{1-\cos 330°}{2}} = \sqrt{\frac{1-\cos 30°}{2}} = \sqrt{\frac{1-\frac{\sqrt{3}}{2}}{2}} = \sqrt{\frac{2-\sqrt{3}}{4}} = \frac{\sqrt{2-\sqrt{3}}}{2}$$

17. To find $\sin 7.5°$, you could use the half-angle formulas for sine and cosine as follows.

$$\sin 7.5° = \sqrt{\frac{1-\cos 15°}{2}} \text{ and } \cos 15° = -\sqrt{\frac{1+\cos 30°}{2}} = \sqrt{\frac{1+\frac{\sqrt{3}}{2}}{2}} = \sqrt{\frac{2+\sqrt{3}}{4}} = \frac{\sqrt{2+\sqrt{3}}}{2}$$

Thus, $$\sin 7.5° = \sqrt{\frac{1-\cos 15°}{2}} = \sqrt{\frac{1-\frac{\sqrt{2+\sqrt{3}}}{2}}{2}} = \sqrt{\frac{2-\sqrt{2+\sqrt{3}}}{4}} = \frac{\sqrt{2-\sqrt{2+\sqrt{3}}}}{2}.$$

18. Show $\sqrt{3-2\sqrt{2}}=\sqrt{2}-1$.

$$3-2\sqrt{2}=3-2\sqrt{2}$$
$$3-2\sqrt{2}=2-2\sqrt{2}+1$$
$$3-2\sqrt{2}=\left(\sqrt{2}\right)^2-2\sqrt{2}+1^2$$
$$\left(\sqrt{3-2\sqrt{2}}\right)^2=\left(\sqrt{2}-1\right)^2$$

If $a^2=b^2$, then $a=b$. Thus, $\sqrt{3-2\sqrt{2}}=\sqrt{2}-1$.

19. Find $\cos\frac{x}{2}$, given $\cos x=\frac{1}{4}$, with $0<x<\frac{\pi}{2}$.

Since $0<x<\frac{\pi}{2}\Rightarrow 0<\frac{x}{2}<\frac{\pi}{4}$, $\cos\frac{x}{2}>0$.

$$\cos\frac{x}{2}=\sqrt{\frac{1+\cos x}{2}}=\sqrt{\frac{1+\frac{1}{4}}{2}}=\sqrt{\frac{4+1}{8}}=\sqrt{\frac{5}{8}}=\frac{\sqrt{5}}{\sqrt{8}}=\frac{\sqrt{5}}{2\sqrt{2}}=\frac{\sqrt{5}}{2\sqrt{2}}\cdot\frac{\sqrt{2}}{\sqrt{2}}=\frac{\sqrt{10}}{4}$$

20. Find $\sin\frac{\theta}{2}$ if $\cos\theta=-\frac{5}{8}$, with $\frac{\pi}{2}<\theta<\pi$.

Since $\frac{\pi}{2}<\theta<\pi\Rightarrow\frac{\pi}{4}<\frac{\theta}{2}<\frac{\pi}{2}$, $\sin\frac{\theta}{2}>0$.

$$\sin\frac{x}{2}=\sqrt{\frac{1-\cos x}{2}}=\sqrt{\frac{1-\left(-\frac{5}{8}\right)}{2}}=\sqrt{\frac{1+\frac{5}{8}}{2}}=\sqrt{\frac{8+5}{16}}=\sqrt{\frac{13}{16}}=\frac{\sqrt{13}}{4}$$

21. Find $\tan\frac{\theta}{2}$, given $\sin\theta=\frac{3}{5}$, with $90°<\theta<180°$.

To find $\tan\frac{\theta}{2}$, we need the values of $\sin\theta$ and $\cos\theta$. We know $\sin\theta=\frac{3}{5}$.

$$\cos\theta=\pm\sqrt{1-\left(\frac{3}{5}\right)^2}=\pm\sqrt{1-\frac{9}{25}}=\pm\sqrt{\frac{16}{25}}=\pm\frac{4}{5}$$

Since $90°<\theta<180°$ (θ is in quadrant II), $\cos\theta<0$. Thus, $\cos\theta=-\frac{4}{5}$.

$$\tan\frac{\theta}{2}=\frac{\sin\theta}{1+\cos\theta}=\frac{\frac{3}{5}}{1+\left(-\frac{4}{5}\right)}=\frac{\frac{3}{5}}{1+\left(-\frac{4}{5}\right)}\cdot\frac{5}{5}=\frac{3}{5-4}=\frac{3}{1}=3$$

22. Find $\cos\dfrac{\theta}{2}$, if $\sin\theta = -\dfrac{1}{5}$, with $180° < \theta < 270°$.

Since $180° < \theta < 270°$, θ is in quadrant III, Thus, $\cos\theta < 0$.

$$\cos\theta = -\sqrt{1-\sin^2\theta} = -\sqrt{1-\left(-\frac{1}{5}\right)^2} = -\sqrt{1-\frac{1}{25}} = -\sqrt{\frac{24}{25}} = -\frac{\sqrt{24}}{5} = -\frac{2\sqrt{6}}{5}$$

Now, find $\cos\dfrac{\theta}{2}$.

Since $180° < \theta < 270° \Rightarrow 90° < \dfrac{\theta}{2} < 135°$, $\dfrac{\theta}{2}$ is in quadrant II. Thus, $\cos\dfrac{\theta}{2} < 0$.

$$\cos\frac{\theta}{2} = -\sqrt{\frac{1+\cos\theta}{2}} = -\sqrt{\frac{1+\left(-\frac{2\sqrt{6}}{5}\right)}{2}}$$

$$= -\sqrt{\frac{1-\frac{2\sqrt{6}}{5}}{2}\cdot\frac{5}{5}} = -\sqrt{\frac{5-2\sqrt{6}}{10}} = -\frac{\sqrt{5-2\sqrt{6}}}{\sqrt{10}} = -\frac{\sqrt{5-2\sqrt{6}}}{\sqrt{10}}\cdot\frac{\sqrt{10}}{\sqrt{10}} = -\frac{\sqrt{50-20\sqrt{6}}}{10}$$

23. Find $\sin\dfrac{x}{2}$, given $\tan x = 2$, with $0 < x < \dfrac{\pi}{2}$.

Since x is in quadrant I, $\sec x > 0$.

$\sec^2 x = \tan^2 x + 1 \Rightarrow \sec^2 x = 2^2 + 1 = 4 + 1 = 5 \Rightarrow \sec x = \sqrt{5}$

$$\cos x = \frac{1}{\sec x} = \frac{1}{\sqrt{5}} = \frac{1}{\sqrt{5}}\cdot\frac{\sqrt{5}}{\sqrt{5}} = \frac{\sqrt{5}}{5}$$

Since $0 < x < \dfrac{\pi}{2} \Rightarrow 0 < \dfrac{x}{2} < \dfrac{\pi}{4}$, $\dfrac{x}{2}$ is in quadrant I, Thus, $\sin\dfrac{x}{2} > 0$.

$$\sin\frac{x}{2} = \sqrt{\frac{1-\cos x}{2}} = \sqrt{\frac{1-\frac{\sqrt{5}}{5}}{2}} = \frac{\sqrt{50-10\sqrt{5}}}{10}$$

24. Find $\cos\dfrac{x}{2}$ if $\cot x = -3$, with $\dfrac{\pi}{2} < x < \pi$.

Use identities to find $\cos x$.

$$\tan x = \frac{1}{\cot x} = -\frac{1}{3}$$

$$\sec^2 x = \tan^2 x + 1 \Rightarrow \sec^2 x = \left(-\frac{1}{3}\right)^2 + 1 \Rightarrow \sec^2 x = \frac{1}{9} + 1 = \frac{10}{9}$$

Since $\dfrac{\pi}{2} < x < \pi, \sec x < 0$. Thus, $\sec x = -\dfrac{\sqrt{10}}{3}$ and $\cos x = -\dfrac{3}{\sqrt{10}} = -\dfrac{3}{\sqrt{10}}\cdot\dfrac{\sqrt{10}}{\sqrt{10}} = -\dfrac{3\sqrt{10}}{10}$.

Since $\dfrac{\pi}{2} < x < \pi \Rightarrow \dfrac{\pi}{4} < \dfrac{x}{2} < \dfrac{\pi}{2}$, $\dfrac{x}{2}$ is in quadrant I. Thus, $\cos\dfrac{x}{2} > 0$.

$$\cos\frac{x}{2} = \sqrt{\frac{1+\cos x}{2}} = \sqrt{\frac{1-\frac{3\sqrt{10}}{10}}{2}} = \sqrt{\frac{10-3\sqrt{10}}{20}} = \frac{\sqrt{10-3\sqrt{10}}}{\sqrt{20}} = \frac{\sqrt{10-3\sqrt{10}}}{2\sqrt{5}}\cdot\frac{\sqrt{5}}{\sqrt{5}} = \frac{\sqrt{50-15\sqrt{10}}}{10}$$

25. Find $\tan\frac{\theta}{2}$, given $\tan\theta = \frac{\sqrt{7}}{3}$, with $180° < \theta < 270°$.

$$\sec^2\theta = \tan^2\theta + 1 \Rightarrow \sec^2\theta = \left(\frac{\sqrt{7}}{3}\right)^2 + 1 = \frac{7}{9} + 1 = \frac{16}{9}$$

Since θ is in quadrant III, $\sec\theta < 0$ and $\sin\theta < 0$.

$$\sec\theta = -\sqrt{\frac{16}{9}} = -\frac{4}{3} \text{ and } \cos\theta = \frac{1}{\sec\theta} = \frac{1}{-\frac{4}{3}} = -\frac{3}{4}$$

$$\sin\theta = -\sqrt{1-\cos^2\theta} = -\sqrt{1-\left(-\frac{3}{4}\right)^2} - \sqrt{1-\frac{9}{16}} = -\frac{\sqrt{7}}{4}$$

$$\tan\frac{\theta}{2} = \frac{\sin\theta}{1+\cos\theta} = \frac{-\frac{\sqrt{7}}{4}}{1+\left(-\frac{3}{4}\right)} = \frac{-\sqrt{7}}{4-3} = -\sqrt{7}$$

26. Find $\cot\frac{\theta}{2}$ if $\tan\theta = -\frac{\sqrt{5}}{2}$, with $90° < \theta < 180°$.
Use identities to find $\sin\theta$ and $\cos\theta$.

$$\sec^2\theta = 1 + \tan^2\theta \Rightarrow \sec^2\theta = 1 + \left(-\frac{\sqrt{5}}{2}\right)^2 = 1 + \frac{5}{4} = \frac{9}{4}$$

Since $90° < \theta < 180°$, $\sec\theta < 0$.

$$\sec\theta = -\sqrt{\frac{9}{4}} = -\frac{3}{2} \Rightarrow \cos\theta = -\frac{2}{3}$$

Since $90° < \theta < 180°$, $\sin\theta > 0$.

$$\sin\theta = \sqrt{1-\cos^2\theta} = \sqrt{1-\left(-\frac{2}{3}\right)^2} = \sqrt{1-\frac{4}{9}} = \sqrt{\frac{5}{9}} = \frac{\sqrt{5}}{3}$$

$$\cot\frac{\theta}{2} = \frac{1}{\tan\frac{\theta}{2}} = \frac{1}{\frac{\sin\theta}{1+\cos\theta}} = \frac{1+\cos\theta}{\sin\theta} = \frac{1+\left(-\frac{2}{3}\right)}{\frac{\sqrt{5}}{3}} = \frac{3-2}{\sqrt{5}} = \frac{1}{\sqrt{5}} = \frac{1}{\sqrt{5}}\cdot\frac{\sqrt{5}}{\sqrt{5}} = \frac{\sqrt{5}}{5}$$

27. Find $\sin\theta$ given $\cos 2\theta = \frac{3}{5}$, θ is in quadrant I.

Since θ is in quadrant I, $\sin\theta > 0$.

$$\sin\theta = \sqrt{\frac{1-\cos 2\theta}{2}} \Rightarrow \sin\theta = \sqrt{\frac{1-\frac{3}{5}}{2}} = \sqrt{\frac{5-3}{10}} = \sqrt{\frac{2}{10}} = \sqrt{\frac{1}{5}} = \frac{1}{\sqrt{5}} = \frac{1}{\sqrt{5}}\cdot\frac{\sqrt{5}}{\sqrt{5}} = \frac{\sqrt{5}}{5}$$

28. Find $\cos\theta$, given $\cos 2\theta = \frac{1}{2}$, θ is in quadrant II.

Since θ is in quadrant II, $\cos\theta < 0$.

$$\cos\theta = -\sqrt{\frac{1+\cos 2\theta}{2}} \Rightarrow \cos\theta = -\sqrt{\frac{1+\frac{1}{2}}{2}} = -\sqrt{\frac{2+1}{4}} = -\sqrt{\frac{3}{4}} = -\frac{\sqrt{3}}{2}$$

29. Find $\cos x$, given $\cos 2x = -\frac{5}{12}$, $\frac{\pi}{2} < x < \pi$.

Since $\frac{\pi}{2} < x < \pi$, $\cos x < 0$.

$$\cos x = -\sqrt{\frac{1+\cos 2x}{2}} \Rightarrow \cos x = -\sqrt{\frac{1+\left(-\frac{5}{12}\right)}{2}} = -\sqrt{\frac{12-5}{24}} = -\sqrt{\frac{7}{24}} = -\frac{\sqrt{7}}{\sqrt{24}} = -\frac{\sqrt{7}}{2\sqrt{6}}\cdot\frac{\sqrt{6}}{\sqrt{6}} = -\frac{\sqrt{42}}{12}$$

30. Find $\sin x$, given $\cos 2x = \frac{2}{3}$, $\pi < x < \frac{3\pi}{2}$.

Since x is in quadrant III, $\sin x < 0$.

$$\sin x = -\sqrt{\frac{1-\cos 2x}{2}} = -\sqrt{\frac{1-\frac{2}{3}}{2}} = -\sqrt{\frac{3-2}{6}} = -\sqrt{\frac{1}{6}} = -\frac{1}{\sqrt{6}} = -\frac{1}{\sqrt{6}}\cdot\frac{\sqrt{6}}{\sqrt{6}} = -\frac{\sqrt{6}}{6}$$

31. $\cos x \approx .9682$ and $\sin x \approx .25$

$$\tan\frac{x}{2} \approx \frac{\sin x}{1+\cos x} = \frac{.25}{1+.9682} = \frac{.25}{1.9682} \approx .127$$

32. $\cos x \approx -.75$ and $\sin x \approx .6614$

$$\tan\frac{x}{2} \approx \frac{\sin x}{1+\cos x} = \frac{.6614}{1+(-.75)} = \frac{.6614}{.25} \approx 2.646$$

33. $\sqrt{\frac{1-\cos 40°}{2}} = \sin\frac{40°}{2} = \sin 20°$

34. $\sqrt{\frac{1+\cos 76°}{2}} = \cos\frac{76°}{2} = \cos 38°$

35. $\sqrt{\frac{1-\cos 147°}{1+\cos 147°}} = \tan\frac{147°}{2} = \tan 73.5°$

36. $\sqrt{\frac{1+\cos 165°}{1-\cos 165°}} = \frac{1}{\tan 82.5°} = \cot 82.5°$

37. $\frac{1-\cos 59.74°}{\sin 59.74°} = \tan\frac{59.74°}{2} = \tan 29.87°$

38. $\frac{\sin 158.2°}{1+\cos 158.2°} = \tan\frac{158.2°}{2} = \tan 79.1°$

39. $\pm\sqrt{\frac{1+\cos 18x}{2}} = \cos\frac{18x}{2} = \cos 9x$

40. $\pm\sqrt{\frac{1+\cos 20\alpha}{2}} = \cos\frac{20\alpha}{2} = \cos 10\alpha$

41. $\pm\sqrt{\frac{1-\cos 8\theta}{1+\cos 8\theta}} = \tan\frac{8\theta}{2} = \tan 4\theta$

42. $\pm\sqrt{\frac{1-\cos 5A}{1+\cos 5A}} = \tan\frac{5A}{2}$

43. $\pm\sqrt{\dfrac{1+\cos\frac{x}{4}}{2}} = \cos\dfrac{\frac{x}{4}}{2} = \cos\dfrac{x}{8}$

44. $\pm\sqrt{\dfrac{1-\cos\frac{3\theta}{5}}{2}} = \sin\dfrac{\frac{3\theta}{5}}{2} = \sin\dfrac{3\theta}{10}$

Exercises 45 – 48 are graphed in the following window.

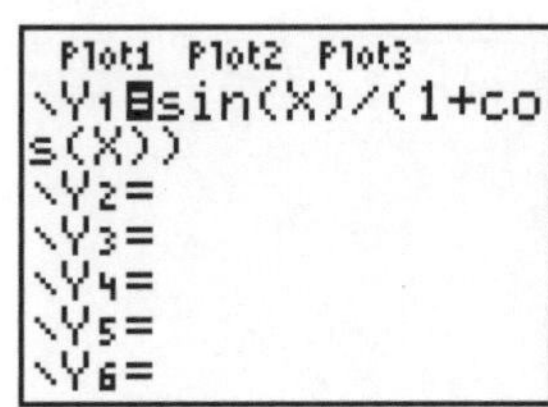

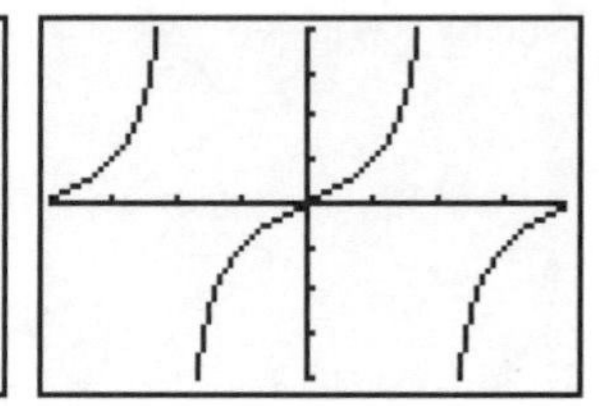

45. $\dfrac{\sin x}{1+\cos x}$ appears to be equivalent to $\tan\dfrac{x}{2}$.

$$\frac{\sin x}{1+\cos x} = \frac{\sin 2\left(\frac{x}{2}\right)}{1+\cos 2\left(\frac{x}{2}\right)} = \frac{2\sin\left(\frac{x}{2}\right)\cos\left(\frac{x}{2}\right)}{1+\left[2\cos^2\left(\frac{x}{2}\right)-1\right]} = \frac{2\sin\left(\frac{x}{2}\right)\cos\left(\frac{x}{2}\right)}{2\cos^2\left(\frac{x}{2}\right)} = \frac{\sin\frac{x}{2}}{\cos\frac{x}{2}} = \tan\frac{x}{2}$$

46. $\dfrac{1-\cos x}{\sin x}$ appears to be equivalent to $\tan\dfrac{x}{2}$.

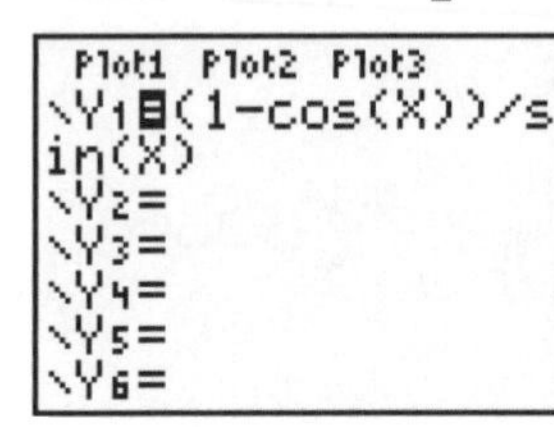

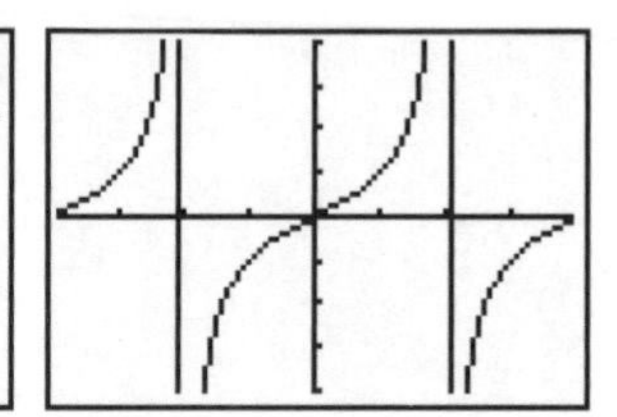

$$\frac{1-\cos x}{\sin x} = \frac{1-\cos 2\left(\frac{x}{2}\right)}{\sin 2\left(\frac{x}{2}\right)} = \frac{1-\left[1-2\sin^2\frac{x}{2}\right]}{2\sin\frac{x}{2}\cos\frac{x}{2}} = \frac{2\sin^2\frac{x}{2}}{2\sin\frac{x}{2}\cos\frac{x}{2}} = \frac{\sin\frac{x}{2}}{\cos\frac{x}{2}} = \tan\frac{x}{2}$$

47. $\dfrac{\tan\frac{x}{2}+\cot\frac{x}{2}}{\cot\frac{x}{2}-\tan\frac{x}{2}}$ appears to be equivalent to $\sec x$.

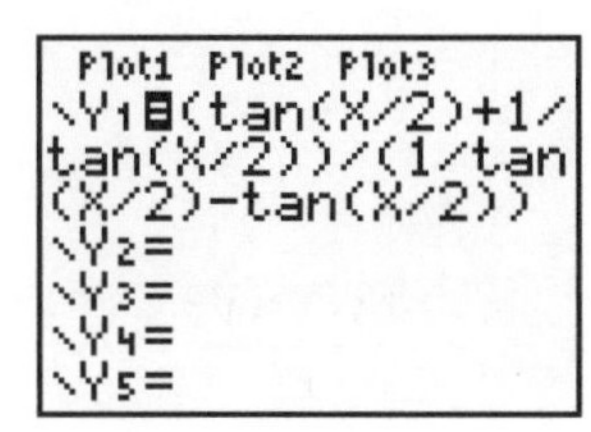

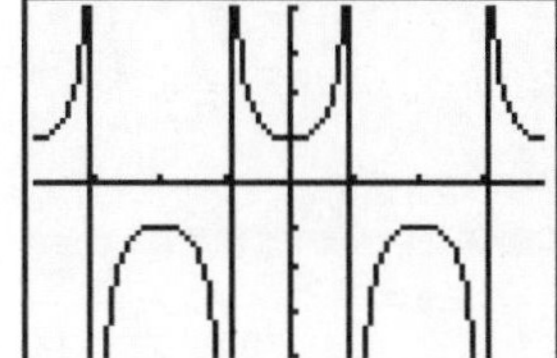

$$\frac{\tan\frac{x}{2}+\cot\frac{x}{2}}{\cot\frac{x}{2}-\tan\frac{x}{2}}=\frac{\frac{\sin\frac{x}{2}}{\cos\frac{x}{2}}+\frac{\cos\frac{x}{2}}{\sin\frac{x}{2}}}{\frac{\cos\frac{x}{2}}{\sin\frac{x}{2}}-\frac{\sin\frac{x}{2}}{\cos\frac{x}{2}}}\cdot\frac{\sin\frac{x}{2}\cos\frac{x}{2}}{\sin\frac{x}{2}\cos\frac{x}{2}}=\frac{\sin^2\frac{x}{2}+\cos^2\frac{x}{2}}{\cos^2\frac{x}{2}-\sin^2\frac{x}{2}}=\frac{1}{\cos 2\left(\frac{x}{2}\right)}=\frac{1}{\cos x}=\sec x$$

48. $1-8\sin^2\frac{x}{2}\cos^2\frac{x}{2}$ appears to be equivalent to $\cos 2x$.

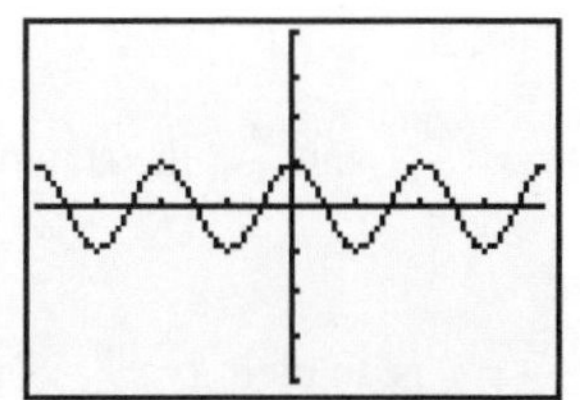

$$1-8\sin^2\frac{x}{2}\cos^2\frac{x}{2}=1-2\left(4\sin^2\frac{x}{2}\cos^2\frac{x}{2}\right)$$
$$=1-2\left(2\sin\frac{x}{2}\cos\frac{x}{2}\right)^2=1-2\left[\sin 2\left(\frac{x}{2}\right)\right]^2=1-2\sin^2 x=\cos 2x$$

49. Verify $\sec^2\frac{x}{2}=\frac{2}{1+\cos x}$ is an identity.

$$\sec^2\frac{x}{2}=\frac{1}{\cos^2\frac{x}{2}}=\frac{1}{\left(\pm\sqrt{\frac{1+\cos x}{2}}\right)^2}=\frac{1}{\frac{1+\cos x}{2}}=\frac{2}{1+\cos x}$$

50. Verify $\cot^2\frac{x}{2}=\frac{(1+\cos x)^2}{\sin^2 x}$ is an identity.

$$\cot^2\frac{x}{2}=\left(\frac{1}{\tan\frac{x}{2}}\right)^2=\frac{1}{\left(\frac{\sin x}{1+\cos x}\right)^2}=\frac{(1+\cos x)^2}{\sin^2 x}$$

51. Verify $\sin^2 \frac{x}{2} = \frac{\tan x - \sin x}{2 \tan x}$ is an identity.

Work with the left side.

$$\sin^2 \frac{x}{2} = \left(\pm\sqrt{\frac{1-\cos x}{2}} \right)^2 = \frac{1-\cos x}{2}$$

Work with the right side.

$$\frac{\tan x - \sin x}{2 \tan x} = \frac{\frac{\sin x}{\cos x} - \sin x}{2 \cdot \frac{\sin x}{\cos x}} = \frac{\frac{\sin x}{\cos x} - \sin x}{2 \cdot \frac{\sin x}{\cos x}} \cdot \frac{\cos x}{\cos x} = \frac{\sin x - \cos x \sin x}{2 \sin x} = \frac{\sin x (1-\cos x)}{2 \sin x} = \frac{1-\cos x}{2}$$

Since $\sin^2 \frac{x}{2} = \frac{1-\cos x}{2} = \frac{\tan x - \sin x}{2 \tan x}$, the statement has been verified.

52. Verify $\frac{\sin 2x}{2 \sin x} = \cos^2 \frac{x}{2} - \sin^2 \frac{x}{2}$ is an identity.

Work with the left side.

$$\frac{\sin 2x}{2 \sin x} = \frac{2 \sin x \cos x}{2 \sin x} = \cos x$$

Work with the right side.

$$\cos^2 \frac{x}{2} - \sin^2 \frac{x}{2} = \frac{1+\cos x}{2} - \frac{1-\cos x}{2} = \frac{2 \cos x}{2} = \cos x$$

Since $\frac{\sin 2x}{2 \sin x} = \cos x = \cos^2 \frac{x}{2} - \sin^2 \frac{x}{2}$, the statement has been verified.

53. Verify $\frac{2}{1+\cos x} - \tan^2 \frac{x}{2} = 1$ is an identity.

$$\frac{2}{1+\cos x} - \tan^2 \frac{x}{2} = \frac{2}{1+\cos x} - \left(\pm\sqrt{\frac{1-\cos x}{1+\cos x}} \right)^2$$

$$= \frac{2}{1+\cos x} - \frac{1-\cos x}{1+\cos x} = \frac{2-1+\cos x}{1+\cos x} = \frac{1+\cos x}{1+\cos x} = 1$$

54. Verify $\tan \frac{\theta}{2} = \csc \theta - \cot \theta$ is an identity.

$$\tan \frac{\theta}{2} = \frac{\sin \theta}{1+\cos \theta} = \frac{\sin \theta}{1+\cos \theta} \cdot \frac{1-\cos \theta}{1-\cos \theta}$$

$$= \frac{\sin \theta - \sin \theta \cos \theta}{1-\cos^2 \theta} = \frac{\sin \theta (1-\cos \theta)}{\sin^2 \theta} = \frac{1-\cos \theta}{\sin \theta} = \frac{1}{\sin \theta} - \frac{\cos \theta}{\sin \theta} = \csc \theta - \cot \theta$$

55. Verify $1-\tan^2\frac{\theta}{2}=\frac{2\cos\theta}{1+\cos\theta}$ is an identity.

$$1-\tan^2\frac{\theta}{2}=1-\left(\frac{\sin\theta}{1+\cos\theta}\right)^2=1-\frac{\sin^2\theta}{(1+\cos\theta)^2}=\frac{(1+\cos\theta)^2-\sin^2\theta}{(1+\cos\theta)^2}$$

$$=\frac{1+2\cos\theta+\cos^2\theta-\sin^2\theta}{(1+\cos\theta)^2}=\frac{1+2\cos\theta+\cos^2\theta-(1-\cos^2\theta)}{(1+\cos\theta)^2}$$

$$=\frac{1+2\cos^2\theta-1+2\cos\theta}{(1+\cos\theta)^2}=\frac{2\cos^2\theta+2\cos\theta}{(1+\cos\theta)^2}=\frac{2\cos\theta(1+\cos\theta)}{(1+\cos\theta)^2}=\frac{2\cos\theta}{1+\cos\theta}$$

56. Verify $\cos x=\dfrac{1-\tan^2\frac{x}{2}}{1+\tan^2\frac{x}{2}}$ is an identity.

Work with the right side.

$$\frac{1-\tan^2\frac{x}{2}}{1+\tan^2\frac{x}{2}}=\frac{1-\frac{1-\cos x}{1+\cos x}}{1+\frac{1-\cos x}{1+\cos x}}=\frac{1-\frac{1-\cos x}{1+\cos x}}{1+\frac{1-\cos x}{1+\cos x}}\cdot\frac{1+\cos x}{1+\cos x}$$

$$=\frac{(1+\cos x)-(1-\cos x)}{(1+\cos x)+(1-\cos x)}=\frac{1+\cos x-1+\cos x}{1+\cos x+1-\cos x}=\frac{2\cos x}{2}=\cos x$$

57. $\tan\frac{A}{2}=\frac{\sin A}{1+\cos A}=\frac{\sin A}{1+\cos A}\cdot\frac{1-\cos A}{1-\cos A}=\frac{\sin A(1-\cos A)}{1-\cos^2 A}=\frac{\sin A(1-\cos A)}{\sin^2 A}=\frac{1-\cos A}{\sin A}$

58. $\sin\frac{\theta}{2}=\frac{1}{m}, m=\frac{3}{2}$

Since $\sin\frac{\theta}{2}=\frac{1}{\frac{3}{2}}=\frac{2}{3}$ and $\sin^2\frac{\theta}{2}=\frac{1-\cos\theta}{2}$, we have the following.

$$\left(\frac{2}{3}\right)^2=\frac{1-\cos\theta}{2}\Rightarrow\frac{4}{9}=\frac{1-\cos\theta}{2}\Rightarrow\frac{8}{9}=1-\cos\theta\Rightarrow-\frac{1}{9}=-\cos\theta\Rightarrow\cos\theta=\frac{1}{9}$$

Thus, we have $\theta=\cos^{-1}\frac{1}{9}\approx 84°$.

59. $\sin\frac{\theta}{2}=\frac{1}{m},\ m=\frac{5}{4}$

Since $\sin\frac{\theta}{2}=\frac{1}{\frac{5}{4}}=\frac{4}{5}$ and $\sin^2\frac{\theta}{2}=\frac{1-\cos\theta}{2}$, we have the following.

$$\left(\frac{4}{5}\right)^2=\frac{1-\cos\theta}{2}\Rightarrow\frac{16}{25}=\frac{1-\cos\theta}{2}\Rightarrow\frac{32}{25}=1-\cos\theta\Rightarrow\frac{7}{25}=-\cos\theta\Rightarrow\cos\theta=-\frac{7}{25}$$

Thus, we have $\theta=\cos^{-1}\left(-\frac{7}{25}\right)\approx 106°$.

60. $\sin\frac{\theta}{2}=\frac{1}{m}, \theta=30°$

$\sin\frac{30°}{2}=\frac{1}{m}\Rightarrow \sin 15°=\frac{1}{m}$

$m=\frac{1}{\sin 15°}\approx 3.9$

61. $\sin\frac{\theta}{2}=\frac{1}{m},\ \theta=60°$

$\sin\frac{60°}{2}=\frac{1}{m}\Rightarrow \sin 30°=\frac{1}{m}$

$\frac{1}{2}=\frac{1}{m}\Rightarrow m=2$

62. **(a)** $\cos\frac{\theta}{2}=\frac{R-b}{R}$

(b) $\tan\frac{\theta}{4}=\frac{1-\cos\frac{\theta}{2}}{\sin\frac{\theta}{2}}=\frac{1-\frac{R-b}{R}}{\frac{50}{R}}=\frac{R-(R-b)}{50}=\frac{R-R+b}{50}=\frac{b}{50}$

(c) If $\tan\frac{\theta}{4}=\frac{12}{50}$, then $\frac{\theta}{4}=\tan^{-1}\frac{12}{50}\approx 13.4957\Rightarrow \theta\approx 53.9828\approx 54°$.

63. They are both radii of the circle.

64. It is the supplement of a $30°$ angle.

65. Their sum is $180°-150°=30°$, and their measures are equal.

66. The length of *DC* is the sum of the lengths of *BC* and *BD*, which is $2+\sqrt{3}$.

67. $(AD)^2=(AC)^2+(CD)^2\Rightarrow (AD)^2=1^2+\left(2+\sqrt{3}\right)^2\Rightarrow (AD)^2=1+4+4\sqrt{3}+3=8+4\sqrt{3}=\left(\sqrt{6}+\sqrt{2}\right)^2$

Thus, $AD=\sqrt{6}+\sqrt{2}$

68. $\cos 15°=\frac{CD}{AD}=\frac{2+\sqrt{3}}{\sqrt{6}+\sqrt{2}}=\frac{2+\sqrt{3}}{\sqrt{6}+\sqrt{2}}\cdot\frac{\sqrt{6}-\sqrt{2}}{\sqrt{6}-\sqrt{2}}=\frac{2\sqrt{6}-2\sqrt{2}+3\sqrt{2}-\sqrt{6}}{6-2}=\frac{\sqrt{6}+\sqrt{2}}{4}$

69. Triangle *ACE* is a $15°-75°$ right triangle. Angle *EAC* measures $15°$.

$\cos 15°=\frac{AC}{AE}=\frac{1}{AE}\Rightarrow AE\cdot\cos 15°=1$

$AE=\frac{1}{\cos 15°}=\frac{1}{\frac{\sqrt{6}+\sqrt{2}}{4}}=\frac{4}{\sqrt{6}+\sqrt{2}}=\frac{4}{\sqrt{6}+\sqrt{2}}\cdot\frac{\sqrt{6}-\sqrt{2}}{\sqrt{6}-\sqrt{2}}=\frac{4\left(\sqrt{6}-\sqrt{2}\right)}{6-2}=\frac{4\left(\sqrt{6}-\sqrt{2}\right)}{4}=\sqrt{6}-\sqrt{2}$

Since the length of *ED* is twice the length of *BD*, *ED* is length 4. In triangle *AED*,

$\sin 15°=\frac{AE}{ED}=\frac{\sqrt{6}-\sqrt{2}}{4}$.

70. $\tan 15°=\frac{AC}{CD}=\frac{1}{2+\sqrt{3}}=\frac{1}{2+\sqrt{3}}\cdot\frac{2-\sqrt{3}}{2-\sqrt{3}}=\frac{2-\sqrt{3}}{4-3}=\frac{2-\sqrt{3}}{1}=2-\sqrt{3}$

Summary Exercises on Verifying Trigonometric Identities

For the following exercises, other solutions are possible.

1. Verify $\tan\theta+\cot\theta=\sec\theta\csc\theta$ is an identity.

Starting on the left side, we have the following.

$$\begin{aligned}\tan\theta+\cot\theta&=\frac{\sin\theta}{\cos\theta}+\frac{\cos\theta}{\sin\theta}\\&=\frac{\sin^2\theta}{\cos\theta\sin\theta}+\frac{\cos^2\theta}{\cos\theta\sin\theta}\\&=\frac{\sin^2\theta+\cos^2\theta}{\cos\theta\sin\theta}\\&=\frac{1}{\cos\theta\sin\theta}\\&=\frac{1}{\cos\theta}\cdot\frac{1}{\sin\theta}\\&=\sec\theta\csc\theta\end{aligned}$$

2. Verify $\csc\theta\cos^2\theta+\sin\theta=\csc\theta$ is an identity.

Starting on the left side, we have the following.

$$\begin{aligned}\csc\theta\cos^2\theta+\sin\theta&=\frac{1}{\sin\theta}\cdot\cos^2\theta+\sin\theta\\&=\frac{\cos^2\theta}{\sin\theta}+\frac{\sin^2\theta}{\sin\theta}\\&=\frac{\cos^2\theta+\sin^2\theta}{\sin\theta}\\&=\frac{1}{\sin\theta}\\&=\csc\theta\end{aligned}$$

3. Verify $\tan\frac{x}{2}=\csc x-\cot x$ is an identity.

Starting on the right side, we have the following.

$$\begin{aligned}\csc x-\cot x&=\frac{1}{\sin x}-\frac{\cos x}{\sin x}\\&=\frac{1-\cos x}{\sin x}\\&=\tan\frac{x}{2}\end{aligned}$$

4. Verify $\sec(\pi-x)=-\sec x$ is an identity.

Starting on the left side, we have the following.

$$\begin{aligned}\sec(\pi-x)&=\frac{1}{\cos(\pi-x)}\\&=\frac{1}{\cos\pi\cos x+\sin\pi\sin x}\\&=\frac{1}{(-1)\cos x+(0)\sin x}\\&=\frac{1}{-\cos x+0}\\&=-\frac{1}{\cos x}\\&=-\sec x\end{aligned}$$

5. Verify $\frac{\sin t}{1+\cos t}=\frac{1-\cos t}{\sin t}$ is an identity.

Starting on the left side, we have the following.

$$\begin{aligned}\frac{\sin t}{1+\cos t}&=\frac{\sin t}{1+\cos t}\cdot\frac{1-\cos t}{1-\cos t}\\&=\frac{\sin t(1-\cos t)}{1-\cos^2 t}\\&=\frac{\sin t(1-\cos t)}{\sin^2 t}\\&=\frac{1-\cos t}{\sin t}\end{aligned}$$

6. Verify $\frac{1-\sin t}{\cos t}=\frac{1}{\sec t+\tan t}$ is an identity.

Starting on the right side, we have the following.

$$\frac{1}{\sec t+\tan t}=\frac{1}{\frac{1}{\cos t}+\frac{\sin t}{\cos t}}=\frac{1}{\frac{1+\sin t}{\cos t}}=\frac{\cos t}{1+\sin t}=\frac{\cos t}{1+\sin t}\cdot\frac{1-\sin t}{1-\sin t}=\frac{\cos t(1-\sin t)}{1-\sin^2 t}=\frac{\cos t(1-\sin t)}{\cos^2 t}=\frac{1-\sin t}{\cos t}$$

7. Verify $\sin 2\theta=\frac{2\tan\theta}{1+\tan^2\theta}$ is an identity.

Starting on the right side, we have the following.

$$\frac{2\tan\theta}{1+\tan^2\theta}=\frac{2\tan\theta}{\sec^2\theta}=\frac{2\cdot\frac{\sin\theta}{\cos\theta}}{\frac{1}{\cos^2\theta}}=2\cdot\frac{\sin\theta}{\cos\theta}\cdot\frac{\cos^2\theta}{1}=2\sin\theta\cos\theta=\sin 2\theta$$

8. Verify $\frac{2}{1+\cos x}-\tan^2\frac{x}{2}=1$ is an identity.

Starting on the left side, we have the following.

$$\begin{aligned}\frac{2}{1+\cos x}-\tan^2\frac{x}{2}&=\\&=\frac{2}{1+\cos x}-\left(\frac{\sin x}{1+\cos x}\right)^2\\&=\frac{2}{1+\cos x}-\frac{\sin^2 x}{(1+\cos x)^2}\\&=\frac{2(1+\cos x)}{(1+\cos x)^2}-\frac{\sin^2 x}{(1+\cos x)^2}\\&=\frac{2+2\cos x-\sin^2 x}{(1+\cos x)^2}\\&=\frac{2+2\cos x-(1-\cos^2 x)}{(1+\cos x)^2}\\&=\frac{2+2\cos x-1+\cos^2 x}{(1+\cos x)^2}\\&=\frac{\cos^2 x+2\cos x+1}{(1+\cos x)^2}\\&=\frac{(1+\cos x)^2}{(1+\cos x)^2}\\&=1\end{aligned}$$

9. Verify $\cot\theta-\tan\theta=\frac{2\cos^2\theta-1}{\sin\theta\cos\theta}$ is an identity.

Starting on the left side, we have the following.

$$\begin{aligned}\cot\theta-\tan\theta&=\frac{\cos\theta}{\sin\theta}-\frac{\sin\theta}{\cos\theta}\\&=\frac{\cos^2\theta}{\sin\theta\cos\theta}-\frac{\sin^2\theta}{\sin\theta\cos\theta}\\&=\frac{\cos^2\theta-\sin^2\theta}{\sin\theta\cos\theta}\\&=\frac{\cos^2\theta-(1-\cos^2\theta)}{\sin\theta\cos\theta}\\&=\frac{\cos^2\theta-1+\cos^2\theta}{\sin\theta\cos\theta}\\&=\frac{2\cos^2\theta-1}{\sin\theta\cos\theta}.\end{aligned}$$

10. Verify $\dfrac{1}{\sec t-1}+\dfrac{1}{\sec t+1}=2\cot t\csc t$ is an identity.

Starting on the left side, we have the following.

$$\frac{1}{\sec t-1}+\frac{1}{\sec t+1}=\frac{1}{\frac{1}{\cos t}-1}+\frac{1}{\frac{1}{\cos t}+1}=\frac{1}{\frac{1}{\cos t}-1}\cdot\frac{\cos t}{\cos t}+\frac{1}{\frac{1}{\cos t}+1}\cdot\frac{\cos t}{\cos t}=\frac{\cos t}{1-\cos t}+\frac{\cos t}{1+\cos t}$$

$$=\frac{\cos t}{1-\cos t}\cdot\frac{1+\cos t}{1+\cos t}+\frac{\cos t}{1+\cos t}\cdot\frac{1-\cos t}{1-\cos t}=\frac{\cos t+\cos^2 t}{1-\cos^2 t}+\frac{\cos t-\cos^2 t}{1-\cos^2 t}$$

$$=\frac{\cos t+\cos^2 t+\cos t-\cos^2 t}{1-\cos^2 t}=\frac{2\cos t}{1-\cos^2 t}=\frac{2\cos t}{\sin^2 t}=2\cdot\frac{\cos t}{\sin t}\cdot\frac{1}{\sin t}=2\cot t\csc t$$

11. Verify $\dfrac{\sin(x+y)}{\cos(x-y)}=\dfrac{\cot x+\cot y}{1+\cot x\cot y}$ is an identity.

Starting on the left side, we have the following.

$$\frac{\sin(x+y)}{\cos(x-y)}=\frac{\sin x\cos y+\cos x\sin y}{\cos x\cos y+\sin x\sin y}=\frac{\sin x\cos y+\cos x\sin y}{\cos x\cos y+\sin x\sin y}\cdot\frac{\frac{1}{\cos x\cos y}}{\frac{1}{\cos x\cos y}}$$

$$=\frac{\frac{\sin x\cos y}{\cos x\cos y}+\frac{\cos x\sin y}{\cos x\cos y}}{\frac{\cos x\cos y}{\cos x\cos y}+\frac{\sin x\sin y}{\cos x\cos y}}=\frac{\frac{\sin x}{\cos x}+\frac{\sin y}{\cos y}}{1+\frac{\sin x}{\cos x}\cdot\frac{\sin y}{\cos y}}=\frac{\cot x+\cot y}{1+\cot x\cot y}$$

12. Verify $1-\tan^2\dfrac{\theta}{2}=\dfrac{2\cos\theta}{1+\cos\theta}$ is an identity.

Starting on the left side, we have the following.

$$1-\tan^2\frac{\theta}{2}=1-\left(\frac{\sin\theta}{1+\cos\theta}\right)^2$$

$$=1-\frac{\sin^2\theta}{(1+\cos\theta)^2}$$

$$=\frac{(1+\cos\theta)^2}{(1+\cos\theta)^2}-\frac{\sin^2\theta}{(1+\cos\theta)^2}$$

$$=\frac{1+2\cos\theta+\cos^2\theta}{(1+\cos\theta)^2}-\frac{\sin^2\theta}{(1+\cos\theta)^2}$$

$$=\frac{1+2\cos\theta+\cos^2\theta-\sin^2\theta}{(1+\cos\theta)^2}$$

$$=\frac{2\cos\theta+\cos^2\theta+(1-\sin^2\theta)}{(1+\cos\theta)^2}$$

$$=\frac{2\cos\theta+\cos^2\theta+\cos^2\theta}{(1+\cos\theta)^2}=\frac{2\cos\theta+2\cos^2\theta}{(1+\cos\theta)^2}$$

$$=\frac{2\cos\theta(1+\cos\theta)}{(1+\cos\theta)^2}=\frac{2\cos\theta}{1+\cos\theta}$$

13. Verify $\dfrac{\sin\theta+\tan\theta}{1+\cos\theta}=\tan\theta$ is an identity.

Starting on the left side, we have the following.

$$\frac{\sin\theta+\tan\theta}{1+\cos\theta}=\frac{\sin\theta+\frac{\sin\theta}{\cos\theta}}{1+\cos\theta}=\frac{\sin\theta+\frac{\sin\theta}{\cos\theta}}{1+\cos\theta}\cdot\frac{\cos\theta}{\cos\theta}=\frac{\sin\theta\cos\theta+\sin\theta}{\cos\theta(1+\cos\theta)}=\frac{\sin\theta(\cos\theta+1)}{\cos\theta(1+\cos\theta)}=\frac{\sin\theta}{\cos\theta}=\tan\theta$$

14. Verify $\csc^4 x-\cot^4 x=\dfrac{1+\cos^2 x}{1-\cos^2 x}$ is an identity.

Starting on the left side, we have the following.

$$\csc^4 x-\cot^4 x=\frac{1}{\sin^4 x}-\frac{\cos^4 x}{\sin^4 x}=\frac{1-\cos^4 x}{\sin^4 x}=\frac{(1+\cos^2 x)(1-\cos^2 x)}{\sin^4 x}=\frac{(1+\cos^2 x)(\sin^2 x)}{\sin^4 x}=\frac{1+\cos^2 x}{\sin^2 x}=\frac{1+\cos^2 x}{1-\cos^2 x}$$

15. Verify $\cos x=\dfrac{1-\tan^2\frac{x}{2}}{1+\tan^2\frac{x}{2}}$ is an identity.

Starting on the right side, we have the following.

$$\frac{1-\tan^2\frac{x}{2}}{1+\tan^2\frac{x}{2}}=\frac{1-\left(\frac{1-\cos x}{\sin x}\right)^2}{1+\left(\frac{1-\cos x}{\sin x}\right)^2}=\frac{1-\frac{(1-\cos x)^2}{\sin^2 x}}{1+\frac{(1-\cos x)^2}{\sin^2 x}}=\frac{1-\frac{(1-\cos x)^2}{\sin^2 x}}{1+\frac{(1-\cos x)^2}{\sin^2 x}}\cdot\frac{\sin^2 x}{\sin^2 x}=\frac{\sin^2 x-(1-\cos x)^2}{\sin^2 x+(1-\cos x)^2}$$

$$=\frac{\sin^2 x-(1-2\cos x+\cos^2 x)}{\sin^2 x+(1-2\cos x+\cos^2 x)}=\frac{\sin^2 x-1+2\cos x-\cos^2 x}{\sin^2 x+1-2\cos x+\cos^2 x}$$

$$=\frac{(1-\cos^2 x)-1+2\cos x-\cos^2 x}{(\sin^2 x+\cos^2 x)+1-2\cos x}=\frac{1-\cos^2 x-1+2\cos x-\cos^2 x}{1+1-2\cos x}$$

$$=\frac{2\cos x-2\cos^2 x}{2-2\cos x}=\frac{2\cos x(1-\cos x)}{2(1-\cos x)}=\cos x$$

16. Verify $\cos 2x=\dfrac{2-\sec^2 x}{\sec^2 x}$ is an identity.

Starting on the right side, we have the following.

$$\frac{2-\sec^2 x}{\sec^2 x}=\frac{2-\frac{1}{\cos^2 x}}{\frac{1}{\cos^2 x}}\cdot\frac{\cos^2 x}{\cos^2 x}=\frac{2\cos^2 x-1}{1}=2\cos^2 x-1=\cos 2x$$

17. Verify $\dfrac{\tan^2 t+1}{\tan t\csc^2 t}=\tan t$ is an identity.

Starting on the left side, we have the following.

$$\frac{\tan^2 t+1}{\tan t\csc^2 t}=\frac{\frac{\sin^2 t}{\cos^2 t}+1}{\frac{\sin t}{\cos t}\cdot\frac{1}{\sin^2 t}}=\frac{\frac{\sin^2 t}{\cos^2 t}+1}{\frac{1}{\cos t\sin t}}=\frac{\frac{\sin^2 t}{\cos^2 t}+1}{\frac{1}{\cos t\sin t}}\cdot\frac{\cos^2 t\sin t}{\cos^2 t\sin t}$$

$$=\frac{\sin^3 t+\cos^2 t\sin t}{\cos t}=\frac{\sin t\left(\sin^2 t+\cos^2 t\right)}{\cos t}=\frac{\sin t(1)}{\cos t}=\frac{\sin t}{\cos t}=\tan t$$

18. Verify $\dfrac{\sin s}{1+\cos s}+\dfrac{1+\cos s}{\sin s}=2\csc s$ is an identity.

Starting on the left side, we have the following.

$$\frac{\sin s}{1+\cos s}+\frac{1+\cos s}{\sin s}=\frac{\sin^2 s}{\sin s(1+\cos s)}+\frac{(1+\cos s)^2}{\sin s(1+\cos s)}=\frac{\sin^2 s+(1+\cos s)^2}{\sin s(1+\cos s)}$$

$$=\frac{\sin^2 s+\left(1+2\cos s+\cos^2 s\right)}{\sin s(1+\cos s)}=\frac{\left(1-\cos^2 s\right)+1+2\cos s+\cos^2 s}{\sin s(1+\cos s)}$$

$$=\frac{2+2\cos s}{\sin s(1+\cos s)}=\frac{2(1+\cos s)}{\sin s(1+\cos s)}=\frac{2}{\sin s}=2\csc s$$

19. Verify $\tan 4\theta=\dfrac{2\tan 2\theta}{2-\sec^2 2\theta}$ is an identity.

Starting on the right side, we have the following.

$$\frac{2\tan 2\theta}{2-\sec^2 2\theta}=\frac{2\cdot\frac{\sin 2\theta}{\cos 2\theta}}{2-\frac{1}{\cos^2 2\theta}}=\frac{2\cdot\frac{\sin 2\theta}{\cos 2\theta}}{2-\frac{1}{\cos^2 2\theta}}\cdot\frac{\cos^2 2\theta}{\cos^2 2\theta}=\frac{2\sin 2\theta\cos 2\theta}{2\cos^2 2\theta-1}=\frac{\sin[2(2\theta)]}{\cos[2(2\theta)]}=\frac{\sin 4\theta}{\cos 4\theta}=\tan 4\theta$$

20. Verify $\tan\left(\dfrac{x}{2}+\dfrac{\pi}{4}\right)=\sec x+\tan x$ is an identity.

Starting on the left side, we have the following.

$$\tan\left(\frac{x}{2}+\frac{\pi}{4}\right)=\frac{\tan\frac{x}{2}+\tan\frac{\pi}{4}}{1-\tan\frac{x}{2}\tan\frac{\pi}{4}}=\frac{\tan\frac{x}{2}+1}{1-\left(\tan\frac{x}{2}\right)(1)}=\frac{\tan\frac{x}{2}+1}{1-\tan\frac{x}{2}}=\frac{\frac{\sin x}{1+\cos x}+1}{1-\frac{\sin x}{1+\cos x}}$$

$$=\frac{\frac{\sin x}{1+\cos x}+1}{1-\frac{\sin x}{1+\cos x}}\cdot\frac{1+\cos x}{1+\cos x}=\frac{\sin x+(1+\cos x)}{(1+\cos x)-\sin x}=\frac{\sin x+1+\cos x}{1+\cos x-\sin x}$$

$$=\frac{\sin x+1+\cos x}{1+\cos x-\sin x}\cdot\frac{\cos x}{\cos x}=\frac{\sin x\cos x+\cos x+\cos^2 x}{\cos x(1+\cos x-\sin x)}$$

$$=\frac{\cos x(1+\sin x)+(1+\sin x)(1-\sin x)}{\cos x(1+\cos x-\sin x)}=\frac{(1+\sin x)(\cos x+1-\sin x)}{\cos x(1+\cos x-\sin x)}$$

$$=\frac{1+\sin x}{\cos x}=\frac{1}{\cos x}+\frac{\sin x}{\cos x}=\sec x+\tan x$$

21. Verify $\frac{\cot s-\tan s}{\cos s+\sin s}=\frac{\cos s-\sin s}{\sin s\cos s}$ is an identity.

Starting on the left side, we have the following.

$$\frac{\cot s-\tan s}{\cos s+\sin s}=\frac{\frac{\cos s}{\sin s}-\frac{\sin s}{\cos s}}{\cos s+\sin s}=\frac{\frac{\cos s}{\sin s}-\frac{\sin s}{\cos s}}{\cos s+\sin s}\cdot\frac{\sin s\cos s}{\sin s\cos s}=\frac{\cos^2 s-\sin^2 s}{(\cos s+\sin s)\sin s\cos s}=\frac{(\cos s+\sin s)(\cos s-\sin s)}{(\cos s+\sin s)\sin s\cos s}=\frac{\cos s-\sin s}{\sin s\cos s}$$

22. Verify $\frac{\tan\theta-\cot\theta}{\tan\theta+\cot\theta}=1-2\cos^2\theta$ is an identity.

Starting on the left side, we have the following.

$$\frac{\tan\theta-\cot\theta}{\tan\theta+\cot\theta}=\frac{\frac{\sin\theta}{\cos\theta}-\frac{\cos\theta}{\sin\theta}}{\frac{\sin\theta}{\cos\theta}+\frac{\cos\theta}{\sin\theta}}=\frac{\frac{\sin\theta}{\cos\theta}-\frac{\cos\theta}{\sin\theta}}{\frac{\sin\theta}{\cos\theta}+\frac{\cos\theta}{\sin\theta}}\cdot\frac{\cos\theta\sin\theta}{\cos\theta\sin\theta}=\frac{\sin^2\theta-\cos^2\theta}{\sin^2\theta+\cos^2\theta}=\frac{\sin^2\theta-\cos^2\theta}{1}$$
$$=\sin^2\theta-\cos^2\theta=\left(1-\cos^2\theta\right)-\cos^2\theta=1-2\cos^2\theta$$

23. Verify $\frac{\tan(x+y)-\tan y}{1+\tan(x+y)\tan y}=\tan x$ is an identity.

Starting on the left side, we have the following.

$$\frac{\tan(x+y)-\tan y}{1+\tan(x+y)\tan y}=\frac{\frac{\tan x+\tan y}{1-\tan x\tan y}-\tan y}{1+\frac{\tan x+\tan y}{1-\tan x\tan y}\cdot\tan y}=\frac{\frac{\tan x+\tan y}{1-\tan x\tan y}-\tan y}{1+\frac{\tan x+\tan y}{1-\tan x\tan y}\cdot\tan y}\cdot\frac{1-\tan x\tan y}{1-\tan x\tan y}$$
$$=\frac{\tan x+\tan y-\tan y(1-\tan x\tan y)}{1-\tan x\tan y+(\tan x+\tan y)\tan y}=\frac{\tan x+\tan x\tan^2 y}{1-\tan x\tan y+\tan x\tan y+\tan^2 y}=\frac{\tan x\left(1+\tan^2 y\right)}{1+\tan^2 y}=\tan x$$

24. Verify $2\cos^2 \frac{x}{2} \tan x = \tan x + \sin x$ is an identity.

Starting on the left side, we have the following.

$$2\cos^2 \frac{x}{2} \tan x = 2\left(\pm\sqrt{\frac{1+\cos x}{2}}\right)^2 \cdot \frac{\sin x}{\cos x}$$
$$= 2\cdot\frac{1+\cos x}{2}\cdot\frac{\sin x}{\cos x} = (1+\cos x)\cdot\frac{\sin x}{\cos x} = \frac{\sin x}{\cos x} + \sin x = \tan x + \sin x$$

25. Verify $\frac{\cos^4 x - \sin^4 x}{\cos^2 x} = 1 - \tan^2 x$ is an identity.

Starting on the left side, we have the following.

$$\frac{\cos^4 x - \sin^4 x}{\cos^2 x} = \frac{(\cos^2 x + \sin^2 x)(\cos^2 x - \sin^2 x)}{\cos^2 x} = \frac{(1)(\cos^2 x - \sin^2 x)}{\cos^2 x}$$
$$= \frac{\cos^2 x - \sin^2 x}{\cos^2 x} = \frac{\cos^2 x}{\cos^2 x} - \frac{\sin^2 x}{\cos^2 x} = 1 - \tan^2 x$$

26. Verify $\frac{\csc t + 1}{\csc t - 1} = (\sec t + \tan t)^2$ is an identity.

Starting on the left side, we have the following.

$$\frac{\csc t + 1}{\csc t - 1} = \frac{\frac{1}{\sin t}+1}{\frac{1}{\sin t}-1} = \frac{\frac{1}{\sin t}+1}{\frac{1}{\sin t}-1}\cdot\frac{\sin t}{\sin t} = \frac{1+\sin t}{1-\sin t} = \frac{1+\sin t}{1-\sin t}\cdot\frac{1+\sin t}{1+\sin t}$$
$$= \frac{(1+\sin t)^2}{1-\sin^2 t} = \frac{(1+\sin t)^2}{\cos^2 t} = \left(\frac{1+\sin t}{\cos t}\right)^2 = \left(\frac{1}{\cos t}+\frac{\sin t}{\cos t}\right)^2 = (\sec t + \tan t)^2$$

27. Verify $\frac{2(\sin x - \sin^3 x)}{\cos x} = \sin 2x$ is an identity.

Starting on the left side, we have the following.

$$\frac{2(\sin x - \sin^3 x)}{\cos x} = \frac{2\sin x(1-\sin^2 x)}{\cos x} = \frac{2\sin x\cos^2 x}{\cos x} = 2\sin x\cos x = \sin 2x$$

28. Verify $\frac{1}{2}\cot\frac{x}{2} - \frac{1}{2}\tan\frac{x}{2} = \cot x$ is an identity.

Starting on the left side, we have the following.

$$\frac{1}{2}\cot\frac{x}{2} - \frac{1}{2}\tan\frac{x}{2} = \frac{1}{2}\cdot\frac{1}{\tan\frac{x}{2}} - \frac{1}{2}\tan\frac{x}{2} = \frac{1}{2}\cdot\frac{1}{\frac{\sin x}{1+\cos x}} - \frac{1}{2}\cdot\frac{1-\cos x}{\sin x}$$
$$= \frac{1+\cos x}{2\sin x} - \frac{1-\cos x}{2\sin x} = \frac{1+\cos x-(1-\cos x)}{2\sin x} = \frac{1+\cos x - 1 + \cos x}{2\sin x}$$
$$= \frac{2\cos x}{2\sin x} = \frac{\cos x}{\sin x} = \cot x$$

Chapter 5: Review Exercises

1. Since $\sec x = \dfrac{1}{\cos x}$, the correct choice is B.

2. Since $\csc x = \dfrac{1}{\sin x}$, the correct choice is A.

3. Since $\tan x = \dfrac{\sin x}{\cos x}$, the correct choice is C.

4. Since $\cot x = \dfrac{\cos x}{\sin x}$, the correct choice is F.

5. Since $\tan^2 x = \dfrac{1}{\cot^2 x}$, the correct choice is D.

6. Since $\sec^2 x = \dfrac{1}{\cos^2 x}$, the correct choice is E.

7. $\sec^2\theta - \tan^2\theta = \dfrac{1}{\cos^2\theta} - \dfrac{\sin^2\theta}{\cos^2\theta} = \dfrac{1-\sin^2\theta}{\cos^2\theta} = \dfrac{\cos^2\theta}{\cos^2\theta} = 1$

8. $\dfrac{\cot\theta}{\sec\theta} = \dfrac{\frac{\cos\theta}{\sin\theta}}{\frac{1}{\cos\theta}} = \dfrac{\cos\theta}{\sin\theta}\cdot\dfrac{\cos\theta}{1} = \dfrac{\cos^2\theta}{\sin\theta}$

9. $\tan^2\theta\left(1+\cot^2\theta\right) = \dfrac{\sin^2\theta}{\cos^2\theta}\left(1+\dfrac{\cos^2\theta}{\sin^2\theta}\right) = \dfrac{\sin^2\theta}{\cos^2\theta}\left(\dfrac{\sin^2\theta+\cos^2\theta}{\sin^2\theta}\right) = \dfrac{\sin^2\theta}{\cos^2\theta}\left(\dfrac{1}{\sin^2\theta}\right) = \dfrac{1}{\cos^2\theta}$

10. $\csc\theta + \cot\theta = \dfrac{1}{\sin\theta} + \dfrac{\cos\theta}{\sin\theta} = \dfrac{1+\cos\theta}{\sin\theta}$

11. $\tan\theta - \sec\theta\csc\theta = \dfrac{\sin\theta}{\cos\theta} - \dfrac{1}{\cos\theta}\cdot\dfrac{1}{\sin\theta} = \dfrac{\sin^2\theta}{\cos\theta\sin\theta} - \dfrac{1}{\cos\theta\sin\theta} = \dfrac{\sin^2\theta - 1}{\cos\theta\sin\theta}$

$= \dfrac{\left(1-\cos^2\theta\right)-1}{\cos\theta\sin\theta} = \dfrac{1-\cos^2\theta-1}{\cos\theta\sin\theta} = \dfrac{-\cos^2\theta}{\sin\theta\cos\theta} = -\dfrac{\cos\theta}{\sin\theta}$

12. $\csc^2\theta + \sec^2\theta = \dfrac{1}{\sin^2\theta} + \dfrac{1}{\cos^2\theta}$

$= \dfrac{1}{\sin^2\theta}\cdot\dfrac{\cos^2\theta}{\cos^2\theta} + \dfrac{1}{\cos^2\theta}\cdot\dfrac{\sin^2\theta}{\sin^2\theta} = \dfrac{\cos^2\theta+\sin^2\theta}{\sin^2\theta\cos^2\theta} = \dfrac{1}{\sin^2\theta\cos^2\theta}$

13. $\cos x = \dfrac{3}{5}$, x is in quadrant IV.

$\sin^2 x = 1-\cos^2 x = 1-\left(\dfrac{3}{5}\right)^2 = 1-\dfrac{9}{25} = \dfrac{16}{25}$

Since x is in quadrant IV, $\sin x < 0$.

$\sin x = -\sqrt{\dfrac{16}{25}} = -\dfrac{4}{5}$; $\tan x = \dfrac{\sin x}{\cos x} = \dfrac{-\frac{4}{5}}{\frac{3}{5}} = -\dfrac{4}{3}$; $\cot(-x) = -\cot x = \dfrac{1}{-\tan x} = \dfrac{1}{-\left(-\frac{4}{3}\right)} = \dfrac{3}{4}$

14. $\tan x = -\frac{5}{4}, \frac{\pi}{2} < x < \pi$

$$\sec^2 x = \tan^2 x + 1 = \left(-\frac{5}{4}\right)^2 + 1 = \frac{25}{16} + 1 = \frac{41}{16}$$

Since x is in quadrant II, $\sec x < 0$, so $\sec x = -\frac{\sqrt{41}}{4}$.

$$\cot x = \frac{1}{\tan x} = \frac{1}{-\frac{5}{4}} = -\frac{4}{5}; \cos x = \frac{1}{\sec x} = \frac{1}{-\frac{\sqrt{41}}{4}} = -\frac{4}{\sqrt{41}} \cdot \frac{\sqrt{41}}{\sqrt{41}} = -\frac{4\sqrt{41}}{41}$$

Since x is in quadrant II, $\sin x > 0$.

$$\sin x = \sqrt{1 - \cos^2 x} = \sqrt{1 - \left(-\frac{4\sqrt{41}}{41}\right)^2} = \sqrt{1 - \frac{656}{1681}} = \sqrt{\frac{1025}{1681}} = \frac{5\sqrt{41}}{41}$$

$$\csc x = \frac{1}{\sin x} = \frac{1}{\frac{5\sqrt{41}}{41}} = \frac{41}{5\sqrt{41}} \cdot \frac{\sqrt{41}}{\sqrt{41}} = \frac{\sqrt{41}}{5}$$

15. Use the fact that $165° = 180° - 15°$.

$$\sin 165° = \sin 180° \cos 15° - \cos 180° \sin 15° = 0 \cdot \cos 15° - (-1)\sin 15° = 0 + \sin 15° = \sin 15°$$

$$= \sin(45° - 30°) = \sin 45° \cos 30° - \cos 45° \sin 30° = \frac{\sqrt{2}}{2} \cdot \frac{\sqrt{3}}{2} - \frac{\sqrt{2}}{2} \cdot \frac{1}{2} = \frac{\sqrt{6} - \sqrt{2}}{4}$$

$$\cos 165° = \cos 180° \cos 15° + \sin 180° \sin 15° = -1 \cdot \cos 15° + 0 \cdot \sin 15° = -\cos 15° + 0 = -\cos 15°$$

$$= -\cos(45° - 30°) = -(\cos 45° \cos 30° + \sin 45° \sin 30°) = -\left(\frac{\sqrt{2}}{2} \cdot \frac{\sqrt{3}}{2} + \frac{\sqrt{2}}{2} \cdot \frac{1}{2}\right) = \frac{-\sqrt{6} - \sqrt{2}}{4}$$

$$\tan 165° = \frac{\tan 180° - \tan 15°}{1 + \tan 180° \tan 15°} = \frac{0 - \tan 15°}{1 + 0 \cdot \tan 15°} = \frac{-\tan 15°}{1 + 0} = -\tan 15°$$

$$= -\frac{\tan 45° - \tan 30°}{1 + \tan 45 \tan 30°} = -\frac{1 - \frac{\sqrt{3}}{3}}{1 + \frac{\sqrt{3}}{3}} = -\frac{1 - \frac{\sqrt{3}}{3}}{1 + \frac{\sqrt{3}}{3}} \cdot \frac{3}{3} = -\frac{3 - \sqrt{3}}{3 + \sqrt{3}} \cdot \frac{3 - \sqrt{3}}{3 - \sqrt{3}}$$

$$= -\frac{9 - 3\sqrt{3} - 3\sqrt{3} + 3}{9 - 3} = -\frac{12 - 6\sqrt{3}}{6} = -(2 - \sqrt{3}) = -2 + \sqrt{3}$$

$$\sec 165° = \frac{1}{\cos 165°} = \frac{1}{\frac{-\sqrt{2} - \sqrt{6}}{4}} = \frac{4}{-\sqrt{2} - \sqrt{6}} \cdot \frac{\sqrt{2} - \sqrt{6}}{\sqrt{2} - \sqrt{6}} = -\frac{4(\sqrt{2} - \sqrt{6})}{2 - 6} = -\frac{4(\sqrt{2} - \sqrt{6})}{-4} = \sqrt{2} - \sqrt{6}$$

$$\csc 165° = \frac{1}{\sin 165°} = \frac{1}{\frac{\sqrt{6} - \sqrt{2}}{4}} = \frac{4}{\sqrt{6} - \sqrt{2}} \cdot \frac{\sqrt{6} + \sqrt{2}}{\sqrt{6} + \sqrt{2}} = \frac{4(\sqrt{6} + \sqrt{2})}{6 - 2} = \frac{4(\sqrt{6} + \sqrt{2})}{4} = \sqrt{6} + \sqrt{2}$$

$$\cot 165° = \frac{1}{\tan 165°} = \frac{1}{-2 + \sqrt{3}} = \frac{1}{-2 + \sqrt{3}} \cdot \frac{-2 - \sqrt{3}}{-2 - \sqrt{3}} = \frac{-2 - \sqrt{3}}{4 - 3} = \frac{-2 - \sqrt{3}}{1} = -2 - \sqrt{3}$$

16. (a) $\sin\frac{\pi}{12}=\sin\left(\frac{\pi}{4}-\frac{\pi}{6}\right)=\sin\frac{\pi}{4}\cos\frac{\pi}{6}-\cos\frac{\pi}{4}\sin\frac{\pi}{6}=\frac{\sqrt{2}}{2}\cdot\frac{\sqrt{3}}{2}-\frac{\sqrt{2}}{2}\cdot\frac{1}{2}=\frac{\sqrt{6}}{4}-\frac{\sqrt{2}}{4}=\frac{\sqrt{6}-\sqrt{2}}{4}$

$\cos\frac{\pi}{12}=\cos\left(\frac{\pi}{4}-\frac{\pi}{6}\right)=\cos\frac{\pi}{4}\cos\frac{\pi}{6}+\sin\frac{\pi}{4}\sin\frac{\pi}{6}=\frac{\sqrt{2}}{2}\cdot\frac{\sqrt{3}}{2}+\frac{\sqrt{2}}{2}\cdot\frac{1}{2}=\frac{\sqrt{6}}{4}+\frac{\sqrt{2}}{4}=\frac{\sqrt{6}+\sqrt{2}}{4}$

$\tan\frac{\pi}{12}=\tan\left(\frac{\pi}{4}-\frac{\pi}{6}\right)=\frac{\tan\frac{\pi}{4}-\tan\frac{\pi}{6}}{1+\tan\frac{\pi}{4}\tan\frac{\pi}{6}}=\frac{1-\frac{\sqrt{3}}{3}}{1+(1)\frac{\sqrt{3}}{3}}=\frac{3-\sqrt{3}}{3+\sqrt{3}}\cdot\frac{3-\sqrt{3}}{3-\sqrt{3}}=\frac{12-6\sqrt{3}}{6}=2-\sqrt{3}$

(b) In this exercise $\frac{\pi}{12}$ is in quadrant I, where all circular functions are positive.

$\sin\frac{\pi}{12}=\sin\frac{\frac{\pi}{6}}{2}=\sqrt{\frac{1-\cos\frac{\pi}{6}}{2}}=\sqrt{\frac{1-\frac{\sqrt{3}}{2}}{2}}=\sqrt{\frac{2-\sqrt{3}}{4}}=\frac{\sqrt{2-\sqrt{3}}}{2}$

$\cos\frac{\pi}{12}=\cos\frac{\frac{\pi}{6}}{2}=\sqrt{\frac{1+\cos\frac{\pi}{6}}{2}}=\sqrt{\frac{1+\frac{\sqrt{3}}{2}}{2}}=\sqrt{\frac{2+\sqrt{3}}{4}}=\frac{\sqrt{2+\sqrt{3}}}{2}$

$\tan\frac{\pi}{12}=\tan\frac{\frac{\pi}{6}}{2}=\frac{\sin\frac{\pi}{6}}{1+\cos\frac{\pi}{6}}=\frac{\frac{1}{2}}{1+\frac{\sqrt{3}}{2}}=\frac{1}{2+\sqrt{3}}\cdot\frac{2-\sqrt{3}}{2-\sqrt{3}}=\frac{2-\sqrt{3}}{4-3}=\frac{2-\sqrt{3}}{1}=2-\sqrt{3}$

17. Since $\cos 210°=\cos(150°+60°)=\cos 150°\cos 60°-\sin 150°\sin 60°$, the correct choice is E.

18. Since $\sin 35°=\cos(90°-35°)=\cos 55°$, the correct choice is B.

19. Since $\tan(-35°)=\cot[90°-(-35°)]=\cot 125°$, the correct choice is J.

20. Since $-\sin 35°=\sin(-35°)$, the correct choice is A.

21. Since $\cos 35°=\cos(-35°)$, the correct choice is I.

22. Since $\cos 75°=\cos\frac{150°}{2}=\sqrt{\frac{1+\cos 150°}{2}}$, the correct choice is C.

23. Since $\sin 75°=\sin(15°+60°)=\sin 15°\cos 60°+\cos 15°\sin 60°$, the correct choice is H.

24. Since $\sin 300°=\sin 2(150°)=2\sin 150°\cos 150°$, the correct choice is D.

25. Since $\cos 300°=\cos 2(150°)=\cos^2 150°-\sin^2 150°$, the correct choice is G.

26. Since $\cos(-55°)=\cos 55°$, the correct choice is B.

27. Find sin $(x + y)$, cos $(x - y)$, and tan $(x + y)$, given $\sin x = -\frac{1}{4}$, $\cos y = -\frac{4}{5}$, x and y are in quadrant III.

Since x and y are in quadrant III, cos x and sin y are negative.

$$\cos x = -\sqrt{1-\sin^2 x} = -\sqrt{1-\left(-\frac{1}{4}\right)^2} = -\sqrt{1-\frac{1}{16}} = -\sqrt{\frac{15}{16}} = -\frac{\sqrt{15}}{4}$$

$$\sin y = -\sqrt{1-\cos^2 y} = -\sqrt{1-\left(-\frac{4}{5}\right)^2} = -\sqrt{1-\frac{16}{25}} = -\sqrt{\frac{9}{25}} = -\frac{3}{5}$$

$$\sin(x+y) = \sin x \cos y + \cos x \sin y = \left(-\frac{1}{4}\right)\left(-\frac{4}{5}\right)+\left(-\frac{\sqrt{15}}{4}\right)\left(-\frac{3}{5}\right) = \frac{4}{20}+\frac{3\sqrt{15}}{20} = \frac{4+3\sqrt{15}}{20}$$

$$\cos(x-y) = \cos x \cos y + \sin x \sin y$$
$$= \left(-\frac{\sqrt{15}}{4}\right)\left(-\frac{4}{5}\right)+\left(-\frac{1}{4}\right)\left(-\frac{3}{5}\right) = \frac{\sqrt{15}}{5}+\frac{3}{20} = \frac{4\sqrt{15}}{20}+\frac{3}{20} = \frac{4\sqrt{15}+3}{20}$$

$$\cos(x+y) = \cos x \cos y - \sin x \sin y$$
$$= \left(-\frac{\sqrt{15}}{4}\right)\left(-\frac{4}{5}\right)-\left(-\frac{1}{4}\right)\left(-\frac{3}{5}\right) = \frac{\sqrt{15}}{5}-\frac{3}{20} = \frac{4\sqrt{15}}{20}-\frac{3}{20} = \frac{4\sqrt{15}-3}{20}$$

$$\tan(x+y) = \frac{\sin(x+y)}{\cos(x+y)} = \frac{\frac{4+3\sqrt{15}}{20}}{\frac{4\sqrt{15}-3}{20}} = \frac{4+3\sqrt{15}}{4\sqrt{15}-3}$$

To find the quadrant of $x + y$, notice that $\sin(x + y) > 0$, which implies $x + y$ is in quadrant I or II. Also $\tan(x + y) > 0$, which implies that $x + y$ is in quadrant I or III. Therefore, $x + y$ is in quadrant I.

28. Find sin $(x + y)$, cos $(x - y)$, and tan $(x + y)$, given $\sin x = \frac{1}{10}$, $\cos y = \frac{4}{5}$, x is in quadrant I, y is in quadrant IV.

Since x is in quadrant I, $\cos x > 0$ and $\cos x = \sqrt{1-\sin^2 x} = \sqrt{1-\left(\frac{1}{10}\right)^2} = \sqrt{1-\frac{1}{100}} = \sqrt{\frac{99}{100}} = \frac{3\sqrt{11}}{10}$.

Since y is in quadrant IV, $\sin y < 0$ and $\sin y = -\sqrt{1-\cos^2 x} = -\sqrt{1-\left(\frac{4}{5}\right)^2} = -\sqrt{1-\frac{16}{25}} = -\sqrt{\frac{9}{25}} = -\frac{3}{5}$.

$$\sin(x+y) = \sin x \cos y + \cos x \sin y = \left(\frac{1}{10}\right)\left(\frac{4}{5}\right)+\frac{3\sqrt{11}}{10}\left(-\frac{3}{5}\right) = \frac{4}{50}-\frac{9\sqrt{11}}{50} = \frac{4-9\sqrt{11}}{50}$$

$$\cos(x-y) = \cos x \cos y + \sin x \sin y = \left(\frac{3\sqrt{11}}{10}\right)\left(\frac{4}{5}\right)+\left(\frac{1}{10}\right)\left(-\frac{3}{5}\right) = \frac{12\sqrt{11}}{50}-\frac{3}{50} = \frac{12\sqrt{11}-3}{50}$$

$$\cos(x+y) = \cos x \cos y - \sin x \sin y = \left(\frac{3\sqrt{11}}{10}\right)\left(\frac{4}{5}\right)-\left(\frac{1}{10}\right)\left(-\frac{3}{5}\right) = \frac{12\sqrt{11}}{50}+\frac{3}{50} = \frac{12\sqrt{11}+3}{50}$$

$$\tan(x+y) = \frac{\sin(x+y)}{\cos(x+y)} = \frac{\frac{4-9\sqrt{11}}{50}}{\frac{12\sqrt{11}+3}{50}} = \frac{4-9\sqrt{11}}{12\sqrt{11}+3}$$

To find the quadrant of $x + y$, notice that $\sin(x + y) < 0$, which implies $x + y$ is in quadrant III or IV. Also $\tan(x + y) < 0$, which implies that $x + y$ is in quadrant II or IV. Therefore, $x + y$ is in quadrant IV.

29. Find sin $(x + y)$, cos $(x - y)$, and tan $(x + y)$, given $\sin x = -\frac{1}{2}$, $\cos y = -\frac{2}{5}$, x and y are in quadrant III.

Since x and y are in quadrant III, cos x and sin y are negative.

$$\cos x = -\sqrt{1-\sin^2 x} = -\sqrt{1-\left(-\frac{1}{2}\right)^2} = -\sqrt{1-\frac{1}{4}} = -\sqrt{\frac{3}{4}} = -\frac{\sqrt{3}}{2}$$

$$\sin y = -\sqrt{1-\cos^2 y} = -\sqrt{1-\left(-\frac{2}{5}\right)^2} = -\sqrt{1-\frac{4}{25}} = -\sqrt{\frac{21}{25}} = -\frac{\sqrt{21}}{5}$$

$$\sin(x+y) = \sin x \cos y + \cos x \sin y = \left(-\frac{1}{2}\right)\left(-\frac{2}{5}\right) + \left(-\frac{\sqrt{3}}{2}\right)\left(-\frac{\sqrt{21}}{5}\right) = \frac{2}{10} + \frac{\sqrt{63}}{10} = \frac{2+3\sqrt{7}}{10}$$

$$\cos(x-y) = \cos x \cos y + \sin x \sin y$$

$$= \left(-\frac{2}{5}\right)\left(-\frac{\sqrt{3}}{2}\right) + \left(-\frac{1}{2}\right)\left(-\frac{\sqrt{21}}{5}\right) = \frac{2\sqrt{3}}{10} + \frac{\sqrt{21}}{10} = \frac{2\sqrt{3}+\sqrt{21}}{10}$$

$$\cos(x+y) = \cos x \cos y - \sin x \sin y$$

$$= \left(-\frac{2}{5}\right)\left(-\frac{\sqrt{3}}{2}\right) - \left(-\frac{1}{2}\right)\left(-\frac{\sqrt{21}}{5}\right) = \frac{2\sqrt{3}}{10} - \frac{\sqrt{21}}{10} = \frac{2\sqrt{3}-\sqrt{21}}{10}$$

$$\tan(x+y) = \frac{\sin(x+y)}{\cos(x+y)} = \frac{\frac{4+3\sqrt{15}}{20}}{\frac{4\sqrt{15}+3}{20}} = \frac{2+3\sqrt{7}}{2\sqrt{3}-\sqrt{21}}$$

To find the quadrant of $x + y$, notice that $\sin(x + y) > 0$, which implies $x + y$ is in quadrant I or II. Also $\tan(x + y) < 0$, which implies that $x + y$ is in quadrant II or IV. Therefore, $x + y$ is in quadrant II.

30. $\sin y = -\frac{2}{3}, \cos x = -\frac{1}{5}$, x is in quadrant II, y is in quadrant III.

First, find $\sin x$ and $\cos y$.

Since x is in quadrant II, $\sin x$ is positive.

$$\sin x = \sqrt{1-\cos^2 x} = \sqrt{1-\left(-\frac{1}{5}\right)^2} = \sqrt{1-\frac{1}{25}} = \sqrt{\frac{24}{25}} = \frac{\sqrt{24}}{5} = \frac{2\sqrt{6}}{5}$$

Since y is in quadrant III, $\cos y$ is negative.

$$\cos y = -\sqrt{1-\sin^2} = -\sqrt{1-\left(-\frac{2}{3}\right)^2} = -\sqrt{1-\frac{4}{9}} = -\sqrt{\frac{5}{9}} = -\frac{\sqrt{5}}{3}$$

$$\sin(x+y) = \sin x \cos y + \cos x \sin y = \left(\frac{2\sqrt{6}}{5}\right)\left(-\frac{\sqrt{5}}{3}\right) + \left(-\frac{1}{5}\right)\left(-\frac{2}{3}\right) = \frac{-2\sqrt{30}}{15} + \frac{2}{15} = \frac{-2\sqrt{30}+2}{15}$$

$$\cos(x-y) = \cos x \cos y + \sin x \sin y = \left(-\frac{1}{5}\right)\left(-\frac{\sqrt{5}}{3}\right) + \left(\frac{2\sqrt{6}}{5}\right)\left(-\frac{2}{3}\right) = \frac{\sqrt{5}}{15} + \frac{-4\sqrt{6}}{15} = \frac{\sqrt{5}-4\sqrt{6}}{15}$$

Continued on next page

30. (continued)

In order to find $\tan(x+y)$ using the formula $\tan(x+y)=\frac{\tan x+\tan y}{1-\tan x\tan y}$, we need to find $\tan x$ and $\tan y$.

$$\tan x=\frac{\sin x}{\cos x}=\frac{\frac{2\sqrt{6}}{5}}{-\frac{1}{5}}=-2\sqrt{6} \text{ and } \tan y=\frac{\sin y}{\cos y}=\frac{-\frac{2}{3}}{-\frac{\sqrt{5}}{3}}=\frac{2}{\sqrt{5}}=\frac{2\sqrt{5}}{5}$$

$$\tan(x+y)=\frac{\tan x+\tan y}{1-\tan x\tan y}=\frac{-2\sqrt{6}+\frac{2\sqrt{5}}{5}}{1-\left(-2\sqrt{6}\right)\left(\frac{2\sqrt{5}}{5}\right)}\cdot\frac{5}{5}=\frac{-10\sqrt{6}+2\sqrt{5}}{5+4\sqrt{30}}$$

To find the quadrant of $x+y$, notice that $\sin(x+y)<0$, which indicates that $x+y$ is in quadrant III or IV. Also, $\tan(x+y)<0$, which indicates that $x+y$ is in quadrant II or IV. Therefore, $x+y$ is in quadrant IV.

31. Find sin $(x+y)$, cos $(x-y)$, and tan $(x+y)$, given $\sin x=\frac{1}{10}$, $\cos y=\frac{4}{5}$, x is in quadrant I, y is in quadrant IV. First, find $\sin x$ and $\cos y$.

Since x is in quadrant I, $\cos x>0$ and $\cos x=\sqrt{1-\sin^2 x}=\sqrt{1-\left(\frac{1}{10}\right)^2}=\sqrt{1-\frac{1}{100}}=\sqrt{\frac{99}{100}}=\frac{3\sqrt{11}}{10}$.

Since y is in quadrant IV, $\sin y<0$ and $\sin y=-\sqrt{1-\cos^2 x}=-\sqrt{1-\left(\frac{4}{5}\right)^2}=-\sqrt{1-\frac{16}{25}}=-\sqrt{\frac{9}{25}}=-\frac{3}{5}$.

$$\sin(x+y)=\sin x\cos y+\cos x\sin y=\left(\frac{1}{10}\right)\left(\frac{4}{5}\right)+\left(\frac{3\sqrt{11}}{10}\right)\left(-\frac{3}{5}\right)=\frac{4}{50}-\frac{9\sqrt{11}}{50}=\frac{4-9\sqrt{11}}{50}$$

$$\cos(x-y)=\cos x\cos y+\sin x\sin y=\left(\frac{3\sqrt{11}}{10}\right)\left(\frac{4}{5}\right)+\left(\frac{1}{10}\right)\left(-\frac{3}{5}\right)=\frac{12\sqrt{11}}{50}-\frac{3}{50}=\frac{12\sqrt{11}-3}{50}$$

In order to find $\tan(x+y)$ using the formula $\tan(x+y)=\frac{\tan x+\tan y}{1-\tan x\tan y}$, we need to find $\tan x$ and $\tan y$.

$$\tan x=\frac{\sin x}{\cos x}=\frac{\frac{1}{10}}{\frac{3\sqrt{11}}{10}}=\frac{1}{3\sqrt{11}}\cdot\frac{\sqrt{11}}{\sqrt{11}}=\frac{\sqrt{11}}{33} \text{ and } \tan y=\frac{\sin y}{\cos y}=\frac{-\frac{3}{5}}{\frac{4}{5}}=-\frac{3}{4}$$

$$\tan(x+y)=\frac{\tan x+\tan y}{1-\tan x\tan y}=\frac{\frac{\sqrt{11}}{33}+\left(-\frac{3}{4}\right)}{1-\left(\frac{\sqrt{11}}{33}\right)\left(-\frac{3}{4}\right)}\cdot\frac{132}{132}=\frac{4\sqrt{11}-99}{132+3\sqrt{11}}, \text{ which is equivalent to } \frac{4-9\sqrt{11}}{12\sqrt{11}+3}.$$

$$\frac{4\sqrt{11}-99}{132+3\sqrt{11}}\cdot\frac{\sqrt{11}}{\sqrt{11}}=\frac{44-99\sqrt{11}}{132\sqrt{11}+33}=\frac{11\left(4-9\sqrt{11}\right)}{11\left(12\sqrt{11}+3\right)}=\frac{4-9\sqrt{11}}{12\sqrt{11}+3}$$

To find the quadrant of $x+y$, notice that $\sin(x+y)<0$, which implies $x+y$ is in quadrant III or IV. Also $\tan(x+y)<0$, which implies that $x+y$ is in quadrant II or IV. Therefore, $x+y$ is in quadrant IV.

32. $\cos x = \frac{2}{9}, \sin y = -\frac{1}{2}$, x is in quadrant IV, y is in quadrant III.

First, find $\sin x$ and $\cos y$.

Since x is in quadrant IV, $\sin x$ is negative.

$$\sin x = -\sqrt{1-\cos^2 x} = -\sqrt{1-\left(\frac{2}{9}\right)^2} = -\sqrt{1-\frac{4}{81}} = -\sqrt{\frac{77}{81}} = -\frac{\sqrt{77}}{9}$$

Since y is in quadrant III, $\cos y$ is negative.

$$\cos y = -\sqrt{1-\sin^2} = -\sqrt{1-\left(-\frac{1}{2}\right)^2} = -\sqrt{1-\frac{1}{4}} = -\sqrt{\frac{3}{4}} = -\frac{\sqrt{3}}{2}$$

$$\sin(x+y) = \sin x \cos y + \cos x \sin y = \left(-\frac{\sqrt{77}}{9}\right)\left(-\frac{\sqrt{3}}{2}\right) + \left(\frac{2}{9}\right)\left(-\frac{1}{2}\right) = \frac{\sqrt{231}}{18} + \frac{-2}{18} = \frac{\sqrt{231}-2}{18}$$

$$\cos(x-y) = \cos x \cos y + \sin x \sin y = \left(\frac{2}{9}\right)\left(-\frac{\sqrt{3}}{2}\right) + \left(-\frac{\sqrt{77}}{9}\right)\left(-\frac{1}{2}\right) = -\frac{2\sqrt{3}}{18} + \frac{\sqrt{77}}{18} = \frac{\sqrt{77}-2\sqrt{3}}{18}$$

In order to find $\tan(x+y)$ using the formula $\tan(x+y) = \frac{\tan x + \tan y}{1-\tan x \tan y}$, we need to find $\tan x$ and $\tan y$.

$$\tan x = \frac{\sin x}{\cos x} = \frac{-\frac{\sqrt{77}}{9}}{\frac{2}{9}} = -\frac{\sqrt{77}}{2} \quad \text{and} \quad \tan y = \frac{\sin y}{\cos y} = \frac{-\frac{1}{2}}{-\frac{\sqrt{3}}{2}} = \frac{1}{\sqrt{3}} \cdot \frac{\sqrt{3}}{\sqrt{3}} = \frac{\sqrt{3}}{3}$$

$$\tan(x+y) = \frac{\tan x + \tan y}{1-\tan x \tan y} = \frac{-\frac{\sqrt{77}}{2} + \frac{\sqrt{3}}{3}}{1-\left(-\frac{\sqrt{77}}{2}\right)\left(\frac{\sqrt{3}}{3}\right)} \cdot \frac{6}{6} = \frac{-3\sqrt{77}+2\sqrt{3}}{6+\sqrt{231}} = \frac{2\sqrt{3}-3\sqrt{77}}{6+\sqrt{231}}$$

To find the quadrant of $x+y$, notice that $\sin(x+y) > 0$, which indicates that $x+y$ is in quadrant I or II. Also, $\tan(x+y) < 0$, which indicates that $x+y$ is in quadrant II or IV. Therefore, $x+y$ is in quadrant II.

33. Find $\sin\theta$ and $\cos\theta$, given $\cos 2\theta = -\frac{3}{4}$, $90° < 2\theta < 180°$.

Since $90° < 2\theta < 180° \Rightarrow 45° < \theta < 90°$, θ is in quadrant I.

$$\cos 2\theta = 1 - 2\sin^2\theta \Rightarrow -\frac{3}{4} = 1 - 2\sin^2\theta \Rightarrow -\frac{7}{4} = -2\sin^2\theta \Rightarrow \sin^2\theta = \frac{7}{8}$$

Since θ is in quadrant I, $\sin\theta > 0$ and $\sin\theta = \sqrt{\frac{7}{8}} = \frac{\sqrt{7}}{2\sqrt{2}} \cdot \frac{\sqrt{2}}{\sqrt{2}} = \frac{\sqrt{14}}{4}$.

Since θ is in quadrant I, $\cos\theta > 0$ and $\cos\theta = \sqrt{1-\sin^2\theta} = \sqrt{1-\frac{7}{8}} = \sqrt{\frac{1}{8}} = \frac{1}{\sqrt{8}} = \frac{1}{2\sqrt{2}} \cdot \frac{\sqrt{2}}{\sqrt{2}} = \frac{\sqrt{2}}{4}$

34. $\cos 2B = \frac{1}{8}$, B is in quadrant IV. Since B is in quadrant IV, $\sin B$ is negative and $\cos B$ is positive.

$$\sin B = -\sqrt{\frac{1-\cos 2B}{2}} = -\sqrt{\frac{1-\frac{1}{8}}{2}} = -\sqrt{\frac{8-1}{16}} = -\sqrt{\frac{7}{16}} = -\frac{\sqrt{7}}{4}$$

$$\cos B = \sqrt{\frac{1+\cos 2B}{2}} = \sqrt{\frac{1+\frac{1}{8}}{2}} = \sqrt{\frac{8+1}{16}} = \sqrt{\frac{9}{16}} = \frac{3}{4}$$

35. Find $\sin 2x$ and $\cos 2x$, given $\tan x = 3$, $\sin x < 0$.

If $\tan x = 3 > 0$ and $\sin x < 0$ then x is in quadrant III and $2x$ is in quadrant I or II.

$$\tan 2x = \frac{2\tan x}{1-\tan^2 x} = \frac{2(3)}{1-3^2} = \frac{6}{1-9} = -\frac{6}{8} = -\frac{3}{4}$$

Since $\tan 2x < 0$, $2x$ is in quadrant II. Thus, $\sec 2x < 0$ and $\sin 2x > 0$.

$$\sec 2x = -\sqrt{1+\tan^2 x} = -\sqrt{1+\left(-\frac{3}{4}\right)^2} = -\sqrt{1+\frac{9}{16}} = -\sqrt{\frac{25}{16}} = -\frac{5}{4}$$

$$\cos 2x = \frac{1}{\sec 2x} = \frac{1}{-\frac{5}{4}} = -\frac{4}{5} \text{ and } \sin 2x = \sqrt{1-\left(-\frac{4}{5}\right)^2} = \sqrt{1-\frac{16}{25}} = \sqrt{\frac{9}{25}} = \frac{3}{5}$$

36. $\sec y = -\frac{5}{3}, \sin y > 0$

$$\cos y = \frac{1}{\sec y} = \frac{1}{-\frac{5}{3}} = -\frac{3}{5};\quad \sin y = \sqrt{1-\cos^2 y} = \sqrt{1-\left(-\frac{3}{5}\right)^2} = \sqrt{1-\frac{9}{25}} = \sqrt{\frac{16}{25}} = \frac{4}{5}$$

$$\sin 2y = 2\sin y\cos y = 2\left(\frac{4}{5}\right)\left(-\frac{3}{5}\right) = -\frac{24}{25}$$

$$\cos 2y = \cos^2 y - \sin^2 y = \left(-\frac{3}{5}\right)^2 - \left(\frac{4}{5}\right)^2 = \frac{9}{25} - \frac{16}{25} = -\frac{7}{25}$$

37. Find $\cos \frac{\theta}{2}$, given $\cos\theta = -\frac{1}{2}$, $90° < \theta < 180°$.

Since $90° < \theta < 180° \Rightarrow 45° < \frac{\theta}{2} < 90°$, $\frac{\theta}{2}$ is in quadrant I and $\cos \frac{\theta}{2} > 0$.

$$\cos\frac{\theta}{2} = \sqrt{\frac{1+\left(-\frac{1}{2}\right)}{2}} = \sqrt{\frac{2-1}{4}} - \sqrt{\frac{1}{4}} = \frac{1}{2}$$

38. Find $\sin\frac{A}{2}$, $\cos A = -\frac{3}{4}$, $90° < A < 180°$.

Since $90° < A < 180° \Rightarrow 45° < \frac{A}{2} < 90°$, $\frac{A}{2}$ is in quadrant I and $\sin\frac{A}{2} > 0$.

$$\sin\frac{A}{2} = \sqrt{\frac{1-\cos A}{2}} = \sqrt{\frac{1-\left(-\frac{3}{4}\right)}{2}} = \sqrt{\frac{4+3}{8}} = \sqrt{\frac{7}{8}} = \frac{\sqrt{7}}{2\sqrt{2}}\cdot\frac{\sqrt{2}}{\sqrt{2}} = \frac{\sqrt{14}}{4}$$

39. Find $\tan x$, given $\tan 2x = 2$, $\pi < x < \frac{3\pi}{2}$.

$$\tan 2x = 2 \Rightarrow \frac{2\tan x}{1-\tan^2 x} = 2 \Rightarrow 2\tan x = 2 - 2\tan^2 x, \text{ if } \tan x \neq \pm 1$$

Thus, we have the equation $\tan^2 x + \tan x - 1 = 0$. We can now use the quadratic formula, $\left(x = \frac{-b \pm \sqrt{b^2 - 4ac}}{2a} \text{ given } ax^2 + bx + c = 0 \right)$ to solve the equation for $\tan x$.

$$\tan x = \frac{-1 \pm \sqrt{1^2 - 4(1)(-1)}}{2} = \frac{-1 \pm \sqrt{1+4}}{2} = \frac{-1 \pm \sqrt{5}}{2}$$

Since $\pi < x < \frac{3\pi}{2}$, x is in quadrant III and $\tan x > 0$. Thus, $\tan x = \frac{-1+\sqrt{5}}{2} = \frac{\sqrt{5}-1}{2}$.

40. Find $\sin y$, given $\cos 2y = -\frac{1}{3}$, with $\frac{\pi}{2} < y < \pi$.

$$\cos 2y = 1 - 2\sin^2 y \Rightarrow -\frac{1}{3} = 1 - 2\sin^2 y \Rightarrow -2\sin^2 y = -\frac{4}{3} \Rightarrow \sin^2 y = \frac{2}{3}$$

Since $\frac{\pi}{2} < y < \pi$, y is in quadrant II and $\sin y > 0$.

Thus, $\sin y = \sqrt{\frac{2}{3}} = \frac{\sqrt{2}}{\sqrt{3}} \cdot \frac{\sqrt{3}}{\sqrt{3}} = \frac{\sqrt{6}}{3}$.

41. Find $\tan \frac{x}{2}$, given $\sin x = .8$, with $0 < x < \frac{\pi}{2}$.

$$\cos^2 x + \sin^2 x = 1 \Rightarrow \cos^2 x + (.8)^2 = 1 \Rightarrow \cos^2 x + .64 = 1 \Rightarrow \cos^2 x = .36$$

Since $0 < x < \frac{\pi}{2}$, x is in quadrant I and $\cos x > 0$.

Thus, $\cos x = \sqrt{.36} = .6$ and $\tan \frac{x}{2} = \frac{1-\cos x}{\sin x} = \frac{1-.6}{.8} = \frac{.4}{.8} = .5$.

42. Find $\sin 2x$, given $\sin x = .6$, with $\frac{\pi}{2} < x < \pi$.

$$\cos^2 x + \sin^2 x = 1 \Rightarrow \cos^2 x + (.6)^2 = 1 \Rightarrow \cos^2 x + .36 = 1 \Rightarrow \cos^2 x = .64$$

Since $\frac{\pi}{2} < x < \pi$, x is in quadrant II and $\cos x < 0$.

Thus, $\cos x = -\sqrt{.64} = -.8$ and $\sin 2x = 2\sin x \cos x = 2(.6)(-.8) = -.96$.

Exercises 43 – 48 are graphed in the following window.

ZOOM MEMORY
1:ZBox
2:Zoom In
3:Zoom Out
4:ZDecimal
5:ZSquare
6:ZStandard
7↓ZTrig

WINDOW
Xmin=-6.152285…
Xmax=6.1522856…
Xscl=1.5707963…
Ymin=-4
Ymax=4
Yscl=1
Xres=1

43. $-\dfrac{\sin 2x+\sin x}{\cos 2x-\cos x}$ appears to be equivalent to $\cot\dfrac{x}{2}$.

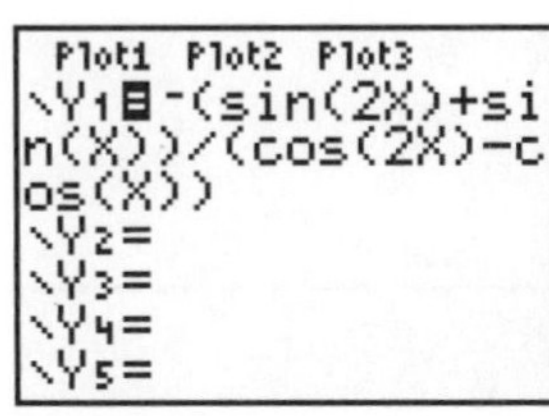

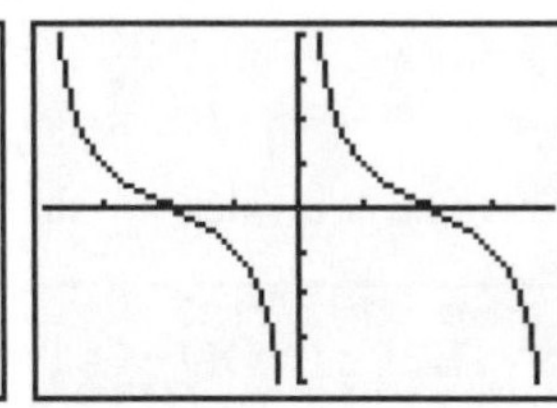

$$-\frac{\sin 2x+\sin x}{\cos 2x-\cos x}=-\frac{2\sin x\cos x+\sin x}{2\cos^2 x-1-\cos x}=-\frac{\sin x(2\cos x+1)}{2\cos^2 x-\cos x-1}$$

$$=-\frac{\sin x(2\cos x+1)}{(2\cos x+1)(\cos x-1)}=-\frac{\sin x}{\cos x-1}=\frac{\sin x}{1-\cos x}=\frac{1}{\frac{1-\cos x}{\sin x}}=\frac{1}{\tan\frac{x}{2}}=\cot\frac{x}{2}$$

44. $\dfrac{1-\cos 2x}{\sin 2x}$ appears to be equivalent $\tan x$.

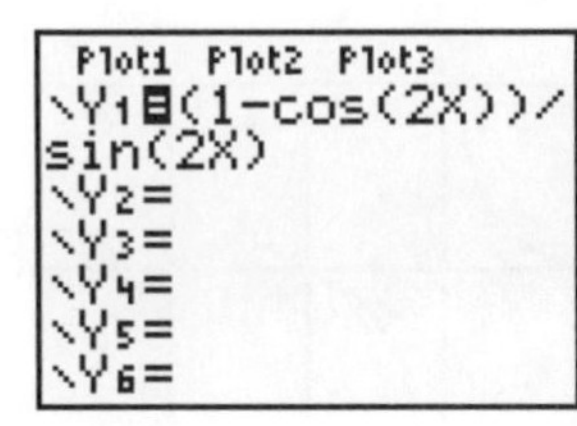

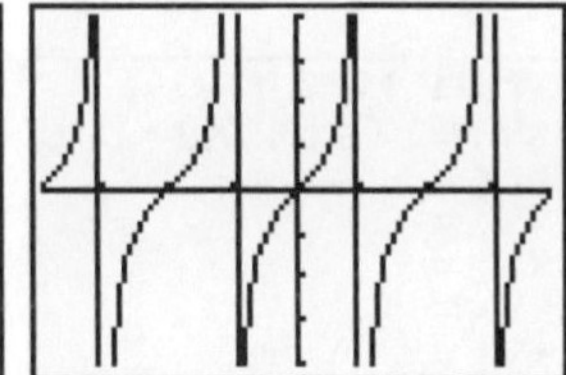

$$\frac{1-\cos 2x}{\sin 2x}=\tan\left(\frac{2x}{2}\right)=\tan x$$

45. $\dfrac{\sin x}{1-\cos x}$ appears to be equivalent to $\cot\dfrac{x}{2}$.

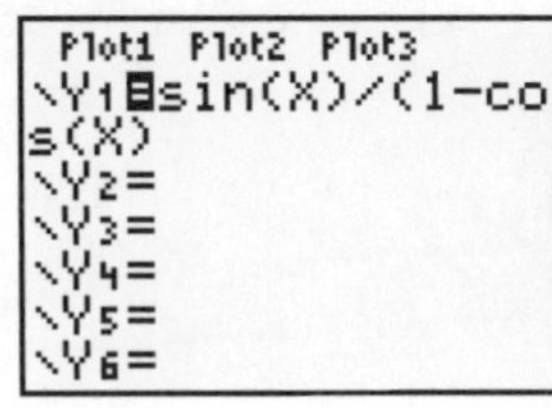

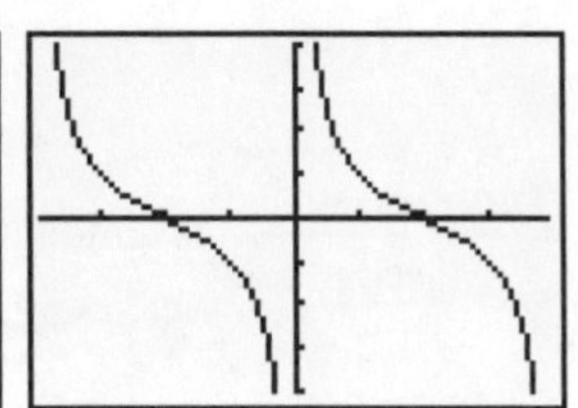

$$\frac{\sin x}{1-\cos x}=\frac{1}{\frac{1-\cos x}{\sin x}}=\frac{1}{\tan\frac{x}{2}}=\cot\frac{x}{2}$$

46. $\dfrac{\cos x \sin 2x}{1+\cos 2x}$ appears to be equivalent to $\sin x$.

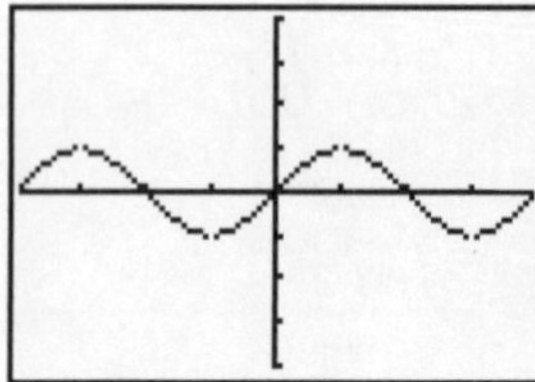

$$\frac{\cos x \sin 2x}{1+\cos 2x} = \frac{\cos x \cdot 2\sin x \cos x}{1+2\cos^2 x-1} = \frac{2\sin x \cos^2 x}{2\cos^2 x} = \sin x$$

47. $\dfrac{2\left(\sin x-\sin^3 x\right)}{\cos x}$ appears to be equivalent to $\sin 2x$.

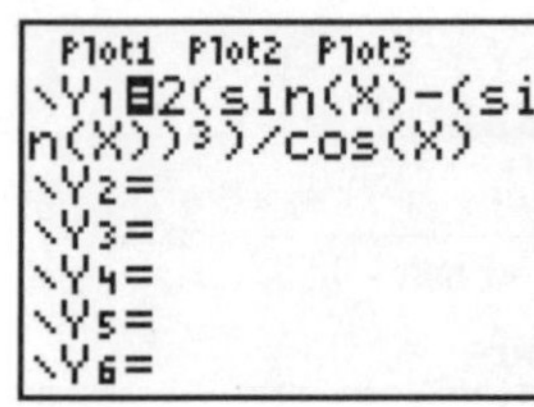

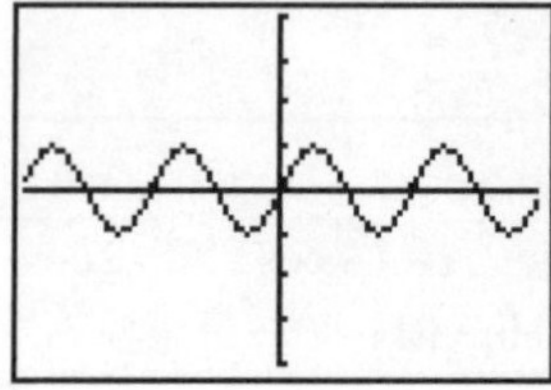

$$\frac{2\left(\sin x-\sin^3 x\right)}{\cos x} = \frac{2\sin x\left(1-\sin^2 x\right)}{\cos x}\cdot\frac{\cos x}{\cos x} = \frac{(2\sin x\cos x)\left(1-\sin^2 x\right)}{\cos^2 x} = \frac{\sin 2x\left(1-\sin^2 x\right)}{1-\sin^2 x} = \sin 2x$$

48. $\csc x-\cot x$ appears to be equivalent to $\tan\dfrac{x}{2}$.

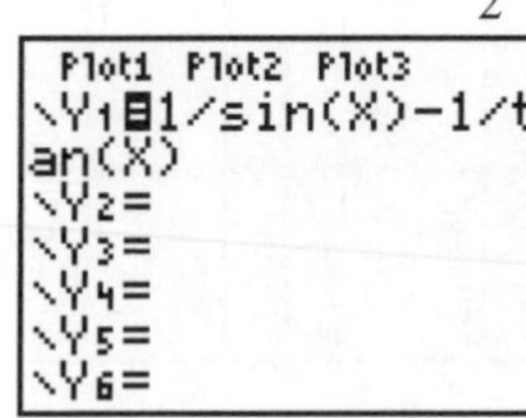

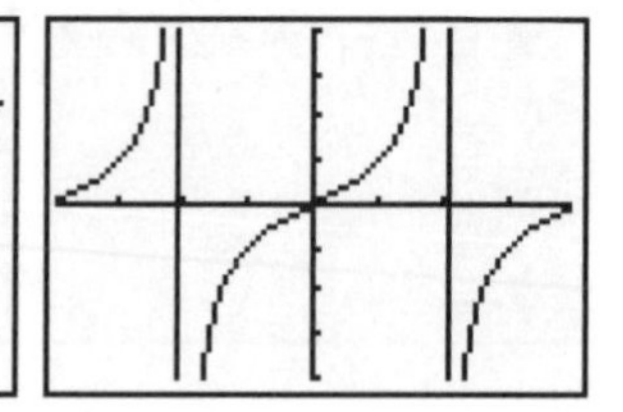

$$\csc x-\cot x = \frac{1}{\sin x}-\frac{\cos x}{\sin x} = \frac{1-\cos x}{\sin x} = \tan\frac{x}{2}$$

49. Verify $\sin^2 x-\sin^2 y=\cos^2 y-\cos^2 x$ is an identity.

$$\sin^2 x-\sin^2 y=\left(1-\cos^2 x\right)-\left(1-\cos^2 y\right)=1-\cos^2 x-1+\cos^2 y=\cos^2 y-\cos^2 x$$

50. Verify $2\cos^3 x-\cos x=\dfrac{\cos^2 x-\sin^2 x}{\sec x}$ is an identity.

Work with the right side.

$$\frac{\cos^2 x-\sin^2 x}{\sec x} = \frac{\cos^2 x-\sin^2 x}{\frac{1}{\cos x}} = \left(\cos^2 x-\sin^2 x\right)\cdot\cos x = \cos^3 x-\sin^2 x\cos x$$

$$= \cos^3 x-\left(1-\cos^2 x\right)\cos x = \cos^3 x-\cos x+\cos^3 x = 2\cos^3 x-\cos x$$

51. Verify $\dfrac{\sin^2 x}{2-2\cos x}=\cos^2\dfrac{x}{2}$ is an identity.

$$\frac{\sin^2 x}{2-2\cos x} = \frac{1-\cos^2 x}{2(1-\cos x)} = \frac{(1-\cos x)(1+\cos x)}{2(1-\cos x)} = \frac{1+\cos x}{2} = \cos^2\frac{x}{2}$$

52. Verify $\dfrac{\sin 2x}{\sin x} = \dfrac{2}{\sec x}$ is an identity.

$$\frac{\sin 2x}{\sin x} = \frac{2\sin x\cos x}{\sin x} = 2\cos x = \frac{2}{\frac{1}{\cos x}} = \frac{2}{\sec x}$$

53. Verify $2\cos A - \sec A = \cos A - \dfrac{\tan A}{\csc A}$ is an identity.

Work with the right side.

$$\cos A - \frac{\tan A}{\csc A} = \cos A - \frac{\frac{\sin A}{\cos A}}{\frac{1}{\sin A}} = \cos A - \frac{\sin^2 A}{\cos A} = \frac{\cos^2 A}{\cos A} - \frac{\sin^2 A}{\cos A} = \frac{\cos^2 A - \sin^2 A}{\cos A}$$

$$= \frac{\cos^2 A - \left(1-\cos^2 A\right)}{\cos A} = \frac{2\cos^2 A - 1}{\cos A} = 2\cos A - \frac{1}{\cos A} = 2\cos A - \sec A$$

54. Verify $\dfrac{2\tan B}{\sin 2B} = \sec^2 B$ is an identity.

$$\frac{2\tan B}{\sin 2B} = \frac{2\cdot\frac{\sin B}{\cos B}}{2\sin B\cos B} = \frac{2\sin B}{2\sin B\cos^2 B} = \frac{1}{\cos^2 B} = \sec^2 B$$

55. Verify $1+\tan^2\alpha = 2\tan\alpha\csc 2\alpha$ is an identity.
Work with the right side.

$$2\tan\alpha\csc 2\alpha = \frac{2\tan\alpha}{\sin 2\alpha} = \frac{2\cdot\frac{\sin\alpha}{\cos\alpha}}{2\sin\alpha\cos\alpha} = \frac{2\sin\alpha}{2\sin\alpha\cos^2\alpha} = \frac{1}{\cos^2\alpha} = \sec^2\alpha = 1+\tan^2\alpha$$

56. Verify $\dfrac{2\cot x}{\tan 2x} = \csc^2 x - 2$ is an identity.

$$\frac{2\cot x}{\tan 2x} = \frac{2}{\tan x\left(\frac{2\tan x}{1-\tan^2 x}\right)} = \frac{2}{\tan x}\cdot\frac{1-\tan^2 x}{2\tan x}$$

$$= \frac{1-\tan^2 x}{\tan^2 x} = \frac{1-\frac{\sin^2 x}{\cos^2 x}}{\frac{\sin^2 x}{\cos^2 x}}\cdot\frac{\cos^2 x}{\cos^2 x} = \frac{\cos^2 x - \sin^2 x}{\sin^2 x} = \frac{1-2\sin^2 x}{\sin^2 x} = \csc^2 x - 2$$

57. Verify $\tan\theta\sin 2\theta = 2-2\cos^2\theta$ is an identity.

$$\tan\theta\sin 2\theta = \tan\theta(2\sin\theta\cos\theta) = \frac{\sin\theta}{\cos\theta}(2\sin\theta\cos\theta) = 2\sin^2\theta = 2\left(1-\cos^2\theta\right) = 2-2\cos^2\theta$$

58. Verify $\csc A\sin 2A - \sec A = \cos 2A\sec A$ is an identity.

$$\csc A\sin 2A - \sec A = \frac{1}{\sin A}(2\sin A\cos A) - \frac{1}{\cos A}$$

$$= 2\cos A - \frac{1}{\cos A} = \frac{2\cos^2 A}{\cos A} - \frac{1}{\cos A} = \frac{2\cos^2 A - 1}{\cos A} = \frac{\cos 2A}{\cos A} = \cos 2A\sec A$$

59. Verify $2\tan x\csc 2x-\tan^2 x=1$ is an identity.

$$2\tan x\csc 2x-\tan^2 x=2\tan x\frac{1}{\sin 2x}-\tan^2 x$$
$$=2\cdot\frac{\sin x}{\cos x}\cdot\frac{1}{2\sin x\cos x}-\frac{\sin^2 x}{\cos^2 x}=\frac{1}{\cos^2 x}-\frac{\sin^2 x}{\cos^2 x}=\frac{1-\sin^2 x}{\cos^2 x}=\frac{\cos^2 x}{\cos^2 x}=1$$

60. Verify $2\cos^2\theta-1=\frac{1-\tan^2\theta}{1+\tan^2\theta}$.

Work with the right side.

$$\frac{1-\tan^2\theta}{1+\tan^2\theta}=\frac{1-\frac{\sin^2\theta}{\cos^2\theta}}{1+\frac{\sin^2\theta}{\cos^2\theta}}\cdot\frac{\cos^2\theta}{\cos^2\theta}=\frac{\cos^2\theta-\sin^2\theta}{\cos^2\theta+\sin^2\theta}=\frac{\cos^2\theta-\sin^2\theta}{1}=\cos^2\theta-\sin^2\theta=\cos 2\theta=2\cos^2\theta-1$$

61. Verify $\tan\theta\cos^2\theta=\frac{2\tan\theta\cos^2\theta-\tan\theta}{1-\tan^2\theta}$ is an identity.

Work with the right side.

$$\frac{2\tan\theta\cos^2\theta-\tan\theta}{1-\tan^2\theta}=\frac{\tan\theta\left(2\cos^2\theta-1\right)}{1-\tan^2\theta}=\frac{\tan\theta\left(2\cos^2\theta-1\right)}{1-\frac{\sin^2\theta}{\cos^2\theta}}\cdot\frac{\cos^2\theta}{\cos^2\theta}$$
$$=\frac{\tan\theta\cos^2\theta\left(2\cos^2\theta-1\right)}{\cos^2\theta-\sin^2\theta}=\frac{\tan\theta\cos^2\theta\left(2\cos^2\theta-1\right)}{2\cos^2\theta-1}=\tan\theta\cos^2\theta$$

62. Verify $\sec^2\alpha-1=\frac{\sec 2\alpha-1}{\sec 2\alpha+1}$ is an identity.

Work with the right side.

$$\frac{\sec 2\alpha-1}{\sec 2\alpha+1}=\frac{\frac{1}{\cos 2\alpha}-1}{\frac{1}{\cos 2\alpha}+1}=\frac{\frac{1}{\cos^2\alpha-\sin^2\alpha}-1}{\frac{1}{\cos^2\alpha-\sin^2\alpha}+1}\cdot\frac{\cos^2\alpha-\sin^2\alpha}{\cos^2\alpha-\sin^2\alpha}=\frac{1-\left(\cos^2\alpha-\sin^2\alpha\right)}{\cos^2\alpha-\sin^2\alpha+1}$$
$$=\frac{\left(1-\cos^2\alpha\right)+\sin^2\alpha}{\cos^2\alpha+\left(1-\sin^2\alpha\right)}=\frac{\sin^2\alpha+\sin^2\alpha}{\cos^2\alpha+\cos^2\alpha}=\frac{2\sin^2\alpha}{2\cos^2\alpha}=\tan^2\alpha=\sec^2\alpha-1$$

63. Verify $2\cos^3 x-\cos x=\frac{\cos^2 x-\sin^2 x}{\sec x}$ is an identity.

Work with the right side.

$$\frac{\cos^2 x-\sin^2 x}{\sec x}=\frac{2\cos^2 x-1}{\frac{1}{\cos x}}\cdot\frac{\cos x}{\cos x}=\left(2\cos^2 x-1\right)\cos x=2\cos^3 x-\cos x$$

64. Verify $\sin^3\theta=\sin\theta-\cos^2\theta\sin\theta$ is an identity.

Work with the right side.

$$\sin\theta-\cos^2\theta\sin\theta=\sin\theta-\left(1-\sin^2\theta\right)\sin\theta=\sin\theta-\sin\theta+\sin^3\theta=\sin^3\theta$$

65. Verify $\tan 4\theta = \dfrac{2\tan 2\theta}{2-\sec^2 2\theta}$.

$$\tan 4\theta = \tan\left[2(2\theta)\right] = \frac{2\tan 2\theta}{1-\tan^2 2\theta} = \frac{2\tan 2\theta}{1-\left(\sec^2 2\theta - 1\right)} = \frac{2\tan 2\theta}{1-\sec^2\theta+1} = \frac{2\tan 2\theta}{2-\sec^2\theta}$$

66. Verify $2\cos^2\dfrac{x}{2}\tan x = \tan x + \sin x$ is an identity.

Work with the right side.

$$\tan x + \sin x = \frac{\sin x}{\cos x} + \sin x = \sin x\left(\frac{1}{\cos x}+1\right) = \sin x\left(\frac{1}{\cos\left[2\left(\frac{x}{2}\right)\right]}+1\right) = \sin x\left(\frac{1}{2\cos^2\frac{x}{2}-1}+1\right)$$

$$= \sin x\left[\frac{1+\left(2\cos^2\frac{x}{2}-1\right)}{2\cos^2\frac{x}{2}-1}\right] = \frac{2\sin x\cos^2\frac{x}{2}}{2\cos^2\frac{x}{2}-1} = \frac{2\sin x\cos^2\frac{x}{2}}{\cos\left[2\left(\frac{x}{2}\right)\right]} = \frac{2\sin x\cos^2\frac{x}{2}}{\cos x}$$

$$= 2\cos^2\frac{x}{2}\cdot\frac{\sin x}{\cos x} = 2\cos^2\frac{x}{2}\tan x$$

67. Verify $\tan\left(\dfrac{x}{2}+\dfrac{\pi}{4}\right) = \sec x + \tan x$ is an identity.

Working with the left side, we have $\tan\left(\dfrac{x}{2}+\dfrac{\pi}{4}\right) = \dfrac{\tan\frac{x}{2}+\tan\frac{\pi}{4}}{1-\tan\frac{x}{2}\tan\frac{\pi}{4}} = \dfrac{\tan\frac{x}{2}+1}{1-\tan\frac{x}{2}}$.

Work with the right side.

$$\sec x + \tan x = \frac{1}{\cos x} + \frac{\sin x}{\cos x} = \frac{1+\sin x}{\cos x} = \frac{\left(\cos^2\frac{x}{2}+\sin^2\frac{x}{2}\right)+\sin\left[2\left(\frac{x}{2}\right)\right]}{\cos\left[2\left(\frac{x}{2}\right)\right]}$$

$$= \frac{\cos^2\frac{x}{2}+\sin^2\frac{x}{2}+2\sin\frac{x}{2}\cos\frac{x}{2}}{\cos^2 x-\sin^2 x} = \frac{\cos^2\frac{x}{2}+2\sin\frac{x}{2}\cos\frac{x}{2}+\sin^2\frac{x}{2}}{\cos^2 x-\sin^2 x}$$

$$= \frac{\left(\cos\frac{x}{2}+\sin\frac{x}{2}\right)^2}{\left(\cos\frac{x}{2}-\sin\frac{x}{2}\right)\left(\cos\frac{x}{2}+\sin\frac{x}{2}\right)} = \frac{\cos\frac{x}{2}+\sin\frac{x}{2}}{\cos\frac{x}{2}-\sin\frac{x}{2}} = \frac{\frac{\cos\frac{x}{2}}{\cos\frac{x}{2}}+\frac{\sin\frac{x}{2}}{\cos\frac{x}{2}}}{\frac{\cos\frac{x}{2}}{\cos\frac{x}{2}}-\frac{\sin\frac{x}{2}}{\cos\frac{x}{2}}} = \frac{1+\tan\frac{x}{2}}{1-\tan\frac{x}{2}}$$

Since $\tan\left(\dfrac{x}{2}+\dfrac{\pi}{4}\right) = \dfrac{\tan\frac{x}{2}+1}{1-\tan\frac{x}{2}} = \sec x + \tan x$, we have verified the identity.

68. Verify $\frac{1}{2}\cot\frac{x}{2}-\frac{1}{2}\tan\frac{x}{2}=\cot x$ is an identity.

$$\frac{1}{2}\cot\frac{x}{2}-\frac{1}{2}\tan\frac{x}{2}=\frac{1}{2}\left(\frac{1+\cos x}{\sin x}\right)-\frac{1}{2}\left(\frac{1-\cos x}{\sin x}\right)=\frac{1+\cos x}{2\sin x}-\frac{1-\cos x}{2\sin x}$$

$$=\frac{(1+\cos x)-(1-\cos x)}{2\sin x}=\frac{1+\cos x-1+\cos x}{2\sin x}=\frac{2\cos x}{2\sin x}=\cot x$$

69. Verify $-\cot\frac{x}{2}=\frac{\sin 2x+\sin x}{\cos 2x-\cos x}$ is an identity.

Work with the right side.

$$\frac{\sin 2x+\sin x}{\cos 2x-\cos x}=\frac{2\sin x\cos x+\sin x}{\left(2\cos^2 x-1\right)-\cos x}=\frac{2\sin x\cos x+\sin x}{2\cos^2 x-\cos x-1}$$

$$=\frac{\sin x(2\cos x+1)}{(2\cos x+1)(\cos x-1)}=\frac{\sin x}{\cos x-1}=-\frac{\sin x}{1-\cos x}=-\frac{1}{\tan\frac{x}{2}}=-\cot\frac{x}{2}$$

70. Verify $\frac{\sin 3t+\sin 2t}{\sin 3t-\sin 2t}=\frac{\tan\frac{5t}{2}}{\tan\frac{t}{2}}$ is an identity. Using sum-to-product identities, we have the following.

$$\frac{\sin 3t+\sin 2t}{\sin 3t-\sin 2t}=\frac{2\sin\left(\frac{3t+2t}{2}\right)\cos\left(\frac{3t-2t}{2}\right)}{2\cos\left(\frac{3t+2t}{2}\right)\sin\left(\frac{3t-2t}{2}\right)}=\frac{\sin\frac{5t}{2}\cos\frac{t}{2}}{\cos\frac{5t}{2}\sin\frac{t}{2}}=\frac{\sin\frac{5t}{2}}{\cos\frac{5t}{2}}\cdot\frac{\cos\frac{t}{2}}{\sin\frac{t}{2}}=\tan\frac{5t}{2}\cot\frac{t}{2}=\frac{\tan\frac{5t}{2}}{\tan\frac{t}{2}}$$

71. **(a)** When $h=0$,

$$D=\frac{v^2\sin\theta\cos\theta+v\cos\theta\sqrt{(v\sin\theta)^2+64\cdot 0}}{32}=\frac{v^2\sin\theta\cos\theta+v\cos\theta\sqrt{v^2\sin^2\theta}}{32}$$

$$=\frac{v^2\sin\theta\cos\theta+(v\cos\theta)(v\sin\theta)}{32}=\frac{v^2\sin\theta\cos\theta+v^2\sin\theta\cos\theta}{32}=\frac{2v^2\sin\theta\cos\theta}{32}=\frac{v^2\sin(2\theta)}{32},$$

which is dependent on both the velocity and angle at which the object is thrown.

(b) $D=\frac{36^2\cdot\sin(2\cdot 30)}{32}=\frac{1296\cdot\sin(60)}{32}=\frac{81\cdot\frac{\sqrt{3}}{2}}{2}=\frac{81\sqrt{3}}{4}\approx 35$ ft

72. **(a)** The period is equal to $\frac{2\pi}{b}=\frac{2\pi}{2\pi\omega}=\frac{1}{\omega}$.

(b) $V=a\sin 2\pi\omega t$ and $I=b\sin 2\pi\omega t$

$W=VI=(a\sin 2\pi\omega t)(b\sin 2\pi\omega t)=ab\sin^2 2\pi\omega t$

Since $\cos 2A=1-2\sin^2 A\Rightarrow\cos 2A-1=-2\sin^2 A\Rightarrow 2\sin^2 A=1-\cos 2A\Rightarrow\sin^2 A=\frac{1-\cos 2A}{2}$,

we have the following.

$$W=ab\sin^2 2\pi\omega t=ab\cdot\frac{1-\cos 2(2\pi\omega t)}{2}=ab\cdot\frac{1-\cos 4\pi\omega t}{2}$$

Thus, the period of V is $\frac{2\pi}{b}=\frac{2\pi}{4\pi\omega}=\frac{1}{2}\cdot\frac{1}{\omega}$. This is half the period of the voltage.

Additional answers will vary.

Chapter 5: Test

1. Given $\tan x = -\frac{5}{6}, \frac{3\pi}{2} < x < 2\pi,$ use trigonometric identities to find $\sin x$ and $\cos x$.

$$\sec^2 x = \tan^2 x + 1 \Rightarrow \sec^2 x = \left(-\frac{5}{6}\right)^2 + 1 = \frac{25}{36} + 1 = \frac{61}{36}$$

Since $\frac{3\pi}{2} < x < 2\pi,$ x is in quadrant IV. Thus, $\sec x > 0$.

Therefore, $\sec x = \sqrt{\frac{61}{36}} = \frac{\sqrt{61}}{6} \Rightarrow \cos x = \frac{1}{\sec x} = \frac{1}{\frac{\sqrt{61}}{6}} = \frac{6}{\sqrt{61}} \cdot \frac{\sqrt{61}}{\sqrt{61}} = \frac{6\sqrt{61}}{61}.$

Since $\cos^2 x + \sin^2 x = 1,$ we have the following.

$$\left(\frac{6\sqrt{61}}{61}\right)^2 + \sin^2 x = 1 \Rightarrow \frac{2196}{3721} + \sin^2 x = 1 \Rightarrow \sin^2 x = \frac{1525}{3721}$$

Since x is in quadrant IV, $\sin x < 0$ and $\sin x = -\sqrt{\frac{1525}{3721}} = -\frac{5\sqrt{61}}{61}.$

2. $\tan^2 x - \sec^2 x = \frac{\sin^2 x}{\cos^2 x} - \frac{1}{\cos^2 x} = \frac{\sin^2 x - 1}{\cos^2 x} = \frac{\left(1-\cos^2 x\right)-1}{\cos^2 x} = -\frac{\cos^2 x}{\cos^2 x} = -1$

3. Find sin $(x + y)$, cos $(x - y)$, and tan $(x + y)$, given $\sin x = -\frac{1}{3}$, $\cos y = -\frac{2}{5}$, x is in quadrant III, y is in quadrant II. First, find $\sin x$ and $\cos y$.
Since x is in quadrant III, $\cos x < 0$ and

$\cos x = -\sqrt{1-\sin^2 x} = -\sqrt{1-\left(-\frac{1}{3}\right)^2} = -\sqrt{1-\frac{1}{9}} = -\sqrt{\frac{8}{9}} = -\frac{2\sqrt{2}}{3}.$

Since y is in quadrant II, $\sin y > 0$ and $\sin y = \sqrt{1-\cos^2 x} = \sqrt{1-\left(-\frac{2}{5}\right)^2} = \sqrt{1-\frac{4}{25}} = \sqrt{\frac{21}{25}} = \frac{\sqrt{21}}{5}.$

$\sin\left(x+y\right) = \sin x \cos y + \cos x \sin y = \left(-\frac{1}{3}\right)\left(-\frac{2}{5}\right) + \left(-\frac{2\sqrt{2}}{3}\right)\left(\frac{\sqrt{21}}{5}\right) = \frac{2}{15} - \frac{2\sqrt{42}}{15} = \frac{2-2\sqrt{42}}{15}$

$\cos\left(x-y\right) = \cos x \cos y + \sin x \sin y = \left(-\frac{2\sqrt{2}}{3}\right)\left(-\frac{2}{5}\right) + \left(-\frac{1}{3}\right)\left(\frac{\sqrt{21}}{5}\right) = \frac{4\sqrt{2}}{15} - \frac{\sqrt{21}}{15} = \frac{4\sqrt{2}-\sqrt{21}}{15}$

In order to find $\tan\left(x+y\right)$ using the formula $\tan\left(x+y\right) = \frac{\tan x + \tan y}{1 - \tan x \tan y}$, we need to find $\tan x$ and $\tan y$.

$$\tan x = \frac{\sin x}{\cos x} = \frac{-\frac{1}{3}}{-\frac{2\sqrt{2}}{3}} = \frac{1}{2\sqrt{2}} \cdot \frac{\sqrt{2}}{\sqrt{2}} = \frac{\sqrt{2}}{4} \text{ and } \tan y = \frac{\sin y}{\cos y} = \frac{\frac{\sqrt{21}}{5}}{-\frac{2}{5}} = -\frac{\sqrt{21}}{2}$$

$$\tan\left(x+y\right) = \frac{\tan x + \tan y}{1 - \tan x \tan y} = \frac{\frac{\sqrt{2}}{4} + \left(-\frac{\sqrt{21}}{2}\right)}{1 - \left(\frac{\sqrt{2}}{4}\right)\left(-\frac{\sqrt{21}}{2}\right)} \cdot \frac{8}{8} = \frac{2\sqrt{2} - 4\sqrt{21}}{8 + \sqrt{42}}$$

4. $\sin(-22.5°) = \pm\sqrt{\dfrac{1-\cos(-45°)}{2}} = \pm\sqrt{\dfrac{1-\frac{\sqrt{2}}{2}}{2}} = \pm\sqrt{\dfrac{2-\sqrt{2}}{4}} = \pm\dfrac{\sqrt{2-\sqrt{2}}}{2}$

Since $-22.5°$ is in quadrant IV, $\sin(-22.5°)$ is negative. Thus, $\sin(-22.5°) = \dfrac{-\sqrt{2-\sqrt{2}}}{2}$.

Exercises 5 –6 are graphed in the following window.

5. $\sec x - \sin x \tan x$ appears to be equivalent to $\cos x$.

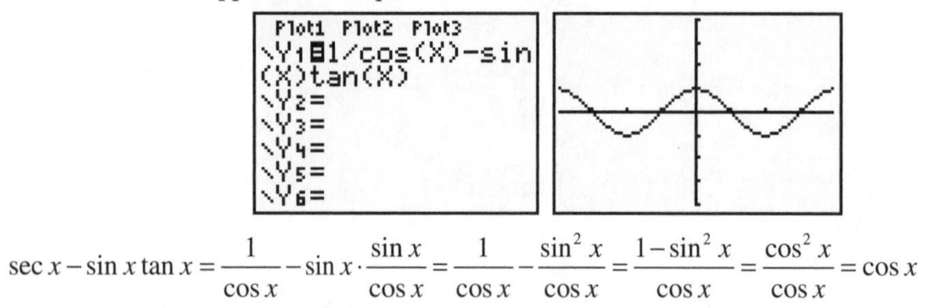

$$\sec x - \sin x \tan x = \frac{1}{\cos x} - \sin x \cdot \frac{\sin x}{\cos x} = \frac{1}{\cos x} - \frac{\sin^2 x}{\cos x} = \frac{1-\sin^2 x}{\cos x} = \frac{\cos^2 x}{\cos x} = \cos x$$

6. $\cot\dfrac{x}{2} - \cot x$ appears to be equivalent to $\csc x$.

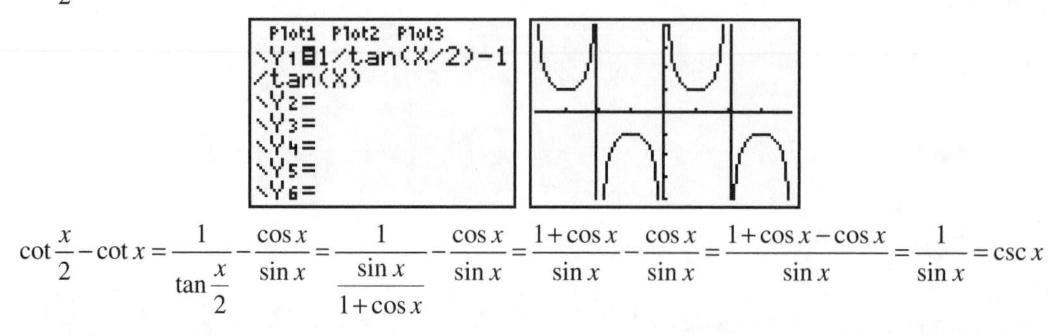

$$\cot\frac{x}{2} - \cot x = \frac{1}{\tan\frac{x}{2}} - \frac{\cos x}{\sin x} = \frac{1}{\frac{\sin x}{1+\cos x}} - \frac{\cos x}{\sin x} = \frac{1+\cos x}{\sin x} - \frac{\cos x}{\sin x} = \frac{1+\cos x - \cos x}{\sin x} = \frac{1}{\sin x} = \csc x$$

7. Verify $\sec^2 B = \dfrac{1}{1-\sin^2 B}$ is an identity.

Work with the right side.

$$\frac{1}{1-\sin^2 B} = \frac{1}{\cos^2 B} = \sec^2 B$$

8. Verify $\cos 2A = \dfrac{\cot A - \tan A}{\csc A \sec A}$ is an identity.

Work with the right side.

$$\frac{\cot A - \tan A}{\csc A \sec A} = \frac{\frac{\cos A}{\sin A} - \frac{\sin A}{\cos A}}{\left(\frac{1}{\sin A}\right)\left(\frac{1}{\cos A}\right)} \cdot \frac{\sin A \cos A}{\sin A \cos A} = \cos^2 A - \sin^2 A = \cos^2 2A$$

9. Verify $\dfrac{\tan x - \cot x}{\tan x + \cot x} = 2\sin^2 x - 1$ is an identity.

$$\frac{\tan x - \cot x}{\tan x + \cot x} = \frac{\dfrac{\sin x}{\cos x} - \dfrac{\cos x}{\sin x}}{\dfrac{\sin x}{\cos x} + \dfrac{\cos x}{\sin x}} = \frac{\dfrac{\sin x}{\cos x} - \dfrac{\cos x}{\sin x}}{\dfrac{\sin x}{\cos x} + \dfrac{\cos x}{\sin x}} \cdot \frac{\cos x \sin x}{\cos x \sin x} = \frac{\sin^2 x - \cos^2 x}{\sin^2 x + \cos^2 x}$$

$$= \frac{\sin^2 x - \cos^2 x}{1} = \sin^2 x - \cos^2 x = \sin^2 x - \left(1 - \sin^2 x\right) = \sin^2 x - 1 + \sin^2 x = 2\sin^2 x - 1$$

10. Verify $\tan^2 - \sin^2 x = \left(\tan x \sin x\right)^2$ is an identity.

$$\tan^2 x - \sin^2 x = \frac{\sin^2 x}{\cos^2 x} - \sin^2 x = \frac{\sin^2 x}{\cos^2 x} - \frac{\sin^2 x \cos^2 x}{\cos^2 x} = \frac{\sin^2 x - \sin^2 x \cos^2 x}{\cos^2 x}$$

$$= \frac{\sin^2 x\left(1 - \cos^2 x\right)}{\cos^2 x} = \frac{\sin^2 x \sin^2 x}{\cos^2 x} = \frac{\sin^2 x}{\cos^2 x} \cdot \sin^2 x = \tan^2 x \sin^2 x = \left(\tan x \sin x\right)^2$$

11. **(a)** $\cos(270° - \theta) = \cos 270° \cos\theta + \sin 270° \sin\theta = 0 \cdot \cos\theta + (-1)\sin\theta = 0 - \sin\theta = -\sin\theta$

(b) $\sin(\pi + \theta) = \sin\pi \cos\theta + \cos\pi \sin\theta = 0 \cdot \cos\theta + (-1)\sin\theta = 0 - \sin\theta = -\sin\theta$

12. **(a)** $V = 163 \sin \omega t$. Since $\sin x = \cos\left(\dfrac{\pi}{2} - x\right)$, $V = 163\cos\left(\dfrac{\pi}{2} - \omega t\right)$.

(b) If $V = 163\sin\omega t = 163\sin 120\pi t = 163\cos\left(\dfrac{\pi}{2} - 120\pi t\right)$, the maximum voltage occurs when $\cos\left(\dfrac{\pi}{2} - 120\pi t\right) = 1$. Thus, the maximum voltage is $V = 163$ volts.

$\cos\left(\dfrac{\pi}{2} - 120\pi t\right) = 1$ when $\dfrac{\pi}{2} - 120\pi t = 2k\pi$, where k is any integer. The first maximum occurs when $\dfrac{\pi}{2} - 120\pi t = 0 \Rightarrow \dfrac{\pi}{2} = 120\pi t \Rightarrow \dfrac{1}{120\pi} \cdot \dfrac{\pi}{2} = t \Rightarrow t = \dfrac{1}{240}$.

The maximum voltage will first occur at $\dfrac{1}{240}$ sec.

Chapter 5: Quantitative Reasoning

$$y' = r\cos\left(\theta + R\right) = r\left[\cos\theta\cos R - \sin\theta\sin R\right] = \left(r\cos\theta\right)\cos R - \left(r\sin\theta\right)\sin R = y\cos R - z\sin R$$

$$z' = r\sin\left(\theta + R\right) = r\cos\left(\frac{\pi}{2} - (\theta + R)\right) = r\cos\left(\left(\frac{\pi}{2} - \theta\right) - R\right) = r\left[\cos\left(\frac{\pi}{2} - \theta\right)\cos R + \sin\left(\frac{\pi}{2} - \theta\right)\sin R\right]$$

$$= r\left[\sin\theta\cos R + \cos\theta\sin R\right] = \left(r\sin\theta\right)\cos R + \left(r\cos\theta\right)\sin R = z\cos R + y\sin R$$

Chapter 6

INVERSE CIRCULAR FUNCTIONS AND TRIGONOMETRIC EQUATIONS

Section 6.1: Inverse Circular Functions

1. For a function to have an inverse, it must be one-to-one.

2. The domain of $y = \arcsin x$ equals the range of $y = \sin x$.

3. The range of $y = \cos^{-1} x$ equals the domain of $y = \cos x$.

4. The point $\left(\frac{\pi}{4}, 1\right)$ lies on the graph of $y = \tan x$. Therefore, the point $\left(1, \frac{\pi}{4}\right)$ lies on the graph of $y = \tan^{-1} x$.

5. If a function *f* has an inverse and $f(\pi) = -1$, then $f^{1}(-1) = \underline{\pi}$.

6. Sketch the reflection of the graph of *f* across the line $y = x$.

7. **(a)** $[-1, 1]$
(b) $\left[-\frac{\pi}{2}, \frac{\pi}{2}\right]$
(c) increasing
(d) -2 is not in the domain.

8. **(a)** $[-1, 1]$
(b) $[0, \pi]$
(c) decreasing
(d) $-\frac{4\pi}{3}$ is not in the range.

9. **(a)** $(-\infty, \infty)$
(b) $\left(-\frac{\pi}{2}, \frac{\pi}{2}\right)$
(c) increasing
(d) no

10. **(a)** $(-\infty, -1] \cup [1, \infty); \left[-\frac{\pi}{2}, 0\right) \cup \left(0, \frac{\pi}{2}\right]$
(b) $(-\infty, -1] \cup [1, \infty); \left[0, \frac{\pi}{2}\right) \cup \left(\frac{\pi}{2}, \pi\right]$
(c) $(-\infty, \infty); (0, \pi)$

11. $\cos^{-1} \frac{1}{a}$

12. Find $\tan^{-1} \frac{1}{a} + \pi\,(\text{ or } 180°)$.

13. $y = \sin^{-1} 0$

$\sin y = 0, \ -\frac{\pi}{2} \le y \le \frac{\pi}{2}$

Since $\sin 0 = 0, \ y = 0$.

14. $y = \tan^{-1} 1$

$\tan y = 1, \ -\frac{\pi}{2} < y < \frac{\pi}{2}$

Since $\tan \frac{\pi}{4} = 1, \ y = \frac{\pi}{4}$.

15. $y = \cos^{-1}(-1)$

$\cos y = -1, \ 0 \le y \le \pi$

Since $\cos \pi = -1, \ y = \pi$.

16. $y = \arctan(-1)$

$\tan y = -1, -\frac{\pi}{2} < y < \frac{\pi}{2}$

Since $\tan\left(-\frac{\pi}{4}\right) = -1$, $y = -\frac{\pi}{4}$.

17. $y = \sin^{-1}(-1)$

$\sin y = -1, -\frac{\pi}{2} \le y \le \frac{\pi}{2}$

Since $\sin\frac{\pi}{2} = 1$, $y = -\frac{\pi}{2}$.

18. $y = \cos^{-1}\frac{1}{2}$

$\cos y = \frac{1}{2}, 0 \le y \le \pi$

Since $\cos\frac{\pi}{3} = \frac{1}{2}$, $y = \frac{\pi}{3}$.

19. $y = \arctan 0$

$\tan y = 0, -\frac{\pi}{2} < y < \frac{\pi}{2}$

Since $\tan 0 = 0$, $y = 0$.

20. $y = \arcsin\left(-\frac{\sqrt{3}}{2}\right)$

$\sin y = -\frac{\sqrt{3}}{2}, -\frac{\pi}{2} \le y \le \frac{\pi}{2}$

Since $\sin\left(-\frac{\pi}{3}\right) = -\frac{\sqrt{3}}{2}$, $y = -\frac{\pi}{3}$.

21. $y = \arccos 0$

$\cos y = 0, 0 \le y \le \pi$

Since $\cos\frac{\pi}{2} = 0$, $y = \frac{\pi}{2}$.

22. $y = \tan^{-1}(-1)$

$\tan y = -1, -\frac{\pi}{2} < y < \frac{\pi}{2}$

Since $\tan\left(-\frac{\pi}{4}\right) = -1$, $y = -\frac{\pi}{4}$.

23. $y = \sin^{-1}\frac{\sqrt{2}}{2}$

$\sin y = \frac{\sqrt{2}}{2}, -\frac{\pi}{2} \le y \le \frac{\pi}{2}$

Since $\sin\frac{\pi}{4} = \frac{\sqrt{2}}{2}$, $y = \frac{\pi}{4}$.

24. $y = \cos^{-1}\left(-\frac{1}{2}\right)$

$\cos y = -\frac{1}{2}, 0 \le y \le \pi$

Since $\cos\frac{2\pi}{3} = -\frac{1}{2}$, $y = \frac{2\pi}{3}$.

25. $y = \arccos\left(-\frac{\sqrt{3}}{2}\right)$

$\cos y = -\frac{\sqrt{3}}{2}, 0 \le y \le \pi$

Since $\cos\frac{5\pi}{6} = -\frac{\sqrt{3}}{2}$, $y = \frac{5\pi}{6}$.

26. $y = \arcsin\left(-\frac{\sqrt{2}}{2}\right)$

$\sin y = -\frac{\sqrt{2}}{2}, -\frac{\pi}{2} \le y \le \frac{\pi}{2}$

Since $\sin\left(-\frac{\pi}{4}\right) = -\frac{\sqrt{2}}{2}$, $y = -\frac{\pi}{4}$.

27. $y = \cot^{-1}(-1)$

$\cot y = -1, 0 < y < \pi$

y is in quadrant II.

The reference angle is $\frac{\pi}{4}$.

Since $\cot\frac{3\pi}{4} = 1$, $y = \frac{3\pi}{4}$.

28. $y = \sec^{-1}\left(-\sqrt{2}\right)$

$\sec y = -\sqrt{2}, 0 \le y \le \pi, y \ne \frac{\pi}{2}$

y is in quadrant II.

The reference angle is $\frac{\pi}{4}$.

Since $\sec\frac{3\pi}{4} = -\sqrt{2}$, $y = \frac{3\pi}{4}$.

29. $y = \csc^{-1}(-2)$

$\csc y = -2,\ -\frac{\pi}{2} \le y \le \frac{\pi}{2},\ y \ne 0$

y is in quadrant IV.

The reference angle is $\frac{\pi}{6}$.

Since $\csc\left(-\frac{\pi}{6}\right) = -2,\ y = -\frac{\pi}{6}$.

30. $y = \text{arc}\cot\left(-\sqrt{3}\right)$

$\cot y = -\sqrt{3},\ 0 < y < \pi$

y is in quadrant II.

The reference angle is $\frac{\pi}{6}$.

Since $\cot\frac{5\pi}{6} = -\sqrt{3},\ y = \frac{5\pi}{6}$.

31. $y = \text{arc}\sec\left(\frac{2\sqrt{3}}{3}\right)$

$\sec y = \frac{2\sqrt{3}}{3},\ 0 \le y \le \pi,\ y \ne \frac{\pi}{2}$

Since $\sec\frac{\pi}{6} = \frac{2\sqrt{3}}{3},\ y = \frac{\pi}{6}$.

32. $y = \csc^{-1}\sqrt{2}$

$\csc\ y = \sqrt{2},\ -\frac{\pi}{2} \le y \le \frac{\pi}{2},\ y \ne 0$

Since $\csc\frac{\pi}{4} = \sqrt{2},\ y = \frac{\pi}{4}$.

33. $\theta = \arctan(-1)$

$\tan\ \theta = -1,\ -90° < \theta < 90°$

is in quadrant IV.

The reference angle is 45°.

Thus, $\theta = -45°$.

34. $\theta = \arccos\left(-\frac{1}{2}\right)$

$\cos\theta = -\frac{1}{2},\ 0° \le \theta \le 180°$

θ is in quadrant II.

The reference angle is 60°.

Thus, $\theta = 180° - 60° = 120°$.

35. $\theta = \arcsin\left(-\frac{\sqrt{3}}{2}\right)$

$\sin\ \theta = -\frac{\sqrt{3}}{2}, -90° \le \theta \le 90°$

θ is in quadrant IV.

The reference angle is 60°.

$\theta = -60°$.

36. $\theta = \arcsin\left(-\frac{\sqrt{2}}{2}\right)$

$\sin\ \theta = -\frac{\sqrt{2}}{2},\ -90° \le \theta \le 90°$

θ is in quadrant IV.

The reference angle is 45°.

$\theta = -45°$

37. $\theta = \cot^{-1}\left(-\frac{\sqrt{3}}{3}\right)$

$\cot\ \theta = -\frac{\sqrt{3}}{3},\ 0° < \theta < 180°$

θ is in quadrant II.

The reference angle is 60°.

$\theta = 180° - 60° = 120°$

38. $\theta = \csc^{-1}(-2)$

$\csc\ \theta = -2$ and $-90° < \theta < 90°,\ \theta \ne 0°$

θ is in quadrant IV.

The reference angle is 30°.

$\theta = -30°$

39. $\theta = \sec^{-1}(-2)$

$\sec\ \theta = -2,\ 0° \le \theta \le 180°, \theta \ne 90°$

θ is in quadrant II.

The reference angle is 60°.

$\theta = 180° - 60° = 120°$.

40. $\theta = \csc^{-1}(-1)$

$\csc\ \theta = -1,\ -90° \le \theta \le 90°,\ \theta \ne 0°$

Since the terminal side of θ lies on the y-axis, there is no reference angle.

$\theta = -90°$

For Exercises 41 – 46, be sure that your calculator is in degree mode. Keystroke sequences may vary based on the type and/or model of calculator being used.

41. $\theta = \sin^{-1}(-.13349122)$

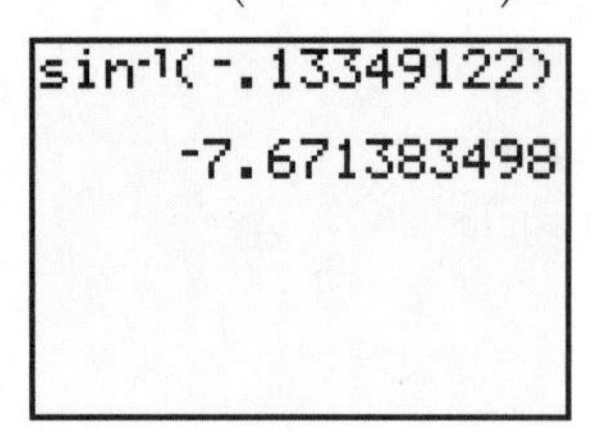

$\sin^{-1}(-.13349122) = -7.6713835°$

42. $\theta = \cos^{-1}(-.13348816)$

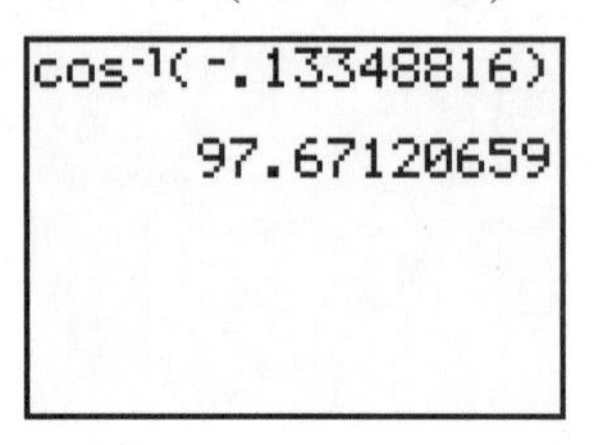

$\cos^{-1}(-.13348816) \approx 97.671207°$

43. $\theta = \arccos(-.39876459)$

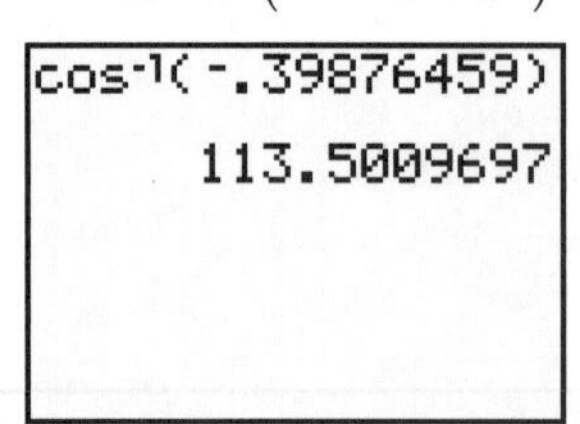

$\arccos(-.39876459) \approx 113.500970°$

44. $\theta = \arcsin .77900016$

$\arcsin .77900016 \approx 51.1691219°$

45. $\theta = \csc^{-1} 1.9422833$

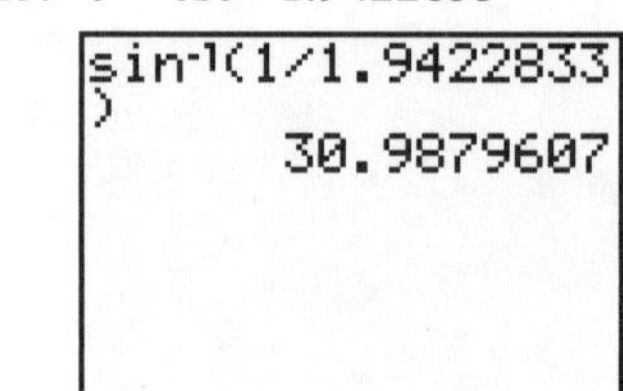

$\csc^{-1} 1.9422833 \approx 30.987961°$

46. $\theta = \cot^{-1} 1.7670492$

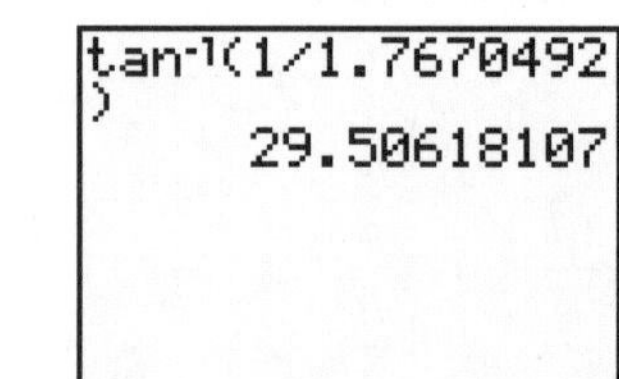

$\cot^{-1} 1.7670492 \approx 29.506181°$

For Exercises 47–52, be sure that your calculator is in radian mode. Keystroke sequences may vary based on the type and/or model of calculator being used.

47. $y = \arctan 1.1111111$

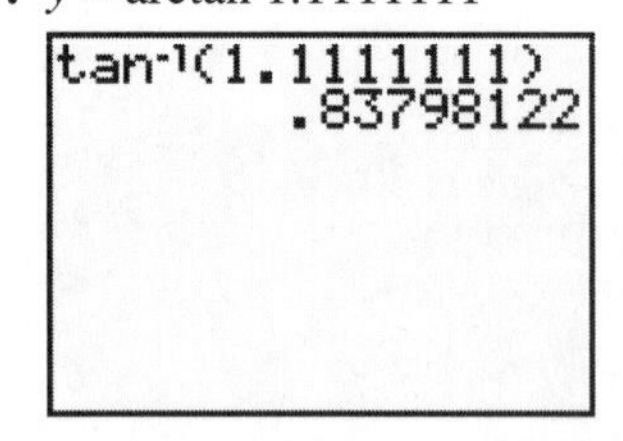

$\arctan 1.1111111 \approx .83798122$

48. $y = \arcsin .81926439$

$\arcsin .81926439 \approx .96012698$

49. $y = \cot^{-1}(-.92170128)$

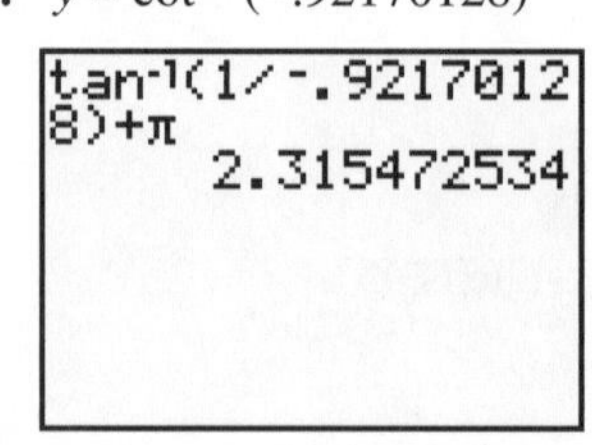

$\cot^{-1}(-.91270128) \approx 2.3154725$

50. $y = \sec^{-1}(-1.2871684)$

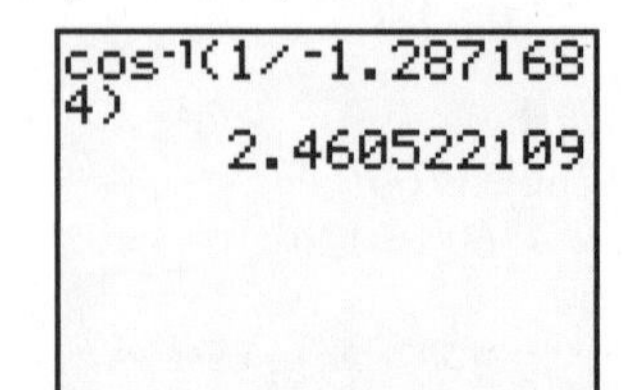

$\sec^{-1}(-1.2871684) \approx 2.4605221$

51. $y = \arcsin .92837781$

sin⁻¹(.92837781)
1.19002379

$\arcsin .92837781 \approx 1.1900238$

52. $y = \arccos .44624593$

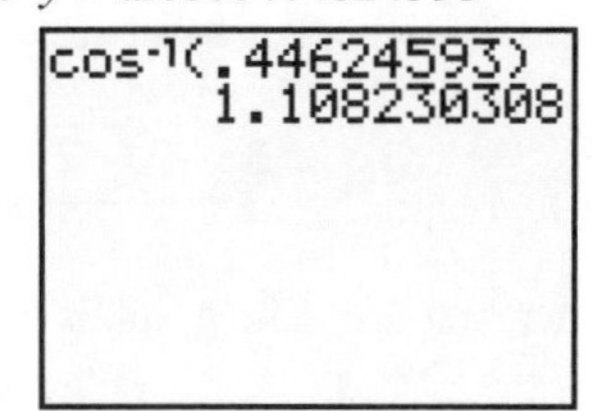

$\arccos .44624593 \approx 1.1082303$

53.

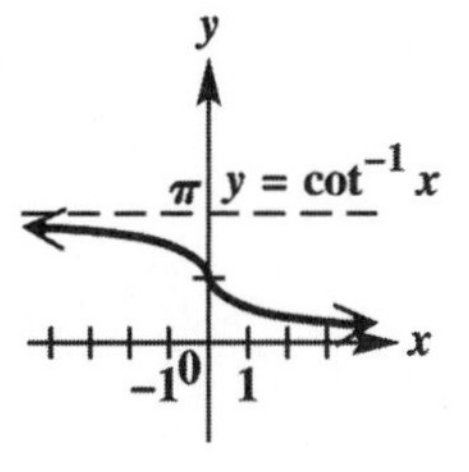

Domain: $(-\infty, \infty)$; Range: $(0, \pi)$

54.

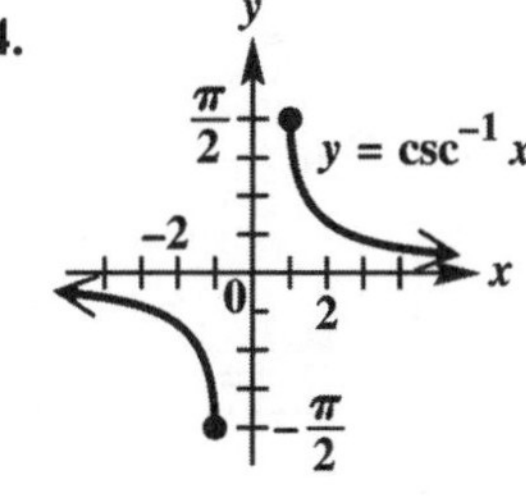

Domain: $(-\infty, -1] \cup [1, \infty)$;

Range: $\left[-\frac{\pi}{2}, 0\right) \cup \left(0, \frac{\pi}{2}\right]$

55.

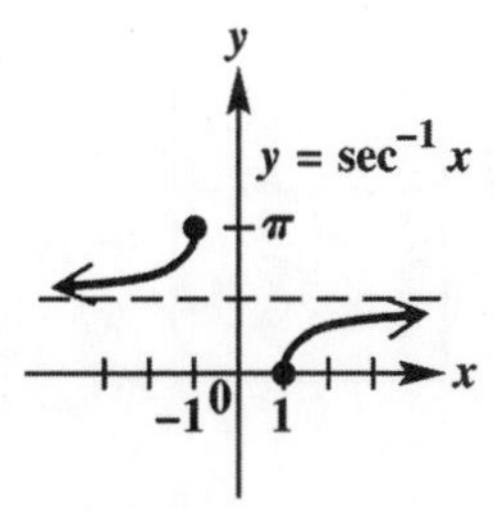

Domain: $(-\infty, -1] \cup [1, \infty)$;

Range: $\left[0, \frac{\pi}{2}\right) \cup \left(\frac{\pi}{2}, \pi\right]$

56.

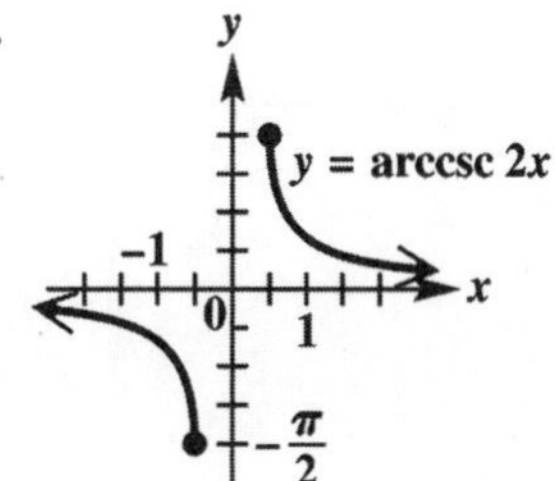

Domain: $\left(-\infty, -\frac{1}{2}\right] \cup \left[\frac{1}{2}, \infty\right)$;

Range: $\left[-\frac{\pi}{2}, 0\right) \cup \left(0, \frac{\pi}{2}\right]$

57.

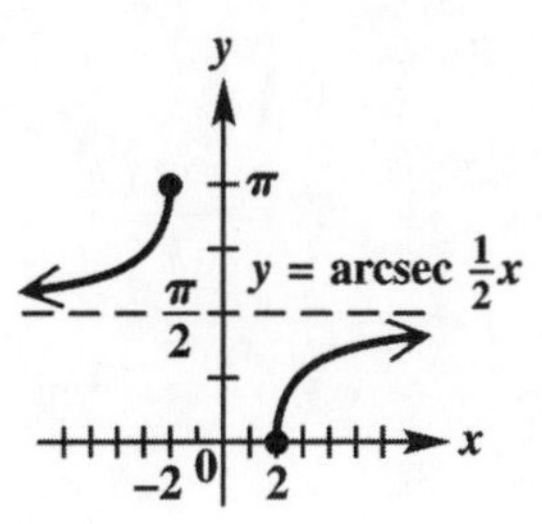

Domain: $(-\infty, -2] \cup [2, \infty)$; Range: $\left[0, \frac{\pi}{2}\right) \cup \left(\frac{\pi}{2}, \pi\right]$

58. 1.003 is not in the domain of $y = \sin^{-1} x$. (Alternatively, you could state that 1.003 is not in the range of $y = \sin x$.)

59. The domain of $y = \tan^{-1} x$ is $(-\infty, \infty)$. (Alternatively, you could state that the range of $y = \tan x$ is $(-\infty, \infty)$.)

60. $f\left[f^{-1}(x)\right]=f\left[\dfrac{x+2}{3}\right]=3\left(\dfrac{x+2}{3}\right)-2=x+2-2=x$

$f^{-1}\left[f(x)\right]=f^{-1}[3x-2]=\dfrac{(3x-2)+2}{3}=\dfrac{3x}{3}=x$

In each case the result is x. The graph is a straight line bisecting the first and third quadrants I and III (i.e., $y=x$).

The graphs for Exercises 61 and 62 are in the standard window. Exercise 62 is graphed in the dot mode to avoid the appearance of vertical line segments.

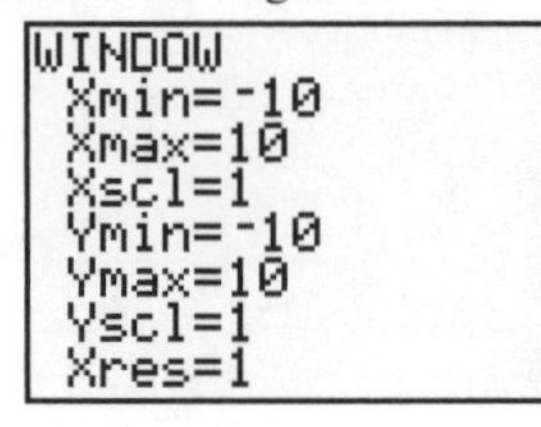

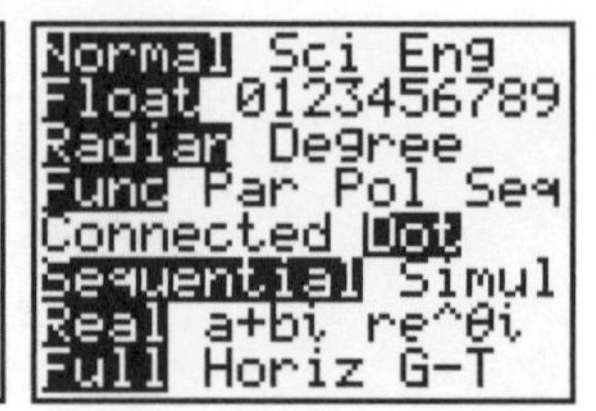

61. It is the graph of $y=x$.

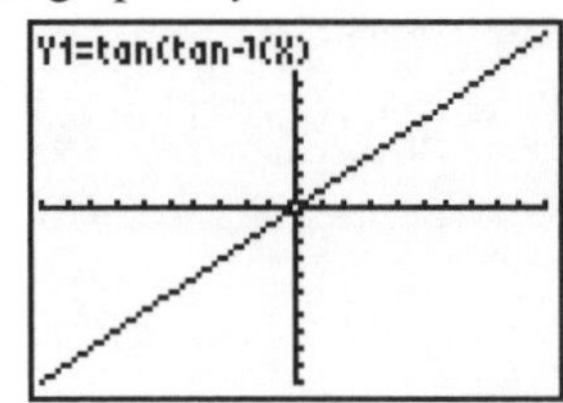

62. It does not agree because the range of the inverse tangent function is $\left(-\dfrac{\pi}{2},\dfrac{\pi}{2}\right)$, not $(-\infty,\infty)$, as was the case in Exercise 61.

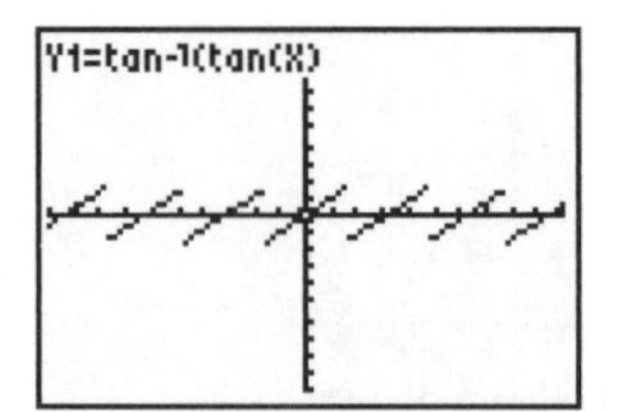

63. $\tan\left(\arccos\dfrac{3}{4}\right)$

Let $\omega=\arccos\dfrac{3}{4}$, so that $\cos\omega=\dfrac{3}{4}$. Since arccos is defined only in quadrants I and II, and $\dfrac{3}{4}$ is positive, ω is in quadrant I. Sketch ω and label a triangle with the side opposite ω equal to $\sqrt{4^2-3^2}=\sqrt{16-9}=\sqrt{7}$.

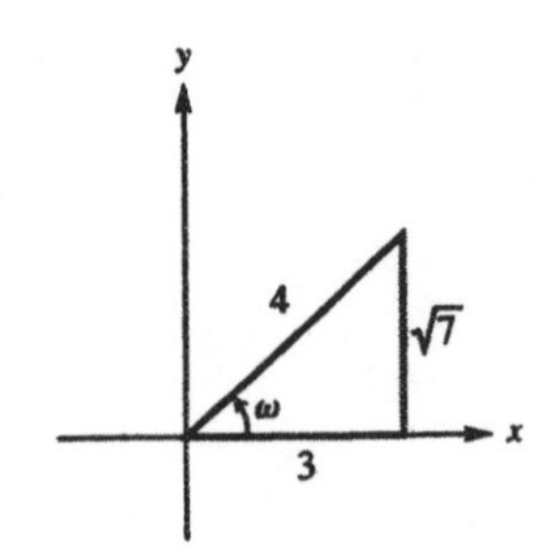

$\tan\left(\arccos\dfrac{3}{4}\right)=\tan\omega=\dfrac{\sqrt{7}}{3}$

64. $\sin\left(\arccos\frac{1}{4}\right)$

Let $\theta = \arccos\frac{1}{4}$, so that $\cos\theta = \frac{1}{4}$. Since arccos is defined only in quadrants I and II, and $\frac{1}{4}$ is positive, θ is in quadrant I. Sketch θ and label a triangle with the side opposite θ equal to $\sqrt{4^2 - 1^2} = \sqrt{16-1} = \sqrt{15}$.

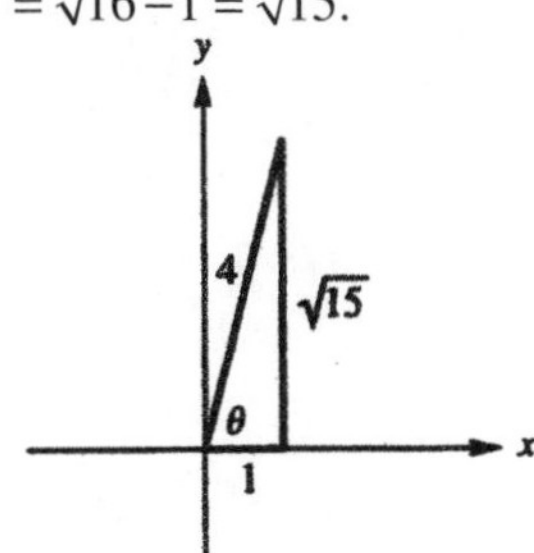

$$\sin\left(\arccos\frac{1}{4}\right) = \sin\theta = \frac{\sqrt{15}}{4}$$

65. $\cos(\tan^{-1}(-2))$

Let $\omega = \tan^{-1}(-2)$, so that $\tan\omega = -2$. Since $\tan^{-1}$ is defined only in quadrants I and IV, and -2 is negative, ω is in quadrant IV. Sketch ω and label a triangle with the hypotenuse equal to $\sqrt{(-2)^2 + 1} = \sqrt{4+1} = \sqrt{5}$.

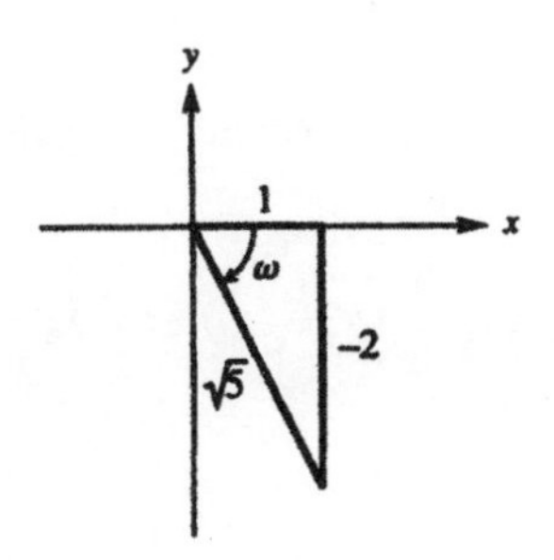

$$\cos(\tan^{-1}(-2)) = \cos\omega = \frac{\sqrt{5}}{5}$$

66. $\sec\left(\sin^{-1}\left(-\frac{1}{5}\right)\right)$

Let $\theta = \sin^{-1}\left(-\frac{1}{5}\right)$, so that $\sin\theta = -\frac{1}{5}$. Since arcsin is defined only in quadrants I and IV, and $-\frac{1}{5}$ is negative, θ is in quadrant IV. Sketch θ and label a triangle with the side adjacent to θ equal to $\sqrt{5^2 - (-1)^2} = \sqrt{25-1} = \sqrt{24} = 2\sqrt{6}$.

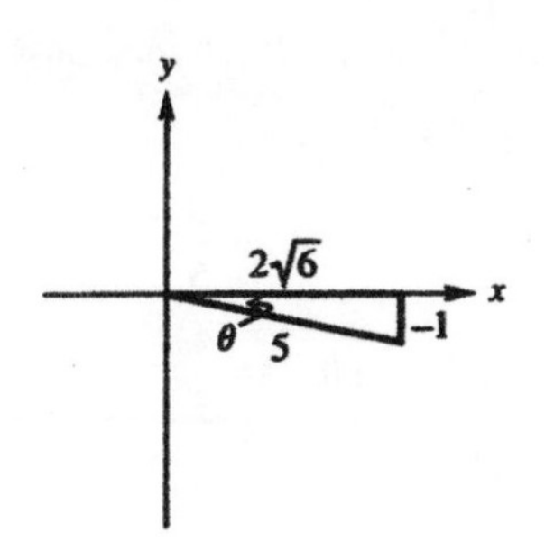

$$\sec\left(\sin^{-1}\left(-\frac{1}{5}\right)\right) = \sec\theta = \frac{5}{2\sqrt{6}} = \frac{5\sqrt{6}}{12}$$

67. $\sin\left(2\tan^{-1}\frac{12}{5}\right)$

Let $\omega=\tan^{-1}\frac{12}{5}$, so that $\tan\omega=\frac{12}{5}$. Since $\tan^{-1}$ is defined only in quadrants I and IV, and $\frac{12}{5}$ is positive, ω is in quadrant I. Sketch ω and label a right triangle with the hypotenuse equal to $\sqrt{12^2+5^2}=\sqrt{144+25}=\sqrt{169}=13.$

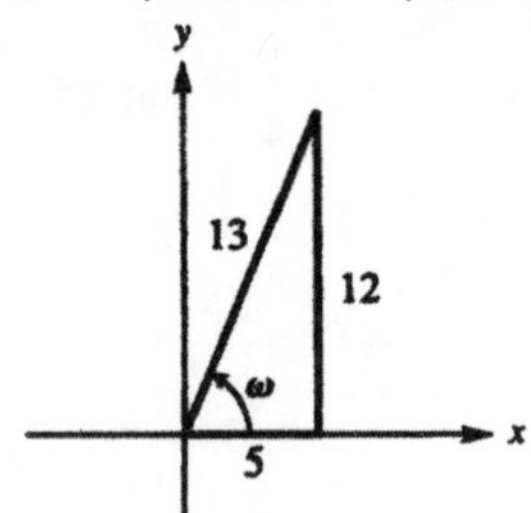

$$\sin\omega=\frac{12}{13}$$

$$\cos\omega=\frac{5}{13}$$

$$\sin\left(2\tan^{-1}\frac{12}{5}\right)=\sin(2\omega)=2\sin\omega\cos\omega=2\left(\frac{12}{13}\right)\left(\frac{5}{13}\right)=\frac{120}{169}$$

68. $\cos\left(2\sin^{-1}\frac{1}{4}\right)$

Let $\theta=\sin^{-1}\frac{1}{4}$, so that $\sin\theta=\frac{1}{4}$. Since $\sin^{-1}$ is defined only in quadrants I and IV, and $\frac{1}{4}$ is positive, θ is in quadrant I. Sketch θ and label a triangle with the side adjacent to θ equal to $\sqrt{4^2-1^2}=\sqrt{16-1}=\sqrt{15}.$

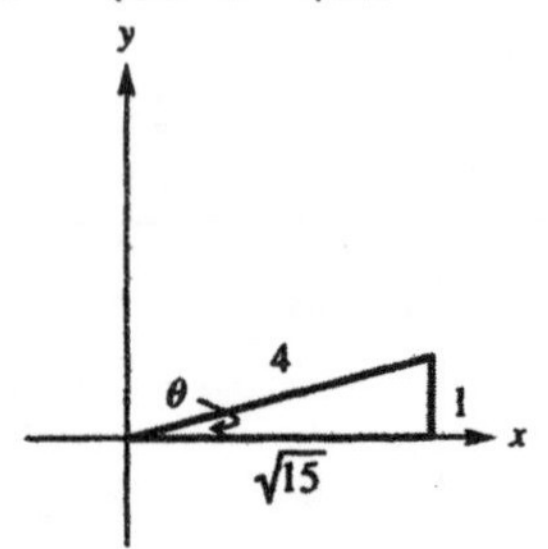

$$\cos\left(2\sin^{-1}\frac{1}{4}\right)=\cos 2\theta=1-2\sin^2\theta$$
$$=1-2\left(\frac{1}{4}\right)^2=1-2\left(\frac{1}{16}\right)$$
$$=1-\frac{1}{8}=\frac{7}{8}$$

69. $\cos\left(2\arctan\frac{4}{3}\right)$

Let $\omega=\arctan\frac{4}{3}$, so that $\tan\omega=\frac{4}{3}$. Since arctan is defined only in quadrants I and IV, and $\frac{4}{3}$ is positive, ω is in quadrant I. Sketch ω and label a triangle with the hypotenuse equal to $\sqrt{4^2+3^2}=\sqrt{16+9}=\sqrt{25}=5.$

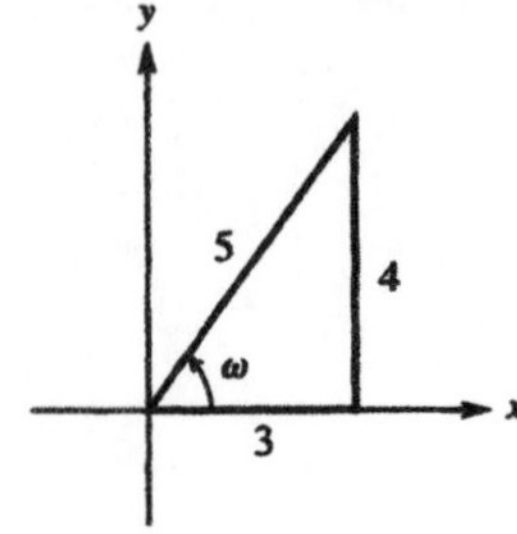

$$\cos\omega=\frac{3}{5}$$

$$\sin\omega=\frac{4}{5}$$

$$\cos\left(2\arctan\frac{4}{3}\right)=\cos(2\omega)=\cos^2\omega-\sin^2\omega=\left(\frac{3}{5}\right)^2-\left(\frac{4}{5}\right)^2=\frac{9}{25}-\frac{16}{25}=-\frac{7}{25}$$

70. $\tan\left(2\cos^{-1}\frac{1}{4}\right)$

Let $\theta = \cos^{-1}\frac{1}{4}$, so that $\cos\theta = \frac{1}{4}$. Since $\cos^{-1}$ is defined only in quadrants I and II, and $\frac{1}{4}$ is positive, θ is in quadrant I. Sketch θ and label a triangle with the side opposite θ equal to $\sqrt{4^2 - 1^2} = \sqrt{16-1} = \sqrt{15}$.

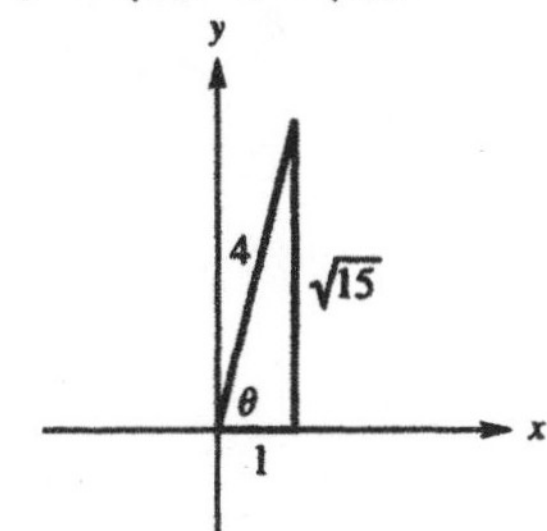

$$\tan\theta = \frac{\sqrt{15}}{1} = \sqrt{15}$$

$$\tan\left(2\cos^{-1}\frac{1}{4}\right) = \tan 2\theta = \frac{2\tan\theta}{1-\tan^2\theta} = \frac{2\sqrt{15}}{1-\left(\sqrt{15}\right)^2} = \frac{2\sqrt{15}}{-14} = -\frac{\sqrt{15}}{7}$$

71. $\sin\left(2\cos^{-1}\frac{1}{5}\right)$

Let $\theta = \cos^{-1}\frac{1}{5}$, so that $\cos\theta = \frac{1}{5}$. The inverse cosine function yields values only in quadrants I and II, and since $\frac{1}{5}$ is positive, θ is in quadrant I. Sketch θ and label the sides of the right triangle. By the Pythagorean theorem, the length opposite to θ will be $\sqrt{5^2 - 1^2} = \sqrt{24} = 2\sqrt{6}$.

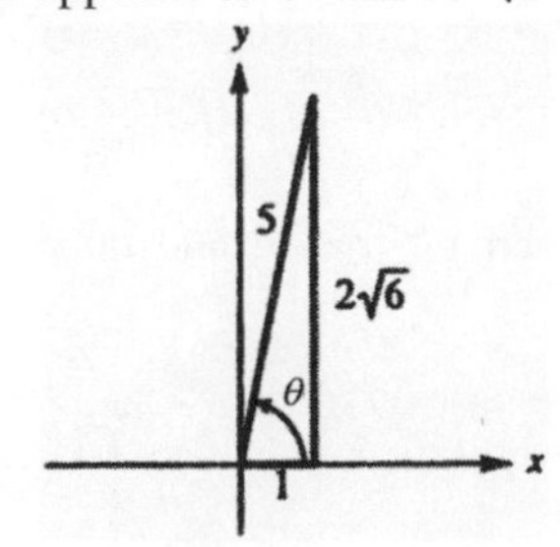

From the figure, $\sin\theta = \frac{2\sqrt{6}}{5}$. Then, $\sin\left(2\cos^{-1}\frac{1}{5}\right) = \sin 2\theta = 2\sin\theta\cos\theta = 2\left(\frac{2\sqrt{6}}{5}\right)\left(\frac{1}{5}\right) = \frac{4\sqrt{6}}{25}$.

72. $\cos\left(2\tan^{-1}(-2)\right)$

Let $\theta = \arctan(-2)$, so that $\tan\theta = -2$. Since arctan is defined only in quadrants I and IV, and -2 is negative, θ is in quadrant IV. Sketch θ and label a triangle with the hypotenuse equal to $\sqrt{(-2)^2+1^2} = \sqrt{4+1} = \sqrt{5}$.

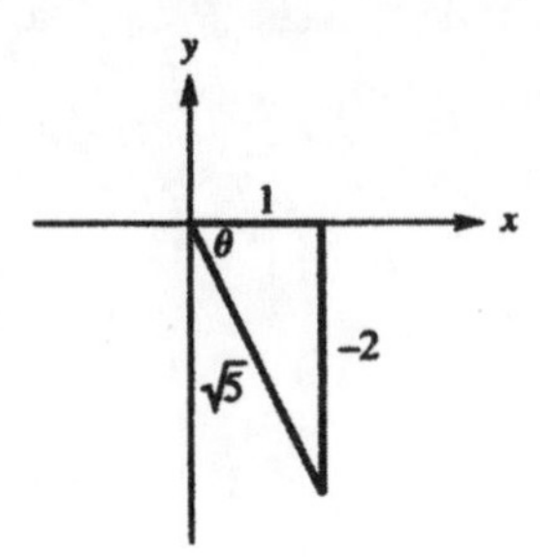

$$\cos\theta = \frac{1}{\sqrt{5}} = \frac{\sqrt{5}}{5}$$

$$\cos\left(2\tan^{-1}(-2)\right) = \cos 2\theta = 2\cos^2\theta - 1 = 2\left(\frac{1}{\sqrt{5}}\right)^2 - 1 = \frac{2}{5} - 1 = -\frac{3}{5}$$

73. $\sec(\sec^{-1} 2)$

Since secant and inverse secant are inverse functions, $\sec\left(\sec^{-1} 2\right) = 2$.

74. $\csc\left(\csc^{-1}\sqrt{2}\right) = \csc\frac{\pi}{4} = \sqrt{2}$

Also, since cosecant and inverse cosecant are inverse functions, $\csc\left(\csc^{-1}\sqrt{2}\right) = \sqrt{2}$.

75. $\cos\left(\tan^{-1}\frac{5}{12} - \cot^{-1}\frac{3}{4}\right)$

Let $\alpha = \tan^{-1}\frac{5}{12}$ and $\beta = \tan^{-1}\frac{4}{3}$. Then $\tan\alpha = \frac{5}{12}$ and $\tan\beta = \frac{3}{4}$. Sketch angles α and β, both in quadrant I.

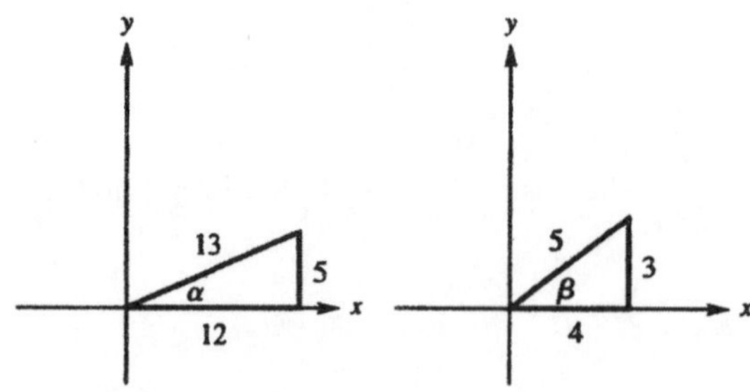

We have $\sin\alpha = \frac{5}{13}$, $\cos\alpha = \frac{12}{13}$, $\sin\beta = \frac{3}{5}$, and $\cos\beta = \frac{4}{5}$.

$$\cos\left(\tan^{-1}\frac{5}{12} - \tan^{-1}\frac{3}{4}\right) = \cos(\alpha - \beta) = \cos\alpha\cos\beta + \sin\alpha\sin\beta$$

$$= \left(\frac{12}{13}\right)\left(\frac{4}{5}\right) + \left(\frac{5}{13}\right)\left(\frac{3}{5}\right) = \frac{48}{65} + \frac{15}{65} = \frac{63}{65}$$

76. $\cos\left(\sin^{-1}\frac{3}{5}+\cos^{-1}\frac{5}{13}\right)$

Let $\omega_1=\sin^{-1}\frac{3}{5}$ and $\omega_2=\cos^{-1}\frac{5}{13}$. Then $\sin\omega_1=\frac{3}{5}$ and $\cos\omega_2=\frac{5}{13}$. Sketch ω_1 and ω_2 in quadrant I.

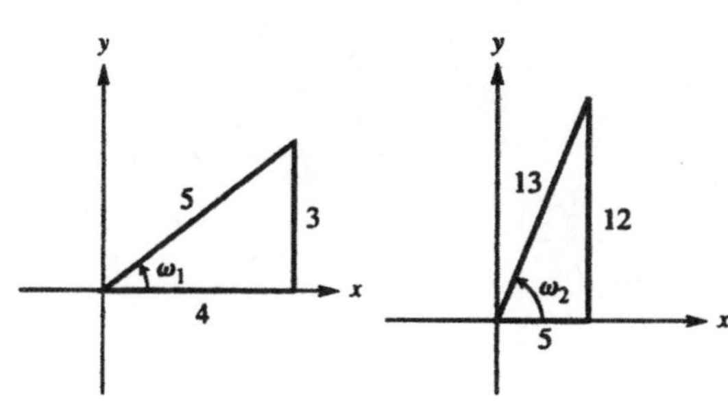

We have $\sin\omega_1=\frac{3}{5}$, $\cos\omega_1=\frac{4}{5}$, $\cos\omega_2=\frac{5}{13}$, and $\sin\omega_2=\frac{12}{13}$.

$$\cos\left(\arcsin\frac{3}{5}+\arccos\frac{5}{13}\right)=\cos(\omega_1+\omega_2)=\cos\omega_1\cos\omega_2-\sin\omega_1\sin\omega_2$$
$$=\left(\frac{4}{5}\right)\left(\frac{5}{13}\right)-\left(\frac{3}{5}\right)\left(\frac{12}{13}\right)=\frac{20}{65}-\frac{36}{65}=-\frac{16}{65}$$

77. $\sin\left(\sin^{-1}\frac{1}{2}+\tan^{-1}(-3)\right)$

Let $\sin^{-1}\frac{1}{2}=A$ and $\tan^{-1}(-3)=B$. Then $\sin A=\frac{1}{2}$ and $\tan B=-3$. Sketch angle A in quadrant I and angle B in quadrant IV.

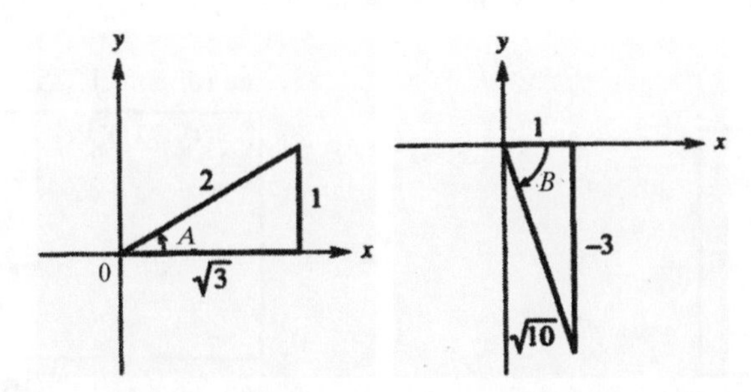

We have $\cos A=\frac{\sqrt{3}}{2}$, $\sin A=\frac{1}{2}$, $\cos B=\frac{1}{\sqrt{10}}=\frac{\sqrt{10}}{10}$, and $\sin B=\frac{-3}{\sqrt{10}}=-\frac{3\sqrt{10}}{10}$.

$$\sin\left(\sin^{-1}\frac{1}{2}+\tan^{-1}(-3)\right)=\sin(A+B)=\sin A\cos B+\cos A\sin B$$
$$=\frac{1}{2}\cdot\frac{1}{\sqrt{10}}+\frac{\sqrt{3}}{2}\cdot\frac{-3}{\sqrt{10}}=\frac{1-3\sqrt{3}}{2\sqrt{10}}=\frac{\sqrt{10}-3\sqrt{30}}{20}$$

78. $\tan\left(\cos^{-1}\frac{\sqrt{3}}{2}-\sin^{-1}\left(-\frac{3}{5}\right)\right)$

Let $\alpha=\cos^{-1}\frac{\sqrt{3}}{2}$, $\beta=\sin^{-1}\left(-\frac{3}{5}\right)$. Sketch angle α in quadrant I and angle β in quadrant IV.

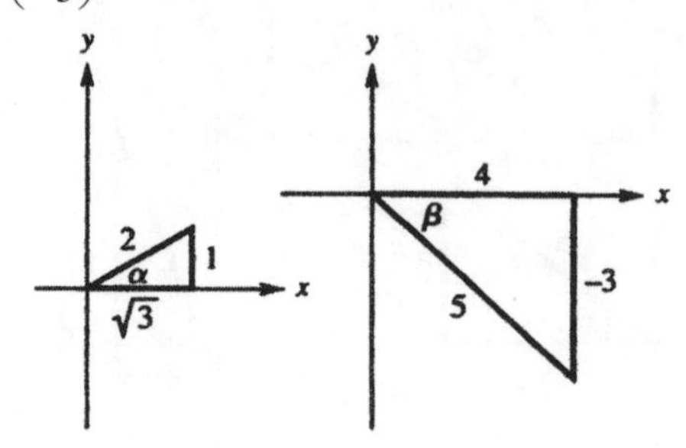

We have $\tan\alpha=\frac{1}{\sqrt{3}}$ and $\tan\beta=-\frac{3}{4}$.

$$\tan\left(\cos^{-1}\frac{\sqrt{3}}{2}-\sin^{-1}\left(-\frac{3}{5}\right)\right)=\tan(\alpha-\beta)=\frac{\tan\alpha-\tan\beta}{1+\tan\alpha\tan\beta}=\frac{\frac{1}{\sqrt{3}}-\left(-\frac{3}{4}\right)}{1+\left(\frac{1}{\sqrt{3}}\right)\left(-\frac{3}{4}\right)}=\frac{\frac{4+3\sqrt{3}}{4\sqrt{3}}}{\frac{4\sqrt{3}-3}{4\sqrt{3}}}$$

$$=\frac{4+3\sqrt{3}}{4\sqrt{3}-3}=\frac{4+3\sqrt{3}}{-3+4\sqrt{3}}\cdot\frac{-3-4\sqrt{3}}{-3-4\sqrt{3}}$$

$$=\frac{-12-25\sqrt{3}-36}{9-48}=\frac{-48-25\sqrt{3}}{-39}=\frac{48+25\sqrt{3}}{39}$$

For Exercises 79–82, your calculator could be in either degree or radian mode. Keystroke sequences may vary based on the type and/or model of calculator being used.

79. $\cos\left(\tan^{-1}.5\right)$

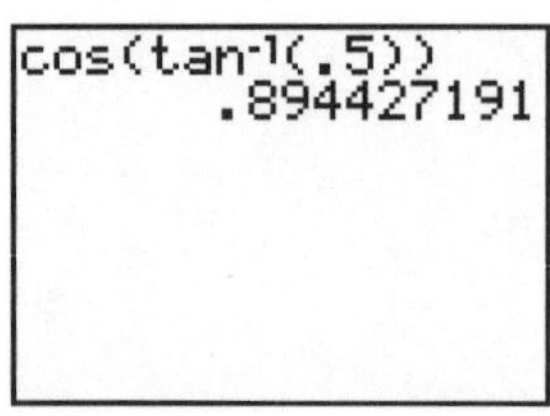

$\cos(\tan^{-1}.5)\approx .894427191$

80. $\sin\left(\cos^{-1}.25\right)$

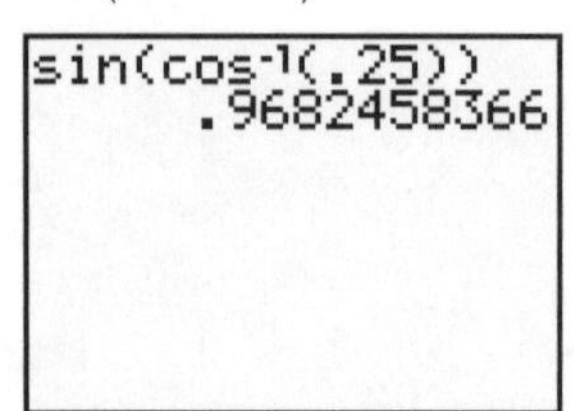

$\sin\left(\cos^{-1}.25\right)\approx .9682458366$

81. tan (arcsin .12251014)

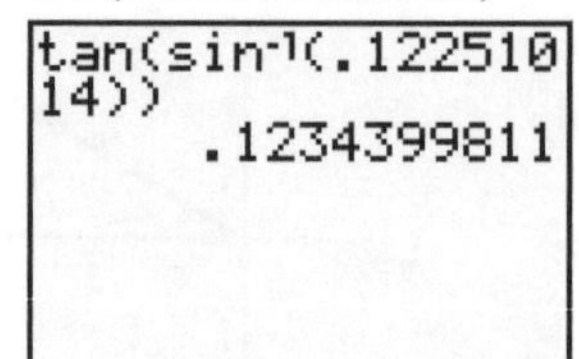

tan (arcsin .12251014) $\approx$ .1234399811

82. cot (arccos .58236841)

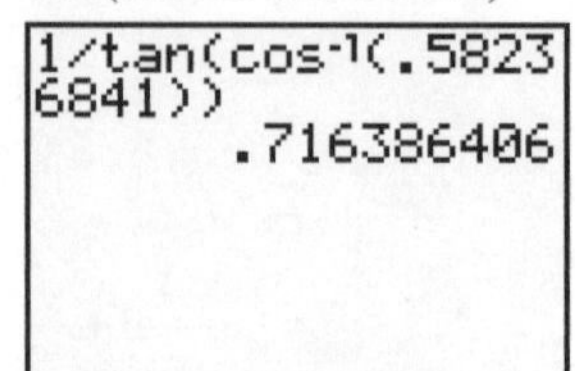

cot (arccos .58236841) $\approx$ .716386406

83. $\sin(\arccos u)$

Let $\theta = \arccos u$, so $\cos\theta = u = \frac{u}{1}$. Since $u > 0$, $0 < \theta < \frac{\pi}{2}$.

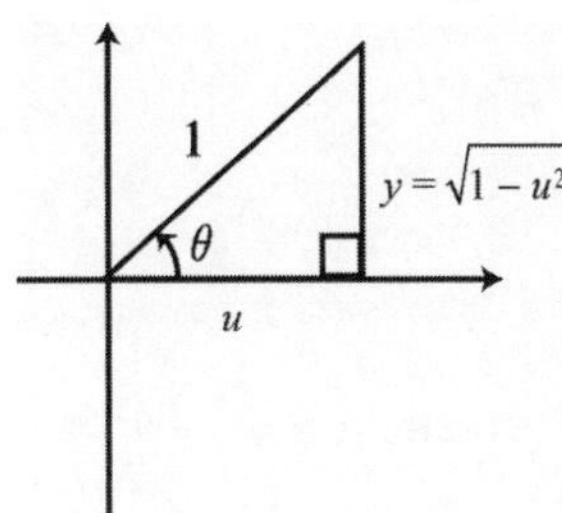

Since $y > 0$, from the Pythagorean theorem, $y = \sqrt{1^2 - u^2} = \sqrt{1 - u^2}$.

Therefore, $\sin\theta = \frac{\sqrt{1-u^2}}{1} = \sqrt{1-u^2}$. Thus, $\sin(\arccos u) = \sqrt{1-u^2}$.

84. $\tan(\arccos u)$

Let $\theta = \arccos u$, so $\cos\theta = u = \frac{u}{1}$. Since $u > 0$, $0 < \theta < \frac{\pi}{2}$.

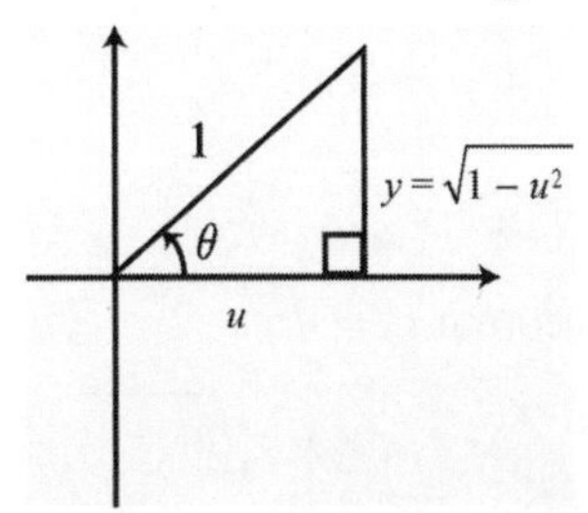

Since $y > 0$, from the Pythagorean theorem, $y = \sqrt{1^2 - u^2} = \sqrt{1 - u^2}$.

Therefore, $\tan\theta = \frac{\sqrt{1-u^2}}{u}$. Thus, $\tan(\arccos u) = \frac{\sqrt{1-u^2}}{u}$.

85. $\cos(\arcsin u)$

Let $\theta = \arcsin u$, so $\sin\theta = u = \frac{u}{1}$. Since $u > 0$, $0 < \theta < \frac{\pi}{2}$.

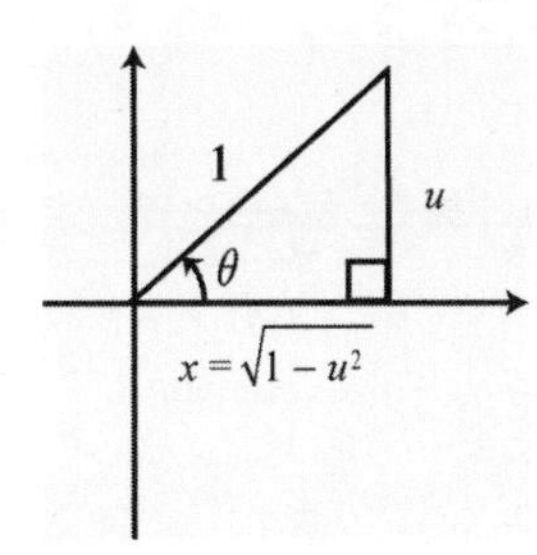

Since $x > 0$, from the Pythagorean theorem, $x = \sqrt{1^2 - u^2} = \sqrt{1 - u^2}$.

Therefore, $\cos\theta = \frac{\sqrt{1-u^2}}{1} = \sqrt{1-u^2}$. Thus, $\cos(\arcsin u) = \sqrt{1-u^2}$

86. $\cot(\arcsin u)$

Let $\theta = \arcsin u$, so $\sin\theta = u = \frac{u}{1}$. Since $u > 0$, $0 < \theta < \frac{\pi}{2}$.

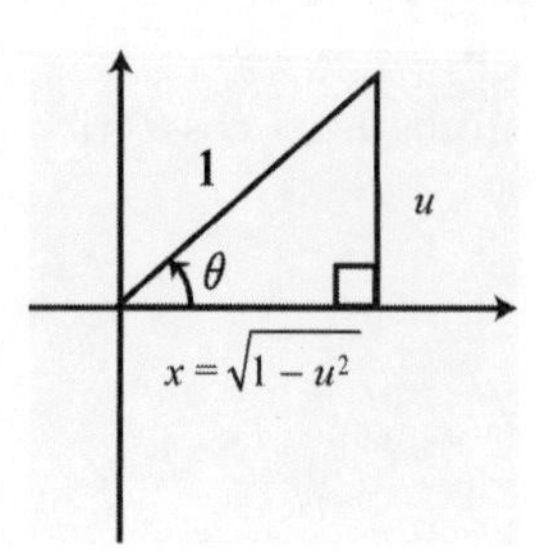

Since $x > 0$, from the Pythagorean theorem, $x = \sqrt{1^2 - u^2} = \sqrt{1-u^2}$.

Therefore, $\cot\theta = \frac{\sqrt{1-u^2}}{u}$. Thus, $\cot(\arcsin u) = \frac{\sqrt{1-u^2}}{u}$.

87. $\sin\left(\sec^{-1}\frac{u}{2}\right)$

Let $\theta = \sec^{-1}\frac{u}{2}$, so $\sec\theta = \frac{u}{2}$. Since $u > 0$, $0 < \theta < \frac{\pi}{2}$.

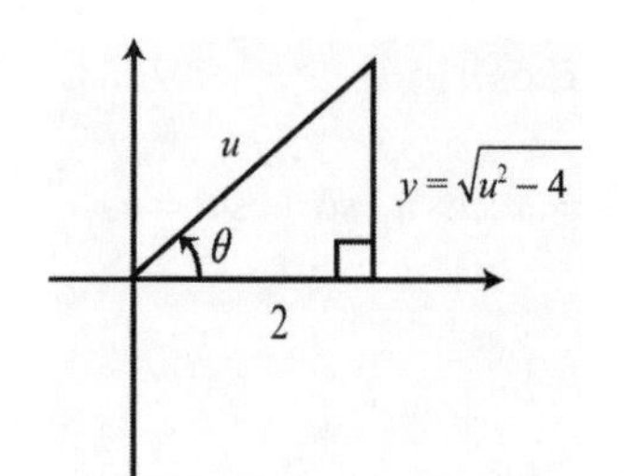

Since $y > 0$, from the Pythagorean theorem, $y = \sqrt{u^2 - 2^2} = \sqrt{u^2 - 4}$.

Therefore, $\sin\theta = \frac{\sqrt{u^2-4}}{u}$. Thus, $\sin\left(\sec^{-1}\frac{u}{2}\right) = \frac{\sqrt{u^2-4}}{u}$.

88. $\cos\left(\tan^{-1}\frac{3}{u}\right)$

Let $\theta = \tan^{-1}\frac{3}{u}$, so $\tan\theta = \frac{3}{u}$. Since $u > 0$, $0 < \theta < \frac{\pi}{2}$.

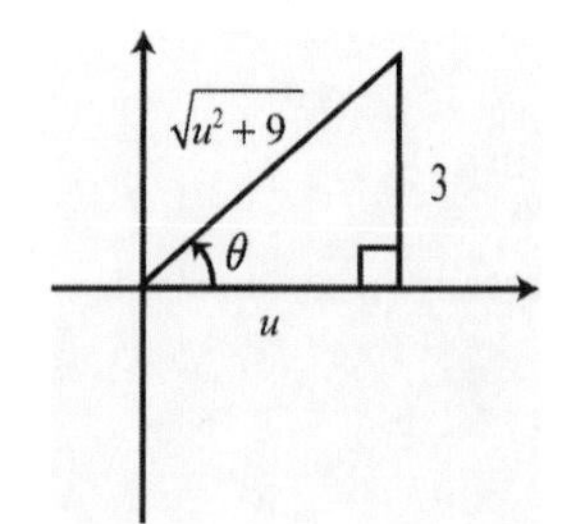

From the Pythagorean theorem, $r = \sqrt{u^2 + 3^2} = \sqrt{u^2 + 9}$.

Therefore, $\cos\theta = \frac{u}{\sqrt{u^2+9}} = \frac{u\sqrt{u^2+9}}{u^2+9}$. Thus, $\cos\left(\tan^{-1}\frac{3}{u}\right) = \frac{u\sqrt{u^2+9}}{u^2+9}$.

89. $\tan\left(\sin^{-1}\dfrac{u}{\sqrt{u^2+2}}\right)$

Let $\theta = \sin^{-1}\dfrac{u}{\sqrt{u^2+2}}$, so $\sin\theta = \dfrac{u}{\sqrt{u^2+2}}$. Since $u > 0$, $0 < \theta < \dfrac{\pi}{2}$.

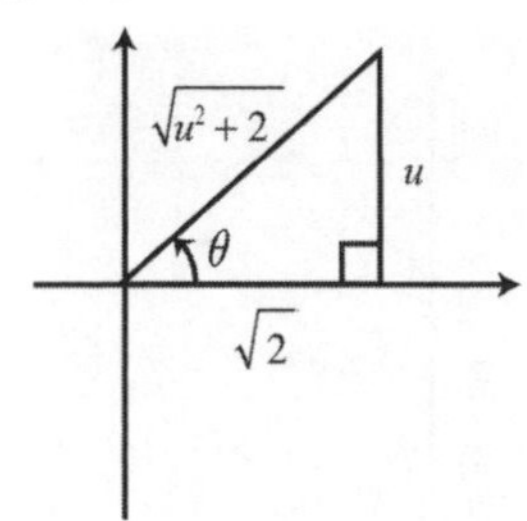

Since $x > 0$, from the Pythagorean theorem, $x = \sqrt{\left(\sqrt{u^2+2}\right)^2 - u^2} = \sqrt{u^2+2-u^2} = \sqrt{2}.$

Therefore, $\tan\theta = \dfrac{u}{\sqrt{2}} = \dfrac{u\sqrt{2}}{2}$. Thus, $\tan\left(\sin^{-1}\dfrac{u}{\sqrt{u^2+2}}\right) = \dfrac{u\sqrt{2}}{2}.$

90. $\sec\left(\cos^{-1}\dfrac{u}{\sqrt{u^2+5}}\right)$

Let $\theta = \cos^{-1}\dfrac{u}{\sqrt{u^2+5}}$, so $\cos\theta = \dfrac{u}{\sqrt{u^2+5}}$. Since $u > 0$, $0 < \theta < \dfrac{\pi}{2}$.

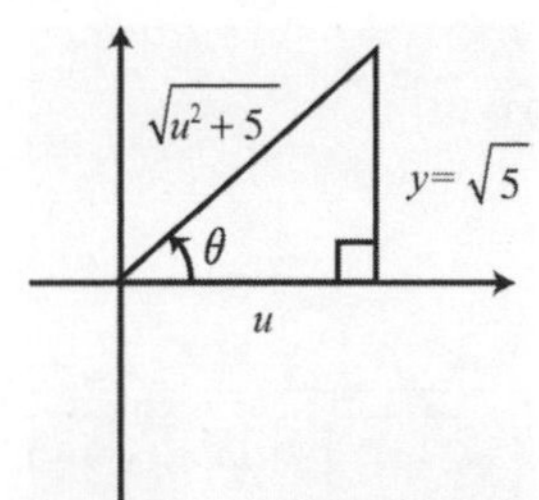

Since $y > 0$, from the Pythagorean theorem, $x = \sqrt{\left(\sqrt{u^2+5}\right)^2 - u^2} = \sqrt{u^2+5-u^2} = \sqrt{5}$. Therefore,

$\sec\theta = \dfrac{\sqrt{u^2+5}}{u}$. Thus, $\sec\left(\cos^{-1}\dfrac{u}{\sqrt{u^2+5}}\right) = \dfrac{\sqrt{u^2+5}}{u}.$

Also note, $\sec\left(\cos^{-1}\dfrac{u}{\sqrt{u^2+5}}\right) = \dfrac{1}{\cos\left(\cos^{-1}\dfrac{u}{\sqrt{u^2+5}}\right)} = \dfrac{1}{\dfrac{u}{\sqrt{u^2+5}}} = \dfrac{\sqrt{u^2+5}}{u}.$

91. $\sec\left(\text{arc}\cot\frac{\sqrt{4-u^2}}{u}\right)$

Let $\theta = \text{arc}\cot\frac{\sqrt{4-u^2}}{u}$, so $\cot\theta = \frac{\sqrt{4-u^2}}{u}$. Since $u>0$, $0<\theta<\frac{\pi}{2}$.

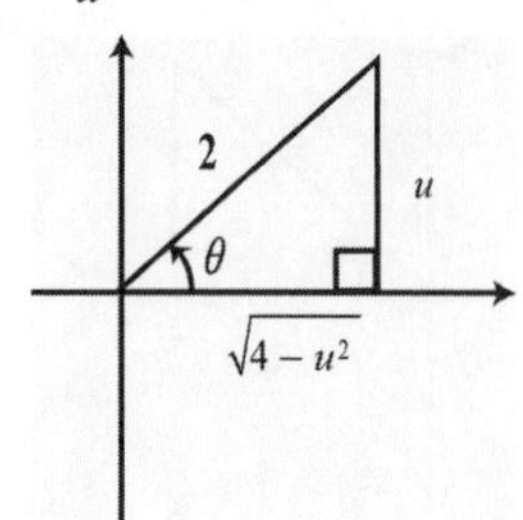

From the Pythagorean theorem, $r = \sqrt{\left(\sqrt{4-u^2}\right)^2 + u^2} = \sqrt{4-u^2+u^2} = \sqrt{4} = 2.$

Therefore, $\sec\theta = \frac{2}{\sqrt{4-u^2}} = \frac{2\sqrt{4-u^2}}{4-u^2}$. Thus, $\sec\left(\text{arc}\cot\frac{\sqrt{4-u^2}}{u}\right) = \frac{2\sqrt{4-u^2}}{4-u^2}$.

92. $\csc\left(\arctan\frac{\sqrt{9-u^2}}{u}\right)$

Let $\theta = \arctan\frac{\sqrt{9-u^2}}{u}$, so

$\tan\theta = \frac{\sqrt{9-u^2}}{u}$. Since $u>0$, $0<\theta<\frac{\pi}{2}$.

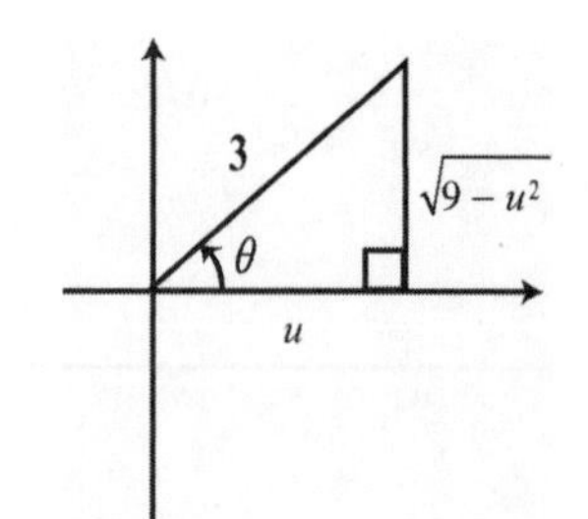

From the Pythagorean theorem, $r = \sqrt{\left(\sqrt{9-u^2}\right)^2 + u^2} = \sqrt{9-u^2+u^2} = \sqrt{9} = 3$. Therefore,

$\csc\theta = \frac{3}{\sqrt{9-u^2}} = \frac{3\sqrt{9-u^2}}{9-u^2}$. Thus, $\csc\left(\arctan\frac{\sqrt{9-u^2}}{u}\right) = \frac{3\sqrt{9-u^2}}{9-u^2}$.

93. (a) $\theta = \arcsin\sqrt{\frac{42^2}{2(42^2)+64(0)}}$

$= \arcsin\sqrt{\frac{42^2}{2(42^2)+0}} = \arcsin\sqrt{\frac{42^2}{2(42^2)}} = \arcsin\sqrt{\frac{1}{2}} = \arcsin\frac{1}{\sqrt{2}} = \arcsin\frac{\sqrt{2}}{2} = 45°$

(b) $\theta = \arcsin\sqrt{\frac{v^2}{2v^2+64(6)}} = \arcsin\sqrt{\frac{v^2}{2v^2+384}}$

As v gets larger and larger, $\sqrt{\frac{v^2}{2v^2+384}} \approx \sqrt{\frac{1}{2}} = \frac{\sqrt{2}}{2}$. Thus, $\theta \approx \arcsin\frac{\sqrt{2}}{2} = 45°$.

The equation of the asymptote is $\theta = 45°$.

94. Let A be the triangle to the "right" of θ and let B be the angle to the "left" of θ. Then $A+\theta+B=\pi$ (since the angles sum to π radians) and $\theta=\pi-A-B$.

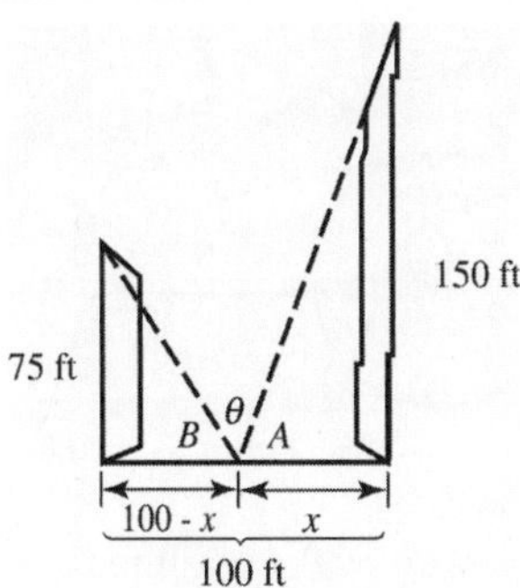

$$\tan A=\frac{150}{x},\text{ so } A=\arctan\left(\frac{150}{x}\right) \text{ and } \tan B=\frac{75}{100-x},\text{ so } B=\arctan\left(\frac{75}{100-x}\right).$$

As a result, $\theta=\pi-B-A \Rightarrow \theta=\pi-\arctan\left(\frac{75}{100-x}\right)-\arctan\left(\frac{150}{x}\right).$

95. $\theta=\tan^{-1}\left(\frac{x}{x^2+2}\right)$

(a) $x=1,\ \theta=\tan^{-1}\left(\frac{1}{1^2+2}\right)=\tan^{-1}\left(\frac{1}{3}\right)\approx 18^\circ$

(b) $x=2,\ \theta=\tan^{-1}\left(\frac{2}{2^2+2}\right)=\tan^{-1}\frac{2}{6}=\tan^{-1}\frac{1}{3}\approx 18^\circ$

(c) $x=3,\ \theta=\tan^{-1}\left(\frac{3}{3^2+2}\right)=\tan^{-1}\frac{3}{11}\approx 15^\circ$

(d) $\tan(\theta+\alpha)=\frac{1+1}{x}=\frac{2}{x}$ and $\tan\alpha=\frac{1}{x}$

$$\tan(\theta+\alpha)=\frac{\tan\theta+\tan\alpha}{1-\tan\theta\tan\alpha}\Rightarrow\frac{2}{x}=\frac{\tan\theta+\frac{1}{x}}{1-\tan\theta\left(\frac{1}{x}\right)}\Rightarrow\frac{2}{x}=\frac{x\tan\theta+1}{x-\tan\theta}$$

$$2(x-\tan\theta)=x(x\tan\theta+1)\Rightarrow 2x-2\tan\theta=x^2\tan\theta+x\Rightarrow 2x-x=x^2\tan\theta+2\tan\theta$$

$$x=\tan\theta\left(x^2+2\right)\Rightarrow\tan\theta=\frac{x}{x^2+2}\Rightarrow\theta=\tan^{-1}\left(\frac{x}{x^2+2}\right)$$

(e) If we graph $y_1=\tan^{-1}\left(\frac{x}{x^2+2}\right)$ using a graphing calculator, the maximum value of the function occurs when x is 1.4142151 m. (Note: Due to the computational routine, there may be a discrepancy in the last few decimal places.)

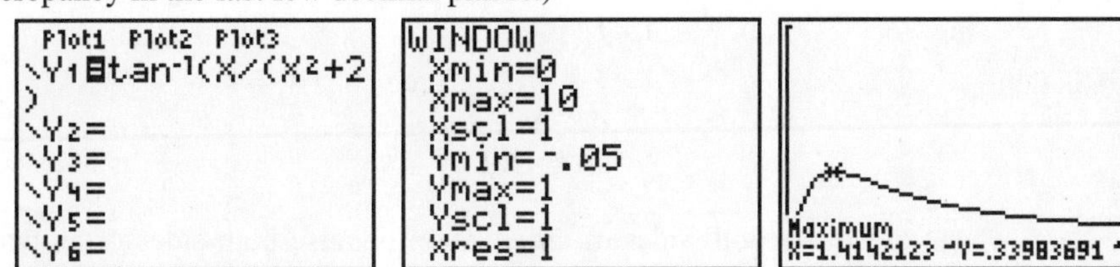

(f) $x=\sqrt{(1)(2)}=\sqrt{2}$

96. Since the diameter of the earth is 7927 miles at the equator, the radius of the earth is 3963.5 miles. Then $\cos\theta = \dfrac{3963.5}{20,000+3963.5} = \dfrac{3963.5}{23,963.5}$ and $\theta = \arccos\left(\dfrac{3963.5}{23,963.5}\right) \approx 80.48°$.

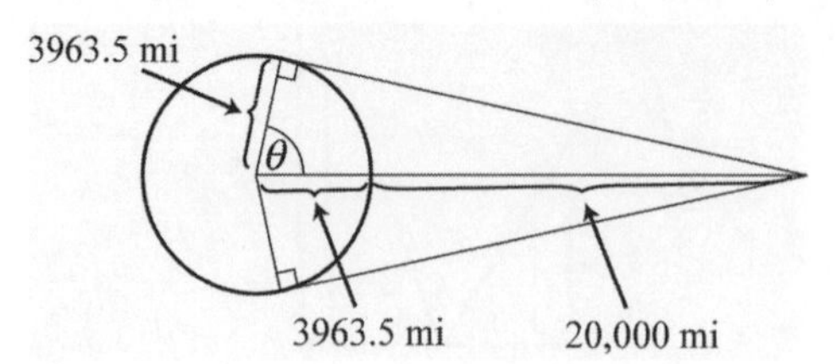

The percent of the equator that can be seen by the satellite is $\dfrac{2\theta}{360}\cdot 100 = \dfrac{2(80.48)}{360} \approx 44.7\%$.

Section 6.2: Trigonometric Equations I

1. Solve the linear equation for $\cot x$.

2. Solve the linear equation for $\sin x$.

3. Solve the quadratic equation for $\sec x$ by factoring.

4. Solve the quadratic equation for $\cos x$ by the zero factor property.

5. Solve the quadratic equation for $\sin x$ using the quadratic formula.

6. Solve the quadratic for $\tan x$ using the quadratic formula.

7. Use the identity to rewrite as an equation with one trigonometric function.

8. Use an identity to rewrite as an equation with one trigonometric function.

9. $-30°$ is not in the interval $[0°, 360°)$.

10. To show that $\left\{0, \dfrac{\pi}{2}, \dfrac{3\pi}{2}\right\}$ is not the correct solution set to the equation $\sin x = 1-\cos x$, you must show that at least one element of the set is not a solution.

Check $x = 0$.

$\sin x = 1-\cos x$
$\sin 0 = 1-\cos 0$?
$0 = 1-1$?
$0 = 0$ True
$x = 0$ is a solution

Check $x = \dfrac{\pi}{2}$.

$\sin x = 1-\cos x$
$\sin\dfrac{\pi}{2} = 1-\cos\dfrac{\pi}{2}$?
$1 = 1-0$?
$1 = 1$ True
$x = \dfrac{\pi}{2}$ is a solution

Check $x = \dfrac{3\pi}{2}$.

$\sin x = 1-\cos x$
$\sin\dfrac{3\pi}{2} = 1-\cos\dfrac{3\pi}{2}$?
$-1 = 1-0$?
$-1 = 1$ False
$x = \dfrac{3\pi}{2}$ is not a solution

Note: In general when you square both sides of an equation or raise both sides of an equation to an even power, you must check you solutions in order to eliminate any extraneous solutions that may occur.

11. $2\cot x+1=-1\Rightarrow 2\cot x=-2\Rightarrow \cot x=-1$

Over the interval $[0, 2\pi)$, the equation $\cot x=-1$ has two solutions, the angles in quadrants II and IV that have a reference angle of $\frac{\pi}{4}$. These are $\frac{3\pi}{4}$ and $\frac{7\pi}{4}$.

Solution set: $\left\{\frac{3\pi}{4}, \frac{7\pi}{4}\right\}$

12. $\sin x+2=3\Rightarrow \sin x=1$

Over the interval $[0, 2\pi)$, the equation $\sin x=1$ has one solution. This solution is $\frac{\pi}{2}$.

Solution set: $\left\{\frac{\pi}{2}\right\}$

13. $2\sin x+3=4\Rightarrow 2\sin x=1\Rightarrow \sin x=\frac{1}{2}$

Over the interval $[0, 2\pi)$, the equation $\sin x=\frac{1}{2}$ has two solutions, the angles in quadrants I and II that have a reference angle of $\frac{\pi}{6}$. These are $\frac{\pi}{6}$ and $\frac{5\pi}{6}$.

Solution set: $\left\{\frac{\pi}{6}, \frac{5\pi}{6}\right\}$

14. $2\sec x+1=\sec x+3\Rightarrow \sec x=2$

Over the interval $[0, 2\pi)$, the equation $\sec x=2$ has two solutions, the angles in quadrants I and IV that have a reference angle of $\frac{\pi}{3}$. These are $\frac{\pi}{3}$ and $\frac{5\pi}{3}$.

Solution set: $\left\{\frac{\pi}{3}, \frac{5\pi}{3}\right\}$

15. $\tan^2 x+3=0\Rightarrow \tan^2 x=-3$

The square of a real number cannot be negative, so this equation has no solution.

Solution set: $\varnothing$

16. $\sec^2 x+2=-1\Rightarrow \sec^2 x=-3$

The square of a real number cannot be negative, so this equation has no solution.

Solution set: $\varnothing$

17. $(\cot x-1)(\sqrt{3}\cot x+1)=0$

$$\cot x-1=0\Rightarrow\cot x=1 \text{ or } \sqrt{3}\cot x+1=0\Rightarrow\sqrt{3}\cot x=-1\Rightarrow\cot x=-\frac{1}{\sqrt{3}}\Rightarrow\cot x=-\frac{\sqrt{3}}{3}$$

Over the interval $[0,2\pi)$, the equation $\cot x=1$ has two solutions, the angles in quadrants I and III that have a reference angle of $\frac{\pi}{4}$. These are $\frac{\pi}{4}$ and $\frac{5\pi}{4}$. In the same interval, $\cot x=-\frac{\sqrt{3}}{3}$ also has two solutions. The angles in quadrants II and IV that have a reference angle of $\frac{\pi}{3}$ are $\frac{2\pi}{3}$ and $\frac{5\pi}{3}$.

Solution set: $\left\{\frac{\pi}{4},\frac{2\pi}{3},\frac{5\pi}{4},\frac{5\pi}{3}\right\}$

18. $(\csc x+2)(\csc x-\sqrt{2})=0$

$$\csc x+2=0\Rightarrow\csc x=-2 \text{ or } \csc x-\sqrt{2}=0\Rightarrow\csc x=\sqrt{2}$$

Over the interval $[0,2\pi)$, the equation $\csc x=-2$ has two solutions, the angles in quadrants III and IV that have a reference angle of $\frac{\pi}{6}$. These are $\frac{7\pi}{6}$ and $\frac{11\pi}{6}$. In the same interval, $\csc x=\sqrt{2}$ also has two solutions. The angles in quadrants I and II that have a reference angle of $\frac{\pi}{4}$ are $\frac{\pi}{4}$ and $\frac{3\pi}{4}$.

Solution set: $\left\{\frac{\pi}{4},\frac{3\pi}{4},\frac{7\pi}{6},\frac{11\pi}{6}\right\}$

19. $\cos^2 x+2\cos x+1=0$

$$\cos^2 x+2\cos x+1=0\Rightarrow(\cos x+1)^2=0\Rightarrow\cos x+1=0\Rightarrow\cos x=-1$$

Over the interval $[0,2\pi)$, the equation $\cos x=-1$ has one solution. This solution is π.

Solution set: $\{\pi\}$

20. $2\cos^2 x-\sqrt{3}\cos x=0$

$$2\cos^2 x-\sqrt{3}\cos x=0\Rightarrow\cos x(2\cos x-\sqrt{3})=0$$

$$\cos x=0 \text{ or } 2\cos x-\sqrt{3}=0\Rightarrow 2\cos x=\sqrt{3}\Rightarrow\cos x=\frac{\sqrt{3}}{2}$$

Over the interval $[0,2\pi)$, the equation $\cos x=0$ has two solutions. These solutions are $\frac{\pi}{2}$ and $\frac{3\pi}{2}$.

In the same interval, $\cos x=\frac{\sqrt{3}}{2}$ also has two solutions. The angles in quadrants I and IV that have a reference angle of $\frac{\pi}{6}$ are $\frac{\pi}{6}$ and $\frac{11\pi}{6}$.

Solution set: $\left\{\frac{\pi}{6},\frac{\pi}{2},\frac{3\pi}{2},\frac{11\pi}{6}\right\}$

21. $-2\sin^2 x = 3\sin x + 1$

$$-2\sin^2 x = 3\sin x + 1 \Rightarrow 2\sin^2 x + 3\sin x + 1 = 0 \Rightarrow (2\sin x + 1)(\sin x + 1) = 0$$

$$2\sin x + 1 = 0 \Rightarrow \sin x = -\frac{1}{2} \text{ or } \sin x + 1 = 0 \Rightarrow \sin x = -1$$

Over the interval $[0, 2\pi)$, the equation $\sin x = -\frac{1}{2}$ has two solutions. The angles in quadrants III and IV that have a reference angle of $\frac{\pi}{6}$ are $\frac{7\pi}{6}$ and $\frac{11\pi}{6}$. In the same interval, $\sin x = -1$ when the angle is $\frac{3\pi}{2}$.

Solution set: $\left\{\frac{7\pi}{6}, \frac{3\pi}{2}, \frac{11\pi}{6}\right\}$

22. $2\cos^2 x - \cos x = 1$

$$2\cos^2 x - \cos x = 1 \Rightarrow 2\cos^2 x - \cos x - 1 = 0 \Rightarrow (2\cos x + 1)(\cos x - 1) = 0$$

$$2\cos x + 1 = 0 \Rightarrow 2\cos x = -1 \Rightarrow \cos x = -\frac{1}{2} \text{ or } \cos x - 1 = 0 \Rightarrow \cos x = 1$$

Over the interval $[0, 2\pi)$, the equation $\cos x = -\frac{1}{2}$ has two solutions. The angles in quadrants II and III that have a reference angle of $\frac{\pi}{3}$ are $\frac{2\pi}{3}$ and $\frac{4\pi}{3}$. In the same interval, $\cos x = 1$ when the angle is 0.

Solution set: $\left\{0, \frac{2\pi}{3}, \frac{4\pi}{3}\right\}$

23. $\left(\cot\theta - \sqrt{3}\right)\left(2\sin\theta + \sqrt{3}\right) = 0$

$$\cot\theta - \sqrt{3} = 0 \Rightarrow \cot\theta = \sqrt{3} \text{ or } 2\sin\theta + \sqrt{3} = 0 \Rightarrow 2\sin\theta = -\sqrt{3} \Rightarrow \sin\theta = -\frac{\sqrt{3}}{2}$$

Over the interval $[0°, 360°)$, the equation $\cot\theta = \sqrt{3}$ has two solutions, the angles in quadrants I and III that have a reference angle of 30°. These are 210° and 330°. In the same interval, the equation $\sin\theta = -\frac{\sqrt{3}}{2}$ has two solutions, the angles in quadrants III and IV that have a reference angle of 60°. These are 240° and 300°.

Solution set: $\{30°, 210°, 240°, 300°\}$

24. $(\tan\theta - 1)(\cos\theta - 1) = 0$

$$\tan\theta - 1 = 0 \Rightarrow \tan\theta = 1 \text{ or } \cos\theta - 1 = 0 \Rightarrow \cos\theta = 1$$

Over the interval $[0°, 360°)$, the equation $\tan\theta = 1$ has two solutions, the angles in quadrants I and III that have a reference angle of 45°. These are 45° and 225°. In the same interval, the equation $\cos\theta = 1$ has one solution. The angle is 0°.

Solution set: $\{0°, 45°, 225°\}$

25. $2\sin\theta - 1 = \csc\theta$

$$2\sin\theta - 1 = \csc\theta \Rightarrow 2\sin\theta - 1 = \frac{1}{\sin\theta} \Rightarrow 2\sin^2\theta - \sin\theta = 1$$
$$\sin^2\theta - \sin\theta - 1 = 0 \Rightarrow (2\sin\theta + 1)(\sin\theta - 1) = 0$$
$$2\sin\theta + 1 = 0 \Rightarrow \sin\theta = -\frac{1}{2} \text{ or } \sin\theta - 1 = 0 \Rightarrow \sin\theta = 1$$

Over the interval $[0°, 360°)$, the equation $\sin\theta = -\frac{1}{2}$ has two solutions, the angles in quadrants III and IV that have a reference angle of 30°. These are 210° and 330°. In the same interval, the only angle θ for which $\sin\theta = 1$ is 90°.

Solution set: $\{90°, 210°, 330°\}$

26. $\tan\theta + 1 = \sqrt{3} + \sqrt{3}\cot\theta$

$$\tan\theta + 1 = \sqrt{3} + \sqrt{3}\cot\theta \Rightarrow \tan\theta + 1 = \sqrt{3} + \frac{\sqrt{3}}{\tan\theta} \Rightarrow \tan^2\theta + \tan\theta = \sqrt{3}\tan\theta + \sqrt{3}$$
$$\tan^2\theta + \left(1 - \sqrt{3}\right)\tan\theta - \sqrt{3} = 0 \Rightarrow \left(\tan\theta - \sqrt{3}\right)(\tan\theta + 1) = 0$$
$$\tan\theta - \sqrt{3} = 0 \Rightarrow \tan\theta = \sqrt{3} \text{ or } \tan\theta + 1 = 0 \Rightarrow \tan\theta = -1$$

Over the interval $[0°, 360°)$, the equation $\tan\theta = \sqrt{3}$ has two solutions, the angles in quadrants I and III that have a reference angle of 60°. These are 60° and 240°. In the same interval, the equation $\tan\theta = -1$ has two solutions, the angles in quadrants II and IV that have a reference angle of 45°. These are 135° and 315°.

Solution set: $\{60°, 135°, 240°, 315°\}$

27. $\tan\theta - \cot\theta = 0$

$$\tan\theta - \cot\theta = 0 \Rightarrow \tan\theta - \frac{1}{\tan\theta} = 0 \Rightarrow \tan^2\theta - 1 = 0 \Rightarrow \tan^2\theta = 1 \Rightarrow \tan\theta = \pm 1$$

Over the interval $[0°, 360°)$, the equation $\tan\theta = 1$ has two solutions, the angles in quadrants I and III that have a reference angle of 45°. These are 45° and 225°. In the same interval, the equation $\tan\theta = -1$ has two solutions, the angles in quadrants II and IV that have a reference angle of 45°. These are 135° and 315°.

Solution set: $\{45°, 135°, 225°, 315°\}$

28. $\cos^2\theta = \sin^2\theta + 1$

$$\cos^2\theta = \sin^2\theta + 1 \Rightarrow 1 - \sin^2\theta = \sin^2\theta + 1 \Rightarrow 1 = 2\sin^2\theta + 1$$
$$2\sin^2\theta = 0 \Rightarrow \sin^2\theta = 0 \Rightarrow \sin\theta = 0$$

Over the interval $[0°, 360°)$, the equation $\sin\theta = 0$ has two solutions. These are 0° and 180°.

Solution set: $\{0°, 180°\}$

29. $\csc^2\theta - 2\cot\theta = 0$

$$\csc^2\theta - 2\cot\theta = 0 \Rightarrow \left(1+\cot^2\theta\right) - 2\cot\theta = 0 \Rightarrow \cot^2\theta - 2\cot\theta + 1 = 0$$
$$\left(\cot\theta - 1\right)^2 = 0 \Rightarrow \cot\theta - 1 = 0 \Rightarrow \cot\theta = 1$$

Over the interval $[0°, 360°)$, the equation $\cot\theta = 1$ has two solutions, the angles in quadrants I and III that have a reference angle of $45°$. These are $45°$ and $225°$.

Solution set: $\{45°, 225°\}$

30. $\sin^2\theta\cos\theta = \cos\theta$

$$\sin^2\theta\cos\theta = \cos\theta \Rightarrow \sin^2\theta\cos\theta - \cos\theta = 0 \Rightarrow \cos\theta\left(\sin^2\theta - 1\right) = 0$$
$$\cos\theta = 0 \text{ or } \sin^2\theta - 1 = 0 \Rightarrow \sin^2\theta = 1 \Rightarrow \sin\theta = \pm 1$$

Over the interval $[0°, 360°)$, the equation $\cos\theta = 0$ has two solutions. These are $90°$ and $270°$. In the same interval, the equation $\sin\theta = 1$ has one solution, namely $90°$. Finally, $\sin\theta = -1$ has one solution, namely $270°$.

Solution set: $\{90°, 270°\}$

31. $2\tan^2\theta\sin\theta - \tan^2\theta = 0$

$$2\tan^2\theta\sin\theta - \tan^2\theta = 0 \Rightarrow \tan^2\theta\left(2\sin\theta - 1\right) = 0$$
$$\tan^2\theta = 0 \Rightarrow \tan\theta = 0 \text{ or } 2\sin\theta - 1 = 0 \Rightarrow 2\sin\theta = 1 \Rightarrow \sin\theta = \frac{1}{2}$$

Over the interval $[0°, 360°)$, the equation $\tan\theta = 0$ has two solutions. These are $0°$ and $180°$. In the same interval, the equation $\sin\theta = \frac{1}{2}$ has two solutions, the angles in quadrants I and II that have a reference angle of $30°$. These are $30°$ and $150°$.

Solution set: $\{0°, 30°, 150°, 180°\}$

32. $\sin^2\theta\cos^2\theta = 0$

$$\sin^2\theta\cos^2\theta = 0$$
$$\sin^2\theta = 0 \Rightarrow \sin\theta = 0 \text{ or } \cos^2\theta = 0 \Rightarrow \cos\theta = 0$$

Over the interval $[0°, 360°)$, the equation $\sin\theta = 0$ has two solutions. These are $0°$ and $180°$. In the same interval, the equation $\cos\theta = 0$ has two solutions. These are $90°$ and $270°$.

Solution set: $\{0°, 90°, 180°, 270°\}$

33. $\sec^2\theta\tan\theta = 2\tan\theta$

$$\sec^2\theta\tan\theta = 2\tan\theta \Rightarrow \sec^2\theta\tan\theta - 2\tan\theta = 0 \Rightarrow \tan\theta\left(\sec^2\theta - 2\right) = 0$$
$$\tan\theta = 0 \text{ or } \sec^2\theta - 2 = 0 \Rightarrow \sec^2\theta = 2 \Rightarrow \sec\theta = \pm\sqrt{2}$$

Over the interval $[0°, 360°)$, the equation $\tan\theta = 0$ has two solutions. These are $0°$ and $180°$. In the same interval, the equation $\sec\theta = \sqrt{2}$ has two solutions, the angles in quadrants I and IV that have a reference angle of $45°$. These are $45°$ and $315°$. Finally, the equation $\sec\theta = -\sqrt{2}$ has two solutions, the angles in quadrants II and III that have a reference angle of $45°$. These are $135°$ and $225°$.

Solution set: $\{0°, 45°, 135°, 180°, 225°, 315°\}$

34. $\cos^2\theta - \sin^2\theta = 0$

$$\cos^2\theta - \sin^2\theta = 0 \Rightarrow \cos^2\theta - \left(1-\cos^2\theta\right) = 0 \Rightarrow 2\cos^2\theta - 1 = 0 \Rightarrow 2\cos^2\theta = 1$$

$$\cos^2\theta = \frac{1}{2} \Rightarrow \cos\theta = \pm\sqrt{\frac{1}{2}} \Rightarrow \cos\theta = \pm\frac{\sqrt{2}}{2}$$

Over the interval $[0°, 360°)$, the equation $\cos\theta = \frac{\sqrt{2}}{2}$ has two solutions, the angles in quadrants I and III that have a reference angle of 45°. These are 45° and 225°. In the same interval, the equation $\cos\theta = -\frac{\sqrt{2}}{2}$ has two solutions, the angles in quadrants II and IV that have a reference angle of 45°. These are 135° and 315°.

Solution set: $\{45°, 135°, 225°, 315°\}$

For Exercises 35 -42, make sure your calculator is in degree mode.

35. $9\sin^2\theta - 6\sin\theta = 1$

$$9\sin^2\theta - 6\sin\theta = 1 \Rightarrow 9\sin^2\theta - 6\sin\theta - 1 = 0$$

We use the quadratic formula with $a = 9$, $b = -6$, and $c = -1$.

$$\sin\theta = \frac{6 \pm \sqrt{36-4(9)(-1)}}{2(9)} = \frac{6 \pm \sqrt{36+36}}{18} = \frac{6 \pm \sqrt{72}}{18} = \frac{6 \pm 6\sqrt{2}}{18} = \frac{1 \pm \sqrt{2}}{3}$$

Since $\sin\theta = \frac{1+\sqrt{2}}{3} > 0$ (and less than 1), we will obtain two angles. One angle will be in quadrant I and the other will be in quadrant II. Using a calculator, if $\sin\theta = \frac{1+\sqrt{2}}{3} \approx .80473787$, the quadrant I angle will be approximately 53.6°. The quadrant II angle will be approximately $180° - 53.6° = 126.4°$.

Since $\sin\theta = \frac{1-\sqrt{2}}{3} < 0$ (and greater than $-$ 1), we will obtain two angles. One angle will be in quadrant III and the other will be in quadrant IV. Using a calculator, if $\sin\theta = \frac{1-\sqrt{2}}{3} \approx -.13807119$, then $\theta \approx -7.9°$. Since this solution is not in the interval [0°, 360°), we must use it as a reference angle to find angles in the interval. Our reference angle will be 7.9°. The angle in quadrant III will be approximately $180° + 7.9° = 187.9°$. The angle in quadrant IV will be approximately $360° - 7.9° = 352.1°$.

Solution set: $\{53.6°, 126.4°, 187.9°, 352.1°\}$

36. $4\cos^2\theta + 4\cos\theta = 1$

$$4\cos^2\theta + 4\cos\theta = 1 \Rightarrow 4\cos^2\theta + 4\cos\theta - 1 = 0$$

We use the quadratic formula with $a = 4$, $b = 4$, and $c = -1$.

$$\cos\theta = \frac{-4 \pm \sqrt{4^2 - 4(4)(-1)}}{2(4)} = \frac{-4 \pm \sqrt{32}}{8} = \frac{-4 + 4\sqrt{2}}{8} = \frac{-1 \pm \sqrt{2}}{2}$$

$\frac{-1-\sqrt{2}}{2}$ is less than -1, which is an impossible value for the cosine function.

Since $\cos\theta = \frac{-1+\sqrt{2}}{2} > 0$ (and less than 1), we will obtain two angles. One angle will be in quadrant I and the other will be in quadrant IV. Using a calculator, if $\cos\theta = \frac{-1+\sqrt{2}}{2} \approx .20710678$, the quadrant I angle will be approximately $78.0°$. The quadrant IV angle will be approximately $360° - 78.0° = 282.0°$.

Solution set: $\{78.0°, 282.0°\}$

37. $\tan^2\theta + 4\tan\theta + 2 = 0$

We use the quadratic formula with $a = 1$, $b = 4$, and $c = 2$.

$$\tan\theta = \frac{-4 \pm \sqrt{16 - 4(1)(2)}}{2(1)} = \frac{-4 \pm \sqrt{16-8}}{2} = \frac{-4 \pm \sqrt{8}}{2} = \frac{-4 \pm 2\sqrt{2}}{2} = -2 \pm \sqrt{2}$$

Since $\tan\theta = -2 + \sqrt{2} < 0$, we will obtain two angles. One angle will be in quadrant II and the other will be in quadrant IV. Using a calculator, if $\tan\theta = -2 + \sqrt{2} = -.5857864$, then $\theta \approx -30.4°$. Since this solution is not in the interval $[0°, 360°)$, we must use it as a reference angle to find angles in the interval. Our reference angle will be $30.4°$. The angle in quadrant II will be approximately $180° - 30.4° = 149.6°$. The angle in quadrant IV will be approximately $360° - 30.4° = 329.6°$.

Since $\tan\theta = -2 - \sqrt{2} < 0$, we will obtain two angles. One angle will be in quadrant II and the other will be in quadrant IV. Using a calculator, if $\tan\theta = -2 - \sqrt{2} = -3.4142136$, then $\theta \approx -73.7°$. Since this solution is not in the interval $[0°, 360°)$, we must use it as a reference angle to find angles in the interval. Our reference angle will be $73.7°$. The angle in quadrant II will be approximately $180° - 73.7° = 106.3°$. The angle in quadrant IV will be approximately $360° - 73.7° = 286.3°$.

Solution set: $\{106.3°, 149.6°, 286.3°, 329.6°\}$

38. $3\cot^2\theta - 3\cot\theta - 1 = 0$

We use the quadratic formula with $a = 3$, $b = -3$, and $c = -1$.

$$\cot\theta = \frac{-(-3) \pm \sqrt{(-3)^2 - 4(3)(-1)}}{2(3)} = \frac{3 \pm \sqrt{9+12}}{6} = \frac{3 \pm \sqrt{21}}{6}$$

Since $\cot\theta = \frac{3+\sqrt{21}}{6} > 0$, we will obtain two angles. One angle will be in quadrant I and the other will be in quadrant III. Using a calculator, if $\cot\theta = \frac{3+\sqrt{21}}{6} \approx 1.2637626$, the quadrant I angle will be approximately 38.4°. The quadrant III angle will be approximately $180° + 38.4° = 218.4°$.

Since $\cot\theta = \frac{3-\sqrt{21}}{6} < 0$, we will obtain two angles. One angle will be in quadrant II and the other will be in quadrant IV. Using a calculator, if $\cot\theta = \frac{3-\sqrt{21}}{6} \approx -.26376262$, the quadrant II angle will be approximately 104.8°. (Note: You need to calculate $\tan^{-1}\left(\frac{1}{\frac{3-\sqrt{21}}{6}}\right) + 180$ to obtain this angle.)

The reference angle is $180° - 104.8° = 75.2°$. Thus, the quadrant IV angle will be approximately $360° - 75.2° = 284.8°$.

Solution set: $\{38.4°, 104.8°, 218.4°, 284.8°\}$

39. $\sin^2\theta - 2\sin\theta + 3 = 0$

We use the quadratic formula with $a = 1$, $b = -2$, and $c = 3$.

$$\sin\theta = \frac{2 \pm \sqrt{4-(4)(1)(3)}}{2(1)} = \frac{2 \pm \sqrt{4-12}}{2} = \frac{2 \pm \sqrt{-8}}{2} = \frac{2 \pm 2i\sqrt{2}}{2} = 1 \pm i\sqrt{2}$$

Since $1 \pm i\sqrt{2}$ is not a real number, the equation has no real solutions.

Solution set: $\varnothing$

40. $2\cos^2\theta + 2\cos\theta - 1 = 0$

We use the quadratic formula with $a = 2$, $b = 2$, and $c = -1$.

$$\cos\theta = \frac{-2 \pm \sqrt{2^2 - 4(2)(-1)}}{2(2)} = \frac{-2 \pm \sqrt{4+8}}{4} = \frac{-2 \pm \sqrt{12}}{4} = \frac{-2 \pm 2\sqrt{3}}{4} = \frac{-1 \pm \sqrt{3}}{2}$$

$\frac{-1-\sqrt{3}}{2}$ is less than –1, which is an impossible value for the cosine function.

Since $\cos\theta = \frac{-1+\sqrt{3}}{2} > 0$ (and less than 1), we will obtain two angles. One angle will be in quadrant I and the other will be in quadrant IV. Using a calculator, if $\cos\theta = \frac{-1+\sqrt{3}}{2} \approx .36602540$, the quadrant I angle will be approximately 68.5°. The quadrant IV angle will be approximately $360° - 68.5° = 291.5°$.

Solution set: $\{68.5°, 291.5°\}$

41. $\cot\theta + 2\csc\theta = 3$

$$\cot\theta + 2\csc\theta = 3 \Rightarrow \frac{\cos\theta}{\sin\theta} + \frac{2}{\sin\theta} = 3 \Rightarrow \cos\theta + 2 = 3\sin\theta \Rightarrow (\cos\theta + 2)^2 = (3\sin\theta)^2$$

$$\cos^2\theta + 4\cos\theta + 4 = 9\sin^2\theta \Rightarrow \cos^2\theta + 4\cos\theta + 4 = 9(1 - \cos^2\theta)$$

$$\cos^2\theta + 4\cos\theta + 4 = 9 - 9\cos^2\theta \Rightarrow 10\cos^2\theta + 4\cos\theta - 5 = 0$$

We use the quadratic formula with $a = 10$, $b = 4$, and $c = -5$.

$$\cos\theta = \frac{-4 \pm \sqrt{4^2 - 4(10)(-5)}}{2(10)} = \frac{-4 \pm \sqrt{16 + 200}}{20} = \frac{-4 \pm \sqrt{216}}{20} = \frac{-4 \pm 6\sqrt{6}}{20} = \frac{-2 \pm 3\sqrt{6}}{10}$$

Since $\cos\theta = \frac{-2 + 3\sqrt{6}}{10} > 0$ (and less than 1), we will obtain two angles. One angle will be in quadrant I and the other will be in quadrant IV. Using a calculator, if $\cos\theta = \frac{-2 + 3\sqrt{6}}{10} \approx .53484692$, the quadrant I angle will be approximately 57.7°. The quadrant IV angle will be approximately $360° - 57.7° = 302.3°$.

Since $\cos\theta = \frac{-2 - 3\sqrt{6}}{10} < 0$ (and greater than – 1), we will obtain two angles. One angle will be in quadrant II and the other will be in quadrant III. Using a calculator, if $\cos\theta = \frac{-2 - 3\sqrt{6}}{10} \approx -.93484692$, the quadrant II angle will be approximately 159.2°. The reference angle is $180° - 159.2° = 20.8°$. Thus, the quadrant III angle will be approximately $180° + 20.8° = 200.8°$.

Since the solution was found by squaring both sides of an equation, we must check that each proposed solution is a solution of the original equation. 302.3° and 200.8° do not satisfy our original equation. Thus, they are not elements of the solution set.

Solution set: $\{57.7°, 159.2°\}$

42. $2\sin\theta = 1 - 2\cos\theta$

$$2\sin\theta = 1 - 2\cos\theta \Rightarrow (2\sin\theta)^2 = (1 - 2\cos\theta)^2 \Rightarrow 4\sin^2\theta = 1 - 4\cos\theta + 4\cos^2\theta$$

$$4(1 - \cos^2\theta) = 1 - 4\cos\theta + 4\cos^2\theta \Rightarrow 4 - 4\cos^2\theta = 1 - 4\cos\theta + 4\cos^2\theta$$

$$0 = -3 - 4\cos\theta + 8\cos^2\theta$$

We use the quadratic formula with $a = 8$, $b = -4$, and $c = -3$.

$$\cos\theta = \frac{-(-4) \pm \sqrt{(-4)^2 - 4(8)(-3)}}{2(8)} = \frac{4 \pm \sqrt{16 + 96}}{16} = \frac{4 \pm \sqrt{112}}{16} = \frac{4 \pm 4\sqrt{7}}{16} = \frac{1 \pm \sqrt{7}}{4}$$

Continued on next page

42. (continued)

Since $\cos\theta = \frac{1+\sqrt{7}}{4} > 0$ (and less than 1), we will obtain two angles. One angle will be in quadrant I and the other will be in quadrant IV. Using a calculator, if $\cos\theta = \frac{1+\sqrt{7}}{4} \approx .91143783$, the quadrant I angle will be approximately 24.3°. The quadrant IV angle will be approximately $360° - 24.3° = 335.7°$.

Since $\cos\theta = \frac{1-\sqrt{7}}{4} < 0$ (and greater than − 1), we will obtain two angles. One angle will be in quadrant II and the other will be in quadrant III. Using a calculator, if $\cos\theta = \frac{1-\sqrt{7}}{4} \approx -.41143783$, the quadrant II angle will be approximately 114.3°. The reference angle is $180° - 114.3° = 65.7°$. Thus, the quadrant III angle will be approximately $180° + 65.7° = 245.7°$.

Since the solution was found by squaring both sides of an equation, we must check that each proposed solution is a solution of the original equation. 24.3° and 245.7° do not satisfy our original equation. Thus, they are not elements of the solution set.

Solution set: $\{114.3°, 335.7°\}$

In Exercises 43 – 46, if you are using a calculator, make sure it is in radian mode.

43. $3\sin^2 x - \sin x - 1 = 0$

We use the quadratic formula with $a = 3$, $b = -1$, and $c = -1$.

$$\sin x = \frac{-(-1) \pm \sqrt{(-1)^2 - 4(3)(-1)}}{2(3)} = \frac{1 \pm \sqrt{1+12}}{6} = \frac{1 \pm \sqrt{13}}{6}$$

Since $\sin x = \frac{1+\sqrt{13}}{6} > 0$ (and less than 1), we will obtain two angles. One angle will be in quadrant I and the other will be in quadrant II. Using a calculator, if $\sin x = \frac{1+\sqrt{13}}{6} \approx .76759188$, the quadrant I angle will be approximately .88. The quadrant II angle will be approximately $\pi - .88 \approx 2.26$.

Since $\sin x = \frac{1-\sqrt{13}}{6} < 0$ (and greater than − 1), we will obtain two angles. One angle will be in quadrant III and the other will be in quadrant IV. Using a calculator, if $\sin x = \frac{1-\sqrt{13}}{6} \approx -.43425855$, then $x \approx -.45$. Since this solution is not in the interval $[0, 2\pi)$, we must use it as a reference angle to find angles in the interval. Our reference angle will be .45. The angle in quadrant III will be approximately $\pi + .45 \approx 3.59$. The angle in quadrant IV will be approximately $2\pi - .45 \approx 5.83$.

Thus, the solutions are approximately $.9 + 2n\pi$, $2.3 + 2n\pi$, $3.6 + 2n\pi$, and $5.8 + 2n\pi$, where n is any integer.

44. $2\cos^2 x+\cos x=1$

$$2\cos^2 x+\cos x=1\Rightarrow 2\cos^2 x+\cos x-1=0\Rightarrow(2\cos x-1)(\cos x+1)=0$$

$$2\cos x-1=0\Rightarrow\cos x=\frac{1}{2}\text{ or }\cos x+1=0\Rightarrow\cos x=-1$$

Over the interval $[0,2\pi)$, the equation $\cos x=\frac{1}{2}$ has two solutions. The angles in quadrants I and IV that have a reference angle of $\frac{\pi}{3}$ are $\frac{\pi}{3}$ and $\frac{5\pi}{3}$. In the same interval, $\cos x=-1$ when the angle is π.

Thus, the solutions are $\frac{\pi}{3}+2n\pi,\ \pi+2n\pi$, and $\frac{5\pi}{3}+2n\pi$, , where n is any integer.

45. $4\cos^2 x-1=0$

$$4\cos^2 x-1=0\Rightarrow\cos^2 x=\frac{1}{4}\Rightarrow\cos x=\pm\frac{1}{2}$$

Over the interval $[0,2\pi)$, the equation $\cos x=\frac{1}{2}$ has two solutions. The angles in quadrants I and IV that have a reference angle of $\frac{\pi}{3}$ are $\frac{\pi}{3}$ and $\frac{5\pi}{3}$. In the same interval, $\cos x=-\frac{1}{2}$ has two solutions. The angles in quadrants II and III that have a reference angle of $\frac{\pi}{3}$ are $\frac{2\pi}{3}$ and $\frac{4\pi}{3}$.

Thus, the solutions are $\frac{\pi}{3}+2n\pi,\frac{2\pi}{3}+2n\pi,\ \frac{4\pi}{3}+2n\pi$, and $\frac{5\pi}{3}+2n\pi$, where n is any integer. This can also be written as $\frac{\pi}{3}+n\pi$ and $\frac{2\pi}{3}+n\pi$, where n is any integer.

46. $2\cos^2 x+5\cos x+2=0$

$$2\cos^2 x+5\cos x+2=0\Rightarrow(2\cos x+1)(\cos x+2)=0$$

$\cos x=-2$ is less than –1, which is an impossible value for the cosine function.

Over the interval $[0,2\pi)$, the equation $\cos x=-\frac{1}{2}$ has two solutions. The angles in quadrants II and III that have a reference angle of $\frac{\pi}{3}$ are $\frac{2\pi}{3}$ and $\frac{4\pi}{3}$.

Thus, the solutions are $\frac{2\pi}{3}+2n\pi$ and $\frac{4\pi}{3}+2n\pi$, where n is any integer.

In Exercises 47 – 50, if you are using a calculator, make sure it is in degree mode.

47. $5\sec^2\theta = 6\sec\theta$

$$5\sec^2\theta = 6\sec\theta \Rightarrow 5\sec^2\theta - 6\sec\theta = 0 \Rightarrow \sec\theta(5\sec\theta - 6) = 0$$

$$\sec\theta = 0 \text{ or } 5\sec\theta - 6 = 0 \Rightarrow \sec\theta = \frac{6}{5}$$

$\sec\theta = 0$ is an impossible values since the secant function must be either ≥ 1 or ≤ -1.

Since $\sec\theta = \frac{6}{5} > 1$, we will obtain two angles. One angle will be in quadrant I and the other will be in quadrant IV. Using a calculator, if $\sec\theta = \frac{6}{5} = 1.2$, the quadrant I angle will be approximately $33.6°$.

The quadrant IV angle will be approximately $360° - 33.6° = 326.4°$.

Thus, the solutions are approximately $33.6° + 360°n$ and $326.4° + 360°n$, where n is any integer.

48. $3\sin^2\theta - \sin\theta = 2$

$$3\sin^2\theta - \sin\theta = 2 \Rightarrow 3\sin^2\theta - \sin\theta - 2 = 0 \Rightarrow (3\sin\theta + 2)(\sin\theta - 1) = 0$$

$$3\sin\theta + 2 = 0 \Rightarrow \sin\theta = -\frac{2}{3} \text{ or } \sin\theta - 1 = 0 \Rightarrow \sin\theta = 1$$

$\sin\theta = 1$ when $\theta = 90°$.

Since $\sin\theta = -\frac{2}{3} < 0$ (and greater than – 1), we will obtain two angles. One angle will be in quadrant III and the other will be in quadrant IV. Using a calculator, if $\sin\theta = -\frac{2}{3} \approx -.66666667$, then $\theta \approx -41.8°$. Since this solution is not in the interval $[0°, 360°)$, we must use it as a reference angle to find angles in the interval. Our reference angle will be $41.8°$. The angle in quadrant III will be approximately $180° + 41.8° = 221.8°$. The angle in quadrant IV will be approximately $360° - 41.8° = 318.2°$.

Thus, the solutions are approximately $90° + 360°n$, $221.8° + 360°n$, and $318.2° + 360°n$, where n is any integer.

49. $\dfrac{2\tan\theta}{3 - \tan^2\theta} = 1$

$$\frac{2\tan\theta}{3 - \tan^2\theta} = 1 \Rightarrow 2\tan\theta = 3 - \tan^2\theta \Rightarrow \tan^2\theta + 2\tan\theta - 3 = 0 \Rightarrow (\tan\theta - 1)(\tan\theta + 3) = 0$$

$$\tan\theta - 1 = 0 \Rightarrow \tan\theta = 1 \text{ or } \tan\theta + 3 = 0 \Rightarrow \tan\theta = -3$$

Over the interval $[0°, 360°)$, the equation $\tan\theta = 1$ has two solutions $45°$ and $225°$. Over the same interval, the equation $\tan\theta = -3$ has two solutions that are approximately $-71.6° + 180° = 108.4°$ and $-71.6° + 360° = 288.4°$.

Thus, the solutions are $45° + 360°n$, $108.4° + 360°n$, $225° + 360°n$ and $288.4° + 360°n$, where n is any integer. Since the period of the tangent function is $180°$, the solutions can also be written as $45° + n \cdot 180°$ and $108° + n \cdot 180°$, where n is any integer.

50. $\sec^2\theta = 2\tan\theta + 4$

$$\sec^2\theta = 2\tan\theta + 4 \Rightarrow \tan^2\theta + 1 = 2\tan\theta + 4 \Rightarrow \tan^2\theta - 2\tan\theta - 3 = 0 \Rightarrow (\tan\theta - 3)(\tan\theta + 1) = 0$$

$$\tan\theta - 3 = 0 \Rightarrow \tan\theta = 3 \text{ or } \tan\theta + 1 = 0 \Rightarrow \tan\theta = -1$$

Over the interval $[0°, 360°)$, the equation $\tan\theta = -1$ has two solutions 135° and 315°. Over the same interval, the equation $\tan\theta = 3$ has two solutions that are approximately 71.6° and $180° + 71.6° = 251.6°$.

Thus, the solutions are $71.6° + 360°n$, $135° + 360°n$, $251.6° + 360°n$ and $315° + 360°n$, where n is any integer. Since the period of the tangent function is 180°, the solutions can also be written as $71.6° + n \cdot 180°$ and $135° + n \cdot 180°$, where n is any integer.

51. The x-intercept method is shown in the following windows.

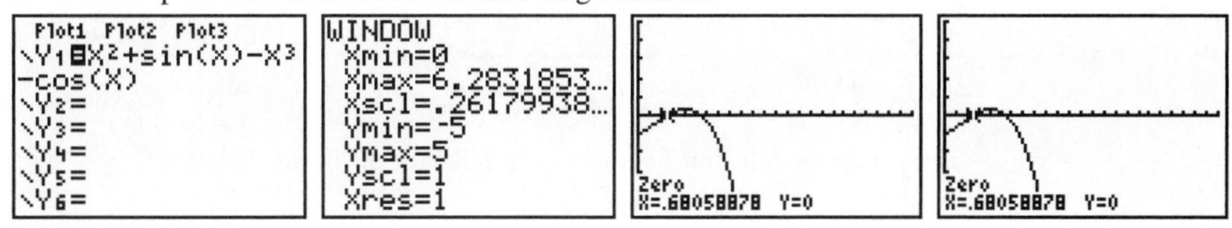

Solution set: $\{.6806, 1.4159\}$

52. The intersection method is shown in the following screens.

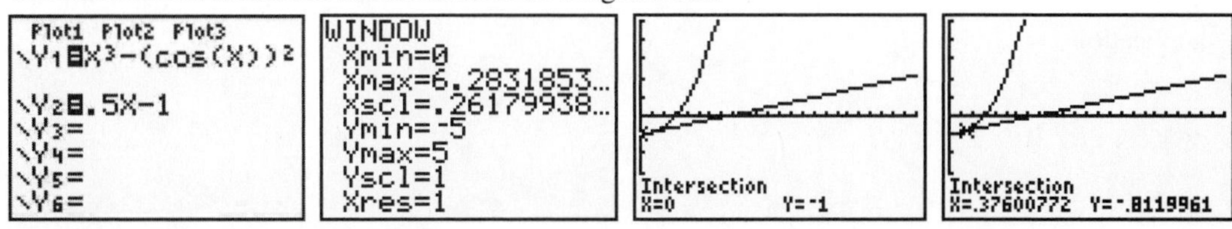

Solution set: $\{0, .3760\}$

53. $P = A\sin(2\pi ft + \phi)$

(a) $0 = .004\sin\left[2\pi(261.63)t + \frac{\pi}{7}\right] \Rightarrow 0 = \sin(1643.87t + .45)$

Since $1643.87t + .45 = n\pi$, we have $t = \frac{n\pi - .45}{1643.87}$, where n is any integer.

If $n = 0$, then $t \approx .000274$. If $n = 1$, then $t \approx .00164$. If $n = 2$, then $t \approx .00355$. If $n = 3$, then $t \approx .00546$. The only solutions for t in the interval [0, .005] are .00164 and .00355.

(b) We must solve the trigonometric equation $P = 0$ to determine when $P \leq 0$. From the graph we can estimate that $P \leq 0$ on the interval [.00164, .00355].

(c) $P < 0$ implies that there is a decrease in pressure so an eardrum would be vibrating outward.

54. $.342D\cos\theta + h\cos^2\theta = \dfrac{16D^2}{V_0^2}$

$V_0 = 60, D = 80, h = 2$

$$.342(80)\cos\theta + 2\cos^2\theta = \frac{16\cdot 80^2}{60^2} \Rightarrow 2\cos^2\theta + 27.36\cos\theta = \frac{256}{9} = 0 \Rightarrow \cos^2\theta + 13.68\cos\theta - \frac{128}{9} = 0$$

We use the quadratic formula with $a = 1$, $b = 13.68$, and $c = -\dfrac{128}{9}$.

$$\cos\theta = \frac{-13.68 \pm \sqrt{13.68^2 - 4\left(-\frac{128}{9}\right)}}{2(1)} = \frac{-13.68 \pm \sqrt{187.1424 + \frac{512}{9}}}{2} \approx \frac{-13.68 \pm 15.6215}{2}$$

$\dfrac{-13.68 - 15.6215}{2}$ is less than -1, which is an impossible value for the cosine function.

Since $\cos\theta = \dfrac{-13.68 + 15.6215}{2} > 0$ (and less than 1), we can obtain two angles. One angle will be in quadrant I and the other will be in quadrant IV. Using a calculator, if $\cos\theta \approx .97075$, the quadrant I angle will be approximately 14°. The quadrant IV angle, however, is not meaningful in this application.

55. $V = \cos 2\pi t, 0 \le t \le \dfrac{1}{2}$

(a) $V = 0,\ \cos 2\pi t = 0 \Rightarrow 2\pi t = \cos^{-1} 0 \Rightarrow 2\pi t = \dfrac{\pi}{2} \Rightarrow t = \dfrac{\frac{\pi}{2}}{2\pi} = \dfrac{1}{4}$ sec

(b) $V = .5,\ \cos 2\pi t = .5 \Rightarrow 2\pi t = \cos^{-1}(.5) \Rightarrow 2\pi t = \dfrac{\pi}{3} \Rightarrow t = \dfrac{\frac{\pi}{3}}{2\pi} = \dfrac{1}{6}$ sec

(c) $V = .25,\ \cos 2\pi t = .25 \Rightarrow 2\pi t = \cos^{-1}(.25) \Rightarrow 2\pi t \approx 1.3181161 \Rightarrow t \approx \dfrac{1.3181161}{2\pi} \approx .21$ sec

56. $e = 20\sin\left(\dfrac{\pi t}{4} - \dfrac{\pi}{2}\right)$

(a) $e = 0 \Rightarrow 0 = 20\sin\left(\dfrac{\pi t}{4} - \dfrac{\pi}{2}\right)$

Since $\arcsin 0 = 0$, solve the following equation.

$$\frac{\pi t}{4} - \frac{\pi}{2} = 0 \Rightarrow \frac{\pi t}{4} = \frac{\pi}{2} \Rightarrow t = 2 \text{ sec}$$

(b) $e = 10\sqrt{3} \Rightarrow 10\sqrt{3} = 20\sin\left(\dfrac{\pi t}{4} - \dfrac{\pi}{2}\right) \Rightarrow \dfrac{\sqrt{3}}{2} = \sin\left(\dfrac{\pi t}{4} - \dfrac{\pi}{2}\right)$

Since $\arcsin\dfrac{\sqrt{3}}{2} = \dfrac{\pi}{3}$, solve the following equation.

$$\frac{\pi t}{4} - \frac{\pi}{2} = \frac{\pi}{3} \Rightarrow \frac{\pi t}{4} = \frac{10\pi}{12} \Rightarrow t = \frac{10}{3} \text{ sec} = 3\frac{1}{3} \text{ sec}$$

57. $s(t) = \sin t + 2\cos t$

(a) $s(t)=\frac{2+\sqrt{3}}{2} \Rightarrow s(t)=\frac{2}{2}+\frac{\sqrt{3}}{2}=2\left(\frac{1}{2}\right)+\frac{\sqrt{3}}{2}=2\cos\left(\frac{\pi}{3}\right)+\sin\left(\frac{\pi}{3}\right)$

One such value is $\frac{\pi}{3}$.

(b) $s(t)=\frac{3\sqrt{2}}{2} \Rightarrow s(t)=\frac{2\sqrt{2}}{2}+\frac{\sqrt{2}}{2}=2\left(\frac{\sqrt{2}}{2}\right)+\frac{\sqrt{2}}{2}=2\cos\left(\frac{\pi}{4}\right)+\sin\left(\frac{\pi}{4}\right)$

One such value is $\frac{\pi}{4}$.

58. In the second line of the "solution", both sides of the equation were divided by $\sin x$. Instead of dividing by $\sin x$, one should have factored $\sin x$ from $\sin^2 x - \sin x$. In the process of dividing both sides by $\sin x$, the solutions of $x = 0$ and $x = \pi$ were eliminated.

Section 6.3: Trigonometric Equations II

1. Since $2x=\frac{2\pi}{3}, 2\pi, \frac{8\pi}{3} \Rightarrow x=\frac{2\pi}{6}, \frac{2\pi}{2}, \frac{8\pi}{6} \Rightarrow x=\frac{\pi}{3}, \pi, \frac{4\pi}{3}$, the solution set is $\left\{\frac{\pi}{3}, \pi, \frac{4\pi}{3}\right\}$.

2. Since $\frac{1}{2}x=\frac{\pi}{16}, \frac{5\pi}{12}, \frac{5\pi}{8} \Rightarrow x=\frac{2\pi}{16}, \frac{10\pi}{12}, \frac{10\pi}{8} \Rightarrow x=\frac{\pi}{8}, \frac{5\pi}{6}, \frac{5\pi}{4}$ the solution set is $\left\{\frac{\pi}{8}, \frac{5\pi}{6}, \frac{5\pi}{4}\right\}$.

3. Since $3\theta = 180°, 630°, 720°, 930° \Rightarrow \theta = \frac{180°}{3}, \frac{630°}{3}, \frac{720°}{3}, \frac{930°}{3} \Rightarrow \theta = 60°, 210°, 240°, 310°$ the solution set is $\{60°, 210°, 240°, 310°\}$.

4. Since $\frac{1}{3}\theta = 45°, 60°, 75°, 90° \Rightarrow \theta = 135°, 180°, 225°, 270°$, the solution set is $\{135°, 180°, 225°, 270°\}$.

5. $\frac{\tan 2\theta}{2} \neq \tan\theta$ for all values of θ.

6. If $\cot\frac{x}{2} - \csc\frac{x}{2} - 1 = 0$ has no solutions in the interval $[0, 2\pi)$, then the graph of $y = \cot\frac{x}{2} - \csc\frac{x}{2} - 1$ will have no x-intercepts in this same interval.

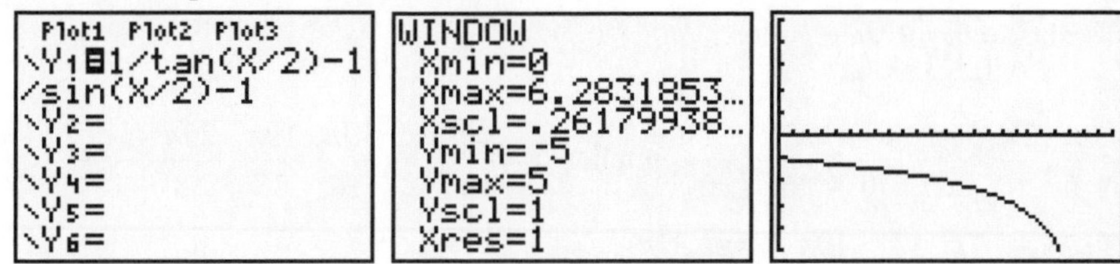

7. $\cos 2x = \frac{\sqrt{3}}{2}$

Since $0 \le x < 2\pi$, $0 \le 2x < 4\pi$. Thus, $2x = \frac{\pi}{6}, \frac{11\pi}{6}, \frac{13\pi}{6}, \frac{23\pi}{6} \Rightarrow x = \frac{\pi}{12}, \frac{11\pi}{12}, \frac{13\pi}{12}, \frac{23\pi}{12}$.

Solution set: $\frac{\pi}{12}, \frac{11\pi}{12}, \frac{13\pi}{12}, \frac{23\pi}{12}$

8. $\cos 2x = -\frac{1}{2}$

Since $0 \le x < 2\pi$, $0 \le 2x < 4\pi$. Thus, $2x = \frac{2\pi}{3}, \frac{4\pi}{3}, \frac{8\pi}{3}, \frac{10\pi}{3} \Rightarrow x = \frac{\pi}{3}, \frac{2\pi}{3}, \frac{4\pi}{3}, \frac{5\pi}{3}$.

Solution set: $\left\{\frac{\pi}{3}, \frac{2\pi}{3}, \frac{4\pi}{3}, \frac{5\pi}{3}\right\}$

9. $\sin 3x = -1$

Since $0 \le x < 2\pi$, $0 \le 3x < 6\pi$. Thus, $3x = \frac{3\pi}{2}, \frac{7\pi}{2}, \frac{11\pi}{2} \Rightarrow x = \frac{\pi}{2}, \frac{7\pi}{6}, \frac{11\pi}{6}$.

Solution set: $\left\{\frac{\pi}{2}, \frac{7\pi}{6}, \frac{11\pi}{6}\right\}$

10. $\sin 3x = 0$

Since $0 \le x < 2\pi$, $0 \le 3x < 6\pi$. Thus, $3x = 0, \pi, 2\pi, 3\pi, 4\pi, 5\pi \Rightarrow x = 0, \frac{\pi}{3}, \frac{2\pi}{3}, \pi, \frac{4\pi}{3}, \frac{5\pi}{3}$.

Solution set: $\left\{0, \frac{\pi}{3}, \frac{2\pi}{3}, \pi, \frac{4\pi}{3}, \frac{5\pi}{3}\right\}$

11. $3\tan 3x = \sqrt{3} \Rightarrow \tan 3x = \frac{\sqrt{3}}{3}$

Since $0 \le x < 2\pi$, $0 \le 3x < 6\pi$.

Thus, $3x = \frac{\pi}{6}, \frac{7\pi}{6}, \frac{13\pi}{6}, \frac{19\pi}{6}, \frac{25\pi}{6}, \frac{31\pi}{6}$ implies $x = \frac{\pi}{18}, \frac{7\pi}{18}, \frac{13\pi}{18}, \frac{19\pi}{18}, \frac{25\pi}{18}, \frac{31\pi}{18}$.

Solution set: $\left\{\frac{\pi}{18}, \frac{7\pi}{18}, \frac{13\pi}{18}, \frac{19\pi}{18}, \frac{25\pi}{18}, \frac{31\pi}{18}\right\}$

12. $\cot 3x = \sqrt{3}$

Since $0 \le x < 2\pi$, $0 \le 3x < 6\pi$.

Thus, $3x = \frac{\pi}{6}, \frac{7\pi}{6}, \frac{13\pi}{6}, \frac{19\pi}{6}, \frac{25\pi}{6}, \frac{31\pi}{6}$ implies $x = \frac{\pi}{18}, \frac{7\pi}{18}, \frac{13\pi}{18}, \frac{19\pi}{18}, \frac{25\pi}{18}, \frac{31\pi}{18}$.

Solution set: $\left\{\frac{\pi}{18}, \frac{7\pi}{18}, \frac{13\pi}{18}, \frac{19\pi}{18}, \frac{25\pi}{18}, \frac{31\pi}{18}\right\}$

13. $\sqrt{2}\cos 2x = -1 \Rightarrow \cos 2x = \frac{-1}{\sqrt{2}} = -\frac{\sqrt{2}}{2}$

Since $0 \le x < 2\pi$, $0 \le 2x < 4\pi$. Thus, $2x = \frac{3\pi}{4}, \frac{5\pi}{4}, \frac{11\pi}{4}, \frac{13\pi}{4} \Rightarrow x = \frac{3\pi}{8}, \frac{5\pi}{8}, \frac{11\pi}{8}, \frac{13\pi}{8}$.

Solution set: $\left\{\frac{3\pi}{8}, \frac{5\pi}{8}, \frac{11\pi}{8}, \frac{13\pi}{8}\right\}$

14. $2\sqrt{3}\sin 2x = \sqrt{3} \Rightarrow \sin 2x = \frac{1}{2}$

Since $0 \le x < 2\pi$, $0 \le 2x < 4\pi$. Thus, $2x = \frac{\pi}{6}, \frac{5\pi}{6}, \frac{13\pi}{6}, \frac{17\pi}{6} \Rightarrow x = \frac{\pi}{12}, \frac{5\pi}{12}, \frac{13\pi}{12}, \frac{17\pi}{12}$.

Solution set: $\left\{\frac{\pi}{12}, \frac{5\pi}{12}, \frac{13\pi}{12}, \frac{17\pi}{12}\right\}$

15. $\sin\frac{x}{2} = \sqrt{2} - \sin\frac{x}{2}$

$$\sin\frac{x}{2} = \sqrt{2} - \sin\frac{x}{2} \Rightarrow \sin\frac{x}{2} + \sin\frac{x}{2} = \sqrt{2} \Rightarrow 2\sin\frac{x}{2} = \sqrt{2} \Rightarrow \sin\frac{x}{2} = \frac{\sqrt{2}}{2}$$

Since $0 \le x < 2\pi$, $0 \le \frac{x}{2} < \pi$. Thus, $\frac{x}{2} = \frac{\pi}{4}, \frac{3\pi}{4} \Rightarrow x = \frac{\pi}{2}, \frac{3\pi}{2}$.

Solution set: $\left\{\frac{\pi}{2}, \frac{3\pi}{2}\right\}$

16. $\tan 4x = 0$

Since $0 \le x < 2\pi$, $0 \le 4x < 8\pi$.

Thus, $4x = 0, \pi, 2\pi, 3\pi, 4\pi, 5\pi, 6\pi, 7\pi$ implies $x = 0, \frac{\pi}{4}, \frac{\pi}{2}, \frac{3\pi}{4}, \pi, \frac{5\pi}{4}, \frac{3\pi}{2}, \frac{7\pi}{4}$.

Solution set: $\left\{0, \frac{\pi}{4}, \frac{\pi}{2}, \frac{3\pi}{4}, \pi, \frac{5\pi}{4}, \frac{3\pi}{2}, \frac{7\pi}{4}\right\}$

17. $\sin x = \sin 2x$

$$\sin x = \sin 2x \Rightarrow \sin x = 2\sin x\cos x \Rightarrow \sin x - 2\sin x\cos x = 0 \Rightarrow \sin x(1 - 2\cos x) = 0$$

Over the interval $[0, 2\pi)$, we have the following.

$$1 - 2\cos x = 0 \Rightarrow -2\cos x = -1 \Rightarrow \cos x = \frac{1}{2} \Rightarrow x = \frac{\pi}{3} \text{ or } \frac{5\pi}{3}$$

$$\sin x = 0 \Rightarrow x = 0 \text{ or } \pi$$

Solution set: $\left\{0, \frac{\pi}{3}, \pi, \frac{5\pi}{3}\right\}$

18. $\cos 2x - \cos x = 0$

We choose an identity for $\cos 2x$ that involves only the cosine function.

$$\cos 2x - \cos x = 0 \Rightarrow \left(2\cos^2 x - 1\right) - \cos x = 0 \Rightarrow 2\cos^2 x - \cos x - 1 = 0 \Rightarrow \left(2\cos x + 1\right)\left(\cos x - 1\right) = 0$$

$$2\cos x + 1 = 0 \text{ or } \cos x - 1 = 0$$

Over the interval $[0, 2\pi)$, we have the following.

$$2\cos x + 1 = 0 \Rightarrow 2\cos x = -1 \Rightarrow \cos x = -\frac{1}{2} \Rightarrow x = \frac{2\pi}{3} \text{ or } \frac{4\pi}{3}$$

$$\cos x - 1 = 0 \Rightarrow \cos x = 1 \Rightarrow x = 0$$

Solution set: $\left\{0, \frac{2\pi}{3}, \frac{4\pi}{3}\right\}$

19. $8\sec^2 \frac{x}{2} = 4 \Rightarrow \sec^2 \frac{x}{2} = \frac{1}{2} \Rightarrow \sec \frac{x}{2} = \pm\frac{\sqrt{2}}{2}$

Since $-\frac{\sqrt{2}}{2}$ is not in the interval $(-\infty, -1]$ and $\frac{\sqrt{2}}{2}$ is not in the interval $[1, \infty)$, this equation has no solution.

Solution set: $\varnothing$

20. $\sin^2 \frac{x}{2} - 2 = 0 \Rightarrow \sin^2 \frac{x}{2} = 2 \Rightarrow \sin \frac{x}{2} = \pm\sqrt{2}$

Since neither $-\sqrt{2}$ nor $\sqrt{2}$ are in the interval $[-1, 1]$, this equation has no solution.

Solution set: $\varnothing$

21. $\sin \frac{x}{2} = \cos \frac{x}{2}$

$$\sin \frac{x}{2} = \cos \frac{x}{2} \Rightarrow \sin^2 \frac{x}{2} = \cos^2 \frac{x}{2} \Rightarrow \sin^2 \frac{x}{2} = 1 - \sin^2 \frac{x}{2} \Rightarrow 2\sin^2 \frac{x}{2} = 1$$

$$\sin^2 \frac{x}{2} = \frac{1}{2} \Rightarrow \sin \frac{x}{2} = \pm\sqrt{\frac{1}{2}} \Rightarrow \sin \frac{x}{2} = \pm\frac{\sqrt{2}}{2}$$

Since $0 \le x < 2\pi$, $0 \le \frac{x}{2} < \pi$. If $\sin \frac{x}{2} = \frac{\sqrt{2}}{2}$, $\frac{x}{2} = \frac{\pi}{4}, \frac{3\pi}{4} \Rightarrow x = \frac{\pi}{2}, \frac{3\pi}{2}$. If $\sin \frac{x}{2} = -\frac{\sqrt{2}}{2}$, there are no solutions in the interval $[0, \pi)$.

Since the solution was found by squaring an equation, the proposed solutions must be checked.

Check $x = \frac{\pi}{2}$

$$\sin \frac{x}{2} = \cos \frac{x}{2}$$

$$\sin \frac{\frac{\pi}{2}}{2} = \cos \frac{\frac{\pi}{2}}{2} \ ?$$

$$\sin \frac{\pi}{4} = \cos \frac{\pi}{4} \ ?$$

$$\frac{\sqrt{2}}{2} = \frac{\sqrt{2}}{2} \quad \text{True}$$

$\frac{\pi}{2}$ is a solution.

Check $x = \frac{3\pi}{2}$

$$\sin \frac{x}{2} = \cos \frac{x}{2}$$

$$\sin \frac{\frac{3\pi}{2}}{2} = \cos \frac{\frac{3\pi}{2}}{2} \ ?$$

$$\sin \frac{3\pi}{4} = \cos \frac{3\pi}{4} \ ?$$

$$\frac{\sqrt{2}}{2} = -\frac{\sqrt{2}}{2} \quad \text{False}$$

$\frac{3\pi}{2}$ is not a solution.

Solution set: $\left\{\frac{\pi}{2}\right\}$

22. $\sec\frac{x}{2}=\cos\frac{x}{2}$

$$\sec\frac{x}{2}=\cos\frac{x}{2}\Rightarrow\frac{1}{\cos\frac{x}{2}}=\cos\frac{x}{2}\Rightarrow\cos^2\frac{x}{2}=1\Rightarrow\cos\frac{x}{2}=\pm 1$$

Since $0\le x<2\pi,\ 0\le\frac{x}{2}<\pi.$ Thus, $\frac{x}{2}=0\Rightarrow x=0.$

Solution set: $\{0\}$

23. $\cos 2x+\cos x=0$

We choose an identity for $\cos 2x$ that involves only the cosine function.

$$\cos 2x+\cos x=0\Rightarrow\left(2\cos^2 x-1\right)+\cos x=0\Rightarrow 2\cos^2 x+\cos x-1=0\Rightarrow\left(2\cos x-1\right)\left(\cos x+1\right)=0$$

$$2\cos x-1=0 \text{ or } \cos x+1=0$$

Over the interval $[0,2\pi)$, we have the following.

$$2\cos x-1=0\Rightarrow 2\cos x=1\Rightarrow\cos x=\frac{1}{2}\Rightarrow x=\frac{\pi}{3}\text{ or }\frac{5\pi}{3}$$

$$\cos x+1=0\Rightarrow\cos x=-1\Rightarrow x=\pi$$

Solution set: $\left\{\frac{\pi}{3},\ \pi,\ \frac{5\pi}{3}\right\}$

24. $\sin x\cos x=\frac{1}{4}$

$$\sin x\cos x=\frac{1}{4}\Rightarrow 2\sin x\cos x=\frac{1}{2}\Rightarrow\sin 2x=\frac{1}{2}$$

Since $0\le x<2\pi,\ 0\le 2x<4\pi.$ Thus, $2x=\frac{\pi}{6},\frac{5\pi}{6},\frac{13\pi}{6},\frac{17\pi}{6}\Rightarrow x=\frac{\pi}{12},\frac{5\pi}{12},\frac{13\pi}{12},\frac{17\pi}{12}.$

Solution set: $\left\{\frac{\pi}{12},\frac{5\pi}{12},\frac{13\pi}{12},\frac{17\pi}{12}\right\}$

25. $\sqrt{2}\sin 3\theta-1=0$

$$\sqrt{2}\sin 3\theta-1=0\Rightarrow\sqrt{2}\sin 3\theta=1\Rightarrow\sin 3\theta=\frac{1}{\sqrt{2}}\Rightarrow\sin 3\theta=\frac{\sqrt{2}}{2}$$

Since $0\le\theta<360°,\ \ 0°\le 3\theta<1080°.$ In quadrant I and II, sine is positive. Thus,

$3\theta=45°,135°,405°,495°,765°,855°\Rightarrow\theta=15°,45°,135°,165°,255°,285°.$

Solution set: $\{15°,45°,135°,165°,255°,285°\}$

26. $-2\cos 2\theta=\sqrt{3}$

$$-2\cos 2\theta=\sqrt{3}\Rightarrow\cos 2\theta=-\frac{\sqrt{3}}{2}$$

Since $0\le\theta<360°,\ \ 0°\le 2\theta<720°.$ In quadrant II and III, cosine is negative.

Thus, $2\theta=150°,210°,510°,570°\ \Rightarrow\theta=75°,105°,255°,285°.$

Solution set: $\{75°,105°,255°,285°\}$

27. $\cos\frac{\theta}{2}=1$

Since $0\le\theta<360°$, $0°\le\frac{\theta}{2}<180°$. Thus, $\frac{\theta}{2}=0°\Rightarrow\theta=0°$.

Solution set: $\{0°\}$

28. $\sin\frac{\theta}{2}=1$

Since $0\le\theta<360°$, $0°\le\frac{\theta}{2}<180°$. Thus, $\frac{\theta}{2}=90°\Rightarrow\theta=180°$.

Solution set: $\{180°\}$

29. $2\sqrt{3}\sin\frac{\theta}{2}=3$

$$2\sqrt{3}\sin\frac{\theta}{2}=3\Rightarrow\sin\frac{\theta}{2}=\frac{3}{2\sqrt{3}}\Rightarrow\sin\frac{\theta}{2}=\frac{3\sqrt{3}}{6}\Rightarrow\sin\frac{\theta}{2}=\frac{\sqrt{3}}{2}$$

Since $0\le\theta<360°$, $0°\le\frac{\theta}{2}<180°$. Thus, $\frac{\theta}{2}=60°,120°\Rightarrow\theta=120°,240°$.

Solution set: $\{120°,240°\}$

30. $2\sqrt{3}\cos\frac{\theta}{2}=-3$

$$2\sqrt{3}\cos\frac{\theta}{2}=-3\Rightarrow\cos\frac{\theta}{2}=\frac{-3}{2\sqrt{3}}\Rightarrow\cos\frac{\theta}{2}=-\frac{\sqrt{3}}{2}$$

Since $0\le\theta<360°$, $0°\le\frac{\theta}{2}<180°$. Thus, $\frac{\theta}{2}=150°\Rightarrow\theta=300°$.

Solution set: $\{300°\}$

31. $2\sin\theta=2\cos2\theta$

$$2\sin\theta=2\cos2\theta\Rightarrow\sin\theta=\cos2\theta\Rightarrow\sin\theta=1-2\sin^2\theta$$
$$2\sin^2\theta+\sin\theta-1=0\Rightarrow(2\sin\theta-1)(\sin\theta+1)=0$$
$$2\sin\theta-1=0\text{ or }\sin\theta+1=0$$

Over the interval $[0°,360°)$, we have the following.

$$2\sin\theta-1=0\Rightarrow2\sin\theta=1\Rightarrow\sin\theta=\frac{1}{2}\Rightarrow\theta=30°\text{ or }150°$$
$$\sin\theta+1=0\Rightarrow\sin\theta=-1\Rightarrow\theta=270°$$

Solution set: $\{30°,150°,270°\}$

32. $\cos\theta-1=\cos2\theta$

$$\cos\theta-1=\cos2\theta\Rightarrow\cos\theta-1=2\cos^2\theta-1\Rightarrow2\cos^2\theta-\cos\theta=0\Rightarrow\cos\theta(2\cos\theta-1)=0$$

Over the interval $[0°,360°)$, we have the following.

$$\cos\theta=0\Rightarrow\theta=90°\text{ or }270°$$
$$2\cos\theta-1=0\Rightarrow2\cos\theta=1\Rightarrow\cos\theta=\frac{1}{2}\Rightarrow\theta=60°\text{ or }300°$$

Solution set: $\{60°,90°,270°,300°\}$

In Exercises 33 – 40, we are to find all solutions.

33. $1-\sin\theta=\cos 2\theta$

$$1-\sin\theta=\cos 2\theta \Rightarrow 1-\sin\theta=1-2\sin^2\theta \Rightarrow 2\sin^2\theta-\sin\theta=0 \Rightarrow \sin\theta(2\sin\theta-1)=0$$

Over the interval $[0°, 360°)$, we have the following.

$$\sin\theta=0 \Rightarrow \theta=0° \text{ or } 180°$$

$$2\sin\theta-1=0 \Rightarrow \sin\theta=\frac{1}{2} \Rightarrow \theta=30° \text{ or } 150°$$

Solution set: $\{0°+360°n,\ 30°+360°n,\ 150°+360°n,\ 180°+360°n,\text{ where } n \text{ is any integer}\}$ or $\{180°n,\ 30°+360°n,\ 150°+360°n,\text{ where } n \text{ is any integer}\}$

34. $\sin 2\theta=2\cos^2\theta$

$$\sin 2\theta=2\cos^2\theta \Rightarrow 2\sin\theta\cos\theta=2\cos^2\theta \Rightarrow \cos^2\theta-\sin\theta\cos\theta=0 \Rightarrow \cos\theta(\cos\theta-\sin\theta)=0$$

Over the interval $[0°, 360°)$, we have the following.

$$\cos\theta=0 \Rightarrow \theta=90° \text{ or } 270°$$

$$\cos\theta-\sin\theta=0 \Rightarrow \cos\theta=\sin\theta \Rightarrow \frac{\sin\theta}{\cos\theta}=1 \Rightarrow \tan\theta=1 \Rightarrow \theta=45° \text{ or } 225°$$

Solution set: $\{45°+360°n,\ 90°+360°n,\ 225°+360°n,\ 270°+360°n,\text{ where } n \text{ is any integer}\}$ or $\{45°+360°n,\ 90°+180°n,\ 225°+360°n,\text{ where } n \text{ is any integer}\}$

35. $\csc^2\frac{\theta}{2}=2\sec\theta$

$$\csc^2\frac{\theta}{2}=2\sec\theta \Rightarrow \frac{1}{\sin^2\frac{\theta}{2}}=\frac{2}{\cos\theta} \Rightarrow 2\sin^2\frac{\theta}{2}=\cos\theta$$

$$2\left(\frac{1-\cos\theta}{2}\right)=\cos\theta \Rightarrow 1-\cos\theta=\cos\theta \Rightarrow 1=2\cos\theta \Rightarrow \cos\theta=\frac{1}{2}$$

Over the interval $[0°, 360°)$, we have the following.

$$\cos\theta=\frac{1}{2} \Rightarrow \theta=60° \text{ or } 300°$$

Solution set: $\{60°+360°n,\ 300°+360°n,\text{ where } n \text{ is any integer}\}$

36. $\cos\theta=\sin^2\frac{\theta}{2}$

$$\cos\theta=\sin^2\frac{\theta}{2} \Rightarrow \cos\theta=\frac{1-\cos\theta}{2} \Rightarrow 2\cos\theta=1-\cos\theta \Rightarrow 3\cos\theta=1 \Rightarrow \cos\theta=\frac{1}{3}$$

In quadrant I and IV, cosine is positive. Over the interval $[0°, 360°)$, we have the following.

$$\cos\theta=\frac{1}{3} \Rightarrow \theta\approx 70.5° \text{ or } 289.5°$$

Solution set: $\{70.5°+360°n,\ 289.5°+360°n,\text{ where } n \text{ is any integer}\}$

37. $2-\sin 2\theta = 4\sin 2\theta$

$$2-\sin 2\theta = 4\sin 2\theta \Rightarrow 2 = 5\sin 2\theta \Rightarrow \sin 2\theta = \frac{2}{5} \Rightarrow \sin 2\theta = .4$$

Since $0 \le \theta < 360°$, $0° \le 2\theta < 720°$. In quadrant I and II, sine is positive.

$$\sin 2\theta = .4 \Rightarrow 2\theta = 23.6°, 156.4°, 383.6°, 516.4°$$

Thus, $\theta = 11.8°, 78.2°, 191.8°, 258.2°$.

Solution set: $\{11.8° + 360°n,\ 78.2° + 360°n,\ 191.8° + 360°n,\ 258.2° + 360°n,$ where n is any integer$\}$
or $\{11.8° + 180°n,\ 78.2° + 180°n,$ where n is any integer$\}$

38. $4\cos 2\theta = 8\sin\theta\cos\theta$

$$4\cos 2\theta = 8\sin\theta\cos\theta \Rightarrow 4\cos 2\theta = 4(2\sin\theta\cos\theta) \Rightarrow 4\cos 2\theta = 4\sin 2\theta \Rightarrow \tan 2\theta = 1$$

Since $0 \le \theta < 360°$, $0° \le 2\theta < 720°$. In quadrant I and III, tangent is positive.

Thus, $2\theta = 45°, 225°, 405°, 585° \Rightarrow \theta = 22.5°, 112.5°, 202.5°, 292.5°$.

Solution set: $\{22.5° + 360°n,\ 112.5° + 360°n,\ 202.5° + 360°n,\ 292.5° + 360°n,$ where n is any integer$\}$
or $\{22.5° + 180°n,\ 112.5° + 180°n,$ where n is any integer$\}$

39. $2\cos^2 2\theta = 1 - \cos 2\theta$

$$2\cos^2 2\theta = 1-\cos 2\theta \Rightarrow 2\cos^2 2\theta + \cos 2\theta - 1 = 0 \Rightarrow (2\cos 2\theta - 1)(\cos 2\theta + 1) = 0$$

Since $0 \le \theta < 360° \Rightarrow 0° \le 2\theta < 720°$, we have the following.

$$2\cos 2\theta - 1 = 0 \Rightarrow 2\cos 2\theta = 1 \Rightarrow \cos 2\theta = \frac{1}{2}$$

$$2\theta = 60°,\ 300°,\ 420°,\ 660° \Rightarrow \theta = 30°,\ 150°,\ 210°,\ 330°$$

or

$$\cos 2\theta + 1 = 0 \Rightarrow \cos 2\theta = -1$$

$$2\theta = 180°,\ 540° \Rightarrow \theta = 90°,\ 270°$$

Solution set: $\{30° + 360°n,\ 90° + 360°n,\ 150° + 360°n,\ 210° + 360°n,\ 270° + 360°n,$
$330° + 360°n,$ where n is any integer$\}$ or
$\{30° + 180°n,\ 90° + 180°n,\ 150° + 180°n,$ where n is any integer$\}$

40. $\sin\theta - \sin 2\theta = 0$

$$\sin\theta - \sin 2\theta = 0 \Rightarrow \sin\theta - 2\sin\theta\cos\theta = 0 \Rightarrow \sin\theta(1 - 2\cos\theta) = 0$$

Over the interval $[0°, 360°)$, we have the following.

$$\sin\theta = 0 \Rightarrow \theta = 0° \text{ or } 180°$$

$$1 - 2\cos\theta = 0 \Rightarrow 2\cos\theta = 1 \Rightarrow \cos\theta = \frac{1}{2} \Rightarrow \theta = 60° \text{ or } 300°$$

Solution set: $\{0° + 360°n,\ 60° + 360°n,\ 180° + 360°n,\ 300° + 360°n,$ where n is any integer$\}$ or
$\{180°n,\ 60° + 360°n,\ 300° + 360°n,$ where n is any integer$\}$

41. The x-intercept method is shown in the following windows.

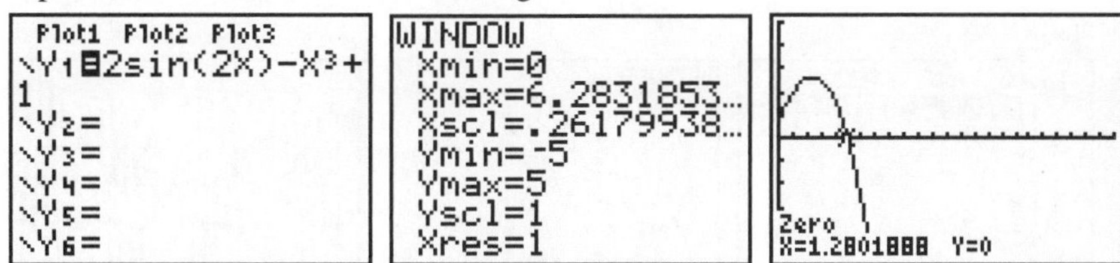

Solution set: $\{1.2802\}$

42. The intersection method is shown in the following screens.

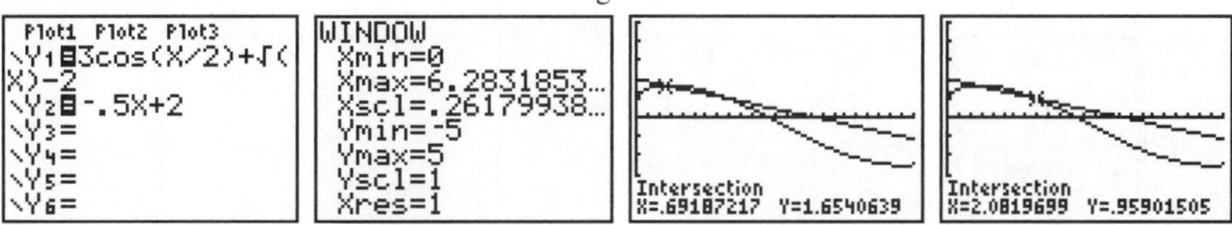

Solution set: $\{.6919, 2.0820\}$

43. **(a)**

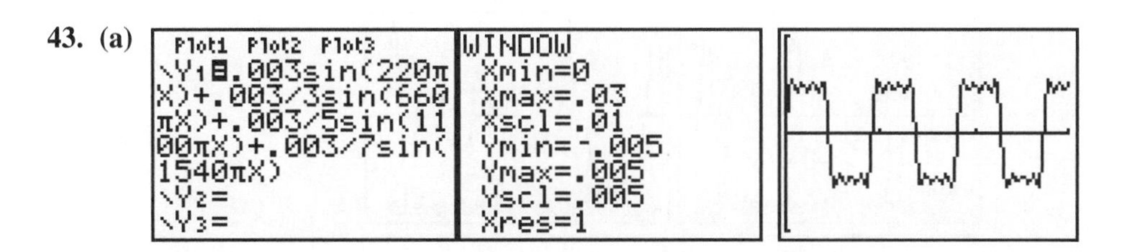

(b) The graph is periodic, and the wave has "jagged square" tops and bottoms.

(c) The eardrum is moving outward when $P < 0$.

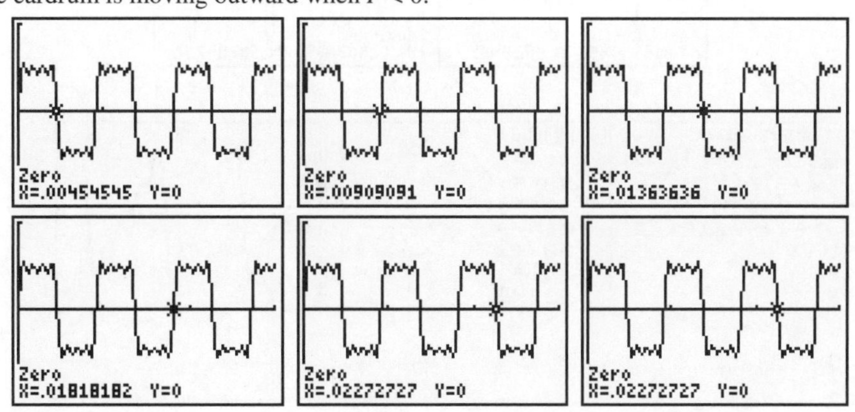

This occurs for the time intervals $(.0045,.0091),(.0136,.0182),(.0227,.0273)$.

44. **(a)** 3 beats per sec

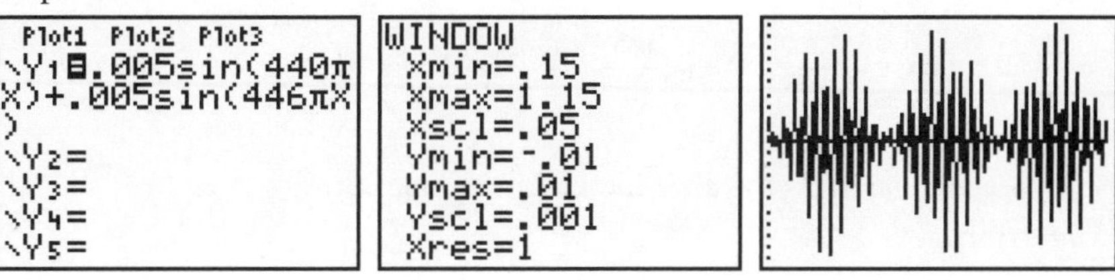

Continued on next page

44. (continued)

(b) 4 beats per sec

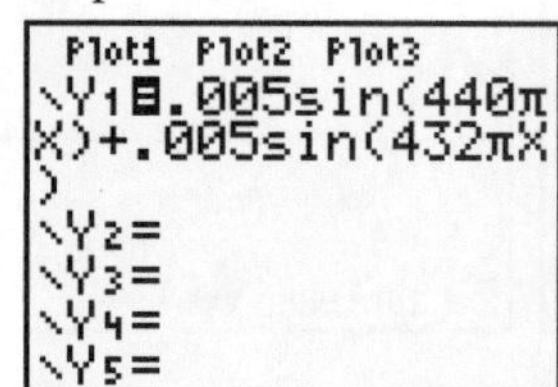

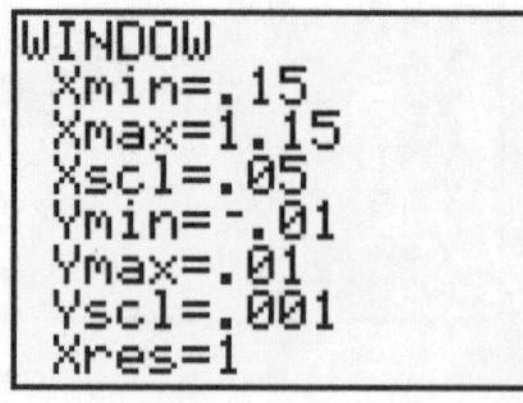

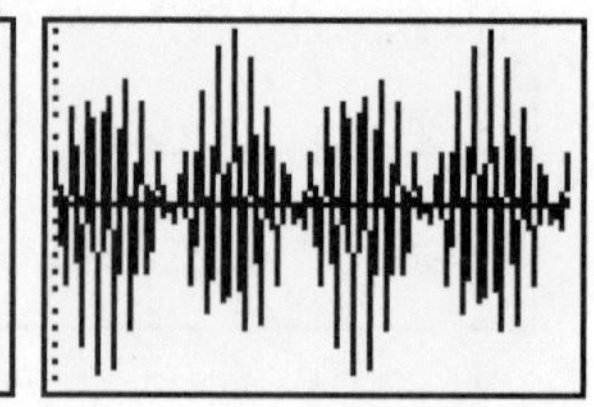

(c) The number of beats is equal to the absolute value of the difference in the frequencies of the two tones.

45. (a)

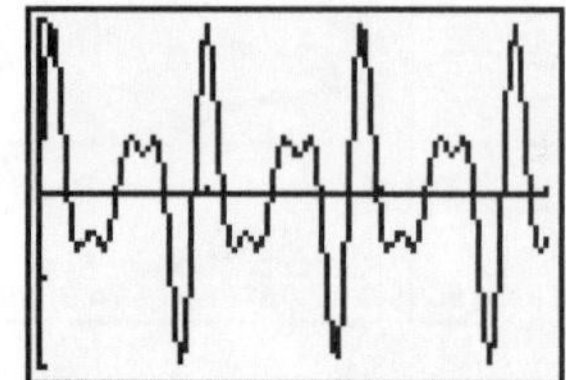

```
Plot1 Plot2 Plot3
\Y1■(1/2)sin(2π*
220X)+(1/3)sin(2
π*330X)+(1/4)sin
(2π*440X)
\Y2=
\Y3=
\Y4=
```

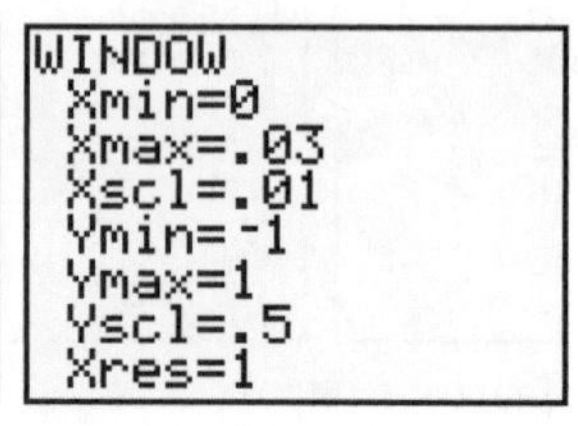

(b) .0007586, .009850, .01894, .02803

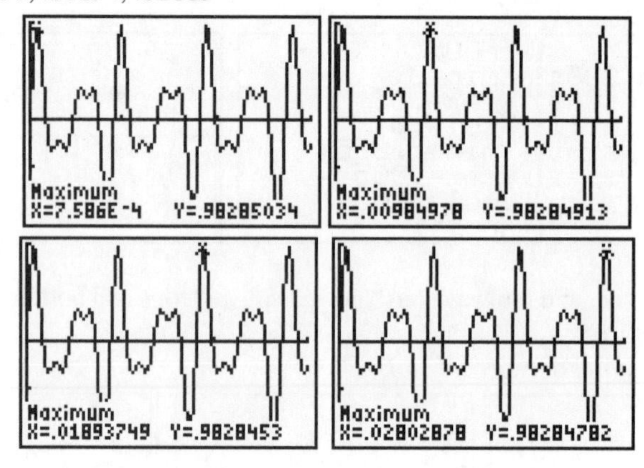

(c) 110 Hz

(d)

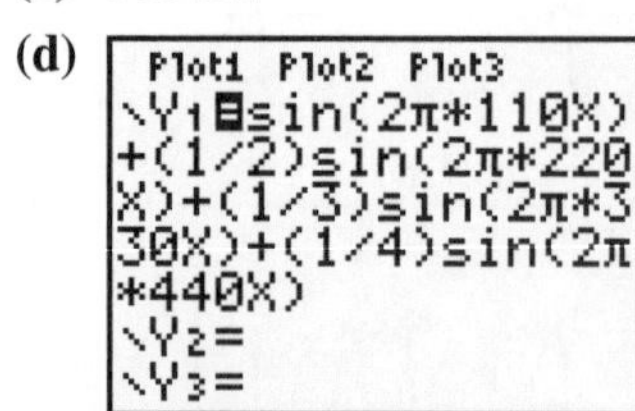


```
WINDOW
Xmin=0
Xmax=.03
Xscl=.01
Ymin=-2
Ymax=2
Yscl=1
Xres=1
```

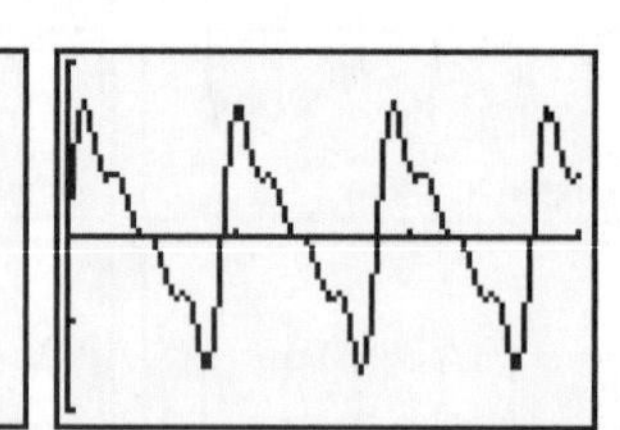

46. $h = \frac{35}{3} + \frac{7}{3}\sin\frac{2\pi x}{365}$

(a) Find x such that $h = 14$.

$$14 = \frac{35}{3} + \frac{7}{3}\sin\frac{2\pi x}{365} \Rightarrow 14 - \frac{35}{3} = \frac{7}{3}\sin\frac{2\pi x}{365} \Rightarrow \frac{7}{3} = \frac{7}{3}\sin\frac{2\pi x}{365} \Rightarrow \sin\frac{2\pi x}{365} = 1$$

$$\frac{2\pi x}{365} = \frac{\pi}{2} \pm 2\pi n \Rightarrow x = \left(\frac{\pi}{2} \pm 2\pi n\right)\left(\frac{365}{2\pi}\right) = \frac{365}{4} \pm 365n = 91.25 \pm 365n, \ n \text{ is any integer.}$$

$x = 91.25$ means about 91.3 days after March 21, on June 20.

Continued on next page

46. (continued)

(b) h assumes its least value when $\sin\frac{2\pi x}{365}$ takes on its least value, which is –1.

$$\sin\frac{2\pi x}{365} = -1 \Rightarrow \frac{2\pi x}{365} = \frac{3\pi}{2} \pm 2\pi n$$

$$x = \left(\frac{3\pi}{2} \pm 2\pi n\right)\left(\frac{365}{2\pi}\right) = \frac{3(365)}{4} \pm 365n = 273.75 \pm 365n, \text{ } n \text{ is any integer.}$$

$x = 273.75$ means about 273.8 days after March 21, on December 19.

(c) Let $h = 10$.

$$10 = \frac{35}{3} + \frac{7}{3}\sin\frac{2\pi x}{365} \Rightarrow 30 = 35 + 7\sin\frac{2\pi x}{365} \Rightarrow -5 = 7\sin\frac{2\pi x}{365}$$

$$-\frac{5}{7} = \sin\frac{2\pi x}{365} \Rightarrow -.71428571 = \sin\left(\frac{2\pi x}{365}\right)$$

In quadrant III and IV, sine is negative. In quadrant III, we have the following.

$$\frac{2\pi x}{365} = \pi + .79560295 \approx 3.9371956 \Rightarrow x \approx \frac{365}{2\pi}(3.9371956) \approx 228.7$$

$x = 228.7$ means 228.7 days after March 21, on November 4.
In quadrant IV, we have the following.

$$\frac{2\pi x}{365} \approx 2\pi - .79560295 \approx 5.4875823 \Rightarrow x \approx \frac{365}{2\pi}(5.4875823) \approx 318.8$$

$x = 318.8$ means about 318.8 days after March 21, on February 2.

47. $i = I_{max} \sin 2\pi ft$

Let $i = 40$, $I_{max} = 100$, $f = 60$.

$$40 = 100\sin\left[2\pi(60)t\right] \Rightarrow 40 = 100\sin 120\pi t \Rightarrow .4 = \sin 120\pi t$$

Using calculator, $120\pi t \approx .4115168 \Rightarrow t \approx \frac{.4115168}{120\pi} \Rightarrow t \approx .0010916 \Rightarrow t \approx .001\text{sec}$.

48. $i = I_{max} \sin 2\pi ft$

Let $i = 50$, $I_{max} = 100$, $f = 120$.

$$50 = 100\sin\left[2\pi(120)t\right] \Rightarrow 50 = 100\sin 240\pi t \Rightarrow \sin 240\pi t = \frac{1}{2}$$

$$240\pi t = \frac{\pi}{6} \Rightarrow t = \frac{1}{1440} \approx .0007 \text{ sec}$$

49. $i = I_{max} \sin 2\pi ft$

Let $i = I_{max}$, $f = 60$.

$$I_{max} = I_{max} \sin\left[2\pi(60)t\right] \Rightarrow 1 = \sin 120\pi t \Rightarrow 120\pi t = \frac{\pi}{2} \Rightarrow 120t = \frac{1}{2} \Rightarrow t = \frac{1}{240} \approx .004\text{sec}$$

50. $i = I_{max} \sin 2\pi ft$

Let $i = \frac{1}{2}I_{max}$, $f = 60$.

$$\frac{1}{2}I_{max} = I_{max} \sin\left[2\pi(60)t\right] \Rightarrow \frac{1}{2} = \sin 120\pi t \Rightarrow 120\pi t = \frac{\pi}{6} \Rightarrow t = \frac{1}{720} \approx .0014 \text{ sec}$$

Section 6.4: Equations Involving Inverse Trigonometric Functions

1. Since $\arcsin 0 = 0$, the correct choice is C.

2. Since $\arctan 1 = \frac{\pi}{4}$, the correct choice is A.

3. Since $\arccos\left(-\frac{\sqrt{2}}{2}\right) = \frac{3\pi}{4}$, the correct choice is C.

4. Since $\arcsin\left(-\frac{1}{2}\right) = -\frac{\pi}{6}$, the correct choice is C.

5. $y = 5\cos x \Rightarrow \frac{y}{5} = \cos x \Rightarrow x = \arccos\frac{y}{5}$

6. $4y = \sin x \Rightarrow x = \arcsin 4y$

7. $2y = \cot 3x \Rightarrow 3x = \text{arccot } 2y \Rightarrow x = \frac{1}{3}\text{arccot } 2y$

8. $6y = \frac{1}{2}\sec x \Rightarrow 12y = \sec x \Rightarrow x = \text{arcsec } 12y$

9. $y = 3\tan 2x \Rightarrow \frac{y}{3} = \tan 2x \Rightarrow 2x = \arctan\frac{y}{3} \Rightarrow x = \frac{1}{2}\arctan\frac{y}{3}$

10. $y = 3\sin\frac{x}{2} \Rightarrow \frac{y}{3} = \sin\frac{x}{2} \Rightarrow \frac{x}{2} = \arcsin\frac{y}{3} \Rightarrow x = 2\arcsin\frac{y}{3}$

11. $y = 6\cos\frac{x}{4} \Rightarrow \frac{y}{6} = \cos\frac{x}{4} \Rightarrow \frac{x}{4} = \arccos\frac{y}{6} \Rightarrow x = 4\arccos\frac{y}{6}$

12. $y = -\sin\frac{x}{3} \Rightarrow \sin\frac{x}{3} = -y \Rightarrow \frac{x}{3} = \arcsin(-y) \Rightarrow x = 3\arcsin(-y)$

13. $y = -2\cos 5x \Rightarrow -\frac{y}{2} = \cos 5x \Rightarrow 5x = \arccos\left(-\frac{y}{2}\right) \Rightarrow x = \frac{1}{5}\arccos\left(-\frac{y}{2}\right)$

14. $y = 3\cot 5x \Rightarrow \cot 5x = \frac{y}{3} \Rightarrow 5x = \text{arccot}\frac{y}{3} \Rightarrow x = \frac{1}{5}\text{arccot}\frac{y}{3}$

15. $y = \cos(x+3) \Rightarrow x+3 = \arccos y \Rightarrow x = -3 + \arccos y$

16. $y = \tan(2x-1) \Rightarrow 2x-1 = \arctan y \Rightarrow 2x = 1 + \arctan y \Rightarrow x = \frac{1}{2}(1 + \arctan y)$

17. $y = \sin x - 2 \Rightarrow y + 2 = \sin x \Rightarrow x = \arcsin(y+2)$

18. $y=\cot x+1 \Rightarrow \cot x = y-1 \Rightarrow x=\text{arccot}\left(y-1\right)$

19. $y=2\sin x-4 \Rightarrow y+4=2\sin x \Rightarrow \dfrac{y+4}{2}=\sin x \Rightarrow x=\arcsin\left(\dfrac{y+4}{2}\right)$

20. $y=4+3\cos x \Rightarrow y-4=3\cos x \Rightarrow \dfrac{y-4}{3}=\cos x \Rightarrow x=\arccos\left(\dfrac{y-4}{3}\right)$

21. Firstly, $\sin x-2 \neq \sin\left(x-2\right)$. If you think of the graph of $y=\sin x-2$, this represents the graph of $f\left(x\right)=\sin x$, shifted 2 units down. If you think of the graph of $y=\sin\left(x-2\right)$, this represents the graph of $f\left(x\right)=\sin x$, shifted 2 units right.

22. $\cos^{-1} 2$ doesn't exist since there is no value x such that $\cos x = 2$.

23. $\dfrac{4}{3}\cos^{-1}\dfrac{y}{4}=\pi \Rightarrow \cos^{-1}\dfrac{y}{4}=\dfrac{3\pi}{4} \Rightarrow \dfrac{y}{4}=\cos\dfrac{3\pi}{4} \Rightarrow \dfrac{y}{4}=-\dfrac{\sqrt{2}}{2} \Rightarrow y=-2\sqrt{2}$

Solution set: $\left\{-2\sqrt{2}\right\}$

24. $4\pi+4\tan^{-1} y=\pi \Rightarrow 4\tan^{-1} y=-3\pi \Rightarrow \tan^{-1} y=-\dfrac{3\pi}{4}$

The range of $\tan^{-1} y$ is $-\dfrac{\pi}{2}<y<\dfrac{\pi}{2}$. Since $-\dfrac{3\pi}{4}$ is not in this interval, the equation has no solution.

Solution set: $\varnothing$

25. $2\arccos\left(\dfrac{y-\pi}{3}\right)=2\pi$

$$2\arccos\left(\frac{y-\pi}{3}\right)=2\pi \Rightarrow \arccos\left(\frac{y-\pi}{3}\right)=\pi \Rightarrow \frac{y-\pi}{3}=\cos\pi$$

$$\frac{y-\pi}{3}=-1 \Rightarrow y-\pi=-3 \Rightarrow y=\pi-3$$

Solution set: $\left\{\pi-3\right\}$

26. $\arccos\left(y-\dfrac{\pi}{3}\right)=\dfrac{\pi}{6} \Rightarrow y-\dfrac{\pi}{3}=\cos\dfrac{\pi}{6} \Rightarrow y=\dfrac{\sqrt{3}}{2}+\dfrac{\pi}{3} \Rightarrow y=\dfrac{3\sqrt{3}+2\pi}{6}$

Solution set: $\left\{\dfrac{3\sqrt{3}+2\pi}{6}\right\}$

27. $\arcsin x = \arctan \frac{3}{4}$

Let $\arctan \frac{3}{4} = u$, so $\tan u = \frac{3}{4}$, u is in quadrant I. Sketch a triangle and label it. The hypotenuse is $\sqrt{3^2 + 4^2} = \sqrt{9+16} = \sqrt{25} = 5.$

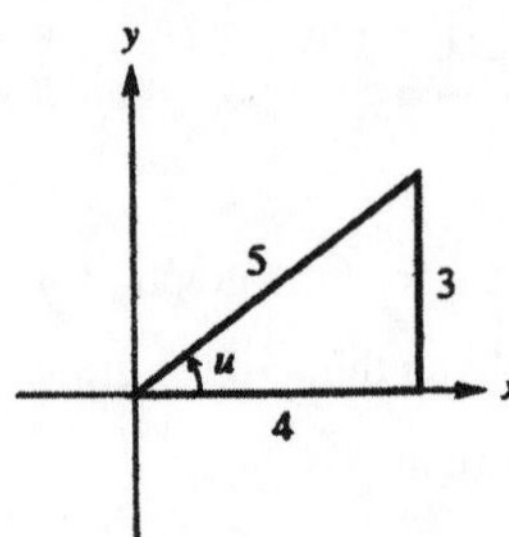

Therefore, $\sin u = \frac{3}{r} = \frac{3}{5}$. This equation becomes $\arcsin x = u$, or $x = \sin u$. Thus, $x = \frac{3}{5}$.

Solution set: $\left\{\frac{3}{5}\right\}$

28. $\arctan x = \arccos \frac{5}{13}$

Let $\arccos \frac{5}{13} = u$, so $\cos u = \frac{5}{13}$. Sketch a triangle and label it. The side opposite angle u is $\sqrt{13^2 - 5^2} = \sqrt{169 - 25} = \sqrt{144} = 12.$

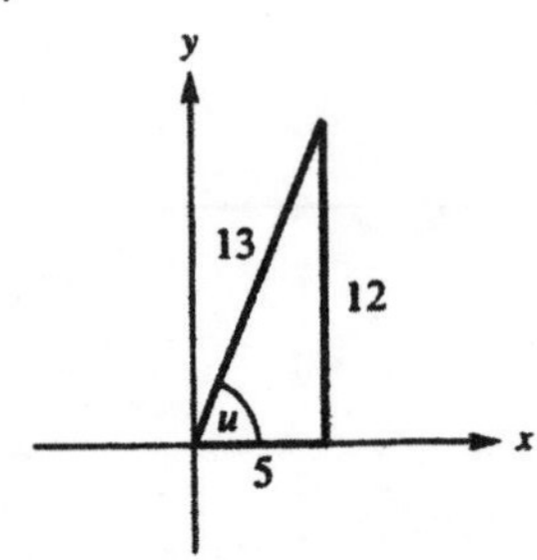

Therefore, $\tan u = \frac{12}{5}$. The equation becomes $\arctan x = u$, or $x = \tan u$. Thus, $x = \frac{12}{5}$.

Solution set: $\left\{\frac{12}{5}\right\}$

29. $\cos^{-1} x = \sin^{-1} \frac{3}{5}$

Let $\sin^{-1} \frac{3}{5} = u$, so $\sin u = \frac{3}{5}$. Sketch a triangle and label it. The hypotenuse is $\sqrt{3^2 + 4^2} = \sqrt{9+16} = \sqrt{25} = 5.$

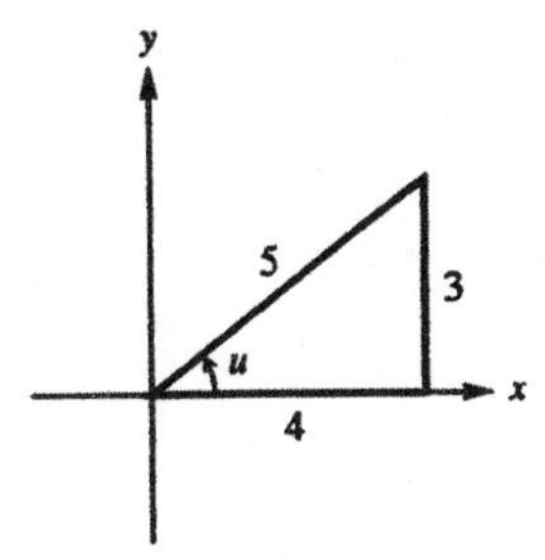

Therefore, $\cos u = \frac{4}{5}$. The equation becomes $\cos^{-1} x = u$, or $x = \cos u$. Thus, $x = \frac{4}{5}$.

Solution set: $\left\{\frac{4}{5}\right\}$

30. $\cot^{-1} x = \tan^{-1} \frac{4}{3}$

Let $\tan^{-1} \frac{4}{3} = u$, so $\tan u = \frac{4}{3}$. Sketch a triangle and label it. The hypotenuse is $\sqrt{3^2 + 4^2} = \sqrt{9+16} = \sqrt{25} = 5.$

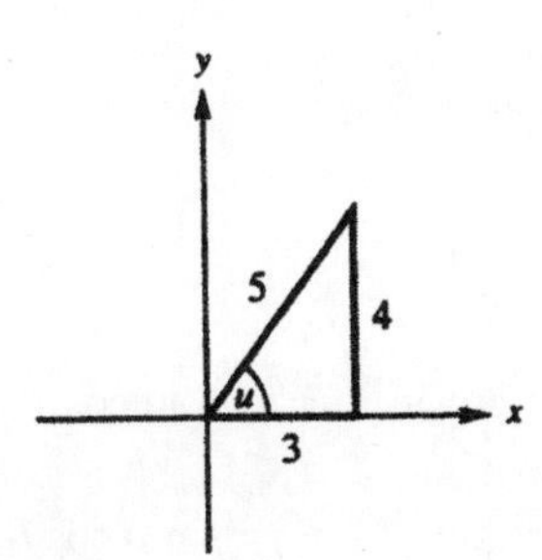

Therefore, $\cot u = \frac{3}{4}$. The equation becomes $\cot^{-1} x = u$, or $x = \cot u$. Thus, $x = \frac{3}{4}$.

Solution set: $\left\{\frac{3}{4}\right\}$

31. $\sin^{-1} x - \tan^{-1} 1 = -\frac{\pi}{4}$

$\sin^{-1} x - \tan^{-1} 1 = -\frac{\pi}{4} \Rightarrow \sin^{-1} x = \tan^{-1} 1 - \frac{\pi}{4} \Rightarrow \sin^{-1} x = \frac{\pi}{4} - \frac{\pi}{4} \Rightarrow \sin^{-1} x = 0 \Rightarrow \sin 0 = x \Rightarrow x = 0$

Solution set: $\{0\}$

32. $\sin^{-1} x + \tan^{-1} \sqrt{3} = \frac{2\pi}{3}$

$$\sin^{-1} x + \tan^{-1} \sqrt{3} = \frac{2\pi}{3} \Rightarrow \sin^{-1} x + \frac{\pi}{3} = \frac{2\pi}{3} \Rightarrow \sin^{-1} x = \frac{\pi}{3} \Rightarrow x = \sin \frac{\pi}{3} \Rightarrow x = \frac{\sqrt{3}}{2}$$

Solution set: $\left\{ \frac{\sqrt{3}}{2} \right\}$

33. $\arccos x + 2 \arcsin \frac{\sqrt{3}}{2} = \pi$

$$\arccos x + 2 \arcsin \frac{\sqrt{3}}{2} = \pi \Rightarrow \arccos x = \pi - 2 \arcsin \frac{\sqrt{3}}{2} \Rightarrow \arccos x = \pi - 2\left(\frac{\pi}{3} \right)$$

$$\arccos x = \pi - \frac{2\pi}{3} \Rightarrow \arccos x = \frac{\pi}{3} \Rightarrow x = \cos \frac{\pi}{3} \Rightarrow x = \frac{1}{2}$$

Solution set: $\left\{ \frac{1}{2} \right\}$

34. $\arccos x + 2 \arcsin \frac{\sqrt{3}}{2} = \frac{\pi}{3}$

$$\arccos x + 2 \arcsin \frac{\sqrt{3}}{2} = \frac{\pi}{3} \Rightarrow \arccos x + 2\left(\frac{\pi}{3} \right) = \frac{\pi}{3} \Rightarrow \arccos x = -\frac{\pi}{3}$$

$-\frac{\pi}{3}$ is not in the range of $\arccos x$. Therefore, the equation has no solution.

Solution set: $\varnothing$

35. $\arcsin 2x + \arccos x = \frac{\pi}{6}$

$$\arcsin 2x + \arccos x = \frac{\pi}{6} \Rightarrow \arcsin 2x = \frac{\pi}{6} - \arccos x \Rightarrow 2x = \sin\left(\frac{\pi}{6} - \arccos x \right)$$

Use the identity $\sin(A - B) = \sin A \cos B - \cos A \sin B$.

$$2x = \sin \frac{\pi}{6} \cos(\arccos x) - \cos \frac{\pi}{6} \sin(\arccos x)$$

Let $u = \arccos x$. Thus, $\cos u = x = \frac{x}{1}$.

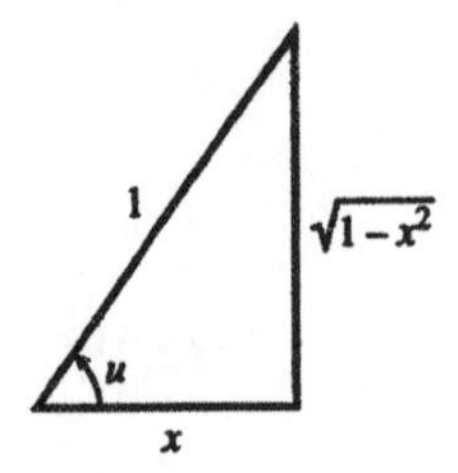

$\sin u = \sqrt{1 - x^2}$

Continued on next page

35. (continued)

$$2x = \sin\frac{\pi}{6}\cdot\cos u - \cos\frac{\pi}{6}\sin u \Rightarrow 2x = \frac{1}{2}x - \frac{\sqrt{3}}{2}\left(\sqrt{1-x^2}\right) \Rightarrow 4x = x - \sqrt{3}\cdot\sqrt{1-x^2}$$

$$3x = -\sqrt{3}\cdot\sqrt{1-x^2} \Rightarrow (3x)^2 = \left(-\sqrt{3}\cdot\sqrt{1-x^2}\right)^2 \Rightarrow 9x^2 = 3\left(1-x^2\right)$$

$$9x^2 = 3 - 3x^2 \Rightarrow 12x^2 = 3 \Rightarrow x^2 = \frac{3}{12} = \frac{1}{4} \Rightarrow x = \pm\frac{1}{2}$$

Check these proposed solutions since they were found by squaring both side of an equation.

Check $x = \frac{1}{2}$.

$$\arcsin 2x + \arccos x = \frac{\pi}{6}$$

$$\arcsin\left(2\cdot\frac{1}{2}\right) + \arccos\left(\frac{1}{2}\right) = \frac{\pi}{6}\ ?$$

$$\frac{\pi}{2} + \frac{\pi}{3} = \frac{\pi}{6}\ ?$$

$$\frac{5\pi}{6} = \frac{\pi}{6} \quad \text{False}$$

$\frac{1}{2}$ is not a solution.

Check $x = -\frac{1}{2}$.

$$\arcsin 2x + \arccos x = \frac{\pi}{6}$$

$$\arcsin\left(2\cdot -\frac{1}{2}\right) + \arccos\left(-\frac{1}{2}\right) = \frac{\pi}{6}\ ?$$

$$-\frac{\pi}{2} + \frac{2\pi}{3} = \frac{\pi}{6}\ ?$$

$$\frac{\pi}{6} = \frac{\pi}{6} \quad \text{True}$$

$-\frac{1}{2}$ is a solution.

Solution set: $\left\{-\frac{1}{2}\right\}$

36. $\arcsin 2x + \arcsin x = \frac{\pi}{2}$

$$\arcsin 2x + \arcsin x = \frac{\pi}{2} \Rightarrow \arcsin 2x = \frac{\pi}{2} - \arcsin x \Rightarrow 2x = \sin\left(\frac{\pi}{2} - \arcsin x\right)$$

Use the identity $\sin(A - B) = \sin A\cos B - \cos A\sin B$.

$$2x = \sin\frac{\pi}{2}\cos(\arcsin x) - \cos\frac{\pi}{2}\sin(\arcsin x)$$

$$2x = 1\cdot\cos(\arcsin x) - 0\cdot\sin(\arcsin x) \Rightarrow 2x = \cos(\arcsin x)$$

Let $\arcsin x = u$, so $\sin u = x = \frac{x}{1}$.

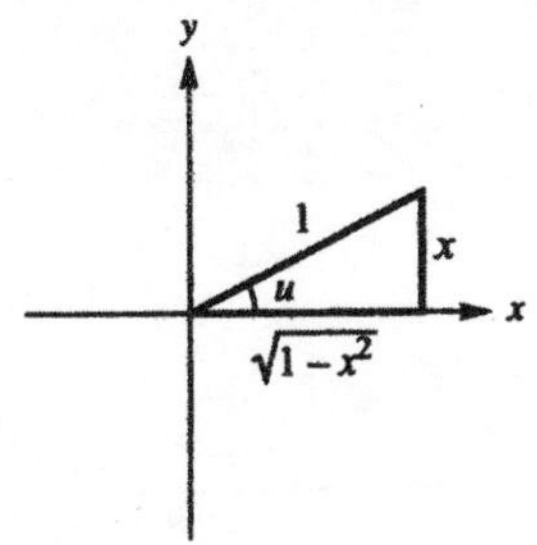

$$\cos u = \sqrt{1-x^2}$$

Continued on next page

36. (continued)

Substitute $\sqrt{1-x^2}$ for $\cos(\arcsin x)$ to obtain the following.

$$2x=\sqrt{1-x^2} \Rightarrow 4x^2=1-x^2 \Rightarrow 5x^2=1 \Rightarrow x^2=\frac{1}{5} \Rightarrow x=\pm\frac{\sqrt{5}}{5}$$

Check these proposed solutions since they were found by squaring an equation.

$$x=\frac{\sqrt{5}}{5}\text{: Since } \arcsin\left(2\cdot\frac{\sqrt{5}}{5}\right)+\arcsin\left(\frac{\sqrt{5}}{5}\right)=\frac{\pi}{2},\ \frac{\sqrt{5}}{5} \text{ is a solution.}$$

$$x=-\frac{\sqrt{5}}{5}\text{: Since } \arcsin\left(2\cdot\frac{-\sqrt{5}}{5}\right)+\arcsin\left(-\frac{\sqrt{5}}{5}\right)=-\frac{\pi}{2},\ \frac{-\sqrt{5}}{5} \text{ is not a solution}$$

Solution set: $\left\{\frac{\sqrt{5}}{5}\right\}$

37. $\cos^{-1} x+\tan^{-1} x=\frac{\pi}{2}$

$$\cos^{-1} x+\tan^{-1} x=\frac{\pi}{2} \Rightarrow \cos^{-1} x=\frac{\pi}{2}-\tan^{-1} x \Rightarrow x=\cos\left(\frac{\pi}{2}-\tan^{-1} x\right)$$

Use the identity cos(*A* – *B*) = cos *A* cos *B* + sin *A* sin *B*.

$$x=\cos\frac{\pi}{2}\cos\left(\tan^{-1} x\right)+\sin\frac{\pi}{2}\sin\left(\tan^{-1} x\right) \Rightarrow x=0\cdot\cos\left(\tan^{-1} x\right)+1\cdot\sin\left(\tan^{-1} x\right) \Rightarrow x=\sin\left(\tan^{-1} x\right)$$

Let $u=\tan^{-1} x$. So, $\tan u=x$.

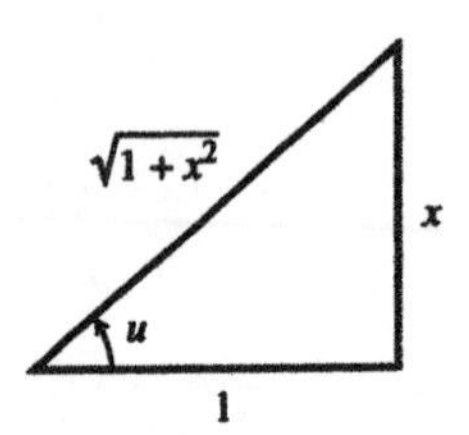

From the triangle, we find $\sin u=\frac{x}{\sqrt{1+x^2}}$, so the equation $x=\sin\left(\tan^{-1} x\right)$ becomes $x=\frac{x}{\sqrt{1+x^2}}$.

Solve this equation.

$$x=\frac{x}{\sqrt{1+x^2}} \Rightarrow x\sqrt{1+x^2}=x \Rightarrow x\sqrt{1+x^2}-x=0 \Rightarrow x\left(\sqrt{1+x^2}-1\right)=0$$

$$x=0 \text{ or } \sqrt{1+x^2}-1=0 \Rightarrow \sqrt{1+x^2}=1 \Rightarrow 1+x^2=1 \Rightarrow x^2=0 \Rightarrow x=0$$

Solution set: $\{0\}$

38. $\sin^{-1} x + \tan^{-1} x = 0$

$$\sin^{-1} x + \tan^{-1} x = 0 \Rightarrow \sin^{-1} x = -\tan^{-1} x \Rightarrow x = \sin\left(-\tan^{-1} x\right) \Rightarrow x = -\sin\left(\tan^{-1} x\right)$$

Let $u = \tan^{-1} x$. So, $\tan u = x$.

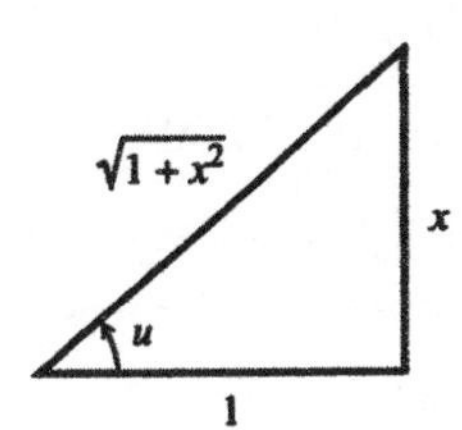

From the triangle, we find $\sin u = \dfrac{x}{\sqrt{1+x^2}}$, so the equation $x = -\sin\left(\tan^{-1} x\right)$ becomes $x = -\dfrac{x}{\sqrt{1+x^2}}$.

Solve this equation.

$$x = -\frac{x}{\sqrt{1+x^2}} \Rightarrow x\sqrt{1+x^2} = -x \Rightarrow x\sqrt{1+x^2} + x = 0 \Rightarrow x\left(\sqrt{1+x^2} + 1\right) = 0$$

$$x = 0 \text{ or } \sqrt{1+x^2} + 1 = 0 \Rightarrow \sqrt{1+x^2} = -1 \Rightarrow 1 + x^2 = 1 \Rightarrow x^2 = 0 \Rightarrow x = 0$$

Solution set: $\{0\}$

39.

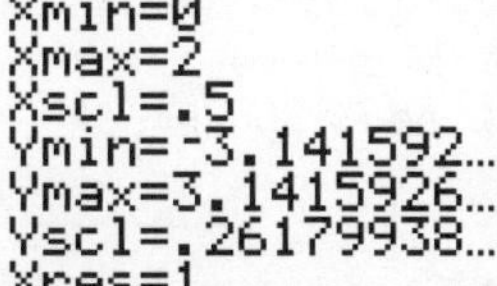

Zero X=.8660254 Y=0

40.

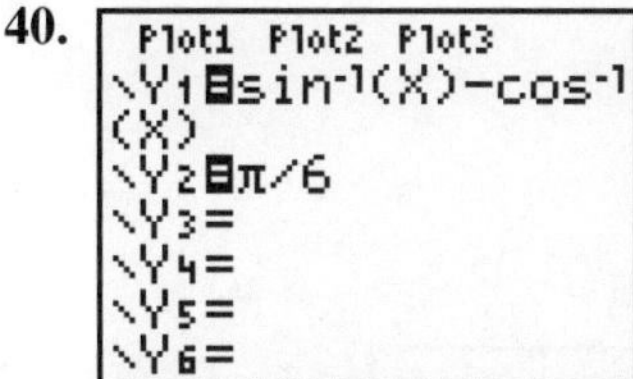

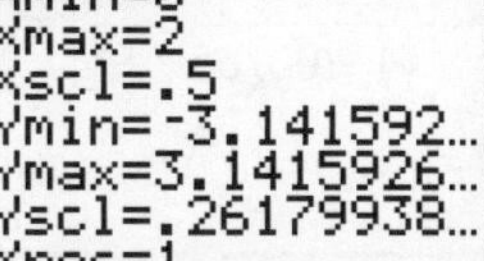

Intersection X=.8660254 Y=.52359878

41. The x-intercept method is shown in the following windows.

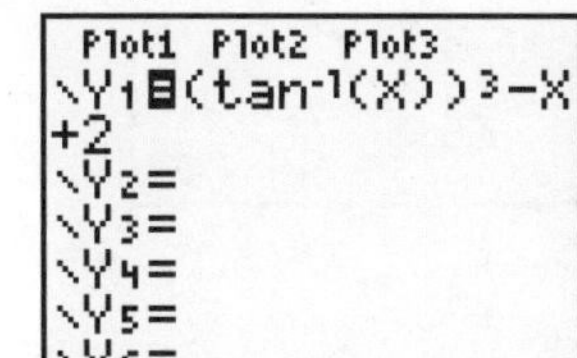

WINDOW Xmin=0 Xmax=6 Xscl=1 Ymin=-3.141592... Ymax=3.1415926... Yscl=.26179938... Xres=1

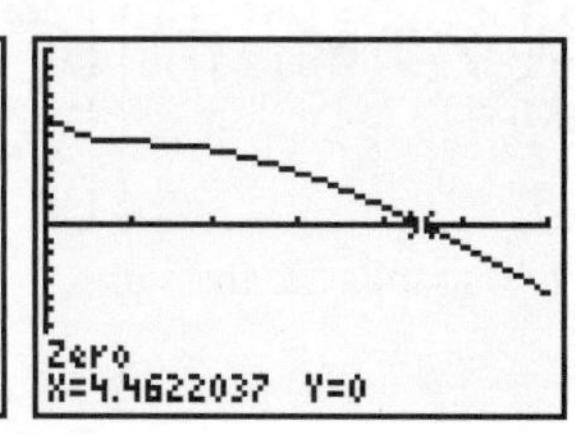

Solution set: $\{4.4622\}$

42. The intersection method is shown in the following screens.

Plot1 Plot2 Plot3 \Y1■πsin⁻¹(.2X)−3 \Y2■−√(X) \Y3= \Y4= \Y5= \Y6=

WINDOW Xmin=0 Xmax=6 Xscl=1 Ymin=-3.141592... Ymax=3.1415926... Yscl=.26179938... Xres=1

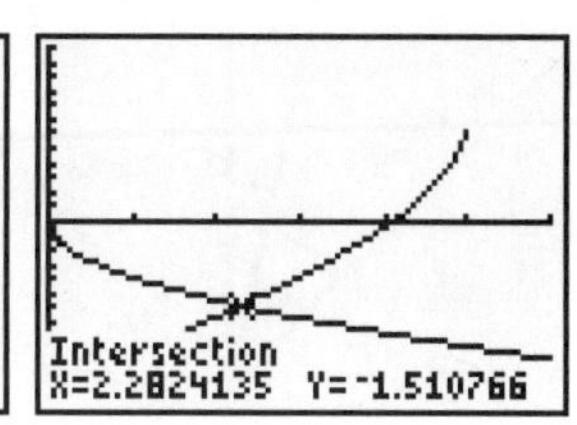

Solution set: $\{2.2824\}$

43. $A=\sqrt{(A_1\cos\phi_1+A_2\cos\phi_2)^2+(A_1\sin\phi_1+A_2\sin\phi_2)^2}$ and $\phi=\arctan\left(\dfrac{A_1\sin\phi_1+A_2\sin\phi_2}{A_1\cos\phi+A_2\cos\phi_2}\right)$

Make sure your calculator is in radian mode.

(a) Let $A_1=.0012$, $\phi_1=.052$, $A_2=.004$, and $\phi_2=.61$.

$$A=\sqrt{(.0012\cos .052+.004\cos .61)^2+(.0012\sin .052+.004\sin .61)^2}\approx .00506$$

$$\phi=\arctan\left(\frac{.0012\sin .052+.004\sin .61}{.0012\cos .052+.004\cos .61}\right)\approx .484$$

If $f=220$, then $P=A\sin(2\pi ft+\phi)$ becomes $P=.00506\sin(440\pi t+.484)$.

(b)

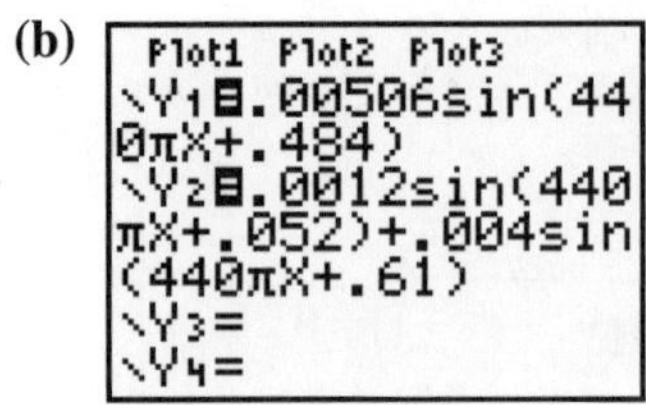

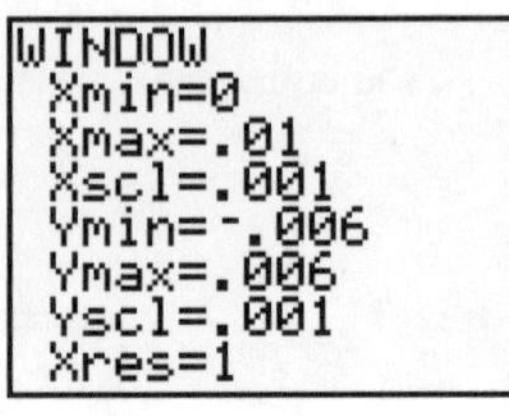

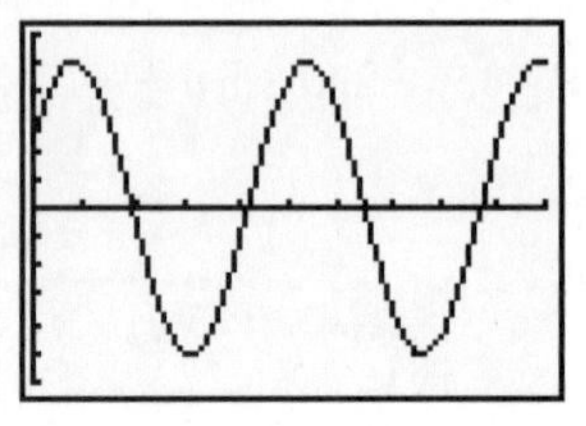

The two graphs are the same.

44. $A=\sqrt{(A_1\cos\phi_1+A_2\cos\phi_2)^2+(A_1\sin\phi_1+A_2\sin\phi_2)^2}$ and $\phi=\arctan\left(\dfrac{A_1\sin\phi_1+A_2\sin\phi_2}{A_1\cos\phi+A_2\cos\phi_2}\right)$

Make sure your calculator is in radian mode.

(a) Let $A_1=.0025$, $\phi_1=\dfrac{\pi}{7}$, $A_2=.001$, $\phi_2=\dfrac{\pi}{6}$, and $f=300$.

$$A=\sqrt{\left[.0025\cos\left(\frac{\pi}{7}\right)+.001\cos\left(\frac{\pi}{6}\right)\right]^2+\left[.0025\sin\left(\frac{\pi}{7}\right)+.001\sin\left(\frac{\pi}{6}\right)\right]^2}\approx .0035$$

$$\phi=\arctan\left[\frac{.0025\sin\left(\frac{\pi}{7}\right)+.001\sin\left(\frac{\pi}{6}\right)}{.0025\cos\left(\frac{\pi}{7}\right)+.001\cos\left(\frac{\pi}{6}\right)}\right]\approx .470$$

If $f=300$, then $P=A\sin(2\pi ft+\phi)$ becomes $P=.0035\sin(600\pi t+.47)$.

(b)

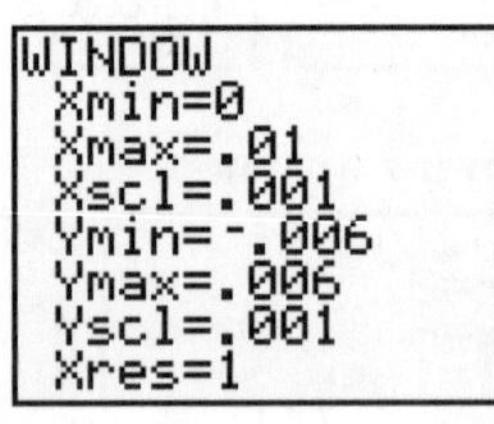

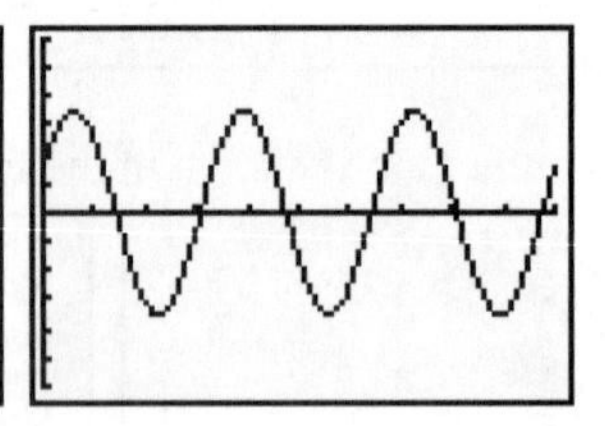

The two graphs are the same.

45. (a) $\tan\alpha=\dfrac{x}{z}$ and $\tan\beta=\dfrac{x+y}{z}$

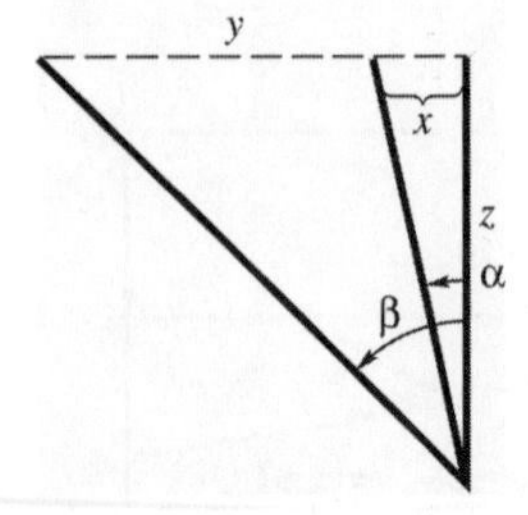

(b) Since $\tan\alpha=\dfrac{x}{z}\Rightarrow z\tan\alpha=x\Rightarrow z=\dfrac{x}{\tan\alpha}$

and

$$\tan\beta=\frac{x+y}{z}\Rightarrow z\tan\beta=x+y\Rightarrow z=\frac{x+y}{\tan\beta},$$

we have $\dfrac{x}{\tan\alpha}=\dfrac{x+y}{\tan\beta}$

Continued on next page

45. (continued)

(c) $(x+y)\tan\alpha = x\tan\beta \Rightarrow \tan\alpha = \dfrac{x\tan\beta}{x+y} \Rightarrow \alpha = \arctan\left(\dfrac{x\tan\beta}{x+y}\right)$

(d) $x\tan\beta = (x+y)\tan\alpha \Rightarrow \tan\beta = \dfrac{(x+y)\tan\alpha}{x} \Rightarrow \beta = \arctan\left(\dfrac{(x+y)\tan\alpha}{x}\right)$

46. (a) $u = \arcsin x \Rightarrow x = \sin u, -\dfrac{\pi}{2} \le u \le \dfrac{\pi}{2}$

(b)

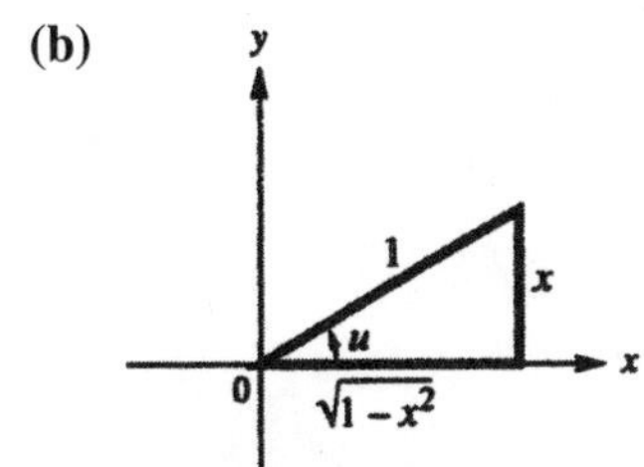

(c) $\tan u = \dfrac{x}{\sqrt{1-x^2}} = \dfrac{x\sqrt{1-x^2}}{1-x^2}$

(d) $u = \arctan\left(\dfrac{x\sqrt{1-x^2}}{1-x^2}\right)$

47. (a) $e = E_{\max}\sin 2\pi ft \Rightarrow \dfrac{e}{E_{\max}} = \sin 2\pi ft \Rightarrow 2\pi ft = \arcsin\dfrac{e}{E_{\max}} \Rightarrow t = \dfrac{1}{2\pi f}\arcsin\dfrac{e}{E_{\max}}$

(b) Let $E_{\max} = 12, e = 5,$ and $f = 100.$

$$t = \frac{1}{2\pi(100)}\arcsin\frac{5}{12} = \frac{1}{200\pi}\arcsin\frac{5}{12} \approx .00068\,\text{sec}$$

48. (a) $\theta = \alpha - \beta$

Since $\tan\alpha = \dfrac{4}{x} \Rightarrow \alpha = \tan^{-1}\left(\dfrac{4}{x}\right)$ and $\tan\beta = \dfrac{1}{x} \Rightarrow \beta = \tan^{-1}\left(\dfrac{1}{x}\right)$, we have the following.

$$\theta = \alpha - \beta \Rightarrow \theta = \tan^{-1}\left(\frac{4}{x}\right) - \tan^{-1}\left(\frac{1}{x}\right)$$

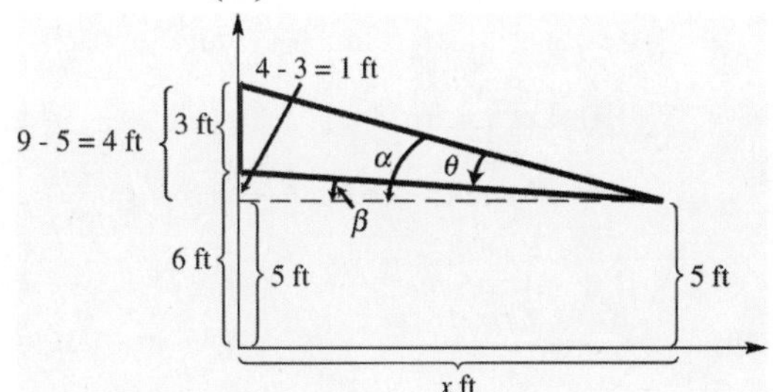

(b) $\theta = \tan^{-1}\left(\dfrac{4}{x}\right) - \tan^{-1}\left(\dfrac{1}{x}\right)$

(i) Let $\theta = \dfrac{\pi}{6}.$

$$\frac{\pi}{6} = \tan^{-1}\frac{4}{x} - \tan^{-1}\frac{1}{x} \Rightarrow \tan\frac{\pi}{6} = \tan\left(\tan^{-1}\frac{4}{x} - \tan^{-1}\frac{1}{x}\right)$$

$$\frac{\sqrt{3}}{3} = \frac{\frac{4}{x} - \frac{1}{x}}{1 + \frac{4}{x}\cdot\frac{1}{x}} \Rightarrow \frac{\sqrt{3}}{3} = \frac{\frac{3}{x}}{1 + \frac{4}{x^2}} \Rightarrow \frac{\sqrt{3}}{3} = \frac{3x}{x^2+4}$$

$$\sqrt{3}x^2 + 4\sqrt{3} = 9x \Rightarrow \sqrt{3}x^2 - 9x + 4\sqrt{3} = 0 \Rightarrow x^2 - 3\sqrt{3}x + 4 = 0$$

Continued on next page

48. (b) (i) (continued)

Using the quadratic formula with $a=1$, $b=-3\sqrt{3}$, and $c=4$.

$$x=\frac{-\left(-3\sqrt{3}\right)\pm\sqrt{\left(-3\sqrt{3}\right)^2-4(1)(4)}}{2(1)}$$

$$x=\frac{3\sqrt{3}\pm\sqrt{27-16}}{2}=\frac{3\sqrt{3}\pm\sqrt{11}}{2}$$

$$x\approx 4.26 \text{ ft or } .94 \text{ ft}$$

(ii) Let $\theta=\frac{\pi}{8}$.

$$\frac{\pi}{8}=\tan^{-1}\frac{4}{x}-\tan^{-1}\frac{1}{x}\Rightarrow \tan\frac{\pi}{8}=\tan\left(\tan^{-1}\frac{4}{x}-\tan^{-1}\frac{1}{x}\right)$$

From part (a), we have the following.

$$\tan\frac{\pi}{8}=\frac{3x}{x^2+4}\Rightarrow\left(\tan\frac{\pi}{8}\right)x^2+4\left(\tan\frac{\pi}{8}\right)=3x$$

$$\left(\tan\frac{\pi}{8}\right)x^2-3x+4\left(\tan\frac{\pi}{8}\right)=0\Rightarrow x^2-3\left(\cot\frac{\pi}{8}\right)x+4=0$$

Using the quadratic formula with $a=1$, $b=-3\left(\cot\frac{\pi}{8}\right)$, and $c=4$.

$$x=\frac{-\left(-3\cot\frac{\pi}{8}\right)\pm\sqrt{\left(-3\cot\frac{\pi}{8}\right)^2-4(1)(4)}}{2(1)}\Rightarrow x\approx 6.64 \text{ ft or } .60 \text{ ft}$$

(c) $\theta=\tan^{-1}\left(\frac{4}{x}\right)-\tan^{-1}\left(\frac{1}{x}\right)$

(i) Let $x=4$.

$$\theta=\tan^{-1}\frac{4}{4}-\tan^{-1}\frac{1}{4}\Rightarrow\theta=\tan^{-1}1-\tan^{-1}\frac{1}{4}\Rightarrow\theta\approx\frac{\pi}{4}-.245\approx .54$$

(ii) Let $x=3$.

$$\theta=\tan^{-1}\frac{4}{3}-\tan^{-1}\frac{1}{3}\Rightarrow\theta\approx .60$$

49. $y=\frac{1}{3}\sin\frac{4\pi t}{3}$

(a) $3y=\sin\frac{4\pi t}{3}\Rightarrow\frac{4\pi t}{3}=\arcsin 3y\Rightarrow 4\pi t=3\arcsin 3y\Rightarrow t=\frac{3}{4\pi}\arcsin 3y$

(b) If $y=.3$ radian, $t=\frac{3}{4\pi}\arcsin .9\Rightarrow t\approx .27\text{ sec}.$

50.

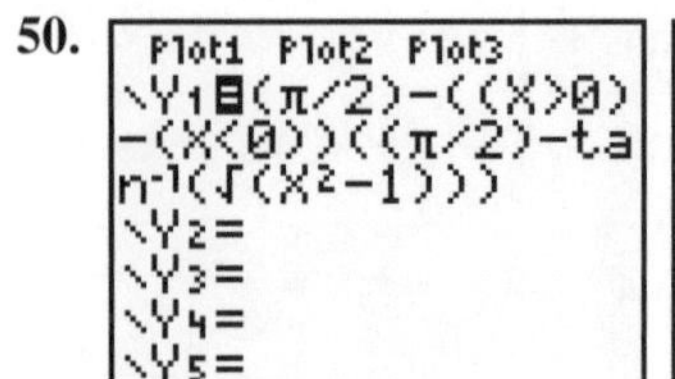

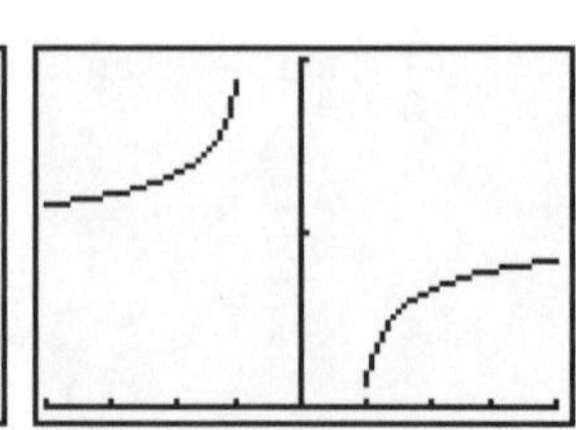

Chapter 6: Review Exercises

1. False; the range of the inverse sine function is $\left[-\frac{\pi}{2}, \frac{\pi}{2}\right]$, while that of the inverse cosine is $[0, \pi]$.

2. False; the range of the inverse tangent function is $\left(-\frac{\pi}{2}, \frac{\pi}{2}\right)$, while that of the inverse cotangent is $(0, \pi)$.

3. False; $\arcsin\left(-\frac{1}{2}\right) = -\frac{\pi}{6}$, not $\frac{11\pi}{6}$.

4. True

5. $y = \sin^{-1}\frac{\sqrt{2}}{2} \Rightarrow \sin y = \frac{\sqrt{2}}{2}$

Since $-\frac{\pi}{2} \le y \le \frac{\pi}{2}$, $y = \frac{\pi}{4}$.

6. $y = \arccos\left(-\frac{1}{2}\right) \Rightarrow \cos y = -\frac{1}{2}$

Since $0 \le y \le \pi$, $y = \frac{2\pi}{3}$.

7. $y = \tan^{-1}\left(-\sqrt{3}\right) \Rightarrow \tan y = -\sqrt{3}$

Since $-\frac{\pi}{2} < y < \frac{\pi}{2}$, $y = -\frac{\pi}{3}$.

8. $y = \arcsin(-1) \Rightarrow \sin y = -1$

Since $-\frac{\pi}{2} \le y \le \frac{\pi}{2}$, $y = -\frac{\pi}{2}$.

9. $y = \cos^{-1}\left(-\frac{\sqrt{2}}{2}\right) \Rightarrow \cos y = -\frac{\sqrt{2}}{2}$

Since $0 \le y \le \pi$, $y = \frac{3\pi}{4}$.

10. $y = \arctan\frac{\sqrt{3}}{3} \Rightarrow \tan y = \frac{\sqrt{3}}{3}$

Since $-\frac{\pi}{2} < y < \frac{\pi}{2}$, $y = \frac{\pi}{6}$.

11. $y = \sec^{-1}(-2) \Rightarrow \sec y = -2$

Since $0 \le y \le \pi$, $y \ne \frac{\pi}{2} \Rightarrow y = \frac{2\pi}{3}$.

12. $y = \operatorname{arccsc}\frac{2\sqrt{3}}{3} \Rightarrow \csc y = \frac{2\sqrt{3}}{3}$

Since $-\frac{\pi}{2} \le y \le \frac{\pi}{2}$ and $y \ne 0$, $y = \frac{\pi}{3}$.

13. $y = \operatorname{arccot}(-1) \Rightarrow \cot y = -1$

Since $0 < y < \pi$, $y = \frac{3\pi}{4}$.

14. $\theta = \arccos\frac{1}{2} \Rightarrow \cos\theta = \frac{1}{2}$

Since $0° \le \theta \le 180°$, $\theta = 60°$.

15. $\theta = \arcsin\left(-\frac{\sqrt{3}}{2}\right) \Rightarrow \sin\theta = -\frac{\sqrt{3}}{2}$

Since $-90° \le \theta \le 90°$, $\theta = -60°$.

16. $\theta = \tan^{-1} 0 \Rightarrow \tan\theta = 0$

Since $-90° < \theta < 90°$, $\theta = 0°$.

For Exercises 17–22, be sure that your calculator is in degree mode. Keystroke sequences may vary based on the type and/or model of calculator being used.

17. $\theta = \arctan 1.7804675$

$\theta = 60.67924514°$

18. $\theta = \sin^{-1}(-.66045320)$

sin⁻¹(⁻.66045320)
⁻41.33444556

$\theta \approx -41.33444556°$

19. $\theta = \cos^{-1} .80396577$

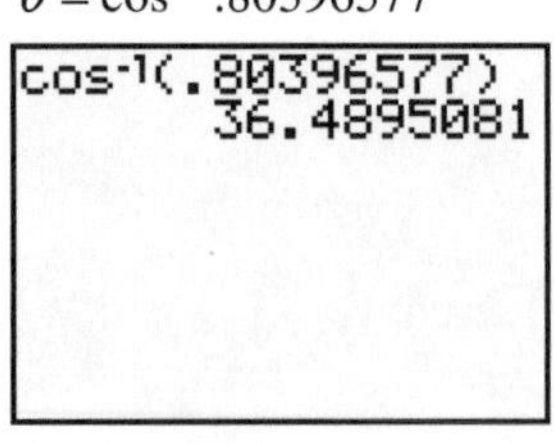

$\theta \approx 36.4895081°$

20. $\theta = \cot^{-1} 4.5046388$

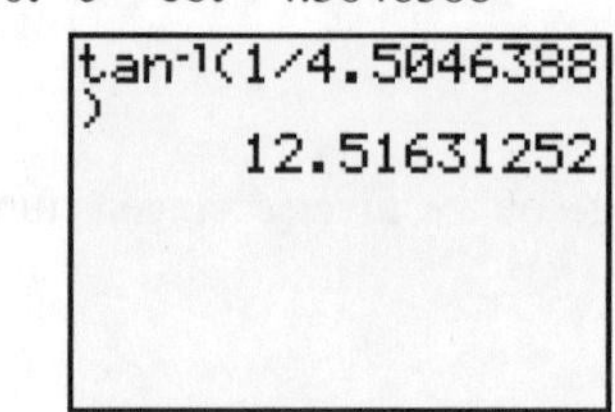

$\theta \approx 12.51631252°$

21. $\theta = \text{arc sec } 3.4723155$

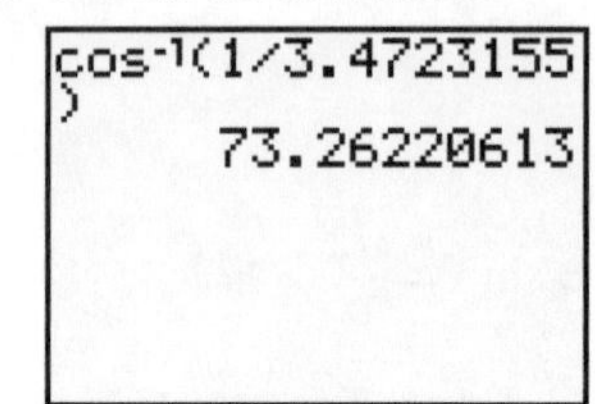

$\theta \approx 73.26220613°$

22. $\theta = \csc^{-1} 7.4890096$

sin⁻¹(1/7.4890096
)
7.673567973

$\theta \approx 7.673567973°$

23. $\cos(\arccos(-1)) = \cos \pi = -1$ or $\cos(\arccos(-1)) = \cos 180° = -1$

24. $\sin\left(\arcsin\left(-\frac{\sqrt{3}}{2}\right)\right) = \sin\left(-\frac{\pi}{3}\right) = -\frac{\sqrt{3}}{2}$ or $\sin\left(\arcsin\left(-\frac{\sqrt{3}}{2}\right)\right) = \sin(-60°) = -\frac{\sqrt{3}}{2}$

25. $\arccos\left(\cos\frac{3\pi}{4}\right) = \arccos\left(-\frac{\sqrt{2}}{2}\right) = \frac{3\pi}{4}$

26. $\text{arcsec}(\sec \pi) = \text{arcsec}(-1) = \pi$

27. $\tan^{-1}\left(\tan\frac{\pi}{4}\right) = \tan^{-1}\frac{\sqrt{2}}{2} = \frac{\pi}{4}$

28. $\cos^{-1}(\cos 0) = \cos^{-1} 1 = 0$

29. $\sin\left(\arccos\frac{3}{4}\right)$

Let $\omega = \arccos\frac{3}{4}$, so that $\cos\omega = \frac{3}{4}$. Since arccos is defined only in quadrants I and II, and $\frac{3}{4}$ is positive, ω is in quadrant I. Sketch ω and label a triangle with the side opposite ω equal to $\sqrt{4^2 - 3^2} = \sqrt{16-9} = \sqrt{7}$.

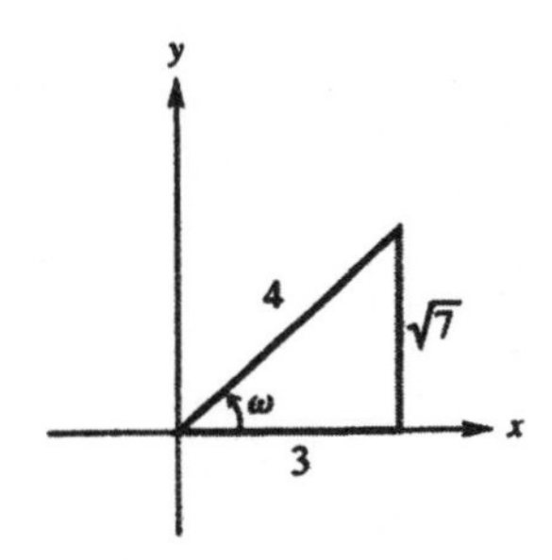

$$\sin\left(\arccos\frac{3}{4}\right) = \sin\omega = \frac{\sqrt{7}}{4}$$

30. $\cos(\arctan 3)$

Let $\theta = \arctan 3$, so that $\tan\theta = 3 = \frac{3}{1}$. Since arctan is defined only in quadrants I and IV, and 3 is positive, θ is in quadrant I. Sketch θ and label a triangle with the hypotenuse equal to $\sqrt{3^2 + 1^2} = \sqrt{10}$.

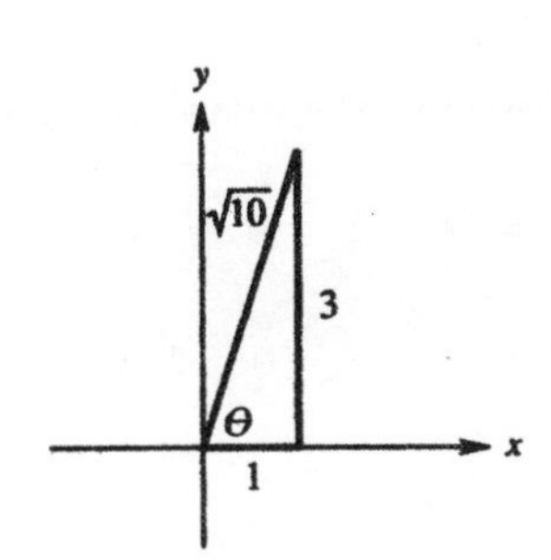

$$\cos(\arctan 3) = \cos\theta = \frac{1}{\sqrt{10}} = \frac{\sqrt{10}}{10}$$

31. $\cos\left(\csc^{-1}(-2)\right)$

Let $\omega = \csc^{-1}(-2)$, so that $\csc\omega = -2$. Since $-\frac{\pi}{2} \le \omega \le \frac{\pi}{2}$ and $\omega \ne 0$, and $\csc\omega = -2$ (negative), ω is in quadrant IV. Sketch ω and label a triangle with side adjacent to ω equal to $\sqrt{2^2 - (-1)^2} = \sqrt{4-1} = \sqrt{3}$.

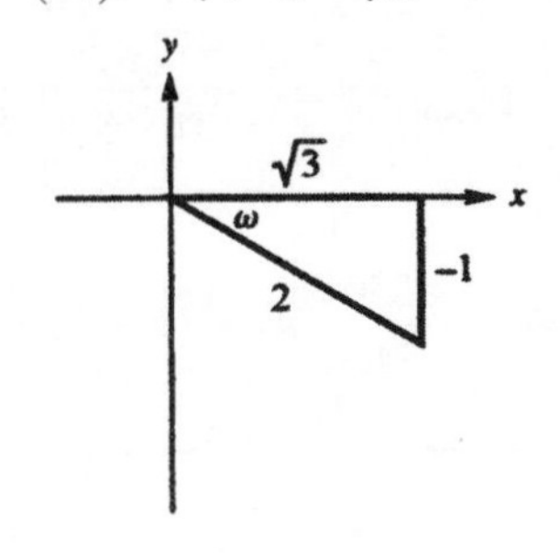

$$\cos\left(\csc^{-1}(-2)\right) = \cos\omega = \frac{\sqrt{3}}{2}$$

32. $\sec\left(2\sin^{-1}\left(-\dfrac{1}{3}\right)\right)$

Let $\theta=\sin^{-1}\left(-\dfrac{1}{3}\right)$, so that $\sin\theta=-\dfrac{1}{3}$. Since arcsin is defined only in quadrants I and IV, and $-\dfrac{1}{3}$ is negative, θ is in quadrant IV. Sketch θ and label a triangle with the side adjacent to θ equal to $\theta=\sqrt{3^2-(-1)^2}=\sqrt{9-1}=\sqrt{8}=2\sqrt{2}$.

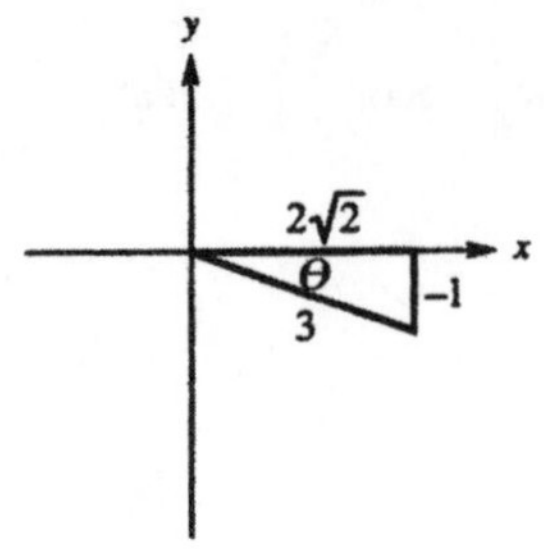

$$\cos\theta=\frac{2\sqrt{2}}{3}$$

$$\sec\left(2\sin^{-1}\left(-\frac{1}{3}\right)\right)=\sec 2\theta$$

$$\sec 2\theta=\frac{1}{\cos 2\theta}=\frac{1}{2\cos^2\theta-1}=\frac{1}{2\left(\frac{2\sqrt{2}}{3}\right)^2-1}=\frac{1}{2\left(\frac{8}{9}\right)-1}=\frac{1}{\frac{16}{9}-1}=\frac{9}{16-9}=\frac{9}{7}$$

33. $\tan\left(\arcsin\dfrac{3}{5}+\arccos\dfrac{5}{7}\right)$

Let $\omega_1=\arcsin\dfrac{3}{5}$, $\omega_2=\arccos\dfrac{5}{7}$. Sketch angles ω_1 and ω_2 in quadrant I. The side adjacent to ω_1 is $\sqrt{5^2-3^2}=\sqrt{25-9}=\sqrt{16}=4$. The side opposite ω_2 is $\sqrt{7^2-5^2}=\sqrt{49-25}=\sqrt{24}=2\sqrt{6}$.

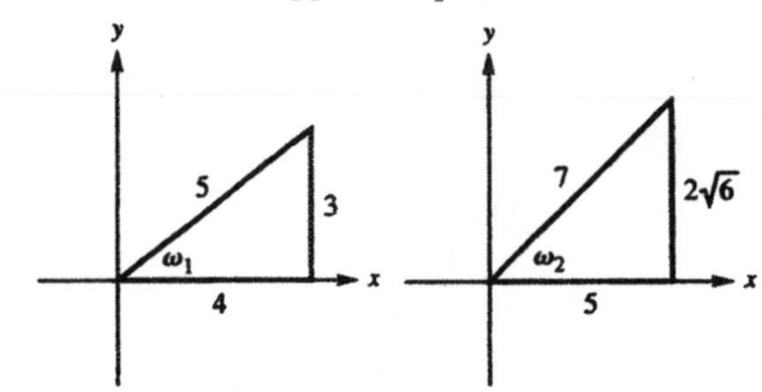

We have $\tan\omega_1=\dfrac{3}{4}$ and $\tan\omega_2=\dfrac{2\sqrt{6}}{5}$.

$$\tan\left(\arcsin\frac{3}{5}+\arccos\frac{5}{7}\right)=\tan(\omega_1+\omega_2)=\frac{\tan\omega_1+\tan\omega_2}{1-\tan\omega_1\tan\omega_2}=\frac{\frac{3}{4}+\frac{2\sqrt{6}}{5}}{1-\left(\frac{3}{4}\right)\left(\frac{2\sqrt{6}}{5}\right)}=\frac{\frac{15+8\sqrt{6}}{20}}{\frac{20-6\sqrt{6}}{20}}$$

$$=\frac{15+8\sqrt{6}}{20-6\sqrt{6}}=\frac{15+8\sqrt{6}}{20-6\sqrt{6}}\cdot\frac{20+6\sqrt{6}}{20+6\sqrt{6}}=\frac{588+250\sqrt{6}}{184}=\frac{294+125\sqrt{6}}{92}$$

34. $\cos\left(\arctan \frac{u}{\sqrt{1-u^2}}\right)$

Let $\theta = \arctan \frac{u}{\sqrt{1-u^2}}$, so $\tan\theta = \frac{u}{\sqrt{1-u^2}}$.

If $u > 0$, $0 < \theta < \frac{\pi}{2}$.

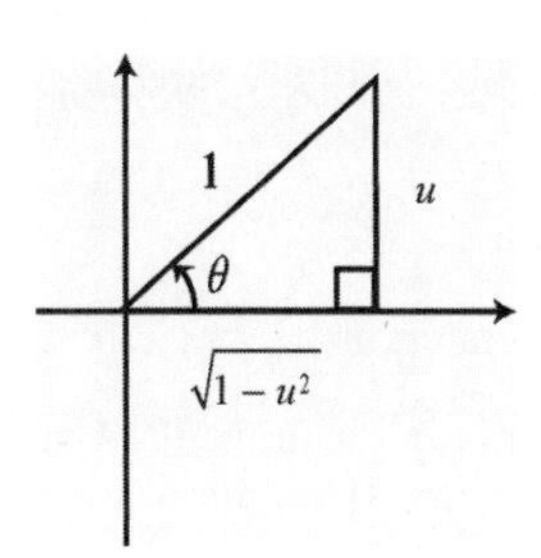

From the Pythagorean theorem, $r = \sqrt{\left(\sqrt{1-u^2}\right)^2 + u^2} = \sqrt{1-u^2+u^2} = \sqrt{1} = 1$. Therefore

$\cos\theta = \frac{\sqrt{1-u^2}}{1} = \sqrt{1-u^2}$. Thus, $\cos\left(\arctan \frac{u}{\sqrt{1-u^2}}\right) = \sqrt{1-u^2}$.

35. $\tan\left(\operatorname{arcsec} \frac{\sqrt{u^2+1}}{u}\right)$

Let $\theta = \operatorname{arcsec} \frac{\sqrt{u^2+1}}{u}$, so $\sec\theta = \frac{\sqrt{u^2+1}}{u}$.

If $u > 0$, $0 < \theta < \frac{\pi}{2}$.

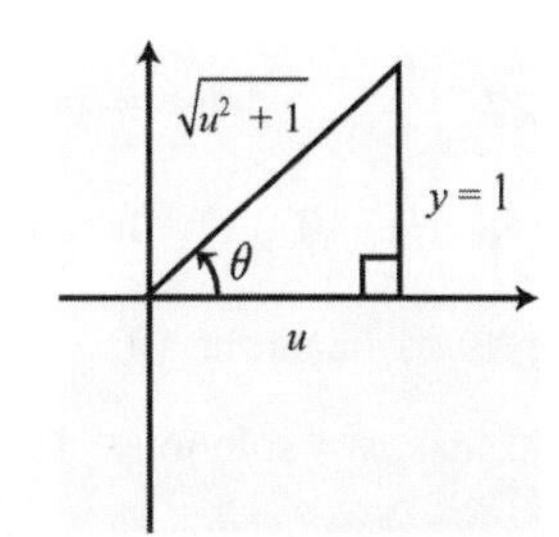

From the Pythagorean theorem, $y = \sqrt{\left(\sqrt{u^2+1}\right)^2 - u^2} = \sqrt{u^2+1-u^2} = \sqrt{1} = 1$. Therefore $\tan\theta = \frac{1}{u}$.

Thus, $\tan\left(\operatorname{arcsec} \frac{\sqrt{u^2+1}}{u}\right) = \frac{1}{u}$.

36. $\sin^2 x = 1$

$$\sin^2 x = 1 \Rightarrow \sin x = \pm 1$$

Over the interval $[0, 2\pi)$, the equation $\sin x = 1$ has one solution. This solution is $\frac{\pi}{2}$. Over the same interval, the equation $\sin x = -1$ has one solution. This solution is $\frac{3\pi}{2}$.

Solution set: $\left\{\frac{\pi}{2}, \frac{3\pi}{2}\right\}$

37. $2\tan x - 1 = 0$

$$2\tan x - 1 = 0 \Rightarrow 2\tan x = 1 \Rightarrow \tan x = \frac{1}{2}$$

Over the interval $[0, 2\pi)$, the equation $\tan x = \frac{1}{2}$ has two solutions. One solution is in quadrant I and the other is in quadrant III. Using a calculator, the quadrant I solution is approximately .463647609. The quadrant III solution would be approximately $.463647609 + \pi \approx 3.605240263$.

Solution set: $\{.463647609, 3.605240263\}$

38. $3\sin^2 x - 5\sin x + 2 = 0$

$$3\sin^2 x - 5\sin x + 2 = 0 \Rightarrow (3\sin x - 2)(\sin x - 1) = 0$$

$$3\sin x - 2 \Rightarrow 3\sin x = 2 \Rightarrow \sin x = \frac{2}{3}$$

$$\sin x - 1 = 0 \Rightarrow \sin x = 1$$

Over the interval $[0, 2\pi)$, the equation $\sin x = \frac{2}{3}$ has two solutions. One solution is in quadrant I and the other is in quadrant II. Using a calculator, the quadrant I solution is approximately .7297276562. The quadrant II solution would be approximately $\pi - .7297276562 \approx 2.411864997$. In the same interval, the equation $\sin x = 1$ has one solution. That solution is $\frac{\pi}{2}$.

Solution set: $\left\{.7297276562, \frac{\pi}{2}, 2.411864997\right\}$

39. $\tan x = \cot x$

Use the identity $\cot x = \frac{1}{\tan x}$, $\tan x \neq 0$.

$$\tan x = \cot x \Rightarrow \tan x = \frac{1}{\tan x} \Rightarrow \tan^2 x = 1 \Rightarrow \tan x = \pm 1$$

Over the interval $[0, 2\pi)$, the equation $\tan x = 1$ has two solutions. One solution is in quadrant I and the other is in quadrant III. These solutions are $\frac{\pi}{4}$ and $\frac{5\pi}{4}$. In the same interval, the equation $\tan x = -1$ has two solutions. One solution is in quadrant II and the other is in quadrant IV. These solutions are $\frac{3\pi}{4}$ and $\frac{7\pi}{4}$.

Solution set: $\left\{\frac{\pi}{4}, \frac{3\pi}{4}, \frac{5\pi}{4}, \frac{7\pi}{4}\right\}$

40. $\sec^4 2x = 4$

$$\sec^4 2x = 4 \Rightarrow \sec^2 2x = 2 \Rightarrow \sec 2x = \pm\sqrt{2}$$

Since $0 \le x < 2\pi, 0 \le 2x < 4\pi$. Thus, $2x = \frac{\pi}{4}, \frac{3\pi}{4}, \frac{5\pi}{4}, \frac{7\pi}{4}, \frac{9\pi}{4}, \frac{11\pi}{4}, \frac{13\pi}{4}, \frac{15\pi}{4}$ implies $x = \frac{\pi}{8}, \frac{3\pi}{8}, \frac{5\pi}{8}, \frac{7\pi}{8}, \frac{9\pi}{8}, \frac{11\pi}{8}, \frac{13\pi}{8}, \frac{15\pi}{8}$.

Solution set: $\left\{\frac{\pi}{8}, \frac{3\pi}{8}, \frac{5\pi}{8}, \frac{7\pi}{8}, \frac{9\pi}{8}, \frac{11\pi}{8}, \frac{13\pi}{8}, \frac{15\pi}{8}\right\}$

41. $\tan^2 2x - 1 = 0$

$$\tan^2 2x - 1 = 0 \Rightarrow \tan^2 2x = 1 \Rightarrow \tan 2x = \pm 1$$

Since $0 \le x < 2\pi, 0 \le 2x < 4\pi$. Thus, $2x = \frac{\pi}{4}, \frac{3\pi}{4}, \frac{5\pi}{4}, \frac{7\pi}{4}, \frac{9\pi}{4}, \frac{11\pi}{4}, \frac{13\pi}{4}, \frac{15\pi}{4}$ implies $x = \frac{\pi}{8}, \frac{3\pi}{8}, \frac{5\pi}{8}, \frac{7\pi}{8}, \frac{9\pi}{8}, \frac{11\pi}{8}, \frac{13\pi}{8}, \frac{15\pi}{8}$.

Solution set: $\left\{\frac{\pi}{8}, \frac{3\pi}{8}, \frac{5\pi}{8}, \frac{7\pi}{8}, \frac{9\pi}{8}, \frac{11\pi}{8}, \frac{13\pi}{8}, \frac{15\pi}{8}\right\}$

42. $\sec\frac{x}{2}=\cos\frac{x}{2}$

$$\sec\frac{x}{2}=\cos\frac{x}{2}\Rightarrow\frac{1}{\cos\frac{x}{2}}=\cos\frac{x}{2}\Rightarrow\cos^2\frac{x}{2}=1\Rightarrow\cos\frac{x}{2}=\pm1$$

Since $0\le x<2\pi, 0\le\frac{x}{2}<\pi$. Thus, the only solution to $\cos\frac{x}{2}=\pm1$ is $x=0$.

Solution set: $\{0+2n\pi, \text{where } n \text{ is any integer}\}$ or $\{2n\pi, \text{where } n \text{ is any integer}\}$

43. $\cos 2x+\cos x=0$

$$\cos 2x+\cos x=0\Rightarrow 2\cos^2 x-1+\cos x=0\Rightarrow 2\cos^2 x+\cos x-1=0\Rightarrow(2\cos x-1)(\cos x+1)=0$$

$$2\cos x-1=0\Rightarrow 2\cos x=1\Rightarrow\cos x=\frac{1}{2}\quad\text{or}\quad\cos x+1=0\Rightarrow\cos x=-1$$

Over the interval $[0,2\pi)$, the equation $\cos x=\frac{1}{2}$ has two solutions. The angles in quadrants I and IV that have a reference angle of $\frac{\pi}{3}$ are $\frac{\pi}{3}$ and $\frac{5\pi}{3}$. In the same interval, $\cos x=-1$ when the angle is π.

Solution set: $\left\{\frac{\pi}{3}+2n\pi,\ \pi+2n\pi,\ \frac{5\pi}{3}+2n\pi, \text{ where } n \text{ is any integer}\right\}$

44. $4\sin x\cos x=\sqrt{3}$

$$4\sin x\cos x=\sqrt{3}\Rightarrow 2\sin x\cos x=\frac{\sqrt{3}}{2}\Rightarrow\sin 2x=\frac{\sqrt{3}}{2}$$

Since $0\le x<2\pi, 0\le 2x<4\pi$. Thus, $2x=\frac{\pi}{3},\frac{2\pi}{3},\frac{7\pi}{3},\frac{8\pi}{3}\Rightarrow x=\frac{\pi}{6},\frac{\pi}{3},\frac{7\pi}{6},\frac{4\pi}{3}$.

Solution set: $\left\{\frac{\pi}{6}+2n\pi,\frac{\pi}{3}+2n\pi,\frac{7\pi}{6}+2n\pi,\frac{4\pi}{3}+2n\pi, \text{ where } n \text{ is any integer}\right\}$ or

$\left\{\frac{\pi}{6}+n\pi,\frac{\pi}{3}+n\pi, \text{ where } n \text{ is any integer}\right\}$

45. $\sin^2\theta+3\sin\theta+2=0$

$$\sin^2\theta+3\sin\theta+2=0\Rightarrow(\sin\theta+2)(\sin\theta+1)=0$$

In the interval $[0°,360°)$, we have the following.

$$\sin\theta+1=0\Rightarrow\sin\theta=-1\Rightarrow\theta=270°\ \text{and}\ \sin\theta+2=0\Rightarrow\sin\theta=-2<-1\Rightarrow\text{no solution}$$

Solution set: $\{270°\}$

46. $2\tan^2\theta=\tan\theta+1$

$$2\tan^2\theta=\tan\theta+1\Rightarrow 2\tan^2\theta-\tan\theta-1=0\Rightarrow(2\tan\theta+1)(\tan\theta-1)=0$$

In the interval $[0°,360°)$, we have the following.

$$2\tan\theta+1=0\Rightarrow\tan\theta=-\frac{1}{2}\Rightarrow\theta\approx 153.4°\ \text{or}\ 333.4°\ \text{(using a calculator)}$$

$$\tan\theta-1=0\Rightarrow\tan\theta=1\Rightarrow\theta=45°\ \text{and}\ 225°$$

Solution set: $\{45°,153.4°,225°,333.4°\}$

47. $\sin 2\theta = \cos 2\theta + 1$

$$\sin 2\theta = \cos 2\theta + 1 \Rightarrow (\sin 2\theta)^2 = (\cos 2\theta + 1)^2 \Rightarrow \sin^2 2\theta = \cos^2 2\theta + 2\cos 2\theta + 1$$

$$1 - \cos^2 2\theta = \cos^2 2\theta + 2\cos 2\theta + 1 \Rightarrow 2\cos^2 2\theta + 2\cos 2\theta = 0$$

$$\cos^2 2\theta + \cos 2\theta = 0 \Rightarrow \cos 2\theta(\cos 2\theta + 1) = 0$$

Since $0° \le \theta < 360°$, $0° \le 2\theta < 720°$.

$$\cos 2\theta = 0 \Rightarrow 2\theta = 90°,\ 270°,\ 450°,\ 630° \Rightarrow \theta = 45°,\ 135°,\ 225°,\ 315°$$

$$\cos 2\theta + 1 = 0 \Rightarrow \cos 2\theta = -1 \Rightarrow 2\theta = 180°,\ 540° \Rightarrow \theta = 90°,\ 270°$$

Possible values for θ are $\theta = 45°, 90°, 135°, 225°, 270°, 315°$.

All proposed solutions must be checked since the solutions were found by squaring an equation.

A value for θ will be a solution if $\sin 2\theta - \cos 2\theta = 1$.

$$\theta = 45°, 2\theta = 90° \Rightarrow \sin 90° - \cos 90° = 1 - 0 = 1$$

$$\theta = 90°, 2\theta = 180° \Rightarrow \sin 180° - \cos 180° = 0 - (-1) = 1$$

$$\theta = 135°, 2\theta = 270° \Rightarrow \sin 270° - \cos 270° = -1 - 0 \ne 1$$

$$\theta = 225°, 2\theta = 450° \Rightarrow \sin 450° - \cos 450° = 1 - 0 = 1$$

$$\theta = 270°, 2\theta = 540° \Rightarrow \sin 540° - \cos 540° = 0 - (-1) = 1$$

$$\theta = 315°, 2\theta = 630° \Rightarrow \sin 630° - \cos 630° = -1 - 0 \ne 1$$

Thus, $\theta = 45°, 90°, 225°, 270°$.

Solution set: $\{45°, 90°, 225°, 270°\}$

48. $2\sin 2\theta = 1$

$$2\sin 2\theta = 1 \Rightarrow \sin 2\theta = \frac{1}{2}$$

Since $0° \le \theta < 360°$, $0° \le 2\theta < 720°$. Thus, $2\theta = 30°, 150°, 390°, 510° \Rightarrow \theta = 15°, 75°, 195°, 255°$.

Solution set: $\{15°, 75°, 195°, 255°\}$

49. $3\cos^2\theta + 2\cos\theta - 1 = 0$

$$3\cos^2\theta + 2\cos\theta - 1 = 0 \Rightarrow (3\cos\theta - 1)(\cos\theta + 1) = 0$$

In the interval $[0°, 360°)$, we have the following.

$$3\cos\theta - 1 = 0 \Rightarrow \cos\theta = \frac{1}{3} \Rightarrow \theta \approx 70.5° \text{ and } 289.5° \text{ (using a calculator)}$$

$$\cos\theta + 1 = 0 \Rightarrow \cos\theta = -1 \Rightarrow \theta = 180°$$

Solution set: $\{70.5°, 180°, 289.5°\}$

50. $5\cot^2\theta - \cot\theta - 2 = 0$

We use the quadratic formula with $a = 5$, $b = -1$, and $c = -2$.

$$\cot\theta = \frac{-(-1) \pm \sqrt{(-1)^2 - 4(5)(-2)}}{2(5)} = \frac{1 \pm \sqrt{1+40}}{10} = \frac{1 \pm \sqrt{41}}{10}$$

Since $\cot\theta = \frac{1+\sqrt{41}}{10} > 0$, we will obtain two angles. One angle will be in quadrant I and the other will be in quadrant III. Using a calculator, if $\cot\theta = \frac{1+\sqrt{41}}{10} \approx .7303124$, the quadrant I angle will be approximately 53.5°. The quadrant III angle will be approximately $180° + 53.5° = 233.5°$.

Since $\cot\theta = \frac{1-\sqrt{41}}{10} < 0$, we will obtain two angles. One angle will be in quadrant II and the other will be in quadrant IV. Using a calculator, if $\cot\theta = \frac{1-\sqrt{41}}{10} \approx -.26376262$, the quadrant II angle will be approximately 118.4°. (Note: You need to calculate $\tan^{-1}\left(\frac{1}{\frac{1-\sqrt{41}}{10}}\right) + 180$ to obtain this angle.)

The reference angle is $180° - 118.4° = 61.6°$. Thus, the quadrant IV angle will be approximately $360° - 61.675.2° = 298.4°$.

Solution set: $\{53.5°, 118.4°, 233.5°, 298.4°\}$

51. $4y = 2\sin x \Rightarrow 2y = \sin x \Rightarrow x = \arcsin 2y$

52. $y = 3\cos\frac{x}{2} \Rightarrow \frac{y}{3} = \cos\frac{x}{2} \Rightarrow \frac{x}{2} = \arccos\frac{y}{3} \Rightarrow x = 2\arccos\frac{y}{3}$

53. $2y = \tan(3x+2) \Rightarrow 3x + 2 = \arctan 2y \Rightarrow 3x = \arctan 2y - 2 \Rightarrow x = \left(\frac{1}{3}\arctan 2y\right) - \frac{2}{3}$

54. $5y = 4\sin x - 3 \Rightarrow \frac{5y+3}{4} = \sin x \Rightarrow x = \arcsin\left(\frac{5y+3}{4}\right)$

55. $\frac{4}{3}\arctan\frac{x}{2} = \pi \Rightarrow \arctan\frac{x}{2} = \frac{3\pi}{4}$

But, by definition, the range of arctan is $\left(-\frac{\pi}{2}, \frac{\pi}{2}\right)$. So, this equation has no solution.

Solution set: $\varnothing$

56. $\arccos x = \arcsin \frac{2}{7}$

Let $u = \arcsin \frac{2}{7}$. Then $\sin u = \frac{2}{7}$, u is in quadrant I since $\sin u = \frac{2}{7} > 0$. The side adjacent to u is $\sqrt{7^2 - 2^2} = \sqrt{49-4} = \sqrt{45} = 3\sqrt{5}$.

$\cos u = \frac{3\sqrt{5}}{7}$

This equation becomes $\arccos x = u$, or $x = \cos u$. Thus, $x = \frac{3\sqrt{5}}{7}$.

Solution set: $\left\{\frac{3\sqrt{5}}{7}\right\}$

57. $\arccos x + \arctan 1 = \frac{11\pi}{12}$

$$\arccos x + \arctan 1 = \frac{11\pi}{12} \Rightarrow \arccos x = \frac{11\pi}{12} - \arctan 1 \Rightarrow \arccos x = \frac{11\pi}{12} - \frac{\pi}{4}$$

$$\arccos x = \frac{11\pi}{12} - \frac{3\pi}{12} = \frac{8\pi}{12} \Rightarrow \arccos x = \frac{2\pi}{3} \Rightarrow \cos \frac{2\pi}{3} = x \Rightarrow x = -\frac{1}{2}$$

Solution set: $\left\{-\frac{1}{2}\right\}$

58. $d = 550 + 450 \cos \frac{\pi}{50} t$

$$d = 550 + 450 \cos \frac{\pi}{50} t \Rightarrow 450 \cos \frac{\pi}{50} t = d - 550 \Rightarrow \cos \frac{\pi}{50} t = \frac{d-550}{450}$$

$$\frac{\pi}{50} t = \arccos\left(\frac{d-550}{450}\right) \Rightarrow t = \frac{50}{\pi} \arccos\left(\frac{d-550}{450}\right)$$

59. (a) Let α be the angle to the left of θ.

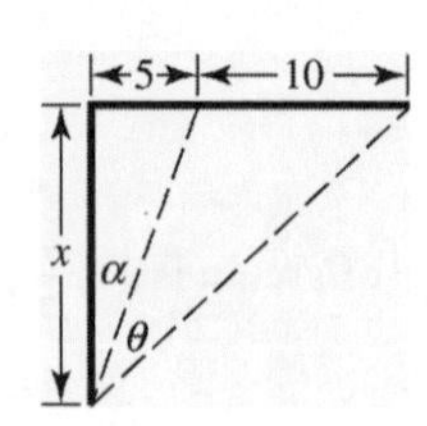

Thus, we have the following.

$$\tan(\alpha+\theta) = \frac{5+10}{x} \Rightarrow \alpha+\theta = \arctan\left(\frac{15}{x}\right) \Rightarrow \theta = \arctan\left(\frac{15}{x}\right) - \alpha \Rightarrow \theta = \arctan\left(\frac{15}{x}\right) - \arctan\left(\frac{5}{x}\right)$$

Continued on next page

59. (continued)

(b) The maximum occurs at approximately =8.66026 ft. There may be a discrepancy in the final digits.

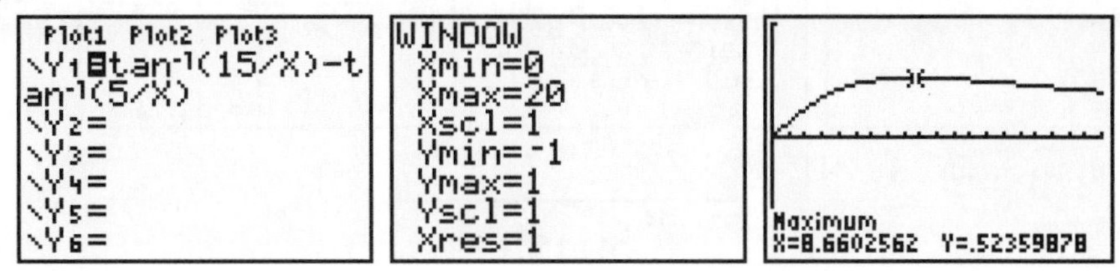

60. $\dfrac{c_1}{c_2}=\dfrac{\sin\theta_1}{\sin\theta_2}$

$$\frac{c_1}{c_2}=\frac{\sin\theta_1}{\sin\theta_2}\Rightarrow .752=\frac{\sin\theta_1}{\sin\theta_2}$$

If $\theta_2=90°$, then $.752=\dfrac{\sin\theta_1}{\sin 90°}=\dfrac{\sin\theta_1}{1}=\sin\theta_1\Rightarrow\theta_1=\sin^{-1}.752\approx 48.8°.$

61. If $\theta_1>48.8°$, then $\theta_2>90°$ and the light beam is completely underwater.

62. $L=6077-31\cos 2\theta$

(a) Solve the equation for θ.

$$L-6077=-31\cos 2\theta\Rightarrow\cos 2\theta=\frac{L-6077}{-31}\Rightarrow 2\theta=\cos^{-1}\left(\frac{L-6077}{-31}\right)$$

$$\theta=\frac{1}{2}\cos^{-1}\left(\frac{L-6077}{-31}\right)\Rightarrow\theta=\frac{1}{2}\cos^{-1}\left(\frac{6074-6077}{-31}\right)\approx 42.2°$$

(b) Substitute L = 6108 in the equation from part (a).

$$\theta=\frac{1}{2}\cos^{-1}\left(\frac{6108-6077}{-31}\right)\Rightarrow\theta=\frac{1}{2}\cos^{-1}\left(\frac{6108-6077}{-31}\right)$$

$$\theta=\frac{1}{2}\cos^{-1}\left(\frac{31}{-31}\right)\Rightarrow\theta=\frac{1}{2}\cos^{-1}(-1)\Rightarrow\theta=\frac{1}{2}(180°)\Rightarrow\theta=90°$$

(c) Substitute L = 6080.2 in the equation from part (a).

$$\theta=\frac{1}{2}\cos^{-1}\left(\frac{6080.2-6077}{-31}\right)\Rightarrow\theta=\frac{1}{2}\cos^{-1}\left(\frac{3.2}{-31}\right)\Rightarrow\theta\approx 48.0°$$

63.

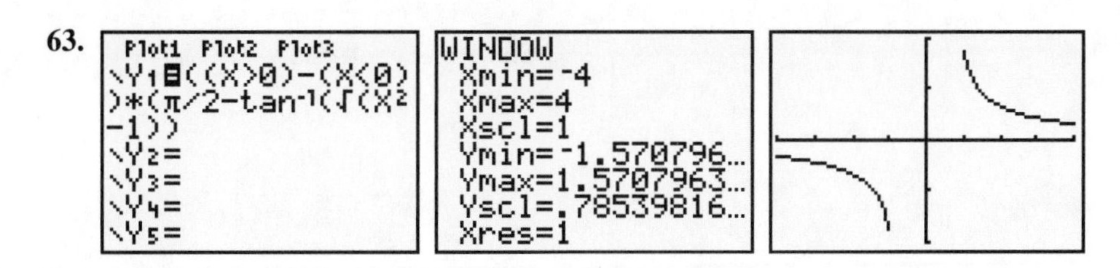

64. (a)

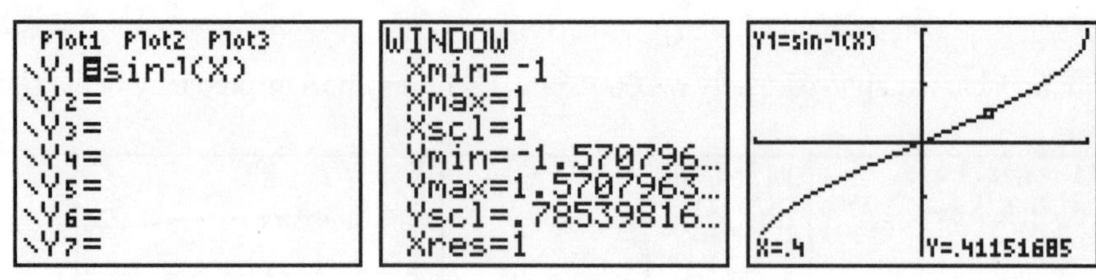

(b) In both cases, $\sin^{-1} .4 \approx .41151685$

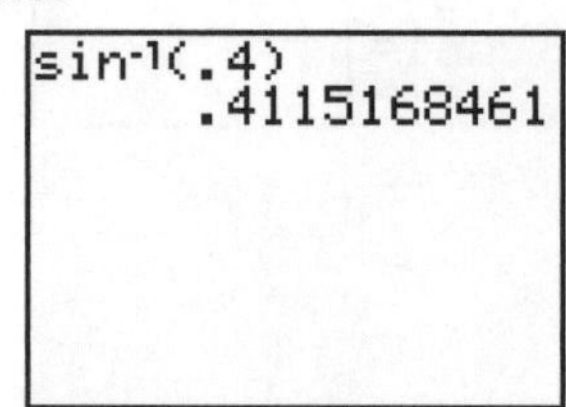

To obtain the screen in the annotated instructor edition, change the mode of the calculator to Horizontal.

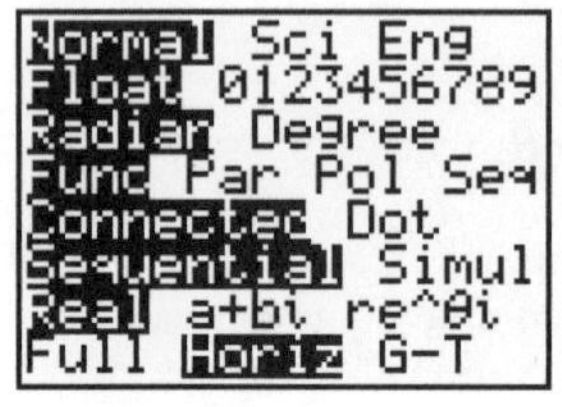

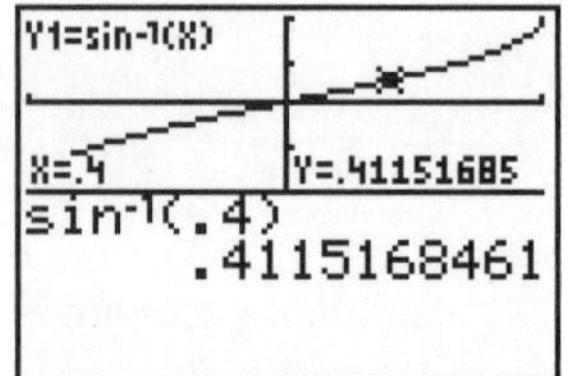

Chapter 6: Test

1.

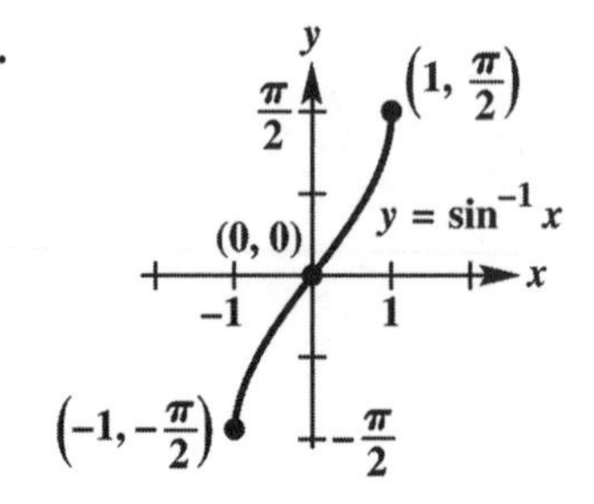

Domain: [–1, 1]; Range: $\left[-\frac{\pi}{2}, \frac{\pi}{2}\right]$

2. (a) $y = \arccos\left(-\frac{1}{2}\right) \Rightarrow \cos y = -\frac{1}{2}$

Since $0 \le y \le \pi$, $y = \frac{2\pi}{3}$.

(b) $y = \sin^{-1}\left(-\frac{\sqrt{3}}{2}\right) \Rightarrow \sin y = -\frac{\sqrt{3}}{2}$

Since $-\frac{\pi}{2} \le y \le \frac{\pi}{2}$, $y = -\frac{\pi}{3}$.

(c) $y = \tan^{-1} 0 \Rightarrow \tan y = 0$

Since $-\frac{\pi}{2} < y < \frac{\pi}{2}$, $y = 0$.

(d) $y = \text{arcsec}(-2) \Rightarrow \sec y = -2$

Since $0 \le y \le \pi$ and $y \ne \frac{\pi}{2}$, $y = \frac{2\pi}{3}$.

3. **(a)** $\cos\left(\arcsin\frac{2}{3}\right)$

Let $\arcsin\frac{2}{3}=u$, so that $\sin u=\frac{2}{3}$. Since arcsin is defined only in quadrants I and IV, and $\frac{2}{3}$ is positive, u is in quadrant I. Sketch u and label a triangle with the side adjacent u equal to $\sqrt{3^2-2^2}=\sqrt{9-4}=\sqrt{5}$.

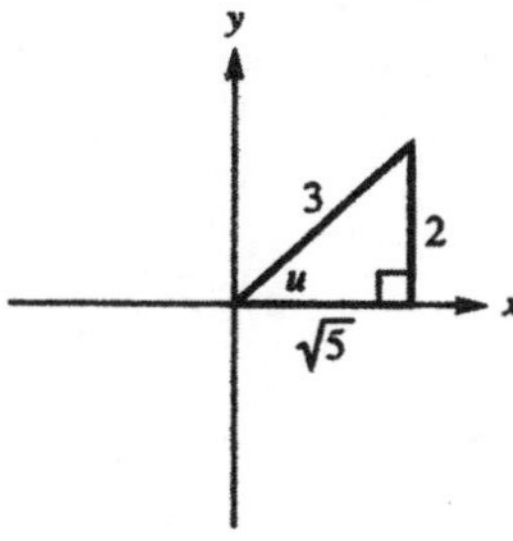

$$\cos\left(\arcsin\frac{2}{3}\right)=\cos u=\frac{\sqrt{5}}{3}$$

(b) $\sin\left(2\cos^{-1}\frac{1}{3}\right)$

Let $\theta=\cos^{-1}\frac{1}{3}$, so that $\cos\theta=\frac{1}{3}$. Since arccos is defined only in quadrants I and II, and $\frac{1}{3}$ is positive, θ is in quadrant I. Sketch θ and label a triangle with the side opposite to θ equal to $\theta=\sqrt{3^2-(-1)^2}=\sqrt{9-1}=\sqrt{8}=2\sqrt{2}$.

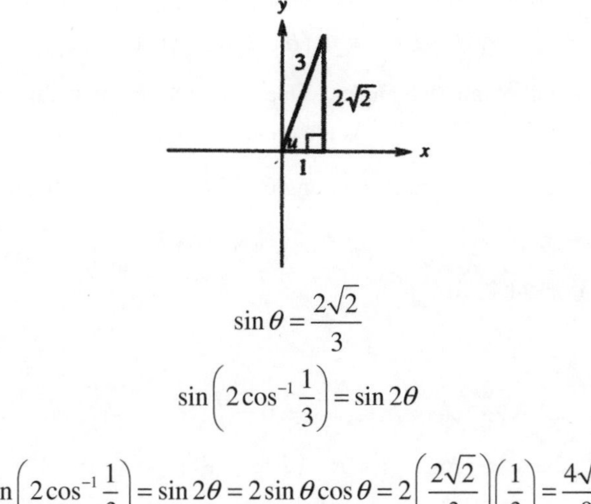

$$\sin\theta=\frac{2\sqrt{2}}{3}$$

$$\sin\left(2\cos^{-1}\frac{1}{3}\right)=\sin 2\theta$$

$$\sin\left(2\cos^{-1}\frac{1}{3}\right)=\sin 2\theta=2\sin\theta\cos\theta=2\left(\frac{2\sqrt{2}}{3}\right)\left(\frac{1}{3}\right)=\frac{4\sqrt{2}}{9}$$

4. Since $-1\le\sin\theta\le 1$, there is no value of θ such that $\sin\theta=3$. Thus, $\sin^{-1}3$ cannot be defined.

5. $\arcsin\left(\sin\frac{5\pi}{6}\right)=\arcsin\frac{1}{2}=\frac{\pi}{6}\neq\frac{5\pi}{6}$

6. $\tan(\arcsin u)$

Let $\theta = \arcsin u$, so $\sin\theta = u = \frac{u}{1}$. If $u > 0$, $0 < \theta < \frac{\pi}{2}$.

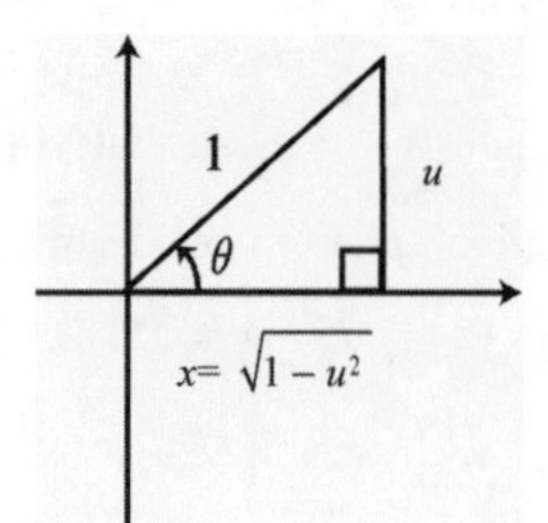

From the Pythagorean theorem, $x = \sqrt{1^2 - u^2} = \sqrt{1-u^2}$. Therefore, $\tan\theta = \frac{u}{\sqrt{1-u^2}} = \frac{u\sqrt{1-u^2}}{1-u^2}$. Thus, $\tan(\arcsin u) = \frac{u\sqrt{1-u^2}}{1-u^2}$.

7. $\sin^2\theta = \cos^2\theta + 1$

$$\sin^2\theta = \cos^2\theta + 1 \Rightarrow \cos^2\theta - \sin^2\theta = -1 \Rightarrow \cos 2\theta = -1$$

Since $0° \le \theta < 360°$, then $0° \le 2\theta < 720°$. Thus, $2\theta = 180°, 540° \Rightarrow \theta = 90°, 270°$.

Solution set: $\{90°, 270°\}$

8. $\csc^2\theta - 2\cot\theta = 4$

$$\csc^2\theta - 2\cot\theta = 4 \Rightarrow 1 + \cot^2\theta - 2\cot\theta = 4 \Rightarrow \cot^2\theta - 2\cot - 3 = 0 \Rightarrow (\cot\theta - 3)(\cot\theta + 1) = 0$$

$$\cot\theta + 1 = 0 \Rightarrow \cot\theta = -1 \Rightarrow \theta = 135° \text{ or } 315°$$

$$\cot\theta - 3 = 0 \Rightarrow \cot\theta = 3 \Rightarrow \theta = \cot^{-1} 3$$

Since $\cot\theta = 3 > 0$, we will obtain two angles. One angle will be in quadrant I and the other will be in quadrant III. Using a calculator, if $\cot\theta = 3$, the quadrant I angle will be approximately 18.4°. (Note: You need to calculate $\tan^{-1}\frac{1}{3}$ to obtain this angle.) The quadrant III angle will be approximately $180° + 18.4° = 198.4°$.

Solution set: $\{18.4°, 135°, 198.4°, 315°\}$

9. $\cos x = \cos 2x$

$$\cos x = \cos 2x \Rightarrow \cos 2x - \cos x = 0 \Rightarrow (2\cos^2 x - 1) - \cos x = 0$$

$$2\cos^2 x - \cos x - 1 = 0 \Rightarrow (2\cos x + 1)(\cos x - 1) = 0$$

$$2\cos x + 1 = 0 \text{ or } \cos x - 1 = 0$$

Over the interval $[0, 2\pi)$, we have the following.

$$2\cos x + 1 = 0 \Rightarrow 2\cos x = -1 \Rightarrow \cos x = -\frac{1}{2} \Rightarrow x = \frac{2\pi}{3} \text{ or } \frac{4\pi}{3}$$

$$\cos x - 1 = 0 \Rightarrow \cos x = 1 \Rightarrow x = 0$$

Solution set: $\left\{0, \frac{2\pi}{3}, \frac{4\pi}{3}\right\}$

10. $2\sqrt{3}\sin\frac{x}{2}=3\Rightarrow\sin\frac{x}{2}=\frac{3}{2\sqrt{3}}\Rightarrow\frac{x}{2}=\sin^{-1}\frac{3}{2\sqrt{3}}\Rightarrow\frac{x}{2}=\sin^{-1}\frac{\sqrt{3}}{2}$

Since $0\le x<2\pi\Rightarrow 0°\le\frac{x}{2}<\pi$, we have the following.

$$\frac{x}{2}=\frac{\pi}{3},\frac{2\pi}{3}\Rightarrow x=\frac{2\pi}{3},\frac{4\pi}{3}$$

Solution set: $\left\{\frac{2\pi}{3}+2\pi n,\ \frac{4\pi}{3}+2\pi n,\text{ where } n \text{ is any integer}\right\}$

11. (a) $y=\cos 3x\Rightarrow 3x=\arccos y\Rightarrow x=\frac{1}{3}\arccos y$

(b) $\arcsin x=\arctan\frac{4}{3}$

Let $\omega=\arctan\frac{4}{3}$. Then $\tan\omega=\frac{4}{3}$. Sketch ω in quadrant I and label a triangle with the hypotenuse equal to $\sqrt{4^2+3^2}=\sqrt{16+9}=\sqrt{25}=5$.

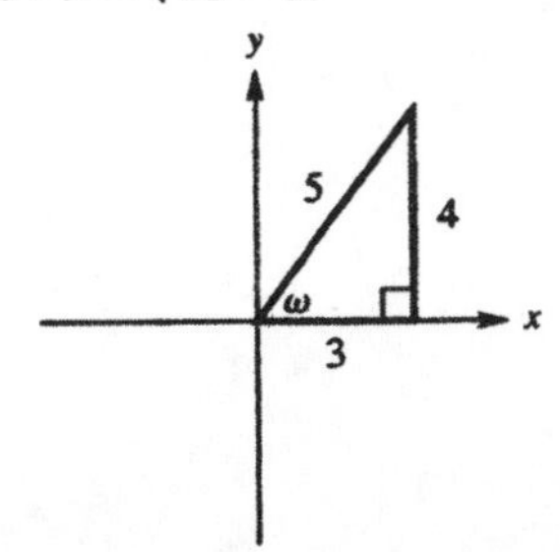

Thus, we have the following.

$$\arcsin x=\arctan\frac{4}{3}\Rightarrow\arcsin x=\omega\Rightarrow x=\sin\omega=\frac{4}{5}$$

12. $y=\frac{\pi}{8}\cos\left[\pi\left(t-\frac{1}{3}\right)\right]$

Solve $0=\frac{\pi}{8}\cos\left[\pi\left(t-\frac{1}{3}\right)\right]$.

$$0=\frac{\pi}{8}\cos\left[\pi\left(t-\frac{1}{3}\right)\right]\Rightarrow 0=\cos\left[\pi\left(t-\frac{1}{3}\right)\right]$$

$$\pi\left(t-\frac{1}{3}\right)=\arccos 0\Rightarrow\pi\left(t-\frac{1}{3}\right)=\frac{\pi}{2}+n\pi$$

$$t-\frac{1}{3}=\frac{1}{2}+n\Rightarrow t=\frac{1}{2}+\frac{1}{3}+n\Rightarrow t=\frac{5}{6}+n,\text{ where } n \text{ is any integer}$$

In the interval $[0,\pi)$, $n=0$, 1, and 2 provide valid values for t. Thus, we have the following.

$$t=\frac{5}{6},\ \frac{5}{6}+1,\ \frac{5}{6}+2\Rightarrow t=\frac{5}{6}\text{sec},\ \frac{11}{6}\text{sec},\ \frac{17}{6}\text{sec}$$

Chapter 6: Quantitative Reasoning

The amount of oil in the tank is 20 times the area shaded in the cross-sectional representation. The shaded area is the area of the sector of the circle defined by 2θ (let 2θ be the measure of the angle depicted in the figure) minus the area of the triangles.

Since $\tan\theta = \frac{\sqrt{8}}{1} = \sqrt{8}$, $\theta = \arctan\sqrt{8}$. Using the Pythagorean theorem, we have the following.

$$b^2 + 1^2 = 3^2 \Rightarrow b^2 + 1 = 9 \Rightarrow b^2 = 8 \Rightarrow b = \sqrt{8}, \text{ where } b \text{ is the base of one triangle}$$

The area of both triangles is $A_T = 2 \cdot \frac{1}{2} bh = 2 \cdot \frac{1}{2}\left(\sqrt{8}\right)(1) = \sqrt{8}$.

Then the sector area is $A_s = \frac{1}{2}(2\theta)r^2 = \left[\arctan\left(\sqrt{8}\right)\right](3)^2 = 9\arctan\sqrt{8}$ (with θ in radians). Thus, the volume of oil in the tank is the following.

$$V_{\text{oil}} = 20(A_s - A_T) = 20\left[9\arctan\left(\sqrt{8}\right) - \sqrt{8}\right] \approx 165 \text{ cubic ft}$$

Chapter 7

APPLICATIONS OF TRIGONOMETRY AND VECTORS

Section 7.1: Oblique Triangles and the Law of Sines

Connections (page 281)

$$X=\frac{(a-h)x}{f\sec\theta-y\sin\theta},\ Y=\frac{(a-h)y\cos\theta}{f\sec\theta-y\sin\theta}$$

1. House: $(x_H, y_H)=(.9,3.5)$; elevation 150 ft

Forest Fire: $(x_F, y_F)=(2.1,-2.4)$; elevation 690 ft

$a=7400$ ft; $f=6$ in.; $\theta=4.1°$

Coordinates of house:

$$X=\frac{(a-h)x}{f\sec\theta-y\sin\theta}\Rightarrow X=\frac{(7400-150)\cdot .9}{6\sec 4.1°-3.5\sin 4.1°}=\frac{6525}{6\sec 4.1°-3.5\sin 4.1°}\approx 1131.8\text{ ft}$$

$$Y=\frac{(a-h)y\cos\theta}{f\sec\theta-y\sin\theta}\Rightarrow Y=\frac{(7400-150)\cdot 3.5\cdot\cos 4.1°}{6\sec 4.1°-3.5\sin 4.1°}=\frac{25,375\cos 4.1°}{6\sec 4.1°-3.5\sin 4.1°}\approx 4390.2\text{ ft}$$

Coordinates of forest fire:

$$X=\frac{(a-h)x}{f\sec\theta-y\sin\theta}\Rightarrow X=\frac{(7400-690)\cdot 2.1}{6\sec 4.1°-(-2.4)\sin 4.1°}=\frac{14,091}{6\sec 4.1°+2.4\sin 4.1°}\approx 2277.5\text{ ft}$$

$$Y=\frac{(a-h)y\cos\theta}{f\sec\theta-y\sin\theta}\Rightarrow Y=\frac{(7400-690)\cdot(-2.4)\cdot\cos 4.1°}{6\sec 4.1°-(-2.4)\sin 4.1°}=\frac{-16,104\cos 4.1°}{6\sec 4.1°+2.4\sin 4.1°}\approx -2596.2\text{ ft}$$

2. The points we need to find the distance between are $(1131.8,4390.2)$ and $(2277.5,-2596.2)$.

Using the distance formula $d=\sqrt{(x_2-x_1)^2+(y_2-y_1)^2}$, we have the following.

$$d=\sqrt{(2277.5-1131.8)^2+(-2596.2-4390.2)^2}$$
$$=\sqrt{1145.7^2+(-6986.4)^2}=\sqrt{1,312,628.49+48,809,784.96}=\sqrt{50122413.45}\approx 7079.7\text{ ft}$$

Exercises

1. A: $\frac{a}{b}=\frac{\sin A}{\sin B}$ can be rewritten as $\frac{a}{\sin A}=\frac{b}{\sin B}$. This is a valid proportion.

B: $\frac{a}{\sin A}=\frac{b}{\sin B}$ is a valid proportion.

C: $\frac{\sin A}{a}=\frac{b}{\sin B}$ cannot be rewritten as $\frac{a}{\sin A}=\frac{b}{\sin B}$. $\frac{\sin A}{a}=\frac{b}{\sin B}$ is not a valid proportion.

D: $\frac{\sin A}{a}=\frac{\sin B}{b}$ is a valid proportion.

2. A: With two angles and the side included between them, we could have $\frac{a}{\underbrace{\sin A}} = \frac{b}{\underbrace{\sin B}} = \frac{\overbrace{c}}{\sin C}$. If you know the measure of angles A and B, you can determine the measure of angle C. This provides enough information to solve the triangle.

B: With two angles and the side opposite one of them, we could have $\frac{a}{\underbrace{\sin A}} = \frac{\overbrace{b}}{\underbrace{\sin B}} = \frac{c}{\sin C}$. If you know the measure of angles A and B, you can determine the measure of angle C. This provides enough information to solve the triangle.

C: With two sides and the angle included between them, we could have $\frac{a}{\underbrace{\sin A}} = \frac{\overbrace{b}}{\sin B} = \frac{\overbrace{c}}{\sin C}$. This does not provide enough information to solve the triangle.

D: With three sides we have, $\frac{\overbrace{a}}{\sin A} = \frac{\overbrace{b}}{\sin B} = \frac{\overbrace{c}}{\sin C}$. This does not provide enough information to solve the triangle.

3. The measure of angle C is $180° - (60° + 75°) = 180° - 135° = 45°$.

$$\frac{a}{\sin A} = \frac{c}{\sin C} \Rightarrow \frac{a}{\sin 60°} = \frac{\sqrt{2}}{\sin 45°} \Rightarrow a = \frac{\sqrt{2}\sin 60°}{\sin 45°} = \frac{\sqrt{2} \cdot \frac{\sqrt{3}}{2}}{\frac{\sqrt{2}}{2}} = \sqrt{2} \cdot \frac{\sqrt{3}}{2} \cdot \frac{2}{\sqrt{2}} = \sqrt{3}$$

4. The measure of angle B is $180° - (45° + 105°) = 180° - 150° = 30°$.

$$\frac{a}{\sin A} = \frac{b}{\sin b} \Rightarrow \frac{a}{\sin 45°} = \frac{10}{\sin 30°} \Rightarrow a = \frac{10\sin 45°}{\sin 30°} = \frac{10 \cdot \frac{\sqrt{2}}{2}}{\frac{1}{2}} = 10 \cdot \frac{\sqrt{2}}{2} \cdot \frac{2}{1} = 10\sqrt{2}$$

5. $A = 37°$, $B = 48°$, $c = 18$ m

$C = 180° - A - B \Rightarrow C = 180° - 37° - 48° = 95°$

$$\frac{b}{\sin B} = \frac{c}{\sin C} \Rightarrow \frac{b}{\sin 48°} = \frac{18}{\sin 95°} \Rightarrow b = \frac{18\sin 48°}{\sin 95°} \approx 13 \text{ m}$$
$$\frac{a}{\sin A} = \frac{c}{\sin C} \Rightarrow \frac{a}{\sin 37°} = \frac{18}{\sin 95°} \Rightarrow a = \frac{18\sin 37°}{\sin 95°} \approx 11 \text{ m}$$

6. $B = 52°$, $C = 29°$, $a = 43$ cm

$A = 180° - B - C \Rightarrow A = 180° - 52° - 29° = 99°$

$$\frac{a}{\sin A} = \frac{b}{\sin B} \Rightarrow \frac{43}{\sin 99°} = \frac{b}{\sin 52°} \Rightarrow b = \frac{43\sin 52°}{\sin 99°} \approx 34 \text{ cm}$$
$$\frac{a}{\sin A} = \frac{c}{\sin C} \Rightarrow \frac{43}{\sin 99°} = \frac{c}{\sin 29°} \Rightarrow c = \frac{43\sin 29°}{\sin 99°} \approx 21 \text{ cm}$$

7. $A = 27.2°$, $C = 115.5°$, $c = 76.0$ ft

$B = 180° - A - C \Rightarrow B = 180° - 27.2° - 115.5° = 37.3°$

$$\frac{a}{\sin A} = \frac{c}{\sin C} \Rightarrow \frac{a}{\sin 27.2°} = \frac{76.0}{\sin 115.5°} \Rightarrow a = \frac{76.0 \sin 27.2°}{\sin 115.5°} \approx 38.5 \text{ ft}$$
$$\frac{b}{\sin B} = \frac{c}{\sin C} \Rightarrow \frac{b}{\sin 37.3°} = \frac{76.0}{\sin 115.5°} \Rightarrow b = \frac{76.0 \sin 37.3°}{\sin 115.5°} \approx 51.0 \text{ ft}$$

8. $C = 124.1°$, $B = 18.7°$, $b = 94.6$ m

$A = 180° - B - C \Rightarrow A = 180° - 18.7° - 124.1° = 37.2°$

$$\frac{a}{\sin A} = \frac{b}{\sin B} \Rightarrow \frac{a}{\sin 37.2°} = \frac{94.6}{\sin 18.7°} \Rightarrow a = \frac{94.6 \sin 37.2°}{\sin 18.7°} \approx 178 \text{ m}$$
$$\frac{c}{\sin C} = \frac{b}{\sin B} \Rightarrow \frac{c}{\sin 124.1°} = \frac{94.6}{\sin 18.7°} \Rightarrow c = \frac{94.6 \sin 124.1°}{\sin 18.7°} \approx 244 \text{ m}$$

9. $A = 68.41°$, $B = 54.23°$, $a = 12.75$ ft

$C = 180° - A - B - C = 180° - 68.41° - 54.23° = 57.36°$

$$\frac{a}{\sin A} = \frac{b}{\sin B} \Rightarrow \frac{12.75}{\sin 68.41°} = \frac{b}{\sin 54.23°} \Rightarrow b = \frac{12.75 \sin 54.23°}{\sin 68.41°} \approx 11.13 \text{ ft}$$
$$\frac{a}{\sin A} = \frac{c}{\sin C} \Rightarrow \frac{12.75}{\sin 68.41°} = \frac{c}{\sin 57.36°} \Rightarrow c = \frac{12.75 \sin 57.36°}{\sin 68.41°} \approx 11.55 \text{ ft}$$

10. $C = 74.08°$, $B = 69.38°$, $c = 45.38$ m

$A = 180° - B - C \Rightarrow A = 180° - 69.38° - 74.08° = 36.54°$

$$\frac{a}{\sin A} = \frac{c}{\sin C} \Rightarrow \frac{a}{\sin 36.54°} = \frac{45.38}{\sin 74.08°} \Rightarrow a = \frac{45.38 \sin 36.54°}{\sin 74.08°} \approx 28.10 \text{ m}$$
$$\frac{b}{\sin B} = \frac{c}{\sin C} \Rightarrow \frac{b}{\sin 69.38°} = \frac{45.38}{\sin 74.08°} \Rightarrow b = \frac{45.38 \sin 69.38°}{\sin 74.08°} \approx 44.17 \text{ m}$$

11. $A = 87.2°, b = 75.9\text{yd}, C = 74.3°$

$B = 180° - A - C \Rightarrow B = 180° - 87.2° - 74.3° = 18.5°$

$$\frac{a}{\sin A} = \frac{b}{\sin B} \Rightarrow \frac{a}{\sin 87.2°} = \frac{75.9}{\sin 18.5°} \Rightarrow a = \frac{75.9 \sin 87.2°}{\sin 18.5°} \approx 239 \text{yd}$$
$$\frac{b}{\sin B} = \frac{c}{\sin C} \Rightarrow \frac{75.9}{\sin 18.5°} = \frac{c}{\sin 74.3°} \Rightarrow c = \frac{75.9 \sin 74.3°}{\sin 18.5°} \approx 230 \text{yd}$$

12. $B = 38°40', a = 19.7\text{cm}, C = 91°40'$

$A = 180° - B - C \Rightarrow A = 180° - 38°40' - 91°40' = 180° - 130°20' = 179°60' - 130°20' = 49°40'$

$$\frac{a}{\sin A} = \frac{b}{\sin B} \Rightarrow \frac{19.7}{\sin 49°40'} = \frac{b}{\sin 38°40'} \Rightarrow b = \frac{19.7 \sin 38°40'}{\sin 49°40'} \approx 16.1 \text{cm}$$
$$\frac{c}{\sin C} = \frac{a}{\sin A} \Rightarrow \frac{c}{\sin 91°40'} = \frac{19.7}{\sin 49°40'} \Rightarrow c = \frac{19.7 \sin 91°40'}{\sin 49°40'} \approx 25.8 \text{ cm}$$

13. $B = 20°50'$, $AC = 132$ ft, $C = 103°10'$

$A = 180° - B - C \Rightarrow A = 180° - 20°50' - 103°10' \Rightarrow A = 56°00'$

$$\frac{AC}{\sin B} = \frac{AB}{\sin C} \Rightarrow \frac{132}{\sin 20°50'} = \frac{AB}{\sin 103°10'} \Rightarrow AB = \frac{132 \sin 103°10'}{\sin 20°50'} \approx 361 \text{ ft}$$

$$\frac{BC}{\sin A} = \frac{AC}{\sin B} \Rightarrow \frac{BC}{\sin 56°00'} = \frac{132}{\sin 20°50'} \Rightarrow BC = \frac{132 \sin 56°00'}{\sin 20°50'} \approx 308 \text{ ft}$$

14. $A = 35.3°$, $B = 52.8°$, $AC = 675$ ft

$C = 180° - A - B \Rightarrow C = 180° - 35.3° - 52.8° = 91.9°$

$$\frac{BC}{\sin A} = \frac{AC}{\sin B} \Rightarrow \frac{BC}{\sin 35.3°} = \frac{675}{\sin 52.8°} \Rightarrow BC = \frac{675 \sin 35.3°}{\sin 52.8°} \approx 490 \text{ ft}$$

$$\frac{AB}{\sin C} = \frac{AC}{\sin B} \Rightarrow \frac{AB}{\sin 91.9°} = \frac{675}{\sin 52.8°} \Rightarrow AB = \frac{675 \sin 91.9°}{\sin 52.8°} \approx 847 \text{ ft}$$

15. $A = 39.70°, C = 30.35°$, $b = 39.74$ m

$B = 180° - A - C \Rightarrow B = 180° - 39.70° - 30.35° \Rightarrow B = 109.95° \approx 110.0°$ (rounded)

$$\frac{a}{\sin A} = \frac{b}{\sin B} \Rightarrow \frac{a}{\sin 39.70°} = \frac{39.74}{\sin 109.95°} \Rightarrow a = \frac{39.74 \sin 39.70°}{\sin 110.0°} \approx 27.01 \text{ m}$$

$$\frac{b}{\sin B} = \frac{c}{\sin C} \Rightarrow \frac{39.74}{\sin 109.95°} = \frac{c}{\sin 30.35°} \Rightarrow c = \frac{39.74 \sin 30.35°}{\sin 110.0°} \approx 21.37 \text{ m}$$

16. $C = 71.83°$, $B = 42.57°$, $a = 2.614$ cm

$A = 180° - B - C \Rightarrow A = 180° - 42.57° - 71.83° = 65.60°$

$$\frac{b}{\sin B} = \frac{a}{\sin A} \Rightarrow \frac{b}{\sin 42.57°} = \frac{2.614}{\sin 65.60°} \Rightarrow b = \frac{2.614 \sin 42.57°}{\sin 65.60°} \approx 1.942 \text{ cm}$$

$$\frac{c}{\sin C} = \frac{a}{\sin A} \Rightarrow \frac{c}{\sin 71.83°} = \frac{2.614}{\sin 65.60°} \Rightarrow c = \frac{2.614 \sin 71.83°}{\sin 65.60°} \approx 2.727 \text{ cm}$$

17. $B = 42.88°$, $C = 102.40°$, $b = 3974$ ft

$A = 180° - B - C \Rightarrow A = 180° - 42.88° - 102.40° = 34.72°$

$$\frac{a}{\sin A} = \frac{b}{\sin B} \Rightarrow \frac{a}{\sin 34.72°} = \frac{3974}{\sin 42.88°} \Rightarrow a = \frac{3974 \sin 34.72°}{\sin 42.88°} \approx 3326 \text{ ft}$$

$$\frac{b}{\sin B} = \frac{c}{\sin C} \Rightarrow \frac{3974}{\sin 42.88°} = \frac{c}{\sin 102.40°} \Rightarrow c = \frac{3974 \sin 102.40°}{\sin 42.88°} \approx 5704 \text{ ft}$$

18. $A = 18.75°$, $B = 51.53°$, $c = 2798$ yd

$C = 180° - A - B \Rightarrow C = 180° - 18.75° - 51.53° = 109.72°$

$$\frac{a}{\sin A} = \frac{c}{\sin C} \Rightarrow \frac{a}{\sin 18.75°} = \frac{2798}{\sin 109.72°} \Rightarrow a = \frac{2798 \sin 18.75°}{\sin 109.72°} \approx 955.4 \text{ yd}$$

$$\frac{b}{\sin B} = \frac{c}{\sin C} \Rightarrow \frac{b}{\sin 51.53°} = \frac{2798}{\sin 109.72°} \Rightarrow b = \frac{2798 \sin 51.53°}{\sin 109.72°} \approx 2327 \text{ yd}$$

19. $A = 39°54', a = 268.7\text{m}, B = 42°32'$

$C = 180° - A - B \Rightarrow C = 180° - 39°54' - 42°32' = 179°60' - 82°26' = 97°34'$

$$\frac{a}{\sin A} = \frac{b}{\sin B} \Rightarrow \frac{268.7}{\sin 39°54'} = \frac{b}{\sin 42°32'} \Rightarrow b = \frac{268.7 \sin 42°32'}{\sin 39°54'} \approx 283.2\text{m}$$

$$\frac{a}{\sin A} = \frac{c}{\sin C} \Rightarrow \frac{268.7}{\sin 39°54'} = \frac{c}{\sin 97°34'} \Rightarrow c = \frac{268.7 \sin 97°34'}{\sin 39°54'} \approx 415.2\text{m}$$

20. $C = 79°18', c = 39.81\text{mm}, A = 32°57'$

$B = 180° - A - C \Rightarrow B = 180° - 32°57' - 79°18' = 179°60' - 112°15' = 67°45'$

$$\frac{a}{\sin A} = \frac{c}{\sin C} \Rightarrow \frac{a}{\sin 32°57'} = \frac{39.81}{\sin 79°18'} \Rightarrow a = \frac{39.81 \sin 32°57'}{\sin 79°18'} \approx 22.04\,\text{mm}$$

$$\frac{b}{\sin B} = \frac{c}{\sin C} \Rightarrow \frac{b}{\sin 67°45'} = \frac{39.81}{\sin 79°18'} \Rightarrow b = \frac{39.81 \sin 67°45'}{\sin 79°18'} \approx 37.50\,\text{mm}$$

21. With three sides we have, $\frac{\overbrace{a}}{\sin A} = \frac{\overbrace{b}}{\sin B} = \frac{\overbrace{c}}{\sin C}$. This does not provide enough information to solve the triangle. Whenever you choose two out of the three ratios to create a proportion, you are missing two pieces of information.

22. In Example 1, the Pythagorean theorem does not apply because triangle *ABC* is not a right triangle. The measure of angle *A* is 32.0°, *B* = 81.8°, therefore *C* = 66.2°. One of the angles has to be 90° for *ABC* to be a right triangle.

23. This is not a valid statement. It is true that if you have the measures of two angles, the third can be found. However, if you do not have at least one side, the triangle can not be uniquely determined. If one considers the congruence axiom involving two angles, ASA, an included side must be considered.

24. No. If *a* is twice as long as *b*, then we have the following.

$$\frac{a}{\sin A} = \frac{b}{\sin B} \Rightarrow \frac{2b}{\sin A} = \frac{b}{\sin B} \Rightarrow 2b \sin B = b \sin A \Rightarrow 2 \sin B = \sin A$$

$\sin A = 2 \sin B$, but *A* is not necessarily twice as large as *B*.

25. $B = 112°10'; C = 15°20'; BC = 354$ m

$A = 180° - B - C$

$A = 180° - 112°10' - 15°20'$

$= 179°60' - 127°30' = 52°30'$

$$\frac{BC}{\sin A} = \frac{AB}{\sin C}$$

$$\frac{354}{\sin 52°30'} = \frac{AB}{\sin 15°20'}$$

$$AB = \frac{354 \sin 15°20'}{\sin 52°30'}$$

$AB \approx 118$ m

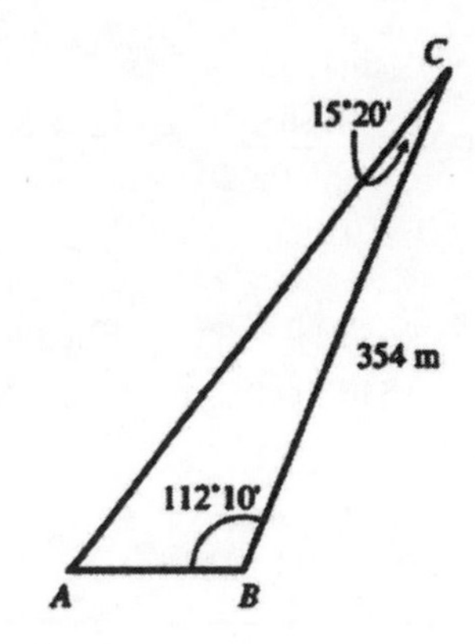

26. $T = 32°50'$; $R = 102°20'$; $TR = 582$ yd

$$S = 180° - 32°50' - 102°20'$$
$$= 179°60' - 135°10' = 44°50'$$

$$\frac{RS}{\sin T} = \frac{TR}{\sin S}$$
$$\frac{RS}{\sin 32°50'} = \frac{582}{\sin 44°50'}$$
$$RS = \frac{582 \sin 32°50'}{\sin 44°50'}$$
$$RS \approx 448 \text{ yd}$$

27. Let d = the distance the ship traveled between the two observations;
L = the location of the lighthouse.

$$L = 180° - 38.8° - 44.2$$
$$= 97.0°$$

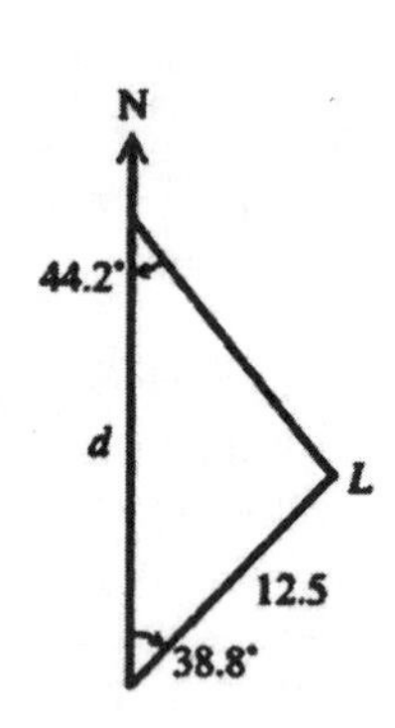

$$\frac{d}{\sin 97°} = \frac{12.5}{\sin 44.2°}$$
$$d = \frac{12.5 \sin 97°}{\sin 44.2°}$$
$$d \approx 17.8 \text{ km}$$

28. Let C = the transmitter

Since side AB is on an east-west line, the angle between it and any north-south line is 90°.

$$A = 90° - 47.7° = 42.3°$$
$$B = 302.5° - 270° = 32.5°$$
$$C = 180° - A - B$$
$$= 180° - 42.3° - 32.5° = 105.2°$$

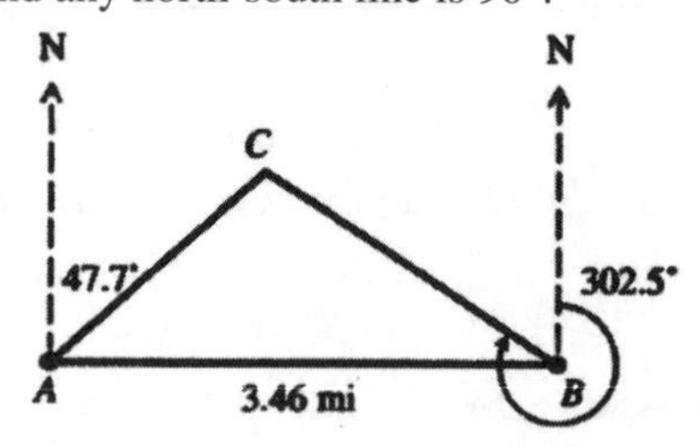

$$\frac{AC}{\sin 32.5°} = \frac{3.46}{\sin 105.2°}$$
$$AC = \frac{3.46 \sin 32.5°}{\sin 105.2°}$$
$$AC \approx 1.93 \text{ mi}$$

29. $\frac{x}{\sin 54.8°} = \frac{12.0}{\sin 70.4°} \Rightarrow x = \frac{12.0 \sin 54.8°}{\sin 70.4°} \approx 10.4$ in.

30. Let M = Mark's location;
L = Lisa's location;
T = the tree's location;
R = a point across the river from Mark so that MR is the distance across the river.

$\theta = 180° - 115.45° = 64.55°$

$\alpha = 180° - 45.47° - 64.55° = 69.98°$

$\beta = \theta = 64.55°$ (Alternate interior angles)

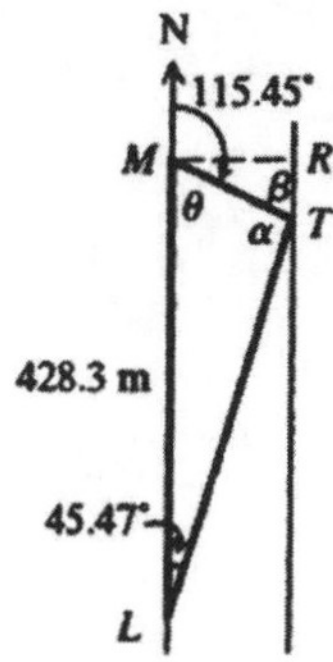

In triangle MTL, $\dfrac{MT}{\sin 45.47°} = \dfrac{428.3}{\sin 69.98°} \Rightarrow MT = \dfrac{428.3 \sin 45.47°}{\sin 69.98°} \approx 324.9645.$

In right triangle MTR, we have the following.

$$\frac{MR}{MT} = \sin\beta \Rightarrow \frac{MR}{324.9645} = \sin 64.55° \Rightarrow MR = 324.9645 \sin 64.55° \approx 293.4 \text{ m}$$

31. We cannot find θ directly because the length of the side opposite angle θ is not given. Redraw the triangle shown in the figure and label the third angle as α.

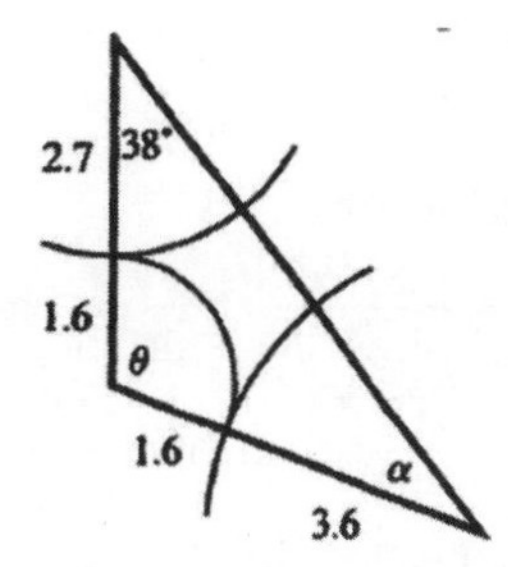

$$\frac{\sin\alpha}{1.6+2.7} = \frac{\sin 38°}{1.6+3.6}$$

$$\frac{\sin\alpha}{4.3} = \frac{\sin 38°}{5.2}$$

$$\sin\alpha = \frac{4.3 \sin 38°}{5.2} \approx .50910468$$

$$\alpha \approx \sin^{-1}(.50910468) \approx 31°$$

$$\theta = 180° - 38° - \alpha \approx 180° - 38° - 31° = 111°$$

32. Label the centers of the atoms A, B, and C.

$a = 2.0 + 3.0 = 5.0$

$c = 3.0 + 4.5 = 7.5$

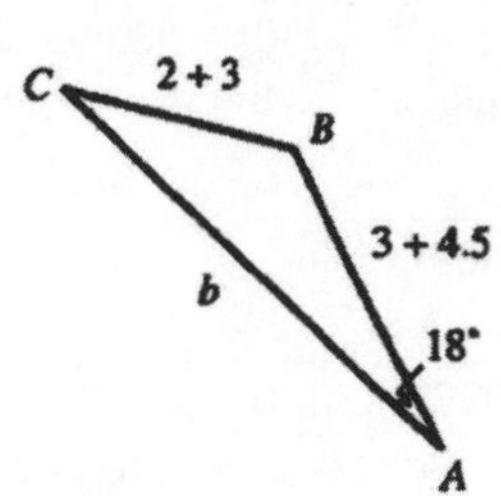

$$\frac{\sin C}{c} = \frac{\sin A}{a} \Rightarrow \frac{\sin C}{7.5} = \frac{\sin 18°}{5} \Rightarrow \sin C = \frac{7.5 \sin 18°}{5} \Rightarrow \sin C \approx .46352549$$

$C \approx \sin^{-1}(.46352549) \approx 28°$

$B \approx 180° - 18° - 28° = 134°$

$$\frac{b}{\sin B} = \frac{a}{\sin A} \Rightarrow \frac{b}{\sin 134°} = \frac{5}{\sin 18°} \Rightarrow b = \frac{5.0 \sin 134°}{\sin 18°} \approx 12$$

The distance between the centers of atoms A and C is 12.

33. Let x = the distance to the lighthouse at bearing N 37° E;
y = the distance to the lighthouse at bearing N 25° E.

$\theta = 180° - 37° = 143°$

$\alpha = 180° - \theta - 25°$
$= 180° - 143° - 25° = 12°$

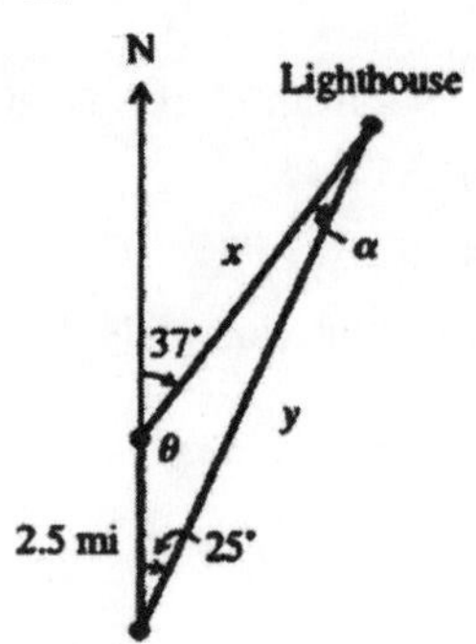

$$\frac{2.5}{\sin\alpha} = \frac{x}{\sin 25°} \Rightarrow x = \frac{2.5\sin 25°}{\sin 12°} \approx 5.1 \text{ mi}$$

$$\frac{2.5}{\sin\alpha} = \frac{y}{\sin\theta} \Rightarrow \frac{2.5}{\sin 12°} = \frac{y}{\sin 143°} \Rightarrow y = \frac{2.5\sin 143°}{\sin 12°} \approx 7.2 \text{ mi}$$

34. Let A = the location of the balloon;
B = the location of the farther town;
C = the location of the closer town.

Angle ABC = 31° and angle ACB = 35° because the angles of depression are alternate interior angles with the angles of the triangle.

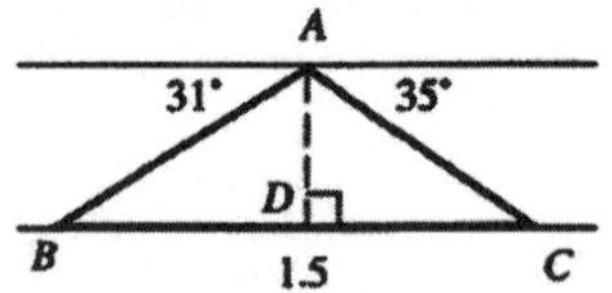

Angle BAC = 180° – 31° – 35° = 114°

$$\frac{1.5}{\sin BAC} = \frac{AB}{\sin ACB} \Rightarrow \frac{1.5}{\sin 114°} = \frac{AB}{\sin 35°} \Rightarrow AB = \frac{1.5\sin 35°}{\sin 114°} \approx .94$$

$$\sin ABC = \frac{AD}{AB} \Rightarrow \sin 31° = \frac{AD}{.94} \Rightarrow AD = .94 \cdot \sin 31° \approx .49$$

The balloon is .49 mi above the ground.

35. Angle C is equal to the difference between the angles of elevation.
$C = B - A = 52.7430° - 52.6997° = .0433°$
The distance BC to the moon can be determined using the law of sines.

$$\frac{BC}{\sin A} = \frac{AB}{\sin C}$$

$$\frac{BC}{\sin 52.6997°} = \frac{398}{\sin .0433°}$$

$$BC = \frac{398\sin 52.6997°}{\sin .0433°}$$

$$BC \approx 418{,}930 \text{ km}$$

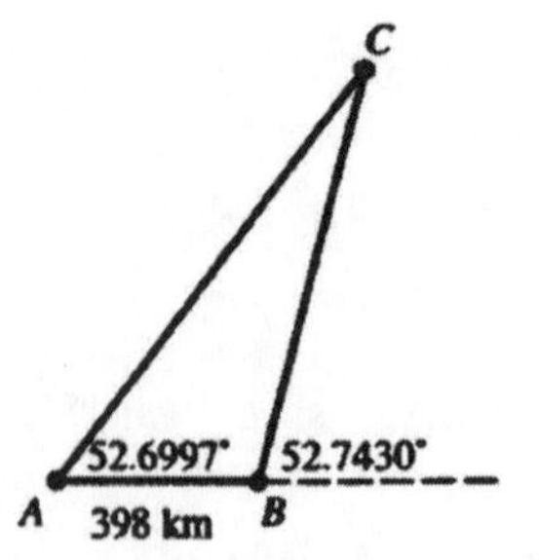

If one finds distance AC, then we have the following.

$$\frac{AC}{\sin B} = \frac{AB}{\sin C} \Rightarrow \frac{AC}{\sin(180° - 52.7430°)} = \frac{398}{\sin .0433°} \Rightarrow \frac{AC}{\sin 127.2570°} = \frac{398}{\sin .0433°}$$

$$AC = \frac{398\sin 127.2570°}{\sin .0433°} \approx 419{,}171 \text{ km}$$

In either case the distance is approximately 419,000 km compared to the actual value of 406,000 km.

36. We must find the length of *CD*

Angle *BAC* is equal to 35° - 30° = 5°.

Side *AB* = 5000ft. Since triangle *ABC* is a right triangle,

$$\cos 5° = \frac{5000}{AC} \Rightarrow AC = \frac{5000}{\cos 5°} \approx 5019\text{ft.}$$

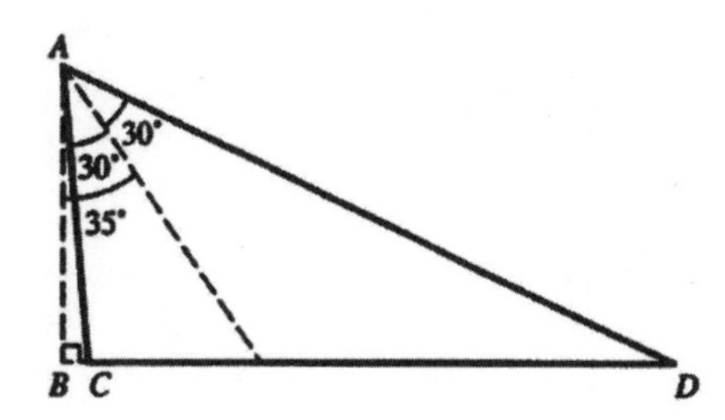

The angular coverage of the lens is 60°, so angle *CAD* = 60°. From geometry, angle *ACB* = 85°, angle *ACD* = 95°, and angle *ADC* = 25°. We now know three angles and one side in triangle ACD and can now use the law of sines to solve for the length of CD.

$$\frac{CD}{\sin 60°} = \frac{AC}{\sin 25°} \Rightarrow \frac{CD}{\sin 60°} = \frac{5019}{\sin 25°} \Rightarrow CD = \frac{5019 \sin 60°}{\sin 25°} \approx 10,285 \text{ ft}$$

The photograph would cover a horizontal distance of approximately 10,285 ft or $\frac{10,285}{5280} \approx 1.95$ mi

37. Determine the length of *CB*, which is the sum of *CD* and *DB*.

Triangle *ACE* is isosceles, so angles *ACE* and *AEC* are both equal to 47°. It follows that angle *ACD* is equal to $47° - 5° = 42°$. Angles *CAD* and *DAB* both equal 43° since the photograph was taken with no tilt and *AD* is the bisector of angle *CAB*. The length of *AD* is 3500 ft. Since the measures of the angles in triangle *ABC* total 180°, angle *ABC* equals $180° - (86° + 42°) = 52°$.

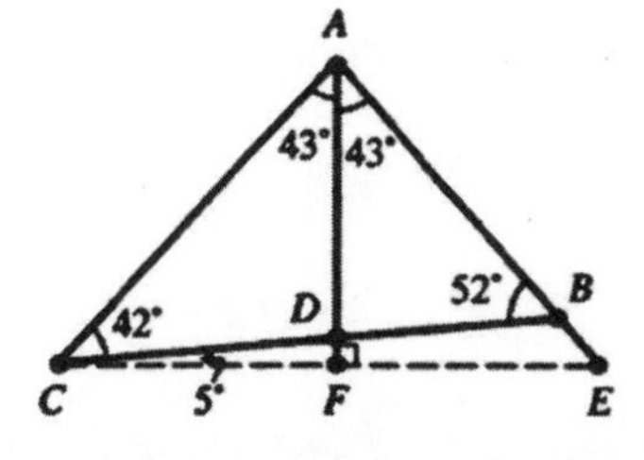

Using the law of sines, we have $CD = \frac{3500 \sin 43°}{\sin 42°} \approx 3567.3$ ft and $DB = \frac{3500 \sin 43°}{\sin 52°} = 3029.1$ ft.

The ground distance in the resulting photograph will equal the length of segment *CB*.

Thus, $CB = CD + DB = 3567.3 + 3029.1 = 6596.4$ ft or approximately 6600 ft.

38. Determine the length of *CB*, which is the sum of *CD* and *DB*.

Triangle *ACE* is isosceles, so angles *ACE* and *AEC* are both equal to 54°. It follows that angle *ACB* is equal to $54° - 5° = 49°$. Angles *CAD* and *DAB* both equal 36° since the photograph is taken with no tilt, and *AD* is the bisector of *CAB* with length 3500 ft. Since the measures of the angles in the triangle *ABC* total 180°, the angle *ABC* equals $180° - (72° + 49°) = 59°$.

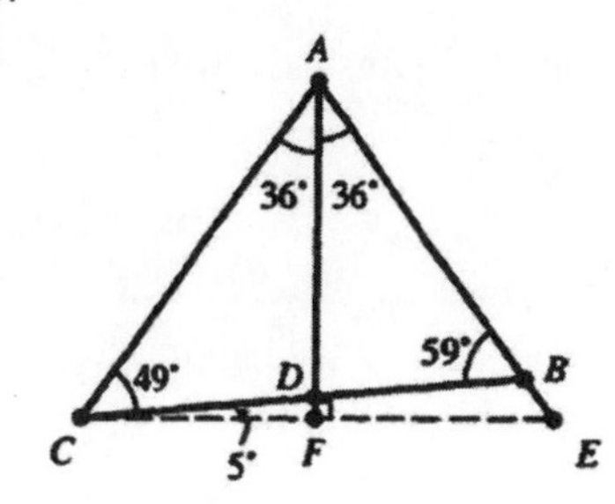

In the figure, angle *CAB* is 72°. Using the law of sines we have the following.

$$CD = \frac{3500 \sin 36°}{\sin 49°} = 2725.9 \text{ ft and } DB = \frac{3500 \sin 36°}{\sin 59°} = 2400.1 \text{ ft.}$$

The ground distance in the resulting photograph will equal the length of segment *CB*.

Thus, $CB = CD + DB = 2725.9 + 2400.1 = 5126$ ft or approximately 5100 ft.

39. To find the area of the triangle, use $\mathcal{A} = \frac{1}{2}bh,$ with $b = 1$ and $h = \sqrt{3}.$

$$\mathcal{A} = \frac{1}{2}(1)\left(\sqrt{3}\right) = \frac{\sqrt{3}}{2}$$

Now use $\mathcal{A} = \frac{1}{2}ab\sin C,$ with $a = \sqrt{3},\ b = 1,$ and $C = 90°.$

$$\mathcal{A} = \frac{1}{2}\left(\sqrt{3}\right)(1)\sin 90° = \frac{1}{2}\left(\sqrt{3}\right)(1)(1) = \frac{\sqrt{3}}{2}$$

40. To find the area of the triangle, use $\mathcal{A} = \frac{1}{2}bh,$ with $b = 2$ and $h = \sqrt{3}.$

$$\mathcal{A} = \frac{1}{2}(2)\left(\sqrt{3}\right) = \sqrt{3}.$$

Now use $\mathcal{A} = \frac{1}{2}ab\sin C,$ with $a = 2,\ b = 2,$ and $C = 60°.$

$$\mathcal{A} = \frac{1}{2}(2)(2)\sin 60° = \frac{1}{2}(2)(2)\left(\frac{\sqrt{3}}{2}\right) = \sqrt{3}.$$

41. To find the area of the triangle, use $\mathcal{A} = \frac{1}{2}bh,$ with $b = 1$ and $h = \sqrt{2}.$

$$\mathcal{A} = \frac{1}{2}(1)\left(\sqrt{2}\right) = \frac{\sqrt{2}}{2}$$

Now use $\mathcal{A} = \frac{1}{2}ab\sin C,$ with $a = 2,\ b = 1,$ and $C = 45°.$

$$\mathcal{A} = \frac{1}{2}(2)(1)\sin 45° = \frac{1}{2}(2)(1)\left(\frac{\sqrt{2}}{2}\right) = \frac{\sqrt{2}}{2}$$

42. To find the area of the triangle, use $\mathcal{A} = \frac{1}{2}bh,$ with $b = 2$ and $h = 1.$

$$\mathcal{A} = \frac{1}{2}(2)(1) = 1$$

Now use $\mathcal{A} = \frac{1}{2}ab\sin C,$ with $a = 2,\ b = \sqrt{2},$ and $C = 45°.$

$$\mathcal{A} = \frac{1}{2}(2)\left(\sqrt{2}\right)\sin 45° = \frac{1}{2}(2)\left(\sqrt{2}\right)\left(\frac{\sqrt{2}}{2}\right) = 1$$

43. $A = 42.5°,\ b = 13.6$ m, $c = 10.1$ m
Angle A is included between sides b and c. Thus, we have the following.

$$\mathcal{A} = \frac{1}{2}bc\sin A = \frac{1}{2}(13.6)(10.1)\sin 42.5° \approx 46.4 \text{ m}^2$$

44. $C = 72.2°,\ b = 43.8$ ft, $a = 35.1$ ft
Angle C is included between sides a and b. Thus, we have the following.

$$\mathcal{A} = \frac{1}{2}ab\sin C = \frac{1}{2}(35.1)(43.8)\sin 72.2° \approx 732 \text{ ft}^2$$

45. $B = 124.5°$, $a = 30.4$ cm, $c = 28.4$ cm
Angle B is included between sides a and c. Thus, we have the following.

$$\mathcal{A} = \frac{1}{2}ac\sin B = \frac{1}{2}(30.4)(28.4)\sin 124.5° \approx 356 \text{ cm}^2$$

46. $C = 142.7°$, $a = 21.9$ km, $b = 24.6$ km
Angle C is included between sides a and b. Thus, we have the following.

$$\mathcal{A} = \frac{1}{2}ab\sin C = \frac{1}{2}(21.9)(24.6)\sin 142.7° \approx 163 \text{ km}^2$$

47. $A = 56.80°$, $b = 32.67$ in., $c = 52.89$ in.
Angle A is included between sides b and c. Thus, we have the following.

$$\mathcal{A} = \frac{1}{2}bc\sin A = \frac{1}{2}(32.67)(52.89)\sin 56.80° \approx 722.9 \text{ in.}^2$$

48. $A = 34.97°$, $b = 35.29$ m, $c = 28.67$ m
Angle A is included between sides b and c. Thus, we have the following.

$$\mathcal{A} = \frac{1}{2}bc\sin A = \frac{1}{2}(35.29)(28.67)\sin 34.97° \approx 289.9 \text{ m}^2$$

49. $A = 30.50°$, $b = 13.00$ cm, $C = 112.60°$
In order to use the area formula, we need to find either a or c.
$B = 180° - A - C \Rightarrow B = 180° - 30.50° - 112.60° = 36.90°$
Finding a:

$$\frac{a}{\sin A} = \frac{b}{\sin B} \Rightarrow \frac{a}{\sin 30.5°} = \frac{13.00}{\sin 36.90°} \Rightarrow a = \frac{13.00\sin 30.5°}{\sin 36.90°} \approx 10.9890\text{cm}$$

$$\mathcal{A} = \frac{1}{2}ab\sin C = \frac{1}{2}(10.9890)(13.00)\sin 112.60° \approx 65.94 \text{ cm}^2$$

Finding c:

$$\frac{b}{\sin B} = \frac{c}{\sin C} \Rightarrow \frac{13.00}{\sin 36.9°} = \frac{c}{\sin 112.6°} \Rightarrow c = \frac{13.00\sin 112.6°}{\sin 36.9°} \approx 19.9889\text{cm}$$

$$\mathcal{A} = \frac{1}{2}bc\sin A = \frac{1}{2}(19.9889)(13.00)\sin 30.5° \approx 65.94 \text{ cm}^2$$

50. $A = 59.80°$, $b = 15.00$ cm, $C = 53.10°$
In order to use the area formula, we need to find either a or c.
$B = 180° - A - C \Rightarrow B = 180° - 59.80° - 53.10° = 67.10°$
Finding a:

$$\frac{a}{\sin A} = \frac{b}{\sin B} \Rightarrow \frac{a}{\sin 59.80°} = \frac{15.00}{\sin 67.10°} \Rightarrow a = \frac{15.00\sin 59.80°}{\sin 67.10°} \approx 14.0733\text{cm}$$

$$\mathcal{A} = \frac{1}{2}ab\sin C = \frac{1}{2}(14.0733)(15.00)\sin 53.10° \approx 84.41 \text{ m}^2$$

Finding c:

$$\frac{b}{\sin B} = \frac{c}{\sin C} \Rightarrow \frac{15.00}{\sin 67.10°} = \frac{c}{\sin 53.10°} \Rightarrow c = \frac{15.00\sin 53.10°}{\sin 67.10°} \approx 13.0216\text{m}$$

$$\mathcal{A} = \frac{1}{2}bc\sin A = \frac{1}{2}(13.0216)(15.00)\sin 59.8° \approx 84.41 \text{ m}^2$$

51. $\mathcal{A} = \frac{1}{2}ab\sin C = \frac{1}{2}(16.1)(15.2)\sin 125° \approx 100 \text{ m}^2$

52. $\text{A} = \frac{1}{2}(52.1)(21.3)\sin 42.2° \approx 373 \text{ m}^2$

53. Since $\frac{a}{\sin A} = \frac{b}{\sin B} = \frac{c}{\sin C} = 2r$ and $r = \frac{1}{2}$ (since the diameter is 1), we have the following.

$$\frac{a}{\sin A} = \frac{b}{\sin B} = \frac{c}{\sin C} = 2\left(\frac{1}{2}\right) = 1$$

Then, $a = \sin A$, $b = \sin B$, and $c = \sin C$.

54. Answers will vary.

55. Since triangles ACD and BCD are right triangles, we have the following.

$$\tan\alpha = \frac{x}{d + BC} \text{ and } \tan\beta = \frac{x}{BC}$$

Since $\tan\beta = \frac{x}{BC} \Rightarrow BC = \frac{x}{\tan\beta}$, we can substitute into $\tan\alpha = \frac{x}{d + BC}$ and solve for x.

$$\tan\alpha = \frac{x}{d + BC} \Rightarrow \tan\alpha = \frac{x}{d + \frac{x}{\tan\beta}} \Rightarrow \frac{\sin\alpha}{\cos\alpha} = \frac{x}{d + \frac{x\cos\beta}{\sin\beta}} \cdot \frac{\sin\beta}{\sin\beta} \Rightarrow \frac{\sin\alpha}{\cos\alpha} = \frac{x\sin\beta}{d\sin\beta + x\cos\beta}$$

$$\sin\alpha(d\sin\beta + x\cos\beta) = x\sin\beta(\cos\alpha) \Rightarrow d\sin\alpha\sin\beta + x\sin\alpha\cos\beta = x\cos\alpha\sin\beta$$

$$d\sin\alpha\sin\beta = x\cos\alpha\sin\beta - x\sin\alpha\cos\beta \Rightarrow d\sin\alpha\sin\beta = x(\cos\alpha\sin\beta - \sin\alpha\cos\beta)$$

$$d\sin\alpha\sin\beta = -x(\sin\alpha\cos\beta - \cos\alpha\sin\beta)$$

$$d\sin\alpha\sin\beta = -x\sin(\alpha - \beta) \Rightarrow d\sin\alpha\sin\beta = x\sin[-(\alpha - \beta)]$$

$$d\sin\alpha\sin\beta = x\sin(\beta - \alpha) \Rightarrow x = \frac{d\sin\alpha\sin\beta}{\sin(\beta - \alpha)}$$

Section 7.2: The Ambiguous Case of the Law of Sines

1. Having three angles will not yield a unique triangle. The correct choice is A.

2. Having three angles will not yield a unique triangle. So, choices A nor C cannot be correct. A triangle can be uniquely determined by three sides, assuming that triangle exists. In the case of choice B, no such triangle can be created with lengths 3, 5, and 20. The correct choice is D.

3. The vertical distance from the point (3, 4) to the x-axis is 4.
   (a) If h is more than 4, two triangles can be drawn. But h must be less than 5 for both triangles to be on the positive x-axis. So, $4 < h < 5$.
   (b) If $h = 4$, then exactly one triangle is possible. If $h > 5$, then only one triangle is possible on the positive x-axis.
   (c) If $h < 4$, then no triangle is possible, since the side of length h would not reach the x-axis.

4. **(a)** Since the side must be drawn to the positive x-axis, no value of h would produce two triangles.

(b) Since the distance from the point (–3, 4) to the origin is 5, any value of h greater than 5 would produce exactly one triangle.

(c) Likewise, any value of h less than or equal to 5 would produce no triangle.

5. $a = 50, b = 26, A = 95°$

$$\frac{\sin A}{a} = \frac{\sin B}{b} \Rightarrow \frac{\sin 95°}{50} = \frac{\sin B}{26} \Rightarrow \sin B = \frac{26 \sin 95°}{50} \approx .51802124 \Rightarrow B \approx 31.2°$$

Another possible value for B is $180° - 31.2° = 148.8°$. This measure, combined with the angle of 95°, is too large, however. Therefore, only one triangle is possible.

6. $b = 60, a = 82, B = 100°$

$$\frac{\sin A}{a} = \frac{\sin B}{b} \Rightarrow \frac{\sin A}{82} = \frac{\sin 100°}{60} \Rightarrow \sin A = \frac{82 \sin 100°}{60} \approx 1.34590393$$

Since $\sin A > 1$ is impossible, no triangle can be drawn with these parts.

7. $a = 31, b = 26, B = 48°$

$$\frac{\sin A}{a} = \frac{\sin B}{b} \Rightarrow \frac{\sin A}{31} = \frac{\sin 48°}{26} \Rightarrow \sin A = \frac{31 \sin 48°}{26} \approx .88605729 \Rightarrow A \approx 62.4°$$

Another possible value for A is $180° - 62.4° = 117.6°$. Therefore, two triangles are possible.

8. $a = 35, b = 30, A = 40°$

$$\frac{\sin A}{a} = \frac{\sin B}{b} \Rightarrow \frac{\sin 40°}{35} = \frac{\sin B}{30} \Rightarrow \sin B = \frac{30 \sin 40°}{35} \approx .55096081 \Rightarrow B \approx 33.4°$$

Another possible value for B is $180° - 33.4° = 146.6°$, but this is too large to be in a triangle that also has a 40° angle. Therefore, only one triangle is possible.

9. $a = 50, b = 61, A = 58°$

$$\frac{\sin A}{a} = \frac{\sin B}{b} \Rightarrow \frac{\sin 58°}{50} = \frac{\sin B}{61} \Rightarrow \sin B = \frac{61 \sin 58°}{50} \approx 1.03461868$$

Since $\sin B > 1$ is impossible, no triangle is possible for the given parts.

10. $B = 54°, c = 28, b = 23$

$$\frac{\sin B}{b} = \frac{\sin C}{c} \Rightarrow \frac{\sin 54°}{23} = \frac{\sin C}{28} \Rightarrow \sin C = \frac{28 \sin 54°}{23} \approx .98489025 \Rightarrow C \approx 80°$$

Another possible value for C is $180° - 80° = 100°$. Therefore, two triangles are possible.

11. $a = \sqrt{6}, b = 2, A = 60°$

$$\frac{\sin B}{b} = \frac{\sin A}{a} \Rightarrow \frac{\sin B}{2} = \frac{\sin 60°}{\sqrt{6}} \Rightarrow \sin B = \frac{2 \sin 60°}{\sqrt{6}} = \frac{2 \cdot \frac{\sqrt{3}}{2}}{\sqrt{6}} = \frac{\sqrt{3}}{\sqrt{6}} = \sqrt{\frac{3}{6}} = \sqrt{\frac{1}{2}} = \frac{1}{\sqrt{2}} \cdot \frac{\sqrt{2}}{2}$$

$$\sin B = \frac{\sqrt{2}}{2} \Rightarrow B = 45°$$

There is another angle between 0° and 180° whose sine is $\frac{\sqrt{2}}{2}$: $180° - 45° = 135°$. However, this is too large because $A = 60°$ and $60° + 135° = 195° > 180°$, so there is only one solution, $B = 45°$.

12. $A = 45°,\ a = 3\sqrt{2},\ b = 3$

$$\frac{\sin B}{b} = \frac{\sin A}{a} \Rightarrow \frac{\sin B}{3} = \frac{\sin 45°}{3\sqrt{2}} \Rightarrow \sin B = \frac{3\sin 45°}{3\sqrt{2}} = \frac{3\cdot\frac{\sqrt{2}}{2}}{3\sqrt{2}} = \frac{1}{2} \Rightarrow B = 30°$$

There is another angle between $0°$ and $180°$ whose sine is $\frac{1}{2}$: $180° - 30° = 150°$. However, this is too large because $A = 45°$ and $45° + 150° = 195° > 180°$, so there is only one solution, $B = 30°$.

13. $A = 29.7°$, $b = 41.5$ ft, $a = 27.2$ ft

$$\frac{\sin B}{b} = \frac{\sin A}{a} \Rightarrow \frac{\sin B}{41.5} = \frac{\sin 29.7°}{27.2} \Rightarrow \sin B = \frac{41.5\sin 29.7°}{27.2} \approx .75593878$$

There are two angles B between $0°$ and $180°$ that satisfy the condition. Since $\sin B \approx .75593878$, to the nearest tenth value of B is $B_1 = 49.1°$. Supplementary angles have the same sine value, so another possible value of B is $B_2 = 180° - 49.1° = 130.9°$. This is a valid angle measure for this triangle since $A + B_2 = 29.7° + 130.9° = 160.6° < 180°$.

Solving separately for angles C_1 and C_2 we have the following.

$$C_1 = 180° - A - B_1 = 180° - 29.7° - 49.1° = 101.2°$$
$$C_2 = 180° - A - B_2 = 180° - 29.7° - 130.9° = 19.4°$$

14. $B = 48.2°$, $a = 890$ cm, $b = 697$ cm

$$\frac{\sin A}{a} = \frac{\sin B}{b} \Rightarrow \frac{\sin A}{890} = \frac{\sin 48.2°}{697} \Rightarrow \sin A = \frac{890\sin 48.2°}{697} = .95189905$$

There are two angles A between $0°$ and $180°$ that satisfy the condition. Since $\sin A \approx .95189905$, to the nearest tenth value of A is $A_1 = 72.2°$. Supplementary angles have the same sine value, so another possible value of A is $A_2 = 180° - 72.2° = 107.8°$. This is a valid angle measure for this triangle since $B + A_2 = 48.2° + 107.8° = 156.0° < 180°$.

Solving separately for angles C_1 and C_2 we have the following.

$$C_1 = 180° - A_1 - B = 180° - 72.2° - 48.2° = 59.6°$$
$$C_2 = 180° - A_2 - B = 180° - 107.8° - 48.2° = 24.0°$$

15. $C = 41°20', b = 25.9$ m, $c = 38.4$ m

$$\frac{\sin B}{b} = \frac{\sin C}{c} \Rightarrow \frac{\sin B}{25.9} = \frac{\sin 41°20'}{38.4} \Rightarrow \sin B = \frac{25.9\sin 41°20'}{38.4} \approx .44545209$$

There are two angles B between $0°$ and $180°$ that satisfy the condition. Since $\sin B \approx .44545209$, to the nearest tenth value of B is $B_1 = 26.5° = 26°30'$. Supplementary angles have the same sine value, so another possible value of B is $B_2 = 180° - 26°30' = 179°60' - 26°30' = 153°30'$. This is not a valid angle measure for this triangle since $C + B_2 = 41°20' + 153°30' = 194°50' > 180°$.

Thus, $A = 180° - 26°30' - 41°20' = 112°10'$.

16. $B = 48°50', a = 3850$ in.,$b = 4730$ in.

$$\frac{\sin A}{a} = \frac{\sin B}{b} \Rightarrow \frac{\sin A}{3850} = \frac{\sin 48°50'}{4730} \Rightarrow \sin A = \frac{3850 \sin 48°50'}{4730} \approx .61274255$$

There are two angles A between $0°$ and $180°$ that satisfy the condition. Since $\sin A \approx .61274255$, to the nearest tenth value of A is $A_1 = 37.8° \approx 37°50'$. Supplementary angles have the same sine value, so another possible value of A is $A_2 = 180° - 37°50' = 179°60' - 37°50' = 142°10'$. This is not a valid angle measure for this triangle since $B + A_2 = 48°50' + 142°10' = 191° > 180°$.

Thus, $C = 180° - A - B = 180° - 37°50' - 48°50' = 179°60' - 86°40' = 93°20'$.

17. $B = 74.3°$, $a = 859$ m, $b = 783$ m

$$\frac{\sin A}{a} = \frac{\sin B}{b} \Rightarrow \frac{\sin A}{859} = \frac{\sin 74.3°}{783} \Rightarrow \sin A = \frac{859 \sin 74.3°}{783} \approx 1.0561331$$

Since $\sin A > 1$ is impossible, no such triangle exists.

18. $C = 82.2°$, $a = 10.9$ km, $c = 7.62$ km

$$\frac{\sin A}{a} = \frac{\sin C}{c} \Rightarrow \frac{\sin A}{10.9} = \frac{\sin 82.2°}{7.62} \Rightarrow \sin A = \frac{10.9 \sin 82.2°}{7.62} \approx 1.4172115$$

Since $\sin A > 1$ is impossible, no such triangle exists.

19. $A = 142.13°$, $b = 5.432$ ft, $a = 7.297$ ft

$$\frac{\sin B}{b} = \frac{\sin A}{a} \Rightarrow \sin B = \frac{b \sin A}{a} \Rightarrow \sin B = \frac{5.432 \sin 142.13°}{7.297} \approx .45697580 \Rightarrow B \approx 27.19°$$

Because angle A is obtuse, angle B must be acute, so this is the only possible value for B and there is one triangle with the given measurements.

$$C = 180° - A - B \Rightarrow C = 180° - 142.13° - 27.19° \Rightarrow C = 10.68°$$

Thus, $B \approx 27.19°$ and $C \approx 10.68°$.

20. $B = 113.72°$, $a = 189.6$ yd, $b = 243.8$ yd

$$\frac{\sin A}{a} = \frac{\sin B}{b} \Rightarrow \frac{\sin A}{189.6} = \frac{\sin 113.72°}{243.8} \Rightarrow \sin A = \frac{189.6 \sin 113.72°}{243.8} \approx .71198940$$

Because angle B is obtuse, angle A must be acute, so this is the only possible value for A and there is one triangle with the given measurements.

$$C = 180° - A - B = 180° - 45.40° - 113.72° = 20.88°$$

Thus, $A = 45.40°$ and $C = 20.88°$.

21. $A = 42.5°$, $a = 15.6$ ft, $b = 8.14$ ft

$$\frac{\sin B}{b} = \frac{\sin A}{a} \Rightarrow \frac{\sin B}{8.14} = \frac{\sin 42.5°}{15.6} \Rightarrow \sin B = \frac{8.14 \sin 42.5°}{15.6} \approx .35251951$$

There are two angles B between $0°$ and $180°$ that satisfy the condition. Since $\sin B \approx .35251951$, to the nearest tenth value of B is $B_1 = 20.6°'$. Supplementary angles have the same sine value, so another possible value of B is $B_2 = 180° - 20.6° = 159.4°$. This is not a valid angle measure for this triangle since $A + B_2 = 42.5° + 159.4° = 201.9° > 180°$.

Thus, $C = 180° - 42.5° - 20.6° = 116.9°$. Solving for c, we have the following.

$$\frac{c}{\sin C} = \frac{a}{\sin A} \Rightarrow \frac{c}{\sin 116.9°} = \frac{15.6}{\sin 42.5°} \Rightarrow c = \frac{15.6 \sin 116.9°}{\sin 42.5°} \approx 20.6 \text{ ft}$$

22. $C = 52.3°$, $a = 32.5$ yd, $c = 59.8$ yd

$$\frac{\sin A}{a} = \frac{\sin C}{c} \Rightarrow \frac{\sin A}{32.5} = \frac{\sin 52.3°}{59.8} \Rightarrow \sin A = \frac{32.5 \sin 52.3°}{59.8} \approx .43001279$$

There are two angles A between $0°$ and $180°$ that satisfy the condition. Since $\sin A \approx .43001279$, to the nearest tenth value of A is $A_1 = 25.5°$. Supplementary angles have the same sine value, so another possible value of A is $A_2 = 180° - 25.5° = 154.5°$. This is not a valid angle measure for this triangle since $C + A_2 = 52.3° + 154.5° = 206.8° > 180°$.

Thus, $B = 180° - 52.3° - 25.5° = 102.2°$. Solving for b, we have the following.

$$\frac{b}{\sin B} = \frac{c}{\sin C} \Rightarrow \frac{b}{\sin 102.2°} = \frac{59.8}{\sin 52.3°} \Rightarrow b = \frac{59.8 \sin 102.2°}{\sin 52.3°} \approx 73.9 \text{ yd}$$

23. $B = 72.2°$, $b = 78.3$ m, $c = 145$ m

$$\frac{\sin C}{c} = \frac{\sin B}{b} \Rightarrow \frac{\sin C}{145} = \frac{\sin 72.2°}{78.3} \Rightarrow \sin C = \frac{145 \sin 72.2°}{78.3} \approx 1.7632026$$

Since $\sin C > 1$ is impossible, no such triangle exists.

24. $C = 68.5°$, $c = 258$ cm, $b = 386$ cm

$$\frac{\sin B}{b} = \frac{\sin C}{c} \Rightarrow \frac{\sin B}{386} = \frac{\sin 68.5°}{258} \Rightarrow \sin B = \frac{386 \sin 68.5°}{258} \approx 1.39202008$$

Since $\sin B > 1$ is impossible, no such triangle exists.

25. $A = 38°40'$, $a = 9.72$ km, $b = 11.8$ km

$$\frac{\sin B}{b} = \frac{\sin A}{a} \Rightarrow \frac{\sin B}{11.8} = \frac{\sin 38°40'}{9.72} \Rightarrow \sin B = \frac{11.8 \sin 38°40'}{9.72} \approx .75848811$$

There are two angles B between $0°$ and $180°$ that satisfy the condition. Since $\sin B \approx .75848811$, to the nearest tenth value of B is $B_1 = 49.3° \approx 49°20'$. Supplementary angles have the same sine value, so another possible value of B is $B_2 = 180° - 49°20' = 179°60' - 49°20' = 130°40'$. This is a valid angle measure for this triangle since $A + B_2 = 38°40' + 130°40' = 169°20' < 180°$.

Continued on next page

25. (continued)

Solving separately for triangles AB_1C_1 and AB_2C_2 we have the following.

AB_1C_1:

$$C_1 = 180° - A - B_1 = 180° - 38°40' - 49°20' = 180° - 88°00' = 92°00'$$

$$\frac{c_1}{\sin C_1} = \frac{a}{\sin A} \Rightarrow \frac{c_1}{\sin 92°00'} = \frac{9.72}{\sin 38°40'} \Rightarrow c_1 = \frac{9.72 \sin 92°00'}{\sin 38°40'} \approx 15.5 \text{ km}$$

AB_2C_2:

$$C_2 = 180° - A - B_2 = 180° - 38°40' - 130°40' = 10°40'$$

$$\frac{c_2}{\sin C_2} = \frac{a}{\sin A} \Rightarrow \frac{c_2}{\sin 10°40'} = \frac{9.72}{\sin 38°40'} \Rightarrow c_2 = \frac{9.72 \sin 10°40'}{\sin 38°40'} \approx 2.88 \text{ km}$$

26. $C = 29°50'$, $a = 8.61$ m, $c = 5.21$ m

$$\frac{\sin A}{a} = \frac{\sin C}{c} \Rightarrow \frac{\sin A}{8.61} = \frac{\sin 29°50'}{5.21} \Rightarrow \sin A = \frac{8.61 \sin 29°50'}{5.21} \approx .82212894$$

There are two angles A between $0°$ and $180°$ that satisfy the condition. Since $\sin A \approx .82212894$, to the nearest tenth value of A is $A_1 = 55.3° \approx 55°20'$. Supplementary angles have the same sine value, so another possible value of A is $A_2 = 180° - 55°20' = 179°60' - 55°20' = 124°40'$. This is a valid angle measure for this triangle since $C + A_2 = 29°50' + 124°40' = 154°10' < 180°$.

Solving separately for triangles A_1B_1C and A_2B_2C, we have the following.

A_1B_1C:

$$B_1 = 180° - C - A_1 = 180° - 55°20' - 29°50' = 179°60' - 85°10' = 94°50'$$

$$\frac{b_1}{\sin B_1} = \frac{c}{\sin C} \Rightarrow \frac{b_1}{\sin 94°50'} = \frac{5.21}{\sin 29°50'} \Rightarrow b_1 = \frac{5.21 \sin 94°50'}{\sin 29°50'} \approx 10.4 \text{ m}$$

A_2B_2C:

$$B_2 = 180° - C - A_2 = 180° - 124°40' - 29°50' = 179°60' - 154°30' = 25°30'$$

$$\frac{b_2}{\sin B_2} = \frac{c}{\sin C} \Rightarrow \frac{b_2}{\sin 25°30'} = \frac{5.21}{\sin 29°50'} \Rightarrow b_2 = \frac{5.21 \sin 25°30'}{\sin 29°50'} \approx 4.51 \text{ m}$$

27. $A = 96.80°, b = 3.589$ ft, $a = 5.818$ ft

$$\frac{\sin B}{b} = \frac{\sin A}{a} \Rightarrow \frac{\sin B}{3.589} = \frac{\sin 96.80°}{5.818} \Rightarrow \sin B = \frac{3.589 \sin 96.80°}{5.818} \approx .61253922$$

There are two angles B between $0°$ and $180°$ that satisfy the condition. Since $\sin B \approx .61253922$, to the nearest hundredth value of B is $B_1 = 37.77°$. Supplementary angles have the same sine value, so another possible value of B is $B_2 = 180° - 37.77° = 142.23°$. This is not a valid angle measure for this triangle since $A + B_2 = 96.80° + 142.23° = 239.03° > 180°$.

Thus, $C = 180° - 96.80° - 37.77° = 45.43°$. Solving for c, we have the following.

$$\frac{c}{\sin C} = \frac{a}{\sin A} \Rightarrow \frac{c}{\sin 45.43°} = \frac{5.818}{\sin 96.80°} \Rightarrow c = \frac{5.818 \sin 45.43°}{\sin 96.80°} \approx 4.174 \text{ ft}$$

28. $C = 88.70°, b = 56.87$ yd, $c = 112.4$ yd

$$\frac{\sin B}{b} = \frac{\sin C}{c} \Rightarrow \frac{\sin B}{56.87} = \frac{\sin 88.70°}{112.4} \Rightarrow \sin B = \frac{56.87 \sin 88.70°}{112.4} \approx .50583062$$

There are two angles B between $0°$ and $180°$ that satisfy the condition. Since $\sin B \approx .50583062$, to the nearest hundredth value of B is $B_1 = 30.39°$. Supplementary angles have the same sine value, so another possible value of B is $B_2 = 180° - 30.39° = 149.61°$. This is not a valid angle measure for this triangle since $C + B_2 = 88.70° + 149.61° = 238.31° > 180°$.

Thus, $A = 180° - C - B = 180° - 88.70° - 30.39° = 60.91°$. Solving for a, we have the following.

$$\frac{a}{\sin A} = \frac{c}{\sin C} \Rightarrow \frac{a}{\sin 60.91°} = \frac{112.4}{\sin 88.70°} \Rightarrow a = \frac{112.4 \sin 60.91°}{\sin 88.70°} \approx 98.25 \text{ yd}$$

29. $B = 39.68°$, $a = 29.81$ m, $b = 23.76$ m

$$\frac{\sin A}{a} = \frac{\sin B}{b} \Rightarrow \frac{\sin A}{29.81} = \frac{\sin 39.68°}{23.76} \Rightarrow \sin A = \frac{29.81 \sin 39.68°}{23.76} \approx .80108002$$

There are two angles A between $0°$ and $180°$ that satisfy the condition. Since $\sin A \approx .80108002$, to the nearest hundredth value of A is $A_1 = 53.23°$. Supplementary angles have the same sine value, so another possible value of A is $A_2 = 180° - 53.23° = 126.77°$. This is a valid angle measure for this triangle since $B + A_2 = 39.68° + 126.77° = 166.45° < 180°$.

Solving separately for triangles A_1BC_1 and A_2BC_2 we have the following.

A_1BC_1 :

$$C_1 = 180° - A_1 - B = 180° - 53.23° - 39.68° = 87.09°$$

$$\frac{c_1}{\sin C_1} = \frac{b}{\sin B} \Rightarrow \frac{c_1}{\sin 87.09°} = \frac{23.76}{\sin 39.68°} \Rightarrow c_1 = \frac{23.76 \sin 87.09°}{\sin 39.68°} \approx 37.16 \text{ m}$$

A_2BC_2:

$$C_2 = 180° - A_2 - B = 180° - 126.77° - 39.68° = 13.55°$$

$$\frac{c_2}{\sin C_2} = \frac{b}{\sin B} \Rightarrow \frac{c_2}{\sin 13.55°} = \frac{23.76}{\sin 39.68°} \Rightarrow c_1 = \frac{23.76 \sin 13.55°}{\sin 39.68°} \approx 8.719 \text{ m}$$

30. $A = 51.20°$, $c = 7986$ cm, $a = 7208$ cm

$$\frac{\sin C}{c} = \frac{\sin A}{a} \Rightarrow \frac{\sin C}{7986} = \frac{\sin 51.20°}{7208} \Rightarrow \sin C = \frac{7986 \sin 51.20°}{7208} \approx .86345630$$

There are two angles C between $0°$ and $180°$ that satisfy the condition. Since $\sin C \approx .86345630$, to the nearest tenth value of C is $C_1 = 59.71°$. Supplementary angles have the same sine value, so another possible value of C is $C_2 = 180° - 59.71° = 120.29°$. This is a valid angle measure for this triangle since $A + C_2 = 51.20° + 120.29° = 171.49° < 180°$.

Solving separately for triangles AB_1C_1 and AB_2C_2 we have the following.

AB_1C_1 :

$$B_1 = 180° - C_1 - A = 180° - 59.71° - 51.20° = 69.09°$$

$$\frac{b_1}{\sin B_1} = \frac{a}{\sin A} \Rightarrow \frac{b_1}{\sin 69.09°} = \frac{7208}{\sin 51.20°} \Rightarrow b_1 = \frac{7208 \sin 69.09°}{\sin 51.20°} \approx 8640 \text{ cm}$$

AB_2C_2:

$$B_2 = 180° - C_2 - A = 180° - 120.29° - 51.20° = 8.51°$$

$$\frac{b_2}{\sin B_2} = \frac{a}{\sin A} \Rightarrow \frac{b_2}{\sin 8.51°} = \frac{7208}{\sin 51.20°} \Rightarrow b_2 = \frac{7208 \sin 8.51°}{\sin 51.20°} \approx 1369 \text{ cm}$$

31. $a=\sqrt{5}, c=2\sqrt{5},\ A=30°$

$$\frac{\sin C}{c}=\frac{\sin A}{a}\Rightarrow\frac{\sin C}{2\sqrt{5}}=\frac{\sin 30°}{\sqrt{5}}\Rightarrow\sin C=\frac{2\sqrt{5}\sin 30°}{\sqrt{5}}=\frac{2\sqrt{5}\cdot\frac{1}{2}}{\sqrt{5}}=\frac{\sqrt{5}}{\sqrt{5}}=1\Rightarrow C=90°$$

This is a right triangle.

32. – 34. Answers will vary.

35. Let $A=38°50'$, $a=21.9$, $b=78.3$.

$$\frac{\sin B}{b}=\frac{\sin A}{a}\Rightarrow\frac{\sin B}{78.3}=\frac{\sin 38°50'}{21.9}\Rightarrow\sin B=\frac{78.3\sin 38°50'}{21.9}\approx 2.2419439$$

Since $\sin B>1$ is impossible, no such piece of property exists.

36. Let $A=28°10'$, $a=21.2$, $b=26.5$.

$$\frac{\sin B}{b}=\frac{\sin A}{a}\Rightarrow\frac{\sin B}{26.5}=\frac{\sin 28°10'}{21.2}\Rightarrow\sin B=\frac{26.5\sin 28°10'}{21.2}\approx .59004745$$

There are two angles B between $0°$ and $180°$ that satisfy the condition. Since $\sin B\approx .59004745$, to the nearest tenth one value of B is $B_1=36.2°\approx 36°10'$. Supplementary angles have the same sine value, so another possible value of B is $B_2=180°-36°10'=179°60'-36°10'=143°50'$. This is a valid angle measure for this triangle since $A+B_2=28°10'+143°50'=172°<180°$.

Solving separately for triangles AB_1C_1 and AB_2C_2, we have the following.

AB_1C_1:

$$C_1=180°-A-B_1=180°-28°10'-36°10'=179°60'-64°20'=115°40'$$

$$\frac{c_1}{\sin C_1}=\frac{a}{\sin A}\Rightarrow\frac{c_1}{\sin 115°40'}=\frac{21.2}{\sin 28°10'}\Rightarrow c_1=\frac{21.2\sin 115°40'}{\sin 28°10'}\approx 40.5\text{ yd}$$

AB_2C_2:

$$C_2=180°-A-B_2=180°-28°10'-143°50'=180°-172°=8°00'$$

$$\frac{c_2}{\sin C_2}=\frac{a}{\sin A}\Rightarrow\frac{c_2}{\sin 8°00'}=\frac{21.2}{\sin 28°10'}\Rightarrow c_2=\frac{21.2\sin 8°00'}{\sin 28°10'}\approx 6.25\text{ yd}$$

37. Prove that $\dfrac{a+b}{b}=\dfrac{\sin A+\sin B}{\sin B}$

Start with the law of sines.

$$\frac{a}{\sin A}=\frac{b}{\sin B}\Rightarrow a=\frac{b\sin A}{\sin B}$$

Substitute for a in the expression $\dfrac{a+b}{b}$.

$$\frac{a+b}{b}=\frac{\frac{b\sin A}{\sin B}+b}{b}=\frac{\frac{b\sin A}{\sin B}+b}{b}\cdot\frac{\sin B}{\sin B}=\frac{b\sin A+b\sin B}{b\sin B}=\frac{b(\sin A+\sin B)}{b\sin B}=\frac{\sin A+\sin B}{\sin B}$$

38. Prove that $\dfrac{a-b}{a+b}=\dfrac{\sin A-\sin B}{\sin A+\sin B}$

Start with the law of sines.

$$\frac{a}{\sin A}=\frac{b}{\sin B}\Rightarrow a=\frac{b\sin A}{\sin B}$$

Substitute for a in the expression $\dfrac{a-b}{a+b}$.

$$\frac{a-b}{a+b}=\frac{\frac{b\sin A}{\sin B}-b}{\frac{b\sin A}{\sin B}+b}=\frac{\frac{b\sin A}{\sin B}-b}{\frac{b\sin A}{\sin B}+b}\cdot\frac{\sin B}{\sin B}=\frac{\sin A-\sin B}{\sin A+\sin B}$$

Section 7.3: The Law of Cosines

Connections (page 298)

$P(2,5), Q(-1,3),$ and $R(4,0)$

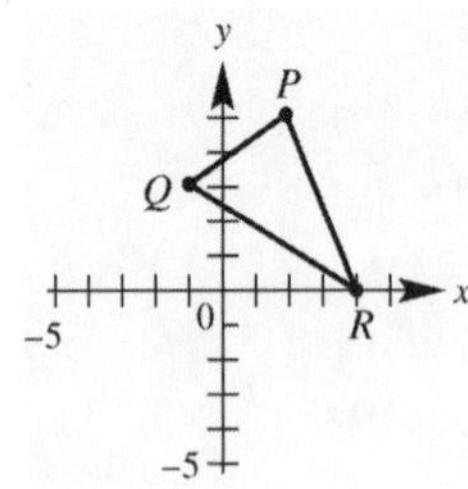

1. $$A=\frac{1}{2}|(x_1y_2-y_1x_2+x_2y_3-y_2x_3+x_3y_1-y_3x_1)|$$
$$=\frac{1}{2}|(2\cdot 3-5(-1)+(-1)\cdot 0-3\cdot 4+4\cdot 5-0\cdot 2)|$$
$$=\frac{1}{2}|(6+5-12+20)|=\frac{1}{2}|19|=\frac{1}{2}\cdot 19=9.5 \text{ sq units}$$

2. distance between $P(2,5)$ and $Q(-1,3)$:

$$a=\sqrt{[2-(-1)]^2+(5-3)^2}=\sqrt{3^2+2^2}=\sqrt{9+4}=\sqrt{13}\approx 3.60555$$

distance between $P(2,5)$ and $R(4,0)$:

$$b=\sqrt{(2-4)^2+(5-0)^2}=\sqrt{(-2)^2+5^2}=\sqrt{4+25}=\sqrt{29}\approx 5.38516$$

distance between $Q(-1,3)$ and $R(4,0)$:

$$c=\sqrt{(-1-4)^2+(3-0)^2}=\sqrt{(-5)^2+3^2}=\sqrt{25+9}=\sqrt{34}\approx 5.83095$$

$$s=\frac{1}{2}(a+b+c)=\frac{1}{2}\left(\sqrt{13}+\sqrt{29}+\sqrt{34}\right)=\frac{\sqrt{13}+\sqrt{29}+\sqrt{34}}{2}\approx 7.41083$$

$$A=\sqrt{s(s-a)(s-b)(s-c)}\approx\sqrt{(7.41083)(3.80528)(2.02567)(1.57988)}\approx 9.5 \text{ sq units}$$

or more precisely,

$$A=\sqrt{s(s-a)(s-b)(s-c)}$$
$$=\sqrt{\frac{\sqrt{13}+\sqrt{29}+\sqrt{34}}{2}\left(\frac{\sqrt{13}+\sqrt{29}+\sqrt{34}}{2}-\sqrt{13}\right)\left(\frac{\sqrt{13}+\sqrt{29}+\sqrt{34}}{2}-\sqrt{29}\right)\left(\frac{\sqrt{13}+\sqrt{29}+\sqrt{34}}{2}-\sqrt{34}\right)}$$
$$=\frac{1}{4}\sqrt{\left(\sqrt{13}+\sqrt{29}+\sqrt{34}\right)\left(\sqrt{13}+\sqrt{29}+\sqrt{34}-2\sqrt{13}\right)\left(\sqrt{13}+\sqrt{29}+\sqrt{34}-2\sqrt{29}\right)\left(\sqrt{13}+\sqrt{29}+\sqrt{34}-2\sqrt{34}\right)}$$
$$=\frac{1}{4}\sqrt{\left(\sqrt{13}+\sqrt{29}+\sqrt{34}\right)\left(\sqrt{29}+\sqrt{34}-\sqrt{13}\right)\left(\sqrt{13}+\sqrt{34}-\sqrt{29}\right)\left(\sqrt{13}+\sqrt{29}-\sqrt{34}\right)}$$
$$=\frac{1}{4}\sqrt{\left[\left(\sqrt{13}+\sqrt{29}+\sqrt{34}\right)\left(\sqrt{29}+\sqrt{34}-\sqrt{13}\right)\right]\left[\left(\sqrt{13}+\sqrt{34}-\sqrt{29}\right)\left(\sqrt{13}+\sqrt{29}-\sqrt{34}\right)\right]}$$
$$=\frac{1}{4}\sqrt{\left(2\sqrt{986}+50\right)\left(2\sqrt{986}-50\right)}=\frac{1}{4}\sqrt{\left(2\sqrt{986}\right)^2-50^2}=\frac{1}{4}\sqrt{3944-2500}=\frac{1}{4}\sqrt{1444}=\frac{1}{4}\cdot 38=9.5 \text{ sq units}$$

3. By the law of cosines, $c^2=a^2+b^2-2ab\cos C\Rightarrow 2ab\cos C=a^2+b^2-c^2\Rightarrow \cos C=\dfrac{a^2+b^2-c^2}{2ab}$.

Using the lengths from part 2, we have the following.

$$\cos C=\frac{13+29-34}{2\sqrt{13}\cdot\sqrt{29}}\approx 0.20601\Rightarrow C\approx 78.1°$$

$$A=\frac{1}{2}ab\sin C=\frac{1}{2}\sqrt{13}\cdot\sqrt{29}\sin 78.1°\approx 9.5 \text{ sq units}$$

Exercises

1. a, b, and C

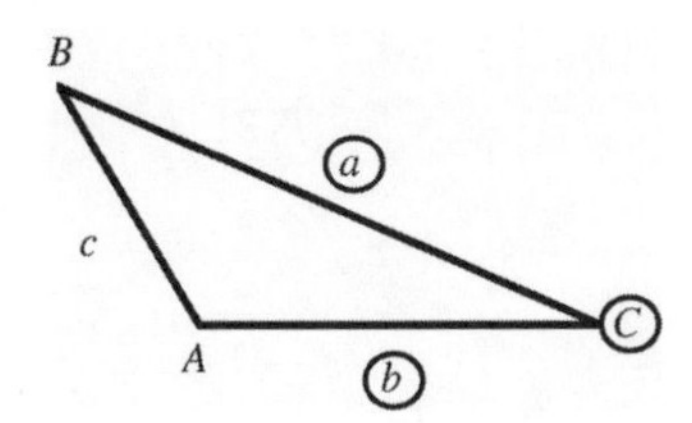

(a) SAS
(b) law of cosines

2. A, C, and c

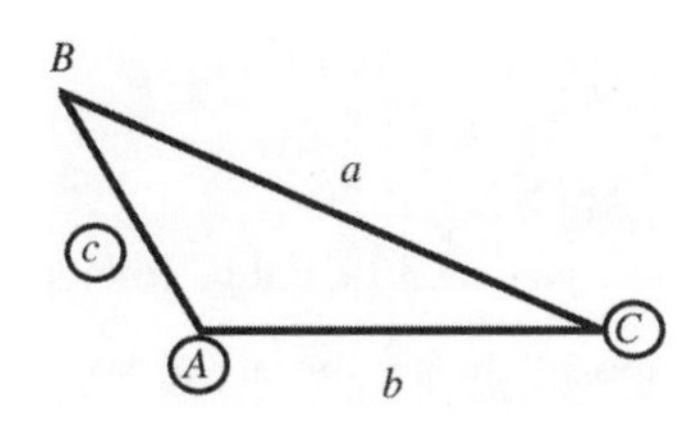

(a) SAA
(b) law of sines

3. a, b, and A

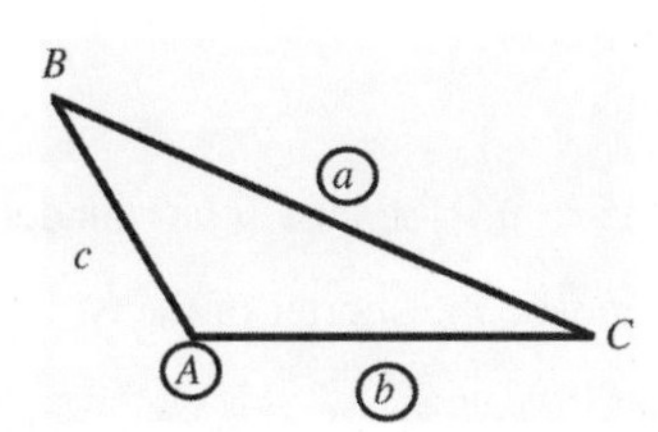

(a) SSA
(b) law of sines

4. a, b, and c

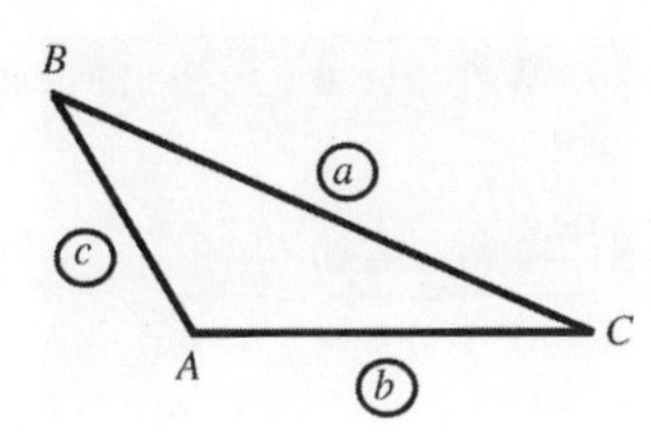

(a) SSS
(b) law of cosines

5. A, B, and c

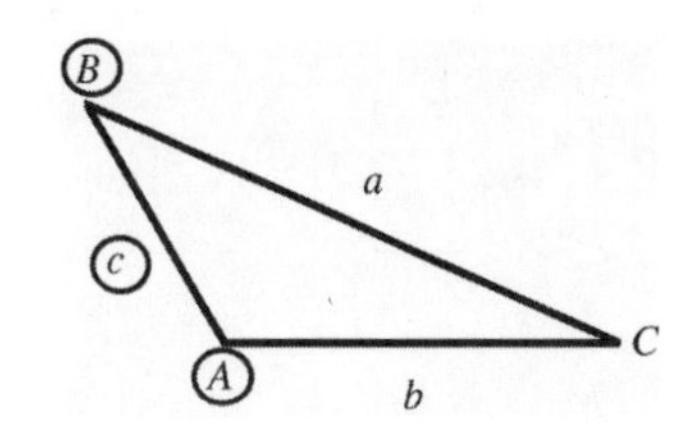

(a) ASA
(b) law of sines

6. a, c, and A

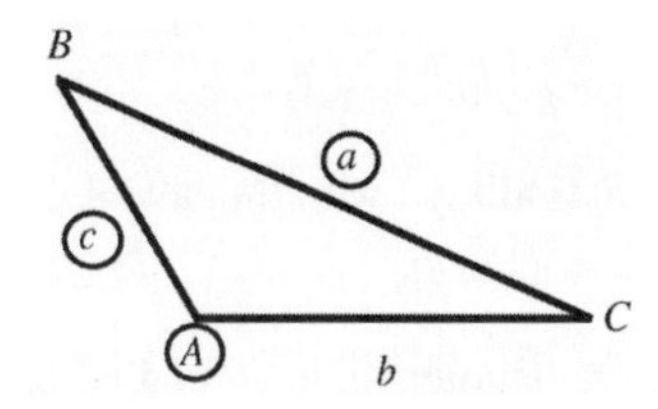

(a) SSA
(b) law of sines

7. a, B, and C

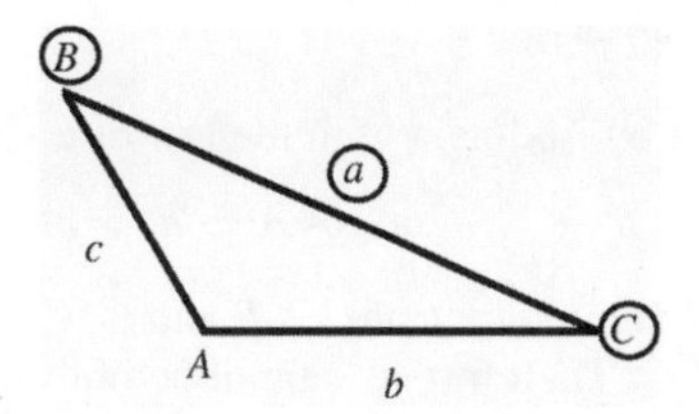

(a) ASA
(b) law of sines

8. b, c, and A

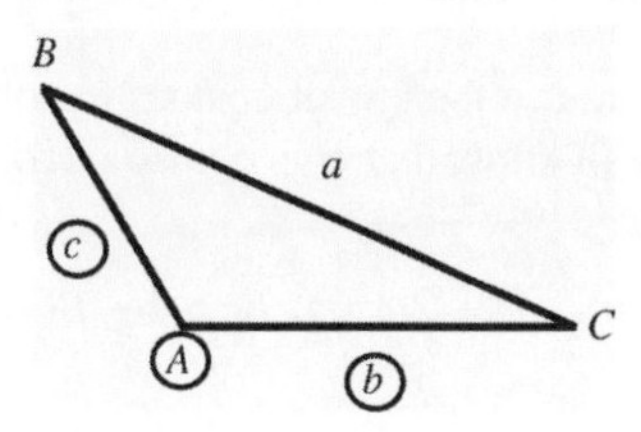

(a) SAS
(b) law of cosines

9. $a^2 = 3^2 + 8^2 - 2(3)(8)\cos 60° = 9 + 64 - 48\left(\frac{1}{2}\right) = 73 - 24 = 49 \Rightarrow a = \sqrt{49} = 7$

10. $a^2 = 1^2 + \left(4\sqrt{2}\right)^2 - 2(1)\left(4\sqrt{2}\right)\cos 45° = 1 + 32 - 8\sqrt{2}\left(\frac{\sqrt{2}}{2}\right) = 33 - 8 = 25 \Rightarrow a = \sqrt{25} = 5$

11. $\cos\theta = \frac{b^2 + c^2 - a^2}{2bc} = \frac{1^2 + \left(\sqrt{3}\right)^2 - 1^2}{2(1)\left(\sqrt{3}\right)} = \frac{1+3-1}{2\sqrt{3}} = \frac{3}{2\sqrt{3}} \cdot \frac{\sqrt{3}}{\sqrt{3}} = \frac{\sqrt{3}}{2} \Rightarrow \theta = 30°$

12. $\cos\theta = \frac{b^2 + c^2 - a^2}{2bc} = \frac{3^2 + 5^2 - 7^2}{2(3)(5)} = \frac{9 + 25 - 49}{30} = -\frac{1}{2} \Rightarrow \theta = 120°$

13. $A = 61°, b = 4, c = 6$

Start by finding a with the law of cosines.

$a^2 = b^2 + c^2 - 2bc\cos A \Rightarrow a^2 = 4^2 + 6^2 - 2(4)(6)\cos 61° \approx 28.729 \Rightarrow a \approx 5.36$ (will be rounded as 5.4)

Of the remaining angles B and C, B must be smaller since it is opposite the shorter of the two sides b and c. Therefore, B cannot be obtuse.

$$\frac{\sin A}{a} = \frac{\sin B}{b} \Rightarrow \frac{\sin 61°}{5.36} = \frac{\sin B}{4} \Rightarrow \sin B = \frac{4\sin 61°}{5.36} \approx .65270127 \Rightarrow B \approx 40.7°$$

Thus, $C = 180° - 61° - 40.7° = 78.3°$.

14. $A = 121°, b = 5, c = 3$

Start by finding a with the law of cosines.

$a^2 = b^2 + c^2 - 2bc\cos A \Rightarrow a^2 = 5^2 + 3^2 - 2(5)(3)\cos 121° \approx 49.5 \Rightarrow a \approx 7.04$ (will be rounded as 7.0)

Of the remaining angles B and C, C must be smaller since it is opposite the shorter of the two sides b and c. Therefore, C cannot be obtuse.

$$\frac{\sin A}{a} = \frac{\sin C}{C} \Rightarrow \frac{\sin 121°}{7.04} = \frac{\sin C}{3} \Rightarrow \sin C = \frac{3\sin 121°}{7.04} \approx .36527016 \Rightarrow C \approx 21.4°$$

Thus, $B = 180° - 121° - 21.4° = 37.6°$.

15. $a = 4, b = 10, c = 8$

We can use the law of cosines to solve for any angle of the triangle. We solve for B, the largest angle. We will know that B is obtuse if $\cos B < 0$.

$$b^2 = a^2 + c^2 - 2ac\cos B \Rightarrow \cos B = \frac{a^2 + c^2 - b^2}{2ac} \Rightarrow \cos B = \frac{4^2 + 8^2 - 10^2}{2(4)(8)} = \frac{-20}{64} = -\frac{5}{16} \Rightarrow B \approx 108.2°$$

We can now use the law of sines or the law of cosines to solve for either A or C. Using the law of cosines to solve for A, we have the following.

$$a^2 = b^2 + c^2 - 2bc\cos A \Rightarrow \cos A = \frac{b^2 + c^2 - a^2}{2bc} \Rightarrow \cos A = \frac{10^2 + 8^2 - 4^2}{2(10)(8)} = \frac{148}{160} = \frac{37}{40} \Rightarrow A \approx 22.3°$$

Thus, $C = 180° - 108.2° - 22.3° = 49.5°$.

16. $a=12,\ b=10,\ c=10$

We can use the law of cosines to solve for any angle of the triangle. Since b and c have the same measure, so do B and C since this would be an isosceles triangle.

If we solve for B, we obtain the following.

$$b^2=a^2+c^2-2ac\cos B\Rightarrow\cos B=\frac{a^2+c^2-b^2}{2ac}\Rightarrow\cos B=\frac{12^2+10^2-10^2}{2(12)(10)}=\frac{144}{240}=\frac{3}{5}\Rightarrow B\approx 53.1^\circ$$

Therefore, $C=B\approx 53.1^\circ$ and $A=180^\circ-53.1^\circ-53.1^\circ=73.8^\circ$.

If we solve for A directly, however, we obtain the following.

$$a^2=b^2+c^2-2bc\cos A\Rightarrow\cos A=\frac{b^2+c^2-a^2}{2bc}\Rightarrow\cos A=\frac{10^2+10^2-12^2}{2(10)(10)}=\frac{56}{200}=\frac{7}{5}\Rightarrow A\approx 73.7^\circ$$

The angles may not sum to 180° due to rounding.

17. $a=5,\ b=7,\ c=9$

We can use the law of cosines to solve for any angle of the triangle. We solve for C, the largest angle. We will know that C is obtuse if $\cos C<0$.

$$c^2=a^2+b^2-2ab\cos C\Rightarrow\cos C=\frac{a^2+b^2-c^2}{2ab}\Rightarrow\cos C=\frac{5^2+7^2-9^2}{2(5)(7)}=\frac{-7}{70}=-\frac{1}{10}\Rightarrow C\approx 95.7^\circ$$

We can now use the law of sines or the law of cosines to solve for either A or B. Using the law of sines to solve for A, we have the following.

$$\frac{\sin A}{a}=\frac{\sin C}{c}\Rightarrow\frac{\sin A}{5}=\frac{\sin 95.7^\circ}{9}\Rightarrow\sin A=\frac{5\sin 95.7^\circ}{9}\approx .55280865\Rightarrow A\approx 33.6^\circ$$

Thus, $B=180^\circ-95.7^\circ-33.6^\circ=50.7^\circ$.

18. $B=55^\circ,\ a=90,\ c=100$

Start by finding b with the law of cosines.

$b^2=a^2+c^2-2ac\cos B\Rightarrow b^2=90^2+100^2-2(90)(100)\cos 55^\circ\approx 7775.6\Rightarrow b\approx 88.18$ (will be rounded as 88.2)

Of the remaining angles A and C, A must be smaller since it is opposite the shorter of the two sides a and c. Therefore, A cannot be obtuse.

$$\frac{\sin A}{a}=\frac{\sin B}{b}\Rightarrow\frac{\sin A}{90}=\frac{\sin 55^\circ}{88.18}\Rightarrow\sin A=\frac{90\sin 55^\circ}{88.18}\approx .83605902\Rightarrow A\approx 56.7^\circ$$

Thus, $C=180^\circ-55^\circ-56.7^\circ=68.3^\circ$.

19. $A=41.4^\circ,\ b=2.78$ yd, $c=3.92$ yd

First find a.

$a^2=b^2+c^2-2bc\cos A\Rightarrow a^2=2.78^2+3.92^2-2(2.78)(3.92)\cos 41.4^\circ\approx 6.7460\Rightarrow a\approx 2.597$ yd (will be rounded as 2.60)

Find B next, since angle B is smaller than angle C (because $b<c$), and thus angle B must be acute.

$$\frac{\sin B}{b}=\frac{\sin A}{a}\Rightarrow\frac{\sin B}{2.78}=\frac{\sin 41.4^\circ}{2.597}\Rightarrow\sin B=\frac{2.78\sin 41.4^\circ}{2.597}\approx .707091182\Rightarrow B\approx 45.1^\circ$$

Finally, $C=180^\circ-41.4^\circ-45.1^\circ=93.5^\circ$.

20. $C = 28.3°$, $b = 5.71$ in., $a = 4.21$ in.
First find c.

$c^2 = a^2 + b^2 - 2ab\cos C \Rightarrow c^2 = 4.21^2 + 5.71^2 - 2(4.21)(5.71)\cos 28.3° \approx 7.9964 \Rightarrow c \approx 2.828$ in. (will be rounded as 2.83)

Find A next, since angle A is smaller than angle B (because $a < b$), and thus angle A must be acute.

$$\frac{\sin A}{a} = \frac{\sin C}{C} \Rightarrow \frac{\sin A}{4.21} = \frac{\sin 28.3°}{2.828} \Rightarrow \sin A = \frac{4.21 \sin 28.3°}{2.828} \approx .70576781 \Rightarrow A \approx 44.9°$$

Finally, $B = 180° - 28.3° - 44.9° = 106.8°$.

21. $C = 45.6°$, $b = 8.94$ m, $a = 7.23$ m
First find c.

$c^2 = a^2 + b^2 - 2ab\cos C \Rightarrow c^2 = 7.23^2 + 8.94^2 - 2(7.23)(8.94)\cos 45.6° \approx 41.7493 \Rightarrow c \approx 6.461$ m (will be rounded as 6.46)

Find A next, since angle A is smaller than angle B (because $a < b$), and thus angle A must be acute.

$$\frac{\sin A}{a} = \frac{\sin C}{c} \Rightarrow \frac{\sin A}{7.23} = \frac{\sin 45.6°}{6.461} \Rightarrow \sin A = \frac{7.23 \sin 45.6°}{6.461} \approx .79951052 \Rightarrow A \approx 53.1°$$

Finally, $B = 180° - 53.1° - 45.6° = 81.3°$.

22. $A = 67.3°$, $b = 37.9$ km, $c = 40.8$ km
First find a.

$a^2 = b^2 + c^2 - 2bc\cos A \Rightarrow 37.9^2 + 40.8^2 - 2(37.9)(40.8)\cos 67.3° \approx 1907.5815 \Rightarrow a \approx 43.68$ km (will be rounded as 43.7)

Find B next, since angle B is smaller than angle C (because $b < c$), and thus angle B must be acute.

$$\frac{\sin B}{b} = \frac{\sin A}{a} \Rightarrow \frac{\sin B}{37.9} = \frac{\sin 67.3°}{43.68} \Rightarrow \sin B = \frac{37.9 \sin 67.3°}{43.68} \approx .80046231 \Rightarrow B \approx 53.2°$$

Finally, $C = 180° - 67.3° - 53.2° = 59.5°$.

23. $a = 9.3$ cm, $b = 5.7$ cm, $c = 8.2$ cm
We can use the law of cosines to solve for any of angle of the triangle. We solve for A, the largest angle. We will know that A is obtuse if $\cos A < 0$.

$$a^2 = b^2 + c^2 - 2bc\cos A \Rightarrow \cos A = \frac{5.7^2 + 8.2^2 - 9.3^2}{2(5.7)(8.2)} \approx .14163457 \Rightarrow A \approx 82°$$

Find B next, since angle B is smaller than angle C (because $b < c$), and thus angle B must be acute.

$$\frac{\sin B}{b} = \frac{\sin A}{a} \Rightarrow \frac{\sin B}{5.7} = \frac{\sin 82°}{9.3} \Rightarrow \sin B = \frac{5.7 \sin 82°}{9.3} \approx .60693849 \Rightarrow B \approx 37°$$

Thus, $C = 180° - 82° - 37° = 61°$.

24. $a = 28$ ft, $b = 47$ ft, $c = 58$ ft
Angle C is the largest, so find it first.

$$c^2 = a^2 + b^2 - 2ab\cos C \Rightarrow \cos C = \frac{28^2 + 47^2 - 58^2}{2(28)(47)} \approx -.14095745 \Rightarrow C \approx 98°$$

Find A next, since angle A is smaller than angle B (because $a < b$), and thus angle A must be acute.

$$\frac{\sin A}{a} = \frac{\sin C}{c} \Rightarrow \frac{\sin A}{28} = \frac{\sin 98°}{58} \Rightarrow \sin A = \frac{28\sin 98°}{58} \approx .47806045 \Rightarrow A \approx 29°$$

Thus, $B = 180° - 29° - 98° = 53°$.

25. $a = 42.9$ m, $b = 37.6$ m, $c = 62.7$ m
Angle C is the largest, so find it first.

$$c^2 = a^2 + b^2 - 2ab\cos C \Rightarrow \cos C = \frac{42.9^2 + 37.6^2 - 62.7^2}{2(42.9)(37.6)} \approx -.20988940 \Rightarrow C \approx 102.1° \approx 102°10'$$

Find B next, since angle B is smaller than angle A (because $b < a$), and thus angle B must be acute.

$$\frac{\sin B}{b} = \frac{\sin C}{c} \Rightarrow \frac{\sin B}{37.6} = \frac{\sin 102.1°}{62.7} \Rightarrow \sin B = \frac{37.6\sin 102.1°}{62.7} \approx .58635805 \Rightarrow B \approx 35.9° \approx 35°50'$$

Thus, $A = 180° - 35°50' - 102°10' = 180° - 138° = 42°00'$.

26. $a = 189$ yd, $b = 214$ yd, $c = 325$ yd
Angle C is the largest, so find it first.

$$c^2 = a^2 + b^2 - 2ab\cos C \Rightarrow \cos C = \frac{189^2 + 214^2 - 325^2}{2(189)(214)} = -.29802700 \Rightarrow C \approx 107.3° \approx 107°20'$$

Find B next, since angle B is smaller than angle A (because $b < a$), and thus angle B must be acute.

$$\frac{\sin B}{b} = \frac{\sin C}{c} \Rightarrow \frac{\sin B}{214} = \frac{\sin 107°20'}{325} \Rightarrow \sin B = \frac{214\sin 107.3°}{325} \approx .62867326 \Rightarrow B \approx 39.0° \approx 39°00'$$

Thus, $A = 180° - 39°00' - 107°20' = 179°60' - 146°20' = 33°40'$.

27. $AB = 1240$ ft, $AC = 876$ ft, $BC = 965$ ft
Let $AB = c$, $AC = b$, $BC = a$
Angle C is the largest, so find it first.

$$c^2 = a^2 + b^2 - 2ab\cos C \Rightarrow \cos C = \frac{965^2 + 876^2 - 1240^2}{2(965)(876)} \approx .09522855 \Rightarrow C \approx 84.5° \text{ or } 84°30'$$

Find B next, since angle B is smaller than angle A (because $b < a$), and thus angle B must be acute.

$$\frac{\sin B}{b} = \frac{\sin C}{c} \Rightarrow \frac{\sin B}{876} = \frac{\sin 84°30'}{1240} \Rightarrow \sin B = \frac{876\sin 84°30'}{1240} \approx .70319925 \Rightarrow B \approx 44.7° \text{ or } 44°10'$$

Thus, $A = 180° - 44°40' - 84°30' = 179°60' - 129°10' = 50°50'$.

28. $AB = 298$ m, $AC = 421$ m, $BC = 324$ m
Let $AB = c$, $AC = b$, $BC = a$.

Angle B is the largest, so find it first.

$$b^2 = a^2 + c^2 - 2ac\cos B \Rightarrow \cos B = \frac{324^2 + 298^2 - 421^2}{2(324)(298)} \approx .08564815 \Rightarrow B \approx 85.1° \approx 85°10'$$

Find C next, since angle C is smaller than angle A (because $c < a$), and thus angle B must be acute.

$$\frac{\sin C}{c} = \frac{\sin B}{b} \Rightarrow \frac{\sin C}{298} = \frac{\sin 85.1°}{421} \Rightarrow \sin C = \frac{298 \sin 85.1°}{421} \approx .70525154 \Rightarrow C \approx 44.8° \approx 44°50'$$

Thus, $A = 180° - 85°10' - 44°50' = 180° - 130° = 50°00'$.

29. $A = 80°\,40'$ $b = 143$ cm, $c = 89.6$ cm

First find a.

$a^2 = b^2 + c^2 - 2bc\cos A \Rightarrow a^2 = 143^2 + 89.6^2 - 2(143)(89.6)\cos 80°\,40' \approx 24{,}321.25 \Rightarrow a \approx 156.0$ cm
(will be rounded as 156)

Find C next, since angle C is smaller than angle B (because $c < b$), and thus angle C must be acute.

$$\frac{\sin C}{c} = \frac{\sin A}{a} \Rightarrow \frac{\sin C}{89.6} = \frac{\sin 80°40'}{156.0} \Rightarrow \sin C = \frac{89.6 \sin 80°40'}{156.0} \approx .56675534 \Rightarrow C \approx 34.5° = 34°30'$$

Finally, $B = 180° - 80°\,40' - 34°\,30' = 179°60' - 115°10' = 64°\,50'$

30. $C = 72°\,40'$, $a = 327$ ft, $b = 251$ ft

First find c.

$c^2 = a^2 + b^2 - 2ab\cos C \Rightarrow c^2 = 327^2 + 251^2 - 2(327)(251)\cos 72°\,40' \approx 121{,}023.55 \Rightarrow c \approx 347.9$ ft
(will be rounded as 348)

Find B next, since angle B is smaller than angle A (because $b < a$), and thus angle B must be acute.

$$\frac{\sin B}{b} = \frac{\sin C}{C} \Rightarrow \frac{\sin B}{251} = \frac{\sin 72°\,40'}{347.9} \Rightarrow \sin B = \frac{251 \sin 72°\,40'}{347.9} \approx .68870795 \Rightarrow B \approx 43.5° = 43°30'$$

Finally, $A = 180° - 72°\,40' - 43°30' = 179°60' - 116°10' = 63°\,50'$.

31. $B = 74.80°$, $a = 8.919$ in., $c = 6.427$ in.
First find b.

$b^2 = a^2 + c^2 - 2ac\cos B \Rightarrow b^2 = 8.919^2 + 6.427^2 - 2(8.919)(6.427)\cos 74.80° \approx 90.7963 \Rightarrow b \approx 9.5287$ in.

(will be rounded as 9.529)

Find C next, since angle C is smaller than angle A (because $c < a$), and thus angle C must be acute.

$$\frac{\sin C}{c} = \frac{\sin B}{b} \Rightarrow \frac{\sin C}{6.427} = \frac{\sin 74.80°}{9.5287} \Rightarrow \sin C = \frac{6.427 \sin 74.80°}{9.5287} \approx .65089267 \Rightarrow C \approx 40.61°$$

Thus, $A = 180° - 74.80° - 40.61° = 64.59°$.

32. $C = 59.70°$, $a = 3.725$ mi, $b = 4.698$ mi
First find c.

$$c^2 = a^2 + b^2 - 2ab\cos C \Rightarrow c^2 = 3.725^2 + 4.698^2 - 2(3.725)(4.698)\cos 59.70° \approx 18.28831 \Rightarrow c \approx 4.27648 \text{ mi}$$

(This will be rounded as 4.276. If we rounded it initially to be 4.2765, then this will cause the final answer to then round to 4.277, which is not correct.)

Find A next, since angle A is smaller than angle B (because $a < b$), and thus angle A must be acute.

$$\frac{\sin A}{a} = \frac{\sin C}{c} \Rightarrow \frac{\sin A}{3.725} = \frac{\sin 59.70°}{4.27648} \Rightarrow \sin A = \frac{3.725\sin 59.70°}{4.27648} \approx .75205506 \Rightarrow A \approx 48.77°$$

Thus, $B = 180° - 48.77° - 59.70° = 71.53°$.

33. $A = 112.8°$, $b = 6.28$ m, $c = 12.2$ m
First find a.

$a^2 = b^2 + c^2 - 2bc\cos A \Rightarrow a^2 = 6.28^2 + 12.2^2 - 2(6.28)(12.2)\cos 112.8° \approx 247.658 \Rightarrow a \approx 15.74$ m (will be rounded as 15.7)

Find B next, since angle B is smaller than angle C (because $b < c$), and thus angle B must be acute.

$$\frac{\sin B}{b} = \frac{\sin A}{a} \Rightarrow \frac{\sin B}{6.28} = \frac{\sin 112.8°}{15.74} \Rightarrow \sin B = \frac{6.28\sin 112.8°}{15.74} \approx .36780817 \Rightarrow B \approx 21.6°$$

Finally, $C = 180° - 112.8° - 21.6° = 45.6°$.

34. $B = 168.2°$, $a = 15.1$ cm, $c = 19.2$ cm
First find b.

$$b^2 = a^2 + c^2 - 2ac\cos B \Rightarrow b^2 = 15.1^2 + 19.2^2 - 2(15.1)(19.2)\cos 168.2° \approx 1164.236 \Rightarrow b \approx 34.12 \text{ cm}$$

(will be rounded as 34.1)

Find A next, since angle A is smaller than angle C (because $a < c$), and thus angle A must be acute.

$$\frac{\sin A}{a} = \frac{\sin B}{b} \Rightarrow \frac{\sin A}{15.1} = \frac{\sin 168.2°}{34.12} \Rightarrow \sin A = \frac{15.1\sin 168.2°}{34.12} \approx .09050089 \Rightarrow A \approx 5.2°$$

Thus, $C = 180° - 5.2° - 168.2° = 6.6°$.

35. $a = 3.0$ ft, $b = 5.0$ ft, $c = 6.0$ ft
Angle C is the largest, so find it first.

$$c^2 = a^2 + b^2 - 2ab\cos C \Rightarrow \cos C = \frac{3.0^2 + 5.0^2 - 6.0^2}{2(3.0)(5.0)} = -\frac{2}{30} = -\frac{1}{15} \approx -.06666667 \Rightarrow C \approx 94°$$

Find A next, since angle A is smaller than angle B (because $a < b$), and thus angle A must be acute.

$$\frac{\sin A}{a} = \frac{\sin C}{c} \Rightarrow \frac{\sin A}{3} = \frac{\sin 94°}{6} \Rightarrow \sin A = \frac{3\sin 94°}{6} \approx .49878203 \Rightarrow A \approx 30°$$

Thus, $B = 180° - 30° - 94° = 56°$.

36. $a = 4.0$ ft, $b = 5.0$ ft, $c = 8.0$ ft
Angle C is the largest, so find it first.

$$c^2 = a^2 + b^2 - 2ab\cos C \Rightarrow \cos C = \frac{4.0^2 + 5.0^2 - 8.0^2}{2(4.0)(5.0)} = -\frac{23}{40} \approx -.57500000 \Rightarrow C \approx 125°$$

Find A next, since angle A is smaller than angle B (because $a < b$), and thus angle A must be acute.

$$\frac{\sin A}{a} = \frac{\sin C}{c} \Rightarrow \frac{\sin A}{4} = \frac{\sin 94°}{8} \Rightarrow \sin A = \frac{4\sin 125°}{8} \approx .40957602 \Rightarrow A \approx 24°$$

Thus, $B = 180° - 24° - 125° = 31°$.

37. There are three ways to apply the law of cosines when $a = 3, b = 4$, and $c = 10$.

Solving for A: $a^2 = b^2 + c^2 - 2bc\cos A \Rightarrow \cos A = \dfrac{4^2 + 10^2 - 3^2}{2(4)(10)} = \dfrac{107}{80} = 1.3375$

Solving for B: $b^2 = a^2 + c^2 - 2ac\cos B \Rightarrow \cos B = \dfrac{3^2 + 10^2 - 4^2}{2(3)(10)} = \dfrac{93}{60} = \dfrac{31}{20} = 1.55$

Solving for C: $c^2 = a^2 + b^2 - 2ab\cos C \Rightarrow \cos C = \dfrac{3^2 + 4^2 - 10^2}{2(3)(4)} = \dfrac{-75}{24} = -\dfrac{25}{8} = -3.125$

Since the cosine of any angle of a triangle must be between –1 and 1, a triangle cannot have sides 3, 4, and 10.

38. Answers will vary.

39. Find AB, or c, in the following triangle.

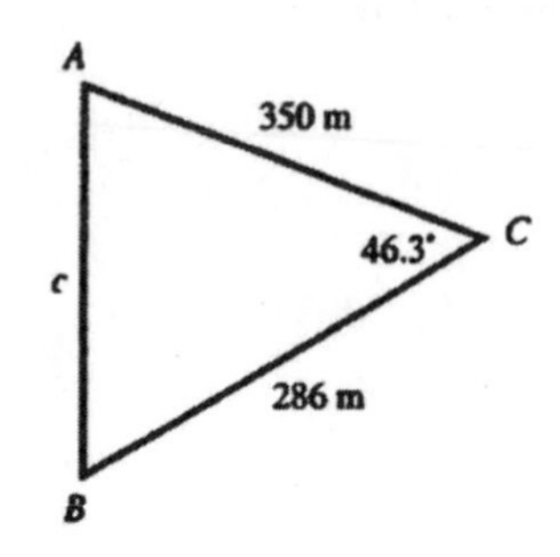

$$c^2 = a^2 + b^2 - 2ab\cos C$$
$$c^2 = 286^2 + 350^2 - 2(286)(350)\cos 46.3°$$
$$c^2 \approx 65{,}981.3$$
$$c \approx 257$$

The length of AB is 257 m.

40. Find the diagonals, BD and AC, of the following parallelogram.

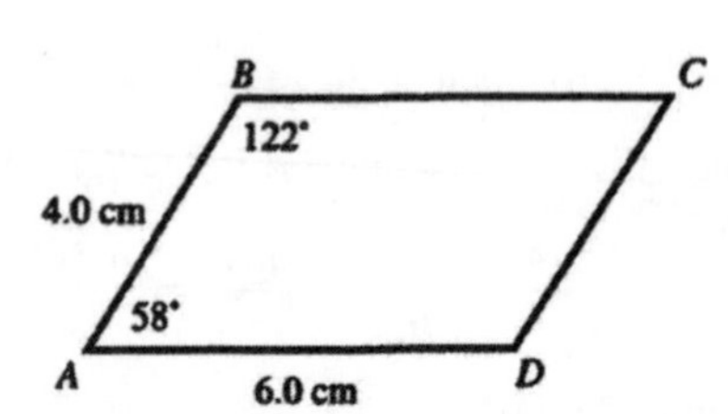

$$BD^2 = AB^2 + AD^2 - 2(AB)(AD)\cos A$$
$$BD^2 = 4^2 + 6^2 - 2(4)(6)\cos 58°$$
$$BD^2 \approx 26.563875$$
$$BD \approx 5.2 \text{ cm}$$

and

$$AC^2 = AB^2 + BC^2 - 2(AB)(BC)\cos B$$
$$AC^2 = 4^2 + 6^2 - 2(4)(6)\cos 122°$$
$$AC^2 \approx 77.436125$$
$$AC \approx 8.8 \text{ cm}$$

The lengths of the diagonals are 5.2 cm and 8.8 cm.

41. Find AC, or b, in the following triangle.

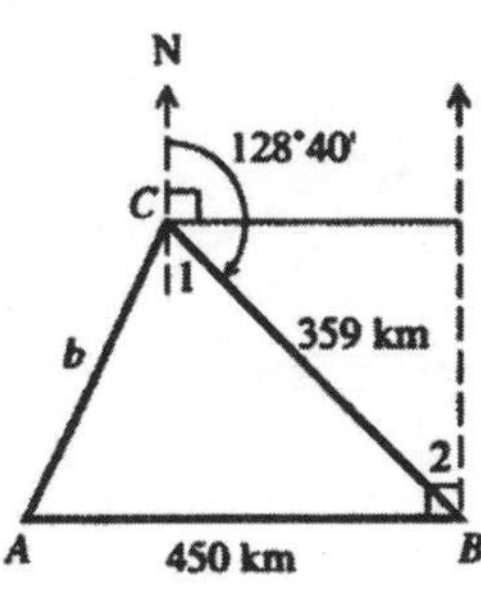

Angle 1 $= 180° - 128° 40' = 179°60' - 128° 40' = 51° 20'$

Angles 1 and 2 are alternate interior angles formed when parallel lines (the north lines) are cut by a transversal, line BC, so angle 2 = angle 1 = $51° 20'$.

Angle ABC = $90°$ – angle 2 $= 89°60' - 51° 20' = 38° 40'$.

$b^2 = a^2 + c^2 - 2ac\cos B \Rightarrow b^2 = 359^2 + 450^2 - 2(359)(450)\cos 38° 40' \approx 79,106 \Rightarrow b \approx 281$ km

C is about 281 km from A.

42. Let B be the harbor, AB is the course of one ship, BC is the course of the other ship. Thus, b = the distance between the ships.

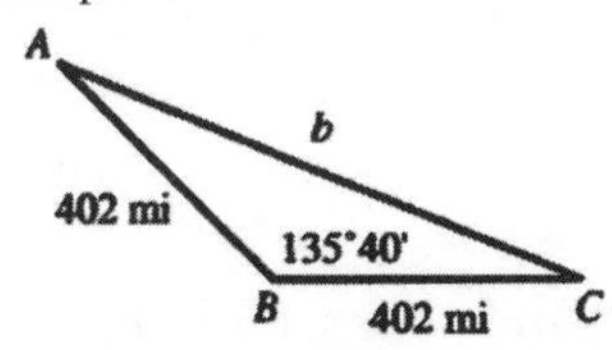

$b^2 = 402^2 + 402^2 - 2(402)(402)\cos 135° 40' \Rightarrow b^2 \approx 554394.25 \Rightarrow b = 745$ mi

43. Sketch a triangle showing the situation as follows.

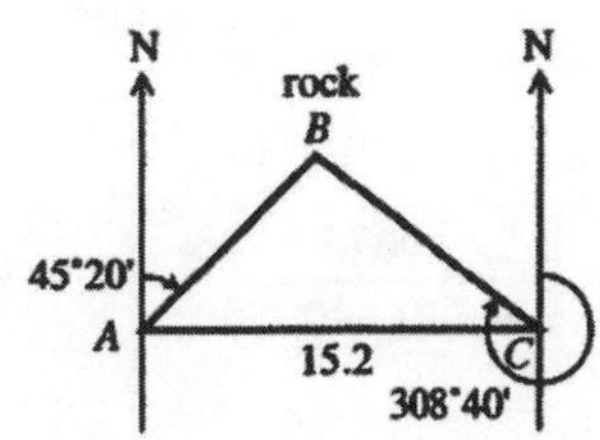

Angle $A = 90° - 45°20' = 89°60' - 45°20' = 44°40'$

Angle $C = 308°40' - 270° = 38°40'$

Angle $B = 180° - A - C = 180° - 44°40' - 38°40' = 179°60' - 83°20' = 96° 40'$

Since we have only one side of a triangle, use the law of sines to find $BC = a$.

$$\frac{a}{\sin A} = \frac{b}{\sin B} \Rightarrow \frac{a}{\sin 44° 40'} = \frac{15.2}{\sin 96° 40'} \Rightarrow a = \frac{15.2 \sin 44° 40'}{\sin 96° 40'} \approx 10.8$$

The distance between the ship and the rock is 10.8 miles.

44. Let d = the distance between the submarine and the battleship.

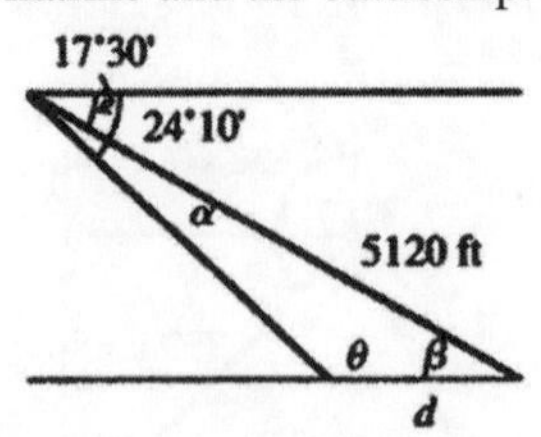

$\alpha = 24°\ 10' - 17°\ 30' = 23°\ 70' - 17°\ 30' = 6°\ 40'$

$\beta = 17°\ 30'$ since the angle of depression to the battleship equals the angle of elevation from the battleship (They are alternate interior angles.).

$\theta = 180° - 6°\ 40' - 17°\ 30' = 179°60' - 14°\ 10' = 155°\ 50'$

Since we have only one side of a triangle, use the law of sines to find d.

$$\frac{d}{\sin 6°\ 40'} = \frac{5120}{\sin 155°\ 50'} \Rightarrow d = \frac{5120 \sin 6°\ 40'}{\sin 155°\ 50'} \approx 1451.9$$

The distance between the submarine and the battleship is 1450 ft. (rounded to three significant digits)

45. Use the law of cosines to find the angle, θ.

$$\cos\theta = \frac{20^2 + 16^2 - 13^2}{2(20)(16)} = \frac{487}{640} \approx .76093750 \Rightarrow \theta \approx 40.5°$$

46. AB is the horizontal distance between points A and B. Using the laws of cosines, we have the following.

$$AB^2 = 10^2 + 10^2 - 2(10)(10)\cos 128° \Rightarrow AB^2 \approx 323.1 \Rightarrow AB \approx 18 \text{ ft}$$

47. Let A = the angle between the beam and the 45-ft cable.

$$\cos A = \frac{45^2 + 90^2 - 60^2}{2(45)(90)} = \frac{6525}{8100} = \frac{29}{36} \approx .80555556 \Rightarrow A \approx 36.3°$$

Let B = the angle between the beam and the 60-ft cable.

$$\cos B = \frac{90^2 + 60^2 - 45^2}{2(90)(60)} = \frac{9675}{10{,}800} = \frac{43}{48} \approx .89583333 \Rightarrow B \approx 26.4°$$

48. Let c = the length of the tunnel.

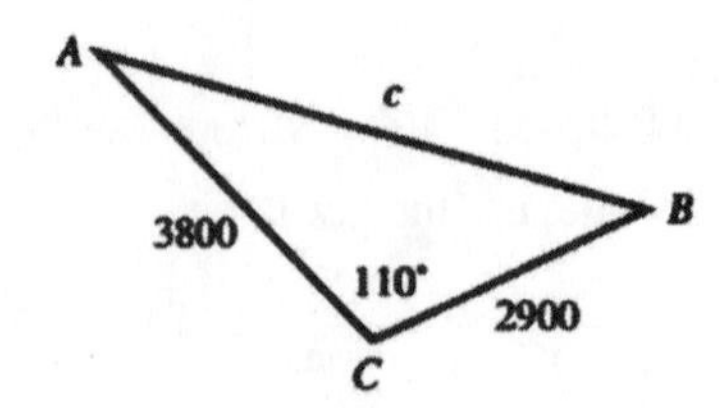

Use the law of cosines to find c.

$c^2 = 3800^2 + 2900^2 - 2(3800)(2900)\cos 110° \Rightarrow c^2 \approx 30{,}388{,}124 \Rightarrow c \approx 5512.5$

The tunnel is 5500 meters long. (rounded to two significant digits)

49. Let A = home plate; B = first base; C = second base; D = third base; P = pitcher's rubber.
Draw AC through P, draw PB and PD.

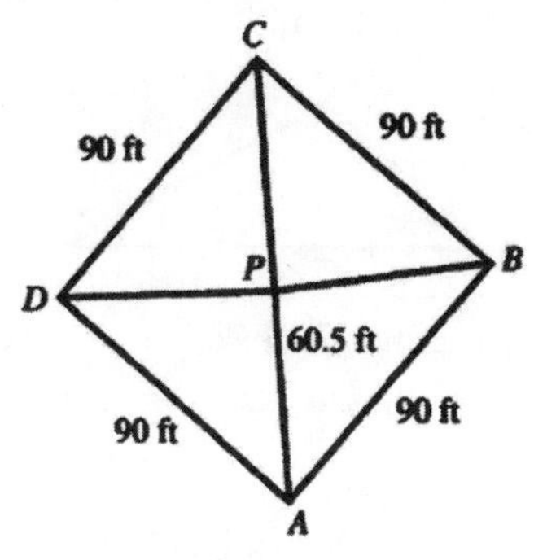

In triangle ABC, angle $B = 90°$, angle A = angle $C = 45°$

$$AC = \sqrt{90^2 + 90^2} = \sqrt{2 \cdot 90^2} = 90\sqrt{2} \text{ and } PC = 90\sqrt{2} - 60.5 \approx 66.8 \text{ ft}$$

In triangle APB, angle $A = 45°$.

$$PB^2 = AP^2 + AB^2 - 2(AP)(AB)\cos A$$
$$PB^2 = 60.5^2 + 90^2 - 2(60.5)(90)\cos 45°$$
$$PB^2 \approx 4059.86$$
$$PB \approx 63.7 \text{ ft}$$

Since triangles APB and APD are congruent, $PB = PD = 63.7$ ft.

The distance to second base is 66.8 ft and the distance to both first and third base is 63.7 ft.

50. Let x = the distance between the ends of the two equal sides.

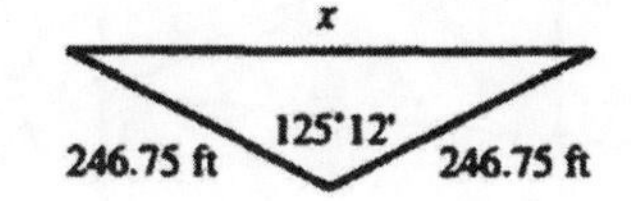

Use the law of cosines to find x.

$$x^2 = 246.75^2 + 246.75^2 - 2(246.75)(246.75)\cos 125°\ 12' \approx 191{,}963.937 \Rightarrow x \approx 438.14$$

The distance between the ends of the two equal sides is 438.14 feet.

51. Find the distance of the ship from point A.

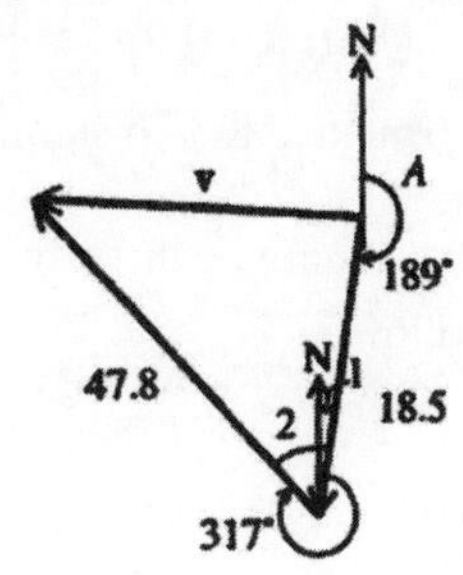

Angle 1 = 189° – 180° = 9°
Angle 2 = 360° – 317° = 43°
Angle 1 + Angle 2 = 9° + 43° = 52°
Use the law of cosines to find v.

$$v^2 = 47.8^2 + 18.5^2 - 2(47.8)(18.5)\cos 52° \approx 1538.23 \Rightarrow v \approx 39.2 \text{ km}$$

52. Let A = the man's location;
B = the factory whistle heard at 3 sec after 5:00;
C = the factory whistle heard at 6 sec after 5:00.

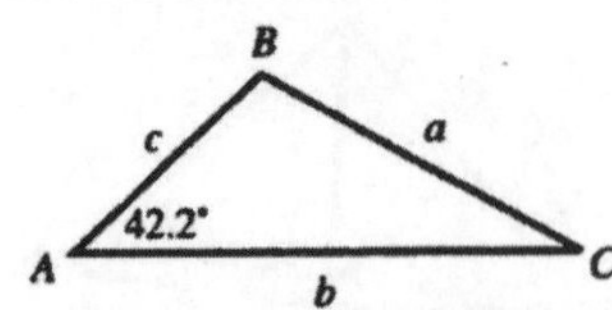

Since sound travels at 344 m per sec and the man hears the whistles in 3 sec and 6 sec, the factories are $c = 3(344) = 1032$ m and $b = 6(344) = 2064$ m from the man.

Using the law of cosines we have the following.

$$a^2 = 1032^2 + 2064^2 - 2(1032)(2064)\cos 42.2° \approx 2{,}169{,}221.3 \Rightarrow a \approx 1472.8$$

The factories are 1473 m apart. (rounded to four significant digits)

53. $\cos A = \dfrac{17^2 + 21^2 - 9^2}{2(17)(21)} = \dfrac{649}{714} \approx .90896359 \Rightarrow A \approx 25°$

Thus, the bearing of B from is $325° + 25° = 350°$.

54. Sketch a triangle showing the situation as follows.

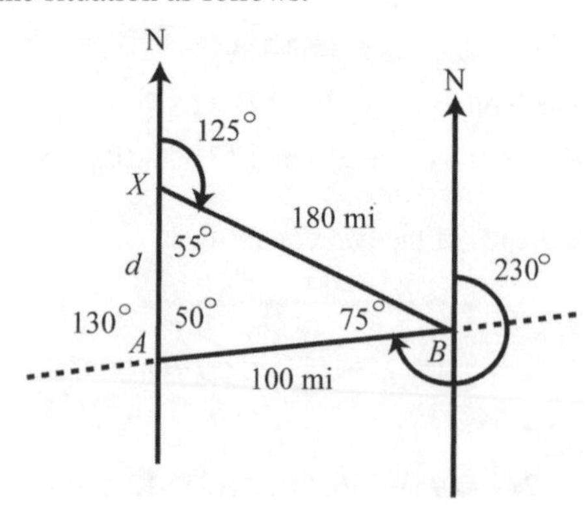

The angle marked $130°$ is the corresponding angle that measures $360° - 230° = 130°$. The angle marked $55°$ is the supplement of the $125°$ angle. Finally, the $75°$ angle is marked as such because $180° - 55° - 50° = 75°$. We can use the law of cosines to solve for the side of the triangle marked d.

$$d^2 = 180^2 + 100^2 - 2(180)(100)\cos 75° \approx 33{,}082 \Rightarrow d \approx 181.9$$

The distance is approximately 180 mi. (rounded to two significant digits)

55. Let c = the length of the property line that cannot be directly measured.

Using the law of cosines, we have the following.

$c^2 = 14.0^2 + 13.0^2 - 2(14.0)(13.0)\cos 70° \approx 240.5 \Rightarrow c \approx 15.5$ ft (rounded to three significant digits)

The length of the property line is approximately 18.0 + 15.5 + 14.0 = 47.5 feet

56. Let A = the point where the ship changes to a bearing of 62°;
C = the point where it changes to a bearing of 115°.

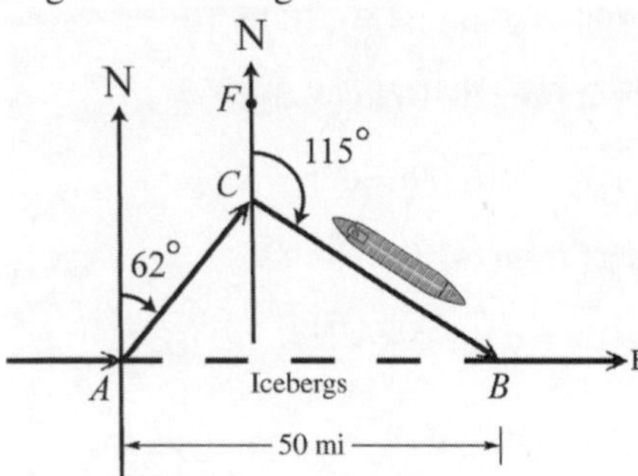

Angle $CAB = 90° - 62° = 28°$

Angle $FCA = 180° - \text{angle } DAC = 180° - 62° = 118°$

Angle $ACB = 360° - \text{ angle } FCB - \text{ angle } FCA = 360° - 115° - 118° = 127°$

Angle $CBA = 180° - \text{ angle } ACB - \text{ angle } CAB = 180° - 127° - 28° = 25°$

Since we have only one side of a triangle, use the law of sines to find CB.

$$\frac{CB}{\sin 28°} = \frac{50}{\sin 127°} \Rightarrow CB = \frac{50 \sin 28°}{\sin 127°} \approx 29.4 \text{ and } \frac{AC}{\sin 25°} = \frac{50}{\sin 127°} \Rightarrow AC = \frac{50 \sin 25°}{\sin 127°} \approx 26.5$$

The ship traveled 26.5 + 29.4 = 55.9 mi.

To avoid the iceberg, the ship had to travel 55.9 – 50 = 5.9 mi.

57. Using the law of cosines we can solve for the measure of angle A.

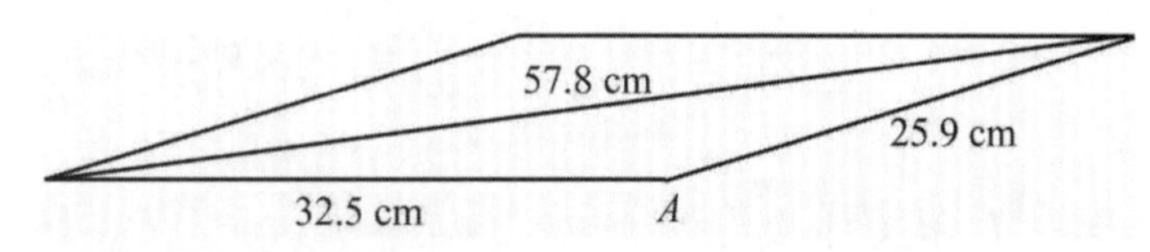

$$\cos A = \frac{25.9^2 + 32.5^2 - 57.8^2}{2(25.9)(32.5)} \approx -.95858628 \Rightarrow A \approx 163.5°$$

58. Let x = the distance from the plane to the mountain when the second bearing is taken.

$$\theta = 180° - 32.7° = 147.3°$$

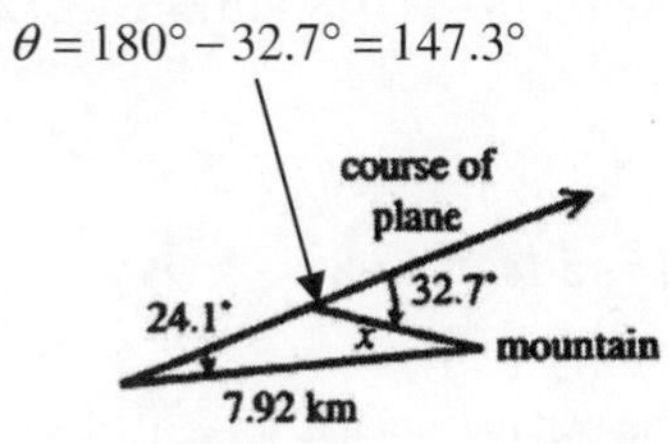

Since we have only one side of a triangle, use the law of sines to find x.

$$\frac{x}{\sin 24.1°} = \frac{7.92}{\sin 147.3°} \Rightarrow x = \frac{7.92 \sin 24.1°}{\sin 147.3°} \approx 5.99$$

The plane is 5.99 km from the mountain. (rounded to three significant digits)

59. Find x using the law of cosines.

$$x^2 = 25^2 + 25^2 - 2(25)(25)\cos 52° \approx 480 \Rightarrow x \approx 22 \text{ ft}$$

60. To find the distance between the towns, d, use the law of cosines.

$$d^2 = 3428^2 + 5631^2 - 2(3428)(5631)\cos 43.33° \approx 15,376,718 \Rightarrow d \approx 3921.3$$

The distance between the two towns is 3921 m. (rounded to four significant digits)

61. Let a be the length of the segment from (0, 0) to (6, 8). Use the distance formula.

$a=\sqrt{(6-0)^2+(8-0)^2}=\sqrt{6^2+8^2}=\sqrt{36+64}=\sqrt{100}=10$

Let b be the length of the segment from (0, 0) to (4, 3).

$b=\sqrt{(4-0)^2+(3-0)^2}=\sqrt{4^2+3^2}=\sqrt{16+9}=\sqrt{25}=5$

Let c be the length of the segment from (4, 3) to (6, 8).

$c=\sqrt{(6-4)^2+(8-3)^2}=\sqrt{2^2+5^2}=\sqrt{4+25}=\sqrt{29}$

$$\cos\theta=\frac{a^2+b^2-c^2}{2ab}\Rightarrow\cos\theta=\frac{10^2+5^2-\left(\sqrt{29}\right)^2}{2(10)(5)}=\frac{100+25-29}{100}=.96\Rightarrow\theta\approx16.26°$$

62. Let a be the length of the segment from (0, 0) to (8, 6). Use the distance formula.

$a=\sqrt{(8-0)^2+(6-0)^2}=\sqrt{8^2+6^2}=\sqrt{64+36}=\sqrt{100}=10$

Let b be the length of the segment from (0, 0) to (12, 5).

$b=\sqrt{(12-0)^2+(5-0)^2}=\sqrt{12^2+5^2}=\sqrt{144+25}=\sqrt{169}=13$

Let c be the length of the segment from (8, 6) to (12, 5).

$c=\sqrt{(12-8)^2+(5-6)^2}=\sqrt{4^2+(-1)^2}=\sqrt{16+1}=\sqrt{17}$

$$\cos\theta=\frac{a^2+b^2-c^2}{2ab}\Rightarrow\cos\theta=\frac{10^2+13^2-\left(\sqrt{17}\right)^2}{2(10)(13)}=\frac{100+169-17}{260}\approx.96923077\Rightarrow\theta\approx14.25°$$

63. Using $\mathsf{A}=\frac{1}{2}bh\Rightarrow\mathsf{A}=\frac{1}{2}(16)\left(3\sqrt{3}\right)=24\sqrt{3}\approx41.57.$

To use Heron's Formula, first find the semiperimeter, $s=\frac{1}{2}(a+b+c)=\frac{1}{2}(6+14+16)=\frac{1}{2}\cdot36=18.$

Now find the area of the triangle.

$\mathsf{A}=\sqrt{s(s-a)(s-b)(s-c)}=\sqrt{18(18-6)(18-14)(18-16)}=\sqrt{18(12)(4)(2)}=\sqrt{1728}=24\sqrt{3}\approx41.57$

Both formulas give the same area.

64. Using $\mathsf{A}=\frac{1}{2}bh\Rightarrow\mathsf{A}=\frac{1}{2}(10)\left(3\sqrt{3}\right)=15\sqrt{3}\approx25.98.$

To use Heron's Formula, first find the semiperimeter, $s=\frac{1}{2}(a+b+c)=\frac{1}{2}(10+6+14)=\frac{1}{2}\cdot30=15.$

Now find the area of the triangle.

$\mathsf{A}=\sqrt{s(s-a)(s-b)(s-c)}=\sqrt{15(15-10)(15-6)(15-14)}=\sqrt{15(5)(9)(1)}=\sqrt{675}=15\sqrt{3}\approx25.98$

Both formulas give the same result.

65. a = 12 m, b = 16 m, c = 25 m

$s=\frac{1}{2}(a+b+c)=\frac{1}{2}(12+16+25)=\frac{1}{2}\cdot53=26.5$

$\mathsf{A}=\sqrt{s(s-a)(s-b)(s-c)}=\sqrt{26.5(26.5-12)(26.5-16)(26.5-25)}=\sqrt{26.5(14.5)(10.5)(1.5)}\approx78\text{ m}^2$

(rounded to two significant digits)

66. $a = 22$ in., $b = 45$ in., $c = 31$ in.

$$s = \frac{1}{2}(a+b+c) = \frac{1}{2}(22+45+31) = \frac{1}{2}\cdot 98 = 49$$

$$\mathsf{A} = \sqrt{s(s-a)(s-b)(s-c)} = \sqrt{49(49-22)(49-45)(49-31)} = \sqrt{49(27)(4)(18)} \approx 310 \text{ in.}^2$$ (rounded to two significant digits)

67. $a = 154$ cm, $b = 179$ cm, $c = 183$ cm

$$s = \frac{1}{2}(a+b+c) = \frac{1}{2}(154+179+183) = \frac{1}{2}\cdot 516 = 258$$

$$\mathsf{A} = \sqrt{s(s-a)(s-b)(s-c)}$$

$$= \sqrt{258(258-154)(258-179)(258-183)} = \sqrt{258(104)(79)(75)} \approx 12{,}600 \text{ cm}^2$$ (rounded to three significant digits)

68. $a = 25.4$ yd, $b = 38.2$ yd, $c = 19.8$ yd

$$s = \frac{1}{2}(a+b+c) = \frac{1}{2}(25.4+38.2+19.8) = \frac{1}{2}\cdot 83.4 = 41.7$$

$$\mathsf{A} = \sqrt{s(s-a)(s-b)(s-c)}$$

$$= \sqrt{41.7(41.7-25.4)(41.7-38.2)(41.7-19.8)} = \sqrt{41.7(16.3)(3.5)(21.9)} \approx 228 \text{ yd}^2$$

(rounded to three significant digits)

69. $a = 76.3$ ft, $b = 109$ ft, $c = 98.8$ ft

$$s = \frac{1}{2}(a+b+c) = \frac{1}{2}(76.3+109+98.8) = \frac{1}{2}\cdot 284.1 = 142.05$$

$$\mathsf{A} = \sqrt{s(s-a)(s-b)(s-c)}$$

$$= \sqrt{142.05(142.05-76.3)(142.05-109)(142.05-98.8)} = \sqrt{142.05(65.75)(33.05)(43.25)} \approx 3650 \text{ ft}^2$$

(rounded to three significant digits)

70. $a = 15.89$ in., $b = 21.74$ in., $c = 10.92$ in.

$$s = \frac{1}{2}(a+b+c) = \frac{1}{2}(15.89+21.74+10.92) = \frac{1}{2}\cdot 48.55 = 24.275$$

$$\mathsf{A} = \sqrt{s(s-a)(s-b)(s-c)} = \sqrt{24.275(24.275-15.89)(24.275-21.74)(24.275-10.92)}$$

$$= \sqrt{24.275(8.385)(2.535)(13.355)} \approx 83.01 \text{ in.}^2$$ (rounded to four significant digits)

71. $AB = 22.47928$ mi, $AC = 28.14276$ mi, $A = 58.56989°$

58.56989°
28.14276 mi
A
C
22.47928 mi
B

This is SAS, so use the law of cosines.

$BC^2 = AC^2 + AB^2 - 2(AC)(AB)\cos A$

$BC^2 = 28.14276^2 + 22.47928^2 - 2(28.14276)(22.47928)\cos 58.56989°$

$BC^2 \approx 637.55393$

$BC \approx 25.24983$

BC is approximately 25.24983 mi. (rounded to seven significant digits)

72. $AB = 22.47928$ mi, $BC = 25.24983$ mi, $A = 58.56989°$

58.56989°
A
C
22.47928 mi
25.24983 mi
B

This is SSA, so use the law of sines.

$$\frac{\sin C}{c} = \frac{\sin A}{a} \Rightarrow \frac{\sin C}{22.47928} = \frac{\sin 58.56989°}{25.24983} \Rightarrow \sin C = \frac{22.47928 \sin 58.56989°}{25.24983} \approx .75965065$$

Thus, $C \approx 49.43341°$ and $B = 180° - A - C = 180° - 58.56989° - 49.43341° = 71.99670°$.

73. Find the area of the Bermuda Triangle using Heron's Formula.

$$s = \frac{1}{2}(a+b+c) = \frac{1}{2}(850+925+1300) = \frac{1}{2}\cdot 3075 = 1537.5$$

$$\text{A} = \sqrt{s(s-a)(s-b)(s-c)} = \sqrt{1537.5(1537.5-850)(1537.5-925)(1537.5-1300)}$$

$$= \sqrt{1537.5(687.5)(612.5)(237.5)} \approx 392{,}128.82$$

The area of the Bermuda Triangle is 392,000 mi^2. (rounded to three significant digits)

74. Find the area of the region using Heron's Formula.

$$s = \frac{1}{2}(a+b+c) = \frac{1}{2}(75+68+85) = \frac{1}{2}\cdot 228 = 114$$

$$\text{A} = \sqrt{s(s-a)(s-b)(s-c)}$$

$$= \sqrt{114(114-75)(114-68)(114-85)} = \sqrt{(114)(39)(46)(29)} \approx 2435.3571 \text{ m}^2$$

$$\text{Number of cans needed } = \frac{(\text{area in m}^2)}{(\text{m}^2 \text{per can})} = \frac{2435.3571}{75} = 32.471428 \text{ cans}$$

She will need to open 33 cans.

75. Perimeter: 9 + 10 + 17 = 36 feet, so the semi-perimeter is $\frac{1}{2}\cdot 36 = 18$ feet.

Use Heron's Formula to find the area.

$$\text{A} = \sqrt{s(s-a)(s-b)(s-c)} = \sqrt{18(18-9)(18-10)(18-17)} = \sqrt{18(9)(8)(1)} = \sqrt{1296} = 36 \text{ ft}$$

Since the perimeter and area both equal 36 feet, the triangle is a *perfect triangle.*

76. **(a)** $s = \frac{1}{2}(a+b+c) = \frac{1}{2}(11+13+20) = \frac{1}{2}\cdot 44 = 22$

$$\text{A} = \sqrt{s(s-a)(s-b)(s-c)}$$

$$= \sqrt{22(22-11)(22-13)(22-20)} = \sqrt{22(11)(9)(2)} = \sqrt{4356} = 66, \text{ which is an integer}$$

(b) $s = \frac{1}{2}(a+b+c) = \frac{1}{2}(13+14+15) = \frac{1}{2}\cdot 42 = 21$

$$\text{A} = \sqrt{s(s-a)(s-b)(s-c)}$$

$$= \sqrt{21(21-13)(21-14)(21-15)} = \sqrt{21(8)(7)(6)} = \sqrt{7056} = 84, \text{ which is an integer}$$

(c) $s = \frac{1}{2}(a+b+c) = \frac{1}{2}(7+15+20) = \frac{1}{2}\cdot 42 = 21$

$$\text{A} = \sqrt{s(s-a)(s-b)(s-c)}$$

$$= \sqrt{21(21-7)(21-15)(21-20)} = \sqrt{21(14)(6)(1)} = \sqrt{1764} = 42, \text{ which is an integer}$$

77. (a) Using the law of sines, we have the following.

$$\frac{\sin C}{c}=\frac{\sin A}{a}\Rightarrow\frac{\sin C}{15}=\frac{\sin 60°}{13}\Rightarrow\sin C=\frac{15\sin 60°}{13}=\frac{15}{13}\cdot\frac{\sqrt{3}}{2}\approx .99926008$$

There are two angles C between $0°$ and $180°$ that satisfy the condition. Since $\sin C\approx .99926008$, to the nearest tenth value of C is $C_1=87.8°$. Supplementary angles have the same sine value, so another possible value of C is $B_2=180°-87.8°=92.2°$.

(b) By the law of cosines, we have the following.

$$\cos C=\frac{a^2+b^2-c^2}{2ab}\Rightarrow\cos C=\frac{13^2+7^2-15^2}{2(13)(7)}=\frac{-7}{182}=-\frac{1}{26}\approx -.03846154\Rightarrow C\approx 92.2°$$

(c) With the law of cosines, we are required to find the inverse cosine of a negative number; therefore; we know angle C is greater than $90°$.

78. Using the law of cosines, we have the following.

$$\cos B=\frac{a^2+c^2-b^2}{2ac}\Rightarrow\cos B=\frac{6^2+5^2-4^2}{2(6)(5)}=\frac{36+25-16}{60}=\frac{3}{4}$$

$$\cos A=\frac{b^2+c^2-a^2}{2bc}\Rightarrow\cos A=\frac{4^2+5^2-6^2}{2(4)(5)}=\frac{16+25-36}{40}=\frac{1}{8}$$

Since $2\cos^2 B-1=2\left(\frac{3}{4}\right)^2-1=2\left(\frac{9}{16}\right)-\frac{16}{16}=\frac{2}{16}=\frac{1}{8}=\cos A$, A is twice the size of B.

79. Given point D is on side AB of triangle ABC such that CD bisects angle C, angle ACD = angle DCB.

Show that $\frac{AD}{DB}=\frac{b}{a}$.

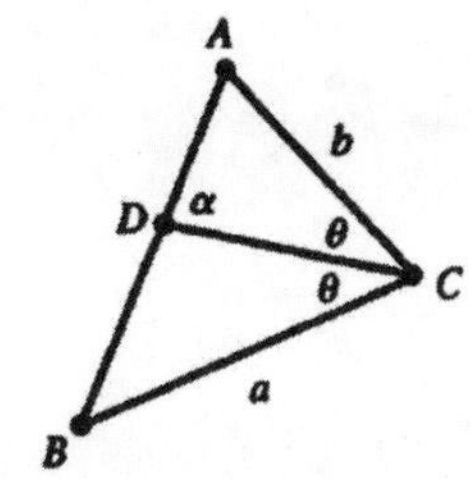

Let $\theta=$ the measure of angle ACD;

$\alpha=$ the measure of angle ADC.

Then $\theta=$ the measure of angle DCB and angle BDC $=180-\alpha$.

By the law of sines, we have the following.

$$\frac{\sin\theta}{AD}=\frac{\sin\alpha}{b}\Rightarrow\sin\theta=\frac{AD\sin\alpha}{b}\text{ and }\frac{\sin\theta}{DB}=\frac{\sin(180°-\alpha)}{a}\Rightarrow\sin\theta=\frac{DB\sin(180°-\alpha)}{a}$$

By substitution, we have the following.

$$\frac{AD\sin\alpha}{b}=\frac{DB\sin(180°-\alpha)}{a}$$

Since $\sin\alpha=\sin(180°-\alpha)$, $\frac{AD}{b}=\frac{DB}{a}$.

Multiplying both sides by $\frac{b}{DB}$, we get $\frac{AD}{b}\cdot\frac{b}{DB}=\frac{DB}{a}\cdot\frac{b}{DB}\Rightarrow\frac{AD}{DB}=\frac{b}{a}$.

80. Let $a = 2,\ b = 2\sqrt{3}, A = 30°, B = 60°$.

Verify $\dfrac{\tan\frac{1}{2}(A-B)}{\tan\frac{1}{2}(A+B)} = \dfrac{a-b}{a+b}$.

$$\frac{\tan\frac{1}{2}(A-B)}{\tan\frac{1}{2}(A+B)} = \frac{\tan\frac{1}{2}(30°-60°)}{\tan\frac{1}{2}(30°+60°)} = \frac{\tan(-15°)}{\tan 45°} \approx -.26794919$$

$$\frac{a-b}{a+b} = \frac{2-2\sqrt{3}}{2+2\sqrt{3}} \cdot \frac{2-2\sqrt{3}}{2-2\sqrt{3}} = \frac{4-8\sqrt{3}+12}{4-12} = \frac{16-8\sqrt{3}}{-8} = -2+\sqrt{3} \approx -.26794919$$

Section 7.4: Vectors, Operations, and the Dot Product

1. Equal vectors have the same magnitude and direction. Equal vectors are **m** and **p**; **n** and **r**.

2. Opposite vectors have the same magnitude but opposite direction. Opposite vectors are **m** and **q**, **p** and **q**, **n** and **s**, **r** and **s**.

3. One vector is a positive scalar multiple of another if the two vectors point in the same direction; they may have different magnitudes.

$$\mathbf{m} = 1\mathbf{p};\ \mathbf{m} = 2\mathbf{t};\ \mathbf{n} = 1\mathbf{r};\ \mathbf{p} = 2\mathbf{t} \text{ or } \mathbf{p} = 1\mathbf{m};\ \mathbf{t} = \frac{1}{2}\mathbf{m};\ \mathbf{r} = 1\mathbf{n};\ \mathbf{t} = \frac{1}{2}\mathbf{p}$$

4. One vector is a negative scalar multiple of another if the two vectors point in the opposite direction; they may have different magnitudes.

$$\mathbf{m} = -1\mathbf{q};\ \mathbf{p} = -1\mathbf{q};\ \mathbf{r} = -1\mathbf{s};\ \mathbf{q} = -2\mathbf{t};\ \mathbf{n} = -1\mathbf{s}$$

5.

6.

-g

7.

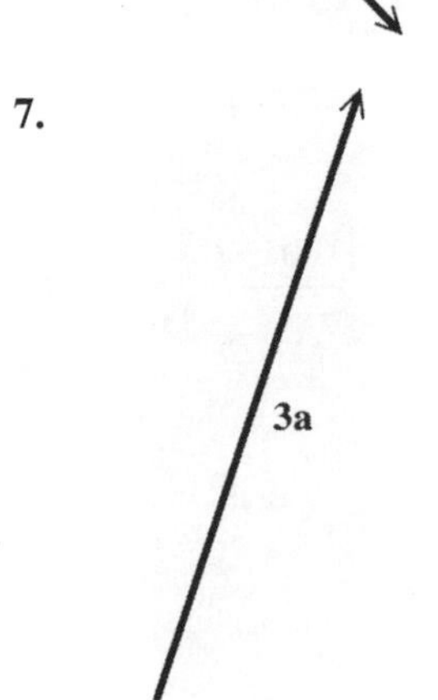

8.

2h

9.

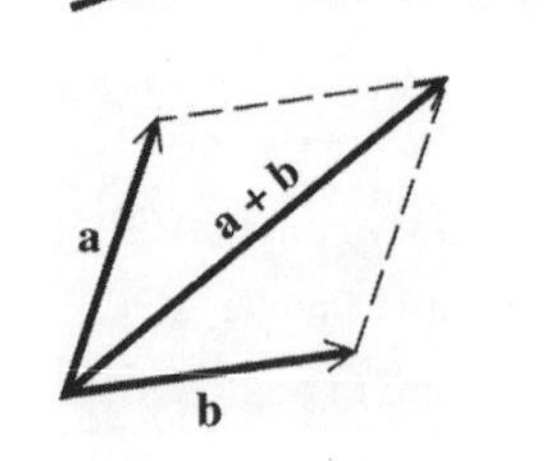

10.

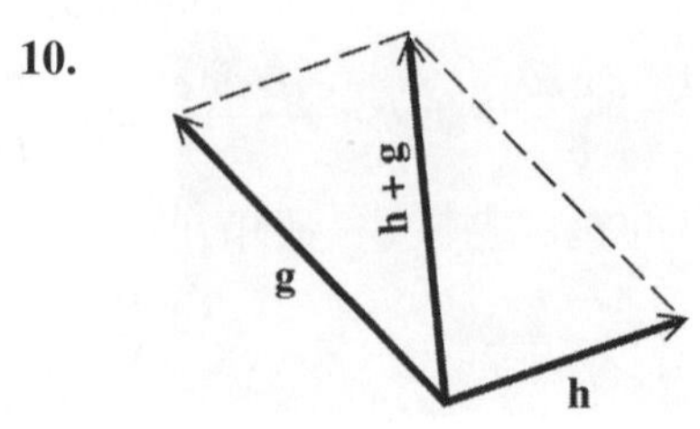

11.

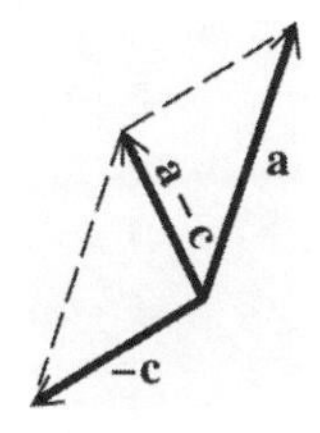

12.

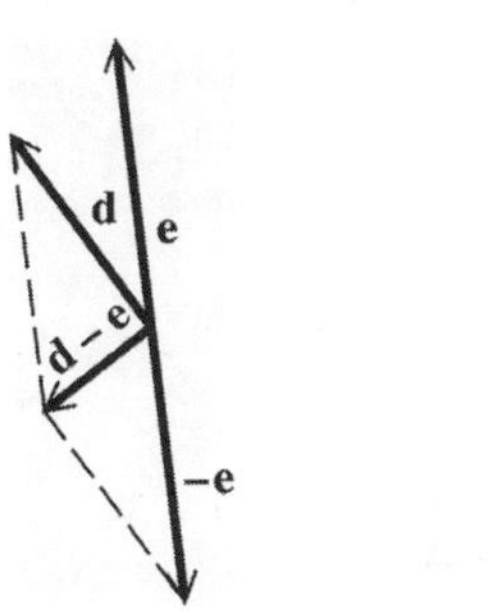

13.

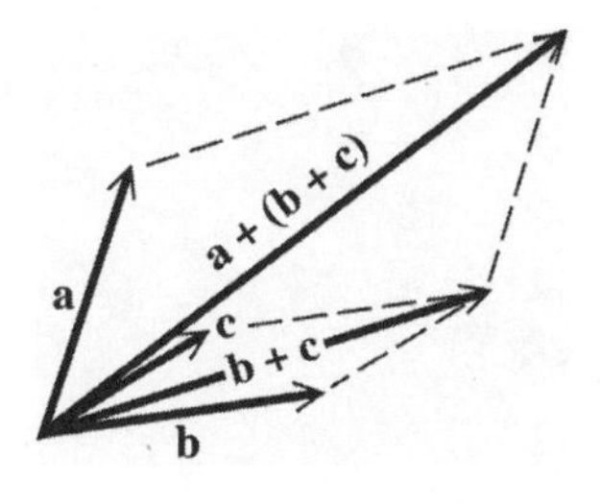

14.

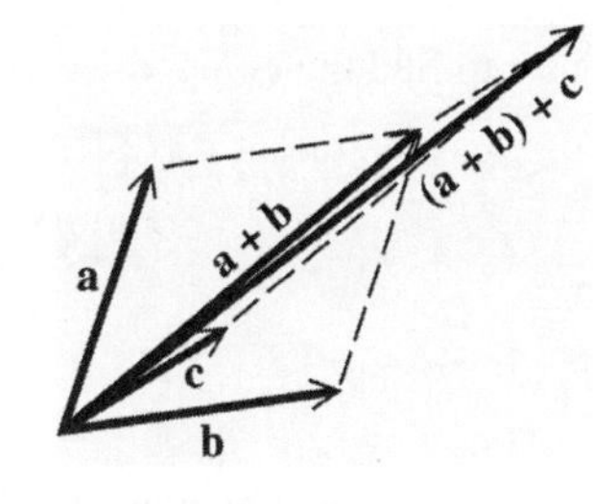

15.

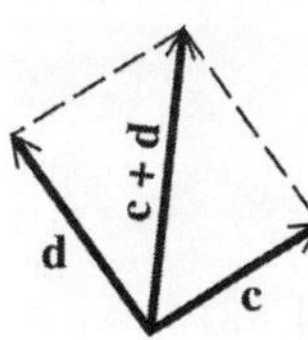

16.

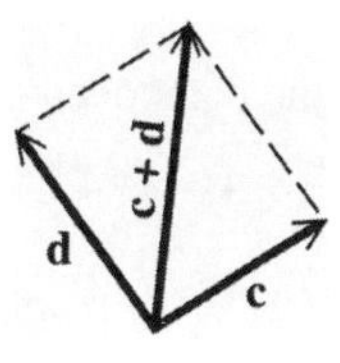

17. $\mathbf{a} + (\mathbf{b} + \mathbf{c}) = (\mathbf{a} + \mathbf{b}) + \mathbf{c}$

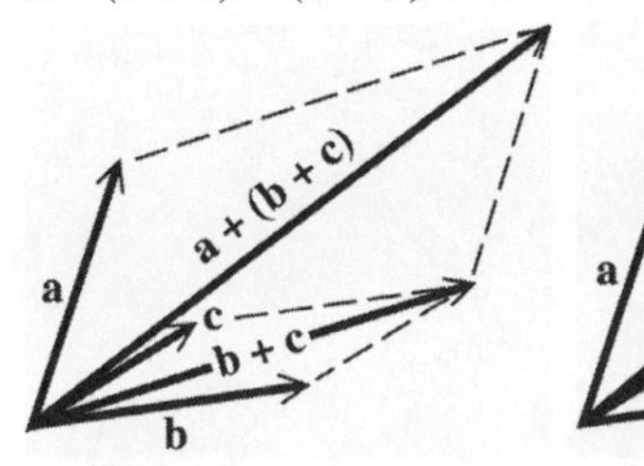

Yes, vector addition is associative.

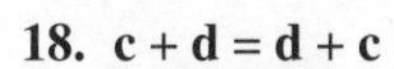

18. $\mathbf{c} + \mathbf{d} = \mathbf{d} + \mathbf{c}$

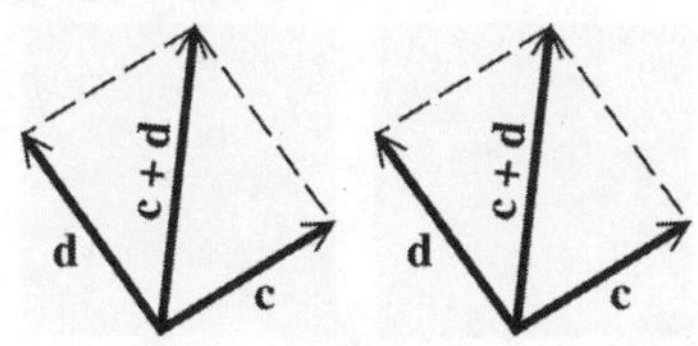

Yes, vector addition is commutative.

19. Use the figure to find the components of **a** and **b**: $\mathbf{a} = \langle -8, 8 \rangle$ and $\mathbf{b} = \langle 4, 8 \rangle$.

(a) $\mathbf{a} + \mathbf{b} = \langle -8, 8 \rangle + \langle 4, 8 \rangle = \langle -8 + 4, 8 + 8 \rangle = \langle -4, 16 \rangle$

(b) $\mathbf{a} - \mathbf{b} = \langle -8, 8 \rangle - \langle 4, 8 \rangle = \langle -8 - 4, 8 - 8 \rangle = \langle -12, 0 \rangle$

(c) $-\mathbf{a} = -\langle -8, 8 \rangle = \langle 8, -8 \rangle$

20. Use the figure to find the components of **a** and **b**: $\mathbf{a} = \langle 4, -4 \rangle$ and $\mathbf{b} = \langle -8, -4 \rangle$.

(a) $\mathbf{a} + \mathbf{b} = \langle 4, -4 \rangle + \langle -8, -4 \rangle = \langle 4 - 8, -4 - 4 \rangle = \langle -4, -8 \rangle$

(b) $\mathbf{a} - \mathbf{b} = \langle 4, -4 \rangle - \langle -8, -4 \rangle = \langle 4 + 8, -4 + 4 \rangle = \langle 12, 0 \rangle$

(c) $-\mathbf{a} = -\langle 4, -4 \rangle = \langle -4, 4 \rangle$

21. Use the figure to find the components of **a** and **b**: $\mathbf{a} = \langle 4, 8 \rangle$ and $\mathbf{b} = \langle 4, -8 \rangle$.

(a) $\mathbf{a} + \mathbf{b} = \langle 4, 8 \rangle + \langle 4, -8 \rangle = \langle 4 + 4, 8 - 8 \rangle = \langle 8, 0 \rangle$

(b) $\mathbf{a} - \mathbf{b} = \langle 4, 8 \rangle - \langle 4, -8 \rangle = \langle 4 - 4, 8 - (-8) \rangle = \langle 0, 16 \rangle$

(c) $-\mathbf{a} = -\langle 4, 8 \rangle = \langle -4, -8 \rangle$

22. Use the figure to find the components of **a** and **b**: $\mathbf{a} = \langle -4, -4 \rangle$ and $\mathbf{b} = \langle 8, 4 \rangle$.

(a) $\mathbf{a} + \mathbf{b} = \langle -4, -4 \rangle + \langle 8, 4 \rangle = \langle -4 + 8, -4 + 4 \rangle = \langle 4, 0 \rangle$

(b) $\mathbf{a} - \mathbf{b} = \langle -4, -4 \rangle - \langle 8, 4 \rangle = \langle -4 - 8, -4 - 4 \rangle = \langle -12, -8 \rangle$

(c) $-\mathbf{a} = -\langle -4, -4 \rangle = \langle 4, 4 \rangle$

23. Use the figure to find the components of **a** and **b**: $\mathbf{a} = \langle -8, 4 \rangle$ and $\mathbf{b} = \langle 8, 8 \rangle$.

(a) $\mathbf{a} + \mathbf{b} = \langle -8, 4 \rangle + \langle 8, 8 \rangle = \langle -8 + 8, 4 + 8 \rangle = \langle 0, 12 \rangle$

(b) $\mathbf{a} - \mathbf{b} = \langle -8, 4 \rangle - \langle 8, 8 \rangle = \langle -8 - 8, 4 - 8 \rangle = \langle -16, -4 \rangle$

(c) $-\mathbf{a} = -\langle -8, 4 \rangle = \langle 8, -4 \rangle$

24. Use the figure to find the components of **a** and **b**: $\mathbf{a} = \langle 8, -4 \rangle$ and $\mathbf{b} = \langle -4, 8 \rangle$.

(a) $\mathbf{a} + \mathbf{b} = \langle 8, -4 \rangle + \langle -4, 8 \rangle = \langle 8 - 4, -4 + 8 \rangle = \langle 4, 4 \rangle$

(b) $\mathbf{a} - \mathbf{b} = \langle 8, -4 \rangle - \langle -4, 8 \rangle = \langle 8 - (-4), -4 - 8 \rangle = \langle 12, -12 \rangle$

(c) $-\mathbf{a} = -\langle 8, -4 \rangle = \langle -8, 4 \rangle$

25. **(a)** $2\mathbf{a} = 2(2\mathbf{i}) = 4\mathbf{i}$

(b) $2\mathbf{a} + 3\mathbf{b} = 2(2\mathbf{i}) + 3(\mathbf{i} + \mathbf{j}) = 4\mathbf{i} + 3\mathbf{i} + 3\mathbf{j} = 7\mathbf{i} + 3\mathbf{j}$

(c) $\mathbf{b} - 3\mathbf{a} = \mathbf{i} + \mathbf{j} - 3(2\mathbf{i}) = \mathbf{i} + \mathbf{j} - 6\mathbf{i} = -5\mathbf{i} + \mathbf{j}$

26. **(a)** $2\mathbf{a} = 2(-\mathbf{i} + 2\mathbf{j}) = -2\mathbf{i} + 4\mathbf{j}$

(b) $2\mathbf{a} + 3\mathbf{b} = 2(-\mathbf{i} + 2\mathbf{j}) + 3(\mathbf{i} - \mathbf{j}) = -2\mathbf{i} + 4\mathbf{j} + 3\mathbf{i} - 3\mathbf{j} = \mathbf{i} + \mathbf{j}$

(c) $\mathbf{b} - 3\mathbf{a} = \mathbf{i} - \mathbf{j} - 3(-\mathbf{i} + 2\mathbf{j}) = \mathbf{i} - \mathbf{j} + 3\mathbf{i} - 6\mathbf{j} = 4\mathbf{i} - 7\mathbf{j}$

27. **(a)** $2\mathbf{a} = 2\langle -1, 2 \rangle = \langle -2, 4 \rangle$

(b) $2\mathbf{a} + 3\mathbf{b} = 2\langle -1, 2 \rangle + 3\langle 3, 0 \rangle = \langle -2, 4 \rangle + \langle 9, 0 \rangle = \langle -2 + 9, 4 + 0 \rangle = \langle 7, 4 \rangle$

(c) $\mathbf{b} - 3\mathbf{a} = \langle 3, 0 \rangle - 3\langle -1, 2 \rangle = \langle 3, 0 \rangle - \langle -3, 6 \rangle = \langle 3 - (-3), 0 - 6 \rangle = \langle 6, -6 \rangle$

28. **(a)** $2\mathbf{a} = 2\langle -2, -1\rangle = \langle -4, -2\rangle$

(b) $2\mathbf{a} + 3\mathbf{b} = 2\langle -2, -1\rangle + 3\langle -3, 2\rangle = \langle -4, -2\rangle + \langle -9, 6\rangle = \langle -4-9, -2+6\rangle = \langle -13, 4\rangle$

(c) $\mathbf{b} - 3\mathbf{a} = \langle -3, 2\rangle - 3\langle -2, -1\rangle = \langle -3, 2\rangle - \langle -6, -3\rangle = \langle -3-(-6), 2-(-3)\rangle = \langle 3, 5\rangle$

29. $|\mathbf{u}| = 12, |\mathbf{w}| = 20, \theta = 27°$

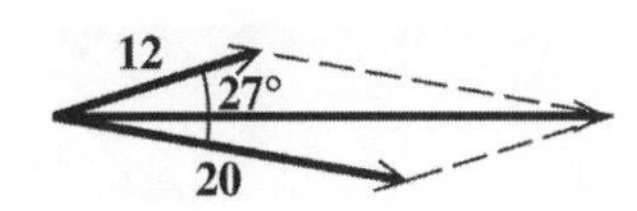

30. $|\mathbf{u}| = 8, |\mathbf{w}| = 12,\ \theta = 20°$

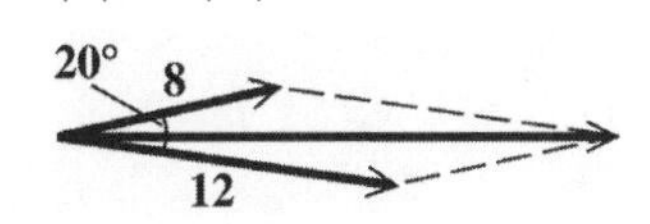

31. $|\mathbf{u}| = 20, |\mathbf{w}| = 30, \theta = 30°$

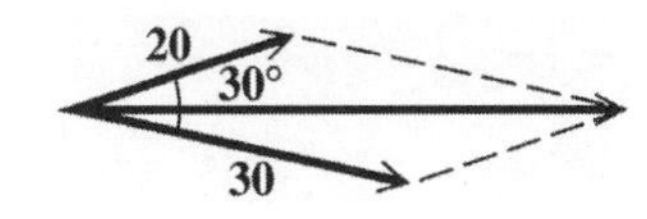

32. $|\mathbf{u}| = 50, |\mathbf{w}| = 70, \theta = 40°$

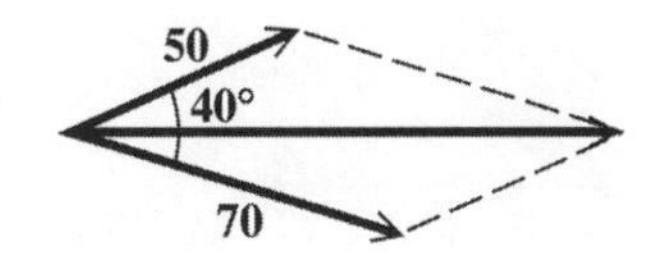

33. Magnitude: $\sqrt{15^2 + (-8)^2} = \sqrt{225 + 64} = \sqrt{289} = 17$

Angle: $\tan\theta' = \dfrac{b}{a} \Rightarrow \tan\theta' = \dfrac{-8}{15} \Rightarrow \theta' = \tan^{-1}\left(-\dfrac{8}{15}\right) \approx -28.1° \Rightarrow \theta = -28.1° + 360° = 331.9°$

(θ lies in quadrant IV)

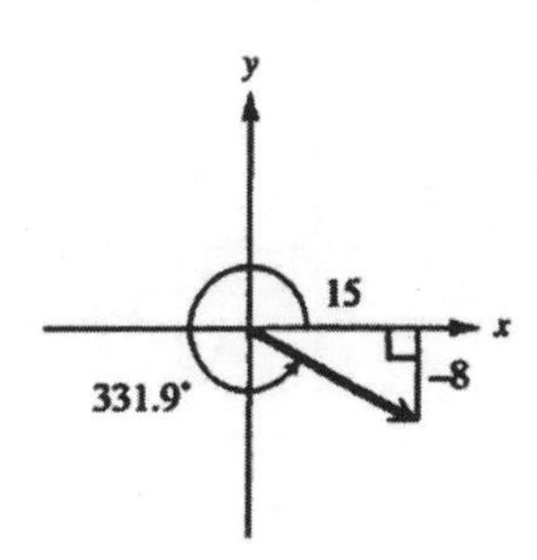

34. Magnitude: $\sqrt{(-7)^2 + 24^2} = \sqrt{49 + 576} = \sqrt{625} = 25$

Angle: $\tan\theta' = \dfrac{b}{a} \Rightarrow \tan\theta' = \dfrac{24}{-7} \Rightarrow \theta' = \tan^{-1}\left(-\dfrac{24}{7}\right) \approx -73.7° \Rightarrow \theta = -73.7° + 180° = 106.3°$

(θ lies in quadrant II)

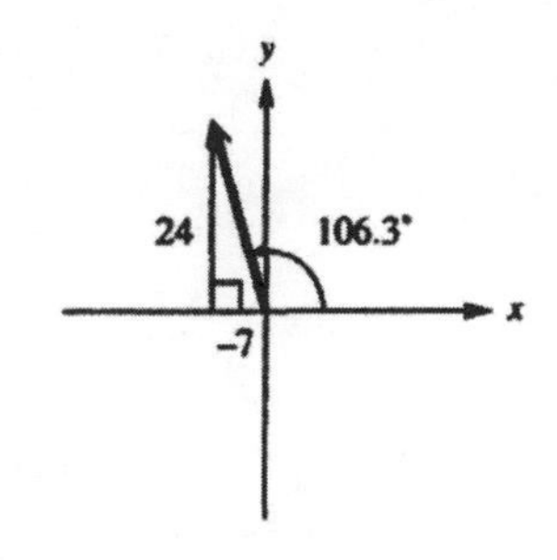

35. Magnitude: $\sqrt{(-4)^2+(4\sqrt{3})^2}=\sqrt{16+48}=\sqrt{64}=8$

Angle: $\tan\theta'=\frac{b}{a}\Rightarrow\tan\theta'=\frac{4\sqrt{3}}{-4}\Rightarrow\theta'=\tan^{-1}\left(-\sqrt{3}\right)=-60°\Rightarrow\theta=-60°+180°=120°$

(θ lies in quadrant II)

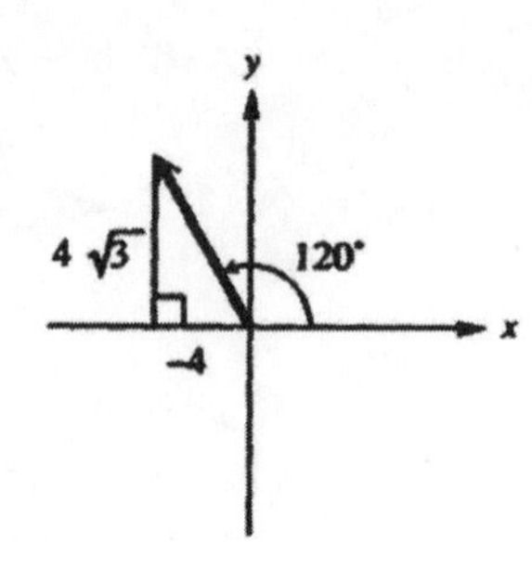

36. Magnitude: $\sqrt{(8\sqrt{2})^2+(-8\sqrt{2})^2}=\sqrt{128+128}=\sqrt{256}=16$

Angle: $\tan\theta'=\frac{b}{a}\Rightarrow\tan\theta=\frac{-8\sqrt{2}}{8\sqrt{2}}\Rightarrow\theta'=\tan^{-1}(-1)\approx-45°\Rightarrow\theta=-45°+360°=315°$

(θ lies in quadrant IV)

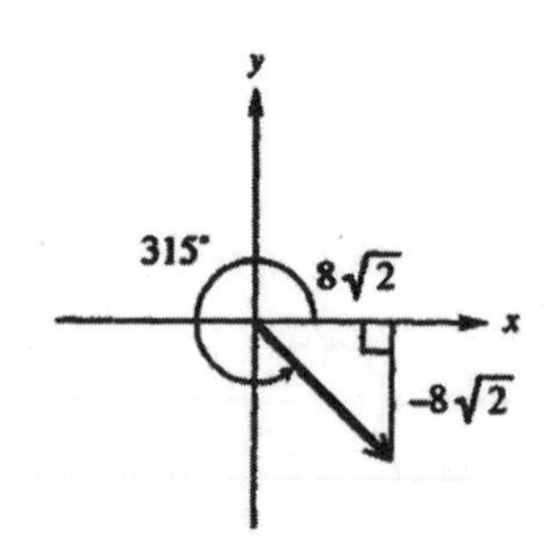

In Exercises 37 – 42, **x** is the horizontal component of **v**, and **y** is the vertical component of **v**. Thus, $|\mathbf{x}|$ is the magnitude of **x** and $|\mathbf{y}|$ is the magnitude of **y**.

37. $\alpha=20°,\ |\mathbf{v}|=50$

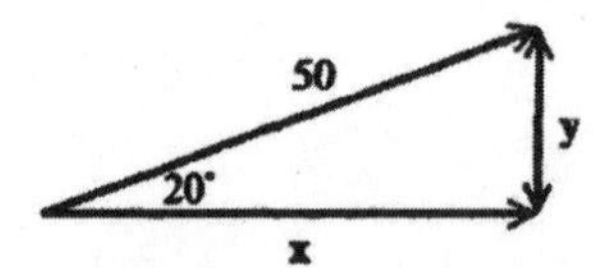

$\mathbf{x}=50\cos 20°\approx 47\Rightarrow|\mathbf{x}|\approx 47$

$\mathbf{y}=50\sin 20°\approx 17\Rightarrow|\mathbf{y}|\approx 17$

38. $\alpha=50°,\ |\mathbf{v}|=26$

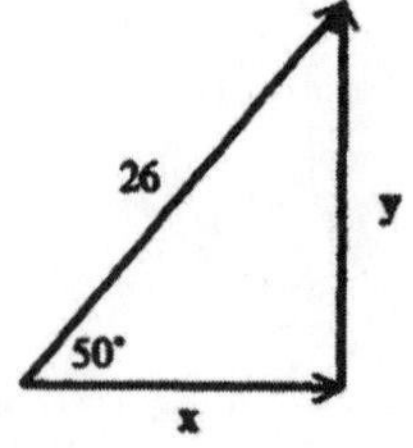

$\mathbf{x}=26\cos 50°\approx 17\Rightarrow|\mathbf{x}|\approx 17$

$\mathbf{y}=26\sin 50°\approx 20\Rightarrow|\mathbf{y}|\approx 20$

39. $\alpha = 35°50'$, $|\mathbf{v}| = 47.8$

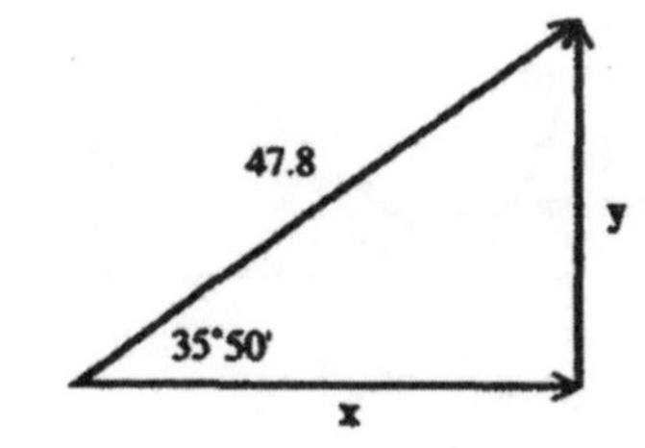

$\mathbf{x} = 47.8\cos 35°50' \approx 38.8 \Rightarrow |\mathbf{x}| \approx 38.8$

$\mathbf{y} = 47.8\sin 35°50' \approx 28.0 \Rightarrow |\mathbf{y}| \approx 28.0$

40. $\alpha = 27°30'$, $|\mathbf{v}| = 15.4$

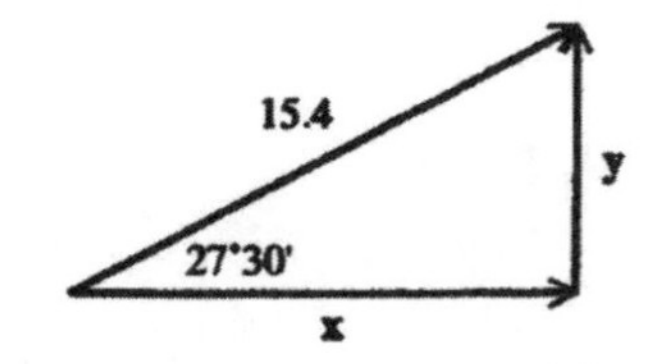

$\mathbf{x} = 15.4\cos 27°30' \approx 13.7 \Rightarrow |\mathbf{x}| \approx 13.7$

$\mathbf{y} = 15.4\sin 27°30' = 7.11 \Rightarrow |\mathbf{y}| \approx 7.11$

41. $\alpha = 128.5°$, $|\mathbf{v}| = 198$

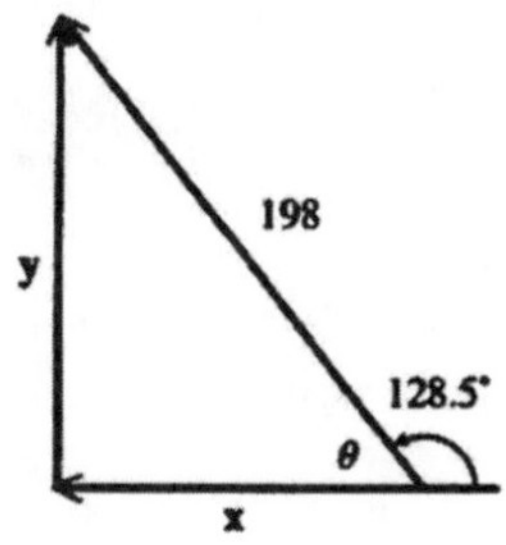

$\mathbf{x} = 198\cos 128.5° \approx -123 \Rightarrow |\mathbf{x}| \approx 123$

$\mathbf{y} = 198\sin 128.5° \approx 155 \Rightarrow |\mathbf{y}| \approx 155$

42. $\alpha = 146.3°$, $|\mathbf{v}| = 238$

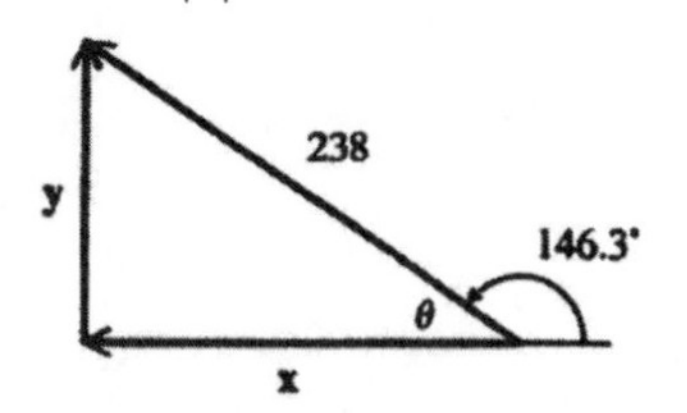

$\mathbf{x} = 238\cos 146.3° \approx -198 \Rightarrow |\mathbf{x}| \approx 198$

$\mathbf{y} = 238\sin 146.3° \approx 132 \Rightarrow |\mathbf{y}| \approx 132$

43. $\mathbf{u} = \langle a, b\rangle = \langle 5\cos(30°), 5\sin(30°)\rangle = \left\langle \frac{5\sqrt{3}}{2}, \frac{5}{2}\right\rangle$

44. $\mathbf{u} = \langle a, b\rangle = \langle 8\cos(60°), 8\sin(60°)\rangle = \langle 4, 4\sqrt{3}\rangle$

45. $\mathbf{v} = \langle a, b\rangle = \langle 4\cos(40°), 4\sin(40°)\rangle \approx \langle 3.0642, 2.5712\rangle$

46. $\mathbf{v} = \langle a, b\rangle = \langle 3\cos(130°), 3\sin(130°)\rangle \approx \langle -1.9284, 2.2981\rangle$

47. $\mathbf{v} = \langle a, b\rangle = \langle 5\cos(-35°), 5\sin(-35°)\rangle \approx \langle 4.0958, -2.8679\rangle$

48. $\mathbf{v} = \langle a, b\rangle = \langle 2\cos(220°), 2\sin(220°)\rangle \approx \langle -1.5321, -1.2856\rangle$

49. Forces of 250 newtons and 450 newtons, forming an angle of 85°

$\alpha = 180° - 85° = 95°$

$|\mathbf{v}|^2 = 250^2 + 450^2 - 2(250)(450)\cos 95°$

$|\mathbf{v}|^2 \approx 284{,}610.04$

$|\mathbf{v}| \approx 533.5$

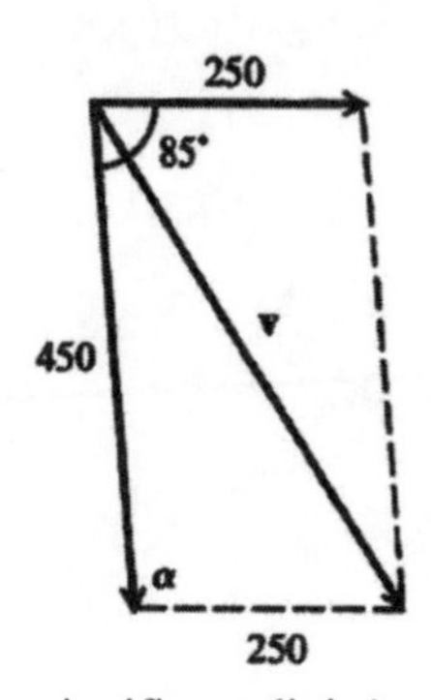

The magnitude of the resulting force is 530 newtons. (rounded to two significant digits)

50. Forces of 19 newtons and 32 newtons, forming an angle of 118°

$\alpha = 180° - 118° = 62°$

$|\mathbf{v}|^2 = 19^2 + 32^2 - 2(19)(32)\cos 62°$

$|\mathbf{v}|^2 \approx 814.12257$

$|\mathbf{v}| \approx 28.53$ newtons

The magnitude of the resulting force is 29 newtons. (rounded to two significant digits)

51. Forces of 116 lb and 139 lb, forming an angle of 140° 50′

$\alpha = 180° - 140°50'$

$= 179°60' - 140°50' = 39°10'$

$|\mathbf{v}|^2 = 139^2 + 116^2 - 2(139)(116)\cos 39°10'$

$|\mathbf{v}|^2 \approx 7774.7359$

$|\mathbf{v}| \approx 88.174$

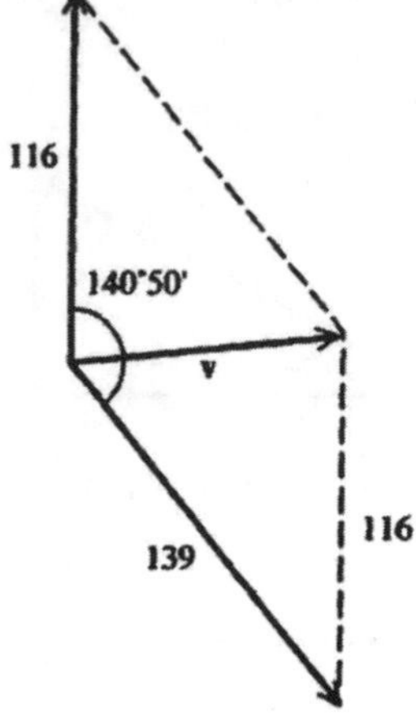

The magnitude of the resulting force is 88.2 lb. (rounded to three significant digits)

52. Forces of 37.8 lb and 53.7 lb, forming an angle of 68.5°

$\alpha = 180° - 68.5° = 111.5°$

$|\mathbf{v}|^2 = 37.8^2 + 53.7^2 - 2(37.8)(53.7)\cos 111.5°$

$|\mathbf{v}|^2 = 5800.4224$

$|\mathbf{v}| = 76.161$

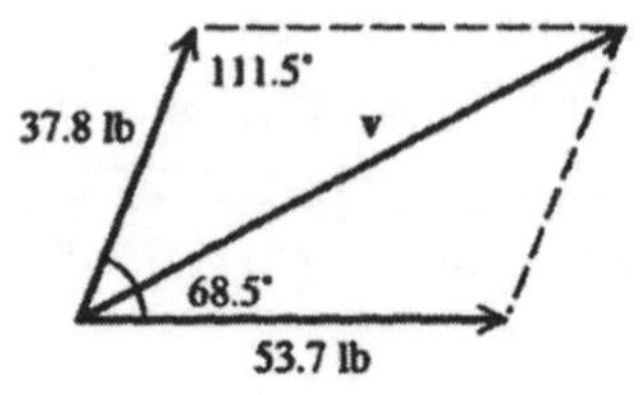

The magnitude of the resulting force is 76.2 lb. (rounded to three significant digits)

53. $\alpha = 180° - 40° = 140°$

$|\mathbf{v}|^2 = 40^2 + 60^2 - 2(40)(60)\cos 140°$

$|\mathbf{v}|^2 \approx 8877.0133$

$|\mathbf{v}| \approx 94.2$ lb

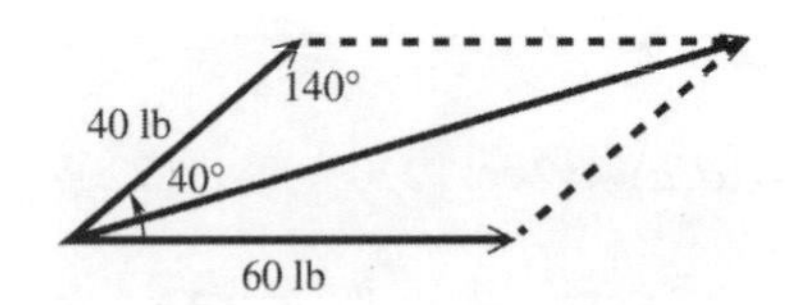

54. $\alpha = 180° - 65° = 115°$

$|\mathbf{v}|^2 = 85^2 + 102^2 - 2(85)(102)\cos 115°$

$|\mathbf{v}|^2 \approx 24{,}957.201$

$|\mathbf{v}| \approx 158.0$ lb

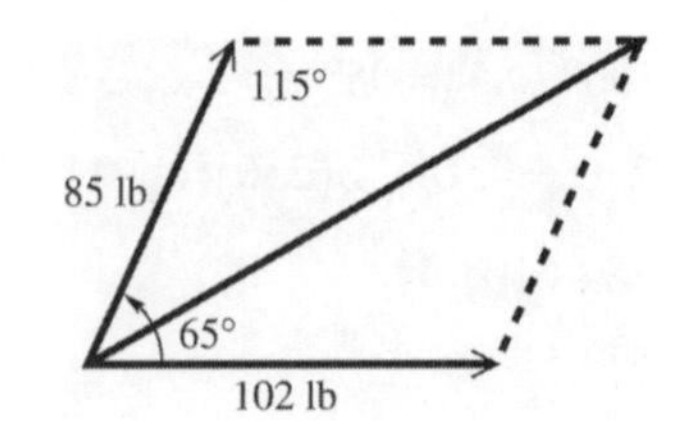

55. $\alpha = 180° - 110° = 70°$

$|\mathbf{v}|^2 = 15^2 + 25^2 - 2(15)(25)\cos 70°$

$|\mathbf{v}|^2 \approx 593.48489$

$|\mathbf{v}| \approx 24.4 \text{ lb}$

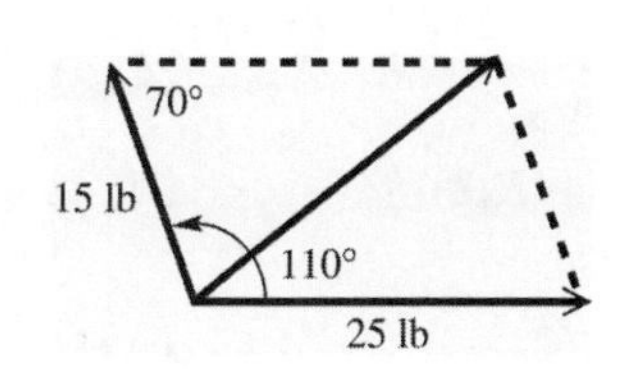

56. $\alpha = 180° - 140° = 40°$

$|\mathbf{v}|^2 = 1500^2 + 2000^2 - 2(1500)(2000)\cos 40°$

$|\mathbf{v}|^2 \approx 1,653,733.3$

$|\mathbf{v}| \approx 1286.0 \text{ lb}$

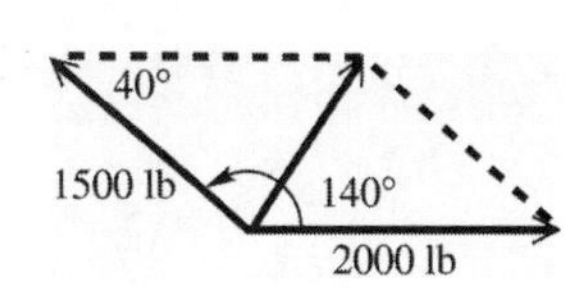

57. If $\mathbf{u} = \langle a, b\rangle$ and $\mathbf{v} = \langle c, d\rangle$, then $\mathbf{u} + \mathbf{v} = \langle a + c, b + d\rangle$.

58. With complex numbers, if $z_1 = a + bi$ and $z_2 = c + di$, then we have the following.

$$z_1 + z_2 = (a + bi) + (c + di) = (a + c) + (b + d)i$$

Additional answers will vary.

For Exercises 59 – 66, $\mathbf{u} = \langle -2, 5\rangle$ and $\mathbf{v} = \langle 4, 3\rangle$.

59. $\mathbf{u} + \mathbf{v} = \langle -2, 5\rangle + \langle 4, 3\rangle = \langle -2 + 4, 5 + 3\rangle = \langle 2, 8\rangle$

60. $\mathbf{u} - \mathbf{v} = \langle -2, 5\rangle - \langle 4, 3\rangle = \langle -2 - 4, 5 - 3\rangle = \langle -6, 2\rangle$

61. $-4\mathbf{u} = -4\langle -2, 5\rangle = \langle -4(-2), -4(5)\rangle = \langle 8, -20\rangle$

62. $-5\mathbf{v} = -5\langle 4, 3\rangle = \langle -5\cdot 4, -5\cdot 3\rangle = \langle -20, -15\rangle$

63. $3\mathbf{u} - 6\mathbf{v} = 3\langle -2, 5\rangle - 6\langle 4, 3\rangle = \langle -6, 15\rangle - \langle 24, 18\rangle = \langle -6 - 24, 15 - 18\rangle = \langle -30, -3\rangle$

64. $-2\mathbf{u} + 4\mathbf{v} = -2\langle -2, 5\rangle + 4\langle 4, 3\rangle$

$= \langle (-2)(-2), (-2)(5)\rangle + \langle 4\cdot 4, 4\cdot 3\rangle = \langle 4, -10\rangle + \langle 16, 12\rangle = \langle 4 + 16, -10 + 12\rangle = \langle 20, 2\rangle$

65. $\mathbf{u} + \mathbf{v} - 3\mathbf{u} = \langle -2, 5\rangle + \langle 4, 3\rangle - 3\langle -2, 5\rangle = \langle -2, 5\rangle + \langle 4, 3\rangle - \langle 3(-2), 3(5)\rangle = \langle -2, 5\rangle + \langle 4, 3\rangle - \langle -6, 15\rangle$

$= \langle -2 + 4, 5 + 3\rangle - \langle -6, 15\rangle = \langle 2, 8\rangle - \langle -6, 15\rangle = \langle 2 - (-6), 8 - 15\rangle = \langle 8, -7\rangle$

66. $2\mathbf{u} + \mathbf{v} - 6\mathbf{v} = 2\langle -2, 5\rangle + \langle 4, 3\rangle - 6\langle 4, 3\rangle = \langle 2(-2), 2\cdot 5\rangle + \langle 4, 3\rangle - \langle 6\cdot 4, 6\cdot 3\rangle$

$= \langle -4, 10\rangle + \langle 4, 3\rangle - \langle 24, 18\rangle = \langle -4 + 4 - 24, 10 + 3 - 18\rangle = \langle -24, -5\rangle$

67. $\langle -5, 8\rangle = -5\mathbf{i} + 8\mathbf{j}$

68. $\langle 6, -3\rangle = 6\mathbf{i} - 3\mathbf{j}$

69. $\langle 2, 0\rangle = 2\mathbf{i} + 0\mathbf{j} = 2\mathbf{i}$

70. $\langle 0, -4\rangle = 0\mathbf{i} - 4\mathbf{j} = -4\mathbf{j}$

71. $\langle 6, -1\rangle \cdot \langle 2, 5\rangle = 6(2)+(-1)(5) = 12-5 = 7$

72. $\langle -3, 8\rangle \cdot \langle 7, -5\rangle = -3(7)+8(-5) = -21-40 = -61$

73. $\langle 2, -3\rangle \cdot \langle 6, 5\rangle = 2(6)+(-3)(5) = 12-15 = -3$

74. $\langle 1, 2\rangle \cdot \langle 3, -1\rangle = 1(3)+2(-1) = 3-2 = 1$

75. $4\mathbf{i} = \langle 4,0\rangle; 5\mathbf{i}-9\mathbf{j} = \langle 5,-9\rangle$

$\langle 4,0\rangle \cdot \langle 5,-9\rangle = 4(5)+0(-9) = 20-0 = 20$

76. $2\mathbf{i}+4\mathbf{j} = \langle 2,4\rangle; -\mathbf{j} = \langle 0,-1\rangle$

$\langle 2,4\rangle \cdot \langle 0,-1\rangle = 2(0)+4(-1) = 0-4 = -4$

77. $\langle 2, 1\rangle \cdot \langle -3, 1\rangle$

$$\cos\theta = \frac{\langle 2, 1\rangle \cdot \langle -3, 1\rangle}{\sqrt{2^2+1^2}\cdot\sqrt{(-3)^2+1^2}} = \frac{-6+1}{\sqrt{5}\cdot\sqrt{10}} = \frac{-5}{5\sqrt{2}} = \frac{-1}{\sqrt{2}} = -\frac{\sqrt{2}}{2} \Rightarrow \theta = 135°$$

78. $\langle 1, 7\rangle \cdot \langle 1, 1\rangle$

$$\cos\theta = \frac{\langle 1, 7\rangle \cdot \langle 1, 1\rangle}{\sqrt{1^2+7^2}\cdot\sqrt{1^2+1^2}} = \frac{1+7}{\sqrt{50}\cdot\sqrt{2}} = \frac{8}{10} = .8 \Rightarrow \theta \approx 36.87°$$

79. $\langle 1, 2\rangle \cdot \langle -6, 3\rangle$

$$\cos\theta = \frac{\langle 1, 2\rangle \cdot \langle -6, 3\rangle}{\sqrt{1^2+2^2}\cdot\sqrt{(-6)^2+3^2}} \frac{-6+6}{\sqrt{5}\sqrt{45}} = \frac{0}{15} = 0 \Rightarrow \theta = 90°$$

80. $\langle 4, 0\rangle \cdot \langle 2, 2\rangle$

$$\cos\theta = \frac{\langle 4, 0\rangle \cdot \langle 2, 2\rangle}{\sqrt{4^2+0^2}\cdot\sqrt{2^2+2^2}} = \frac{8+0}{\sqrt{16}\cdot\sqrt{8}} = \frac{8}{8\sqrt{2}} = \frac{1}{\sqrt{2}} = \frac{\sqrt{2}}{2} \Rightarrow \theta = 45°$$

81. First write the given vectors in component form.

$$3\mathbf{i}+4\mathbf{j} = \langle 3,4\rangle \text{ and } \mathbf{j} = \langle 0,1\rangle$$

$$\cos\theta = \frac{\langle 3, 4\rangle \cdot \langle 0, 1\rangle}{|\langle 3, 4\rangle||\langle 0, 1\rangle|} = \frac{\langle 3, 4\rangle \cdot \langle 0, 1\rangle}{\sqrt{3^2+4^2}\cdot\sqrt{0^2+1^2}} = \frac{0+4}{\sqrt{25}\cdot\sqrt{1}} = \frac{4}{5\cdot 1} = \frac{4}{5} = .8 \Rightarrow \theta = \cos^{-1}.8 \approx 36.87°$$

82. First write the given vectors in component form.

$$-5\mathbf{i}+12\mathbf{j} = \langle -5,12\rangle \text{ and } 3\mathbf{i}+2\mathbf{j} = \langle 3,2\rangle$$

$$\cos\theta = \frac{\langle -5, 12\rangle \cdot \langle 3, 2\rangle}{\sqrt{(-5)^2+12^2}\cdot\sqrt{3^2+2^2}} = \frac{-15+24}{\sqrt{169}\sqrt{13}} = \frac{9}{13\sqrt{13}} = \frac{9\sqrt{13}}{169} \Rightarrow \theta \approx 78.93°$$

For Exercises 83 – 86, $\mathbf{u} = \langle -2,1 \rangle$, $\mathbf{v} = \langle 3,4 \rangle$, and $\mathbf{w} = \langle -5,12 \rangle$.

83. $(3\mathbf{u}) \cdot \mathbf{v} = (3\langle -2, 1 \rangle) \cdot \langle 3, 4 \rangle = \langle -6, 3 \rangle \cdot \langle 3, 4 \rangle = -18 + 12 = -6$

84. $\mathbf{u} \cdot (\mathbf{v} - \mathbf{w}) = \langle -2,1 \rangle \cdot (\langle 3,4 \rangle - \langle -5,12 \rangle) = \langle -2,1 \rangle \cdot \langle 3-(-5), 4-12 \rangle = \langle -2,1 \rangle \cdot \langle 8,-8 \rangle = -16 - 8 = -24$

85. $\mathbf{u} \cdot \mathbf{v} - \mathbf{u} \cdot \mathbf{w} = \langle -2, 1 \rangle \cdot \langle 3,4 \rangle - \langle -2,1 \rangle \cdot \langle -5, 12 \rangle = (-6+4) - (10+12) = -2 - 22 = -24$

86. $\mathbf{u} \cdot (3\mathbf{v}) = \langle -2,1 \rangle \cdot (3\langle 3,4 \rangle) = \langle -2,1 \rangle \cdot \langle 9,12 \rangle = -18 + 12 = -6$

87. Since $\langle 1,2 \rangle \cdot \langle -6,3 \rangle = -6 + 6 = 0$, the vectors are orthogonal.

88. Since $\langle 3,4 \rangle \cdot \langle 6,8 \rangle = 18 + 32 = 50 \neq 0$, the vectors are not orthogonal.

89. Since $\langle 1,0 \rangle \cdot \langle \sqrt{2},0 \rangle = \sqrt{2} + 0 = \sqrt{2} \neq 0$, the vectors are not orthogonal

90. Since $\langle 1,1 \rangle \cdot \langle 1,-1 \rangle = 1 - 1 = 0$, the vectors are orthogonal.

91. $\sqrt{5}\mathbf{i} - 2\mathbf{j} = \langle \sqrt{5}, -2 \rangle; -5\mathbf{i} + 2\sqrt{5}\mathbf{j} = \langle -5, 2\sqrt{5} \rangle$

Since $\langle \sqrt{5}, -2 \rangle \cdot \langle -5, 2\sqrt{5} \rangle = -5\sqrt{5} - 4\sqrt{5} = -9\sqrt{5} \neq 0$, the vectors are not orthogonal.

92. $-4\mathbf{i} + 3\mathbf{j} = \langle -4,3 \rangle; 8\mathbf{i} - 6\mathbf{j} = \langle 8,-6 \rangle$

Since $\langle -4,3 \rangle \cdot \langle 8,-6 \rangle = -32 - 18 = -50 \neq 0$, the vectors are not orthogonal

93. Draw a line parallel to the x-axis and the vector **u** + **v** (shown as a dashed line). Since $\theta_1 = 110°$, its supplementary angle is 70°. Further, since $\theta_2 = 260°$, the angle α is $260° - 180° = 80°$. Then the angle CBA becomes 180 – (80 + 70) = 180 – 150 = 30°.

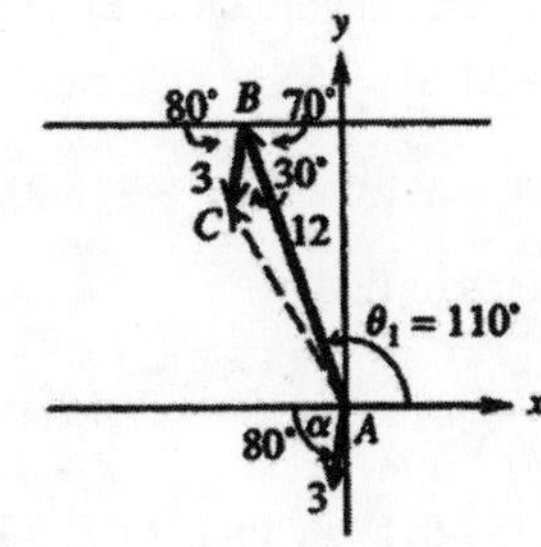

Using the law of cosines, the magnitude of **u** + **v** is the following.

$$|\mathbf{u} + \mathbf{v}|^2 = a^2 + c^2 - 2ac\cos B$$

$$|\mathbf{u} + \mathbf{v}|^2 = 3^2 + 12^2 - 2(3)(12)\cos 30° = 9 + 144 - 72 \cdot \frac{\sqrt{3}}{2} = 153 - 36\sqrt{3} \approx 90.646171$$

Thus, $|\mathbf{u} + \mathbf{v}| \approx 9.5208$.

Using the law of sines, we have the following.

$$\frac{\sin A}{a} = \frac{\sin B}{b} \Rightarrow \frac{\sin A}{3} = \frac{\sin 30°}{9.5208} \Rightarrow \sin A = \frac{3\sin 30°}{9.5208} = \frac{3 \cdot \frac{1}{2}}{9.5208} \approx .15754979 \Rightarrow A \approx 9.0647°$$

The direction angle of **u** + **v** is $110° + 9.0647° = 119.0647°$.

94. Since $a = 12\cos 110° \approx -4.10424172$ and $b = 12\sin 110° \approx 11.27631145$, $\langle a, b\rangle \approx \langle -4.1042, 11.2763\rangle$.

95. Since $c = 3\cos 260° \approx -.52094453$ and $d = 3\sin 260° \approx -2.95442326$, $\langle c, d\rangle \approx \langle -.5209, -2.9544\rangle$.

96. If $\mathbf{u} = \langle a, b\rangle = \langle -4.10424, 11.27631\rangle$ and $\mathbf{v} = \langle c, d\rangle = \langle -.52094, -2.95442\rangle$, then $\mathbf{u}+\mathbf{v}$ is the following.

$$\begin{aligned}\mathbf{u}+\mathbf{v} &= \langle -4.10424172 + (-.52094453), 11.27631145 + (-2.95442326)\rangle \\ &= \langle -4.62518625, 8.32188819\rangle \approx \langle -4.6252, 8.3219\rangle\end{aligned}$$

97. Magnitude: $\sqrt{(-4.62518625)^2 + 8.32188819^2} \approx 9.5208$

Angle: $\tan\theta' = \dfrac{8.32188819}{-4.625186258} \Rightarrow \theta' \approx -60.9353° \Rightarrow \theta = -60.9353° + 180° = 119.0647°$
(θ lies in quadrant II)

98. They are the same. Preference of method is an individual choice.

Section 7.5: Applications of Vectors

1. Find the direction and magnitude of the equilibrant.
Since $A = 180° - 28.2° = 151.8°$, we can use the law of cosines to find the magnitude of the resultant, **v**.

$|\mathbf{v}|^2 = 1240^2 + 1480^2 - 2(1240)(1480)\cos 151.8° \approx 6962736.2 \Rightarrow |\mathbf{v}| \approx 2639$ lb (will be rounded as 2640)

Use the law of sines to find α.

$$\frac{\sin\alpha}{1240} = \frac{\sin 151.8°}{2639} \Rightarrow \sin\alpha = \frac{1240\sin 151.8°}{2639} \approx .22203977 \Rightarrow \alpha \approx 12.8°$$

Thus, we have 2640 lb at an angle of $\theta = 180° - 12.8° = 167.2°$ with the 1480-lb force.

2. Find the direction and magnitude of the equilibrant.
Since $A = 180° - 24.5° = 155.5°$, we can use the law of cosines to find the magnitude of the resultant, **v**.

$$|\mathbf{v}|^2 = 840^2 + 960^2 - 2(840)(960)\cos 155.5° \approx 3094785.5 \Rightarrow |\mathbf{v}| \approx 1759 \text{ lb}$$

Use the law of sines to find α.

$$\frac{\sin\alpha}{960} = \frac{\sin 155.5°}{1759} \Rightarrow \sin\alpha = \frac{960\sin 155.5°}{1759} \approx .22632491 \Rightarrow \alpha \approx 13.1°$$

Thus, we have 1759 lb at an angle of $\theta = 180° - 13.1° = 166.9°$ with the 840-lb force.

3. Let α = the angle between the forces.

To find α, use the law of cosines to find θ.

$$786^2 = 692^2 + 423^2 - 2(692)(423)\cos\theta$$

$$\cos\theta = \frac{692^2 + 423^2 - 786^2}{2(692)(423)} \approx .06832049$$

$$\theta \approx 86.1°$$

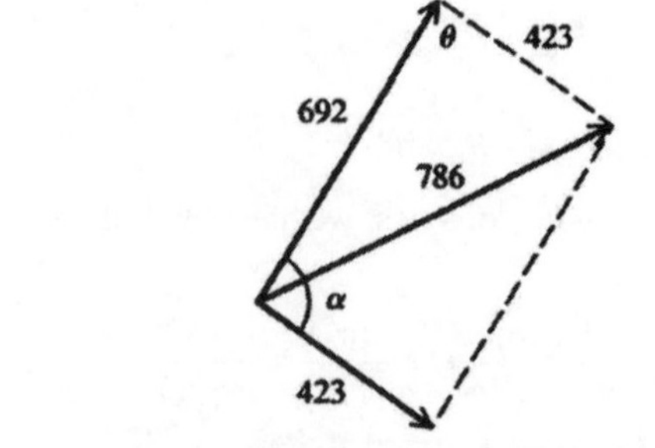

Thus, $\alpha = 180° - 86.1° = 93.9°$.

4. Let θ = the angle between the forces.

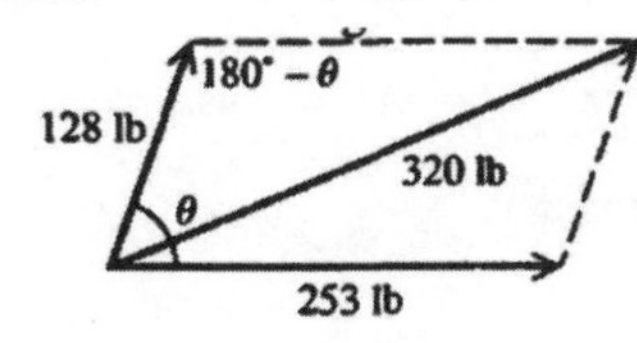

$$320^2 = 128^2 + 253^2 - 2(128)(253)\cos(180° - \theta)$$

$$\cos(180° - \theta) = \frac{128^2 + 253^2 - 320^2}{2(128)(253)} \approx -.33978199$$

$$180° - \theta = 109.9°$$

$$\theta = 70.1°$$

5. Use the parallelogram rule. In the figure, **x** represents the second force and **v** is the resultant.

$$\alpha = 180° - 78°50'$$
$$= 179°60' - 78°50' = 101°10'$$

and

$$\beta = 78°50' - 41°10' = 37°40'$$

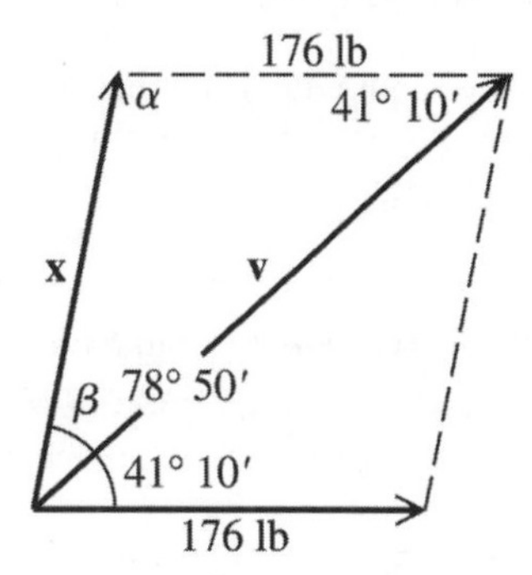

Using the law of sines, we have the following.

$$\frac{|\mathbf{x}|}{\sin 41°10'} = \frac{176}{\sin 37°40'} \Rightarrow |\mathbf{x}| = \frac{176 \sin 41°10'}{\sin 37°40'} \approx 190$$

$$\frac{|\mathbf{v}|}{\sin \alpha} = \frac{176}{\sin 37°40'} \Rightarrow |\mathbf{v}| = \frac{176 \sin 101°10'}{\sin 37°40'} \approx 283$$

Thus, the magnitude of the second force is about 190 lb and the magnitude of the resultant is about 283 lb.

6. Let $|\mathbf{f}|$ = the second force and $|\mathbf{r}|$ = the magnitude of the resultant.

$$\theta = 180° - 42°10'$$
$$= 179°60' - 42°10' = 137°50'$$

and

$$\beta = 42°10' - 32°40'$$
$$= 41°70' - 32°40' = 9°30'$$

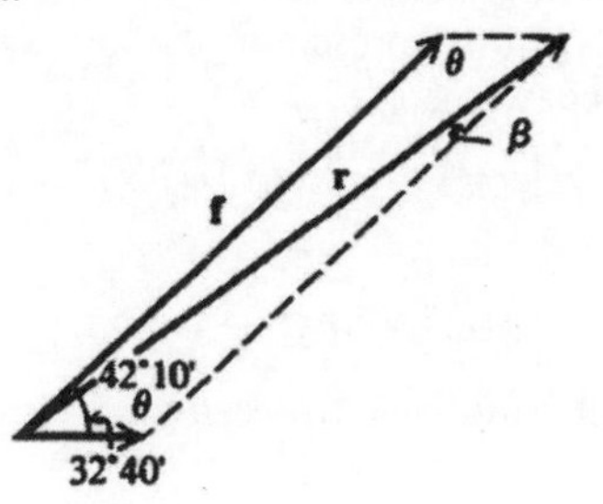

Using the law of sines, we have the following.

$$\frac{|\mathbf{r}|}{\sin \theta} = \frac{28.7}{\sin 9°30'} \Rightarrow |\mathbf{r}| = \frac{28.7 \sin 9°30'}{\sin 37°40'} \approx 116.73 \text{ lb (will be rounded as 117)}$$

$$\frac{|\mathbf{r}|}{\sin \theta} = \frac{28.7}{\sin 9°30'} \Rightarrow |\mathbf{r}| = \frac{28.7 \sin 137°50'}{\sin 9°30'} \approx 116.73 \text{ lb}$$

$$\frac{|\mathbf{f}|}{\sin 32°40'} = \frac{|\mathbf{r}|}{\sin 137°50'} \Rightarrow |\mathbf{f}| = \frac{116.73 \sin 32°40'}{\sin 137°50'} \approx 93.9 \text{ lb}$$

The magnitude of the resultant is 117 lb; the second force is 93.9 lb.

7. Let θ = the angle that the hill makes with the horizontal.

The 80-lb downward force has a 25-lb component parallel to the hill. The two right triangles are similar and have congruent angles.

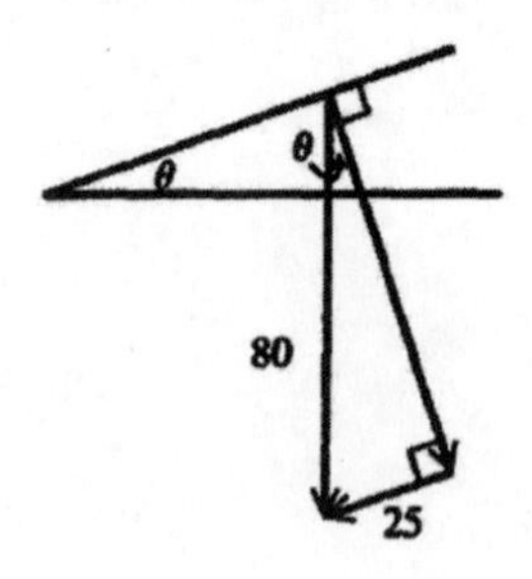

$$\sin\theta = \frac{25}{80} = \frac{5}{16} = .3125 \Rightarrow \theta \approx 18^\circ$$

8. Let $|\mathbf{f}|$ = the force required to keep the car parked on hill.

$$\frac{|\mathbf{f}|}{3000} = \sin 15^\circ \Rightarrow |\mathbf{f}| = 3000\sin 15^\circ = 776.5 \text{ lb}$$

The force required to keep the car parked on the hill is approximately 780 lb. (rounded to two significant digits)

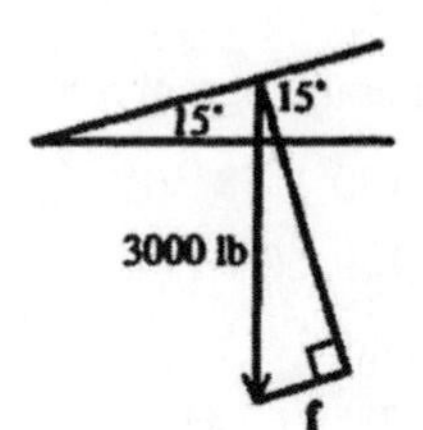

9. Find the force needed to pull a 60-ton monolith along the causeway.

The force needed to pull 60 tons is equal to the magnitude of **x**, the component parallel to the causeway.

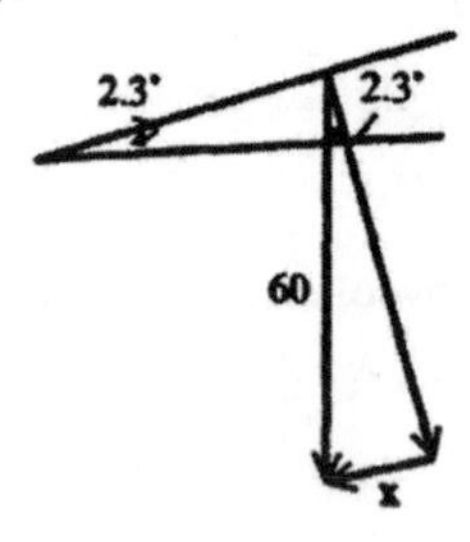

$$\sin 2.3^\circ = \frac{|\mathbf{x}|}{60} \Rightarrow |\mathbf{x}| = 60\sin 2.3^\circ \approx 2.4 \text{ tons}$$

The force needed is 2.4 tons.

10. Let **r** = the vertical component of the person exerting a 114-lb force;

s = the vertical component of the person exerting a 150-lb force.

The weight of the box is the sum of the magnitudes of the vertical components of the two vectors representing the forces exerted by the two people.

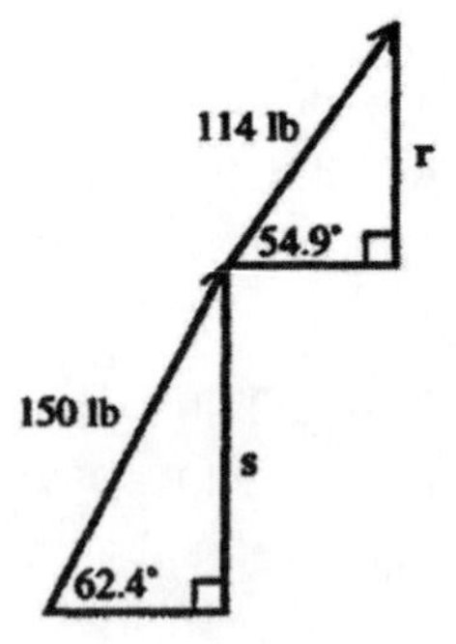

$$|\mathbf{r}| = 114\sin 54.9^\circ \approx 93.27$$

and

$$|\mathbf{s}| = 150\sin 62.4^\circ \approx 132.93$$

Thus, the weight of box is $|\mathbf{r}| + |\mathbf{s}| \approx 93.27 + 132.93 = 226.2 \approx 226$ lb.

11. Like Example 3 on page 316 or your text, angle B equals angle θ and here the magnitude of vector **BA** represents the weight of the stump grinder. The vector **AC** equals vector **BE**, which represents the force required to hold the stump grinder on the incline. Thus, we have the following.

$$\sin B = \frac{18}{60} = \frac{3}{10} = .3 \Rightarrow B \approx 17.5^\circ$$

12. Like Example 3 on page 316 or your text, angle B equals angle θ and here the magnitude of vector **BA** represents the weight of the pressure washer. The vector **AC** equals vector **BE**, which represents the force required to hold the pressure washer on the incline. Thus we have the following.

$$\sin B = \frac{30}{80} = \frac{3}{8} = .375 \Rightarrow B \approx 22.0°$$

13. Find the weight of the crate and the tension on the horizontal rope.

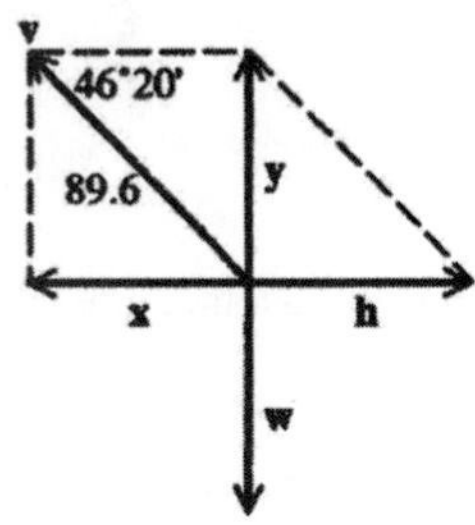

v has horizontal component **x** and vertical component **y**. The resultant **v** + **h** also has vertical component **y**. The resultant balances the weight of the crate, so its vertical component is the equilibrant of the crate's weight.

$$|\mathbf{w}| = |\mathbf{y}| = 89.6 \sin 46° 20' \approx 64.8 \text{ lb}$$

Since the crate is not moving side-to-side, **h**, the horizontal tension on the rope, is the opposite of **x**.

$$|\mathbf{h}| = |\mathbf{x}| = 89.6 \cos 46° 20' \approx 61.9 \text{ lb}$$

The weight of the crate is 64.8 lb; the tension is 61.9 lb.

14. Draw a vector diagram showing the three forces acting at a point in equilibrium. Then, arrange the forces to form a triangle. The angles between the forces are the supplements of the angles of the triangle.

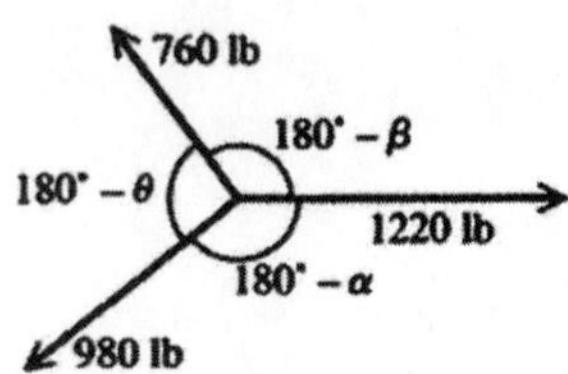

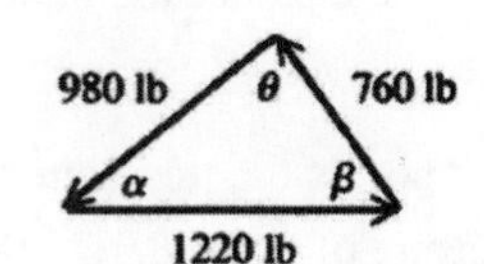

(1) $\cos\theta = \dfrac{980^2 + 760^2 - 1220^2}{2(980)(760)} = \dfrac{49,600}{1,489,600} = \dfrac{31}{931} \approx .03329753 \Rightarrow \theta \approx 88.1°$

The angle between the 760-lb and 980-lb forces is $180° - \theta = 180° - 88.1° = 91.9°$

(2) $\dfrac{\sin\alpha}{760} = \dfrac{\sin\theta}{1220} \Rightarrow \sin\alpha = \dfrac{760 \sin 88.1°}{1220} \approx .62260833 \Rightarrow \alpha \approx 38.5°$

The angle between the 1220-lb and 980-lb forces is $180° - \alpha = 180° - 38.5° = 141.5°$

(3) $\beta = 180° - \alpha - \theta = 180° - 38.5° - 88.1° = 53.4°$

The angle between the 760-lb and 1220-lb forces is $180° - \beta = 180° - 53.4° = 126.6°$.

15. Refer to the diagram on the right.

In order for the ship to turn due east, the ship must turn the measure of angle *CAB*, which is $90° - 34° = 56°$. Angle *DAB* is therefore $180° - 56° = 124°$.

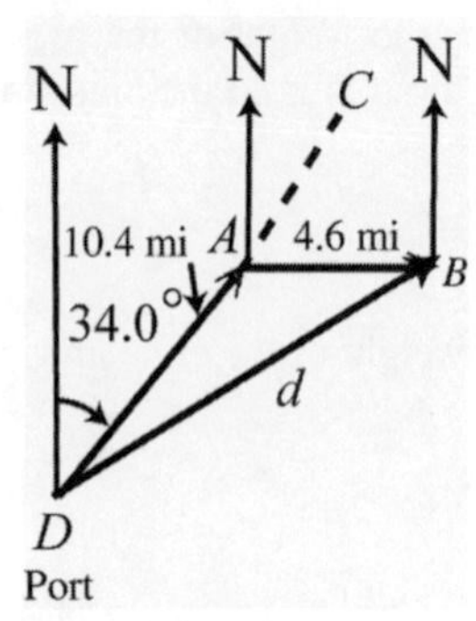

Using the law of cosines, we can solve for the distance the ship is from port as follows.

$$d^2 = 10.4^2 + 4.6^2 - 2(10.4)(4.6)\cos 124° \approx 182.824 \Rightarrow d \approx 13.52$$

Thus, the distance the ship is from port is 13.5 mi. (rounded to three significant digits)

To find the bearing, we first seek the measure of angle *ADB*, which we will refer to as angle *D*. Using the law of cosines we have the following.

$$\cos D = \frac{13.52^2 + 10.4^2 - 4.6^2}{2(13.52)(10.4)} \approx .95937073 \Rightarrow D \approx 16.4°$$

Thus, the bearing is $34.0° + D = 34.0° + 16.4° = 50.4°$.

16. Refer to the diagram on the right.

In order for the luxury liner to turn due west, the measure of angle *ABD* must be $180° - 110° = 70°$. Thus, angle *DBC* is $90° - 70° = 20°$.

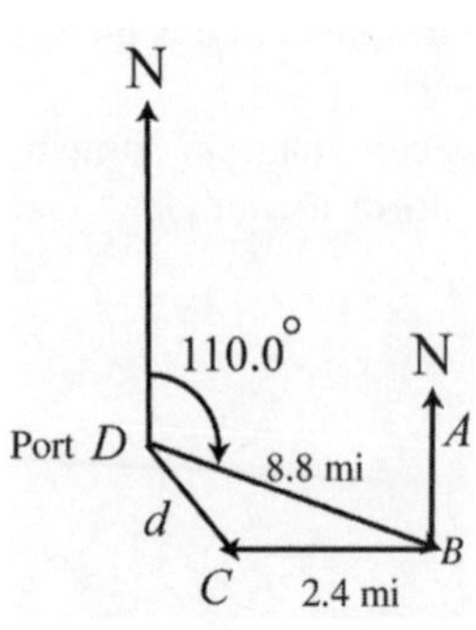

Using the law of cosines, we can solve for the distance the luxury liner is from port as follows.

$$d^2 = 8.8^2 + 2.4^2 - 2(8.8)(2.4)\cos 20° \approx 43.507 \Rightarrow d \approx 6.596$$

Thus, the distance the luxury liner is from port is 6.6 mi. (rounded to two significant digits)

To find the bearing, we first seek the measure of angle *BDC*, which we will refer to as angle *D*. Using the law of cosines we have the following.

$$\cos D = \frac{6.596^2 + 8.8^2 - 2.4^2}{2(6.596)(8.8)} \approx .99222683 \Rightarrow D \approx 7.1°$$

Thus, the bearing is $110.0° + D = 110.0° + 7.1° = 117.1°$.

17. Find the distance of the ship from point A.

Angle 1 = 189° – 180° = 9°

Angle 2 = 360° – 317° = 43°

Angle 1 + Angle 2 = 9° + 43° = 52°

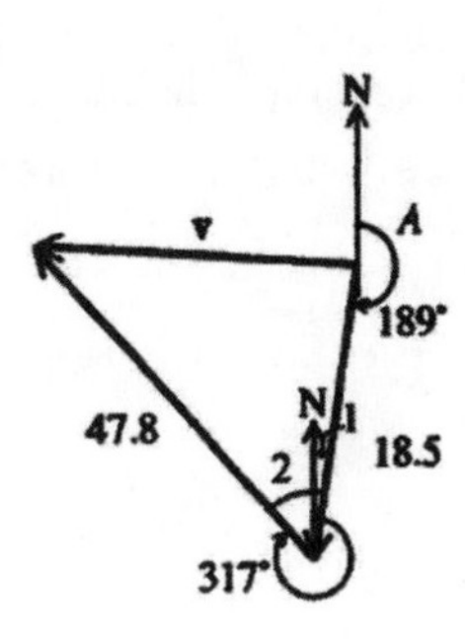

Use the law of cosines to find $|\mathbf{v}|$.

$$|\mathbf{v}|^2 = 47.8^2 + 18.5^2 - 2(47.8)(18.5)\cos 52° \approx 1538.23 \Rightarrow |\mathbf{v}| \approx 39.2 \text{ km}$$

18. Find the distance of the ship from point X.

Angle 1 = 200° – 180° = 20°

Angle 2 = 360° – 320° = 40°

Angle 1 + Angle 2 = 20° + 40° = 60°

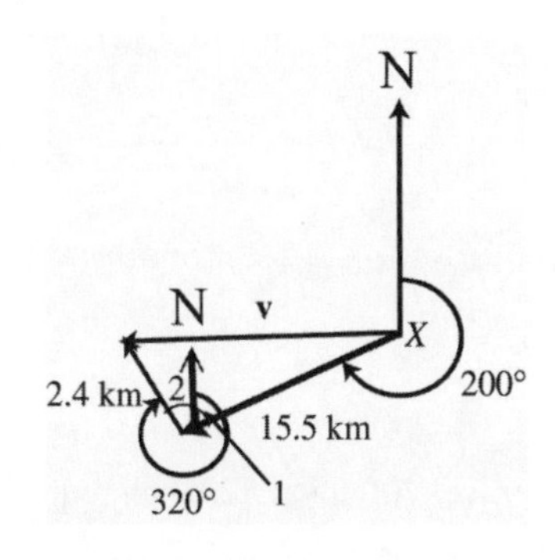

Use the law of cosines to find $|\mathbf{v}|$.

$$|\mathbf{v}|^2 = 2.4^2 + 15.5^2 - 2(2.4)(15.5)\cos 60° = 208.81 \Rightarrow |\mathbf{v}| \approx 14.5 \text{ km}$$

19. Let x = be the actual speed of the motorboat;
y = the speed of the current.

$$\sin 10° = \frac{y}{20.0} \Rightarrow y = 20.0 \sin 10° \approx 3.5$$

$$\cos 10° = \frac{x}{20.0} \Rightarrow x = 20.0 \cos 10° \approx 19.7$$

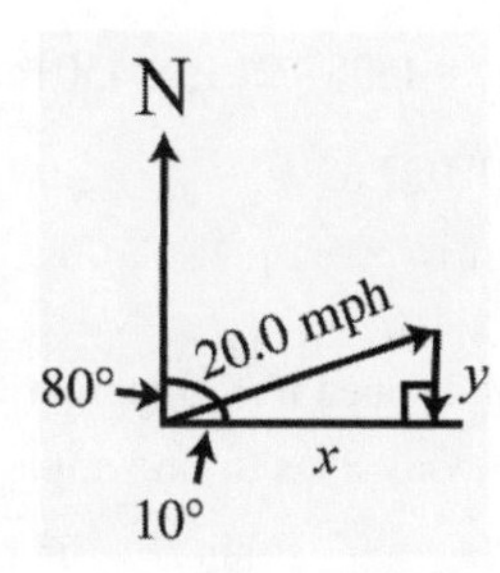

The speed of the current is 3.5 mph and the actual speed of the motorboat is 19.7 mph.

20. At what time did the pilot turn? The plane will fly 2.6 hr before it runs out of fuel. In 2.6 hr, the carrier will travel (2.6)(32) = 83.2 mi and the plane will travel a total of (2.6)(520) = 1352 mi. Suppose it travels x mi on its initial bearing; then it travels $1352 - x$ mi after having turned.

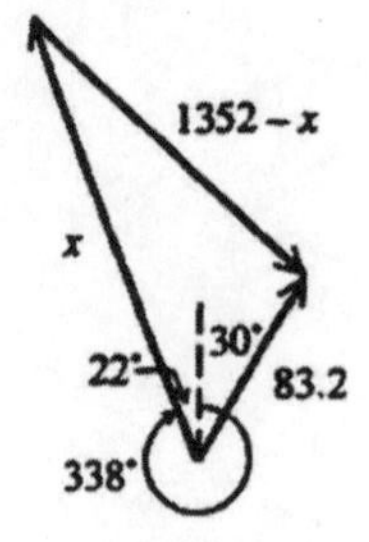

Use the law of cosines to get an equation in x and solve.

$$\begin{aligned}
(1352-x)^2 &= x^2 + 83.2^2 - 2(x)(83.2)\cos 52° \\
1,827,904 - 2704x + x^2 &= x^2 + 6922.24 - (166.4\cos 52°)x \\
1,827,904 - 2704x &= 6922.24 - (166.4\cos 52°)x \\
1,820,982 - 2704x &= -(166.4\cos 52°)x \\
1,820,982 &= 2704x - (166.4\cos 52°)x \\
1,820,982 &= 2(2704 - 166.4\cos 52°)x \\
x &= \frac{1,820,982}{2704 - 166.4\cos 52°} \\
x &\approx 700
\end{aligned}$$

To travel 700 mi at 520 mph requires $\frac{700}{520} = 1.35$ hr, or 1 hr and 21 min.

Thus, the pilot turned at 1 hr 21 min after 2 P.M., or at 3:21 P.M.

21. Let $\mathbf{v}$ = the ground speed vector.
Find the bearing and ground speed of the plane.
Angle $A = 233° - 114° = 119°$
Use the law of cosines to find $|\mathbf{v}|$.

$$|\mathbf{v}|^2 = 39^2 + 450^2 - 2(39)(450)\cos 119°$$
$$|\mathbf{v}|^2 \approx 221,037.82$$
$$|\mathbf{v}| \approx 470.1$$

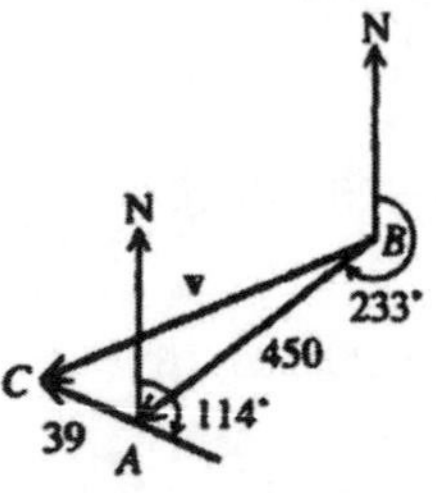

The ground speed is 470 mph. (rounded to two significant digits)
Use the law of sines to find angle B.

$$\frac{\sin B}{39} = \frac{\sin 119°}{470.1} \Rightarrow \sin B = \frac{39\sin 119°}{470.1} \approx .07255939 \Rightarrow B \approx 4°$$

Thus, the bearing is $B + 233° = 4° + 233° = 237°$.

22. **(a)** Relative to the banks, the motorboat will be traveling at the following speed.

$x^2+3^2=7^2 \Rightarrow x^2+9=49 \Rightarrow x^2=40 \Rightarrow x=\sqrt{40}\approx 6.325$ mph (will be rounded as 6.32)

(b) At 6.325 mph, the motorboat travels the following rate.

$$\frac{6.325\text{miles}}{\text{hour}}\cdot\frac{5280\text{ feet}}{1\text{ mile}}\cdot\frac{1\text{ hour}}{60\text{ minutes}}\cdot\frac{1\text{ minute}}{60\text{ seconds}}=9.277\text{ ft/sec}$$

Crossing a 132-foot wide river would take $132\text{ feet}\cdot\dfrac{1\text{ second}}{9.277\text{ feet}}\approx 14.23$ seconds

(c) $\sin\theta=\dfrac{3}{7}\approx .42857143 \Rightarrow \theta\approx 25.38°$

23. Let $|\mathbf{x}|$ = the airspeed and $|\mathbf{d}|$ = the ground speed.

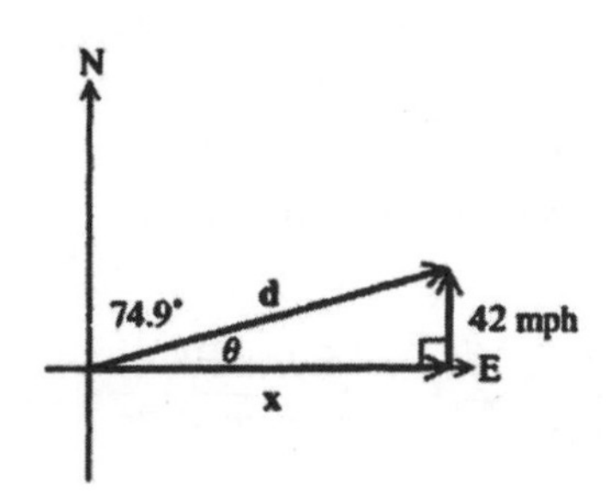

$$\theta=90°-74.9°=15.1°$$

$$\frac{|\mathbf{x}|}{42}=\cot 15.1° \Rightarrow |\mathbf{x}|=42\cot 15.1°=\frac{42}{\tan 15.1°}\approx 156\text{ mph}$$

$$\frac{|\mathbf{d}|}{42}=\csc 15.1° \Rightarrow |\mathbf{d}|=42\csc 15.1°=\frac{42}{\sin 15.1°}\approx 161\text{ mph}$$

24. Let $\mathbf{v}$ = the ground speed vector of the plane. Find the actual bearing of the plane.

Angle $A=266.6°-175.3°=91.3°$

Use the law of cosines to find $|\mathbf{v}|$.

$|\mathbf{v}|^2=25^2+650^2-2(25)(650)\cos 91.3°$

$|\mathbf{v}|^2=423,862.33$

$|\mathbf{v}|=651.0$

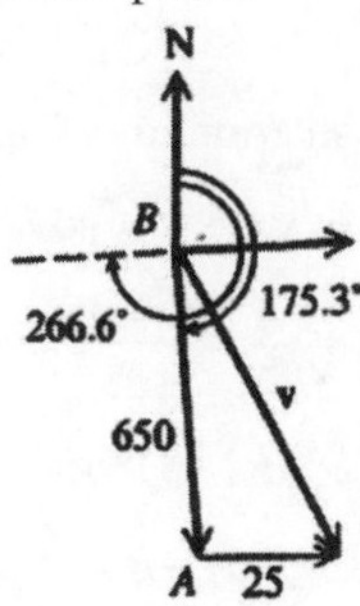

Use the law of sines to find, B the angle that $\mathbf{v}$ makes with the airspeed vector of the plane.

$$\frac{\sin B}{25}=\frac{\sin A}{651.0} \Rightarrow \sin B=\frac{25\sin 91.3°}{651.0}\approx .03839257 \Rightarrow B\approx 2.2°$$

Thus, the bearing is $175.1°-B=175.3°-2.2°=173.1°$.

25. Let **c** = the ground speed vector.

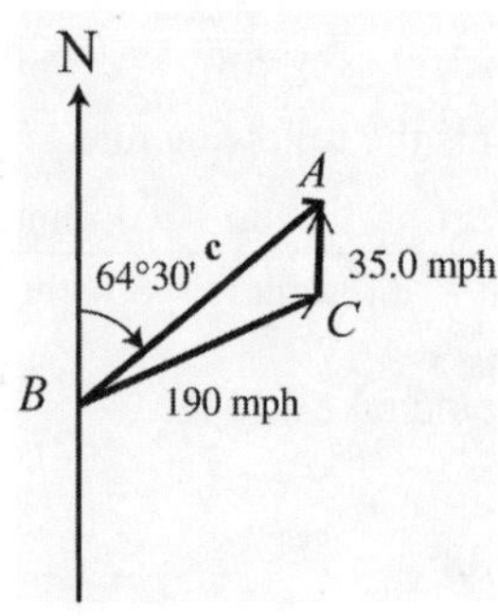

By alternate interior angles, angle $A = 64°30'$.

Use the law of sines to find B.

$$\frac{\sin B}{35.0} = \frac{\sin A}{190} \Rightarrow \sin B = \frac{35.0 \sin 64°30'}{190} \approx .16626571 \Rightarrow B \approx 9.57° \approx 9°30'$$

Thus, the bearing is $64°30' + B = 64°30' + 9°30' = 74°00'$.

Since $C = 180° - A - B = 180° - 64.50° - 9.57° = 105.93°$, we use the law of sine to find the ground speed.

$$\frac{|\mathbf{c}|}{\sin C} = \frac{35.0}{\sin B} \Rightarrow |\mathbf{c}| = \frac{35.0 \sin 105.93°}{\sin 9.57°} \approx 202 \text{ mph}$$

The bearing is $74°00'$; the ground speed is 202 mph.

26. Let **c** = the ground speed vector.

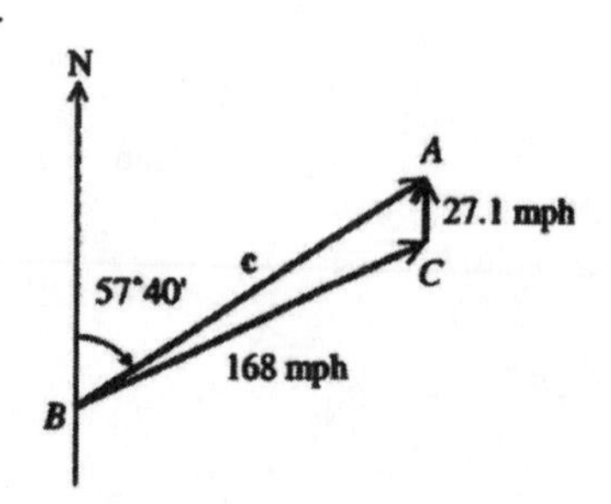

By alternate interior angles, angle $A = 57°\,40'$.

Use the law of sines to find B.

$$\frac{\sin B}{27.1} = \frac{\sin A}{168} \Rightarrow \sin B = \frac{27.1 \sin 57°\,40'}{168} \approx .13629861 \Rightarrow B \approx 7.83° \approx 7°\,50'$$

Thus, the bearing is $57°\,40' + B = 57°\,40' + 7°\,50' = 65°\,30'$.

Since $C = 180° - A - B = 180° - 57°\,40' - 7.83° \approx 114°\,30'$, we use the law of sine to find the ground speed.

$$\frac{|\mathbf{c}|}{\sin C} = \frac{27.1}{\sin B} \Rightarrow |\mathbf{c}| = \frac{27.1 \sin 114°\,30'}{\sin 7°\,50'} \approx 181 \text{ mph}$$

The bearing is $65°\,30'$; the ground speed is 181 mph.

27. Let **v** = the airspeed vector.

The ground speed is $\frac{400 \text{ mi}}{2.5 \text{ hr}} = 160$ mph.

angle $BAC = 328° - 180° = 148°$

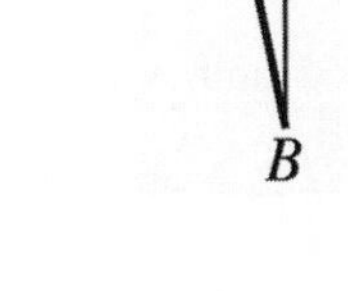

Using the law of cosines to find $|\mathbf{v}|$, we have the following.

$$|\mathbf{v}|^2 = 11^2 + 160^2 - 2(11)(160)\cos 148°$$
$$|\mathbf{v}|^2 \approx 28,706.1$$
$$|\mathbf{v}| \approx 169.4$$

The airspeed must be 170 mph. (rounded to two significant digits)

Use the law of sines to find B.

$$\frac{\sin B}{11} = \frac{\sin 148°}{169.4} \Rightarrow \sin B = \frac{11 \sin 148°}{169.4} \Rightarrow \sin B \approx .03441034 \Rightarrow B \approx 2.0°$$

The bearing must be approximately $360° - 2.0° = 358°$.

28. Let **v** = the ground speed vector.
angle $C = 180° - 78° = 102°$

$$|\mathbf{v}|^2 = 23^2 + 192^2 - 2(23)(192)\cos 102°$$
$$|\mathbf{v}|^2 \approx 39,229.28$$
$$|\mathbf{v}| \approx 198.1$$

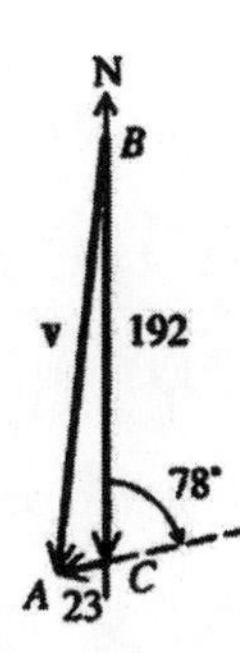

Thus, the ground speed is 198 mph. (rounded to three significant digits)

$$\frac{\sin B}{23} = \frac{\sin 102°}{198.1} \Rightarrow \sin B = \frac{23 \sin 102°}{198.1} \approx .11356585 \Rightarrow B \approx 6.5°$$

Thus, the bearing is $180° + B = 180° + 6.5° = 186.5°$.

29. Find the ground speed and resulting bearing.
Angle $A = 245° - 174° = 71°$
Use the law of cosines to find $|\mathbf{v}|$.

$$|\mathbf{v}|^2 = 30^2 + 240^2 - 2(30)(240)\cos 71°$$
$$|\mathbf{v}|^2 \approx 53,811.8$$
$$|\mathbf{v}| \approx 232.1$$

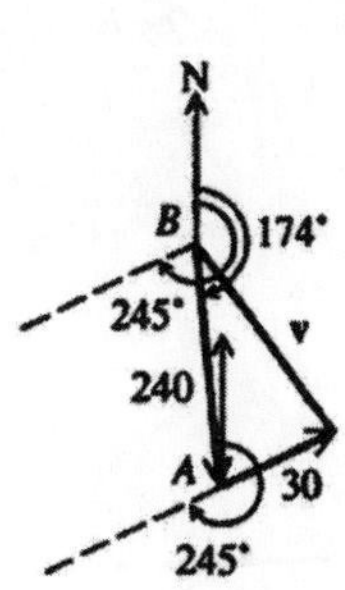

The ground speed is 230 km per hr. (rounded to two significant digits)
Use the law of sines to find angle B.

$$\frac{\sin B}{30} = \frac{\sin 71°}{230} \Rightarrow \sin B = \frac{30 \sin 71°}{230} \approx .12332851 \Rightarrow B \approx 7°$$

Thus, the bearing is $174° - B = 174° - 7° = 167°$.

30. (a) First change 10.34″ to radians in order to use the length of arc formula.

$$10.34'' \cdot \frac{1^\circ}{3600''} \cdot \frac{\pi}{180^\circ} \approx 5.013 \times 10^{-5} \text{ radian.}$$

In one year Barnard's Star will move in the tangential direction the following distance.

$$s = r\theta = \left(35 \times 10^{12}\right)\left(5.013 \times 10^{-5}\right) = 175.455 \times 10^{7} = 1{,}754{,}550{,}000 \text{ mi.}$$

In one second Barnard's Star moves $\frac{1{,}754{,}550{,}000}{60 \cdot 60 \cdot 24 \cdot 365} \approx 55.6$ mi tangentially.

Thus, $\mathbf{v}_t \approx 56$ mi/sec. (rounded to two significant digits)

(b) The magnitude of **v** is given by $|\mathbf{v}|^2 = |\mathbf{v}_r|^2 + |\mathbf{v}_t|^2 - 2|\mathbf{v}_r||\mathbf{v}_t|\cos 90^\circ$.
Since $|\mathbf{v}_r| = 67$ and $|\mathbf{v}_t| \approx 55.6$, we have the following.

$$|\mathbf{v}|^2 = 67^2 + 55.6^2 - 2(67)(55.6)(0) = 4489 + 3091.36 - 0 = 7580.36 \Rightarrow |\mathbf{v}| = \sqrt{7580.36} \approx 87.1$$

Since the magnitude or length of **v** is 87, **v** represents a velocity of 87 mi/sec. (both rounded to two significant digits)

31. $\mathbf{R} = \mathbf{i} - 2\mathbf{j}$ and $\mathbf{A} = .5\mathbf{i} + \mathbf{j}$

(a) Write the given vector in component form. $\mathbf{R} = \mathbf{i} - 2\mathbf{j} = \langle 1, -2 \rangle$ and $\mathbf{A} = .5\mathbf{i} + \mathbf{j} = \langle .5, 1 \rangle$

$$|\mathbf{R}| = \sqrt{1^2 + (-2)^2} = \sqrt{1+4} = \sqrt{5} \approx 2.2 \text{ and } |\mathbf{A}| = \sqrt{.5^2 + 1^2} = \sqrt{.25 + 1} = \sqrt{1.25} \approx 1.1$$

About 2.2 in. of rain fell. The area of the opening of the rain gauge is about 1.1 in.^{2}.

(b) $V = |\mathbf{R} \cdot \mathbf{A}| = |\langle 1, -2 \rangle \cdot \langle .5, 1 \rangle| = |.5 + (-2)| = |-1.5| = 1.5$

The volume of rain was 1.5 in.^{3}.

(c) **R** and **A** should be parallel and point in opposite directions.

32. $\mathbf{a} = \langle a_1, a_2 \rangle, \mathbf{b} = \langle b_1, b_2 \rangle$, and $\mathbf{a} - \mathbf{b} = \langle a_1 - b_1, a_2 - b_2 \rangle$

$$\begin{aligned}
|\mathbf{a} - \mathbf{b}|^2 &= |\mathbf{a}|^2 + |\mathbf{b}|^2 - 2|\mathbf{a}||\mathbf{b}|\cos\theta \\
(a_1 - b_1)^2 + (a_2 - b_2)^2 &= a_1^2 + a_2^2 + b_1^2 + b_2^2 - 2|\mathbf{a}||\mathbf{b}|\cos\theta \\
a_1^2 - 2a_1b_1 + b_1^2 + a_2^2 - 2a_2b_2 + b_2^2 &= a_1^2 + a_2^2 + b_1^2 + b_2^2 - 2|\mathbf{a}||\mathbf{b}|\cos\theta \\
-2a_1b_1 - 2a_2b_2 &= -2|\mathbf{a}||\mathbf{b}|\cos\theta \\
a_1b_1 + a_2b_2 &= |\mathbf{a}||\mathbf{b}|\cos\theta \\
\mathbf{a} \cdot \mathbf{b} &= |\mathbf{a}||\mathbf{b}|\cos\theta
\end{aligned}$$

Chapter 7: Review Exercises

1. Find b, given $C = 74.2°$, $c = 96.3$ m, $B = 39.5°$.
Use the law of sines to find b.

$$\frac{b}{\sin B} = \frac{c}{\sin C} \Rightarrow \frac{b}{\sin 39.5°} = \frac{96.3}{\sin 74.2°} \Rightarrow b = \frac{96.3 \sin 39.5°}{\sin 74.2°} \approx 63.7 \text{ m}$$

2. Find B, given $A = 129.7°$, $a = 127$ ft, $b = 69.8$ ft.
Use the law of sines to find B.

$$\frac{\sin B}{b} = \frac{\sin A}{a} \Rightarrow \frac{\sin B}{69.8} = \frac{\sin 129.7°}{127} \Rightarrow \sin B = \frac{69.8 \sin 129.7°}{127} \approx .42286684$$

Since angle A is obtuse, angle B is acute. Thus, $B \approx 25.0°$.

3. Find B, given $C = 51.3°$, $c = 68.3$ m, $b = 58.2$ m.
Use the law of sines to find B.

$$\frac{\sin B}{b} = \frac{\sin C}{c} \Rightarrow \frac{\sin B}{58.2} = \frac{\sin 51.3°}{68.3} \Rightarrow \sin B = \frac{58.2 \sin 51.3°}{68.3} \approx .66502269$$

There are two angles B between 0° and 180° that satisfy the condition. Since $\sin B \approx .66502269$, to the nearest tenth value of B is $B_1 = 41.7°$. Supplementary angles have the same sine value, so another possible value of B is $B_2 = 180° - 41.7° = 138.3°$. This is not a valid angle measure for this triangle since $C + B_2 = 51.3° + 138.3° = 189.6° > 180°$.

Thus, $B = 41.7°$.

4. Find b, given $a = 165$ m, $A = 100.2°$, $B = 25.0°$.
Use the law of sines to find b.

$$\frac{b}{\sin B} = \frac{a}{\sin A} \Rightarrow \frac{b}{\sin 25.0°} = \frac{165}{\sin 100.2°} \Rightarrow b = \frac{165 \sin 25.0°}{\sin 100.2°} \approx 70.9 \text{ m}$$

5. Find A, given $B = 39° 50'$, $b = 268$ m, $a = 340$ m.
Use the law of sines to find A.

$$\frac{\sin A}{a} = \frac{\sin B}{b} \Rightarrow \frac{\sin A}{340} = \frac{\sin 39° 50'}{268} \Rightarrow \sin A = \frac{340 \sin 39° 50'}{268} \approx .81264638$$

There are two angles A between 0° and 180° that satisfy the condition. Since $\sin A \approx .81264638$, to the nearest tenth value of A is $A_1 = 54.4° \approx 54°20'$. Supplementary angles have the same sine value, so another possible value of A is $A_2 = 180° - 54°20' = 179°60' - 54°20' = 125°40'$. This is a valid angle measure for this triangle since $B + A_2 = 39°50' + 125°40' = 165°30' < 180°$.

$A = 54° 20'$ or $A = 125° 40'$

6. Find A, given $C = 79° 20'$, $c = 97.4$ mm, $a = 75.3$ mm.
Use the law of sines to find A.

$$\frac{\sin A}{a} = \frac{\sin C}{c} \Rightarrow \frac{\sin A}{75.3} = \frac{\sin 79° 20'}{97.4} \Rightarrow \sin A = \frac{75.3 \sin 79° 20'}{97.4} \approx .75974194$$

There are two angles A between $0°$ and $180°$ that satisfy the condition. Since $\sin A \approx .75974194$, to the nearest hundredth value of A is $A_1 = 49.44° \approx 49°30'$. Supplementary angles have the same sine value, so another possible value of A is $A_2 = 180° - 49°30' = 179°60' - 49°30' = 130°30'$. This is not a valid angle measure for this triangle since $C + A_2 = 79°20' + 130°30' = 209°50' > 180°$.

Thus, $A = 49° 30'$.

7. No; If you are given two angles of a triangle, then the third angle is known since the sum of the measures of the three angles is $180°$. Since you are also given one side, there will only be one triangle that will satisfy the conditions.

8. No; the sum of a and b do not exceed c.

9. $a = 10, B = 30°$

(a) The value of b that forms a right triangle would yield exactly one value for A. That is, $b = 10 \sin 30° = 5$. Also, any value of b greater than or equal to 10 would yield a unique value for A.

(b) Any value of b between 5 and 10, would yield two possible values for A.

(c) If b is less than 5, then no value for A is possible.

10. $A = 140°, a = 5, \text{and } b = 7$
With these conditions, we can try to solve the triangle with the law of sines.

$$\frac{\sin B}{b} = \frac{\sin A}{a} \Rightarrow \frac{\sin B}{7} = \frac{\sin 140°}{5} \Rightarrow \sin B = \frac{7 \sin 140°}{5} \approx .89990265 \Rightarrow B \approx 64°$$

Since $A + B = 140° + 64° = 204° > 180°$, no such triangle exists.

11. Find A, given $a = 86.14$ in., $b = 253.2$ in., $c = 241.9$ in.
Use the law of cosines to find A.

$$a^2 = b^2 + c^2 - 2bc \cos A \Rightarrow \cos A = \frac{b^2 + c^2 - a^2}{2bc} = \frac{253.2^2 + 241.9^2 - 86.14^2}{2(253.2)(241.9)} \approx .94046923$$

Thus, $A \approx 19.87°$ or $19° 52'$.

12. Find b, given $B = 120.7°$, $a = 127$ ft, $c = 69.8$ ft.
Use the law of cosines to find b.

$$b^2 = a^2 + c^2 - 2ac \cos B \Rightarrow b^2 = 127^2 + 69.8^2 - 2(127)(69.8) \cos 120.7° \approx 30,052.6 \Rightarrow b \approx 173 \text{ ft}$$

13. Find a, given $A = 51° 20'$, $c = 68.3$ m, $b = 58.2$ m.
Use the law of cosines to find a.

$$a^2 = b^2 + c^2 - 2bc \cos A \Rightarrow a^2 = 58.2^2 + 68.3^2 - 2(58.2)(68.3) \cos 51° 20' \approx 3084.99 \Rightarrow a \approx 55.5 \text{ m}$$

14. Find B, given $a = 14.8$ m, $b = 19.7$ m, $c = 31.8$ m.
Use the law of cosines to find B.

$$b^2 = a^2 + c^2 - 2ac\cos B \Rightarrow \cos B = \frac{a^2 + c^2 - b^2}{2ac} = \frac{14.8^2 + 31.8^2 - 19.7^2}{2(14.8)(31.8)} \approx .89472845$$

Thus, $B \approx 26.5°$ or $26°\ 30'$.

15. Find a, given $A = 60°$, $b = 5$cm, $c = 21$ cm.
Use the law of cosines to find a.

$$a^2 = b^2 + c^2 - 2bc\cos A \Rightarrow a^2 = 5^2 + 21^2 - 2(5)(21)\cos 60° = 361 \Rightarrow a = 19\text{ cm}$$

16. Find A, given $a = 13$ ft, $b = 17$ ft, $c = 8$ ft.
Use the law of cosines to find A.

$$a^2 = b^2 + c^2 - 2bc\cos A \Rightarrow \cos A = \frac{b^2 + c^2 - a^2}{2bc} = \frac{17^2 + 8^2 - 13^2}{2(17)(8)} = \frac{184}{272} = \frac{23}{34} \approx .67647059 \Rightarrow A \approx 47°$$

17. Solve the triangle, given $A = 25.2\ °$, $a = 6.92$ yd, $b = 4.82$ yd.

$$\frac{\sin B}{b} = \frac{\sin A}{a} \Rightarrow \sin B = \frac{b\sin A}{a} \Rightarrow \sin B = \frac{4.82\sin 25.2°}{6.92} \approx .29656881$$

There are two angles B between $0°$ and $180°$ that satisfy the condition. Since $\sin B \approx .29656881$, to the nearest tenth value of B is $B_1 = 17.3°$. Supplementary angles have the same sine value, so another possible value of B is $B_2 = 180° - 17.3° = 162.7°$. This is not a valid angle measure for this triangle since $A + B_2 = 25.2° + 162.7° = 187.9° > 180°$.

$$C = 180° - A - B \Rightarrow C = 180° - 25.2° - 17.3° \Rightarrow C = 137.5°$$

Use the law of sines to find c.

$$\frac{c}{\sin C} = \frac{a}{\sin A} \Rightarrow \frac{c}{\sin 137.5°} = \frac{6.92}{\sin 25.2°} \Rightarrow c = \frac{6.92\sin 137.5°}{\sin 25.2°} \approx 11.0\text{ yd}$$

18. Solve the triangle, given $A = 61.7°$, $a = 78.9$ m, $b = 86.4$ m.

$$\frac{\sin B}{b} = \frac{\sin A}{a} \Rightarrow \sin B = \frac{b\sin A}{a} \Rightarrow \sin B = \frac{86.4\sin 61.7°}{78.9} \approx .96417292$$

There are two angles B between $0°$ and $180°$ that satisfy the condition. Since $\sin B \approx .96417292$, to the nearest tenth value of B is $B_1 = 74.6°$. Supplementary angles have the same sine value, so another possible value of B is $B_2 = 180° - 74.6° = 105.4°$. This is a valid angle measure for this triangle since $A + B_2 = 61.7° + 105.4° = 167.1° < 180°$.

Solving separately for triangles AB_1C_1 and AB_2C_2 we have the following.

AB_1C_1:

$$C_1 = 180° - A - B_1 = 180° - 61.7° - 74.6° = 43.7°$$

$$\frac{c_1}{\sin C_1} = \frac{a}{\sin A} \Rightarrow c_1 = \frac{a\sin C_1}{\sin A} \Rightarrow c_1 = \frac{78.9\sin 43.7°}{\sin 61.7°} \approx 61.9\text{ m}$$

AB_2C_2:

$$C_2 = 180° - A - B_2 = 180° - 61.7° - 105.4° = 12.9°$$

$$\frac{c_2}{\sin C_2} = \frac{a}{\sin A} \Rightarrow c_2 = \frac{a\sin C_2}{\sin A} \Rightarrow c_2 = \frac{78.9\sin 12.9°}{\sin 61.7°} \approx 20.0\text{ m}$$

19. Solve the triangle, given $a = 27.6$ cm, $b = 19.8$ cm, $C = 42° 30'$.
This is a SAS case, so using the law of cosines.

$$c^2 = a^2 + b^2 - 2ab\cos C \Rightarrow c^2 = 27.6^2 + 19.8^2 - 2(27.6)(19.8)\cos 42° 30' \approx 347.985 \Rightarrow c \approx 18.65 \text{cm}$$

(will be rounded as 18.7)

Of the remaining angles A and B, B must be smaller since it is opposite the shorter of the two sides a and b. Therefore, B cannot be obtuse.

$$\frac{\sin B}{b} = \frac{\sin C}{c} \Rightarrow \frac{\sin B}{19.8} = \frac{\sin 42°30'}{18.65} \Rightarrow \sin B = \frac{19.8 \sin 42°30'}{18.65} \approx .717124859 \Rightarrow B \approx 45.8° \approx 45°50'$$

Thus, $A = 180° - B - C = 180° - 45°\ 50' - 42°\ 30' = 179°60' - 88°20' = 91°\ 40'$.

20. Solve the triangle, given $a = 94.6$ yd, $b = 123$ yd, $c = 109$ yd.
We can use the law of cosines to solve for any angle of the triangle. We solve for B, the largest angle. We will know that B is obtuse if $\cos B < 0$.

$$\cos B = \frac{a^2 + c^2 - b^2}{2ac} \Rightarrow \cos B = \frac{94.6^2 + 109^2 - 123^2}{2(94.6)(109)} = .27644937 \Rightarrow B \approx 74.0°$$

Of the remaining angles A and C, A must be smaller since it is opposite the shorter of the two sides a and c. Therefore, A cannot be obtuse.

$$\frac{\sin A}{a} = \frac{\sin B}{b} \Rightarrow \frac{\sin A}{94.6} = \frac{\sin 74.0°}{123} \Rightarrow \sin A = \frac{94.6 \sin 74.0°}{123} \approx .73931184 \Rightarrow A \approx 47.7°$$

Thus, $C = 180° - A - B = 180° - 47.7° - 74.0° = 58.3°$.

21. Given $b = 840.6$ m, $c = 715.9$ m, $A = 149.3°$, find the area.
Angle A is included between sides b and c. Thus, we have the following.

$$\text{A} = \frac{1}{2} bc \sin A = \frac{1}{2}(840.6)(715.9)\sin 149.3° \approx 153{,}600 \text{ m}^2$$ (rounded to four significant digits)

22. Given $a = 6.90$ ft, $b = 10.2$ ft, $C = 35° 10'$, find the area.
Angle C is included between sides a and b. Thus, we have the following.

$$\text{A} = \frac{1}{2} ab \sin C = \frac{1}{2}(6.90)(10.2)\sin 35°\ 10' \approx 20.3 \text{ ft}^2$$ (rounded to three significant digits)

23. Given $a = .913$ km, $b = .816$ km, $c = .582$ km, find the area.
Use Heron's formula to find the area.

$$s = \frac{1}{2}(a+b+c) = \frac{1}{2}(.913 + .816 + .582) = \frac{1}{2} \cdot 2.311 = 1.1555$$

$$\begin{aligned} \text{A} &= \sqrt{s(s-a)(s-b)(s-c)} \\ &= \sqrt{1.1555(1.1555 - .913)(1.1555 - .816)(1.1555 - .582)} \\ &= \sqrt{1.1555(.2425)(.3395)(.5735)} \approx .234 \text{ km}^2 \end{aligned}$$ (rounded to three significant digits)

24. Given $a = 43$ m, $b = 32$ m, $c = 51$ m, find the area.
Use Heron's formula to find the area.

$$s = \frac{1}{2}(a+b+c) = \frac{1}{2}(43 + 32 + 51) = \frac{1}{2} \cdot 126 = 63$$

$$\begin{aligned} \text{A} &= \sqrt{s(s-a)(s-b)(s-c)} \\ &= \sqrt{63(63-43)(63-32)(63-51)} = \sqrt{63(20)(31)(12)} \approx 680 \text{ m}^2 \end{aligned}$$ (rounded to two significant digits)

25. Since $B = 58.4°$ and $C = 27.9°$, $A = 180° - B - C = 180° - 58.4° - 27.9° = 93.7°$.
Using the law of sines, we have the following.

$$\frac{AB}{\sin C} = \frac{125}{\sin A} \Rightarrow \frac{AB}{\sin 27.9°} = \frac{125}{\sin 93.7°} \Rightarrow AB = \frac{125 \sin 27.9°}{\sin 93.7°} \approx 58.61$$

The canyon is 58.6 feet across. (rounded to three significant digits)

26. The angle opposite the 8.0 ft flagpole is $180° - 115° - 22° = 43°$.

Using the law of sines, we have the following.

$$\frac{8}{\sin 43°} = \frac{x}{\sin 115°}$$
$$x = \frac{8 \sin 115°}{\sin 43°}$$
$$x \approx 10.63$$

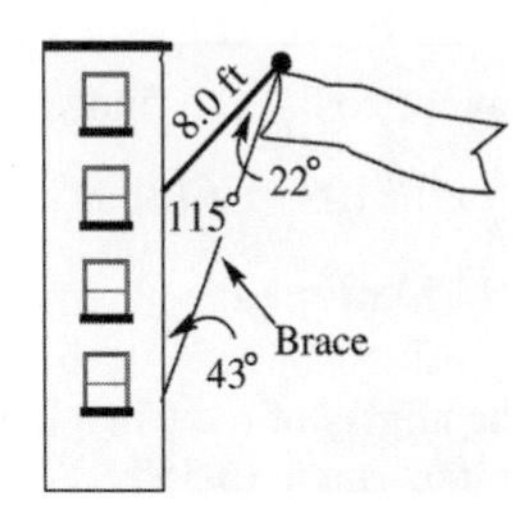

The brace is 11 feet long. (rounded to two significant digits)

27. Let AC = the height of the tree.
Angle $A = 90° - 8.0° = 82°$
Angle $C = 180° - B - A = 30°$
Use the law of sines to find $AC = b$.

$$\frac{b}{\sin B} = \frac{c}{\sin C}$$
$$\frac{b}{\sin 68°} = \frac{7.0}{\sin 30°}$$
$$b = \frac{7.0 \sin 68°}{\sin 30°}$$
$$b \approx 12.98$$

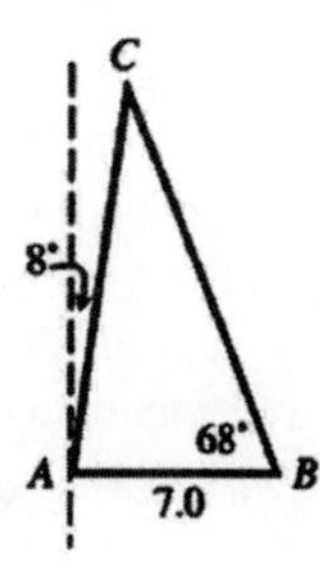

The tree is 13 meters tall. (rounded to two significant digits)

28. Let d = the distance between the ends of the wire.
This situation is SAS, so we should use the law of cosines.

$$d^2 = 15.0^2 + 12.2^2 - 2(15.0)(12.2)\cos 70.3° \approx 250.463 \Rightarrow d \approx 15.83 \text{ ft}$$

The ends of the wire should be places 15.8 ft apart. (rounded to three significant digits)

29. Let h = the height of tree.

$$\theta = 27.2° - 14.3° = 12.9°$$
$$\alpha = 90° - 27.2° = 62.8°$$
$$\frac{h}{\sin\theta} = \frac{212}{\sin\alpha}$$
$$\frac{h}{\sin 12.9°} = \frac{212}{\sin 62.8°}$$
$$h = \frac{212 \sin 12.9°}{\sin 62.8°}$$
$$h \approx 53.21$$

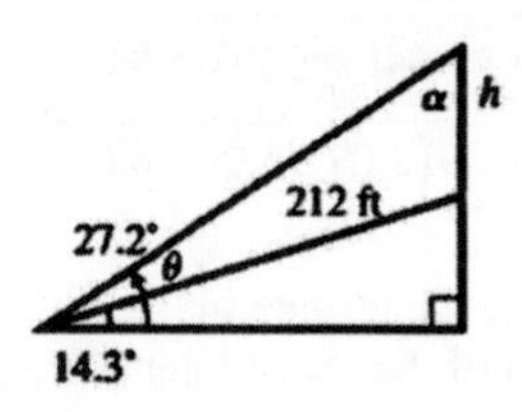

The height of the tree is 53.2 ft. (rounded to three significant digits)

30. $AB = 150$ km, $AC = 102$ km, $BC = 135$ km

Use the law of cosines to find the measure of angle *C*.

$$(AB)^2 = (AC)^2 + (BC)^2 - 2(AC)(BC)\cos C \Rightarrow 150^2 = 102^2 + 135^2 - 2(102)(135)\cos C$$

$$\cos C = \frac{102^2 + 135^2 - 150^2}{2(102)(135)} \approx .22254902 \Rightarrow C \approx 77.1°$$

31. Let x = the distance between the boats.

In 3 hours the first boat travels 3(36.2) = 108.6 km and the second travels 3(45.6)=136.8 km.

Use the law of cosines to find x.

$$x^2 = 108.6^2 + 136.8^2 - 2(108.6)(136.8)\cos 54°10' \approx 13{,}113.359 \Rightarrow x \approx 115 \text{ km}$$

They are 115 km apart.

32. To find the angles of the triangle formed by the ship's positions with the lighthouse, we find the supplementary angle to 55°: 180° – 55° = 125°. The third angle in the triangle is the following.

$$180° - 125° - 30° = 25°$$

Using the law of sines, we have the following.

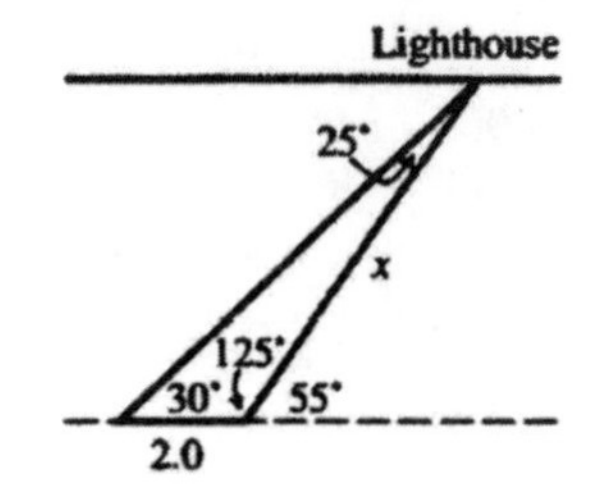

$$\frac{2}{\sin 25°} = \frac{x}{\sin 30°}$$

$$x = \frac{2\sin 30°}{\sin 25°}$$

$$\approx 2.4 \text{ mi}$$

The ship is 2.4 miles from the lighthouse.

33. Use the distance formula to find the distances between the points.

Distance between (–8, 6) and (0, 0):

$$\sqrt{(-8-0)^2 + (6-0)^2} = \sqrt{(-8)^2 + 6^2} = \sqrt{64+36} = \sqrt{100} = 10$$

Distance between (–8, 6) and (3, 4):

$$\sqrt{(-8-3)^2 + (6-4)^2} = \sqrt{(-11)^2 + 2^2} = \sqrt{121+4} = \sqrt{125} \approx 11.18$$

Distance between (3, 4) and (0, 0):

$$\sqrt{(3-0)^2 + (4-0)^2} = \sqrt{3^2 + 4^2} = \sqrt{9+16} = \sqrt{25} = 5$$

$$s \approx \frac{1}{2}(10 + 11.18 + 5) = \frac{1}{2}\cdot 26.18 = 13.09$$

$$\text{A} = \sqrt{s(s-a)(s-b)(s-c)} = \sqrt{13.09(13.09-10)(13.09-11.18)(13.09-5)}$$

$$= \sqrt{13.09(3.09)(1.91)(8.09)} \approx 25 \text{ sq units (rounded to two significant digits)}$$

34. Divide the quadrilateral into two triangles.

The area of the quadrilateral is the sum of the areas of the two triangles.

$$\text{Area of quadrilateral} = \frac{1}{2}(65)(130)\sin 80° + \frac{1}{2}(120)(120)\sin 70° \approx 10{,}926.6$$

The area of the quadrilateral is $11{,}000 \text{ ft}^2$. (rounded to two significant digits)

35. a − b

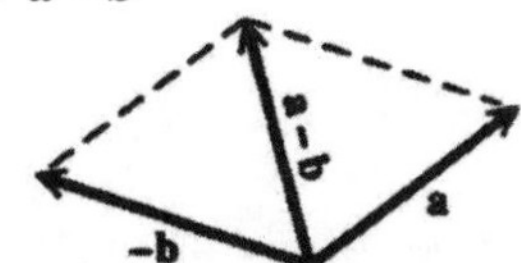

36. a + 3c

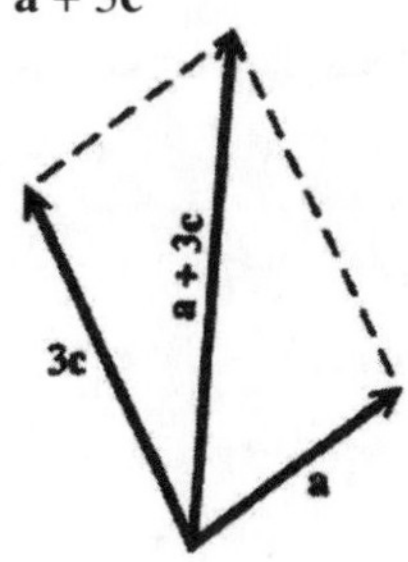

37. **(a)** true
(b) false

38. Forces of 142 newtons and 215 newtons, forming an angle of 112°

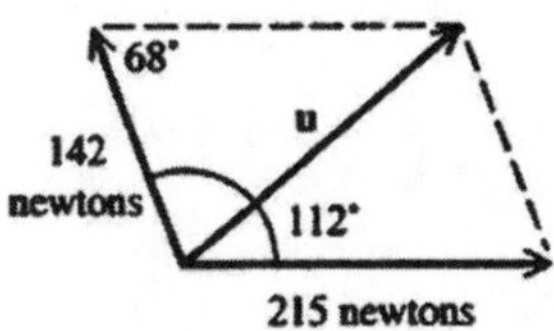

$180° - 112° = 68°$

$|\mathbf{u}|^2 = 215^2 + 142^2 - 2(215)(142)\cos 68°$

$|\mathbf{u}|^2 \approx 43515.5$

$|\mathbf{u}| \approx 209$ newtons

39. $\alpha = 180° - 52° = 128°$

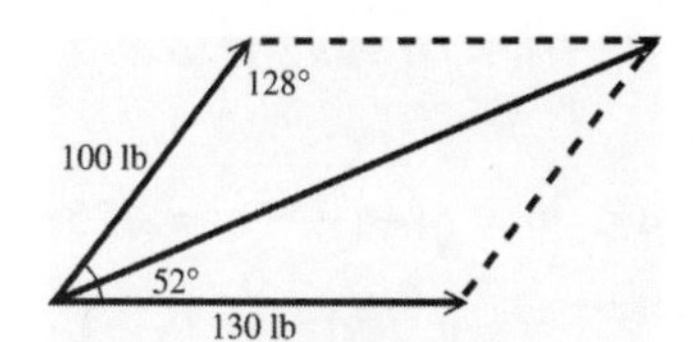

$|\mathbf{v}|^2 = 100^2 + 130^2 - 2(100)(130)\cos 128°$

$|\mathbf{v}|^2 \approx 42907.2$

$|\mathbf{v}| \approx 207$ lb

40. $|\mathbf{v}| = 50, \theta = 45°$

horizontal: $x = |\mathbf{v}|\cos\theta = 50\cos 45° = \dfrac{50\cdot\sqrt{2}}{2} = 25\sqrt{2}$

vertical: $y = |\mathbf{v}|\sin\theta = 50\sin 45° = \dfrac{50\cdot\sqrt{2}}{2} = 25\sqrt{2}$

41. $|\mathbf{v}| = 964, \theta = 154°20'$

horizontal: $x = |\mathbf{v}|\cos\theta = 964\cos 154°\ 20' \approx 869$

vertical: $y = |\mathbf{v}|\sin\theta = 964\sin 154°\ 20' \approx 418$

42. $\mathbf{u} = \langle 21, -20\rangle$

magnitude: $|\mathbf{u}| = \sqrt{21^2 + (-20)^2} = \sqrt{441 + 400} = \sqrt{841} = 29$

Angle: $\tan\theta' = \dfrac{b}{a} \Rightarrow \tan\theta' = \dfrac{-20}{21} \Rightarrow \theta' = \tan^{-1}\left(-\dfrac{20}{21}\right) \approx -43.6° \Rightarrow \theta = -43.6° + 360° = 316.4°$

(θ lies in quadrant IV)

43. $\mathbf{u} = \langle -9, 12\rangle$

magnitude: $|\mathbf{u}| = \sqrt{(-9)^2 + 12^2} = \sqrt{81 + 144} = \sqrt{225} = 15$

Angle: $\tan\theta' = \dfrac{b}{a} \Rightarrow \tan\theta' = \dfrac{12}{-9} \Rightarrow \theta' = \tan^{-1}\left(-\dfrac{4}{3}\right) \approx -53.1° \Rightarrow \theta = -53.1° + 180° = 126.9°$

(θ lies in quadrant II)

44. $\mathbf{v} = 2\mathbf{i} - \mathbf{j},\ \ \mathbf{u} = -3\mathbf{i} + 2\mathbf{j}$

First write the given vectors in component form.

$$\mathbf{v} = 2\mathbf{i} - \mathbf{j} = \langle 2, -1 \rangle \text{ and } \mathbf{u} = -3\mathbf{i} + 2\mathbf{j} = \langle -3, 2 \rangle$$

(a) $2\mathbf{v} + \mathbf{u} = 2\langle 2,-1\rangle + \langle -3,2\rangle = \langle 2\cdot 2, 2(-1)\rangle + \langle -3,2\rangle = \langle 4,-2\rangle + \langle -3,2\rangle = \langle 4+(-3), -2+2\rangle = \langle 1,0\rangle = \mathbf{i}$

(b) $2\mathbf{v} = 2\langle 2,-1\rangle = \langle 2\cdot 2, 2(-1)\rangle = \langle 4,-2\rangle = 4\mathbf{i} - 2\mathbf{j}$

(c) $\mathbf{v} - 3\mathbf{u} = \langle 2,-1\rangle - 3\langle -3,2\rangle$

$$= \langle 2,-1\rangle - \langle 3(-3), 3\cdot 2\rangle = \langle 2,-1\rangle - \langle -9,6\rangle = \langle 2-(-9), -1-6\rangle = \langle 11,-7\rangle = 11\mathbf{i} - 7\mathbf{j}$$

45. $\mathbf{a} = \langle 3,-2\rangle,\ \mathbf{b} = \langle -1,3\rangle$

$\mathbf{a}\cdot\mathbf{b} = \langle 3,-2\rangle \cdot \langle -1,3\rangle = 3(-1) + (-2)\cdot 3 = -3 - 6 = -9$

$$\cos\theta = \frac{\mathbf{a}\cdot\mathbf{b}}{|\mathbf{a}||\mathbf{b}|} \Rightarrow \cos\theta = \frac{-9}{|\langle 3,-2\rangle||\langle -1,3\rangle|}$$

$$= \frac{-9}{\sqrt{3^2+(-2)^2}\sqrt{(-1)^2+3^2}} = -\frac{9}{\sqrt{9+4}\sqrt{1+9}} = -\frac{9}{\sqrt{13}\sqrt{10}} = -\frac{9}{\sqrt{130}} \approx -.78935222$$

Thus, $\theta \approx 142.1°$.

46. $|\mathbf{u}| = \sqrt{(-4)^2 + 3^2} = \sqrt{25} = 5$

$$\mathbf{v} = \frac{\mathbf{u}}{|\mathbf{u}|} = \frac{\langle -4,3\rangle}{5} = \left\langle -\frac{4}{5}, \frac{3}{5}\right\rangle$$

47. $|\mathbf{u}| = \sqrt{5^2 + 12^2} = \sqrt{169} = 13$

$$\mathbf{v} = \frac{\mathbf{u}}{|\mathbf{u}|} = \frac{\langle 5,12\rangle}{13} = \left\langle \frac{5}{13}, \frac{12}{13}\right\rangle$$

48. Let $\mathbf{x}$ = the resultant vector.

$\alpha = 180° - 45° = 135°$

$|\mathbf{x}|^2 = 200^2 + 100^2 - 2(200)(100)\cos 135°$

$|\mathbf{x}|^2 = 40{,}000 + 10{,}000 - 40{,}000\left(-\frac{\sqrt{2}}{2}\right)$

$|\mathbf{x}|^2 = 50{,}000 + 20{,}000\sqrt{2}$

$|\mathbf{x}|^2 \approx 78{,}284.27$

$|\mathbf{x}| \approx 279.8$

Using the law of cosines again, we get the following.

$$\cos\theta = \frac{100^2 + 279.8^2 - 200^2}{2(100)(279.8)} \approx .86290279 \Rightarrow \theta \approx 30.4°$$

The force is 280 newtons (rounded) at an angle of 30.4° with the first rope.

49. Let $|\mathbf{x}|$ be the resultant force.

$\theta = 180° - 15° - 10° = 155°$

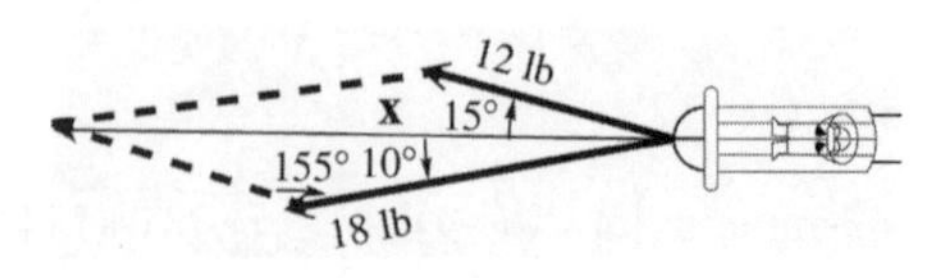

$|\mathbf{x}|^2 = 12^2 + 18^2 - 2(12)(18)\cos 155°$

$|\mathbf{x}|^2 \approx 859.5$

$|\mathbf{x}| \approx 29$

The magnitude of the resultant force on Jessie and the sled is 29 lb.

50. Let θ = the angle that the hill makes with the horizontal.

The downward force of 2800 lb has component AC perpendicular to the hill and component CB parallel to the hill. AD represents the force of 186 lb that keeps the car from rolling down the hill. Since vectors AD and BC are equal, $|BC| = 186$. Angle B = angle EAB because they are alternate interior angles to the two right triangles are similar. Hence angle θ = angle BAC.

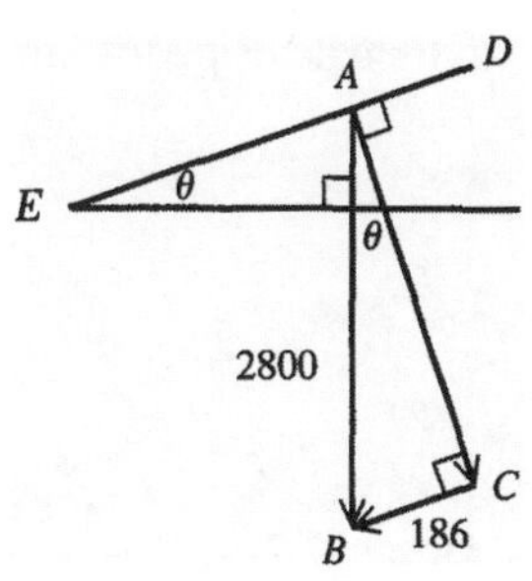

Since $\sin BAC = \frac{186}{2800} = \frac{93}{1400} \Rightarrow BAC = \theta \approx 3.8° \approx 3°50'$, the angle that the hill makes with the horizontal is $3°50'$.

51. Let **v** = the ground speed vector.

$\alpha = 212° - 180° = 32°$ and $\beta = 50°$ because they are alternate interior angles. Angle opposite to 520 is $\alpha + \beta = 82°$.

Using the law of sines, we have the following.

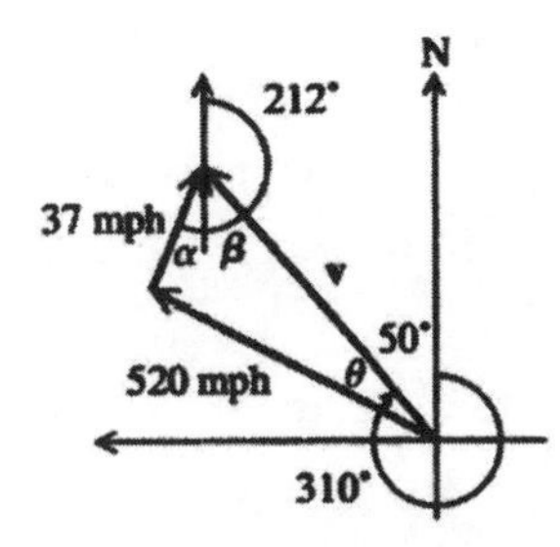

$$\frac{\sin\theta}{37} = \frac{\sin 82°}{520}$$
$$\sin\theta = \frac{37\sin 82°}{520}$$
$$\sin\theta \approx .07046138$$
$$\theta \approx 4°$$

Thus, the bearing is $360° - 50° - \theta = 306°$. The angle opposite **v** is $180° - 82° - 4° = 94°$. Using the laws of sines, we have the following.

$$\frac{|\mathbf{v}|}{\sin 94°} = \frac{520}{\sin 82°} \Rightarrow |\mathbf{v}| = \frac{520\sin 94°}{\sin 82°} \approx 524 \text{ mph}$$

The pilot should fly on a bearing of 306°. Her actual speed is 524 mph.

52. Let **v** = the resultant vector.

Angle $A = 180° - (130° - 90°) = 140°$

Use the law of cosines to find the magnitude of the resultant **v**.

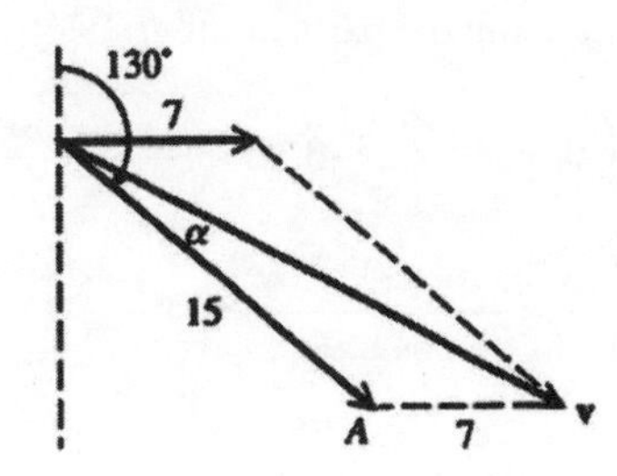

$$|\mathbf{v}|^2 = 15^2 + 7^2 - 2(15)(7)\cos 140°$$
$$|\mathbf{v}|^2 \approx 434.87$$
$$|\mathbf{v}| \approx 20.9$$

Use the law of sines to find α.

$$\frac{\sin\alpha}{7} = \frac{\sin 140°}{20.9} \Rightarrow \sin\alpha = \frac{7\sin 140°}{20.9} \approx .21528772 \Rightarrow \alpha \approx 12°$$

The resulting speed is 21 km per hr (rounded) and bearing is $130° - 12° = 118°$.

53. Refer to the diagram below. In each of the triangles *ABP* and *PBC,* we know two angles and one side. Solve each triangle using the law of sines.

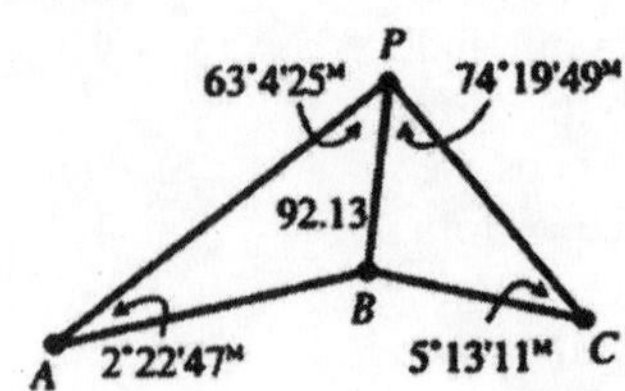

$$AB=\frac{92.13\sin 63°\ 4'\ 25''}{\sin 2°\ 22'\ 47''}\approx 1978.28 \text{ ft and } BC=\frac{92.13\sin 74°\ 19'\ 49''}{\sin 5°\ 13'\ 11''}\approx 975.05 \text{ ft}$$

54. $A=30°, B=60°, C=90°, a=7, b=7\sqrt{3}, c=14$; Newton's formula: $\frac{a+b}{c}=\frac{\cos\frac{1}{2}(A-B)}{\sin\frac{1}{2}C}$

Verify using the parts of the triangle.

We will need $\cos\frac{1}{2}\cdot 30°=\sqrt{\frac{1+\cos 30°}{2}}=\sqrt{\frac{1+\frac{\sqrt{3}}{2}}{2}}=\sqrt{\frac{2+\sqrt{3}}{4}}=\frac{\sqrt{2+\sqrt{3}}}{2}$ by a half-angle formula.

$$\frac{7+7\sqrt{3}}{14}\overset{?}{=}\frac{\cos\frac{1}{2}(30°-60°)}{\sin\frac{1}{2}\cdot 90°}\Rightarrow\frac{1+\sqrt{3}}{2}\overset{?}{=}\frac{\cos\frac{1}{2}(-30°)}{\sin 45°}\Rightarrow\frac{1+\sqrt{3}}{2}\overset{?}{=}\frac{\cos(-15°)}{\sin 45°}$$

$$\frac{1+\sqrt{3}}{2}\overset{?}{=}\frac{\cos 15°}{\sin 45°}\Rightarrow\frac{1+\sqrt{3}}{2}\overset{?}{=}\frac{\cos\frac{1}{2}\cdot 30°}{\sin 45°}\Rightarrow\frac{1+\sqrt{3}}{2}\overset{?}{=}\frac{\frac{\sqrt{2+\sqrt{3}}}{2}}{\frac{\sqrt{2}}{2}}\Rightarrow\frac{1+\sqrt{3}}{2}\overset{?}{=}\frac{\sqrt{2+\sqrt{3}}}{\sqrt{2}}$$

$$\frac{\sqrt{\left(1+\sqrt{3}\right)^2}}{\sqrt{2^2}}\overset{?}{=}\sqrt{\frac{2+\sqrt{3}}{2}}\Rightarrow\sqrt{\frac{\left(1+\sqrt{3}\right)^2}{2^2}}\overset{?}{=}\sqrt{\frac{2+\sqrt{3}}{2}}\Rightarrow\sqrt{\frac{1+2\sqrt{3}+3}{4}}\overset{?}{=}\sqrt{\frac{2+\sqrt{3}}{2}}$$

$$\sqrt{\frac{4+2\sqrt{3}}{4}}\overset{?}{=}\sqrt{\frac{2+\sqrt{3}}{2}}\Rightarrow\sqrt{\frac{2+\sqrt{3}}{2}}=\sqrt{\frac{2+\sqrt{3}}{2}}$$

55. $A=30°, B=60°, C=90°, a=7, b=7\sqrt{3}, c=14$; Mollweide's formula: $\frac{a-b}{c}=\frac{\sin\frac{1}{2}(A-B)}{\cos\frac{1}{2}C}$

Verify using the parts of the triangle.

We will need $\sin\frac{1}{2}\cdot 30°=\sqrt{\frac{1-\cos 30°}{2}}=\sqrt{\frac{1-\frac{\sqrt{3}}{2}}{2}}=\sqrt{\frac{2-\sqrt{3}}{4}}=\frac{\sqrt{2-\sqrt{3}}}{2}$ by a half-angle formula.

$$\frac{7-7\sqrt{3}}{14}\overset{?}{=}\frac{\sin\frac{1}{2}(30°-60°)}{\cos\frac{1}{2}\cdot 90°}\Rightarrow\frac{1-\sqrt{3}}{2}\overset{?}{=}\frac{\sin\frac{1}{2}(-30°)}{\cos 45°}\Rightarrow-\frac{\sqrt{3}-1}{2}\overset{?}{=}\frac{\sin\frac{1}{2}(-30°)}{\cos 45°}\Rightarrow-\frac{\sqrt{3}-1}{2}\overset{?}{=}\frac{\sin(-15°)}{\cos 45°}$$

$$-\frac{\sqrt{3}-1}{2}\overset{?}{=}\frac{-\sin 15°}{\sin 45°}\Rightarrow-\frac{\sqrt{3}-1}{2}\overset{?}{=}-\frac{\sin\frac{1}{2}\cdot 30°}{\sin 45°}\Rightarrow-\frac{\sqrt{3}-1}{2}\overset{?}{=}-\frac{\frac{\sqrt{2-\sqrt{3}}}{2}}{\frac{\sqrt{2}}{2}}\Rightarrow-\frac{\sqrt{3}-1}{2}\overset{?}{=}-\frac{\sqrt{2-\sqrt{3}}}{\sqrt{2}}$$

$$-\frac{\sqrt{\left(\sqrt{3}-1\right)^2}}{\sqrt{2^2}}\overset{?}{=}-\sqrt{\frac{2-\sqrt{3}}{2}}\Rightarrow-\sqrt{\frac{\left(\sqrt{3}-1\right)^2}{2^2}}\overset{?}{=}-\sqrt{\frac{2-\sqrt{3}}{2}}\Rightarrow-\sqrt{\frac{3-2\sqrt{3}+1}{4}}\overset{?}{=}-\sqrt{\frac{2-\sqrt{3}}{2}}$$

$$-\sqrt{\frac{4-2\sqrt{3}}{4}}\overset{?}{=}-\sqrt{\frac{2-\sqrt{3}}{2}}\Rightarrow-\sqrt{\frac{2-\sqrt{3}}{2}}=-\sqrt{\frac{2-\sqrt{3}}{2}}$$

Chapter 7: Test

1. Find C, given $A = 25.2°$, $a = 6.92$ yd, $b = 4.82$ yd.
Use the law of sines to first find the measure of angle B.

$$\frac{\sin 25.2°}{6.92} = \frac{\sin B}{4.82} \Rightarrow \sin B = \frac{4.82 \sin 25.2°}{6.92} \approx .29656881 \Rightarrow B \approx 17.3°$$

Use the fact that the angles of a triangle sum to 180° to find the measure of angle C.

$$C = 180 - A - B = 180° - 25.2° - 17.3° = 137.5°$$

Angle C measures 137.5°.

2. Find c, given $C = 118°$, $b = 130$ km, $a = 75$ km.
Using the law of cosines to find the length of c.

$$c^2 = a^2 + b^2 - 2ab\cos C \Rightarrow c^2 = 75^2 + 130^2 - 2(75)(130)\cos 118° \approx 31679.70 \Rightarrow c \approx 178.0 \text{ km}$$

c is approximately 180 km. (rounded to two significant digits)

3. Find B, given $a = 17.3$ ft, $b = 22.6$ ft, $c = 29.8$ ft.
Using the law of cosines, find the measure of angle B.

$$b^2 = a^2 + c^2 - 2ac\cos B \Rightarrow \cos B = \frac{a^2 + c^2 - b^2}{2ac} = \frac{17.3^2 + 29.8^2 - 22.6^2}{2(17.3)(29.8)} \approx .65617605 \Rightarrow B \approx 49.0°$$

B is approximately 49.0°.

4. Find the area of triangle ABC, given $C = 118°$, $b = 130$ km, $a = 75$ km.

$$A = \frac{1}{2}ab\sin C = \frac{1}{2}(75)(130)\sin 118° \approx 4304.4$$

The area of the triangle is approximately 4300 km^2. (rounded to two significant digits)
(*Note*: Since c was found in Exercise 2, Heron's formula can also be used.)

5. Since $B > 90°$, b must be the longest side of the triangle.

(a) $b > 10$

(b) none

(c) $b \le 10$

6. The semi-perimeter s is $s = \frac{1}{2}(a+b+c) = \frac{1}{2}(22+26+40) = \frac{1}{2} \cdot 88 = 44.$
Using Heron's formula, we have the following.

$$A = \sqrt{s(s-a)(s-b)(s-c)} = \sqrt{44(44-22)(44-26)(44-40)} = \sqrt{44(22)(18)(4)} = \sqrt{69{,}696} = 264$$

The area of the triangle is 264 square units.

7. magnitude: $|\mathbf{v}| = \sqrt{(-6)^2 + 8^2} = \sqrt{36+64} = \sqrt{100} = 10$

angle: $\tan\theta' = \frac{y}{x} \Rightarrow \tan\theta' = \frac{8}{-6} = -\frac{4}{3} \approx -1.33333333 \Rightarrow \theta' \approx -53.1° \Rightarrow \theta = -53.1° + 180° = 126.9°$

(θ lies in quadrant II)

The magnitude $|\mathbf{v}|$ is 10 and θ = 126.9°.

8. **a + b**

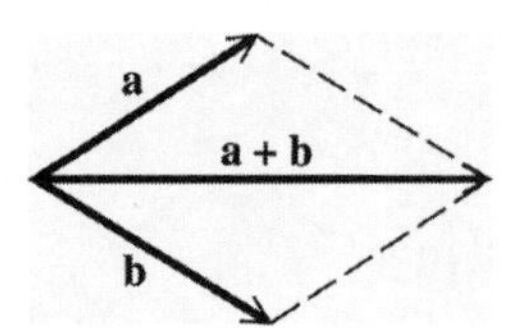

9. $\mathbf{u} = \langle -1, 3\rangle, \mathbf{v} = \langle 2, -6\rangle$

(a) $\mathbf{u}+\mathbf{v} = \langle -1, 3\rangle + \langle 2, -6\rangle = \langle -1+2, 3+(-6)\rangle = \langle 1, -3\rangle$

(b) $-3\mathbf{v} = -3\langle 2, -6\rangle = \langle -3\cdot 2, -3(-6)\rangle = \langle -6, 18\rangle$

(c) $\mathbf{u}\cdot\mathbf{v} = \langle -1, 3\rangle \cdot \langle 2, -6\rangle = -1(2)+3(-6) = -2-18 = -20$

(d) $|\mathbf{u}| = \sqrt{(-1)^2+3^2} = \sqrt{1+9} = \sqrt{10}$

10. Given $A = 24°\ 50', B = 47°\ 20'$ and $AB = 8.4$ mi, first find the measure of angle C.

$$\begin{aligned} C &= 180° - 47°\ 20' - 24°\ 50' \\ &= 179°60' - 72°\ 10' \\ &= 107°\ 50' \end{aligned}$$

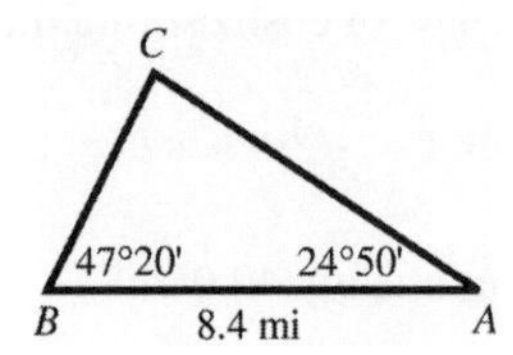

Use this information and the law of sines to find AC.

$$\frac{AC}{\sin 47°\ 20'} = \frac{8.4}{\sin 107°\ 50'} \Rightarrow AC = \frac{8.4 \sin 47°\ 20'}{\sin 107°50'} \approx 6.49 \text{ mi}$$

Drop a perpendicular line from C to segment AB.

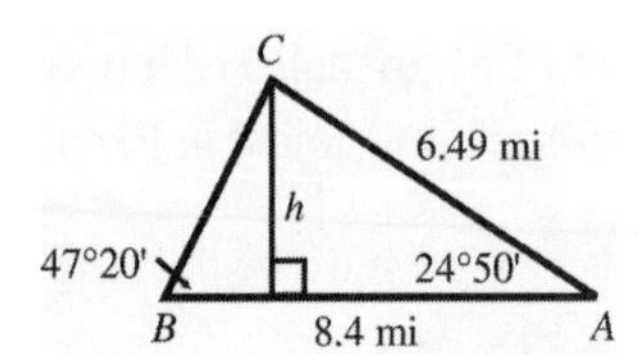

Thus, $\sin 24°50' = \frac{h}{6.49} \Rightarrow h \approx 6.49 \sin 24°50' \approx 2.7$ mi. The balloon is 2.7 miles off the ground.

11. horizontal: $x = |\mathbf{v}|\cos\theta = 569\cos 127.5° \approx -346$ and vertical: $y = |\mathbf{v}|\sin\theta = 569\sin 127.5° \approx 451$

The vector is $\langle -346, 451\rangle$.

12. Consider the figure below.

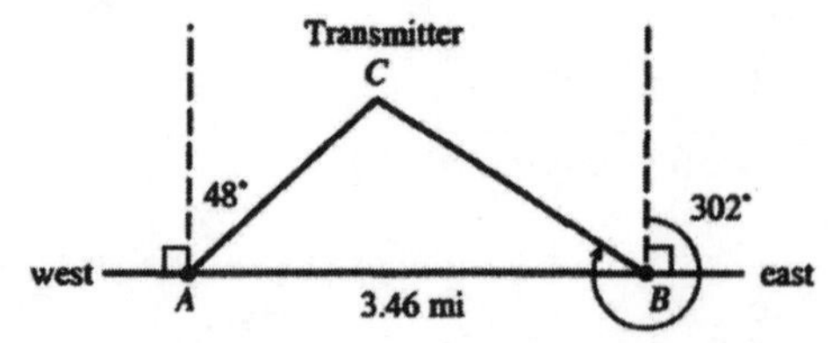

Since the bearing is 48° from A, angle A in ABC must be 90° – 48° = 42°. Since the bearing is 302° from B, angle B in ABC must be 302° – 270° = 32°. The angles of a triangle sum to 180°, so $C = 180° - A - B = 180° - 42° - 32° = 106°$. Using the law of sines, we have the following.

$$\frac{b}{\sin B} = \frac{c}{\sin C} \Rightarrow \frac{b}{\sin 32°} = \frac{3.46}{\sin 106°} \Rightarrow b = \frac{3.46 \sin 32°}{\sin 106°} \approx 1.91 \text{ mi}$$

The distance from A to the transmitter is 1.91 miles. (rounded to two significant digits)

13. Consider the diagram below.

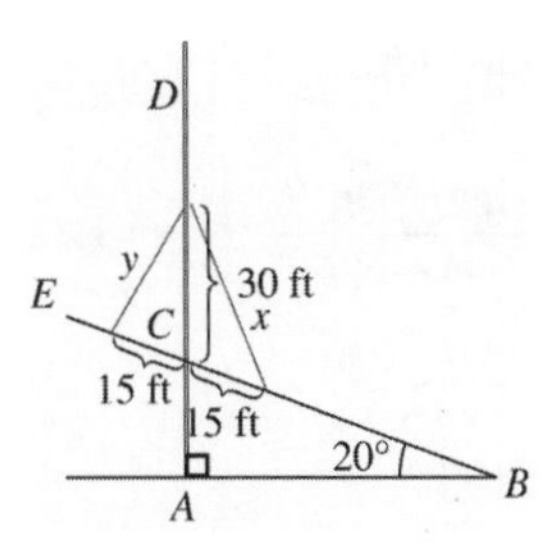

If we extend the flagpole, a right triangle *CAB* is formed. Thus, the measure of angle *BCA* is $90° - 20° = 70°$. Since angle *DCB* and *BCA* are supplementary, the measure of angle *DBC* is $180° - 70° = 110°$. We can now use the law of cosines to find the measure of the support wire on the right, *x*.

$$x^2 = 30^2 + 15^2 - 2(30)(15)\cos 110° \approx 1432.818 \approx 37.85 \text{ ft}$$

Now, to find the length of the support wire on the left, we have different ways to find it. One way would be to use the approximation for *x* and use the law of cosines. To avoid using the approximate value, we will find *y* with the same method as for *x*.

Since angle *DCB* and *DCE* are supplementary, the measure of angle *DCE* is $180° - 110° = 70°$. We can now use the law of cosines to find the measure of the support wire on the left, *y*.

$$y^2 = 30^2 + 15^2 - 2(30)(15)\cos 70° \approx 817.182 \approx 28.59 \text{ ft}$$

The length of the two wires are 28.59 ft and 37.85 ft.

14. Let $|\mathbf{x}|$ be the equilibrant force.

$\theta = 180° - 35° - 25° = 120°$

$|\mathbf{x}|^2 = 15^2 + 20^2 - 2(15)(20)\cos 120°$

$|\mathbf{x}|^2 = 925$

$|\mathbf{x}| \approx 30.4$

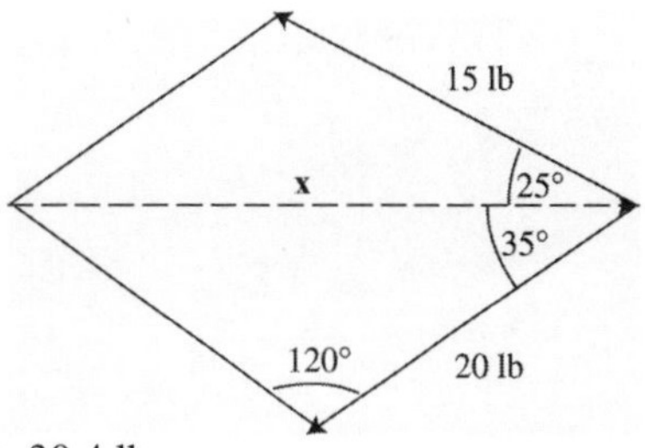

The magnitude of the equilibrant force Michael must apply is 30.4 lb.

Chapter 7: Quantitative Reasoning

1. We can use the area formula $A = \frac{1}{2} rR\sin B$ for this triangle. By the law of sines, we have the following.

$$\frac{r}{\sin A} = \frac{R}{\sin C} \Rightarrow r = \frac{R\sin A}{\sin C}$$

Siince $\sin C = \sin\left[180° - (A+B)\right] = \sin(A+B)$, we have the following.

$$r = \frac{R\sin A}{\sin C} \Rightarrow r = \frac{R\sin A}{\sin(A+B)}$$

By substituting into our area formula, we have the following.

$$A = \frac{1}{2} rR\sin B \Rightarrow A = \frac{1}{2}\left[\frac{R\sin A}{\sin(A+B)}\right] R\sin B \Rightarrow A = \frac{1}{2}\cdot\frac{\sin A\sin B}{\sin(A+B)} R^2$$

Since there are a total of 10 stars, the total area covered by the stars is the following.

$$A = 10\left[\frac{1}{2}\cdot\frac{\sin A\sin B}{\sin(A+B)} R^2\right] = \left[5\frac{\sin A\sin B}{\sin(A+B)}\right] R^2$$

Continued on next page

Quantitative Reasoning (continued)

2. If $A = 18°$ and $B = 36°$ we have the following.

$$A = \left[\frac{5\sin 18°\sin 36°}{\sin(18+36)°}\right]R^2 \approx 1.12257R^2$$

3. **(a)** $11.4 \text{ in.} \cdot \frac{10}{13} \text{ in.} \approx 8.77 \text{ in.}^2$

(b) $A = 50\left[5\frac{\sin 18°\sin 36°}{\sin(18°+36°)}\right] \cdot .308^2 \approx 5.32 \text{ in.}^2$

(c) red

Chapter 8

COMPLEX NUMBERS, POLAR EQUATIONS, AND PARAMETRIC EQUATIONS

Section 8.1: Complex Numbers

1. true

2. true

3. true

4. true

5. false (Every real number is a complex number.)

6. true

7. -6 is real and complex.

8. 0 is real and complex.

9. $10i$ is pure imaginary and complex.

10. $-8i$ is pure imaginary and complex.

11. $2+i$ is complex.

12. $-5-2i$ is complex.

13. π is real and complex.

14. $\sqrt{8}$ is real and complex.

15. $\sqrt{-9}=3i$ is pure imaginary and complex.

16. $\sqrt{-16}=4i$ is pure imaginary and complex.

17. $\sqrt{-25}=i\sqrt{25}=5i$

18. $\sqrt{-36}=i\sqrt{36}=6i$

19. $\sqrt{-10}=i\sqrt{10}$

20. $\sqrt{-15}=i\sqrt{15}$

21. $\sqrt{-288}=i\sqrt{288}=i\sqrt{144\cdot 2}=12i\sqrt{2}$

22. $\sqrt{-500}=i\sqrt{500}=i\sqrt{100\cdot 5}=10i\sqrt{5}$

23. $-\sqrt{-18}=-i\sqrt{18}=-i\sqrt{9\cdot 2}=-3i\sqrt{2}$

24. $-\sqrt{-80}=-i\sqrt{80}=-i\sqrt{16\cdot 5}=-4i\sqrt{5}$

25. $x^2=-16$

$x=\pm\sqrt{-16}=\pm 4i$

Solution set: $\{\pm 4i\}$

26. $x^2=-36$

$x=\pm\sqrt{-36}=\pm 6i$

Solution set: $\{\pm 6i\}$

27. $x^2+12=0$

$x^2=-12$

$x=\pm\sqrt{-12}$

$x=\pm 2i\sqrt{3}$

Solution set: $\left\{\pm 2i\sqrt{3}\right\}$

28. $x^2+48=0$

$x^2=-48$

$x=\sqrt{-48}$

$x=\pm 4i\sqrt{3}$

Solution set: $\left\{\pm 4i\sqrt{3}\right\}$

In Exercises 29 – 36, we will be using the quadratic formula.

$$\text{Given } ax^2+bx+c=0, \text{ where } a\neq 0, \ x=\frac{-b\pm\sqrt{b^2-4ac}}{2a}.$$

Equations must be place in standard form before applying the quadratic formula.

29. $3x^2+2=-4x \Rightarrow 3x^2+4x+2=0$

$$x=\frac{-4\pm\sqrt{4^2-4(3)(2)}}{2(3)}=\frac{-4\pm\sqrt{-8}}{6}=\frac{-4\pm 2i\sqrt{2}}{6}=\frac{2\left(-2\pm i\sqrt{2}\right)}{6}=\frac{-2\pm i\sqrt{2}}{3}=-\frac{2}{3}\pm\frac{i\sqrt{2}}{3}$$

Solution set: $\left\{-\frac{2}{3}\pm\frac{i\sqrt{2}}{3}\right\}$

30. $2x^2+3x=-2 \Rightarrow 2x^2+3x+2=0$

$$x=\frac{-3\pm\sqrt{3^2-4(2)(2)}}{2(2)}=\frac{-3\pm\sqrt{9-16}}{4}=\frac{-3\pm\sqrt{-7}}{4}=\frac{-3\pm i\sqrt{7}}{4}=-\frac{3}{4}\pm\frac{\sqrt{7}}{4}i$$

Solution set: $\left\{-\frac{3}{4}\pm\frac{\sqrt{7}}{4}i\right\}$

31. $x^2-6x+14=0$

$$x=\frac{-(-6)\pm\sqrt{(-6)^2-4(1)(14)}}{2(1)}=\frac{6\pm\sqrt{36-56}}{2}=\frac{6\pm\sqrt{-20}}{2}=\frac{6\pm 2i\sqrt{5}}{2}=\frac{2\left(3\pm i\sqrt{5}\right)}{2}=3\pm i\sqrt{5}$$

Solution set: $\left\{3\pm i\sqrt{5}\right\}$

32. $x^2+4x+11=0$

$$x=\frac{-4\pm\sqrt{4^2-4(1)(11)}}{2(1)}=\frac{-4\pm\sqrt{16-44}}{2}=\frac{-4\pm\sqrt{-28}}{2}=\frac{-4\pm 2i\sqrt{7}}{2}=\frac{2\left(-2\pm i\sqrt{7}\right)}{2}=-2\pm i\sqrt{7}$$

Solution set: $\left\{-2\pm i\sqrt{7}\right\}$

33. $4\left(x^2-x\right)=-7 \Rightarrow 4x^2-4x=-7 \Rightarrow 4x^2-4x+7=0$

$$x=\frac{-(-4)\pm\sqrt{(-4)^2-4(4)(7)}}{2(4)}=\frac{4\pm\sqrt{16-112}}{8}=\frac{4\pm\sqrt{-96}}{8}=\frac{4\pm 4i\sqrt{6}}{8}=\frac{4\left(1\pm i\sqrt{6}\right)}{8}=\frac{1\pm i\sqrt{6}}{2}=\frac{1}{2}\pm\frac{\sqrt{6}}{2}i$$

Solution set: $\left\{\frac{1}{2}\pm\frac{\sqrt{6}}{2}i\right\}$

34. $3\left(3x^2-2x\right)=-7 \Rightarrow 9x^2-6x=-7$

$$9x^2-6x+7=0$$

$$x=\frac{-(-6)\pm\sqrt{(-6)^2-4(9)(7)}}{2(9)}$$

$$=\frac{6\pm\sqrt{36-252}}{18}$$

$$=\frac{6\pm\sqrt{-216}}{18}$$

$$=\frac{6\pm 6i\sqrt{6}}{18}$$

$$=\frac{6\left(1\pm i\sqrt{6}\right)}{18}$$

$$=\frac{1\pm i\sqrt{6}}{3}$$

$$=\frac{1}{3}\pm\frac{\sqrt{6}}{3}i$$

Solution set: $\left\{\frac{1}{3}\pm\frac{\sqrt{6}}{3}i\right\}$

35. $x^2+1=-x \Rightarrow x^2+x+1=0$

$$x=\frac{-1\pm\sqrt{1^2-4(1)(1)}}{2(1)}$$

$$=\frac{-1\pm\sqrt{1-4}}{2}$$

$$=\frac{-1\pm\sqrt{-3}}{2}$$

$$=\frac{-1\pm i\sqrt{3}}{2}$$

$$=-\frac{1}{2}\pm\frac{\sqrt{3}}{2}i$$

Solution set: $\left\{-\frac{1}{2}\pm\frac{\sqrt{3}}{2}i\right\}$

36. $x^2+2=2x \Rightarrow x^2-2x+2=0$

$$x=\frac{-(-2)\pm\sqrt{(-2)^2-4(1)(2)}}{2(1)}=\frac{2\pm\sqrt{4-8}}{2}=\frac{2\pm 2i}{2}=\frac{2(1\pm i)}{2}=1\pm i$$

Solution set: $\{1\pm i\}$

37. $\sqrt{-13}\cdot\sqrt{-13}=i\sqrt{13}\cdot i\sqrt{13}$

$$=i^2\left(\sqrt{13}\right)^2=-1\cdot 13=-13$$

38. $\sqrt{-17}\cdot\sqrt{-17}=i\sqrt{17}\cdot i\sqrt{17}$

$$=i^2\left(\sqrt{17}\right)^2=-1\cdot 17=-17$$

39. $\sqrt{-3}\cdot\sqrt{-8}=i\sqrt{3}\cdot i\sqrt{8}=i^2\sqrt{3\cdot 8}$

$$=-1\cdot\sqrt{24}=-\sqrt{4\cdot 6}=-2\sqrt{6}$$

40. $\sqrt{-5}\cdot\sqrt{-15}=i\sqrt{5}\cdot i\sqrt{15}=i^2\sqrt{5\cdot 15}$

$$=-1\cdot\sqrt{75}=-\sqrt{25\cdot 3}=-5\sqrt{3}$$

41. $\frac{\sqrt{-30}}{\sqrt{-10}}=\frac{i\sqrt{30}}{i\sqrt{10}}=\sqrt{\frac{30}{10}}=\sqrt{3}$

42. $\frac{\sqrt{-70}}{\sqrt{-7}}=\frac{i\sqrt{70}}{i\sqrt{7}}=\sqrt{\frac{70}{7}}=\sqrt{10}$

43. $\frac{\sqrt{-24}}{\sqrt{8}}=\frac{i\sqrt{24}}{\sqrt{8}}=i\sqrt{\frac{24}{8}}=i\sqrt{3}$

44. $\frac{\sqrt{-54}}{\sqrt{27}}=\frac{i\sqrt{54}}{\sqrt{27}}=i\sqrt{\frac{54}{27}}=i\sqrt{2}$

45. $\frac{\sqrt{-10}}{\sqrt{-40}}=\frac{i\sqrt{10}}{i\sqrt{40}}=\sqrt{\frac{10}{40}}=\sqrt{\frac{1}{4}}=\frac{1}{2}$

46. $\frac{\sqrt{-40}}{\sqrt{20}}=\frac{i\sqrt{40}}{\sqrt{20}}=i\sqrt{\frac{40}{20}}=i\sqrt{2}$

47. $\dfrac{\sqrt{-6}\cdot\sqrt{-2}}{\sqrt{3}}=\dfrac{i\sqrt{6}\cdot i\sqrt{2}}{\sqrt{3}}=i^2\sqrt{\dfrac{6\cdot 2}{3}}=-1\cdot\sqrt{\dfrac{12}{3}}=-\sqrt{4}=-2$

48. $\dfrac{\sqrt{-12}\cdot\sqrt{-6}}{\sqrt{8}}=\dfrac{i\sqrt{12}\cdot i\sqrt{6}}{\sqrt{8}}=i^2\sqrt{\dfrac{12\cdot 6}{8}}=-1\cdot\sqrt{\dfrac{72}{8}}=-\sqrt{9}=-3$

49. $(3+2i)+(4-3i)=(3+4)+[2+(-3)]i=7+(-1)i=7-i$

50. $(4-i)+(2+5i)=(4+2)+(-1+5)i=6+4i$

51. $(-2+3i)-(-4+3i)=[-2-(-4)]+(3-3)i=2+0i=2$

52. $(-3+5i)-(-4+5i)=[-3-(-4)]+(5-5)i=1+0i=1$

53. $(2-5i)-(3+4i)-(-2+i)=[2-3-(-2)]+(-5-4-1)i=1+(-10)i=1-10i$

54. $(-4-i)-(2+3i)+(-4+5i)=[-4-2+(-4)]+(-1-3+5)i=-10+i$

55. $-i-2-(6-4i)-(5-2i)=(-2-6-5)+[-1-(-4)-(-2)]i=-13+5i$

56. $3-(4-i)-4i+(-2+5i)=[3-4+(-2)]+[-(-1)-4+5]i=-3+2i$

57. $(2+i)(3-2i)=2(3)+2(-2i)+i(3)+i(-2i)=6-4i+3i-2i^2=6-i-2(-1)=6-i+2=8-i$

58. $(-2+3i)(4-2i)=-2(4)-2(-2i)+3i(4)+3i(-2i)$
$=-8+4i+12i-6i^2=-8+16i-6(-1)=-8+16i+6=-2+16i$

59. $(2+4i)(-1+3i)=2(-1)+2(3i)+4i(-1)+4i(3i)$
$=-2+6i-4i+12i^2=-2+2i+12(-1)=-2+2i-12=-14+2i$

60. $(1+3i)(2-5i)=1(2)+1(-5i)+3i(2)+3i(-5i)=2-5i+6i-15i^2=2+i-15(-1)=2+i+15=17+i$

61. $(-3+2i)^2=(-3)^2+2(-3)(2i)+(2i)^2$ Square of a binomial
$=9-12i+4i^2=9-12i+4(-1)=9-12i-4=5-12i$

62. $(2+i)^2=2^2+2(2)(i)+i^2=4+4i+i^2=4+4i+(-1)=3+4i$

63. $(3+i)(-3-i)=3(-3)+3(-i)+i(-3)+i(-i)=-9-3i-3i-i^2=-9-6i-(-1)=-9-6i+1=-8-6i$

64. $(-5-i)(5+i)=-5(5)-5i-i(5)-i(i)=-25-5i-5i-i^2=-25-10i-(-1)=-25-10i+1=-24-10i$

65. $(2+3i)(2-3i)=2^2-(3i)^2=4-9i^2=4-9(-1)=4+9=13$

66. $(6-4i)(6+4i)=6^2-(4i)^2=36-16i^2$
$=36-16(-1)=36+16=52$

67. $(\sqrt{6}+i)(\sqrt{6}-i)=(\sqrt{6})^2-i^2$
$=6-(-1)=6+1=7$

68. $(\sqrt{2}-4i)(\sqrt{2}+4i)=(\sqrt{2})^2-(4i)^2=2-16i^2=2-16(-1)=2+16=18$

69. $i(3-4i)(3+4i)=i[(3-4i)(3+4i)]=i[3^2-(4i)^2]=i[9-16i^2]=i[9-16(-1)]=i(9+16)=25i$

70. $i(2+7i)(2-7i)=i[(2+7i)(2-7i)]=i[2^2-(7i)^2]=i[4-49i^2]=i[4-49(-1)]=i(4+49)=53i$

71. $3i(2-i)^2=3i[2^2-2(2)(i)+i^2]=3i[4-4i+(-1)]=3i(3-4i)=9i-12i^2=9i-12(-1)=12+9i$

72. $-5i(4-3i)^2=-5i[4^2-2(4)(3i)+(3i)^2]=-5i[16-24i+9i^2]$
$=-5i[16-24i+9(-1)]=-5i(16-24i-9)=-5i(7-24i)$
$=-35i+120i^2=-35i+120(-1)=-35i-120=-120-35i$

73. $(2+i)(2-i)(4+3i)=[(2+i)(2-i)](4+3i)$
$=[2^2-i^2](4+3i)=[4-(-1)](4+3i)=5(4+3i)=20+15i$

74. $(3-i)(3+i)(2-6i)=[(3-i)(3+i)](2-6i)$
$=[3^2-i^2](2-6i)=[9-(-1)](2-6i)=10(2-6i)=20-60i$

75. $i^{21}=i^{20}\cdot i=(i^4)^5\cdot i=1^5\cdot i=i$

76. $i^{25}=i^{24}\cdot i=(i^4)^6\cdot i=1^6\cdot i=i$

77. $i^{22}=i^{20}\cdot i^2=(i^4)^5\cdot(-1)=1^5\cdot(-1)=-1$

78. $i^{26}=i^{24}\cdot i^2=(i^4)^6\cdot(-1)=1^6\cdot(-1)=-1$

79. $i^{23}=i^{20}\cdot i^3=(i^4)^5\cdot i^3=1^5\cdot(-i)=-i$

80. $i^{27}=i^{24}\cdot i^3=(i^4)^6\cdot i^3=1^6\cdot(-i)=-i$

81. $i^{24}=(i^4)^6=1^6=1$

82. $i^{32}=(i^4)^8=1^8=1$

83. $i^{-9}=i^{-12}\cdot i^3=(i^4)^{-3}\cdot i^3=1^{-3}\cdot(-i)=-i$

84. $i^{-10}=i^{-12}\cdot i^2=(i^4)^{-3}\cdot i^2=1^{-3}\cdot(-1)=-1$

85. $\frac{1}{i^{-11}}=i^{11}=i^8\cdot i^3=(i^4)^2\cdot i^3=1^2\cdot(-i)=-i$

86. $\frac{1}{i^{12}}=i^{-12}=(i^4)^{-3}=1^{-3}=1$

87. – 88. Answers will vary.

89. $\dfrac{6+2i}{1+2i}=\dfrac{(6+2i)(1-2i)}{(1+2i)(1-2i)}$

$=\dfrac{6-12i+2i-4i^2}{1^2-(2i)^2}=\dfrac{6-10i-4(-1)}{1-4i^2}=\dfrac{6-10i+4}{1-4(-1)}=\dfrac{10-10i}{1+4}=\dfrac{10-10i}{5}=\dfrac{10}{5}-\dfrac{10}{5}i=2-2i$

90. $\dfrac{14+5i}{3+2i}=\dfrac{(14+5i)(3-2i)}{(3+2i)(3-2i)}=\dfrac{42-28i+15i-10i^2}{3^2-(2i)^2}$

$=\dfrac{42-13i-10(-1)}{9-4i^2}=\dfrac{42-13i+10}{9-4(-1)}=\dfrac{52-13i}{9+4}=\dfrac{52-13i}{13}=\dfrac{52}{13}-\dfrac{13}{13}i=4-i$

91. $\dfrac{2-i}{2+i}=\dfrac{(2-i)(2-i)}{(2+i)(2-i)}=\dfrac{2^2-2(2i)+i^2}{2^2-i^2}=\dfrac{4-4i+(-1)}{4-(-1)}=\dfrac{3-4i}{5}=\dfrac{3}{5}-\dfrac{4}{5}i$

92. $\dfrac{4-3i}{4+3i}=\dfrac{(4-3i)(4-3i)}{(4+3i)(4-3i)}$

$=\dfrac{4^2-2(4)(3i)+(3i)^2}{4^2-(3i)^2}=\dfrac{16-24i+9i^2}{16-9i^2}=\dfrac{16-24i+9(-1)}{16-9(-1)}=\dfrac{16-24i-9}{16+9}=\dfrac{7-24i}{25}=\dfrac{7}{25}-\dfrac{24}{25}i$

93. $\dfrac{1-3i}{1+i}=\dfrac{(1-3i)(1-i)}{(1+i)(1-i)}=\dfrac{1-i-3i+3i^2}{1^2-i^2}=\dfrac{1-4i+3(-1)}{1-(-1)}=\dfrac{1-4i-3}{2}=\dfrac{-2-4i}{2}=\dfrac{-2}{2}-\dfrac{4}{2}i=-1-2i$

94. $\dfrac{-3+4i}{2-i}=\dfrac{(-3+4i)(2+i)}{(2-i)(2+i)}$

$=\dfrac{-6-3i+8i+4i^2}{2^2-i^2}=\dfrac{-6+5i+4(-1)}{4-(-1)}=\dfrac{-6+5i-4}{5}=\dfrac{-10+5i}{5}=\dfrac{-10}{5}+\dfrac{5}{5}i=-2+i$

95. $\dfrac{5}{i}=\dfrac{5(-i)}{i(-i)}=\dfrac{-5i}{-i^2}$

$=\dfrac{-5i}{-(-1)}=\dfrac{-5i}{1}=-5i$ or $0-5i$

96. $\dfrac{6}{i}=\dfrac{6(-i)}{i(-i)}=\dfrac{-6i}{-i^2}$

$=\dfrac{-6i}{-(-1)}=\dfrac{-6i}{1}=-6i$ or $0-6i$

97. $\dfrac{-8}{-i}=\dfrac{-8\cdot i}{-i\cdot i}=\dfrac{-8i}{-i^2}$

$=\dfrac{-8i}{-(-1)}=\dfrac{-8i}{1}=-8i$ or $0-8i$

98. $\dfrac{-12}{-i}=\dfrac{-12\cdot i}{-i\cdot i}=\dfrac{-12i}{-i^2}$

$=\dfrac{-12i}{-(-1)}=\dfrac{-12i}{1}=-12i$ or $0-12i$

99. $\dfrac{2}{3i}=\dfrac{2(-3i)}{3i\cdot(-3i)}=\dfrac{-6i}{-9i^2}=\dfrac{-6i}{-9(-1)}=\dfrac{-6i}{9}=-\dfrac{2}{3}i$ or $0-\dfrac{2}{3}i$

Note: In the above solution, we multiplied the numerator and denominator by the complex conjugate of $3i$, namely $-3i$. Since there is a reduction in the end, the same results can be achieved by multiplying the numerator and denominator by $-i$.

100. $\frac{5}{9i}=\frac{5(-9i)}{9i\cdot(-9i)}=\frac{-45i}{-81i^2}=\frac{-45i}{-81(-1)}=\frac{-45i}{81}=-\frac{5}{9}i$ or $0-\frac{5}{9}i$

101. We need to show that $\left(\frac{\sqrt{2}}{2}+\frac{\sqrt{2}}{2}i\right)^2=i.$

$$\left(\frac{\sqrt{2}}{2}+\frac{\sqrt{2}}{2}i\right)^2=\left(\frac{\sqrt{2}}{2}\right)^2+2\cdot\frac{\sqrt{2}}{2}\cdot\frac{\sqrt{2}}{2}i+\left(\frac{\sqrt{2}}{2}i\right)^2$$

$$=\frac{2}{4}+2\cdot\frac{2}{4}i+\frac{2}{4}i^2=\frac{1}{2}+i+\frac{1}{2}i^2=\frac{1}{2}+i+\frac{1}{2}(-1)=\frac{1}{2}+i-\frac{1}{2}=i$$

102. We need to show that $\left(\frac{\sqrt{3}}{2}+\frac{1}{2}i\right)^3=i.$

$$\left(\frac{\sqrt{3}}{2}+\frac{1}{2}i\right)^3=\left(\frac{\sqrt{3}}{2}+\frac{1}{2}i\right)\left(\frac{\sqrt{3}}{2}+\frac{1}{2}i\right)^2=\left(\frac{\sqrt{3}}{2}+\frac{1}{2}i\right)\left[\left(\frac{\sqrt{3}}{2}\right)^2+2\cdot\frac{\sqrt{3}}{2}\cdot\frac{1}{2}i+\left(\frac{1}{2}i\right)^2\right]$$

$$=\left(\frac{\sqrt{3}}{2}+\frac{1}{2}i\right)\left[\frac{3}{4}+\frac{\sqrt{3}}{2}i+\frac{1}{4}i^2\right]=\left(\frac{\sqrt{3}}{2}+\frac{1}{2}i\right)\left[\frac{3}{4}+\frac{\sqrt{3}}{2}i+\frac{1}{4}(-1)\right]$$

$$=\left(\frac{\sqrt{3}}{2}+\frac{1}{2}i\right)\left[\frac{3}{4}+\frac{\sqrt{3}}{2}i-\frac{1}{4}\right]=\left(\frac{\sqrt{3}}{2}+\frac{1}{2}i\right)\left(\frac{2}{4}+\frac{\sqrt{3}}{2}i\right)=\left(\frac{\sqrt{3}}{2}+\frac{1}{2}i\right)\left(\frac{1}{2}+\frac{\sqrt{3}}{2}i\right)$$

$$=\frac{\sqrt{3}}{2}\cdot\frac{1}{2}+\frac{\sqrt{3}}{2}\cdot\frac{\sqrt{3}}{2}i+\frac{1}{2}\cdot\frac{1}{2}i+\frac{1}{2}\frac{\sqrt{3}}{2}i^2=\frac{\sqrt{3}}{4}+\frac{3}{4}i+\frac{1}{4}i+\frac{\sqrt{3}}{4}i^2$$

$$=\frac{\sqrt{3}}{4}+\frac{4}{4}i+\frac{\sqrt{3}}{4}(-1)=\frac{\sqrt{3}}{4}+i+\left(-\frac{\sqrt{3}}{4}\right)=i$$

103. $I=8+6i,\ Z=6+3i$

$$E=IZ\Rightarrow E=(8+6i)(6+3i)=48+24i+36i+18i^2=48+60i+18(-1)=48+60i-18=30+60i$$

104. $I=10+6i,\ Z=8+5i$

$$E=IZ\Rightarrow(10+6i)(8+5i)=80+50i+48i+30i^2=80+98i+30(-1)=80+98i-30=50+98i$$

105. $I=7+5i,\ E=28+54i$

$$E=IZ\Rightarrow(28+54i)=(7+5i)Z$$

$$Z=\frac{28+54i}{7+5i}\cdot\frac{7-5i}{7-5i}=\frac{196-140i+378i-270i^2}{7^2-(5i)^2}$$

$$Z=\frac{196+238i-270(-1)}{49-25i^2}=\frac{196+238i+270}{49-25(-1)}=\frac{466+238i}{49+25}=\frac{2(233+119i)}{74}=\frac{233+119i}{37}=\frac{233}{37}+\frac{119}{37}i$$

106. $E = 35 + 55i,\ Z = 6 + 4i$

$$E = IZ \Rightarrow (35+55i) = I(6+4i)$$

$$I = \frac{35+55i}{6+4i} \cdot \frac{6-4i}{6-4i} = \frac{210-140i+330i-220i^2}{6^2-(4i)^2}$$

$$I = \frac{210+190i-220(-1)}{36-16i^2} = \frac{210+190i+220}{36-16(-1)} = \frac{430+190i}{36+16} = \frac{2(215+95i)}{52} = \frac{215+95i}{26} = \frac{215}{26} + \frac{95}{26}i$$

107. The total impedance would be $Z = 50 + 60 + 17i + 15i = 110 + 32i$.

108. For $Z = 110 + 32i, a = 110$ and $b = 32$. Thus, $\tan\theta = \frac{b}{a} \Rightarrow \tan\theta = \frac{32}{110} \approx .29090909 \Rightarrow \theta \approx 16.22°$.

Section 8.2: Trigonometric (Polar) Form of Complex Numbers

1. The absolute value of a complex number represents the length (or magnitude) of the vector representing it in the complex plane.

2. It is the angle formed by the vector and the positive x - axis.

3.

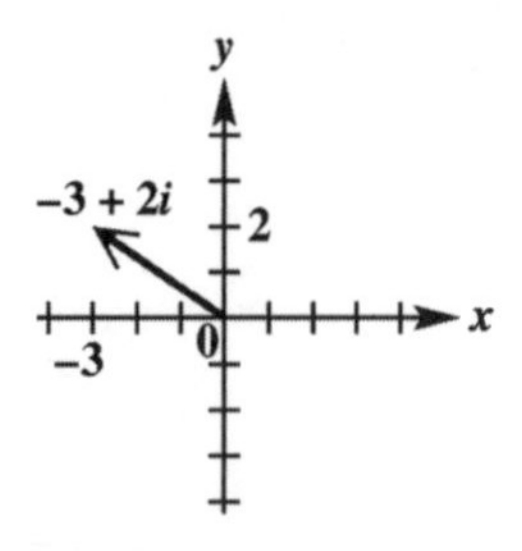

6.

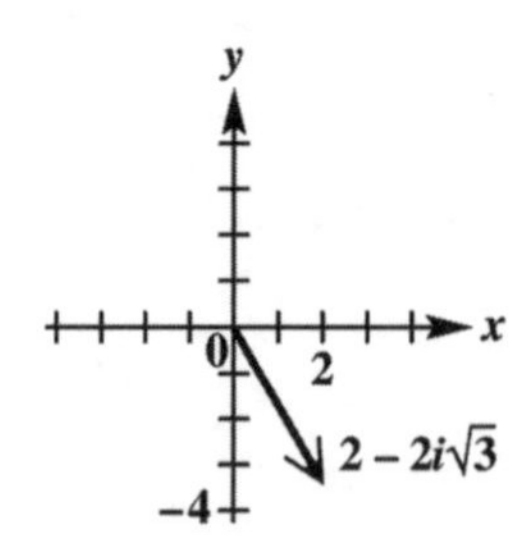

4.

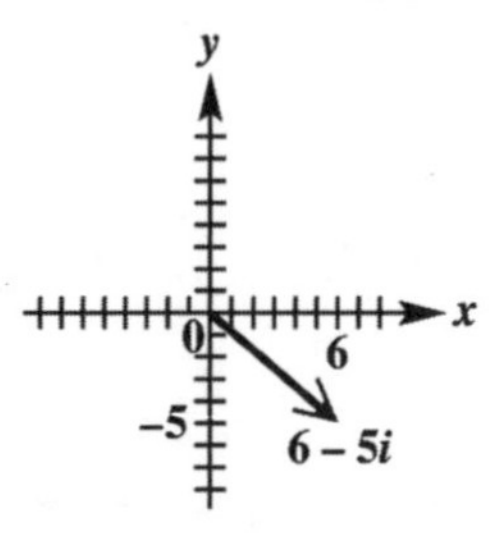

7.

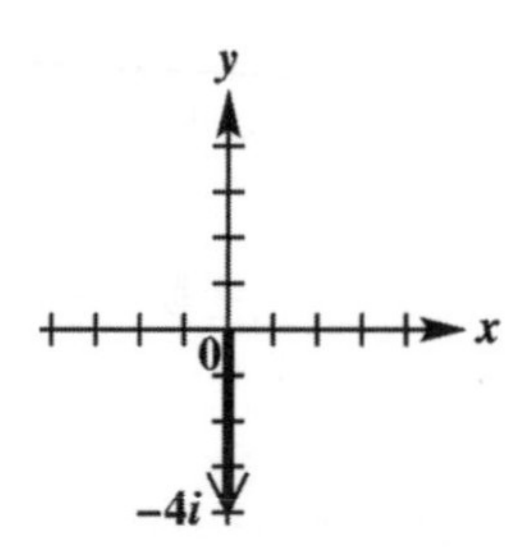

5.

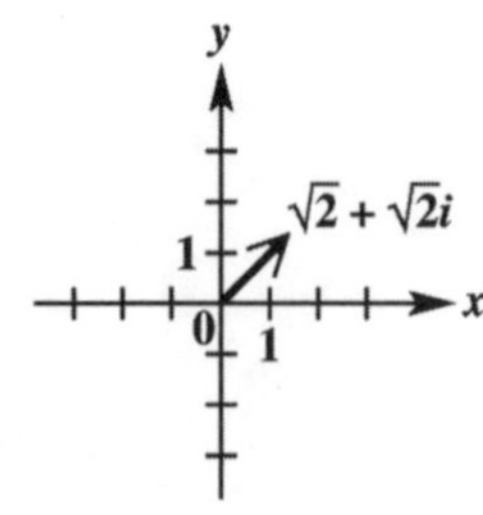

8.

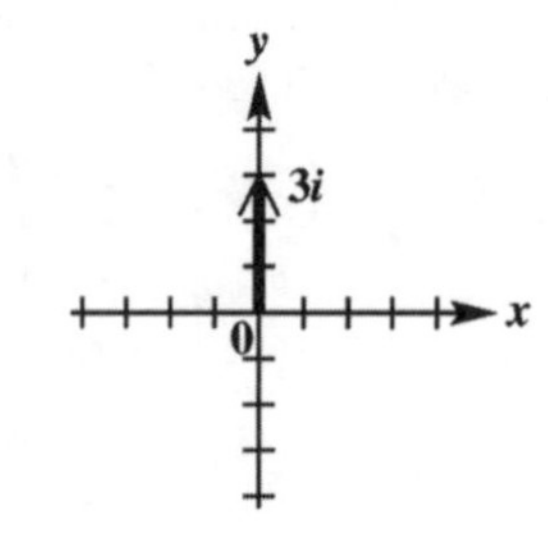

9.

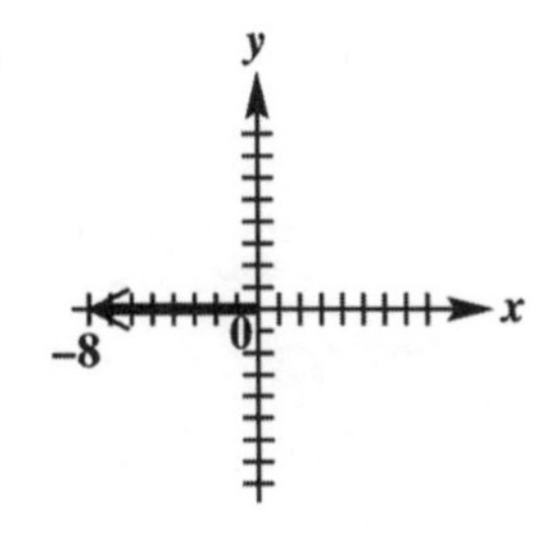

10.

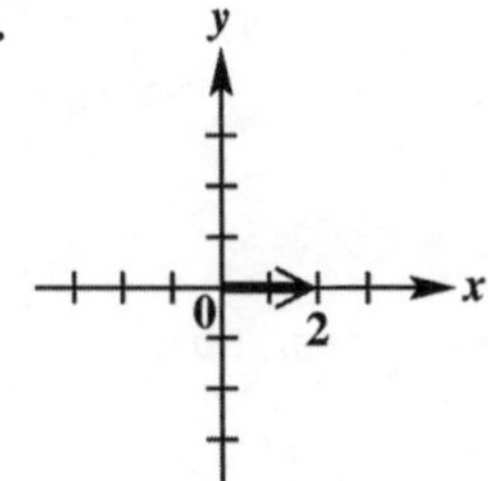

11. $1-4i$

12. $-4-i$

13. $4-3i,\ -1+2i$
$(4-3i)+(-1+2i)=3-i$

14. $2+3i,\ -4-i$
$(2+3i)+(-4-i)=-2+2i$

15. $5-6i,\ -2+3i$
$(5-6i)+(-2+3i)=3-3i$

16. $7-3i,\ -4+3i$
$(7-3i)+(-4+3i)=3+0i$ or 3

17. $-3,\ 3i$
$(-3+0i)+(0+3i)=-3+3i$

18. $6,\ -2i$
$(6+0i)+(0-2i)=6-2i$

19. $2+6i,\ -2i$
$(2+6i)+(0-2i)=2+4i$

20. $4-2i,\ 5$
$(4-2i)+(5+0i)=9-2i$

21. $7+6i,\ 3i$
$(7+6i)+(0+3i)=7+9i$

22. $-5-8i,\ -1$
$(-5-8i)+(-1+0i)=-6-8i$

23. $2(\cos 45°+i\ \sin 45°)=2\left(\frac{\sqrt{2}}{2}+i\frac{\sqrt{2}}{2}\right)=\sqrt{2}+i\sqrt{2}$

24. $4(\cos\ 60°+i\ \sin\ 60°)=4\left(\frac{1}{2}+i\frac{\sqrt{3}}{2}\right)=2+2i\sqrt{3}$

25. $10(\cos 90°+i\ \sin 90°)=10(0+i)=0+10i=10i$

26. $8(\cos 270°+i\sin 270°)=8(0-1i)=-8i$

27. $4(\cos 240°+i\ \sin 240°)=4\left(-\frac{1}{2}-i\frac{\sqrt{3}}{2}\right)=-2-2i\sqrt{3}$

28. $2(\cos 330°+i\sin\ 330°)=2\left(\frac{\sqrt{3}}{2}-\frac{1}{2}i\right)=\sqrt{3}-i$

29. $3\text{ cis }150°=3(\cos\ 150°+i\ \sin 150°)=3\left(-\frac{\sqrt{3}}{2}+\frac{1}{2}i\right)=-\frac{3\sqrt{3}}{2}+\frac{3}{2}i$

30. $2\text{cis}30° = 2(\cos 30° + i\sin 30°) = 2\left(\frac{\sqrt{3}}{2} + \frac{1}{2}i\right) = \sqrt{3} + i$

31. $5\text{cis}300° = 5(\cos 300° + i\sin 300°) = 5\left[\frac{1}{2} + i\left(-\frac{\sqrt{3}}{2}\right)\right] = \frac{5}{2} - \frac{5\sqrt{3}}{2}i$

32. $6 \text{ cis } 135° = (\cos 135° + i \sin 135°) = 6\left(-\frac{\sqrt{2}}{2} + \frac{\sqrt{2}}{2}i\right) = -3\sqrt{2} + 3i\sqrt{2}$

33. $\sqrt{2} \text{ cis } 180° = \sqrt{2}(\cos 180° + i \sin 180°) = \sqrt{2}(-1 + i \cdot 0) = -\sqrt{2} + 0i = -\sqrt{2}$

34. $\sqrt{3} \text{ cis } 315° = \sqrt{3}(\cos 315° + i \sin 315°) = \sqrt{3}\left[\frac{\sqrt{2}}{2} + i\left(-\frac{\sqrt{2}}{2}\right)\right] = \frac{\sqrt{6}}{2} - \frac{\sqrt{6}}{2}i$

35. $4(\cos(-30°) + i\sin(-30°)) = 4(\cos 30° - i\sin 30°) = 4\left(\frac{\sqrt{3}}{2} - \frac{1}{2}i\right) = 2\sqrt{3} - 2i$

36. $\sqrt{2}(\cos(-60°) + i\sin(-60°)) = \sqrt{2}(\cos 60° - i\sin 60°) = \sqrt{2}\left(\frac{1}{2} - \frac{\sqrt{3}}{2}i\right) = \frac{\sqrt{2}}{2} - \frac{\sqrt{6}}{2}i$

37. $3 - 3i$

Sketch a graph of $3 - 3i$ in the complex plane.

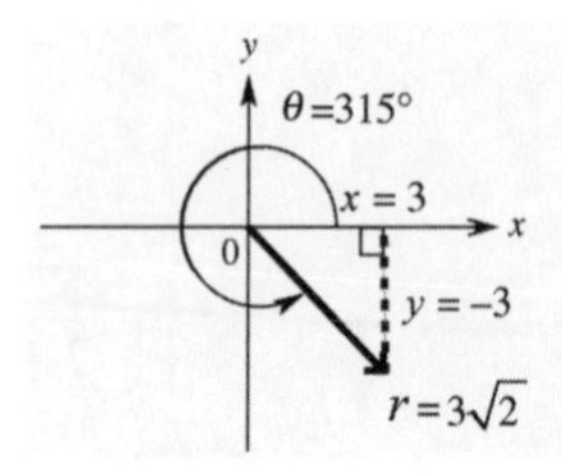

Since $x = 3$ and $y = -3$, $r = \sqrt{3^2 + (-3)^2} = \sqrt{9+9} = \sqrt{18} = 3\sqrt{2}$ and $\tan\theta = \frac{y}{x} = \frac{-3}{3} = -1.$ Since $\tan\theta = -1,$ the reference angle for θ is $45°$. The graph shows that θ is in quadrant IV, so $\theta = 360° - 45° = 315°$. Therefore, $3 - 3i = 3\sqrt{2}(\cos 315° + i\sin 315°)$

38. $-2 + 2i\sqrt{3}$

Sketch a graph of $-2 + 2i\sqrt{3}$ in the complex plane.

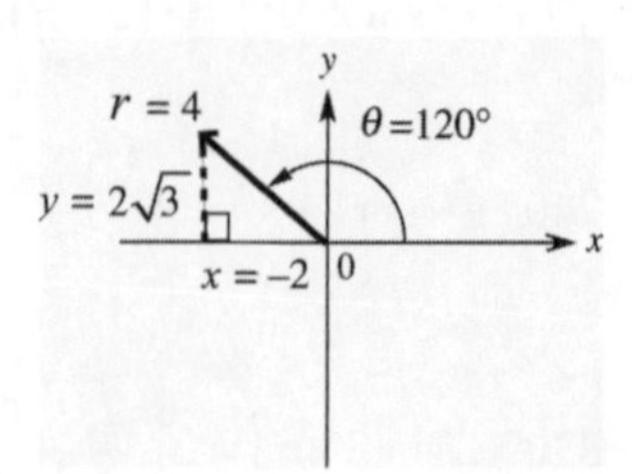

Since $x = -2$ and $y = 2\sqrt{3}$, $r = \sqrt{(-2)^2 + (2\sqrt{3})^2} = \sqrt{4+12} = \sqrt{16} = 4$ and $\tan\theta = \frac{2\sqrt{3}}{-2} = -\sqrt{3}.$ Since $\tan\theta = -\sqrt{3},$ the reference angle for θ is $60°$. The graph shows that θ is in quadrant II, so $\theta = 180° - 60° = 120°$. Therefore, $-2 + 2i\sqrt{3} = 4(\cos 120° + i\sin 120°)$.

39. $-3-3i\sqrt{3}$

Sketch a graph of $-3-3i\sqrt{3}$ in the complex plane.

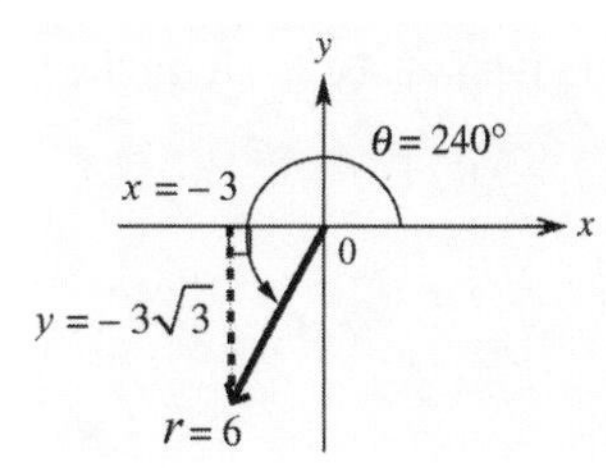

Since $x=-3$ and $y=-3\sqrt{3}$, $r=\sqrt{(-3)^2+\left(-3\sqrt{3}\right)^2}=\sqrt{9+27}=\sqrt{36}=6$ and $\tan\theta=\frac{-3\sqrt{3}}{-3}=\sqrt{3}$. Since $\tan\theta=\sqrt{3}$, the reference angle for θ is 60°. The graph shows that θ is in quadrant III, so $\theta=180°+60°=240°$. Therefore, $-3-3i\sqrt{3}=6\left(\cos 240°+i\sin 240°\right)$.

40. $1+i\sqrt{3}$

Sketch a graph of $1+i\sqrt{3}$ in the complex plane.

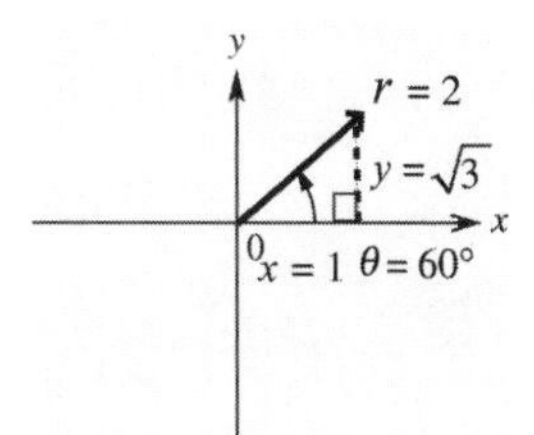

Since $x=1$ and $y=\sqrt{3}$, $r=\sqrt{1^2+\left(\sqrt{3}\right)^2}=\sqrt{1+3}=\sqrt{4}=2$ and $\tan\theta=\frac{\sqrt{3}}{1}=\sqrt{3}$. Since $\tan\theta=\sqrt{3}$, the reference angle for θ is 60°. The graph shows that θ is in quadrant I, so $\theta=60°$. Therefore, $1+i\sqrt{3}=2\left(\cos 60°+i\sin 60°\right)$.

41. $\sqrt{3}-i$

Sketch a graph of $\sqrt{3}-i$ in the complex plane.

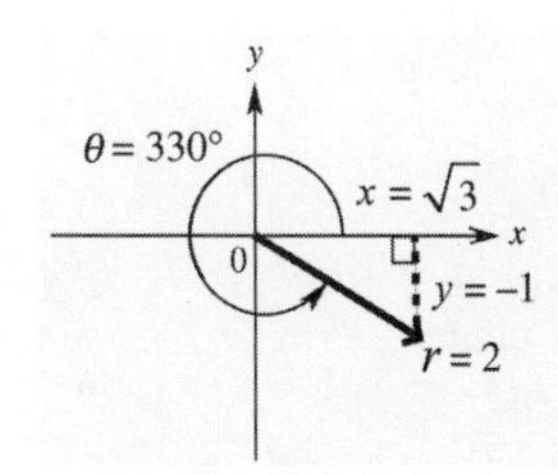

Since $x=\sqrt{3}$ and $y=-1$, $r=\sqrt{\left(\sqrt{3}\right)^2+(-1)^2}=\sqrt{3+1}=\sqrt{4}=2$ and $\tan\theta=\frac{-1}{\sqrt{3}}=-\frac{\sqrt{3}}{3}$. Since $\tan\theta=-\frac{\sqrt{3}}{3}$, the reference angle for θ is 30°. The graph shows that θ is in quadrant IV, so $\theta=360°-30°=330°$. Therefore, $\sqrt{3}-i=2\left(\cos 330°+i\sin 330°\right)$.

42. $4\sqrt{3}+4i$

Sketch a graph of $4\sqrt{3}+4i$ in the complex plane.

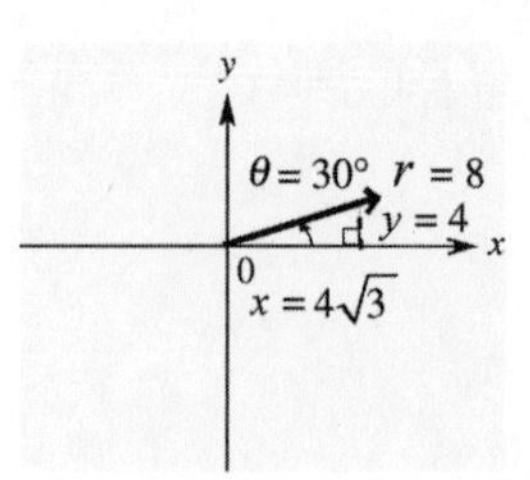

Since $x=4\sqrt{3}$ and $y=4$, $r=\sqrt{\left(4\sqrt{3}\right)^2+4^2}=\sqrt{48+16}=\sqrt{64}=8$ and $\tan\theta=\frac{4}{4\sqrt{3}}=\frac{1}{\sqrt{3}}=\frac{\sqrt{3}}{3}$.

Since $\tan\theta=\frac{\sqrt{3}}{3}$, the reference angle for θ is 30°. The graph shows that θ is in quadrant I, so $\theta=30°$. Therefore, $4\sqrt{3}+4i=8\left(\cos 30°+i\sin 30°\right)$.

43. $-5-5i$

Sketch a graph of $-5-5i$ in the complex plane.

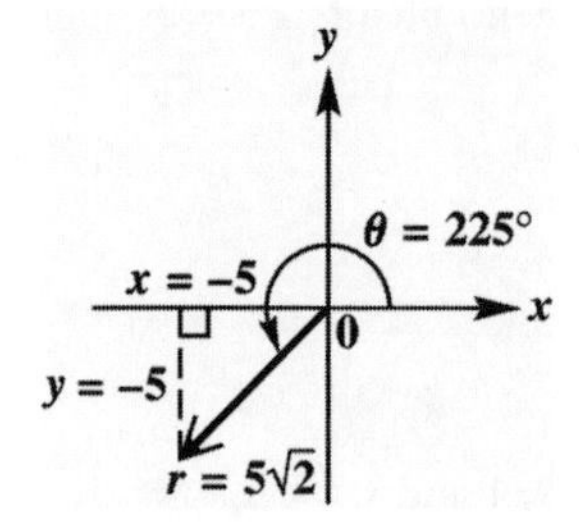

Since $x=-5$ and $y=-5$, $r=\sqrt{(-5)^2+(-5)^2}=\sqrt{25+25}=\sqrt{50}=5\sqrt{2}$ and $\tan\theta=\frac{y}{x}=\frac{-5}{-5}=1$.
Since $\tan\theta=1$, the reference angle for θ is 45°. The graph shows that θ is in quadrant III, so $\theta=180°+45°=225°$. Therefore, $-5-5i=5\sqrt{2}\left(\cos 225°+i\sin 225°\right)$.

44. $-\sqrt{2}+i\sqrt{2}$

Sketch a graph of $-\sqrt{2}+i\sqrt{2}$ in the complex plane.

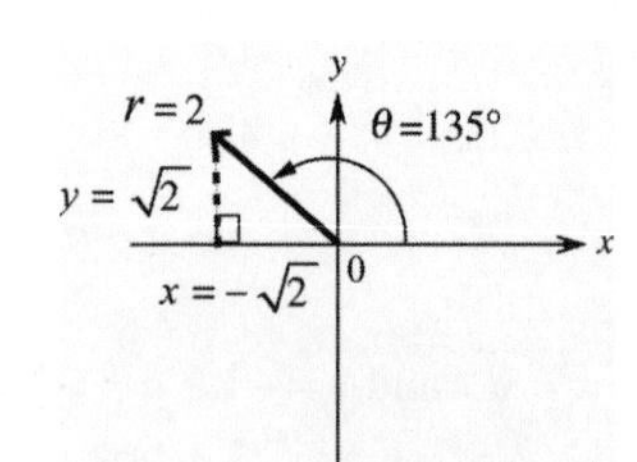

Since $x=-\sqrt{2}$ and $y=\sqrt{2}$, $r=\sqrt{\left(-\sqrt{2}\right)^2+\left(\sqrt{2}\right)^2}=\sqrt{2+2}=\sqrt{4}=2$ and $\tan\theta=\frac{\sqrt{2}}{-\sqrt{2}}=-1$. Since $\tan\theta=-1$, the reference angle for θ is 45°. The graph shows that θ is in quadrant II, so $\theta=180°-45°=135°$. Therefore, $-\sqrt{2}+i\sqrt{2}=2\left(\cos 135°+i\sin 135°\right)$.

45. $2+2i$

Sketch a graph of $2+2i$ in the complex plane.

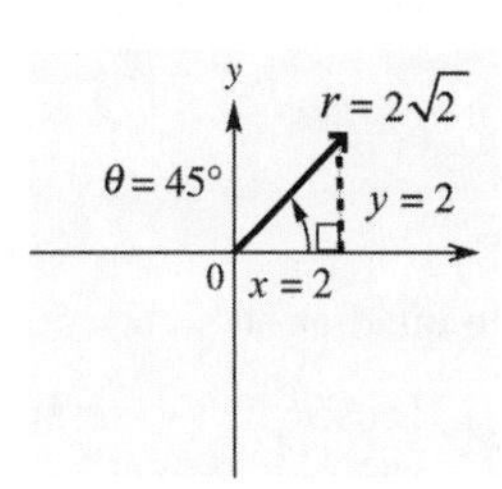

Since $x=2$ and $y=2$, $r=\sqrt{2^2+2^2}=\sqrt{4+4}=\sqrt{8}=2\sqrt{2}$ and $\tan\theta=\frac{2}{2}=1$. Since $\tan\theta=1$, the reference angle for θ is 45°. The graph shows that θ is in quadrant I, so $\theta=45°$. Therefore, $2+2i=2\sqrt{2}\left(\cos 45°+i\sin 45°\right)$.

46. $-\sqrt{3}+i$

Sketch a graph of $-\sqrt{3}+i$ in the complex plane.

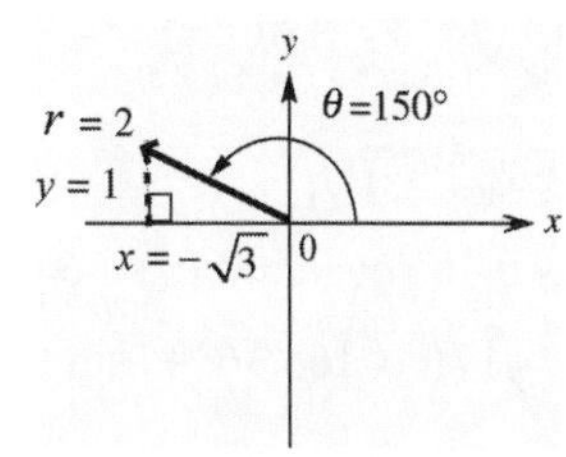

Since $x=-\sqrt{3}$ and $y=1$, $r=\sqrt{\left(-\sqrt{3}\right)^2+1^2}=\sqrt{3+1}=\sqrt{4}=2$ and $\tan\theta=\frac{1}{-\sqrt{3}}=-\frac{\sqrt{3}}{3}$. Since $\tan\theta=-\frac{\sqrt{3}}{3}$, the reference angle for θ is 30°. The graph shows that θ is in quadrant II, so $\theta=180°-30°=150°$. Therefore, $-\sqrt{3}+i=2\left(\cos 150°+i\sin 150°\right)$.

47. $5i=0+5i$

$0+5i$ is on the positive y-axis, so $\theta=90°$ and $x=0,\ y=5\Rightarrow r=\sqrt{0^2+5^2}=\sqrt{0+25}=5$

Thus, $5i=5\left(\cos 90°+i\sin 90°\right)$.

48. $-2i=0-2i$

$0-2i$ is on the negative y-axis, so $\theta=270°$ and $x=0,\ y=-2\Rightarrow r=\sqrt{(-2)^2+0^2}=\sqrt{4}=2$

Thus, $-2i=2\left(\cos 270°+i\sin 270°\right)$.

49. $-4=-4+0i$

$-4+0i$ is on the negative x-axis, so $\theta=180°$ and $x=-4,\ y=0\Rightarrow r=\sqrt{(-4)^2+0^2}=\sqrt{16}=4$

Thus, $-4=4\left(\cos 180°+i\sin 180°\right)$.

50. $7=7+0i$

$7+0i$ is on the positive x-axis, so $\theta=0°$ and $x=7,\ y=0\Rightarrow r=\sqrt{7^2+0^2}=\sqrt{49+0}=\sqrt{49}=7$

Thus, $7=7\left(\cos 0°+i\sin 0°\right)$.

51. $2+3i$

$x=2, y=3 \Rightarrow r=\sqrt{2^2+3^2}=\sqrt{4+9}=\sqrt{13}$

$\tan\theta=\frac{3}{2}$

$2+3i$ is in quadrant I, so $\theta=56.31°$

$2+3i=\sqrt{13}\left(\cos 56.31°+i\sin 56.31°\right)$

52. $\cos 35°+i\sin 35°=.8192+.5736i$

53. $3\left(\cos 250°+i\sin 250°\right)=-1.0261-2.8191i$

54. $-4+i$

$x=-4,\ y=1 \Rightarrow r=\sqrt{(-4)^2+1}=\sqrt{16+1}=\sqrt{17}$

$\tan\theta=\frac{1}{-4}=-\frac{1}{4}$

$-4+i$ is in the quadrant II, so $\theta=165.96°$.

$-4+i=\sqrt{17}(\cos 165.96°+i\sin 165.96°)$

55. $12i=0+12i$

$x=0,\ y=12 \Rightarrow r=\sqrt{0^2+12^2}=\sqrt{0+144}=\sqrt{144}=12$

$0+12i$ is on the positive y-axis, so $0=90°$.

$12i=12(\cos 90°+i\sin 90°)$

56. $3\text{ cis }180°=3\left(\cos 180°+i\sin 180°\right)=3\left(-1+0i\right)=-3+0i$ or -3

57. $3+5i$

$x=3,\ y=5 \Rightarrow r=\sqrt{3^2+5^2}=\sqrt{9+25}=\sqrt{34}$

$\tan\theta=\frac{5}{3}$

$3+5i$ is in the quadrant I, so $\theta=59.04°$.

$3+5i=\sqrt{34}\left(\cos 59.04°+i\sin 59.04°\right)$

58. $\text{cis }110.5°=\cos 110.5°+i\sin 110.5°\approx -.3502+.9367i$

59. Since the modulus represents the magnitude of the vector in the complex plane, $r=1$ would represent a circle of radius one centered at the origin.

60. When graphing $x+yi$ in the plane as (x, y), if x and y are equal, we are graphing points in the form $\left(x, x\right)$. These points make up the line $y=x$.

61. Since the real part of $z=x+yi$ is 1, the graph of $1+yi$ would be the vertical line $x=1$.

62. When graphing $x+yi$ in the plane as (x, y), since the imaginary part is 1, the points are of the form $(x, 1)$. These points constitute the horizontal line $y=1$.

63. $z = -.2i$

$z^2 - 1 = (-.2i)^2 - 1 = .04i^2 - 1 = .04(-1) - 1 = -.04 - 1 = -1.04$

The modulus is 1.04.

$(z^2 - 1)^2 - 1 = (-1.04)^2 - 1 = 1.0816 - 1 = .0816$

The modulus is .0816.

$\left[(z^2 - 1)^2 - 1\right]^2 - 1 = (.0816)^2 - 1 = .0665856 - 1 = -.99334144$

The modulus is .99334144.

The moduli do not exceed 2. Therefore, z is in the Julia set.

64. **(a)** Let $z_1 = a + bi$ and its complex conjugate be $z_2 = a - bi$.

$|z_1| = \sqrt{a^2 + b^2}$ and $|z_2| = \sqrt{a^2 + (-b)^2} = \sqrt{a^2 + b^2} = |z_1|$.

(b) Let $z_1 = a + bi$ and $z_2 = a - bi$.

$$z_1^2 - 1 = (a + bi)^2 - 1$$
$$= (a^2 + 2abi + b^2i^2) - 1 = a^2 + 2abi + b^2(-1) - 1 = a^2 - b^2 + 2abi - 1 = (a^2 - b^2 - 1) + (2ab)i$$
$$= c + di$$

$$z_2^2 - 1 = (a - bi)^2 - 1$$
$$= (a^2 - 2abi + b^2i^2) - 1 = a^2 - 2abi + b^2(-1) - 1 = a^2 - b^2 - 2abi - 1 = (a^2 - b^2 - 1) - (2ab)i$$
$$= c - di$$

(c) – (d) The results are again complex conjugates of each other. At each iteration, the resulting values from z_1 and z_2 will always be complex conjugates. Graphically, these represent points that are symmetric with respect to the x-axis, namely points such as (a, b) and $(a, -b)$.

(e) Answers will vary.

65. Let $z = r(\cos\theta + i\sin\theta)$. This would represent a vector with a magnitude of r and argument θ. Now, the conjugate of z should be the vector with magnitude r pointing in the $-\theta$ direction. Thus, $r\left[\cos(360° - \theta) + i\sin(360° - \theta)\right]$ satisfies these conditions as shown.

$$r\left[\cos(360° - \theta) + i\sin(360° - \theta)\right] = r\left[\cos 360°\cos\theta + \sin 360°\sin\theta + i\sin 360°\cos\theta - i\cos 360°\sin\theta\right]$$
$$= r\left[1\cdot\cos\theta + 0\cdot\sin\theta + i(0)\cos\theta - i(1)\sin\theta\right]$$
$$= r(\cos\theta - i\sin\theta) = r\left[\cos(-\theta) + i\sin(-\theta)\right]$$

This represents the conjugate of z, which is a reflection over the x-axis.

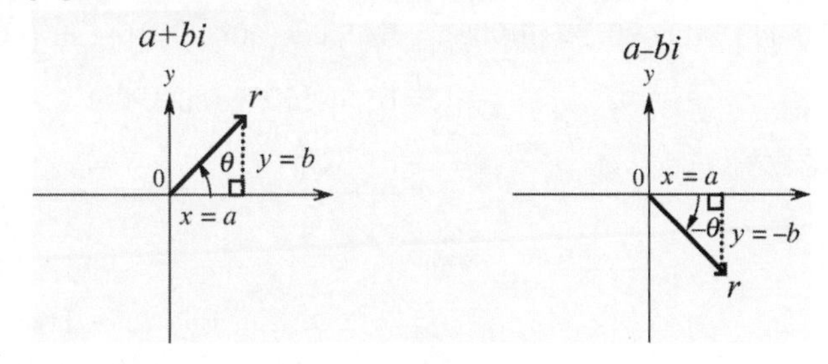

66. Let $z = r(\cos\theta + i\sin\theta)$. This would represent a vector with a magnitude of r and angle θ. Now, $-z$ should be the vector with magnutide of r pointing in the opposite direction. An angle that would make this vector point in the opposite direction would be $\theta + \pi$. Thus, $-z$ would be the following.

$$r\left[\cos(\theta+\pi) + i\sin(\theta+\pi)\right]$$

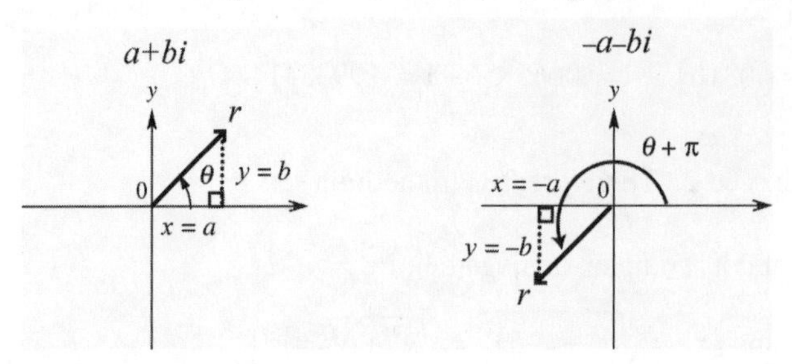

67. $(a+bi)-(c+di) = e+0i$, so, $b-d=0$ or $b=d$.

Therefore, the terminal points of the vectors corresponding to $a+bi$ and $c+di$ lie on a horizontal line. (B)

68. The absolute value of a complex number is equal to the length of the vector representing it in the complex plane. Therefore, the only way for the absolute value of the sum of two complex numbers to be equal to the sum of their absolute value is for the two vectors representing the numbers to have the same direction. (C)

69. The difference of the vectors is equal to the sum of one of the vectors plus the opposite of the other vector. If this sum is represented by two parallel vectors, the sum of their absolute value will equal to the absolute value of their sum. In order for this sum to be represented by two parallel vectors their difference must be represented by two vectors with opposite directions. (A)

70. Let $z = a+bi$. Thus, $|z| = \sqrt{a^2+b^2}$. Now, let $iz = i(a+bi) = ai + bi^2 = ai + b(-1) = ai - b = -b + ai$. Therefore, we have the following.

$$|iz| = \sqrt{(-b)^2 + a^2} = \sqrt{b^2 + a^2} = \sqrt{a^2+b^2} = |z|.$$

Section 8.3: The Product and Quotient Theorems

1. When multiplying two complex numbers in trigonometric form, we multiply their absolute values and add their arguments.

2. When dividing two complex numbers in trigonometric form, we divide their absolute values and subtract their arguments.

3. $$\begin{aligned}\left[3(\cos 60^\circ + i\sin 60^\circ)\right]\cdot\left[2(\cos 90^\circ + i\sin 90^\circ)\right] &= 3\cdot 2\left[\cos(60^\circ+90^\circ) + i\sin(60^\circ+90^\circ)\right]\\ &= 6(\cos 150^\circ + i\sin 150^\circ)\\ &= 6\left(-\frac{\sqrt{3}}{2} + \frac{1}{2}i\right) = -3\sqrt{3} + 3i\end{aligned}$$

4. $\left[4(\cos 30°+i\sin 30°)\right]\cdot\left[5(\cos 120°+i\sin 120°)\right]=4\cdot 5\left[\cos(30°+120°)+i\sin(30°+120°)\right]$

$$=20(\cos 150°+i\sin 150°)$$

$$=20\left(-\frac{\sqrt{3}}{2}+\frac{1}{2}i\right)=-10\sqrt{3}+10i$$

5. $\left[2(\cos 45°+i\sin 45°)\right]\cdot\left[2(\cos 225°+i\sin 225°)\right]=2\cdot 2\left[\cos(45°+225°)+i\sin(45°+225°)\right]$

$$=4(\cos 270°+i\sin 270°)=4(0-i)=0-4i \text{ or } -4i$$

6. $\left[8(\cos 300°+i\sin 300°)\cdot[5(\cos 120°+i\sin 120°)\right]=8\cdot 5\left[\cos(300°+120°)+i\sin(300°+120°)\right]$

$$=40(\cos 420°+i\sin 420°)=40(\cos 60°+i\sin 60°)$$

$$=40\left(\frac{1}{2}+i\frac{\sqrt{3}}{2}\right)=20+20i\sqrt{3}$$

7. $\left[4(\cos 60°+i\sin 60°)\right]\cdot\left[6(\cos 330°+i\sin 330°)\right]=4\cdot 6\left[(\cos(60°+330°)+i\sin(60°+330°)\right]$

$$=24(\cos 390°+i\sin 390°)=24(\cos 30°+i\sin 30°)$$

$$=24\left(\frac{\sqrt{3}}{2}+i\frac{1}{2}\right)=12\sqrt{3}+12i$$

8. $\left[8(\cos 210°+i\sin 210°)\right]\cdot\left[2(\cos 330°+i\sin 330°)\right]=8\cdot 2\left[\cos(210°+330°)+i\sin(210°+330°)\right]$

$$=16(\cos 540°+i\sin 540°)$$

$$=16(\cos 180°+i\sin 180°)$$

$$=16(-1+0\cdot i)=-16+0i \text{ or } -16$$

9. $[5 \text{ cis } 90°][3 \text{ cis } 45°]=5\cdot 3\left[\text{cis}(90°+45°)\right]=15 \text{ cis } 135°$

$$=15(\cos 135°+i\sin 135°)=15\left(-\frac{\sqrt{2}}{2}+\frac{\sqrt{2}}{2}i\right)=-\frac{15\sqrt{2}}{2}+\frac{15\sqrt{2}}{2}i$$

10. $[6 \text{ cis } 120°][5 \text{ cis }(-30°)]=6\cdot 5\left[\text{cis}(120°+(-30°))\right]$

$$=30 \text{ cis } 90°=30(\cos 90°+i\sin 90°)=30(0+1i)=0+30i=30i$$

11. $\left[\sqrt{3}\text{ cis } 45°\right]\left[\sqrt{3}\text{ cis } 225°\right]=\sqrt{3}\cdot\sqrt{3}\left[\text{cis}(45°+225°)\right]$

$$=3 \text{ cis } 270°=3(\cos 270°+i\sin 270°)=3(0-i)=0-3i \text{ or } -3i$$

12. $\left[\sqrt{2}\text{ cis } 300°\right]\left[\sqrt{2}\text{ cis } 270°\right]=\sqrt{2}\cdot\sqrt{2}\left[\text{cis}(300°+270°)\right]=2 \text{ cis } 570°$

$$=2 \text{ cis } 210°=2(\cos 210°+i\sin 210°)=2\left(-\frac{\sqrt{3}}{2}-\frac{1}{2}i\right)=-\sqrt{3}-i$$

13. $\dfrac{4(\cos 120° + i \sin 120°)}{2(\cos 150° + i \sin 150°)} = \dfrac{4}{2}\left[\cos(120° - 150°) + i \sin(120° - 150°)\right] = 2\left[\cos(-30°) + i \sin(-30°)\right]$

$$= 2\left[\cos 30° - i \sin 30°\right] = 2\left(\frac{\sqrt{3}}{2} - \frac{1}{2}i\right) = \sqrt{3} - i$$

14. $\dfrac{24(\cos 150° + i \sin 150°)}{2(\cos 30° + i \sin 30°)} = \dfrac{24}{2}\left[\cos(150° - 30°) + i \sin(150° - 30°)\right]$

$$= 12\left[\cos(120°) + i \sin(120°)\right] = 12\left(-\frac{1}{2} + \frac{\sqrt{3}}{2}i\right) = -6 + 6i\sqrt{3}$$

15. $\dfrac{10(\cos 225° + i \sin 225°)}{5(\cos 45° + i \sin 45°)} = \dfrac{10}{5}\left[\cos(225° - 45°) + i \sin(225° - 45°)\right]$

$$= 2(\cos 180° + i \sin 180°) = 2(-1 + 0 \cdot i) = -2 + 0i \text{ or } -2$$

16. $\dfrac{16(\cos 300° + i \sin 300°)}{8(\cos 60° + i \sin 60°)} = \dfrac{16}{8}\left[\cos(300° - 60°) + i \sin(300° - 60°)\right]$

$$= 2(\cos 240° + i \sin 240°) = 2\left(-\frac{1}{2} - \frac{\sqrt{3}}{2}i\right) = -1 - i\sqrt{3}$$

17. $\dfrac{3 \text{ cis } 305°}{9 \text{ cis } 65°} = \dfrac{1}{3}\text{cis}(305° - 65°) = \dfrac{1}{3}(\text{cis } 240°) = \dfrac{1}{3}(\cos 240° + i \sin 240°) = \dfrac{1}{3}\left(-\dfrac{1}{2} - \dfrac{\sqrt{3}}{2}i\right) = -\dfrac{1}{6} - \dfrac{\sqrt{3}}{6}i$

18. $\dfrac{12 \text{ cis } 293°}{6 \text{ cis } 23°} = 2 \text{ cis}(293° - 23°) = 2(\text{cis } 270°) = 2(\cos 270° + i \sin 270°) = 2(0 - 1 \cdot i) = 0 - 2i = -2i$

19. $\dfrac{8}{\sqrt{3} + i}$

numerator: $8 = 8 + 0 \cdot i$ and $r = 8$

$\theta = 0°$ since $\cos 0° = 1$ and $\sin 0° = 0$, so $8 = 8 \text{ cis } 0°$.

denominator: $\sqrt{3} + i$ and $r = \sqrt{\left(\sqrt{3}\right)^2 + 1^2} = \sqrt{3 + 1} = \sqrt{4} = 2$

$\tan\theta = \dfrac{1}{\sqrt{3}} = \dfrac{\sqrt{3}}{3}$

Since x and y are both positive, θ is in quadrant I, so $\theta = 30°$. Thus, $\sqrt{3} + i = 2 \text{ cis } 30°$.

$$\frac{8}{\sqrt{3} + i} = \frac{8 \text{ cis } 0°}{2 \text{ cis } 30°}$$

$$= \frac{8}{2}\text{cis}(0 - 30°) = 4 \text{ cis}(-30°) = 4\left[\cos(-30°) + i \sin(-30°)\right] = 4\left(\frac{\sqrt{3}}{2} - i\frac{1}{2}\right) = 2\sqrt{3} - 2i$$

20. $\dfrac{2i}{-1+i\sqrt{3}}$

numerator: $2i = 0 + 2i$ and $r = 2$

$\theta = 90°$ since $\cos 90° = 0$ and $\sin 90° = 1$, so $2i = 2 \text{ cis } 90°$.

denominator: $-1 + i\sqrt{3}$

$$r = \sqrt{(-1)^2 + \left(-\sqrt{3}\right)^2} = \sqrt{1+3} = \sqrt{4} = 2$$

$$\tan\theta = \frac{-\sqrt{3}}{-1} = \sqrt{3}$$

Since x is negative and y is positive, θ is in quadrant II, so $\theta = 60° + 180° = 240°$.

Thus, $-1 + i\sqrt{3} = 2 \text{ cis } 240°$

$$\frac{2i}{-1-i\sqrt{3}} = \frac{2\text{cis } 90°}{2\text{cis } 240°} = \frac{2}{2}\text{cis}\left(90° - 240°\right) = 1\text{cis}\left(-150°\right) = \cos\left(-150°\right) + i\sin\left(-150°\right) = -\frac{\sqrt{3}}{2} - \frac{1}{2}i$$

21. $\dfrac{-i}{1+i}$

numerator: $-i = 0 - i$ and $r = \sqrt{0^2 + (-1)^2} = \sqrt{0+1} = \sqrt{1} = 1$

$\theta = 270°$ since $\cos 270° = 0$ and $\sin 270° = -1$, so $-i = 1 \text{ cis} 270°$.

denominator: $1 + i$

$r = \sqrt{1^2 + 1^2} = \sqrt{1+1} = \sqrt{2}$ and $\tan\theta = \dfrac{y}{x} = \dfrac{1}{1} = 1$

Since x and y are both positive, θ is in quadrant I, so $\theta = 45°$. Thus,

$1 + i = \sqrt{2} \text{ cis } 45°$

$$\frac{-i}{1+i} = \frac{\text{cis } 270°}{\sqrt{2}\text{ cis } 45°} = \frac{1}{\sqrt{2}}\text{cis }\left(270° - 45°\right)$$

$$= \frac{\sqrt{2}}{2}\text{ cis } 225° = \frac{\sqrt{2}}{2}\left(\cos 225° + i\sin 225°\right) = \frac{\sqrt{2}}{2}\left(-\frac{\sqrt{2}}{2} - i\cdot\frac{\sqrt{2}}{2}\right) = -\frac{1}{2} - \frac{1}{2}i$$

22. $\dfrac{1}{2-2i}$

numerator: $1 = 1 + 0\cdot i$ and $r = 1$

$\theta = 0°$ since $\cos 0° = 1$ and $\sin 0° = 0$, so $1 = 1 \text{ cis } 0°$.

denominator: $2 - 2i$ and $r = \sqrt{2^2 + (-2)^2} = \sqrt{4+4} = \sqrt{8} = 2\sqrt{2}$

$$\tan\theta = \frac{y}{x} = \frac{-2}{2} = -1$$

Since x is positive and y is negative, θ is in quadrant IV, so $\theta = -45°$. Thus,

$2 - 2i = 2\sqrt{2}\text{ cis}\left(-45°\right)$.

$$\frac{1}{2-2i} = \frac{1\text{ cis } 0°}{2\sqrt{2}\text{ cis}\left(-45°\right)} = \frac{1}{2\sqrt{2}}\text{cis}\left[0 - (-45°)\right]$$

$$= \frac{\sqrt{2}}{4}\text{cis } 45° = \frac{\sqrt{2}}{4}\left(\cos 45° + i\sin 45°\right) = \frac{\sqrt{2}}{4}\left(\frac{\sqrt{2}}{2} + i\frac{\sqrt{2}}{2}\right) = \frac{1}{4} + \frac{1}{4}i$$

23. $\dfrac{2\sqrt{6}-2i\sqrt{2}}{\sqrt{2}-i\sqrt{6}}$

numerator: $2\sqrt{6}-2i\sqrt{2}$ and $r=\sqrt{\left(2\sqrt{6}\right)^2+\left(-2\sqrt{2}\right)^2}=\sqrt{24+8}=\sqrt{32}=4\sqrt{2}$

$\tan\theta=\dfrac{-2\sqrt{2}}{2\sqrt{6}}=-\dfrac{1}{\sqrt{3}}=-\dfrac{\sqrt{3}}{3}$

Since x is positive and y is negative, θ is in quadrant IV, so $\theta=-30°$. Thus, $2\sqrt{6}-2i\sqrt{2}=4\sqrt{2}\text{ cis}\left(-30°\right)$.

denominator: $\sqrt{2}-i\sqrt{6}$ and $r=\sqrt{\left(\sqrt{2}\right)^2+\left(-\sqrt{6}\right)^2}=\sqrt{2+6}=\sqrt{8}=2\sqrt{2}$

$\tan\theta=\dfrac{-\sqrt{6}}{\sqrt{2}}=-\sqrt{3}$

Since x is positive and y is negative, θ is in quadrant IV, so $\theta=-60°$. Thus, $\sqrt{2}-i\sqrt{6}=2\sqrt{2}\text{ cis}\left(-30°\right)$

$$\frac{2\sqrt{6}-2i\sqrt{2}}{\sqrt{2}-i\sqrt{6}}=\frac{4\sqrt{2}\text{ cis}\left(-30°\right)}{2\sqrt{2}\text{ cis}\left(-60°\right)}$$

$$=\frac{4\sqrt{2}}{2\sqrt{2}}\text{cis}\left[-30°-\left(-60°\right)\right]=2\text{cis }30°=2\left(\cos 30°+i\sin 30°\right)=2\left(\frac{\sqrt{3}}{2}+i\frac{1}{2}\right)=\sqrt{3}+i$$

24. $\dfrac{4+4i}{2-2i}$

numerator: $4+4i$ and $r=\sqrt{4^2+4^2}=\sqrt{16+16}=\sqrt{32}=4\sqrt{2}$

$\tan\theta=\dfrac{-2}{2}=-1$

Since x and y are both positive, θ is in quadrant I, so $\theta=45°$. Thus, $2+2i=4\sqrt{2}\text{ cis }45°$

denominator: $2-2i$ and $r=\sqrt{2^2+\left(-2\right)^2}=\sqrt{4+4}=\sqrt{8}=2\sqrt{2}$

$\tan\theta=\dfrac{y}{x}=\dfrac{-2}{2}=-1$

Since x is positive and y is negative, θ is in quadrant IV, so $\theta=-45°$. Thus, $2-2i=2\sqrt{2}\text{ cis}\left(-45°\right)$

$$\frac{4+4i}{2-2i}=\frac{4\sqrt{2}\text{ cis }45°}{2\sqrt{2}\text{ cis}\left(-45°\right)}$$

$$=\frac{4\sqrt{2}}{2\sqrt{2}}\text{cis}\left[45°-\left(-45°\right)\right]=2\text{ cis }90°=2\left(\cos 90°+i\sin 90°\right)=2\left(0+i\right)=0+2i=2i$$

25. $\left[2.5\left(\cos 35°+i\sin 35°\right)\right]\cdot\left[3.0\left(\cos 50°+i\sin 50°\right)\right]=2.5\cdot 3.0\left[\cos\left(35°+50°\right)+i\sin\left(35°+50°\right)\right]$

$$=7.5\left(\cos 85°+i\sin 85°\right)\approx .6537+7.4715i$$

26. $\left[4.6\left(\cos 12°+i\sin 12°\right)\right]\cdot\left[2.0\left(\cos 13°+i\sin 13°\right)\right]=4.6\cdot 2.0\left[\cos\left(12°+13°\right)+i\sin\left(12°+13°\right)\right]$

$$=9.2\left(\cos 25°+i\sin 25°\right)\approx 8.3380+3.8881i$$

27. $(12\text{ cis } 18.5^\circ)(3\text{ cis } 12.5^\circ) = 12 \cdot 3\text{ cis } (18.5^\circ + 12.5^\circ)$
$$= 36\text{ cis } 31^\circ = 36(\cos 31^\circ + i \sin 31^\circ) \approx 30.8580 + 18.5414i$$

28. $(4\text{ cis } 19.25^\circ)(7\text{ cis } 41.75^\circ) = 4 \cdot 7\text{ cis } (19.25^\circ + 41.75^\circ)$
$$= 28\text{ cis } 61^\circ = 28(\cos 61^\circ + i \sin 61^\circ) \approx 13.5747 + 24.4894i$$

29. $\dfrac{45(\cos 127^\circ + i \sin 127^\circ)}{22.5(\cos 43^\circ + i \sin 43^\circ)} = \dfrac{45}{22.5}\left[\cos(127^\circ - 43^\circ) + i \sin(127^\circ - 43^\circ)\right]$
$$= 2(\cos 84^\circ + i \sin 84^\circ) \approx .2091 + 1.9890i$$

30. $\dfrac{30(\cos 130^\circ + i \sin 130^\circ)}{10(\cos 21^\circ + i \sin 21^\circ)} = \dfrac{30}{10}\left[\cos(130^\circ - 21^\circ) + i \sin(130^\circ - 21^\circ)\right]$
$$= 3(\cos 109^\circ + i \sin 109^\circ) \approx -.9767 + 2.8366i$$

31. $\left[2\text{cis}\dfrac{5\pi}{9}\right]^2 = \left[2\text{cis}\dfrac{5\pi}{9}\right]\left[2\text{cis}\dfrac{5\pi}{9}\right]$
$$= 2 \cdot 2\text{ cis}\left(\frac{5\pi}{9} + \frac{5\pi}{9}\right) = 4\text{cis}\frac{10\pi}{9} = 4\left(\cos\frac{10\pi}{9} + i \sin\frac{10\pi}{9}\right) \approx -3.7588 - 1.3681i$$

32. $\left[24.3\text{cis}\dfrac{7\pi}{12}\right]^2 = \left[24.3\text{cis}\dfrac{7\pi}{12}\right]\left[24.3\text{cis}\dfrac{7\pi}{12}\right] = 24.3 \cdot 24.3\text{ cis}\left(\dfrac{7\pi}{12} + \dfrac{7\pi}{12}\right)$
$$= 590.49\text{cis}\frac{7\pi}{6} = 590.49\left(\cos\frac{7\pi}{6} + i \sin\frac{7\pi}{6}\right) \approx -511.3793 - 295.2450i$$

In Exercises 33–39, $w = -1 + i$ and $z = -1 - i$.

33. $w \cdot z = (-1+i)(-1-i) = -1(-1) + (-1)(-i) + i(-1) + i(-i) = 1 + i - i - i^2 = 1 - (-1) = 2$

34. $w = -1 + i$:

$r = \sqrt{(-1)^2 + 1^2} = \sqrt{1+1} = \sqrt{2}$ and $\tan\theta = \dfrac{1}{-1} = -1$

Since x is negative and y is positive, θ is in quadrant II, so $\theta = 135^\circ$. Thus, $w = \sqrt{2}\text{ cis } 135^\circ$.

$z = -1 - i$:

$r = \sqrt{(-1)^2 + (-1)^2} = \sqrt{1+1} = \sqrt{2}$ and $\tan\theta = \dfrac{-1}{-1} = 1$

Since x and y are negative, θ is in quadrant III, so $\theta = 225^\circ$. Thus, $z = \sqrt{2}\text{ cis } 225^\circ$.

35. $w \cdot z = (\sqrt{2}\text{ cis } 135^\circ)(\sqrt{2}\text{ cis } 225^\circ) = \sqrt{2} \cdot \sqrt{2}\left[\text{ cis } (135^\circ + 225^\circ)\right] = 2\text{ cis } 360^\circ = 2\text{ cis } 0^\circ$

36. $2\text{ cis } 0^\circ = 2(\cos 0^\circ + i \sin 0^\circ) = 2(1 + 0 \cdot i) = 2 \cdot 1 = 2$; It is the same.

37. $\dfrac{w}{z} = \dfrac{-1+i}{-1-i} = \dfrac{-1+i}{-1-i} \cdot \dfrac{-1+i}{-1+i} = \dfrac{1-i-i+i^2}{1-i^2} = \dfrac{1-2i+(-1)}{1-(-1)} = \dfrac{-2i}{2} = -i$

38. $\frac{w}{z}=\frac{\sqrt{2}\text{ cis }135°}{\sqrt{2}\text{ cis }225°}=\frac{\sqrt{2}}{\sqrt{2}}\text{ cis }(135°-225°)=\text{ cis }(-90°)$

39. $\text{cis }(-90°)=\cos(-90°)+i\sin(-90°)=0+i(-1)=0-i=-i$; It is the same.

40. Answers will vary.

41. The two results will have the same magnitude because in both cases you are finding the product of 2 and 5. We must now determine if the arguments are the same. In the first product, the argument of the product will be $45°+90°=135°$. In the second product, the argument of the product will be $-315°+(-270°)=-585°$. Now, $-585°$ is coterminal with $-585°+2\cdot 360°=135°$. Thus, the two products are the same since they have the same magnitude and argument.

42. $z=r(\cos\theta+i\sin\theta)$

$1=1+0\cdot i=1(\cos 0°+i\sin 0°),\ \frac{1}{z}=\frac{1(\cos 0°+i\sin 0°)}{r(\cos\theta+i\sin\theta)}=\frac{1}{r}[\cos(0°-\theta°)+i\sin(0°-\theta°)]$

$=\frac{1}{r}[\cos(0°-\theta°)+i\sin(0°-\theta°)]=\frac{1}{r}[\cos(-\theta)+i\sin(-\theta°)]=\frac{1}{r}(\cos\theta-i\sin\theta)$

43. $E=8(\cos 20°+i\sin 20°), R=6, X_L=3,\ I=\frac{E}{Z}, Z=R+X_L i$

Write $Z=6+3i$ in trigonometric form.

$x=6$, and $y=3\Rightarrow r=\sqrt{6^2+3^2}=\sqrt{36+9}=\sqrt{45}=3\sqrt{5}$.

$\tan\theta=\frac{3}{6}=\frac{1}{2}$, so $\theta\approx 26.6°$

$Z=3\sqrt{5}\text{ cis }26.6°$

$I=\frac{8\text{ cis }20°}{3\sqrt{5}\text{ cis }26.6°}=\frac{8}{3\sqrt{5}}\text{ cis }(20°-26.6°)=\frac{8\sqrt{5}}{15}\text{ cis }(-6.6°)=\frac{8\sqrt{5}}{15}[\cos(-6.6°)+i\sin(-6.6°)]\approx 1.18-.14i$

44. $E=12(\cos 25°+i\sin 25°), R=3,\ X_L=4$, and $X_c=6$

$I=\frac{E}{R+(X_L-X_c)i}=\frac{12\text{ cis }25°}{3+(4-6)i}=\frac{12\text{ cis }25°}{3-2i}$

Write $3-2i$ in trigonometric form. $x=3$, $y=-2$, so $r=\sqrt{3^2+(-2)^2}=\sqrt{9+4}=\sqrt{13}$.

$\tan\theta=-\frac{2}{3}$, so θ=326.31° since θ is in quadrant IV.

Continuing to find the current, we have the following.

$I=\frac{12\text{ cis }25°}{\sqrt{13}\text{ cis }326.31°}=\frac{12}{\sqrt{13}}\text{ cis }(25°-326.3°)$

$=\frac{12\sqrt{13}}{13}\text{ cis }(-301.3°)=\frac{12\sqrt{13}}{13}[\cos(-301.3°)+i\sin(-301.3°)]\approx 1.7+2.8i$

45. Since $Z_1 = 50 + 25i$ and $Z_2 = 60 + 20i$, we have the following.

$$\frac{1}{Z_1} = \frac{1}{50+25i}\cdot\frac{50-25i}{50-25i} = \frac{50-25i}{50^2-25^2i^2} = \frac{50-25i}{2500-625(-1)} = \frac{50-25i}{2500+625} = \frac{50-25i}{3125} = \frac{2}{125} - \frac{1}{125}i$$

and

$$\frac{1}{Z_2} = \frac{1}{60+20i}\cdot\frac{60-20i}{60-20i} = \frac{60-20i}{60^2-20^2i^2} = \frac{60-20i}{3600-400(-1)} = \frac{60-20i}{3600+400} = \frac{60-20i}{4000} = \frac{3}{200} - \frac{1}{200}i$$

and

$$\frac{1}{Z_1} + \frac{1}{Z_2} = \left(\frac{2}{125} - \frac{1}{125}i\right) + \left(\frac{3}{200} - \frac{1}{200}i\right) = \left(\frac{2}{125} + \frac{3}{200}\right) - \left(\frac{1}{125} + \frac{1}{200}\right)i$$
$$= \left(\frac{16}{1000} + \frac{15}{1000}\right) - \left(\frac{8}{1000} + \frac{5}{1000}\right)i = \frac{31}{1000} - \frac{13}{1000}i$$

$$Z = \frac{1}{\frac{1}{Z_1} + \frac{1}{Z_2}} = \frac{1}{\frac{31}{1000} - \frac{13}{1000}i} = \frac{1000}{31-13i}\cdot\frac{31+13i}{31+13i} = \frac{1000(31+13i)}{31^2-13^2i^2}$$
$$= \frac{31,000+13,000}{961-169(-1)} = \frac{31,000+13,000}{961+169} = \frac{31,000+13,000i}{1130} = \frac{3100}{113} + \frac{1300}{113}i \approx 27.43 + 11.5i$$

46. $\tan\theta = \frac{11.5}{27.43} \Rightarrow \theta = \tan^{-1}\frac{11.5}{27.43} \approx 22.75°$ since θ is in quadrant I.

Section 8.4: Demoivre's Theorem; Powers and Roots of Complex Numbers

1. $$[3(\cos 30° + i\sin 30°)]^3 = 3^3[\cos(3\cdot 30°) + i\sin(3\cdot 30°)]$$
$$= 27(\cos 90° + i\sin 90°) = 27(0 + 1\cdot i) = 0 + 27i \text{ or } 27i$$

2. $$[2(\cos 135° + i\sin 135°)]^4 = 2^4[\cos(4\cdot 135°) + i\sin(4\cdot 135°)]$$
$$= 16(\cos 540° + i\sin 540°) = 16(\cos 180° + i\sin 180°)$$
$$= 16(-1 + 0\cdot i) = -16 + 0i \text{ or } -16$$

3. $$(\cos 45° + i\sin 45°)^8 = [\cos(8\cdot 45°) + i\sin(8\cdot 45°)] = \cos 360° + i\sin 360° = 1 + 0\cdot i \text{ or } 1$$

4. $$[2(\cos 120° + i\sin 120°)]^3 = 2^3[\cos(3\cdot 120°) + i\sin(3\cdot 120°)]$$
$$= 8(\cos 360° + i\sin 360°) = 8(1 + 0\cdot i) = 8 + 0i \text{ or } 8$$

5. $$[3 \text{ cis } 100°]^3 = 3^3 \text{ cis } (3\cdot 100°)$$
$$= 27 \text{ cis } 300° = 27(\cos 300° + i\sin 300°) = 27\left(\frac{1}{2} - \frac{\sqrt{3}}{2}i\right) = \frac{27}{2} - \frac{27\sqrt{3}}{2}i$$

6. $[3 \text{ cis } 40°]^3 = 3^3 \text{ cis } (3 \cdot 40°)$

$$= 27 \text{ cis } 120° = 27(\cos 120° + i \sin 120°) = 27\left(-\frac{1}{2} + \frac{\sqrt{3}}{2}i\right) = -\frac{27}{2} + \frac{27\sqrt{3}}{2}i$$

7. $\left(\sqrt{3} + i\right)^5$

First write $\sqrt{3} + i$ in trigonometric form.

$$r = \sqrt{\left(\sqrt{3}\right)^2 + 1^2} = \sqrt{3+1} = \sqrt{4} = 2 \text{ and } \tan\theta = \frac{1}{\sqrt{3}} = \frac{\sqrt{3}}{3}$$

Because x and y are both positive, θ is in quadrant I, so $\theta = 30°$.

$$\sqrt{3} + i = 2(\cos 30° + i \sin 30°)$$

$$\left(\sqrt{3} + i\right)^5 = \left[2(\cos 30° + i \sin 30°)\right]^5 = 2^5\left[\cos(5 \cdot 30°) + i \sin(5 \cdot 30°)\right]$$

$$= 32(\cos 150° + i \sin 150°) = 32\left(-\frac{\sqrt{3}}{2} + i\frac{1}{2}\right) = -16\sqrt{3} + 16i$$

8. $\left(2\sqrt{2} - 2i\sqrt{2}\right)^6$

First write $2\sqrt{2} - 2i\sqrt{2}$ in trigonometric form.

$$r = \sqrt{\left(2\sqrt{2}\right)^2 + \left(-2\sqrt{2}\right)^2} = \sqrt{8+8} = \sqrt{16} = 4 \text{ and } \tan\theta = \frac{-2\sqrt{2}}{2\sqrt{2}} = -1$$

Because x is positive and y is negative, θ is in quadrant IV, so $\theta = 315°$.

$$2\sqrt{2} - 2i\sqrt{2} = 4(\cos 315° + i \sin 315°)^6$$

$$\left(2\sqrt{2} - 2i\sqrt{2}\right)^6 = \left[4(\cos 315° + i \sin 315°)\right]^6 = 4^6\left[\cos(6 \cdot 315°) + i \sin(6 \cdot 315°)\right]$$

$$= 4096\left[\cos 1890° + i \sin 1890°\right] = 4096(\cos 90° + i \sin 90°)$$

$$= 4096(0 + 1 \cdot i) = 0 + 4096i \text{ or } 4096i$$

9. $\left(2 - 2i\sqrt{3}\right)^4$

First write $2 - 2i\sqrt{3}$ in trigonometric form.

$$r = \sqrt{2^2 + \left(-2\sqrt{3}\right)^2} = \sqrt{4+12} = \sqrt{16} = 4 \text{ and } \tan\theta = \frac{-2\sqrt{3}}{2} = -\sqrt{3}$$

Because x is positive and y is negative, θ is in quadrant IV, so $\theta = 300°$.

$$2\text{–}2\text{i}\sqrt{3} = 4(\cos 300° + i \sin 300°)$$

$$\left(2\text{–}2\text{i}\sqrt{3}\right)^4 = \left[4(\cos 300° + i \sin 300°)\right]^4 = 4^4\left[\cos(4 \cdot 300°) + i \sin(4 \cdot 300°)\right]$$

$$= 256(\cos 1200° + i \sin 1200°) = 256\left(-\frac{1}{2} + \frac{i\sqrt{3}}{2}\right) = -128 + 128i\sqrt{3}$$

10. $\left(\frac{\sqrt{2}}{2}-\frac{\sqrt{2}}{2}i\right)^8$

First write $\frac{\sqrt{2}}{2}-\frac{\sqrt{2}}{2}i$ in trigonometric form.

$$r=\sqrt{\left(\frac{\sqrt{2}}{2}\right)^2+\left(-\frac{\sqrt{2}}{2}\right)^2}=\sqrt{\frac{2}{4}+\frac{2}{4}}=\sqrt{1}=1 \text{ and } \tan\theta=\frac{-\frac{\sqrt{2}}{2}}{\frac{\sqrt{2}}{2}}=-1$$

Because x is positive and y is negative, θ is in quadrant IV, so $\theta=315°$.

$$\frac{\sqrt{2}}{2}-\frac{\sqrt{2}}{2}i=\cos 315°+i\sin 315°$$

$$\begin{aligned}\left(\frac{\sqrt{2}}{2}-\frac{\sqrt{2}}{2}i\right)^8&=(\cos 315°+i\sin 315°)^8\\&=\cos(8\cdot 315)°+i\sin(8\cdot 315)°\\&=\cos 2520°+i\sin 2520°\\&=\cos 0°+i\sin 0°=1+0i \text{ or } 1\end{aligned}$$

11. $(-2-2i)^5$

First write $-2-2i$ in trigonometric form.

$$r=\sqrt{(-2)^2+(-2)^2}=\sqrt{4+4}=\sqrt{8}=2\sqrt{2} \text{ and } \tan\theta=\frac{-2}{-2}=1$$

Because x and y are both negative, θ is in quadrant III, so $\theta=225°$.

$$-2-2i=2\sqrt{2}(\cos 225°+i\sin 225°)$$

$$\begin{aligned}(-2-2i)^5&=\left[2\sqrt{2}(\cos 225°+i\sin 225°)\right]^5=\left(2\sqrt{2}\right)^5\left[\cos(5\cdot 225°)+i\sin(5\cdot 225°)\right]\\&=32\sqrt{32}(\cos 1125°+i\sin 1125°)=128\sqrt{2}(\cos 45°+i\sin 45°)\\&=128\sqrt{2}\left(\frac{\sqrt{2}}{2}+\frac{\sqrt{2}}{2}i\right)=128+128i\end{aligned}$$

12. $(-1+i)^7$

First write $-2-2i$ in trigonometric form.

$$r=\sqrt{(-1)^2+1^2}=\sqrt{1+1}=\sqrt{2}=\sqrt{2} \text{ and } \tan\theta=\frac{1}{-1}=-1$$

Because x is negative and y is positive, θ is in quadrant II, so $\theta=135°$.

$$-1+i=\sqrt{2}(\cos 135°+i\sin 135°)$$

$$\begin{aligned}(-1+i)^7&=\left[\sqrt{2}(\cos 135°+i\sin 135°)\right]^7=\left(\sqrt{2}\right)^7\left[\cos(7\cdot 135°)+i\sin(7\cdot 135°)\right]\\&=8\sqrt{2}(\cos 945°+i\sin 945°)=8\sqrt{2}(\cos 225°+i\sin 225°)\\&=8\sqrt{2}\left(-\frac{\sqrt{2}}{2}-i\frac{\sqrt{2}}{2}\right)=-8-8i\end{aligned}$$

13. (a) $\cos 0° + i\sin 0° = 1(\cos 0° + i\sin 0°)$

We have $r = 1$ and $\theta = 0°$.

Since $r^3(\cos 3\alpha + i\sin 3\alpha) = 1(\cos 0° + i\sin 0°)$, then we have the following.

$$r^3 = 1 \Rightarrow r = 1 \text{ and } 3\alpha = 0° + 360° \cdot k \Rightarrow \alpha = \frac{0° + 360° \cdot k}{3} = 0° + 120° \cdot k = 120° \cdot k, \ k \text{ any integer.}$$

If $k = 0$, then $\alpha = 0°$. If $k = 1$, then $\alpha = 120°$. If $k = 2$, then $\alpha = 240°$.

So, the cube roots are $\cos 0° + i\sin 0°$, $\cos 120° + i\sin 120°$, and $\cos 240° + i\sin 240°$.

(b)

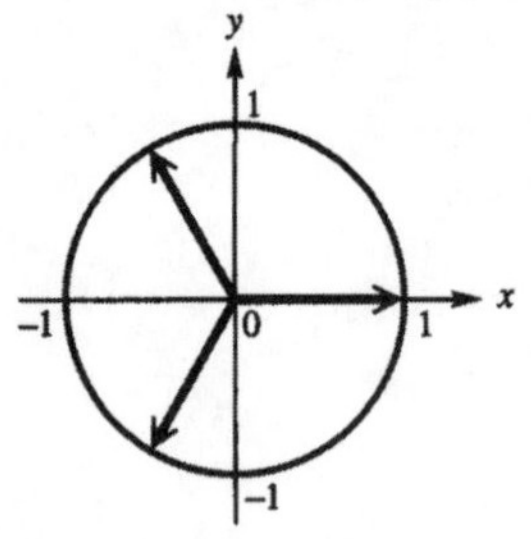

14. (a) Find the cube roots of $\cos 90° + i\sin 90° = 1(\cos 90° + i\sin 90°)$.

We have $r = 1$ and $\theta = 90°$.

Since $r^3(\cos 3\alpha + i\sin 3\alpha) = 1(\cos 90° + i\sin 90°)$, then we have the following.

$$r^3 = 1 \Rightarrow r = 1 \text{ and } 3\alpha = 90° + 360° \cdot k \Rightarrow \alpha = \frac{90° + 360° \cdot k}{3} = 30° + 120° \cdot k, \ k \text{ any integer.}$$

If $k = 0$, then $\alpha = 30° + 0° = 30°$. If $k = 1$, then $\alpha = 30° + 120° = 150°$.

If $k = 2$, then $\alpha = 30° + 240° = 270°$.

So, the cube roots are $\cos 30° + i\sin 30°$, $\cos 150° + i\sin 150°$, and $\cos 270° + i\sin 270°$.

(b)

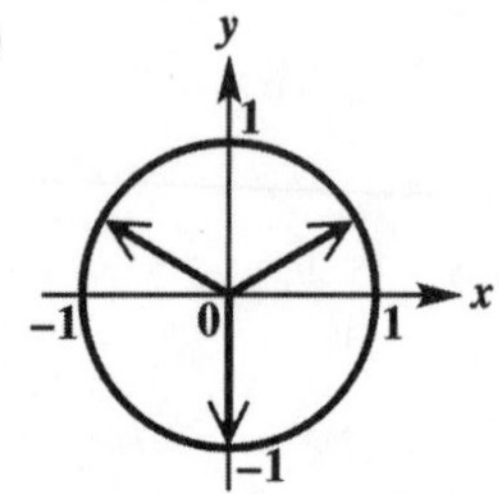

15. (a) Find the cube roots of 8 cis 60°

We have $r = 8$ and $\theta = 60°$.

Since $r^3(\cos 3\alpha + i\sin 3\alpha) = 8(\cos 60° + i\sin 60°)$, then we have the following.

$$r^3 = 1 \Rightarrow r = 1 \text{ and } 3\alpha = 60° + 360° \cdot k \Rightarrow \alpha = \frac{60° + 360° \cdot k}{3} = 20° + 120° \cdot k, \ k \text{ any integer.}$$

If $k = 0$, then $\alpha = 20° + 0° = 20°$. If $k = 1$, then $\alpha = 20° + 120° = 140°$.

If $k = 2$, then $\alpha = 20° + 240° = 260°$.

So, the cube roots are 2 cis 20°, 2 cis 140°, and 2 cis 260°.

(b)

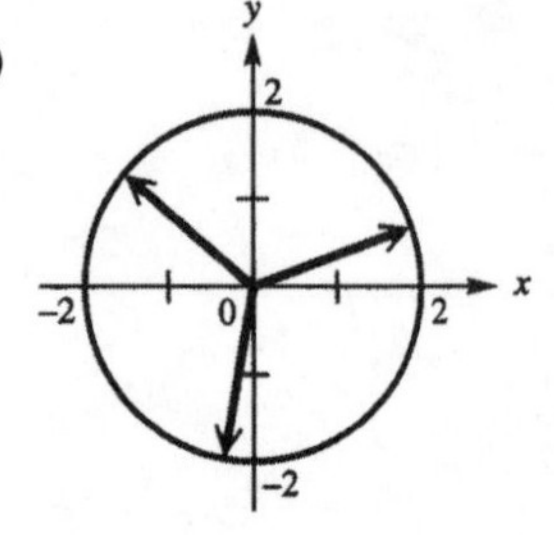

16. (a) Find the cube roots of 27 cis 300°.

We have $r = 27$ and $\theta = 300°$.

Since $r^3(\cos 3\alpha + i\sin 3\alpha) = 27(\cos 300° + i\sin 300°)$, then we have the following.

$$r^3 = 27 \Rightarrow r = 3 \text{ and } 3\alpha = 300° + 360° \cdot k \Rightarrow \alpha = \frac{300° + 360° \cdot k}{3} = 100° + 120° \cdot k, \; k \text{ any integer.}$$

If $k = 0$, then $\alpha = 100° + 0° = 100°$. If $k = 1$, then $\alpha = 100° + 120° = 220°$.

If $k = 2$, then $\alpha = 100° + 240° = 340°$.

So, the cube roots are $3 \text{ cis} 100°, 3 \text{ cis } 220°,$ and $3 \text{ cis } 340°$.

(b)

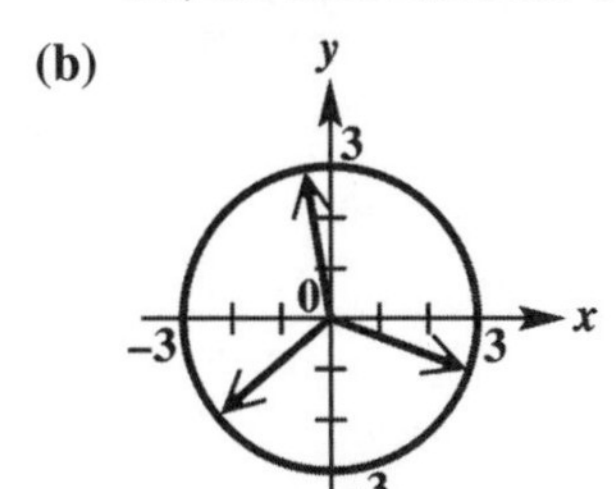

17. (a) Find the cube roots of $-8i = 8(\cos 270° + i \sin 270°)$

We have $r = 8$ and $\theta = 270°$.

Since $r^3(\cos 3\alpha + i\sin 3\alpha) = 8(\cos 270° + i\sin 270°)$, then we have the following.

$$r^3 = 8 \Rightarrow r = 2 \text{ and } 3\alpha = 270° + 360° \cdot k \Rightarrow \alpha = \frac{270° + 360° \cdot k}{3} = 90° + 120° \cdot k, \; k \text{ any integer.}$$

If $k = 0$, then $\alpha = 90° + 0° = 90°$. If $k = 1$, then $\alpha = 90° + 120° = 210°$.

If $k = 2$, then $\alpha = 90° + 240° = 330°$.

So, the cube roots are $2(\cos 90° + i\sin 90°)$, $2(\cos 210° + i\sin 210°)$, and $2(\cos 330° + i\sin 330°)$.

(b)

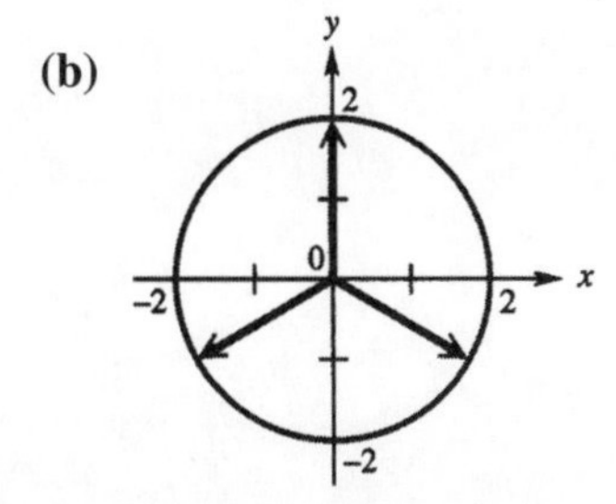

18. (a) Find the cube roots of $27i = 27(\cos 90° + i \sin 90°)$.

We have $r = 27$ and $\theta = 90°$.

Since $r^3(\cos 3\alpha + i\sin 3\alpha) = 27(\cos 90° + i\sin 90°)$, then we have the following.

$$r^3 = 27 \Rightarrow r = 3 \text{ and } 3\alpha = 90° + 360° \cdot k \Rightarrow \alpha = \frac{90° + 360° \cdot k}{3} = 30° + 120° \cdot k, \; k \text{ any integer.}$$

If $k = 0$, then $\alpha = 30° + 0° = 30°$. If $k = 1$, then $\alpha = 30° + 120° = 150°$.

If $k = 2$, then $\alpha = 30° + 240° = 270°$.

So, the cube roots are $3(\cos 30° + i\sin 30°)$, $3(\cos 150° + i\sin 150°)$, and $3(\cos 270° + i\sin 270°)$.

(b)

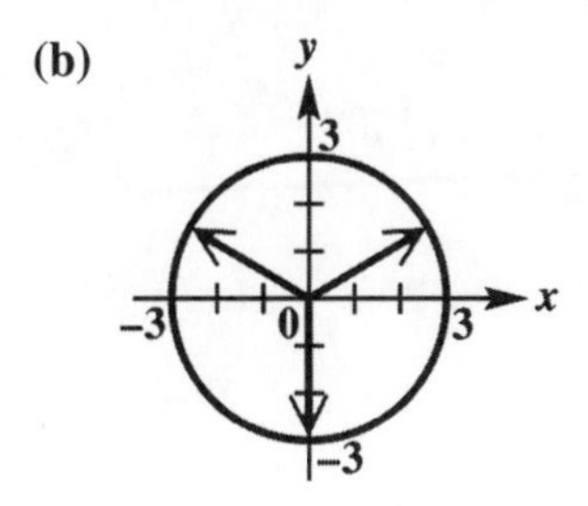

19. (a) Find the cube roots of $-64 = 64(\cos 180° + i \sin 180°)$

We have $r = 64$ and $\theta = 180°$.

Since $r^3(\cos 3\alpha + i\sin 3\alpha) = 64(\cos 180° + i\sin 180°)$, then we have the following.

$$r^3 = 64 \Rightarrow r = 4 \text{ and } 3\alpha = 180° + 360° \cdot k \Rightarrow \alpha = \frac{180° + 360° \cdot k}{3} = 60° + 120° \cdot k, \ k \text{ any integer.}$$

If $k = 0$, then $\alpha = 60° + 0° = 60°$. If $k = 1$, then $\alpha = 60° + 120° = 180°$.

If $k = 2$, then $\alpha = 60° + 240° = 300°$.

So, the cube roots are $4(\cos 60° + i\sin 60°)$, $4(\cos 180° + i\sin 180°)$, and $4(\cos 300° + i\sin 300°)$.

(b)

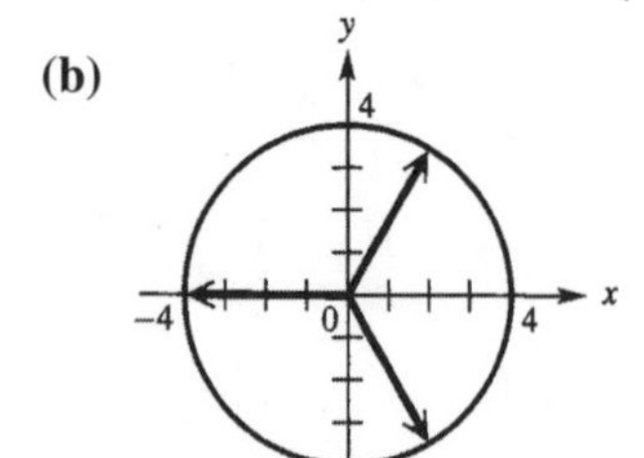

20. (a) Find the cube roots of $27 = 27(\cos 0° + i \sin 0°)$.

We have $r = 27$ and $\theta = 0°$.

Since $r^3(\cos 3\alpha + i\sin 3\alpha) = 27(\cos 0° + i\sin 0°)$, then we have the following.

$$r^3 = 27 \Rightarrow r = 3 \text{ and } 3\alpha = 0° + 360° \cdot k \Rightarrow \alpha = \frac{0° + 360° \cdot k}{3} = 0° + 120° \cdot k = 120° \cdot k, \ k \text{ any integer.}$$

If $k = 0$, then $\alpha = 0°$. If $k = 1$, then $\alpha = 120°$. If $k = 2$, then $\alpha = 240°$.

So, the cube roots are $3(\cos 0° + i\sin 0°)$, $3(\cos 120° + i\sin 120°)$, and $3(\cos 240° + i\sin 240°)$.

(b)

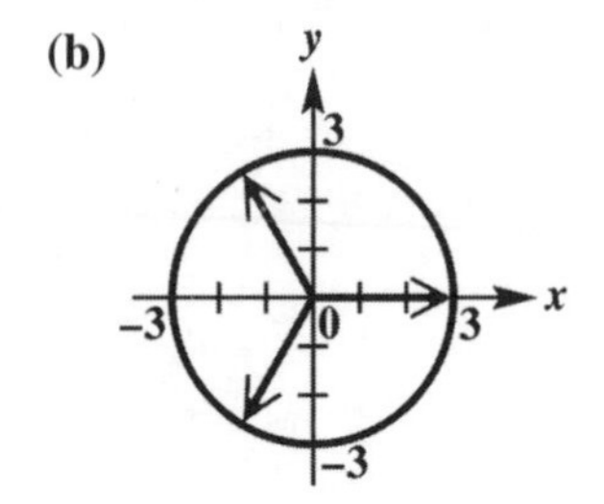

21. (a) Find the cube roots of $1 + i\sqrt{3}$.

We have $r = \sqrt{1^2 + (\sqrt{3})^2} = \sqrt{1+3} = \sqrt{4} = 2$ and $\tan\theta = \frac{\sqrt{3}}{1} = \sqrt{3}$. Since θ is in quadrant I, $\theta = 60°$.

Thus, $1 + i\sqrt{3} = 2\left(\frac{1}{2} + i\frac{\sqrt{3}}{2}\right) = 2(\cos 60° + i\sin 60°)$.

Since $r^3(\cos 3\alpha + i\sin 3\alpha) = 2(\cos 60° + i\sin 60°)$, then we have the following.

$$r^3 = 2 \Rightarrow r = \sqrt[3]{2} \text{ and } 3\alpha = 60° + 360° \cdot k \Rightarrow \alpha = \frac{60° + 360° \cdot k}{3} = 20° + 120° \cdot k, \ k \text{ any integer.}$$

If $k = 0$, then $\alpha = 20° + 0° = 20°$. If $k = 1$, then $\alpha = 20° + 120° = 140°$.

If $k = 2$, then $\alpha = 20° + 240° = 260°$.

So, the cube roots are $\sqrt[3]{2}(\cos 20° + i\sin 20°)$, $\sqrt[3]{2}(\cos 140° + i\sin 140°)$, and $\sqrt[3]{2}(\cos 260° + i\sin 260°)$.

Continued on next page

21. (continued)

(b)

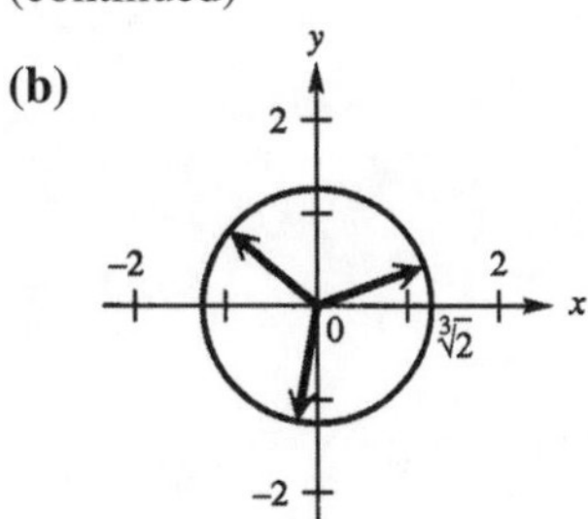

22. (a) Find the cube roots of $2-2i\sqrt{3}$.

We have $r=\sqrt{2^2+\left(-2\sqrt{3}\right)^2}=\sqrt{4+12}=\sqrt{16}=4$ and $\tan\theta=\dfrac{\sqrt{3}}{1}=\sqrt{3}$. Since θ is in quadrant IV, $\theta=300°$.

Thus, $2-2i\sqrt{3}=4\left(\dfrac{1}{2}-i\dfrac{\sqrt{3}}{2}\right)=4\left(\cos 300°+i\sin 300°\right)$.

Since $r^3\left(\cos 3\alpha+i\sin 3\alpha\right)=4\left(\cos 300°+i\sin 300°\right)$, then we have the following.

$r^3=4\Rightarrow r=\sqrt[3]{4}$ and $3\alpha=300°+360°\cdot k\Rightarrow\alpha=\dfrac{300°+360°\cdot k}{3}=100°+120°\cdot k$, k any integer.

If $k=0$, then $\alpha=100°+0°=100°$. If $k=1$, then $\alpha=100°+120°=220°$.

If $k=2$, then $\alpha=100°+240°=340°$.

So, the cube roots are $\sqrt[3]{4}\left(\cos 100°+i\sin 100°\right)$, $\sqrt[3]{4}\left(\cos 220°+i\sin 220°\right)$, and $\sqrt[3]{4}\left(\cos 340°+i\sin 340°\right)$.

(b)

23. (a) Find the cube roots of $-2\sqrt{3}+2i$.

We have $r=\sqrt{\left(-2\sqrt{3}\right)^2+2^2}=\sqrt{12+4}=\sqrt{16}=4$ and $\tan\theta=\dfrac{2}{-2\sqrt{3}}=-\dfrac{\sqrt{3}}{3}$. Since θ is in quadrant II, $\theta=150°$.

Thus, $-2\sqrt{3}+2i=4\left(-\dfrac{\sqrt{3}}{2}+\dfrac{1}{2}i\right)=4\left(\cos 150°+i\sin 150°\right)$.

Since $r^3\left(\cos 3\alpha+i\sin 3\alpha\right)=4\left(\cos 150°+i\sin 150°\right)$, then we have the following.

$r^3=4\Rightarrow r=\sqrt[3]{4}$ and $3\alpha=150°+360°\cdot k\Rightarrow\alpha=\dfrac{150°+360°\cdot k}{3}=50°+120°\cdot k$, k any integer.

If $k=0$, then $\alpha=50°+0°=50°$. If $k=1$, then $\alpha=50°+120°=170°$.

If $k=2$, then $\alpha=50°+240°=290°$.

So, the cube roots are $\sqrt[3]{4}\left(\cos 50°+i\sin 50°\right)$, $\sqrt[3]{4}\left(\cos 170°+i\sin 170°\right)$, and $\sqrt[3]{4}\left(\cos 290°+i\sin 290°\right)$.

Continued on next page

23. (continued)

(b)

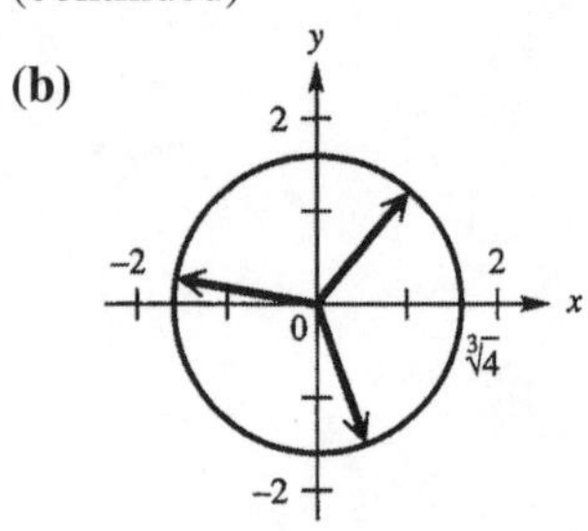

24. **(a)** Find the cube roots of $\sqrt{3}-i$.

We have $r=\sqrt{\left(-\sqrt{3}\right)^2+(-1)^2}=\sqrt{3+1}=\sqrt{4}=2$ and $\tan\theta=-\frac{\sqrt{3}}{3}$. Since θ is in quadrant IV, $\theta=330°$.

Thus, $\sqrt{3}-i=2\left(\frac{\sqrt{3}}{2}-\frac{1}{2}i\right)=2(\cos 330°+i\sin 330°)$.

Since $r^3(\cos 3\alpha+i\sin 3\alpha)=2(\cos 330°+i\sin 330°)$, then we have the following.

$$r^3=2\Rightarrow r=\sqrt[3]{2} \text{ and } 3\alpha=330°+360°\cdot k\Rightarrow \alpha=\frac{330°+360°\cdot k}{3}=110°+120°\cdot k,\ k \text{ any integer.}$$

If $k=0$, then $\alpha=110°+0°=110°$. If $k=1$, then $\alpha=110°+120°=230°$.
If $k=2$, then $\alpha=110°+240°=350°$.

So, the cube roots are $\sqrt[3]{2}(\cos 110°+i\sin 110°)$, $\sqrt[3]{2}(\cos 230°+i\sin 230°)$, and $\sqrt[3]{2}(\cos 350°+i\sin 350°)$.

(b)

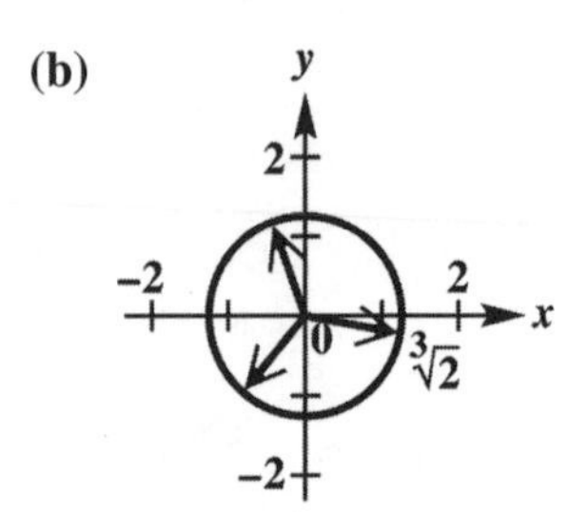

25. Find all the second (or square) roots of $1=1(\cos 0°+i\sin 0°)$.

Since $r^2(\cos 2\alpha+i\sin 2\alpha)=1(\cos 0°+i\sin 0°)$, then we have the following.

$$r^2=1\Rightarrow r=1 \text{ and } 2\alpha=0°+360°\cdot k\Rightarrow \alpha=\frac{0°+360°\cdot k}{2}=0°+180°\cdot k=180°\cdot k,\ k \text{ any integer.}$$

If $k=0$, then $\alpha=0°$. If $k=1$, then $\alpha=180°$.

So, the second roots of 1 are $\cos 0°+i\sin 0°$, and $\cos 180°+i\sin 180°$. (or 1 and −1)

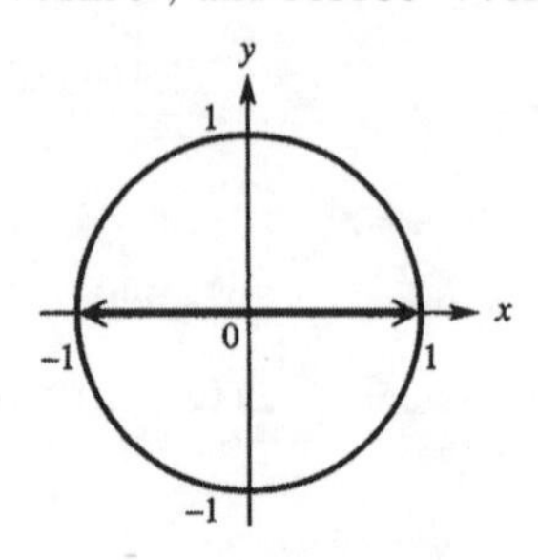

26. Find all the fourth roots of $1 = 1(\cos 0° + i\sin 0°)$.

Since $r^4(\cos 4\alpha + i\sin 4\alpha) = 1(\cos 0° + i\sin 0°)$, then we have the following.

$$r^4 = 1 \Rightarrow r = 1 \text{ and } 4\alpha = 0° + 360°\cdot k \Rightarrow \alpha = \frac{0° + 360°\cdot k}{4} = 0° + 90°\cdot k = 90°\cdot k,\ k \text{ any integer.}$$

If $k = 0$, then $\alpha = 0°$. If $k = 1$, then $\alpha = 90°$. If $k = 2$, then $\alpha = 180°$. If $k = 3$, then $\alpha = 270°$.

So, the fourth roots of 1 are as follows.

$\cos 0° + i\sin 0°$, $\cos 90° + i\sin 90°$, $\cos 180° + i\sin 180°$, and $\cos 270° + i\sin 270°$ (or 1, i, –1 and $-i$)

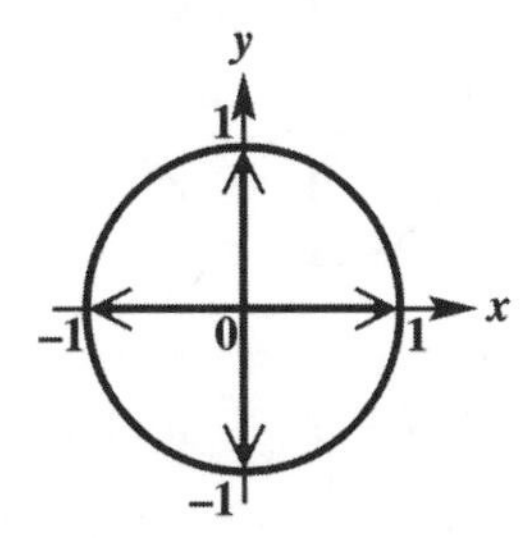

27. Find all the sixth roots of $1 = 1(\cos 0° + i\sin 0°)$.

Since $r^6(\cos 6\alpha + i\sin 6\alpha) = 1(\cos 0° + i\sin 0°)$, then we have the following.

$$r^6 = 1 \Rightarrow r = 1 \text{ and } 6\alpha = 0° + 360°\cdot k \Rightarrow \alpha = \frac{0° + 360°\cdot k}{6} = 0° + 60°\cdot k = 60°\cdot k,\ k \text{ any integer.}$$

If $k = 0$, then $\alpha = 0°$. If $k = 1$, then $\alpha = 60°$. If $k = 2$, then $\alpha = 120°$.
If $k = 3$, then $\alpha = 180°$. If $k = 4$, then $\alpha = 240°$. If $k = 5$, then $\alpha = 300°$.

So, the sixth roots of 1 are $\cos 0° + i\sin 0°$, $\cos 60° + i\sin 60°$, $\cos 120° + i\sin 120°$, $\cos 180° + i\sin 180°$, $\cos 240° + i\sin 240°$, and $\cos 300° + i\sin 300°$. $\left(\text{or } 1,\ \frac{1}{2} + \frac{\sqrt{3}}{2}i,\ -\frac{1}{2} + \frac{\sqrt{3}}{2}i,\ -1,\ -\frac{1}{2} - \frac{\sqrt{3}}{2}i, \text{ and } \frac{1}{2} - \frac{\sqrt{3}}{2}i\right)$

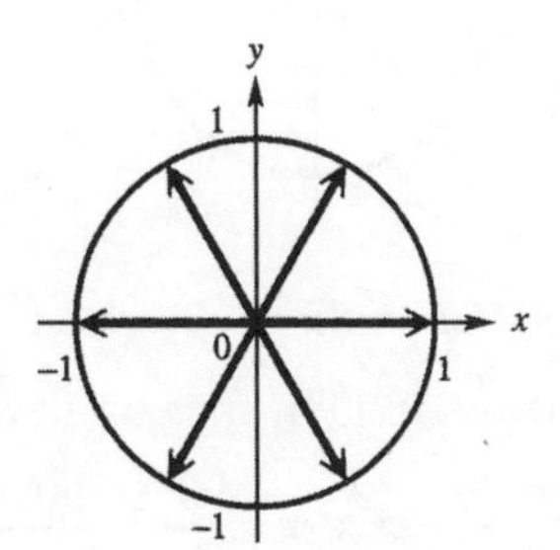

28. Find all the eighth roots of $1 = 1(\cos 0° + i\sin 0°)$.

Since $r^8(\cos 8\alpha + i\sin 8\alpha) = 1(\cos 0° + i\sin 0°)$, then we have the following.

$$r^8 = 1 \Rightarrow r = 1 \text{ and } 8\alpha = 0° + 360°\cdot k \Rightarrow \alpha = \frac{0° + 360°\cdot k}{8} = 0° + 45°\cdot k = 45°\cdot k,\ k \text{ any integer.}$$

If $k = 0$, then $\alpha = 0°$. If $k = 1$, then $\alpha = 45°$. If $k = 2$, then $\alpha = 90°$. If $k = 3$, then $\alpha = 135°$.
If $k = 4$, then $\alpha = 180°$. If $k = 5$, then $\alpha = 225°$. If $k = 6$, then $\alpha = 270°$. If $k = 7$, then $\alpha = 315°$.

Continued on next page

28. (continued)

So, the sixth roots of 1 are $\cos 0° + i\sin 0°$, $\cos 45° + i\sin 45°$, $\cos 90° + i\sin 90°$, $\cos 135° + i\sin 135°$, $\cos 180° + i\sin 180°$, $\cos 225° + i\sin 225°$, $\cos 270° + i\sin 270°$, and $\cos 315° + i\sin 315°$.

$$\left(\text{or } 1, \frac{\sqrt{2}}{2}+\frac{\sqrt{2}}{2}i, i, -\frac{\sqrt{2}}{2}+\frac{\sqrt{2}}{2}, -1, -\frac{\sqrt{2}}{2}-\frac{\sqrt{2}}{2}i, -i, \text{ and } \frac{\sqrt{2}}{2}-\frac{\sqrt{2}}{2}i\right)$$

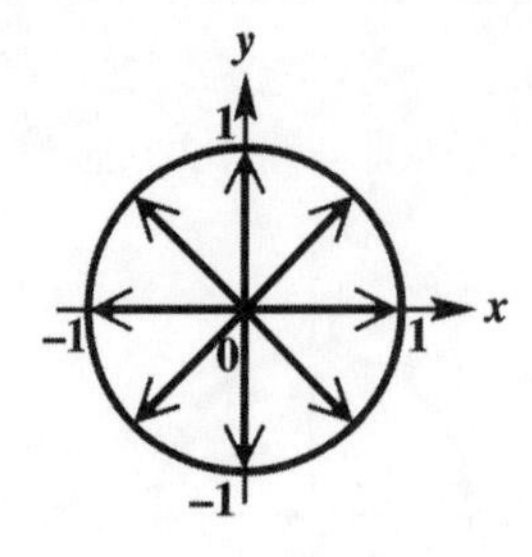

29. Find all the second (square) roots of $i = 1(\cos 90° + i\sin 90°)$.

Since $r^2(\cos 2\alpha + i\sin 2\alpha) = 1(\cos 90° + i\sin 90°)$, then we have the following.

$$r^2 = 1 \Rightarrow r = 1 \text{ and } 2\alpha = 90° + 360° \cdot k \Rightarrow \alpha = \frac{90° + 360° \cdot k}{2} = 45° + 180° \cdot k, \ k \text{ any integer.}$$

If $k = 0$, then $\alpha = 45° + 0° = 45°$. If $k = 1$, then $\alpha = 45° + 180° = 225°$.

So, the second roots of i are $\cos 45° + i\sin 45°$, and $\cos 225° + i\sin 225°$.

$$\left(\text{or } \frac{\sqrt{2}}{2}+\frac{\sqrt{2}}{2}i \text{ and } -\frac{\sqrt{2}}{2}-\frac{\sqrt{2}}{2}i\right)$$

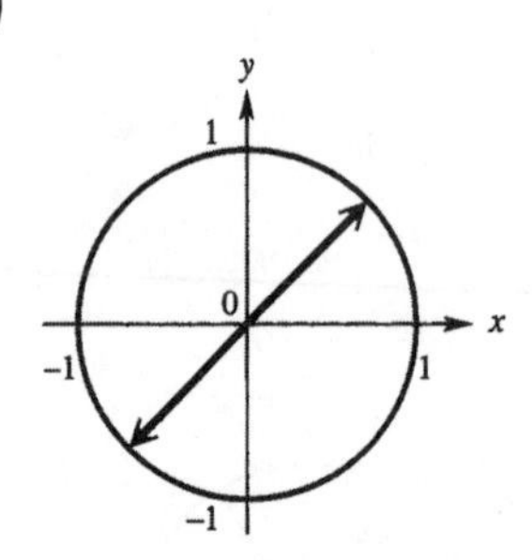

30. Find all the fourth roots of $i = 1(\cos 90° + i\sin 90°)$.

Since $r^4(\cos 4\alpha + i\sin 4\alpha) = 1(\cos 90° + i\sin 90°)$, then we have the following.

$$r^4 = 1 \Rightarrow r = 1 \text{ and } 4\alpha = 90° + 360° \cdot k \Rightarrow \alpha = \frac{90° + 360° \cdot k}{4} = 22.5° + 90° \cdot k, \ k \text{ any integer.}$$

If $k = 0$, then $\alpha = 22.5° + 0° = 22.5°$. If $k = 1$, then $\alpha = 22.5° + 90° = 112.5°$.

If $k = 2$, then $\alpha = 22.5° + 180° = 202.5°$. If $k = 3$, then $\alpha = 22.5° + 270° = 292.5°$.

So, the fourth roots of i are $\cos 22.5° + i\sin 22.5°$, $\cos 112.5° + i\sin 112.5°$, $\cos 202.5° + i\sin 202.5°$, and $\cos 292.5° + i\sin 292.5°$.

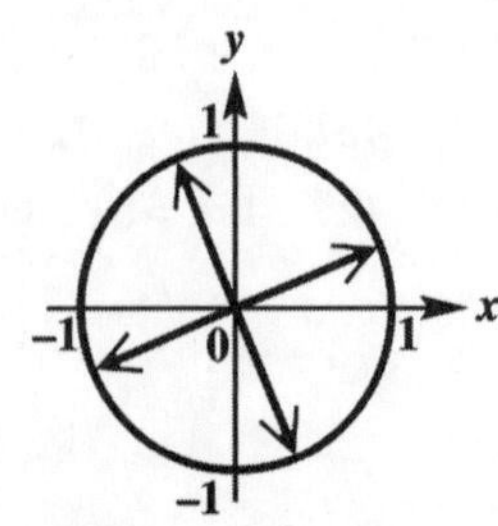

31. $x^3 - 1 = 0 \Rightarrow x^3 = 1$
We have $r = 1$ and $\theta = 0°$.

$$x^3 = 1 = 1 + 0i = 1(\cos 0° + i\sin 0°)$$

Since $r^3(\cos 3\alpha + i\sin 3\alpha) = 1(\cos 0° + i\sin 0°)$, then we have the following.

$$r^3 = 1 \Rightarrow r = 1 \text{ and } 3\alpha = 0° + 360°\cdot k \Rightarrow \alpha = \frac{0° + 360°\cdot k}{3} = 0° + 120°\cdot k = 120°\cdot k,\ k \text{ any integer.}$$

If $k = 0$, then $\alpha = 0°$. If $k = 1$, then $\alpha = 120°$. If $k = 2$, then $\alpha = 240°$.

Solution set: $\{\cos 0° + i\sin 0°,\ \cos 120° + i\sin 120°,\ \cos 240° + i\sin 240°\}$ or

$$\left\{1,\ -\frac{1}{2} + \frac{\sqrt{3}}{2}i,\ -\frac{1}{2} - \frac{\sqrt{3}}{2}i\right\}$$

32. $x^3 + 1 = 0 \Rightarrow x^3 = -1$
We have $r = 1$ and $\theta = 180°$.

$$x^3 = -1 = -1 + 0i = 1(\cos 180° + i\sin 180°)$$

Since $r^3(\cos 3\alpha + i\sin 3\alpha) = 1(\cos 180° + i\sin 180°)$, then we have the following.

$$r^3 = 1 \Rightarrow r = 1 \text{ and } 3\alpha = 180° + 360°\cdot k \Rightarrow \alpha = \frac{180° + 360°\cdot k}{3} = 60° + 120°\cdot k,\ k \text{ any integer.}$$

If $k = 0$, then $\alpha = 60° + 0° = 60°$. If $k = 1$, then $\alpha = 60° + 120° = 180°$.
If $k = 2$, then $\alpha = 60° + 240° = 300°$.

Solution set: $\{\cos 60° + i\sin 60°,\ \cos 180° + i\sin 180°,\ \cos 300° + i\sin 300°\}$ or

$$\left\{\frac{1}{2} + \frac{\sqrt{3}}{2}i,\ -1,\ \frac{1}{2} - \frac{\sqrt{3}}{2}i\right\}$$

33. $x^3 + i = 0 \Rightarrow x^3 = -i$
We have $r = 1$ and $\theta = 270°$.

$$x^3 = -i = 0 - i = 1(\cos 270° + i\sin 270°)$$

Since $r^3(\cos 3\alpha + i\sin 3\alpha) = 1(\cos 270° + i\sin 270°)$, then we have the following.

$$r^3 = 1 \Rightarrow r = 1 \text{ and } 3\alpha = 270° + 360°\cdot k \Rightarrow \alpha = \frac{270° + 360°\cdot k}{3} = 90° + 120°\cdot k,\ k \text{ any integer.}$$

If $k = 0$, then $\alpha = 90° + 0° = 90°$. If $k = 1$, then $\alpha = 90° + 120° = 210°$.
If $k = 2$, then $\alpha = 90° + 240° = 330°$.

Solution set: $\{\cos 90° + i\sin 90°,\ \cos 210° + i\sin 210°,\ \cos 330° + i\sin 330°\}$ or

$$\left\{0,\ -\frac{\sqrt{3}}{2} - \frac{1}{2}i,\ \frac{\sqrt{3}}{2} - \frac{1}{2}i\right\}$$

34. $x^4 + i = 0 \Rightarrow x^4 = -i$
We have $r = 1$ and $\theta = 270°$.

$$x^4 = -i = 0 - i = 1(\cos 270° + i\sin 270°)$$

Since $r^4(\cos 4\alpha + i\sin 4\alpha) = 1(\cos 270° + i\sin 270°)$, then we have the following.

$$r^4 = 1 \Rightarrow r = 1 \text{ and } 4\alpha = 270° + 360°\cdot k \Rightarrow \alpha = \frac{270° + 360°\cdot k}{4} = 67.5° + 90°\cdot k,\ k \text{ any integer.}$$

If $k = 0$, then $\alpha = 67.5° + 0° = 67.5°$. If $k = 1$, then $\alpha = 67.5° + 90° = 157.5°$.
If $k = 2$, then $\alpha = 67.5° + 180° = 247.5°$. If $k = 3$, then $\alpha = 67.5° + 270° = 337.5°$.

Solution set: $\{\cos 67.5° + i\sin 67.5°,\ \cos 157.5° + i\sin 157.5°,\ \cos 247.5° + i\sin 247.5°,\ \cos 337.5° + i\sin 337.5°\}$

35. $x^3 - 8 = 0 \Rightarrow x^3 = 8$

We have $r = 8$ and $\theta = 0°$.

$$x^3 = 8 = 8 + 0i = 8(\cos 0° + i \sin 0°)$$

Since $r^3(\cos 3\alpha + i \sin 3\alpha) = 8(\cos 0° + i \sin 0°)$, then we have the following.

$$r^3 = 8 \Rightarrow r = 2 \text{ and } 3\alpha = 0° + 360° \cdot k \Rightarrow \alpha = \frac{0° + 360° \cdot k}{3} = 0° + 120° \cdot k = 120° \cdot k, \ k \text{ any integer.}$$

If $k = 0$, then $\alpha = 0°$. If $k = 1$, then $\alpha = 120°$. If $k = 2$, then $\alpha = 240°$.

Solution set: $\{2(\cos 0° + i \sin 0°),\ 2(\cos 120° + i \sin 120°),\ 2(\cos 240° + i \sin 240°)\}$ or

$\{2,\ -1 + \sqrt{3}i,\ -1 - \sqrt{3}i\}$

36. $x^3 + 27 = 0 \Rightarrow x^3 = -27$

We have $r = 27$ and $\theta = 180°$.

$$x^3 = -27 = -27 + 0i = 27(\cos 180° + i \sin 180°)$$

Since $r^3(\cos 3\alpha + i \sin 3\alpha) = 27(\cos 180° + i \sin 180°)$, then we have the following.

$$r^3 = 27 \Rightarrow r = 3 \text{ and } 3\alpha = 180° + 360° \cdot k \Rightarrow \alpha = \frac{180° + 360° \cdot k}{3} = 60° + 120° \cdot k, \ k \text{ any integer.}$$

If $k = 0$, then $\alpha = 60° + 0° = 60°$. If $k = 1$, then $\alpha = 60° + 120° = 180°$.

If $k = 2$, then $\alpha = 60° + 240° = 300°$.

Solution set: $\{3(\cos 60° + i \sin 60°),\ 3(\cos 180° + i \sin 180°),\ 3(\cos 300° + i \sin 300°)\}$ or

$$\left\{\frac{3}{2} + \frac{3\sqrt{3}}{2}i,\ -3,\ \frac{3}{2} - \frac{3\sqrt{3}}{2}i\right\}$$

37. $x^4 + 1 = 0 \Rightarrow x^4 = -1$

We have $r = 1$ and $\theta = 180°$.

$$x^4 = -1 = -1 + 0i = 1(\cos 180° + i \sin 180°)$$

Since $r^4(\cos 4\alpha + i \sin 4\alpha) = 1(\cos 180° + i \sin 180°)$, then we have the following.

$$r^4 = 1 \Rightarrow r = 1 \text{ and } 4\alpha = 180° + 360° \cdot k \Rightarrow \alpha = \frac{180° + 360° \cdot k}{4} = 45° + 90° \cdot k, \ k \text{ any integer.}$$

If $k = 0$, then $\alpha = 45° + 0° = 45°$. If $k = 1$, then $\alpha = 45° + 90° = 135°$.

If $k = 2$, then $\alpha = 45° + 180° = 225°$. If $k = 3$, then $\alpha = 45° + 270° = 315°$.

Solution set: $\{\cos 45° + i \sin 45°,\ \cos 135° + i \sin 135°,\ \cos 225° + i \sin 225°,\ \cos 315° + i \sin 315°\}$ or

$$\left\{\frac{\sqrt{2}}{2} + \frac{\sqrt{2}}{2}i,\ -\frac{\sqrt{2}}{2} + \frac{\sqrt{2}}{2}i,\ -\frac{\sqrt{2}}{2} - \frac{\sqrt{2}}{2}i,\ \frac{\sqrt{2}}{2} - \frac{\sqrt{2}}{2}i\right\}$$

38. $x^4+16=0 \Rightarrow x^4=-16$

We have $r=16$ and $\theta=180°$.

$$x^4=-16=-16+0i=16(\cos 180°+i\sin 180°)$$

Since $r^4(\cos 4\alpha+i\sin 4\alpha)=16(\cos 180°+i\sin 180°)$, then we have the following.

$$r^4=16 \Rightarrow r=2 \text{ and } 4\alpha=180°+360°\cdot k \Rightarrow \alpha=\frac{180°+360°\cdot k}{4}=45°+90°\cdot k,\ k \text{ any integer.}$$

If $k=0$, then $\alpha=45°+0°=45°$. If $k=1$, then $\alpha=45°+90°=135°$.

If $k=2$, then $\alpha=45°+180°=225°$. If $k=3$, then $\alpha=45°+270°=315°$.

Solution set:

$\{2(\cos 45°+i\sin 45°),\ 2(\cos 135°+i\sin 135°),\ 2(\cos 225°+i\sin 225°),\ 2(\cos 315°+i\sin 315°)\}$ or

$\{\sqrt{2}+i\sqrt{2},\ -\sqrt{2}+i\sqrt{2},\ -\sqrt{2}-i\sqrt{2},\ \sqrt{2}-i\sqrt{2}\}$

39. $x^4-i=0 \Rightarrow x^4=i$

We have $r=1$ and $\theta=90°$.

$$x^4=i=0+i=1(\cos 90°+i\sin 90°)$$

Since $r^4(\cos 4\alpha+i\sin 4\alpha)=1(\cos 90°+i\sin 90°)$, then we have the following.

$$r^4=1 \Rightarrow r=1 \text{ and } 4\alpha=90°+360°\cdot k \Rightarrow \alpha=\frac{90°+360°\cdot k}{4}=22.5°+90°\cdot k,\ k \text{ any integer.}$$

If $k=0$, then $\alpha=22.5°+0°=22.5°$. If $k=1$, then $\alpha=22.5°+90°=112.5°$.

If $k=2$, then $\alpha=22.5°+180°=202.5°$. If $k=3$, then $\alpha=22.5°+270°=292.5°$.

Solution set: $\{\cos 22.5°+i\sin 22.5°,\ \cos 112.5°+i\sin 112.5°,\ \cos 202.5°+i\sin 202.5°,\ \cos 292.5°+i\sin 292.5°\}$

40. $x^5-i=0 \Rightarrow x^5=i$

We have $r=1$ and $\theta=90°$.

$$x^5=i=0+i=1(\cos 90°+i\sin 90°)$$

Since $r^5(\cos 5\alpha+i\sin 5\alpha)=1(\cos 90°+i\sin 90°)$, then we have the following.

$$r^5=1 \Rightarrow r=1 \text{ and } 5\alpha=90°+360°\cdot k \Rightarrow \alpha=\frac{90°+360°\cdot k}{5}=18°+72°\cdot k,\ k \text{ any integer.}$$

If $k=0$, then $\alpha=18°+0°=18°$. If $k=1$, then $\alpha=18°+72°=90°$.

If $k=2$, then $\alpha=18°+144°=162°$. If $k=3$, then $\alpha=18°+216°=234°$.

If $k=4$, then $\alpha=18°+288°=306°$.

Solution set: $\{\cos 18°+i\sin 18°,\ \cos 90°+i\sin 90° \text{ (or 0)},$

$\cos 162°+i\sin 162°,\ \cos 234°+i\sin 234°,\ \cos 306°+i\sin 306°\}$

41. $x^3 - (4 + 4i\sqrt{3}) = 0 \Rightarrow x^3 = 4 + 4i\sqrt{3}$

We have $r = \sqrt{4^2 + (4\sqrt{3})^2} = \sqrt{16+48} = \sqrt{64} = 8$ and $\tan\theta = \frac{4\sqrt{3}}{4} = \sqrt{3}$. Since θ is in quadrant I, $\theta = 60°$.

$$x^3 = 4 + 4i\sqrt{3} = 8\left(\frac{1}{2} + i\frac{\sqrt{3}}{2}\right) = 8(\cos 60° + i\sin 60°)$$

Since $r^3(\cos 3\alpha + i\sin 3\alpha) = 8(\cos 60° + i\sin 60°)$, then we have the following.

$$r^3 = 8 \Rightarrow r = 2 \text{ and } 3\alpha = 60° + 360°\cdot k \Rightarrow \alpha = \frac{60° + 360°\cdot k}{3} = 20° + 120°\cdot k,\ k \text{ any integer.}$$

If $k = 0$, then $\alpha = 20° + 0° = 20°$. If $k = 1$, then $\alpha = 20° + 120° = 140°$.
If $k = 2$, then $\alpha = 20° + 240° = 260°$.

Solution set: $\{2(\cos 20° + i\sin 20°), 2(\cos 140° + i\sin 140°), 2(\cos 260° + i\sin 260°)\}$

42. $x^4 - (8 + 8i\sqrt{3}) = 0 \Rightarrow x^4 = 8 + 8i\sqrt{3}$

We have $r = \sqrt{8^2 + (8\sqrt{3})^2} = \sqrt{64+192} = \sqrt{256} = 16$ and $\tan\theta = \frac{8\sqrt{3}}{8} = \sqrt{3}$. Since θ is in quadrant I, $\theta = 60°$.

$$x^4 = 8 + 8i\sqrt{3} = 16\left(\frac{1}{2} + i\frac{\sqrt{3}}{2}\right) = 16(\cos 60° + i\sin 60°)$$

Since $r^4(\cos 4\alpha + i\sin 4\alpha) = 16(\cos 60° + i\sin 60°)$, then we have the following.

$$r^4 = 16 \Rightarrow r = 2 \text{ and } 4\alpha = 60° + 360°\cdot k \Rightarrow \alpha = \frac{60° + 360°\cdot k}{4} = 15° + 90°\cdot k,\ k \text{ any integer.}$$

If $k = 0$, then $\alpha = 15° + 0° = 15°$. If $k = 1$, then $\alpha = 15° + 90° = 105°$.
If $k = 2$, then $\alpha = 15° + 180° = 195°$. If $k = 3$, then $\alpha = 15° + 270° = 285°$.

Solution set:

$$\{2(\cos 15° + i\sin 15°),\ 2(\cos 105° + i\sin 105°),\ 2(\cos 195° + i\sin 195°),\ 2(\cos 285° + i\sin 285°)\}$$

43. $x^3 - 1 = 0 \Rightarrow (x-1)(x^2 + x + 1) = 0$

Setting each factor equal to zero, we have the following.

$$x - 1 = 0 \Rightarrow x = 1 \text{ and } x^2 + x + 1 = 0 \Rightarrow x = \frac{-1 \pm \sqrt{1^2 - 4\cdot 1\cdot 1}}{2\cdot 1} = \frac{-1 \pm \sqrt{-3}}{2} = \frac{-1 \pm i\sqrt{3}}{2} = -\frac{1}{2} \pm \frac{\sqrt{3}}{2}i$$

Thus, $x = 1, -\frac{1}{2} + \frac{\sqrt{3}}{2}i, -\frac{1}{2} - \frac{\sqrt{3}}{2}i$. We see that the solutions are the same as Exercise 31.

44. $x^3 + 27 = 0 \Rightarrow (x+3)(x^2 - 3x + 9) = 0$

Setting each factor equal to zero, we have the following.

$$x + 3 = 0 \Rightarrow x = -3 \text{ and } x^2 - 3x + 9 = 0 \Rightarrow x = \frac{-(-3) \pm \sqrt{(-3)^2 - 4\cdot 1\cdot 9}}{2\cdot 1} = \frac{3 \pm \sqrt{-27}}{2} = \frac{3 \pm 3i\sqrt{3}}{2} = \frac{3}{2} \pm \frac{3\sqrt{3}}{2}i$$

Thus, $x = -3, \frac{3}{2} + \frac{3\sqrt{3}}{2}i, \frac{3}{2} - \frac{3\sqrt{3}}{2}i$. We see that the solutions are the same as Exercise 36.

45. $(\cos\theta + i\sin\theta)^2 = 1^2(\cos 2\theta + i\sin 2\theta) = \cos 2\theta + i\sin 2\theta$

46. $(\cos\theta + i\sin\theta)^2 = \cos^2\theta + 2i\sin\theta\cos\theta + i^2\sin^2\theta = \cos^2\theta + 2i\sin\theta\cos\theta + (-1)\sin^2\theta$

$= \cos^2\theta + 2i\sin\theta\cos\theta - \sin^2\theta = (\cos^2\theta - \sin^2\theta) + i(2\sin\theta\cos\theta) = \cos 2\theta + i\sin 2\theta$

47. Two complex numbers $a + bi$ and $c + di$ are equal only if $a = c$ and $b = d$.
Thus, $a = c$ implies $\cos^2\theta - \sin^2\theta = \cos 2\theta$.

48. Two complex numbers $a + bi$ and $c + di$ are equal only if $a = c$ and $b = d$.
Thus, $b = d$ implies $2\sin\theta\cos\theta = \sin 2\theta$.

49. **(a)** If $z = 0 + 0i$, then $z = 0$, $0^2 + 0 = 0$, $0^2 + 0 = 0$, and so on. The calculations repeat as 0, 0, 0, . . ., and will never exceed a modulus of 2. The point $(0,0)$ is part of the Mandelbrot set. The pixel at the origin should be turned on.

(b) If $z = 1 - 1i$, then $(1-i)^2 + (1-i) = 1 - 3i$. The modulus of $1 - 3i$ is $\sqrt{1^2 + (-3)^2} = \sqrt{1+9} = \sqrt{10}$, which is greater than 2. Therefore, $1 - 1i$ is not part of the Mandelbrot set, and the pixel at $(1,-1)$ should be left off.

(c) If $z = -.5i$, $(-.5i)^2 - .5i = -.25 - .5i$; $(-.25 - .5i)^2 + (-.25 - .5i) = -.4375 - .25i$;
$(-.4375 - .25i)^2 + (-.4375 - .25i) = -.308593 - .03125i$;
$(-.308593 - .03125i)^2 + (-.308593 - .03125i) = -.214339 - .0119629i$;
$(-.214339 - .0119629i)^2 + (-.214339 - .0119629i) = -.16854 - .00683466i$.
This sequence appears to be approaching the origin, and no number has a modulus greater than 2. Thus, $-.5i$ is part of the Mandelbrot set, and the pixel at $(0,-.5)$ should be turned on.

50. **(a)** Let $f(z) = \dfrac{2z^3+1}{3z^2}$ and $z_1 = i$. Then, $z_2 = f(z_1) = \dfrac{2i^3+1}{3i^2} = \dfrac{2(-i)+1}{3(-1)} = \dfrac{-2i+1}{-3} = -\dfrac{1}{3} + \dfrac{2}{3}i$.

Similarly, $z_3 = f(z_2) = \dfrac{2z_2^{\ 3}+1}{3z_2^{\ 2}} \approx -.58222 + .92444i$ and $z_4 = f(z_3) = \dfrac{2z_3+1}{3z_3^{\ 2}} \approx -.50879 + .868165i$.

The values of z seem to approach $w_2 = -\dfrac{1}{2} + \dfrac{\sqrt{3}}{2}i$. Color the pixel at $(0,1)$ blue.

(b) Let $f(z) = \dfrac{2z^3+1}{3z^2}$ and $z_1 = 2 + i$. Then, $z_2 = f(z_1) = \dfrac{2(2+i)^3+1}{3(2+i)^2} \approx 1.37333 + .61333i$.

Similarly, $z_3 = f(z_2) = \dfrac{2z_2^{\ 3}+1}{3z_2^{\ 2}} \approx 1.01389 + .299161i$, $z_4 = f(z_3) = \dfrac{2z_3^{\ 3}+1}{3z_3^{\ 2}} \approx .926439 + .0375086i$,

and $z_5 = f(z_4) = \dfrac{2z_4^{\ 3}+1}{3z_4^{\ 2}} \approx 1.00409 - .00633912i$. The values of z seem to approach $w_1 = 1$.
Color the pixel at $(2,1)$ red.

(c) Let $f(z) = \dfrac{2z^3+1}{3z^2}$ and $z_1 = -1 - i$. Then, $z_2 = f(z_1) = \dfrac{2(-1-i)^3+1}{3(-1-i)^2} = -\dfrac{2}{3} - \dfrac{5}{6}i$. Similarly,

$z_3 = f(z_2) = \dfrac{2z_2^{\ 3}+1}{3z_2^{\ 2}} \approx .508691 - .841099i$ and $z_4 = f(z_3) = \dfrac{2z_3^{\ 3}+1}{3z_3^{\ 2}} \approx -.499330 - .866269i$. The

values of z seem to approach $w_3 = -\dfrac{1}{2} - \dfrac{\sqrt{3}}{2}i$. Color the pixel at $(-1, -1)$ yellow.

51. Using the trace function, we find that the other four fifth roots of 1 are: .30901699 + .95105652i, –.809017 + .58778525i, –.809017 – .5877853i, .30901699 – .9510565i.

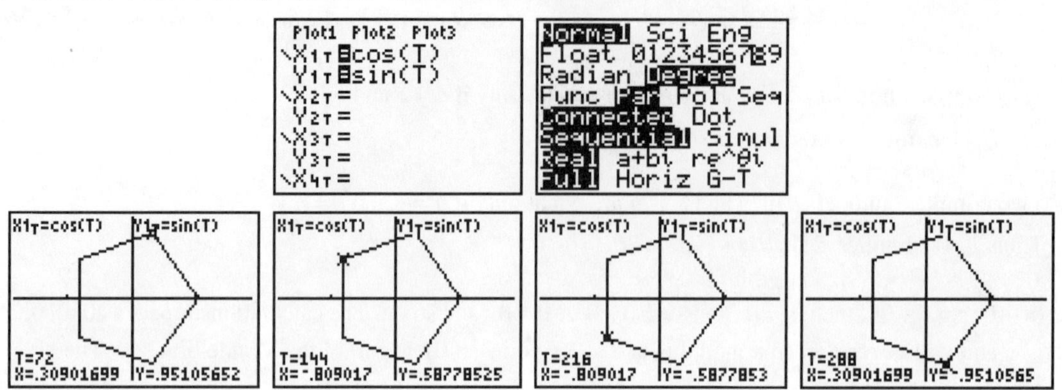

52. Using the trace function, we find that three of the tenth roots of 1 are: 1, .80901699 + .58778525i, .30901699 + .95105652i.

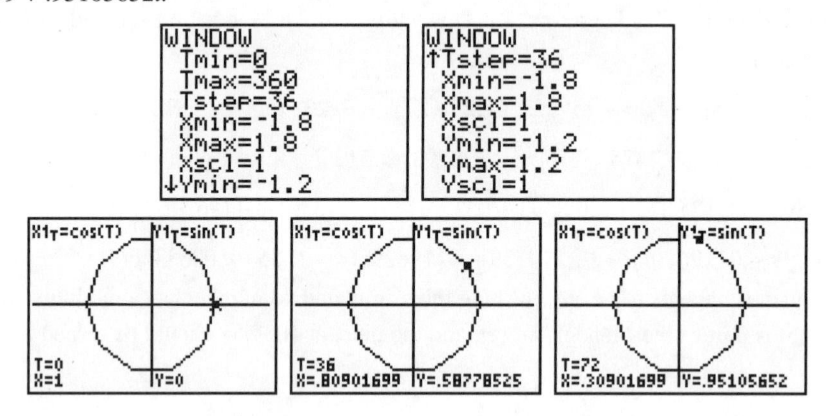

53. $2+2i\sqrt{3}$ is one cube root of a complex number.

$$r=\sqrt{2^2+\left(2\sqrt{3}\right)^2}=\sqrt{4+12}=4 \text{ and } \tan\theta=\frac{2\sqrt{3}}{2}=\sqrt{3}$$

Because x and y are both positive, θ is in quadrant I, so $\theta=60°$.

$$2+2i\sqrt{3}=4(\cos 60°+i\sin 60°)$$

$$\left(2+2i\sqrt{3}\right)^3=\left[4(\cos 60°+i\sin 60°)\right]^3=4^3\left[\cos(3\cdot 60°)+i\sin(3\cdot 60°)\right]$$
$$=64(\cos 180°+i\sin 180°)=64(-1+0\cdot i)=-64+0\cdot i$$

Since the graphs of the other roots of – 64 must be equally spaced around a circle and the graphs of these roots are all on a circle that has center at the origin and radius 4, the other roots are as follows.

$$4\left[\cos(60°+120°)+i\sin(60°+120°)\right]=4(\cos 180°+i\sin 180°)=4(-1+i\cdot 0)=-4$$

$$4\left[\cos(60°+240°)+i\sin(60°+240°)\right]=4(\cos 300°+i\sin 300°)=4\left(\frac{1}{2}-\frac{\sqrt{3}}{2}i\right)=2-2i\sqrt{3}$$

54. $x^3+4-5i=0 \Rightarrow x^3=-4+5i$

$$r=\sqrt{(-4)^2+5^2}=\sqrt{16+25}=\sqrt{41} \Rightarrow r^{1/n}=r^{1/3}=\left(\sqrt{41}\right)^{1/3}\approx 1.85694 \text{ and } \tan\theta=-\frac{5}{4}=-1.25$$

Since θ is in quadrant II, $\theta \approx 128.6598°$ and $\alpha=\dfrac{128.6598°+360°\cdot k}{3}\approx 42.887°+120°\cdot k,$ where k is an integer.

$$x \approx 1.85694(\cos 42.887°+i\sin 42.887°), 1.85694(\cos 162.887°+i\sin 162.887°),$$
$$1.85694(\cos 282.887°+i\sin 282.887°)$$

Solution set: $\{1.3606+1.2637i, -1.7747+.5464i, .4141-1.8102i\}$

55. $x^5+2+3i=0 \Rightarrow x^5=-2-3i$

$$r=\sqrt{(-2)^2+(-3)^2}=\sqrt{4+9}=\sqrt{13} \Rightarrow r^{1/n}=r^{1/5}=\left(\sqrt{13}\right)^{1/5}=13^{1/10}\approx 1.2924 \text{ and } \tan\theta=\frac{-3}{-2}=1.5$$

Since θ is in quadrant III, $\theta \approx 236.310°$ and $\alpha=\dfrac{236.310°+360°\cdot k}{5}=47.262°+72°\cdot k,$ where k is an integer.

$$x \approx 1.29239(\cos 47.262°+i\sin 47.262°), 1.29239(\cos 119.262°+i\sin 119.2622°),$$
$$1.29239(\cos 191.262°+i\sin 191.2622°), 1.29239(\cos 263.262°+i\sin 263.2622°),$$
$$1.29239(\cos 335.262°+i\sin 335.2622°)$$

Solution set: $\{.87708+.94922i, -.63173+1.1275i, -1.2675-.25240i, -.15164-1.28347i, 1.1738-.54083i\}$

56. The number 1 has 64 complex 64th roots. Two of them are real, 1 and –1, and 62 of them are not real.

57. The statement, "Every real number must have two real square roots," is false. Consider, for example, the real number –4. Its two square roots are $2i$ and $-2i$, which are not real.

58. The statement, "Some real numbers have three real cube roots," is false. Every real number has only one cube root that is also a real number. Then there are two cube roots, which are conjugates that are not real.

59. If z is an nth root of 1, then $z^n=1$. Since $1=\frac{1}{1}=\frac{1}{z^n}=\left(\frac{1}{z}\right)^n$, then $\frac{1}{z}$ is also an nth root of 1.

60. – 62. Answers will vary.

Section 8.5: Polar Equations and Graphs

1. **(a)** II (since $r>0$ and $90°<\theta<180°$)

(b) I (since $r>0$ and $0°<\theta<90°$)

(c) IV (since $r>0$ and $-90°<\theta<0°$)

(d) III (since $r>0$ and $180°<\theta<270°$)

2. **(a)** positive x-axis
(b) negative x-axis
(c) negative y-axis
(d) positive y-axis (since $450° - 360° = 90°$)

For Exercises 3–12, answers may vary.

3. **(a)**

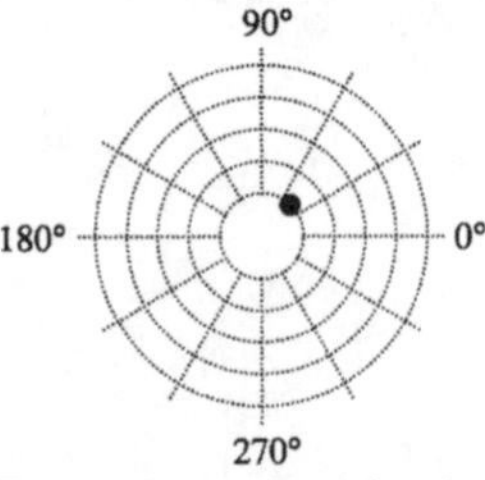

(b) Two other pairs of polar coordinates for (1, 45°) are (1, 405°) and (–1, 225°).

(c) Since $x = r\cos\theta \Rightarrow x = 1 \cdot \cos 45° = \frac{\sqrt{2}}{2}$ and $y = r\sin\theta \Rightarrow y = 1 \cdot \sin 45° = \frac{\sqrt{2}}{2}$, the point is $\left(\frac{\sqrt{2}}{2}, \frac{\sqrt{2}}{2}\right)$.

4. **(a)**

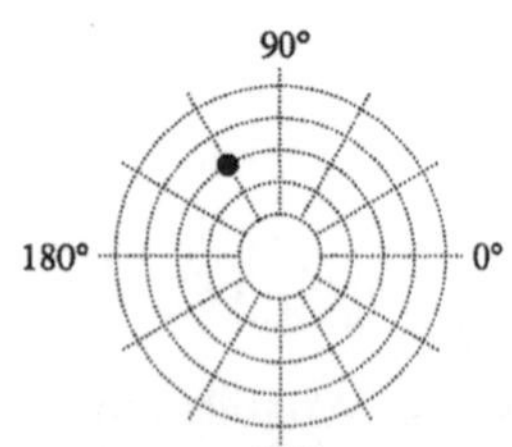

(b) Two other pairs of polar coordinates for (3, 120°) are (3, 480°) and (–3, 300°).

(c) Since $x = r\cos\theta \Rightarrow x = 3\cos 120 = -\frac{3}{2}$ and $y = r\sin\theta \Rightarrow y = 3\sin 120° = \frac{3\sqrt{3}}{2}$, the point is $\left(-\frac{3}{2}, \frac{3\sqrt{3}}{2}\right)$.

5. **(a)**

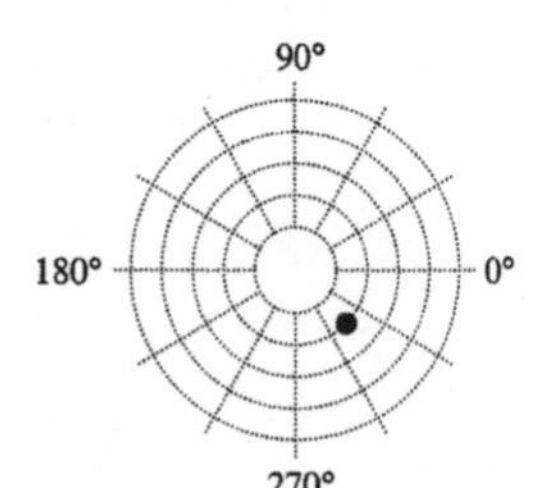

(b) Two other pairs of polar coordinates for (–2, 135°) are (–2, 495°) and (2, 315°).

(c) Since $x = r\cos\theta \Rightarrow x = (-2)\cos 135° = \sqrt{2}$ and $y = r\sin\theta \Rightarrow y = (-2)\sin 135° = \sqrt{2}$, the point is $\left(\sqrt{2}, -\sqrt{2}\right)$.

6. **(a)**

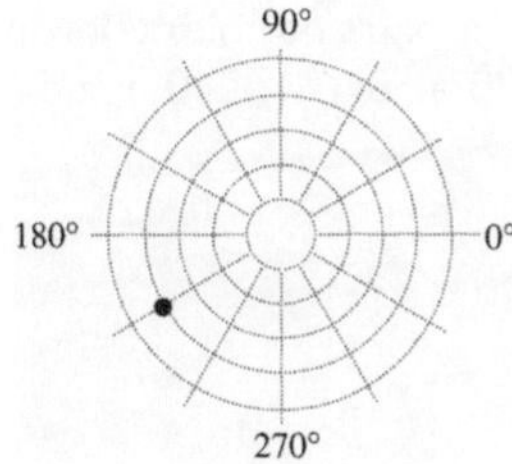

(b) Two other pairs of polar coordinates for (–4, 30°) are $(-4, 390°)$ and $(4, 210°)$.

(c) Since $x = r\cos\theta \Rightarrow x = (-4)\cos 30° = -2\sqrt{3}$ and $y = r\sin\theta \Rightarrow y = (-4)\sin 30° = -2,$ the point is $\left(-2\sqrt{3}, -2\right)$.

7. **(a)**

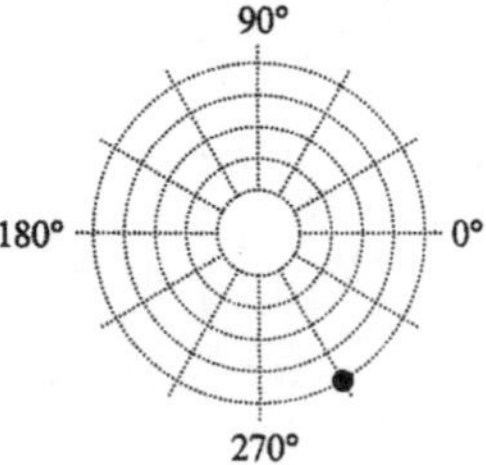

(b) Two other pairs of polar coordinates for (5, – 60°) are (5, 300°) and (–5, 120°).

(c) Since $x = r\cos\theta \Rightarrow x = 5\cos(-60°) = \frac{5}{2}$ and $y = r\sin\theta \Rightarrow y = 5\sin(-60°) = -\frac{5\sqrt{3}}{2}$, the point is $\left(\frac{5}{2}, -\frac{5\sqrt{3}}{2}\right)$.

8. **(a)**

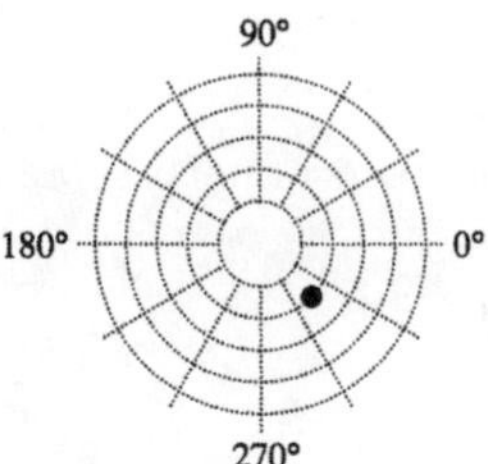

(b) Two other pairs of polar coordinates for (2, – 45°) are (2, 315°) and (–2, 135°).

(c) Since $x = r\cos\theta \Rightarrow x = 2\cos(-45°) = \sqrt{2}$ and $y = r\sin\theta \Rightarrow y = 2\sin(-45°) = -\sqrt{2},$ the point is $\left(\sqrt{2}, -\sqrt{2}\right)$.

9. **(a)**

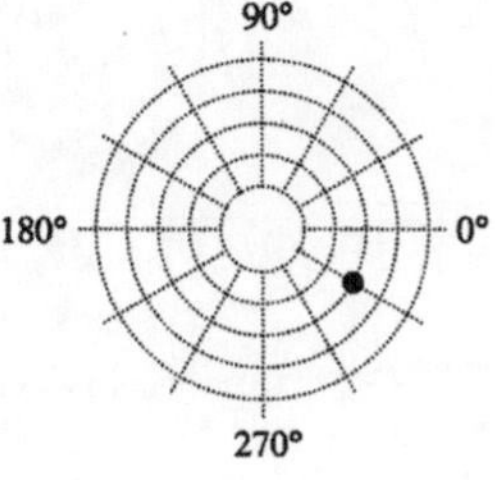

(b) Two other pairs of polar coordinates for (–3, –210°) are (–3, 150°) and (3, –30°).

(c) Since $x = r\cos\theta \Rightarrow x = (-3)\cos(-210°) = \frac{3\sqrt{3}}{2}$ and $y = r\sin\theta \Rightarrow y = (-3)\sin(-210°) = -\frac{3}{2}$, the point is $\left(\frac{3\sqrt{3}}{2}, -\frac{3}{2}\right)$.

10. (a)

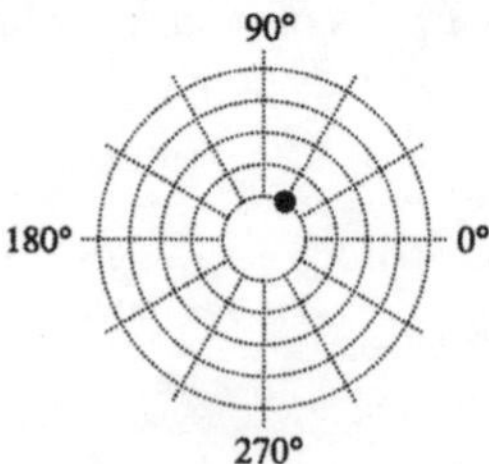

(b) Two other pairs of polar coordinates for $(-1, -120°)$ are $(-1, 240°)$ and $(1, 60°)$.

(c) Since $x = r\cos\theta \Rightarrow x = (-1)\cos(-120°) = \frac{1}{2}$ and $y = r\sin\theta \Rightarrow y = (-1)\sin(-120°) = \frac{\sqrt{3}}{2}$, the point is $\left(\frac{1}{2}, \frac{\sqrt{3}}{2}\right)$.

11. (a)

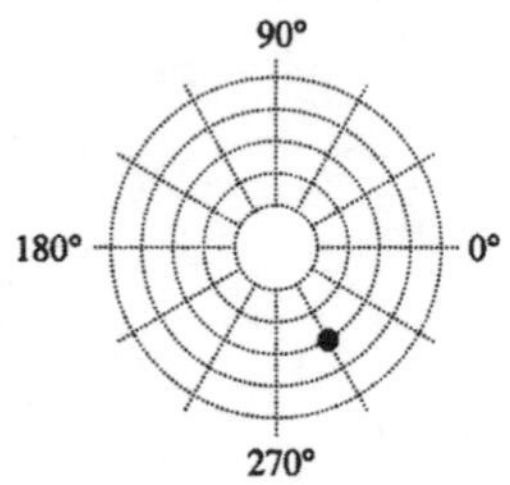

(b) Two other pairs of polar coordinates for $\left(3, \frac{5\pi}{3}\right)$ are $\left(3, \frac{11\pi}{3}\right)$ and $\left(-3, \frac{2\pi}{3}\right)$.

(c) Since $x = r\cos\theta \Rightarrow x = 3\cos\frac{5\pi}{3} = \frac{3}{2}$ and $y = r\sin\theta \Rightarrow y = 3\sin\frac{5\pi}{3} = -\frac{3\sqrt{3}}{2}$, the point is $\left(\frac{3}{2}, -\frac{3\sqrt{3}}{2}\right)$.

12. (a)

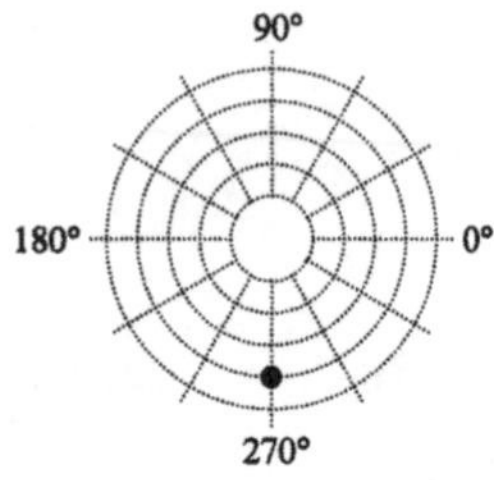

(b) Two other pairs of polar coordinates for $\left(4, \frac{3\pi}{2}\right)$ are $\left(4, -\frac{\pi}{2}\right)$ and $\left(-4, \frac{\pi}{2}\right)$.

(c) Since $x = r\cos\theta \Rightarrow x = 4\cos\frac{3\pi}{2} = 0$ and $y = r\sin\theta \Rightarrow y = 4\sin\frac{3\pi}{2} = -4$, the point is $(0, -4)$.

For Exercises 13–21, answers may vary.

13. (a)

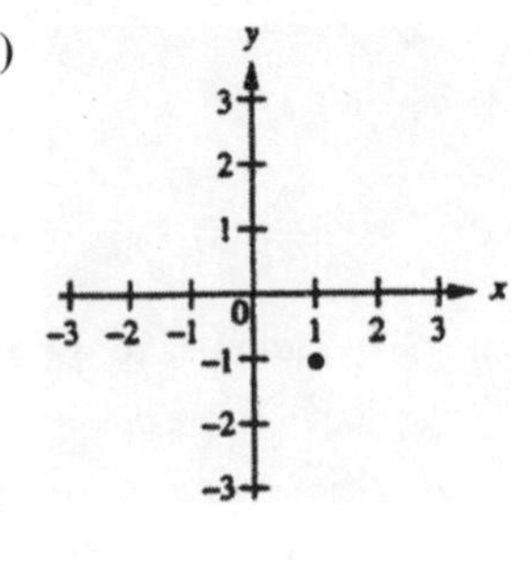

(b) $r = \sqrt{1^2 + (-1)^2} = \sqrt{1+1} = \sqrt{2}$ and $\theta = \tan^{-1}\left(\frac{-1}{1}\right) = \tan^{-1}(-1) = -45°$, since θ is in quadrant IV. Since $360° - 45° = 315°$, one possibility is $(\sqrt{2}, 315°)$. Alternatively, if $r = -\sqrt{2}$, then $\theta = 315° - 180° = 135°$. Thus, a second possibility is $(-\sqrt{2}, 135°)$.

14. (a)

(b) $r=\sqrt{1^2+1^2}=\sqrt{1+1}=\sqrt{2}$ and $\theta=\tan^{-1}\left(\frac{1}{1}\right)=\tan^{-1}1=45°$, since θ is in quadrant I. So, one possibility is $\left(\sqrt{2},45°\right)$. Alternatively, if $r=-\sqrt{2}$, then $\theta=45°+180°=225°$. Thus, a second possibility is $\left(-\sqrt{2},225°\right)$.

15. (a)

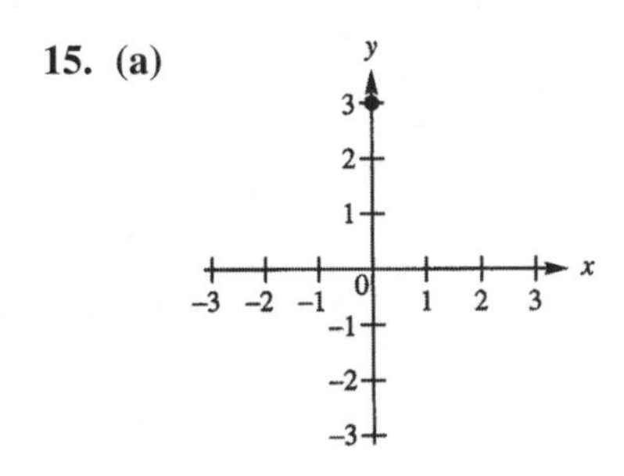

(b) $r=\sqrt{0^2+3^2}=\sqrt{0+9}=\sqrt{9}=3$ and $\theta=90°$, since $(0,3)$ is on the positive y-axis. So, one possibility is $(3,90°)$. Alternatively, if $r=-3$, then $\theta=90°+180°=270°$. Thus, a second possibility is $(-3,270°)$.

16. (a)

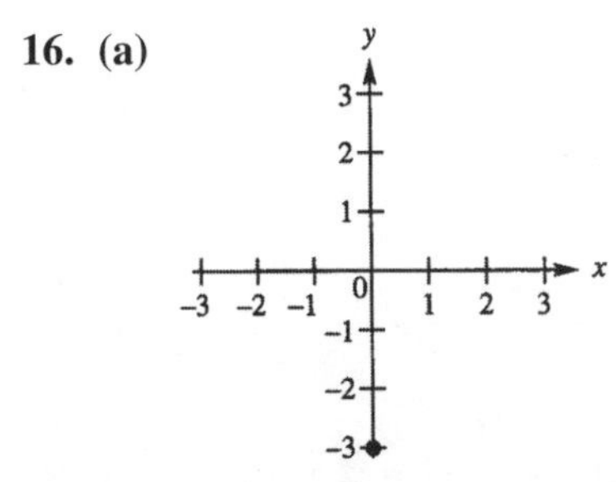

(b) $r=\sqrt{0^2+(-3)^2}=\sqrt{0+9}=\sqrt{9}=3$ and $\theta=270°$, since $(0,3)$ is on the negative y-axis. So, one possibility is $(3,270°)$. Alternatively, if $r=-3$, then $\theta=270°-180°=90°$. Thus, a second possibility is $(-3,90°)$.

17. (a)

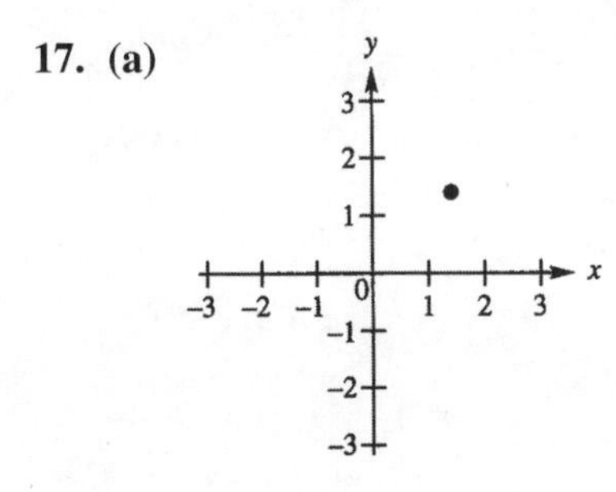

(b) $r=\sqrt{\left(\sqrt{2}\right)^2+\left(\sqrt{2}\right)^2}=\sqrt{2+2}=\sqrt{4}=2$ and $\theta=\tan^{-1}\left(\frac{\sqrt{2}}{\sqrt{2}}\right)=\tan^{-1}1=45°$, since θ is in quadrant I. So, one possibility is $(2,45°)$. Alternatively, if $r=-2$, then $\theta=45°+180°=225°$. Thus, a second possibility is $(-2,225°)$.

18. (a)

(b) $r=\sqrt{\left(-\sqrt{2}\right)^2+\left(\sqrt{2}\right)^2}=\sqrt{2+2}=\sqrt{4}=2$ and $\theta=\tan^{-1}\left(\frac{\sqrt{2}}{-\sqrt{2}}\right)=\tan^{-1}(-1)$, since θ is in quadrant II we have $\theta=135°$. So, one possibility is $(2,135°)$. Alternatively, if $r=-2$, then $\theta=135°+180°=315°$. Thus, a second possibility is $(-2,315°)$.

19. (a)

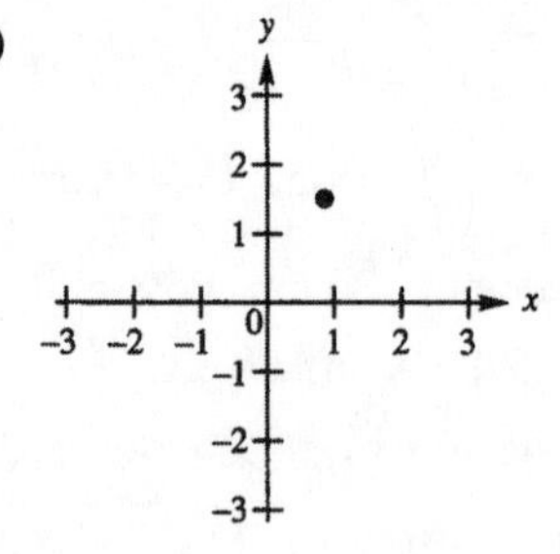

(b) $r=\sqrt{\left(\frac{\sqrt{3}}{2}\right)^2+\left(\frac{3}{2}\right)^2}=\sqrt{\frac{3}{4}+\frac{9}{4}}=\sqrt{\frac{12}{4}}=\sqrt{3}$ and $\theta=\arctan\left(\frac{3}{2}\cdot\frac{2}{\sqrt{3}}\right)=\tan^{-1}\left(\sqrt{3}\right)=60°$, since θ is in quadrant I. So, one possibility is $\left(\sqrt{3},60°\right)$. Alternatively, if $r=-\sqrt{3}$, then $\theta=60°+180°=\ 240°$. Thus, a second possibility is $\left(-\sqrt{3},240°\right)$.

20. (a)

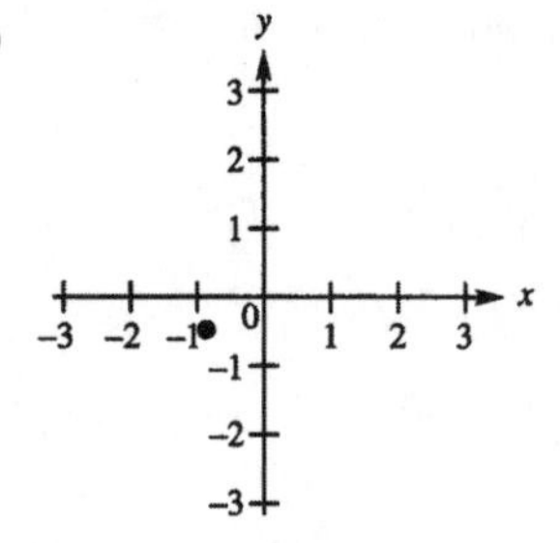

(b) $r=\sqrt{\left(-\frac{\sqrt{3}}{2}\right)^2+\left(-\frac{1}{2}\right)^2}=\sqrt{\frac{3}{4}+\frac{1}{4}}=\sqrt{\frac{4}{4}}=1$ and $\theta=\arctan\left(\frac{-1}{2}\cdot\left(\frac{-2}{\sqrt{3}}\right)\right)=\tan^{-1}\left(\frac{1}{\sqrt{3}}\right)=$ $\tan^{-1}\left(\frac{\sqrt{3}}{3}\right)=210°$ since θ is in quadrant III. So, one possibility is $(1,210°)$. Alternatively, if $r=-1$, then $\theta=210°-180°=30°$. Thus, a second possibility is $(-1,30°)$.

21. (a)

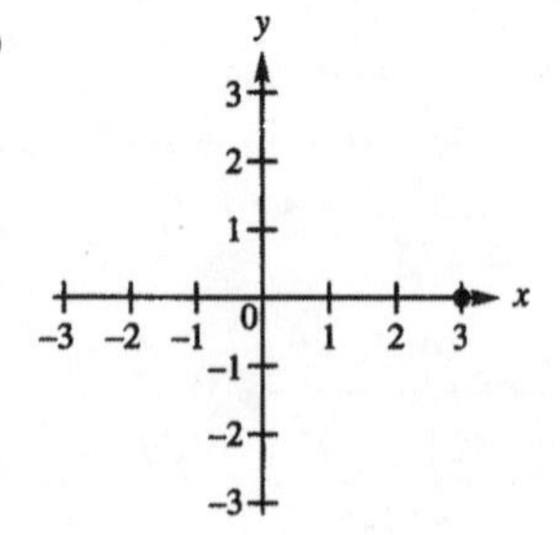

(b) $r=\sqrt{3^2+0^2}=\sqrt{9+0}=\sqrt{9}=3$ and $\theta=0°$, since $(3,0)$ is on the positive *x*-axis. So, one possibility is $(3,0°)$. Alternatively, if $r=-3$, then $\theta=0°+180°=180°$. Thus, a second possibility is $(-3,180°)$.

22. (a)

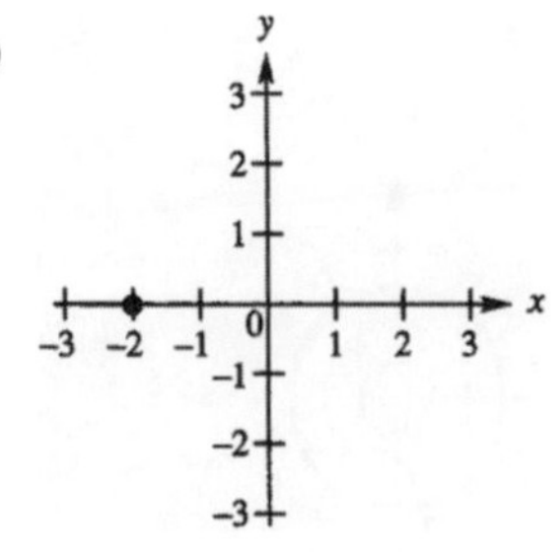

(b) $r=\sqrt{(-2)^2+0^2}=\sqrt{4+0}=\sqrt{4}=2$ and $\theta=180°$, since $(-2,0)$ is on the negative x-axis. So, one possibility is $(2,180°)$. Alternatively, if $r=-2$, then $\theta=180°-180°=0°$. Thus, a second possibility is $(-2,0°)$.

23. $x-y=4$

Using the general form for the polar equation of a line, $r=\dfrac{c}{a\cos\theta+b\sin\theta}$, with $a=1, b=-1$, and $c=4$, the polar equation is $r=\dfrac{4}{\cos\theta-\sin\theta}$.

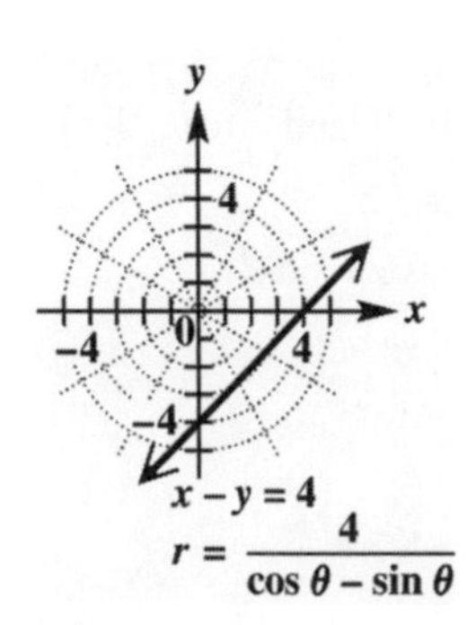

24. $x+y=-7$

Using the general form for the polar equation of a line, $r=\dfrac{c}{a\cos\theta+b\sin\theta}$, with $a=1, b=1$, and $c=-7$, the polar equation is $r=\dfrac{-7}{\cos\theta+\sin\theta}$.

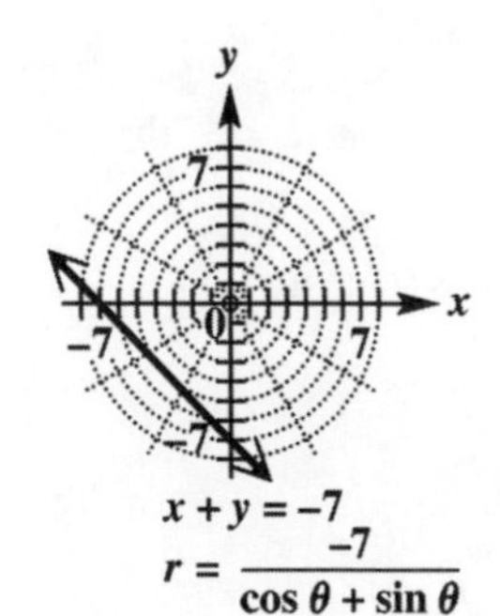

25. $x^2+y^2=16 \Rightarrow r^2=16 \Rightarrow r=\pm 4$

The equation of the circle in polar form is $r=4$ or $r=-4$.

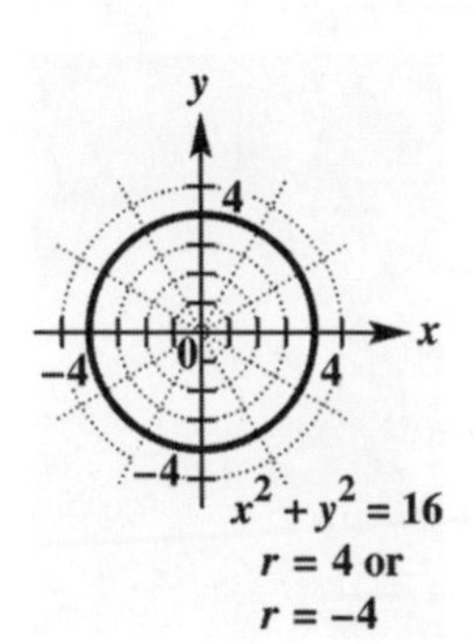

26. $x^2 + y^2 = 9 \Rightarrow r^2 = 9 \Rightarrow r = \pm 3$
The equation of the circle in polar form is $r = 3$ or $r = -3$.

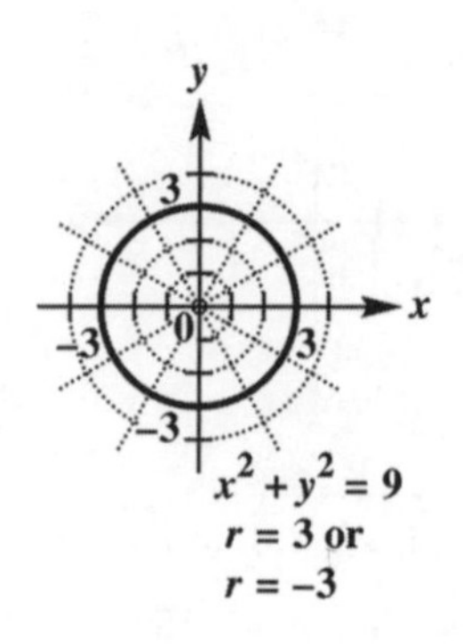

27. $2x + y = 5$
Using the general form for the polar equation of a line, $r = \dfrac{c}{a\cos\theta + b\sin\theta}$, with $a = 2, b = 1$, and $c = 5$, the polar equation is $r = \dfrac{5}{2\cos\theta + \sin\theta}$.

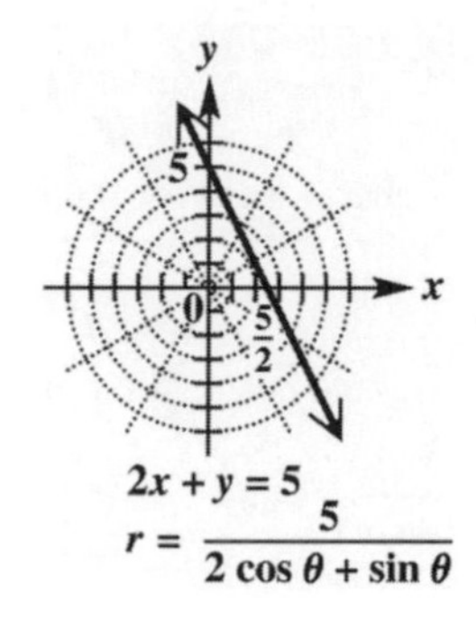

28. $3x - 2y = 6$
Using the general form for the polar equation of a line, $r = \dfrac{c}{a\cos\theta + b\sin\theta}$, with $a = 3, b = -2$, and $c = 6$, the polar equation is $r = \dfrac{6}{3\cos\theta - 2\sin\theta}$.

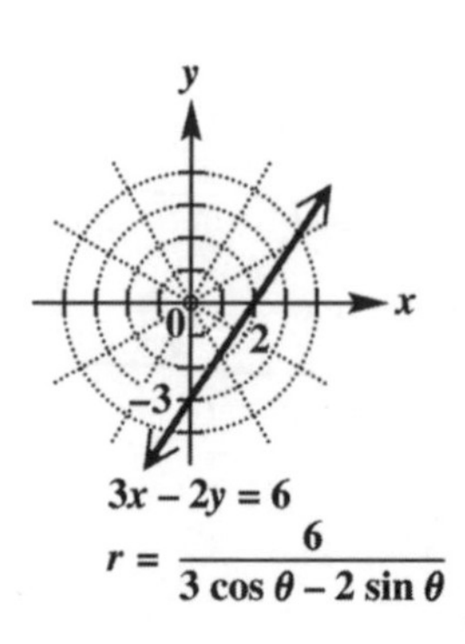

29. $r\sin\theta = k$

30. $r = \dfrac{k}{\sin\theta}$

31. $r = \dfrac{k}{\sin\theta} \Rightarrow r = k\csc\theta$

32. $y = 3$

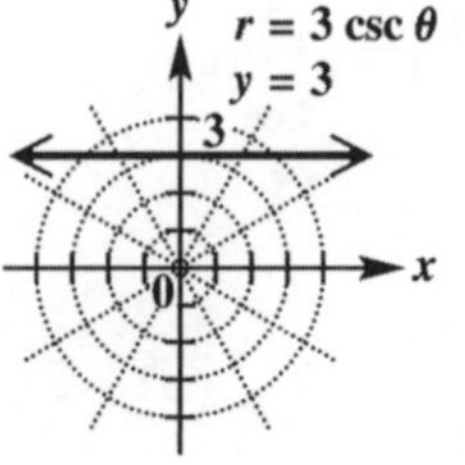

33. $r\cos\theta = k$

34. $r = \dfrac{k}{\cos\theta}$

35. $r = \dfrac{k}{\cos\theta} \Rightarrow r = k\sec\theta$

36. $x = 3$

37. $r = 3$ represents the set of all points 3 units from the pole. The correct choice is C.

38. $r = \cos 3\theta$ is a rose curve with 3 petals. The correct choice is D.

39. $r = \cos 2\theta$ is a rose curve with $2 \cdot 2 = 4$ petals. The correct choice is A

40. The general form for the polar equation of a line is $r = \dfrac{c}{a\cos\theta + b\sin\theta}$, where the standard form of a line $ax + by = c$. $r = \dfrac{2}{\cos\theta + \sin\theta}$ is a line. The correct choice is B.

41. $r = 2 + 2\cos\theta$ (cardioid)

θ	0°	30°	60°	90°	120°	150°
$\cos\theta$	1	.9	.5	0	–.5	–.9
$r = 2 + 2\cos\theta$	4	3.8	3	2	1	.2

θ	180°	210°	240°	270°	300°	330°
$\cos\theta$	–1	–.9	–.5	0	.5	.9
$r = 2 + 2\cos\theta$	0	.3	1	2	3	3.7

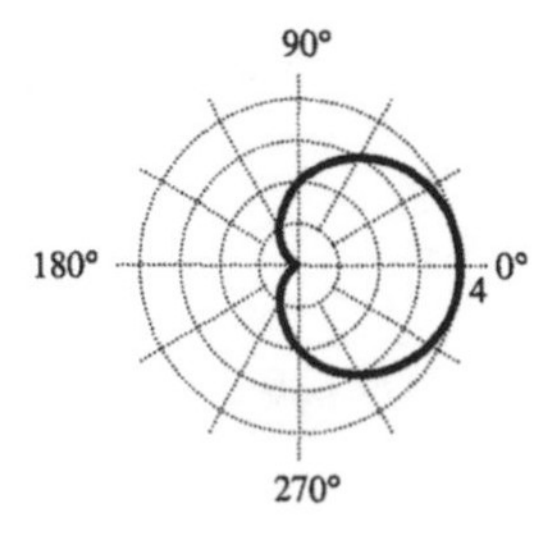

42. $r = 8 + 6\cos\theta$ (limaçon)

θ	0°	30°	60°	90°	120°	150°
$\cos\theta$	1	.9	.5	0	–.5	–.9
$r = 8 + 6\cos\theta$	14	13.2	11	8	5	2.8

θ	180°	210°	240°	270°	300°	330°
$\cos\theta$	–1	–.9	–.5	0	.5	.9
$r = 8 + 6\cos\theta$	2	2.8	5	8	11	13.2

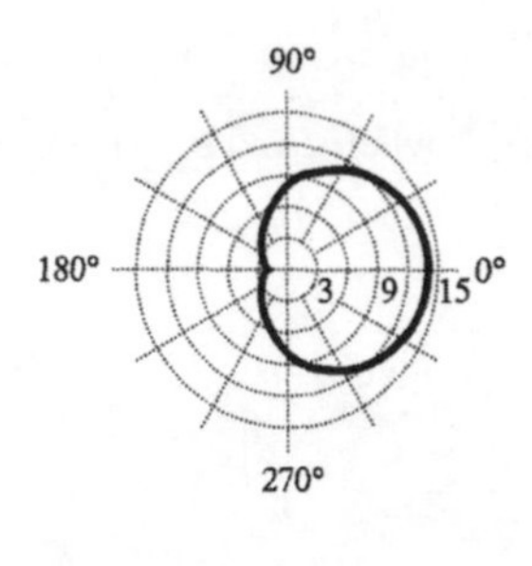

43. $r = 3 + \cos\theta$ (limaçon)

θ	0°	30°	60°	90°	120°	150°
$r = 3 + \cos\theta$	4	3.9	3.5	3	2.5	2.1

θ	180°	210°	240°	270°	300°	330°
$r = 3 + \cos\theta$	2	2.1	2.5	3	3.5	3.9

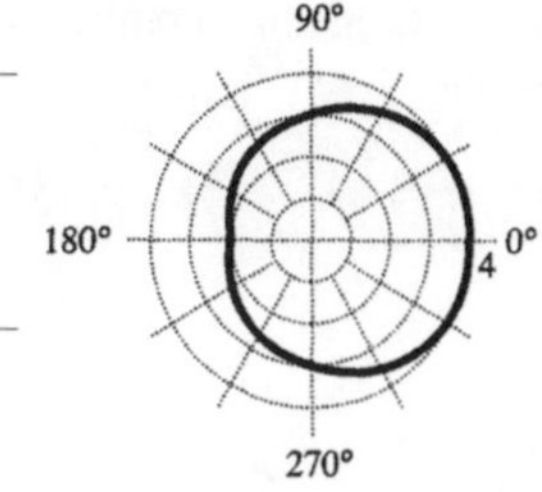

44. $r = 2 - \cos\theta$ (limaçon)

θ	0°	30°	60°	90°	135°	180°	225°	270°	315°
$r = 2 - \cos\theta$	1	1.1	1.5	2	2.7	3	2.7	2	1.3

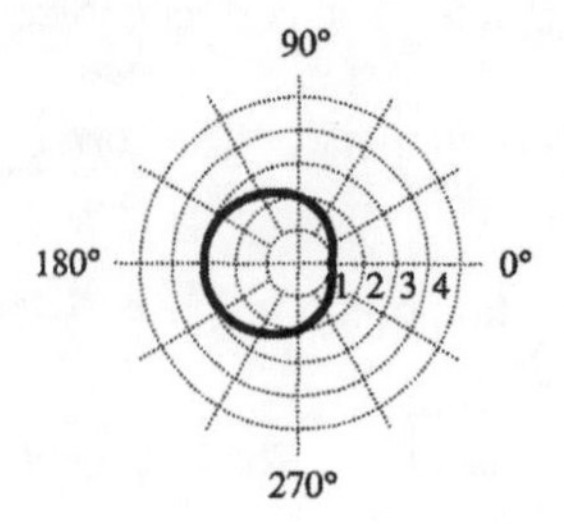

45. $r = 4\cos 2\theta$ (four-leaved rose)

θ	0°	30°	45°	60°	90°	120°	135°	150°
$r = 4\cos 2\theta$	4	2	0	−2	−4	−2	0	2

θ	180°	210°	225°	240°	270°	300°	315°	330°
$r = 4\cos 2\theta$	4	2	0	−2	−4	−2	0	2

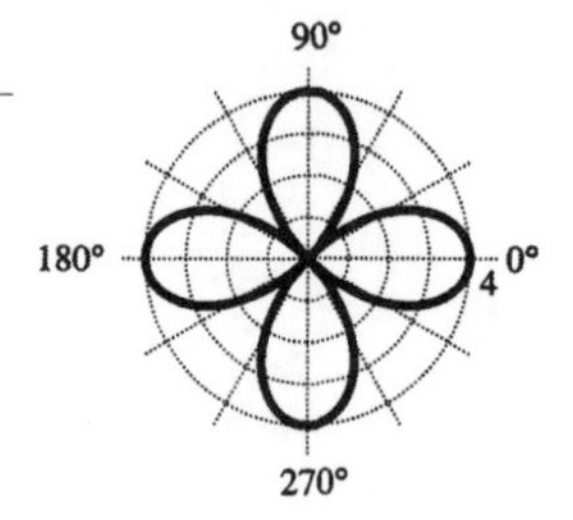

46. $r = 3\cos 5\theta$ (five-leaved rose)

$r = 0$ when $\cos 5\theta = 0$, or $5\theta = 90° + 360° \cdot k = 18° + 72° \cdot k$, where k is an integer, or $\theta = 18°, 90°, 162°, 234°$.

θ	0°	18°	36°	54°	72°	90°	108°	162°
$r = 3\cos 5\theta$	3	0	−3	0	3	0	−3	0

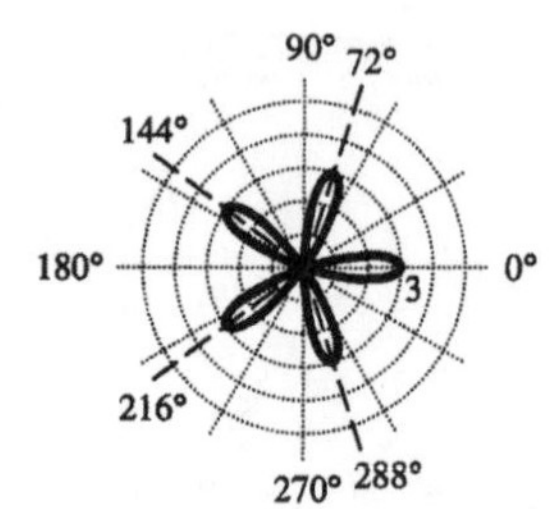

Pattern 3, 0, −3, 0, 3 continues for every 18°.

47. $r^2 = 4\cos 2\theta \Rightarrow r = \pm 2\sqrt{\cos 2\theta}$ (lemniscate)

Graph only exists for [0°, 45°], [135°, 225°], and [315°, 360°] because $\cos 2\theta$ must be positive.

θ	0°	30°	45°	135°	150°
$r = \pm 2\sqrt{\cos 2\theta}$	±2	±1.4	0	0	±1.4

θ	180°	210°	225°	315°	330°
$r = \pm 2\sqrt{\cos 2\theta}$	±2	±1.4	0	0	±1.4

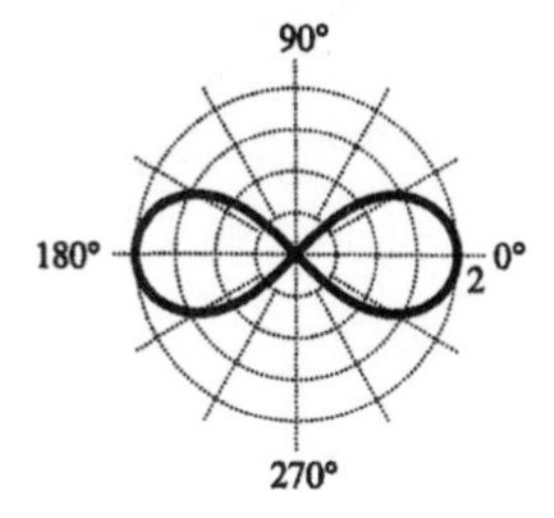

48. $r^2 = 4\sin 2\theta \Rightarrow r = \pm 2\sqrt{\sin 2\theta}$ (lemniscate)

Graph only exists for [0°, 90°] and [180°, 270°] because sin θ must be positive.

θ	0°	30°	45°	60°	90°	180°	225°	270°
$r = \pm 2\sqrt{\sin 2\theta}$	0	±1.86	±2	±1.86	0	0	±2	0

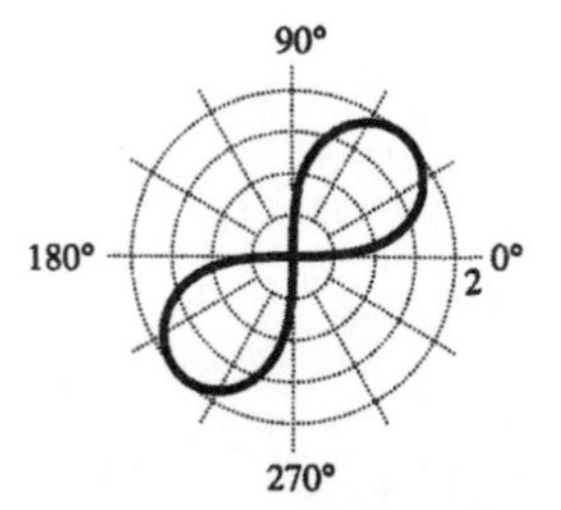

49. $r = 4 - 4\cos\theta$ (cardioid)

θ	0°	30°	60°	90°	120°	150°
$r = 4 - 4\cos\theta$	0	.5	2	4	6	7.5

θ	180°	210°	240°	270°	300°	330°
$r = 4 - 4\cos\theta$	8	7.5	6	4	2	.5

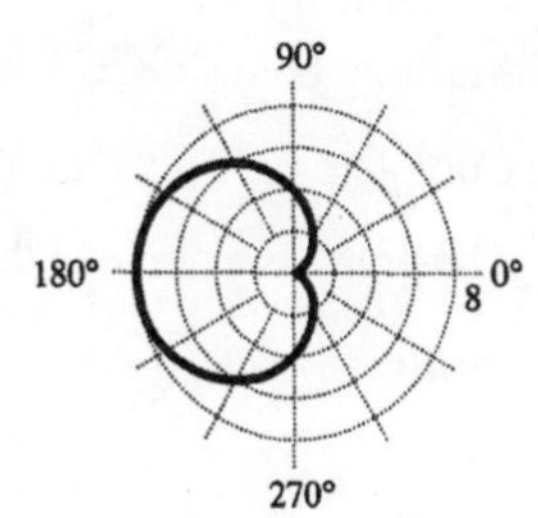

50. $r = 6 - 3\cos\theta$ (limaçon)

θ	0°	45°	90°	135°	180°	270°	360°
$r = 6 - 3\cos\theta$	3	3.9	6	8.1	9	6	3

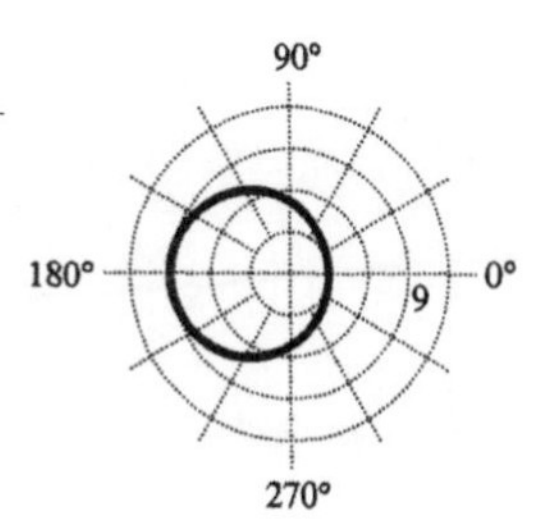

51. $r = 2\sin\theta\tan\theta$ (cissoid)

r is undefined at $\theta = 90°$ and $\theta = 270°$.

θ	0°	30°	45°	60°	90°	120°	135°	150°	180°
$r = 2\sin\theta\tan\theta$	0	.6	1.4	3	–	–3	–1.4	–.6	0

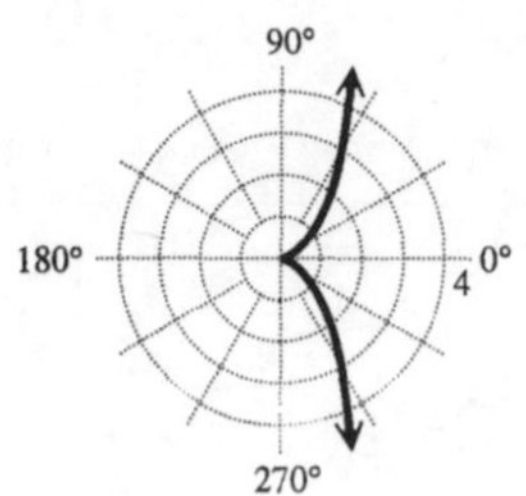

Notice that for [180°, 360°), the graph retraces the path traced for [0°, 180°).

52. $r = \dfrac{\cos 2\theta}{\cos\theta}$ (cissoid with a loop)

r is undefined at $\theta = 90°$ and $\theta = 270°$ and $r = 0$ at 45°, 135°, 225°, and 315°.

θ	0°	45°	60°	70°	80°
$r = \dfrac{\cos 2\theta}{\cos\theta}$	1	0	–1	–2.2	–5.4

θ	90°	100°	110°	135°	180°
$r = \dfrac{\cos 2\theta}{\cos\theta}$	–	5.4	2.2	0	–1

Notice that for [180°, 360°), the graph retraces the path traced for [0°, 180°).

53. $r = 2\sin\theta$

Multiply both sides by r to obtain $r^2 = 2r\sin\theta$. Since $r^2 = x^2 + y^2$ and $y = r\sin\theta$, $x^2 + y^2 = 2y$.

Complete the square on y to obtain $x^2 + y^2 - 2y + 1 = 1 \Rightarrow x^2 + (y-1)^2 = 1$.

The graph is a circle with center at $(0,1)$ and radius 1.

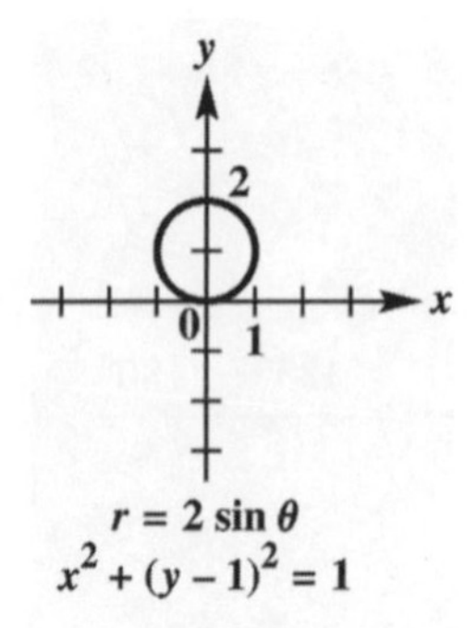

54. $r = 2\cos\theta$

Multiply both sides by r to obtain $r^2 = 2r\cos\theta$. Since $r^2 = x^2 + y^2$ and $x = r\cos\theta$, $x^2 + y^2 = 2x$.

Complete the square on x to get the equation of a circle to obtain $x^2 - 2x + y^2 = 0 \Rightarrow (x-1)^2 + y^2 = 1$.

The graph is a circle with center at $(1,0)$ and radius 1.

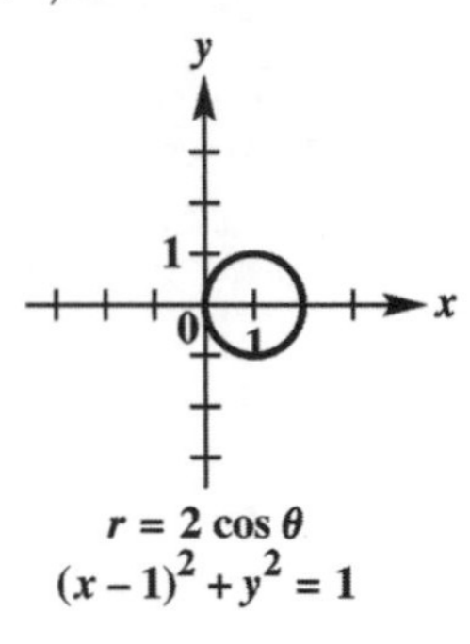

$r = 2\cos\theta$
$(x-1)^2 + y^2 = 1$

55. $r = \dfrac{2}{1-\cos\theta}$

Multiply both sides by $1 - \cos\theta$ to obtain $r - r\cos\theta = 2$. Substitute $r = \sqrt{x^2 + y^2}$ to obtain the following.

$\sqrt{x^2+y^2} - x = 2 \Rightarrow \sqrt{x^2+y^2} = 2 + x \Rightarrow x^2 + y^2 = (2+x)^2 \Rightarrow x^2 + y^2 = 4 + 4x + x^2 \Rightarrow y^2 = 4(1+x)$

The graph is a parabola with vertex at $(-1,0)$ and axis $y = 0$.

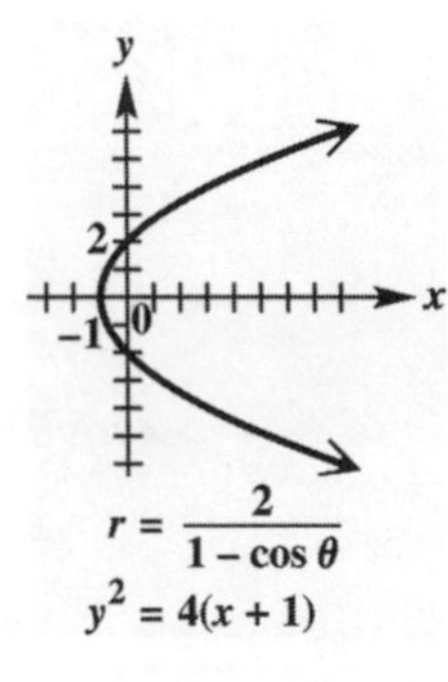

$r = \dfrac{2}{1-\cos\theta}$
$y^2 = 4(x+1)$

56. $r = \dfrac{3}{1-\sin\theta}$

$$r = \frac{3}{1-\sin\theta} \Rightarrow r - r\sin\theta = 3 \Rightarrow r = r\sin\theta + 3 \Rightarrow \sqrt{x^2+y^2} = y + 3$$

$$x^2 + y^2 = y^2 + 6y + 9 \Rightarrow x^2 = 6y + 9 \Rightarrow x^2 = 6\left(y + \frac{3}{2}\right)$$

The graph is a parabola with axis $x = 0$ and vertex $\left(0, -\dfrac{3}{2}\right)$.

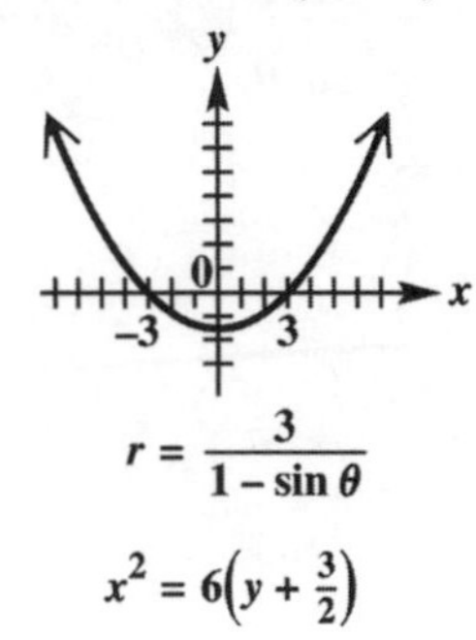

$r = \dfrac{3}{1-\sin\theta}$

$x^2 = 6\left(y + \frac{3}{2}\right)$

57. $r + 2\cos\theta = -2\sin\theta$

$$r + 2\cos\theta = -2\sin\theta \Rightarrow r^2 = -2r\sin\theta - 2r\cos\theta \Rightarrow x^2 + y^2 = -2y - 2x$$

$$x^2 + 2x + y^2 + 2y = 0 \Rightarrow x^2 + 2x + 1 + y^2 + 2y + 1 = 2 \Rightarrow (x+1)^2 + (y+1)^2 = 2$$

The graph is a circle with center $(-1, -1)$ and radius $\sqrt{2}$.

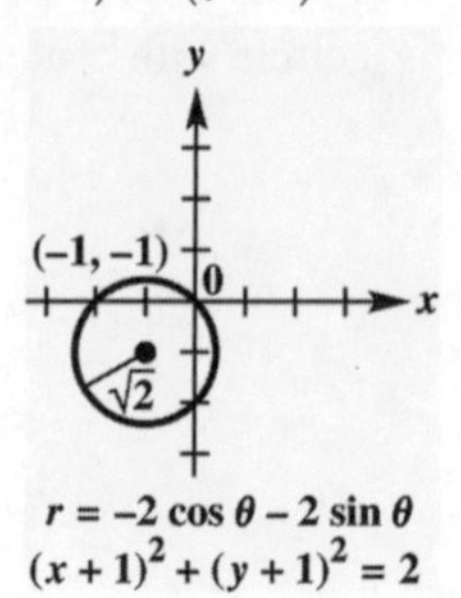

58. $r = \dfrac{3}{4\cos\theta - \sin\theta}$

Using the general form for the polar equation of a line, $r = \dfrac{c}{a\cos\theta + b\sin\theta}$, with $a = 4$, $b = -1$, and $c = 3$, we have $4x - y = 3$.

The graph is a line with intercepts $(0, -3)$ and $\left(\frac{3}{4}, 0\right)$.

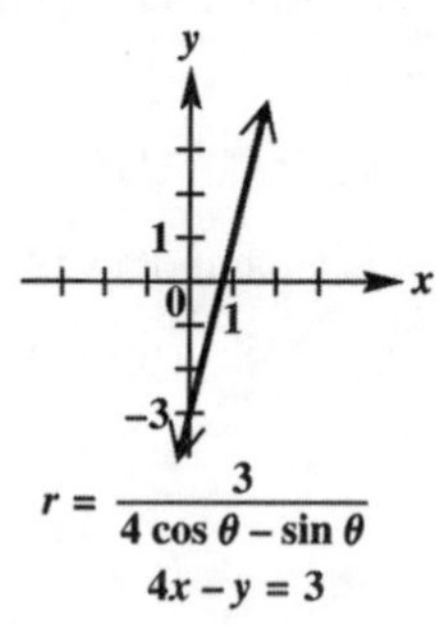

59. $r = 2\sec\theta$

$$r = 2\sec\theta \Rightarrow r = \frac{2}{\cos\theta} \Rightarrow r\cos\theta = 2 \Rightarrow x = 2$$

The graph is a vertical line, intercepting the x-axis at 2.

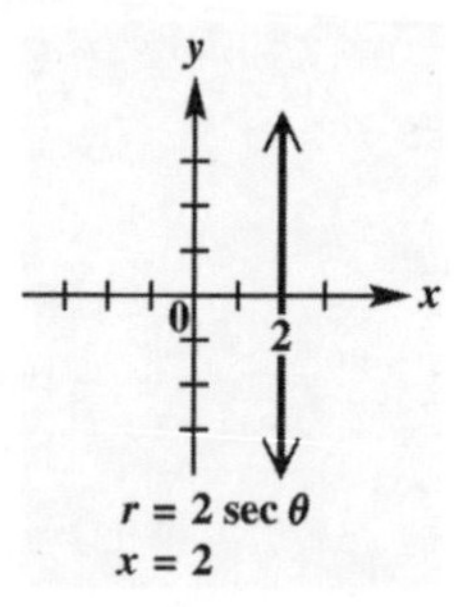

60. $r = -5\csc\theta$

$$r = -5\csc\theta \Rightarrow r = -\frac{5}{\sin\theta} \Rightarrow r\sin\theta = -5 \Rightarrow y = -5$$

The graph is a horizontal line, intercepting the y-axis at -5.

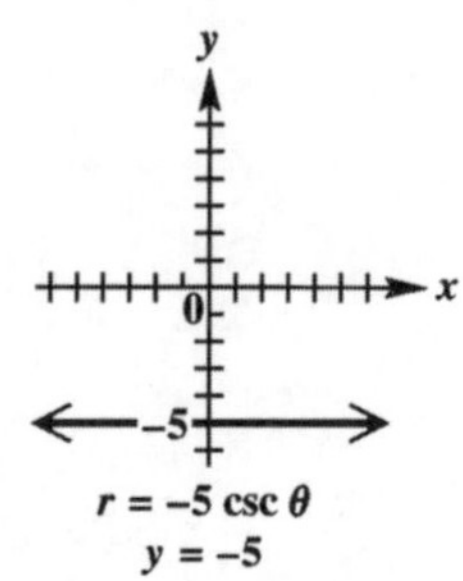

61. $r = \dfrac{2}{\cos\theta + \sin\theta}$

Using the general form for the polar equation of a line, $r = \dfrac{c}{a\cos\theta + b\sin\theta}$, with $a = 1,\ b = 1,$ and $c = 2,$ we have $x + y = 2.$

The graph is a line with intercepts $(0, 2)$ and $(2, 0)$.

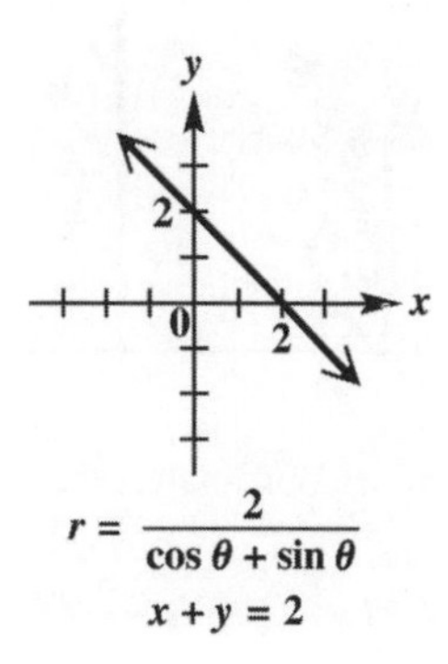

62. $r = \dfrac{2}{2\cos\theta + \sin\theta}$

Using the general form for the polar equation of a line, $r = \dfrac{c}{a\cos\theta + b\sin\theta}$, with $a = 2,\ b = 1,$ and $c = 2,$ we have $2x + y = 2.$

The graph is a line with intercepts $(0, 2)$ and $(1, 0)$.

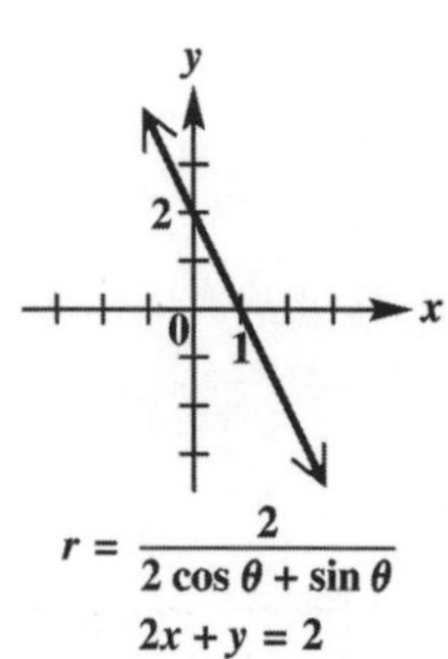

63. Graph $r = \theta$, a spiral of Archimedes.

θ	–360°	–270°	–180°	–90°
θ (radians)	–6.3	–4.7	–3.1	–1.6
$r = \theta$	–6.3	–4.7	–3.1	–1.6

θ	0°	90°	180°	270°	360°
θ (radians)	0	1.6	3.1	4.7	6.3
$r = \theta$	0	1.6	3.1	4.7	6.3

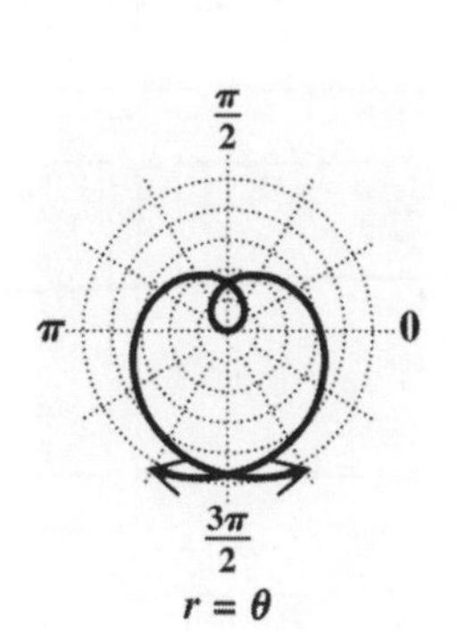

64.

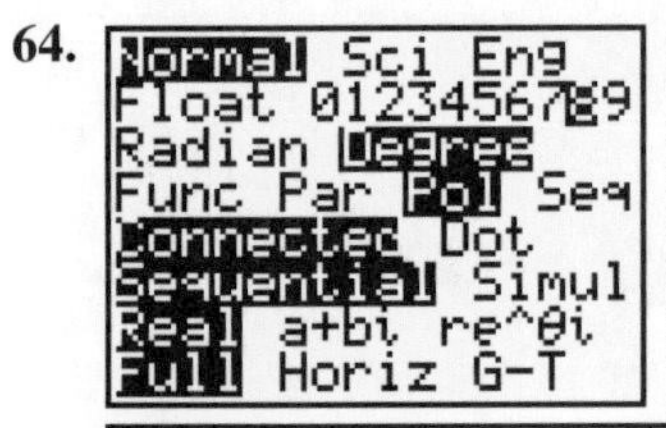

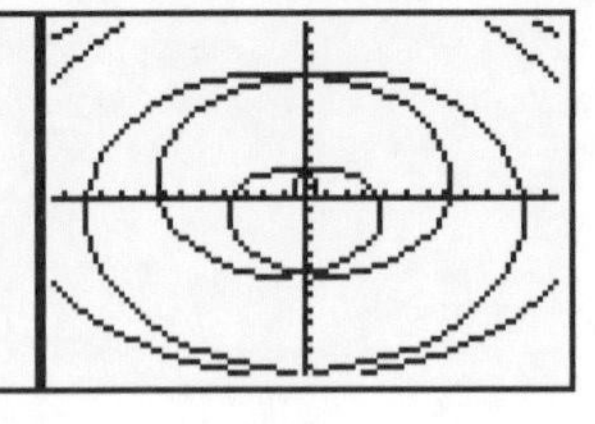

65. In rectangular coordinates, the line passes through $(1,0)$ and $(0,2)$. So $m=\frac{2-0}{0-1}=\frac{2}{-1}=-2$ and $(y-0)=-2(x-1)\Rightarrow y=-2x+2\Rightarrow 2x+y=2$. Converting to polar form $r=\frac{c}{a\cos\theta+b\sin\theta}$, we have: $r=\frac{2}{2\cos\theta+\sin\theta}$.

66. Answers will vary.

67. (a) $(r,-\theta)$

(b) $(r,\pi-\theta)$ or $(-r,-\theta)$

(c) $(r,\pi+\theta)$ or $(-r,\theta)$

68. (a) $-\theta$

(b) $\pi-\theta$

(c) $-r$; $-\theta$

(d) $-r$

(e) $\pi+\theta$

(f) the polar axis

(g) the line $\theta=\frac{\pi}{2}$

69.

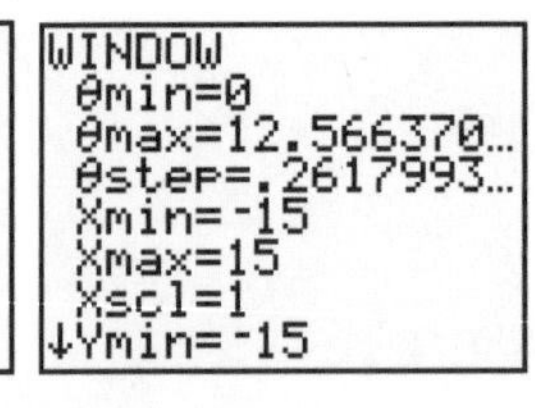

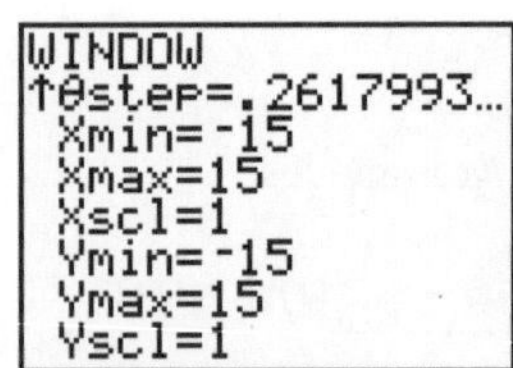

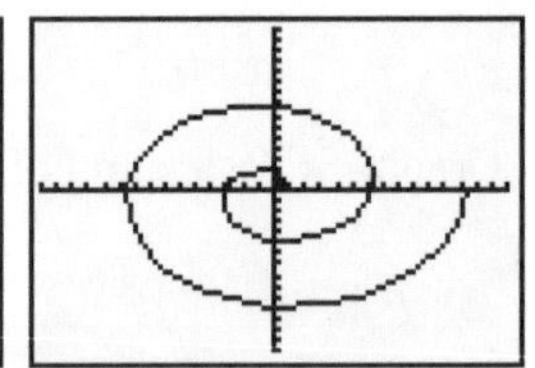

70.

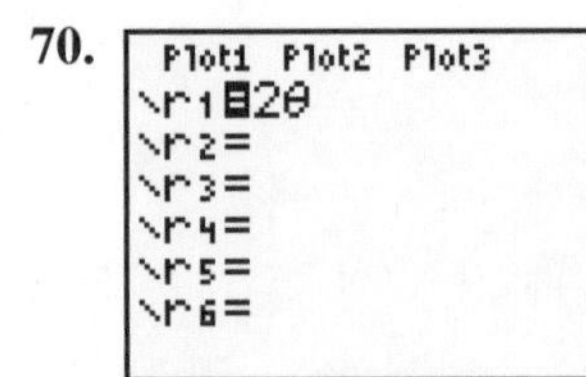

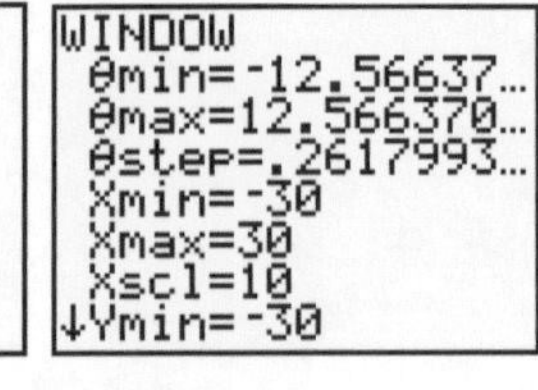

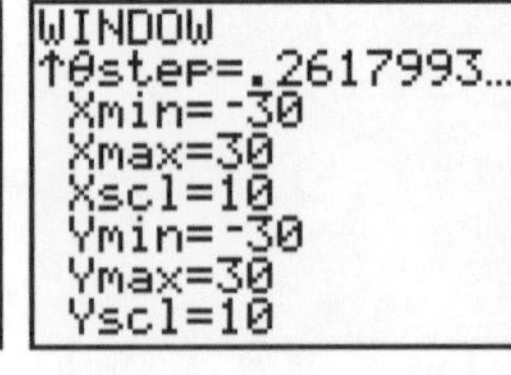

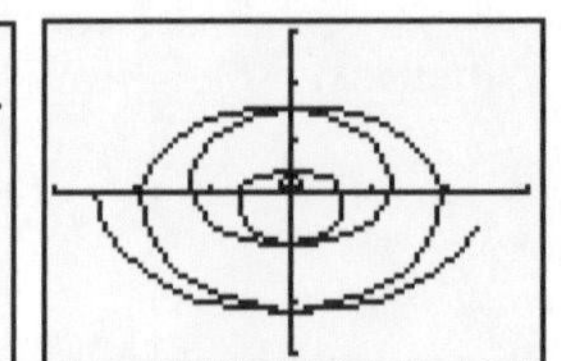

71.

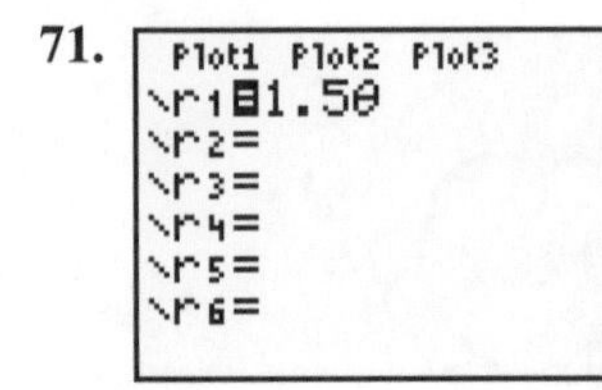

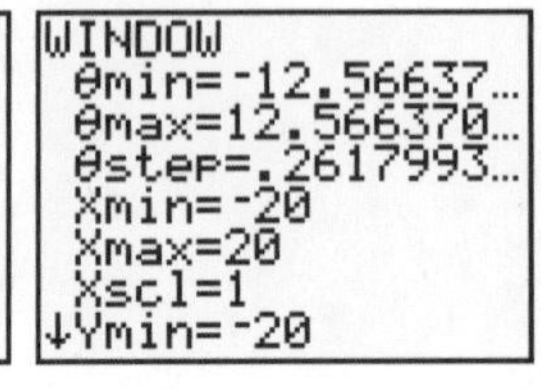

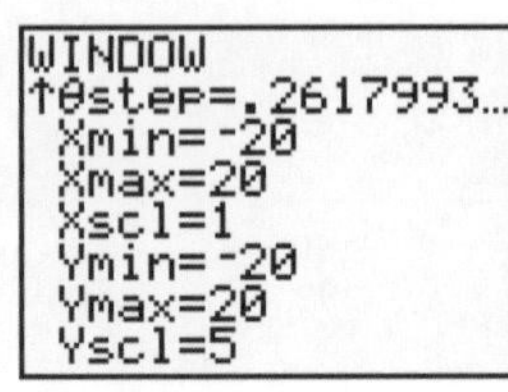

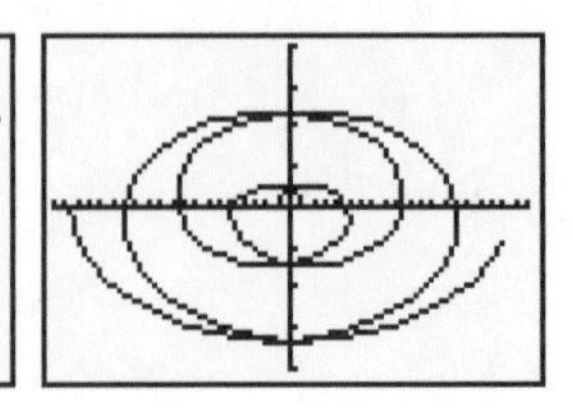

72.

Plot1 Plot2 Plot3
\r1 = -θ
\r2=
\r3=
\r4=
\r5=
\r6=

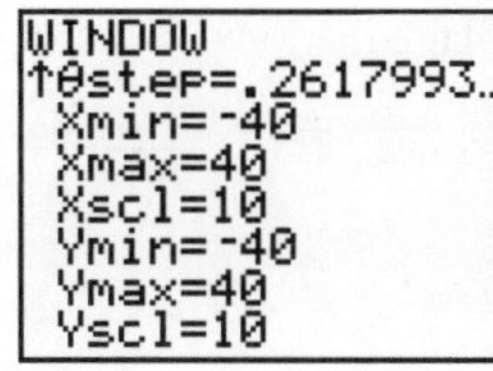

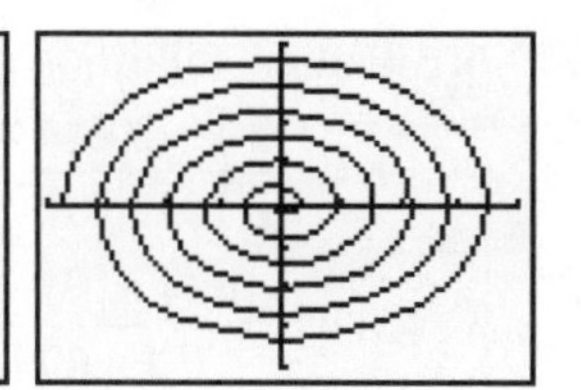

73. $r = 4\sin\theta,\ r = 1 + 2\sin\theta,\ 0 \le \theta < 2\pi$

$$4\sin\theta = 1 + 2\sin\theta \Rightarrow 2\sin\theta = 1 \Rightarrow \sin\theta = \frac{1}{2} \Rightarrow \theta = \frac{\pi}{6} \text{ or } \frac{5\pi}{6}$$

The points of intersection are $\left(4\sin\frac{\pi}{6}, \frac{\pi}{6}\right) = \left(2, \frac{\pi}{6}\right)$ and $\left(4\sin\frac{5\pi}{6}, \frac{5\pi}{6}\right) = \left(2, \frac{5\pi}{6}\right)$.

74. $r = 3,\ r = 2 + 2\cos\theta;\ 0° \le \theta < 360°$

$$3 = 2 + 2\cos\theta \Rightarrow 1 = 2\cos\theta \Rightarrow \cos\theta = \frac{1}{2} \Rightarrow \theta = 60° \text{ or } 300°$$

The points of intersection are $(3, 60°), (3, 300°)$

75. $r = 2 + \sin\theta,\ r = 2 + \cos\theta,\ 0 \le \theta < 2\pi$

$$2 + \sin\theta = 2 + \cos\theta \Rightarrow \sin\theta = \cos\theta \Rightarrow \theta = \frac{\pi}{4} \text{ or } \frac{5\pi}{4}$$

$$r = 2 + \sin\frac{\pi}{4} = 2 + \frac{\sqrt{2}}{2} = \frac{4+\sqrt{2}}{2} \quad \text{and} \quad r = 2 + \sin\frac{5\pi}{4} = 2 - \frac{\sqrt{2}}{2} = \frac{4-\sqrt{2}}{2}$$

The points of intersection are $\left(\frac{4+\sqrt{2}}{2}, \frac{\pi}{4}\right)$ and $\left(\frac{4-\sqrt{2}}{2}, \frac{5\pi}{4}\right)$.

76. $r = \sin 2\theta,\ r = \sqrt{2}\cos\theta,\ 0 \le \theta < \pi$

$\sin 2\theta = \sqrt{2}\cos\theta \Rightarrow 2\sin\theta\cos\theta = \sqrt{2}\cos\theta \Rightarrow 2\sin\theta\cos\theta - \sqrt{2}\cos\theta = 0 \Rightarrow \cos\theta\left(2\sin\theta - \sqrt{2}\right) = 0$

$$\cos\theta = 0 \text{ or } 2\sin\theta - \sqrt{2} = 0 \Rightarrow 2\sin\theta = \sqrt{2} \Rightarrow \sin\theta = \frac{\sqrt{2}}{2}$$

Thus, $\theta = \frac{\pi}{2}$ or $\theta = \frac{\pi}{4}$ or $\frac{3\pi}{4}$. The points of intersection are the following.

$$\left(\sin 2\cdot\frac{\pi}{2}, \frac{\pi}{2}\right) = \left(0, \frac{\pi}{2}\right),\ \left(\sin 2\cdot\frac{\pi}{4}, \frac{\pi}{4}\right) = \left(1, \frac{\pi}{4}\right), \text{ and } \left(\sin 2\cdot\frac{3\pi}{4}, \frac{3\pi}{4}\right) = \left(-1, \frac{3\pi}{4}\right).$$

77. (a) Plot the following polar equations on the same polar axis in radian mode: Mercury: $r=\dfrac{.39(1-.206^2)}{1+.206\cos\theta}$; Venus: $r=\dfrac{.78(1-.007^2)}{1+.007\cos\theta}$; Earth: $r=\dfrac{1(1-.017^2)}{1+.017\cos\theta}$;

Mars: $r=\dfrac{1.52(1-.093^2)}{1+.093\cos\theta}$.

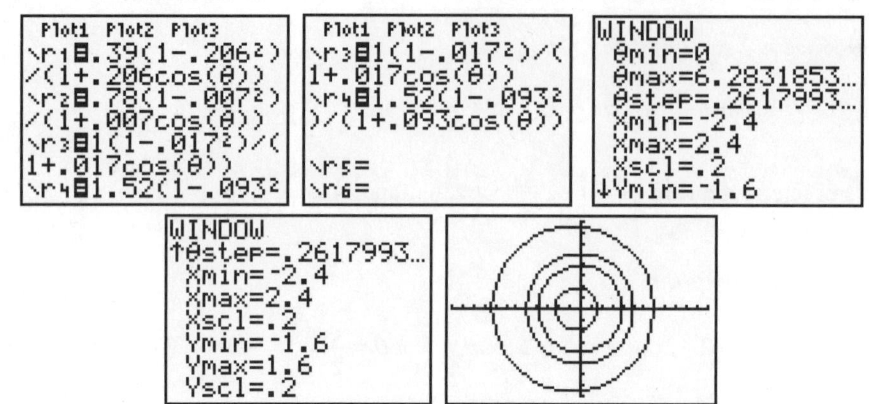

(b) Plot the following polar equations on the same polar axis: Earth: $r=\dfrac{1(1-.017^2)}{1+.017\cos\theta}$;

Jupiter: $r=\dfrac{5.2(1-.048^2)}{1+.048\cos\theta}$; Uranus: $r=\dfrac{19.2(1-.047^2)}{1+.047\cos\theta}$; Pluto: $r=\dfrac{39.4(1-.249^2)}{1+.249\cos\theta}$.

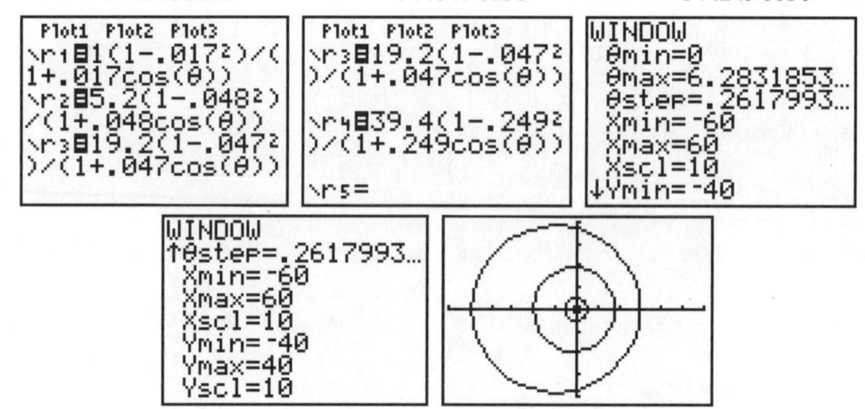

Continued on next page

77. (continued)

(c) We must determine if the orbit of Pluto is always outside the orbits of the other planets. Since Neptune is closest to Pluto, plot the orbits of Neptune and Pluto on the same polar axes.

$$\text{Neptune: } r=\frac{30.1(1-.009^2)}{1+.009\cos\theta};\quad \text{Pluto: } r=\frac{39.4(1-.249^2)}{1+.249\cos\theta}$$

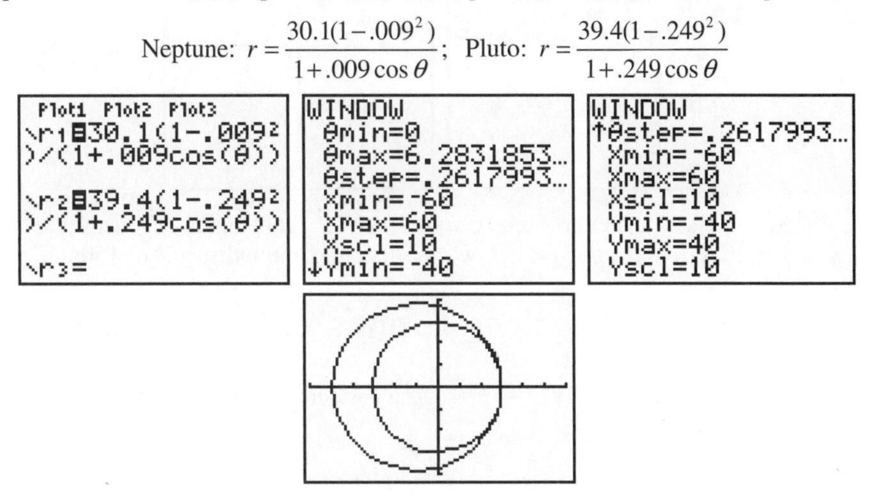

The graph shows that their orbits are very close near the polar axis.

Use ZOOM or change your window to see that the orbit of Pluto does indeed pass inside the orbit of Neptune. Therefore, there are times when Neptune, not Pluto, is the farthest planet from the sun. (However, Pluto's average distance from the sun is considerably greater than Neptune's average distance.)

78. **(a)** In degree mode, graph $r^2=40{,}000\cos 2\theta$.

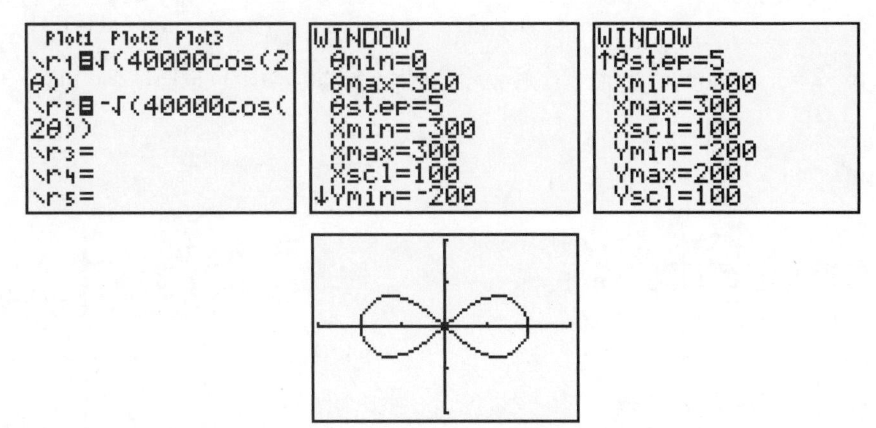

Inside the "figure eight" the radio signal can be received. This region is generally in an east-west direction from the two radio towers with a maximum distance of 200 mi.

Continued on next page

78. (continued)

(b) In degree mode, graph $r^2 = 22,500 \sin 2\theta$.

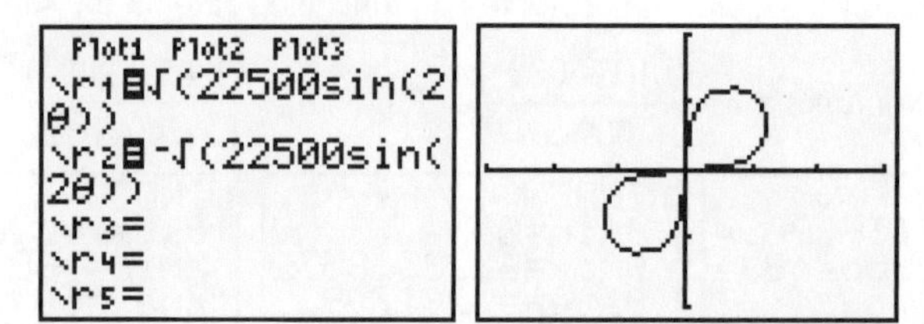

Inside the "figure eight" the radio signal can be received. This region is generally in a northeast-southwest direction from the two radio towers with a maximum distance of 150 mi.

Section 8.6: Parametric Equations, Graphs, and Applications

1. At $t = 2$, $x = 3(2) + 6 = 12$ and $y = -2(2) + 4 = 0$. The correct choice is C.

2. At $t = \frac{\pi}{4}$, $x = \cos\left(\frac{\pi}{4}\right) = \frac{\sqrt{2}}{2}$ and $y = \sin\left(\frac{\pi}{4}\right) = \frac{\sqrt{2}}{2}$. The correct choice is D.

3. At $t = 5$, $x = 5$ and $y = 5^2 = 25$. The correct choice is A.

4. At $t = 2$, $x = 2^2 + 3 = 7$ and $y = 2^2 - 2 = 2$. The correct choice is B.

5. **(a)** $x = t + 2$, $y = t^2$, for t in $[-1, 1]$

t	$x = t + 2$	$y = t^2$
-1	$-1 + 2 = 1$	$(-1)^2 = 1$
0	$0 + 2 = 2$	$0^2 = 0$
1	$1 + 2 = 3$	$1^2 = 1$

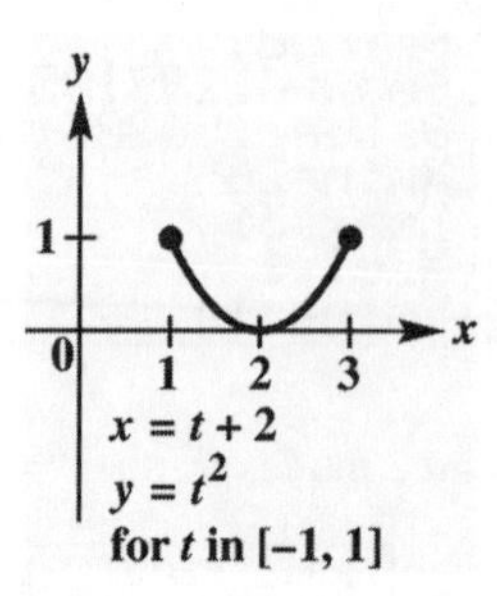

(b) $x - 2 = t$, therefore $y = (x - 2)^2$ or $y = x^2 - 4x + 4$. Since t is in $[-1, 1]$, x is in $[-1 + 2, 1 + 2]$ or $[1, 3]$.

6. **(a)** $x = 2t$, $y = t + 1$, for t in $[-2, 3]$

t	$x = 2t$	$y = t + 1$
-2	$2(-2) = -4$	$-2 + 1 = -1$
-1	$2(-1) = -2$	$-1 + 1 = 0$
0	$2(0) = 0$	$0 + 1 = 1$
1	$2(1) = 2$	$1 + 1 = 2$
2	$2(2) = 4$	$2 + 1 = 3$
3	$2(3) = 6$	$3 + 1 = 4$

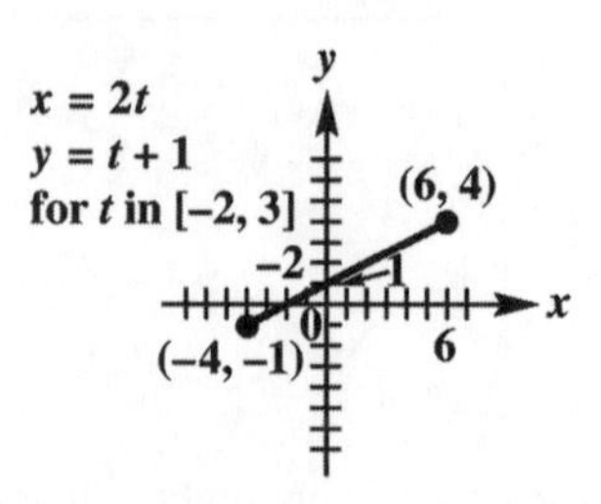

(b) Since $x = 2t \Rightarrow \frac{x}{2} = t$, we have $y = \frac{x}{2} + 1$. Since t is in $[-2, 3]$, x is in $[2(-2), 2(3)]$ or $[-4, 6]$.

7. **(a)** $x = \sqrt{t}$, $y = 3t - 4$, for t in [0, 4].

t	$x = \sqrt{t}$	$y = 3t - 4$
0	$\sqrt{0} = 0$	$3(0) - 4 = -4$
1	$\sqrt{1} = 1$	$3(1) - 4 = -1$
2	$\sqrt{2} = 1.4$	$3(2) - 4 = 2$
3	$\sqrt{3} = 1.7$	$3(3) - 4 = 5$
4	$\sqrt{4} = 2$	$3(4) - 4 = 8$

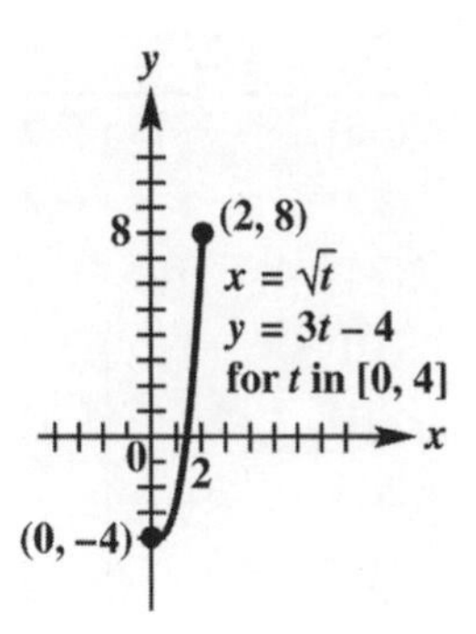

(b) $x = \sqrt{t}$, $y = 3t - 4$

Since $x = \sqrt{t} \Rightarrow x^2 = t$, we have $y = 3x^2 - 4$. Since t is in [0, 4], x is in $[\sqrt{0}, \sqrt{4}]$ or [0, 2].

8. **(a)** $x = t^2$, $y = \sqrt{t}$, for t in [0, 4]

t	$x = t^2$	$y = \sqrt{t}$
0	$0^2 = 0$	$\sqrt{0} = 0$
1	$1^2 = 1$	$\sqrt{1} = 1$
2	$2^2 = 4$	$\sqrt{2} = 1.414$
3	$3^2 = 9$	$\sqrt{3} = 1.732$
4	$4^2 = 16$	$\sqrt{4} = 2$

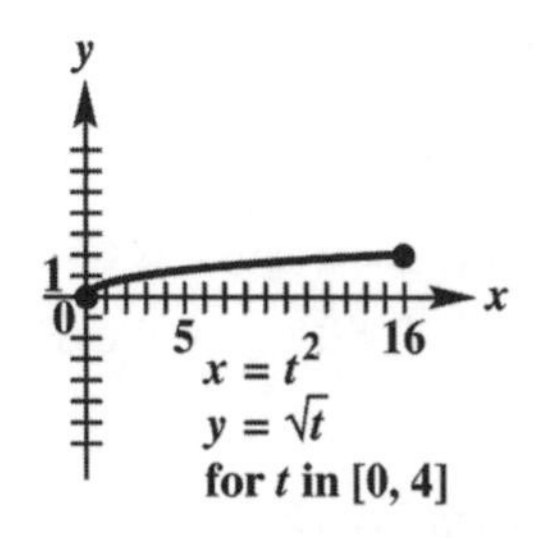

(b) Since $y = \sqrt{t} \Rightarrow y^2 = t$, we have $x = t^2 = \left(y^2\right)^2 = y^4$ or $y = \sqrt[4]{x}$. Since t is in [0, 4], x is in $[0^2, 4^2]$, or [0, 16].

9. **(a)** $x = t^3 + 1$, $y = t^3 - 1$, for t in $(-\infty, \infty)$

t	$x = t^3 + 1$	$y = t^3 - 1$
-2	$(-2)^3 + 1 = -7$	$(-2)^3 - 1 = -9$
-1	$(-1)^3 + 1 = 0$	$(-1)^3 - 1 = -2$
0	$0^3 + 1 = 1$	$0^3 - 1 = -1$
1	$1^3 + 1 = 2$	$1^3 - 1 = 0$
2	$2^3 + 1 = 9$	$2^3 - 1 = 7$
3	$3^3 + 1 = 28$	$3^3 - 1 = 26$

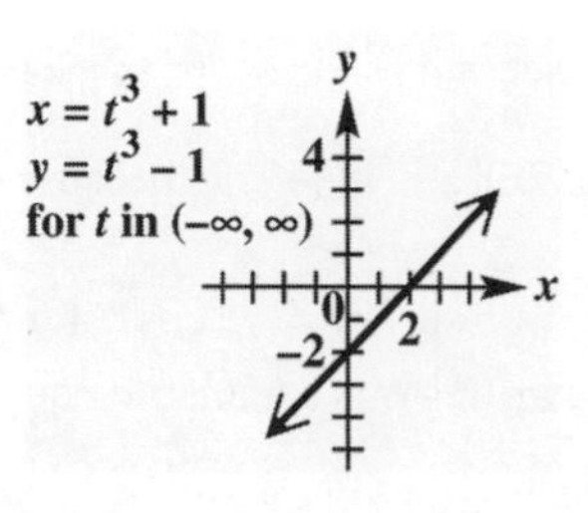

(b) Since $x = t^3 + 1$, we have $x - 1 = t^3$. Since $y = t^3 - 1$, we have $y = (x-1) - 1 = x - 2$. Since t is in $(-\infty, \infty)$, x is in $(-\infty, \infty)$.

10. (a) $x = 2t - 1,\ y = t^2 + 2$, for t in $(-\infty, \infty)$

t	$x = 2t - 1$	$y = t^2 + 2$
−3	$2(-3) - 1 = -7$	$(-3)^2 + 2 = 11$
−2	$2(-2) - 1 = -5$	$(-2)^2 + 2 = 6$
−1	$2(-1) - 1 = -3$	$(-1)^2 + 2 = 3$
0	$2(0) - 1 = -1$	$0^2 + 2 = 2$
1	$2(1) - 1 = 1$	$1^2 + 2 = 3$
2	$2(2) - 1 = 3$	$2^2 + 2 = 6$
3	$2(3) - 1 = 5$	$3^2 + 2 = 11$

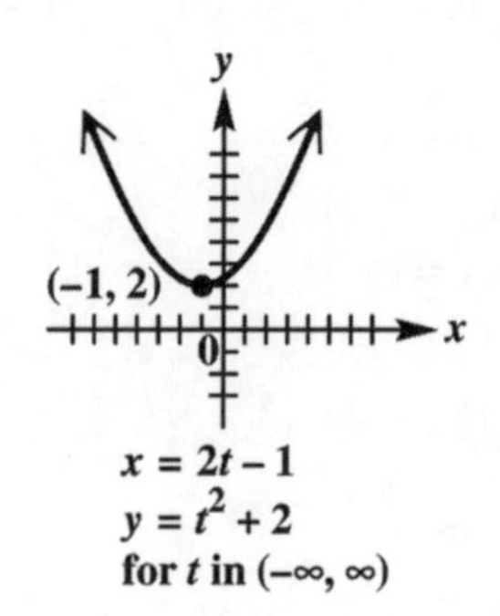

(b) Since $x = 2t - 1 \Rightarrow x + 1 = 2t$, we have $\frac{x+1}{2} = t$. Since $y = t^2 + 2$, we have the following.

$$y = \left(\frac{x+1}{2}\right)^2 + 2 = \frac{1}{4}(x+1)^2 + 2$$

Since t is in $(-\infty, \infty)$ and $x = 2t - 1$, x is in $(-\infty, \infty)$.

11. (a) $x = 2\sin t,\ y = 2\cos t$, for t in $[0, 2\pi]$

t	$x = 2\sin t$	$y = 2\cos t$
0	$2\sin 0 = 0$	$2\cos\theta = 2$
$\frac{\pi}{6}$	$2\sin\frac{\pi}{6} = 1$	$2\cos\frac{\pi}{6} = \sqrt{3}$
$\frac{\pi}{4}$	$2\sin\frac{\pi}{4} = \sqrt{2}$	$2\cos\frac{\pi}{4} = \sqrt{2}$
$\frac{\pi}{3}$	$2\sin\frac{\pi}{3} = \sqrt{3}$	$2\cos\frac{\pi}{3} = 1$
$\frac{\pi}{2}$	$2\sin\frac{\pi}{2} = 2$	$2\cos\frac{\pi}{2} = 0$

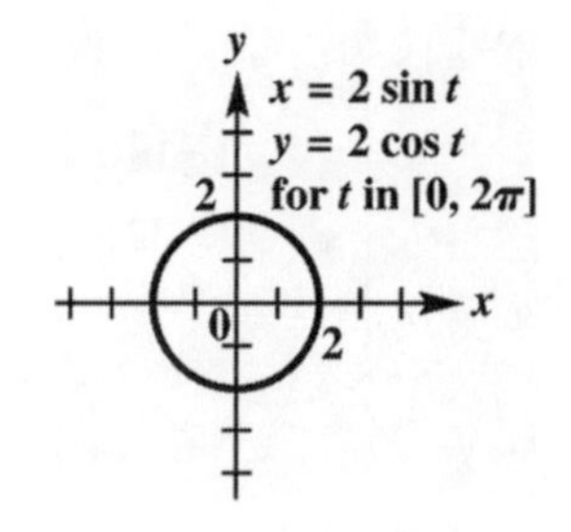

(b) Since $x = 2\sin t$ and $y = 2\cos t$, we have $\frac{x}{2} = \sin t$ and $\frac{y}{2} = \cos t$.

Since $\sin^2 t + \cos^2 t = 1$, we have $\left(\frac{x}{2}\right)^2 + \left(\frac{y}{2}\right)^2 = 1 \Rightarrow \frac{x^2}{4} + \frac{y^2}{4} = 1 \Rightarrow x^2 + y^2 = 4$. Since t is in $[0, 2\pi]$, x is in $[-2, 2]$ because the graph is a circle, centered at the origin, with radius 2.

12. (a) $x = \sqrt{5}\sin t,\ y = \sqrt{3}\cos t$, for t in $[0, 2\pi]$

Rewriting our parametric equations, we have $\frac{x}{\sqrt{5}} = \sin t$ and $\frac{y}{\sqrt{3}} = \cos t$. Since $-1 \le \sin t \le 1$, we have the following.

$$-\sqrt{5} \le \sqrt{5}\sin t \le \sqrt{5}$$

Therefore, x is in $[-\sqrt{5}, \sqrt{5}]$. The graph is an ellipse, centered at the origin, with vertices $(\sqrt{5}, 0), (-\sqrt{5}, 0), (0, \sqrt{3}), (0, -\sqrt{3})$.

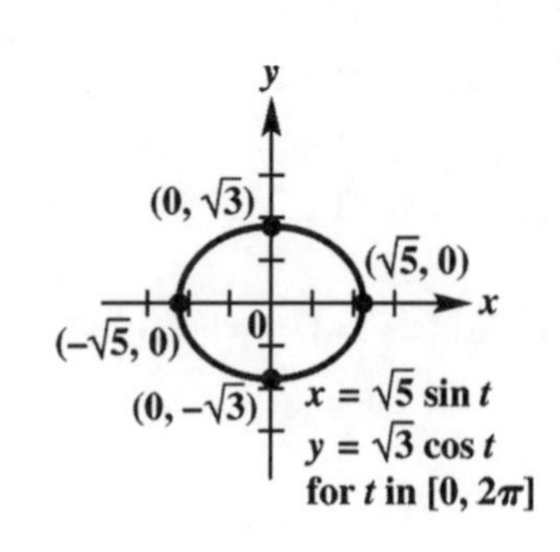

(b) Since $\sin^2 t + \cos^2 t = 1$, we have $\left(\frac{x}{\sqrt{5}}\right)^2 + \left(\frac{y}{\sqrt{3}}\right)^2 = 1 \Rightarrow \frac{x^2}{5} + \frac{y^2}{3} = 1$.

13. (a) $x = 3\tan t,\ y = 2\sec t,$ for t in $\left(-\frac{\pi}{2}, \frac{\pi}{2}\right)$

t	$x = 3\tan t$	$y = 2\sec t$
$-\frac{\pi}{3}$	$3\tan\left(-\frac{\pi}{3}\right) = -3\sqrt{3}$	$2\sec\left(-\frac{\pi}{3}\right) = 4$
$-\frac{\pi}{6}$	$3\tan\left(-\frac{\pi}{6}\right) = -\sqrt{3}$	$2\sec\left(-\frac{\pi}{6}\right) = \frac{4\sqrt{3}}{3}$
0	$3\tan 0 = 0$	$2\sec 0 = 2$
$\frac{\pi}{6}$	$3\tan\frac{\pi}{6} = \sqrt{3}$	$2\sec\frac{\pi}{6} = \frac{4\sqrt{3}}{3}$
$\frac{\pi}{3}$	$3\tan\frac{\pi}{3} = 3\sqrt{3}$	$2\sec\frac{\pi}{3} = 4$

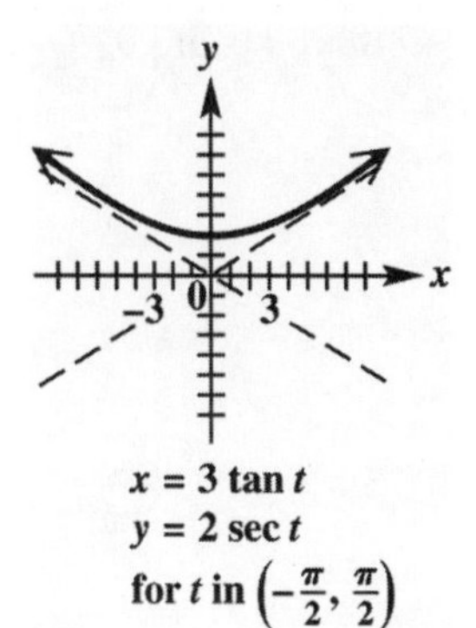

(b) Since $\frac{x}{3} = \tan t,\ \frac{y}{2} = \sec t,$ and $1 + \tan^2 t = \left(\frac{y}{2}\right)^2 = \sec^2 t,$ we have the following.

$$1 + \left(\frac{x}{3}\right)^2 = \left(\frac{y}{2}\right)^2 \Rightarrow 1 + \frac{x^2}{9} = \frac{y^2}{4} \Rightarrow y^2 = 4\left(1 + \frac{x^2}{9}\right) \Rightarrow y = 2\sqrt{1 + \frac{x^2}{9}}$$

Since this graph is the top half of a hyperbola, x is in $(-\infty, \infty)$.

14. (a) $x = \cot t,\ y = \csc t,$ for t in $(0, \pi)$

Since t is in $(0, \pi)$ and the value of the cotangent of a value close to 0 is very large and the value of the cotangent of a value close to π is very small, x is in $(-\infty, \infty)$. The graph is the top half of a hyperbola with vertex $(0, 1)$.

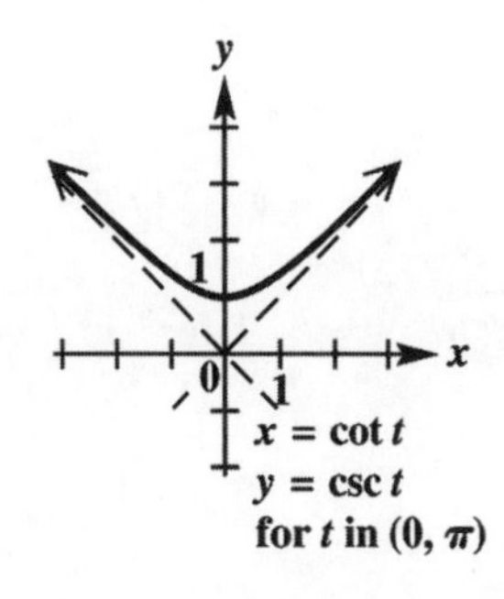

(b) Since $1 + \cot^2 t = \csc^2 t,$ we have $1 + x^2 = y^2 \Rightarrow y = \sqrt{1 + x^2}$.

15. (a) $x = \sin t,\ y = \csc t$ for t in $(0, \pi)$

Since t is in $(0, \pi)$ and $x = \sin t$, x is in $(0, 1]$.

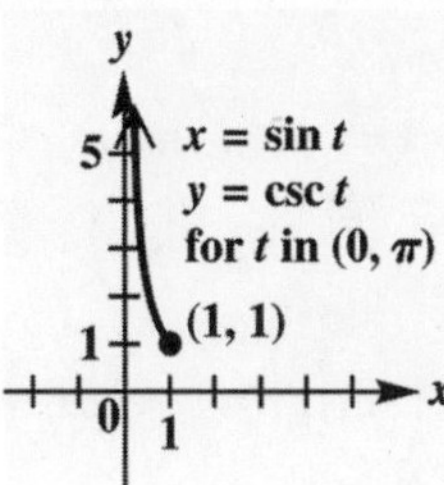

(b) Since $x = \sin t$ and $y = \csc t = \frac{1}{\sin t}$, we have $y = \frac{1}{x}$, where x is in $(0, 1]$.

16. (a) $x = \tan t, y = \cot t$, for t in $\left(0, \frac{\pi}{2}\right)$

Since t is in $\left(0, \frac{\pi}{2}\right)$, x is in $(0, \infty)$.

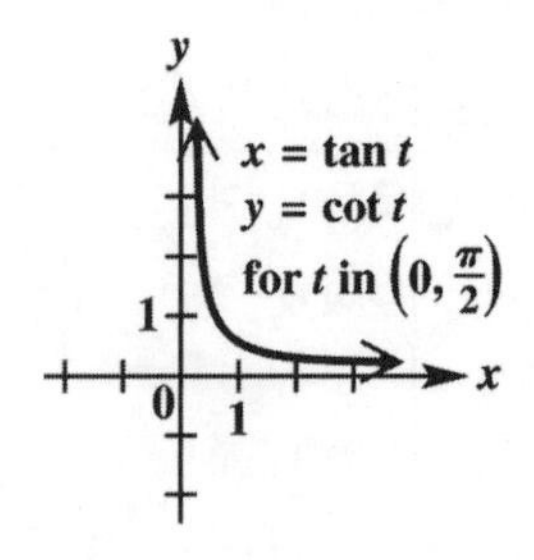

(b) Since $\cot t = \frac{1}{\tan t}$, $y = \frac{1}{x}$, where x is in $(0, \infty)$.

17. (a) $x = t, y = \sqrt{t^2 + 2}$, for t in $(-\infty, \infty)$

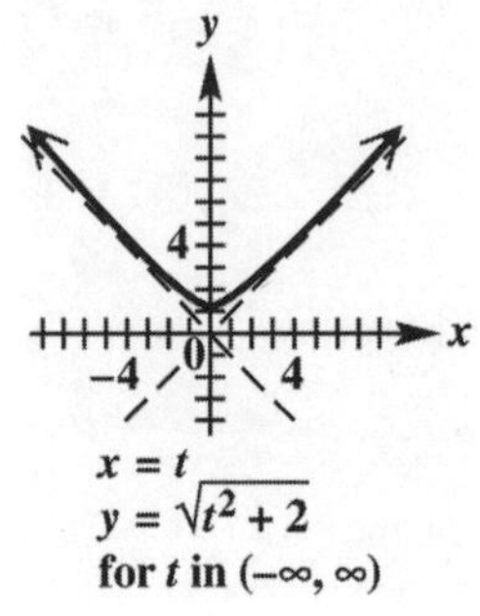

(b) Since $x = t$ and $y = \sqrt{t^2 + 2}$, $y = \sqrt{x^2 + 2}$. Since t is in $(-\infty\ \infty)$ and $x = t$, x is in $(-\infty, \infty)$.

18. (a) $x = \sqrt{t}$, $y = t^2 - 1$, for t in $[0, \infty)$

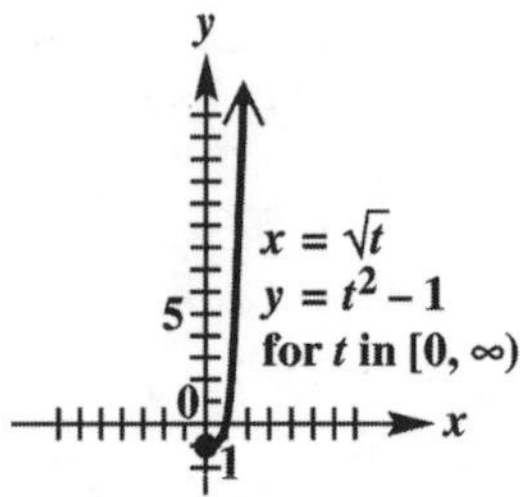

(b) Since $x = \sqrt{t}$, we have $x^2 = t$. Therefore, $y = t^2 - 1 = (x^2)^2 - 1 = x^4 - 1$. Since t is in $[0, \infty)$, x is in $\left[\sqrt{0}, \sqrt{\infty}\right)$ or $[0, \infty)$.

19. (a) $x = 2 + \sin t, y = 1 + \cos t$, for t in $[0, 2\pi]$

Since this is a circle centered at $(2,1)$ with radius 1, and t is in $[0, 2\pi]$, x is in $[1, 3]$.

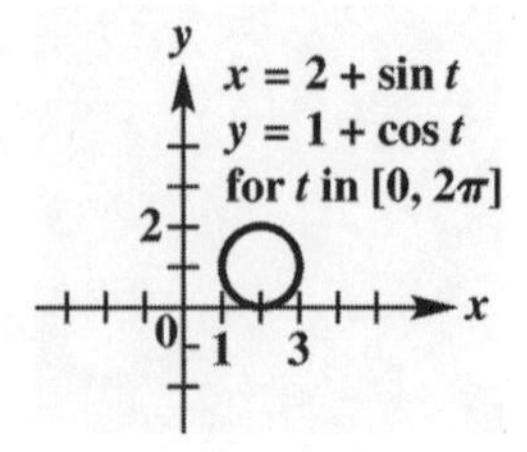

(b) Since $x = 2 + \sin t$ and $y = 1 + \cos t$, $x - 2 = \sin t$ and $y - 1 = \cos t$. Since $\sin^2 t + \cos^2 t = 1$, we have $(x-2)^2 + (y-1)^2 = 1$.

20. (a) $x = 1 + 2\sin t,\ y = 2 + 3\cos t$, for t in $[0, 2\pi]$

Since $x - 1 = 2\sin t \Rightarrow \dfrac{x-1}{2} = \sin t$ and $y - 2 = 3\cos t \Rightarrow \dfrac{y-2}{3} = \cos t$, x is in $[-1, 3]$. The graph is an ellipse with center $(1, 2)$ and axes endpoints $(3,2)$, $(-1,2)$, $(1,5)$, $(1,-1)$.

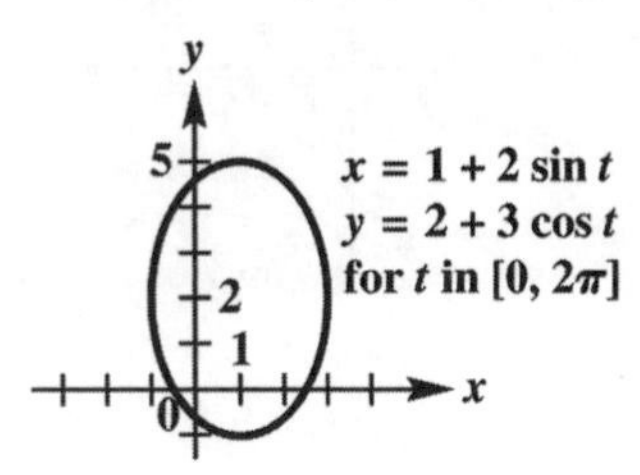

(b) Since $\sin^2 t + \cos^2 t = 1$, $\left(\dfrac{x-1}{2}\right)^2 + \left(\dfrac{y-2}{3}\right)^2 = 1 \Rightarrow \dfrac{(x-1)^2}{4} + \dfrac{(y-2)^2}{9} = 1$.

Also, $-1 \le \sin t \le 1, \Rightarrow -2 \le 2\sin t \le 2 \Rightarrow -1 \le 1 + 2\sin t \le 3$.

21. (a) $x = t + 2,\ y = \dfrac{1}{t+2}$, for $t \ne 2$

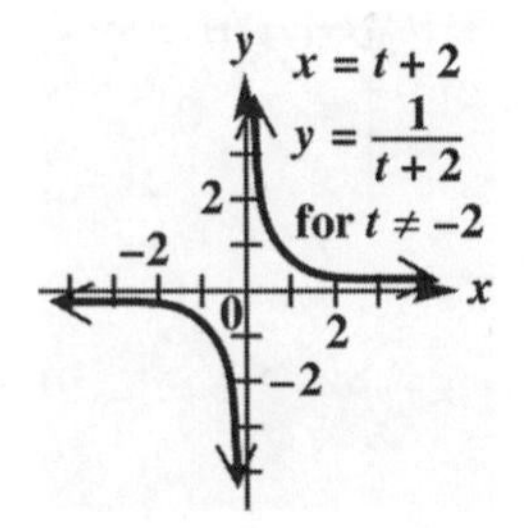

(b) Since $x = t + 2$ and $y = \dfrac{1}{t+2}$, we have $y = \dfrac{1}{x}$. Since $t \ne -2$, $x \ne -2 + 2$, $x \ne 0$. Therefore, x is in $(-\infty, 0) \cup (0, \infty)$.

22. (a) $x = t - 3,\ y = \dfrac{2}{t-3}$, for $t \ne 3$

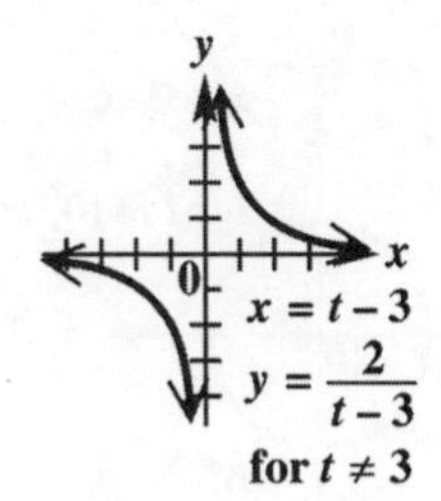

(b) Since $y = \dfrac{2}{t-3}$, we have $y = \dfrac{2}{x}$. Since $t \ne 3$, $x \ne 3 - 3 = 0$. Therefore, x is in $(-\infty, 0) \cup (0, \infty)$.

23. (a) $x = t + 2,\ y = t - 4$, for t in $(-\infty, \infty)$

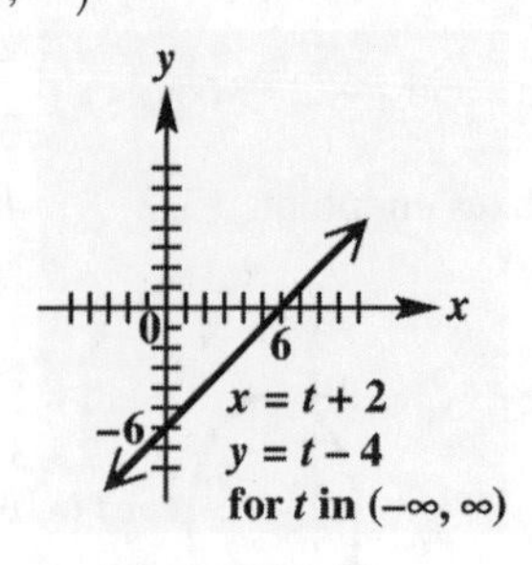

(b) Since $x = t + 2$, we have $t = x - 2$. Since $y = t - 4$, we have $y = (x - 2) - 4 = x - 6$. Since t is in $(-\infty, \infty)$, x is in $(-\infty, \infty)$.

24. (a) $x = t^2 + 2,\ y = t^2 - 4$, for t in $(-\infty, \infty)$

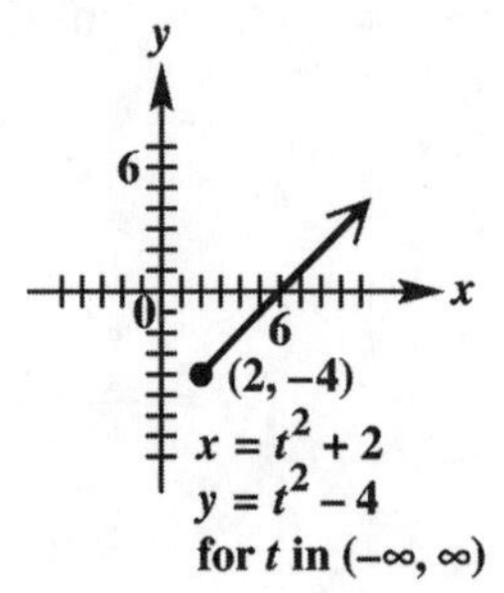

(b) Since $x = t^2 + 2$, we have $t^2 = x - 2$. Since $y = t^2 - 4$, we have $y = (x - 2) - 4 = x - 6$. Since t is in $(-\infty, \infty)$, x is in $[2, \infty)$.

25. $x = 3\cos t,\ y = 3\sin t$

Since $x = 3\cos t \Rightarrow \cos t = \frac{x}{3}$, $y = 3\sin t \Rightarrow \sin t = \frac{y}{3}$, and $\sin^2 t + \cos^2 t = 1$, we have the following.

$$\left(\frac{y}{3}\right)^2 + \left(\frac{x}{3}\right)^2 = 1 \Rightarrow \frac{y^2}{9} + \frac{x^2}{9} = 1 \Rightarrow x^2 + y^2 = 9$$

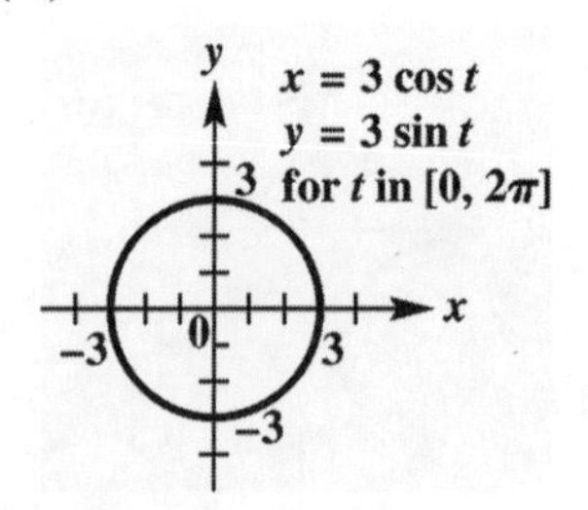

This is a circle centered at the origin with radius 3.

26. $x = 2\cos t,\ y = 2\sin t$

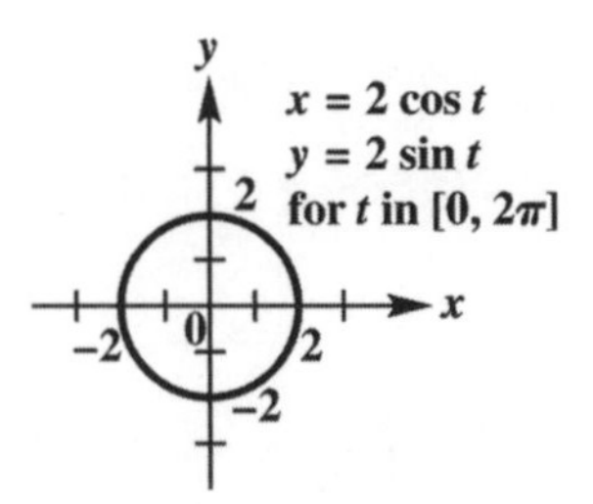

Since $x = 2\cos t \Rightarrow \cos t = \frac{x}{2}$, $y = 2\sin t \Rightarrow \sin t = \frac{y}{2}$, and $\sin^2 t + \cos^2 t = 1$, we have the following.

$$\left(\frac{y}{2}\right)^2 + \left(\frac{x}{2}\right)^2 = 1 \Rightarrow \frac{y^2}{4} + \frac{x^2}{4} = 1 \Rightarrow x^2 + y^2 = 4$$

This is a circle centered at the origin with radius 2.

27. $x = 3\sin t,\ y = 2\cos t$

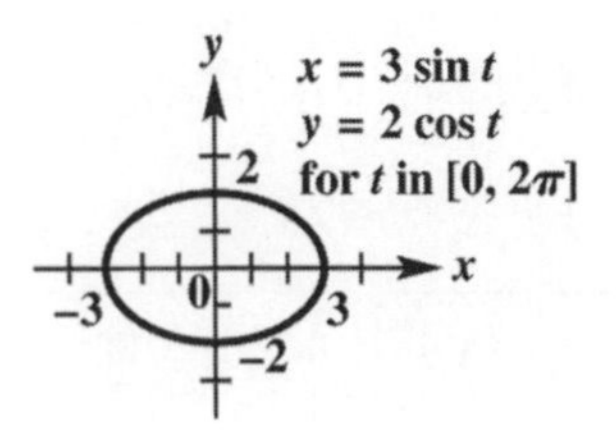

Since $x = 3\sin t \Rightarrow \sin t = \frac{x}{3}$, $y = 2\cos t \Rightarrow \cos t = \frac{y}{2}$, and $\sin^2 t + \cos^2 t = 1$, we have the following.

$$\left(\frac{x}{3}\right)^2 + \left(\frac{y}{2}\right)^2 = 1 \Rightarrow \frac{x^2}{9} + \frac{y^2}{4} = 1$$

This is an ellipse centered at the origin with axes endpoints $(-3,0), (3,0), (0,-2), (0,2)$.

28. $x = 4\sin t,\ y = 3\cos t$

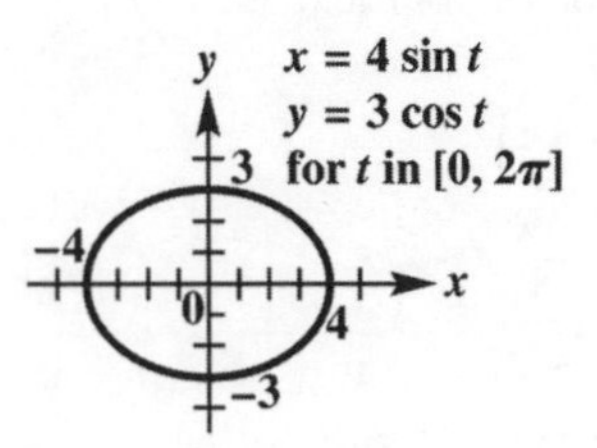

Since $x = 4\sin t \Rightarrow \sin t = \frac{x}{4}$, $y = 3\cos t \Rightarrow \cos t = \frac{y}{3}$, and $\sin^2 t + \cos^2 t = 1$, we have the following.

$$\left(\frac{x}{4}\right)^2 + \left(\frac{y}{3}\right)^2 = 1 \Rightarrow \frac{x^2}{16} + \frac{y^2}{9} = 1$$

This is an ellipse centered at the origin axes endpoints $(-4,0), (4,0), (0,-3), (0,3)$.

In Exercises 29 – 32, answers may vary.

29. $y=(x+3)^2-1$

$x=t,\ y=(t+3)^2-1$ for t in $(-\infty,\infty)$; $x=t-3,\ y=t^2-1$ for t in $(-\infty,\infty)$

30. $y=(x+4)^2+2$

$x=t,\ y=(t+4)^2+2$ for t in $(-\infty,\infty)$; $x=t-4,\ y=t^2+2$ for t in $(-\infty,\infty)$

31. $y=x^2-2x+3=(x-1)^2+2$

$x=t,\ y=(t-1)^2+2$ for t in $(-\infty,\infty)$; $x=t+1,\ y=t^2+2$ for t in $(-\infty,\infty)$

32. $y=x^2-4x+6=(x-2)^2+2$

$x=t,\ y=(t-2)^2+2$ for t in $(-\infty,\infty)$; $x=t+2,\ y=t^2+2$ for t in $(-\infty,\infty)$

33. $x=2t-2\sin t, y=2-2\cos t,$ for t in $[0,8\pi]$

t	0	$\frac{\pi}{2}$	π	$\frac{3\pi}{2}$	2π	3π	4π	5π	6π	7π	8π
$x=2t-2\sin t$	0	1.4	2π	11.4	4π	6π	8π	10π	12π	14π	16π
$y=2-2\cos t$	0	2	4	2	0	4	0	4	0	4	0

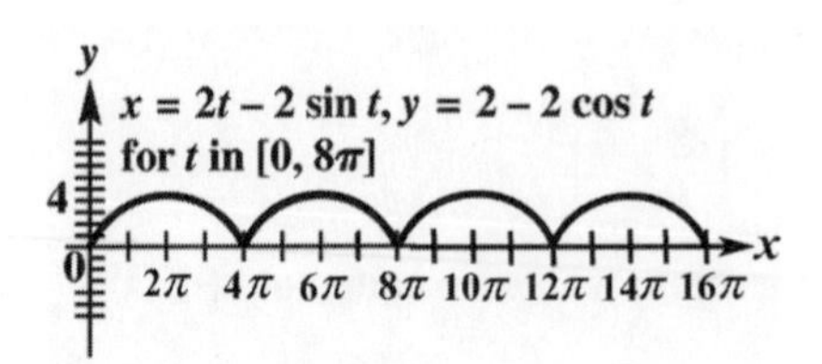

Note: The graph in the answer section of your text shows up to 8π on the x-axis.

34. $x=t-\sin t, y=1-\cos t,$ for t in $[0,4\pi]$

t	0	$\frac{\pi}{2}$	π	$\frac{3\pi}{2}$	2π	$\frac{5\pi}{2}$	3π	$\frac{7\pi}{2}$	4π
$x=t-\sin t$	0	.6	π	5.7	2π	6.8	3π	12.0	4π
$y=1-\cos t$	0	1	2	1	0	1	2	1	0

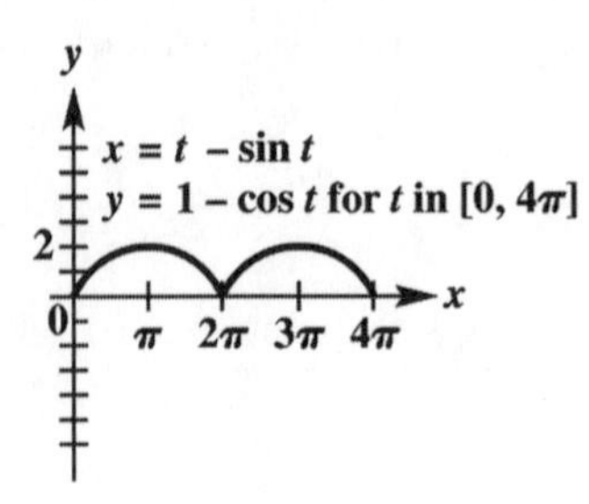

Exercises 35 – 38 are graphed in parametric mode in the following window.

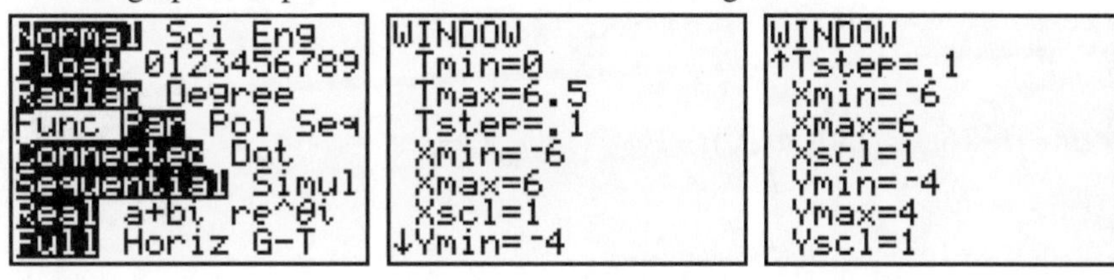

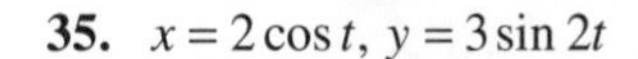

35. $x = 2\cos t,\ y = 3\sin 2t$

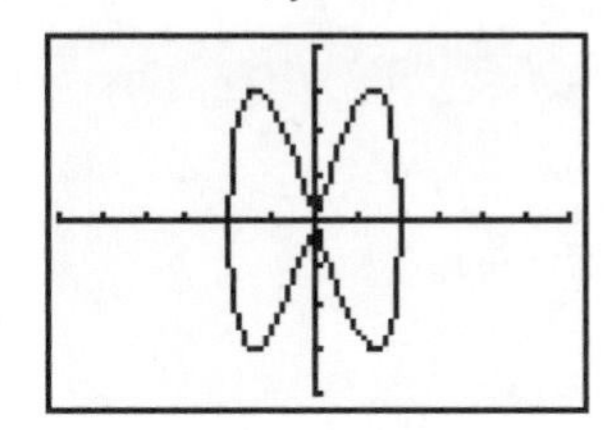

36. $x = 3\cos 2t,\ y = 3\sin 3t$

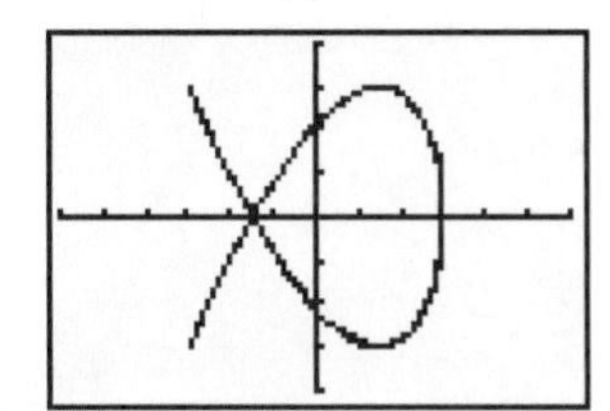

37. $x = 3\sin 4t,\ y = 3\cos 3t$

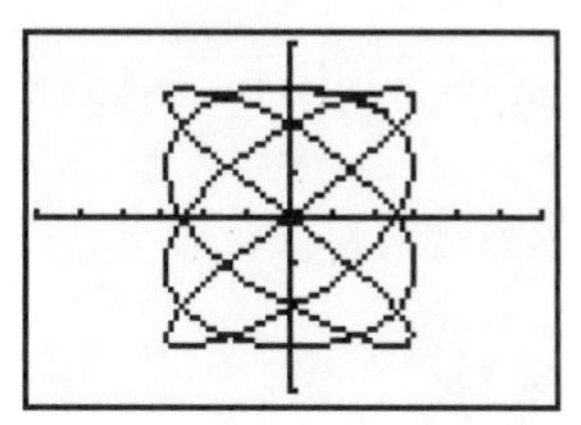

38. $x = 4\sin 4t,\ y = 3\sin 5t$

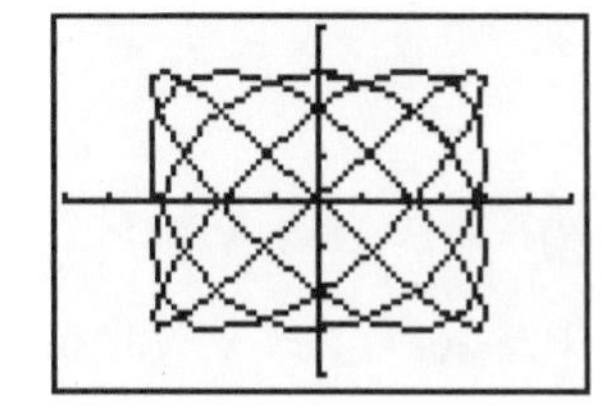

For Exercises 39–42, recall that the motion of a projectile (neglecting air resistance) can be modeled by: $x = (v_0 \cos\theta)t,\ y = (v_0 \sin\theta)t - 16t^2$ for t in $[0, k]$.

39. (a) $x = (v\cos\theta)t \Rightarrow x = (48\cos 60°)t = 48\left(\frac{1}{2}\right)t = 24t$

$$y = (v\sin\theta)t - 16t^2 \Rightarrow y = (48\sin 60°)t - 16t^2 = 48 \cdot \frac{\sqrt{3}}{2}t - 16t^2 = -16t^2 + 24\sqrt{3}t$$

(b) $t = \frac{x}{24}$, so $y = -16\left(\frac{x}{24}\right)^2 + 24\sqrt{3}\left(\frac{x}{24}\right) = -\frac{x^2}{36} + \sqrt{3}x$

(c) $y = -16t^2 + 24\sqrt{3}t$

When the rocket is no longer in flight, $y = 0$. Solve $0 = -16t^2 + 24\sqrt{3}t \Rightarrow 0 = t\left(-16t + 24\sqrt{3}\right)$.

$$t = 0 \text{ or } -16t + 24\sqrt{3} = 0 \Rightarrow -16t = -24\sqrt{3} \Rightarrow t = \frac{24\sqrt{3}}{16} \Rightarrow t = \frac{3\sqrt{3}}{2} \approx 2.6$$

The flight time is about 2.6 seconds. The horizontal distance at $t = \frac{3\sqrt{3}}{2}$ is the following.

$$x = 24t = 24\left(\frac{3\sqrt{3}}{2}\right) \approx 62 \text{ ft}$$

40. (a) $x=(v\cos\theta)t \Rightarrow x=(150\cos 60°)t=150\left(\frac{1}{2}\right)t=75t$

$$y=(v\sin\theta)t-16t^2=(150\sin 60°)t-16t^2=150\frac{\sqrt{3}}{2}t-16t^2=-16t^2+75\sqrt{3}t$$

(b) $t=\frac{x}{75}$, so $y=-16\left(\frac{x}{75}\right)^2+75\sqrt{3}\left(\frac{x}{75}\right)=-\frac{16}{5625}x^2+\sqrt{3}x$

(c) $y=-16t^2+75\sqrt{3}t$

When the golf ball is no longer in flight, $y=0$. Solve $0=-16t^2+75\sqrt{3}t \Rightarrow 0=t\left(-16t+75\sqrt{3}\right)$.

$$t=0 \text{ or } -16t+75\sqrt{3}=0 \Rightarrow -16t=-75\sqrt{3} \Rightarrow t=\frac{75\sqrt{3}}{16}\approx 8.1$$

The flight time is about 8.1 seconds. The horizontal distance at $t=\frac{75\sqrt{3}}{16}$ is the following.

$$x=75t=75\left(\frac{75\sqrt{3}}{16}\right)\approx 609 \text{ ft}$$

41. (a) $x=(v\cos\theta)t \Rightarrow x=(88\cos 20°)t$

$$y=(v\sin\theta)t-16t^2+2 \Rightarrow y=(88\sin 20°)t-16t^2+2$$

(b) Since $t=\frac{x}{88\cos 20°}$, we have the following.

$$y=88\sin 20°\left(\frac{x}{88\cos 20°}\right)-16\left(\frac{x}{88\cos 20°}\right)^2+2=(\tan 20°)x-\frac{x^2}{484\cos^2 20°}+2$$

(c) Solving $0=-16t^2+(88\sin 20°)t+2$ by the quadratic formula, we have the following.

$$t=\frac{-88\sin 20°\pm\sqrt{(88\sin 20°)^2-4(-16)(2)}}{(2)(-16)}=\frac{-30.098\pm\sqrt{905.8759+128}}{-32}\Rightarrow t\approx -0.064 \text{ or } 1.9$$

Discard $t=-0.064$ since it is an unacceptable answer. At $t\approx 1.9$ sec, $x=(88\cos 20°)t\approx 161$ ft. The softball traveled 1.9 sec and 161 feet.

42. (a) $x=(v\cos\theta)t \Rightarrow x=(136\cos 29°)t$

$$y=(v\sin\theta)t-16t^2+2.5 \Rightarrow y=(136\sin 29°)t-16t^2+2.5=2.5-16t^2+(136\sin 29°)t$$

(b) Since $t=\frac{x}{136\cos 29°}$, we have the following.

$$y=(136\sin 29°)\left(\frac{x}{136\cos 29°}\right)-16\left(\frac{x}{136\cos 29°}\right)^2+2.5=(\tan 29°)x-\frac{x^2}{1156\cos^2 29°}+2.5$$

(c) Solving $0=-16t^2+(136\sin 29°)t+2.5$ by the quadratic formula, we have the following.

$$t=\frac{-136\sin 29°\pm\sqrt{(136\sin 29°)^2-4(-16)(2.5)}}{2(-16)}=\frac{-65.934\pm\sqrt{4347.3066+160}}{-32}\Rightarrow t\approx -.04,\ 4.2$$

Discard $t=-.04$ since it gives an unacceptable answer. At $t\approx 4.2$ sec, $x=(136\cos 29°)t\approx 495$ ft. The baseball traveled 4.2 sec and 495 feet.

43. (a) $x=(v\cos\theta)t \Rightarrow x=(128\cos 60°)t=128\left(\frac{1}{2}\right)t=64t$

$y=(v\sin\theta)t-16t^2+2.5 \Rightarrow y=(128\sin 60°)t-16t^2+8=128\left(\frac{\sqrt{3}}{2}\right)t-16t^2+8=64\sqrt{3}t-16t^2+8$

Since $t=\frac{x}{64}$, $y=64\sqrt{3}\left(\frac{x}{64}\right)-16\left(\frac{x}{64}\right)^2+8=y=-\frac{1}{256}x^2+\sqrt{3}x+8$. This is a parabolic path.

(b) Solving $0=-16t^2+64\sqrt{3}t+8$ by the quadratic formula, we have the following.

$$t=\frac{-64\sqrt{3}\pm\sqrt{\left(64\sqrt{3}\right)^2-4(-16)(8)}}{2(-16)}=\frac{-64\sqrt{3}\pm\sqrt{12{,}800}}{-32}=\frac{-64\sqrt{3}\pm 80\sqrt{2}}{-32}\Rightarrow t\approx -.07,\ 7.0$$

Discard $t=-.07$ since it gives an unacceptable answer. At $t\approx 7.0$ sec, $x=64t=448$ ft. The rocket traveled approximately 7 sec and 448 feet.

44. $x=(v\cos\theta)t \Rightarrow x=(88\cos 45°)t=88\left(\frac{\sqrt{2}}{2}\right)t=44\sqrt{2}t$

$y=(v\sin\theta)t-2.66t^2+h \Rightarrow y=(88\sin 45°)t-2.66t^2+0=88\left(\frac{\sqrt{2}}{2}\right)t-2.66t^2=44\sqrt{2}t-2.66t^2$

Solving $0=44\sqrt{2}t-2.66t^2$ by factoring, we have the following.

$$0=44\sqrt{2}t-2.66t^2 \Rightarrow 0=\left(44\sqrt{2}-2.66t\right)t \Rightarrow t=0 \text{ or } 44\sqrt{2}-2.66t=0$$

$t=0$ implies that the golf ball was initially on the ground, which is true. Solving $44\sqrt{2}-2.66t=0$, we obtain the following.

$$44\sqrt{2}-2.66t=0 \Rightarrow 44\sqrt{2}=2.66t \Rightarrow t=\frac{44\sqrt{2}}{2.66}\approx 23.393 \text{ sec}$$

At $t\approx 23.393$ sec, $x=44\sqrt{2}t\approx 1456$ ft. The golf ball traveled approximately 1456 feet.

45. (a) $x=(v\cos\theta)t \Rightarrow x=(64\cos 60°)t=64\left(\frac{1}{2}\right)t=32t$

$y=(v\sin\theta)t-16t^2+3 \Rightarrow y=(64\sin 60°)t-16t^2+3=64\left(\frac{\sqrt{3}}{2}\right)t-16t^2+3=32\sqrt{3}t-16t^2+3$

(b) Solving $0=-16t^2+32\sqrt{3}t+3$ by the quadratic formula, we have the following.

$$t=\frac{-32\sqrt{3}\pm\sqrt{\left(32\sqrt{3}\right)^2-4(-16)(3)}}{2(-16)}=\frac{-32\sqrt{3}\pm\sqrt{3264}}{-32}=\frac{-32\sqrt{3}\pm 8\sqrt{51}}{-32}\Rightarrow t\approx -.05,\ 3.52$$

Discard $t=-.07$ since it gives an unacceptable answer. At $t\approx 3.52$ sec, $x=32t\approx 112.6$ ft. The ball traveled approximately 112.6 feet.

Continued on next page

45. (continued)

(c) To find the maximum height, find the vertex of $y=-16t^2+32\sqrt{3}t+3$.

$$y=-16t^2+32\sqrt{3}t+3=-16\left(t^2-2\sqrt{3}t\right)+3=-16\left(t^2-2\sqrt{3}t+3-3\right)+3$$
$$y=-16\left(t-\sqrt{3}\right)^2+48+3=-16\left(t-\sqrt{3}\right)^2+51$$

The maximum height of 51 ft is reached at $\sqrt{3}\approx 1.73$ sec. Since $x=32t$, the ball has traveled horizontally $32\sqrt{3}\approx 55.4$ ft.

(d) To determine if the ball would clear a 5-ft-high fence that is 100 ft from the batter, we need to first determine at what time is the ball 100 ft from the batter. Since $x=32t$, the time the ball is 100 ft from the batter is $t=\frac{100}{32}=3.125$ sec. We next need to determine how high off the ground the ball is at this time. Since $y=32\sqrt{3}t-16t^2+3$, the height of the ball is the following.

$$y=32\sqrt{3}(3.125)-16(3.125)^2+3\approx 20.0 \text{ ft}$$

Since this height exceeds 5 ft, the ball will clear the fence.

46. (a)

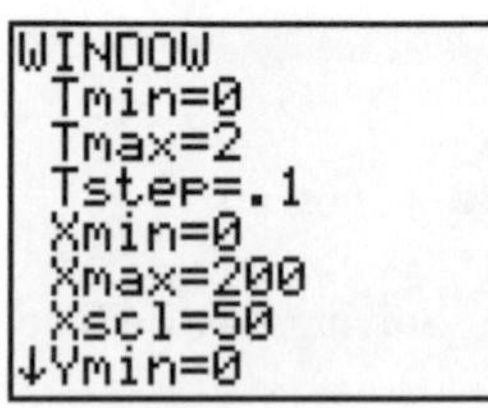

WINDOW
↑Tstep=.1
Xmin=0
Xmax=200
Xscl=50
Ymin=0
Ymax=30
Yscl=10

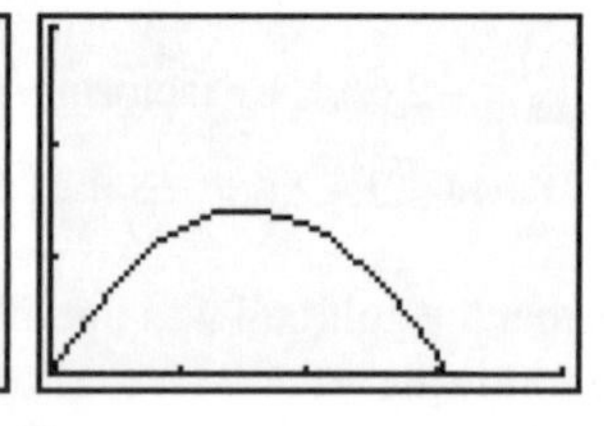

(b) $x=82.69265063t=v(\cos\theta)t\Rightarrow 82.69265063=v\cos\theta$

$y=-16t^2+30.09777261t=v(\sin\theta)t-16t^2\Rightarrow 30.09777261=v\sin\theta$

Thus, $\frac{30.09777261}{82.69265063}=\frac{v\sin\theta}{v\cos\theta}\Rightarrow 0.3697=\tan\theta\Rightarrow\theta\approx 20.0°$

(c) $30.09777261=v\sin 20.0°\Rightarrow v\approx 88.0$ ft/sec

Thus, the parametric equations are $x=88(\cos 20.0°)t,\ y=-16t^2+88(\sin 20.0°)t$.

47. (a)

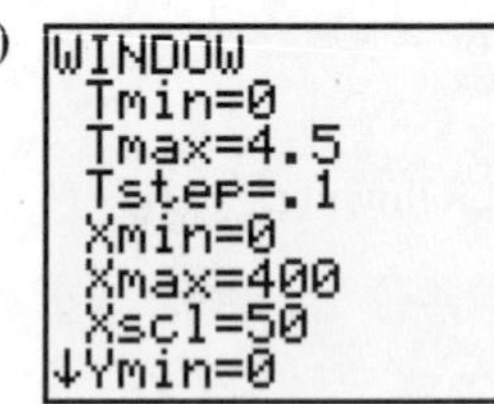

WINDOW
↑Tstep=.1
Xmin=0
Xmax=400
Xscl=50
Ymin=0
Ymax=100
Yscl=10

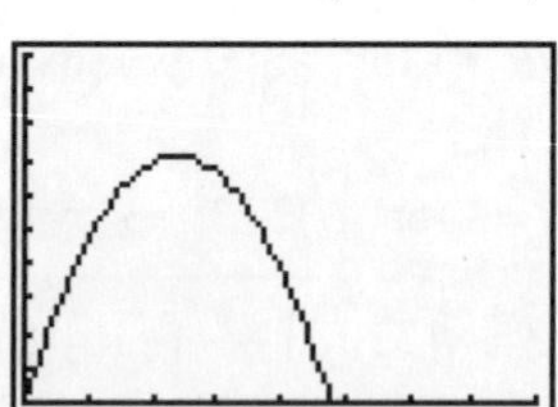

(b) $x=56.56530965t=v(\cos\theta)t\Rightarrow 56.56530965=v\cos\theta$

$y=-16t^2+67.41191099t=-16t^2+v(\sin\theta)t\Rightarrow 67.41191099=v\sin\theta$

Thus, $\frac{67.41191099}{56.56530965}=\frac{v\sin\theta}{v\cos\theta}\Rightarrow 1.1918=\tan\theta\Rightarrow\theta\approx 50.0°$.

(c) $67.41191099=v\sin 50.0°\Rightarrow v\approx 88.0$ ft/sec

Thus, the parametric equations are $x=88(\cos 50.0°)t,\ y=-16t^2+88(\sin 50.0°)t$.

For Exercises 48–51, many answers are possible.

48. The equation of a line with slope m through (x_1, y_1) is $y - y_1 = m(x - x_1)$.
To find two parametric representations, let $x = t$. We therefore have the following.

$$y - y_1 = m(t - x_1) \Rightarrow y = m(t - x_1) + y_1$$

For another representation, let $x = t^2$. We therefore have the following.

$$y - y_1 = m(t^2 - x_1) \Rightarrow y = m(t^2 - x_1) + y_1$$

49. $y = a(x - h)^2 + k$

To find one parametric representation, let $x = t$. We therefore have, $y = a(t - h)^2 + k$.

For another representation, let $x = t + h$. We therefore have $y = a(t + h - h)^2 + k = at^2 + k$.

50. $\frac{x^2}{a^2} - \frac{y^2}{b^2} = 1$

To find a parametric representation, let $x = a\sec\theta$. We therefore have the following.

$$\frac{(a\sec\theta)^2}{a^2} - \frac{y^2}{b^2} = 1 \Rightarrow \frac{a^2\sec^2\theta}{a^2} - \frac{y^2}{b^2} = 1 \Rightarrow \sec^2\theta - \frac{y^2}{b^2} = 1 \Rightarrow \sec^2\theta - 1 = \frac{y^2}{b^2}$$

$$\tan^2\theta = \frac{y^2}{b^2} \Rightarrow b^2\tan^2\theta = y^2 \Rightarrow b\tan\theta = y$$

51. $\frac{x^2}{a^2} + \frac{y^2}{b^2} = 1$

To find a parametric representation, let $x = a\sin t$. We therefore have the following.

$$\frac{(a\sin t)^2}{a^2} + \frac{y^2}{b^2} = 1 \Rightarrow \frac{a^2\sin^2 t}{a^2} + \frac{y^2}{b^2} = 1 \Rightarrow \sin^2 t + \frac{y^2}{b^2} = 1 \Rightarrow \frac{y^2}{b^2} = 1 - \sin^2 t$$

$$y^2 = b^2(1 - \sin^2 t) \Rightarrow y^2 = b^2\cos^2 t \Rightarrow y = b\cos t$$

52. To show that $r = a\theta$ is given parmetrically by $x = a\theta\cos\theta,\ y = a\theta\sin\theta$, for θ in $(-\infty, \infty)$,

we must show that the parametric equations yield $r = a\theta$, where $r^2 = x^2 + y^2$.

$$r^2 = x^2 + y^2 \Rightarrow r^2 = (a\theta\cos\theta)^2 + (a\theta\sin\theta)^2 \Rightarrow r^2 = a^2\theta^2\cos^2\theta + a^2\theta^2\sin^2\theta$$

$$r^2 = a^2\theta^2\cos^2\theta + a^2\theta^2\sin^2\theta \Rightarrow r^2 = a^2\theta^2(\cos^2\theta + \sin^2\theta) \Rightarrow r^2 = a^2\theta^2 \Rightarrow r = \pm a\theta \text{ or just } r = a\theta$$

This implies that the parametric equations satisfy $r = a\theta$.

53. To show that $r\theta = a$ is given parmetrically by $x = \frac{a\cos\theta}{\theta},\ y = \frac{a\sin\theta}{\theta}$, for θ in $(-\infty, 0) \cup (0, \infty)$,

we must show that the parametric equations yield $r\theta = a$, where $r^2 = x^2 + y^2$.

$$r^2 = x^2 + y^2 \Rightarrow r^2 = \left(\frac{a\cos\theta}{\theta}\right)^2 + \left(\frac{a\sin\theta}{\theta}\right)^2 \Rightarrow r^2 = \frac{a^2\cos^2\theta}{\theta^2} + \frac{a^2\sin^2\theta}{\theta^2}$$

$$r^2 = \frac{a^2}{\theta^2}\cos^2\theta + \frac{a^2}{\theta^2}\sin^2\theta \Rightarrow r^2 = \frac{a^2}{\theta^2}(\cos^2\theta + \sin^2\theta) \Rightarrow r^2 = \frac{a^2}{\theta^2} \Rightarrow r = \pm\frac{a}{\theta} \text{ or just } r = \frac{a}{\theta}$$

This implies that the parametric equations satisfy $r\theta = a$.

54. The second set of equations $x = \cos t$, $y = -\sin t$, t in $[0, 2\pi]$ trace the circle out clockwise. A table of values confirms this.

t	$x = \cos t$	$y = -\sin t$
0	$\cos 0 = 1$	$-\sin 0 = 0$
$\frac{\pi}{6}$	$\cos \frac{\pi}{6} = \frac{\sqrt{3}}{2}$	$-\sin \frac{\pi}{6} = -\frac{1}{2}$
$\frac{\pi}{4}$	$\cos \frac{\pi}{4} = \frac{\sqrt{2}}{2}$	$-\sin \frac{\pi}{4} = -\frac{\sqrt{2}}{2}$
$\frac{\pi}{3}$	$\cos \frac{\pi}{3} = \frac{1}{2}$	$-\sin \frac{\pi}{3} = -\frac{\sqrt{3}}{2}$
$\frac{\pi}{2}$	$\cos \frac{\pi}{2} = 0$	$-\sin \frac{\pi}{2} = -1$

55. If $x = f(t)$ is replaced by $x = c + f(t)$, the graph will be translated c units horizontally.

56. If $y = g(t)$ is replaced by $y = d + g(t)$, the graph is translated vertically d units.

Chapter 8: Review Exercises

1. $\sqrt{-9} = i\sqrt{9} = 3i$

2. $\sqrt{-12} = 2i\sqrt{3}$

3. $x^2 = -81 \Rightarrow x = \pm\sqrt{-81} \Rightarrow x = \pm i\sqrt{81} \Rightarrow x = \pm i(9) \Rightarrow x = \pm 9i$

Solution set: $\{\pm 9i\}$

4. $x(2x+3) = -4 \Rightarrow 2x^2 + 3x = -4 \Rightarrow 2x^2 + 3x + 4 = 0$

Use the quadratic formula with $a = 2$, $b = 3$, and $c = 4$.

$$x = \frac{-3 \pm \sqrt{3^2 - 4(2)(4)}}{2(2)} = \frac{-3 \pm \sqrt{9-32}}{4} = \frac{-3 \pm \sqrt{-23}}{4} = \frac{-3 \pm i\sqrt{23}}{4}$$

Solution set: $\left\{\frac{-3 \pm i\sqrt{23}}{4}\right\}$

5. $(1-i)-(3+4i)+2i = (1-3)+(-1-4+2)i = -2-3i$

6. $(2-5i)+(9-10i)-3 = (2+9-3)+(-5-10)i = 8-15i$

7. $(6-5i)+(2+7i)-(3-2i) = (6+2-3)+(-5+7+2)i = 5+4i$

8. $(4-2i)-(6+5i)-(3-i) = (4-6-3)+(-2-5+1)i = -5-6i$

9. $(3+5i)(8-i) = 24-3i+40i-5i^2 = 24+37i-5(-1) = 29+37i$

10. $(4-i)(5+2i) = 20+8i-5i-2i^2 = 22+3i$

11. $(2+6i)^2 = 4+24i+36i^2 = 4+24i+36(-1) = -32+24i$

12. $(6-3i)^2 = 36-36i+9i^2 = 27-36i$

12. $(6-3i)^2 = 36-36i+9i^2$
$= 27-36i$

13. $(1-i)^3 = (1-i)^2(1-i)$
$= (1-2i+i^2)(1-i)$
$= -2i(1-i) = -2i+2i^2$
$= -2i-2 \text{ or } -2-2i$

14. $(2+i)^3 = (2+i)^2(2+i)$
$= (4+4i+i^2)(2+i)$
$= [4+4i+(-1)](2+i)$
$= (3+4i)(2+i)$
$= 6+3i+8i+4i^2 = 2+11i$

15. $\dfrac{6+2i}{3-i} = \dfrac{6+2i}{3-i} \cdot \dfrac{3+i}{3+i}$
$= \dfrac{18+12i+2i^2}{9-i^2}$
$= \dfrac{18+12i+2(-1)}{9-(-1)} = \dfrac{16+12i}{10}$
$= \dfrac{8+6i}{5} = \dfrac{8}{5}+\dfrac{6}{5}i$

16. $\dfrac{2-5i}{1+i} = \dfrac{(2-5i)(1-i)}{(1+i)(1-i)}$
$= \dfrac{2-2i-5i+5i^2}{1-i^2}$
$= \dfrac{-3-7i}{2} = -\dfrac{3}{2}-\dfrac{7}{2}i$

17. $\dfrac{2+i}{1-5i} = \dfrac{2+i}{1-5i} \cdot \dfrac{1+5i}{1+5i}$
$= \dfrac{2+11i+5i^2}{1-25i^2}$
$= \dfrac{-3+11i}{26}$
$= -\dfrac{3}{36}+\dfrac{11}{26}i$

18. $\dfrac{3+2i}{i} = \dfrac{(3+2i)(-i)}{i(-i)}$
$= \dfrac{-3i-2i^2}{-i^2}$
$= 2-3i$

19. $i^{53} = i^{52} \cdot i = (i^4)^{13} \cdot i$
$= 1^{13} \cdot i = 1 \cdot i$
$= i \text{ or } 0+i$

20. $i^{-41} = i^{-44} \cdot i^3 = (i^4)^{-11}(-i) = 1^{-11}(-i) = 1 \cdot (-i) = -i \text{ or } 0-i$

21. $[5(\cos 90° + i \sin 90°)] \cdot [6(\cos 180° + i \sin 180°)] = 5 \cdot 6[\cos(90°+180°) + i \sin(90°+180°)]$
$= 30(\cos 270° + i \sin 270°)$
$= 30(0-i)$
$= 0\text{–}30i \text{ or } \text{–}30i$

22. $[3 \operatorname{cis} 135°][2 \operatorname{cis} 105°] = 3 \cdot 2 \operatorname{cis}(135°+105°)$
$= 6 \operatorname{cis} 240° = 6(\cos 240° + i \sin 240°) = 6\left(-\dfrac{1}{2}-\dfrac{\sqrt{3}}{2}i\right) = -3-3\sqrt{3}i$

23. $\dfrac{2(\cos 60° + i \sin 60°)}{8(\cos 300° + \sin 300°)} = \dfrac{2}{8}[\cos(60°-300°) + i \sin(60°-300°)]$
$= \dfrac{1}{4}[\cos(-240°) + i \sin(-240°)] = \dfrac{1}{4}[\cos(240°) - i \sin(240°)]$
$= \dfrac{1}{4}[-\cos 60° + i \sin 60°] = \dfrac{1}{4}\left(-\dfrac{1}{2}+\dfrac{\sqrt{3}}{2}\right) = -\dfrac{1}{8}+\dfrac{\sqrt{3}}{8}i$

24. $\frac{4 \text{ cis } 270°}{2 \text{ cis } 90°} = \frac{4}{2} \text{cis}(270° - 90°) = 2 \text{ cis } 180° = 2(\cos 180° + i \sin 180°) = 2(-1 + 0i) = -2 + 0i$ or -2

25. $\left(\sqrt{3} + i\right)^3$

$r = \sqrt{\left(\sqrt{3}\right)^2 + 1^2} = \sqrt{3+1} = \sqrt{4} = 2$ and since θ is in quadrant I, $\tan\theta = \frac{1}{\sqrt{3}} = \frac{\sqrt{3}}{3} \Rightarrow \theta = 30°$.

$$\begin{aligned}\left(\sqrt{3} + i\right)^3 &= \left[2(\cos 30° + i \sin 30°)\right]^3 \\ &= 2^3\left[\cos(3 \cdot 30°) + i \sin(3 \cdot 30°)\right] = 8(\cos 90° + i \sin 90°) = 0 + 8i = 8i\end{aligned}$$

26. $(2 - 2i)^5$

$r = \sqrt{2^2 + (-2)^2} = \sqrt{4+4} = \sqrt{8} = 2\sqrt{2}$ and $\tan\theta = \frac{-2}{2} = -1 \Rightarrow \theta = 315°$, since θ is in quadrant IV,

$$\begin{aligned}(2-2i)^5 &= \left[2\sqrt{2}(\cos 315° + i \sin 315°)\right]^5 = \left(2\sqrt{2}\right)^5\left[\cos(5 \cdot 315°) + i \sin(5 \cdot 315°)\right] \\ &= 128\sqrt{2}(\cos 1575° + i \sin 1575°) = 128\sqrt{2}(\cos 135° + i \sin 135°) \\ &= 128\sqrt{2}(-\cos 45° + i \sin 45°) = 128\sqrt{2}\left(-\frac{\sqrt{2}}{2} + i\frac{\sqrt{2}}{2}\right) = -128 + 128i\end{aligned}$$

27. $(\cos 100° + i \sin 100°)^6 = \cos(6 \cdot 100°) + i \sin(6 \cdot 100°)$

$$= \cos 600° + i \sin 600° = \cos 240° + i \sin 240° = -\cos 60° - i \sin 60° = -\frac{1}{2} - \frac{\sqrt{3}}{2}i$$

28. The vector representing a real number will lie on the x - axis in the complex plane.

29.

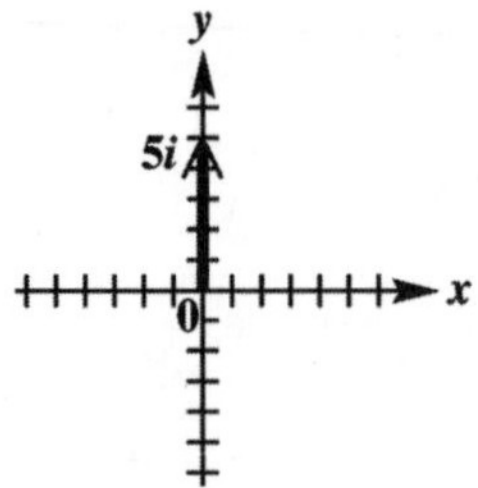

30.

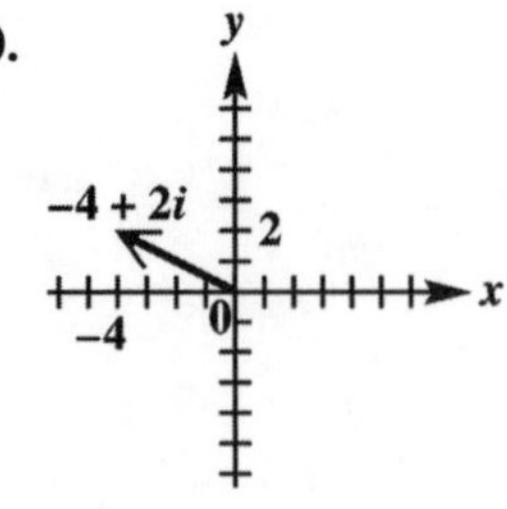

31.

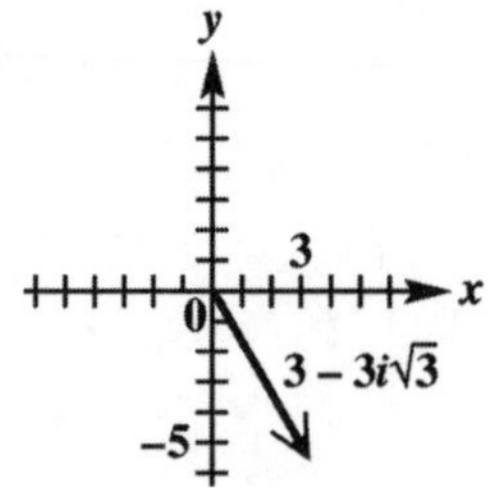

32. The resultant of $7 + 3i$ and $-2 + i$ is $(7 + 3i) + (-2 + i) = 5 + 4i$.

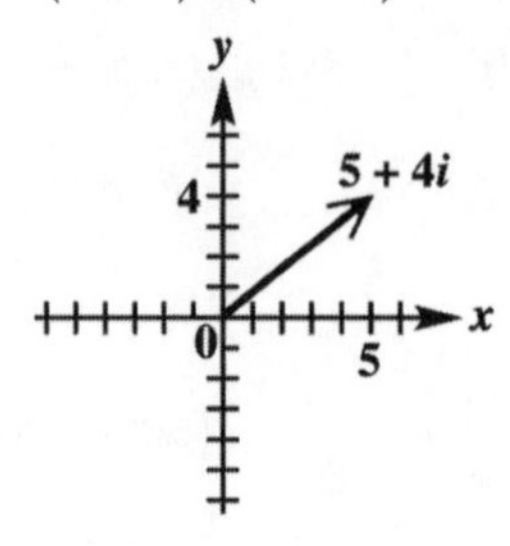

33. $-2 + 2i$

$r = \sqrt{(-2)^2 + 2^2} = \sqrt{4+4} = \sqrt{8} = 2\sqrt{2}$

Since θ is in quadrant II, $\tan\theta = \frac{2}{-2} = -1 \Rightarrow \theta = 135°$. Thus, $-2 + 2i = 2\sqrt{2}(\cos 135° + i \sin 135°)$.

34. $3(\cos 90° + i\sin 90°) = 3(0+i) = 0+3i$ or $3i$

35. $2(\cos 225° + i\sin 225°) = 2(-\cos 45° - i\sin 45°) = 2\left(-\frac{\sqrt{2}}{2} - \frac{i\sqrt{2}}{2}\right) = -\sqrt{2} - i\sqrt{2}$

36. $-4+4i\sqrt{3}$

$r = \sqrt{(-4)^2 + (4\sqrt{3})^2} = \sqrt{16+48} = 8$

Since θ is in quadrant II, $\tan\theta = \frac{4\sqrt{3}}{-4} = -\sqrt{3} \Rightarrow \theta = 120°$.

Thus, $-4+4i\sqrt{3} = 8(\cos 120° + i\sin 120°)$.

37. $1-i$

$r = \sqrt{1^2 + (-1)^2} = \sqrt{1+1} = \sqrt{2}$ and $\tan\theta = \frac{-1}{1} = -1 \Rightarrow \theta = 315°$, since θ is in quadrant IV.

Thus, $1-i = \sqrt{2}(\cos 315° + i\sin 315°)$.

38. $4\text{ cis } 240° = 4(\cos 240° + i\sin 240°) = 4(-\cos 60° - i\sin 60°) = 4\left(-\frac{1}{2} - i\frac{\sqrt{3}}{2}\right) = -2-2i\sqrt{3}$

39. $-4i$

Since $r=4$ and the point $(0,-4)$ intersects the negative y-axis, $\theta = 270°$ and $-4i = 4(\cos 270° + i\sin 270°)$.

40. Since the modulus of z is 2, the graph would be a circle, centered at the origin, with radius 2.

41. $z = x + yi$

Since the imaginary part of z is the negative of the real part of z, we are saying $y = -x$. This is a line.

42. Convert $-2 + 2i$ to polar form.

$r = \sqrt{(-2)^2 + 2^2} = \sqrt{4+4} = \sqrt{8}$ and $\theta = \tan^{-1}\left(\frac{2}{-2}\right) = \tan^{-1}(-1) = 135°$, since θ is in quadrant II.

Thus, $-2+2i = \sqrt{8}(\cos 135° + i\sin 135°)$.

Since $r^5(\cos 5\alpha + i\sin 5\alpha) = \sqrt{8}(\cos 135° + i\sin 135°)$, then we have the following.

$r^5 = \sqrt{8} \Rightarrow r = \sqrt[10]{8}$ and $5\alpha = 135° + 360°\cdot k \Rightarrow \alpha = \frac{135° + 360°\cdot k}{5} = 27° + 72°\cdot k$, k any integer.

If $k=0$, then $\alpha = 27° + 0° = 27°$. If $k=1$, then $\alpha = 27° + 72° = 99°$.
If $k=2$, then $\alpha = 27° + 144° = 171°$. If $k=3$, then $\alpha = 27° + 216° = 243°$.
If $k=4$, then $\alpha = 27° + 288° = 315°$.

So, the fifth roots of $-2 + 2i$ are $\sqrt[10]{8}(\cos 27° + i\sin 27°)$, $\sqrt[10]{8}(\cos 99° + i\sin 99°)$, $\sqrt[10]{8}(\cos 171° + i\sin 171°)$, $\sqrt[10]{8}(\cos 243° + i\sin 243°)$, and $\sqrt[10]{8}(\cos 315° + i\sin 315°)$..

43. Convert $1 - i$ to polar form

$r = \sqrt{1^2 + (-1)^2} = \sqrt{1+1} = \sqrt{2}$ and $\tan\theta = \frac{-1}{1} = -1 \Rightarrow \theta = 315°$, since θ is in quadrant IV.

Thus, $1 - i = \sqrt{2}(\cos 315° + i \sin 315°)$.

Since $r^3(\cos 3\alpha + i \sin 3\alpha) = \sqrt{2}(\cos 315° + i \sin 315°)$, then we have the following.

$$r^3 = \sqrt{2} \Rightarrow r = \sqrt[6]{2} \text{ and } 3\alpha = 315° + 360° \cdot k \Rightarrow \alpha = \frac{315° + 360° \cdot k}{3} = 105° + 120° \cdot k, \ k \text{ any integer.}$$

If $k = 0$, then $\alpha = 105° + 0° = 105°$. If $k = 1$, then $\alpha = 105° + 120° = 225°$.
If $k = 2$, then $\alpha = 105° + 240° = 345°$.

So, the cube roots of $1 - i$ are $\sqrt[6]{2}(\cos 105° + i \sin 105°)$, $\sqrt[6]{2}(\cos 225° + i \sin 225°)$, and $\sqrt[6]{2}(\cos 345° + i \sin 345°)$.

44. The real number –32 has one real fifth root. The one real fifth root is –2, and all other fifth roots are not real.

45. The number –64 has no real sixth roots because a real number raised to the sixth power will never be negative.

46. $x^3 + 125 = 0 \Rightarrow x^3 = -125$

We have, $r = 125$ and $\theta = 180°$.

$$x^3 = -125 = -125 + 0i = 125(\cos 180° + i \sin 180°)$$

Since $r^3(\cos 3\alpha + i \sin 3\alpha) = 125(\cos 180° + i \sin 180°)$, then we have the following.

$$r^3 = 125 \Rightarrow r = 5 \text{ and } 3\alpha = 180° + 360° \cdot k \Rightarrow \alpha = \frac{180° + 360° \cdot k}{3} = 60° + 120° \cdot k, \ k \text{ any integer.}$$

If $k = 0$, then $\alpha = 60° + 0° = 60°$. If $k = 1$, then $\alpha = 60° + 120° = 180°$.
If $k = 2$, then $\alpha = 60° + 240° = 300°$.

Solution set: $\{5(\cos 60° + i \sin 60°),\ 5(\cos 180° + i \sin 180°),\ 5(\cos 300° + i \sin 300°)\}$

47. $x^4 + 16 = 0 \Rightarrow x^4 = -16$

We have, $r = 16$ and $\theta = 180°$.

$$x^4 = -16 = -16 + 0i = 16(\cos 180° + i \sin 180°)$$

Since $r^4(\cos 4\alpha + i \sin 4\alpha) = 16(\cos 180° + i \sin 180°)$, then we have the following.

$$r^4 = 16 \Rightarrow r = 2 \text{ and } 4\alpha = 180° + 360° \cdot k \Rightarrow \alpha = \frac{180° + 360° \cdot k}{4} = 45° + 90° \cdot k, \ k \text{ any integer.}$$

If $k = 0$, then $\alpha = 45° + 0° = 45°$. If $k = 1$, then $\alpha = 45° + 90° = 135°$.
If $k = 2$, then $\alpha = 45° + 180° = 225°$. If $k = 3$, then $\alpha = 45° + 270° = 315°$.

Solution set:
$\{2(\cos 45° + i \sin 45°),\ 2(\cos 135° + i \sin 135°),\ 2(\cos 225° + i \sin 225°),\ 2(\cos 315° + i \sin 315°)\}$

48. $x^2 + i = 0 \Rightarrow x^2 = -i$

We have, $r = 1$ and $\theta = 270°$.

$$x^2 = -i = 0 - i = 1(\cos 270° + i \sin 270°)$$

Since $r^2(\cos 2\alpha + i \sin 2\alpha) = 1(\cos 270° + i \sin 270°)$, then we have the following.

$$r^2 = 1 \Rightarrow r = 1 \text{ and } 2\alpha = 270° + 360° \cdot k \Rightarrow \alpha = \frac{270° + 360° \cdot k}{2} = 135° + 180° \cdot k, \ k \text{ any integer.}$$

If $k = 0$, then $\alpha = 135° + 0° = 135°$. If $k = 1$, then $\alpha = 135° + 180° = 315°$.

Solution set: $\{\cos 135° + i \sin 135°, \ \cos 315° + i \sin 315°\}$

49. $\left(-1, \sqrt{3}\right)$

$r = \sqrt{(-1)^2 + \left(\sqrt{3}\right)^2} = \sqrt{1+3} = \sqrt{4} = 2$ and $\theta = \tan^{-1}\left(-\frac{\sqrt{3}}{1}\right) = \tan^{-1}\left(-\sqrt{3}\right) = 120°$, since θ is in quadrant II. Thus, the polar coordinates are $(2, 120°)$.

50. $(5, 315°)$

$x = r\cos\theta \Rightarrow x = 5\cos 315° = 5\left(\frac{\sqrt{2}}{2}\right) = \frac{5\sqrt{2}}{2}$ and $y = r\sin\theta \Rightarrow y = 5\sin 315° = 5\left(-\frac{\sqrt{2}}{2}\right) = -\frac{5\sqrt{2}}{2}$.

Thus, the rectangular coordinates are $\left(\frac{5\sqrt{2}}{2}, -\frac{5\sqrt{2}}{2}\right)$.

51. The angle must be quadrantal.

52. Since r is constant, the graph will be a circle.

53. $r = 4\cos\theta$ is a circle.

θ	0°	30°	45°	60°	90°	120°	135°	150°	180°
$r = 4\cos\theta$	4	3.5	2.8	2	0	−2	−2.8	−3.5	−4

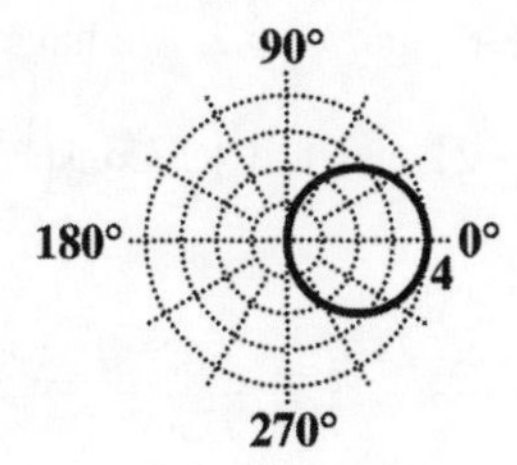

$r = 4\cos\theta$

Graph is retraced in the interval (180°, 360°).

54. $r = -1 + \cos\,\theta$ is a cardioid.

θ	0°	30°	45°	60°	90°
$r = -1 + \cos\theta$	0	−.7	−.3	−.5	0

θ	120°	135°	150°	180°	270°	315°
$r = -1 + \cos\theta$	−1.5	−1.7	−1.9	−2	−1	−.3

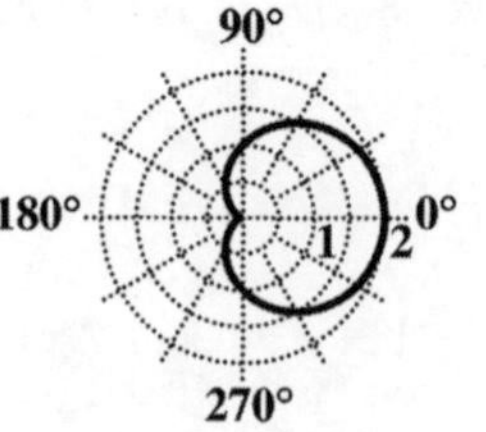

$r = -1 + \cos\theta$

55. $r = 2\sin 4\theta$ is an eight-leaved rose.

θ	0°	7.5°	15°	22.5°	0°	37.5°	45°
$r = 2\sin 4\theta$	0	1	$\sqrt{3}$	2	$\sqrt{3}$	1	0

θ	52.5°	60°	67.5°	75°	82.5°	90°	52.5°
$r = 2\sin 4\theta$	−1	$-\sqrt{3}$	−2	$-\sqrt{3}$	−1	0	−1

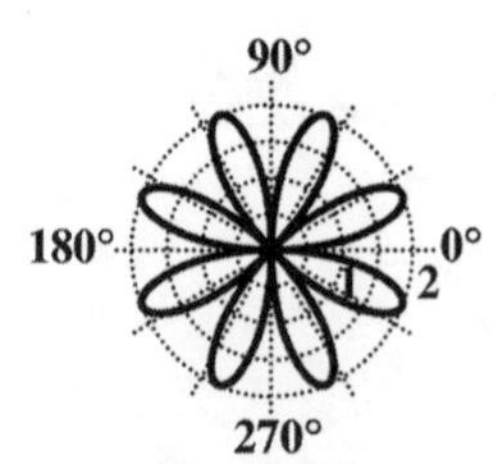

$r = 2\sin 4\theta$

The graph continues to form eight petals for the interval [0°, 360°).

56. Since $r = \dfrac{2}{2\cos\theta - \sin\theta}$, we can use the general form for the polar equation of a line,

$r = \dfrac{c}{a\cos\theta + b\sin\theta}$, with $a = 2,\ b = -1,$ and $c = 2,$ we have $2x - y = 2$.

The graph is a line with intercepts $(0,-2)$ and $(1,0)$. Constructing a table of values, will also result in the graph.

θ	0°	30°	45°	60°	90°
$r = \dfrac{2}{2\cos\theta - \sin\theta}$	1	1.6	2.8	15.9	−2

θ	120°	135°	150°	180°	270°	315°
$r = \dfrac{2}{2\cos\theta - \sin\theta}$	−1.1	−.9	−1.9	−1	2	.9

90°
180° 0°
1 2
270°

$r = \dfrac{2}{2\cos\theta - \sin\theta}$

57. $r=\dfrac{3}{1+\cos\theta}$

$$r=\frac{3}{1+\cos\theta}\Rightarrow r(1+\cos\theta)=3\Rightarrow r+r\cos\theta=3\Rightarrow\sqrt{x^2+y^2}+x=3\Rightarrow\sqrt{x^2+y^2}=3-x$$

$$x^2+y^2=(3-x)^2\Rightarrow x^2+y^2=9-6x+x^2\Rightarrow y^2=9-6x\Rightarrow y^2+6x-9=0\Rightarrow y^2=-6x+9$$

$$y^2=-6\left(x-\frac{3}{2}\right)\text{ or }y^2+6x-9=0$$

58. $r=\sin\theta+\cos\theta$

$$r=\sin\theta+\cos\theta\Rightarrow r^2=r\sin\theta+r\cos\theta\Rightarrow x^2+y^2=x+y$$

$$x^2+y^2-x-y=0\Rightarrow\left(x^2-x\right)+\left(y^2-y\right)=0\Rightarrow\left(x^2-x+\frac{1}{4}\right)+\left(y^2-y+\frac{1}{4}\right)=\frac{1}{4}+\frac{1}{4}$$

$$\left(x-\frac{1}{2}\right)^2+\left(y-\frac{1}{2}\right)^2=\frac{1}{2}\text{ or }x^2+y^2-x-y=0$$

59. $r=2\Rightarrow\sqrt{x^2+y^2}=2\Rightarrow x^2+y^2=4$

60. $y=x\Rightarrow r\sin\theta=r\cos\theta\Rightarrow\sin\theta=\cos\theta\text{ or }\tan\theta=1$

61. $y=x^2\Rightarrow r\sin\theta=r^2\cos^2\theta\Rightarrow\sin\theta=r\cos^2\theta\Rightarrow r=\dfrac{\sin\theta}{\cos^2\theta}\Rightarrow r=\dfrac{\sin\theta}{\cos\theta}\cdot\dfrac{1}{\cos\theta}=\tan\theta\sec\theta$

$$r=\tan\theta\sec\theta\text{ or }r=\frac{\tan\theta}{\cos\theta}$$

62. Suppose (r,θ) is on the graph, $-\theta$ reflects this point with respect to the x-axis, and $-r$ reflects the resulting point with respect to the origin. The net result is that the original point is reflected with respect to the y-axis. The correct choice is B.

63. If (r,θ) lies on the graph, $(-r,\theta)$ would reflect that point into the quadrant diagonal to the original quadrant. The correct choice is A.

64. If (r,θ) lies on the graph, $(r,-\theta)$ would reflect that point across the x-axis. Therefore, there is symmetry about the x-axis. The correct choice is C.

65. If (r,θ) lies on the graph, $(r,\pi-\theta)$ would reflect that point into the other quadrant on the same side of the x-axis as the original quadrant. Therefore, there is symmetry about the y-axis. The correct choice is B.

66. $y=2$

$y=r\sin\theta\Rightarrow r\sin\theta=2\Rightarrow r=\dfrac{2}{\sin\theta}$ or $r=2\csc\theta$

67. $x=2$

$x=r\cos\theta\Rightarrow r\cos\theta=2\Rightarrow r=\dfrac{2}{\cos\theta}$ or $r=2\sec\theta$

68. $x^2 + y^2 = 4$

Since $x = r\cos\theta$ and $y = r\sin\theta$, we have the following.

$$(r\cos\theta)^2 + (r\sin\theta)^2 = 4 \Rightarrow r^2\cos\theta^2 + r^2\sin^2\theta = 4 \Rightarrow r^2\left(\cos\theta^2 + \sin^2\theta\right) = 4 \Rightarrow r^2 = 4$$

$$r = -2 \text{ or } r = -2$$

69. $x + 2y = 4$

Since $x = r\cos\theta$ and $y = r\sin\theta$, we have the following.

$$r\cos\theta + 2(r\sin\theta) = 4 \Rightarrow r(\cos\theta + 2\sin\theta) = 4 \Rightarrow r = \frac{4}{\cos\theta + 2\sin\theta}$$

70. To show that the distance between (r_1, θ_1) and (r_2, θ_2) is given by $d = \sqrt{r_1^2 + r_2^2 - 2r_1r_2\cos(\theta_1 - \theta_2)}$, we can convert the polar coordinates to rectangular coordinates, apply the distance formula.

(r_1, θ_1) in rectangular coordinates is $(r_1\cos\theta_1, r_1\sin\theta_1)$.

(r_2, θ_2) in rectangular coordinates is $(r_2\cos\theta_2, r_2\sin\theta_2)$.

$$\begin{aligned}
d &= \sqrt{(r_1\cos\theta_1 - r_2\cos\theta)^2 + (r_1\sin\theta_1 - r_2\sin\theta)^2} \\
&= \sqrt{\left(r_1^2\cos^2\theta_1 - 2r_1r_2\cos\theta_1\cos\theta_2 + r_2^2\cos^2\theta\right) + \left(r_1^2\sin^2\theta_1 - 2r_1r_2\sin\theta_1\sin\theta_2 + r_2^2\sin^2\theta\right)} \\
&= \sqrt{r_1^2\left(\cos^2\theta_1 + \sin^2\theta_1\right) - 2r_1r_2\left(\cos\theta_1\cos\theta_2 - \sin\theta_1\sin\theta_2\right) + r_2^2\left(\cos^2\theta + \sin^2\theta\right)} \\
&= \sqrt{r_1^2 \cdot 1 - 2r_1r_2\cos(\theta_1 - \theta_2) + r_2^2 \cdot 1} \\
&= \sqrt{r_1^2 + r_2^2 - 2r_1r_2\cos(\theta_1 - \theta_2)}
\end{aligned}$$

71. $x = t + \cos t,\ y = \sin t$ for t in $[0, 2\pi]$

t	0	$\frac{\pi}{6}$	$\frac{\pi}{3}$	$\frac{\pi}{2}$	$\frac{3\pi}{4}$	π
$x = t + \cos t$	0	$\frac{\pi}{6} + \frac{\sqrt{3}}{2}$ ≈ 1.4	$\frac{\pi}{3} + \frac{1}{2}$ ≈ 1.5	$\frac{\pi}{2}$ ≈ 1.6	$\frac{3\pi}{4} - \frac{\sqrt{2}}{2}$ ≈ 1.6	$\pi - 1$ ≈ 2.1
$y = \sin t$	0	$\frac{1}{2} = .5$	$\frac{\sqrt{3}}{2} \approx 1.7$	1	$\frac{\sqrt{2}}{2} \approx .7$	0

t	$\frac{7\pi}{6}$	$\frac{5\pi}{4}$	$\frac{4\pi}{3}$	$\frac{3\pi}{2}$	$\frac{7\pi}{4}$	2π
$x = t + \cos t$	$\frac{7\pi}{6} - \frac{\sqrt{3}}{2}$ ≈ 2.8	$\frac{5\pi}{4} - \frac{\sqrt{2}}{2}$ ≈ 3.2	$\frac{4\pi}{3} - \frac{1}{2}$ ≈ 3.7	$\frac{3\pi}{2}$ ≈ 4.7	$\frac{7\pi}{4} + \frac{\sqrt{2}}{2}$ ≈ 6.2	$2\pi + 1$ ≈ 7.3
$y = \sin t$	$-\frac{1}{2} = -.5$	$-\frac{\sqrt{2}}{2} \approx -.7$	$-\frac{\sqrt{3}}{2} \approx -1.7$	-1	$-\frac{\sqrt{2}}{2} \approx -.7$	0

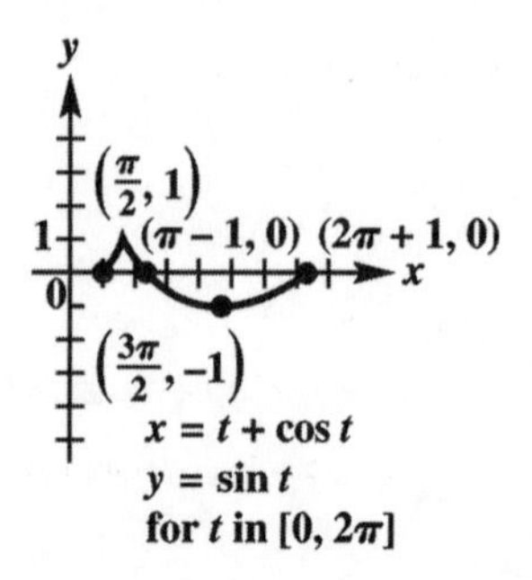

72. $x = 3t + 2$, $y = t - 1$, for t in $[-5, 5]$

Since $t = y + 1$, substitute $y + 1$ for t in the equation for x.

$$x = 3(y+1)+2 \Rightarrow x = 3y+3+2 \Rightarrow x = 3y+5 \Rightarrow x-3y = 5$$

Since t is in $[-5, 5]$, x is in $[3(-5) + 2, 3(5) + 2]$ or $[-13, 17]$.

73. $x = \sqrt{t-1}$, $y = \sqrt{t}$, for t in $[1, \infty)$

Since $x = \sqrt{t-1} \Rightarrow x^2 = t-1 \Rightarrow t = x^2+1$, substitute x^2+1 for t in the equation for y to obtain $y = \sqrt{x^2+1}$. Since t is in $[1, \infty)$, x is in $[\sqrt{1-1}, \infty)$ or $[0, \infty)$.

74. $x = t^2+5$, $y = \dfrac{1}{t^2+1}$, for t in $(-\infty, \infty)$

Since $t^2 = x-5$, substitute $x - 5$ for t^2 in the equation for y.

$$y = \frac{1}{t^2+1} \Rightarrow y = \frac{1}{(x-5)+1} \Rightarrow y = \frac{1}{x-4}$$

Since $x = t^2+5$ and $t^2 \ge 0$, $x \ge 0+5 = 5$. Therefore, x is in $[5, \infty)$.

75. $x = 5 \tan t$, $y = 3 \sec t$, for t in $\left(-\dfrac{\pi}{2}, \dfrac{\pi}{2}\right)$

Since $\dfrac{x}{5} = \tan t$, $\dfrac{y}{3} = \sec t$, and $1+\tan^2 t = \sec^2 t$, we have the following.

$$1+\left(\frac{x}{5}\right)^2 = \left(\frac{y}{3}\right)^2 \Rightarrow 1+\frac{x^2}{25} = \frac{y^2}{9} \Rightarrow 9\left(1+\frac{x^2}{25}\right) = y^2 \Rightarrow y = \sqrt{9\left(1+\frac{x^2}{25}\right)} \Rightarrow y = 3\sqrt{1+\frac{x^2}{25}}$$

y is positive since $y = 3 \sec t > 0$ for t in $\left(-\dfrac{\pi}{2}, \dfrac{\pi}{2}\right)$.

Since t is in $\left(-\dfrac{\pi}{2}, \dfrac{\pi}{2}\right)$ and $x = 5 \tan t$ is undefined at $-\dfrac{\pi}{2}$ and $\dfrac{\pi}{2}$, x is in $(-\infty, \infty)$.

76. $x = \cos 2t$, $y = \sin t$ for t in $(-\pi, \pi)$

$\cos 2t = \cos^2 t - \sin^2 t$ (double angle formula)

Since $\cos^2 t + \sin^2 t = 1$, we have the following.

$$\cos^2 t+\sin^2 t = 1 \Rightarrow (\cos^2 t-\sin^2 t)+2\sin^2 t = 1 \Rightarrow x+2y^2 = 1 \Rightarrow 2y^2 = -x+1 \Rightarrow 2y^2 = -(x-1)$$

$$y^2 = -\frac{1}{2}(x-1) \text{ or } 2y^2+x-1 = 0$$

Since t is in $(-\pi, \pi)$ and $\cos 2t$ is in $[-1, 1]$, x is in $[-1, 1]$.

77. The radius of the circle that has center $(3,4)$ and passes through the origin is the following.

$$r = \sqrt{(3-0)^2+(4-0)^2} = \sqrt{3^2+4^2} = \sqrt{9+16} = \sqrt{25} = 5$$

Thus, the equation of this circle is $(x-3)^2+(y-4)^2 = 5^2$

Since $\cos^2 t+\sin^2 y = 1 \Rightarrow 25\cos^2 t+25\sin^2 t = 25 \Rightarrow (5\cos t)^2+(5\sin t)^2 = 5^2$, we can have $5\cos t = x-3$ and $5\sin t = y-4$. Thus, a pair of parametric equations can be the following.

$$x = 3+5\cos t,\ y = 4+5\sin t, \text{ where } t \text{ in } [0, 2\pi]$$

78. **(a)** Let $z_1 = a+bi$ and its complex conjugate be $z_2 = a-bi$.

$|z_1| = \sqrt{a^2+b^2}$ and $|z_2| = \sqrt{a^2+(-b)^2} = \sqrt{a^2+b^2} = |z_1|$.

(b) Let $z_1 = a+bi$ and $z_2 = a-bi$.

$$\begin{aligned} z_1^2 + z_1 &= (a+bi)^2 + (a+bi) \\ &= (a^2+2abi+b^2i^2)+(a+bi) = a^2+2abi+b^2(-1)+a+bi = a^2-b^2+2abi+a+bi \\ &= (a^2-b^2+a)+(2ab+b)i = c+di \end{aligned}$$

$$\begin{aligned} z_2^2 + z_2 &= (a-bi)^2 + (a-bi) \\ &= (a^2-2abi+b^2i^2)+(a-bi) = a^2-2abi+b^2(-1)+a-bi = a^2-b^2-2abi+a-bi \\ &= (a^2-b^2+a)-(2ab+b)i = c-di \end{aligned}$$

(c) – (d) The results are again complex conjugates of each other. At each iteration, the resulting values from z_1 and z_2 will always be complex conjugates. Graphically, these represent points that are symmetric with respect to the *x*-axis, namely points such as (a,b) and $(a,-b)$.

79. **(a)** $x = (v\cos\theta)t \Rightarrow x = (118\cos 27°)t$ and $y = (v\sin\theta)t - 16t^2 + h \Rightarrow y = (118\sin 27°)t - 16t^2 + 3.2$

(b) Since $t = \dfrac{x}{118\cos 27°}$, we have the following.

$$y = 118\sin 27° \cdot \frac{x}{118\cos 27°} - 16\left(\frac{x}{118\cos 27°}\right)^2 + 3.2 = 3.2 - \frac{4}{3481\cos^2 27°}x^2 + (\tan 27°)x$$

(c) Solving $0 = -16t^2 + (118\sin 27°)t + 3.2$ by the quadratic formula, we have the following.

$$t = \frac{-118\sin 27° \pm \sqrt{(118\sin 27°)^2 - 4(-16)(3.2)}}{2(-16)} \Rightarrow t \approx -.06,\ 3.406$$

Discard $t = -0.06$ sec since it is an unacceptable answer. At $t = 3.4$ sec, the baseball traveled $x = (118\cos 27°)(3.406) \approx 358$ ft.

Chapter 8: Test

1. **(a)** $w + z = (2-4i)+(5+i) = (2+5)+(-4+1)i = 7-3i$

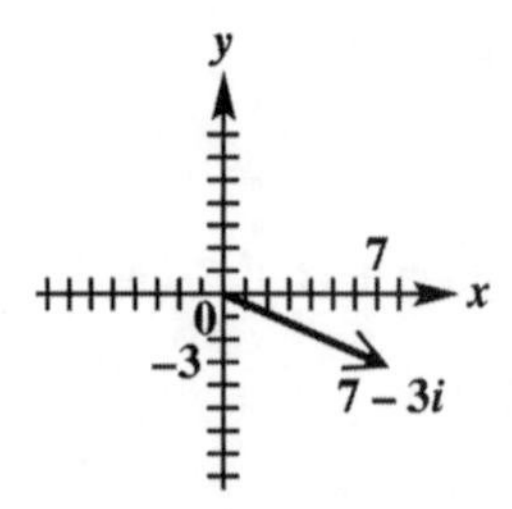

(b) $w - z = (2-4i)-(5+i) = (2-5)+(-4-1)i = -3-5i$

(c) $wz = (2-4i)(5+i) = 10+2i-20i-4i^2 = 10-4(-1)-18i = 14-18i$

(d) $\dfrac{w}{z} = \dfrac{2-4i}{5+i}\cdot\dfrac{5-i}{5-i} = \dfrac{10-2i-20i+4i^2}{25-5i+5i-i^2} = \dfrac{10+4(-1)-22i}{25-(-1)} = \dfrac{6-22i}{26} = \dfrac{3}{13}-\dfrac{11}{13}i$

2. **(a)** $i^{15} = i^{12+3} = i^{12} \cdot i^3 = \left(i^4\right)^3 \cdot i^3 = 1(-i) = -i$

(b) $(1+i)^2 = (1+i)(1+i) = 1+i+i+i^2 = 1+2i+(-1) = 2i$

3. $2x^2 - x + 4 = 0$

Using the quadratic formula we have the following.

$$x = \frac{-(-1) \pm \sqrt{(-1)^2 - 4(2)(4)}}{2(2)} = \frac{1 \pm \sqrt{1-32}}{4} = \frac{1 \pm \sqrt{-31}}{4} = \frac{1 \pm i\sqrt{31}}{4} = \frac{1}{4} \pm \frac{\sqrt{31}}{4}i$$

Solution set: $\left\{ \frac{1}{4} \pm \frac{\sqrt{31}}{4}i \right\}$

4. **(a)** $3i$

$r = \sqrt{0^2 + 3^2} = \sqrt{0+9} = \sqrt{9} = 3$

The point $(0, 3)$ is on the positive y-axis, so, $\theta = 90°$. Thus, $3i = 3(\cos 90° + i \sin 90°)$.

(b) $1+2i$

$r = \sqrt{1^2 + 2^2} = \sqrt{1+4} = \sqrt{5}$

Since θ is in quadrant I, $\theta = \tan^{-1}\left(\frac{2}{1}\right) = \tan^{-1} 2 \approx 63.43°$.

Thus, $1+2i = \sqrt{5}(\cos 63.43° + i \sin 63.43°)$.

(c) $-1 - \sqrt{3}i$

$r = \sqrt{(-1)^2 + (-\sqrt{3})^2} = \sqrt{1+3} = \sqrt{4} = 2$

Since θ is in quadrant III, $\theta = \tan^{-1}\left(\frac{-\sqrt{3}}{-1}\right) = \tan^{-1}\sqrt{3} = 240°$.

Thus, $-1 - \sqrt{3}i = 2(\cos 240° + i \sin 240°)$

5. **(a)** $3(\cos 30° + i \sin 30°) = 3\left(\frac{\sqrt{3}}{2} + \frac{1}{2}i\right) = \frac{3\sqrt{3}}{2} + \frac{3}{2}i$

(b) $4 \text{ cis } 4° = 3.06 + 2.57i$

(c) $3(\cos 90° + i \sin 90°) = 3(0 + 1 \cdot i) = 0 + 3i = 3i$

6. $w = 8(\cos 40° + i \sin 40°), z = 2(\cos 10° + i \sin 10°)$

(a) $wz = 8 \cdot 2\left[\cos(40° + 10°) + i \sin(40° + 10°)\right] = 16(\cos 50° + i \sin 50°)$

(b) $\frac{w}{z} = \frac{8}{2}\left[\cos(40° - 10°) + i \sin(40° - 10°)\right] = 4(\cos 30° + i \sin 30°) = 4\left(\frac{\sqrt{3}}{2} + \frac{1}{2}i\right) = 2\sqrt{3} + 2i$

(c) $z^3 = \left[2(\cos 10° + i \sin 10°)\right]^3$

$= 2^3 (\cos 3 \cdot 10° + i \sin 3 \cdot 10°) = 8(\cos 30° + i \sin 30°) = 8\left(\frac{\sqrt{3}}{2} + \frac{1}{2}i\right) = 4\sqrt{3} + 4i$

7. Find all the fourth roots of $-16i = 16(\cos 270° + i\sin 270°)$.

Since $r^4(\cos 4\alpha + i\sin 4\alpha) = 16(\cos 270° + i\sin 270°)$, then we have the following.

$$r^4 = 16 \Rightarrow r = 2 \text{ and } 4\alpha = 270° + 360° \cdot k \Rightarrow \alpha = \frac{270° + 360° \cdot k}{4} = 67.5° + 90° \cdot k,\ k \text{ any integer.}$$

If $k = 0$, then $\alpha = 67.5°$. If $k = 1$, then $\alpha = 157.5°$.
If $k = 2$, then $\alpha = 247.5°$. If $k = 3$, then $\alpha = 337.5°$.

The fourth roots of -16i are $2(\cos 67.5° + \sin 67.5), 2(\cos 157.5° + i\sin 157.5°)$, $2(\cos 247.5° + i\sin 247.5°)$, and $2(\cos 337.5° + i\sin 337.5°)$.

For Exercise 8, answers may vary.

8. **(a)** $(0,5)$

$r = \sqrt{0^2 + 5^2} = \sqrt{0+25} = \sqrt{25} = 5$

The point $(0,5)$ is on the positive y-axis. Thus, $\theta = 90°$. One possibility is (5, 90°).
Alternatively, if $\theta = 90° - 360° = -270°$, a second possibility is (5, –270°).

(b) $(-2,-2)$

$r = \sqrt{(-2)^2 + (-2)^2} = \sqrt{4+4} = \sqrt{8} = 2\sqrt{2}$

Since θ is in quadrant III, $\theta = \tan^{-1}\left(\frac{-2}{-2}\right) = \tan^{-1} 1 = 225°$. One possibility is $(2\sqrt{2}, 225°)$.

Alternatively, if $\theta = 225° - 360° = -135°$, a second possibility is $(2\sqrt{2}, -135°)$.

9. **(a)** $(3, 315°)$

$x = r\cos\theta \Rightarrow x = 3\cos 315° = 3 \cdot \frac{\sqrt{2}}{2} = \frac{3\sqrt{2}}{2}$ and $y = r\sin\theta \Rightarrow y = 3\sin 315° = 3\left(-\frac{\sqrt{2}}{2}\right) = \frac{-3\sqrt{2}}{2}$

The rectangular coordinates are $\left(\frac{3\sqrt{2}}{2}, \frac{-3\sqrt{2}}{2}\right)$.

(b) $(-4, 90°)$

$x = r\cos\theta \Rightarrow x = -4\cos 90° = 0$ and $y = r\sin\theta \Rightarrow y = -4\sin 90° = -4$

The rectangular coordinates are $(0,-4)$.

10. $r = 1 - \cos\theta$ is a cardioid.

θ	0°	30°	45°	60°	90°	135°
$r = 1 - \cos\theta$	0	.1	.3	.5	1	1.7

θ	180°	225°	270°	315°	360°
$r = 1 - \cos\theta$	2	1.7	1	.3	0

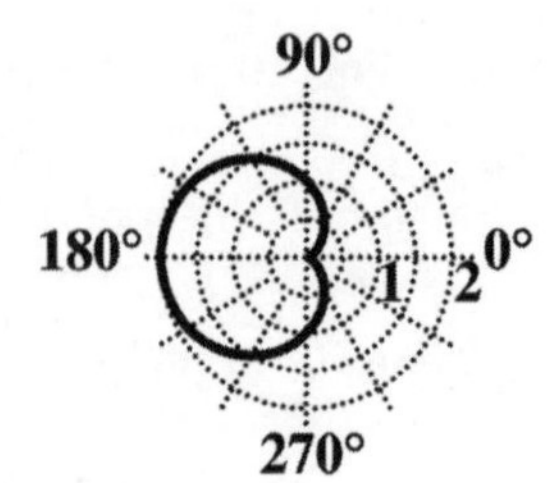

$r = 1 - \cos\theta$

11. $r = 3\cos 3\theta$ is a three-leaved rose.

θ	0°	30°	45°	60°	90°	120°	135°	150°	180°
$r = 3\cos 3\theta$	3	0	–2.1	–3	0	3	2.1	0	–3

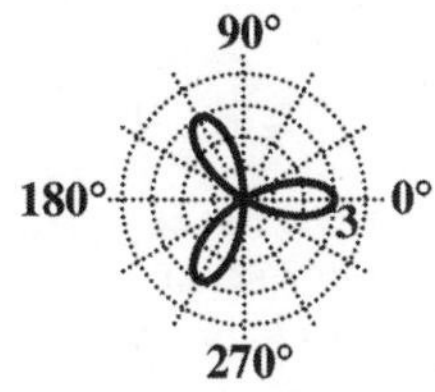

Graph is retraced in the interval (180°, 360°).

12. (a) Since $r = \dfrac{4}{2\sin\theta - \cos\theta} = \dfrac{4}{-1\cdot\cos\theta + 2\sin\theta}$, we can use the general form for the polar equation of a line, $r = \dfrac{c}{a\cos\theta + b\sin\theta}$, with $a = -1$, $b = 2$, and $c = 4$, we have the following.

$$-x + 2y = 4 \text{ or } x - 2y = -4$$

The graph is a line with intercepts $(-4,0)$ and $(0,2)$.

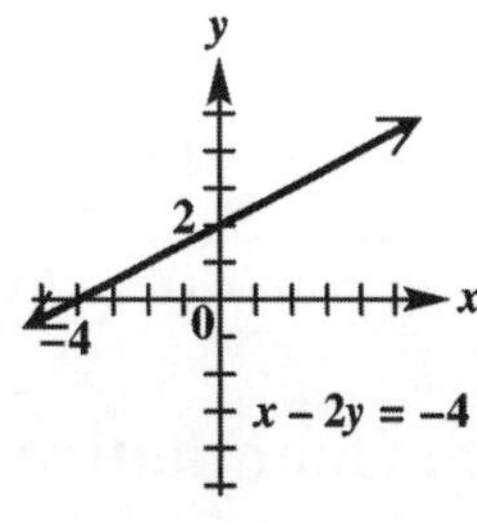

(b) $r = 6$ represents the equation of a circle centered at the origin with radius 6, namely $x^2 + y^2 = 36$.

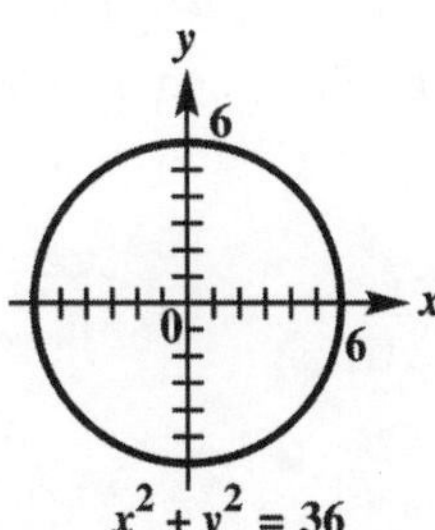

13. $x = 4t - 3,\ y = t^2$ for t in [–3, 4]

t	x	y
–3	–15	9
–1	–7	1
0	–3	0
1	1	1
2	5	4
4	13	16

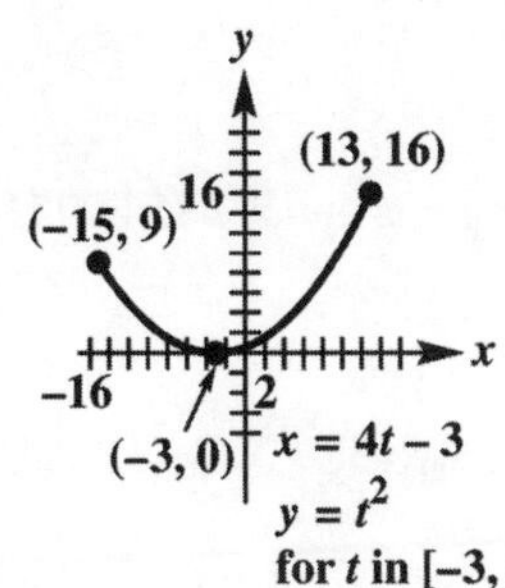

Since $x = 4t - 3 \Rightarrow t = \dfrac{x+3}{4}$ and $y = t^2$ we have $y = \left(\dfrac{x+3}{4}\right)^2 = \dfrac{1}{4}(x+3)^2$, where x is in [–15, 13]

14. $x = 2\cos 2t$, $y = 2\sin 2t$ for t in $[0, 2\pi]$

t	0	$\frac{\pi}{8}$	$\frac{\pi}{4}$	$\frac{3\pi}{8}$	$\frac{\pi}{2}$	$\frac{5\pi}{8}$	$\frac{3\pi}{4}$	π	$\frac{5\pi}{4}$	$\frac{3\pi}{2}$	$\frac{7\pi}{4}$	2π
x	2	$\sqrt{2}$	0	$-\sqrt{2}$	–2	$-\sqrt{2}$	0	2	0	–2	0	2
y	0	$\sqrt{2}$	2	$\sqrt{2}$	0	$-\sqrt{2}$	–2	0	2	0	–2	0

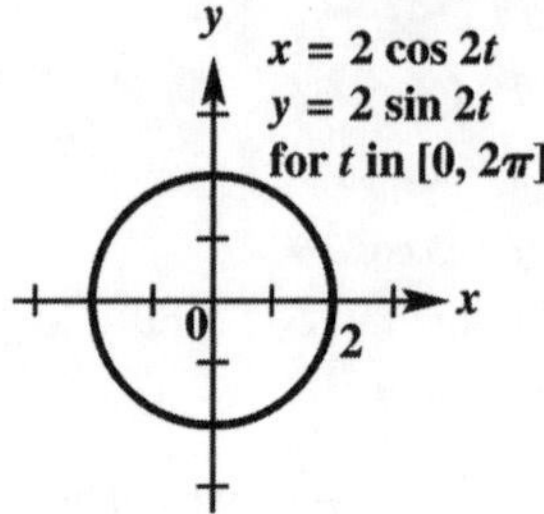

Since $x = 2\cos 2t \Rightarrow \cos 2t = \dfrac{x}{2}$, $y = 2\sin 2t \Rightarrow \sin 2t = \dfrac{y}{2}$, and $\cos^2(2t) + \sin^2(2t) = 1$, we have $\left(\dfrac{x}{2}\right)^2 + \left(\dfrac{y}{2}\right)^2 = 1 \Rightarrow \dfrac{x^2}{4} + \dfrac{y^2}{4} = 1 \Rightarrow x^2 + y^2 = 4$, where x is in $[-1, 1]$.

15. $z = -1 + i$

$z^2 - 1 = (-1+i)^2 - 1 = 1 - i - i + i^2 - 1 = -2i - 1 = -1 - 2i$

Since $r = \sqrt{(-1)^2 + (-2)^2} = \sqrt{1+4} = \sqrt{5} > 2$, z is not in the Julia set.

Chapter 8: Quantitative Reasoning

Since $y = \dfrac{k}{x^n}$; $k = 23.5$, and $n = 1.153$, $y = \dfrac{23.5}{x^{1.153}}$.

1. $x = 6$ in.

Since $y = \dfrac{23.5}{6^{1.153}} \approx 2.9776$, the total length is about $2.9776 \cdot 6 \approx 17.9$ in.

2. $x = .1$ in.

Since $y = \dfrac{23.5}{.1^{1.153}} \approx 334.2473$, the total length is about $334.2473 \cdot .1 \approx 33.4$ in.

3. $x = .01$ in.

Since $y = \dfrac{23.5}{.01^{1.153}} \approx 4754.0951$, the total length is about $4754.0951 \cdot .01 \approx 47.5$ in.

Chapter 9
EXPONENTIAL AND LOGARITHMIC FUNCTIONS

Section 9.1: Exponential Functions

1. $f(x) = 3^x$
$f(2) = 3^2 = 9$

2. $f(x) = 3^x$
$f(3) = 3^3 = 27$

3. $f(x) = 3^x$
$f(-2) = 3^{-2} = \frac{1}{3^2} = \frac{1}{9}$

4. $f(x) = 3^x$
$f(-3) = 3^{-3} = \frac{1}{3^3} = \frac{1}{27}$

5. $g(x) = \left(\frac{1}{4}\right)^x$
$g(2) = \left(\frac{1}{4}\right)^2 = \frac{1}{16}$

6. $g(x) = \left(\frac{1}{4}\right)^x$
$g(3) = \left(\frac{1}{4}\right)^3 = \frac{1}{64}$

7. $g(x) = \left(\frac{1}{4}\right)^x$
$g(-2) = \left(\frac{1}{4}\right)^{-2} = 4^2 = 16$

8. $g(x) = \left(\frac{1}{4}\right)^x$
$g(-3) = \left(\frac{1}{4}\right)^{-3} = 4^3 = 64$

9. $f(x) = 3^x$
$f\left(\frac{3}{2}\right) = 3^{3/2} \approx 5.196$

10. $g(x) = \left(\frac{1}{4}\right)^x$
$g\left(\frac{3}{2}\right) = \left(\frac{1}{4}\right)^{3/2} = \frac{1}{\left(\sqrt{4}\right)^3} = \frac{1}{2^3} = \frac{1}{8}$

11. $g(x) = \left(\frac{1}{4}\right)^x$
$g(2.34) = \left(\frac{1}{4}\right)^{2.34} \approx .039$

12. $f(x) = 3^x$
$f(1.68) = 3^{1.68} \approx 6.332$

13. The y-intercept of $f(x) = 3^x$ is 1, and the x-axis is a horizontal asymptote. Make a table of values.

x	$f(x)$
-2	$\frac{1}{9} \approx .1$
-1	$\frac{1}{3} \approx .3$
$-\frac{1}{2}$	$\approx .6$
0	1
$\frac{1}{2}$	≈ 1.7
1	3
2	9

Plot these points and draw a smooth curve through them. This is an increasing function. The domain is $(-\infty, \infty)$ and the range is $(0, \infty)$ and is one-to-one.

14. The y-intercept of $f(x)=4^x$ is 1, and the x-axis is a horizontal asymptote. Make a table of values.

x	$f(x)$
-2	.0625
-1	.25
$-\frac{1}{2}$	.5
0	1
$\frac{1}{2}$	2
1	4
2	16

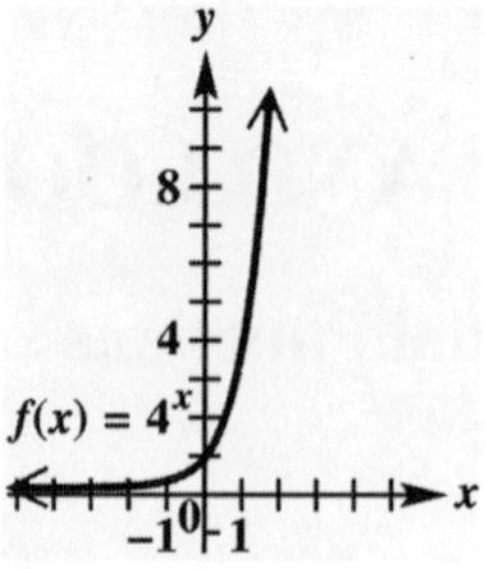

y
8
4
$f(x) = 4^x$
x
−1 0 1

Plot these points and draw a smooth curve through them. This is an increasing function. The domain is $(-\infty,\infty)$ and the range is $(0,\infty)$ and is one-to-one.

15. The y-intercept of $f(x)=\left(\frac{1}{3}\right)^x$ is 1, and the x-axis is a horizontal asymptote. Make a table of values.

x	$f(x)$
-2	9
-1	3
$-\frac{1}{2}$	≈ 1.7
0	1
$\frac{1}{2}$	$\approx .6$
1	$\frac{1}{3} \approx .3$
2	$\frac{1}{9} \approx .1$

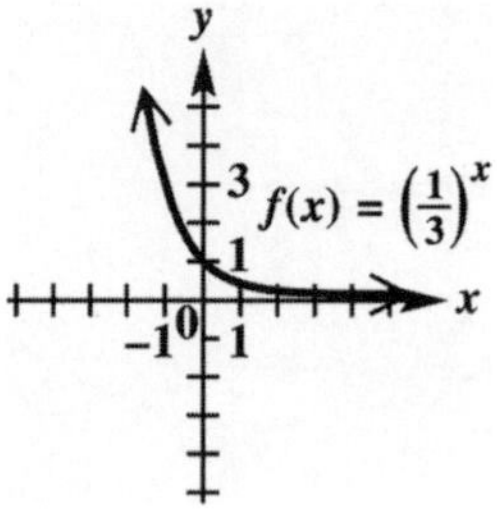

y
3
$f(x) = \left(\frac{1}{3}\right)^x$
1
x
−1 0 1

Plot these points and draw a smooth curve through them. This is a decreasing function. The domain is $(-\infty,\infty)$ and the range is $(0,\infty)$ and is one-to-one. Note: Since $f(x)=\left(\frac{1}{3}\right)^x=\left(3^{-1}\right)^x=3^{-x}$, the graph of $f(x)=\left(\frac{1}{3}\right)^x$ is the reflection of the graph of $f(x)=3^x$ (Exercise 13) about the y-axis.

16. The y-intercept of $f(x)=\left(\frac{1}{4}\right)^x$ is 1, and the x-axis is a horizontal asymptote. Make a table of values.

x	$f(x)$
-2	16
-1	4
$-\frac{1}{2}$	2
0	1
$\frac{1}{2}$	.5
1	.25
2	.0625

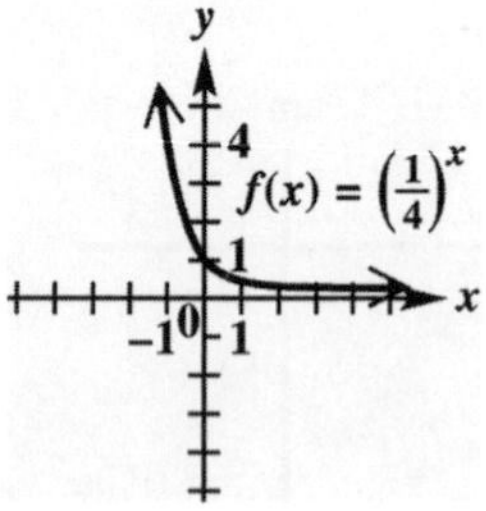

y
4
$f(x) = \left(\frac{1}{4}\right)^x$
1
x
−1 0 1

Plot these points and draw a smooth curve through them. This is a decreasing function. The domain is $(-\infty,\infty)$ and the range is $(0,\infty)$ and is one-to-one. Note: Since $f(x)=\left(\frac{1}{4}\right)^x=\left(4^{-1}\right)^x=4^{-x}$, the graph of $f(x)=\left(\frac{1}{4}\right)^x$ is the reflection of the graph of $f(x)=4^x$ (Exercise 14) about the y-axis.

17. The y-intercept of $f(x)=\left(\frac{3}{2}\right)^x$ is 1, and the x-axis is a horizontal asymptote. Make a table of values.

x	$f(x)$
-2	$\approx .4$
-1	$\approx .7$
$-\frac{1}{2}$	$\approx .8$
0	1
$\frac{1}{2}$	≈ 1.2
1	1.5
2	2.25

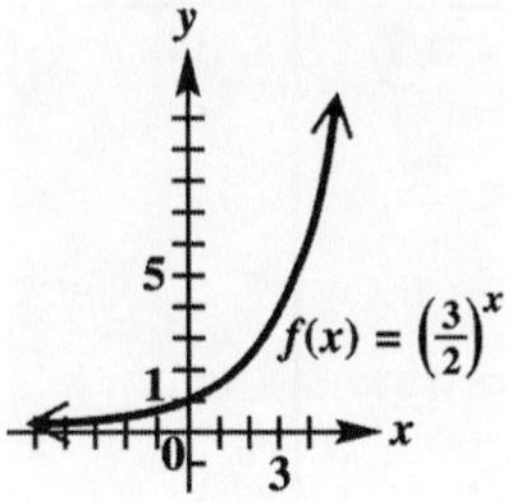

y
5
1
0
3
x
$f(x) = \left(\frac{3}{2}\right)^x$

Plot these points and draw a smooth curve through them. This is an increasing function. The domain is $(-\infty,\infty)$ and the range is $(0,\infty)$ and is one-to-one.

18. The y-intercept of $f(x)=\left(\frac{2}{3}\right)^x$ is 1, and the x-axis is a horizontal asymptote. Make a table of values.

x	$f(x)$
-2	2.25
-1	1.5
$-\frac{1}{2}$	≈ 1.2
0	1
$\frac{1}{2}$	$\approx .8$
1	$\approx .7$
2	$\approx .4$

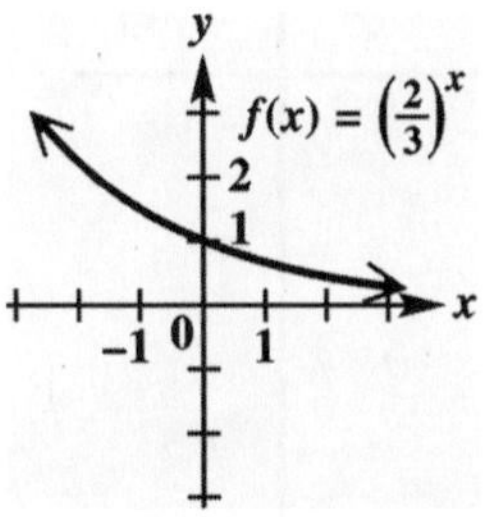

y
$f(x) = \left(\frac{2}{3}\right)^x$
2
1
-1
0
1
x

Plot these points and draw a smooth curve through them. This is a decreasing function. The domain is $(-\infty,\infty)$ and the range is $(0,\infty)$ and is one-to-one. Note: Since $f(x)=\left(\frac{2}{3}\right)^x=\left[\left(\frac{3}{2}\right)^{-1}\right]^x=\left(\frac{3}{2}\right)^{-x}$, the graph of $f(x)=\left(\frac{2}{3}\right)^x$ is the reflection of the graph of $f(x)=\left(\frac{3}{2}\right)^x$ (Exercise 17) about the y-axis.

19. The y-intercept of $f(x)=10^x$ is 1, and the x-axis is a horizontal asymptote. Make a table of values.

x	$f(x)$
-2	$.01$
-1	$.1$
$-\frac{1}{2}$	$\approx .3$
0	1
$\frac{1}{2}$	≈ 3.2
1	10
2	100

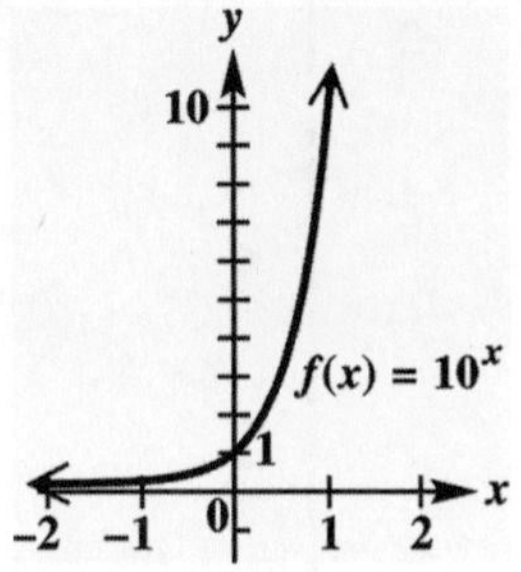

y
10
$f(x) = 10^x$
1
-2
-1
0
1
2
x

Plot these points and draw a smooth curve through them. This is an increasing function. The domain is $(-\infty,\infty)$ and the range is $(0,\infty)$ and is one-to-one.

20. The y-intercept of $f(x)=10^{-x}$ is 1, and the x-axis is a horizontal asymptote. Make a table of values.

x	$f(x)$
-2	100
-1	10
$-\frac{1}{2}$	≈ 3.2
0	1
$\frac{1}{2}$	$\approx .3$
1	.1
2	.01

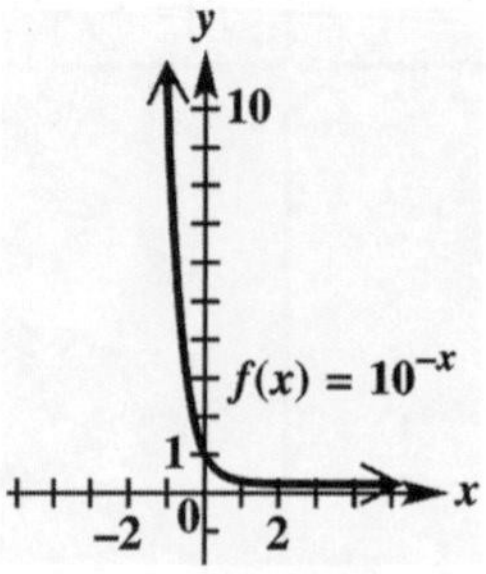

Plot these points and draw a smooth curve through them. This is a decreasing function. The domain is $(-\infty,\infty)$ and the range is $(0,\infty)$ and is one-to-one. Note: The graph of $f(x)=10^{-x}$ is the reflection of the graph of $f(x)=10^{x}$ (Exercise 19) about the y-axis.

21. The y-intercept of $f(x)=4^{-x}$ is 1, and the x-axis is a horizontal asymptote. Make a table of values.

x	$f(x)$
-2	16
-1	4
$-\frac{1}{2}$	2
0	1
$\frac{1}{2}$	.5
1	.25
2	.0625

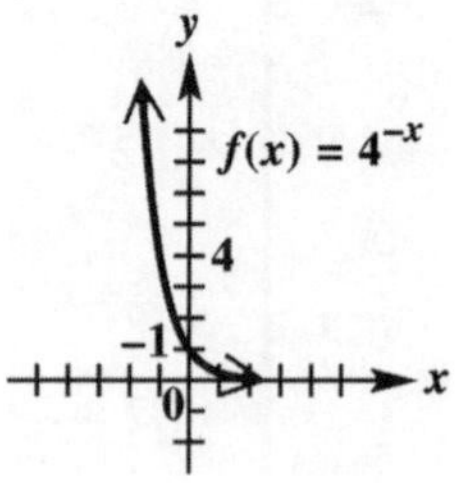

Plot these points and draw a smooth curve through them. This is a decreasing function. The domain is $(-\infty,\infty)$ and the range is $(0,\infty)$ and is one-to-one. Note: The graph of $f(x)=4^{-x}$ is the reflection of the graph of $f(x)=4^{x}$ (Exercise 14) about the y-axis.

22. The y-intercept of $f(x)=6^{-x}$ is 1, and the x-axis is a horizontal asymptote. Make a table of values.

x	$f(x)$
-2	36
-1	6
$-\frac{1}{2}$	≈ 2.4
0	1
$\frac{1}{2}$	$\approx .4$
1	$\frac{1}{6}\approx .2$
2	$\frac{1}{36}\approx .03$

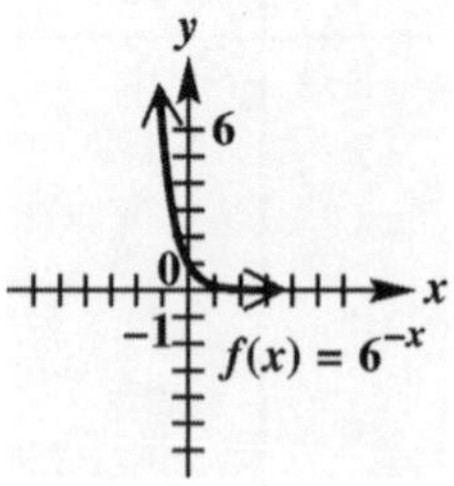

Plot these points and draw a smooth curve through them. This is an increasing function. The domain is $(-\infty,\infty)$ and the range is $(0,\infty)$ and is one-to-one.

23. The y-intercept of $f(x)=2^{|x|}$ is 1, and the x-axis is a horizontal asymptote. Make a table of values.

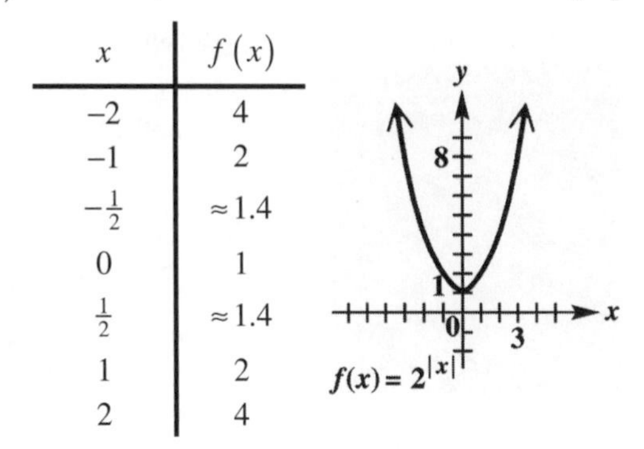

x	$f(x)$
-2	4
-1	2
$-\frac{1}{2}$	≈ 1.4
0	1
$\frac{1}{2}$	≈ 1.4
1	2
2	4

Plot these points and draw a smooth curve through them. The domain is $(-\infty,\infty)$ and the range is $[1,\infty)$ and is not one-to-one. Note: For $x < 0$, $|x|=-x$, so the graph is the same as that of $f(x)=2^{-x}$. For $x\geq 0$, we have $|x|=x$, so the graph is the same as that of $f(x)=2^{x}$. Since $|-x|=|x|$, the graph is symmetric with respect to the y-axis.

24. The y-intercept of $f(x)=2^{-|x|}$ is 1, and the x-axis is a horizontal asymptote. Make a table of values.

x	$f(x)$
-2	$\frac{1}{4}=.25$
-1	$\frac{1}{2}=.5$
$-\frac{1}{2}$	$\approx .7$
0	1
$\frac{1}{2}$	$\approx .7$
1	$\frac{1}{2}=.5$
2	$\frac{1}{4}=.25$

Plot these points and draw a smooth curve through them. The domain is $(-\infty,\infty)$ and the range is $(0,1]$ and is not one-to-one. Note: For $x < 0$, $|x|=-x$, so the graph is the same as that of $f(x)=2^{-(-x)}=2^{x}$. For $x\geq 0$, we have $|x|=x$, so the graph is the same as that of $f(x)=2^{-x}$. Since $|-x|=|x|$, the graph is symmetric with respect to the y-axis.

For Exercises 25 – 28, refer to the following graph of $f(x)=2^{x}$.

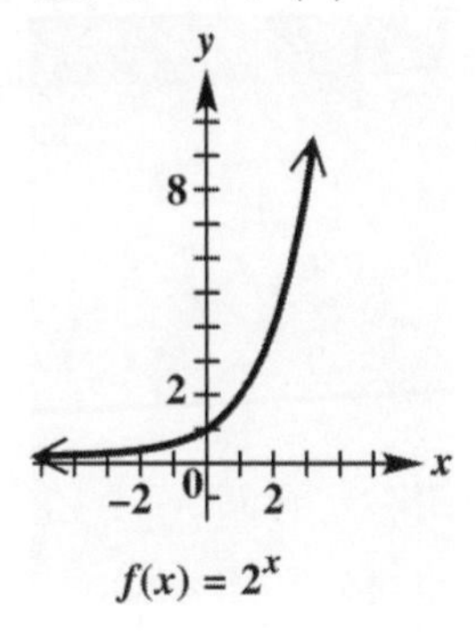

25. The graph of $f(x)=2^x+1$ is obtained by translating the graph of $f(x)=2^x$ up 1 unit.

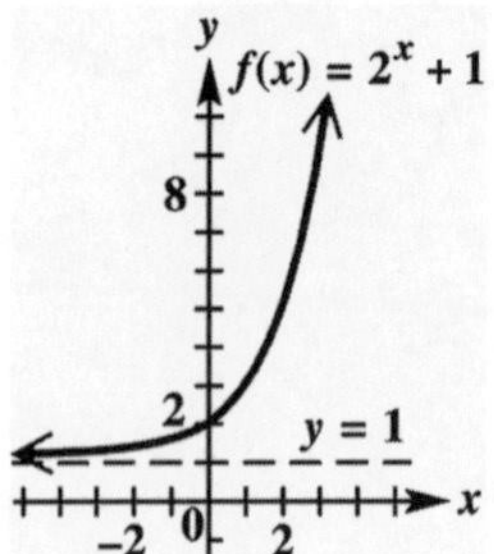

26. The graph of $f(x)=2^x-4$ is obtained by translating the graph of $f(x)=2^x$ down 4 units.

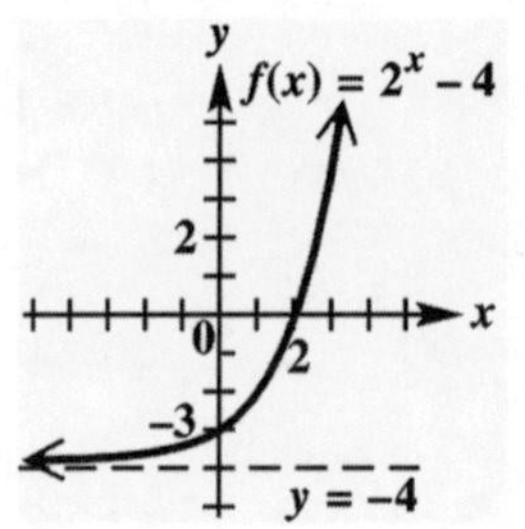

27. Since $f(x)=2^{x+1}=2^{x-(-1)}$, the graph is obtained by translating the graph of $f(x)=2^x$ to the left 1 unit.

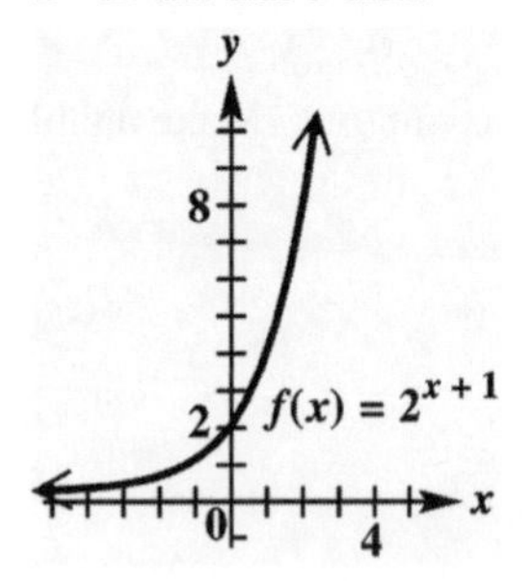

28. The graph of $f(x)=2^{x-4}$ is obtained by translating the graph of $f(x)=2^x$ to the right 4 units.

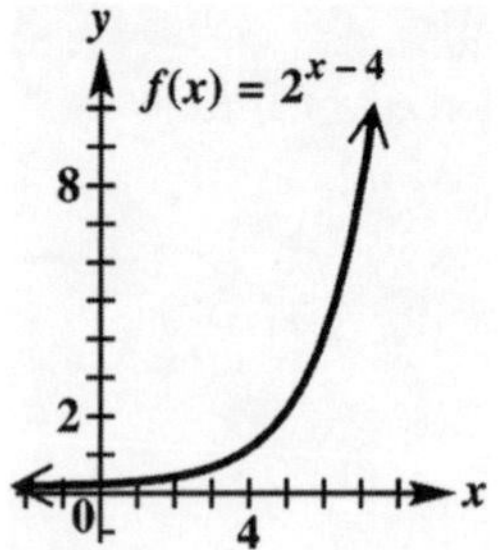

For Exercises 29 –32, refer to the following graph of $f(x)=\left(\frac{1}{3}\right)^x$.

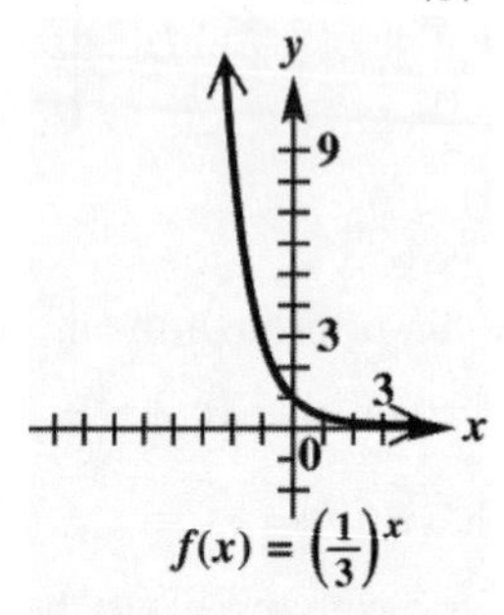

29. The graph of $f(x)=\left(\frac{1}{3}\right)^x-2$ obtained by translating the graph of $f(x)=\left(\frac{1}{3}\right)^x$ down 2 units.

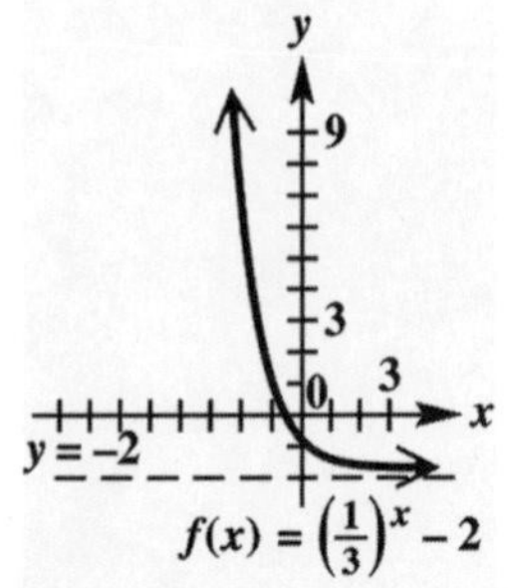

30. The graph of $f(x)=\left(\frac{1}{3}\right)^x+4$ is obtained by translating the graph of $f(x)=\left(\frac{1}{3}\right)^x$ up 4 units.

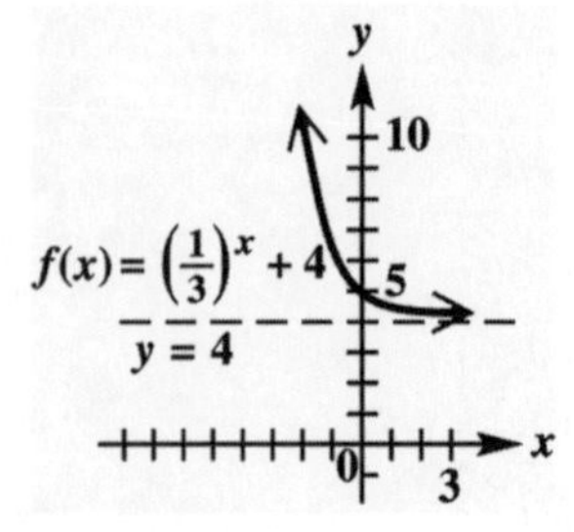

31. Since $f(x)=\left(\frac{1}{3}\right)^{x+2}=\left(\frac{1}{3}\right)^{x-(-2)}$, the graph is obtained by translating the graph of $f(x)=\left(\frac{1}{3}\right)^{x}$ 2 units to the left.

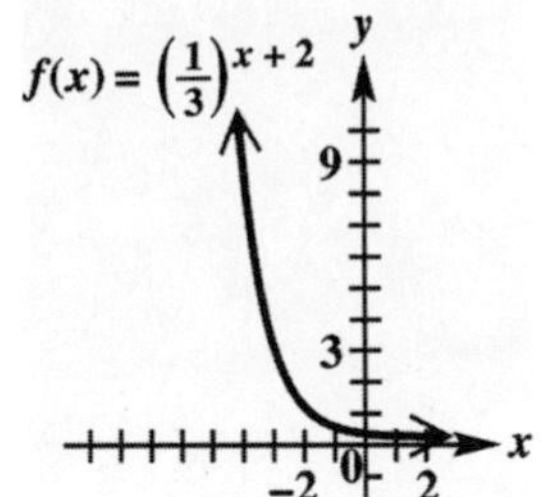

32. The graph of $f(x)=\left(\frac{1}{3}\right)^{x-4}$ is obtained by translating the graph of $f(x)=\left(\frac{1}{3}\right)^{x}$ 4 units to the right.

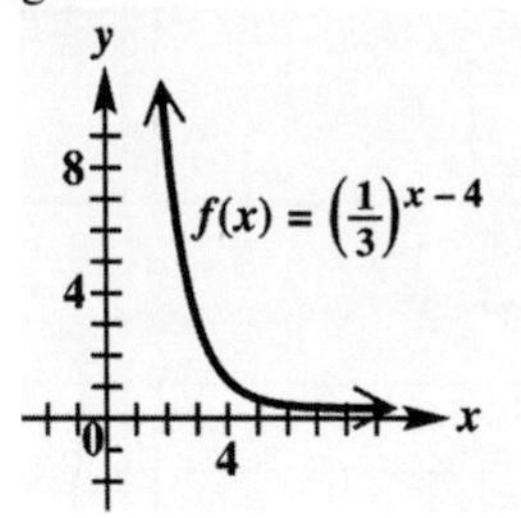

33. $a=2.3$
Since graph A is increasing, $a > 1$. Since graph A is the middle of the three increasing graphs, the value of a must be the middle of the three values of a greater than 1.

34. $a=1.8$
Since graph B is increasing, $a > 1$. Since graph B increases at the slowest rate of the three increasing graphs, the value of a must be the smallest of the three values of a greater than 1.

35. $a=.75$
Since graph C is decreasing, $0 < a < 1$. Since graph C decreases at the slowest rate of the three decreasing graphs, the value of a must be the closest to 1 of the three values of less than 1.

36. $a=.4$
Since graph D is decreasing, $0 < a < 1$. Since graph D is the middle of the three decreasing graphs, the value of a must be the middle of the three values of a less than 1.

37. $a=.31$
Since graph E is decreasing, $0 < a < 1$. Since graph E decreases at the fastest rate of the three decreasing graphs, the value of a must be the closest to 0 of the three values of a less than 1.

38. $a=3.2$
Since graph F is increasing, $a > 1$. Since graph F increases at the fastest rate of the three increasing graphs, the value of a must be the largest of the three values of a greater than 1.

39. $f(x)=\frac{e^{x}-e^{-x}}{2}$

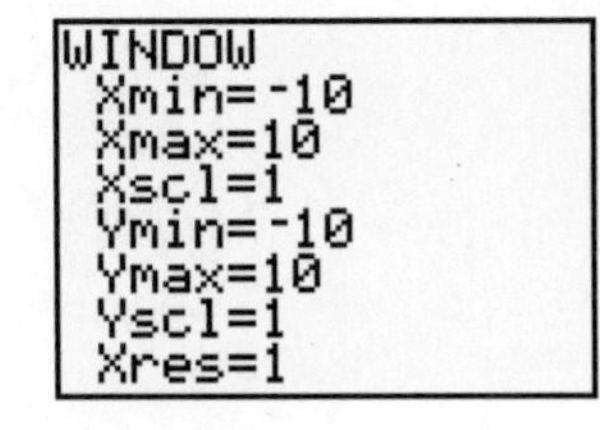

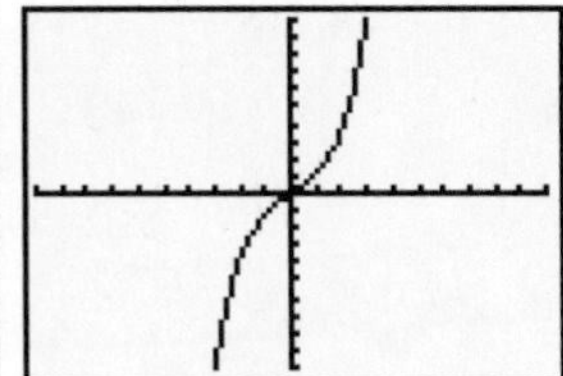

40. $f(x)=\frac{e^{x}+e^{-x}}{2}$

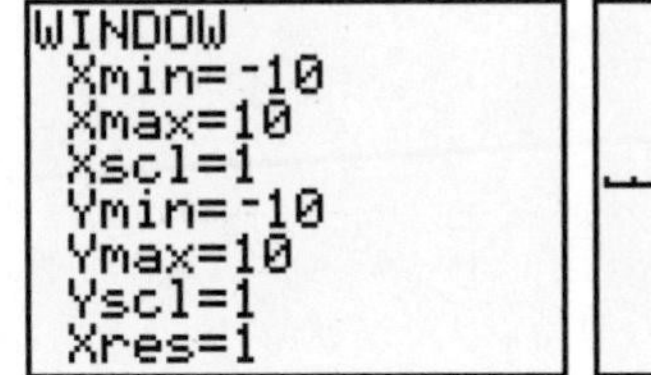

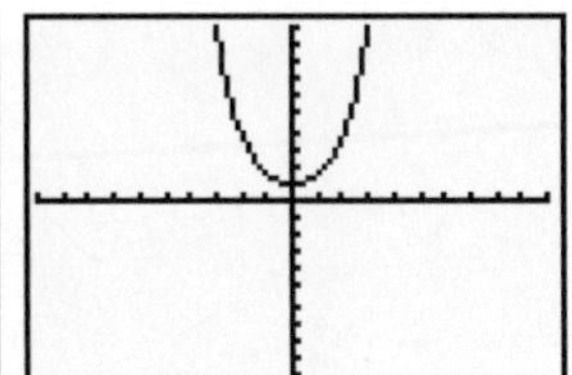

41. $f(x) = x \cdot 2^x$

```
WINDOW
 Xmin=-5
 Xmax=5
 Xscl=1
 Ymin=-2
 Ymax=2
 Yscl=1
 Xres=1
```

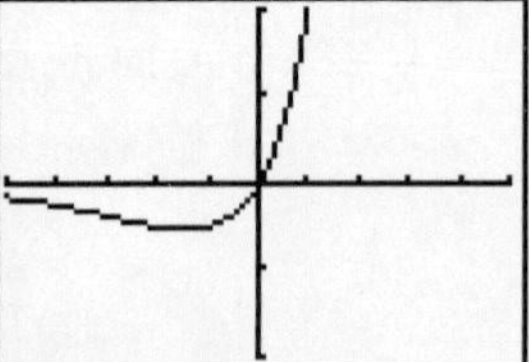

42. $f(x) = x^2 \cdot 2^{-x}$

```
WINDOW
 Xmin=-2
 Xmax=8
 Xscl=1
 Ymin=-2
 Ymax=5
 Yscl=1
 Xres=1
```

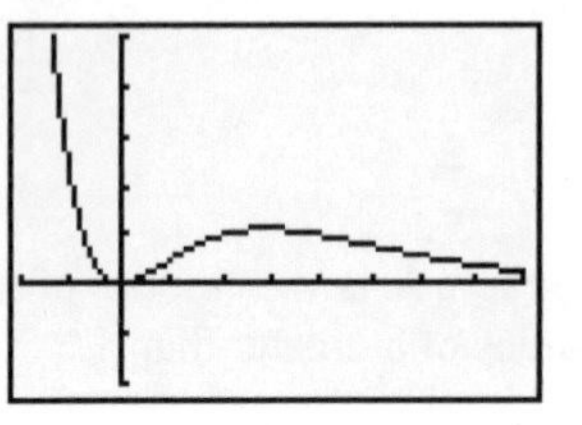

43.
$$4^x = 2$$
$$\left(2^2\right)^x = 2^1$$
$$2^{2x} = 2^1$$
$$2x = 1$$
$$x = \tfrac{1}{2}$$

Solution set: $\left\{\tfrac{1}{2}\right\}$

44.
$$125^r = 5$$
$$\left(5^3\right)^r = 5^1$$
$$5^{3r} = 5^1$$
$$3r = 1$$
$$r = \tfrac{1}{3}$$

Solution set: $\left\{\tfrac{1}{3}\right\}$

45.
$$\left(\tfrac{1}{2}\right)^k = 4$$
$$\left(2^{-1}\right)^k = 2^2$$
$$2^{-k} = 2$$
$$-k = 2$$
$$k = -2$$

Solution set: $\{-2\}$

46.
$$\left(\tfrac{2}{3}\right)^x = \tfrac{9}{4}$$
$$\left(\tfrac{2}{3}\right)^x = \left(\tfrac{3}{2}\right)^2$$
$$\left(\tfrac{2}{3}\right)^x = \left[\left(\tfrac{2}{3}\right)^{-1}\right]^2$$
$$\left(\tfrac{2}{3}\right)^x = \left(\tfrac{2}{3}\right)^{-2}$$
$$x = -2$$

Solution set: $\{-2\}$

47.
$$2^{3-y} = 8$$
$$2^{3-y} = 2^3$$
$$3 - y = 3$$
$$-y = 0$$
$$y = 0$$

Solution set: $\{0\}$

48.
$$5^{2p+1} = 25$$
$$5^{2p+1} = 5^2$$
$$2p + 1 = 2$$
$$2p = 1$$
$$p = \tfrac{1}{2}$$

Solution set: $\left\{\tfrac{1}{2}\right\}$

49. $e^{4x-1} = \left(e^2\right)^x$

$e^{4x-1} = e^{2x}$

$4x-1 = 2x$

$-1 = -2x$

$\frac{1}{2} = x$

Solution set: $\left\{\frac{1}{2}\right\}$

50. $e^{3-x} = \left(e^3\right)^{-x}$

$e^{3-x} = e^{-3x}$

$3-x = -3x$

$3 = -2x$

$-\frac{3}{2} = x$

Solution set: $\left\{-\frac{3}{2}\right\}$

51. $27^{4z} = 9^{z+1}$

$\left(3^3\right)^{4z} = \left(3^2\right)^{z+1}$

$3^{3(4z)} = 3^{2(z+1)}$

$3^{12z} = 3^{2z+2}$

$12z = 2z+2$

$10z = 2$

$z = \frac{1}{5}$

Solution set: $\left\{\frac{1}{5}\right\}$

52. $32^t = 16^{1-t}$

$\left(2^5\right)^t = \left(2^4\right)^{1-t}$

$2^{5t} = 2^{4(1-t)}$

$2^{5t} = 2^{4-4t}$

$5t = 4-4t$

$9t = 4$

$t = \frac{4}{9}$

Solution set: $\left\{\frac{4}{9}\right\}$

53. $4^{x-2} = 2^{3x+3}$

$\left(2^2\right)^{x-2} = 2^{3x+3}$

$2^{2(x-2)} = 2^{3x+3}$

$2^{2x-4} = 2^{3x+3}$

$2x-4 = 3x+3$

$-4 = x+3$

$-7 = x$

Solution set: $\{-7\}$

54. $\left(\frac{1}{2}\right)^{3x-6} = 8^{x+1}$

$\left(2^{-1}\right)^{3x-6} = \left(2^3\right)^{x+1}$

$2^{-(3x-6)} = 2^{3(x+1)}$

$2^{-3x+6} = 2^{3x+3}$

$-3x+6 = 3x+3$

$6 = 6x+3$

$3 = 6x$

$\frac{1}{2} = x$

Solution set: $\left\{\frac{1}{2}\right\}$

55. $\left(\frac{1}{e}\right)^{-x} = \left(\frac{1}{e^2}\right)^{x+1}$

$\left(e^{-1}\right)^{-x} = \left(e^{-2}\right)^{x+1}$

$e^x = e^{-2(x+1)}$

$e^x = e^{-2x-2}$

$x = -2x-2$

$3x = -2$

$x = -\frac{2}{3}$

Solution set: $\left\{-\frac{2}{3}\right\}$

56. $e^{k-1} = \left(\frac{1}{e^4}\right)^{k+1}$

$e^{k-1} = \left(e^{-4}\right)^{k+1}$

$e^{k-1} = e^{-4(k+1)}$

$e^{k-1} = e^{-4k-4}$

$k-1 = -4k-4$

$5k-1 = -4$

$5k = -3 \Rightarrow k = -\frac{3}{5}$

Solution set: $\left\{-\frac{3}{5}\right\}$

57. $\left(\sqrt{2}\right)^{x+4}=4^x$

$\left(2^{1/2}\right)^{x+4}=\left(2^2\right)^x$

$2^{(1/2)(x+4)}=2^{2x}$

$2^{(1/2)x+2}=2^{2x}$

$\frac{1}{2}x+2=2x$

$2=\frac{3}{2}x$

$x=\frac{2}{3}\cdot 2=\frac{4}{3}$

Solution set: $\left\{\frac{4}{3}\right\}$

58. $\left(\sqrt[3]{5}\right)^{-x}=\left(\frac{1}{5}\right)^{x+2}$

$\left(5^{1/3}\right)^{-x}=\left(5^{-1}\right)^{x+2}$

$5^{-(1/3)x}=5^{-(x+2)}$

$5^{-(1/3)x}=5^{-x-2}$

$-\frac{1}{3}x=-x-2$

$\frac{2}{3}x=-2$

$x=-2\cdot\frac{3}{2}=-3$

Solution set: $\{-3\}$

59. $\frac{1}{27}=b^{-3}$

$3^{-3}=b^{-3}$

$b=3$

Alternate solution:

$\frac{1}{27}=b^{-3}$

$\frac{1}{27}=\frac{1}{b^3}$

$27=b^3$

$b=\sqrt[3]{27}=3$

Solution set: $\{3\}$

60. $\frac{1}{81}=k^{-4}$

$\frac{1}{81}=\frac{1}{k^4}$

$81=k^4$

$k=\pm\sqrt[4]{81}$

$k=\pm 3$

Solution set: $\{-3,3\}$

61. $4=r^{2/3}\Rightarrow 4^{3/2}=\left(r^{2/3}\right)^{3/2}\Rightarrow\left(\pm\sqrt{4}\right)^3=r\Rightarrow r=(\pm 2)^3\Rightarrow r=\pm 8$

One should check all proposed solutions in the original equation when you raise both sides to a power.

Check $r=-8$.

$4=r^{2/3}$

$4=(-8)^{2/3}$?

$4=\left(\sqrt[3]{-8}\right)^2$

$4=(-2)^2$

$4=4$

This is a true statement.
-8 is a solution.

Check $r=8$.

$4=r^{2/3}$

$4=8^{2/3}$?

$4=\left(\sqrt[3]{8}\right)^2$

$4=2^2$

$4=4$

This is a true statement.
8 is a solution.

Solution set: $\{-8,8\}$

62. $z^{5/2}=32\Rightarrow\left(z^{5/2}\right)^{2/5}=32^{2/5}\Rightarrow z=32^{2/5}\Rightarrow z=\left(\sqrt[5]{32}\right)^2\Rightarrow z=2^2\Rightarrow z=4$

One should check all proposed solutions in the original equation when you raise both sides to a power.

Check $z=4$.

$$z^{5/2}=32$$
$$4^{5/2}=32\quad ?$$
$$\left(\sqrt{4}\right)^5=32$$
$$2^5=32$$
$$32=32$$

This is a true statement.
4 is a solution.

Solution set: $\{4\}$

63. **(a)** Use the compound interest formula to find the future value, $A=P\left(1+\frac{r}{m}\right)^{tm}$, given $m=2$, $P=8906.54$, $r=.05$, and $t=9$.

$$A=P\left(1+\tfrac{r}{m}\right)^{tm}=(8906.54)\left(1+\tfrac{.05}{2}\right)^{9(2)}=(8906.54)(1+.025)^{18}\approx 13{,}891.16276$$

Rounding to the nearest cent, the future value is \$13,891.16.
The amount of interest would be \$13,891.16 − \$8906.54 = \$4984.62.

(b) Use the continuous compounding interest formula to find the future value, $A=Pe^{rt}$, given $P=8906.54$, $r=.05$, and $t=9$.

$$A=Pe^{rt}=8906.54e^{.05(9)}=8906.54e^{.45}\approx 8906.54(1.568312)\approx 13{,}968.23521$$

Rounding to the nearest cent, the future value is \$13,968.24.
The amount of interest would be \$13,968.24 − \$8906.54 = \$5061.70.

64. **(a)** Use the compound interest formula to find the future value, $A=P\left(1+\frac{r}{m}\right)^{tm}$, given $m=4$, $P=56{,}780$, $r=.053$, and $t=\frac{23}{4}$.

$$A=P\left(1+\tfrac{r}{m}\right)^{tm}=(56{,}780)\left(1+\tfrac{.053}{4}\right)^{23}=(56{,}780)(1+.01325)^{23}\approx 76{,}855.9462$$

Rounding to the nearest cent, the future value is \$76,855.95.
The amount of interest would be \$76,855.95 − \$56,780 = \$20,075.95.

(b) Use the continuous compounding interest formula to find the future value, $A=Pe^{rt}$, given $P=56{,}780$, $r=.053$, and $t=15$.

$$A=Pe^{rt}=56{,}780e^{.053(15)}=56{,}780e^{.795}\approx 56{,}780(2.214441)=125{,}735.9598$$

Rounding to the nearest cent, the future value is \$125,735.96.
The amount of interest would be \$125,735.96 − \$56,780 = \$68,955.96.

65. Use the compound interest formula to find the present amount, $A = P\left(1+\frac{r}{m}\right)^{tm}$, given $m = 4$, $A = 25{,}000$, $r = .06$, and $t = \frac{11}{4}$.

$$A = P\left(1+\tfrac{r}{m}\right)^{tm}$$

$$25{,}000 = P\left(1+\tfrac{.06}{4}\right)^{(11/4)(4)} \Rightarrow 25{,}000 = P(1.015)^{11} \Rightarrow P = \frac{25{,}000}{(1.015)^{11}} \approx \$21{,}223.33083$$

Rounding to the nearest cent, the present value is \$21,223.33.

66. Use the compound interest formula to find the present amount, $A = P\left(1+\frac{r}{m}\right)^{tm}$, given $m = 12$, $A = 45{,}000$, $r = .036$, and $t = 1$.

$$A = P\left(1+\tfrac{r}{m}\right)^{tm}$$

$$45{,}000 = P\left(1+\tfrac{.036}{12}\right)^{1(12)} \Rightarrow 45{,}000 = P(1.003)^{12} \Rightarrow P = \frac{45{,}000}{(1.003)^{12}} \approx \$43{,}411.15267$$

Rounding to the nearest cent, the present value is \$43,411.15.

67. Use the compound interest formula to find the present value, $A = P\left(1+\frac{r}{m}\right)^{tm}$, given $m = 4$, $A = 5{,}000$, $r = .035$, and $t = 10$.

$$A = P\left(1+\tfrac{r}{m}\right)^{tm}$$

$$5{,}000 = P\left(1+\tfrac{.035}{4}\right)^{10(4)} \Rightarrow 5{,}000 = P(1.00875)^{40} \Rightarrow P = \frac{5{,}000}{(1.00875)^{40}} \approx \$3528.808535$$

Rounding to the nearest cent, the present value is \$3528.81.

68. Use the compound interest formula to find the interest rate, $A = P\left(1+\frac{r}{m}\right)^{tm}$, given $m = 12$, $A = 65{,}325$, $P = 65{,}000$, and $t = \frac{6}{12} = \frac{1}{2}$.

$$A = P\left(1+\tfrac{r}{m}\right)^{tm}$$

$$65{,}325 = 65{,}000\left(1+\tfrac{r}{12}\right)^{(1/2)(12)} \Rightarrow 65{,}325 = 65{,}000\left(1+\tfrac{r}{12}\right)^{6}$$

$$1.005 = \left(1+\tfrac{r}{12}\right)^{6} \Rightarrow (1.005)^{\frac{1}{6}} = 1+\tfrac{r}{12} \Rightarrow (1.005)^{\frac{1}{6}} - 1 = \tfrac{r}{12} \Rightarrow 12\left[(1.005)^{\frac{1}{6}} - 1\right] = r$$

$$r \approx .0099792301$$

The interest rate, to the nearest tenth, is 1.0%.

69. Use the compound interest formula to find the interest rate, $A = P\left(1+\frac{r}{m}\right)^{tm}$, given $m = 4$, $A = 1500$, $P = 1200$, and $t = 5$.

$$A = P\left(1+\tfrac{r}{m}\right)^{tm}$$

$$1500 = 1200\left(1+\tfrac{r}{4}\right)^{5(4)} \Rightarrow 1500 = 1200\left(1+\tfrac{r}{4}\right)^{20} \Rightarrow 1.25 = \left(1+\tfrac{r}{4}\right)^{20}$$

$$(1.25)^{\frac{1}{20}} = 1+\tfrac{r}{4} \Rightarrow (1.25)^{\frac{1}{20}} - 1 = \tfrac{r}{4} \Rightarrow 4\left[(1.25)^{\frac{1}{20}} - 1\right] = r$$

$$r \approx .044878604$$

The interest rate, to the nearest tenth, is 4.5%.

70. Use the compound interest formula to find the interest rate, $A = P\left(1+\frac{r}{m}\right)^{tm}$, given $m = 4$, $A = 8400$, $P = 5000$, and $t = 8$.

$$A = P\left(1+\frac{r}{m}\right)^{tm}$$

$$8400 = 5000\left(1+\frac{r}{4}\right)^{8(4)} \Rightarrow 8400 = 5000\left(1+\frac{r}{4}\right)^{32} \Rightarrow 1.68 = \left(1+\frac{r}{4}\right)^{32}$$

$$(1.68)^{\frac{1}{32}} = 1+\frac{r}{4} \Rightarrow (1.68)^{\frac{1}{32}} - 1 = \frac{r}{4} \Rightarrow 4\left[(1.68)^{\frac{1}{32}} - 1\right] = r$$

$$r \approx .0653777543$$

The interest rate, to the nearest tenth, is 6.5%.

71. For each bank we need to calculate $\left(1+\frac{r}{m}\right)^{m}$. Since the base, $1+\frac{r}{m}$, is greater than 1, we need only compare the three values calculated to determine which bank will yield the least amount of interest. It is understood that the amount of time, t, and the principal, P, are the same for all three banks.

Bank A: Calculate $\left(1+\frac{r}{m}\right)^{m}$ where $m = 1$ and $r = .064$.

$$\left(1+\frac{.064}{1}\right)^{1} = (1+.064)^{1} = (1.064)^{1} = 1.064$$

Bank B: Calculate $\left(1+\frac{r}{m}\right)^{m}$ where $m = 12$ and $r = .063$.

$$\left(1+\frac{.063}{12}\right)^{12} = (1+.00525)^{12} = (1.00525)^{12} \approx 1.064851339$$

Bank C: Calculate $\left(1+\frac{r}{m}\right)^{m}$ where $m = 4$ and $r = .0635$.

$$\left(1+\frac{.0635}{4}\right)^{4} = (1+.015875)^{4} = (1.015875)^{4} \approx 1.06502816$$

Bank A will charge you the least amount of interest, even though it has the highest stated rate.

72. Given $P = 10{,}000$, $r = .05$, and $t = 10$, the compound interest formula, $A = P\left(1+\frac{r}{m}\right)^{tm}$, becomes $A = 10{,}000\left(1+\frac{.05}{m}\right)^{10m}$.

(a) $m = 1 \Rightarrow A = 10{,}000\left(1+\frac{.05}{1}\right)^{10(1)} = 10{,}000(1+.05)^{10} \approx 16{,}288.94627$

Rounding to the nearest cent, the future value is \$16,288.95.

(b) $m = 4 \Rightarrow A = 10{,}000\left(1+\frac{.05}{4}\right)^{10(4)} = 10{,}000(1+.0125)^{40} \approx 16{,}436.19463$

Rounding to the nearest cent, the future value is \$16,436.19.

(c) $m = 12 \Rightarrow A = 10{,}000\left(1+\frac{.05}{12}\right)^{10(12)} = 10{,}000\left(1+\frac{.05}{12}\right)^{120} \approx 16{,}470.09498$

Rounding to the nearest cent, the future value is \$16,470.09.

(d) $m = 365 \Rightarrow A = 10{,}000\left(1+\frac{.05}{365}\right)^{10(365)} = 10{,}000\left(1+\frac{.05}{365}\right)^{3650} \approx 16{,}486.64814$

Rounding to the nearest cent, the future value is \$16,486.65.

73. (a)

```
WINDOW
 Xmin=-1000
 Xmax=11000
 Xscl=1000
 Ymin=0
 Ymax=1200
 Yscl=100
 Xres=1
```

(b) From the graph above, we can see that the data are not linear but exponentially decreasing.

(c)

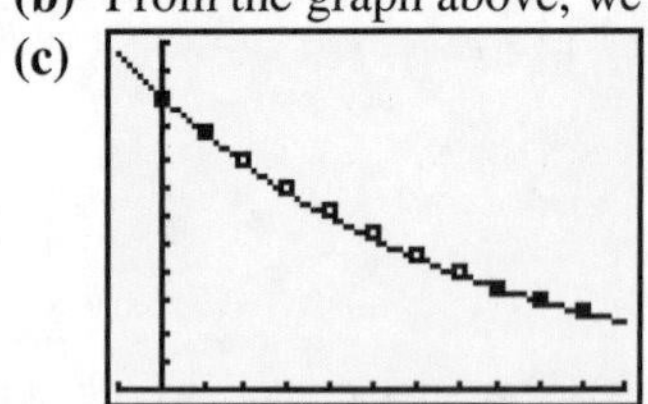

(d)
$$P(x) = 1013e^{-.0001341x}$$
$$P(1500) = 1013e^{-.0001341(1500)} \approx 1013(.817790) \approx 828$$
$$P(11{,}000) = 1013e^{-.0001341(11{,}000)} \approx 1013(.228756) \approx 232$$

When the altitude is 1500 m, the function P gives a pressure of 828 mb, which is less than the actual value of 846 mb. When the altitude is 11,000 m, the function P gives a pressure of 232 mb, which is more than the actual value of 227 mb.

74. $y = 6073e^{.0137x}$

(a) 2000 is 10 years after 1990, so use $x = 10$.

$$y = 6073e^{.0137(10)} = 6073e^{.137} \approx 6073(1.146828) \approx 6965$$

The function gives a population of about 6965 million, which differs from the actual value by $6965 - 6079 = 886$ million.

(b) 2005 is 15 years after 1990, so use $x = 15$.

$$y = 6073e^{.0137(15)} = 6073e^{.2055} \approx 6073(1.228139) \approx 7458$$

The function gives a population of about 7458 million.

(c) 2015 is 25 years after 1990, so use $x = 25$.

$$y = 6073e^{.0137(25)} = 6073e^{.3425} \approx 6073(1.408464) \approx 8554$$

The function gives a population of about 8554 million.

(d) Answers will vary.

75. (a) Evaluate $T = 50{,}000(1+.06)^n$ where $n = 4$.

$$T = 50{,}000(1+.06)^4 = 50{,}000(1.06)^4 \approx 63{,}123.848$$

Total population after 4 years is about 63,000.

(b) Evaluate $T = 30{,}000(1+.12)^n$ where $n = 3$.

$$T = 30{,}000(1+.12)^3 = 30{,}000(1.12)^3 = 42{,}147.84$$

There would be about 42,000 deer after 3 years.

(c) Evaluate $T = 45{,}000(1+.08)^n$ where $n = 5$.

$$T = 45{,}000(1+.08)^5 = 45{,}000(1.08)^5 \approx 66{,}119.76346$$

There would be about 66,000 deer after 5 years. Thus, we can expect about $66{,}000 - 45{,}000 = 21{,}000$ additional deer after 5 years.

76. $p(t)=250-120(2.8)^{-.5t}$

(a) $p(2)=250-120(2.8)^{-.5(2)}=250-120(2.8)^{-1}\approx 207.1428571$
After 2 months, a person will type about 207 symbols per minute.

(b) $p(4)=250-120(2.8)^{-.5(4)}=250-120(2.8)^{-2}\approx 234.6938776$
After 4 months, a person will type about 235 symbols per minute.

(c) $p(10)=250-120(2.8)^{-.5(10)}=250-120(2.8)^{-5}\approx 249.3027459$
After 10 months, a person will type about 249 symbols per minute.

(d) The number of symbols approaches 250.

Section 9.2: Logarithmic Functions

1. **(a)** C; $\log_2 16=4$ because $2^4=16$.

(b) A; $\log_3 1=0$ because $3^0=1$.

(c) E; $\log_{10}.1=-1$ because $10^{-1}=.1$.

(d) B; $\log_2\sqrt{2}=\frac{1}{2}$ because $2^{1/2}=\sqrt{2}$.

(e) F; $\log_e\left(\frac{1}{e^2}\right)=-2$ because $e^{-2}=\frac{1}{e^2}$.

(f) D; $\log_{1/2}8=-3$ because $\left(\frac{1}{2}\right)^{-3}=8$.

2. **(a)** F; $\log_3 81=4$ because $3^4=16$.

(b) B; $\log_3\frac{1}{3}=-1$ because $3^{-1}=\frac{1}{3}$.

(c) A; $\log_{10}.01=-2$ because $10^{-2}=.01$.

(d) D; $\log_6\sqrt{6}=\frac{1}{2}$ because $6^{1/2}=\sqrt{6}$.

(e) C; $\log_e 1=0$ because $e^0=1$.

(f) E; $\log_3 27^{3/2}=\frac{9}{2}$
because $3^{9/2}=\left(3^3\right)^{3/2}=27^{3/2}$.

3. $3^4=81$ is equivalent to $\log_3 81=4$.

4. $2^5=32$ is equivalent to $\log_2 32=5$.

5. $\left(\frac{2}{3}\right)^{-3}=\frac{27}{8}$ is equivalent to $\log_{2/3}\frac{27}{8}=-3$.

6. $10^{-4}=.0001$ is equivalent to $\log_{10}.0001=-4$.

7. $\log_6 36=2$ is equivalent to $6^2=36$.

8. $\log_5 5=1$ is equivalent to $5^1=5$.

9. $\log_{\sqrt{3}}81=8$ is equivalent to $\left(\sqrt{3}\right)^8=81$.

10. $\log_4\frac{1}{64}=-3$ is equivalent to $4^{-3}=\frac{1}{64}$.

11. Answers will vary.

12. $\log_a 1=0$ for all real numbers a, because $a^0=1$, $(a\neq 0)$ for all real numbers a.

13. $x=\log_5\frac{1}{625}$

$5^x=\frac{1}{625}\Rightarrow 5^x=\dfrac{1}{5^4}$

$5^x=5^{-4}\Rightarrow x=-4$

Solution set: $\{-4\}$

14. $x=\log_3\frac{1}{81}$

$3^x=\frac{1}{81}\Rightarrow 3^x=\frac{1}{3^4}$

$3^x=3^{-4}\Rightarrow x=-4$

Solution set: $\{-4\}$

15. $x=\log_{10}.001$

$10^x=.001\Rightarrow 10^x=10^{-3}$

$x=-3$

Solution set: $\{-3\}$

16. $x=\log_6\frac{1}{216}$

$6^x=\frac{1}{216}\Rightarrow 6^x=\frac{1}{6^3}$

$6^x=6^{-3}\Rightarrow x=-3$

Solution set: $\{-3\}$

17. $x = \log_8 \sqrt[4]{8}$

$8^x = \sqrt[4]{8} \Rightarrow 8^x = 8^{1/4}$

$x = \frac{1}{4}$

Solution set: $\left\{\frac{1}{4}\right\}$

18. $x = 8\log_{100} 10$

$x = \log_{100} 10^8$

$100^x = 10^8 \Rightarrow 100^x = 10^{2(4)}$

$100^x = \left(10^2\right)^4 \Rightarrow 100^x = 100^4$

$x = 4$

Solution set: $\{4\}$

19. $x = 3^{\log_3 8}$; Writing as a logarithmic equation we have $\log_3 8 = \log_3 x$ which implies $x = 8$. Using the Theorem of Inverses on page 410, we can directly state that $x = 8$.
Solution set: $\{8\}$

20. $x = 12^{\log_{12} 5}$; Writing as a logarithmic equation we have $\log_{12} 5 = \log_{12} x$ which implies $x = 5$. Using the Theorem of Inverses on page 410, we can directly state that $x = 5$.
Solution set: $\{5\}$

21. $x = 2^{\log_2 9}$ Writing as a logarithmic equation we have $\log_2 9 = \log_2 x$ which implies $x = 9$. Using the Theorem of Inverses on page 410, we can directly state that $x = 9$.
Solution set: $\{9\}$

22. $x = 8^{\log_8 11}$ Writing as a logarithmic equation we have $\log_8 11 = \log_8 x$ which implies $x = 11$. Using the Theorem of Inverses on page 410, we can directly state that $x = 11$.
Solution set: $\{11\}$

23. $\log_x 25 = -2$

$x^{-2} = 25 \Rightarrow x^{-2} = 5^2 \Rightarrow \left(x^{-2}\right)^{-1/2} = \left(5^2\right)^{-1/2}$

$x = 5^{-1}$ You do not include a $\pm$ since the base, x, cannot be negative.

$x = \frac{1}{5}$

Solution set: $\left\{\frac{1}{5}\right\}$

24. $\log_x \frac{1}{16} = -2$

$x^{-2} = \frac{1}{16} \Rightarrow x^{-2} = \frac{1}{2^4} \Rightarrow x^{-2} = 2^{-4} \Rightarrow \left(x^{-2}\right)^{-1/2} = \left(2^{-4}\right)^{-1/2}$

$x = 2^2$ You do not include a $\pm$ since the base, x, cannot be negative.

$x = 4$

Solution set: $\{4\}$

25. $\log_4 x = 3$

$4^3 = x$

$64 = x$

Solution set: $\{64\}$

26. $\log_2 x = -1$

$2^{-1} = x$

$\frac{1}{2} = x$

Solution set: $\left\{\frac{1}{2}\right\}$

27. $x = \log_4 \sqrt[3]{16}$

$4^x = \sqrt[3]{16}$

$4^x = (16)^{1/3}$

$4^x = (4^2)^{1/3}$

$4^x = 4^{2/3}$

$x = \frac{2}{3}$

Solution set: $\left\{\frac{2}{3}\right\}$

28. $x = \log_5 \sqrt[4]{25}$

$5^x = \sqrt[4]{25} \Rightarrow 5^x = (25)^{1/4}$

$5^x = (5^2)^{1/4} \Rightarrow 5^x = 5^{2/4}$

$5^x = 5^{1/2} \Rightarrow x = \frac{1}{2}$

Solution set: $\left\{\frac{1}{2}\right\}$

29. $\log_x 3 = -1$

$x^{-1} = 3 \Rightarrow (x^{-1})^{-1} = 3^{-1}$

$x = \frac{1}{3}$

Solution set: $\left\{\frac{1}{3}\right\}$

30. $\log_x 1 = 0$

$x^0 = 1$

This statement is true for any valid base, x.

Solution set: $(0,1) \cup (1,\infty)$

31. Answers will vary.

32. y (2.2319281) represents the exponent to which 2 must be raised in order to obtain x (5).

For Exercises 33 – 35, refer to the following graph of $f(x) = \log_2 x$.

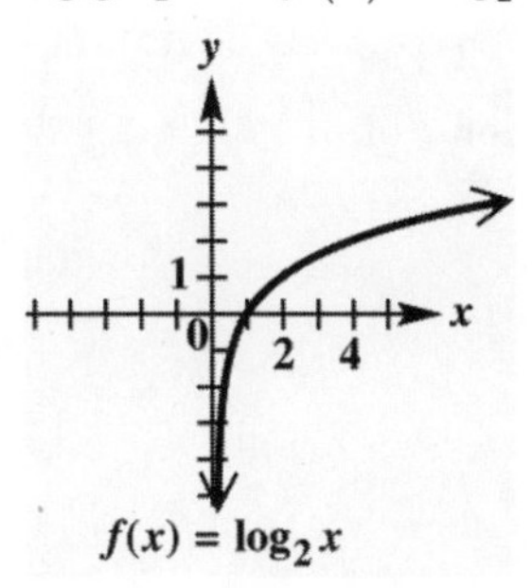

33. The graph of $f(x) = (\log_2 x) + 3$ is obtained by translating the graph of $f(x) = \log_2 x$ up 3 units.

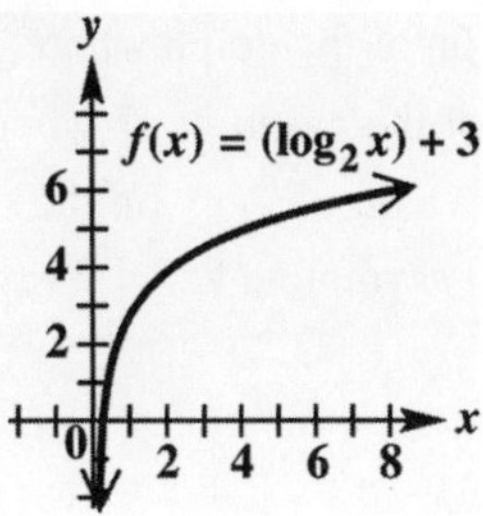

34. The graph of $f(x) = \log_2(x+3)$ is obtained by translating the graph of $f(x) = \log_2 x$ to the left 3 units. The graph has a vertical asymptote at $x = -3$.

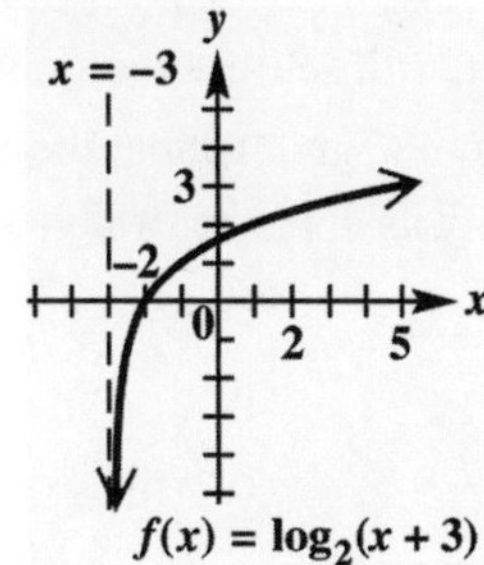

35. To find the graph of $f(x)=|\log_2(x+3)|$, translate the graph of $f(x)=\log_2 x$ to the left 3 units to obtain the graph of $\log_2(x+3)$. (See Exercise 34.) For the portion of the graph where $f(x)\geq 0$, that is, where $x\geq -2$, use the same graph as in 34. For the portion of the graph in 34 where $f(x)<0$, $-3<x<-2$, reflect the graph about the x-axis. In this way, each negative value of $f(x)$ on the graph in 34 is replaced by its opposite, which is positive. The graph has a vertical asymptote at $x = -3$.

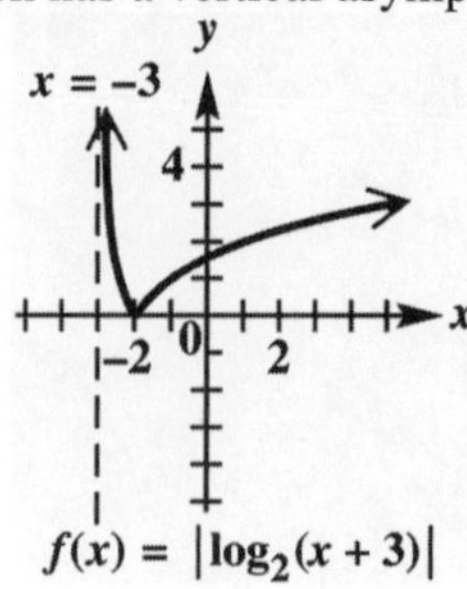

For Exercises 36 – 38, refer to the following graph of $f(x)=\log_{1/2} x$.

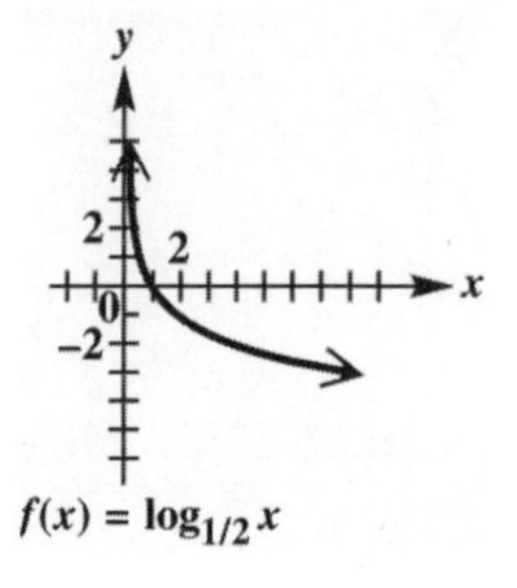

36. The graph of $f(x)=(\log_{1/2} x)-2$ is obtained by translating the graph of $f(x)=\log_{1/2} x$ down 2 units.

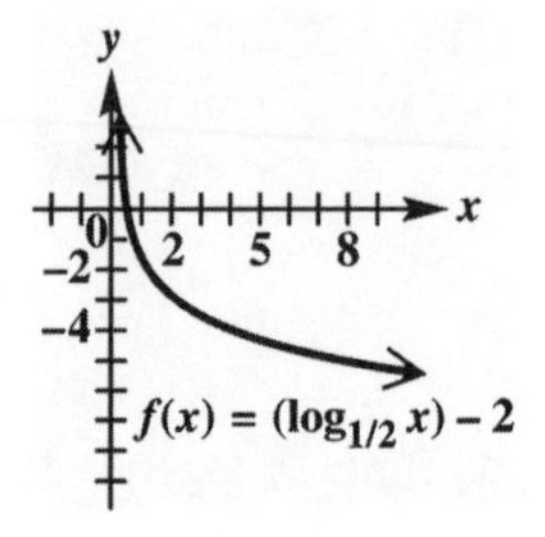

37. The graph of $f(x)=\log_{1/2}(x-2)$ is obtained by translating the graph of $f(x)=\log_{1/2} x$ to the right 2 units. The graph has a vertical asymptote at $x = 2$.

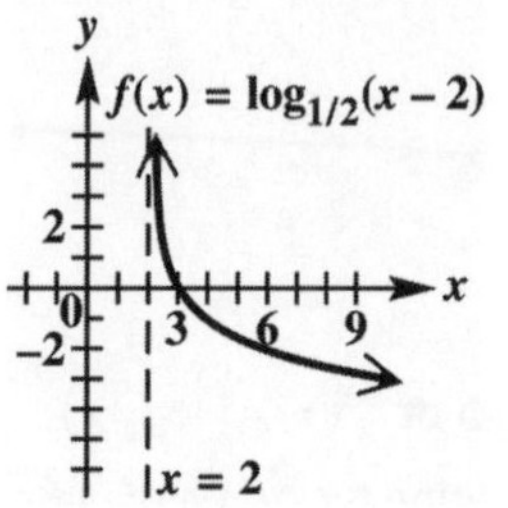

38. To find the graph of $f(x)=|\log_{1/2}(x-2)|$ translate the graph of $f(x)=\log_{1/2} x$ to the right 2 units to obtain the graph of $\log_{1/2}(x-2)$. (See Exercise 37.) For the portion of the graph where $f(x)\geq 0$, that is, where $2<x\leq 3$, use the same graph as in 37. For the portion of the graph in 37 where $f(x)<0$, $x > 3$, reflect the graph about the x-axis. In this way, each negative value of $f(x)$ on the graph in 37 is replaced by its opposite, which is positive. The graph has a vertical asymptote at $x = 2$.

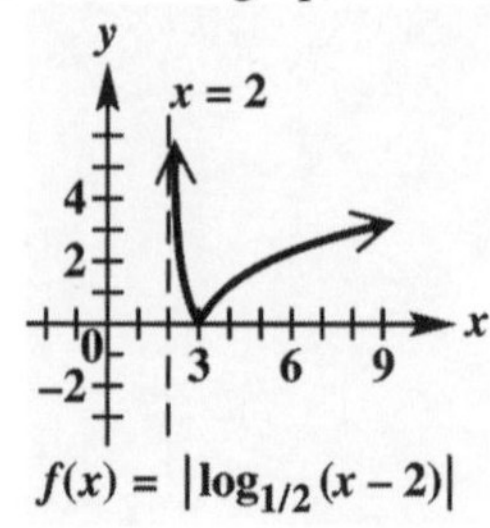

39. Because $f(x)=\log_2 x$ has a vertical asymptote, which is the y-axis (the line $x=0$), x-intercept of 1, and is increasing, the correct choice is the graph in E.

40. Because $f(x)=\log_2 2x$ has a vertical asymptote, which is the y-axis (the line $x=0$), has an x-intercept when $2x=1 \Rightarrow x=\frac{1}{2}$, and is increasing, the correct choice is the graph in D.

41. Because $f(x)=\log_2 \frac{1}{x}=\log_2 x^{-1}=-\log_2 x$, it has a vertical asymptote, which is the y-axis (the line $x=0$), has an x-intercept 1, and is the reflection of $f(x)=\log_2 x$ across the x-axis, it is decreasing and the correct choice is the graph in B

42. Because $f(x)=\log_2 \frac{x}{2}$ has a vertical asymptote, which is the y-axis (the line $x=0$), has an x-intercept when $\frac{x}{2}=1 \Rightarrow x=2$, and is increasing, the correct choice is the graph in C.

43. Because $f(x)=\log_2(x-1)$ represents the horizontal shift of $f(x)=\log_2 x$ to the right 1 unit, the function has a vertical asymptote which is the line $x=1$, has an x-intercept when $x-1=1 \Rightarrow x=2$, and is increasing, the correct choice is the graph in F.

44. Because $f(x)=\log_2(-x)$ represents a reflection of $f(x)=\log_2 x$ over the y-axis, it has a vertical asymptote, which is the y-axis (the line $x=0$), and passes through $(-1,0)$, the correct choice is the graph in A.

45. $f(x)=\log_5 x$

Since $f(x)=y=\log_5 x$, we can write the exponential form as $x=5^y$ to find ordered pairs that satisfy the equation. It is easier to choose values for y and find the corresponding values of x. Make a table of values.

x	$y=\log_5 x$
$\frac{1}{25}=.04$	-2
$\frac{1}{5}=.2$	-1
1	0
5	1
25	2

The graph can also be found by reflecting the graph of $f(x)=5^x$ about the line $y=x$. The graph has the y-axis as a vertical asymptote.

46. $f(x)=\log_{10} x$

Since $f(x)=y=\log_{10} x$, we can write the exponential form as $x=10^y$ to find ordered pairs that satisfy the equation. It is easier to choose values for y and find the corresponding values of x. Make a table of values.

x	$y=\log_{10} x$
$\frac{1}{100}=.01$	-2
$\frac{1}{10}=.1$	-1
1	0
10	1
100	2

The graph can also be found by reflecting the graph of $f(x)=10^x$ about the line $y=x$. The graph has the y-axis as a vertical asymptote.

47. $f(x)=\log_{1/2}(1-x)$

Since $f(x)=y=\log_{1/2}(1-x)$, we can write the exponential form as $1-x=\left(\frac{1}{2}\right)^y \Rightarrow x=1-\left(\frac{1}{2}\right)^y$ to find ordered pairs that satisfy the equation. It is easier to choose values for y and find the corresponding values of x. Make a table of values.

x	$y=\log_{1/2}(1-x)$
-3	-2
-1	-1
0	0
$\frac{1}{2}=.5$	1
$\frac{3}{4}=.75$	2

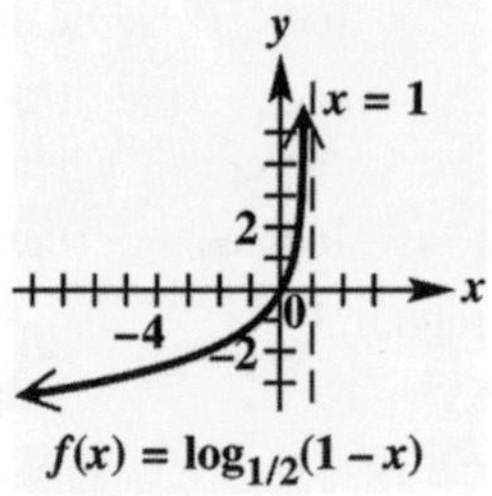

The graph has the line $x=1$ as a vertical asymptote.

48. $f(x)=\log_{1/3}(3-x)$

Since $f(x)=y=\log_{1/3}(3-x)$, we can write the exponential form as $3-x=\left(\frac{1}{3}\right)^y \Rightarrow x=3-\left(\frac{1}{3}\right)^y$ to find ordered pairs that satisfy the equation. It is easier to choose values for y and find the corresponding values of x. Make a table of values.

x	$y=\log_{1/3}(3-x)$
-6	-2
0	-1
2	0
$\frac{8}{3}\approx 2.7$	1
$\frac{26}{9}\approx 2.9$	2

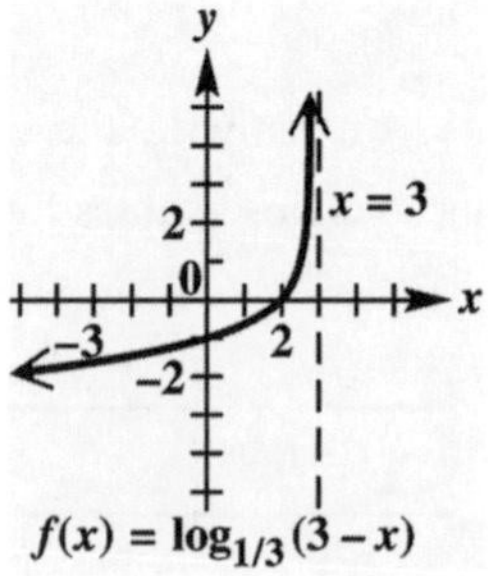

The graph has the line $x=3$ as a vertical asymptote.

49. $f(x)=\log_3(x-1)$

Since $f(x)=y=\log_3(x-1)$, we can write the exponential form as $x-1=3^y \Rightarrow x=3^y+1$ to find ordered pairs that satisfy the equation. It is easier to choose values for y and find the corresponding values of x. Make a table of values.

x	$y=\log_3(x-1)$
$\frac{10}{9}\approx 1.1$	-2
$\frac{4}{3}\approx 1.3$	-1
2	0
4	1
10	2

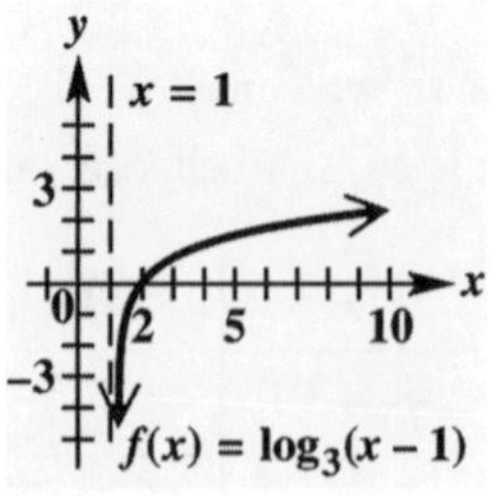

The vertical asymptote will be $x=1$.

50. $f(x) = \log_2 (x^2)$

Since $f(x) = y = \log_2 (x^2)$, we can write the exponential form as $x^2 = 2^y$ to find ordered pairs that satisfy the equation. It is easier to choose values for *y* and find the corresponding values of *x*. Make a table of values.

x	$y = \log_5 x$
$\pm\sqrt{\frac{1}{4}} = \pm\frac{1}{2} = \pm.5$	-2
$\pm\sqrt{\frac{1}{2}} \approx \pm.7$	-1
± 1	0
$\pm\sqrt{2} \approx \pm 1.4$	1
± 2	2

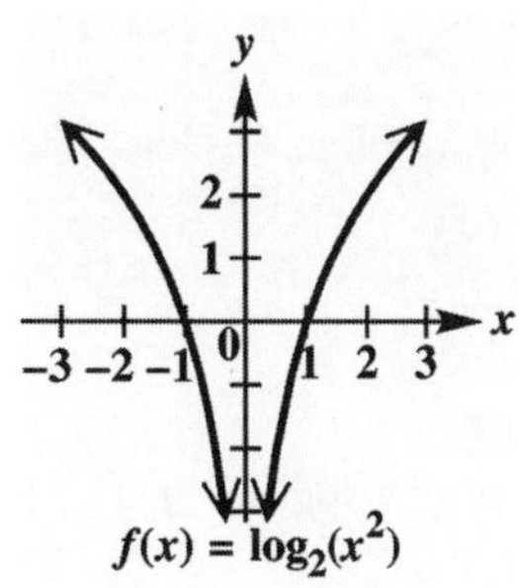

The graph has the *y*-axis as a vertical asymptote and is symmetric with respect to the *y*-axis.

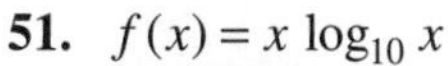

51. $f(x) = x \log_{10} x$

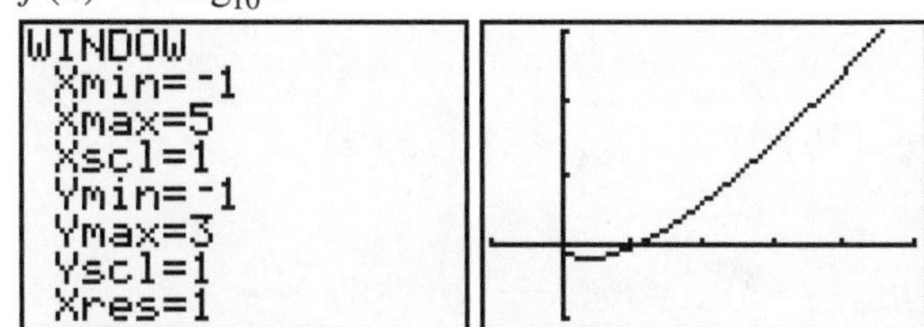

52. $f(x) = x^2 \log_{10} x$

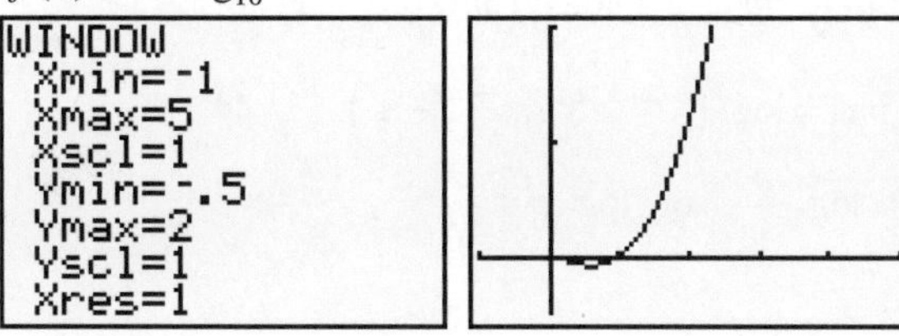

53. $\log_2 \frac{6x}{y} = \log_2 6x - \log_2 y = \log_2 6 + \log_2 x - \log_2 y$

54. $\log_3 \frac{4p}{q} = \log_3 4p - \log_3 q = \log_3 4 + \log_3 p - \log_3 q$

55. $\log_5 \frac{5\sqrt{7}}{3} = \log_5 5\sqrt{7} - \log_5 3 = \log_5 5 + \log_5 \sqrt{7} - \log_5 3 = 1 + \log_5 7^{1/2} - \log_5 3 = 1 + \frac{1}{2}\log_5 7 - \log_5 3$

56. $\log_2 \frac{2\sqrt{3}}{5} = \log_2 2\sqrt{3} - \log_2 5 = \log_2 2 + \log_2 \sqrt{3} - \log_2 5 = 1 + \log_2 3^{1/2} - \log_2 5 = 1 + \frac{1}{2}\log_2 3 - \log_2 5$

57. $\log_4 (2x + 5y)$

Since this is a sum, none of the logarithm properties apply, so this expression cannot be simplified.

58. $\log_6 (7m + 3q)$

Since this is a sum, none of the logarithm properties apply, so this expression cannot be simplified.

59. $\log_m \sqrt{\frac{5r^3}{z^5}} = \log_m \left(\frac{5r^3}{z^5}\right)^{1/2} = \frac{1}{2}\log_m \frac{5r^3}{z^5} = \frac{1}{2}\left(\log_m 5r^3 - \log_m z^5\right)$

$= \frac{1}{2}\left(\log_m 5 + \log_m r^3 - \log_m z^5\right) = \frac{1}{2}\left(\log_m 5 + 3\log_m r - 5\log_m z\right)$

60. $\log_p \sqrt[3]{\frac{m^5n^4}{t^2}} = \log_p\left(\frac{m^5n^4}{t^2}\right)^{1/3} = \frac{1}{3}\log_p\left(\frac{m^5n^4}{t^2}\right)$
$= \frac{1}{3}\left(\log_p m^5 + \log_p n^4 - \log_p t^2\right) = \frac{1}{3}\left(5\log_p m + 4\log_p n - 2\log_p t\right)$

61. $\log_a x + \log_a y - \log_a m = \log_a xy - \log_a m = \log_a \frac{xy}{m}$

62. $(\log_b k - \log_b m) - \log_b a = \log_b \frac{k}{m} - \log_b a = \log_b \frac{k}{ma}$

63. $2\log_m a - 3\log_m b^2 = \log_m a^2 - \log_m\left(b^2\right)^3 = \log_m a^2 - \log_m b^6 = \log_m \frac{a^2}{b^6}$

64. $\frac{1}{2}\log_y p^3q^4 - \frac{2}{3}\log_y p^4q^3 = \log_y\left(p^3q^4\right)^{1/2} - \log_y\left(p^4q^3\right)^{2/3} = \log_y \frac{\left(p^3q^4\right)^{1/2}}{\left(p^4q^3\right)^{2/3}}$
$= \log_y \frac{p^{3/2}q^2}{p^{8/3}q^2} = \log_y \frac{p^{9/6}}{p^{16/6}} = \log_y\left(p^{(9/6)-(16/6)}\right) = \log_y\left(p^{-7/6}\right)$

65. $2\log_a(z-1) + \log_a(3z+2),\ z>1$

$2\log_a(z-1) + \log_a(3z+2) = \log_a(z-1)^2 + \log_a(3z+2) = \log_a\left[(z-1)^2(3z+2)\right]$

66. $\log_b(2y+5) - \frac{1}{2}\log_b(y+3) = \log_b(2y+5) - \log_b(y+3)^{1/2} = \log_b \frac{2y+5}{(y+3)^{1/2}} = \log_b \frac{2y+5}{\sqrt{y+3}}$

67. $-\frac{2}{3}\log_5 5m^2 + \frac{1}{2}\log_5 25m^2 = \log_5\left(5m^2\right)^{-2/3} + \log_5\left(25m^2\right)^{1/2} = \log_5\left[\left(5m^2\right)^{-2/3}\cdot\left(25m^2\right)^{1/2}\right]$
$= \log_5\left(5^{-2/3}m^{-4/3}\cdot 5m\right) = \log_5\left(5^{-2/3}\cdot 5^1\cdot m^{-4/3}\cdot m^1\right)$
$= \log_5\left(5^{1/3}\cdot m^{-1/3}\right) = \log_5 \frac{5^{1/3}}{m^{1/3}}$ or $\log_5 \sqrt[3]{\frac{5}{m}}$

68. $-\frac{3}{4}\log_3 16p^4 - \frac{2}{3}\log_3 8p^3 = \log_3\left(16p^4\right)^{-3/4} - \log_3\left(8p^3\right)^{2/3} = \log_3\left[\frac{\left(16p^4\right)^{-3/4}}{\left(8p^3\right)^{2/3}}\right]$
$= \log_3 \frac{16^{-3/4}p^{-3}}{4p^2} = \log_3 \frac{2^{-3}}{4p^2\cdot p^3} = \log_3 \frac{1}{2^3\cdot 4p^{2+3}} = \log_3 \frac{1}{8\cdot 4p^5} = \log_3 \frac{1}{32p^5}$

69. $\log_{10} 6 = \log_{10}(2\cdot 3) = \log_{10} 2 + \log_{10} 3 = .3010 + .4771 = .7781$

70. $\log_{10} 12 = \log_{10}\left(3\cdot 2^2\right) = \log_{10} 3 + 2\log_{10} 2 = .4771 + 2(.3010) = .4771 + .6020 = 1.0791$

71. $\log_{10} \frac{9}{4} = \log_{10} 9 - \log_{10} 4 = \log_{10} 3^2 - \log_{10} 2^2$
$= 2\log_{10} 3 - 2\log_{10} 2 = 2(.4771) - 2(.3010) = .9542 - .6020 = .3522$

72. $\log_{10} \frac{20}{27} = \log_{10} 20 - \log_{10} 27 = \log_{10}(2\cdot 10) - \log_{10} 3^3 = \log_{10} 2 + \log_{10} 10 - 3\log_{10} 3$
$= .3010 + 1 - 3(.4771) = .3010 + 1 - 1.4313 = -.1303$

73. $\log_{10}\sqrt{30} = \log_{10} 30^{1/2} = \frac{1}{2}\log_{10} 30 = \frac{1}{2}\log_{10}(10\cdot 3)$
$= \frac{1}{2}(\log_{10} 10 + \log_{10} 3) = \frac{1}{2}(1 + .4771) = \frac{1}{2}(1.4771) \approx .7386$

74. $\log_{10} 36^{1/3} = \frac{1}{3}\log_{10} 36 = \frac{1}{3}\log_{10} 6^2 = \frac{2}{3}\log_{10}(2\cdot 3)$

$= \frac{2}{3}(\log_{10} 2 + \log_{10} 3) = \frac{2}{3}(.3010 + .4771) = \frac{2}{3}(.7781) \approx .5187$

75. **(a)** If the x-values are representing years, 3 months is $\frac{3}{12} = \frac{1}{4} = .25$ yr and 6 months is $\frac{6}{12} = \frac{1}{2} = .5$ yr.

L1	L2	L3
.25	.83	------
.5	.91	
2	1.35	
5	2.46	
10	3.54	
30	4.58	
------	------	

L3 =

```
WINDOW
 Xmin=0
 Xmax=35
 Xscl=5
 Ymin=0
 Ymax=5
 Yscl=1
 Xres=1
```

(b) Answers will vary.

76. **(a)** From the graph, $\log_3 .3 = -1.1$

(b) From the graph, $\log_3 .8 = -.2$

77. $f(x) = \log_a x$ and $f(3) = 2$

$2 = \log_a 3 \Rightarrow a^2 = 3 \Rightarrow (a^2)^{1/2} = 3^{1/2} \Rightarrow a = \sqrt{3}$

(There is no $\pm$ because a must be positive and not equal to 1.)

We now have $f(x) = \log_{\sqrt{3}} x$.

(a)

$f\left(\frac{1}{9}\right) = \log_{\sqrt{3}} \frac{1}{9}$

$y = \log_{\sqrt{3}} \frac{1}{9}$

$(\sqrt{3})^y = \frac{1}{9}$

$(3^{1/2})^y = \frac{1}{3^2}$

$3^{y/2} = 3^{-2}$

$\frac{y}{2} = -2$

$y = -4$

(b)

$f(27) = \log_{\sqrt{3}} 27$

$y = \log_{\sqrt{3}} 27$

$(\sqrt{3})^y = 27$

$(3^{1/2})^y = 3^3$

$3^{y/2} = 3^3$

$\frac{y}{2} = 3$

$y = 6$

78. (5, 4) is equivalent to stating that when x = 5 we have y = 4. So for $f(x) = \log_a x$ or $y = \log_a x$, we have $4 = \log_a 5$.

Section 9.3: Evaluating Logarithms; Equations and Applications

1. For $f(x) = a^x$, where $a > 0$, the function is *increasing* over its entire domain.

2. For $g(x) = \log_a x$, where $a > 1$, the function is *increasing* over its entire domain.

3. $f(x) = 5^x$

This function is one-to-one.

Step 1 Replace $f(x)$ with y and interchange x and y.

$$y = 5^x$$
$$x = 5^y$$

Step 2 Solve for y.

$$x = 5^y$$
$$y = \log_5 x$$

Step 3 Replace y with $f^{-1}(x)$.

$$f^{-1}(x) = \log_5 x$$

4. Since $4^{\log_4 11} = 11$, the exponent to which 4 must be raised to obtain 11 is $\log_4 11$.

5. A base e logarithm is called a natural logarithm, while a base 10 logarithm is called a common logarithm.

6. $\log_3 12 = \frac{\ln 12}{\ln 3}$

7. $\log_2 0$ is undefined because there is no power of 2 that yields a result of 0. In other words, the equation $2^x = 0$ has no solution.

8. Let $\log_2 12 = x$. This implies $2^x = 12$. Since $2^3 = 8$ and $2^4 = 16$, $\log_2 12$ must lie between 3 and 4.

9. $\log 8 = .90308999$

10. $\ln 2.75 = 1.0116009$

11. $\log 36 \approx 1.5563$

12. $\log 72 \approx 1.8573$

13. $\log .042 \approx -1.3768$

14. $\log .319 \approx -.4962$

15. $\log\left(2\times10^4\right) \approx 4.3010$

16. $\log\left(2\times10^{-4}\right) \approx -3.6990$

17. $\ln 36 \approx 3.5835$

18. $\ln 72 \approx 4.2767$

19. $\ln .042 \approx 3.1701$

20. $\ln .319 \approx -1.1426$

21. $\ln\left(2\times e^4\right) \approx 4.6931$

22. $\ln\left(2\times e^{-4}\right) \approx -3.3069$

23. Grapefruit, 6.3×10^{-4}

$$\text{pH} = -\log\left[H_3 0^+\right] = -\log\left(6.3\times10^{-4}\right) = -\left(\log 6.3 + \log 10^{-4}\right) = -(.7793-4) = -.7993+4 \approx 3.2$$

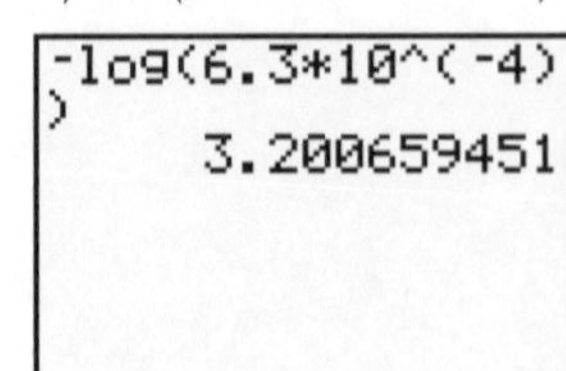

The answer is rounded to the nearest tenth because it is customary to round pH values to the nearest tenth. The pH of grapefruit is 3.2.

24. Crackers, 3.9×10^{-9}

$$pH=-\log\left[H_30^+\right]=-\log\left(3.9\times10^{-9}\right)=-\left(\log 3.9+\log 10^{-9}\right)=-(.59106-9)=-(-8.409)\approx 8.4$$

```
-log(3.9*10^(-9)
)
          8.408935393
```

The answer is rounded to the nearest tenth because it is customary to round pH values to the nearest tenth. The pH of crackers is 8.4.

25. Limes, 1.6×10^{-2}

$$pH=-\log\left[H_30^+\right]=-\log\left(1.6\times10^{-2}\right)=-\left(\log 1.6+\log 10^{-2}\right)=-(.2041-2)=-(-1.7959)\approx 1.8$$

```
-log(1.6*10^(-2)
)
          1.795880017
```

The pH of limes is 1.8.

26. Sodium hydroxide (lye), 3.2×10^{-4}

$$pH=-\log\left[H_30^+\right]=-\log\left(3.2\times10^{-14}\right)=-\left(\log 3.2+\log 10^{-14}\right)=-(.5052-14)=-(-13.49)\approx 13.5$$

```
-log(3.2*10^(-4)
)
          3.494850022
```

The pH of sodium hydroxide is 13.5.

27. Soda pop, 2.7

$$pH=-\log\left[H_30^+\right]$$
$$2.7=-\log\left[H_30^+\right]$$
$$-2.7=\log\left[H_30^+\right]$$
$$\left[H_30^+\right]=10^{-2.7}$$

```
10^(-2.7)
         .0019952623
```

$$\left[H_30^+\right]\approx 2.0\times10^{-3}$$

28. Wine, 3.4

$$pH=-\log\left[H_30^+\right]$$
$$3.4=-\log\left[H_30^+\right]$$
$$-3.4=\log\left[H_30^+\right]$$
$$\left[H_30^+\right]=10^{-3.4}$$

```
10^(-3.4)
     3.981071706E-4
```

$$\left[H_30^+\right]\approx 4.0\times10^{-4}$$

29. Beer, 4.8

$$\text{pH} = -\log\left[H_3 0^+\right]$$
$$4.8 = -\log\left[H_3 0^+\right]$$
$$-4.8 = \log\left[H_3 0^+\right]$$
$$\left[H_3 0^+\right] = 10^{-4.8}$$

```
10^(-4.8)
  1.584893192E-5
```

$$\left[H_3 0^+\right] \approx 1.6 \times 10^{-5}$$

30. Drinking water, 6.5

$$\text{pH} = -\log\left[H_3 0^+\right]$$
$$6.5 = -\log\left[H_3 0^+\right]$$
$$-6.5 = \log\left[H_3 0^+\right]$$
$$\left[H_3 0^+\right] = 10^{-6.5}$$

```
10^(-6.5)
  3.16227766E-7
```

$$\left[H_3 0^+\right] \approx 3.2 \times 10^{-7}$$

31. Wetland, 2.49×10^{-5}

$$\begin{aligned} \text{pH} &= -\log\left[H_3 0^+\right] \\ &= -\log\left(2.49 \times 10^{-5}\right) \\ &= -\left(\log 2.49 + \log 10^{-5}\right) \\ &= -\log 2.49 - (-5) \\ &= -\log 2.49 + 5 \end{aligned}$$
$$\text{pH} \approx 4.6$$

Since the pH is between 4.0 and 6.0, it is a poor fen.

32. Wetland, 2.49×10^{-2}

$$\begin{aligned} \text{pH} &= -\log\left[H_3 0^+\right] \\ &= -\log\left(2.49 \times 10^{-2}\right) \\ &= -\left(\log 2.49 + \log 10^{-2}\right) \\ &= -\log 2.49 - (-2) \\ &= -\log 2.49 + 2 \end{aligned}$$
$$\text{pH} \approx 1.6$$

Since the pH is 3.0 or less, it is a bog.

33. Wetland, 2.49×10^{-7}

$$\begin{aligned} \text{pH} &= -\log\left[H_3 0^+\right] \\ &= -\log\left(2.49 \times 10^{-7}\right) \\ &= -\left(\log 2.49 + \log 10^{-7}\right) \\ &= -\log 2.49 - (-7) \\ &= -\log 2.49 + 7 \end{aligned}$$
$$\text{pH} \approx 6.6$$

Since the pH is greater than 6.0, it is a rich fen.

34. **(a)** $\log 398.4 \approx 2.60031933$

(b) $\log 39.84 \approx 1.60031933$

(c) $\log 3.984 \approx .6003193298$

(d) The whole number parts will vary, but the decimal parts will be the same.

For Exercises 35 – 42, as noted on page 437 of the text, the solutions will be evaluated at the intermediate steps to four decimal places. However, the final answers are obtained without rounding the intermediate steps.

35. $\log_2 5 = \frac{\ln 5}{\ln 2} \approx \frac{1.6094}{.6931} \approx 2.3219$

We could also have used the common logarithm. $\log_2 5 = \frac{\log 5}{\log 2} \approx \frac{.6990}{.3010} \approx 2.3219$

36. $\log_2 9 = \frac{\ln 9}{\ln 2} \approx \frac{2.1972}{.6931} \approx 3.1699$

We could also have used the common logarithm. $\log_2 9 = \frac{\log 9}{\log 2} \approx \frac{.9542}{.3010} \approx 3.1699$

37. $\log_8 .59 = \frac{\log .59}{\log 8} \approx \frac{-.2291}{.9031} \approx -.2537$

We could also have used the natural logarithm. $\log_8 .59 = \frac{\ln .59}{\ln 8} \approx \frac{-.5276}{2.0794} \approx -.2537$

38. $\log_8 .71 = \frac{\log .71}{\log 8} \approx \frac{-.1487}{.9031} \approx -.1647$

We could also have used the natural logarithm. $\log_8 .71 = \frac{\ln .71}{\ln 8} \approx \frac{-.3425}{2.0794} \approx -.1647$

39. Since $\sqrt{13} = 13^{1/2}$, we have $\log_{\sqrt{13}} 12 = \frac{\ln 12}{\ln \sqrt{13}} = \frac{\ln 12}{\frac{1}{2}\ln 13} \approx \frac{2.4849}{1.2825} \approx 1.9376.$

The required logarithm can also be found by entering $\ln\sqrt{13}$ directly into the calculator.

We could also have used the common logarithm. $\log_{\sqrt{13}} 12 = \frac{\log 12}{\log\sqrt{13}} = \frac{\log 12}{\frac{1}{2}\log 13} \approx \frac{1.0792}{.5570} \approx 1.9376$

40. Since $\sqrt{19} = 19^{1/2}$, we have $\log_{\sqrt{19}} 5 = \frac{\ln 5}{\ln\sqrt{19}} = \frac{\ln 5}{\frac{1}{2}\ln 19} \approx \frac{1.6094}{1.4722} \approx 1.0932$

The required logarithm can also be found by entering $\ln\sqrt{19}$ directly into the calculator.

We could also have used the common logarithm. $\log_{\sqrt{19}} 5 = \frac{\log 5}{\log\sqrt{19}} = \frac{\log 5}{\frac{1}{2}\log 19} \approx \frac{.6990}{.6394} \approx 1.0932$

41. $\log_{.32} 5 = \frac{\log 5}{\log .32} \approx \frac{.6990}{-.4949} \approx -1.4125$

We could also have used the natural logarithm. $\log_{.32} 5 = \frac{\ln 5}{\ln .32} \approx \frac{1.6094}{-1.1394} \approx -1.4125$

42. $\log_{.91} 8 = \frac{\log 8}{\log .91} \approx \frac{.9031}{-.0410} \approx -22.0488$

We could also have used the natural logarithm. $\log_{.91} 8 = \frac{\ln 8}{\ln .91} \approx \frac{2.0794}{-.0943} \approx -22.0488$

43. $g(x) = e^x$

(a) $g(\ln 3) = e^{\ln 3} = 3$

(b) $g\left[\ln\left(5^2\right)\right] = e^{\ln 5^2} = 5^2$ or 25

(c) $g\left[\ln\left(\frac{1}{e}\right)\right] = e^{\ln(1/e)} = \frac{1}{e}$

44. $f(x) = 3^x$

(a) $f(\log_3 7) = 3^{\log_3 7} = 7$

(b) $f\left[\log_3(\ln 3)\right] = 3^{\log_3(\ln 3)} = \ln 3$

(c) $f\left[\log_3(2\ln 3)\right] = 3^{\log_3(2\ln 3)}$

$= 2\ln 3$ or $\ln 9$

45. $f(x) = \ln x$

(a) $f\left(e^5\right) = \ln e^5 = 5$

(b) $f\left(e^{\ln 3}\right) = \ln e^{\ln 3} = \ln 3$

(c) $f\left(e^{2\ln 3}\right) = \ln e^{2\ln 3} = 2\ln 3$ or $\ln 9$.

46. $f(x) = \log_2 x$

(a) $f\left(2^3\right) = \log_2 2^3 = 3$

(b) $f\left(2^{\log_2 2}\right) = \log_2 2^{\log_2 2} = \log_2 2 = 1$

(c) $f\left(2^{2\log_2 2}\right) = \log_2 2^{2\log_2 2}$

$= 2\log_2 2 = 2 \cdot 1 = 2$

47. $d = 10\log\frac{I}{I_0}$, where d is the decibel rating.

(a) $d = 10\log\frac{100I_0}{I_0}$
$= 10\log_{10}100 = 10(2) = 20$

(b) $d = 10\log\frac{1000I_0}{I_0}$
$= 10\log_{10}1000 = 10(3) = 30$

(c) $d = 10\log\frac{100{,}000I_0}{I_0}$
$= 10\log_{10}100{,}000 = 10(5) = 50$

(d) $d = 10\log\frac{1{,}000{,}000I_0}{I_0}$
$= 10\log_{10}1{,}000{,}000 = 10(6) = 60$

(e) $I = 2I_0$
$d = 10\log\frac{2I_0}{I_0} = 10\log 2 \approx 3.0103$
The described rating is increased by about 3 decimals.

48. $d = 10\log\frac{I}{I_0}$, where d is the decibel rating.

(a) $d = 10\log\frac{115I_0}{I_0}$
$= 10\log 115 \approx 21$

(b) $d = 10\log\frac{9{,}500{,}000I_0}{I_0}$
$= 10\log 9{,}500{,}000 \approx 70$

(c) $d = 10\log\frac{1{,}200{,}000{,}000I_0}{I_0}$
$= 10\log 1{,}200{,}000{,}000 \approx 91$

(d) $d = 10\log\frac{895{,}000{,}000{,}000I_0}{I_0}$
$= 10\log 895{,}000{,}000{,}000 \approx 120$

(e) $d = 10\log\frac{109{,}000{,}000{,}000{,}000I_0}{I_0}$
$= 10\log 109{,}000{,}000{,}000{,}000 \approx 140$

49. $r = \log_{10}\frac{I}{I_0}$, where r is the Richter scale rating of an earthquake.

(a) $r = \log_{10}\frac{1000I_0}{I_0} = \log_{10}1000 = 3$

(b) $r = \log_{10}\frac{1{,}000{,}000I_0}{I_0} = \log_{10}1{,}000{,}000 = 6$

(c) $r = \log_{10}\frac{100{,}000{,}000I_0}{I_0} = \log_{10}100{,}000{,}000 = 8$

50. $r = \log_{10}\frac{I}{I_0}$

$6.7 = \log_{10}\frac{I}{I_0} \Rightarrow 10^{6.7} = \frac{I}{I_0} \Rightarrow I_0 10^{6.7} = I \Rightarrow I \approx 5{,}011{,}872 I_0$

The reading was about 5,000,000 I_0.

51. $r = \log_{10}\frac{I}{I_0}$

$8.1 = \log_{10}\frac{I}{I_0} \Rightarrow 10^{8.1} = \frac{I}{I_0} \Rightarrow I_0 10^{8.1} = I \Rightarrow I \approx 125{,}892{,}541 I_0$

The reading was about 126,000,000 I_0.

52. To find out how much greater the force of the 1985 earthquake than the 1999 earthquake, we need to find the ratio of their reading in terms of I_0. $\frac{10^{8.1}I_0}{10^{6.7}I_0} = \frac{10^{8.1}}{10^{6.7}} = 10^{8.1-6.7} = 10^{1.4} \approx 25$

It was about 25 times greater.

53. $f(x) = -269 + 73\ln x$ and in the year 2004, $x = 104$.

$f(104) = -269 + 73\ln 104 \approx 70$

Thus, the number of visitors in the year 2004 will be about 70 million. We must assume that the rate of increase continues to be logarithmic.

54. **(a)** $f(t) = 74.61 + 3.84\ln t,\ t \ge 1$

In the year 2002, $t = 8$

$f(8) = 74.61 + 3.84\ln 8 \approx 82.5951$

Thus, the percent of freshmen entering college in 2004 who performed volunteer work during their last year of high school will be about 82.5951%. This is exceptionally close to the percent shown in the graph and to the actual percent of 82.6%.

(b) Answers will vary.

55. If $a = .36$, then $S(n) = a\ln\left(1+\frac{n}{a}\right) = .36\ln\left(1+\frac{n}{.36}\right)$.

(a) $S(100) = .36\ln\left(1+\frac{100}{.36}\right) \approx 2.0269 \approx 2$

(b) $S(200) = .36\ln\left(1+\frac{200}{.36}\right) \approx 2.2758 \approx 2$

(c) $S(150) = .36\ln\left(1+\frac{150}{.36}\right) \approx 2.1725 \approx 2$

(d) $S(10) = .36\ln\left(1+\frac{10}{.36}\right) \approx 1.2095 \approx 1$

56. If $a = .88$, then $S(n) = a\ln\left(1+\frac{n}{a}\right) = .88\ln\left(1+\frac{n}{.88}\right)$.

(a) $S(50) = .88\ln\left(1+\frac{50}{.88}\right) \approx 3.5704 \approx 4$

(b) $S(100) = .88\ln\left(1+\frac{100}{.88}\right) \approx 4.1728 \approx 4$

(c) $S(250) = .88\ln\left(1+\frac{250}{.88}\right) \approx 4.9745 \approx 5$

57. The index of diversity H for 2 species is given by $H = -\left[P_1\log_2 P_1 + P_2\log_2 P_2\right]$. When $P_1 = \frac{50}{100} = .5$ and $P_2 = \frac{50}{100} = .5$ we have $H = -\left[.5\log_2 .5 + .5\log_2 .5\right]$. Since $\log_2 .5 = \log_2 \frac{1}{2} = \log_2 2^{-1} = -1$, we have $H = -\left[.5(-1) + .5(-1)\right] = -(-1) = 1$. Thus, the index of diversity is 1.

58. $H = \left[P_1\log_2 P_1 + P_2\log_2 P_2 + P_3\log_2 P_3 + P_4\log_2 P_4\right]$

$= -\left[.521\log_2 .521 + .324\log_2 .324 + .081\log_2 .081 + .074\log_2 .074\right]$

We need the change-of-base theorem to calculate each term. Using the natural log we have the following.

$$H = -\left[.521\log_2 .521 + .324\log_2 .324 + .081\log_2 .081 + .074\log_2 .074\right]$$
$$= -\left[.521\frac{\ln .521}{\ln 2} + .324\frac{\ln .324}{\ln 2} + .081\frac{\ln .081}{\ln 2} + .074\frac{\ln .074}{\ln 2}\right]$$
$$= -\left[.521\ln .521 + .324\ln .324 + .081\ln .081 + .074\ln .074\right]/\ln 2$$

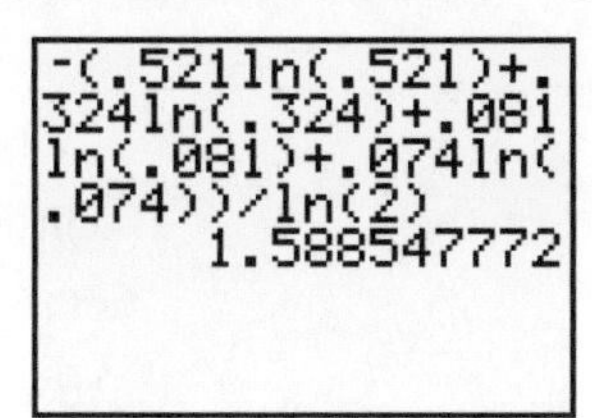

The index of diversity approximately1.59.

59. Since x is the exponent to which 7 must be raised in order to obtain 19, the solution is $\log_7 19$ or $\frac{\log 19}{\log 7}$ or $\frac{\ln 19}{\ln 7}$.

60. Since x is the exponent to which 3 must be raised in order to obtain 10, the solution is $\log_3 10$ or $\frac{\log 10}{\log 3}$ or $\frac{\ln 10}{\ln 3}$.

61. Since x is the exponent to which $\frac{1}{2}$ must be raised in order to obtain 12, the solution is $\log_{1/2} 12$ or $\frac{\log 12}{\log\left(\frac{1}{2}\right)}$ or $\frac{\ln 12}{\ln\left(\frac{1}{2}\right)}$.

62. Since x is the exponent to which $\frac{1}{3}$ must be raised in order to obtain 4, the solution is $\log_{1/3} 4$ or $\frac{\log 4}{\log\left(\frac{1}{3}\right)}$ or $\frac{\ln 4}{\ln\left(\frac{1}{3}\right)}$.

63.
$$\begin{aligned} 3^x &= 6 \\ \ln 3^x &= \ln 6 \\ x \ln 3 &= \ln 6 \\ x &= \tfrac{\ln 6}{\ln 3} \approx 1.6309 \end{aligned}$$
Solution set: $\{1.6309\}$

64.
$$\begin{aligned} 4^x &= 12 \\ \ln 4^x &= \ln 12 \\ x \ln 4 &= \ln 12 \\ x &= \tfrac{\ln 12}{\ln 4} \approx 1.7925 \end{aligned}$$
Solution set: $\{1.7925\}$

65.
$$\begin{aligned} 6^{1-2x} &= 8 \\ \ln 6^{1-2x} &= \ln 8 \\ (1-2x)\ln 6 &= \ln 8 \\ 1-2x &= \tfrac{\ln 8}{\ln 6} \\ 2x &= 1-\tfrac{\ln 8}{\ln 6} \\ x &= \tfrac{1}{2}\left(1-\tfrac{\ln 8}{\ln 6}\right) \approx -.0803 \end{aligned}$$
or
$$\begin{aligned} \ln 6^{1-2x} &= \ln 8 \\ (1-2x)\ln 6 &= \ln 8 \\ \ln 6 - 2x\ln 6 &= \ln 8 \\ -2x\ln 6 &= \ln 8 - \ln 6 \\ -x\ln 6^2 &= \ln \tfrac{8}{6} \\ -x\ln 36 &= \ln \tfrac{4}{3} \\ x &= -\frac{\ln \frac{4}{3}}{\ln 36} \approx -.0803 \end{aligned}$$
Solution set: $\{-.0803\}$

66.
$$\begin{aligned} 3^{2x-5} &= 13 \\ \ln 3^{2x-5} &= \ln 13 \\ (2x-5)\ln 3 &= \ln 13 \\ 2x-5 &= \tfrac{\ln 13}{\ln 3} \\ 2x &= 5+\tfrac{\ln 13}{\ln 3} \\ x &= \tfrac{1}{2}\left(5+\tfrac{\ln 13}{\ln 3}\right) \approx 3.6674 \end{aligned}$$
or
$$\begin{aligned} 3^{2x-5} &= 13 \\ \ln 3^{2x-5} &= \ln 13 \\ (2x-5)\ln 3 &= \ln 13 \\ 2x\ln 3 - 5\ln 3 &= \ln 13 \\ x\ln 3^2 - \ln 3^5 &= \ln 13 \\ x\ln 9 - \ln 243 &= \ln 13 \\ x\ln 9 &= \ln 13 + \ln 243 \\ x\ln 9 &= \ln 3159 \\ x &= \tfrac{\ln 3159}{\ln 9} \approx 3.6674 \end{aligned}$$
Solution set: $\{3.6674\}$

67.
$$\begin{aligned} 2^{x+3} &= 5^x \\ \ln 2^{x+3} &= \ln 5^x \\ (x+3)\ln 2 &= x\ln 5 \\ x\ln 2 + 3\ln 2 &= x\ln 5 \\ x\ln 2 - x\ln 5 &= -3\ln 2 \\ x(\ln 2 - \ln 5) &= -\ln 2^3 \\ x\left(\ln \tfrac{2}{5}\right) &= -\ln 8 \\ x &= \frac{-\ln 8}{\ln \frac{2}{5}} \approx 2.2694 \end{aligned}$$
Solution set: $\{2.2694\}$

68.
$$\begin{aligned} 6^{x+3} &= 4^x \\ \ln 6^{x+3} &= \ln 4^x \\ (x+3)\ln 6 &= x\ln 4 \\ x\ln 6 + 3\ln 6 &= x\ln 4 \\ x\ln 6 - x\ln 4 &= -3\ln 6 \\ x(\ln 6 - \ln 4) &= -3\ln 6 \\ x\left(\ln \tfrac{6}{4}\right) &= -\ln 6^3 \\ x\left(\ln \tfrac{3}{2}\right) &= -\ln 216 \\ x &= \frac{-\ln 216}{\ln \frac{3}{2}} \approx -13.2571 \end{aligned}$$
Solution set: $\{-13.2571\}$

69. $e^{x-1} = 4$

$\ln e^{x-1} = \ln 4$

$x - 1 = \ln 4$

$x = \ln 4 + 1 \approx 2.3863$

Solution set: $\{2.3863\}$

70. $e^{2-x} = 12$

$\ln e^{2-x} = \ln 12$

$2 - x = \ln 12$

$-x = -2 + \ln 12$

$x = 2 - \ln 2 \approx -.4849$

Solution set: $\{-.4849\}$

71. $2e^{5x+2} = 8$

$e^{5x+2} = 4$

$\ln e^{5x+2} = \ln 4$

$5x + 2 = \ln 4$

$5x = \ln 4 - 2$

$x = \frac{1}{5}(\ln 4 - 2) \approx -.1227$

Solution set: $\{-.1227\}$

72. $10e^{3x-7} = 5$

$e^{3x-7} = \frac{1}{2}$

$\ln e^{3x-7} = \ln \frac{1}{2}$

$3x - 7 = \ln \frac{1}{2}$

$3x = \ln \frac{1}{2} + 7$

$x = \frac{1}{3}\left(\ln \frac{1}{2} + 7\right) \approx 2.1023$

Solution set: $\{2.1023\}$

73. $2^x = -3$ has no solution since 2 raised to any power is positive.

Solution set: $\varnothing$

74. $3^x = -6$ has no solution since 3 raised to any power is positive.

Solution set: $\varnothing$

75. $e^{8x} \cdot e^{2x} = e^{20}$

$e^{10x} = e^{20}$

$10x = 20$

$x = 2$

Solution set: $\{2\}$

76. $e^{6x} \cdot e^{x} = e^{21}$

$e^{7x} = e^{21}$

$7x = 21$

$x = 3$

Solution set: $\{3\}$

77. $\ln(6x+1) = \ln 3$

$6x + 1 = 3$

$6x = 2$

$x = \frac{2}{6} = \frac{1}{3}$

Solution set: $\left\{\frac{1}{3}\right\}$

78. $\ln(7-x) = \ln 12$

$7 - x = 12$

$-x = 5$

$x = -5$

Solution set: $\{-5\}$

79. $\log 4x - \log(x-3) = \log 2$

$\log \frac{4x}{x-3} = \log 2 \Rightarrow \frac{4x}{x-3} = 2 \Rightarrow 4x = 2(x-3) \Rightarrow 4x = 2x - 6 \Rightarrow 2x = -6 \Rightarrow x = -3$

Since x is negative, $4x$ is negative. Therefore, $\log 4x$ is not defined. Thus, the proposed solution must be discarded.

Solution set: $\varnothing$

80. $\ln(-x) + \ln 3 = \ln(2x-15) \Rightarrow \ln(-3x) = \ln(2x-15) \Rightarrow -3x = 2x - 15 \Rightarrow -5x = -15 \Rightarrow x = 3$

Since x is positive, $-x$ is negative. Therefore, $\ln(-x)$ is not defined. Thus, the proposed solution must be discarded.

Solution set: $\varnothing$

81. $\log(2x-1)+\log 10x = \log 10 \Rightarrow \log[(2x-1)(10x)] = \log 10 \Rightarrow (2x-1)(10x) = 10$

$20x^2 - 10x = 10 \Rightarrow 20x^2 - 10x - 10 = 0 \Rightarrow 2x^2 - x - 1 = 0 \Rightarrow (2x+1)(x-1) = 0$

$$2x+1=0 \text{ or } x-1=0$$
$$x=-\tfrac{1}{2} \qquad x=1$$

Since $x=-\frac{1}{2}$ is negative, $10x$ is negative. Therefore, $\log 10x$ is not defined. This proposed solution must be discarded.

Solution set: $\{1\}$

82. $\ln 5x - \ln(2x-1) = \ln 4$

$\ln \frac{5x}{2x-1} = \ln 4 \Rightarrow \frac{5x}{2x-1} = 4 \Rightarrow 5x = 4(2x-1) \Rightarrow 5x = 8x - 4 \Rightarrow -3x = -4 \Rightarrow x = \frac{4}{3}$

Solution set: $\left\{\frac{4}{3}\right\}$

83. $\log(x+25) = 1 + \log(2x-7) \Rightarrow \log(x+25) - \log(2x-7) = 1 \Rightarrow \log_{10} \frac{x+25}{2x-7} = 1$

$\frac{x+25}{2x-7} = 10^1 \Rightarrow x+25 = 10(2x-7) \Rightarrow x+25 = 20x-70 \Rightarrow 25 = 19x-70 \Rightarrow 95 = 19x \Rightarrow 5 = x$

Solution set: $\{5\}$

84. $\ln(5+4x) - \ln(3+x) = \ln 3$

$\ln \frac{5+4x}{3+x} = \ln 3 \Rightarrow \frac{5+4x}{3+x} = 3 \Rightarrow 5+4x = 3(3+x) \Rightarrow 5+4x = 9+3x \Rightarrow x+5 = 9 \Rightarrow x = 4$

Solution set: $\{4\}$

85. $\log x + \log(3x-13) = 1 \Rightarrow \log_{10}[x(3x-13)] = 1$

$x(3x-13) = 10^1 \Rightarrow 3x^2 - 13x = 10 \Rightarrow 3x^2 - 13x - 10 = 0 \Rightarrow (3x+2)(x-5) = 0$

$$3x+2=0 \text{ or } x-5=0$$
$$x=-\tfrac{2}{3} \qquad x=5$$

Since the negative solution $\left(x=-\frac{2}{3}\right)$ is not in the domain of $\log x$, it must be discarded.

Solution set: $\{5\}$

86. $\ln(2x+5) + \ln x = \ln 7$

$\ln[x(2x+5)] = \ln 7 \Rightarrow x(2x+5) = 7 \Rightarrow 2x^2 + 5x - 7 = 0 \Rightarrow (2x+7)(x-1) = 0$

$$2x+7=0 \text{ or } x-1=0$$
$$x=-\tfrac{7}{2} \qquad x=1$$

Since the negative solution $\left(x=-\frac{7}{2}\right)$ is not in the domain of $\log x$, it must be discarded.

Solution set: $\{1\}$

87. $\log_6 4x - \log_6(x-3) = \log_6 12$

$\log_6 \frac{4x}{x-3} = \log_6 12 \Rightarrow \frac{4x}{x-3} = 12 \Rightarrow 4x = 12(x-3) \Rightarrow 4x = 12x - 36 \Rightarrow 36 = 8x \Rightarrow \frac{36}{8} = x \Rightarrow x = \frac{9}{2}$

Solution set: $\left\{\frac{9}{2}\right\}$

88. $\log_2 3x + \log_2 3 = \log_2(2x+15)$

$\log_2[3x(3)] = \log_2(2x+15) \Rightarrow 3x(3) = 2x+15 \Rightarrow 9x = 2x+15 \Rightarrow 7x = 15 \Rightarrow x = \frac{15}{7}$

Solution set: $\left\{\frac{15}{7}\right\}$

89.

$$5^{x+2} = 2^{2x-1}$$
$$\ln 5^{x+2} = \ln 2^{2x-1}$$
$$(x+2)\ln 5 = (2x-1)\ln 2$$
$$x\ln 5 + 2\ln 5 = 2x\ln 2 - \ln 2$$
$$x\ln 5 + \ln 5^2 = x\ln 2^2 - \ln 2$$
$$x\ln 5 + \ln 25 = x\ln 4 - \ln 2$$
$$\ln 25 + \ln 2 = x\ln 4 - x\ln 5$$
$$\ln 25 + \ln 2 = x(\ln 4 - \ln 5)$$
$$\frac{\ln 25+\ln 2}{\ln 4-\ln 5} = x$$
$$x = \frac{\ln 50}{\ln \frac{4}{5}}$$
$$x \approx -17.5314$$

Solution set: $\{-17.5314\}$

90.

$$6^{x-3} = 3^{4x+1}$$
$$\ln 6^{x-3} = \ln 3^{4x+1}$$
$$(x-3)\ln 6 = (4x+1)\ln 3$$
$$x\ln 6 - 3\ln 6 = 4x\ln 3 + \ln 3$$
$$x\ln 6 - \ln 6^3 = x\ln 3^4 + \ln 3$$
$$x\ln 6 - \ln 216 = x\ln 81 + \ln 3$$
$$x\ln 6 - x\ln 81 = \ln 3 + \ln 216$$
$$x(\ln 6 - \ln 81) = \ln 3 + \ln 216$$
$$x = \frac{\ln 3+\ln 216}{\ln 6-\ln 81}$$
$$x = \frac{\ln(3\cdot 216)}{\ln \frac{6}{81}}$$
$$x = \frac{\ln 648}{\ln \frac{2}{27}}$$
$$x \approx -2.4874$$

Solution set: $\{-2.4874\}$

91. $\ln e^x - \ln e^3 = \ln e^5$

$$x - 3 = 5$$
$$x = 8$$

Solution set: $\{8\}$

92. $\ln e^x - 2\ln e = \ln e^4$

$$x - 2 = 4$$
$$x = 6$$

Solution set: $\{6\}$

93. $\log_2(\log_2 x) = 1 \Rightarrow \log_2 x = 2^1 \Rightarrow \log_2 x = 2 \Rightarrow x = 2^2 \Rightarrow x = 4$

Solution set: $\{4\}$

94. $\log x = \sqrt{\log x}$

$$(\log x)^2 = \left(\sqrt{\log x}\right)^2 \Rightarrow (\log x)^2 = \log x \Rightarrow (\log x)^2 - \log x = 0 \Rightarrow \log x(\log x - 1) = 0$$

$$\log_{10} x = 0 \quad \text{or} \quad \log_{10} x - 1 = 0$$
$$x = 10^0 \qquad \log_{10} x = 1$$
$$x = 1 \qquad x = 10^1 = 10$$

Since the work involves squaring both sides, both proposed solutions must be checked in the original equation.

Check $x = 1$.

$$\log x = \sqrt{\log x}$$
$$\log 1 = \sqrt{\log 1}\ ?$$
$$0 = \sqrt{0}$$
$$0 = 0$$

This is a true statement.
1 is a solution.

Check $x = 10$.

$$\log x = \sqrt{\log x}$$
$$\log 10 = \sqrt{\log 10}\ ?$$
$$1 = \sqrt{1}$$
$$1 = 1$$

This is a true statement.
10 is a solution.

Solution set: $\{1, 10\}$

95. $\log x^2 = (\log x)^2 \Rightarrow 2\log x = (\log x)^2 \Rightarrow (\log x)^2 - 2\log x = 0 \Rightarrow \log x(\log x - 2) = 0$

$$\log_{10} x = 0 \quad \text{or} \quad \log_{10} x - 2 = 0$$
$$x = 10^0 \qquad \log_{10} x = 2$$
$$x = 1 \qquad x = 10^2 = 100$$

Solution set: $\{1, 100\}$

96. $\log_2 \sqrt{2x^2} = \frac{3}{2} \Rightarrow \sqrt{2x^2} = 2^{3/2} \Rightarrow \left(\sqrt{2x^2}\right)^2 = \left(2^{3/2}\right)^2 \Rightarrow 2x^2 = 2^3 \Rightarrow 2x^2 = 8 \Rightarrow x^2 = 4 \Rightarrow x = \pm 2$

Since the solution involves squaring both sides, both proposed solutions must be checked in the original equation.

Check $x = -2$.

$$\log_2 \sqrt{2x^2} = \tfrac{3}{2}$$
$$\log_2 \sqrt{2(-2)^2} = \tfrac{3}{2} \ ?$$
$$\log_2 \sqrt{2(4)} = \tfrac{3}{2}$$
$$\log_2 \sqrt{8} = \tfrac{3}{2}$$
$$\log_2 \sqrt{2^3} = \tfrac{3}{2}$$
$$\log_2 2^{3/2} = \tfrac{3}{2}$$
$$\tfrac{3}{2} = \tfrac{3}{2}$$

This is a true statement.
-2 is a solution.

Check $x = 2$.

$$\log_2 \sqrt{2x^2} = \tfrac{3}{2}$$
$$\log_2 \sqrt{2(2)^2} = \tfrac{3}{2} \ ?$$
$$\log_2 \sqrt{2(4)} = \tfrac{3}{2}$$
$$\log_2 \sqrt{8} = \tfrac{3}{2}$$
$$\log_2 \sqrt{2^3} = \tfrac{3}{2}$$
$$\log_2 2^{3/2} = \tfrac{3}{2}$$
$$\tfrac{3}{2} = \tfrac{3}{2}$$

This is a true statement.
2 is a solution.

Solution set: $\{-2, 2\}$

97. Double the 2000 value is $2(7{,}990) = 15{,}980$.

$$f(x) = 8160(1.06)^x$$
$$15{,}980 = 8160(1.06)^x$$
$$\tfrac{15{,}980}{8160} = 1.06^x$$
$$\tfrac{47}{24} = 1.06^x$$
$$\ln \tfrac{47}{24} = \ln 1.06^x$$
$$\ln \tfrac{47}{24} = x \ln 1.06$$
$$\frac{\ln \frac{47}{24}}{\ln 1.06} = x$$
$$x \approx 11.53$$

During 2011, the cost of a year's tuition, room and board, and fees at a public university will be double the cost in 2000.

98. $f(t) = 11.65\left(1 - e^{-t/1.27}\right)$

(a) At the finish line $t = 9.86$.

$$f(9.86) = 11.65\left(1 - e^{-9.86/1.27}\right) \approx 11.6451$$

He was running approximately 11.6451 m per sec as he crossed the finish line.

(b)
$$10 = 11.65\left(1 - e^{-t/1.27}\right)$$
$$\tfrac{10}{11.65} = 1 - e^{-t/1.27}$$
$$e^{-t/1.27} = 1 - \tfrac{10}{11.65}$$
$$-\tfrac{t}{1.27} = \ln\left(1 - \tfrac{10}{11.65}\right)$$
$$t = -1.27 \ln\left(1 - \tfrac{10}{11.65}\right) \approx 2.4823$$

After 2.4823 sec, he was running at a rate of 10 m per sec.

99. $f(x)=\dfrac{25}{1+1364.3e^{-x/9.316}}$

(a) In 1997, $x = 97$.

$$f(97)=\dfrac{25}{1+1364.3e^{-97/9.316}}\approx 24$$

In 1997, about 24% of U.S. children lived in a home without a father.

(b) $10=\dfrac{25}{1+1364.3e^{-x/9.316}}\Rightarrow 10\left(1+1364.3e^{-x/9.316}\right)=25\Rightarrow 10+13{,}643e^{-x/9.316}=25$

$13{,}643e^{-x/9.316}=15\Rightarrow e^{-x/9.316}=\frac{15}{13{,}643}\Rightarrow -\frac{x}{9.316}=\ln\frac{15}{13{,}643}\Rightarrow x=-9.316\ln\frac{15}{13{,}643}\approx 63.47$

During 1963, 10% of U.S. children lived in a home without a father.

100. $f(x)=-301\ln\frac{x}{207}$

(a) The left side is a reflection of the right side across the vertical axis of the tower; the graph of $f(-x)$ is exactly the reflection of the graph of $f(x)$ across the y-axis.

(b) x is half the length. We have $x=\frac{15.7488}{2}=7.8744$. The height is $f(7.8744)=-301\ln\frac{7.8744}{207}\approx 984$ ft.

(c) Let $y = 500$ and solve for x.

$$500=-301\ln\frac{x}{207}\Rightarrow -\frac{500}{301}=\ln\frac{x}{207}\Rightarrow e^{-500/301}=\frac{x}{207}\Rightarrow x=207e^{-500/301}\approx 39$$

The height is 500 feet, about 39 feet from the center.

Chapter 9: Review Exercises

1. $y=\log_{.3}x$

The point (1, 0) is on the graph of every function of the form $y=\log_a x$, so the correct choice must be either B or C. Since the base is $a = .3$ and $0 < .3 < 1$, $y=\log_{.3}x$ is a decreasing function, and so the correct choice must be B.

2. $y=e^x$

The point (0, 1) is on the graph since $e^0=1$, so the correct choice must be either A or D. Since the base is e and $e > 1$, $y=e^x$ is an increasing function, and so the correct choice must be A.

3. $y=\ln x=\log_e x$

The point (1, 0) is on the graph of every function of the form $y=\log_a x$, so the correct choice must be either B or C. Since the base is $a = e$ and $e > 1$, $y = \ln x$ is an increasing function, and so the correct choice must be C.

4. $y=(.3)^x$

The point (0, 1) is on the graph since $(.3)^0=1$, so the correct choice must be either A or D. Since the base is .3 and $0 < .3 < 1$, $y=(.3)^x$ is a decreasing function, and so the correct choice must be D.

5. $2^5=32$ is written in logarithmic form as $\log_2 32=5$.

6. $100^{1/2}=10$ is written in logarithmic form as $\log_{100}10=\frac{1}{2}$.

7. $\left(\frac{3}{4}\right)^{-1}=\frac{4}{3}$ is written in logarithmic form as $\log_{3/4}\frac{4}{3}=-1$.

8. The y-intercept of $f(x)=(1.5)^{x+2}$ is $f(0)=(1.5)^{0+2}=1.5^2=2.25$ and the x-axis is a horizontal asymptote. Make a table of values, plot the points, and draw a smooth curve through them.

x	$f(x)$
−4	$\approx .44$
−3	$\approx .67$
−2.5	$\approx .82$
−2	1
−1.5	≈ 1.22
−1	1.5
0	2.25

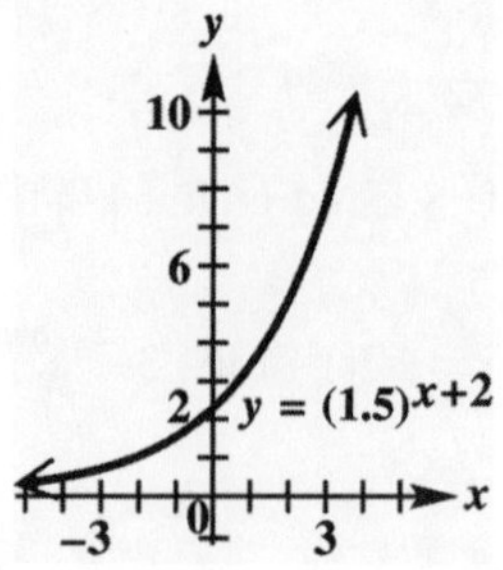

9. $\log_3 4$ is the logarithm to the base 3 of 4. ($\log_4 3$ would be the logarithm to the base 4 of 3.)

10. The exact value of $\log_3 9$ is 2 since $3^2=9$.

11. The exact value of $\log_3 27$ is 3 since $3^3=27$.

12. $\log_3 16$ must lie between 2 and 3. Because the function defined by $y=\log_3 x$ is increasing and $9<16<27$, we have $\log_3 9<\log_3 16<\log_3 27$.

13. By the change-of-base theorem, $\log_3 16=\frac{\log 16}{\log 3}=\frac{\ln 16}{\ln 3}\approx 2.523719014$. This value is between 2 and 3, as predicted in Exercise 12.

14. The exact value of $\log_5 \frac{1}{5}$ is -1 since $5^{-1}=\frac{1}{5}$. The exact value of $\log_5 1$ is 0 since $5^0=1$.

15. $\log_5 .68$ must lie between -1 and 0. Since the function defined by $y=\log_5 x$ is increasing and $\frac{1}{5}=.2<.68<1$, we must have $\log_5 .2<\log_5 .68<\log_5 1$. By the change-of-base theorem, we have $\log_5 .68=\frac{\log .68}{\log 5}=\frac{\ln .68}{\ln 5}\approx -.2396255723$. This value is between -1 and 0, as predicted above.

16. $\log_9 27=\frac{3}{2}$ is written in exponential form as $9^{3/2}=27$.

17. $\log 3.45\approx .5378$ is written in exponential form as $10^{.5378}\approx 3.45$.

18. $\ln 45\approx 3.8067$ is written in exponential form as $e^{3.8067}\approx 45$.

19. Let $f(x)=\log_a x$ be the required function.
Then $f(81)=4\Rightarrow \log_a 81=4\Rightarrow a^4=81\Rightarrow a^4=3^4\Rightarrow a=3$. The base is 3.

20. Let $f(x)=a^x$ be the required function.
Then $f(-4)=\frac{1}{16}\Rightarrow a^{-4}=\frac{1}{16}\Rightarrow a^{-4}=2^{-4}\Rightarrow a=2$. The base is 2.

21. $\log_3 \frac{mn}{5r}=\log_3 mn-\log_3 5r=\log_3 m+\log_3 n-(\log_3 5+\log_3 r)=\log_3 m+\log_3 n-\log_3 5-\log_3 r$

22. $\log_5\left(x^2y^4\sqrt[5]{m^3p}\right)=\log_5 x^2y^4\left(m^3p\right)^{1/5}$

$$=\log_5 x^2+\log_5 y^4+\log_5\left(m^3p\right)^{1/5}=2\log_5 x+4\log_5 y+\tfrac{1}{5}\left(\log_5 m^3p\right)$$

$$=2\log_5 x+4\log_5 y+\tfrac{1}{5}\left(\log_5 m^3+\log_5 p\right)=2\log_5 x+4\log_5 y+\tfrac{1}{5}\left(3\log_5 m+\log_5 p\right)$$

23. $\log_7(7k+5r^2)$

Since this is the logarithm of a sum, this expression cannot be simplified.

24. $\log 45.6 \approx 1.6590$

25. $\log .0411 \approx -1.3862$

26. $\ln 470 \approx 6.1527$

27. $\ln 144{,}000 \approx 11.8776$

28. To find $\log_3 769$, use the change-of-base theorem. We have $\log_3 769=\frac{\log 769}{\log 3}=\frac{\ln 769}{\ln 3}\approx 6.0486.$

29. To find $\log_{2/3}\frac{5}{8}$, use the change-of-base theorem. We have $\log_{2/3}\frac{5}{8}=\frac{\log\frac{5}{8}}{\log\frac{2}{3}}=\frac{\ln\frac{5}{8}}{\ln\frac{2}{3}}\approx 1.1592.$

30.

$$8^x=32$$
$$\left(2^3\right)^x=2^5$$
$$2^{3x}=2^5$$
$$3x=5\Rightarrow x=\tfrac{5}{3}$$

Solution set: $\left\{\frac{5}{3}\right\}$

31.

$$x^{-3}=\tfrac{8}{27}$$
$$\left(\tfrac{1}{x}\right)^3=\left(\tfrac{2}{3}\right)^3$$
$$\tfrac{1}{x}=\tfrac{2}{3}$$
$$x=\tfrac{3}{2}$$

Solution set: $\left\{\frac{3}{2}\right\}$

32. $10^{2x-3}=17$

Take common logarithms of both sides since the base of the exponential is 10.

$$\log 10^{2x-3}=\log 17\Rightarrow(2x-3)\log 10=\log 17\Rightarrow 2x-3=\log 17$$
$$2x=\log 17+3\Rightarrow x=\tfrac{\log 17+3}{2}\approx 2.1152$$

Solution set: $\{2.1152\}$

33. $4^{x+3}=5^{2-x}$

Take the natural logarithms of both sides. Note: You could also solve by taking the common logarithm of both sides.

$$\ln 4^{x+3}=\ln 5^{2-x}$$
$$(x+3)\ln 4=(2-x)\ln 5$$
$$x\ln 4+3\ln 4=2\ln 5-x\ln 5$$
$$x\ln 4+x\ln 5=2\ln 5-3\ln 4$$
$$x(\ln 4+\ln 5)=\ln 5^2-\ln 4^3$$
$$x(\ln 4+\ln 5)=\ln 25-\ln 64$$
$$x=\tfrac{\ln 25-\ln 64}{\ln 4+\ln 5}$$
$$x=\frac{\ln\frac{25}{64}}{\ln 20}\approx -.3138$$

Solution set: $\{-.3138\}$

34. $e^{x+1}=10$

Take the natural logarithms of both sides.

$$e^{x+1}=10$$
$$\ln e^{x+1}=\ln 10$$
$$(x+1)\ln e=\ln 10$$
$$x+1=\ln 10$$
$$x=\ln 10-1\approx 1.3026$$

Solution set: $\{1.3026\}$

35. $\log_{64} x=\frac{1}{3}$

$$64^{1/3}=x$$
$$x=\sqrt[3]{64}=4$$

Solution set: $\{4\}$

36. $\ln(6x)-\ln(x+1)=\ln 4$

$$\ln\frac{6x}{x+1}=\ln 4$$
$$\frac{6x}{x+1}=4$$
$$6x=4(x+1)$$
$$6x=4x+4$$
$$2x=4$$
$$x=2$$

Solution set: $\{2\}$

37. $\log_{16}\sqrt{x+1}=\frac{1}{4}$

$$\sqrt{x+1}=16^{1/4}$$
$$\left(\sqrt{x+1}\right)^2=\left(16^{1/4}\right)^2$$
$$x+1=16^{1/2}$$
$$x+1=\sqrt{16}$$
$$x+1=4$$
$$x=3$$

Since the solution involves squaring both sides, the proposed solution must be checked in the original equation.

Check $x=3$.

$$\log_{16}\sqrt{x+1}=\frac{1}{4}$$
$$\log_{16}\sqrt{3+1}=\frac{1}{4}\ ?$$
$$\log_{16}\sqrt{4}=\frac{1}{4}$$
$$\log_{16}4^{1/2}=\frac{1}{4}$$
$$\frac{1}{2}\log_{16}4=\frac{1}{4}$$
$$\frac{1}{2}\log_{16}\sqrt{16}=\frac{1}{4}$$
$$\frac{1}{2}\log_{16}16^{1/2}=\frac{1}{4}$$
$$\frac{1}{2}\cdot\frac{1}{2}=\frac{1}{4}$$
$$\frac{1}{4}=\frac{1}{4}$$

This is a true statement.
3 is a solution.

Solution set: $\{3\}$

38. $\ln x+3\ln 2=\ln\frac{2}{x}$

$$\ln x+\ln 2^3=\ln\frac{2}{x}$$
$$\ln(x\cdot 2^3)=\ln\frac{2}{x}$$
$$\ln 8x=\ln\frac{2}{x}$$
$$8x=\frac{2}{x}$$
$$8x^2=2$$
$$x^2=\frac{1}{4}$$
$$x=\pm\frac{1}{2}.$$

Since the negative solution $\left(x=-\frac{1}{2}\right)$ is not in the domain of $\ln x$, it must be discarded.

Solution set: $\left\{\frac{1}{2}\right\}$

39. $\log x+\log(x-3)=1$

$$\log_{10}x(x-3)=1$$
$$10^1=x(x-3)$$
$$10=x^2-3x$$
$$0=x^2-3x-10$$
$$0=(x+2)(x-5)$$
$$x+2=0 \quad\text{or}\quad x-5=0$$
$$x=-2 \qquad\qquad x=5$$

Since the negative solution $(x=-2)$ is not in the domain of $\log x$, it must be discarded.

Solution set: $\{5\}$

40. **(a)** $6.6=\log_{10}\frac{I}{I_0}$

$$\frac{I}{I_0}=10^{6.6}$$
$$I=I_0 10^{6.6}\approx 3{,}981{,}071.71I_0$$

The magnitude was about $4{,}000{,}000I_0$.

(b) $6.5=\log_{10}\frac{I}{I_0}$

$$10^{6.5}=\frac{I}{I_0}$$
$$I_0 10^{6.5}=I$$
$$I\approx 3{,}162{,}277.66I_0$$

The magnitude was about $3{,}200{,}000I_0$.

(c) Consider the ratio of the magnitudes.

$$\frac{4{,}000{,}000\ I_0}{3{,}200{,}000\ I_0}=\frac{40}{32}=\frac{5}{4}=1.25$$

The earthquake with a measure of 6.6 was about 1.25 times as great.

41. **(a)** $8.3 = \log_{10} \frac{I}{I_0}$

$\frac{I}{I_0} = 10^{8.3}$

$I = 10^{8.3} I_0 \approx 199{,}526{,}231.5 I_0$

The magnitude was about $200{,}000{,}000 I_0$.

(b) $7.1 = \log_{10} \frac{I}{I_0}$

$\frac{I}{I_0} = 10^{7.1}$

$I = 10^{7.1} I_0 \approx 12{,}589{,}254.12 I_0$

The magnitude was about $13{,}000{,}000 I_0$.

(c) $\frac{200{,}000{,}000 I_0}{13{,}000{,}000 I_0} = \frac{200}{13} \approx 15.38$

The 1906 earthquake had a magnitude more than 15 times greater than the 1989 earthquake. Note: If the more precise values found in parts a and b were used, the 1906 earthquake had a magnitude of almost 16 times greater than the 1989 earthquake.

42. For 89 decibels, we have

$89 = 10 \log \frac{I}{I_0}$

$8.9 = \log \frac{I}{I_0}$

$\frac{I}{I_0} = 10^{8.9} \Rightarrow I = 10^{8.9} I_0.$

For 86 decibels, we have

$86 = 10 \log \frac{I}{I_0}$

$\frac{I}{I_0} = 10^{8.6} \Rightarrow I = 10^{8.6} I_0.$

To compare these intensities, find their ratio.

$$\frac{10^{8.9} I_0}{10^{8.6} I_0} = 10^{8.9-8.6} = 10^{.3} \approx 2$$

From this calculation, we see that 89 decibels is about twice as loud as 86 decibels. This is a 100% increase.

43. Substitute $A = 5760$, $P = 3500$, $t = 10$, $m = 1$ into the formula $A = P\left(1+\frac{r}{m}\right)^{tm}$.

$$5760 = 3500\left(1+\frac{r}{1}\right)^{10(1)} \Rightarrow \frac{288}{175} = (1+r)^{10} \Rightarrow \left(\frac{288}{175}\right)^{1/10} = 1+r \Rightarrow \left(\frac{288}{175}\right)^{1/10} - 1 = r \Rightarrow r \approx .051$$

The annual interest rate, to the nearest tenth, is 5.1%.

44. Substitute $P = 48{,}000$, $A = 58{,}344$, $r = .05$, and $m = 2$ into the formula $A = P\left(1+\frac{r}{m}\right)^{tm}$.

$$58{,}344 = 48{,}000\left(1+\frac{.05}{2}\right)^{t(2)}$$

$$58{,}344 = 48{,}000(1.025)^{2t}$$

$$1.2155 = (1.025)^{2t}$$

$$\ln 1.2155 = \ln (1.025)^{2t}$$

$$\ln 1.2155 = 2t \ln 1.025$$

$$t = \frac{\ln 1.2155}{2 \ln 1.025} \approx 4.0$$

$48,000 will increase to $58,344 in about 4.0 yr.

45. First, substitute $P = 10{,}000$, $r = .08$, $t = 12$, and $m = 1$ into the formula $A = P\left(1+\frac{r}{m}\right)^{tm}$.

$$A = 10{,}000\left(1+\frac{.08}{1}\right)^{12(1)} = 10{,}000(1.08)^{12} \approx 25{,}181.70$$

After the first 12 yr, there would be about $25,181.70 in the account. To finish off the 21-year period, substitute $P = 25{,}181.70$, $r = .10$, $t = 9$, and $m = 2$ into the formula $A = P\left(1+\frac{r}{m}\right)^{tm}$.

$$A = 25{,}181.70\left(1+\frac{.10}{2}\right)^{9(2)} = 25{,}181.70(1+.05)^{18} = 25{,}181.70(1.05)^{18} \approx 60{,}602.76$$

At the end of the 21-year period, about $60,606.76 would be in the account. Note: If it was possible to transfer the money accrued after the 12 years to the new account without rounding, the amount after the 21-year period would be $60,606.77. The difference is not significant.

46. First, substitute $P = 12{,}000$, $r = .05$, $t = 8$, and $m = 1$ into the formula $A = P\left(1+\frac{r}{m}\right)^{tm}$.

$$A = 12{,}000\left(1+\tfrac{.05}{1}\right)^{8(1)} = 12{,}000(1.05)^{8} \approx 17{,}729.47$$

After the first 8 yr, there would be \$17,729.47 in the account. To finish off the 14-year period, substitute $P = 17{,}729.47$, $r = .06$, $t = 6$, and $m = 1$ into the formula $A = P\left(1+\frac{r}{m}\right)^{tm}$.

$$A = 17{,}729.47\left(1+\tfrac{.06}{1}\right)^{6(1)} = 17{,}729.47(1.06)^{6} \approx 25{,}149.59$$

At the end of the 14-year period, \$25,149.59 would be in the account.

47. To find t, substitute $a = 2$, $P = 1$, and $r = .04$ into $A = Pe^{rt}$ and solve.

$$2 = 1 \cdot e^{.04t}$$
$$2 = e^{.04t}$$
$$\ln 2 = \ln e^{.04t}$$
$$\ln 2 = .04t$$
$$t = \tfrac{\ln 2}{.04} \approx 17.3$$

It would take about 17.3 yr.

48. For each of the following parts the window is as follows.

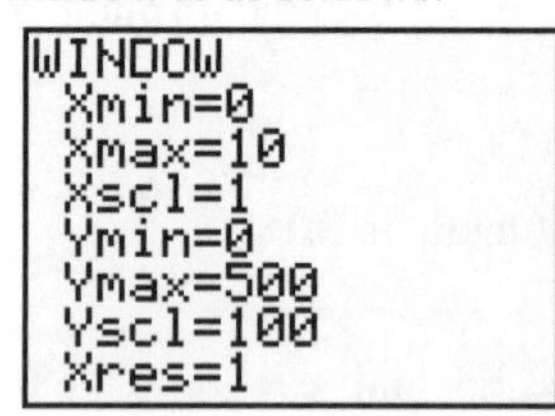

(a) $A(t) = t^2 - t + 350$

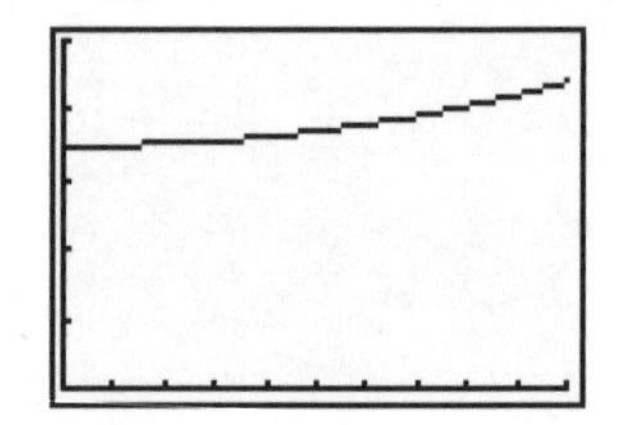

(b) $A(t) = 350\log(t+1)$

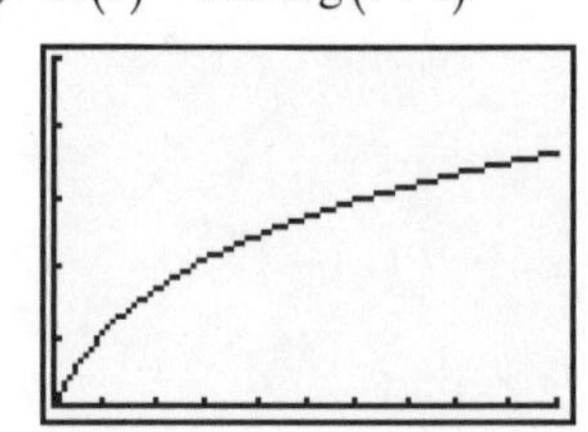

(c) $A(t) = 350(.75)^t$

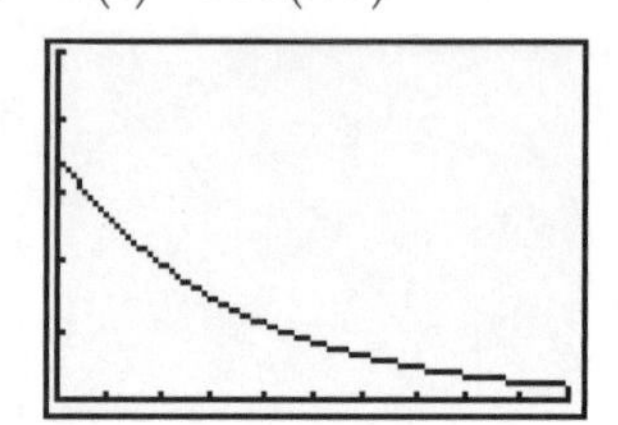

(d) $A(t) = 100(.95)^t$

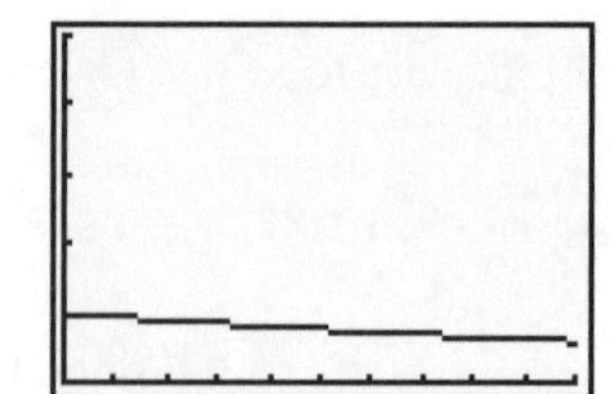

Function (c) best describes $A(t)$.

49. Double the 2003 total payoff value is $2(152.7) = 305.4$. Using the function $f(x) = 93.54e^{.16x}$, we solve for x when $f(x) = 305.4$.

$$93.54e^{.16x} = 305.4 \Rightarrow e^{.16x} = \frac{305.4}{93.54} \Rightarrow \ln e^{.16x} = \ln \frac{305.4}{93.54} \Rightarrow .16x = \ln \frac{305.4}{93.54} \Rightarrow x = \frac{\ln \frac{305.4}{93.54}}{.16} \approx 7.40$$

Since x represents the number of years since 2000, in 2007 the total payoff value will be double of 2003.

50. **(a)** Plot the year on the x-axis and the number of processors on the y-axis. Let $x = 0$ correspond to the year 1971.

Year	Transistors
$1971 - 1971 = 0$	2300
$1986 - 1971 = 15$	275,000
$1989 - 1971 = 18$	1,200,000
$1993 - 1971 = 22$	3,300,000
$1995 - 1971 = 24$	5,500,000
$1997 - 1971 = 26$	9,500,000
$2000 - 1971 = 29$	42,000,000

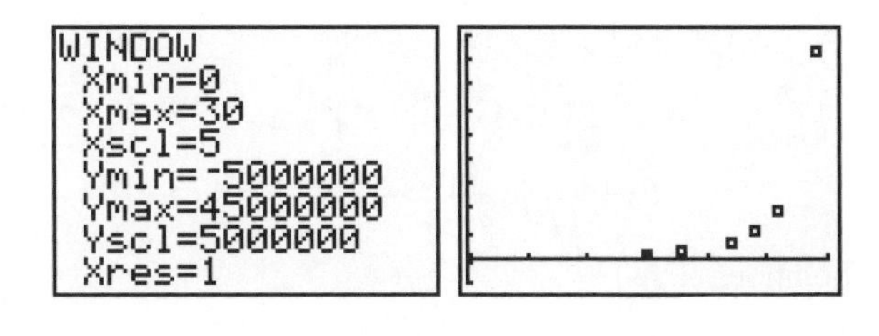

(b) The data are clearly not linear and do not level off like a logarithmic function. The data are increasing at a faster rate as x increases. Of the three choices, an exponential function will describe this data best.

(c) Using the exponential regression feature on the TI graphing calculator, we have if $f(x) = a(b)^x$, then $f(x) \approx 2278(1.392)^x$. Other answers are possible using the techniques described in this chapter.

(d) Since $2205 - 1971 = 34$, we can predict the number of transistors on a chip in the year 2005 by evaluating $f(34)$.

$$f(34) \approx 2278(1.392)^{34} \approx 174{,}296{,}119.9$$

There will be approximately 174,000,000 transistors on a chip in the year 2005.

Chapter 9: Test

1. **(a)** $y = \log_{1/3} x$

The point (1, 0) is on the graph of every function of the form $y = \log_a x$, so the correct choice must be either B or C. Since the base is $a = \frac{1}{3}$ and $0 < \frac{1}{3} < 1$, $y = \log_{1/3} x$ is a decreasing function, and so the correct choice must be B.

(b) $y = e^x$

The point (0, 1) is on the graph since $e^0 = 1$, so the correct choice must be either A or D. Since the base is e and $e > 1$, $y = e^x$ is an increasing function, and so the correct choice must be A.

Continued on next page

1. (continued)

(c) $y = \ln x$ or $y = \log_e x$

The point (1, 0) is on the graph of every function of the form $y = \log_a x$, so the correct choice must be B or C. Since the base is $a = e$ and $e > 1$, $y = \ln x$ is an increasing function, and the correct choice must be C.

(d) $y = \left(\frac{1}{3}\right)^x$

The point (0, 1) is on the graph since $\left(\frac{1}{3}\right)^0 = 1$, so the correct choice must be either A or D. Since the base is $\frac{1}{3}$ and $0 < \frac{1}{3} < 1$, $y = \left(\frac{1}{3}\right)^x$ is a decreasing function, and so the correct choice must be D.

2. $\left(\frac{1}{8}\right)^{2x-3} = 16^{x+1} \Rightarrow \left(2^{-3}\right)^{2x-3} = \left(2^4\right)^{x+1} \Rightarrow 2^{-3(2x-3)} = 2^{4(x+1)} \Rightarrow 2^{-6x+9} = 2^{4x+4}$

$$-6x + 9 = 4x + 4 \Rightarrow -10x + 9 = 4 \Rightarrow -10x = -5 \Rightarrow x = \frac{1}{2}$$

Solution set: $\left\{\frac{1}{2}\right\}$

3. $4^{3/2} = 8$ is written in logarithmic form as $\log_4 8 = \frac{3}{2}$.

4. $\log_8 4 = \frac{2}{3}$ is written in exponential form as $8^{2/3} = 4$.

5. They are inverses of each other.

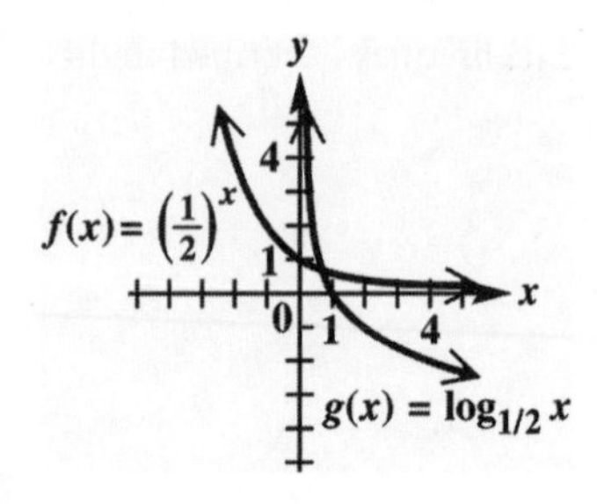

6. $\log_7 \frac{x^2 \sqrt[4]{y}}{z^3} = \log_7 x^2 + \log_7 \sqrt[4]{y} - \log_7 z^3 = \log_7 x^2 + \log_7 y^{1/4} - \log_7 z^3 = 2\log_7 x + \frac{1}{4}\log_7 y - 3\log_7 z$

7. $\log 237.4 \approx 2.3755$

8. $\ln .0467 \approx -3.0640$

9. $\log_9 13 = \frac{\ln 13}{\ln 9} = \frac{\log 13}{\log 9} \approx 1.1674$

10. $\log\left(2.49 \times 10^{-3}\right) \approx -2.6038$

11. $\log_x 25 = 2$

$$x^2 = 25$$
$$x^2 = 5^2$$
$$x = 5$$

Solution set: $\{5\}$

12. $\log_4 32 = x$

$$4^x = 32$$
$$\left(2^2\right)^x = 2^5$$
$$2^{2x} = 2^5$$
$$2x = 5$$
$$x = \frac{5}{2}$$

Solution set: $\left\{\frac{5}{2}\right\}$

13. $\log_2 x + \log_2(x+2) = 3 \Rightarrow \log_2 x(x+2) = 3 \Rightarrow 2^3 = x(x+2)$

$$8 = x^2 + 2x \Rightarrow 0 = x^2 + 2x - 8 \Rightarrow 0 = (x+4)(x-2)$$

$$x+4=0 \quad \text{or} \quad x-2=0$$

$$x=-4 \qquad\qquad x=2$$

Since the negative solution $(x=-4)$ is not in the domain of $\log_2 x$, it must be discarded.

Solution set: $\{2\}$

14.

$$5^{x+1} = 7^x$$
$$\ln 5^{x+1} = \ln 7^x$$
$$(x+1)\ln 5 = x\ln 7$$
$$x\ln 5 + \ln 5 = x\ln 7$$
$$\ln 5 = x\ln 7 - x\ln 5$$
$$\ln 5 = x(\ln 7 - \ln 5)$$
$$x = \frac{\ln 5}{\ln 7 - \ln 5} = \frac{\ln 5}{\ln \frac{7}{5}} \approx 4.7833$$

Solution set: $\{4.7833\}$

15. $\ln x - 4\ln 3 = \ln \frac{5}{x}$

$$\ln x - \ln 3^4 = \ln \frac{5}{x}$$
$$\ln x - \ln 81 = \ln \frac{5}{x}$$
$$\ln \frac{x}{81} = \ln \frac{5}{x}$$
$$\frac{x}{81} = \frac{5}{x}$$
$$x^2 = 405$$
$$x = \pm\sqrt{405} \approx \pm 20.1246$$

Since the negative solution $(x=-20.1246)$ is not in the domain of $\ln x$, it must be discarded.

Solution set: $\{20.1246\}$

16. Answers will vary.

$\log_5 27$ is the exponent to which 5 must be raised in order to obtain 27. To approximate $\log_5 27$ on your calculator, use the change-of-base formula; $\log_5 27 = \frac{\log 27}{\log 5} = \frac{\ln 27}{\ln 5} \approx 2.048$.

17. $v(t) = 176(1 - e^{-.18t})$

Find the time t at which $v(t) = 147$.

$$147 = 176(1 - e^{-.18t}) \Rightarrow \frac{147}{176} = 1 - e^{-.18t} \Rightarrow -e^{-.18t} = \frac{147}{176} - 1 \Rightarrow -e^{-.18t} = -\frac{29}{176}$$

$$e^{-.18t} = \frac{29}{176} \Rightarrow \ln e^{-.18t} = \ln \frac{29}{176} \Rightarrow -.18t = \ln \frac{29}{176} \Rightarrow t = \frac{\ln \frac{29}{176}}{-.18} \approx 10.02$$

It will take the skydiver about 10 sec to attain the speed of 147 ft per sec (100 mph).

18. (a) Substitute $P = 5000$, $A = 18{,}000$, $r = .068$, and $m = 12$ into the formula $A = P\left(1+\frac{r}{m}\right)^{tm}$.

$$18{,}000 = 5000\left(1+\frac{.068}{12}\right)^{t(12)} \Rightarrow 3.6 = \left(1+\frac{.068}{12}\right)^{12t} \Rightarrow \ln 3.6 = \ln\left(1+\frac{.068}{12}\right)^{12t}$$

$$\ln 3.6 = 12t\ln\left(1+\frac{.068}{12}\right) \Rightarrow t = \frac{\ln 3.6}{12\ln\left(1+\frac{.068}{12}\right)} \approx 18.9$$

It will take about 18.9 years.

(b) Substitute $P = 5000$, $A = 18{,}000$, and $r = .068$, and into the formula $A = Pe^{rt}$.

$$18{,}000 = 5000e^{.068t} \Rightarrow 3.6 = e^{.068t} \Rightarrow \ln 3.6 = \ln e^{.068t} \Rightarrow \ln 3.6 = .068t \Rightarrow t = \frac{\ln 3.6}{.068} \approx 18.8$$

It will take about 18.8 years.

Chapter 9: Quantitative Reasoning

1. Since you are taxed on the entire amount when it is withdrawn at 60, calculate 60% (40% goes to taxes) of $A = P\left(1+\frac{r}{m}\right)^{tm}$ when P = 3000, r = .08, m = 1 and $t = 60-25 = 35$.

$$.60\left[(3000)\left(1+\tfrac{.08}{1}\right)^{35(1)}\right] = \left[.60(3000)\right]\left(1+\tfrac{.08}{1}\right)^{35(1)} = 1800(1.08)^{35} \approx 26,613.62$$

$26,613.62 will remain after taxes are paid.

2. Since you are taxed on the money and the annual interest, calculate $A = P\left(1+\frac{r}{m}\right)^{tm}$ when P = 1800, $r = .048$, m = 1 and $t = 35$.

$$1800(1+.048)^{35(1)} = 1800(1.048)^{35} \approx 9287.90$$

$9287.90 will be available at age 60.

3. Since $26,613.62 - 9287.90 = 17,325.72$, to the nearest dollar, $17,326 will be additionally earned with the IRA.

4. Since $.60\left[(3000)\left(1+\tfrac{.08}{1}\right)^{35(1)}\right] = \left[.60(3000)\right]\left(1+\tfrac{.08}{1}\right)^{35(1)}$, this is the same calculation.

Appendix A
EQUATIONS AND INEQUALITIES

1. Solve the equation $2x+3=x-5$.
$$\begin{aligned} 2x+3&=x-5\\ 2x+3-x&=x-5-x\\ x+3&=-5\\ x+3-3&=-5-3\\ x&=-8 \end{aligned}$$
Moreover, replacing x with -8 in the equation $2x+3=x-5$ yields a true statement. Therefore, the given statement is true.

2. The left side can be written as
$$\begin{aligned} 5(x-9)&=5\left[x+(-9)\right]=5x+5(-9)\\ &=5x+(-45)=5x-45, \end{aligned}$$
which is the same as the right side. Therefore, the statement is true.

3. A linear equation can be a contradiction, an identity, or conditional. If it is a contradiction, it has no solution; if it is an identity, it has more than two solutions; and if it is conditional, it has exactly one solution. Therefore, the given statement is false.

4. Answers will vary.

5. B cannot be written in the form $ax+b=0$.
A can be written as $15x-7=0$ or $15x+(-7)=0$, C can be written as $2x=0$ or $2x+0=0$, and D can be written as $-.04x-.4=0$ or $-.04x+(-.4)=0$.

6. The student's answer is not correct. Additional answers will vary.

7. $$\begin{aligned} 5x+2&=3x-6\\ 2x+2&=-6\\ 2x&=-8\\ x&=-4 \end{aligned}$$
Solution set: $\{-4\}$

8. $$\begin{aligned} 9x+1&=7x-9\\ 2x+1&=-9\\ 2x&=-10\\ x&=-5 \end{aligned}$$
Solution set: $\{-5\}$

9. $$\begin{aligned} 6(3x-1)&=8-(10x-14)\\ 18x-6&=8-10x+14\\ 18x-6&=22-10x\\ 28x-6&=22\\ 28x&=28\\ x&=1 \end{aligned}$$
Solution set: $\{1\}$

10. $$\begin{aligned} 4(-2x+1)&=6-(2x-4)\\ -8x+4&=6-2x+4\\ -8x+4&=10-2x\\ 4&=10+6x\\ -6&=6x\\ -1&=x \end{aligned}$$
Solution set: $\{-1\}$

11. $$\begin{aligned} \frac{5}{6}x-2x+\frac{1}{3}&=\frac{2}{3}\\ 6\cdot\left[\frac{5}{6}x-2x+\frac{1}{3}\right]&=6\cdot\frac{2}{3}\\ 5x-12x+2&=4\\ -7x+2&=4\\ -7x&=2\\ x&=-\frac{2}{7} \end{aligned}$$
Solution set: $\{-\frac{2}{7}\}$

12.
$$\frac{3}{4}+\frac{1}{5}x-\frac{1}{2}=\frac{4}{5}x$$
$$20\cdot\left[\frac{3}{4}+\frac{1}{5}x-\frac{1}{2}\right]=20\cdot\frac{4}{5}x$$
$$15+4x-10=16x$$
$$4x+5=16x$$
$$5=12x$$
$$\frac{5}{12}=x$$
Solution set: $\left\{\frac{5}{12}\right\}$

13.
$$3x+2-5(x+1)=6x+4$$
$$3x+2-5x-5=6x+4$$
$$-2x-3=6x+4$$
$$-3=8x+4$$
$$-7=8x$$
$$\frac{-7}{8}=x$$
$$-\frac{7}{8}=x$$
Solution set: $\left\{-\frac{7}{8}\right\}$

14.
$$5(x+3)+4x-5=-(2x-4)$$
$$5x+15+4x-5=-2x+4$$
$$9x+10=-2x+4$$
$$11x+10=4$$
$$11x=-6$$
$$x=\frac{-6}{11}=-\frac{6}{11}$$
Solution set: $\left\{-\frac{6}{11}\right\}$

15.
$$2\left[x-(4+2x)+3\right]=2x+2$$
$$2(x-4-2x+3)=2x+2$$
$$2(-x-1)=2x+2$$
$$-2x-2=2x+2$$
$$-2=4x+2$$
$$-4=4x$$
$$-1=x$$
Solution set: $\{-1\}$

16.
$$4\left[2x-(3-x)+5\right]=-7x-2$$
$$4(2x-3+x+5)=-7x-2$$
$$4(3x+2)=-7x-2$$
$$12x+8=-7x-2$$
$$19x+8=-2$$
$$19x=-10$$
$$x=\frac{-10}{19}=-\frac{10}{19}$$
Solution set: $\left\{-\frac{10}{19}\right\}$

17.
$$\frac{1}{7}(3x-2)=\frac{x+10}{5}$$
$$35\cdot\left[\frac{1}{7}(3x-2)\right]=35\cdot\left[\frac{x+10}{5}\right]$$
$$5(3x-2)=7(x+10)$$
$$15x-10=7x+70$$
$$8x-10=70$$
$$8x=80$$
$$x=10$$
Solution set: $\{10\}$

18.
$$\frac{1}{5}(2x+5)=\frac{x+2}{3}$$
$$15\cdot\left[\frac{1}{5}(2x+5)\right]=15\cdot\left[\frac{x+2}{3}\right]$$
$$3(2x+5)=5(x+2)$$
$$6x+15=5x+10$$
$$x+15=10$$
$$x=-5$$
Solution set: $\{-5\}$

19.
$$.2x-.5=.1x+7$$
$$10(.2x-.5)=10(.1x+7)$$
$$2x-5=x+70$$
$$x-5=70$$
$$x=75$$
Solution set: $\{75\}$

20.
$$.01x+3.1=2.03x-2.96$$
$$100(.01x+3.1)=100(2.03x-2.96)$$
$$x+310=203x-296$$
$$310=202x-296$$
$$606=202x$$
$$3=x$$
Solution set: $\{3\}$

21. $-4(2x-6)+7x=5x+24$
$$-8x+24+7x=5x+24$$
$$-x+24=5x+24$$
$$24=6x+24$$
$$0=6x$$
$$0=x$$
Solution set: $\{0\}$

22. $-8(3x+4)+2x=4(x-8)$
$$-24x-32+2x=4x-32$$
$$-22x-32=4x-32$$
$$-32=26x-32$$
$$0=26x$$
$$0=x$$
Solution set: $\{0\}$

23. $4(2x+7)=2x+25+3(2x+1)$
$$8x+28=2x+25+6x+3$$
$$8x+28=8x+28$$
$$28=28$$
$$0=0$$
identity; $\{\text{all real numbers}\}$

24. $\frac{1}{2}(6x+14)=x+1+2(x+3)$
$$3x+7=x+1+2x+6$$
$$3x+7=3x+7$$
$$7=7$$
$$0=0$$
identity; $\{\text{all real numbers}\}$

25. $2(x-7)=3x-14$
$$2x-14=3x-14$$
$$-14=x-14$$
$$0=x$$
conditional equation; $\{0\}$

26. $-8(x+3)=-8x-5(x+1)$
$$-8x-24=-8x-5x-5$$
$$-8x-24=-13x-5$$
$$5x-24=-5$$
$$5x=19$$
$$x=\frac{19}{5}$$
conditional equation; $\left\{\frac{19}{5}\right\}$

27. $8(x+7)=4(x+12)+4(x+1)$
$$8x+56=4x+48+4x+4$$
$$8x+56=8x+52$$
$$56=52$$
contradiction; $\varnothing$

28. $-6(2x+1)-3(x-4)=-15x+1$
$$-12x-6-3x+12=-15x+1$$
$$-15x+6=-15x+1$$
$$6=1$$
contradiction; $\varnothing$

29. **(a).** $x^2 = 4$

$x = \pm\sqrt{4} = \pm 2$; E

(b). $x^2 - 2 = 0$

$x^2 = 2$

$x = \pm\sqrt{2}$; C

(c). $x^2 = 8$

$x = \pm\sqrt{8} = \pm 2\sqrt{2}$; A

(d). $x - 2 = 0$

$x = 2$; B

(e). $x + 2 = 0$

$x = -2$; D

30. D is the only one set up for direct use of the zero-factor property.

$(3x+1)(x-7) = 0$

$3x + 1 = 0$ or $x - 7 = 0$

$x = -\frac{1}{3}$ or $x = 7$

Solution set: $\left\{-\frac{1}{3}, 7\right\}$

31. B is the only one set up for direct use of the square root property.

$(2x+5)^2 = 7 \Rightarrow 2x + 5 = \pm\sqrt{7}$

$2x = -5 \pm \sqrt{7} \Rightarrow x = \dfrac{-5 \pm \sqrt{7}}{2}$

Solution set: $\left\{\frac{-5\pm\sqrt{7}}{2}\right\}$

32. A is the only set up so that the values of a, b, and c can be determined immediately.

$3x^2 - 17x - 6 = 0$ yields $a = 3$, $b = -17$, and $c = -6$.

$$x = \frac{-b \pm \sqrt{b^2 - 4ac}}{2a} = \frac{-(-17) \pm \sqrt{(-17)^2 - 4(3)(-6)}}{2(3)} = \frac{17 \pm \sqrt{289 - (-72)}}{6} = \frac{17 \pm \sqrt{361}}{6} = \frac{17 \pm 19}{6}$$

$$\frac{17+19}{6} = \frac{36}{6} = 6 \text{ and } \frac{17-19}{6} = \frac{-2}{6} = -\frac{1}{3}$$

Solution set: $\left\{-\frac{1}{3}, 6\right\}$

33. $x^2 - 5x + 6 = 0$

$(x-2)(x-3) = 0$

$x - 2 = 0$ or $x - 3 = 0$

$x = 2$ or $x = 3$

Solution set: $\{2, 3\}$

34. $x^2 + 2x - 8 = 0$

$(x+4)(x-2) = 0$

$x + 4 = 0$ or $x - 2 = 0$

$x = -4$ or $x = 2$

Solution set: $\{-4, 2\}$

35. $5x^2 - 3x - 2 = 0$

$(5x+2)(x-1) = 0$

$5x + 2 = 0$ or $x - 1 = 0$

$x = -\frac{2}{5}$ or $x = 1$

Solution set: $\left\{-\frac{2}{5}, 1\right\}$

36. $2x^2 - x - 15 = 0$

$(2x+5)(x-3) = 0$

$2x + 5 = 0$ or $x - 3 = 0$

$x = -\frac{5}{2}$ or $x = 3$

Solution set: $\left\{-\frac{5}{2}, 3\right\}$

37. $-4x^2 + x = -3$

$0 = 4x^2 - x - 3$

$0 = (4x+3)(x-1)$

$4x + 3 = 0$ or $x - 1 = 0$

$x = -\frac{3}{4}$ or $x = 1$

Solution set: $\left\{-\frac{3}{4}, 1\right\}$

38. $-6x^2 + 7x = -10$

$0 = 6x^2 - 7x - 10 = 0$

$0 = (6x+5)(x-2) = 0$

$6x + 5 = 0$ or $x - 2 = 0$

$x = -\frac{5}{6}$ or $x = 2$

Solution set: $\left\{-\frac{5}{6}, 2\right\}$

39. $x^2 = 16$

$$x = \pm\sqrt{16} = \pm 4$$

Solution set: $\{\pm 4\}$

40. $x^2 = 25$

$$x = \pm\sqrt{25} = \pm 5$$

Solution set: $\{\pm 5\}$

41. $x^2 = 27$

$$x = \pm\sqrt{27} = \pm 3\sqrt{3}$$

Solution set: $\{\pm 3\sqrt{3}\}$

42. $x^2 = 48$

$$x = \pm\sqrt{48} = \pm 4\sqrt{3}$$

Solution set: $\{\pm 4\sqrt{3}\}$

43. $(x+5)^2 = 40$

$$x+5 = \pm\sqrt{40}$$
$$x+5 = \pm 2\sqrt{10}$$
$$x = -5 \pm 2\sqrt{10}$$

Solution set: $\{-5 \pm 2\sqrt{10}\}$

44. $(x-7)^2 = 24$

$$x-7 = \pm\sqrt{24}$$
$$x-7 = \pm 2\sqrt{6}$$
$$x = 7 \pm 2\sqrt{6}$$

Solution set: $\{7 \pm 2\sqrt{6}\}$

45. $(3x-1)^2 = 12$

$$3x-1 = \pm\sqrt{12}$$
$$3x = 1 \pm 2\sqrt{3}$$
$$x = \frac{1 \pm 2\sqrt{3}}{3}$$

Solution set: $\left\{\frac{1\pm 2\sqrt{3}}{3}\right\}$

46. $(4x+1)^2 = 20$

$$4x+1 = \pm\sqrt{20}$$
$$4x = -1 \pm 2\sqrt{5}$$
$$x = \frac{-1 \pm 2\sqrt{5}}{4}$$

Solution set: $\left\{\frac{-1\pm 2\sqrt{5}}{4}\right\}$

47. $x^2 - x - 1 = 0$

Let $a = 1, b = -1,$ and $c = -1$.

$$x = \frac{-b \pm \sqrt{b^2 - 4ac}}{2a}$$
$$= \frac{-(-1) \pm \sqrt{(-1)^2 - 4(1)(-1)}}{2(1)}$$
$$= \frac{1 \pm \sqrt{1+4}}{2} = \frac{1 \pm \sqrt{5}}{2}$$

Solution set: $\left\{\frac{1\pm\sqrt{5}}{2}\right\}$

48. $x^2 - 3x - 2 = 0$

Let $a = 1, b = -3,$ and $c = -2$.

$$x = \frac{-b \pm \sqrt{b^2 - 4ac}}{2a}$$
$$= \frac{-(-3) \pm \sqrt{(-3)^2 - 4(1)(-2)}}{2(1)}$$
$$= \frac{3 \pm \sqrt{9+8}}{2} = \frac{3 \pm \sqrt{17}}{2}$$

Solution set: $\left\{\frac{3\pm\sqrt{17}}{2}\right\}$

49. $x^2 - 6x = -7 \Rightarrow x^2 - 6x + 7 = 0$

Let $a = 1, b = -6,$ and $c = 7$.

$$x = \frac{-b \pm \sqrt{b^2 - 4ac}}{2a}$$
$$= \frac{-(-6) \pm \sqrt{(-6)^2 - 4(1)(7)}}{2(1)}$$
$$= \frac{6 \pm \sqrt{36-28}}{2} = \frac{6 \pm \sqrt{8}}{2}$$
$$= \frac{6 \pm 2\sqrt{2}}{2} = 3 \pm \sqrt{2}$$

Solution set: $\{3 \pm \sqrt{2}\}$

50. $x^2-4x=-1 \Rightarrow x^2-4x+1=0$
Let $a=1, b=-4,$ and $c=1$.

$$x=\frac{-b\pm\sqrt{b^2-4ac}}{2a}$$
$$=\frac{-(-4)\pm\sqrt{(-4)^2-4(1)(1)}}{2(1)}$$
$$=\frac{4\pm\sqrt{16-4}}{2}=\frac{4\pm\sqrt{12}}{2}$$
$$=\frac{4\pm2\sqrt{3}}{2}=2\pm\sqrt{3}$$

Solution set: $\left\{2\pm\sqrt{3}\right\}$

51. $2x^2-4x-3=0$
$$x^2-2x-\tfrac{3}{2}=0$$
$$x^2-2x+1=\tfrac{3}{2}+1$$
$$\text{Note:}\left[\tfrac{1}{2}\cdot(-2)\right]^2=(-1)^2=1$$
$$(x-1)^2=\tfrac{5}{2}$$
$$x-1=\pm\sqrt{\tfrac{5}{2}}$$
$$x-1=\pm\tfrac{\sqrt{10}}{2}$$
$$x=1\pm\tfrac{\sqrt{10}}{2}=\tfrac{2\pm\sqrt{10}}{2}$$

Solution set: $\left\{\frac{2\pm\sqrt{10}}{2}\right\}$

52. $-3x^2+6x+5=0$
$$x^2-2x-\tfrac{5}{3}=0$$
$$x^2-2x+1=\tfrac{5}{3}+1$$
$$\text{Note:}\left[\tfrac{1}{2}\cdot(-2)\right]^2=(-1)^2=1$$
$$(x-1)^2=\tfrac{8}{3}$$
$$x-1=\pm\sqrt{\tfrac{8}{3}}$$
$$x-1=\pm\tfrac{\sqrt{24}}{3}$$
$$x-1=\pm\tfrac{2\sqrt{6}}{3}$$
$$x=1\pm\tfrac{2\sqrt{6}}{3}=\tfrac{3\pm2\sqrt{6}}{3}$$

Solution set: $\left\{\frac{3\pm2\sqrt{6}}{3}\right\}$

53. $\tfrac{1}{2}x^2+\tfrac{1}{4}x-3=0$
$$4\left(\tfrac{1}{2}x^2+\tfrac{1}{4}x-3\right)=4\cdot0$$
$$2x^2+x-12=0$$
Let $a=2, b=1,$ and $c=-12$.

$$x=\frac{-b\pm\sqrt{b^2-4ac}}{2a}$$
$$=\frac{-1\pm\sqrt{1^2-4(2)(-12)}}{2(2)}$$
$$=\frac{-1\pm\sqrt{1+96}}{4}=\frac{-1\pm\sqrt{97}}{4}$$

Solution set: $\left\{\frac{-1\pm\sqrt{97}}{4}\right\}$

54. $\tfrac{2}{3}x^2+\tfrac{1}{4}x=3$
$$12\left(\tfrac{2}{3}x^2+\tfrac{1}{4}x\right)=12\cdot3$$
$$8x^2+3x=36$$
$$8x^2+3x-36=0$$
Let $a=8, b=3,$ and $c=-36$.

$$x=\frac{-b\pm\sqrt{b^2-4ac}}{2a}$$
$$=\frac{-3\pm\sqrt{3^2-4(8)(-36)}}{2(8)}$$
$$=\frac{-3\pm\sqrt{9+1152}}{16}$$
$$=\frac{-3\pm\sqrt{1161}}{16}=\frac{-3\pm3\sqrt{129}}{16}$$

Solution set: $\left\{\frac{-3\pm3\sqrt{129}}{16}\right\}$

55.
$$.2x^2 + .4x - .3 = 0$$
$$10(.2x^2 + .4x - .3) = 10 \cdot 0$$
$$2x^2 + 4x - 3 = 0$$
Let $a = 2, b = 4,$ and $c = -3.$
$$x = \frac{-b \pm \sqrt{b^2 - 4ac}}{2a} = \frac{-4 \pm \sqrt{4^2 - 4(2)(-3)}}{2(2)} = \frac{-4 \pm \sqrt{16 + 24}}{4} = \frac{-4 \pm \sqrt{40}}{4} = \frac{-4 \pm 2\sqrt{10}}{4} = \frac{-2 \pm \sqrt{10}}{2}$$
Solution set: $\left\{\frac{-2 \pm \sqrt{10}}{2}\right\}$

56.
$$.1x^2 - .1x = .3$$
$$10(.1x^2 - .1x) = 10 \cdot .3$$
$$x^2 - x = 3$$
$$x^2 - x - 3 = 0$$
Let $a = 1, b = -1,$ and $c = -3.$
$$x = \frac{-b \pm \sqrt{b^2 - 4ac}}{2a} = \frac{-(-1) \pm \sqrt{(-1)^2 - 4(1)(-3)}}{2(1)} = \frac{1 \pm \sqrt{1 + 12}}{2} = \frac{1 \pm \sqrt{13}}{2}$$
Solution set: $\left\{\frac{1 \pm \sqrt{13}}{2}\right\}$

57. **(a)** $x < -4$
The interval includes all real numbers less than –4, not including –4. The correct interval notation is $(-\infty, -4)$, so the correct choice is F.

(b) $x \le 4$
The interval includes all real numbers less than or equal to 4, so it includes 4. The correct interval notation is $(-\infty, 4]$, so the correct choice is J.

(c) $-2 < x \le 6$
The interval includes all real numbers from –2 to 6, not including –2, but including 6. The correct interval notation is (–2, 6], so the correct choice is A.

(d) $0 \le x \le 8$
The interval includes all real numbers between 0 and 8, including 0 and 8. The correct interval notation is [0, 8], so the correct choice is H.

(e) $x \ge -3$
The interval includes all real numbers greater than or equal to –3, so it includes –3. The correct interval notation is $[-3, \infty)$, so the correct choice is I.

(f) $4 \le x$
The interval includes all real numbers greater than or equal to 4, so it includes 4. The correct interval notation is $[4, \infty)$, so the correct choice is D.

(g) The interval shown on the number line includes all real numbers between –2 and 6, including –2, but not including 6. The correct interval notation is [–2, 6), so the correct choice is B.

(h) The interval shown on the number line includes all real numbers between 0 and 8, not including 0 or 8. The correct interval notation is (0, 8), so the correct choice is G.

(i) The interval shown on the number line includes all real numbers greater than 3, not including 3. The correct interval notation is $(3, \infty)$, so the correct choice is E.

(j) The interval includes all real numbers less than or equal to – 4, so it includes – 4. The correct interval notation is $(-\infty, -4]$, so the correct choice is C.

58. D

$-7 < x < -10$ would mean $-7 < x$ and $x < -10$, which is equivalent to $x > -7$ and $x < -10$. There is no real number that is simultaneously to the right of –7 and to the left of –10 on a number line.

59.

$$2x+1 \le 9$$
$$2x+1-1 \le 9-1$$
$$2x \le 8$$
$$\frac{2x}{2} \le \frac{8}{2}$$
$$x \le 4$$

Solution set: $(-\infty, 4]$

Graph: 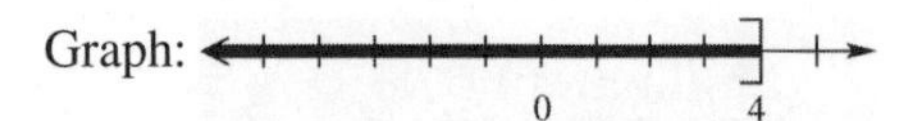

60.

$$3x-2 \le 13$$
$$3x-2+2 \le 13+2$$
$$3x \le 15$$
$$\frac{3x}{3} \le \frac{15}{3}$$
$$x \le 5$$

Solution set: $(-\infty, 5]$

Graph: 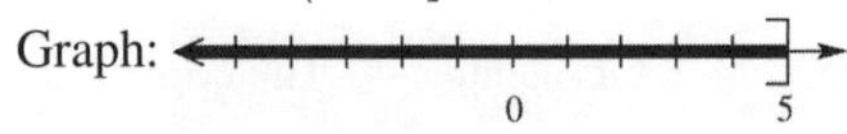

61.

$$-3x-2 \le 1$$
$$-3x-2+2 \le 1+2$$
$$-3x \le 3$$
$$\frac{-3x}{-3} \ge \frac{3}{-3}$$
$$x \ge -1$$

Solution set: $[-1, \infty)$

Graph:

62.

$$-5x+3 \ge -2$$
$$-5x+3-3 \ge -2-3$$
$$-5x \ge -5$$
$$\frac{-5x}{-5} \le \frac{-5}{-5}$$
$$x \le 1$$

Solution set: $(-\infty, 1]$

Graph:

63.

$$2(x+5)+1 \ge 5+3x$$
$$2x+10+1 \ge 5+3x$$
$$2x+11 \ge 5+3x$$
$$2x+11-3x \ge 5+3x-3x$$
$$-x+11 \ge 5$$
$$-x+11-11 \ge 5-11$$
$$\frac{-x}{-1} \le \frac{-6}{-1}$$
$$x \le 6$$

Solution set: $(-\infty, 6]$

Graph:

64.

$$6x-(2x+3) \ge 4x-5$$
$$6x-2x-3 \ge 4x-5$$
$$4x-3 \ge 4x-5$$
$$4x-4x-3 \ge 4x-5-4x$$
$$-3 \ge -5$$

The inequality is true when x is any real number.

Solution set: $(-\infty, \infty)$

Graph:

65.

$$8x-3x+2 < 2(x+7)$$
$$5x+2 < 2x+14$$
$$5x+2-2x < 2x+14-2x$$
$$3x+2 < 14$$
$$3x+2-2 < 14-2$$
$$3x < 12$$
$$\frac{3x}{3} < \frac{12}{3}$$
$$x < 4$$

Solution set: $(-\infty, 4)$

Graph:

66. $2-4x+5(x-1)<-6(x-2)$

$$2-4x+5x-5<-6x+12$$
$$x-3<-6x+12$$
$$x-3+6x<-6x+12+6x$$
$$7x-3<12$$
$$7x-3+3<12+3$$
$$7x<15$$
$$\frac{7x}{7}<\frac{15}{7}$$
$$x<\tfrac{15}{7}$$

Solution set: $\left(-\infty,\tfrac{15}{7}\right)$

Graph:

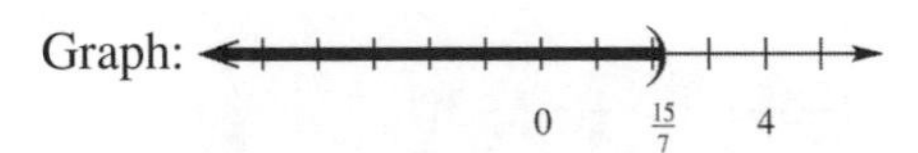

67. $\dfrac{4x+7}{-3}\le 2x+5$

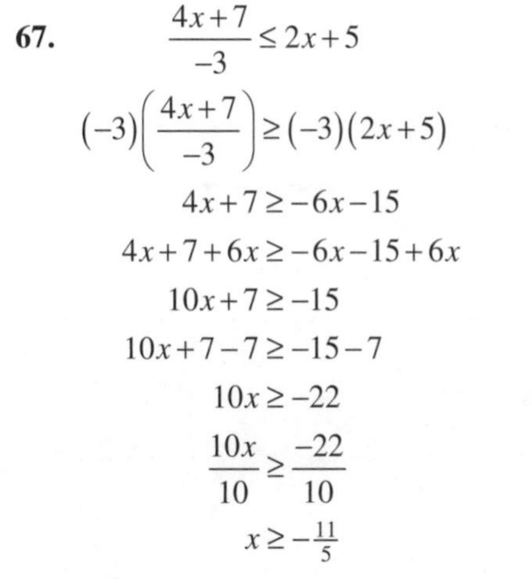

$$(-3)\left(\frac{4x+7}{-3}\right)\ge(-3)(2x+5)$$
$$4x+7\ge -6x-15$$
$$4x+7+6x\ge -6x-15+6x$$
$$10x+7\ge -15$$
$$10x+7-7\ge -15-7$$
$$10x\ge -22$$
$$\frac{10x}{10}\ge\frac{-22}{10}$$
$$x\ge -\tfrac{11}{5}$$

Solution set: $\left[-\tfrac{11}{5},\infty\right)$

Graph: $-\tfrac{11}{5}$ 0

68. $\dfrac{2x-5}{-8}\le 1-x$

$$(-8)\left(\frac{2x-5}{-8}\right)\ge(-8)(1-x)$$
$$2x-5\ge -8+8x$$
$$2x-5-8x\ge -8+8x-8x$$
$$-6x-5\ge -8$$
$$-6x-5+5\ge -8+5$$
$$-6x\ge -3$$
$$\frac{-6x}{-6}\le\frac{-3}{-6}\Rightarrow x\le\tfrac{1}{2}$$

Solution set: $\left(-\infty,\tfrac{1}{2}\right]$

Graph: 0 $\tfrac{1}{2}$

69. $-3<7+2x<13$

$$-3-7<7+2x-7<13-7$$
$$-10<2x<6$$
$$\frac{-10}{2}<\frac{2x}{2}<\frac{6}{2}$$
$$-5<x<3$$

Solution set: $(-5,3)$

Graph:

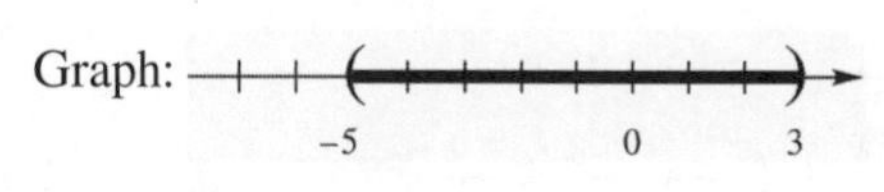

70. $-4<5+3x<8$

$$-4-5<5+3x-5<8-5$$
$$-9<3x<3$$
$$\frac{-9}{3}<\frac{3x}{3}<\frac{3}{3}$$
$$-3<x<1$$

Solution set: $(-3,1)$

Graph:

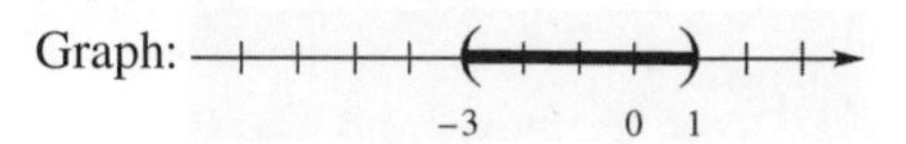

71. $10\le 2x+4\le 16$

$$10-4\le 2x+4-4\le 16-4$$
$$6\le 2x\le 12$$
$$\frac{6}{2}\le\frac{2x}{2}\le\frac{12}{2}$$
$$3\le x\le 6$$

Solution set: $[3,6]$

Graph:

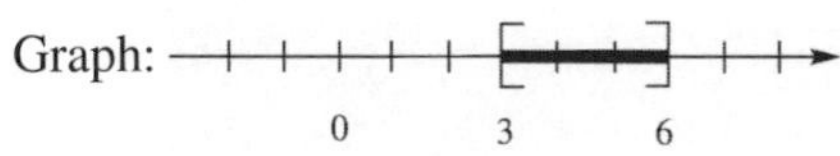

72.
$$-6 \le 6x+3 \le 21$$
$$-6-3 \le 6x+3-3 \le 21-3$$
$$-9 \le 6x \le 18$$
$$\frac{-9}{6} \le \frac{6x}{6} \le \frac{18}{6}$$
$$-\tfrac{3}{2} \le x \le 3$$

Solution set: $\left[-\tfrac{3}{2}, 3\right]$

Graph:

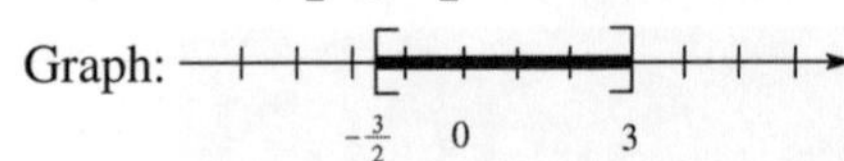

73.

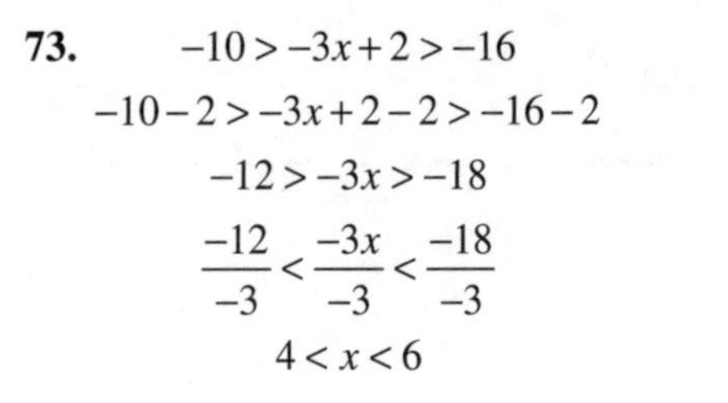

$$-10 > -3x+2 > -16$$
$$-10-2 > -3x+2-2 > -16-2$$
$$-12 > -3x > -18$$
$$\frac{-12}{-3} < \frac{-3x}{-3} < \frac{-18}{-3}$$
$$4 < x < 6$$

Solution set: $(4, 6)$

Graph: (number line with labels 0, 4, 6)

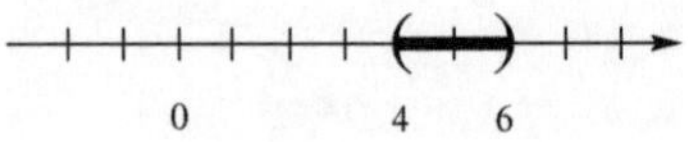

74.
$$4 > -6x+5 > -1$$
$$4-5 > -6x+5-5 > -1-5$$
$$-1 > -6x > -6$$
$$\frac{-1}{-6} < \frac{-6x}{-6} < \frac{-6}{-6}$$
$$\tfrac{1}{6} < x < 1$$

Solution set: $\left(\tfrac{1}{6}, 1\right)$

Graph: (number line with labels 0, $\frac{1}{6}$, 1)

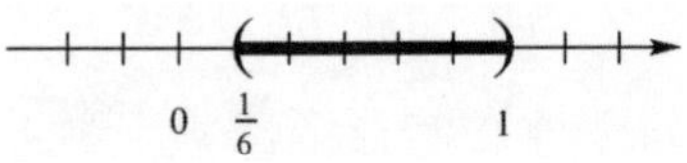

75.
$$-4 \le \frac{x+1}{2} \le 5$$
$$2(-4) \le 2\left(\frac{x+1}{2}\right) \le 2(5)$$
$$-8 \le x+1 \le 10$$
$$-8-1 \le x+1-1 \le 10-1$$
$$-9 \le x \le 9$$

Solution set: $[-9, 9]$

Graph: (number line with labels −9, 0, 9)

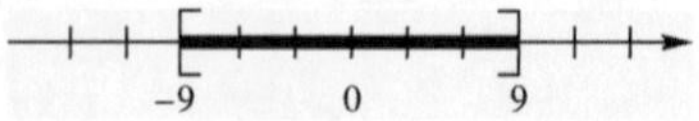

76.

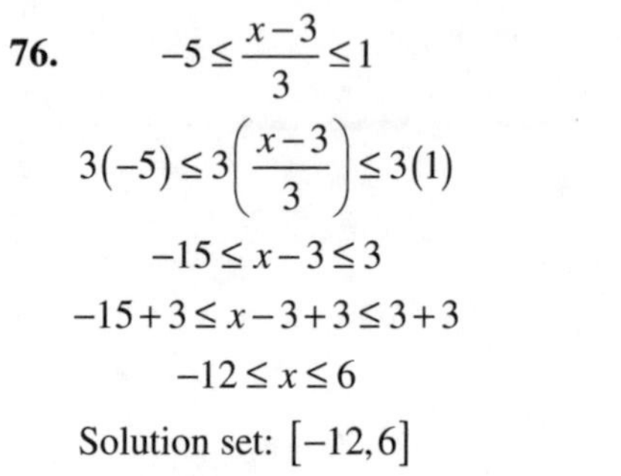

$$-5 \le \frac{x-3}{3} \le 1$$
$$3(-5) \le 3\left(\frac{x-3}{3}\right) \le 3(1)$$
$$-15 \le x-3 \le 3$$
$$-15+3 \le x-3+3 \le 3+3$$
$$-12 \le x \le 6$$

Solution set: $[-12, 6]$

Graph: (number line with labels −12, 0, 6)

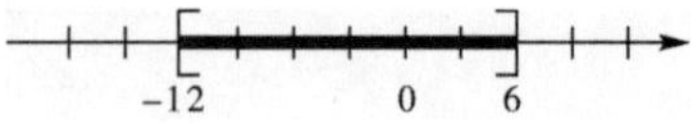

Appendix B
GRAPHS OF EQUATIONS

1. $(3,2)$

2. $(-7,6)$

3. $(-7,-4)$

4. $(8,-5)$

5. $(0,5)$

6. $(-8,0)$

7. $(4.5,7)$

8. $(-7.5,8)$

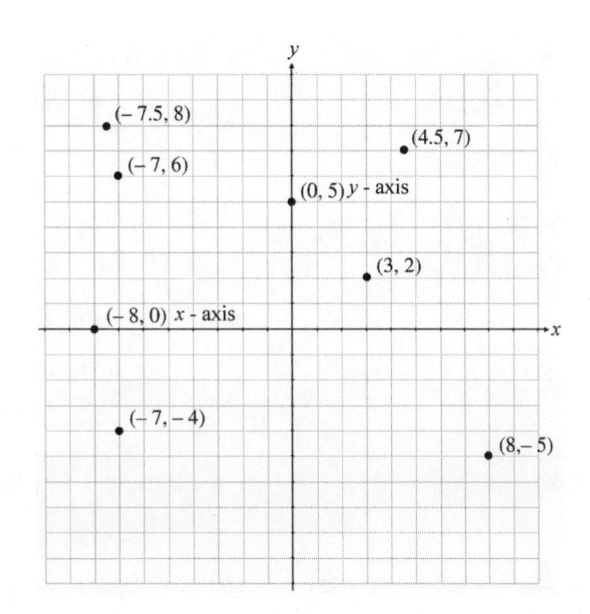

9. $(-5,1.25)$ has a negative x-coordinate and a positive y-coordinate. The point lies is quadrant II.

10. $(\pi,-3)$ has a positive x-coordinate and a negative y-coordinate. The point lies is quadrant IV.

11. $(-1.4,-2.8)$ has a negative x-coordinate and a negative y-coordinate. The point lies is quadrant III.

12. $\left(1+\sqrt{3},\frac{1}{2}\right)$ has a positive x-coordinate and a positive y-coordinate. The point lies is quadrant I.

13. Answers will vary.
If (a,b) is in the first quadrant, then $(a,-b)$ is in the fourth quadrant. To locate this new point, move the point (a,b), $2b$ units downward.

14. Answers will vary.
If (a,b) is in the first quadrant, then $(-a,b)$ is in the second quadrant. To locate this new point, move the point (a,b), $2b$ units to the left.

15. **(a)** Since (a,b) lies in quadrant II, $a<0, b>0$. Therefore $-a>0$ and $(-a,b)$ would lie in quadrant I.

(b) Since (a,b) lies in quadrant II, $a<0, b>0$. Therefore $-a>0, -b<0$, and $(-a,-b)$ would lie in quadrant IV.

(c) Since (a,b) lies in quadrant II, $a<0, b>0$. Therefore $-b<0$ and $(a,-b)$ would lie in quadrant III.

(d) Since (a,b) lies in quadrant II, $a<0, b>0$. The point (b,a) has x-coordinate positive and y-coordinate negative Therefore, the point would lie in quadrant IV.

16. Since $xy=1$, and 1 is a positive number attained by the product of two variables x and y, then either x and y are both positive (quadrant I), or both negative (quadrant III). The graph of $xy=1$ will lie in quadrants I and III.

17. $(a,b,c)=(9,12,15)$

$$\begin{aligned} a^2+b^2 &= 9^2+12^2 \\ &= 81+144 \\ &= 225 = 15^2 = c^2 \end{aligned}$$

The triple is a Pythagorean triple since $a^2+b^2=c^2$.

18. $(a,b,c)=(6,8,10)$

$$\begin{aligned} a^2+b^2 &= 6^2+8^2 \\ &= 36+64 \\ &= 100 = 10^2 = c^2 \end{aligned}$$

The triple is a Pythagorean triple since $a^2+b^2=c^2$.

19. $(a,b,c)=(5,10,15)$

$$\begin{aligned} a^2+b^2 &= 5^2+10^2 \\ &= 25+100 \\ &= 125 \neq 225 = 15^2 = c^2 \end{aligned}$$

The triple is not a Pythagorean triple since $a^2+b^2 \neq c^2$.

20. $(a,b,c)=(7,24,25)$

$$\begin{aligned} a^2+b^2 &= 7^2+24^2 \\ &= 49+576 \\ &= 625 = 25^2 = c^2 \end{aligned}$$

The triple is a Pythagorean triple since $a^2+b^2=c^2$.

21. By the Pythagorean theorem, we have $x^2=\left(\frac{34.5}{2}\right)^2+(55.4)^2$. Solving for x we have the following.

$$x^2=\left(\tfrac{34.5}{2}\right)^2+(55.4)^2 \Rightarrow x^2=(17.25)^2+(55.4)^2 \Rightarrow x^2=297.5625+3069.16$$

$$x^2=3366.7225 \Rightarrow x=\sqrt{3366.7225}\approx 58.02 \text{ ft}$$

22. Let d = the diagonal.

By the Pythagorean theorem, we have $d^2=5^2+12^2$. Solving for d we have the following.

$$d^2=5^2+12^2 \Rightarrow d^2=25+144 \Rightarrow d^2=169 \Rightarrow d=\sqrt{169}=13 \text{ ft}$$

23. $AC^2+BC^2=AB^2$

$$(1000)^2+BC^2=(1000.5)^2 \Rightarrow 1{,}000{,}000+BC^2=1{,}001{,}000.25$$

$$BC^2=1000.25 \Rightarrow BC=\sqrt{1000.25}\approx 31.6 \text{ ft}$$

24. To find the length AB, apply the Pythagorean theorem to triangle ABX.

$$AB^2 = AX^2 + BX^2 \Rightarrow AB^2 = 8^2 + 8^2 \Rightarrow AB^2 = 64 + 64$$
$$AB^2 = 128 \Rightarrow AB = \sqrt{128} \approx 11.3137$$

The distance across the top of the fabric is $20 \cdot AB$ (since there are 20 strips) plus 42 in. for the side of the unused triangle at the end. (Since it is a $45° - 45°$ right triangle, it is isosceles and its two equal sides are 42 in.) Thus, the distance across the top is as follows.

$$20(11.3137) + 42 = 226.274 + 42 = 268.274 \text{ in.}$$

Since the price of the material is \$10 per linear yard, convert from inches to yards to find the number of linear yards. Since there are 36 inches in 1 yard we have $\frac{268.274}{36} = 7.452$ yd.

The cost of the material is therefore, $7.452(10) = \$74.52$.

25. Let $(x_1, y_1) = (5,6)$. Let $(x_2, y_2) = (5,0)$ since we want the point on the x-axis with x-value 5.

$$d = \sqrt{(5-5)^2 + [0-(-6)]^2} = \sqrt{0^2 + 6^2} = \sqrt{0+36} = \sqrt{36} = 6$$

26. Let $(x_1, y_1) = (5,-6)$. Let $(x_2, y_2) = (0,-6)$ since we want the point on the y-axis with y-value –6.

$$d = \sqrt{(0-5)^2 + [-6-(-6)]^2} = \sqrt{(-5)^2 + 0^2} = \sqrt{25+0} = \sqrt{25} = 5$$

27. $P(-5, -7)$, $Q(-13, 1)$

(a) $d(P, Q) = \sqrt{[-13-(-5)]^2 + [1-(-7)]^2} = \sqrt{(-8)^2 + 8^2} = \sqrt{128} = 8\sqrt{2}$

(b) The midpoint M of the segment joining points P and Q has the following coordinates.

$$\left(\frac{-5+(-13)}{2}, \frac{-7+1}{2}\right) = \left(\frac{-18}{2}, \frac{-6}{2}\right) = (-9, -3)$$

28. $P(-4, 3)$, $Q(2, -5)$

(a) $d(P, Q) = \sqrt{[2-(-4)]^2 + (-5-3)^2} = \sqrt{6^2 + (-8)^2} = \sqrt{100} = 10$

(b) The midpoint M of the segment joining points P and Q has the following coordinates.

$$\left(\frac{-4+2}{2}, \frac{3+(-5)}{2}\right) = \left(\frac{-2}{2}, \frac{-2}{2}\right) = (-1, -1)$$

29. $P(8, 2)$, $Q(3, 5)$

(a) $d(P, Q) = \sqrt{(3-8)^2 + (5-2)^2} = \sqrt{(-5)^2 + 3^2} = \sqrt{25+9} = \sqrt{34}$

(b) The midpoint M of the segment joining points P and Q has coordinates $\left(\frac{8+3}{2}, \frac{2+5}{2}\right) = \left(\frac{11}{2}, \frac{7}{2}\right)$.

30. $P(-6, -5)$, $Q(6, 10)$

(a) $d(P, Q) = \sqrt{[6-(-6)]^2 + [10-(-5)]^2} = \sqrt{12^2 + 15^2} = \sqrt{144+225} = \sqrt{369} = 3\sqrt{41}$

(b) The midpoint M of the segment joining points P and Q has the following coordinates.

$$\left(\frac{-6+6}{2}, \frac{-5+10}{2}\right) = \left(\frac{0}{2}, \frac{5}{2}\right) = \left(0, \frac{5}{2}\right)$$

31. $P(-8,4),\ Q(3,-5)$

(a) $d(P,Q)=\sqrt{[3-(-8)]^2+(-5-4)^2}=\sqrt{11^2+(-9)^2}=\sqrt{121+81}=\sqrt{202}$

(b) The midpoint M of the segment joining points P and Q has the following coordinates.

$$\left(\frac{-8+3}{2},\frac{4+(-5)}{2}\right)=\left(-\frac{5}{2},-\frac{1}{2}\right)$$

32. $P(6,-2),\ Q(4,6)$

(a) $d(P,Q)=\sqrt{(4-6)^2+[6-(-2)]^2}=\sqrt{(-2)^2+8^2}=\sqrt{4+64}=\sqrt{68}=2\sqrt{17}$

(b) The midpoint M of the segment joining points P and Q has the following coordinates.

$$\left(\frac{6+4}{2},\frac{-2+6}{2}\right)=\left(\frac{10}{2},\frac{4}{2}\right)=(5,2)$$

33. $P(3\sqrt{2},4\sqrt{5}),\ Q(\sqrt{2},-\sqrt{5})$

(a) $d(P,Q)=\sqrt{\left(\sqrt{2}-3\sqrt{2}\right)^2+\left(-\sqrt{5}-4\sqrt{5}\right)^2}=\sqrt{\left(-2\sqrt{2}\right)^2+\left(-5\sqrt{5}\right)^2}=\sqrt{8+125}=\sqrt{133}$

(b) The midpoint M of the segment joining points P and Q has the following coordinates.

$$\left(\frac{3\sqrt{2}+\sqrt{2}}{2},\frac{4\sqrt{5}+\left(-\sqrt{5}\right)}{2}\right)=\left(\frac{4\sqrt{2}}{2},\frac{3\sqrt{5}}{2}\right)=\left(2\sqrt{2},\frac{3\sqrt{5}}{2}\right)$$

34. $P(-\sqrt{7},8\sqrt{3}),Q(5\sqrt{7},-\sqrt{3})$

(a) $d(P,Q)=\sqrt{\left[5\sqrt{7}-\left(-\sqrt{7}\right)\right]^2+\left(-\sqrt{3}-8\sqrt{3}\right)^2}=\sqrt{(6\sqrt{7})^2+(-9\sqrt{3})^2}=\sqrt{252+243}=\sqrt{495}=3\sqrt{55}$

(b) The midpoint M of the segment joining points P and Q has the following coordinates.

$$\left(\frac{-\sqrt{7}+5\sqrt{7}}{2},\frac{8\sqrt{3}+\left(-\sqrt{3}\right)}{2}\right)=\left(\frac{4\sqrt{7}}{2},\frac{7\sqrt{3}}{2}\right)=\left(2\sqrt{7},\frac{7\sqrt{3}}{2}\right)$$

35. Find x such that the distance between $(x,7)$ and $(2,3)$ is 5. Use the distance formula, and solve for x.

$$5=\sqrt{(2-x)^2+(3-7)^2}$$
$$5=\sqrt{(2-x)^2+(-4)^2}$$
$$5=\sqrt{(2-x)^2+16}$$
$$25=(2-x)^2+16$$
$$9=(2-x)^2$$
$$\pm3=2-x$$
$$3=2-x \quad\text{or}\quad -3=2-x$$
$$1=-x \qquad -5=-x$$
$$x=-1 \qquad x=5$$

Therefore, x is -1 or 5.

36. Find y such that the distance between $(5,y)$ and $(8,-1)$ is 5. Use the distance formula, and solve for y.

$$5=\sqrt{(8-5)^2+(-1-y)^2}$$
$$5=\sqrt{3^2+(-1-y)^2}$$
$$5=\sqrt{9+(-1-y)^2}$$
$$25=9+(-1-y)^2$$
$$16=(-1-y)^2$$
$$\pm4=-1-y$$
$$4=-1-y \quad\text{or}\quad -4=-1-y$$
$$5=-y \qquad -3=-y$$
$$y=-5 \qquad y=3$$

Therefore, y is -5 or 3.

37. Find y such that the distance between $(3, y)$ and $(-2,9)$ is 5. Use the distance formula to solve for y.

$$12=\sqrt{(-2-3)^2+(9-y)^2} \Rightarrow 12=\sqrt{(-5)^2+(-9-y)^2} \Rightarrow 12=\sqrt{25+(-9-y)^2}$$

$$144=25+(-9-y)^2 \Rightarrow 119=(-9-y)^2 \Rightarrow \pm\sqrt{119}=-9-y$$

$$\begin{array}{lcl} \sqrt{119}=9-y & \text{or} & -\sqrt{119}=9-y \\ -9+\sqrt{119}=-y & & -\sqrt{119}-9=-y \\ y=9-\sqrt{119} & & y=9+\sqrt{119} \end{array}$$

Therefore, y is $9-\sqrt{119}$ or $9+\sqrt{119}$.

38. Find x such that the distance between $(x,11)$ and $(5,-4)$ is 17. Use the distance formula, and solve for x.

$$17=\sqrt{(5-x)^2+(-4-11)^2} \Rightarrow 17=\sqrt{(5-x)^2+(-15)^2} \Rightarrow 17=\sqrt{(5-x)^2+225}$$

$$289=(5-x)^2+225 \Rightarrow 64=(5-x)^2 \Rightarrow \pm 8=5-x$$

$$\begin{array}{lcl} 8=5-x & \text{or} & -8=5-x \\ 3=-x & & -13=-x \\ x=-3 & & x=13 \end{array}$$

Therefore, x is -3 or 13.

39. Let $P=(x,y)$ be a point 5 units from $(0,0)$. Then, by the distance formula $5=\sqrt{(x-0)^2+(y-0)^2} \Rightarrow 5=\sqrt{x^2+y^2} \Rightarrow 25=x^2+y^2$. This is the required equation. The graph is a circle with center $(0,0)$ and radius 5.

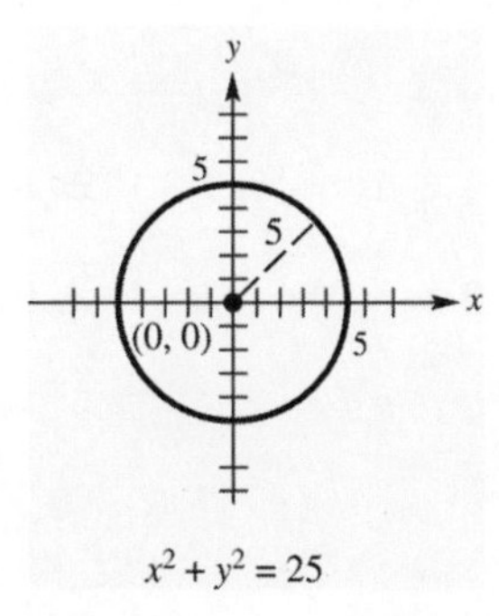

40. Let $P=(x,y)$ be a point 3 units from $(-5,6)$. Then, by the distance formula $3=\sqrt{[x-(-5)]^2+(y-6)^2} \Rightarrow 3=\sqrt{(x+5)^2+(y-6)^2} \Rightarrow 9=(x+5)^2+(y-6)^2$. This is the required equation. The graph is a circle with center $(-5,6)$ and radius 3.

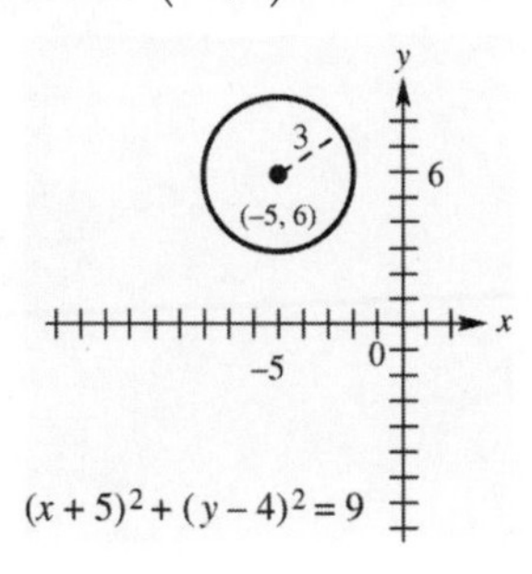

41. The points to use would be (1982, 79.1) and (2002, 69.3). Their midpoint is the following.

$$\left(\frac{1982+2002}{2}, \frac{79.1+69.3}{2}\right) = \left(\frac{3984}{2}, \frac{148.4}{2}\right) = (1992, 74.2)$$

The estimate is 74.2%. This is 1.1% less than the actual percent of 75.3.

42. The midpoint between (1980, 4.5) and (1990, 5.2) is $\left(\frac{1980+1990}{2}, \frac{4.5+5.2}{2}\right) = (1985, 4.85)$. In 1985, the enrollment was 4.85 million. The midpoint between (1990, 5.2) and (2000, 5.8) is $\left(\frac{1990+2000}{2}, \frac{5.2+5.8}{2}\right) = (1995, 5.5)$. In 1995, the enrollment was 5.5 million.

43. **(a)**

x	y	
0	−2	y-intercept: $x=0 \Rightarrow 6y=3(0)-12 \Rightarrow 6y=-12 \Rightarrow y=-2$
4	0	x-intercept: $y=0 \Rightarrow 6(0)=3x-12 \Rightarrow 0=3x-12 \Rightarrow 12=3x \Rightarrow 4=x$
2	−1	additional point

(b)

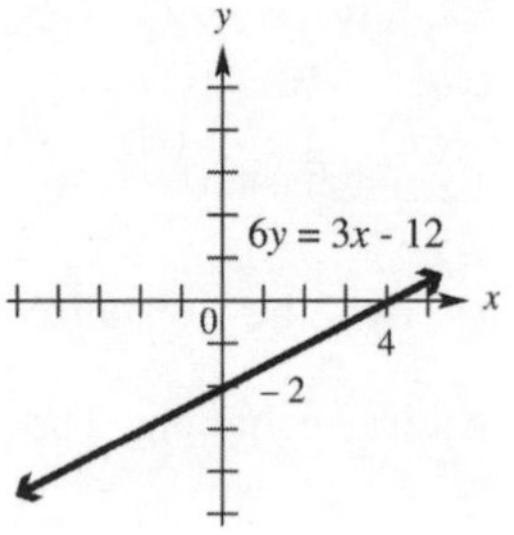

$6y = 3x - 12$

44. **(a)**

x	y	
0	3	y-intercept: $x=0 \Rightarrow 6y=-6(0)+18 \Rightarrow 6y=18 \Rightarrow y=3$
3	0	x-intercept: $y=0 \Rightarrow 6(0)=-6x+18 \Rightarrow 0=-6x+18 \Rightarrow 6x=18 \Rightarrow x=3$
1	2	additional point

(b)

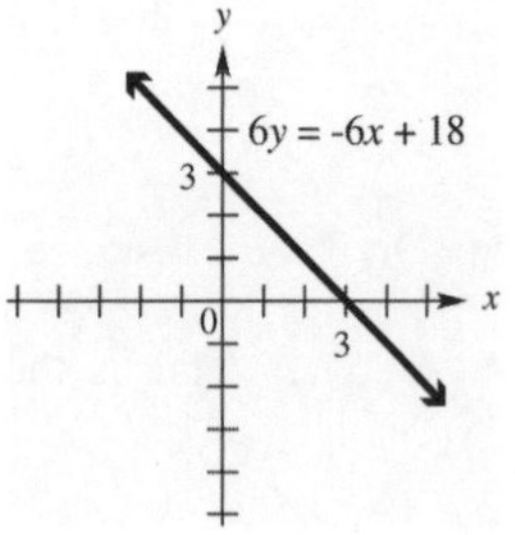

$6y = -6x + 18$

45. (a)

x	y	
0	0	x- and y-intercept: $0 = 0^2$
1	1	additional point
−2	4	additional point

(b)

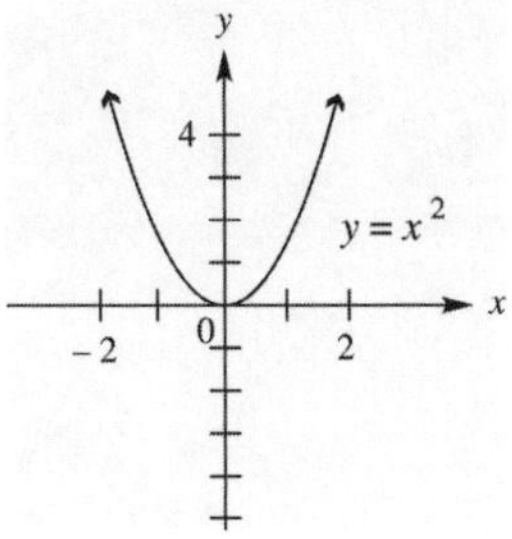

46. (a)

x	y	
0	2	y-intercept: $x = 0 \Rightarrow y = 0^2 + 2 \Rightarrow y = 0 + 2 \Rightarrow y = 2$
−1	3	additional point
2	6	additional point

no x-intercept: $y = 0 \Rightarrow 0 = x^2 + 2 \Rightarrow -2 = x^2 \Rightarrow \pm\sqrt{-2} = x$

(b)

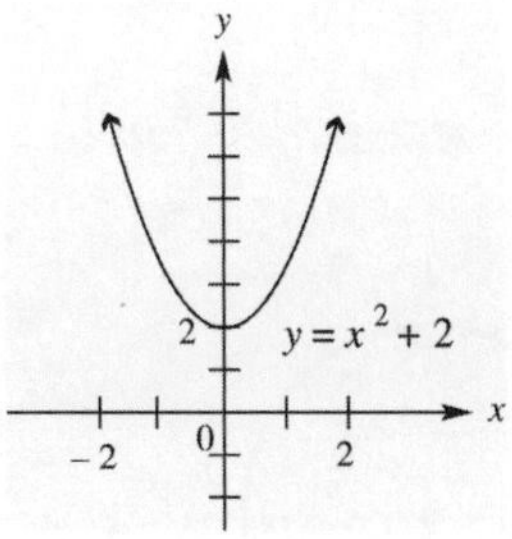

47. (a)

x	y	
0	0	x- and y-intercept: $0 = 0^3$
−1	−1	additional point
2	8	additional point

(b)

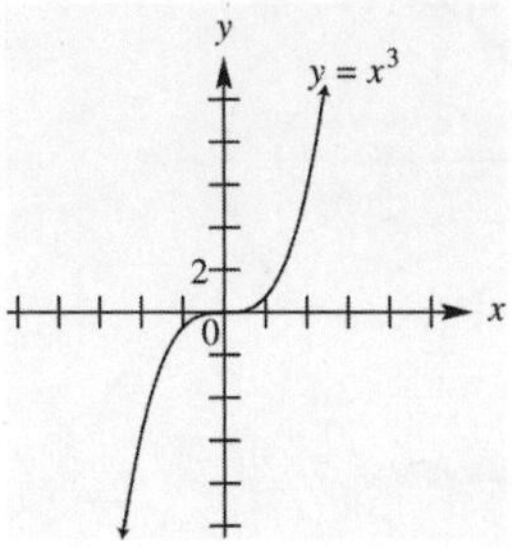

48. (a)

x	y	
0	0	x- and y-intercept: $0 = -0^3$
1	−1	additional point
2	−8	additional point

(b)

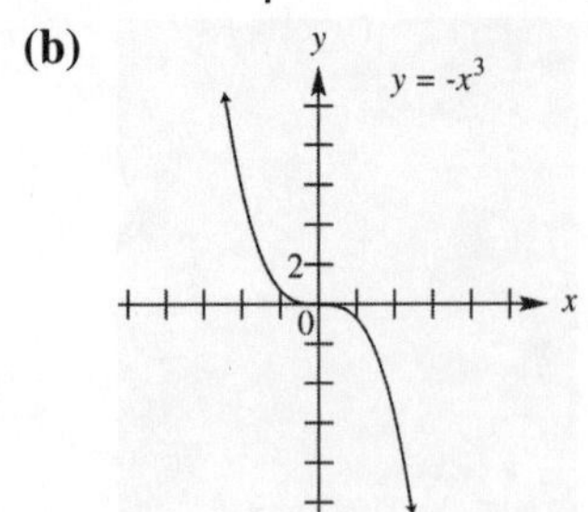

49. (a) Center (0, 0), radius 6

$$\sqrt{(x-0)^2+(y-0)^2}=6$$
$$(x-0)^2+(y-0)^2=6^2$$
$$x^2+y^2=36$$

(b)

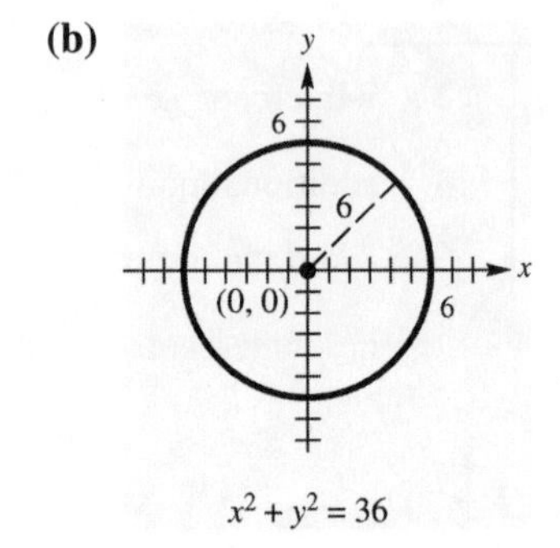

$x^2 + y^2 = 36$

50. (a) Center (0, 0), radius 9

$$\sqrt{(x-0)^2+(y-0)^2}=9$$
$$(x-0)^2+(y-0)^2=9^2$$
$$x^2+y^2=81$$

(b)

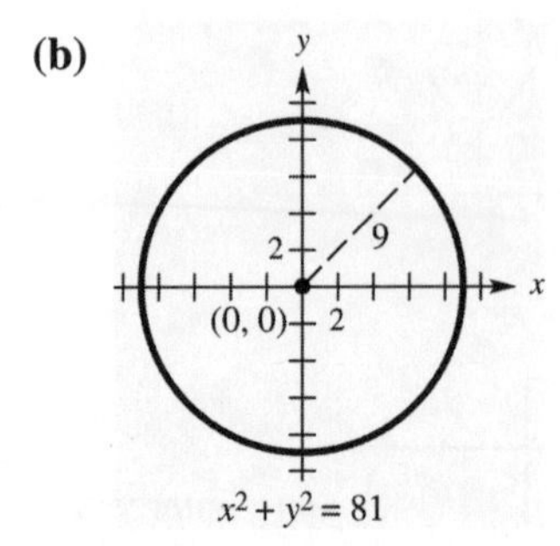

$x^2 + y^2 = 81$

51. (a) Center (2, 0), radius 6

$$\sqrt{(x-2)^2+(y-0)^2}=6$$
$$(x-2)^2+(y-0)^2=6^2$$
$$(x-2)^2+y^2=36$$

(b)

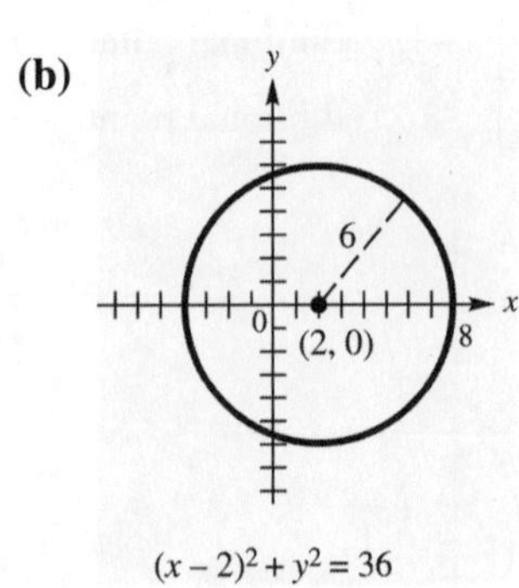

$(x-2)^2 + y^2 = 36$

52. (a) Center (0, –3), radius 7

$$\sqrt{(x-0)^2+[y-(-3)]^2}=7$$
$$(x-0)^2+[y-(-3)]^2=7^2$$
$$x^2+(y+3)^2=49$$

(b)

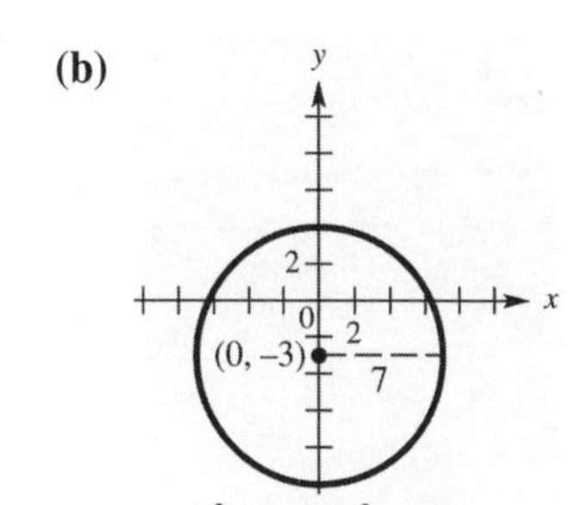

$x^2 + (y + 3)^2 = 49$

53. (a) Center (–2, 5), radius 4

$$\sqrt{[x-(-2)]^2+(y-5)^2}=4$$
$$[x-(-2)]^2+(y-5)^2=4^2$$
$$(x+2)^2+(y-5)^2=16$$

(b)

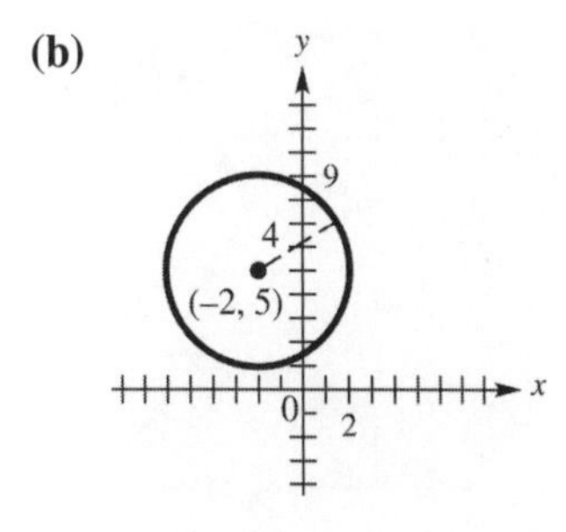

$(x + 2)^2 + (y - 5)^2 = 16$

54. (a) Center (4, 3), radius 5

$$\sqrt{(x-4)^2+(y-3)^2}=5$$
$$(x-4)^2+(y-3)^2=5^2$$
$$(x-4)^2+(y-3)^2=25$$

(b)

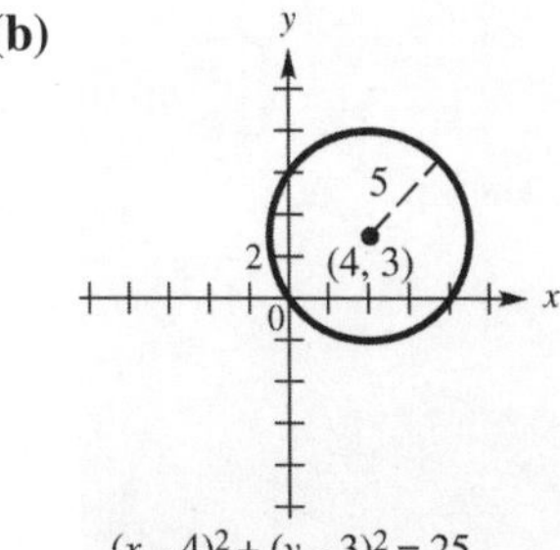

$(x - 4)^2 + (y - 3)^2 = 25$

55. The radius of this circle is the distance from the center $C(3, 2)$ to the x-axis. This distance is 2, so $r = 2$.

$$(x-3)^2+(y-2)^2=2^2 \Rightarrow (x-3)^2+(y-2)^2=4$$

56. The radius is the distance from the center $C(-4, 3)$ to the point $P(5, 8)$.

$$r=\sqrt{[5-(-4)]^2+(8-3)^2}=\sqrt{9^2+5^2}=\sqrt{106}$$

The equation of the circle is $[x-(-4)]^2+(y-3)^2=(\sqrt{106})^2 \Rightarrow (x+4)^2+(y-3)^3=106.$

Appendix C
FUNCTIONS

1. Answers will vary.

2. Answers will vary.

3. The graphs in (a) and (c), and (d) represent function. The graph in (b) fails the vertical line test; it is not the graph of a function.

4. One example is $\{(-3,4),(2,4),(2,6),(6,4)\}$.

5. The relation is a function because for each different x-value there is exactly one y-value. This correspondence can be shown as follows.

{5, 3, 4, 9} x-values

{1, 2, 9, 6} y-values

6. The relation is a function because for each different x-value there is exactly one y-value. This correspondence can be shown as follows.

{8, 5, 9, 3} x-values

{0, 4, 3, 8} y-values

7. Two ordered pairs, namely $(2,4)$ and $(2,5)$, have the same x-value paired with different y-values, so the relation is not a function.

8. Two ordered pairs, namely $(9,-2)$ and $(9,2)$, have the same x-value paired with different y-values, so the relation is not a function.

9. The relation is a function because for each different x-value there is exactly one y-value. This correspondence can be shown as follows.

{-3, 4, -2} x-values

{1, 7} y-values

10. The relation is a function because for each different x-value there is exactly one y-value. This correspondence can be shown as follows.

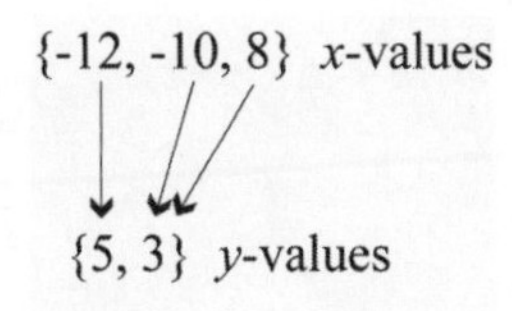

11. Two sets of ordered pairs, namely $(1,1)$ and $(1,-1)$ as well as $(2,4)$ and $(2,-4)$, have the same x-value paired with different y-values, so the relation is not a function.

domain: $\{0,1,2\}$; range: $\{-4,-1,0,1,4\}$

12. The relation is a function because for each different x-value there is exactly one y-value. This correspondence can be shown as follows.

{2, 3, 4, 5} x-values

{5, 7, 9, 11} y-values

domain: $\{2,3,4,5\}$; range: $\{5,7,9,11\}$

13. The relation is a function because for each different x-value there is exactly one y-value. This correspondence can be shown as follows.

{2, 5, 11, 17} x-values

{1, 7, 20} y-values

domain: $\{2,5,11,17\}$; range: $\{1,7,20\}$

14. The relation is not a function because for each different x-value there is not exactly one y-value. This correspondence can be shown as follows.

{1, 2, 3} x-values

{10, 15, 20, 25} y-values

domain: $\{1,2,3\}$; range: $\{10,15,20,25\}$

15. This graph represents a function. If you pass a vertical line through the graph, one x-value corresponds to only one y-value.

domain: $(-\infty,\infty)$; range: $(-\infty,\infty)$

16. This graph represents a function. If you pass a vertical line through the graph, one x-value corresponds to only one y-value.

domain: $(-\infty,\infty)$; range: $(-\infty,4]$

17. This graph does not represent a function. If you pass a vertical line through the graph, there are places where one value of x corresponds to two values of y.

domain: $[3,\infty)$; range: $(-\infty,\infty)$

18. This graph represents a function. If you pass a vertical line through the graph, one x-value corresponds to only one y-value.

domain: $(-\infty,\infty)$; range: $(-\infty,\infty)$

19. This graph does not represent a function. If you pass a vertical line through the graph, there are places where one value of x corresponds to two values of y.

domain: $[-4,4]$; range: $[-3,3]$

20. This graph represents a function. If you pass a vertical line through the graph, one x-value corresponds to only one y-value.

domain: $[-2,2]$; range: $[0,4]$

21. $y = x^2$ represents a function since y is always found by squaring x. Thus, each value of x corresponds to just one value of y. x can be any real number. Since the square of any real number is not negative, the range would be zero or greater.

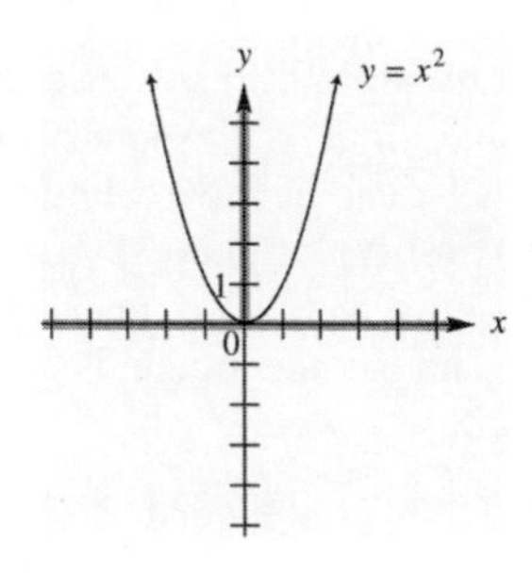

domain: $(-\infty, \infty)$; range: $[0, \infty)$

22. $y = x^3$ represents a function since y is always found by cubing x. Thus, each value of x corresponds to just one value of y. x can be any real number. Since the cube of any real number could be negative, positive, or zero, the range would be any real number.

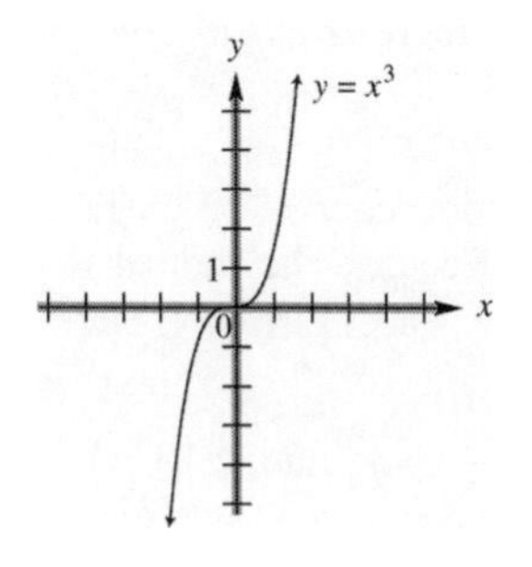

domain: $(-\infty, \infty)$; range: $(-\infty, \infty)$

23. The ordered pairs $(1,1)$ and $(1,-1)$ both satisfy $x = y^6$. This equation does not represent a function. Because x is equal to the sixth power of y, the values of x are nonnegative. Any real number can be raised to the sixth power, so the range of the relation is all real numbers.

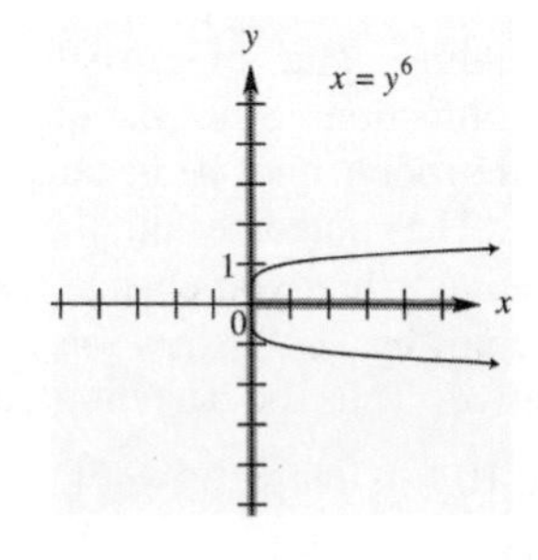

domain: $[0, \infty)$ range: $(-\infty, \infty)$

24. The ordered pairs $(1,1)$ and $(1,-1)$ both satisfy $x = y^4$. This equation does not represent a function. Because x is equal to the fourth power of y, the values of x are nonnegative. Any real number can be raised to the fourth power, so the range of the relation is all real numbers.

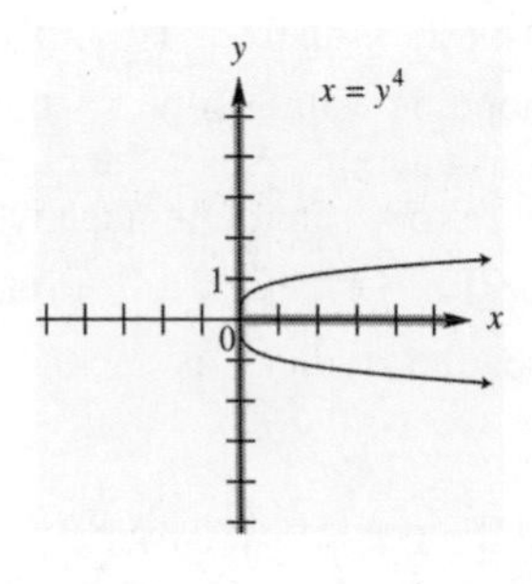

domain: $[0, \infty)$ range: $(-\infty, \infty)$

25. $y = 2x - 6$ represents a function since y is found by multiplying x by 2 and subtracting 6. Each value of x corresponds to just one value of y. x can be any real number, so the domain is all real numbers. Since y is twice x, less 6, y also may be any real number, and so the range is also all real numbers.

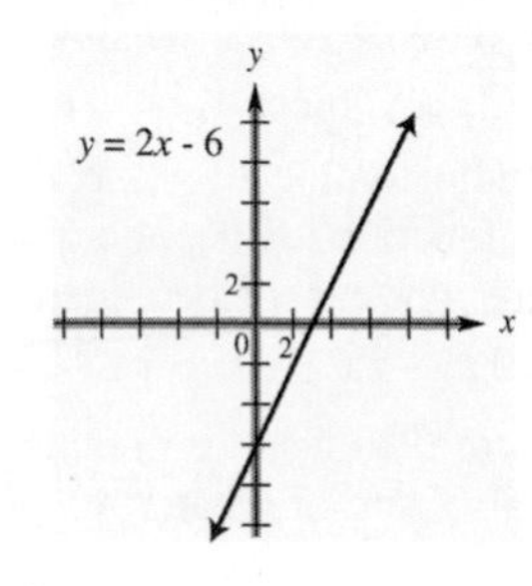

domain: $(-\infty, \infty)$; range: $(-\infty, \infty)$

26. $y = -6x + 8$ represents a function since y is found by multiplying x by -6 and adding 8. Each value of x corresponds to just one value of y. x can be any real number, so the domain is all real numbers. Since y is -6 times x, plus 8, y also may be any real number, and so the range is also all real numbers.

domain: $(-\infty, \infty)$; range: $(-\infty, \infty)$

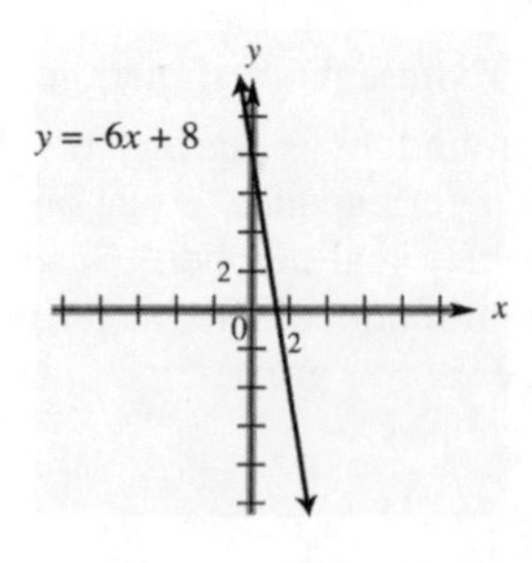

27. For any choice of x in the domain of $y = \sqrt{x}$, there is exactly one corresponding value of y, so this equation defines a function. Since the quantity under the square root cannot be negative, we have $x \geq 0$. Because the radical is nonnegative, the range is also zero or greater.

domain: $[0, \infty)$; range: $[0, \infty)$

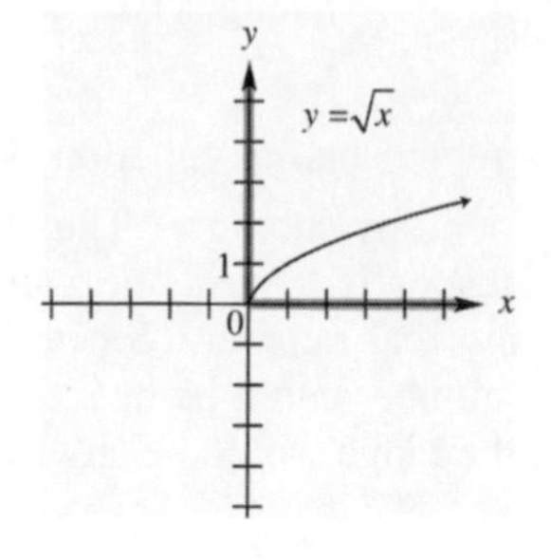

28. For any choice of x in the domain of $y = -\sqrt{x}$, there is exactly one corresponding value of y, so this equation defines a function. Since the quantity under the square root cannot be negative, we have $x \geq 0$. The outcome of the radical is nonnegative, when you change the sign (by multiplying by -1), the range becomes nonpositive. Thus the range is zero or less.

domain: $[0, \infty)$; range: $(-\infty, 0]$

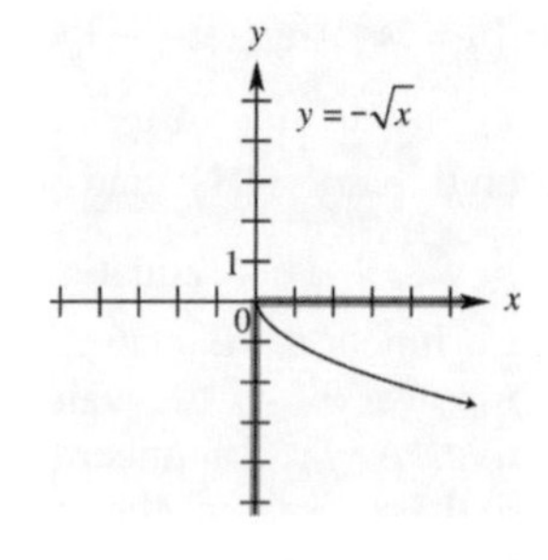

29. For any choice of x in the domain of $y = \sqrt{4x + 2}$, there is exactly one corresponding value of y, so this equation defines a function. Since the quantity under the square root cannot be negative, we have $4x + 2 \geq 0 \Rightarrow 4x \geq -2 \Rightarrow x \geq \frac{-2}{4}$ or $x \geq -\frac{1}{2}$. Because the radical is nonnegative, the range is also zero or greater.

domain: $\left[-\frac{1}{2}, \infty\right)$; range: $[0, \infty)$

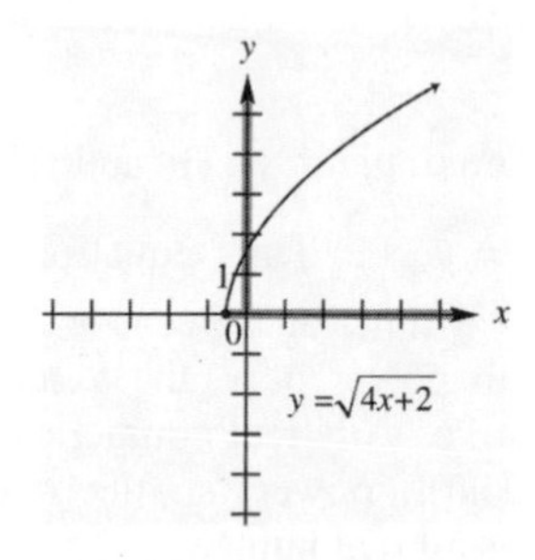

30. For any choice of x in the domain of $y = \sqrt{9 - 2x}$, there is exactly one corresponding value of y, so this equation defines a function. Since the quantity under the square root cannot be negative, we have $9 - 2x \geq 0 \Rightarrow -2x \geq -9 \Rightarrow x \leq \frac{-9}{-2}$ or $x \leq \frac{9}{2}$. Because the radical is nonnegative, the range is also zero or greater.

domain: $\left(-\infty, \frac{9}{2}\right]$; range: $[0, \infty)$

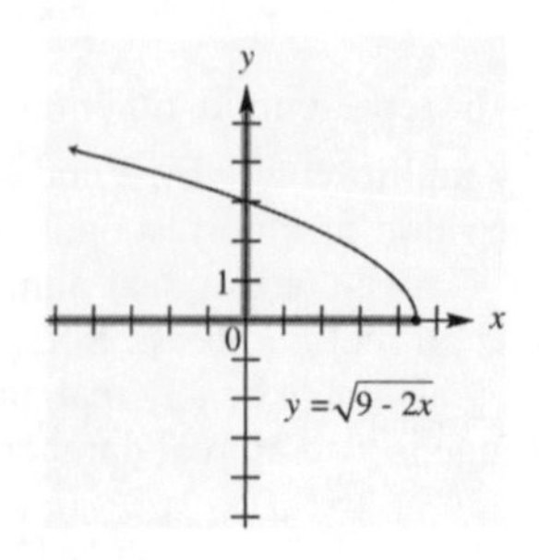

31. Given any value in the domain of $y = \frac{2}{x-9}$, we find y by subtracting 9, then dividing into 2. This process produces one value of y for each value of x in the domain, so this equation is a function. The domain includes all real numbers except those that make the denominator equal to zero, namely $x = 9$. Values of y can be negative or positive, but never zero. Therefore, the range will be all real numbers except zero.

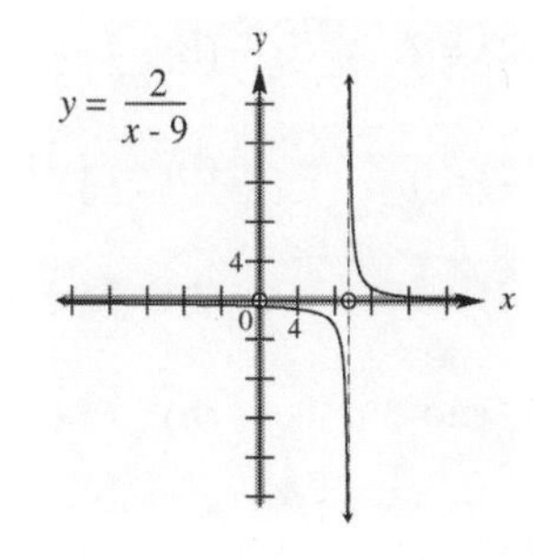

domain: $(-\infty, 9) \cup (9, \infty)$; range: $(-\infty, 0) \cup (0, \infty)$

32. Given any value in the domain of $y = \frac{-7}{x-16}$, we find y by subtracting 16, then dividing into −7. This process produces one value of y for each value of x in the domain, so this equation is a function. The domain includes all real numbers except those that make the denominator equal to zero, namely $x = 16$. Values of y can be negative or positive, but never zero. Therefore, the range will be all real numbers except zero.

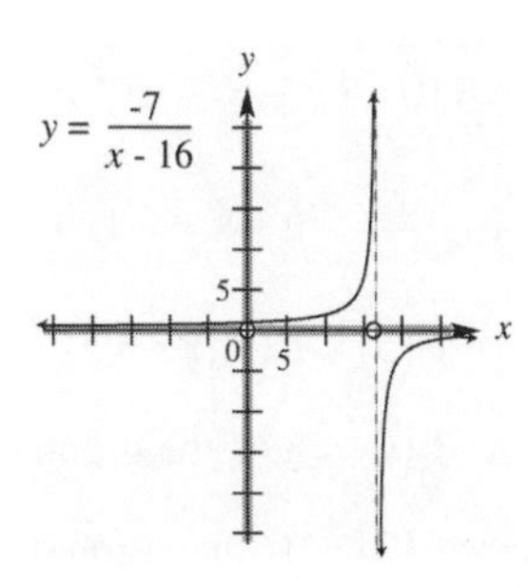

domain: $(-\infty, 16) \cup (16, \infty)$; range: $(-\infty, 0) \cup (0, \infty)$

33. B

34. Answers will vary.
An example is: The cost of gasoline depends on the number of gallons used; so cost is a function of number of gallons.

35. $f(x) = -3x + 4$
$f(0) = -3 \cdot 0 + 4 = 0 + 4 = 4$

36. $f(x) = -3x + 4$
$f(-3) = -3(-3) + 4 = 9 + 4 = 13$

37. $g(x) = -x^2 + 4x + 1$
$g(-2) = -(-2)^2 + 4(-2) + 1$
$= -4 + (-8) + 1 = -11$

38. $g(x) = -x^2 + 4x + 1$
$g(10) = -10^2 + 4 \cdot 10 + 1$
$= -100 + 40 + 1 = -59$

39. $f(x) = -3x + 4$
$f(p) = -3p + 4$

40. $g(x) = -x^2 + 4x + 1$
$g(k) = -k^2 + 4k + 1$

41. $f(x) = -3x + 4$
$f(-x) = -3(-x) + 4 = 3x + 4$

42. $g(x) = -x^2 + 4x + 1$
$g(-x) = -(-x)^2 + 4(-x) + 1$
$= -x^2 - 4x + 1$

43. $f(x) = -3x + 4$
$f(a+4) = -3(a+4) + 4$
$= -3a - 12 + 4 = -3a - 8$

44. $f(x) = -3x + 4$
$f(2m-3) = -3(2m-3) + 4$
$= -6m + 9 + 4 = -6m + 13$

45. **(a)** $f(2)=2$ **(b)** $f(-1)=3$

46. **(a)** $f(2)=5$ **(b)** $f(-1)=11$

47. **(a)** $f(2)=3$ **(b)** $f(-1)=-3$

48. **(a)** $f(2)=-3$ **(b)** $f(-1)=2$

49. **(a)** $f(-2)=0$ **(b)** $f(0)=4$
(c) $f(1)=2$ **(d)** $f(4)=4$

50. **(a)** $f(-2)=5$ **(b)** $f(0)=0$
(c) $f(1)=2$ **(d)** $f(4)=4$

51. **(a)** $f(0)=11$ **(b)** $f(6)=9$
(c) Since $f(-2)=0,\ a=-2$.
(d) Since $f(2)=f(7)=f(8)=10,\ x=2,7,8$.
(e) We have $(8,f(8))=(8,10)$ and $(10,f(10))=(10,0)$. Using the distance formula we have
$d=\sqrt{(10-8)^2+(0-10)^2}=\sqrt{2^2+(-10)^2}=\sqrt{4+100}=\sqrt{104}$.

52. **(a)** $f(2)=1.2$
(b) Since $f(5)=-2.4,\ x=5$.
(c) Since $f(0)=3.6$, the graph of f intersects the y-axis at $(0,3.6)$.
(d) Since $f(3)=0$, the graph of f intersects the x-axis at $(3,0)$.

53. **(a)** $[4,\infty)$ **(b)** $(-\infty,-1]$
(c) $[-1,4]$

54. **(a)** $(-\infty,1]$ **(b)** $[4,\infty)$
(c) $[1,4]$

55. **(a)** $(-\infty,4]$ **(b)** $[4,\infty)$
(c) none

56. **(a)** none **(b)** $(-\infty,\infty)$
(c) none

57. **(a)** none
(b) $(-\infty,-2]$; $[3,\infty)$
(c) $(-2,3)$

58. **(a)** $(3,\infty)$ **(b)** $(-\infty,-3)$
(c) $(-3,3]$

59. **(a)** yes
(b) $[0,24]$
(c) When $t=8,\ y=1200$ from the graph. At 8 A.M., approximately 1200 megawatts is being used.
(d) at 17 hr or 5 P.M.; at 4 A.M.
(e) $f(12)=2000$; At 12 noon, electricity use is 2000 megawatts.
(f) increasing from 4 A.M. to 5 P.M.; decreasing from midnight to 4 A.M. and from 5 P.M. to midnight

60. **(a)** At $t = 8$, $y = 24$ from the graph. Therefore, there are 24 units of the drug in the bloodstream at 8 hours.
(b) Level increases between 0 and 2 hours and decreases between 2 and 12 hours.
(c) The coordinates of the highest point are (2, 64). Therefore, at 2 hours, the level of the drug in the bloodstream reaches its greatest value of 64 units.
(d) After the peak, $y = 16$ at $t = 10$.
10 hours – 2 hours = 8 hours after the peak. 8 additional hours are required for the level to drop to 16 units.

Appendix D
GRAPHING TECHNIQUES

1. **(a)** B; $y=(x-7)^2$ is a shift of $y=x^2$, 7 units to the right.

(b) E; $y=x^2-7$ is a shift of $y=x^2$, 7 units downward.

(c) F; $y=7x^2$ is a vertical stretch of $y=x^2$, by a factor of 7.

(d) A; $y=(x+7)^2$ is a shift of $y=x^2$, 7 units to the left.

(e) D; $y=x^2+7$ is a shift of $y=x^2$, 7 units upward.

(f) C; $y=\frac{1}{7}x^2$ is a vertical shrink of $y=x^2$.

2. **(a)** E; $y=4x^2$ is a vertical stretch of $y=x^2$, by a factor of 4.

(b) C; $y=-x^2$ is a reflection of $y=x^2$, over the x-axis.

(c) D; $y=(-x)^2$ is a reflection of $y=x^2$, over the y-axis.

(d) A; $y=(x+4)^2$ is a shift of $y=x^2$, 4 units to the left.

(e) B; $y=x^2+4$ is a shift of $y=x^2$, 4 units up.

3. **(a)** B; $y=x^2+2$ is a shift of $y=x^2$, 2 units upward.

(b) A; $y=x^2-2$ is a shift of $y=x^2$, 2 units downward.

(c) G; $y=(x+2)^2$ is a shift of $y=x^2$, 2 units to the left.

(d) C; $y=(x-2)^2$ is a shift of $y=x^2$, 2 units to the right.

(e) F; $y=2x^2$ is a vertical stretch of $y=x^2$, by a factor of 2.

(f) D; $y=-x^2$ is a reflection of $y=x^2$, across the x-axis.

(g) H; $y=(x-2)^2+1$ is a shift of $y=x^2$, 2 units to the right and 1 unit upward.

(h) E; $y=(x+2)^2+1$ is a shift of $y=x^2$, 2 units to the left and 1 unit upward.

4. $y=3x^2$

x	$y=x^2$	$y=3x^2$
−2	4	12
−1	1	3
0	0	0
1	1	3
2	4	12

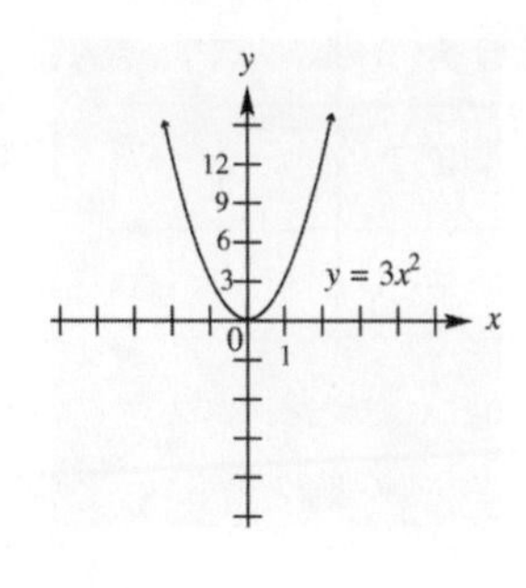

5. $y = 4x^2$

x	$y = x^2$	$y = 4x^2$
–2	4	16
–1	1	4
0	0	0
1	1	4
2	4	16

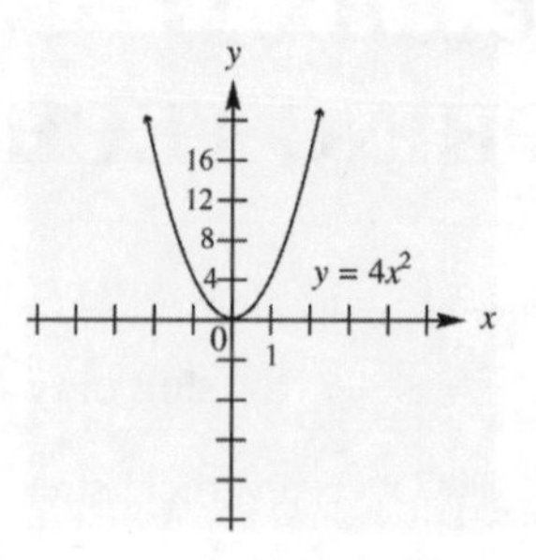

6. $y = \frac{1}{3}x^2$

x	$y = x^2$	$y = \frac{1}{3}x^2$
–3	9	3
–2	4	$\frac{4}{3}$
–1	1	$\frac{1}{3}$
0	0	0
1	1	$\frac{1}{3}$
2	4	$\frac{4}{3}$
3	9	3

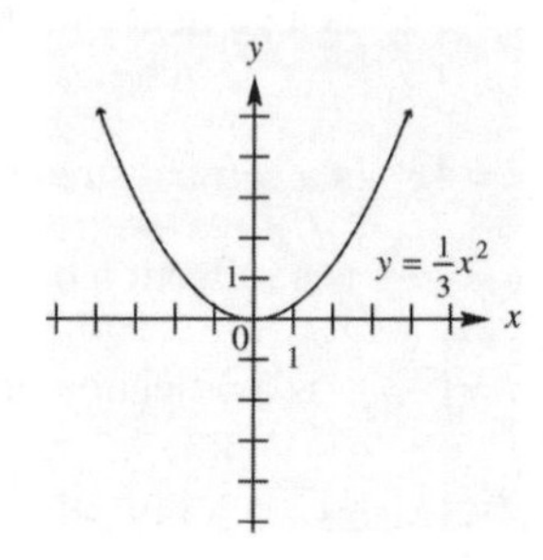

7. $y = \frac{2}{3}x^2$

x	$y = x^2$	$y = \frac{2}{3}x^2$
–3	9	6
–2	4	$\frac{8}{3}$
–1	1	$\frac{2}{3}$
0	0	0
1	1	$\frac{2}{3}$
2	4	$\frac{8}{3}$
3	9	6

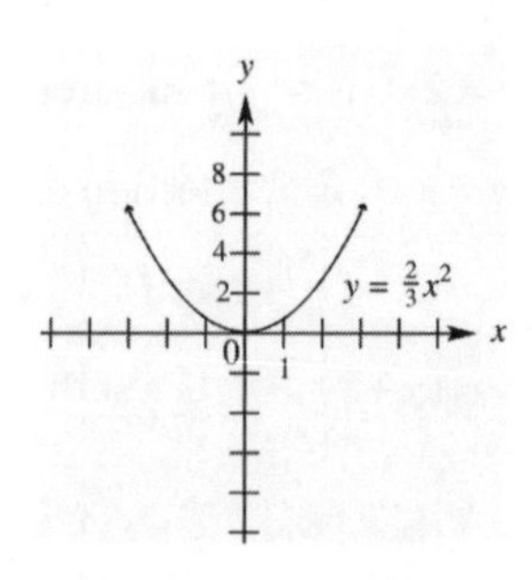

8. $y = -\frac{1}{2}x^2$

x	$y = x^2$	$y = -\frac{1}{2}x^2$
−3	9	$-\frac{9}{2}$
−2	4	−2
−1	1	$-\frac{1}{2}$
0	0	0
1	1	$-\frac{1}{2}$
2	4	−2
3	9	$-\frac{9}{2}$

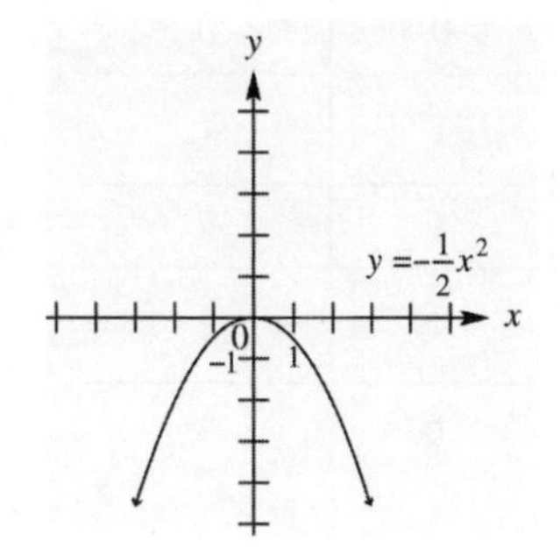

9. $y = -3x^2$

x	$y = x^2$	$y = -3x^2$
−3	9	−27
−2	4	−12
−1	1	−3
0	0	0
1	1	−3
2	4	−12
3	9	−27

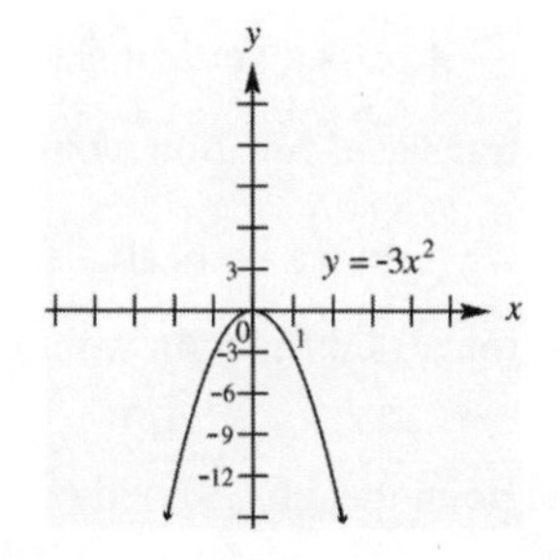

10. $y = 2\sqrt{-x}$

x	$-x$	$y = \sqrt{-x}$	$y = 2\sqrt{-x}$
0	0	0	0
−1	1	1	2
−4	4	2	4
−9	9	3	6
−16	16	4	8

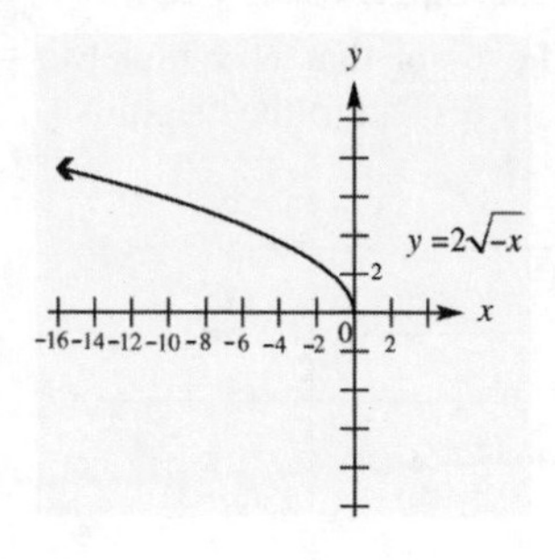

11. $y = \sqrt{-2x}$

x	$-2x$	$y = \sqrt{-2x}$
0	0	0
$-\frac{1}{2}$	1	1
−2	4	2
$-\frac{9}{2}$	9	3
−8	16	4

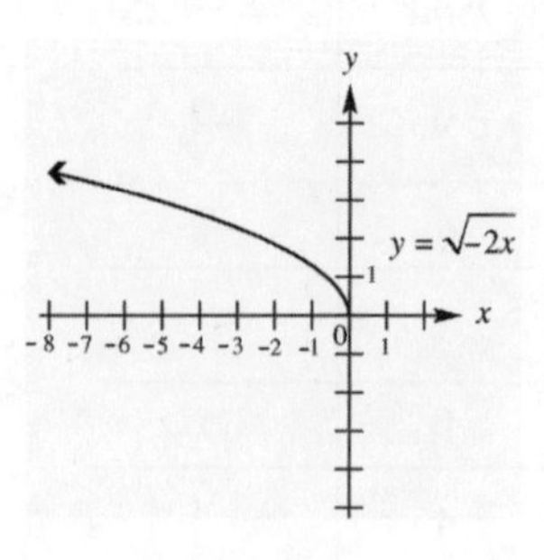

12. **(a)** $y = f(x+4)$ is a horizontal translation of *f*, 4 units to the left. The point that corresponds to $(8,12)$ on this translated function would be $(8-4,12) = (4,12)$.

(b) $y = f(x)+4$ is a vertical translation of *f*, 4 units up. The point that corresponds to $(8,12)$ on this translated function would be $(8,12+4) = (8,16)$.

13. **(a)** $y = \frac{1}{4}f(x)$ is a vertical shrinking of *f*, by a factor of $\frac{1}{4}$. The point that corresponds to $(8,12)$ on this translated function would be $\left(8,\frac{1}{4}\cdot 12\right) = (8,3)$.

(b) $y = 4f(x)$ is a vertical stretching of *f*, by a factor of 4. The point that corresponds to $(8,12)$ on this translated function would be $(8,4\cdot 12) = (8,48)$.

14. **(a)** The point that corresponds to $(8,12)$ when reflected across the *x*-axis would be $(8,-12)$.

(b) The point that corresponds to $(8,12)$ when reflected across the *y*-axis would be $(-8,12)$.

15. **(a)** The point that is symmetric to (5, −3) with respect to the *x*-axis is (5, 3).

(b) The point that is symmetric to (5, −3) with respect to the *y*-axis is (−5, −3).

(c) The point that is symmetric to (5, −3) with respect to the origin is (−5, 3).

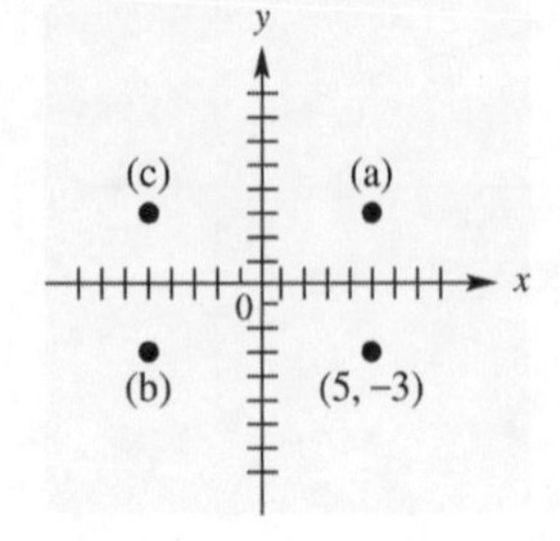

16. **(a)** The point that is symmetric to (−6, 1) with respect to the *x*-axis is (−6, −1).

(b) The point that is symmetric to (−6, 1) with respect to the *y*-axis is (6, 1).

(c) The point that is symmetric to (−6, 1) with respect to the origin is (6, −1).

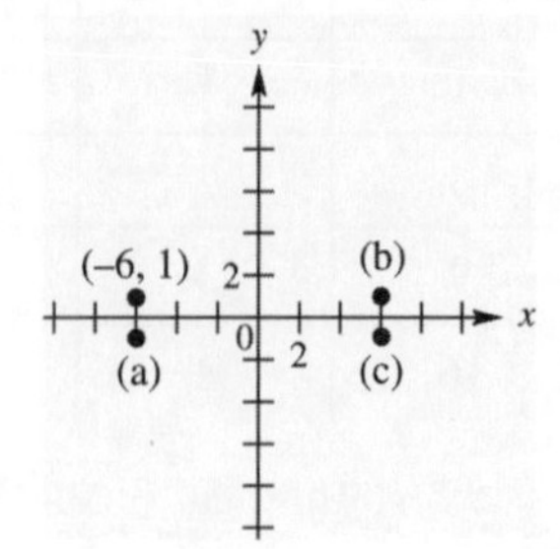

17. (a) The point that is symmetric to (–4, –2) with respect to the x-axis is (–4, 2).

(b) The point that is symmetric to (–4, –2) with respect to the y-axis is (4, –2).

(c) The point that is symmetric to (–4, –2) with respect to the origin is (4, 2).

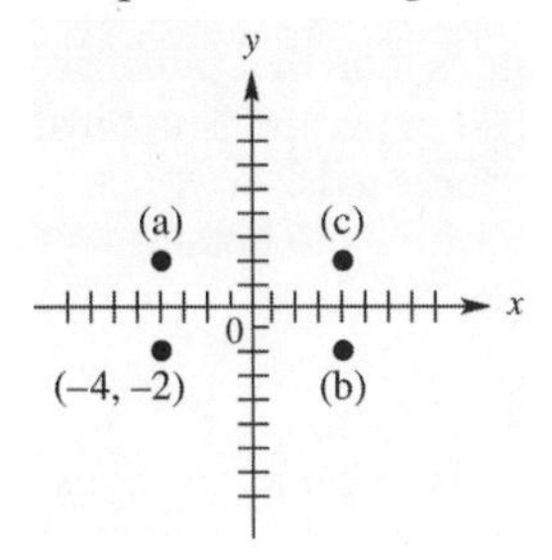

18. (a) The point that is symmetric to (–8, 0) with respect to the x-axis is (–8, 0), since this point lies on the x-axis.

(b) The point that is symmetric to the point (–8, 0) with respect to the y-axis is (8, 0).

(c) The point that is symmetric to the point (–8, 0) with respect to the origin is (8, 0).

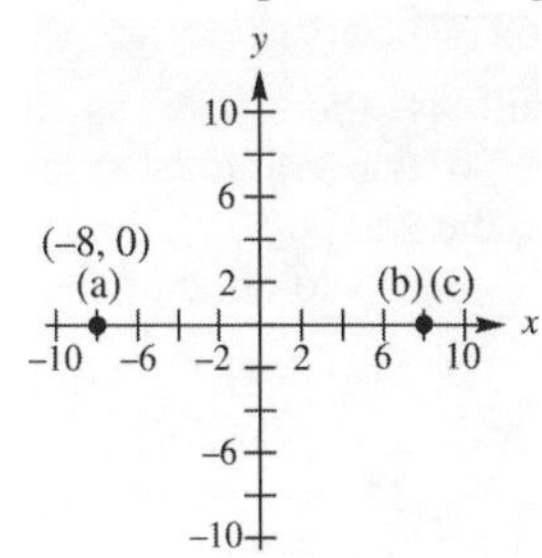

19. $y = x^2 + 2$

Replace x with $-x$ to obtain

$y = (-x)^2 + 2 = x^2 + 2.$

The result is the same as the original equation, so the graph is symmetric with respect to the y-axis. Since y is a function of x, the graph cannot be symmetric with respect to the x-axis.

Replace x with $-x$ and y with $-y$ to obtain

$-y = (-x)^2 + 2 \Rightarrow -y = x^2 + 2 \Rightarrow y = -x^2 - 2.$

The result is not the same as the original equation, so the graph is not symmetric with respect to the origin. Therefore, the graph is symmetric with respect to the y-axis only.

20. $y = 2x^4 - 1$

Replace x with $-x$ to obtain

$y = 2(-x)^4 - 1 = 2x^4 - 1$.

The result is the same as the original equation, so the graph is symmetric with respect to the y-axis. Since y is a function of x, the graph cannot be symmetric with respect to the x-axis.

Replace x with $-x$ and y with $-y$ to obtain

$$-y = 2(-x)^4 - 1$$
$$-y = 2x^4 - 1$$
$$y = -2x^4 + 1.$$

The result is not the same as the original equation, so the graph is not symmetric with respect to the origin.

Therefore, the graph is symmetric with respect to the y-axis only.

21. $x^2 + y^2 = 10$

Replace x with $-x$ to obtain the following.

$$(-x)^2 + y^2 = 10$$
$$x^2 + y^2 = 10$$

The result is the same as the original equation, so the graph is symmetric with respect to the y-axis.

Replace y with $-y$ to obtain the following.

$$x^2 + (-y)^2 = 10$$
$$x^2 + y^2 = 10$$

The result is the same as the original equation, so the graph is symmetric with respect to the x-axis. Since the graph is symmetric with respect to the x-axis and y-axis, it is also symmetric with respect to the origin.

22. $y^2 = \frac{-5}{x^2}$

Replace x with $-x$ to obtain

$y^2 = \frac{-5}{(-x)^2}$

$y^2 = \frac{-5}{x^2}.$

The result is the same as the original equation, so the graph is symmetric with respect to the y-axis.

Replace y with $-y$ to obtain

$(-y)^2 = \frac{-5}{x^2}$

$y^2 = \frac{-5}{x^2}.$

The result is the same as the original equation, so the graph is symmetric with respect to the x-axis. Since the graph is symmetric with respect to the x-axis and y-axis, it is also symmetric with respect to the origin.

Therefore, the graph is symmetric with respect to the x-axis, the y-axis, and the origin.

23. $y = -3x^3$

Replace x with $-x$ to obtain

$y = -3(-x)^3$

$y = -3(-x^3)$

$y = 3x^3.$

The result is not the same as the original equation, so the graph is not symmetric with respect to the y-axis.

Replace y with $-y$ to obtain

$-y = -3x^3$

$y = 3x^3.$

The result is not the same as the original equation, so the graph is not symmetric with respect to the x-axis.

Replace x with $-x$ and y with $-y$ to obtain

$-y = -3(-x)^3$

$-y = -3(-x^3)$

$-y = 3x^3$

$y = -3x^3.$

The result is the same as the original equation, so the graph is symmetric with respect to the origin. Therefore, the graph is symmetric with respect to the origin only.

24. $y = x^3 - x$

Replace x with $-x$ to obtain

$y = (-x)^3 - (-x)$

$y = -x^3 + x.$

The result is not the same as the original equation, so the graph is not symmetric with respect to the y-axis.

Replace y with $-y$ to obtain

$-y = x^3 - x$

$y = -x^3 + x.$

The result is not the same as the original equation, so the graph is not symmetric with respect to the x-axis.

Replace x with $-x$ and y with $-y$ to obtain

$-y = (-x)^3 - (-x)$

$-y = -x^3 + x$

$y = x^3 - x.$

The result is the same as the original equation, so the graph is symmetric with respect to the origin.

Therefore, the graph is symmetric with respect to the origin only.

25. $y = x^2 - x + 7$

Replace x with $-x$ to obtain

$y = (-x)^2 - (-x) + 7$

$y = x^2 + x + 7.$

The result is not the same as the original equation, so the graph is not symmetric with respect to the y-axis.

Since y is a function of x, the graph cannot be symmetric with respect to the x-axis.

Replace x with $-x$ and y with $-y$ to obtain

$-y = (-x)^2 - (-x) + 7$

$-y = x^2 + x + 7$

$y = -x^2 - x - 7.$

The result is not the same as the original equation, so the graph is not symmetric with respect to the origin.

Therefore, the graph has none of the listed symmetries.

26. $y = x + 12$

Replace x with $-x$ to obtain the following.

$$y = (-x) + 12$$
$$y = -x + 12$$

The result is not the same as the original equation, so the graph is not symmetric with respect to the y-axis. Since y is a function of x, the graph cannot be symmetric with respect to the x-axis.

Replace x with $-x$ and y with $-y$ to obtain the following.

$$-y = (-x) + 12$$
$$y = x - 12$$

The result is not the same as the original equation, so the graph is not symmetric with respect to the origin. Therefore, the graph has none of the listed symmetries.

27. $y = x^2 - 1$

This graph may be obtained by translating the graph of $y = x^2$, 1 unit downward.

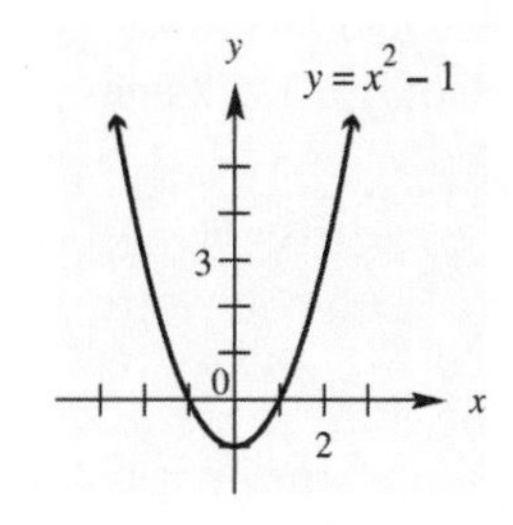

28. $y = x^2 + 1$

This graph may be obtained by translating the graph of $y = x^2$, 1 units upward.

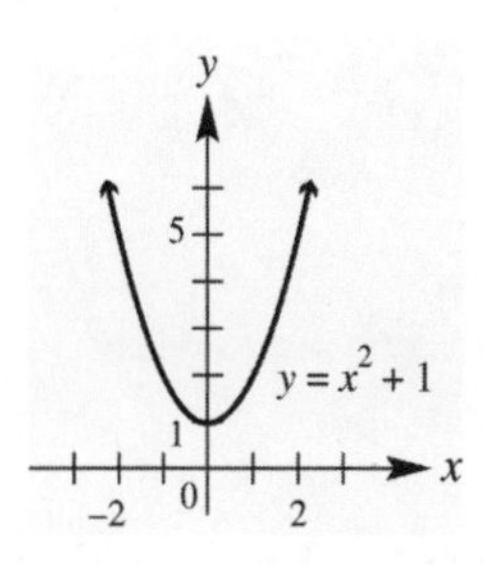

29. $y = x^2 + 2$

This graph may be obtained by translating the graph of $y = x^2$, 2 units upward.

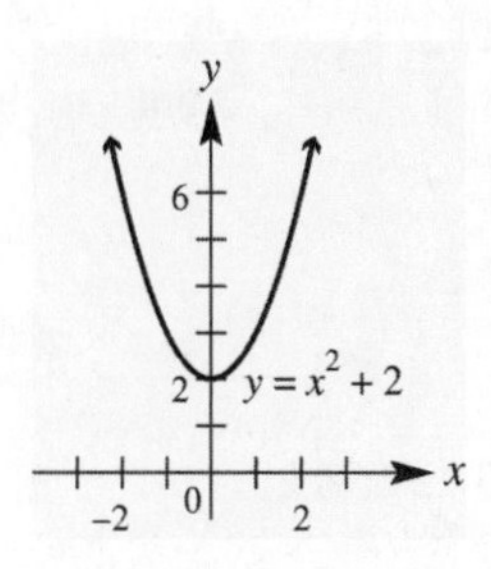

30. $y = x^2 - 2$

This graph may be obtained by translating the graph of $y = x^2$, 2 units downward.

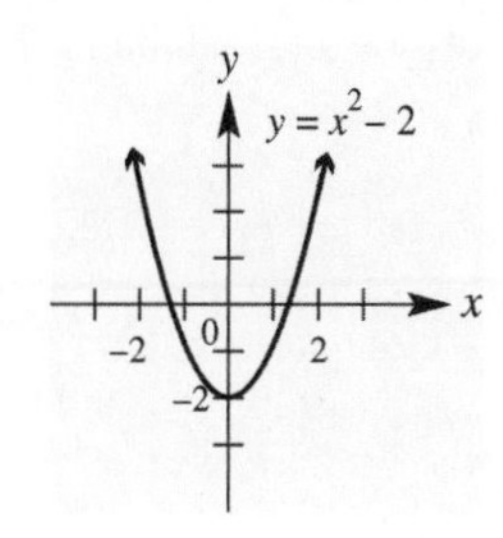

31. $y=(x-1)^2$

This graph may be obtained by translating the graph of $y=x^2$, 1 units to the right.

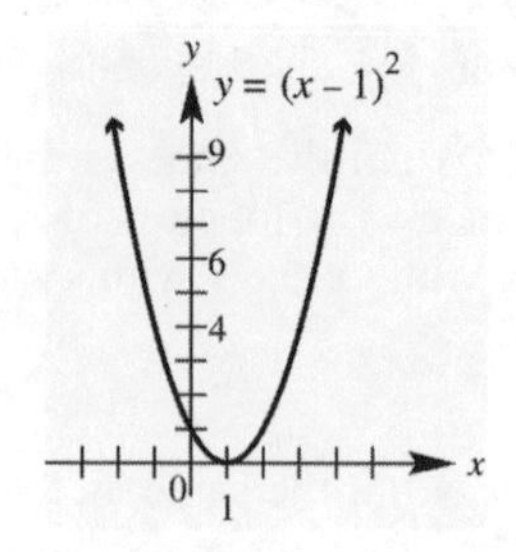

32. $y=(x-2)^2$

This graph may be obtained by translating the graph of $y=x^2$, 2 units to the right.

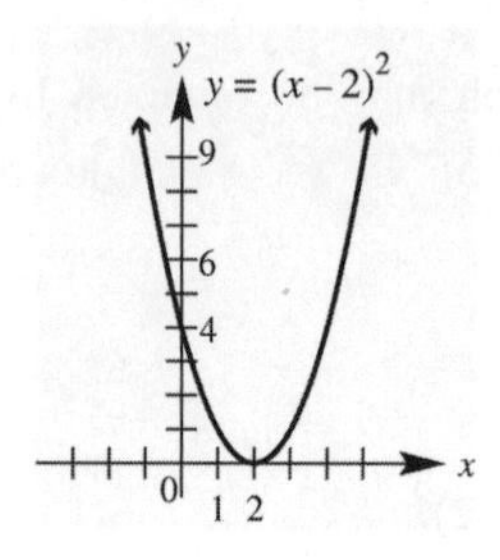

33. $y=(x+2)^2$

This graph may be obtained by translating the graph of $y=x^2$, 2 units to the left.

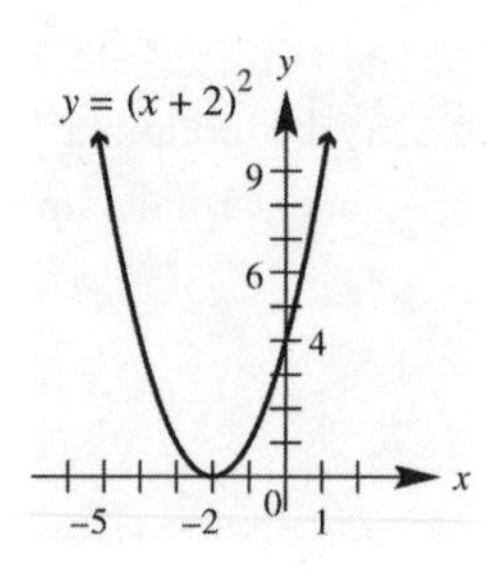

34. $y=(x+3)^2$

This graph may be obtained by translating the graph of $y=x^2$, 3 units to the left.

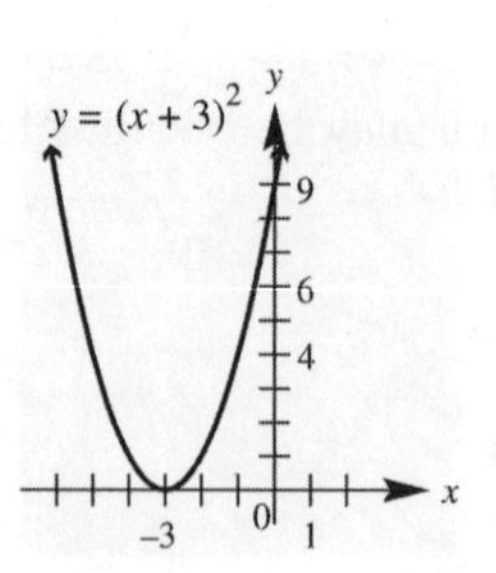

35. $y=(x+3)^2-4$

This graph may be obtained by translating the graph of $y=x^2$, 3 units to the left and 4 units down.

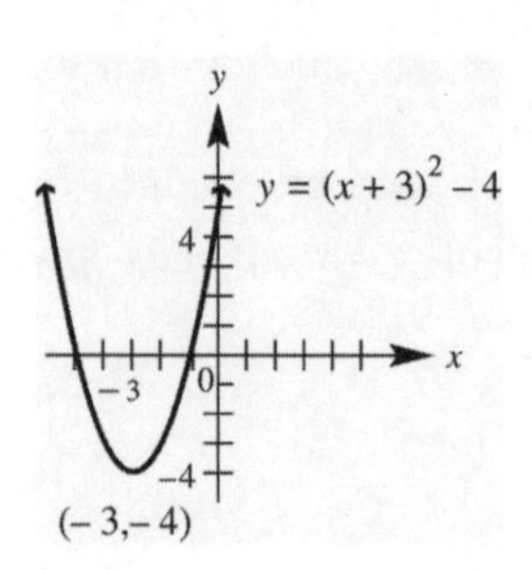

36. $y = (x-5)^3 - 4$
This graph can be obtained by translating the graph of $y = x^3$, 5 units to the right and 4 units down.

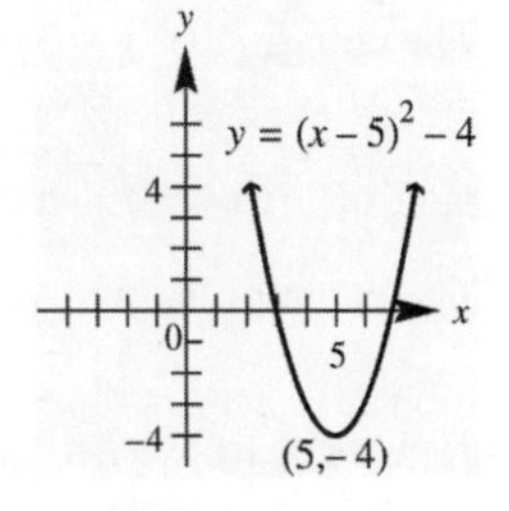

37. $y = 2x^2 - 1$
This graph may be obtained by translating the graph of $y = x^2$, 1 unit down. It is then stretched vertically by a factor of 2.

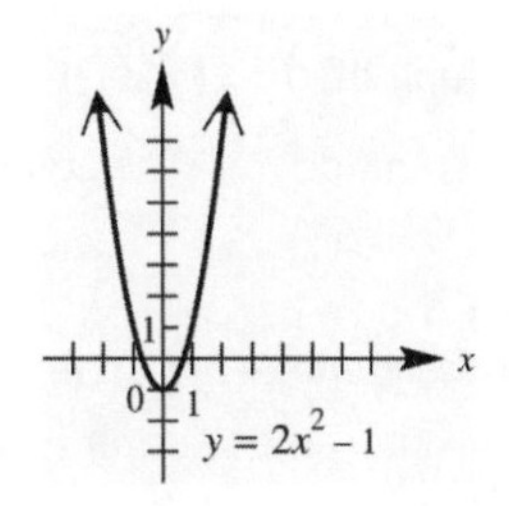

38. $y = \frac{2}{3}(x-2)^2$
This graph may be obtained by translating the graph of $y = x^2$, 2 units to the right. It is then shrunk vertically by a factor of $\frac{2}{3}$.

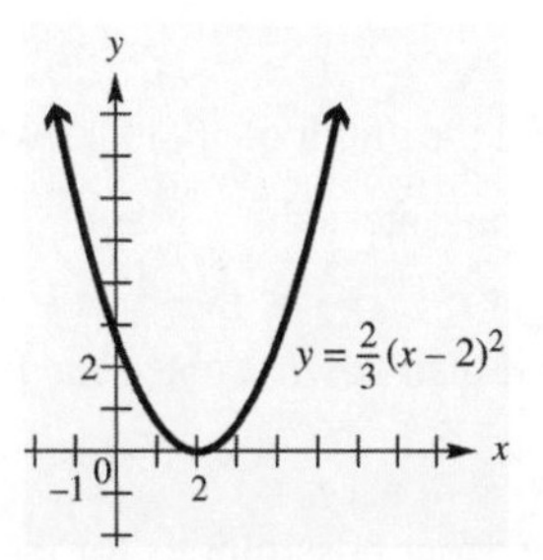

39. $f(x) = 2(x-2)^2 - 4$
This graph may be obtained by translating the graph of $y = x^2$, 2 units to the right and 4 units down. It is then stretched vertically by a factor of 2.

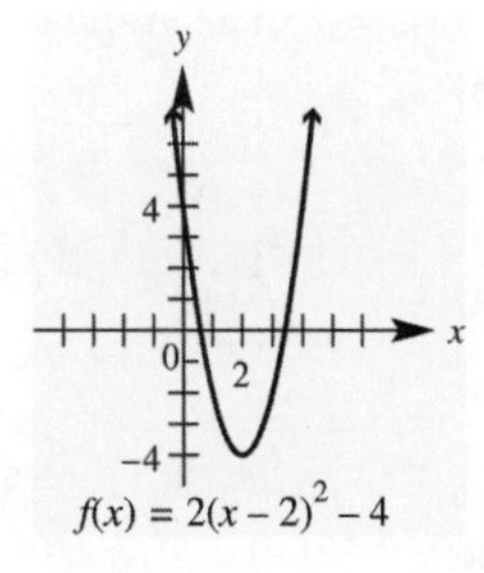

40. $f(x) = -3(x-2)^2 + 1$
This graph may be obtained by translating the graph of $y = x^2$, 2 units to the right and 1 unit up. It is then stretched vertically by a factor of 3 and reflected over the x-axis.

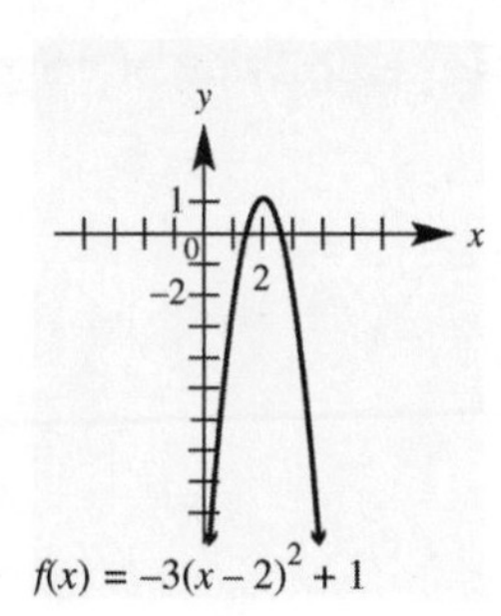

41. It is the graph of $f(x)=|x|$ translated 1 unit to the left, reflected across the x-axis, and translated 3 units up. The equation is $y=-|x+1|+3$.

42. It is the graph of $g(x)=\sqrt{x}$ translated 4 units to the left, reflected across the x-axis, and translated 2 units up. The equation is $y=-\sqrt{x+4}+2$.

43. It is the graph of $g(x)=\sqrt{x}$ translated 4 units to the left, stretched vertically by a factor of 2, and translated 4 units down. The equation is $y=2\sqrt{x+4}-4$.

44. It is the graph of $f(x)=|x|$ translated 2 units to the right, shrunken vertically by a factor of $\frac{1}{2}$, and translated 1 unit down. The equation is $y=\frac{1}{2}|x-2|-1$.

45. $f(x) = 2x + 5$
Translate the graph of $f(x)$ up 2 units to obtain the graph of $t(x)=(2x+5)+2=2x+7$.
Now translate the graph of $t(x) = 2x + 7$ left 3 units to obtain the graph of $g(x)=2(x+3)+7=2x+6+7=2x+13$. (Note that if the original graph is first translated to the left 3 units and then up 2 units, the final result will be the same.)

46. $f(x) = 3 - x$
Translate the graph of $f(x)$ down 2 units to obtain the graph of $t(x)=(3-x)-2=-x+1$.
Now translate the graph of $t(x)=-x+1$ right 3 units to obtain the graph of $g(x)=-(x-3)+1=-x+3+1=-x+4$. (Note that if the original graph is first translated to the right 3 units and then down 2 units, the final result will be the same.)

47. Answers will vary.
There are four possibilities for the constant, c.
i) $c>0$ $|c|>1$ The graph of $F(x)$ is stretched vertically by a factor of c.
ii) $c>0$ $|c|<1$ The graph of $F(x)$ is shrunk vertically by a factor of c.
iii) $c<0$ $|c|>1$ The graph of $F(x)$ is stretched vertically by a factor of $-c$ and reflected over the x-axis.
iv) $c<0$ $|c|<1$ The graph of $F(x)$ is shrunk vertically by a factor of $-c$ and reflected over the x-axis.

48. The graph of $y=F(x+h)$ represents a horizontal shift of the graph of $y=F(x)$. If $h>0$, it is a shift to the left h units. If $h<0$, it is a shift to the left $-h$ units (h is negative).
The graph of $y=F(x)+h$ is not the same as the graph of $y=F(x+h)$. The graph of $y=F(x)+h$ represents a vertical shift of the graph of $y=F(x)$.